Earth Science Today

BOOKS IN THE BROOKS/COLE EARTH SCIENCE AND ASTRONOMY SERIES:

INTERNET AND MEDIA PRODUCTS IN THE BROOKS/COLE EARTH SCIENCE SERIES:

Earth Online: An Internet Guide for Earth Science, RITTER

In-TERRA-Active: Physical Geology Interactive Student CD-ROM, BROWN AND DUNNING

GeoLink: Instructor's Research CD-ROM Library for Geology, WEST

The Geology Workbook for the Web, BLACKERBY

EARTH SCIENCE

Earth Science, DUTCH, MONROE, MORAN

Earth Online: An Internet Guide for Earth Science, RITTER

METEOROLOGY

Meteorology Today, 5th, AHRENS

Essentials of Meteorology, 2nd, AHRENS

Meteorology Today for Scientists and Engineers, STULL

ASTRONOMY

Horizons: Exploring the Universe, 5th, SEEDS

Foundations of Astronomy, 4th, SEEDS

Universe: Origins and Evolutions, SNOW AND BROWNSBERGER

Astronomy: The Cosmic Journey, 5th, HARTMANN AND IMPEY

Moons and Planets, an Introduction to Planetary Science, 3rd, HARTMANN

Astronomy on the Web, KURLAND

Current Perspectives in Astronomy and Physics, POMPEA

PHYSICAL GEOLOGY

Physical Geology: Exploring the Earth, 3rd, MONROE AND WICANDER

Essentials of Geology, WICANDER AND MONROE

Physical Geology, BENTON

HISTORICAL GEOLOGY

Historical Geology: Evolution of the Earth and Life Through Time, 2nd, WICANDER AND MONROE

Earth and Environment Through Time: A Lab Manual for Historical Geology, 2nd, HICKEY

COMBINED PHYSICAL AND HISTORICAL GEOLOGY

The Changing Earth, 2nd, MONROE AND WICANDER

ENVIRONMENTAL GEOLOGY

Geology and the Environment, 2nd, PIPKEN AND TRENT

CURRENT ISSUES READER FOR GEOLOGY

Current Perspectives in Geology, 1998 edition, MCKINNEY/SHARIFF/TOLLIVER

OCEANOGRAPHY

Essentials of Oceanography, GARRISON

Oceanography: An Invitation to Marine Science, 2nd, GARRISON

Oceanography: An Introduction, 5th, INGMANSON AND WALLACE

Introduction to Ocean Science, SEGAR

Marine Life and the Sea, MILNE

Contact your local Brooks/Cole representative for a review copy of any of the above titles or call Faculty Support at 1-800-423-0563.

Earth Science Today

BRENDAN MURPHY
St. Francis Xavier University

DAMIAN NANCE
Ohio University

Brooks/Cole • Wadsworth
I(T)P® an International Thomson Publishing Company

Pacific Grove, CA • Albany, NY • Belmont, CA • Boston • Cincinnati • Johannesburg
London • Madrid • Melbourne • Mexico City • New York • Scottsdale, AZ
Singapore • Tokyo • Toronto

Earth Science Editors: *Stacey Purviance, Nina Horne*
Development Editors: *Heather Dutton, Jody Larson, Laurel Smith*
Assistant Editor: *Marie Carigma-Sambilay*
Editorial Assistant: *Erin Conlon*
Marketing Managers: *Christine Henry, Halee Dinsey, Tami Cueny*
Project Editor: *Tanya Nigh*
Print Buyer: *Stacey Weinberger*
Permissions Editor: *Bob Kauser*
Production: *Electronic Publishing Services Inc., NYC*
Designer: *Cloyce Wall*
Copy Editor: *Electronic Publishing Services Inc., NYC*
Illustrator: *Electronic Publishing Services Inc., NYC*
Cover Design: *Cloyce Wall*
Cover Image: *Bruce Roberts, The Heceta Head Lighthouse in Oregon*
Compositor: *Electronic Publishing Services Inc., NYC*
Printer: *RR Donnelley & Sons, Inc.*

Printed in the United States of America.
3   4   5   6   7   8   9   10

For more information, contact Brooks/Cole Publishing Company, 511 Forest Lodge Road, Pacific Grove, CA 93950

International Thomson Publishing Europe
Berkshire House 168-173
High Holborn
London WC1V 7AA
England

Thomas Nelson Australia
102 Dodds Street
South Melbourne, 3205
Victoria, Australia

Nelson Canada
1120 Birchmount Road
Scarborough, Ontario
Canada M1K 5G4

International Thomson Publishing Southern Africa
Building 18, Constantia Square
138 Sixteenth Road, P.O. Box 2459
Halfway House, 1685 South Africa

International Thomson Editores
Seneca 53
Col. Polanco
11560 México, D. F., México

International Thomson Publishing Asia
221 Henderson Road
#05-10 Henderson Building
Singapore 0315

International Thomson Publishing Japan
Hirakawacho Kyowa Building, 3F
2-2-1 Hirakawacho
Chiyoda-ku, Tokyo 102
Japan

You can request permission to use material from this text through the following phone and fax numbers:
Phone: 1-800-730-2214; Fax: 1-800-730-2215.

Library of Congress Cataloging-in-Publication Data
Murphy, Brendan.
    Earth science today / Brendan Murphy, Damian Nance.
    p.   cm — (The Brooks/Cole earth science and astronomy series)
    Includes index.
    ISBN 0-534-52182-7 (hc.)
    1. Earth sciences.   I. Nance, Damian.   II. Title.   III. Series
QE26.2.M87   1998
550—DC21   98-43312

Dedication

To our parents, our families, and our teachers

Brief Contents

Contents

Chapter 18

Energy Resources 545

Preface

To The Instructor

Our Approach

Today's students are tomorrow's decision-makers, whether their future careers are in politics, finance, technology, medicine, or basic science. It is their decisions that collectively will decide the fate of our planet. As instructors in Earth science or geology, therefore, we have a most important mission. We must ensure that our students have the opportunity to obtain a basic understanding of the Earth so that they are equipped to make informed, environmentally responsible decisions in their future careers.

We hope that our textbook helps you, the instructor, attain these goals, and that the student's appreciation of our amazing planet and its place in the cosmos is enhanced. We attempt to convey the message that understanding the Earth is exciting, and that it enriches and heightens our sense of awareness of the world around us.

To achieve this, we adopt an integrated, interdisciplinary Earth systems approach to planet Earth that involves the study of its geology, biology, chemistry, and physics. Our book emphasizes that Earth systems are an interplay of the traditional sciences and do not respect the artificial boundaries we tend to place between them. If our goal is to understand our planet, therefore, we must break down these barriers. Only through this approach, we believe, can we hope to gain an understanding of global environmental problems and so be in a position to address them.

We recognize that students taking an introductory course in Earth science or geology may come from widely varying academic backgrounds. Many may not have taken a science course in high school, never mind a course in Earth science. We also recognize that for a significant proportion of students, this will be their only formal exposure to Earth science. So, we believe that we have a dual mission. We must not only provide a strong fundamental background for students who wish to pursue geology or Earth science in more advanced courses, but we must also offer all students the opportunity to learn about the processes that shape the planet in which they live. In addition, we hope that some students who are unsure about what they might select as a career will at least consider geology or Earth science as a potential major.

The goal of this text is to provide an understanding of, and appreciation for, the Earth. Our approach is to stress concepts, processes, and principles because these should serve all students well. A thorough grasp of any science requires some basic knowledge of facts and terminology, the language of the science. However, we are careful about overtaxing the student. In our experience, *excess* terminology at an introductory level leads to rote-learning. Students tend to navigate from one term to the next and fall into the trap of thinking that by learning the terminology they will understand the concepts and principles. The trap is sprung at exam time! Furthermore, knowledge of terminology fades with time, especially for those students for whom this is a terminal course in Earth science. A course with an emphasis on the understanding of concepts and principles, we believe, has a much longer shelf-life.

For example, we stress how gases released from volcanoes over the 4.6 billion years of Earth history have had a first order influence on the evolution of our atmosphere, how water pumped from volcanoes contributed to the oceans and the origin of life, and why, in this respect, the Earth is unique in the solar system. We consider the vast array of terminology that describes these volcanoes and their various types of volcanic eruptions to be of secondary importance at an introductory level, the lexicon of terms sometimes distracting the student from the primary message. This terminology, we feel, is better learned in the appropriate upper level courses.

At the same time, we do not wish to sacrifice or dilute the conceptual content of the material. Our challenge, then, is to present concepts and processes in a straightforward manner rather than risk losing student interest by burying them in unnecessary terminology. If, at the end of your course, the students display an appetite for learning more, then we will have achieved our goal.

Organization of the text

The book is divided into seven parts. The first four parts deal with each of the Earth's surface reservoirs in turn. In Part I, we introduce the foundations of modern geology and continue with an examination of the solid Earth. In Part II, we discuss the hydrosphere, followed by the atmosphere in Part III. Since a detailed study of the modern biosphere is more appropriate in an introductory biology course, Part IV looks at the evolution of the biosphere over geologic time in the context of the other Earth systems. This part also serves as a summary of Earth history.

In Part V, we examine the Earth within the context of the solar system and the cosmos. Equipped with this basic background, we then discuss the balance between Earth resources (Part VI) and environmental responsibility (Part VII). In Part VII, we also address specific environmental issues using the principles and concepts developed in the previous chapters. In a way, this part serves as a thematic review and demonstrates how a thorough knowledge of Earth systems provides important insights into modern environmental problems.

Each part is subdivided into chapters. The chapters are designed to be as freestanding as possible, the beginning of each chapter identifying the important portions of the previous text that form the basis for what follows. An overview at the beginning of each chapter also prepares the student for the concepts that follow. Each chapter ends with a summary and a listing of key concepts and terms to facilitate that essential last-minute brush-up before tests! Finally, two sets of questions brings each chapter to a close. The Review Questions focus on the most important parts of the chapter. The Study Questions require further thought on the part of the student and a working knowledge of the preceding chapters in the text.

In dividing the text into parts based on Earth systems, we run the risk of failing to follow our own advice. However, we repeatedly stress the processes that bridge the traditional sciences and link the various reservoirs. This linkage will be apparent in many parts of the text. The most important linkages will be highlighted with "Process in Action Icons" and in the Visual Summaries at the end of each part (see box in "To the Student").

In Part I, we describe the basic principles of modern geology, and then use these to describe the solid Earth. Following an introduction (Chapter 1), Chapter 2 discusses minerals and rocks, the basic materials that compose the solid Earth. This part acknowledges that much of our modern understanding of the Earth is rooted in classical geology. In Chapter 3, we gain an impression of the scale and duration of natural processes. We acquire a sense of geologic time by examining the age of the Earth and the methods by which geologic events are dated. By discussing the basic principles of relative and absolute time, we show how rocks and minerals may be used to piece together geologic history and so reveal the evolution of the Earth through geologic time. In so doing, we gain important insights into the pace of evolutionary change. Finally, in Chapter 4, we set the stage for our treatment of the solid Earth by examining the fascinating story behind the development of plate tectonic theory, a story that we also use to illustrate the scientific method.

In our treatment of the solid Earth (chapters 5 to 7), we focus on the processes that introduce new chemical constituents to the surface of the Earth. Chapters 5 and 6 examine the processes that govern the evolution of the Earth's surface. Here we outline the principles of plate tectonics, the fundamental unifying concept in the Earth sciences that accounts for the origin and evolution of the Earth's major surface features such as the continents and oceans, and explains the distribution of global phenomena such as earthquakes, volcanoes, and mountain belts. Here we get a sense of the dynamic nature of our planet. In Chapter 7, we examine the internal structure of the Earth from its surface to its core. In doing so, we discover the internal heat engine that is the driving force behind plate tectonics and many of the Earth's surface processes. We also learn that some ingenious techniques have given us a surprisingly thorough knowledge of the chemistry of our planet.

In Part II (Chapters 8, 9, and 10), we examine the hydrosphere. We first look at the hydrologic cycle and the factors that control the chemistry of ocean water. Using the material covered in Part I as a background, we then analyze the origin and ultimate fate of the oceans. Chapter 9 deals with the circulation of ocean water on both a global and local scale. Here we find the major influences to include radiant energy from the Sun, the spin of the Earth about its own axis, and the temperature contrast between the poles and the equator. We use the phenomenon of El Niño as a prime example of the importance of understanding global circulation patterns. Chapter 10 is devoted to fresh water on continental landmasses, a vital aspect of life on our planet.

Part III (chapters 11 and 12) deals with the atmosphere and follows a similar approach to that used in Part II. We first examine in Chapter 11 the average composition and temperature of the atmosphere and, once again, find that the material covered in the preceding parts plays a key role in our understanding of the evolution of the Earth's atmosphere and the controls behind past and future climates. In Chapter 12, we look at the motion of air masses and the factors that influence modern climates and our day-to-day weather patterns. We also examine the processes that lead to weather "events," such as hurricanes, tornadoes, and monsoons, and the global disruption of typical weather patterns brought about by an El Niño. We show how satellites are our "eyes in the sky" and how the use of this technology has resulted in amazing advances in our understanding of the dynamic forces behind our weather and climate. We also show how improved weather forecasting now provides advanced warning of dangerous weather and can help predict catastrophes such as widespread drought or the failure of the monsoon rains.

The biosphere is explicitly dealt with in courses in the life sciences. In Part IV (chapters 13 and 14), however, we focus on the origin and evolution of the biosphere and the ways in which it interacts with the solid Earth, hydrosphere, and atmosphere. In a sense, this part serves as a synthesis of Earth history, emphasizing the evolution of the Earth systems and the key events and processes that have governed our planet's development from its origin to the present day. In Chapter 13, we outline the evolution of life and the Earth systems in the Precambrian and follow this, in Chapter 14, with a similar treatment of the Phanerozoic Eon. Chapter 14 also describes our most recent history and deals with the forces that shaped our modern landscape.

In Part V, we broaden our horizons to look at the planets of the Solar System (Chapter 15) and the Sun and stars in the universe beyond it (Chapter 16). The study of other planets in the Solar System provides some very important perspectives on the evolution of our own planet. The uniqueness of planet Earth can be placed in the context of its position in the Solar System, its size and composition, and its distance from the

Sun. We find that these factors, in particular, have caused the Earth to take a different evolutionary path from those of its nearest planetary neighbors. Recent evidence that there may once have been microbial life on Mars adds further fuel to the debate that has pre-occupied us for centuries. Is there anybody out there or are we alone in this vast universe?

Our knowledge of our place in the cosmos is improving rapidly, thanks to amazing advances in both land-based and orbiting telescopes, such as the Hubble space telescope. Exciting breakthroughs like those that suggest there may be other solar systems similar to our own, are typical of the information explosion that confronts us almost daily. No doubt, this trend will continue into the immediate future. Our challenge, then, is to lay the groundwork so that students can both understand and assess this information as each new discovery is made.

We start our examination of the universe with the star we know best, the Sun. Earth systems are greatly influenced by radiant energy from the Sun. Although the Sun played little direct influence in the *origin* of our atmosphere and hydrosphere, it is the major external influence on the global-scale circulation of these two reservoirs and has a profound effect on our weather and climate, as we know from personal experience. Indeed, the Earth's surface is where solar energy and the Earth's internal processes interact. In our treatment of the Sun, we concentrate on solar processes and phenomena that influence the Earth today, and may have influenced the planet in the past. By examining the universe beyond, we gain an appreciation for our position in the cosmos.

In Part VI, we deal with Earth resources and the factors that influence the generation of economic deposits of minerals (Chapter 17) and energy reserves (Chapter 18). Nowhere is the interaction of the Earth's reservoirs more closely linked to the development of human society than it is in our exploitation of the planet's natural resources. It is the demands of society that make a deposit economically feasible to exploit. These demands have changed dramatically over the past 20 years and will continue to do so in the foreseeable future. Our treatment of mineral resources emphasizes our dependence on metallic and industrial minerals, the inequity of their distribution, and the important natural processes that concentrate them. It also examines some of the methods used in mineral exploration and the environmental consequences of mineral exploitation. The underlying principles of a source, a medium of transport, and an environment of deposition that concentrates the commodity, are common themes in the formation of many of our natural resources. Our treatment of energy examines the issues of supply and demand, and looks at the variable contributions of each of the Earth's reservoirs in providing our energy needs. We find that energy sources derived from the solid Earth, such as fossil fuels and nuclear power, are convenient but nonrenewable, whereas those obtained from the hydrosphere, atmosphere, and biosphere, such as hydroelectric power, solar energy, and biomass fuel, are renewable but often difficult to harness. Since questions concerning the future of our finite mineral and energy reserves are likely to become major issues in Earth science as we enter the 21st Century, we end our survey of the Earth's natural resources with a look at what the future may hold.

With this background, we are now equipped to evaluate Earth systems and the environment (Part VII). In fact, just as our current environmental problems provide a test of the extent to which we truly understand our own planet, this part will test the students' understanding of the key concepts developed in each of the preceding chapters.

Chapters 19 and 20 demonstrate how the Earth's reservoirs are linked and discuss the exciting new scientific discipline of biogeochemistry. In Chapter 20, we address key environmental issues in a deliberately provocative fashion to encourage debate among students and between students and the course instructor. Can we really sustain the resource exploitation needed to support modern development? In recognizing rhythms in the natural interaction among Earth systems, can we detect the distortion associated with modern development and our own tampering with the Earth's reservoirs? And in understanding the mechanisms that drive these rhythms, can we learn how best to alleviate some of our current environmental problems?

In the Afterword, we attempt to point out to students the importance of environmental responsibility and some of the important challenges and influences that will affect their lives. We stress that important and fulfilling work in Earth science remains to be done. This may convince students that a career in Earth science is worthwhile, although that is not our main objective. Most students will not have taken a course in Earth science or geology in high school, and this chapter is intended to give them an idea of the basic research that remains to be done. It is also important to impress upon our future decision makers the importance and urgency of basic research and its relationship to society.

Through our emphasis on the concepts and principles of Earth science, we hope to provide students, majors and non-majors alike, with a measure of understanding about the planet on which they live.

Acknowledgments

An undertaking such as this one is impossible to complete without the help and expertise of many people. The team at Brooks/Cole, and before that, at Wadsworth, encouraged and advised us every step of the way. Their expertise and cheerfulness kept us motivated and ensured that we completed the task at hand and made it a pleasurable experience. Heather Dutton, Senior Developmental Editor, navigated us through many revisions, deadlines, and revised deadlines, and her expert editing made substantial improvements to the text and artwork. We greatly appreciate her commitment and the weekly conference calls in which timely words of encouragement and advice kept us to the task at hand. Kim Leistner-Little, Acquisitions Editor for Earth Science, and Gary Carlson, Publisher for Sciences, shared our initial enthusiasm for the project and were largely responsible for us becoming involved with the Brooks/Cole team. Stacey Purviance, who took over from Kim as Acquisitions Editor during the course of the project, provided important guidance as the book began to take shape and headed towards completion. Her expertise and patience were invaluable in keeping the project on track. We greatly acknowledge the impressive skills of Tanya Nigh, Senior Project Editor; Bob

Kauser, Permissions Editor; Erin Conlon, Editorial Assistant; and Marie Carigma-Sambilay, Assistant Editor, without whom this book would not have been completed on schedule.

And for their enthusiastic marketing efforts, we'd like to acknowledge: Halee Dinsey, Christine Henry, Joey Jodar, and Tami Strang. And for media development, we owe great thanks to Claire Masson, our Media Editor.

We are also indebted to Laurel Smith, a freelance art developmental editor; Jody Larson, a freelance developmental editor; and to Anthony Calcara and Justin Menza of Electronic Publishing Services for their tireless help in organizing and editing the text and the illustrations. Several friends and colleagues (Alan Anderson, Chris Fontana, Dan Kontak, Johnny and Leslie Buckland-Nicks, Duffy and Marie MacDonald, Karen Magee, Joe and Kay Martinez, Dorothy Sack, Glen Stockmal, John Waldron, Marc St. Onge, and Hank Williams) kindly provided photographs that helped to illustrate various parts of the text. We are also grateful to Marcie Mealia and Kerry Baruth for their initial interest and enthusiasm. JBM would like to particularly acknowledge the hospitality and support of Ed and Mona Nichols and of Electromagnetic Instruments during sabbatical leave in California in 1996–97. We are also grateful to Nova Scotia Links for partially funding the hiring of undergraduate and graduate students who helped with many aspects of this book.

We thank all the reviewers (listed below) for their time and expertise in providing thorough and constructive reviews, especially those few, including Cindy Murphy, who read virtually the whole text and offered important advice. We also thank our colleagues in our respective departments at St. Francis Xavier University and Ohio University who gave us the breathing room and encouragement to finish "the book." In this regard, we are particularly indebted to Daphne Metts.

To our families, however, we owe our greatest debt. Our wives, Cindy Murphy and Rita Nance, helped us along every step of the way, and our kids, Orla and Declan Murphy, and André, Sarah, and Christopher Nance were most understanding while their daddies completed "the book." The love, patience, understanding, and assistance of our families during all stages of the project were a constant source of inspiration. Without them, this book would not have become a reality. Now it's time to take them to the beach.

List of Reviewers

James R. Albanese
State University of New York
Oneonta, NY

Elizabeth Anthony
University of Texas at El Paso

Sandra Barr
Acadia University
Wolfville, Nova Scotia

Hugo Beltrami
St. Francis Xavier University
Antigonish, Nova Scotia

William C. Cornell
University of Texas at El Paso

J. Warner Cribb
Middle Tennessee State University
Murfreesboro, TN

William Culver
St. Petersburg Jr. College—Gibbs Campus
St. Petersburg, FL

James R. Ebert
State University of New York
Oneonta, NY

Stephen C. Hildreth, Jr.
Lawrence University
Canton, NY

William F. Kean
University of Wisconsin at Milwaukee

Lisa Kellman
St. Francis Xavier University
Antigonish, Nova Scotia

David T. King, Jr.
Auburn University
Auburn, AL

George Kipphut
Murray State University
Murray, KY

Arthur C. Lee
Westark Community College
Ft. Smith, AR

Keenan Lee
Colorado School of Mines
Golden, CO

Ellen P. Metzger
San Jose State University
San Jose, CA

Susan K. Morgan
Utah State University
Logan, Utah

John E. Poling
California Polytechnic State University
San Luis Obispo, CA

Gerry Simila
California State University—Northridge

Alison J. Smith
Kent State University
Kent, OH

Keith Sverdrup
University of Wisconsin at Milwaukee

J. Robert Thompson
Glendale Community College
Glendale, AZ

To The Student

We live in amazing times. In the past 20 years we have learned an enormous amount about our planetary home, and new information confronts us almost daily. We can scarcely watch the news or read a newspaper without learning of some new and exciting discovery that relates to the Earth and our position in the cosmos. At the same time, we are receiving mixed messages about the state of our local and global environment. This information now comes at us at such a bewildering pace, it is difficult to assimilate it all.

As we enter the 21st Century, it is appropriate to remind ourselves that just a century ago the world and our knowledge of it was entirely different. Our forebears could never have visualized the scientific and technological advances of the 20th Century. We now have a comprehensive understanding of the origin of mountains and oceans, and explanations for the origins of earthquakes and volcanic activity. We have devised ingenious methods that allow us to image the Earth's interior, determine its composition, and understand some of the internal processes that go on thousands of kilometers below its surface. Even weather forecasts have become more reliable through the use of satellites as our eyes in the sky. Our forays into space have given us a new appreciation of the tiny portion of the universe we occupy. At the same time, we have come to realize that our planet is fragile and that our own actions may compromise our very existence.

Our goal in this textbook is to convey the wonder and excitement of discovery while heightening your knowledge, appreciation, and interest in the planet on which we live. Indeed, since we all use its resources, we have a certain obligation to learn something of the way in which the Earth works. And who is to say that the technological advances of the 20th Century will not continue or even accelerate into the 21st? So it may well be in your best interest to have a working knowledge of the world around you.

Each one of us has a daily impact on our planet in some small way. Cumulatively, this impact may be enormous. In the near future, your decisions, big and small, may impact the environment on a local, regional, and even global scale. Faced with such decisions, we hope that this book, in some small way, will help you make the right ones.

Believe it or not, it was not so very long ago that we were students like you, and we remember only too well how hectic student life can be. If channeled properly, however, these will be some of the most rewarding years of your life, full of heightened intellectual and social experiences. We also remember how much easier it was to learn if the material we were studying was presented in an interesting and dynamic way. Luckily for us, we both had excellent teachers in our introductory courses. In the classrooms where we now teach, we try to inspire our students through our own fascination and enthusiasm for the science, and by showing them that learning is fun. We hope that this also comes across in our textbook. If, in addition, you are considering a career in geology or Earth science, we encourage you to do so. We have both thoroughly enjoyed our own careers and most geologists would say the same. A healthy appreciation and respect for the working of the world around us has a wonderfully calming influence when the waters of life get a little rough!

As you read the book, you will find that we emphasize concepts and processes over excess terminology. Perhaps this is because we are both plagued with poor memories and we know all too well how one's memory of terminology fades with time! While terminology is important to the language of science, it is through concepts and processes that science is understood. We therefore believe that courses that are process-oriented will have a far longer shelf-life than those based on terminology.

If you choose a career in Earth science, you will have the opportunity to contribute directly to our understanding of planet Earth. Should you choose a different career path, we hope that this textbook will increase your understanding of the planet on which we live. Our challenge is to present the material to you in language that is as jargon-free as possible and to capture the excitement and wonder of our planet. We hope we achieve these goals and, if you are sitting comfortably, we will begin.

Earth Science Today has visual cues to help students really see the interactions described in the text. **Visual Summaries** close each part and cover all the key interactions within that group of chapters. Two **Process in Action** icons visually emphasize the interconnectedness of the chapters and the information within them. Interactions between chapters are shown by the large icon. Those within chapters are marked by the smaller icon.

Each part opens with the large **Process in Action** icon. This icon helps to reinforce the nature of the interactions between the Earth's surface reservoirs—the solid Earth, hydrosphere, atmosphere, and biosphere. Each of the Earth's reservoirs is allocated a segment within the large **Process in Action** icon. The solid Earth is positioned in the center, reflecting its influence on all the reservoirs. Other reservoirs are positioned along the periphery, reflecting their connections to the solid Earth, the Sun, and each other. Textures within each reservoir's segment mark the heads in all the chapters, giving the student another visual cue to where they are.

The small **Process in Action** icon positioned throughout the chapters (and seen at the beginning of this box) is designed to send a signal that two or more reservoirs are interacting. When they come across the icon, students can then ask themselves what reservoirs are working together to make this particular event occur.

Visual Summaries are designed to help the student make a visual connection between the concepts. The **Visual Summaries** also offer the student a way to conceptualize the integration, movement, and change described in each part. They illustrate the "big picture" by showing the interdependence of the processes, concepts, and principles involved in Earth science.

Damian Nance and Brendan Murphy with their families

About the Authors

Brendan Murphy is a Professor of Geology at St. Francis Xavier University in Nova Scotia, Canada. His research interests are related to the many processes involved in mountain building. In order to understand these processes, he has delved into fields such as the deformation and recrystallization of rocks; the chemical signature of igneous, metamorphic, and sedimentary rocks; and the use of isotopic systems to constrain plate tectonic events. In 1993, he received the Presidents Research Award at St. Francis Xavier University and, in 1995, the Distinguished Service Award of the Atlantic Geoscience Society. Murphy is from Birr, Ireland, famous for the Birr Telescope. Indeed, it was on a high school trip to view this telescope that his interest in the sciences was ignited. He obtained his B.Sc. from University College, Dublin, in 1975 and moved to Canada where he completed his M.Sc. at Acadia University in 1977, before moving to McGill University, where he earned his Ph.D degree in 1982. He has taught at St. Francis Xavier University since 1982, including a term as chair of the Department of Geology from 1989 to 1995. He is married to Cindy Murphy, and they have two children—Orla and Declan.

Damian Nance is a Professor of Geology at Ohio University and is currently the chair of the Department of Geological Sciences. His research interests have ranged from the use of Gulf Coast salt domes as repositories for nuclear waste to the recent history of glaciation on Mount Olympus, Greece, and its bearing on climate change. But the focus of much of his research has centered on the early evolution of the Appalachian Mountains and on the global-scale plate tectonic processes responsible for their development. In 1991, he received his college's Outstanding Teaching Award and has twice been nominated for the university's Outstanding Graduate Faculty Award. Nance is from Cornwall, England, obtained his B.Sc. degree from the University of Leicester in 1972, and completed his PhD at the University of Cambridge in 1978. He moved to Canada in 1976 and taught at St. Francis Xavier University before moving to Ohio University in 1980. He is married to Rita Nance, and they have three children—André, Sarah, and Christopher.

Murphy and Nance began collaborating in 1985 because of their shared interest in the origin of ancient mountain belts preserved in the Appalachians. They soon realized that the evolution of these belts reflected global-scale plate tectonic processes, in which they were both keenly interested. In addition to writing textbooks, they have each authored over 75 scientific papers, 30 of which they coauthored. Nance has also co-edited a 400-page volume on Appalachian geology for the Geological Society of America and has published works on another of his interests, that of mining history and the use of steam power in the mining industry.

Both having taught at St. Francis Xavier University, Nance teases Murphy that he had to leave so that Murphy could get a job, to which Murphy responds that he refused to take the job until Nance had left!

1

Introduction

1.1 Earth and Its Environment

Since the dawn of civilization, humans have sought to understand the world around them and the universe beyond. Our thirst for knowledge is almost insatiable. Slowly, through meticulous observations and ingenious insights, has come understanding. With this understanding comes a greater appreciation for the beauty of planet Earth, its place in the solar system and the cosmos, and the processes that resulted in its evolution from a most inhospitable beginning nearly 4.6 billion years ago to the marvelous life-sustaining planet of today. We can now probe its inner depths and behold the manner of its creation in the death of distant stars. We marvel at nature's awesome powers in sculpting our varied landscape, and have come to realize that a serene landscape can mask an inner turmoil that periodically unleashes earthquakes and volcanic eruptions at the planet's surface.

Advances in our understanding of the Earth have generally been accompanied by society's attempts to use this knowledge for what is euphemistically called progress. Those born in the early part of the twentieth century have witnessed such progress at a bewildering rate. In their youth, the most reliable form of communication was the pony express. They have since witnessed the development of radio, television, supersonic travel, space exploration, personal computers, and multimedia. Now, gigabytes of technical information can be beamed around the globe in an instant.

The Earth also provides us with bountiful resources that we use so widely in our day-to-day lives that we often take them for granted. There's an old saying: "If you don't grow it, you have to mine it." Look around your room at all the inanimate objects, each of which has been harvested from the Earth's resources. For example, all of our kitchen appliances together use close to 100 raw materials that were directly or indirectly extracted from the rocks and minerals of the Earth (Fig. 1.1).

However, the pace of technological change and the explosion in global population have placed a strain on these resources, and on the energy supply needed to fuel economic development. Most of this energy supply is extracted from the Earth in the form of fossil fuels (coal, oil, and natural gas), each of which is a finite resource. For example, it has been estimated that we will run out of oil by the middle of the next century.

The use of these resources may come at a higher price than we realize. Our heightened awareness of the world around us has led to an unprecedented concern about the fate of our planet. Over the past 20 years, evidence has been mounting that, for the first time in its 4.6-billion-year history, the Earth may be losing the natural control of its environment because of the activities of humankind. Many scientists view the environmental modification these activities may have caused to be damaging, and perhaps irreversible. If this view is correct, we are faced with the necessity of changing the way we live and fundamentally restructuring our traditional economic base; otherwise, our planet may no longer be able to support our existence. Other scientists are skeptical about the ability of humankind to damage a planet that has survived many natural

Figure 1.1
The most commonplace objects utilize the Earth's resources. Kitchen appliances use close to 100 raw materials that were extracted from the Earth. (Photo Heather Dutton)

catastrophes throughout its history, and point out that the Earth has many buffers or built-in controls to dampen the effects of human activities. Furthermore, they suggest that much of what is called global change can be explained as natural fluctuation.

With such conflicting views, it is difficult for our elected representatives, most of whom have little scientific background, to formulate environmental policy that will not needlessly disadvantage their constituents' ability to compete nationally and internationally. At the same time, economic forces demand more of Earth's resources. Raw materials such as minerals, fossil fuels, and water contribute new wealth to local and national economies in the form of unprocessed goods that are directly derived from these materials. The processing of these raw materials and the manufacture of new products creates additional jobs and wealth. Many communities and nations depend on the exploitation of these resources and the employment this exploitation creates.

If we are to practice responsible management of our planet, it is crucial that we distinguish between the kind of global change induced by human activity and that which is part of a natural cycle. For example, is industrial development and the consumption of energy responsible for global warming, ozone depletion, and pollution of our waters by acid rain? Or are these human activities dwarfed by natural processes such as volcanic eruptions and changing patterns in ocean circulation, both of which bring about dramatic changes in the environment? What kinds of human development can be sustained by our planet, and is sustainable development even possible? Scientists are only beginning to study these problems.

In this context, the study of the record of Earth's evolution offers a unique and clear perspective. During this evolution, natural processes unaffected by the influences of the modern era are recorded in the rocks, sediments, and soils they produced. By understanding these processes, we can more clearly recognize the signals that result from human interference with the natural order, and so formulate policies to deal with this interference.

The Earth has its own pulse, its own natural rhythms, many of which profoundly influence phenomena as all-embracing as the motion and growth of continents, and the birth and destruction of oceans. These rhythms relate to the interaction between the solid Earth beneath our feet, its watery surface (or hydrosphere), its atmosphere, and the realm of life (or biosphere). These rhythms were in operation long before the modern era of industrialization, and may extend back to the very earliest stages of the Earth's development. However, we must be able to recognize them if human interference with the environment is to be distinguished from the natural order of things. For example, if a familiar song is cut off before it ends, we can usually continue to hum the tune because we recognize its rhythmic beat. In a similar way, if we are to gain an understanding of the modern Earth, we must first recognize the rhythms and beats of its geologic past so that we can carry its song into the future.

The accuracy of our predictions depends on how well we have learned the Earth's past rhythms. It may seem that 200 years of industrialization is a trivially short interval in a song lasting 4.6 billion years. However, we seek distortions in this song, and these distortions are instantly recognizable just as they are in a recording of familiar music. Therefore, if human activity truly represents a distortion in the evolution of our planet, it can be readily identified. However, this is possible if, and only if, we know enough about the Earth's history to recognize its natural pulse.

We scientists have long known that the Earth has, at various times during its evolution, witnessed the creation and extinction of lifeforms, as well as the formation of global-scale deserts and ice ages. We now believe that all these processes may be interconnected. It is becoming increasingly clear that the Earth's evolution is driven by cycles that involve a delicate interaction and feedback between biological and geological phenomena. These cycles may control long-term fluctuations in climate, evolution, and the environment. What we need in the Earth sciences to determine is whether human activity enhances, interferes with, overrides, or short-circuits these processes, and if so, how.

In this book, many topics relating to the science of the Earth and its environment will be examined. Its goal is to improve our understanding of the planet we live on. For unless we understand the Earth, how can we assess the effect of human development on it? For example, why do volcanoes occur, and what effects do they have on global climate? What fluctuations have occurred in global temperature over the last billion years and what controlled them? What are the effects of the ever-changing patterns of the Earth's orbit around the Sun? Did global warming begin with the retreat of the ice sheets thousands of years ago, or is it a more recent phenomenon in which we have played a role? Is our climate related to the chemistry of ocean water? And why is ocean water salty when the rivers that drain into the oceans are not? Where did ocean water come from, and why are we the only planet in the solar system with abundant liquid water? Why does the Earth have an atmosphere unlike any in the solar system and how did it evolve? What is ozone and how does it form in our atmosphere? How do fossil fuels such as oil, coal, and natural gas form, and did their formation affect our climate? If it did, will burning fossil fuels have the opposite climatic effect? Why are these and other natural resources located in some places and not in others? The answers to these and many other questions are fundamental to modern society. Just as successful managers must understand what they are managing, so we must understand the Earth if we are to live in harmony with it, and so survive upon it.

If tomorrow's decision makers are to make the best choices for us and our planet, they must have some understanding of global processes. They must also have an appreciation for the processes that have allowed the Earth to evolve over billions of years from cosmic dust to the life-sustaining planet that we take for granted today.

1.2 Earth Science: A Study of Earth Systems

If we had to sum up the driving force behind the Earth's 4.6 billion years of evolution in a single sentence, it would be this: The Earth is cooling down. The Earth loses more heat from its surface than it absorbs from the Sun. This is achieved by the flow of heat (known as **heat flow**) from the hot core of our planet to its cool surface, where the heat escapes (Fig. 1.2). Even today, heat is escaping everywhere at the Earth's surface. Throughout its history, nearly all processes affecting the Earth on

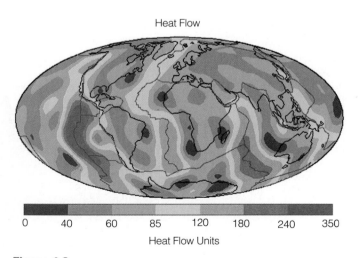

Heat Flow

| 0 | 40 | 60 | 85 | 120 | 180 | 240 | 350 |

Heat Flow Units

Figure 1.2

Heat flow measurements on the Earth's surface depict the escape of heat from the Earth's interior into the atmosphere. Note how they vary significantly from one location to another. Heat escapes far more efficiently beneath the oceans, indicating that the Earth's crust is much hotter in these regions.

a global scale, such as the formation of its rocks and minerals, the evolution of life, the origin and composition of the atmosphere, continents and oceans, and the development of our present-day environment, are all related to this cooling. On the largest scale, the Earth's continents, oceans, and atmosphere have, in effect, been exhumed from the Earth's interior just as the froth in a glass of soda-pop is exhumed to the surface as it is poured.

The heat flow escaping into the atmosphere from the Earth's surface can be measured at any location on Earth regardless of how geologically tranquil the locality may appear to be. However, the heat is escaping more efficiently from some parts of the surface than it is from others. This is illustrated in Fig. 1.2, which shows that the amount of heat flow varies from one location to another. The figure reveals an obvious correlation between regions of high heat flow and the Earth's crust beneath the oceans. This indicates that the Earth is unusually hot in these regions and, as with a hot radiator in a room, heat flow is highest where the surface is hottest.

1.2.1 The Earth's Reservoirs

Planet Earth consists of four main components or **reservoirs** (see the photo at the beginning of this chapter).

- The **solid Earth,** which comprises the Earth beneath our feet with its rocky outer layer or **lithosphere** and its hot, mushy interior.

- The **hydrosphere,** which comprises the water of our planet, whether it is in oceans, lakes, or streams, frozen in ice and snow, or trapped underground in soils and rock fractures.
- The **atmosphere,** which is Earth's gaseous envelope and the air we breathe. It is mostly made up of nitrogen and oxygen, with smaller but important concentrations of other gases such as carbon dioxide, water vapor, and ozone.
- The **biosphere,** which is the realm of life, comprising plants and animals that are either living or in the process of decay.

Just as artificial reservoirs are linked to important stream systems, so too are these natural reservoirs linked together at the Earth's surface. At the surface, these reservoirs also interact with radiant energy from the Sun. Some of these linkages are obvious; the most commonly cited is the hydrologic cycle (Fig. 1.3), which describes the motion of water at or near the Earth's surface. In this cycle, radiant energy from the Sun strikes the ocean surface and initiates the evaporation of ocean water into the atmosphere. Winds then blow this moisture onto land, where it falls as rain or snow and stimulates plant and animal life. Most of the water eventually drains into major stream systems and so is returned to the oceans to close the cycle.

Other examples of linkages between the Earth's surface reservoirs are less obvious, but are equally important. For example, lavas that erupt from volcanoes bring new chemicals from the interior of the Earth to the surface. As the lavas cool, they solidify to form part of the solid Earth. Because of

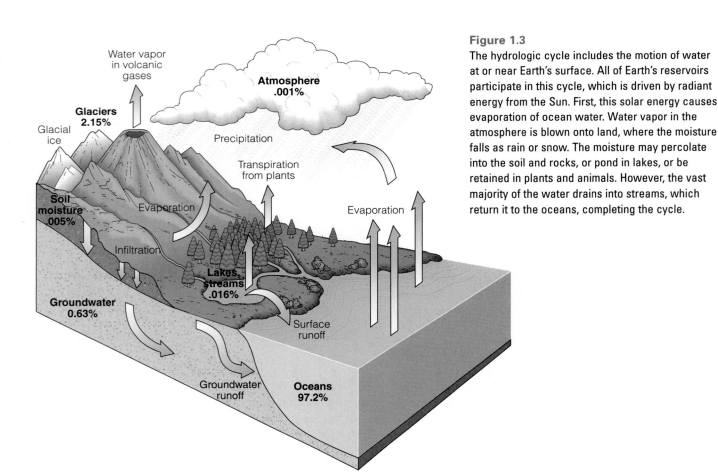

Figure 1.3
The hydrologic cycle includes the motion of water at or near Earth's surface. All of Earth's reservoirs participate in this cycle, which is driven by radiant energy from the Sun. First, this solar energy causes evaporation of ocean water. Water vapor in the atmosphere is blown onto land, where the moisture falls as rain or snow. The moisture may percolate into the soil and rocks, or pond in lakes, or be retained in plants and animals. However, the vast majority of the water drains into streams, which return it to the oceans, completing the cycle.

Figure 1.4

Fields on the slopes of a volcano in Bali, Indonesia, are terraced to retard erosion and retain nutrients. Periodically replenished by volcanic ash rich in elements such as calcium and phosphorus that are essential to plant growth, Balinese soils produce some of the most productive croplands in the world. (Photo Damian Nance)

this chemical enrichment, soils in the vicinity of volcanoes are often very fertile, and support the biosphere by promoting plant life (Fig. 1.4). Exposure of the lavas and their soils to the air and rain causes them to weather and disintegrate, and they are eroded and transported to the sea. This process brings vital chemical nutrients such as calcium to the sea, and also helps marine life to flourish. The same volcano also vents gases that, over the course of Earth history, have profoundly influenced the composition of our atmosphere. One of these gases is water vapor, which cools and condenses to form rain and may ultimately drain to the sea to add to the water in our oceans. Circulation in the atmosphere and the oceans is driven by the Sun's radiant energy. In this way, the products of volcanic eruptions link all four of the Earth's surface reservoirs and are distributed among them by the radiant energy of the Sun.

Besides volcanoes Earth scientists know of few other mechanisms that introduce new chemistry to the surface of the Earth. As our bones, soft tissue, and blood are made up of chemicals, it follows that many of the molecules in our bodies are likely to have arrived at the Earth's surface from its interior by way of a volcano. When or where this happened we will never know. Indeed, even adjacent molecules in our bodies are most unlikely to have arrived at the Earth's surface at the same time, or even from the same volcano. So if you feel a little scrambled today, there's a good, natural reason for it!

More importantly, the above example shows how a geological event results in chemical constituents being transport-ed in response to the laws of physics, first from the interior of the Earth to the exterior, and then across the Earth's surface to the oceans. As this happens, the biosphere flourishes. The natural world does not compartmentalize the scientific disciplines the way university and college courses do. Our ability to understand our planet and its environment, and our willingness to live in harmony with nature, may therefore depend on our ability to break down the artificial barriers between the traditional sciences. Only then can we examine **Earth systems science,** as the study of the interactions between the Earth's surface reservoirs is known.

Figure 1.5 summarizes the nature of these interactions and the connections (shown by arrows) between them. The solid Earth is positioned in the center because all of the reservoirs are directly influenced by the movement of chemical constituents from the interior of the Earth to its surface. The other reservoirs are positioned along the periphery because they are not only connected to the solid Earth and each other, but they also absorb and process radiant energy from the Sun. In contrast, the effects of solar activity on the solid Earth are much more limited. Throughout the rest of the book, a process icon will mark places where the most important processes or interactions between the reservoirs occur.

Other cycles can now be viewed within this global perspective. For example, the hydrologic cycle of Fig. 1.3 is merely a portion, albeit an important one, of Earth system cycles.

1.3 A Revolution in the Earth Sciences

Geology, the geological sciences, the geosciences, and *Earth science* are all terms used, sometimes interchangeably, for the study of planet Earth. As the term *geology* is derived from the Greek words for "the science of the Earth," you might think that these terms are synonymous and that their usage is redundant.

This apparent confusion masks a revolution that has occurred in the study of our planet, a revolution that is not only related to recent technological advances, but is also one of emphasis. When we, the authors, were student geologists (which was not that long ago), one of the first principles we were taught was that "the present is the key to the past." This means that those geologic processes affecting the Earth in the past are best understood by studying those shaping the Earth today (see Chapter 3). Therefore, for every ancient environment there is a similar modern environment that can be used to aid in its interpretation. Although this approach is still very important, our current environmental problems have forced a shift in emphasis. Our knowledge and understanding of Earth's evolution have progressed to the point where Earth scientists can detect the Earth's natural rhythms of global change over geologic time. As these rhythms and the processes responsible for them can be projected into the future, they can also be used to predict the fate of our planet. Therefore, in a sense, we have reversed the principle we were taught as students. It is the past that has become the key to the present and future.

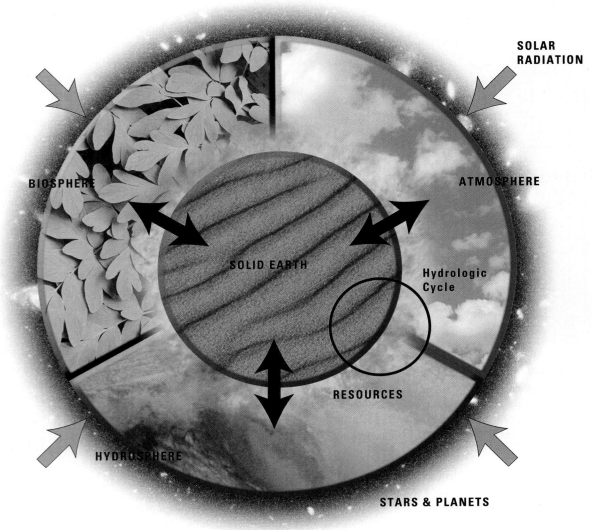

Figure 1.5
The Earth's surface reservoirs interact and mutually influence one another. Processes in the interior of the solid Earth have led to the formation of its rocky outer layer (lithosphere), air (atmosphere), and water (hydrosphere). These reservoirs, in turn, provide the basic prerequisites for life (biosphere). The solid Earth is placed in the center because it has been the main source for the chemicals of the atmosphere and hydrosphere, and has provided nutrients for the biosphere. Arrows to the reservoirs go in both directions, depicting the feedback between them. Interactions between the atmosphere, hydrosphere, and biosphere are also greatly influenced by solar radiation, shown in the outer part of the diagram. The position of the hydrologic cycle of Figure 1.3 is shown from this global perspective by the circle in the bottom right of the diagram.

According to the definitions found in most dictionaries, geology employs physical, chemical, biological, and mathematical methods to aid in the understanding of the Earth, its materials, behavior, history, and environment. A geologist's work consequently involves the study, mapping, and interpretation of the Earth's evolution as revealed by its rocks, minerals, and fossils, and by its atmosphere and oceans, as well as the exploration and development of its valuable resources and the evaluation of the environmental implications of these activities. The physical forces that have acted on the Earth, the processes that govern the chemistry of its minerals and rocks, and the biology of its ancient life, as revealed by fossils, are equally vital elements in our understanding of the Earth's evolution. However, as we shall see, many clues to the Earth's early history have been lost. Just as a detective has difficulty in evaluating the evidence in an old criminal case, so the geologist has difficulty evaluating the Earth's early history when the evidence has been obliterated by younger geological events. To remedy this, the geologist must rely on indirect and circumstantial evidence, much of which comes from other celestial bodies, including the Moon, the terrestrial planets

Mercury, Mars and Venus, and meteorites that are thought to represent the remnants of the early solar system.

Until very recently, much of the evidence related to the evolution of the Earth was obtained from the rocks and minerals that occur at or near the Earth's surface. As a result, the term *geology* became synonymous with the study of rocks and minerals despite the wide-ranging scope of its definition. However, recent technological advances have allowed us to model the chemical, physical, and organic features that these rocks and minerals preserve, and has given us insight into both ancient and modern atmospheres and oceans. Our voyages into space have also allowed us to view the Earth in the context of its neighbors within the solar system.

As a result of these developments, geologists such as ourselves now know that the study of the Earth cannot be compartmentalized into standalone subdisciplines for our own convenience. For example, the volcanic activity that produces new rocks at the Earth's surface has also produced our atmosphere, and has greatly influenced the composition of our oceans. Even a single volcanic eruption can have a dramatic effect. For example, in 1991 the eruption of Mount Pinatubo in the Philippines (Fig. 1.6a) rocketed particles into the atmosphere to heights of up to 30 km (18.6 miles). In only 21 days, the ejected mixtures of gases, or aerosols, encircled the Earth (Fig. 1.6b). These aerosols included an estimated 20 million tons of sulfur dioxide, a gas that combines with rain droplets in the atmosphere to produce acid rain. This is comparable with the entire annual industrial output of sulfur dioxide in the United States. If this is the effect of only a single volcanic eruption, imagine the profound effect on the composition of the atmosphere of all the eruptions that have occurred throughout Earth's history.

Recent technological advances, such as those responsible for the data shown in Figure 1.6b, have brought to the forefront a wide range of important subdisciplines in the study of the Earth, such as geochemistry, geophysics, and planetary geology. As a result, it has become necessary to look beyond rocks and minerals, and synthesize information from all of these sources, if we are to understand our planet more fully. In order to emphasize this new reality, the subdisciplines that relate to the solid Earth are collectively called the geological sciences, or the geosciences for short. The term *Earth science* has an even wider scope, and is analogous in its use to the term *life sciences*. It embraces all disciplines that relate to the modern and ancient Earth, including those that pertain to its liquid and gaseous envelopes, such as oceanography, meteorology, and climatology, and disciplines such as agronomy and soil science.

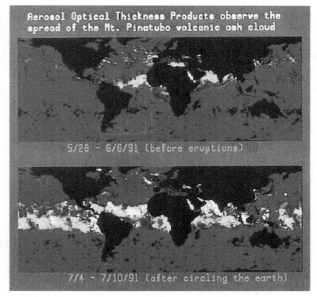

Figure 1.6
(a) The eruption of Mount Pinatubo (left) in the Philippines on June 12, 1991, was the second largest of the twentieth century. It is estimated to have ejected more than 5 cubic kilometers (1.2 cubic miles) of material, five times more than the eruption of Mount St. Helens in 1980. Ash, gas, and aerosols (gas mixtures) were ejected into the atmosphere to heights of up to 30 km (18.6 miles).
(b) Distribution of the aerosols before (top) and after (bottom) the eruption. Within 21 days the ejected aerosols became distributed into a band that encircled the Earth. The ejected gases included an estimated 20 million tons of sulfur dioxide, a gas that combines with rain droplets in the atmosphere to produce acid rain.

1.4 Earth Science and the Scientific Method

The goal of science is to understand the natural world. The scientist works at the frontiers of human knowledge to provide the fundamental cornerstones on which modern society is built. Indeed, the modern world is testament to the practical applications of this basic knowledge.

Earth science, like all sciences, advances by a methodical approach known as the **scientific method.** A scientist, or team of scientists, identifies a gap in our understanding of a given process. In order to correct this situation, the first stage involves the collection of data. This basic information is obtained by measurement, experiment, and observation relevant to the problem. Then a **hypothesis** is formulated as a tentative explanation of the data. The relationship between global warming and the level of greenhouse gases such as carbon dioxide in the atmosphere is a recent example of such an hypothesis. The hypothesis must advance the simplest possible explanation of the available data. However, by definition, it is unproven and must be rigorously tested. To be effective, the test must distinguish between rival hypotheses, and so must be one that could disprove the proposed hypothesis. Similarly, an unbeaten horse must pass the test of the Kentucky Derby before it will be recognized as a true champion. The recent proposal that there may once have been life on Mars is an example of an hypothesis that is now undergoing rigorous testing. This hypothesis has focused the attention of the scientific world and has heightened public awareness of the importance of the Pathfinder mission to Mars. However, even the proponents of this hypothesis do not claim it to be foolproof; they only believe that it is the simplest explanation of their current observations.

A hypothesis that passes a number of tests where rival hypotheses fail becomes a **theory,** which amounts to a vote of confidence by the scientific community. However, theories are not fact, and are still subject to further testing that may possibly disprove them. But, as a theory passes more and more tests, it is regarded with a higher degree of confidence. As we will discover, one such theory is that of plate tectonics, which forms the cornerstone of modern Earth science and provides an explanation for a wide variety of natural phenomena such as the cause of volcanic eruptions and earthquakes, and the origin of mountains, continents, and oceans.

After many tests and numerous observations, measurements, and experiments, a group of scientific concepts may be formulated into a **law.** This is an expression or statement that accounts for a natural process or processes and from which there is no known deviation. Although a law must provide an exact description of a process, it need not provide an explanation for it. For example, when Johannes Kepler proposed the **laws of planetary motion** in the early seventeenth century, he could not provide a satisfactory explanation for them. In fact, there was no explanation for planetary motion until the late seventeenth century, when Sir Isaac Newton proposed the law of universal gravitation.[1]

It is very important that all hypotheses, theories, and laws be subject to reexamination as new data is acquired. The results of these reexaminations are often communicated to the scientific community at conferences and in publications. The thousands of new scientific papers published annually in Earth science alone indicate that even our basic knowledge of planet Earth is increasing. However, much still remains to be done. Just as the first athlete to break the 4-minute mile extended the physical capabilities of humankind and led others to strive for new records, so science in its relentless quest for knowledge, advances the frontiers of our understanding and continues to seek new ones.

It should be noted that the scientific method has its critics. Some point out that its application retards the development of science, that the constant probing and testing is too pedestrian for the rapidly changing modern world. Others point out that it has fundamentally failed to predict the environmental consequences of certain scientific findings.

While acknowledging these criticisms, it is important to realize that the foundations of modern science were built on the application of the scientific method. This connection will be apparent in several parts of this book. In addition, scientists need a philosophical approach to science similar to the one provided by the scientific method in order to communicate effectively.

1.5 Earth Science and Society

The Earth is a dynamic and exciting planet, and has continuously evolved throughout its history. During this time, entire oceans were created and destroyed, mountains were uplifted and eroded away, catastrophic meteorite impacts, volcanic eruptions, and earthquakes occurred repeatedly, climate change produced both global greenhouses and global icehouses, and all these events were interspersed with long periods of relative calm. In addition, each of these events greatly influenced the evolution and extinction of living things. It is the task of the Earth scientist to try and understand the origin, significance, and order of these events so that we may understand more fully our planet and the environmental challenges that confront us.

This understanding is vital if we are to locate, use, and harness Earth's resources. As has been discussed, almost all of the everyday goods that society now requires are derived from Earth materials. This fact, together with the rapidly growing population of many developing nations, puts the Earth scientist in the forefront of trying to provide society with the raw materials on which to build its future, while at the same time

---

[1]The term *scientific method* was not used in the time of Kepler and Newton. Since their deaths, however, their laws have been scrutinized for centuries by scientists who applied this method in their research.

addressing the environmental challenges we are confronted with in the use of the Earth's resources.

With this role comes great responsibility. The relevance of Earth science to resource management and environmental impact assessment cannot be overemphasized. The increasing demands of society for energy and raw materials are now viewed in the context of sustainable development and environmental stewardship. An understanding of Earth systems places the Earth scientist in a unique position to evaluate both the potential environmental impact posed by the development of natural resources and the natural hazards associated with geological phenomena. As a result, Earth scientists play a vital role in balancing some of the conflicting economic pressures that society faces today.

Today's Earth scientists (see Fig. 1.7) make their fundamental contribution to society by:

• Developing an understanding of the Earth's evolution throughout its history, as well as the linkages between its solid crust, its atmosphere, its water, and its life
• The investigation, evaluation, and supply of natural resources, including minerals, fossil fuels, water, and air quality
• The assessment of the environmental effects of resource exploitation and human development on a local to global scale
• The prediction and mitigation of natural hazards, whether from geologic processes such as landslides, earthquakes, and volcanic eruptions, or from severe weather events such as floods, tsunamis, and droughts, or from toxic Earth materials such as asbestos and radon

It is important to realize that many of these roles have come about only recently as a result of society's newly heightened sense of environmental responsibility. As society's demands and concerns evolve, the role of the Earth scientist in society evolves as well.

Because of the interdisciplinary nature of the subject matter, Earth scientists must be educated in the important principles of all the basic sciences. A university degree in Earth science can provide a stepping stone to a wide range of career opportunities. Traditionally, these are considered to be in the resource and environmental industries, in government agencies, and in schools and universities. However, career opportunities also exist in other areas. As an example, financial institutions and law firms increasingly recognize that employees with a knowledge of the Earth's processes offer important insights that affect investments, economic planning, and the responsible development of these resources.

1.6 Goals for the Rest of the Book

As we have learned, the study of many important Earth processes transcends the artificial boundaries between the sciences, and we too must transcend these boundaries if we wish to understand them. Several fundamental goals have guided the writing of this textbook, the material covered, and the sequence in which the material is covered. The goals are:

• To help students understand the nature of Earth systems in the modern world
• To help students see how these interactions have guided Earth's evolution
• To demonstrate that an understanding of Earth systems and their evolution provides important insights into environmental problems, and the necessary background for responsible resource development

The record of Earth's history is fragmentary. Nevertheless, the "geodetective" can piece together sufficient clues, mainly from the rock record, to provide a conceptual overview of Earth processes. A portion of the ancient lithosphere is preserved as rocks up to 4 billion years old. Some of the ancient rocks preserve fossils, and so provide a partial record of the ancient biosphere. Many of these rocks record features similar to modern marine sediments. By analogy, they can be assumed to represent vestiges of ancient marine environments and therefore provide a record of ancient marine processes. In this way the geoscientist gains insights into the history of the hydrosphere. Direct evidence pertaining to the evolution of the atmosphere is more difficult to find. However, indirect evidence, such as that derived from the fossil record and from the other terrestrial planets, provides some remarkably good guidelines.

Our planet is our main benefactor. If humans are to live on the Earth and exploit its resources, each of us has a responsibility to understand the planet and recognize and appreciate both its strengths and its delicate fragility. We hope that when you reach the end of this book you will have a firmer understanding and broader appreciation of the planet on which you live, and will have developed an appetite for learning more.

Figure 1.7
Earth scientists today are at the forefront of developing an understanding of Earth processes and how we can best use and manage its resources.

PART 1

The Solid Earth

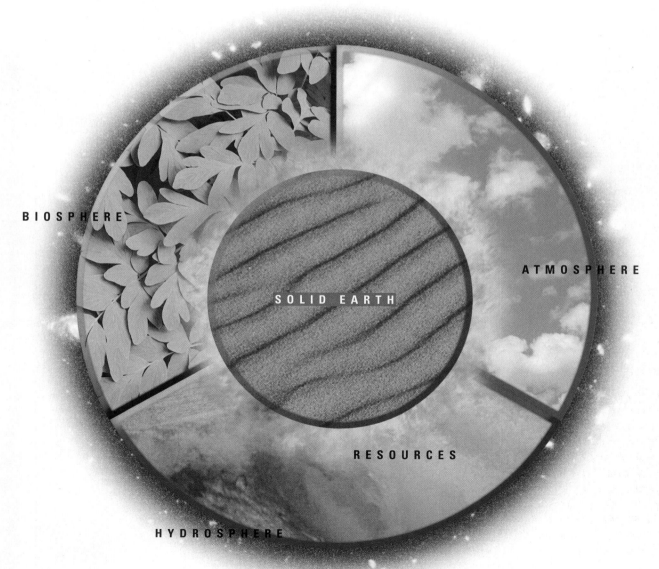

BIOSPHERE

ATMOSPHERE

SOLID EARTH

RESOURCES

HYDROSPHERE

STARS & PLANETS

CHAPTER 2

Earth Materials: Minerals and Rocks

If we are to understand the interactions between the solid Earth and the Earth's other reservoirs, the hydrosphere, atmosphere, and biosphere, we must start by examining the basic ingredients of the solid Earth called minerals and rocks. Minerals and rocks are the building blocks of the Earth's continents and ocean floors, and form the basic materials used to understand the evolution of the planet through geologic time. The composition and distribution of minerals and rocks also tell us much about the inner workings of the Earth, the origin of its lithosphere, and the nature of its outermost rocky layer called the crust. In this chapter, we investigate Earth materials by examining some of the more common minerals and the rocks they combine to produce.

To understand minerals, which are naturally occurring chemical compounds, we first turn to the chemical elements that make up minerals and to the various chemical bonds that hold them together in an orderly arrangement. You will learn that minerals are crystalline solids with a specific internal structure and formula, and that they have a characteristic set of physical properties that are used in their identification.

Although they number in the thousands, you will discover that less than 1% of known minerals occur commonly in rocks. As the most abundant elements in the Earth's crust are oxygen and silicon, the most abundant rock-forming minerals contain both of these elements and are called silicates. Variations in the way oxygen and silicon bond together, and the way they are bonded to other elements, gives rise to large numbers of silicate minerals. The properties of these silicates depend on their composition and structure, and each forms only under suitable chemical and physical conditions. Because of this, a great deal can be learned about the environment in which a rock formed from its constituent silicate minerals.

You will learn that rocks are aggregates of mineral grains, and fall into three broad categories based on the way they form. Igneous rocks represent the products of cooling molten rock, or magma. Sedimentary rocks are the product of surface processes such as erosion and deposition. Metamorphic rocks form within the crust where the pressures and temperatures are high. Within these three groups, rocks are further classified according to their grain size and composition, which is reflected in the minerals they contain.

The grain size of igneous rocks is primarily a function of the speed at which they cool. Igneous rocks that cool slowly within the Earth's crust have large mineral grains, whereas those that cool rapidly on the Earth's surface are fine-grained or even glassy. The composition of igneous rocks is measured by the amount of silicon and oxygen (or silica) they contain, and reveals clues to their origin. You will discover that the composition of a molten rock can change as it cools, and has a dramatic effect on its physical properties. This influence on physical properties will be found to have important implications for the nature and prediction of volcanic eruptions.

You will find that most sedimentary rocks are the product of erosion and deposition, and are named according to the size of their component grains. Yet others result from chemical and biochemical processes, and are named on the basis of their composition. Importantly, we will find that the attributes of sedimentary rocks vary with the environment in which they were deposited. For example, those found in continental environments differ from those deposited in marine or coastal environments. Because of this, we can learn much about the history of the Earth by studying the depositional environments recorded in ancient sedimentary rocks.

Finally we turn to metamorphic rocks, which form when minerals become unstable and react to form new minerals. This occurs below the Earth's surface where changing conditions can cause such instability. The process of transformation is called metamorphism. Metamorphic rocks form in the hidden interior of mountains, in areas adjacent to bodies of molten rock, and along fractures in the Earth's crust. Changing conditions in the Earth's interior are often accompanied by forces that deform the rocks, causing them to flow or break. Because of this, metamorphism is often accompanied by deformation, which imparts a variety of textures to the rock that can be used to classify them. Metamorphic rocks that lack such textures are named on the basis of their composition.

The minerals present in metamorphic rocks depend largely on the conditions of pressure and temperature during metamorphism. The study of metamorphic rocks enables us to learn much about the processes and conditions that exist within the Earth's crust.

The information in this chapter will give you an understanding of minerals and rocks. You will then be ready to explore the ways in which geologists have used them to determined the nature and history of the solid Earth.

2.1 Minerals

The solid Earth is made of rocks and rocks consist of consolidated mixtures of mineral grains in much the same way that stained-glass windows are mosaics of pieces of colored glass. Therefore, minerals are the basic ingredients of rocks and it is with minerals that our examination of the solid Earth should start.

In this section, we will learn that minerals are chemical compounds and that they are built up of chemical elements linked together in scaffoldinglike frameworks by a variety of chemical bonds. We will find that their atoms are arranged in an orderly fashion and that, because of this, minerals have specific chemical compositions. If they are left free to grow, minerals make the beautiful geometric forms we call crystals. We will also learn that most minerals have characteristic crystal shapes that can be used to identify them, along with other physical properties such as color and hardness. We will then see how minerals combine to make rocks, and so form the basic ingredients of the solid Earth and the basis of many of our natural resources.

2.1.1 Elements and Bonding: How Minerals are Built

The building blocks of all minerals, and indeed all chemical compounds, are the chemical **elements.** Elements, either singly or in combination, are the fundamental substance of all matter and each has a name and a chemical symbol. Over 100 elements are known to exist but only 92 occur naturally. The remainder are produced artificially only in nuclear reactions. Some minerals such as gold and sulfur are made from a single element, but most are combinations of elements that form stable chemical compounds. To understand how this occurs, we must first look to the atoms that comprise the elements.

Atoms

Each chemical element is made up of fundamental units known as **atoms.** Atoms vary in size depending on the element, but are typically on the order of 0.0000000001 (10^{-10}) m across. In other words, 10 billion atoms lined up side-by-side would form a chain about one meter in length. Each atom consists of a central unit, called a **nucleus,** where most of the mass (or substance) of the atom is concentrated. It also consists of a cloud of orbiting particles called **electrons,** which have almost no mass and carry a negative electrical charge. Electrons revolve around the central nucleus in a series of orbits that reflect differences in their energy levels. The nucleus contains two types of particles: **protons,** which carry a positive electric charge, and **neutrons,** which are of almost identical mass but carry no charge. All elements are defined by the number of protons contained in their nucleus. In this way, all atoms of the same element contain the same number of protons and this number is different from that of all other elements. For example, the simplest of the elements, hydrogen, has a single proton in its atomic nucleus (Fig. 2.1a), and is therefore assigned an **atomic number** of 1 in the periodic table of elements. Oxygen, on the other hand, has eight protons in its nucleus (Fig. 2.1b), whereas iron has twenty-six. Their atomic numbers are 8 and 26, respectively. Uranium, the heaviest of all naturally occurring elements, has an atomic number of 92.

Types of Bonds

The nucleus of an atom is tiny, about 100,000 times smaller than the atom itself, but it is a relatively stable part of an atom. The orbiting electrons are more loosely held within the atomic structure. As a result, elements such as metals have a tendency to lose electrons whereas others tend to accept them. Yet others share electrons with neighboring atoms. This mobility of electrons is the essence of bonding and allows elements to combine with their neighbors to form stable combinations of atoms called **molecules.** In this way, one atom of carbon (C) will combine with two atoms of oxygen (O) to form a single molecule of carbon dioxide (CO_2).

Atoms combine to form molecules by using one of four types of bonds. These four types are: ionic bonds, covalent bonds, metallic bonds, and van der Waals forces.

Ionic Bonds. **Ionic bonds** are the simplest of the four and involve the transfer of an electron from one atom to another. Metals such as sodium, potassium, and calcium have only a few electrons in the outer orbits of their atoms, and these are detached with relative ease. On the other hand, nonmetals such as chlorine and fluorine have several electrons in the outer orbits of their atoms but can readily accept more. Consequently, if a metal with a tendency to lose an electron such as sodium (Na) encounters a nonmetal with an tendency to accept one such as chlorine (Cl), the two elements will combine by the transfer of an electron (Fig. 2.2a). The result is an ionically-bonded molecule of the compound sodium chloride (NaCl). This compound is also known as table salt (Fig. 2.2b).

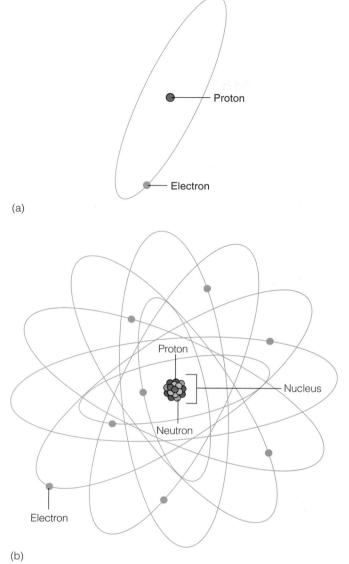

(a)

(b)

Figure 2.1
Atomic structures of (a) hydrogen, with an electron orbiting a single proton, and (b) oxygen, with eight electrons orbiting a dense nucleus containing eight protons and eight neutrons (note that the first two electrons are in a lower orbit than the remaining six).

In a molecule of table salt, the sodium and chlorine are no longer in their simple atomic states, but have become charged atoms called **ions.** The sodium ion has a positive electrical charge because it has lost an electron, whereas the chlorine ion has a negative charge because it has gained one. Because of their opposing charges, the two ions attract each other. The chemical formula for table salt is often written as Na^+Cl^-. A positively charged ion like that of sodium is called a **cation,** whereas a negatively charged ion like that of chlorine is called an **anion.**

Covalent Bonds. Molecules formed by bonds created by sharing electrons are known as **covalent bonds.** Each atom donates an electron to the bond such that the electrons are shared by both atoms. The shared electrons orbit between the adjacent atomic nuclei, providing the attraction that binds the molecule together. Although these bonds are generally more difficult to break than ionic bonds, they are not as rigid and allow the adjacent atoms to vibrate.

Diamond, the crystalline form of carbon, is an example of a mineral held together by shared electrons (Fig. 2.3). Each carbon atom has four nearest neighbors, so that four electrons are shared. This makes the covalent bonds between the carbon atoms very strong, and explains why diamonds are so hard and must be heated to such high temperatures (3500°C at the Earth's surface) before they melt.

Metallic Bonds. When atoms combine to form a piece of metal, each atom gives up an outer electron and becomes a positive ion. As a result, a common cloud of shared electrons is formed. These shared electrons roam throughout the material, holding together the metal cations. This type of bonding is known as a **metallic bond.** The electron cloud is an effective reflector of solar radiation, which is why metals have shiny surfaces. The electrons within the cloud can also carry energy from one end of the piece of metal to the other, which is why metals are generally excellent conductors of heat and electricity.

In contrast to metals, most minerals that make up the rocks of the solid Earth contain a combination of ionic and covalent bonding, and are relatively poor conductors (and so are good insulators) because the electrons are more firmly bound to their adjacent cations. As these are the dominant minerals of the Earth's rocky outermost layer or **crust,** the crust as a whole is a poor conductor. On the other hand, ore bodies containing high concentrations of metals can be good conductors. Because of this, they can sometimes be detected by measuring conductance in the rocks below the Earth's surface.

Van der Waals Forces. A fourth type of bonding, named **van der Waals forces** after the Dutch physicist who first proposed their existence, occurs between adjacent molecules rather than between atoms, and allows the molecules to bind together. Van der Waals forces are weak forces of attraction that exist between the negatively charged electron cloud of one molecule and the positively charged nuclei of an adjacent molecule. For example, the structure of the water molecule (H_2O) is such that the two hydrogens are positioned at one end of the molecule while the oxygen is located at the other. Thus, the charge distribution is not uniform throughout the molecule. Instead, the molecule is electrically polarized. The hydrogen atoms have a tendency to donate electrons and so are positively charged. The oxygen atoms have a tendency to pull the electrons donated by the hydrogen toward them and so are negatively charged. As a result, the positive ends of water molecules are attracted to the negative ends of their neighbors and vice versa. This weak force of attraction is called the van der Waals bond.

Molecules do not need to be permanently polarized to exhibit van der Waals forces. Nonpolarized molecules can also attract each other as electrons are in constant motion. As a result, a temporary polarity can exist at any moment because of the unequal distribution of electrons.

Although van der Waals forces are the weakest of all bonding arrangements, those present in liquids must be overcome

Figure 2.2
Atomic structure of sodium chloride and crystal structure of salt. (a) Sodium atom easily loses its outermost electron to become a positively charged sodium cation. Chlorine atom readily accepts the electron and becomes a negatively charged anion. The opposite charges of the sodium and chlorine ions attract each other to form an ionic bond. (b) Crystal structure of halite (salt) showing the relative sizes of the two ions and their location within the crystal structure.

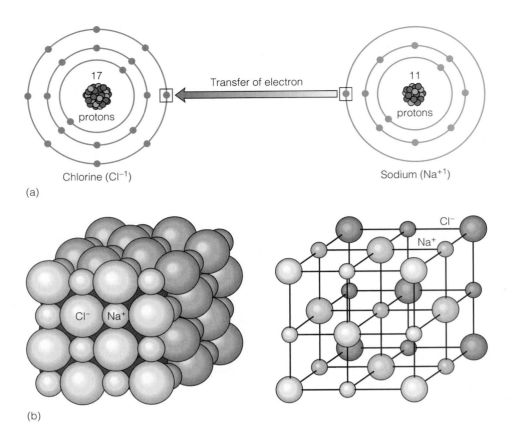

before the liquid can vaporize. Because of this, boiling water requires a significant input of additional energy (known as the latent heat) before each water molecule can become independent of its neighbors and evaporate.

2.1.2 What is a Mineral?

Minerals are the basic components of the rocks that make up the solid Earth. A **mineral** is defined as a naturally occurring, inorganic crystalline solid with a specific chemical structure and formula. That is, a mineral is a chemical that occurs in nature in the absence of living organisms and consists of atoms arranged in an orderly pattern so that it has a particular chemical composition and will grow to form crystals. Although nearly 4000 naturally occurring minerals have been identified, less

than 1% of these are common in rocks and can be referred to as *rock-forming minerals*.

Naturally Occurring and Inorganic

As minerals are defined as having formed in the natural world, artificially manufactured inorganic solids are not usually regarded as minerals. Because of this, synthetic diamond is not a mineral in the strictest sense, although natural diamond is. Similarly, the requirement that the solid be inorganic not only excludes animal and vegetable matter, but also excludes those materials derived from such organic matter like coal and oil. However, some materials produced by living things are considered minerals. For example, many sea shells are minerals composed of the chemical compound calcium carbonate ($CaCO_3$), the mineral name for which is *calcite*.

Figure 2.3
Atomic structure of diamond. (a) Covalent bonds formed by adjacent carbon atoms sharing electrons. (b) Three-dimensional structure of carbon atoms in diamond.

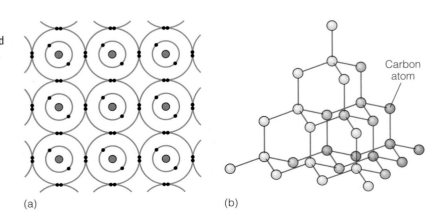

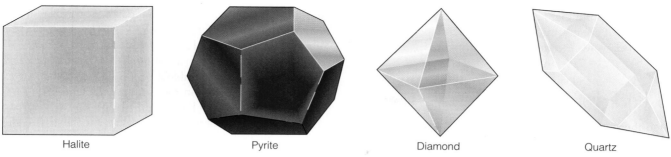

Figure 2.4
Mineral crystals differ in shape for different minerals. Shown here are crystals of the minerals halite, pyrite, diamond, and quartz.

Crystalline Solids

All minerals are **crystalline solids** because their ions are arranged in regular three-dimensional configurations. Left free to grow, minerals will form perfect **crystals** having a regular geometrical form made up of flat crystal faces with straight edges. This crystal form, which differs in shape for different minerals (Fig. 2.4), is an external expression of the orderly internal arrangement of a mineral's ions. Solid materials like glass, which lack a regular atomic structure and are therefore not crystalline, are not considered minerals.

Because they are crystalline solids, minerals will grow only under the appropriate physical conditions. For example, the minerals *graphite* (used as pencil lead) and diamond are both composed of a single essential element, carbon (C). However, the internal arrangement of the carbon atoms within the crystal structure of the two minerals is different, and this difference affects the stability of the two minerals. Whereas the structure of diamond forms naturally only at very high pressures, the structure of graphite is only stable if the pressures are relatively low. Because of this, diamonds only form deep in the Earth's interior where the pressures exerted by the weight of the rocks above are very high. Graphite is found nearer the Earth's surface where the load is less and the pressures are much lower.

Chemical Composition and Formula

Because the ions of minerals are arranged in a regular configuration, all minerals have a specific composition that can be expressed as a **chemical formula.** The chemical formula of a mineral lists its component ions according to the relative proportion in which they occur in the mineral structure. Table salt, for example, is actually the mineral *halite* and has the chemical formula NaCl (sodium chloride). This means that it is made up of equal numbers of sodium (Na) and chlorine (Cl) ions bonded together. The common mineral *quartz* has the chemical formula SiO_2 (silicon dioxide, or silica) and therefore contains twice as many ions of the element oxygen (O) as it does of the element silicon (Si).

In reality, no natural mineral is absolutely pure. Quartz, for example, may contain small amounts of other elements. These impurities are evident in the various colors displayed by different forms of quartz (Fig. 2.5), from the deep purple of *amethyst* produced by trace amounts of iron (Fe), to the pale pink of *rose quartz*

produced by traces of the metal titanium (Ti). It is the availability of silicon and oxygen that is essential to the formation of quartz, because without these two elements the mineral simply could not form. It is the relative proportion of essential elements in a mineral that is expressed in a mineral's chemical formula.

Although each mineral has its own chemical formula, that formula is not necessarily unique to the mineral. In fact, quite different minerals can sometimes have the same chemical formula because they have the same chemical composition. For example, the minerals diamond and graphite are made up of a single essential element, carbon, so that both possess the same chemical formula (C). However, the internal arrangement of the carbon atoms within the crystal structure of the two minerals is quite different. Because of this, the appearance and properties of the two minerals are distinct. Minerals such as

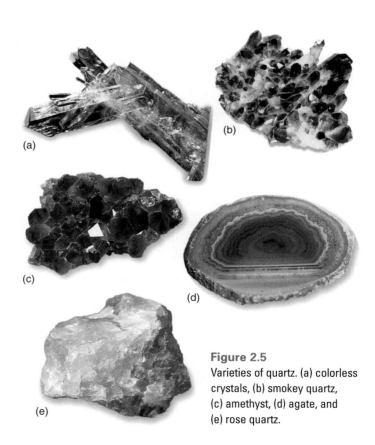

Figure 2.5
Varieties of quartz. (a) colorless crystals, (b) smokey quartz, (c) amethyst, (d) agate, and (e) rose quartz.

diamond and graphite that have the same chemical composition but different crystal structures are known as **polymorphs,** from the Greek words meaning "many formed."

2.1.3 Rock-Forming Silicates

Although the minerals we are most familiar with are the precious ones we call gemstones, most of the common rock-forming minerals are **silicates.** Silicates are minerals in which silicon (Si) and oxygen (O) are chemically bonded to a variety of metallic ions such as sodium (Na), potassium (K), calcium (Ca), aluminum (Al), iron (Fe), magnesium (Mg), and titanium (Ti). For example, silicon and oxygen combined with potassium and aluminum produces the common rock-forming mineral *orthoclase* ($KAlSi_3O_8$) (Table 2-1).

Of the 92 naturally occurring chemical elements, oxygen, silicon, and these seven metals are the most abundant in the Earth's crust (Table 2-2). The vast majority of minerals in crustal rocks consequently represent various combinations of these elements (Table 2-3). In addition, many rock-forming silicates contain water, not in its more familiar liquid state, but incorporated as an integral part of the crystal structure. Its appearance in a mineral generally reflects crystal growth in the presence of dispersed water in the crust, and is shown by the symbol OH in the mineral's chemical formula.

With silicon and oxygen accounting for approximately 75% of crustal rocks by weight (Table 2-2), it is not surprising that silicates are the most common rock-forming minerals. The most abundant of these in the Earth's crust is the silicate, *feldspar,* which is usually white or pink in color (Fig. 2.6). This is an aluminum silicate that, in addition to aluminum, silicon, and oxygen contains either potassium (*orthoclase*) or sodium and/or calcium (*plagioclase*). Two important rock-forming aluminum silicates that, in addition to aluminum, incorporate water into their crystal structure (and so are Al-OH silicates), are *muscovite,* or white mica, and the *clay minerals,* such as *kaolinite.*

Other common rock-forming silicates contain magnesium and iron in addition to silicon and oxygen, and are therefore dark colored (Fig. 2.6). Among these are biotite, hornblende (amphibole), pyroxene, and olivine. *Biotite* (black mica) contains potassium, aluminum, and water and so is a K-Mg-Fe-Al-OH silicate. Dark-green *hornblende* is the most common member of a family of minerals called *amphiboles* and is calcium-bearing. It is a Ca-Mg-Fe-Al-OH silicate. *Pyroxenes* are a group of minerals that generally resemble amphiboles but lack

Table 2-1
Rock-Forming Minerals

MINERAL	COMPOSITION	PRIMARY OCCURRENCE
Ferromagnesian Silicates		
Olivine	$(Mg,Fe)_2SiO_4$	Igneous, metamorphic rocks
Pyroxene group Augite most common	Ca,Mg,Fe,Al silicate	Igneous, metamorphic rocks
Amphibole group Hornblende most common	Hydrous* Na, Ca, Mg, Fe, Al silicate	Igneous, metamorphic rocks
Biotite	Hydrous K, Mg, Fe silicate	All rock types
Non-ferromagnesian Silicates		
Quartz	SiO_2	All rock types
Potassium feldspar group Orthoclase, microcline	$KAlSi_3O_8$	All rock types
Plagioclase feldspar group	Varies from $CaAl_2Si_2O_8$ to $NaAlSi_3O_8$	All rock types
Muscovite	Hydrous K , Al silicate	All rock types
Clay mineral group	Varies	Soils and sedimentary rocks
Carbonates		
Calcite	$CaCo_3$	Sedimentary rocks
Dolomite	$CaMg(CO_3)_2$	Sedimentary rocks
Sulfates		
Anhydrite	$CaSO_4$	Sedimentary rocks
Gypsum	$CaSO_4 \cdot 2H_2O$	Sedimentary rocks
Halides		
Halite	NaCl	Sedimentary rocks

*Contains elements of water.

Table 2-2
Common Elements in Earth's Crust

ELEMENT	SYMBOL	PERCENTAGE OF CRUST	
		(BY WEIGHT)	(BY ATOMS)
Oxygen	O	46.6%	62.6%
Silicon	Si	27.7	21.2
Aluminum	Al	8.1	6.5
Iron	Fe	5.0	1.9
Calcium	Ca	3.6	1.9
Sodium	Na	2.8	2.6
Potassium	K	2.6	1.4
Magnesium	Mg	2.1	1.8
All others		1.5	0.1

Table 2-3
Relative Abundance of Minerals in the Earth's Crust

MINERAL	PERCENTAGE
Plagioclase	39%
Quartz	12%
Orthoclase	12%
Pyroxenes	11%
Micas	5%
Amphiboles	5%
Clay minerals	5%
Other silicates	3%
Nonsilicates	8%

aluminum. They are mostly Ca-Mg-Fe silicates and are usually green, brown, or black in color. *Olivine*, which is thought to be the most abundant mineral in the region of the Earth's interior immediately below the crust, is simply an Mg-Fe silicate and is olive-green like its better-known gemstone form "peridot."

Structure of Silicates

All rock-forming silicates have in common a basic structural unit known as the **silicate tetrahedron,** so called because of its tetrahedral shape, which is a pyramid-like shape with a triangular base (Fig. 2.7). The silicate tetrahedron consists of a central silicon cation with a positive electrical charge bonded with four negatively charged oxygen anions. Individual tetrahedra may be linked together in a variety of forms that are characteristic of the mineral. The ions of metals are then bonded to these tetrahedra. The vast number of naturally occurring silicate minerals is indicative of the many different ways in which this basic tetrahedral unit can combine with various metal ions to form stable crystal structures.

The silicate tetrahedra in quartz and feldspars occur in complex three dimensional networks, whereas those in micas and clay minerals occur in sheetlike forms. In contrast, the silicate tetrahedra in amphiboles are arranged in double chains, whereas those in pyroxenes are arranged in single chains. Those in olivine occur as isolated tetrahedra. These tetrahedral arrangements (Fig. 2.8) form the basic building blocks of silicate mineral families, and each has a characteristic crystalline expression that facilitates its identification.

2.1.4 Nonsilicate Rock-Forming Minerals

Not all rock-forming minerals are silicates. Indeed, a small but significant proportion of the Earth's crust is made up of **nonsilicate** minerals (Fig. 2.9) in which ions of the common metals are bonded to elements other than a combination of silicon and oxygen. The most common of these are the **carbonate** minerals that make up rocks like limestone and marble. In carbonate minerals, common metallic ions such as calcium (Ca) and magnesium (Mg) are bonded to a combination of carbon (C) and oxygen (O) to produce minerals such as *calcite* ($CaCO_3$) and *dolomite* [$CaMg(CO_3)_2$].

Another important group of nonsilicate rock-forming minerals is produced by the evaporation of salt water from ancient seas. These so-called **evaporite** minerals include *halite* (or rock salt), which is the mineral name for common table salt, sodium

Figure 2.6
Some common rock-forming silicate minerals. (a) feldspar, (b) quartz, (c) muscovite, and (d) hornblende.

(a)

(b)

(c)

(d)

Figure 2.7

Silicate tetrahedron, the basic building block of silicate mineral families. (a) Expanded view showing oxygen atoms at the corners of a tetrahedron with a silicon atom at its center. (b) View showing the arrangement of the same atoms as they actually occur. (c) Diagrammatic representation of the silicate tetrahedron.

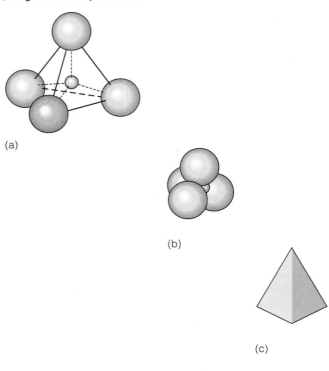

(a)

(b)

(c)

chloride (NaCl), and *gypsum*. Gypsum is formed when calcium (Ca) is bonded to a combination of sulfur (S) and oxygen (O) to form a calcium sulfate ($CaSO_4 \cdot 2H_2O$).

Other nonsilicates are not sufficiently common to be considered rock-forming minerals, However, they do form locally important mineral deposits. Most are relatively simple compounds in which a metal ion is either bonded to oxygen to form an **oxide,** as is the case for the iron ore *hematite* (Fe_2O_3), or is bonded to sulfur to form a **sulfide,** as is the case for the lead ore *galena* (PbS) and the zinc ore *sphalerite* (ZnS). Yet others comprise uncombined or **native** metals, such as gold (Au).

2.1.5 Mineral Identification

Just as a gemologist can discover whether a precious mineral is genuine or fake by studying its properties, so a mineralogist can identify minerals by an examination of their properties. These properties are an expression of the mineral's chemical formula and crystal structure and, when evaluated as a whole, provide a set of diagnostic criteria that distinguish one mineral from another. Like a doctor, a mineralogist performs a routine program of examination and testing that facilitates the diagnosis. Because single tests rarely provide an accurate identification, generally a series of tests must be performed before a confident diagnosis is possible.

Even then, the diagnosis may remain elusive and more sophisticated tests may be necessary. For example, microscopic studies can identify the presence of clay minerals in a sample, but cannot identify which of the many natural clays may be present. In forensic science, this distinction is crucial if samples of soil taken from a suspect's shoes are to be tied to the scene of the crime.

Figure 2.8

Arrangement of silicate tetrahedra in some common rock-forming silicates: (a) Isolated tetrahedra, (b) single and double chains, (c) continuous sheets, and (d) three-dimensional networks.

			Formula of negatively charged ion group	Silicon to oxygen ratio	Example
(a)	Isolated tetrahedra		$(SiO_4)^{-4}$	1:4	Olivine
(b)	Continuous chains of tetrahedra	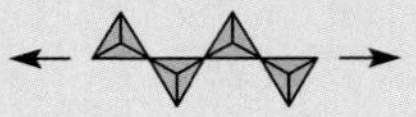 Single chain	$(SiO_3)^{-2}$	1:3	Pyroxene group
		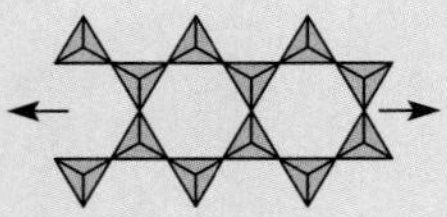 Double chain	$(Si_4O_{11})^{-6}$	4:11	Amphibole group
(c)	Continuous sheets	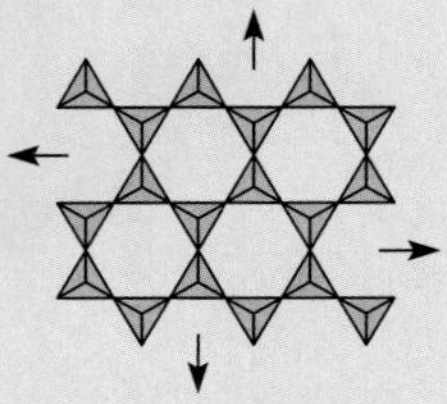	$(Si_4O_{10})^{-4}$	2:5	Micas
(d)	Three-dimensional networks	Too complex to be shown by a simple two-dimensional drawing	$(SiO_2)^0$	1:2	Quartz

(a)

(b)

(c)

Figure 2.9
Some common rock-forming nonsilicate minerals: (a) calcite, (b) gypsum, and (c) halite.

However, the reactions of minerals to high-energy X-rays or electron beams not only provide more sophisticated tests on a mineral identity, but can also resolve crystal structure and composition on a very fine scale. Therefore these tests are used routinely, both to identify unknown minerals and to provide details on the structure and composition of known minerals. By these methods, chemical analyses can also be determined from very small areas of a crystal, so that compositional variations within an individual mineral can be mapped out and important impurities identified.

2.1.6 Properties of Minerals

An accurate diagnosis of an unknown mineral usually starts with the routine examination of a mineral in a hand specimen, either with the naked eye or with a hand lens. This preliminary identification is concerned with those physical properties that reflect the internal structure and composition of the mineral. Obviously, we cannot determine the detailed chemistry of a mineral from this kind of examination, but we can deduce its idealized composition and structure. Diagnostic physical properties that can be examined in hand specimens include a mineral's color and streak, luster, crystal form, hardness, cleavage, and specific gravity. Mineral identification tables based on these properties are provided in Appendix C.

Color and Streak

One of the first things most of us are likely to notice about a crystal is its **color.** The color of a crystal is often related to the overall composition of the mineral. Minerals such as quartz and feldspars tend to be light colored because their compositions are dominated by silicon, aluminum, sodium, potassium, and calcium. On the other hand, minerals rich in iron and magnesium, like the pyroxenes and amphiboles, tend to be dark colored (Fig. 2.6).

Color alone, however, can be badly misleading. For example, the semiprecious minerals amethyst and rose quartz have very different colors (Fig. 2.5) but to a mineralogist they are merely varieties of quartz. Therefore, color in this case cannot be related to overall composition as the chemical formula of both minerals (SiO_2) is the same. Their difference in color is entirely a consequence of minor impurities trapped within the crystal. In fact, a particularly tricky mineral identification quiz can have up to ten different specimens, each with a different color, but all of the same mineral. (One of the authors of this book fell for such a trick in his student days.)

Color, particularly in dark minerals, can also reflect grain size, which is not an inherent property of a mineral. To alleviate this problem, the mineral is ground against an unglazed porcelain plate to produce a **streak,** the color of which is characteristic of the mineral in powdered form (Fig. 2.10).

Luster

If you look around your room, it is obvious that materials respond quite differently to reflected light. A mirror is highly reflective whereas cushions and mattresses are not. The reflectivity of the wall depends on its color and the quality of the paint or wallpaper. Similarly, minerals also appear differently in reflected light. This property is known as **luster.** Earlier in this chapter we described how metals are strong reflectors of light and therefore appear shiny. This is because metals are held together by reflective clouds of electrons above the nuclei of their atoms. In contrast, most silicate minerals are held together by a variety of covalent and ionic bonds and are less reflective. Therefore, two main types of mineral luster are recognized: **metallic** and **nonmetallic** (Fig. 2.11). Among the nonmetals, a variety of mineral lusters occur, including waxy, greasy, vitreous (or glassy), brilliant (like diamond), and dull.

Crystal Form

A mineral's **crystal form** is defined by the geometric arrangement of its crystal faces and is an external expression of the mineral's internal structure. Any two samples of the same mineral will have the same crystal form because they have the same set of crystal faces. They may differ in size, but every face on one crystal will have an equivalent parallel face on the other (Fig. 2.12). This implies that the angles between a mineral's crystal faces can be used to specifically identify the mineral. However, a severe limitation to the application of this property is the fact that minerals rarely grow in an environment where their

Figure 2.10
The color of a powdered mineral, or streak, is sometimes more diagnostic than the color of the mineral. Hematite, the mineral name for iron oxide or rust (Fe_2O_3), occurs in various forms (left and right). On a porcelain plate, however, its streak is always red.

Figure 2.12
Crystal form is the external expression of a mineral's orderly internal structure. Well-developed quartz crystals show a characteristic six-sided crystal form with pyramid-shaped ends.

(a)

(b)

Figure 2.11
Luster is the appearance of a mineral in reflected light. (a) The lead ore galena (PbS) has the appearance of a metal and is said to have a metallic luster. (b) Halite or rock salt (NaCl) has the appearance of glass and its luster is said to be nonmetallic and vitreous.

Figure 2.13
Habit is a characteristic outward appearance of a mineral. Gypsum sometimes grows in fibers and is said to have a fibrous habit.

growth is unhindered by that of their nearest neighbors. This competition for space means that minerals are rarely given the opportunity to develop well-formed crystal faces. However, even in environments where their growth is restricted, minerals often have a characteristic outward appearance or **habit** that aids in their identification. For example, amphiboles such as hornblende tend to be elongate, whereas micas such as muscovite and biotite are platy. The nonsilicate mineral gypsum sometimes grows in fine elongate threads; this type of habit is described as fibrous (Fig. 2.13).

Hardness

Depending on the strength of their chemical bonds, minerals also vary in hardness. The strength of the covalent bonds in the mineral diamond, for example, makes it the hardest naturally occurring mineral. The mineral talc has extremely weak bonds and

therefore is very soft. The contrasting hardness of these two minerals is reflected in their industrial uses; diamond is used as an abrasive whereas talc is the principal ingredient of talcum powder.

Technically, **hardness** is defined as the relative resistance of a mineral to being scratched. Fredrich Mohs, an Austrian mineralogist, devised a scale of relative hardness. The scale is divided into ten steps, each of which is defined by a common mineral (Table 2-4). Diamond, being the hardest mineral, is given a value of 10. Talc, the softest, is given a hardness of 1. A mineral ranked higher on the scale will scratch one ranked lower down. Your fingernail has a hardness between 2 and 3, and will therefore scratch gypsum (hardness 2) but not calcite (hardness 3). A pocketknife or a plate of glass has a hardness between 5 and 6, and so will scratch the mineral apatite but not potassium feldspar. It is important to realize that this scale is not a linear one, but merely ranks minerals in terms of their hardness relative to ten common ones.

Table 2-4
Mohs Scale of Hardness

HARDNESS	MINERAL	HARDNESS OF SOME COMMON OBJECTS
10	Diamond	
9	Corundum	
8	Topaz	
7	Quartz	
		Steel file (6½)
6	Orthoclase	
		Glass (5½–6)
5	Apatite	
4	Fluorite	
3	Calcite	Copper penny (3)
		Fingernail (2½)
2	Gypsum	
1	Talc	

Cleavage and Fracture

An additional property of many minerals is their tendency to break along preferred, rather than random, directions. This tendency is called **cleavage.** Where it occurs, each preferred fracture surface is typically smooth and planar and is termed a **cleavage plane.** Cleavage planes are not crystal faces, because they are not expressions of the outward growth of a mineral. Instead, cleavages reflect systematic weaknesses in the internal structure of a mineral along which the mineral tends to break.

Silicate minerals have a wide variety of cleavage patterns reflecting the wide variation in their mineral structure. Minerals like quartz, for example, which consist of complex three-dimensional networks of silica tetrahedra, exhibit no obvious cleavage. In contrast, micas and clay minerals, which consist of stacked sheets of silica tetrahedra held together by relatively weak bonds, have a well-defined cleavage plane parallel to the sheets (Fig. 2.14). As a result, micas are characteristically flaky and can be easily pried apart along their cleavage surfaces.

Minerals that break along irregular surfaces rather than cleavage planes are said to possess a **fracture.** Any mineral can be broken to produce irregular fractures, but some minerals break along fracture surfaces of characteristic shape. For example, the mineral quartz typically breaks like glass along smoothly curved surfaces and is said to possess a **conchoidal fracture** (Fig. 2.15).

Specific Gravity and Density

The **specific gravity** of a mineral is defined as the ratio of its weight relative to the weight of an equal volume of water. For example, quartz weighs 2.65 times the same volume of water and so has a specific gravity of 2.65. This value is much the same as the mineral's **density,** which is calculated by dividing the mass of the mineral by its volume. The specific gravity and density of a mineral reflect a combination of its chemical composition and mineral structure. As an example, silicates such as quartz and feldspar are dominated by relatively "light" elements and their specific gravities typically range from 2.6 to 2.9. Because these are the dominant minerals in the Earth's continental crust, it is not surprising that the continental crust has an average density of 2.7. Iron- and magnesium-rich silicates, such as the pyroxenes and olivine, have specific gravities that typically range from 2.7 to 3.4. These minerals dominate the Earth's oceanic crust, which consequently has an average density between 2.9 and 3.0.

Minerals that are polymorphs of each other, that is, they have the same composition but different crystal structures, may have significantly different specific gravities. Graphite and diamond, for example, are both composed of carbon but have average specific gravities of 2.2 and 3.5, respectively. The difference reflects the tighter bonding arrangement of diamond, which leads to tighter packing of its carbon atoms.

2.2 Rocks

Just as the chemical elements are the building blocks of minerals, so minerals are the basic ingredients of rocks. Like minerals, rocks are naturally occurring solids, but beyond this the similarity between the two ends. Minerals, as we have learned, are inorganic solids and have crystalline structures and characteristic physical properties that are determined by the regular internal arrangement of their constituent atoms. Rocks, however, are not necessarily inorganic or crystalline and do not possess a specific chemical structure and formula. Instead, a **rock** is an aggregate

Figure 2.14
(a) Mica being pulled apart along its cleavage planes.
(b) Relationship of cleavage to the crystal structure of mica (shown here simplified).

(a)

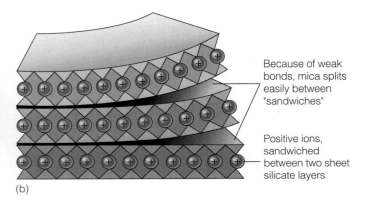
(b)

Because of weak bonds, mica splits easily between "sandwiches"

Positive ions, sandwiched between two sheet silicate layers

Figure 2.15
Quartz displaying the smoothly curved or conchoidal fracture surfaces along which it breaks.

of one or more minerals that have either been cemented together or have grown together in an interlocking mosaic. As a result, most rocks contain several different minerals.

The rock we call granite, for example, always contains quartz and feldspar, often with variable amounts of muscovite and biotite (Fig. 2.16). Other rocks like limestone are predominantly composed of only one mineral; in the case of limestone the mineral is calcite. Although similar in composition to calcite, limestone is not considered a mineral because it does not have a characteristic crystal structure. Instead, it is made up of countless interlocking calcite grains. Likewise coal, while similar in composition to both graphite and diamond, is a rock but not a mineral because it does not have a regular atomic framework and is organic rather than inorganic in origin.

In this section we will learn how minerals combine to form rocks, and how all rocks fall into one of three broad categories, known as **igneous, sedimentary,** and **metamorphic,** based on the way they form. We will discover that igneous rocks form from the cooling of molten liquids whereas sedimentary rocks are created by processes that operate on the Earth's surface, and that metamorphic rocks are produced beneath the Earth's surface by the effects of heat and pressure on preexisting rocks. We will also learn that rocks can be further classified according to the size, shape, and chemistry of their component mineral grains. Finally, we will learn that in examining rocks much can be discovered about the nature of the solid Earth. We will see that rocks and minerals together are the building materials of the solid Earth, and that a knowledge of both is essential if we are to understand the workings of the solid Earth and the ways in which it interacts with Earth's other surface reservoirs. Nowhere is this better illustrated than in the continual interaction that occurs between the three categories of rocks during the process we call the rock cycle.

2.2.1 The Rock Cycle

In the view of James Hutton, the eighteenth century naturalist often considered to be the father of modern geology, the Earth is in continuous but gradual change, constantly wearing down and building up under the power of its own internal heat engine. This insightful view is vividly illustrated in the relationship between the three basic groups of rocks that make up the Earth's continents and ocean floors. As the Earth's ever-changing surface goes through episodes of uplift, erosion, deposition, and burial, each of these rock groups is transformed into one of the others in a continuous process known as the **rock cycle** (Fig. 2.17).

The first of these rock groups is called **igneous rock** (from the Latin word *ignis* for fire), and forms whenever a molten

Figure 2.16
Rocks are made up of minerals. This granite, for example, is an aggregate of quartz, biotite, and feldspar.

Quartz (mineral)

Granite (rock)

Biotite (mineral)

Feldspar (mineral)

rock, or **magma**, freezes. Igneous rocks can form at the Earth's surface, as the product of volcanic eruptions, in which case they cool very rapidly and are immediately subject to erosion by water, wind, or ice. Alternatively, they can solidify within the Earth's crust, in which case they cool slowly and are protected from the elements until such time that uplift and erosion of the overlying rocks exposes them at the surface.

Once exposed at the Earth's surface, all rocks undergo **weathering** during which they slowly disintegrate and decompose as a result of their interaction with water, ice, and air. The weathered materials may first move down slope under the influence of gravity but are eventually picked up and transported by one of several agents of erosion: glaciers, running water, wind, or waves. Eventually the eroded material is deposited as unconsolidated **sediment.** Although some of this sediment may be deposited in river valleys or in lakes or deserts, most is eventually transported to the sea.

Once deposited, various processes combine to turn unconsolidated sediment into **sedimentary rock,** the second of the three basic rock groups. The sediment is slowly buried by the progressive deposition of further sediment, and so becomes compacted by the weight of the overlying material. It may also come into contact with percolating, mineral-charged ground waters which deposit material between the grains of sediment, cementing them together. This conversion of unconsolidated sediment into solid rock is known as **lithification** (from the Greek word, *lithos*, meaning rock), and results in the formation of sedimentary rock. Uplift and erosion of sedimentary rock will produce new unconsolidated sediment. In this way, the process of deposition, burial, and lithification may be repeated many times, each time resulting in a new sedimentary rock.

If sedimentary rocks come into contact with molten rock, or are involved in mountain building and become deeply buried as a result, they may be subjected to intense pressure and heat. As the rock responds to these new conditions, the minerals it contains become unstable and may react to form more stable minerals. This causes the chemical and physical makeup of the rock to alter, and the third basic rock group, known as **metamorphic rock** (from the Greek words for "to change form"), is produced.

What happens next depends on whether burial continues, or whether metamorphism is followed by uplift and erosion. In

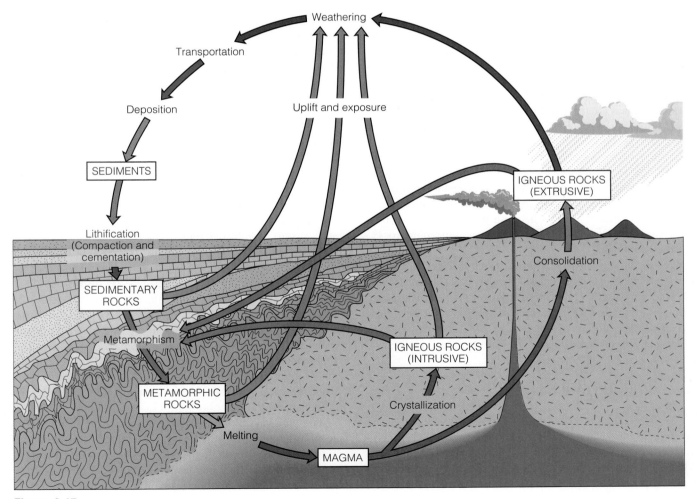

Figure 2.17
The rock cycle shows the relationship between the three main groups of rocks: igneous, sedimentary, and metamorphic. Arrows within the circle show possible shortcuts in the ideal cycle.

the latter case, the metamorphic rocks may ultimately become exposed at the Earth's surface. The product of their weathering then provides the raw material for new sedimentary rocks. Alternatively, if the metamorphic rocks continue to be buried, they may encounter high enough temperatures that they melt, creating the raw material (magma) for new igneous rocks when it cools. If an igneous rock is subjected to the great pressures and temperatures of mountain building, it may undergo alteration and revert to a new metamorphic rock. However, if the same igneous rock finds its way to the Earth's surface so that it is exposed to the agents of erosion, the product of its weathering would provide the raw material for new sedimentary rocks, and the rock cycle will have come full circle.

In this way, igneous rocks ultimately disintegrate into sediment through the processes of weathering, transportation, and deposition; sediment is converted into sedimentary rock through the process of lithification; sedimentary rocks may be transformed into metamorphic rocks through the application of heat and pressure; and metamorphic rocks can provide the raw material for igneous rocks through melting and solidification. Not all rocks follow this full cycle, however, as shortcuts are also possible. These shortcuts occur whenever igneous rocks are transformed directly into metamorphic rocks or when metamorphic rocks become exposed and are subjected to the process of weathering.

So each of the three basic rock groups and the geologic processes that produce them are closely interrelated by the continual pattern of interchange we call the rock cycle. The cycle shows how one rock type can be transformed into another in an endless pattern of creation and destruction. This demonstrates in vivid fashion that the Earth is a planet of gradual but continuous change—a dynamic planet just as Hutton envisaged.

About 95% of the Earth's crust is composed of igneous and metamorphic rocks, although it is thought that most of the metamorphic rocks were originally igneous rocks before they were transformed. Sedimentary rocks form only a thin veneer on the Earth's surface. This is because they form only at the Earth's surface and are transformed to metamorphic rocks when they are buried.

Because of the rock cycle, all rocks can be grouped into one of three varieties—igneous, sedimentary, or metamorphic—on the basis of how they were formed. However, because rocks are made up of minerals, rocks from each of three groups can be further classified according to the texture (size and shape) and composition (chemistry) of their component mineral grains.

2.2.2 Igneous Rocks

Formation

Igneous rocks form when molten rock, or **magma,** which generally ranges from 700°C to 1200°C in temperature, cools and solidifies. The process of solidification is known as **crystallization** because it involves the growth of crystals in the melt as it cools. Just as more sugar will dissolve in hot tea than in cold, mineral components held in solution in the magma at high temperatures

become less soluble as the magma cools. Eventually, the magma cools to a threshold level below which the components can no longer be held in solution, and crystals begin to form.

Volcanic eruptions produce the most dramatic and familiar form of igneous rock (Fig. 2.18). During eruptions, igneous rocks are produced at the Earth's surface and magma may flow across the land to form a **lava.** Most lavas erupt at temperatures between 1000°C and 1200°C, and so cool quite rapidly at the Earth's surface. As a result, crystals have little time to grow and are consequently small. The resulting igneous rock is said to be **fine-grained.** In fact, cooling may be so rapid that crystals do not get a chance to form and the product becomes a **volcanic glass.** This natural process is mimicked in the industrial procedure of manufacturing glass by rapidly chilling molten sand.

Volcanoes are the end product of the ascent of hot, buoyant magma from the interior of the Earth toward its surface. However, much of this molten rock never reaches the surface and is instead trapped in large caverns, or **magma chambers,** in the Earth's interior. Many of these chambers are like storage tanks that are interconnected by an intricate array of pipes and fractures, much like the plumbing system in a house. Magmas within these chambers cool very slowly so that crystals have far more time to grow. As a result, igneous rocks produced deep in the Earth's interior contain large crystals and are described as **coarse-grained.** Such deep-seated igneous rocks are also termed **plutonic** after Pluto, the Greek god of the underworld. The exposure of plutonic rocks at the Earth's surface is testament to the rock cycle (Fig. 2.17), as it requires the subsequent removal by erosion of the rocks that were above them when the plutonic rocks crystallized.

Neighboring plutonic and volcanic rocks at the Earth's surface are often closely related. In these situations, the deep-seated bodies of plutonic rock, or **plutons,** often represent the crystallized remnants of ancient magma chambers that once fed the surface volcanoes that produced the volcanic rock. Because they are related, the two types of rocks would have very similar chemical compositions. Their different rates of cooling, however, would ensure a quite different overall appearance, the coarse grain size of the slowly cooled plutonic rock contrasting with the fine grain size of the more rapidly cooled volcanic rock.

The two groups of rocks might also be distinguished by field inspection as illustrated in Fig. 2.19. Because volcanic rocks are erupted onto the Earth's surface, they are said to be **extrusive** and are often deposited in layers that are generally parallel to those of the underlying rock. Subsequent rock layers deposited on top will continue the parallel succession. In contrast, plutonic rocks are said to be **intrusive** because they intrude into their surroundings and so commonly cut across any layering in the neighboring rocks.

The pathways by which a magma ascends toward the surface or migrates from one magma chamber to another are conduits that cut through the surrounding rock and exploit any weakness that they encounter. Magma trapped in these conduits cools to form sheet-like plutonic bodies known as dikes and sills. **Dikes** cut through existing layers and because of this are said to be **discordant** with respect to the layering in

(a)

(b)

(c)

Figure 2.18
(a) Volcanic eruption producing a flow of hot lava. (b) Fine-grained, and (c) glassy rocks produced by the rapid crystallization of lava flows.

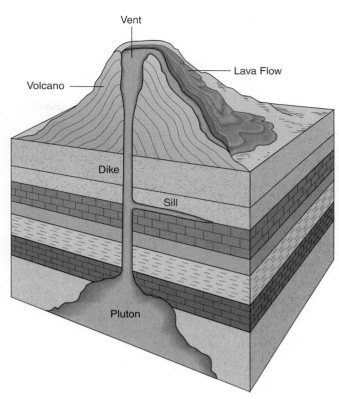

Figure 2.19
Schematic diagram illustrating the relationship between plutons, dikes, sills, and volcanoes.

the surrounding rock, whereas **sills** form along the boundary between layers and so are said to be **concordant** (Fig. 2.20). The formation of sills testifies to the pressure exerted by ascending magma and reflects the exploitation of weak layers which are preferentially pried apart.

Classification: Types of Igneous Rocks

Igneous rocks, whether plutonic or volcanic, are classified according to both their mineral composition and grain size. In turn, the mineral composition and grain size reflect the way they formed (Fig. 2.21). As with all classification schemes there are a few exceptions, but the approach accounts for the vast majority of igneous rocks. The simplest scheme is a compositional one based on the total silica content (SiO_2), which for most igneous rocks ranges between 45% and 80% by weight. This variation in silica content can be used to subdivide igneous rocks into four categories: felsic, mafic, intermediate, and ultra-mafic. Within each category, however, grain size may be coarse or fine depending on the rate at which the magma cooled. Because magma chambers feed volcanos, and dikes and sills reflect the path of magma ascent toward the surface (Fig. 2.19), a single igneous event may produce plutons, dikes, sills, and lavas of similar composition but very different grain size.

(a)

(b)

Figure 2.20
Field photographs of (a) dikes and (b) a sill.

Felsic. Igneous rocks rich in silica, like the coarse-grained plutonic rock **granite** or its fine-grained volcanic equivalent **rhyolite** tend to contain abundant quartz, K-feldspar and Na-rich plagioclase, and so are light in color. Because of the abundance of f̲eldspar and s̲ilica, the rocks are described as **felsic** (Fig. 2.21). Felsic rocks also have low densities because sodium and potassium are relatively light elements.

Granite is perhaps the most familiar igneous rock because of its use in public buildings. It contains quartz, K-feldspar, and Na-rich plagioclase, sometimes with smaller amounts of biotite, muscovite, and hornblende. The individual mineral grains in coarse-grained granite are large (often about the size of a thumbnail), so that it is usually quite easy to distinguish the different minerals on the basis of their appearance. The feldspars are usually tabular and are often the largest grains. They are commonly pinkish in the case of K-feldspar, or white in the case of Na-rich plagioclase. Quartz, on the other hand, is a glassy grey mineral and tends to fill the spaces between the more rectangular feldspars. Biotite and hornblende are black and usually form smaller grains that are platelike or rodlike, respectively. Like biotite, muscovite forms platy grains but is silver rather than black. The effect of this mineral combination is to produce a grey rock with a distinctive, speckled appearance (Fig.2.16).

Mafic. Igneous rocks that are relatively poor in silica, are richer in the heavier elements magnesium (Mg) and iron (F̲e), like the fine-grained volcanic rock **basalt** and its coarse-grained plutonic equivalent **gabbro**. These rocks are consequently dark in color, have quite high densities, and are described as **mafic** (Fig. 2.21). Basalt is probably the best-known volcanic igneous rock and, unlike granite, is almost black and very uniform in color. It is also noticeably heavier than granite with a density of 3.0 compared to granite's 2.7. It is obvious from its appearance that basalt differs from granite in both composition and grain size. Unlike granite, the individual mineral grains are rarely visible to the naked eye and the rock must be magnified if its component minerals (Ca-rich plagioclase and pyroxene, with or without olivine) are to be identified. In fact, basalt contains no quartz or K-feldspar, whereas granite contains no pyroxene or olivine.

Because both basalt and gabbro come from the same type of magma, they each contain the same minerals: Ca-rich plagioclase, pyroxene, and possibly olivine. However, gabbro has crystals that are readily visible to the naked eye because gabbro is a plutonic rock and therefore cools much more slowly than the volcanic lavas that cool to become basalt.

Intermediate. Igneous rocks that lie between the compositional extremes of felsic and mafic rocks are considered **intermediate.** Their mineral composition is dominated by plagioclase containing both sodium and calcium and they often contain hornblende and biotite. As a result, the colors, densities, and compositions of intermediate rocks lie between those of felsic and mafic rocks.

	FELSIC	INTERMEDIATE	MAFIC	ULTRAMAFIC
Dominant Minerals	Quartz Potassium feldspar	Amphibole Intermediate plagioclase feldspar	Pyroxene Calcium-rich plagioclase feldspar	Olivine Pyroxene
% Silica Content	>65%	52–65%	45–52%	<45%
Color	Light-colored Less than 15% dark minerals	Medium-colored 15–40% dark minerals	Dark gray to black More than 40% dark minerals	Dark-green to black Nearly 100% dark minerals
Coarse-grained	Granite	Granodiorite/Diorite	Gabbro	Peridotite
Fine-grained	Rhyolite	Dacite/Andesite	Basalt	

Granite   *Diorite*   *Gabbro*   *Peridotite*

Rhyolite   *Andesite*   *Basalt*

Figure 2.21
Classification of igneous rocks.

Examples of intermediate rocks include fine-grained **dacite** and **andesite,** the two most common volcanic rocks of mountain belts, and their coarse-grained equivalents, **granodiorite,** and **diorite,** the two most common plutonic rocks.

Ultramafic. Ultramafic rocks, such as coarse-grained **peridotite,** are uncommon on the Earth's surface but are thought to dominate the Earth's interior in the region immediately below the crust. They contain the least amount of silica and are made up of the minerals pyroxene and olivine, and so are the richest in iron and magnesium and have the highest densities. They are distinguished from gabbros by the relative lack of plagioclase. Ultramafic lavas are rare in the geologic record and there are no modern volcanoes that produce lavas of this composition. However, there are important examples of such flows early in the Earth's history and, as we shall learn in Chapter 14, their existence has interesting implications for the evolution of the planet.

Evolution of Igneous Rocks

Chemical Evolution: Bowen's Reaction Series. In the early part of the 20th century the Canadian geoscientist, Norman Bowen, proposed that mafic magmas similar in composition to basalt or gabbro may evolve by cooling and crystallization to produce more silica-rich magmas like those represented by granite or rhyolite. Using laboratory experiments, Bowen monitored the cooling history of basaltic liquid and established a regular sequence in the order of crystallization of its component mafic minerals. This sequence reflects the various temperatures under which different mafic minerals are stable. As basaltic liquid cools, olivine is typically the first mineral to crystallize from the melt (Fig. 2.22). However, as the liquid continues to cool a critical temperature is reached where pyroxene becomes the more stable mafic mineral. At that temperature, the olivine crystals react with the remaining liquid to produce new crystals of pyroxene. With further cooling, amphibole becomes the more stable mineral and is produced by a reaction between the pyroxene crystals and the liquid. At still lower temperatures, biotite is produced in a similar fashion by reaction between the melt and the amphibole crystals. This sequence, in which one mineral becomes converted to another at a specific temperature in the cooling history, is known as a Bowen's Discontinuous Reaction Series.

At the same time that this sequence of mafic minerals is forming, plagioclase feldspar is also crystallizing out of the liquid. Plagioclase is the most common mineral in igneous rocks and has a wide range of compositions that vary between calcium-rich and sodium-rich varieties. As the liquid cools, the most calcium-rich variety crystallizes first (Fig. 2.22). With continued cooling, however, progressively more sodium-rich varieties become more stable. Each new plagioclase crystal

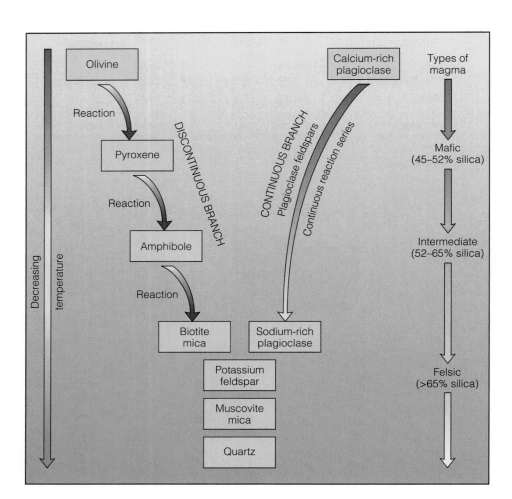

Figure 2.22
Bowen's Reaction Series.

consequently reacts with the remaining liquid as it cools, almost as soon as it forms, in a continuous cycle of crystallization and reaction. This process is known as Bowen's Continuous Reaction Series, and with each reaction progressively more sodium-rich plagioclase is produced.

The process of continuous reaction with falling temperatures displayed by plagioclase is distinct from that which characterizes the sequence of mafic minerals, where reactions occur only at specific temperatures during the cooling history and so are discontinuous. However, both reaction series operate simultaneously and have fundamental implications for the relationship between magmas of various compositions and the rocks they produce as they cool. Basaltic liquids initially crystallize olivine and calcium-rich plagioclase, which are minerals typically found in mafic rocks such as basalt and gabbro. During cooling, however, these early-forming crystals react with the melt and are replaced by amphibole, biotite, and a feldspar intermediate in composition between the calcium- and sodium-rich varieties. These minerals are commonly found in rocks of intermediate composition such as andesite and diorite. At the end of the reaction series, these minerals are replaced by quartz, muscovite, potassium-feldspar, and more sodium-rich plagioclase, all of which are typical of rhyolites and granites.

Bowen proposed that, in most instances, early-forming mafic minerals do not completely react with the melt. Instead, because they are denser than the melt, they sink and accumulate in layers on the floor of the magma chamber in a process known as **crystal settling** (Fig. 2.23). As each layer of crystals forms, it buries the layer beneath so that the crystals are effectively prevented from reacting with the melt above. Because these mafic minerals are richer in iron and magnesium and poorer in silicon than the liquid, the settling and burial of these crystals effectively removes these elements from the melt. As a consequence, the remaining fraction of melt changes in composition, becoming richer in the remaining elements such as silicon, sodium, potassium, and water. This process is known as **fractionation.** The melt consequently evolves from a basaltic toward a rhyolitic composition as crystal settling proceeds.

The concept is similar to eating a box of candies which have a variety of colors (Fig. 2.24). The box may originally contain an equal representation of all colors. However, if you eat your favorite color first, this color will become progressively depleted. If you then eat your second and third favorite colors, the remaining population of candies is very different from (and much less appealing than) the one with which you started.

We now know that there are many ways other than crystal settling in which early-forming crystals can be separated from the remaining magma. For example, in some instances plagioclase crystals will float rather than sink. Crystals can also become caught up in convection currents within large magma

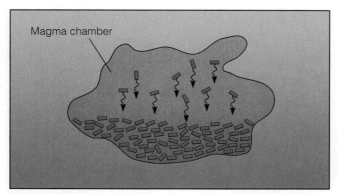

Figure 2.23
Fractionation by crystal settling. Early formed minerals have a higher density than the magma and settle to the bottom of the magma chamber where they accumulate.

chambers or may adhere laterally to the cooler walls of the chamber. However, in each case, the process promotes the chemical evolution of the liquid in the same way as crystal settling. In fact, many experts working on igneous rocks believe that the overall process may be far more dynamic than the isolated system outlined above. For example, magmas may change their compositions by reaction, assimilation, and digestion of the chamber walls. In addition, they may interact with each other. As an example, if mafic and felsic magmas coexist, as is commonly the case in modern volcanically active regions, they may end up exploiting the same conduits and become mixed during their ascent through the crust. This process is

known as **magma mixing** (Fig. 2.25) and commonly produces a magma of intermediate composition.

The chemical evolution of igneous rocks is a dynamic process that may involve one or all of the processes described above in any given area. In addition to these processes, we shall learn in Chapter 5 that differences in the composition and patterns of evolution of igneous rocks also reflects the fact that their respective magmas come from very different areas of melting within the Earth's interior.

Physical Evolution: Explosive Volcanic Eruptions. The contrasting compositions of volcanic rocks produces important variations in their physical properties as well. For example, we have seen that mafic rocks are denser than those of felsic composition. Similarly, a magma's viscosity also depends on its composition. The **viscosity** of a liquid is its resistance to flow. The less viscous a magma, the more easily it flows. Mafic magmas with low silica values have low viscosities whereas more silica-rich felsic to intermediate magmas tend to be highly viscous and gassy. This difference in viscosity profoundly effects the explosiveness of volcanic eruptions, which range from the spectacular lava fountains typical of Hawaii to the devastating pyroclastic eruptions of volcanoes like Mount St. Helens.

Because mafic magmas are fluid, basalt lavas like those of Hawaii tend to flow rapidly and spread out. Any gases trapped as bubbles within such fluid lavas can escape with relative ease, just as they do in a glass of soda pop. As a result, basaltic eruptions are rarely more violent than the familiar jets of incandescent lava we call **lava fountains** (Fig. 2.26). Like the spray from a shaken bot-

(a)   (b)   (c)

Figure 2.24
Three photographs showing the changes in the relative proportion of colored candies as one color is preferentially selected. This example is analogous to the process of fractionation of magma as crystals are effectively removed by settling and the proportion of the remaining components in the magma increases. (Photos/Heather Dutton)

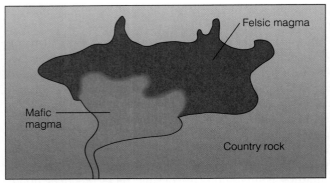

Figure 2.25
Magma mixing. Mafic magma mixes with a felsic magma to produce one of intermediate composition.

tle of champagne, these fountains are produced by the rapid expansion of trapped gas bubbles as the fluid magma reaches the surface and the pressure on it abruptly drops.

As a viscous felsic to intermediate magma ascends toward the surface, the reduced pressure similarly facilitates the formation of gas bubbles, just as it does when you open a can of soda pop. In this case, however, the bubbles are initially held captive by the viscous melt. As the magma continues to rise, a threshold is eventually reached where the pressure from the gas bubbles explodes the magma into hot fragments. At first, the solid rock overlying the magma acts like a pressure cooker, temporarily sealing in the violent cauldron below. However, rocks are an imperfect seal, and when the lid breaks an explosive eruption is inevitable. Explosive or **pyroclastic eruptions** (from the Greek words "pyro" meaning fire and "klastos" meaning broken) vent the gases into the atmosphere and jettison fragments of the hot magma and the surrounding rocks into the air. These eventually fall to the surface to form a **pyroclastic deposit**.

Pyroclastic deposits are classified according to their particle size. In general, the larger particles fall first so that coarser-grained deposits are found closest to the eruptive site. The most finely grained deposits, on the other hand, may be very widely dispersed, eventually forming thin ash beds far from their original source.

The eruption of Mount St. Helens in May, 1980, is a typical example of an pyroclastic eruption (Fig. 2.27). Here the ascent of silica-rich magma to the surface was temporarily impeded by the overlying solid rock. The renewed ascent of magma was indicated by the venting of steam and ash from the volcano's summit and by increasing earthquake activity. Over a period of about one minute, a sequence of events that cascaded like dominoes triggered a series of violent eruptions. The ascent of magma first caused the mountain to bulge, rendering the mountain slope of rock debris and glacial ice unstable. An earthquake caused by the ascent of magma triggered a landslide which removed part of the rock cover holding the magma in. This abruptly reduced the pressure on the magma chamber and caused the gases held within it to rapidly expand, triggering a sideways blast. This blast, in turn, triggered another landslide that further reduced pressure, initiating an eruption from the volcano's central outlet, or **vent**.

Figure 2.26
Lava fountain giving rise to fluid basaltic lava flow, Kilauea volcano, Hawaii.

The most devastating of all violent eruptions is known as a **pyroclastic flow** (Fig. 2.28). This type of flow has some of the properties of lava flows in that it descends down the flanks of volcanoes, but it is also rich in the explosive gases characteristic of pyroclastic eruptions. These flows occur when a mixture of pyroclastic fragments, lubricated by gases and lava, becomes too dense to be vented upwards. The presence of gases significantly reduces the viscosity of these flows. As a result, they are highly mobile, attaining velocities of up to 400 km/hr (250 mph) and spreading as far as 100 km (60 mi) from the eruption. In 1902, a pyroclastic flow destroyed the city of St. Pierre on the island of Martinique in the Caribbean, killing 29,000 people.

Can Explosive Eruptions be Predicted?   Explosive volcanic eruptions present a far greater environmental hazard than eruptions which consist of lava alone, but how can they be predicted? Like most geological phenomena, the inherent difficulty in prediction is that volcanoes are the product of millions of years of geological activity, most of which occurs far too slowly to lend itself to predictions on a time scale of days and weeks. Nevertheless, predictions of volcanic eruptions have had a considerable degree of success because the warning signs, although subtle, can now be identified and interpreted.

Volcanic activity is a phenomena that tends to recur, so a region that has not experienced any in the recent past will probably not experience any in the foreseeable future. Therefore, regions that are prone to eruptions can be readily identified and the sites of previous eruptions can be monitored

(a)

(b)

Figure 2.27
Eruption of Mount St. Helens on May 18, 1980. (a) Initial sideways blast caused by earthquake-triggered collapse of volcano's north face which sharply reduced pressure on the magma within the volcano. (b) Subsequent central eruption from the volcanic vent ejected ash to a height of 19 km (12 mi).

in order to identify any unusual activity. Activities that commonly precede eruptions include inflation of the magma chamber beneath the volcano, increased earthquake activity, and subtle changes in the gas content emanating from the volcanic vent (Fig. 2.29).

Active volcanoes are fed from an intricate plumbing system consisting of magma chambers. Magma usually ascends toward the surface by flowing from one magma chamber to another in response to the local buildup and release of pressure. As magma rises to the chamber immediately below a volcano, the chamber may become inflated. This inflation can be detected by using sensitive tiltmeters and altimeters which respectively measure the slope and elevation (relative to sea level) of the Earth's surface. Satellites, too, can sometimes detect the changes in slope and elevation that accompany inflation of a magma chamber.

Inflation of a magma chamber and the migration of magma from one chamber to another are also accompanied by earthquake activity of low-to-medium intensity. Hence, an unusual amount of earthquake activity usually precedes an eruption. In addition, subtle changes in the gases emanating from the volcano may occur as new magma invades the magma chamber and mixes with the magma already present. For example, the 1991 eruption of Mount Pinatubo in the Philippines was preceded by small but measurable changes in gas emissions which indicated that invasion of the chamber beneath the volcano had occurred. This, together with increased earthquake activity, was used to successfully predict the eruption, and tens of thousands of Filipinos were safely evacuated from the area closest to the volcano.

Figure 2.28
Pyroclastic flow races down the side of Mount St. Helens at speeds in excess of 100 km (60 mi) per hour.

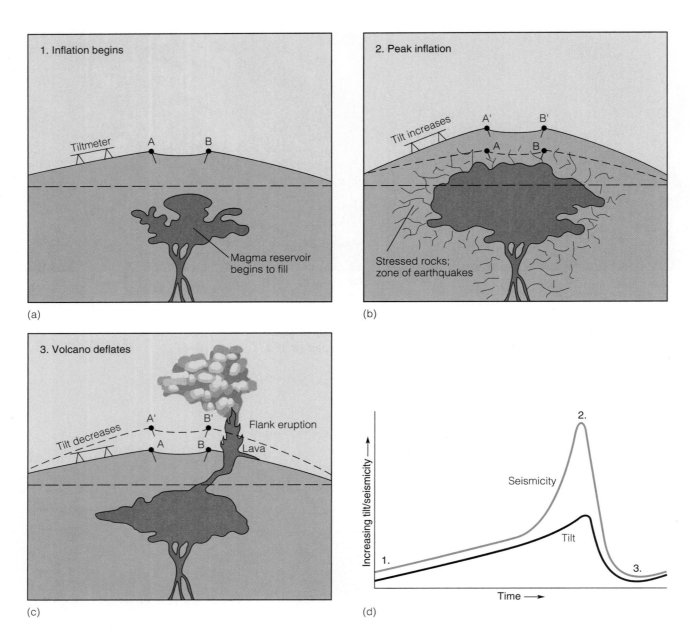

Figure 2.29
Monitoring the three stages in a typical Hawaiian eruption. (a) Volcano begins to inflate and earthquake activity increases as magma reservoir swells. (b) Inflation and earthquake activity peak. (c) Volcano erupts and abruptly deflates as earthquake activity diminishes. (d) Relationship between earthquake activity and tilt from stages 1 to 3.

Igneous rocks, as we have now learned, are rocks produced by the cooling of magmas. The grain sizes of these rocks get larger the longer it takes for the magma to cool. Volcanic rocks erupt at the Earth's surface and therefore cool rapidly and are fine-grained, whereas deep-seated plutonic rocks cool slowly and so are coarse. The minerals in igneous rocks vary with the chemistry of the magma, which can range from silica-rich felsic compositions through progressively more silica-poor intermediate, mafic, and ultramafic varieties. The composition of a magma may change as the magma cools, evolving toward more felsic compositions as crystallization proceeds. A magma's composition also determines its viscosity which increases with increased silica content. Eruptions of low-viscosity mafic lavas are consequently more quiescent than those of felsic magmas, the higher viscosity of which can result in devastating volcanic explosions. Fortunately, the eruption of such volcanoes can often be predicted by the activities that precede them.

2.2.3 Sedimentary Rocks

Formation

In contrast to igneous rocks that form by "hot" processes, sedimentary rocks are formed by "cold" processes that occur only at the Earth's surface. They are the product of the deposition of sediment, which is formed either by erosion of bedrock and soils or by chemical processes. Sand, gravel, and mud are sediments commonly produced by erosion, whereas the salt deposit left behind on a dry lake bed is a common example of a chemical sediment. Other chemical sediments form by organic processes. An example of this is chalk, which is made up of the skeletal remains of microorganisms. These two sources of sediment correspond to the two groups of sedimentary rocks, termed detrital and chemical/biochemical, respectively.

Most sedimentary rocks are **detrital.** Detrital rocks are the product of mechanical erosion by water, wind, and ice, and the transport and deposition of the resulting **detritus** (from the Latin word for "rock waste"). Their texture, in which discrete fragments and particles are either cemented or compacted together, is known as **clastic** (from the Greek word "klastos" for broken).

Detritus generally consists of individual minerals such as quartz and feldspar, or fragments of rocks. Identification of the rock fragments yields important information on the composition of the source region where the erosion took place. In fact, these fragments can sometimes be matched to their original bedrock and can show that detritus may travel considerable distances from its source. For example, some of the detritus currently being deposited at the mouth of the Mississippi River came from the continental interior of North America as far away as Wyoming, Ohio, and Minnesota (Fig. 2.30).

Depending on their stability, the minerals that make up the detritus will behave quite differently as they are transported from the site of erosion to that of their deposition. Quartz, for example, is chemically very stable and so usually survives the journey intact, although its surfaces often become progressively rounded (Fig. 2.33a). Muscovite (or white mica) is also resistant to alteration and tends to survive sediment transport. However, feldspar and many of the mafic minerals are less strongly bonded and tend to break down by reacting with water to form clay minerals. Thus, the presence of feldspar and mafic minerals in sedimentary rocks generally indicates limited transport (Fig. 2.33b). In contrast, the mineral content of detrital sedimentary rocks that travel the farthest typically comprises some combination of quartz, muscovite, and clay minerals.

Less common than detrital sedimentary rocks are those that form through chemical and biochemical processes. **Chemical** sedimentary rocks, such as salt deposits, are entirely inorganic in origin and form by chemical settling, or **precipitation,** when a body of salt water evaporates. **Biochemical** sedimentary rocks, such as coal and limestone, involve organic processes and form from the fossil remains of the organisms themselves or from the remains of their skeletons. The process of deposition for both chemical and biochemical sedimentary rocks tends to produce a texture of interlocking crystals rather than broken fragments, and so is termed **nonclastic.**

Figure 2.30
River deltas, like that of the Mississippi shown here, form where sediment-rich rivers enter the sea at locations where wave energy is limited.

Classification: Types of Sedimentary Rocks

Sedimentary rocks, as we have learned, fall into one of two groups: detrital and chemical/biochemical. These two groups form the basis of the classification of sedimentary rocks.

Detrital Sedimentary Rocks. Detrital sedimentary rocks are classified according to their grain size (Fig. 2.34). **Mudstones** and **siltstones** are very fine-grained varieties with clay- (less than 0.004 mm) and/or silt-sized (from 0.004 to 0.062mm) particles. Thinly layered, or laminated, mudstones that tend to break into flakes are called **shales. Sandstones** are made up of sand-sized (from 0.062 to 2 mm) grains and include the rock types known as **greywacke, arkose,** and **quartz sandstone,** in which the sand grains are dominated by rock fragments, feldspar, and quartz, respectively. Detrital sedimentary rocks containing fragments that are greater than 2 mm in size, such as gravel, pebbles, and boulders, are called either **conglomerates** if the fragments are rounded, or **breccias,** if they are angular.

Chemical and Biochemical Sedimentary Rocks. Other sedimentary rocks are the product of chemical and biochemical deposition so that their component minerals form a nonclastic texture of interlocking crystals. These nonclastic sedimentary rocks are classified according to their composition (Fig. 2.34).

Chemical sediments are formed by inorganic processes. **Evaporites** are rocks composed of salts left behind by the

SALT OF THE EARTH

Precious minerals such as diamond, gold, and silver are glamorous because their worth lies in their appearance as well as their distinctive properties. Many common minerals, however, also have a profound influence on our lives although their roles go virtually unheralded. As a result, these important resources are often not used very efficiently.

The common mineral halite, or rock salt, is an excellent example of a vital but often overlooked resource. It has been produced naturally on several occasions in Earth history by the evaporation of great inland seas which left behind huge deposits of salt. For example, when the Mediterranean Sea dried up some 6 million years ago, the salt deposited on its floor reached thicknesses of over 2000 m (6500 ft).

More recently, the same process operating on a much smaller scale created Great Salt Lake in northern Utah. Starting about 30,000 years ago, during the last Ice Age, greater precipitation and reduced evaporation caused huge quantities of water to pond in northwestern Utah, forming a great freshwater lake that Earth Scientists call Lake Bonneville (Fig. 2.31). Following the Ice Age, as the climate became warmer and drier, the evaporation of its fresh water caused this great lake to shrink rapidly and become saltier. Eventually, large portions of

the lake dried up, leaving behind deposits of salt. Today, one such portion makes up the Bonneville Salt Flats, famed as a raceway for land-speed records. Only remnants of Lake Bonneville, such as Great Salt Lake and Utah Lake, now remain, but its former extent can still be seen from the prominent wave-cut terraces that were produced as its shoreline receded and now lie high in the surrounding uplands.

Lake Bonneville, however, was small in comparison to some of the great inland seas of the past. For example, seas like those that evaporated to produce the Louanne Salt, now buried beneath the Gulf Coast states, or the Salina Salt, now buried beneath the Great Lakes.

Both of these deposits contain huge thicknesses of rock salt that is literally the salt of the Earth. Rock salt from these deposits is mined in both Canada and the United States (Fig. 2.32) and, in one form or another, it impacts each of us every day. Indeed, it is essential to life itself.

Rock salt is made up of interlocking crystals of the mineral halite (NaCl) that we use daily as table salt. In fact, well over 2 million metric tons of salt are used each year by the food industry in North America alone. Salt for human consumption, however,

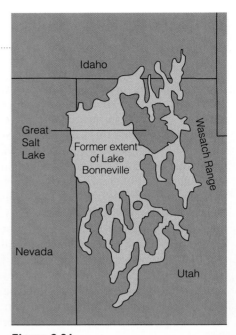

Figure 2.31
Former extent of Lake Bonneville in northern Utah.

is only about 3% of the total domestic demand. Together, the United States and Canada produce over 50 million metric tons of salt each year and the demand is higher still. Most is used by the chemical industry in the production of such chemicals as chlorine, caustic soda, and polyvinyl chloride. More familiar to us is its role in highway de-icing, a use that accounts for some 20 million metric tons a year.

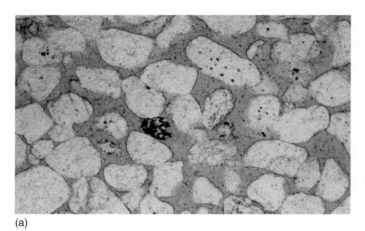

(a)

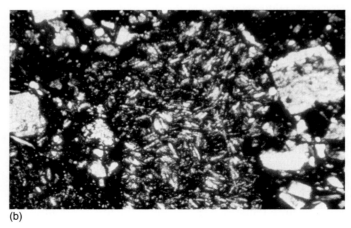

(b)

Figure 2.33
Differences in sand grains with transport distance as seen through the microscope. (a) Far-traveled sand with well-rounded quartz grains of similar size. (b) Near-source sand with angular quartz grains and large angular fragments of unstable volcanic rock.

Figure 2.32

Rising above the swamps of the Atchafalaya Basin in southern Louisiana, Avery Island sits atop a salt dome mined by the Cargill Salt Company. It is also cultivated for peppers and is, perhaps, better known as the home of Tabasco Sauce. (Photo Damian Nance)

in this way are important sources of oil and natural gas in the Gulf Coast region of the United States.

Rock salt is consequently a sedimentary rock of enormous economic importance, both as a resource of its component mineral, halite, and because of its ability to form reservoirs for oil and natural gas. It is a mineral on which our lives depend. However, its use is not without consequences. Salt has been criticized by both the environmental and medical communities. Its use for highway de-icing can damage road-side vegetation and contaminate ground water supplies, and causes automobiles to rust. Although essential to the human diet, overcon-sumption of salt may cause hypertension and significant efforts have been made to reduce the quantity of salt added to foodstuffs. There may also be health risks in using chlorine pro-duced from salt for the treatment of municipal water supplies, and its use in chlorofluorocar-bons (CFCs) is thought to contribute to ozone depletion in the upper atmosphere.

Rock salt is consequently an excellent example of the necessity to maintain a bal-ance between our need to use the Earth's resources and the environmental and/or medical consequences of that use. Rock salt is one of many resources that the Earth provides, which add daily to the quality of our lives. Like most resources, it has a finite lifespan because we exploit it faster than nature can replenish it. It is up to us to use these resources wisely and not take them for granted.

However, rock salt is also a highly mobile material and will flow under the weight of overlying sedimentary rocks just as glacier ice will flow under the influence of gravity. It is also of low density (2.2) and, when buried by clastic sedimentary rocks of higher density (2.6), it can become gravita-tionally unstable. Where this occurs, the salt tends to exploit weaknesses in the overlying rock layers (such as fractures) and pushes its way upward toward the surface in huge fin-gers called salt domes. As salt domes rise, the layers of sedimentary rock they pierce are drawn upward and the rocks above are arched to form structures in which petroleum can accumulate. Petroleum reservoirs formed

---

evaporation of water. As evaporation diminishes the volume of water, salts in solution become saturated and crystals develop. The best known of these evaporites is rock salt, which is made up of grains of the mineral halite (NaCl) (see Living on Earth: Salt of the Earth). Minerals that comprise other evaporites include gypsum ($CaSO_4 \cdot 2H_2O$), which is used for plastering, anhydrite ($CaSO_4$), which is used as "plaster of Paris" in the protection of broken bones, and dolomite ($CaMgCO_3$).

Biochemical sediments on the other hand, involve organic processes. **Coal,** for example, is formed from highly compressed plant remains. Some sediments of this group can be either inor-ganic or biochemical in origin. **Limestone,** which is composed of the mineral calcite ($CaCO_3$), and **chert,** which is made up of microcrystalline quartz (SiO_2), can be the product of the direct chemical precipitation of these minerals from water. Alternatively, organisms may extract these components from sea water. When they die, their shells and other skeletal remains are then deposited to form calcareous or silica-rich beds.

Depositional Environments

The physical settings in which sediments are deposited such as a flood plain, delta, dune, or lake, are known as **depositional environments,** and much can be learned about the history of our planet by studying the ancient depositional environments preserved in the sedimentary rock record.

Most of our understanding of ancient depositional environ-ments is derived from the study of modern settings of sediment deposition. The findings are then applied to ancient sedimenta-ry rocks based on the understanding that the sedimentary char-acteristics of modern depositional environments should be duplicated by those of the past. As we shall learn in Chapter 3, this idea that "the present is the key to the past" is one of the basic principles on which the science of the geology has been built.

In modern settings, there are many environments where sediments are deposited. However, most fall into one of three categories: continental, marine, or the coastal environment lying between these two (Fig. 2.35).

Continental. Continental environments are dominated by the erosion and deposition associated with rivers and streams which dissect the landscape and carve out broad floodplains. In cold climates ice may take the place of water as the principal agent of erosion and deposition and in hot desert regions wind may perform this function.

Because each of these agents of erosion and deposition produce sediment of quite different character, the nature of the sedimentary rock deposited in continental environments is strongly influenced by climate. Desert sands, for example, are typically made up of wind-blown grains. Winds are effective sorters of detritus so these grains are very similar in size, and are said to be "well sorted." When glaciers melt, on the other hand, debris ranging from boulders to mud are deposited rapidly side-by-side. Such deposits are said to be "poorly sorted" (see Figure 2.33).

This difference reflects the varying abilities of these agents of erosion to sort material as it is transported, and is best displayed by flowing water. For example, you may have observed

DETRITAL ROCKS			CHEMICAL/BIOCHEMICAL ROCKS		
SEDIMENT NAME AND PARTICLE SIZE	**DESCRIPTION**		**TEXTURE**	**COMPOSITION**	
Gravel >2 mm	Rounded rock fragments	 *Conglomerate*	Clastic or nonclastic	Calcite, $CaCO_3$	*Limestone*
	Angular rock fragments	 *Breccia*	Clastic or nonclastic	Dolomite, $CaMg(CO_3)_2$	*Dolomite*
Sand 0.062–2 mm	Quartz predominant	 *Sandstone*	Nonclastic	Halite, NaCl	*Rock salt*
	Quartz with >25% feldspar	 *Akrose*	Nonclastic	Gypsum, $CaSO_4 \cdot 2H_2O$	*Gypsum*
	Dark color; quartz with considerable feldspar, clay, and rock fragments	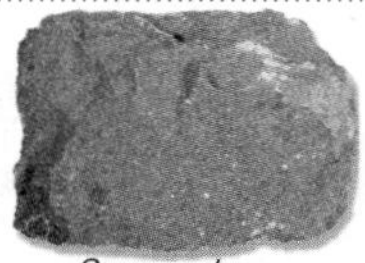 *Graywacke*	Nonclastic	Microcrystalline quartz, SiO_2	*Chert*
Mud 0.004 mm	Splits into thin layers	 *Shale*	Nonclastic	Altered plant remains	*Coal*
	Breaks into clumps or blocks	 *Mudstone*			

Figure 2.34
Classification of sedimentary rocks.

that rivers become muddier following major storms. This reflects the enhanced ability of a river to erode and transport detritus while it is in flood. This tendency is even more pronounced in monsoon climates, like that of India, where there is a strong seasonality in rainfall, and boulders transported downstream during periods of high-energy flooding can be quite large. If a river overflows its banks and spreads across the floodplain, the current slows. As a result, coarser detritus can no longer be carried, and so is deposited in a bank or **natural levee** along the river's edge, while finer-grained material previously held in suspension is deposited on the floodplain. Thus, muds may be deposited on the floodplain at the same time that coarser material is being deposited in the main channel (Fig. 2.36).

Marine. In marine environments, the ongoing erosion of continents may result in the deposition of a thick pile of sediment along their ocean margins. Modern examples of such deposits along the margins of the Atlantic Ocean average 2.5 km (1.6 mi) in thickness. The rapid buildup of sediment renders localized areas of the margin unstable, especially if there is earthquake activity at the same time. These instabilities can cause submarine avalanches, in which sediment hurtles down submarine canyons in muddy flows called **turbidity currents.** The sediment is typically deposited at the mouth of the canyon in a **submarine fan** (Fig. 2.35).

Where sediment supply to marine environments is not overwhelming, either in the amount of deposited material or in the rate at which it is deposited, ecosystems flourish.

Because many organisms in the shallow marine realm utilize calcium carbonate ($CaCO_3$) in their shells and skeletons, which are sometimes used to build massive reefs like those of the Bahamas, the presence of ecosystems is often recorded in the deposition of limestone beds.

In many marine environments, individual layers of sediment are not continuous. In nearshore environments, like beaches, where water movement is commonly more energetic, sand-sized grains are preferentially deposited. At the same time, clay-sized particles are deposited in the lower-energy environments further offshore to form muds. If the offshore environment is conducive to a flourishing ecology, however, limestone sediments will form instead. In tracing a layer of sediment from nearshore to offshore, one may start in a sediment dominated by sand, pass through one dominated by mud, and end with one dominated by limestone (Fig. 2.37), as the proportion of detritus decreases.

Coastal. Coastal regions commonly show some of the characteristics of both continental and marine deposition. Vast volumes of sediments are deposited where the water velocity of a major river slows down as the river meets the sea. Because of this, the river can no longer carry its sediment load, which results in it often being dumped at the river mouth to form a **delta.** At the mouth of the Mississippi River, for example, a delta has formed at the site where more than 500 million tons of sediment are deposited annually at the margin of the Gulf of Mexico (Fig. 2.30). Coastal sediments may also be deposited

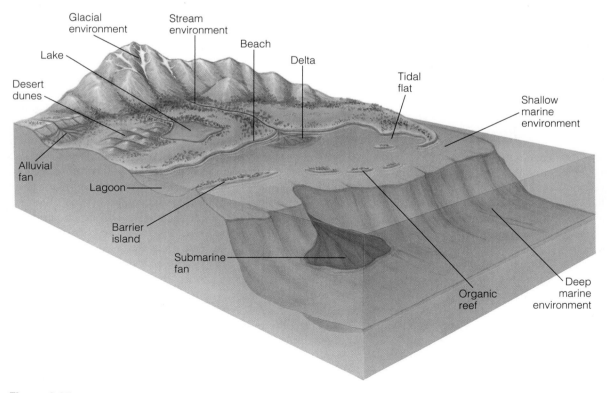

Figure 2.35

Major depositional environments. Sediments can be deposited in continental environments such as rivers, lakes. and deserts, in shallow or deep marine environments, and in coastal environments that lie between the two and include tidal flats, barrier islands, and lagoons.

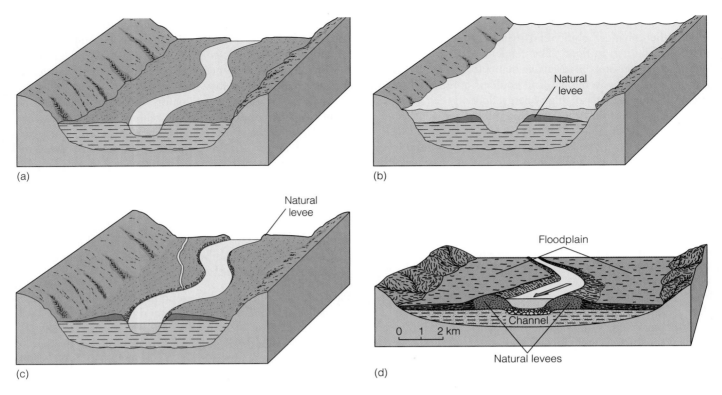

Figure 2.36
Stages in the formation of deposits on a floodplain. (a) Stream at low-water stage. (b) Flooding stream and deposition of natural levees after many episodes of flooding. (c) After flooding. (d) The coarsest sediment is deposited in the stream channel, where the stream's velocity is greatest, while sand is deposited on the levees, and finer silt and clay are deposited on the floodplain as velocity diminishes during overflow.

in muddy tidewater areas known as **tidal flats,** around offshore sand bars called **barrier islands,** or in lagoons, which are long bodies of seawater isolated from the oceans by sand bars or reefs (Fig. 2.35).

Interpreting the Sedimentary Rock Record

As we shall learn in Chapter 3, when this understanding of modern depositional environments is applied to the ancient rock record, the layers of sedimentary rock stacked one on the other can be read like the pages of a book, each recording the setting in which the original sediment was deposited. Because each sedimentary layer must be younger than the one on which it was deposited, the sedimentary rock record documents the history of sedimentation, and all changes in sedimentary rock type record variations in deposition with time. These variations reflect the dynamic nature of the Earth's surface, and the continual change in the sedimentary environments developed on it. The positions of coastal settings, for example, and therefore the distribution of nearshore-to-offshore sedimentary environments, continually change with time. This is because the distribution of sedimentary deposits in coastal regions is very susceptible to the rise and fall of sea level. As we shall learn in Chapter 8, sea level can fluctuate for a wide variety of reasons. For example, today we are concerned that global warming may cause melting of the polar ice caps,

resulting in rising sea levels that may swamp our coastal regions. Similarly, at the onset of the Ice Age approximately 1.6 million years ago, sea level abruptly dropped.

As sea levels rise and fall, coastal regions migrate. Rising sea level produces what is known as a **marine transgression,** during which the relative distribution of sand, mud, and limestone migrates shoreward. Thus limestone is deposited on top of a mudstone, and mudstone is deposited on top of sandstone (Fig. 2.37). Falling sea level results in a **marine regression,** during which the opposite pattern is produced, with sandstone deposition on top of mudstone and mudstone deposition on limestone. Therefore, either fluctuation will be documented in the sedimentary rock record by changes in rock type.

In summary, sedimentary rocks are those produced at the Earth's surface by the deposition of sediment that is either of detrital origin or is formed through chemical and biochemical processes. Detrital sedimentary rocks, such as sandstones and shales, have a clastic texture of broken fragments and are classified according to their grain size. In contrast, chemical and biochemical sedimentary rocks, such as evaporites and limestones, have a nonclastic texture of interlocking grains and are classified according to their composition. Sedimentary rocks can form in a wide variety of depositional environments, each with distinctive sedimentary characteristics and ranging from continental through coastal to marine. The sedimentary rocks preserved in the rock record document ancient depositional

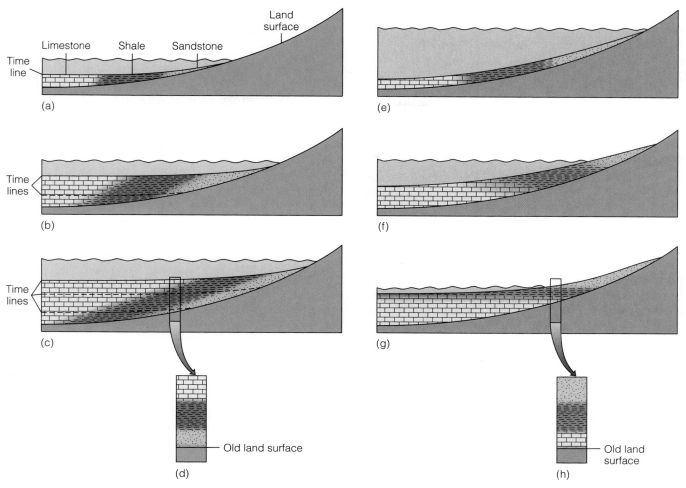

Figure 2.37
Deposition of sedimentary rocks with rising and falling sea level. (a), (b), and (c) Three stages of a marine transgression (sea level rise), and (d) resulting vertical sequence of sedimentary rocks. (e), (f) and (g) Three stages of a marine regression (sea level fall), and (h) resulting vertical sequence of sedimentary rocks.

environments, and much can be learned about the history of the Earth by using a knowledge of modern depositional environments to interpret those of the past.

2.2.4 Metamorphic Rocks

Formation

In contrast to sedimentary rocks, metamorphic rocks are formed by processes that can only occur below the Earth's surface. **Metamorphic rocks** are those that have been subjected to such great pressures and heat that their textures and/or compositions have been profoundly changed from their original state. Metamorphic rocks differ from igneous rocks because the changes involved in their formation take place while the rock is still in the solid state and do not involve melting.

When crustal rocks are deeply buried, they may encounter conditions unlike those under which they formed. They become subjected to the intense pressures and increased temperatures that exist at depth, and may also undergo chemical changes as a result of their interaction with hot circulating fluids. In response to these new conditions, the rocks may experience significant changes in their mineral compositions and textures, and so become transformed into new kinds of rock (Fig. 2.38).

The process of transformation is called **metamorphism** (from the Greek words meaning "to change form"). It is simply the path taken by the rock to reach a state of equilibrium with the new environment. If some of a rock's original minerals are subjected to conditions significantly different from those under which they formed, they may become unstable and break down or react together to form new stable minerals. Because new conditions can come into play anywhere from the base of the crust to within a few kilometers of the Earth's surface, metamorphic rocks provide important clues about the geologic processes that operate at various levels within the Earth's crust. Many minerals, for example, only form under a limited range of pressure or temperature conditions. Therefore, their presence in a metamorphic rock helps to determine the conditions of the rock's metamorphism and, consequently, the kinds of pressures and temperatures that existed within the Earth's crust at the place where it formed (see Dig In: Key Metamorphic Minerals, page 45).

Most metamorphic rocks are the product of one of three types of metamorphism: regional, contact, and dynamic, and are classified according to their texture and mineral content.

Types of Metamorphism

Three types of metamorphism, each producing characteristic metamorphic rocks, are distinguished by the environment in which they occur. **Regional metamorphism** occurs in the roots of mountain belts, **contact metamorphism** occurs around hot igneous bodies, and **dynamic metamorphism** occurs along fault zones.

Regional Metamorphism. Regional metamorphism takes place when large areas of crustal rocks are subjected to the elevated temperatures and pressures associated with mountain building. Regional metamorphic rocks are therefore produced inside mountain ranges and become exposed over wide regions of the Earth's surface following their uplift and erosion. For example, regional metamorphic rocks are being produced today in the hidden interior of the Himalayas and will become exposed when the Himalayas are sufficiently eroded. Ancient mountain chains like the Appalachians have undergone hundreds of millions of years of erosion so that metamorphic rocks, like those forming beneath today's Himalayas, are now widely exposed.

Contact Metamorphism. Metamorphic rocks are also produced by contact metamorphism, which occurs when a rock is heated by the presence of an adjacent body of magma. For example, sedimentary rocks adjacent to an intrusive igneous body commonly show progressive changes within the contact zone that are simply the result of baking. In this case, the metamorphism is largely the result of elevated temperatures and the igneous body is said to possess a halo, or **aureole,** of contact metamorphic rocks (Fig. 2.40). A metamorphic rock formed in this way is called a **hornfels.**

Because contact metamorphism occurs only in the immediate vicinity of an igneous body, it tends to be limited in extent. In regions where mountain building has taken place, however, intense temperatures invariably lead to the generation of magma so that the overall result is an interesting interplay of regional and contact metamorphic effects.

Dynamic Metamorphism. The least-common metamorphic rocks are those produced by dynamic metamorphism. This occurs along fractures in the Earth's crust along which significant movement has taken place. Dynamic metamorphic rocks are largely the product of elevated pressures and form as the result of either crushing or smearing of rocks along the zone of fracturing. Three factors are important in determining the type of rock formed: temperature, pressure, and the intensity of deformation known as the **strain rate.** For an active fracture zone, the strain rate is simply a measure of the speed at which the fracture zone moves. Crushing tends

(a)

(b)

(c)

(d)

Figure 2.38
Progressive metamorphism transforms a sedimentary shale (a) first to a slate (b) then to a schist (c), and finally a gneiss (d) with increasing metamorphic intensity (grade).

Key Metamorphic Minerals

Three minerals which, with few exceptions, form only in metamorphic rocks merit special mention. They are the minerals andalusite, kyanite, and sillimanite. These three minerals are polymorphs, that is, they have the same chemical formula (Al_2SiO_5) but different crystal structures. As implied by their chemical formula, they tend to form in aluminum-rich rocks, typically those that were originally mudstones. At most pressures and temperatures, however, only one of the three is stable. Andalusite and kyanite are both stable at relatively low temperatures. However, andalusite has the lowest density of the three (about 3.2) and is therefore only stable at relatively low pressures. On the other hand, kyanite has the highest density (about 3.6) and is stable at much higher pressures. Sillimanite (density about 3.25) is only stable at relatively high temperatures.

Because of their stability under different conditions of pressure and temperature, which of the three polymorphs develop in a metamorphic rock is an important indicator of the conditions of metamorphism. Their relative stabilities over a range of pressures and temperatures has been the subject of much experimentation and theoretical analysis. The most commonly accepted scheme is shown in Fig. 2.39. Each of the fields in this diagram represent the range in pressures and temperatures over which one of the three polymorphs is stable. At the pressure-temperature conditions represented by the lines, two of the polymorphs can stably coexist, and at one point (near 500°C and 3.5 kbars pressure) all three polymorphs are stable. Therefore, the stable presence of all three minerals in a metamorphic rock uniquely fixes the pressure-temperature conditions of metamorphism.

Because metamorphic rocks do not form at the Earth's surface, we cannot study

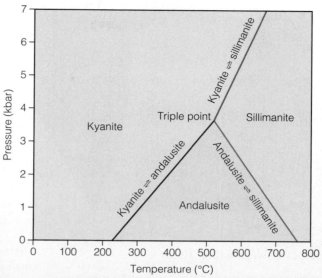

Figure 2.39

Aluminosilicate polymorphs, andalusite, sillimanite and kyanite all have the same chemical composition (Al_2SiO_5) but very different physical properties and crystal structures. This is because they form under different conditions of pressure and temperature. Kyanite, the polymorph with the highest density, is stable at high pressures and low temperatures. Andalusite, the polymorph with the lowest density, is stable at low pressures and low temperatures. Sillimanite is stable at high pressures and high temperatures. The graph shows the range of pressures and temperatures over which each polymorph is stable. The double arrows mean that the transformations are reversible. All three minerals can stably coexist at the triple point.

the modern environments in which they are forming as we do with sedimentary rocks. Those metamorphic rocks that do occur at the surface today are exposed only because the rocks above them have been removed by erosion as part of the rock cycle.

You might ask why metamorphic rocks are exposed on the Earth's surface at all. Why didn't their minerals simply revert back to those that are stable under surface conditions during uplift and erosion? In order to answer this question we need to recall a basic principle of chemical reactions. This principle states that the rate of a chemical reaction increases exponentially with temperature, so the higher the temperature the faster the reaction goes. Chemical reactions are highly efficient at high temperatures and very sluggish at low temperatures. This is because high temperatures increase the energy at the site of the reaction and so facilitate the chemical breakdown. Metamorphic rocks tend to record the maximum temperature that the rocks were subjected to. As these rocks are uplifted and cooled, reaction rates drastically diminish so that the metamorphic minerals formed at elevated temperatures are preserved.

---

to occur in near-surface environments where the rocks are cold and brittle, and so pulverize to form a rock called a **fault breccia.** Smearing, on the other hand, requires higher temperatures and produces a distinctive streaky metamorphic rock called a **mylonite**.

The effect of strain rate on rocks is not obvious but can be illustrated with a simple analogy. If you take a stick of butter from the refrigerator and subject it to a high strain rate by abruptly bending it, the brittle stick will snap. If, on the other hand, you subject it to a low strain rate by flexing it slowly, it will be more compliant and bend or smear without breaking. Rocks are similarly more prone to smear at low strain rates, whereas they tend to snap, and so produce brittle structures when strain rates are high.

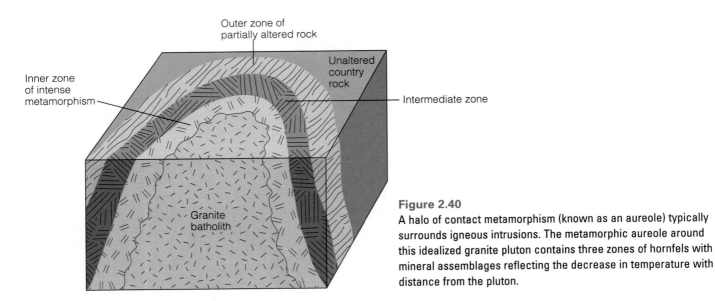

Figure 2.40
A halo of contact metamorphism (known as an aureole) typically surrounds igneous intrusions. The metamorphic aureole around this idealized granite pluton contains three zones of hornfels with mineral assemblages reflecting the decrease in temperature with distance from the pluton.

Classification: Types of Metamorphic Rocks

Metamorphic rocks are classified on the basis of their texture and mineralogy (Fig. 2.41), from which much can be learned about the conditions under which they formed. Metamorphic rocks are divided into two main groups, **foliated** and **nonfoliated,** based on the presence or absence of layering within the rock, known as a **foliation.** Within these two groups, metamorphic rocks are further divided according to their mineral content, which varies with the conditions of pressure and temperature during metamorphism, or **metamorphic grade.**

Foliated Metamorphic Rocks. The texture most commonly found in metamorphic rocks is the closely spaced planar layering known as a foliation, and metamorphic rocks that possess such a feature are termed foliated metamorphic rocks. Foliation is produced by the alignment and segregation of minerals that occurs as a result of their growth under conditions of directed pressure (Fig. 2.42). These conditions are common during mountain-building processes.

Which minerals define the foliation depends on the composition of the rock being metamorphosed and the intensity or grade of metamorphism. The metamorphic grade is considered to increase as the pressures and temperatures of metamorphism increase. Metamorphism that occurs under conditions of increasing pressures and temperatures, like that which accompanies the progressive burial of metamorphic rocks, is known as **prograde metamorphism** because the metamorphic conditions progress from lower to higher metamorphic grade.

Prograde metamorphism is accompanied by two important trends in metamorphic rocks. First, while directed pressure produces a foliation, increasing grades of metamorphism produce coarser-grained minerals because the transport of appropriate chemical elements to the growing minerals is much more efficient at high temperatures. Second, prograde reactions tend to use up some of the fluid stored within the reacting crystals. As a result, low-grade metamorphic rocks tend to contain many minerals with water in their crystal lattices, whereas high-grade metamorphic rocks contain a relatively small amount of these minerals.

For example, consider the prograde metamorphism of a sedimentary rock such as a shale (Fig. 2.43). Shales contain a lot of clay minerals which store abundant water in their crystal lattices. The formula of the important clay mineral kaolinite $[Al_4Si_4O_{10}(OH)_8]$ reflects the abundance of water stored in its crystal lattice. As a shale undergoes metamorphism as part of the rock cycle, these clay minerals become unstable and react to form very fine-grained flaky minerals such as the mica, muscovite $[KAl_3Si_3O_{10}(OH)_2]$. As is evident from the formula, the new mica grains contain less water in their mineral structure than clay minerals. They are also so small that they can only be clearly seen with the aid of a microscope. As a result of directed pressure, these flaky minerals are aligned into foliation planes along which the metamorphic rock tends to split. In this way, sedimentary shale changes to form the fine-grained metamorphic rock we know as **slate,** once widely used in roofing (Fig. 2.38).

With increasing metamorphism, the micas in the slate continually recrystallize and become coarser-grained. Eventually this process produces a rock known as a **schist,** in which the recrystallized micas are visible to the naked eye. Both slates and schists are produced by the regional metamorphism of shale, but schists occur in regions that have experienced higher temperatures and pressures. As the metamorphic grade increases still further, micas are no longer stable and the rocks show no preferential direction of splitting. Instead, their foliation is a mineral banding formed by alternating layers of different mineral composition. The coarse-grained rock produced is called a **gneiss.** Because they form at high grade, gneisses tend to be rich in minerals like garnet, that contain little or no water in their crystal lattices.

Most foliated metamorphic rocks are given individual names that simply reflect the nature of their foliation and the minerals they contain. In most cases, their names are self-explanatory. For example, the names **mica schist,** for a schist made of mica, and **granite gneiss,** for a gneiss formed from granite, need no additional clarification.

TEXTURE	METAMORPHIC ROCK	TYPICAL MINERALS	METAMORPHIC GRADE	CHARACTERISTICS OF ROCKS	PARENT ROCK	
Foliated	Slate	Clays, micas, chlorite	Low	Fine-grained, splits easily into flat pieces	Mudrocks, claystones, volcanic ash	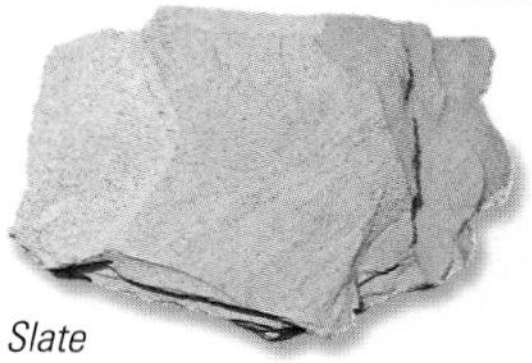 *Slate*
	Schist	Micas, chlorite, quartz talc, hornblende, garnet, staurolite, graphite	Low to high	Distinct foliation, minerals visible	Mudrocks, carbonates, mafic igneous rocks	 *Schist*
	Gneiss	Quartz, feldspars, hornblende, micas	High	Segregated light and dark bands visible	Mudrocks, sandstones, felsic igneous rocks	 *Gneiss*
Nonfoliated	Marble	Calcite, dolomite	Low to high	Interlocking grains of calcite or dolomite, reacts with HCl	Limestone or dolostone	 *Marble*
	Quartzite	Quartz	Medium to high	Interlocking quartz	Quartz sandstone	 *Quartzite*
	Amphibolite	Hornblende, plagioclase	Medium to high	Dark-colored, weakly foliated	Mafic igneous rocks	 *Amphibolite*
	Hornfels	Micas, garnets, andalusite, cordierite, quartz	Low to medium	Fine-grained, equidimensional grains, hard, dense	Mudrocks	 *Hornfels*

Figure 2.41
Classification of metamorphic rocks.

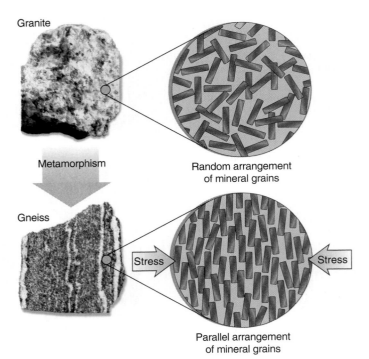

Granite

Metamorphism

Gneiss

Random arrangement
of mineral grains

Stress   Stress

Parallel arrangement
of mineral grains

Figure 2.42
Under directed pressure, planar minerals, such as the micas, become reoriented so that their surfaces are aligned at right angles to the direction of pressure. The resulting mineral alignment gives the rock a planar texture known as a foliation. If the coarse-grained igneous rock (granite) above underwent intense metamorphism, it could end up looking like the foliated metamorphic rock (gneiss) below.

Formula	Mineral	Environment
$Al_4Si_4O_{10}(OH)_8$	Clays	Sedimentary
$(MgAl)_3(SiAl)_2O_5(OH)_4$	Chlorite	Greenschist facies
$KAl_3Si_3O_{10}(OH)_2$	Muscovite	
$KMg_3AlSi_3O_{10}(OH)_2$	Biotite	
$Fe_3Al_2Si_3O_{12}$	Garnet	
$Fe_2Al_9O_6(SiO_4)_4(OH)_2$	Staurolite	Amphibolite facies
Al_2SiO_5	Kyanite	
Al_2SiO_5	Sillimanite	

Increased metamorphism

Figure 2.43
Minerals that develop from clay-rich shales with progressive metamorphism show a decrease in the amount of water in their mineral formulae with increasing metamorphic grade.

Nonfoliated Metamorphic Rocks.   Other metamorphic rocks, because of the minerals they contain or the conditions under which they form, lack a foliation and are termed **nonfoliated metamorphic rocks.** Coarse, crystalline **marbles** produced by the metamorphism of limestone, and coarse, crystalline **quartzite** produced from the metamorphism of quartz sandstone, are common nonfoliated metamorphic rocks. Coarse, hornblende-rich **amphibolites,** produced by the metamorphism of basalt, may also be nonfoliated. All three rocks are largely made up of only one mineral (calcite or dolomite in marble, quartz in quartzite, and hornblende in amphibolite), and each lacks micas. As a result, neither a splitting foliation like that of schists, nor a mineral banding like than of gneisses can form. Instead, all these rocks have a nonfoliated texture of interlocking crystals.

The metamorphic rock **hornfels** (Fig. 2.41) is also nonfoliated. In this case, however, the rock lacks a foliation because it formed in the absence of directed pressure as it is the product of contact metamorphism.

Metamorphic Facies

Much can be learned about the conditions under which metamorphism took place from the mineralogy of the rocks themselves. As different minerals are stable under different conditions of temperature and pressure, the conditions of metamorphism for a given rock are confined to the range of pressures and tem-

peratures over which all of its constituent minerals are stable. This range can be determined experimentally and has been used to constrain the temperature and pressure conditions, and so define what is known as the **facies** of metamorphism. A **metamorphic facies** is an association of metamorphic rocks that were metamorphosed under similar conditions of temperature and pressure. Hence, the temperature and pressure (or grade) of metamorphism of all rocks in a given metamorphic facies are similar.

The most important metamorphic facies are those shown in Figure 2.44. Of these, the most common is the **greenschist facies,** so called because the metamorphism of many rock types at this grade produces schists that contain a variety of green minerals. At slightly more elevated temperatures and pressures, amphiboles such as hornblende become stable, and metamorphism enters the **amphibolite facies.** At still higher temperatures and pressures, such as those found in the lower crust, mineral assemblages of the **granulite facies** become stable. This type of facies is so called because rocks metamorphosed at this grade commonly show a granular texture.

Under certain conditions, very high pressures can be developed at much lower temperatures than those of the granulite facies. The temperatures are, in fact, similar to those of the greenschist facies, and as with this lower-pressure facies, schists are one of the principal metamorphic rocks produced. However, one of the characteristic stable minerals under high-pressure conditions is a blue amphibole known as **glaucophane.** Because this blue amphibole gives its color to the rocks in which it occurs, this low-temperature but high-pressure metamorphism produces rocks of the **blueschist facies.**

The highest grade of metamorphism, the rare **eclogite facies,** develops only under extremes of crustal temperatures and pressures. As we shall learn in Chapters 5 and 6, rocks of the blueschist and eclogite facies are usually associated with subduction zones, along which the Earth's tectonic plates are consumed. This is because rocks encounter the necessary extremes of pressure and temperature only through subduction.

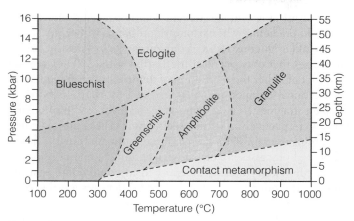

Figure 2.44
Simplified diagram showing the temperature and pressure conditions of the various metamorphic facies. Metamorphic facies are characterized by particular minerals that form under the same general pressure-temperature conditions.

The limit of metamorphism occurs where temperatures exceed the melting point of a rock, in which case a partially molten rock is produced and we enter the realm of igneous rocks.

Metamorphism and Deformation of Rocks

Rocks deform when they are subjected to conditions of directed pressure that substantially differ from those under which they formed. Because of this, deformation is often intimately related to, and accompanied by, metamorphism. Whereas metamorphism refers to changes in the mineral content and texture of rocks, deformation results in changes in their shape and volume. The development of metamorphic foliation, for example, occurs as the result of directed pressure and so is directly related to the deformation that accompanied metamorphism. As we will learn in subsequent chapters, deformation and metamorphism are manifestations of plate tectonic activity related to the processes and events that build the continental crust.

2.3 Revisiting the Rock Cycle

As we have now learned, the solid Earth is made up of rocks, the chemical constituents of which are minerals. Minerals, in turn, are made up of atoms linked together in a regular framework by a variety of chemical bonds. Because of this, the composition of any mineral can be expressed in terms of a chemical formula. Also, all minerals have a characteristic set of physical properties that depend on the nature of their component atoms, the geometry of their arrangement, and the strength of the bonds that link them. Similarly, all minerals exhibit a characteristic range of pressures and temperatures over which they are stable.

Rocks, on the other hand, are mineral aggregates and are grouped according to the way they form. Igneous rocks form from the cooling of molten rock, sedimentary rocks are produced by surface processes, and metamorphic rocks result from the action of pressure and heat in the Earth's interior. Although the origins of the three rock groups vary, the environments in which they form are linked by the rock cycle. Sedimentary rocks are converted to metamorphic rocks by deep burial, and metamorphic rocks convert to igneous rocks if they are heated to the point of melting. Likewise, igneous rocks can be metamorphosed, and both igneous and metamorphic rocks convert to new sedimentary rocks during uplift and erosion.

The cycling of crustal rocks from the Earth's interior to the surface and back again, also cycles the minerals that rocks contain. Because the stability of minerals varies, those stable in the Earth's interior may break down at the surface, and those stable at the surface may break down at depth. Clay minerals, for example, are only found in sedimentary rocks because they are stable only at low temperatures and break down to form other rock-forming silicates in the high-temperature environments of igneous and metamorphic rocks. Conversely, minerals like hornblende and biotite, which are common in igneous and metamorphic rocks because they are stable at high temperatures, tend to break down during low-temperature weathering and are therefore rarely found in sedimentary rocks. Mineral reactions in low-temperature environments, however, are very slow, allowing minerals to survive for long periods at the Earth's surface even when they are unstable. As a result, all three rock types are exposed at the Earth's surface.

This continual cycling of rocks and minerals is possible because the Earth is a dynamic planet. As we shall learn in subsequent chapters, it is because this cycling brings crustal rocks and their constituent minerals from the Earth's inaccessible interior to the surface, that we are able to study the interior of the crust and so have come to understand the dynamic processes that occur within it.

2.4 Chapter Summary

The Earth's continents and ocean floors consist of rocks and their component minerals which form the basic materials we use to understand the Earth's evolution throughout geologic time. Minerals are naturally occurring, inorganic, crystalline solids with a specific chemical structure and formula. The structure and formula are an expression of the manner in which the chemical elements bond together to form the minerals.

The fundamental unit of the elements in minerals is the atom, each of which is made up of a central nucleus of protons and neutrons, and an orbiting cloud of electrons. The mobility of electrons allows atoms to bond to form molecules. Bonds are formed by electron transfer (ionic bonds), by the sharing of electrons (covalent bonds), and by the development of a common electron cloud (metallic bonds). Weaker van der Waals forces occur between electrically polarized molecules.

Of the 4000 known minerals, less than 1% commonly occur in rocks. Because the crust is dominated by silicon and oxygen, most rock-forming minerals are silicates. Silicon and oxygen bond with each other and with other elements to form the silicate minerals. The basic framework of all silicates is the silicate tetrahedron, which is an expression of the bonding arrangement between silicon and oxygen. Individual tetrahedra may be linked to each other and bonded with metal ions in a variety of forms, each characteristic of a different silicate family. The properties of these minerals largely reflect their chemistry and the character of their bonding. Common mineral properties include color, luster, crystal form, hardness, cleavage, and specific gravity, which are expressions of a mineral's internal bonding arrangements. Minerals have different stabilities and will form in a rock only if the rock has the appropriate chemistry and the physical conditions are suitable.

Rocks are classified into three varieties: igneous, sedimentary, and metamorphic, which reflect their mode of origin. As portrayed by the rock cycle, one variety of rock can be transformed into another in vivid affirmation of an ever-changing, dynamic Earth. Igneous rocks form from the cooling of molten magma. Sedimentary rocks are produced either by cycles of erosion and deposition, or by chemical and biochemical precipitation at the Earth's surface. Metamorphic rocks form beneath the Earth's surface when one rock is transformed into another by the action of pressure, temperature, and/or fluids. Each of these rock groups are identified and classified by the relative abundance of certain key minerals (which indicate their composition) and by the grain size of these minerals.

Igneous rocks form by cooling and crystallization of molten rock, or magma. Because magma is less dense than surrounding rocks, it is buoyant and rises toward the surface. If it reaches the surface, volcanoes are produced, and fine-grained or glassy volcanic rocks form from the rapid cooling of lava. Igneous rocks that erupt on the Earth's surface in this way are said to be extrusive. Because they are deposited more-or-less horizontally, they are generally parallel to the sedimentary rock layers on which they sit and to those deposited afterward.

Magmas trapped in chambers beneath the Earth's surface cool more slowly and so form relatively coarse-grained, plutonic igneous rocks. Such deep-seated igneous rocks are said to be intrusive because they cut into preexisting rocks layers as they rise through the Earth's crust.

Igneous rocks are classified on the basis of differences in their texture and chemical composition. However, magmas may change their chemistry during their cooling histories such that mafic (iron- and magnesium-rich) magma may evolve to become more felsic (feldspar- and silica-rich) as a result of a succession of crystal-melt interactions known as Bowen's Reaction Series. A single episode of magmatism can also produce plutons, dikes, sills, and lava flows of similar chemical composition, but very different grain size because of variations in cooling rates.

Silica-poor (basaltic) magmas are typically very fluid and so tend to erupt as fast-flowing lava flows. Silica-rich magmas are generally very viscous and gaseous, and tend to cause explosive volcanic eruptions. Explosive eruptions produce pyroclastic deposits and pose a much greater threat to society than those that consist of lava alone. Recent technological advances have aided in predicting when explosive eruptions will occur.

Unlike igneous rocks, sedimentary rocks form by "cold" processes that only occur at the Earth's surface. They are the product of either the deposition of detritus or the precipitation of chemical compounds, characterized by clastic textures of broken fragments and nonclastic textures of interlocking grains, respectively. Detrital sedimentary rocks are classified according to grain size, whereas those formed by chemical/biochemical processes are classified according to composition.

Sedimentary rocks are deposited in continental, marine, and coastal environments. Continental deposits are predominantly the product of erosion, transport, and deposition by rivers and streams, although under extreme climatic conditions, wind and ice may also be important. Deposition in marine environments is influenced by the ongoing erosion of continents and the accumulation of thick piles of sediment along their ocean margins. Instabilities on these margins can result in submarine avalanches that can carry sediment to the deep ocean floor. In many marine environments, individual sediment layers are not continuous. Near-shore sands, for example, commonly give way to muds, and then to limestone sediments in the offshore environment. As sea level rises and falls, the position of the coastline changes with time. Because of this, the character of sedimentary deposits at any one location also changes, resulting in changes of rock type with time. In coastal environments, river dump their sediment load as they slow down and meet the sea. These environments are characterized by deltas, tidal flats, barrier islands, and lagoons.

Metamorphic rocks form below the Earth's surface as a result of great pressure and heat, which cause significant changes in the texture and mineral composition of preexisting rocks. Increasing pressure generally favors the stability of denser minerals over lighter ones, whereas increasing temperature favors the stability of those without water in their crystal structures over those containing water. The minerals and textures that form depend largely on the chemistry of the rock and the conditions of pressure and temperature. Metamorphic rocks consequently contain important information about the geological processes that operate at various levels within the Earth's crust.

Metamorphic rocks that develop over wide areas in the roots of mountain belts are said to have experienced regional metamorphism. In these situations, metamorphism is intimately linked to rock deformation. Metamorphic rocks are also produced by contact metamorphism adjacent to hot plutonic bodies, and by the crushing and smearing of dynamic metamorphism along active fracture zones.

Based on the presence or absence of layering, metamorphic rocks are subdivided into foliated and nonfoliated varieties that are respectively classified according to their grain size and composition. Metamorphic minerals vary with the pressure and temperature conditions (or grade) of metamorphism and so provide clues to the conditions under which metamorphism took place. An association of metamorphic rocks formed under similar conditions of pressure and temperature define a metamorphic facies.

Key Terms and Concepts

Look for the highlighted items on the web at:
WWW.BROOKSCOLE.COM

- Minerals and rocks are the basic components of the solid Earth
- Minerals are naturally occurring, inorganic crystalline solids with specific chemical structures and formulae
- Elements, the building blocks of minerals, are made up of atoms in which electrons orbit a central nucleus of protons and neutrons
- Chemical bonding, caused by the mobility of electrons, holds the elements of a mineral together in an orderly arrangement
- Most rock-forming minerals are silicates in which a variety of metallic ions are bonded with tetrahedral arrangements of silicon and oxygen
- The physical properties of minerals (color and streak, crystal form, hardness, cleavage and fracture, specific gravity, and density) reflect their internal structure and chemical composition
- Three groups of rocks (igneous, sedimentary, and metamorphic) are distinguished on the basis of their modes of origin
- The rock cycle is a dynamic crustal process by which each rock group can be transformed into one of the others
- Igneous rocks form by the crystallization of a cooling magma and are classified according to their grain size (fine or volcanic, coarse or plutonic) and chemical composition (felsic, intermediate, mafic, ultramafic)
- According to Bowen's Reaction Series, mafic magmas evolve during cooling as crystals react either discontinuously or continuously with the melt. Separation of early-formed minerals by crystal

settling causes the composition of the remaining melt to become more felsic, a process called fractionation
- Basaltic lavas are fluid because silica-poor (mafic) magmas have low viscosities. Silica-rich (felsic) magmas are more viscous, so eruptions are explosive but usually predictable
- Sedimentary rocks are formed by either detrital or chemical/biochemical deposition at the Earth's surface. The former show clastic textures and are classified according to grain size, whereas the latter are nonclastic and classified according to composition
- By understanding modern depositional environments (continental, marine, coastal), the ancient sedimentary rock record can be interpreted
- Metamorphic rocks form beneath the Earth's surface by the action of heat, pressure, and/or fluids on preexisting rocks. Metamorphism can be regional, contact or dynamic, and often accompanies deformation. Foliated metamorphic rocks possess a layering or foliation, and are classified according to the grain size of the metamorphic minerals. Nonfoliated metamorphic rocks are classified on the basis of their composition
- Metamorphic minerals tend to reflect the conditions (or grade) of metamorphism and so reveal processes occurring within the Earth's crust

Review Questions

1. What is the difference between a mineral and a rock?
2. What are atoms and what do they consist of?
3. Distinguish between the four types of chemical bonds that are common in minerals.
4. What is meant by the term "crystalline solid"?
5. What is the difference between silicate and nonsilicate minerals?
6. What is the difference between the mineral properties of color and streak?
7. What is Mohs scale of hardness and how is it used?
8. What is mineral cleavage and how does a cleavage face differ from a crystal face?
9. What is the rock cycle, and what does it tell us about the nature of the crust?
10. What are the three categories of rocks, and how do their modes of origin differ?

11. Why are the crystals in a granite so much larger than those in a basalt?
12. How do felsic igneous rocks differ from those of mafic composition?
13. What are the chemical and textural differences between (a) granite and rhyolite, (b) gabbro and granite, and (c) gabbro and basalt?
14. How are sediments converted into sedimentary rock?
15. How do clastic and nonclastic sedimentary rocks differ in their mode of formation?
16. Which of the following are clastic sedimentary rocks and why? (a) Sandstone, (b) limestone, (c) mudstone, and (d) rock salt.
17. How are metamorphic rocks classified?
18. What is the difference between regional, contact, and dynamic metamorphism?
19. Why are some metamorphic rocks foliated?
20. What is the distinction between a schist and a slate?

Study Questions

Research the highlighted questions using InfoTrac College Edition.

1. How do ionic, covalent and metallic bonds differ in their use of electron mobility?
2. Explain the relationship between metallic bonding and some of the key properties of metals? How may these properties aid in mineral exploration?
3. Why do minerals have specific chemical formulae whereas rocks do not?
4. In what ways do crystal form, hardness, and cleavage reflect a mineral's internal structure?
5. Why are silicates the dominant rock-forming minerals and how are the silicate families related to the structural arrangement of the silicate tetrahedron?
6. Why are silicate minerals typical of igneous rocks whereas nonsilicate minerals are more common in sedimentary rocks?

7. Explain how a metamorphic rocks derived from a sedimentary one can be exposed at the Earth's surface without its minerals reverting back to those that were present in the original sedimentary rock.
8. Rocks are divided into three varieties (igneous, sedimentary, and metamorphic) based on their mode of origin. Distinguish between these varieties.
9. How does Bowen's Reaction Series influence the composition of a cooling mafic magma?
10. Explain how the distribution of sedimentary rocks in coastal regions is sensitive to changes in sea level.
11. Why can sedimentary rocks be produced only at the Earth's surface, while metamorphic rocks only form below the surface?
12. Why is metamorphism commonly accompanied by deformation?

Geologic Time and the Age of the Earth

For most of us the immensity of geologic time is very hard to grasp. Yet a comprehension of its length is essential to any understanding of Earth science. The reason is simple. Although some geologic activities, such as earthquakes and volcanic eruptions, are rapid events that we can experience, most of the processes that shape the Earth operate so slowly that it is only with the passage of vast stretches of time that their effects are realized. The profound changes these processes have wrought on the Earth's surface is therefore testament to the antiquity of the planet. However, during a human lifetime, their effects are often imperceptible. Our purpose in this chapter is to provide a sense of the vastness of geologic time by examining the methods by which the age and history of the Earth have been established.

The story behind our present understanding of the Earth's antiquity, and the history of attempts to determine its age, form a fascinating chapter in the development of modern Earth science. First, methods were established by which the relative age of geologic events could be determined. In revealing whether one episode was older or younger than another, these methods allowed geologic events to be placed in sequence. In this way, rock layers could be arranged in the order in which they formed even though their true age was not known.

Next, methods of establishing the absolute age of geologic events came with the discovery of radioactivity. Knowledge of the fixed rate at which radioactive minerals break down made it possible to actually date rocks in years, and from this has come our modern understanding of geologic time. It is through such radiometric dating, for example, that we now believe the Earth to be 4.6 billion years old. In addition, through radiometric dating we now have the means of timing past geologic events and the rate at which they occurred so that we can study the pace and rhythms of natural processes.

Only with an appreciation of the immense span of geologic time do the true dynamics of the processes that shape the Earth become apparent. Given millions of years, profound change can be realized by the cumulative effects of even the slowest of processes. Thus, movements that seem imperceptible can be recognized as the agents of great transformation, capable of uplifting huge mountain ranges, or of opening and closing entire ocean basins. Once we have grasped the vastness of geologic time, we will be ready to explore the workings of the solid Earth and the dramatic story of the slow but ceaseless movement of its lithosphere.

3.1 The Vastness of Geologic Time

Although a sense of time distinguishes humans from other animals, our comprehension of geologic time is prejudiced by the limited length of a human lifetime. We can visualize historical time involving hundreds or even thousands of years with little difficulty because we can relate to it in terms of human lifespans. However, our ability to comprehend time becomes much more difficult when its length is too long to be viewed from this simple human perspective. Like the vast interstellar distances used by astronomers, the scale of geologic time where the unit is a million years is extremely difficult for us to grasp. Yet we need to comprehend the vast reaches of geological time if we are to understand the processes that have shaped our planet. Because these processes act on our planet very slowly, they have been able to produce the profound changes our planet has experienced only because the Earth is very ancient, and the period of time over which they have operated is immense.

A sense of geologic time and the slowness of natural processes is not just of academic interest. Many of our environmental concerns stem from the fact that we consume resources far faster than nature can replenish them, and we produce pollutants far faster than nature can absorb them. Understanding geologic time also influences our ability to predict catastrophic events such as earthquakes and volcanic eruptions. Even though these geologic events may only take seconds to have their impact, we must recognize that they are the culmination of natural processes that have taken a very long time to build up. Any warning signals may therefore be evident only in the most imperceptible changes.

3.1.1 Relative Time and Absolute Time

How do we measure geologic time and how have we determined the age of the Earth? It is often said that a geologist is like a detective trying to find and evaluate clues at the scene of a crime. Our arena is the **outcrop** where rock formations appear at the surface of the Earth (Fig. 3.1). In examining outcrops we try to recreate the events that produced the exposed rocks in the order in which they occurred so that we can come to understand the final outcome as it is presented before us. The outcrop shown in Figure 3.1, for example, reveals layers of sedimentary rock that have been bent into a U-shape. From this, a history can be recreated in which sedimentary deposition was followed by bending of the sedimentary layers. However, the evidence provided by outcrops is often fragmentary and some on-site investigations must be complemented by laboratory work on rock specimens taken from the scene.

At the outcrop, geologists can often place geologic events, such as the deposition and bending of sedimentary rocks, in sequence. That is, we can determine their **relative age**. More rarely, clues to the actual time that the events took place may be present. For example, distinctive fossils whose age is known may be present in rocks associated with a particular event. In the outcrop shown in Figure 3.1, the time of sedimentary deposition could be determined if fossils were present in some of the sedimentary rock layers.

Alternatively, it may be possible to determine the actual age of the rocks by sophisticated laboratory measurements of the amount of radioactive decay in certain minerals present within them. The more decay, the older the rock. This approach provides a precise determination of the **absolute age** of the rocks, that is, the amount of time that has elapsed since the rock formed.

The final determination of the sequence of a particular set of geologic events is usually achieved by combining clues observed at the outcrop with those provided by the laboratory. However, a vital aspect of this approach is that the physical evidence visible at the outcrop should be consistent with the laboratory evidence. For example, a detective who deduces the time of a crime from a smashed wristwatch needs evidence to ensure that the wristwatch was not tampered with after the event. Similarly, geologists must take care to ensure that the samples used for determining the absolute age of a rock accurately reflect the events they have deduced from observing the outcrop. The physical evidence revealed by the outcrop in Figure 3.1, for example, requires that the sedimentary rocks were deposited before they were bent. If the absolute ages of deposition and bending could be determined from samples collected at the scene, it is essential that the laboratory results also show the age of deposition to be older than the age of bending.

Figure 3.1
An outcrop of folded sedimentary rocks on Interstate 68 at Sideling Hill in western Maryland reveals clues to the history of the Appalachian Mountains (Photo Damian Nance).

3.2 A Short History of Geologic Time

3.2.1 "As Old as the Hills"

In order to develop an appreciation for geological time and gain an understanding of the basic principles on which the modern geologic time scale is based, it is useful to review major deductions made by previous generations of scientists and some of the controversies that beset them in their endeavor to establish the age of the Earth.

The common expression "as old as the hills" is based on the observation, made first by ancient civilizations, that the landscape changes extremely slowly and must therefore be far older than the life-span of many human generations. Yet monuments created of stone, the very material that the hills are made of, do not survive the passage of time without change. Therefore, the hills may be old but they are not unchanging.

The fact that rivers turn muddy after major rainfalls is evidence that material is being eroded from the hills and transported by streams. Much of this material is not deposited until the streams reach the sea. For example, most of the sediment that now resides on the continental shelf of the western seaboard of the North Atlantic once formed part of Appalachian mountain belt of eastern North America before being eroded and transported to the ocean. However, this being the case, the mountains must once have been much higher than they are today. Could they once have been as high as the Himalayas, as geologists believe? If so, they must be ancient indeed.

If we were able to reverse the slow process of erosion, just imagine how long would it take to rebuild these mountains to their former heights. And what of the rocks of which the hills are made? Surely these must be far older than the hills themselves. So how old are the hills and the rocks that form them, and how can we tell?

3.2.2 Theories, Principles, and People

Catastrophism

Attempts to measure the age of the Earth began in the middle of the seventeenth century with the biblical chronology of Anglican Archbishop James Ussher. Based on careful study of the Bible's continuous record of births and deaths, Ussher was able to place the Earth's creation in the year 4004 B.C. The Earth then was a mere 6000 years old. Other scholars subsequently duplicated Ussher's result and even refined it to the day and hour; 9:00 A.M. on October 26.

Other cultures place different dates on the time of creation. In the Mayan calendar, for example, the deepest probings of eternity reach back 400 million years, while in the Hindu calendar, the present world came into existence almost 2 billion years ago.

In the Christian world of Western Europe, however, Ussher's figure of 6000 years became dogma and was taken seriously for over 200 years by scholars who considered the Earth to have formed as a molten ball that cooled to give a lumpy crust with mountains, valleys, and oceans. Thus the hills too were only 6000 years old. All geographic changes on the Earth's surface were believed to be catastrophic in nature, as exemplified by Noah's Flood, which was viewed as one of a series of successive destructions of the Earth's surface by supernatural catastrophies.

In support of this principle of **catastrophism,** eighteenth century scholars who came to be known as the **Catastrophists,** pointed to the existence of terraces in river banks and to great valleys that were far too large to have been eroded by the streams they contained as evidence of Noah's Flood. Further evidence was seen in fossilized marine organisms in rocks exposed far from the sea and in the existence over much of northern Europe of great boulders scattered far from outcrops of similar rock (Fig. 3.2). These **erratics,** so called because they had apparently moved from their original location, are now known to have been deposited by glaciers during the last Ice Age but were used by the Catastrophists as evidence of sudden catastrophic flooding. Earthquakes and volcanic eruptions, like the one that had buried the Roman city of Pompeii on the flanks of Vesuvius in 79 A.D., were also seen as sudden catastrophic events consistent with a 6000-year calendar for Earth history.

Steno's Three Principles

Evidence that the Earth might be much older than Ussher's 6000 years soon started to accumulate as scholars turned their

Figure 3.2

Woodcut from Andrew Ramsay's book "Physical Geology and Geography of Great Britain," first published in 1864, showing erratic boulders in the glaciated valley of Llanberis in North Wales. Lying far from outcrops of similar rock, boulders such as these were used by the Catastrophists as evidence of catastrophic flooding.

attention to rocks and the origin of rock layering. Shortly after Archbishop Ussher's pronouncement, the Italian philosopher Nicholas Steno proposed three fundamental principles that still form the basis of all geologic mapping. The principles were based on three simple arguments, namely that sedimentary rock layers are deposited on top of each other, that the layers are flat-lying when they are deposited, and that they are deposited over wide areas.

If, Steno argued, sedimentary rock layers or **strata** (from the Latin word for bed) are deposited on top of one another, then in any vertical sequence the oldest layers will occur at the base and the youngest at the top. This ordering of layers, which is known as the principle of **superposition,** also applies to the fossils the rock layers contain and is consequently at the heart of relative age dating. Steno also argued that sedimentary strata would have lain flat when they were first deposited, that is, they would have been originally horizontal. The principle is therefore known as **original horizontality.** Finally he argued that the flat-lying sedimentary strata would have initially extended in all directions, that is, they would originally have been laterally continuous. The principle is therefore known as **original lateral continuity.** Steeply inclined rock layers must therefore have suffered later tilting, and otherwise similar layers separated across a valley, for example, must have been originally continuous (Fig. 3.3).

Neptunists

How did these layers get there? By the end of the eighteenth century, scholars who came to be known as the **Neptunists** believed a great globe-engulfing sea laid down all of the Earth's rocky layers, gradually exposing them as the water slowly receded. Crystalline rocks such as granite were thought to have formed from the chemical precipitation of crystals from seawater. In the rock classification scheme we use today, this means that the Neptunists interpreted granites to be chemical sedimentary rocks like rock salt rather than the igneous rocks we now know them to be.

Where true sedimentary rocks, such as sandstone and shale, overlay crystalline material, the Neptunists interpreted these to have been deposited as the ocean withdrew and the ancient crystalline rocks were exposed to erosion. Where basalts occurred in this sequence of sedimentary rocks, they were attributed to the burning of coal seams. The youngest and topmost layers were recent deposits eroded from the older crystalline and sedimentary rocks and washed down from the emerging mountains as the ocean retreated. Thus, mountains were viewed simply as high points in the original surface of the Earth.

Although incorrect in their interpretation, these scholars were right in thinking that the Earth had to be ancient. Six thousand years was insufficient time for their explanation of how the layers formed. This separated them from the Catastrophists, although their differing explanations of Earth history had two aspects in common. The Neptunists shared with the Catastrophists the belief that huge quantities of water once covered the Earth and had deposited the sediments of the Earth's surface on its retreat. Also, like the Catastrophists, they had no explanation as to where this water had gone.

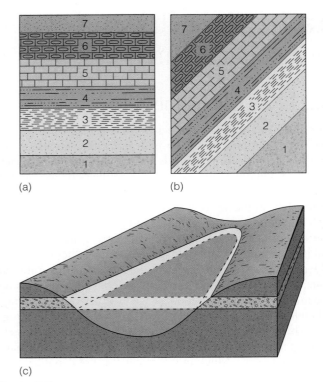

Figure 3.3
Steno's three principles concerning the deposition of sedimentary rock strata. (a) The appearance of a sequence of sedimentary rock strata at the time of deposition, numbered 1 to 7 in the order in which they were deposited. (b) The appearance of the same sequence of sedimentary strata after tilting. (c) The original continuity of a sedimentary rock layer allows it to be traced from one side of a valley to the other.

Plutonists

After a debate that lasted many decades, the Neptunists ultimately lost their case to scholars known as **Plutonists,** who insisted that crystalline rocks such as granite were not chemical precipitates but had crystallized from a molten state. The Plutonists were the first to argue for a dynamic and changing Earth, and in so doing were the harbingers of modern geological thinking.

Foremost amongst these Plutonists, and often considered the "father of modern geology," was the Scotsman James Hutton. As part of an exciting scientific movement in late-eighteenth century Edinburgh known as the "Scottish Enlightenment," Hutton watched the effects of weathering and erosion and became convinced that, given enough time, processes at work in the environment today could have produced all of the features previously attributed to catastrophic flooding.

Hutton's Principle of Gradualism

In his pivotal book "Theory of the Earth" published in 1795, Hutton proposed that the Earth, powered by its own internal heat, is in continuous but gradual change, and is constantly decaying, renewing, and repairing itself. This idea has come to be known as **gradualism.** Rather than 6000 years of Earth history, gradualism required the Earth to be truly ancient with, as Hutton put it, "no vestige of a beginning, no prospect of an end".

In defense of an ancient and slowly changing Earth, Hutton compared the level of erosion in old mountain belts with that observed in ancient monuments which had changed little since they were built. Just south of the Scottish border, Hutton studied Hadrian's Wall. Built by the Romans between Scotland and England in 122 A.D., the wall had suffered little deterioration despite 16 centuries of erosion (see also Fig. 10.1). If, Hutton argued, 16 centuries could barely scar a 6-foot-tall wall, the time needed to wear down mountains and carry sand and gravel to the sea, must be enormous indeed.

Unconformities

Hutton found further evidence for the enormous length of geological time on the banks of the River Jed in Scotland. Here, vertical layers of rock worn down by erosion were covered by horizontal beds, a relationship known as an **unconformity** (Fig. 3.4). Originally, each set of rock layers had been laid down horizontally beneath the sea. In the years between the deposition of the first set and that of the second, earth movements of unimaginable force had tilted the lower layers and uplifted them to form mountains. Erosion then wore down these mountains. The lower layers were once again submerged and then the upper layers were deposited on top. Now both sets of rock layers had been uplifted above sea level and were being eroded (Fig. 3.5).

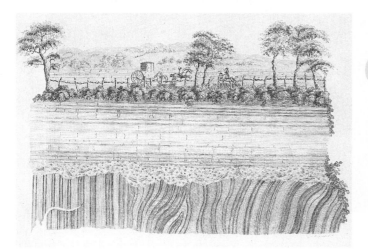

Figure 3.4
In 1787, when most scholars believed the Earth to be only a few thousand years old, this unconformity on the banks of the River Jed in Scotland was used by James Hutton to prove the enormous length of geologic time. Both layers of rock, now at right angles, were laid down horizontally beneath the sea. Yet in the years between, earth movements uplifted the lower layers to form mountains. Erosion wore them down before they were once again submerged and the upper layers deposited on top. Now all stand above sea level again, a calendar of events millions of years long.

The unconformity separating the two sets of layers is therefore an erosion surface on which the upper layers were deposited horizontally. Because of this, the upper layers cut across the tilted lower layers. The unconformity is therefore an example of a **cross-cutting relationship,** in which a younger sequence truncates an older one. As we shall discover in Section 3.3, such relationships are fundamental in determining the relative ages of geological events because they help to establish the sequence of those events. Hutton realized that no single catastrophic event could have produced the sequence of events represented in an unconformity like that on the River Jed. Instead, it required time and enormous lengths of it. Unconformities represent major time gaps in the geological record of layered rocks during which no rocks are deposited. These gaps must involve sufficient time for the lower rock layers to be uplifted, tilted, and eroded before being resubmerged and then buried by younger rocks (Fig. 3.5). On the River Jed, we now know that erosion of the tilted layers alone took 70 million years.

Since Hutton's time many other unconformities have been identified. Indeed, they are now known to be common features of the sedimentary rock record. As we shall see in Section 3.4.1, the apparently continuous succession of sedimentary rocks exposed in the walls of the Grand Canyon in fact contain several unconformities, the largest (at the base of the succession) representing a time gap of some 1.5 billion years.

Smith's Principle of Faunal Succession

Further south in England, the study of rock layers had also become the vocation of a canal engineer called William Smith. In conducting a survey for a proposed canal between London and Bristol, Smith noticed that the sedimentary strata were consistently tilted to the east and that each layer contained characteristic fossils that were distinct from those in the layers above and below. Smith was therefore able to identify the same layer in different parts of the country based on its fossil content. Developing on Steno's concept of original lateral continuity, "Strata Smith" (as he later became known) established the basis for the principle of **faunal succession.** According to this principle, individual rock layers can be "fingerprinted" and **correlated** (or traced) from one place to another on the basis of the fossils they contained (Fig. 3.6).

By plotting the distribution of sedimentary strata containing the same fossils, Smith was able to produce some of the first geologic maps of southern England (Fig. 3.7). Furthermore, by following Steno's principle of superposition and arranging each of the rock layers into a vertical sequence or **stratigraphic column** of younger-upon-older strata, Smith was able to observe the changes in fossils from layer to layer. The results suggested progressive evolution rather than abrupt change, and so supported gradualism over catastrophism.

Of greater importance was Smith's recognition that the collection of fossils in any one layer represented organisms that lived during the time that the layer was deposited. Because the same collection of fossils never occurred in earlier layers and never reappeared in later ones, the fossil content of a layer

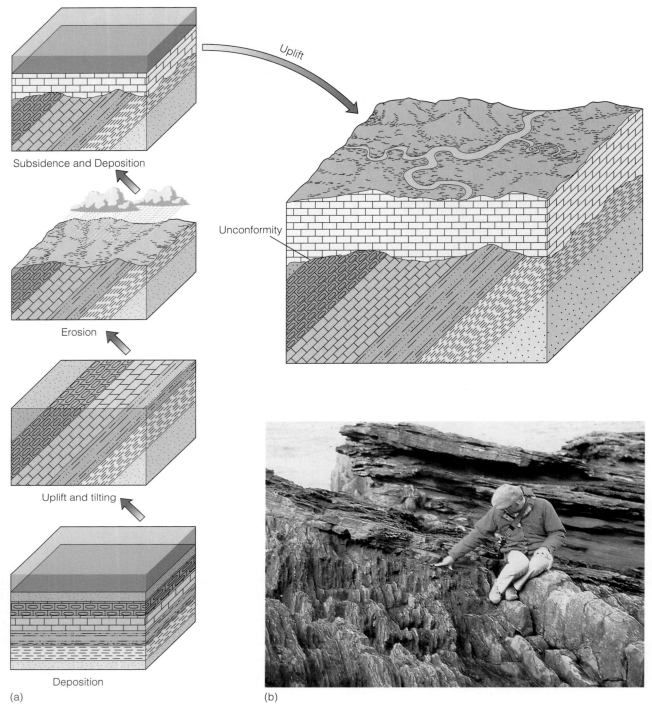

Subsidence and Deposition

Erosion

Uplift and tilting

Deposition

Uplift

Unconformity

(a)

(b)

Figure 3.5
(a) Successive stages in the development of an unconformity. (b) Unconformity at Siccar Point, Scotland, that was to prove the confirmation of Hutton's view of gradualism and the immensity of geologic time. Here, tilted rock layers rest on vertical ones in vivid testament of Hutton's concept of continuous but gradual change.

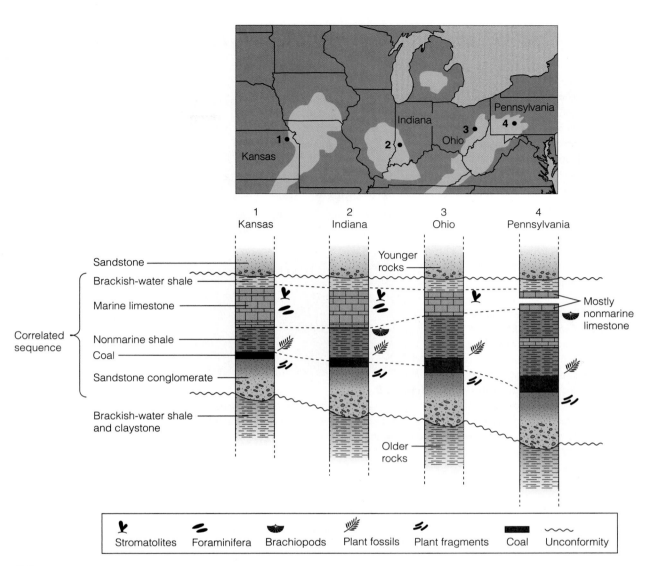

Figure 3.6
Correlation of the sedimentary rock sequences and fossil assemblages above and below a coal horizon in four coal fields in the American Midwest indicates that the coal was deposited at or about the same time at all four sites.

Figure 3.7
Part of William Smith's geologic map of southern England made possible through correlation and the application of his principle of faunal succession. Published in 1815, this was the first large-scale geologic map of any country.

Since the mid-twentieth century, North America has depended on oil and gas for more than half of its energy needs, and today that figure is fast approaching two-thirds. Oil and gas are the lifeblood of modern society, and as the larger, more easily found oil fields are exhausted, the search for new reservoirs becomes ever more important. At the heart of this search is the technique of correlation, a principle that has changed little from the ideas of Nicolas Steno in the seventeenth century and those of William Smith more than a century-and-a-half ago.

However, unlike the insightful observations of Steno and Smith, the search for oil and gas takes place beneath the Earth's surface. Oil and gas are produced by the burial and heating of sedimentary rocks that contain organic matter. Because oil and gas are low-density fluids, they tend to migrate upward toward the surface once they form, through the interconnecting cracks and spaces that exist in most rocks. To form a reservoir, they must be prevented from reaching the surface by an impenetrable rock layer that stops their upward movement and causes them to pond below. The oil and gas can then be extracted by drilling a hole through the impenetrable layer, allowing them to continue their upward migration by providing a path to the surface.

The search for oil and gas involves the detection of areas where such ponding may have occurred. For a viable deposit to form several conditions must be met. First, it is necessary to identify those rocks that might have provided a source for oil and gas. Then a route

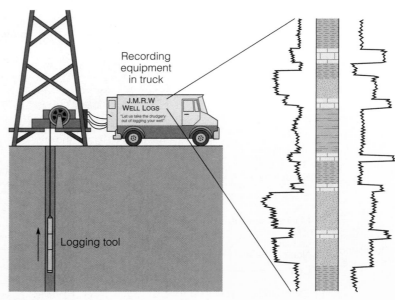

Figure 3.8
Characterizing rock formations in the subsurface by well logging. Measurements made by the logging tool as it is withdrawn from a drill hole are recorded at the surface and printed out as a log that shows how the physical properties of the rocks encountered vary with depth.

must be found that the oil and gas might have followed as they migrated upward. Finally, other rocks must be identified that might have prevented the oil and gas from reaching the surface by acting as impenetrable layers. Each of these steps takes place in the subsurface, in rocks that occur below ground.

Subsurface geology involves the acquisition and interpretation of data on geologic features that occur below the Earth's surface. In the oil industry,

drilling provides one of the main sources of such information. Drill holes, however, are just a few inches in diameter and are merely pinpricks into the subsurface. To obtain a larger picture of the subsurface geology, the data acquired from one drill hole must be correlated with that from others. Therefore, correlation plays a vital role in the interpretation of subsurface data. Where drilling is involved, correlation is based on data obtained either directly,

could be used to establish its position in the stratigraphic column. That is, the fossil content could be used to determine the layer's relative age.

Lyell's Principle of Uniformitarianism

England's Charles Lyell dealt another blow to catastrophism with the publication of the first volume of his highly influential "Principles of Geology" in 1830. In this book, Lyell described in great detail the concept that geological processes at work on the Earth today have acted with the same intensity from the earliest times to the present.

This elaboration of Hutton's idea of gradualism, which came to be known as the principle of **uniformitarianism,** is based on the view that only natural causes can be used to explain nat-

ural events. Thus, past events can only be explained by analogy with modern ones. Furthermore, uniformity of process through time means a uniform rate of change, and therefore, a geologic time scale adequate for the slowness of natural processes.

As proof of the gradual nature of geological processes, Lyell pointed to the Roman temple of Jupiter Serapis, near Naples in Italy (Fig. 3.10). Here fragile columns remain standing despite evidence for considerable sea level change over the past 2000 years in the form of holes bored by marine organisms that are now set high in the columns. Surely, Lyell argued, only gradual processes could have brought about such a significant change in sea level without toppling the columns. Lyell's three-volume work was to revolutionize the natural sciences because it reached a wide audience and, for the first time, brought togeth-

from rocks recovered from the drill hole, or indirectly, from measurements made in the drill hole.

During drilling, rock core or rock fragments are commonly recovered from the drill hole. These provide a wealth of information on the subsurface geology and are examined not only to determine their rock type and its potential for producing or collecting oil and gas, but also for their fossil content. Given the small size of these samples, large fossils are rarely encountered, so fossils of microscopic organisms are examined instead. These microfossils, however, follow the principle of faunal succession in exactly the same way as the larger fossils studied by Smith. They can therefore be used to determine the relative age of the sample and its correlation with samples from other drill holes. In this way, rock formations penetrated by one drill hole can be correlated in the subsurface with those penetrated by another.

A more sophisticated method of characterizing rock formations in the subsurface is that of **well logging** (Fig. 3.8). In this procedure, an instrument package known as a "logging tool" is lowered into the drill hole, and measurements are made of a variety of physical properties in the rocks encountered. Rock properties such as fluid content, density, radioactivity, resistivity (resistance to electrical currents), and permeability (ability to transmit fluids) can be made in this way. Measurements are made as the logging tool is slowly withdrawn from the drill hole, the data being recorded at the surface where it is print-

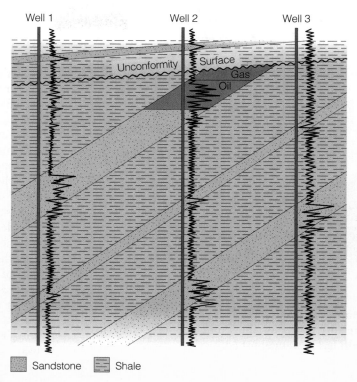

Sandstone   Shale

Figure 3.9

Correlation of well logs from three drill holes reveals an unconformity in the subsurface and, beneath it, a potential reservoir of oil and gas.

ed out in the form of a wavy trace. The pattern of the trace can then be compared with measurements made in other drill holes.

For example, in Figure 3.9 well logs from three drill holes have been correlated to reveal the presence of a subsurface unconformity. Unconformities are important targets in oil and gas exploration because of the potential for oil

and gas to become ponded beneath the unconformity surface. In Figure 3.9, this occurred where oil and gas that migrated upward though a sandstone unit became confined beneath the unconformity by units of impenetrable shale. Only by applying the principle of correlation, however, can the unconformity and the reservoir it produced be revealed.

er all of the important new developments in Earth science. Indeed, it is no accident that just a year after the first volume was published, a copy was to cross the Atlantic on the voyage of "HMS Beagle" in the possession of Charles Darwin.

Unfortunately, under the title of "Uniformitarianism" Lyell's concept of uniformity of process through time is apt to be misunderstood. For example, uniformity of intensity and rate does not require all geologic processes to be gradual. Some, such as volcanic eruptions and earthquakes, are far from gradual as Lyell was well aware. Indeed, many less obvious features, such as rock layering itself, may record quite short-lived events in an otherwise gradual history of deposition. On shorelines subjected to occasional hurricanes, for example, layers of shallow-marine sediments are frequently preserved

only after severe storms. Each storm produces a single "storm deposit" layer after the hurricane abates. All sediment deposited between major storms is stirred up by the new storm and any layering it possessed is consequently destroyed. Therefore Uniformitarianism, by which Lyell meant a uniformity of natural law, should not be taken to preclude catastrophic events, but rather should be taken to mean "the present is the key to the past." Interpreted in this fashion, the principle can be more aptly described as **actualism.**

The remarkable progression of ideas from those of Ussher to those of Lyell (summarized in Table 3-1) brought about a profound change in the way that scholars viewed the planet on which they lived. To this day, the principles of Steno, Hutton, Smith, and Lyell lie at the heart of our understanding of

Figure 3.10
Woodcut from Charles Lyell's book of 1830, showing the Temple of Jupiter Serapis near Naples, Italy. Borings by marine shells on the temple's columns were used by Lyell and other Gradualists as proof of the gradual nature of geologic processes.

geologic time. In addition, those of Steno form the basic principles of relative dating by which much of the geologic history of the Earth has been established.

3.3 Relative Dating

According to Steno's principles of superposition, original horizontality, and original lateral continuity, sedimentary layers are deposited in sequence, one above the other in flat-lying, laterally extensive sheets. Consequently, the relative age of sedimentary layers in a vertical sequence of strata is given by their relative position, the oldest layers occurring at the base while the youngest ones lie at the top. Furthermore, steeply inclined sedimentary layers must have been tilted following deposition,

and layers should be laterally traceable except where they have been removed by later erosion (Fig. 3.3).

For example, in the sequence of rock layers shown in Figure 3.11a, Steno's principles imply that the oldest unit is layer A while the youngest is layer F. Unit D, however, is a buried lava flow rather than a sedimentary layer. Yet Steno's principles still apply. Baking of the underlying bed (layer C) by the lava flow shows that its eruption must have followed the deposition of layer C. Similarly, the presence of fragments of the lava in sedimentary layer E shows that the deposition of this layer occurred after the lava was erupted. Hence, the lava flow must be younger than layer C and older than layer E.

The rule that holds that fragments of one rock in a layer of another must be older than the layer itself, is another principle of relative dating known as the **rule of inclusions.** Figure 3.11b illustrates yet another, known as the **rule of cross-cutting relationships.** This rule, which is attributed to James Hutton, holds that any feature that cuts across a rock must be younger than the rock it cuts. Therefore, an intrusive body is always younger than the rocks it intrudes and an offsetting fracture (or fault) is always younger than the rocks it offsets. In Figure 3.11b, an intrusive sheet of basalt (in this case a dike) cuts across and bakes sedimentary layers A through C but is present as fragments in layer D. Its intrusion must therefore have occurred after the deposition of layer C but before that of layer D. Because uplift and erosion are required to expose the dike, the upper surface of layer C is actually an unconformity that cuts across the dike. The fault, on the other hand, cuts the dike and all layers except layer E. It is therefore younger than the dike and must have formed after the deposition of layer D and before that of layer E.

The application of all of the principles of relative dating is illustrated by the geologic history represented in Figure 3.12.

Table 3-1
Doctrines Behind the Understanding of Geologic Time

PRINCIPLE	DESCRIPTION	ADVOCATE	DATE
1. Catastrophism	Earth created fully formed in 4004 B.C. All changes to surface are result of supernatural catastrophies	Archbishop James Ussher	1654
2. Superposition	In sequence of sedimentary layers, youngest material is at the top	Nicholas Steno	1669
3. Original Horizontality	Sedimentary layers deposited horizontally	Nicholas Steno	1669
4. Original Lateral Continuity	Sedimentary strata deposited in continuous layers	Nicholas Steno	1669
5. Gradualism	Earth ancient. Earth's surface in continuous but gradual change	James Hutton	1795
6. Faunal Succession (Correlation)	Fossil contents of sedimentary layer different from layers above and below, allowing layer to be traced and its relative age determined	William Smith	1817
7. Uniformitarianism (Actualism)	Geologic processes have always acted with the same rate as they do today	Charles Lyell	1830

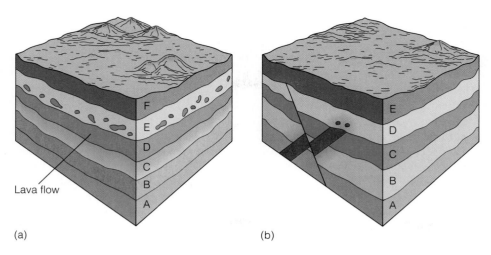

Figure 3.11
Determination of relative age (a) by the superposition of sedimentary layers (A, B, C, E and F) and a lava flow (D), and (b) by cross-cutting relationships between sedimentary layers (A through E), an intrusion, and the offset of both along a fracture.

(a)

Lava flow

(b)

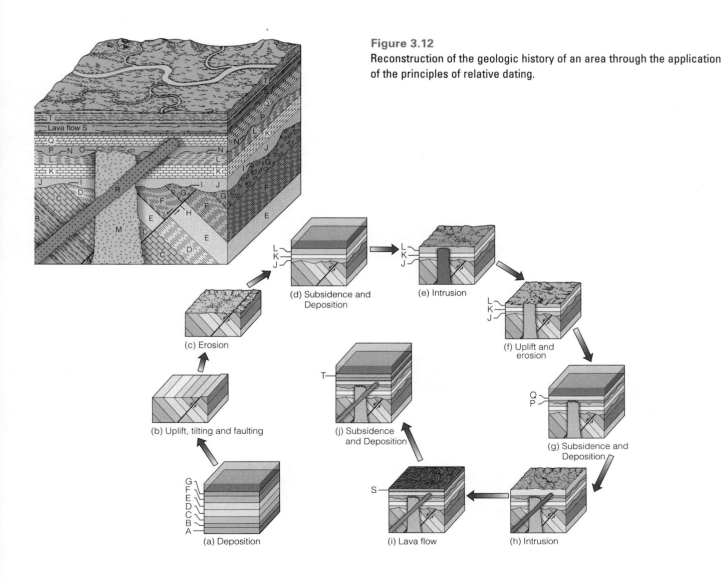

Figure 3.12
Reconstruction of the geologic history of an area through the application of the principles of relative dating.

(c) Erosion

(d) Subsidence and Deposition

(e) Intrusion

(f) Uplift and erosion

(b) Uplift, tilting and faulting

(j) Subsidence and Deposition

(g) Subsidence and Deposition

(a) Deposition

(i) Lava flow

(h) Intrusion

Geologic Maps

A geological map shows the distribution of the various rock types exposed at the Earth's surface. From such a map, the relationships between the exposed rock units and the geological history of the region can be interpreted. This information forms the cornerstone of much of our knowledge of Earth processes, and so is an essential element of resource exploration and environmental assessment.

The detail portrayed on a geological map depends on the **scale** at which the map is drawn. On a map of North America, for example, the distribution of surface rocks is far too complicated and intricate to be represented accurately. Maps of this type necessarily incorporate synthesis and simplification so that information appropriate to the study can be represented.

Most governmental geological survey maps are conducted at a scale of 1:50,000. At this scale, one centimeter on the map represents 50,000 centimeters (or 500 meters) on the ground, and one inch on the map represents 50,000 inches (or 4166.6 feet). In more important areas, maps may be drawn at a more detailed scale, such as 1:10,000 (one centimeter represents 100 meters). Either of these two scales are suitable for resolving the distribution of major rock units.

In reality, many of these rocks are hidden beneath a cover of soil and vegetation.

In showing the distribution of rock units beneath this cover, a geological map must therefore involve a degree of interpretation as to the nature of the bedrock. In a sense, the task of geological mapping is like having to complete a jigsaw with several of the pieces missing. From the overall pattern, an educated guess can be made on the nature of the missing pieces. The final map is an interpretation of all the available data, and as such, is subject to scrutiny and ongoing refinement as new data becomes available.

The actual geological map is the end product of a process that begins with the systematic examination of all known outcrops in the area where the bedrock pokes through its cover of soil and vegetation. First, the outcrop locations are plotted on an available base map, usually one that shows surface elevations in the form of topographic contours. The outcrops are then color-coded according to rock type. For example, in Figure 3.13a blue is used for limestone, yellow for sandstone, and gray for shale. Any exposed contacts between adjacent rock types are also identified as these provide important information on the location of the boundaries between rock units. The orientation of features such as the bedding are then measured and used to interpret the geology hidden beneath the soil and vegetation (Fig. 3.13b).

Two parameters, known as strike and dip (Fig. 3.14a), are used to describe the orientation of bedding planes (see also Chapter 6). Used in combination, these two variables uniquely define the orientation of any flat surface. The **strike** of an individual bed is the orientation of an imaginary horizontal line on the bedding plane measured relative to North. Because sea level is horizontal, the intersection of the sea surface with a bed in coastal regions will be parallel to the strike direction of the bed (Fig. 3.14b). In other regions, the strike direction can best be visualized by bringing a notebook, held horizontally, into contact with the bedding plane. The line of intersection between the two planes is parallel to the strike of the bedding plane.

Dip is a measurement of the maximum slope of a plane relative to the horizontal (Fig. 3.14). All inclined planes have a line of maximum slope. If you were to sprinkle water on the plane, gravity would drive the water down this line. The angle the line makes with the horizontal is the dip. Horizontal beds therefore have a dip of zero degrees, while the dip of vertical beds is 90 degrees.

Orientational data such as strike and dip are portrayed on a geological map as symbols. For strike and dip, a T-shaped symbol is typically used (Fig. 3.13b). The strike is portrayed by a relatively long straight line at the head of the symbol, the orientation of which

Here, several successions of sedimentary strata have been variably offset by faulting and cut by intrusions, and from this the following sequence of events can be deciphered. Layers A through G form a continuous succession of parallel sedimentary strata which, according to Steno's principles, were deposited in sequence and later tilted, faulted, and eroded. Following uplift and erosion, layers J through L were deposited on the erosion surface. The base of layer J cuts across layers A through G, producing an unconformity. It also cuts across the fault, indicating that faulting occurred prior to the deposition of layer J. According to the rule of cross-cutting relationships, the sequence A to L was then intruded by unit M. The area was then uplifted and eroded, and layers P and Q were deposited on the erosion surface to pro-

duce a second unconformity above which fragments of unit M occur in layer P. Next comes unit R, which according to the rule of cross-cutting relationships was intruded at some time following the deposition of layer Q. Lava S then flowed over layer Q, following which layer T was deposited.

In this fashion, the relative chronology of events can be built up for any area. However, it is important to note that the process only establishes the *sequence* of events, not their actual age. We cannot tell from relative age dating how long ago any one of these events took place. Nor can we determine the duration of an event or the time interval between events. Such information did not become available until the advent of absolute age dating (see Section 3.6).

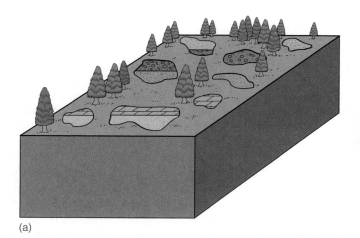

(a)

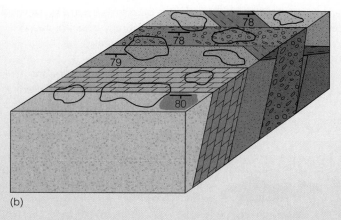

(b)

Figure 3.13

Construction of a geologic map and cross section from surface rock outcrops. (a)Area with scattered outcrops where the bedrock pokes through its cover of soil and vegetation. (b) Data from outcrops, including the strike and dip of the bedding, are used to infer what is present in the covered areas. The lines shown represent boundaries between the different rock types. The result is a geologic map (top surface) showing the bedrock as if the cover had been removed, and a cross section (sideview) constructed from the location and orientation of the rocks exposed.

(continued)

3.4 Relative Time and the Geologic Time Scale

3.4.1 Stratigraphy and the Stratigraphic Column

Throughout the nineteenth century, Steno's principle of superposition and Smith's principle of faunal succession were used by geologists to compile an ever larger and more complete stratigraphic column of younger-upon-older strata arranged in vertical sequence. Clearly, no more than a fraction of the entire succession of sedimentary strata (or **stratigraphy**) can be observed at one outcrop. However, as successions of sedimentary strata observed at one particular outcrop were correlated with those at other locations where portions of the succession either higher or lower in the sequence of strata were exposed, it became possible to construct a composite stratigraphic column in which all sedimentary layers were arranged in the order in which they had been deposited.

The construction of such a composite stratigraphic column is illustrated in Figure 3.16 using the geology of three national parks in the southwestern United States. The Grand Canyon in Arizona exposes one of the Earth's most complete rock sequences, with a vertical thickness of almost 2 km (1.2 mi).

is parallel to the direction of strike as measured in the field. The dip is portrayed as a short tick at the center of this line that points in the downslope direction. A number beside the tick indicates the angle of dip.

The location of individual outcrops, the position of contacts between rock units, and the strike and dip of bedding can then be used to, in effect, "complete the jigsaw puzzle" by filling in the gaps between outcrops. The result is a map that shows the geology of the area as if the cover of soil and vegetation had been removed (Fig. 3.13c).

In portraying the surface distribution of rock units, a geological map is also a two-

dimensional rendition of a three-dimensional geological problem. However, the three-dimensional geometry can often be inferred. Although the map itself is two-dimensional, the orientational data can also be used to construct cross sections from which a three-dimensional picture of the geology can be obtained. In Figure 3.13c, for example, the orientational data reveals a sequence of steeply dipping beds. A more complex situation is shown in Figure 3.15a for an area of outcrops otherwise similar to that of Figure 3.13a. Here, however, two quite different orientations of bedding are indicated by the exposed contacts between adjacent rock

types, and by the symbols for strike and dip (Fig. 3.15b). When this information is used to determine the three-dimensional geometry, the resulting cross section (side view) reveals the presence of an unconformity not unlike that observed by Hutton on the banks of the River Jed.

Geologic maps can provide a wealth of information about the distribution, composition, and disposition of the rocks in a given area, and are fundamental to the interpretation of an area's geologic history. However, they also play a key role in a variety of economic and environmental issues. By revealing the three-dimensional relationships among rock units, maps play a vital

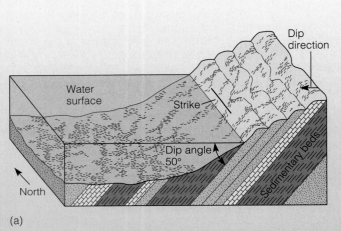

Figure 3.14

Strike and dip. (a) The strike, dip, and direction of dip of a succession of tilted sedimentary beds. In this case, the beds strike north and dip 50° west. (b) The strike of sedimentary beds (dipping to the left) along the coast on North Devon in England, is shown by the intersection of the beds with the sea surface (photo Damian Nance).

From Steno's principles we may conclude that the oldest unit is the Vishnu Schist while the youngest is the Moenkopi Formation. Yet the exposed sequence represents only part of the complete stratigraphy of northern Arizona and neighboring Utah. Only by correlating the Kaibab Limestone at the top of the Grand Canyon sequence with the base of the succession at Zion National Park, and then correlating the Navajo Sandstone at the top of the Zion succession with the base of the sequence at Bryce Canyon, can the complete stratigraphy be constructed.

The exposed sequence at the Grand Canyon also contains some significant time gaps in the form of unconformities. For example, about 150 million years of the rock record is missing at the base of the Temple Butte Limestone, and the unconformity at the base of the Tapeats Sandstone represents a time gap of approximately 1.5 billion years. Because the rocks representing these intervals have either been removed or were never deposited, evidence of a rock record that might fill these missing intervals must be sought elsewhere.

Throughout the nineteenth century, gaps in the rock record at one location were filled by those from another to produce an ever more complete stratigraphic column. By the end of the century, this composite stratigraphic column, into which the jigsaw of worldwide rock strata had been compiled, had become the basis of a sophisticated geologic time scale (Fig. 3.17). Because the stratigraphic column arranged sedimentary strata according to their relative age, subdivisions of the column became subdivisions of geologic time.

Such subdivisions were possible because the strata contained fossils that changed from layer to layer as new species emerged or existing species became extinct. Different portions of the stratigraphic column were consequently found to contain different assemblages of fossils, and were subdivided and named accordingly. In the mid-nineteenth century, this changing fossil record played an important role in the development of Darwin's theory of evolution. The names given to the subdivisions, however, are significant because they

role in the exploration for groundwater, oil, and gas and the search for economic mineral deposits. They are also used during the planning of major engineering projects such as the construction of highways and dams, and in the assessment of earthquake risk and volcanic hazard in regions of active faulting and volcanic activity. Geologic maps are also consulted for many aspects of land-use planning. For example, they are used when choosing a site for a new well field or landfill or when planning development in areas subject to landslides. Geologic maps consequently play an important role in many aspects of society and are used for many purposes by people in many professions.

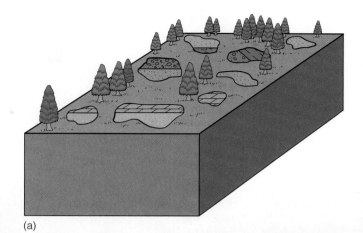

(a)

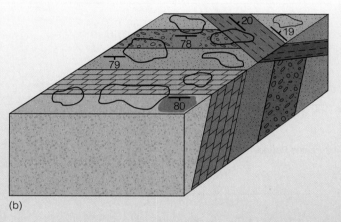

(b)

Figure 3.15

Unconformity revealed from surface rock outcrops. (a) Level area with outcrops similar to those of Figure 3.13. (b) Strike and dip of the bedding and boundaries between different rock types are used to produce a geologic map (top surface) and cross section (side view) that reveal the presence of an unconformity.

remain with us today as the terms we apply to the subdivisions of geologic time.

3.4.2 The Geologic Time Scale

The passage of geologic time as recorded in the stratigraphic column is marked by the emergence and extinction of fossil life-forms. Long intervals of geologic time separated by major extinction events are called **eras**. Early paleontologists discovered that the fossil record was punctuated by two major extinction events that were global in scale. They used these events to define two major boundaries in the geologic time scale, such that each mass extinction can truly be said to define "the end of an era." However, they did not know what had caused these extinctions, nor did they know how many millions of years ago they had occurred. Only with the advent of radiometric dating (see Section 3.6) has the absolute age of these events become clear.

The first of these major boundaries separates the era called the **Paleozoic**, meaning "ancient life," from the era called the **Mesozoic**, which means "middle life." The second separates the Mesozoic Era from the era called the **Cenozoic**, which means "recent life." Abundant fossils first appear at the beginning of the Paleozoic Era. Indeed, the abrupt appearance of plentiful fossils worldwide is used to separate the Paleozoic Era from all rocks deposited beforehand. Although some of the fossils of the Paleozoic Era represent the ancestors of modern lifeforms, many have left no living relatives because they became extinct at the end of the era. In fact, the massive extinction that marks the end of the Paleozoic is estimated to have eliminated nearly 90% of all marine species.

Dinosaur bones, in contrast, are not found in the strata of the Paleozoic Era, but occur only in overlying rocks of the Mesozoic Era. Fossil birds, mammals, and flowering plants are also seen for the first time in rocks of this era. The end of the Mesozoic is marked by another massive extinction,

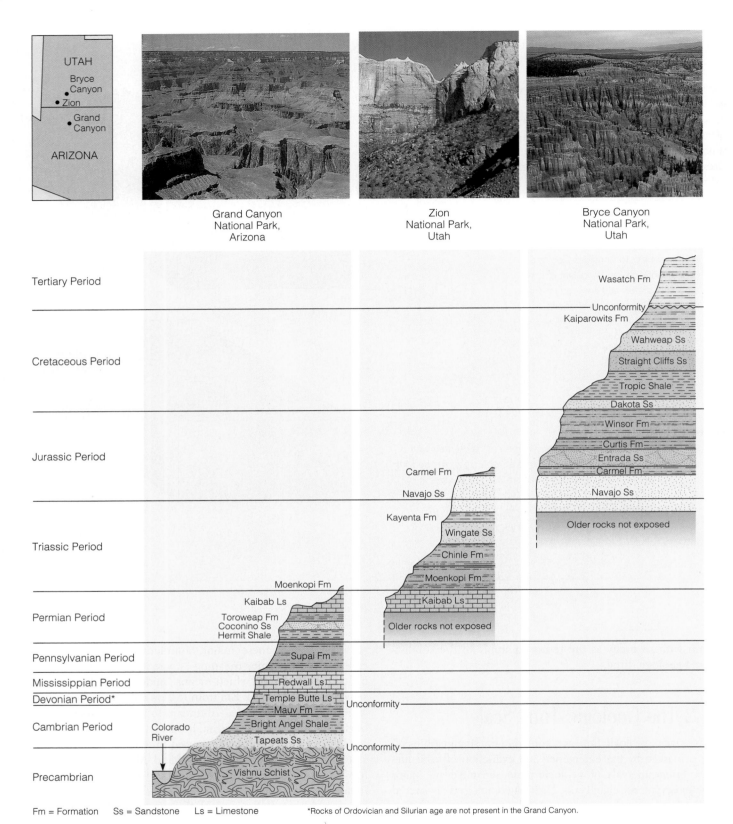

Grand Canyon
National Park,
Arizona

Zion
National Park,
Utah

Bryce Canyon
National Park,
Utah

UTAH

Bryce Canyon
• Zion
Grand Canyon

ARIZONA

Tertiary Period

Wasatch Fm

Unconformity
Kaiparowits Fm

Cretaceous Period

Wahweap Ss
Straight Cliffs Ss
Tropic Shale
Dakota Ss

Jurassic Period

Winsor Fm
Curtis Fm
Entrada Ss
Carmel Fm
Carmel Fm
Navajo Ss
Navajo Ss

Older rocks not exposed

Kayenta Fm

Triassic Period

Wingate Ss
Chinle Fm
Moenkopi Fm

Moenkopi Fm

Permian Period

Kaibab Ls
Toroweap Fm
Coconino Ss
Hermit Shale

Kaibab Ls

Older rocks not exposed

Pennsylvanian Period

Supai Fm

Mississippian Period

Redwall Ls

Devonian Period*

Temple Butte Ls
Mauv Fm

Unconformity

Cambrian Period

Colorado River

Bright Angel Shale
Tapeats Ss

Unconformity

Precambrian

Vishnu Schist

Fm = Formation   Ss = Sandstone   Ls = Limestone   *Rocks of Ordovician and Silurian age are not present in the Grand Canyon.

Figure 3.16

The Grand Canyon in northern Arizona affords one of the Earth's most complete exposed rock sequences: almost 2 vertical km (about 1.2 mi) of rock ranging up to 3 billion years in age. However, even here there are unconformities representing gaps in the local rock record, and correlations must be made with rocks exposed in other parts of the Colorado Plateau, such as Zion and Bryce Canyon, before the geological history of the entire region can be determined.

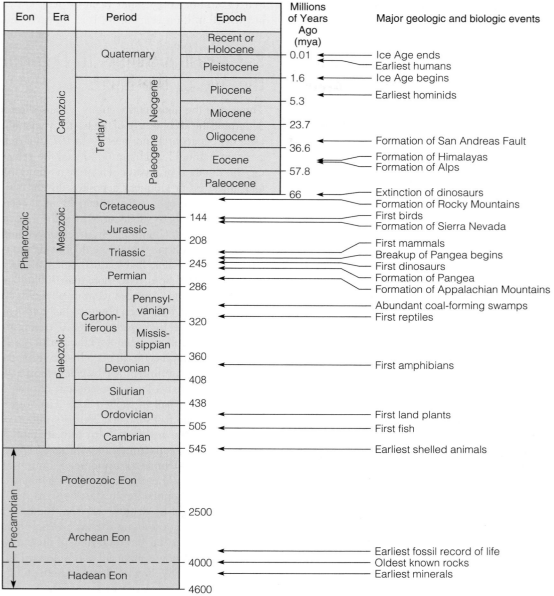

Eon	Era	Period		Epoch	Millions of Years Ago (mya)	Major geologic and biologic events
Phanerozoic	Cenozoic	Quaternary		Recent or Holocene	0.01	Ice Age ends
						Earliest humans
				Pleistocene	1.6	Ice Age begins
		Tertiary	Neogene	Pliocene		Earliest hominids
				Miocene	5.3	
					23.7	
			Paleogene	Oligocene		Formation of San Andreas Fault
					36.6	Formation of Himalayas
				Eocene		Formation of Alps
					57.8	
				Paleocene		
					66	Extinction of dinosaurs
	Mesozoic	Cretaceous				Formation of Rocky Mountains
					144	First birds
		Jurassic				Formation of Sierra Nevada
					208	
		Triassic				First mammals
					245	Breakup of Pangea begins
						First dinosaurs
	Paleozoic	Permian				Formation of Pangea
					286	Formation of Appalachian Mountains
		Carbon-iferous	Pennsyl-vanian			Abundant coal-forming swamps
					320	First reptiles
			Missis-sippian			
					360	
		Devonian				First amphibians
					408	
		Silurian				
					438	
		Ordovician				First land plants
					505	First fish
		Cambrian				
					545	Earliest shelled animals
Precambrian		Proterozoic Eon				
					2500	
		Archean Eon				
						Earliest fossil record of life
					4000	Oldest known rocks
		Hadean Eon				Earliest minerals
					4600	

Figure 3.17

The geological time scale showing the subdivision of geologic time into eons, eras, periods, and epochs. Figures in mya (millions of years ago) approximate the beginning of each time division.

second only in scale to that which brought the Paleozoic to a close. This extinction, which claimed the dinosaurs and many other animal and plant species, heralds the start of the Cenozoic, the era during which fossil mammals first become abundant.

The four largest intervals of geologic time, some with durations of billions of years, are called **eons,** which represent fundamental stages in the planet's evolution. They include the **Archean,** meaning "primeval," which starts with the Earth's oldest known rocks, the **Proterozoic,** meaning "earlier life," which contains the first record of multicellular organisms, and the **Phanerozoic,** meaning "visible life," during which fossils

became abundant. However, it is only during the Phanerozoic that subdivision of geologic time based on plentiful fossils is possible. Subdivision of the Archean and Proterozoic became possible only with the advent of radiometric dating (see Section 3.6), and they remain undivided in Figure 3.17. Indeed, the vast interval of geologic time that precedes the Phanerozoic Eon is often referred to simply as the **Precambrian,** a name that implies that it predates the Cambrian, which is first subdivision of the Paleozoic Era.

With the comings and goings of literally thousands of different fossil life forms it is possible to subdivide each of the three eras of the Phanerozoic into shorter intervals called **periods.** These were named, for the most part, after places where rocks of that age are found. Thus, the Devonian Period of the Paleozoic Era is named after the county of Devon in southwest England, whereas the Jurassic Period of the Mesozoic Era is named for the Jura Mountains in Switzerland. But the names are applied worldwide. It was during the Devonian Period, for example, that the Temple Butte Limestone was deposited in what is now the Grand Canyon, and it was during the Jurassic Period that much of the Navajo Sandstone was deposited at Zion and Bryce Canyon (Fig. 3.16).

Certain fossils in particular were useful for the purpose of subdivision because there were easily recognized, widely distributed around the world, and restricted to a narrow interval of rock strata. Some of these so-called **index fossils** were sufficiently short-lived that they could be used to further subdivide geologic periods into even shorter **epochs.** For example, the oldest *hominids*, the primate family that includes present-day humans, date back to the Early Pliocene Epoch at the end of the Tertiary Period (Fig. 3.17).

Detailed studies of sedimentary strata and the fossils they contained enabled nineteenth century geologists to construct a sophisticated geologic time scale, and a multitude of names that are used to this day were introduced to catalog rocks according to their relative age. Only by dividing geologic time in this fashion did it become possible to examine the evolution of the Earth and compare the geologic history of one part of the world with that of another. This time scale, although constructed over a century ago, is still the calendar in use today.

Despite its refinement into eras, periods, and epochs the scale could only record the sequence of geologic time, not the actual number of years in any given time interval. In other words, the eras, periods, and epochs could be assigned relative ages but not absolute dates. So how were their absolute ages determined?

3.5 How Old is the Earth?: The Quest for an Absolute Age

3.5.1 Accuracy Versus Precision

In dealing with absolute age, it is important to distinguish accuracy from precision. **Precision** refers to the exactness of an estimate whereas **accuracy** refers to its correctness. For example, if when asked for the time you reply "it's just after twelve" when the correct time is 12:01 p.m., your response is accurate but imprecise. On the other hand, if you reply "16 minutes past twelve" because your watch is running 15 minutes fast, your response is precise but inaccurate.

To this day, the most precise determination of the age of the Earth is that based on the biblical chronology pioneered by Archbishop James Ussher, according to which the Earth celebrated its 6000th birthday on October 26, 1997, at exactly 9:00 A.M. The determination is precise because it was given to the nearest minute, a resolution that modern science can not hope to duplicate. Although it is precise, according to modern science it is wildly inaccurate. Modern estimates place the age of the Earth at around 4600 million (or 4.6 billion) years. This determination is nowhere near as precise as

that of Archbishop Ussher because it is known only to the nearest 10 million years or so. It is, however, more accurate.

3.5.2 Past Attempts to Determine the Age of the Earth

The end of the nineteenth century witnessed several attempts to determine the absolute age of the Earth by scientific methods that were neither accurate nor precise. In 1878, for example, the Irish geologist Samuel Haughton introduced the principle of estimating the duration of a particular geologic time interval from the total thickness of sediment deposited during that interval. With the construction of a stratigraphic column, it became possible to estimate the age of its subdivisions from the total thickness of strata they contained. Unfortunately, much uncertainty existed about the rates at which the strata were deposited, and estimates of the amount of time that had elapsed since the beginning of the Paleozoic ranged from 18 to more than 700 million years.

Lord Kelvin, a noted Victorian physicist, used a different approach. It had long been observed that temperatures in mines increased with depth, implying a flow of heat from the interior of the Earth to its surface. Therefore the Earth, Lord Kelvin argued, is losing heat and must have been hotter in the past. In 1897, he attempted to determine its age by calculating how long it would take the planet to cool to its present state of heat loss, assuming it had begun as a ball of molten rock. His answer was that the Earth's crust had first solidified 20 to 40 million years ago.

In 1898, the Irishman John Joly attempted to determine the age of the oceans from their saltiness, a suggestion first proposed by the celebrated astronomer Edmund Halley in 1715. Recognizing that the oceans owe their saltiness to the accumulation of salts supplied by streams, he suggested that the age of the oceans might be determined from the total amount of salt in the sea. By determining the average salt content of stream water and estimating the total flow of all the world's streams, Joly calculated how much salt was delivered to the oceans in any given year. Dividing this figure into the total amount of salt in the sea gave Joly an estimate of how long it had taken to bring the world's ocean water to their present level of saltiness, assuming no loss of salt. Joly's answer was 80 to 90 million years.

Although Lord Kelvin's and Joly's estimates were of similar magnitude (both being in the order of the tens of millions of years), both wildly underestimated the age of the Earth as modern methods were later to prove. The reasons in both cases are similar. Lord Kelvin assumed a steadily cooling planet, unaware that naturally occurring radioactive elements provide the Earth with its own internal heat source. Just as an electric heater loses heat without cooling, so Kelvin's assumption that the Earth must be cooling because it was losing heat was flawed. Joly, on the other hand, assumed that the oceans were getting progressively saltier whereas the world's salt deposits testify to the removal of huge quantities of salt from the oceans by chemical precipitation. Therefore, Joly's answer does not give the age of the oceans, but rather is a crude estimate of the "residence time" of salt in the oceans. This is the average time that salt spends in the ocean after its removal from the land by streams and before its removal from the ocean by precipitation or other processes.

3.6 Radiometric Dating

Modern estimates of absolute time and modern determinations of the age with the Earth did not become possible until 1896, when Henri Bequerel discovered that the atoms of some chemical elements, such as uranium, are inherently unstable and break down in a predictable fashion. This property is known as **radioactivity** and its explanation is highly technical because it concerns the structure of the atom itself (see Dig In: Radioactivity and the Dating of Rocks, pages 72–73). However, in simple terms it occurs when an instability exists in the nucleus of an atom, as would occur, for example, if the nucleus contained too many particles.

As we learned in Chapter 2, the nucleus of an atom can contain two kinds of particles, called protons and neutrons. Protons have a positive charge and their number is the atomic number which determines the element of the atom. Neutrons have no charge and their number in the nucleus of an atom can vary without affecting the atom's chemical properties. However, neutrons do have mass, so their presence in an atomic nucleus will increase an atom's mass as their number rises, and this can cause the nucleus to become unstable. Whenever the number of neutrons in the atomic nucleus of a chemical element varies slightly, each variation is called an **isotope** of that element. Isotopes of elements therefore have the same number of protons but varying numbers of neutrons, and they therefore have the same chemical properties but differing atomic weights. Isotopes are therefore identified, not by their atomic number but by their **mass number,** which is the sum of the number of protons and neutrons in their atomic nuclei. For example, the element chlorine can have either 18 or 20 neutrons in its atomic nuclei in addition to 17 protons. The isotopes are consequently termed chlorine-35 and chlorine-37.

Isotopes are very common in nature. In fact, most elements possess several isotopes, the vast majority of which are perfectly stable. However, isotopes that have too many particles in their nuclei are unstable and spontaneously break down to isotopes with more stable nuclei, releasing energy and particles in the process. If the new isotope produced by this radioactive decay is also radioactive, several breakdown steps from the original radioactive **parent** isotope may be necessary to form the final stable nonradioactive isotope or **daughter.** For example, many steps are involved in the ultimate breakdown of uranium to lead. The energy released through each step of this process is enormous and forms the basis of atomic power. However, it is the pace of the process that is the key to its use for dating.

The rate at which a parent decays to its daughter varies from one radioactive element to another. However, for any given parent, the rate is a fixed one and cannot be changed by any known means. Because of this, naturally occurring radioactive elements form the most reliable of clocks, decaying at a constant rate regardless of any physical or chemical changes in their environment.

The decay process starts with a set number of parent atoms and no daughter product. As the parent decays, the number of parent atoms decreases while the number of daughter atoms grows. At some point in time the amount of parent isotope will have decreased to exactly half its original quantity. The length of time it takes to reach this point is known as the isotope's **half life** (Fig. 3.18) and varies from one isotope to another.

The length of an isotope's half life is a measure of its rate of decay and can be determined experimentally. For example, the half life for the decay of uranium-235 to lead-207 is 704 million years. This means that 704 million years after the decay process starts, 50% of the uranium parent atoms will have decayed to atoms of lead, the daughter product. After another 704 million years, or a total of two half lives, half of the remaining parent atoms will have decayed to atoms of the daughter product and the remaining atoms of uranium will have shrunk to 25% of their original number. After another 704 million years, or three half lives, the atoms of uranium will have shrunk to 12.5% of their original number, and so on. This process whereby the amount of the parent isotope is halved in each successive half life is known as **exponential decay.** Since one daughter is formed by the decay of a single parent atom, the total number of atoms does not change during the decay process. As a result, the number of atoms of the daughter product steadily increases, equaling the number of parent atoms after one half life and reaching 75% of the total number of atoms after two. After three half lives, 87.5% of the total number of atoms will be those of the daughter isotope.

We can draw a parallel with the process of radioactive decay by tossing a set of sixteen coins that originally lay heads up. Because each coin must land as either heads or tails, there is a fifty-fifty chance that it will be either one. So the odds favor

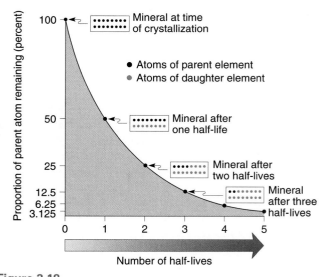

Figure 3.18

Radioactive decay converts a radioactive parent isotope to a stable daughter isotope. With the passage of each half life, the number of atoms of the radioactive isotope is reduced by half, whereas the number of atoms of the daughter isotope increases by the same quantity. By measuring the ratio of the quantities of the parent to the daughter isotopes, geologists can determine the absolute ages of some rocks, providing the half life of the parent isotope is known.

Radioactivity and the Dating of Rocks

To understand the principle of radiometric dating, that is, how the age of a rock is determined from the decay of radioactive elements contained within it, we must turn to the nature of elements themselves and the atoms that comprise them. The atom is the fundamental unit of all elements and consists of a nucleus, containing protons and neutrons, and a negatively charged cloud of orbiting electrons. Whereas the number of positively charged protons in the nucleus of an atom (the atomic number) is fixed in any given element, the number of uncharged neutrons can vary.

Atoms of the same element that have differing numbers of neutrons in their nuclei are called **isotopes.** Carbon, for example, always has six protons but may have six, seven, or eight neutrons. Carbon is therefore said to possess three isotopes which are respectively termed carbon-12, carbon-13, and carbon-14, based on the total number of particles in their atomic nuclei (the **mass number**). Oxygen, with eight protons, may have eight, nine, or ten neutrons, and so is said to have three isotopes termed oxygen-16, oxygen-17, and oxygen-18. Uranium with 92 protons, has two important isotopes, uranium-238 (with 146 neutrons) and uranium-235 (with 143 neutrons).

In 1896, Henri Bequerel discovered that not all isotopes are stable. Instead, some are radioactive and spontaneously break down to more stable isotopes through the release of energy and particles. Through the loss or gain of protons, the breakdown of a radioactive parent isotope produces a stable daughter isotope of a new element. Carbon-12 and carbon-13, for example, are stable isotopes but carbon-14, the basis of radiocarbon dating, is radioactive and decays by the spontaneous breakdown of one neutron to a proton and an electron (which is ejected). The daughter, now with seven protons rather than six, is nitrogen, and the specific isotope, now with seven neutrons instead of eight, is nitrogen-14. This form of radioactive decay involving the ejection of an electron is called **beta particle emission.** Two other forms of decay involve either **alpha particle emission** (the ejection of two protons and two neutrons) or **electron capture** and the conversion of a proton in the nucleus to a neutron (Fig. 3.19). In each case, the number of protons in the atomic nucleus (the atomic number) changes and so a different element is produced.

In nature, radioactive isotopes are commonly incorporated into newly formed crystals

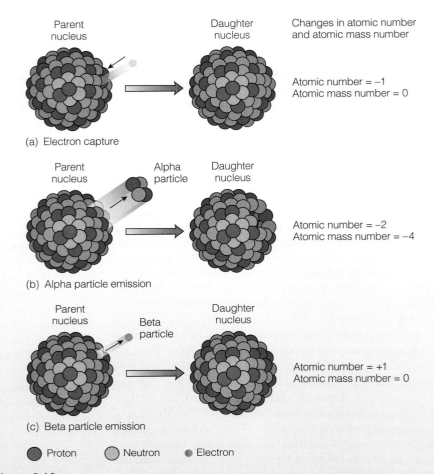

(a) Electron capture

Changes in atomic number and atomic mass number

Atomic number = −1
Atomic mass number = 0

(b) Alpha particle emission

Atomic number = −2
Atomic mass number = −4

(c) Beta particle emission

Atomic number = +1
Atomic mass number = 0

● Proton　◯ Neutron　● Electron

Figure 3.19

Common types of radioactive decay, each resulting in a change in the number of protons in the atomic nucleus and therefore, a change in element.

as impurities. Once incorporated, decay begins and the number of parent atoms of the isotope progressively decreases with time while the number of daughter atoms progressively increases. The decay process is a continuous one that is unaffected by any outside influences. As a result, we can be certain that a particular fraction of all of the atoms of the radioactive parent isotope will decay to atoms of the stable daughter isotope in a given length of time. For example, when a new crystal containing a particular radioactive element first forms in a rock, only the parent isotope will be present. As time passes, however, atoms of the parent isotope decay to the daughter, so that the amount of the parent decreases while the amount of the daughter isotope goes up. By measuring the ratio of parent isotope to daughter isotope, we can determine the length of time over which breakdown of the parent has occurred (providing we know the parent element's rate of radioactive decay)

and so establish the age of the mineral. Measurement of the amount of both parent and daughter isotope is made on an instrument known as a **mass spectrometer**.

In order for a radioactive isotope to be useful for dating purposes, it must occur naturally in rock-forming minerals and must decay at a slow enough rate that both parent and daughter isotopes are present in measurable quantities. The most suitable radioactive decay systems are shown in Table 3-2. Of these, uranium-lead, rubidium-strontium, and potassium-argon are the decay systems most frequently used to date rocks. The breakdown of uranium to lead is used to date rocks such as granite that contain uranium-bearing minerals like zircon and uraninite. The rubidium-strontium and potassium-argon systems, on the other hand, are used to date rocks containing potassium-bearing minerals such as orthoclase, muscovite, biotite, and hornblende.

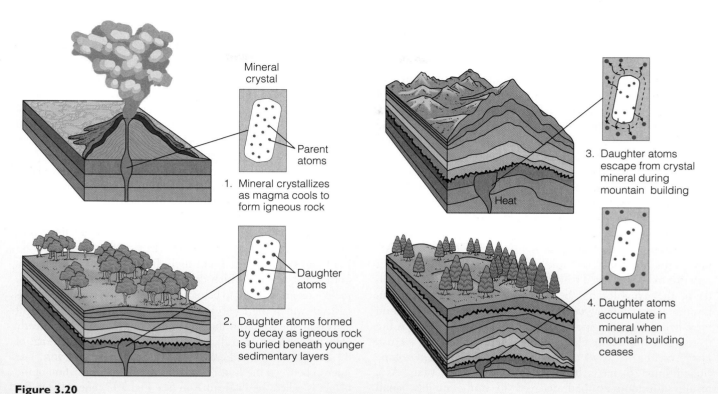

Figure 3.20

Loss of daughter isotopes from a rock heated during mountain building. Following original crystallization of rock, daughter atoms accumulate as atoms of the parent isotope decay. Later, heat causes expansion of minerals, allowing daughter atoms to escape and migrate elsewhere. Although the decay of the parent isotope proceeds normally throughout, and daughter isotopes resume accumulating after this heating episode ceases and the minerals contract, the loss of these daughter isotopes during this period skews the resulting parent-to-daughter ratio such that the age of the rock appears younger than it is. Complete loss of daughter atoms during the heating event resets the radiometric clock to the time of mountain building.

The breakdown of carbon-14 is used to date carbon-bearing material such as shells, bones, and charcoal. However, because of its short half life carbon dating is effective only in materials that are less than 100,000 years old. Beyond this, too little of the parent isotope remains for accurate measurement. Carbon-14 is therefore inappropriate for dating most geologic events but is extensively used in archeology and in the dating of events in the recent geologic past, such as the last Ice Age.

Once a radiometric age has been obtained, determining what that age means differs in different circumstances and may require some interpretation. When dating a detrital sedimentary rock, for example, is one dating the time of deposition or the age of the rocks from which the detritus was derived? Likewise, when dating a metamorphic rock, is one dating the time of metamorphism or the original age of the rock that was metamorphosed? As we will see, the answer depends on the mineral selected for dating.

Recent advances in analytical techniques and instrumentation have made uranium-lead

dating one of the most precise and widely used methods of radiometric age determination. The technique utilizes the highly resistant mineral *zircon*, which contains trace amounts of uranium and is common in very small quantities in granites. In granites, the age obtained from zircon crystals gives the date of their crystallization, and therefore the time at which the granite solidified from a molten liquid. The advantage of zircon in this case is its resistance to any form of alteration. (Note that in the form "cubic zirconia" it is used as a substitute for diamond). Therefore a zircon age will accurately date the granite in which it occurs regardless of the geologic events that may have affected the granite since its crystallization. Like a Timex watch, it can take a licking and keep on ticking! On the other hand, zircon is of no use in dating sedimentary rocks. A zircon crystal taken from a beach in Florida, for example, will not provide the age of the beach. Rather, it will provide the crystallization age of the granite from which the zircon grain was originally derived.

Radiometric techniques such as potassium-argon, on the other hand, are useful specifically because the minerals they are used to

date are susceptible to alteration. If, as a result of some geologic event, the daughter isotope escapes from the crystal in which it was produced, the radiometric clock is effectively reset to the time of that event (Fig. 3.20). In potassium-argon dating, the daughter product is a gas and is readily lost if the rock is reheated. This poses a problem if the method is used to date a particular potassium-bearing mineral, but is an advantage in dating the reheating event.

As an example, the potassium-argon method is widely used for dating potassium-bearing micas such as muscovite and biotite which, like zircon, are common in granites. If the granite has remained unaffected by geologic events since its crystallization, the ages obtained from all three of these minerals will be similar and will date the granite itself. However, if the granite was subsequently reheated as the result of some geologic event, the age obtained by the potassium-argon method is likely to date the most recent reheating. Under these circumstances, the zircon age of the granite will be different from its mica age and care must be taken to ensure that the two ages are correctly interpreted.

eight of each (50% heads and 50% tails). If all those showing tails are removed and the remaining eight are tossed again, the odds remain the same, this time favoring four heads and four tails. Each flip is therefore similar to a half life because it reduces the number of coins showing heads by a factor of two, first from 16 to 8 and then from 8 to 4.

The half life of any one radioactive isotope is fixed, but it varies markedly from one isotope to another. Obtaining an absolute age by radiometric means is therefore a matter of determining the parent-to-daughter ratio of a suitable radioactive element with a known half life. Once the decay rates of the different radioactive isotopes had been determined, a means was finally available for measuring geologic time.

Of the many radioactive isotopes, however, only a few are suitable for dating purposes. To be useful, the parent isotope must occur in minerals commonly found in rocks, neither parent nor daughter isotope must be able to leave the material being dated too easily, and the duration of the parent's half life must be appropriate for measuring purposes. That is, the isotope must not decay so rapidly that too little of the parent remains to be measured with accuracy, or so slowly that too little of the daughter has been produced. For example, aluminum has a radioactive isotope (aluminum-26) that could have been of great value for dating because aluminum is the third most abundant element present in the Earth's crust. However, the decay of aluminum-26 to its daughter isotope (magnesium-26) has a half life on only 720,000 years. As a result, aluminum-26 had decayed to immeasurably small quantities before the Earth's crustal rocks had even formed.

The parent-daughter isotope systems that have proved to be most appropriate for dating purposes because they best meet the criteria listed above, are shown in Table 3-2. Of these, the isotope systems uranium-lead, rubidium-strontium, and potassium-argon are used most frequently. Because they are measured by radioactive decay, dates determined by these isotope systems are called **radiometric** ages.

3.7 Absolute Time and the Geologic Time Scale

As a result of the application of radiometric dating to rocks around the world, numerical ages have been found for each of the major boundaries of the stratigraphic column. We now know for example that the Paleozoic era started some 545 million years ago, that the Mesozoic era began 245 million years ago, and that the demise of the dinosaurs at the beginning of the Cenozoic era occurred 66 million years ago. The oldest rocks yet discovered were found near Great Slave Lake in Canada and have a uranium-lead age of 3962 million years. Individual crystals of the mineral *zircon* have been found in Australia with ages of 4200 million years. However, they are older than the sedimentary rocks in which they occur. Rocks as old as those near Great Slave Lake are extremely rare because the Earth is a dynamic planet and has recycled itself many times over since the oldest rocks were formed. In fact, the oldest rocks on Earth are not from the Earth itself. Instead, they are meteorites from space, most of which date back 4600 million years. These meteorites are thought to be representative of the primitive material from which the Earth first formed. The oldest rocks brought back from the Moon by the Apollo missions also date back nearly 4600 million years, and all indications suggest that this is the age of the Solar System, and therefore, the true age of planet Earth. The interval between the Earth's oldest rocks and the formation of the planet is called the **Hadean Eon** (see Fig. 3.17) after Hades, the Greek god of the underworld.

Assigning absolute ages to the geologic time scale is not as easy as it may seem at first. This is because the major subdivisions are based on fossils, and fossil-bearing sedimentary rocks are not easily dated by radiometric means.

This, in turn, is because most dating techniques measure how long ago a mineral crystallized. The time at which a mineral is incorporated into sediment is not recorded because its

Table 3-2
The Six Radioactive Isotope Systems Most Suitable for Radiometric Dating

ISOTOPES		HALF-LIFE OF PARENT (YEARS)	EFFECTIVE DATING RANGE (YEARS)	MATERIAL THAT CAN BE DATED	
PARENT	DAUGHTER				
Uranium-238	Lead-206	4.5 billion	10 million to 4.6 billion	Zircon Uraninite	
Uranium-235	Lead-207	704 million			
Thorium-232	Lead-208	14 billion			
Rubidium-87	Strontium-87	48.8 billion	10 million to 4.6 billion	Muscovite Biotite Orthoclase Whole metamorphic or igneous rock	
Potassium-40	Argon-40	1.3 billion	100,000 to 4.6 billion	Glauconite Muscovite Biotite	Hornblende Whole volcanic rock
Carbon-14	Nitrogen-14	5730 years	less than 100,000	Shell, bones and charcoal	

radioactive clock is unaffected by this process. So the relationship between the relative time scale (established on the basis of fossils) and absolute time (established through radiometric dating) has had to be constructed by carefully dating minerals of the same relative age as the fossiliferous sediments.

An example of this approach is shown in Figure 3.21. In Section 1, dated volcanic lavas are interlayered with fossiliferous sedimentary rocks. While the sedimentary rocks can be assigned to, say, the Devonian Period on the basis of the fossils they contain, the actual age of the middle layer (layer 2) is bracketed by the absolute age of the volcanic lavas above and below. The lavas can be dated because they contain minerals that crystallized as the lava cooled. In this case, the ages of volcanic lavas A and B indicate that the fossils in the intervening sedimentary rocks of layer 2 are between 400 and 375 million years old.

We can also be sure that the rocks below lava A (layer 1) are more than 400 million years old, and that those above lava B (layer 3) are less than 375 million years old, but we do not know by how much. However, Section 2 exposes fossiliferous sedimentary rocks that correlate with layer 3, and an overlying sedimentary sequence (layer 4) that contains fossils belonging to the Carboniferous Period and a dated volcanic lava (lava C) that is 350 million years old. From this we can tell that the deposition of layers 3 and 4 must have occurred between 375 and 350 million years ago. Because this must also be the age of the intervening fossils, the boundary between the Devonian and Carboniferous periods is similarly restricted to this time interval.

An obvious drawback in this approach is the necessity for conveniently placed datable units such as volcanic lavas. If no lavas were present in Section 1 and 2, for example, it would not have been possible to date the Devonian-Carboniferous boundary. Because of this, the ages assigned to the geologic time scale are under constant scrutiny and refinement. The recent age revision for the start of the Paleozoic era is an excellent example of this. Prior to the revision, an age of about 570 million years for this boundary had long been widely accepted. However, suspicions about the accuracy of this age arose with the dating of volcanic rocks in Newfoundland, Canada. These rocks yielded an age of 565 million years, yet they lay at least 4 km (2.5 mi) beneath those containing the earliest Paleozoic fossils. This suggested a considerably younger age for the base of the Paleozoic. Consequently, attention was focused on similar sequences worldwide, as a result of which new dates have been obtained that more accurately place the age of the boundary at about 545 million years.

3.8 A Sense of Time

With radiometric age dating we now have the means of determining the absolute age of past geologic events and the rate at which they occurred. We can determine when events such as mass extinctions, ice ages, and major episodes of volcanism or mountain building occurred during Earth history, and so examine the pace and rhythms of natural activity in a way that has never before been possible. We can now confirm Hutton's speculation that the time needed to wear down mountains and

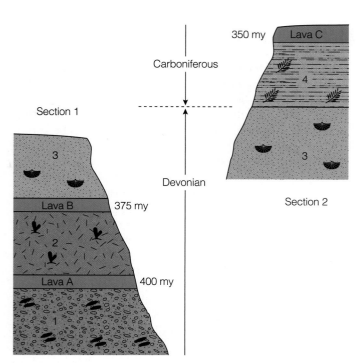

Figure 3.21
Idealized example of how the absolute age of fossiliferous sedimentary rocks is determined by dating associated volcanic lavas.

carry sand and gravel to the sea is enormous. We now know, for example, that the forces that uplifted the Appalachians ceased more than 250 million years ago, yet the mountains still rise to a heights in excess of 2000 m (6600 ft).

We can also identify patterns of global change, such as the comings and goings of ice ages, and the time scale over which they occur. Our ability to examine Earth history in this way is of enormous importance. Today, for example, the possibility of global warming is a major environmental concern, yet we have little understanding of either the impact of such a trend or the time it might take to become noticeable, if it was to occur. With an absolute time scale we can identify natural patterns of climate change in the past and the rates at which they occurred. We can also test scientific ideas or computer models which predict events to occur at certain times or at certain rates.

As we calibrate the geologic time scale in ever greater detail, we should remember that its major subdivisions (Fig. 3.17) are still based on fossils using the principles established by Hutton, Smith, and Lyell over 150 years ago. Before any rock can be dated, for example, its age relative to associated units must first be determined, and the area from which it was obtained must be mapped geologically. These techniques are as important today as they were in the past and still depend to a large extent on the principles of correlation and relative age dating developed by these pioneers of Earth science. Furthermore, the vast majority of rocks do not lend themselves to dating because they do not contain minerals appropriate for the purpose. It is still far easier, for example, to date a sedimentary rock on the basis of its fossil content than it is to date the rock itself. Consequently, the fossil-based subdivisions of the geologic time scale are still very much in use today.

Understanding the immensity of geologic time is essential to any understanding of Earth science. Over 200 years ago the Earth was widely believed to be a mere 6000 years old. All changes to the Earth's surface were attributed to supernatural disasters, like Noah's Flood; a view that came to be known as Catastrophism.

The examination of sedimentary rock layers led to the recognition of several key principles by which they could be placed in sequence. Because the layers were deposited in succession, the principle of superposition placed the oldest at the base and the youngest at the top in the form of a stratigraphic column. In this way, their relative age, or the order in which they formed, could be determined.

With the further recognition that sedimentary rocks were originally deposited horizontally in laterally continuous layers (expressed in the principles of original horizontality and original lateral continuity), and that individual rock layers could be identified on the basis of their fossil content, came the principle of faunal succession. From this, the age equivalence of rock layers in separate successions could be determined, so that age equivalent layers could be correlated with each other. Application of the principle of inclusions and cross-cutting relationships further allowed the relative age of sedimentary rock layers to be determined with respect to other features, such as intrusive rocks and faults.

Placed in succession, sedimentary rock layers and their fossils suggested gradual rather than abrupt change, advancing the principle of gradualism over that of catastrophism. In the further development of gradualism as the principle of uniformitarianism, geological processes were recognized as having acted in the past in the same way as they do today (a doctrine known as actualism). This uniformity of process implied generally uniform rates of change, and so necessitated a length of geologic time adequate for the slowness of natural processes.

Placed in succession, sedimentary strata were compiled into a stratigraphic column, the subdivision of which was based on the fossils the strata contained. Because this arranged strata according to their relative age, the subdivisions of the stratigraphic column became those of the geologic time scale. Subdivision of geologic time into eons, eras, periods, and epochs (as the level of refinement increases), is consequently based on the fossil record, its boundaries coinciding with the emergence and extinction of fossil life forms. Assembled in this way, the time scale recorded the sequence of events but not their age in years.

Early attempts to determine the age of the Earth, based on factors such as the thickness of sedimentary strata in the stratigraphic column, the saltiness of the oceans, and the time needed to permit the planet to cool to its present condition from a molten ball, were imprecise because of great uncertainties in the rates at which geologic processes operate. Only with the discovery of radioactivity did a precise method of dating rocks become available.

Because radioactive parent isotopes decay at fixed rates to non-radioactive daughter isotopes, the absolute age of a rock or mineral can be determined from the ratio of the amount of daughter isotope produced to that of the parent isotope remaining, providing the rate at which the parent isotope decays to its daughter is known. This rate of decay can be determined experimentally, and is expressed in terms of the half life of the parent isotope, that is, the time it takes for a given amount of the parent to decay to half that amount.

Through radiometric dating, absolute ages have been determined for the geologic time scale, and the age of the Earth is now placed at 4.6 billion years. Because the Earth is a dynamic planet that continually recycles the rocks on its surface, its age is based on dates obtained from meteorites thought to be representative of the primitive material from which the planet first formed. Many of the boundaries of the geologic time scale, however, are based on fossils that cannot be dated by radiometric means. Instead, the age of these boundaries must be bracketed by dating suitable rocks immediately above and below.

With absolute ages, not only can we date past geologic events, but we can identify the patterns of natural processes and the rates at which they occur. And finally, with the confirmation of the immense span of geologic time comes an appreciation for the profound changes that can be produced by even the most imperceptible of movements, or the slowest of processes.

Key Terms and Concepts

Look for activities exploring the highlighted items on the web at:
WWW.BROOKSCOLE.COM

- Relative time of older versus younger events and absolute time in years
- Archbishop Ussher's biblical chronology reveals the Earth to have been created in 4004 B.C. Changes to the Earth's surface attributed to supernatural catastrophes (principle of catastrophism)
- Nicolas Steno's principles of superposition, original horizontality and original lateral continuity suggest that sedimentary strata are deposited in sequence, horizontally, and in continuous layers
- Neptunists propose all rocks to have been deposited in water but Plutonists insist that rocks like granite and basalt crystallized from a molten state

- James Hutton advocates an ancient Earth, the surface of which is in continuous but gradual change (principle of gradualism) as demonstrated by unconformities
- William Smith recognizes that the evolving fossil content of sedimentary strata (principle of faunal succession) allows strata to be correlated and their relative age determined
- Charles Lyell proposes that geologic processes act today as they have always done (principle of uniformitarianism or actualism)
- Use of inclusions and cross-cutting relationships in relative age dating

- Construction of a stratigraphic column of younger upon older strata forms the basis of a geological time scale with eons, eras, periods, and epochs
- Accuracy versus precision
- Calculations based on sedimentation rates (Samuel Haughton), planetary cooling (Lord Kelvin), and the saltiness of the oceans (John Joly) underestimate the age of the Earth

- Henri Bequerel discovers radioactivity in which unstable parent isotopes spontaneously break down to stable daughter isotopes
- Exponential decay, half lives, and radiometric dating are used to establish an absolute time scale

Review Questions

1. Why is an appreciation of the length of geologic time important to Earth science?
2. What did Nicolas Steno mean by the superposition, original horizontality, and original lateral continuity of sedimentary successions?
3. How did the Neptunists and Plutonists differ in their interpretation of crystalline rocks such as granite?
4. How did James Hutton use Hadrian's Wall to argue for an ancient Earth?
5. What is meant by the correlation of rock layers and how was this used to establish the stratigraphic column?

6. What is the difference between accuracy and precision?
7. What is an isotope?
8. What is radioactivity and what takes place during radioactive decay?
9. What is meant by the half life of a radioactive isotope?
10. Under what circumstances can the radiometric age of a mineral differ from that of the rock from which it was obtained?
11. What is the absolute age of the Earth now thought to be and how has this age been determined?
12. Why has it been difficult to assign absolute ages to the geologic time scale?

Study Questions

Research the highlighted question using InfoTrac College Edition.

1. What is the essential difference between the uniformitarian and catastrophic approach to geology? Which best explains features found on the Earth's surface?
2. Describe the sequence of events necessary to produce an outcrop in which a sequence of folded strata is seen to be unconformably overlain by a sequence of tilted strata.
3. Use the principles of relative dating to determine the geologic history of the area shown in Figure 3.22.
4. Describe the concepts of relative and absolute time and outline how both are used in the geologic time scale (Fig. 3.18).
5. Lord Kelvin calculated the age of the Earth to be 20 to 40 million years based on its rate of heat loss, whereas John Joly calculated its age to be 80 to 89 million years based on the saltiness of the oceans. Why are both of these calculations seriously flawed?

6. A feldspar crystal analyzed for potassium-argon dating is found to contain 75% of its original potassium-40. Using Fig. 3.18, determine the age of the feldspar crystal, given that the half life of potassium-40 is 1.3 billion years. What assumptions are being made in determining this age?
7. A vertical sequence of Paleozoic sedimentary rocks contains two lava flows with ages of 405 and 415 million years (Fig. 3.23). The older flow is overlain by rocks containing Silurian fossils, whereas the younger flow overlies rocks with Devonian fossils. What are the ages of the rocks above, below, and between the two flows, and how can this be used to constrain the age of the boundary between the Silurian and Devonian periods?

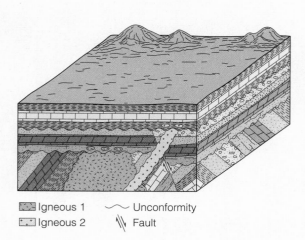

Figure 3.22
Figure to accompany Study Question 3.

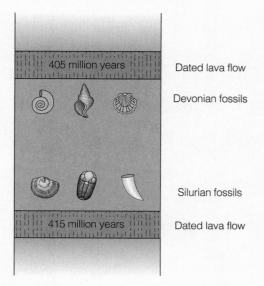

Figure 3.23
Figure to accompany Study Question 7.

CHAPTER 4

Plate Tectonics: Development of a Theory

The Earth is a dynamic planet with an internal power source that continually shifts and reshapes the planet's surface. We now know that the entire surface of the Earth is mobile and that this mobility accounts for many of the geological processes that shape our planet. Like a giant cracked eggshell, the outer rocky layer of the Earth is broken into shifting plates which can cause entire continents to move, and whole oceans to open or close, as they jostle and compete for position. This dynamic view of the Earth forms the basis of a comprehensive theory known as plate tectonics, the introduction of which has revolutionized our understanding of the world we live on.

Yet as recently as the early twentieth century these were radical ideas. The continents and ocean basins were considered fixed and permanent features of the Earth's surface and when the notion of moving continents was first proposed as continental drift, it was met with ridicule from the scientific community of the time. But the evolution of this hypothesis from its proposal and initial rejection, through its modification with the recognition that sea floors also move, to its rebirth in the theory of plate tectonics, is one of the most fascinating chapters in the history of science and forms a uniquely instructive lesson in the scientific method.

In this chapter we explore the remarkable story behind the development of the theory of plate tectonics. We do this by first examining the original lines of evidence used to support the idea of drifting continents and the flaw in the hypothesis that led to its rejection by the scientific community of the day. We then turn to subsequent developments that revitalized the idea of continental drift. We first examine remarkable discoveries in the study of the Earth's magnetic field that provided compelling new evidence for movement of the continents. We learn that when these discoveries were used to interpret strange magnetic patterns found on the floors of the oceans, they suggested that the ocean floors were also moving and that this movement was the result of a process known as sea floor spreading. Finally, we examine the evidence that led to the confirmation of sea floor spreading and the recognition of a process known as subduction, by which ocean floor is destroyed. We learn that with the recognition of these two processes came the revolution in Earth science we call plate tectonics. We end the chapter with a summary of plate tectonics as we understand it today.

Continental Drift

Ever since maps of the Atlantic Ocean were first published, scholars have pondered the remarkable jigsaw fit between the ocean's opposing coastlines (Fig. 4.1). Was this fit significant or was it only coincidence? The precision of the match only got better as maps of the Atlantic improved and, at the beginning of the twentieth century, this jigsaw fit became one of the many lines of evidence that led the German meteorologist Alfred Wegener to propose the hypothesis of **continental drift**. According to this hypothesis, the fit of the coastlines was no coincidence. Rather, Wegener proposed that the match was so striking because the continents of Europe, Africa, and the Americas were once joined together and had subsequently broken up and moved apart over enormous distances.

Wegener's concept of how the Earth's geography had changed with time is dramatically illustrated by his own maps (Fig. 4.2). In these, he reassembles the continents into a single landmass or supercontinent during the Late Carboniferous Period, some 300 million years ago. He called this supercontinent **Pangea (or Pangaea)**, meaning "all lands." Although Wegener presented his ideas in a book first published in 1915, almost half a century would pass before they gained much support. In fact, as recently as the late 1960's many introductory geology textbooks mentioned continental drift only as a curious, discredited idea. When exploring the evidence for the drift of the southern continents, for example, students of the first edition of Bernhard Kumme's well-known textbook "History of the Earth," published in 1961, were cautioned to "keep in mind that many historical geologists (perhaps the great majority) believe in the permanence of the

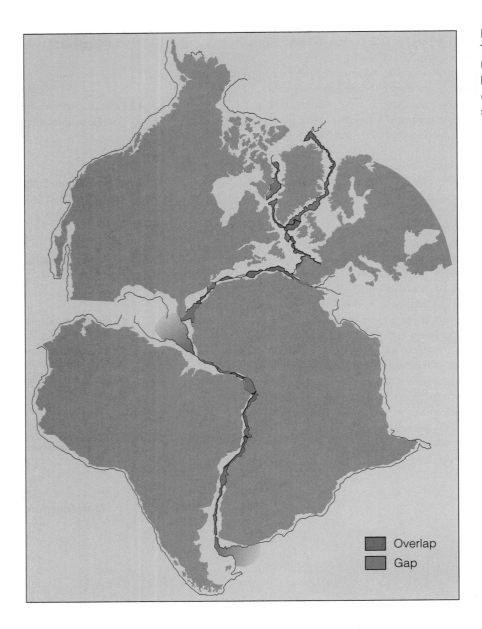

Figure 4.1

The best fit of the continents around the Atlantic Ocean was obtained by Sir Edward Bullard and his colleagues at Cambridge University using a computer. It was found to lie on the continental slope at a depth of about 1000 m (500 fathoms).

Overlap
Gap

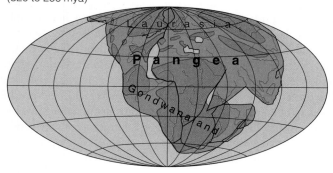

Late Carboniferous
(320 to 286 mya)

Eocene
(55 to 36 mya.)

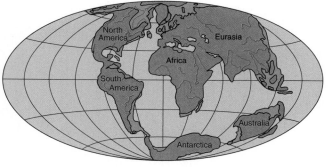

Early Pleistocene
(1.6 to 0.01 mya)

mya = Millions of years ago

Figure 4.2
Wegener's reconstruction of the supercontinent Pangea and the distribution of continents produced by its breakup. Dark blue areas represent shallow seas. Absolute ages of the geologic periods assigned to each map were unknown to Wegener and have been added.

continents and oceans and see nothing in the physical or biological data . . . to justify drastic explanations."

Yet today we know that continents do indeed move, a realization that has revolutionized our picture of the Earth. So what happened to change our understanding of the planet and why did it take so long?

4.1.1 A Controversy Unleashed

Although others had united the continents in much the same way before him, the publication of Wegener's book sparked an international controversy that was to last almost a half century. At the beginning of the twentieth century, the prevailing view among geologists was that major features of the Earth's surface like continents and oceans were fixed and permanent, having been formed during the formation of the planet. Because it challenged established science, Wegener's hypothesis of moving continents was met with fierce opposition from a shocked scientific community. Yet his ideas were not dismissed out of hand because Wegener was a reputable scientist and had amassed an imposing collection of facts and opinion in support of his hypothesis from a large variety of sources. Although some of his evidence was convincing, much was speculative and provoked fierce debate. In the mid-1920s, for example, Wegener was invited to present his case at two major conferences held in London and New York, and did so persuasively. However, he failed to convince the majority of his audience at both venues. Although the mood of the London meeting was sympathetic, a comment from a critic in attendance illustrates the majority view. "In examining ideas so novel as those of Wegener," he said, "it is not easy to avoid bias. A moving continent is as strange to us as a moving Earth was to our ancestors, and we may be as prejudiced as they were. On the other hand, if continents have moved, many former difficulties disappear, and we may be tempted to forget the difficulties of the theory itself and the imperfections of the evidence."

In New York, the audience was more hostile and some of the criticism became intense, as evidenced by an attack from Edward W. Berry, then professor of paleontology at Johns Hopkins University. Wegener's method, he said, "is not scientific, but takes the familiar course of an initial idea, a selective search through the literature for corroborative evidence, ignoring most of the facts that are opposed to the idea, and ending in a state of auto-intoxication in which the subjective idea comes to be considered as an objective fact." And so, following a decade during which continental drift was one of the most hotly debated issues in science, Wegener's hypothesis was believed to have been discredited and fell into an obscurity that was to last for more than 25 years.

4.1.2 The Jigsaw Fit of Continents

So what was Wegener's evidence for continental drift and why did it fail to convince the geologists of his day? Of course, it was easy for the skeptical world of science to dismiss any significance in the jigsaw fit of continents because coastlines were known to be ephemeral features of the Earth's surface that could be changed by a single storm. So how could their shape have any meaning after millions of years of coastal erosion?

In fact, the best fit between the Atlantic continents has since been shown by computer analysis of their geometry to lie, not at their coastlines, but along lines drawn halfway down their continental slopes at an ocean depth of 1000 m (500 fathoms).

Unlike coastlines, which mark only the edge of the land, continental slopes mark the edge of the continents themselves and so are far more fundamental features of the Earth's surface. But the ocean floors had not been mapped in Wegener's day, so he had no way of knowing the shape of the continental slopes.

However, jigsaw puzzles are not put together solely on the basis of the shape of their pieces. A jigsaw is created when an image is cut along irregular lines. So when the puzzle is correctly reassembled, not only do the shapes of the pieces fit together, but the pictures on each of the pieces come together to recreate the image. In other words, the pieces make more sense together than they do apart. Like our severed image, if the continents were fragments of what was once a much larger landmass as Wegener proposed, they might be expected to share certain large-scale features that had formed before the landmass broke up. Wegener found evidence of such features in ancient mountain belts, in major fault lines, and in certain distinctive rock formations, each of which was brought into continuity when the continents were reassembled.

Along the Atlantic seaboard of North America and Europe, several ancient mountain ranges like those of the

Appalachian, Caledonian, and Hercynian belts are abruptly truncated or cut off by coastlines (Fig. 4.3a). Geologists of Wegener's day had no explanation as to why this should be so. For example, the Appalachian mountain belt of eastern North America terminates along the Atlantic coast of Newfoundland. These mountains were known to have formed in two main stages during the mid-Paleozoic (some 350 to 400 million years ago) and Late Paleozoic (some 270 to 300 million years ago), before the breakup of Wegener's single landmass during the Early Mesozoic (some 160 to 230 million years ago). Mountains of very similar age and geology to the mid-Paleozoic stage of the Appalachians are also found in eastern Greenland, but the belt is abruptly truncated along the coast. On the other side of the Atlantic, the same is true of the ancient Caledonian mountain belt of western Europe, which is truncated by the coastlines of western Britain and Scandinavia. Rocks of identical age and geology to those of the late Paleozoic stage of the Appalachians also occur in the Hercynian belt, an ancient mountain range abruptly cut off by the coastlines of southern Europe and northwest Africa.

When the continents are reassembled into Wegener's single landmass, these ancient mountain ranges are brought together to

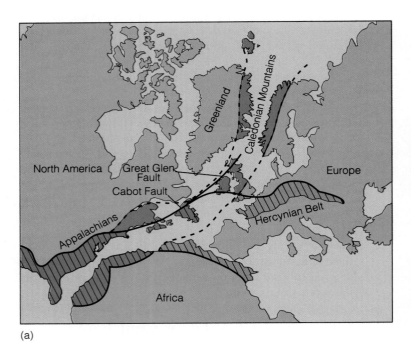

(a)

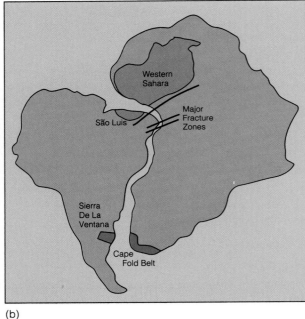

(b)

Figure 4.3
Ancient mountain chains come together to form continuous belts, old crustal fractures line up, and distinctive rock formations match when the continents on either side of (a) the North Atlantic and (b) the South Atlantic are pieced together according to the jigsaw fit of their coastlines. Thus the mid-Paleozoic portion of the Appalachians continues into the Caledonian Mountains of Great Britain, eastern Greenland and Norway, whereas the Late Paleozoic portion of the Appalachians matches the Hercynian belt of northwest Africa and southern Europe. Likewise, the mid- to Late Paleozoic Cabot Fault of eastern Newfoundland aligns with the Great Glen Fault of Scotland and northern Ireland. In a similar fashion, the Early Mesozoic Cape Fold Belt of South Africa continues into the Sierra de la Ventana of Argentina, major fracture zones in the bite of Africa line up with those in eastern Brazil, and the 2 billion-year-old rocks of the Sahara match those of São Luis in northeastern Brazil.

form two continuous mountain belts. Therefore, the patterns on the jigsaw pieces matched, the trend of the mid-Paleozoic Appalachians continuing into Greenland and the Caledonian Mountains of Scotland and Scandinavia, while the Late Paleozoic Appalachians continued into the Hercynian belt of southern Europe. In Wegener's view these mountain belts were once continuous, but were severed by the opening of the Atlantic Ocean.

In addition to the continuity of mountain belts, Wegener's reassembly of the European and North American continents also brought together major crustal fractures or faults (see Chapter 6). In 1962, for example, the eminent Canadian geophysicist, J. Tuzo Wilson, proposed that the Great Glen Fault of Scotland was a continuation of the Cabot Fault of Newfoundland (Fig. 4.3a). Both structures are major crustal fractures similar to the modern San Andreas Fault in California, which were active during the mid- to Late Paleozoic. The continuity and alignment of these two faults when Europe and North America are reassembled is another pattern match on Wegener's continental jigsaw puzzle.

In 1937, a similar argument was made for the southern Atlantic by the celebrated South African geologist, Alexander du Toit. Here the Early Mesozoic Cape Fold Belt of South Africa, which reaches the coast at Capetown, is closely comparable in age and geology to the Sierra de la Ventana of Argentina, which extends into South America just south of Buenos Aires. When these two continents are brought together in the manner proposed by Wegener, the two mountain belts connect up to form another perfect match (Fig. 4.3b). Likewise, major fracture zones that reach the coast in the bite of Africa line up with similar fracture zones in eastern Brazil, and a distinctive region of 2 billion-year-old rocks in the western Sahara continues into the São Luis region of northeastern Brazil.

4.1.3 The Enigma of Ancient Climates

Other matching patterns lay in the rocks themselves and provided one of Wegener's most compelling lines of argument. On many of the southern continents, very distinctive glacial deposits had been found that were formed during the Late Carboniferous, some 300 million years ago. Their widespread distribution across large areas of South America, South Africa, India, and Australia (Fig. 4.4) suggested that these continents had experienced a massive continental glaciation during the Late Paleozoic, the scale of which was comparable to that of the northern continents during the last Ice Age.

If the continents had fixed positions, as scientists of Wegener's day believed, then the distribution of these glacial deposits defied explanation. For example, during the last Ice Age, the Earth's polar ice caps expanded towards lower latitudes as continental glaciers advanced outwards from the poles. Following the doctrine of Actualism ("the present is the key to the past"), Wegener's contemporaries might have expected ancient ice ages to follow a similar pattern. However, the doctrine broke down when the Late Paleozoic glacial deposits were plotted on present-day maps (Fig. 4.4). Rather than a single polar ice cap, the map revealed several ice caps at widely different latitudes. In Africa, for example, the glacial deposits suggested that continental glaciers had reached the equator, whereas they had come only to within 40 degrees of the equator during the last Ice Age. The deposits also suggested the wholesale glaciation of tropical India while the cooler latitudes of Asia to the north showed no such record. If the continents occupied fixed positions, the evidence of glaciation in tropical latitudes and its absence from the polar regions of North America, northern Europe, and Russia, made very little sense.

The problem only got worse as the areas in which these deposits occurred were explored. Detailed field measurements of the flow directions of the ice, as recorded in the grooves and scratch marks carved in the bedrock by the glaciers (see arrows in Fig. 4.4), also defied common sense. In both India and South Africa, the flow of ice was directed toward the north or south pole from equatorial regions rather than the other way around. In South America and Australia, the direction of the glacier movement suggested that the ice had defied gravity, flowing inland and uphill from the sea! Boulders in the glacial deposits of Brazil also contained rock types that could be found nowhere else in South America, but which matched rocks found in South Africa.

Figure 4.4
Directions of ice flow (arrows) and distribution of Late Paleozoic glacial deposits (in white) on a present day map of the southern continents. Note that in India and South Africa, the flow of ice is directed from equatorial regions toward the north and south poles, respectively. Note also that in South America and Australia the direction of ice flow is inland from the sea. Both patterns of ice flow are highly improbable.

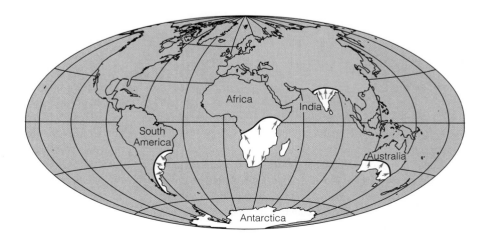

Any attempt to explain this ancient glaciation in a world in which the continents had fixed positions resulted in apparently insurmountable problems. However, each of these apparently contradictory observations could be readily accounted for in Wegener's reconstruction of the southern continents into a portion of Pangea now known as **Gondwanaland** (Fig. 4.5). With the continents reassembled in a jigsaw fit, his reconstruction produced a single ice cap centered on the south pole, then located in southern Africa. Plotted on the same reconstruction, the directions of glacial movement suggested an outward flow of ice from the south pole in a manner typical of a polar ice cap. With the intervening oceans removed, ice no longer needed to flow inland (and uphill) from the sea, and the presence of South African boulders in the glacial deposits of Brazil could easily be accounted for by the transport of material from one continent to the other by the outward flow of ice.

Therefore, according to Wegener, the Late Carboniferous continental glaciation did not take place at tropical latitudes. Rather the southern continents at that time were assembled near the south pole. Long after the ice age was over, continental drift dispersed these continents and moved the glacial deposits toward regions closer to the equator.

Other rock types that could be used as indicators of past climate also suggested either movement of the continents or unaccountable shifts in the Earth's major climatic zones. From the doctrine of Actualism, we know that coal deposits form in subtropical swampy regions whereas major salt deposits are produced in arid marine climates as a result of the evaporation of sea water. So the presence of coal deposits in Antarctica and the Sahara Desert suggested that swampy subtropical conditions had once existed in areas of present-day continental glaciation and desert. Similarly, the distribution across much of northeastern North America and western Europe of desert sandstones and massive deposits of minerals formed by evaporation, such as rock salt and gypsum, suggested that arid conditions had occurred in the past in areas that today enjoy temperate climates. Worse still, the occurrence of reef-building coral limestones in Greenland and northern Canada suggested that tropical seas had existed in the past north of the Arctic Circle! Because the Earth's major climatic belts are predominantly controlled by their latitude, radical shifts in the Earth's latitudes were apparently implied. But the latitude of any given point on the Earth's surface is dictated by the axis about which the Earth spins, and this has not changed. So how could this be?

4.1.4 The Puzzling Fossil Record

Wegener also found evidence for continental drift in the distribution of certain fossils. For example, identical species of the Permian reptile *Mesosaurus* (Fig. 4.6a) had been found on both sides of the South Atlantic. Although it was a marine reptile, Mesosaurus was not thought to be a strong swimmer and, like the modern alligator, would not have been able to swim across an entire ocean. Similarly, fossils of the Triassic reptile *Cynognathus* (Fig. 4.6b) had been found in South America and Africa, while those of the Triassic reptile *Lystrosaurus* (Fig. 4.6c) had been found in Antarctica, Africa, and India. Today these continents are separated by major oceans. While it is conceivable that *Lystrosaurus*, a sheep-sized, land-dwelling reptile, could have walked from Africa to India, it certainly could not have crossed the polar ocean to Antarctica.

Wegener found similar evidence in the distribution of an unusual, apparently stunted assemblage of fossil ferns known as the *Glossopteris* flora (Fig. 4.6d). Characterized by a small number of readily identifiable species, this unique flora of Late Paleozoic ferns had been found in South America, South Africa, India, Australia, and Antarctica, as well as in the Falkland Islands and Madagascar. Because the seeds of *Glossopteris* were several millimeters across, they could not have been carried great distances by the wind and are unlikely to have been able to float. As birds had not yet evolved to carry the seeds, how had the distribution of this flora been achieved?

The seemingly paradoxical distribution of fossil species such as *Mesosaurus, Cynognathus, Lystrosaurus* and the *Glossopteris* flora was also readily accounted for on Wegener's maps which reassembled the southern continents into a single landmass by closing the intervening oceans (Fig. 4.6). Without oceans to cross, the distribution of these and other species posed no dilemmas. But those opposed to continental drift were forced to advocate the improbable existence of former "land bridges" that had at one time connected the present continents but which had now sunk below the surface of the oceans.

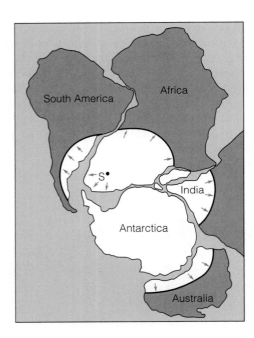

Figure 4.5
Distribution of the ice flow directions (arrows) and Late Paleozoic glacial deposits (in white) shown on Figure 4.4, plotted on Wegener's reconstruction of the southern continents into a single landmass called Gondwanaland. Note the coherent distribution of glacial deposits centered on the south pole and the outward radiating directions of ice flow typical of polar ice caps.

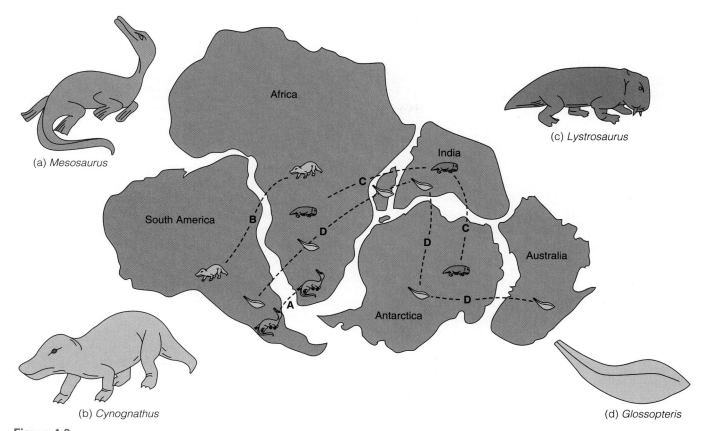

Figure 4.6
The distribution of some of the animals and plants whose fossils are now found on the widely separated continents of Gondwanaland. The 2-foot long, freshwater reptile *Mesosaurus* (a) occurs in sediments of Permian age in South America and Africa. The land reptile *Cynognathus* (b) has been found on the same two continents in sediments of Triassic age. The sheep-sized land reptile *Lystrosaurus* (c) occurs in sediments of Triassic age in Africa, India, and Antarctica. The large-seeded *Glossopteris* (d) flora is found in sediments of Pennsylvanian and Permian age on all five continents. The ranges of the reptiles, none of which were long-distance swimmers, and the large-seeded flora on continents now separated by vast oceans is readily accounted for on this modern reconstruction of Gondwanaland.

4.1.5 Wegener Presents His Case

In 1915, Wegener compiled a wealth of evidence in favor of continental drift in a book he entitled "The Origin of Continents and Oceans." Although the evidence was circumstantial, the case he made was a compelling one. Using criteria that included the fit of coastlines, the continuity of mountain belts and major faults, the record of past climate, and the distribution of fossils, he argued convincingly for the existence of the single landmass or supercontinent he called Pangea during the Late Carboniferous Period (Fig. 4.7). With the south pole placed in the center of the glacial deposits of Gondwanaland, the coal measures of the period took up their expected locations along the reconstructed position of the equator. The reconstruction also produced two continuous belts of wind-blown sands and deposits of rock salt and gypsum, indicating arid conditions at latitudes either side of the equator identical to those of today's desert belts.

Upon the breakup of this supercontinent, new coastlines were established and formerly continuous belts were severed, leading to the development of today's continental jigsaw. As the continental fragments dispersed to their present geographic positions, their latitudes changed and as they did so, their climates changed too. Floras and faunas became isolated from each other and, as a result, followed different evolutionary paths. Continental drift was therefore able to explain many enigmas: the surprising fit of the continents, the dilemma of ancient climates, the puzzling fossil record, and the distribution of today's animals and plants.

4.1.6 The Fatal Flaw

Although Wegener's ideas were compelling, much of his evidence was based on inadequate data, and his views were rejected by most scientists of his day. While appealing to some, especially to paleontologists who found many answers to the dilemmas of the fossil record in the consequences of continental drift, they were dismissed by geophysicists who focused

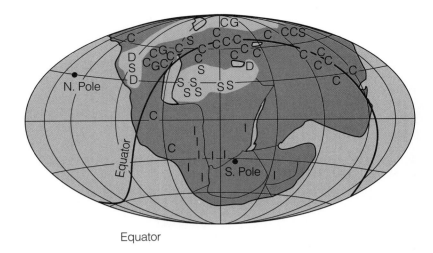

Equator

Figure 4.7

Wegener's reassembly of the continents during the Late Carboniferous based on the paleoclimatic evidence available in 1924. Glaciated regions indicated by I (ice) lie close to the south pole. Coal deposits (C) mainly lie near the Late Carboniferous equator. Arid regions (light brown) with desert sandstone (D), salt (S), and gypsum (G) lie on either side of the equator.

on the process of continental drift and found it in defiance of the laws of physics. As was repeatedly pointed out by Wegener's greatest adversary, the eminent British geophysicist Harold Jeffreys, the chief weakness in the hypothesis of drifting continents was Wegener's inability to provide a viable mechanism, one that would allow the process to operate. What force, the geophysicists argued, could possibly cause the continents to plough their way across the ocean floors?

Although this argument was to prove fatal to the acceptance of Wegener's hypothesis, such criticism plays a fundamental role in the advancement of science. For example, three hundred years earlier, Copernicus and Galileo had encountered similar criticism in advocating that the planets revolved around the Sun rather than the Earth. They too had a wealth of observational data that supported the Sun's position at the center of the Solar System in contradiction to the prevailing view of a stationary Earth at the center of the Universe. But, like Wegener, they could not come up with a mechanism that accounted for their observations. In fact, it was not until Sir Isaac Newton discovered gravity one hundred years later, that a mechanism was finally provided. But in proposing their controversial hypothesis, Copernicus and Galileo set the stage for Newton's discovery of gravity and its application to planetary motion. Indeed, many would argue that it is the criticisms of ingenious insight that drives scientists to discovery, first by testing new hypotheses, and then by deciphering the mechanisms upon which they are based. As we shall see, we still do not fully understand the driving forces behind continental drift, or its companion hypothesis, plate tectonics, and this must be considered a serious weakness. Nevertheless, the weight of observational data is overwhelming and has convinced all but a few skeptics of the validity of moving continents. Yet, these skeptics continue to perform an important function by keeping the convinced on their toes and driving them toward further insights into the underlying forces involved.

Wegener described the drift of the southern continents away from the south pole as **polflucht**, or "flight from the poles," which he attributed to the gravitational attraction of the Earth's

equatorial bulge. Such forces certainly exist, but as Jeffreys was able to demonstrate, they are far too weak to drag the southern continents northward. Wegener also suggested that the westward movement of the Americas had occurred as a result of tidal forces in the Earth's crust produced by the gravitational attraction between the continents and the Moon. Again, such forces exist but are hopelessly inadequate for the task. Indeed, since tidal friction acts like a break on the spinning Earth, tidal forces strong enough to move continents would long ago have brought the Earth's eastward rotation to a halt, as Jeffreys was quick to point out. So Wegener's provocative ideas found little acceptance among scientists of his day because it appeared no natural process stood even the remotest chance of being able to account for it. His hypothesis was rejected and his ideas fell into obscurity until strange discoveries on the floor of the ocean revived his challenging vision.

4.2 Paleomagnetism and Sea Floor Spreading

Ironically, the revival and eventual vindication of Wegener's hypothesis of continental drift was to come from the study of geophysics, the very field of science that had so effectively discredited the idea during Wegener's lifetime. The first key to how our planet works lay in the phenomenon of magnetism and its application to studies of the Earth's magnetic field.

The magnetic properties of the iron-rich mineral magnetite, or lodestone, were known to the Ancient Greeks, but the magnetic compass was not invented until the 13th century when the Chinese floated lodestone on wood and found that it caused the wood to rotate to a fixed position. The use of a compass in navigation and pathfinding lies in the fact that a magnetized needle that is free to rotate about a vertical axis always comes to rest with one end pointing in the direction of the north magnetic pole. The reason why a compass needle con-

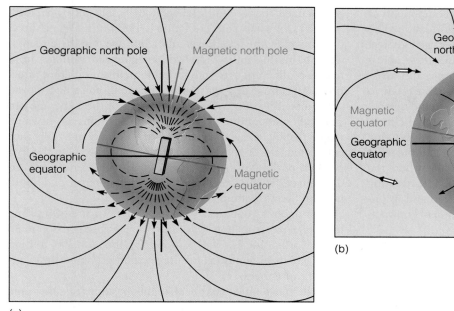

(a)

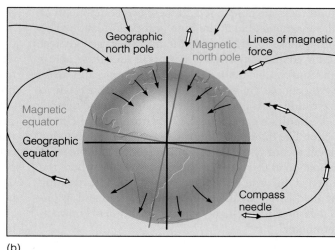

(b)

Figure 4.8

(a) The Earth's present magnetic field in which lines of magnetic force form a pattern of closed loops that emerge from the planet at the magnetic south pole and re-enter at the magnetic north pole. (b) Compass needles align themselves with the magnetic lines of force and so are parallel to the Earth's surface at the equator, but tilt at ever-increasing angles as the magnetic poles are approached.

sistently points north-south in this fashion was first explained by the 17th century British physicist William Gilbert, who suggested that the Earth itself acted as a huge magnet, the force of which controlled the orientation of smaller ones.

Although the exact relationship is still a matter of intense debate, scientists now think that this magnetism is related to the motion of a liquid layer deep within the Earth's interior. This layer is rich in iron and spins like a dynamo under the influence of the Earth's rotation. The spinning dynamo, in turn, generates the planet's magnetic field (see Dig In: The Earth's Magnetism and Its Dynamic Core, in Chapter 7). This field behaves as if a simple bar magnet were aligned north-south at the Earth's center, producing lines of magnetic force which emerge from the southern hemisphere and loop through space to return in the northern hemisphere (Fig. 4.8a). Therefore, the lines of magnetic force are parallel to the Earth's surface at the equator but plunge ever more steeply as the poles are approached, pointing up out of the ground in the southern hemisphere and down into the ground in the northern hemisphere.

The orientation of a compass needle pivoted on its center of gravity is controlled by these invisible lines of magnetic force (Fig. 4.8b). The lines cause the compass needle to point north but also cause the needle to tilt, the north end tilting downward in the northern hemisphere and upward in the southern hemisphere. So a compass reading really has two components—a horizontal component, or swing, used in navigation and pathfinding, and a

vertical component, or tilt, known as the **magnetic inclination.** Because the amount of inclination progressively increases with distance from the equator, and the direction of tilt is different in the northern and southern hemispheres, the inclination of a compass needle can be used to determine latitude (Fig. 4.9).

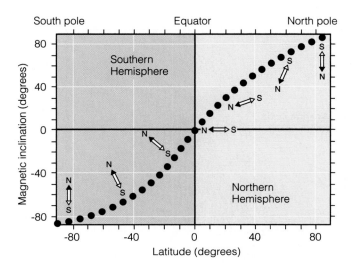

Figure 4.9

Relationship between latitude and magnetic inclination and its effect on the tilt of a compass needle.

Certain rocks are able to record these two components of the Earth's magnetic field at the time of their formation. The records from these rocks provided the first direct geophysical evidence for continental drift. When basalt lava flows across the Earth's surface and starts to cool, for example, tiny magnetically susceptible grains of iron oxide (magnetite) start to crystallize within it. As they cool, these grains become magnetized by the Earth's magnetic field. Once frozen into the rock, they function like thousands of tiny compass needles, creating a permanent record of the north-south direction and the magnetic inclination at the point on the Earth's surface where the lava erupted at the time it solidified (Fig. 4.10). The resulting magnetic field is very weak but it can be detected and its orientation measured with a highly sensitive instrument known as a **magnetometer.**

The study and measurement of the Earth's magnetic field from ancient rocks is called **paleomagnetism.** When the history of the Earth's magnetic field was first examined in this way, the results were startling. Rather than pointing to the present day magnetic poles as expected, the tiny magnetite compass needles in ancient basalt lavas were found to point in many different directions. Could the poles have moved about?

Yet more startling were results that suggested the north and south poles had in the past repeatedly swapped positions. Could this be true? It was in trying to answer these puzzling questions that some of the old ideas of continental drift were revived.

4.2.1 Apparent Polar Wander

When modern basalt lava flows, like those of Hawaii, are examined paleomagnetically, their tiny magnetite compass needles are found to point toward the present day magnetic poles. But this is not the case for ancient basalt lavas.

Magnetized rock samples taken from basalts all over North America, for example, were found to point in the same direction providing the basalts were the same age, but the direction was not north-south. Moreover, magnetized samples from basalts of different age were found to point in different directions, suggesting that the Earth's magnetic poles had moved with time. Yet harder to explain, basalts of the same age but from different continents gave different positions for the same magnetic pole. Only the geologically most recent basalts gave calculated "north pole" positions that coincided with the true position of the North Pole. If the continents were fixed, this data suggested that the Earth had several north and south poles at the same time!

In order to obtain a systematic view of these puzzling magnetic patterns, pole positions were calculated for rocks from North America and Europe for the last 300 million years. The data compiled for the position of the north pole is shown in Figure 4.11.

The line connecting the changing positions of either pole suggested by successively older basalt samples from a given continent is called a **polar wander curve,** and traces the pole's apparent movement (Fig. 4.11a). Note that the north polar path for the last 300 million years based on North American rocks starts at the North Pole because this is the calculated pole position determined from the most recent basalt lavas. However, it diverges progressively further from the north pole as the lavas get older, ending up off the coast of China near Shanghai 300 million years ago.

A corresponding analysis of the European data produced similar results. European basalts of the same age define the same "north pole," but basalts of different ages have different "north poles" whose positions can be connected to define a polar wander curve for Europe (Fig. 4.11a). However, this polar wander curve has a different path from that of North America.

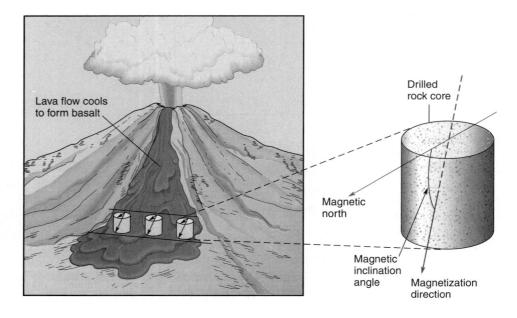

Figure 4.10
In a study of paleomagnetism, rock cores drilled from a basalt lava flow are used to determine the direction in which the magnetic minerals within the rock were aligned at the time the lava solidified. The magnetic inclination angle is the tilt of the magnetization direction relative to the horizontal.

Lava flow cools to form basalt

Drilled rock core

Magnetic north

Magnetic inclination angle

Magnetization direction

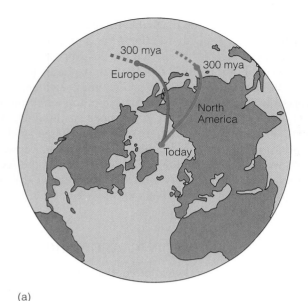

(a)

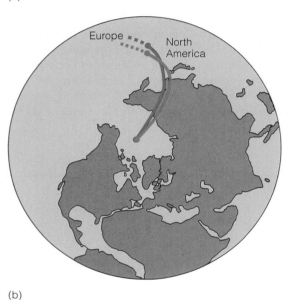

(b)

Figure 4.11
(a) Apparent polar wander paths for North America and Europe for the past 300 million years define similar but separate curves indicating that the two continents have moved relative to both the poles and each other during this time interval. (b) The separation of the two paths is attributed to the opening of the Atlantic Ocean and is reconciled if the two continents are reassembled in their Pangea configurations.

If the continents had fixed positions, this would be very difficult to explain.

But the position of the magnetic poles is thought to be geographically fixed by the Earth's axis of rotation. If so, how could the poles "wander" or change their positions, and how could several of them exist at the same time? The answers are: they don't wander and there can't be several of them at once.

So how can we explain these data? There are only two realistic options. Either the magnetic poles have moved in some unknown manner, or the continents have moved. If the continents moved, the magnetic records frozen into ancient basalts when they erupted on continents millions of years ago would become reoriented by the movement. As the positions of the magnetic poles are constrained by the Earth's rotational axis, it is most unlikely that they have moved to the extent suggested by the data. By eliminating the first option we are left with the second, that is, the movement of the continents. Furthermore, when the continents are reassembled into a Pangea "fit" like the one proposed by Wegener, the two polar wander paths coincide (Fig. 4.11b). This suggests that the two continents were once joined and that their polar wander paths diverged with the opening of the North Atlantic Ocean.

Taken together, the data suggest that what actually moves are the continents on which the lavas erupted. Thus, polar wander curves only trace *apparent* (rather than actual) polar movement and are more correctly termed **apparent polar wander curves.** But their discovery provided dramatic geophysical evidence for continental drift. Apparent polar wander curves for individual continents showed them to have moved relative to the poles, whereas different polar wander curves for different continents showed that the continents had moved relative to each other. In this way, the present separation of the North American and European paths can be attributed to the progressive opening of the Atlantic Ocean, such that the paths coincide if the Atlantic is closed in the manner Wegener proposed.

The paleomagnetic record was heralded as conclusive evidence for continental drift. Wegener's proposal that the continents had moved relative to the poles based on the evidence for Late Carboniferous glaciation, was now supported by geophysical data. Indeed, Late Carboniferous paleomagnetic data from the southern continents proved to be consistent with the existence of Gondwanaland and placed the south pole near the position proposed by Wegener. However, while the paleomagnetic data provided strong support for continental drift, the quest to find a viable mechanism still remained.

4.2.2 Paleomagnetic Reversals

A key piece of the puzzle was provided by an even more startling paleomagnetic discovery. Magnetized samples of basalt from the same locality were repeatedly found in which the north and south poles had apparently been interchanged. That is, the north magnetic pole suddenly swapped position with the south magnetic pole so that the tiny compass needles produced in basalt lavas erupted at the time pointed south rather than north. Did this mean that rocks could somehow reverse their magnetization or, more incredibly, did the Earth's magnetic field periodically reverse itself? To test these rival possibilities, Allan Cox of Stanford University and Brent Dalrymple of the U.S. Geological Survey set about collecting magnetized samples from all over the world in order to compare their mineral content, their direction of magnetization, and their age. If, they argued, rocks were reversely magnetized because they contained a certain mineral, then all reversely magnetized samples would contain that mineral. If, on the other hand, the Earth's magnetic field had reversed, then samples from all over the world would record the magnetic reversal at precisely the same time. Amazingly, their study showed that it was the Earth's magnetic field that periodically reversed itself during an event known as a **magnetic polarity reversal.**

Although the reason why such magnetic reversals should take place is still poorly understood, they are now known to have occurred quite frequently in the recent geologic past, with major reversals (**magnetic epochs**) occurring every million years or so and shorter flips (**magnetic events**) lasting a few thousand to a few tens of thousands of years. We are presently in a time of **normal magnetic polarity,** and have been for the past 700,000 years. During this time, the north ends of all compass needles have pointed towards the north magnetic pole. Prior to this, however, the Earth experienced an interval of **reversed magnetic polarity** that lasted almost 2 million years. During this interval, the magnetic field everywhere on Earth had the opposite orientation to that which it has now, so that the north end of a compass needle would have rotated toward the present South Pole. Some unknown motion of the Earth's liquid outer core is believed to produce these reversals. But whatever their cause, their discovery was destined to provide what is for most geoscientists, proof of continental drift and would unlock the dramatic mechanism by which it was accomplished.

As more data on the magnetic orientation of dated samples accumulated from all over the world, a precise time scale could be constructed that charted the duration of each magnetic reversal. From this, the Earth's magnetic field was found to have reversed itself many times, eighteen such reversals having occurred in the past four and a half million years alone (Fig. 4.12). Like a bar code on produce in a supermarket, the pattern of these reversals is unique, with little order in the timing or duration of any particular episode. But as we shall now see, the pattern would reveal the secret to a dynamic Earth history with the discovery of the very same magnetic bar code in rocks on the ocean floor.

4.2.3 Marine Magnetic Anomalies

Ironically, the main evidence for continental drift was not to come from the continents. Instead, it was to come from paleomagnetic studies of the oceans, and with the recognition that the ocean floor, hidden from view by ocean water and a veneer of unconsolidated sediment, was actually part of a dynamic crust with concealed mountain chains, active volcanoes, and frequent earthquakes. This new understanding was to provide the missing element in Wegener's original hypothesis.

After the Second World War, the strategic importance of the oceans greatly increased. With the advent of nuclear sub-

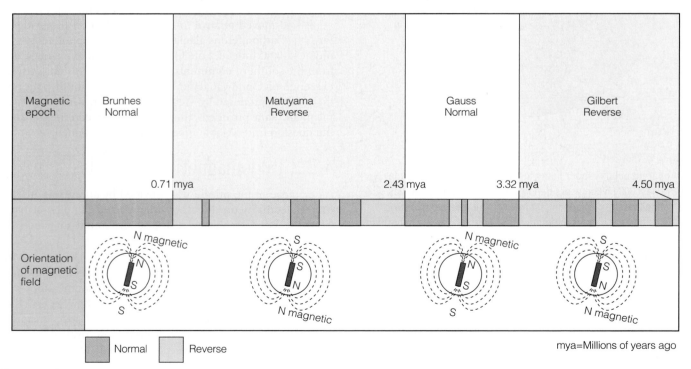

Figure 4.12
Times of normal and reversed polarity in the Earth's magnetic field during the past 4.5 million years established from the study of magnetized samples from basalt lava flows of known age.

marines, a major effort was made to chart the ocean floors. So, while the mystery of magnetic reversals was being unraveled on land, scientists were also charting magnetic rocks on the sea floor by towing sensitive magnetometers across the world's oceans. In a remarkable survey of the floor of the Atlantic Ocean southwest of Iceland, an odd magnetic pattern or **magnetic anomaly** was identified in the basaltic rocks of the seabed. The survey was centered on part of the "Mid-Atlantic Ridge", a vast underwater mountain range rising 3000 meters (10,000 feet) above the ocean floor in the central Atlantic (see Fig. 5.25). Echo sounding revealed this hidden mountain range to be part of the greatest mountain chain on Earth. True, the Himalayas are higher, but they are confined to a relatively small region of southern Asia. The Mid-Atlantic Ridge, in contrast, runs the entire length of the Atlantic Ocean along a line that mirrors its coastlines. The ridge is split lengthwise by a great rift valley and is frequently shaken by shallow earthquakes. It also links up with a similar mountain chain in the Indian Ocean to form part of a system of **mid-ocean ridges,** 65,000 kilometers (40,000 miles) in length, that girdles the Earth like the seam on a giant baseball.

Until the early 1960's, however, no theory existed that could account for the origin of mid-ocean ridges. But when the magnetic properties of the Mid-Atlantic Ridge were charted southwest of Iceland, the survey revealed long, zebra-like stripes of normal and reversed magnetic polarity in the basaltic ocean floor that alternated in a symmetrical fashion on either side of the ridge axis (Fig. 4.13a).

4.2.4 A Pattern Explained

The pattern went unexplained until 1963, when Fred Vine and Drummond Matthews at Cambridge University showed that it could be accounted for by a theory known as **sea floor spreading.** According to this theory, first proposed by Harry Hess at Princeton University, circulating currents under the ocean floor cause molten rock or magma to ooze up beneath mid-ocean ridges as the sea floor spreads away to either side as if it were carried on a giant conveyor belt. If, Vine and Matthews reasoned, lava flowed out along the center of mid-ocean ridges, then the iron minerals in the lava would become magnetized in the prevailing direction of the Earth's magnetic field. Because magma is continually being inserted into the ridge, previously crystallized rock would be pushed progressively sideways. The process is similar to that of repeatedly inserting books into the middle of a bookshelf. The other books on the shelf become progressively pushed to the side. The magma, of course, is inserted from below rather than the side, but the overall effect is similar.

As the newly generated, magnetically imprinted rock was then carried away from the ridge at a uniform rate in both directions, more lava would be erupted along the ridge and would, in turn, be magnetized. As the process was continuous, any reversal in the Earth's magnetic polarity would be recorded as a strip of rock displaying the reversed magnetization. And because the process is symmetric with respect to the ridge axis, the ridge

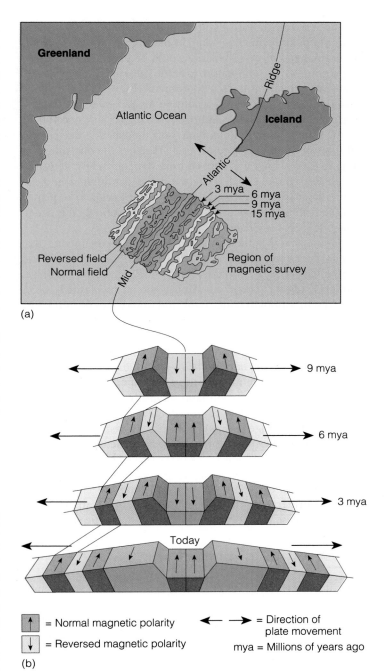

Figure 4.13

(a) Symmetric pattern of magnetic variation across part of the Mid-Atlantic Ridge southwest of Iceland surveyed by magnetometers towed behind an ocean research vessel. Light blue stripes identify ocean floor rocks showing matching reverse magnetization. Intervening grey stripes correspond to rocks with normal polarity.
(b) The origin of the magnetic anomaly pattern across the Mid-Atlantic Ridge. Basalt lava erupting along the crest of the ridge records the Earth's prevailing magnetic polarity as it cools and solidifies. Sea floor spreading then carries the rock away from the ridge in both directions as new lava erupts along the ridge axis. Occurring continuously, this process captures each reversal in the Earth's magnetic field, leading to alternating bands of normal and reversed polarity distributed symmetrically about the ridge axis.

would resemble a mirror with matching bands of normal and reversed polarity on either side. In this way, symmetrical bands of normal and reversed polarity would be produced, each having formed as a continuous strip at the ridge axis only to be torn in two lengthwise as the sea floor spread (Fig. 4.13b). Thus, the magnetic stripes provided a continuous record of the Earth's history of magnetic reversals frozen into the basaltic ocean floor at the time of its formation. Note how the barcode-like pattern of reversals on the sea bed precisely matches the one shown in Figure 4.12, which was derived from the continents. This provided the connecting link between the continental and sea floor magnetic records, and permitted a time scale to be placed on the stripes of the ocean floor since the continental magnetic record had been dated.

Vine and Matthews were able to correctly predict that ocean floor rocks at the ridge crest would be very young and would have today's normal polarity. On either side of the ridge, however, ocean floor rocks would become increasingly older with distance from the ridge crest, and would have polarities that matched the Earth's recent history of magnetic reversals found on land. This is because the rocks at the center of the ridge are those emplaced most recently like the books on our bookshelf, whereas those at the edge of the ocean are the oldest.

The remarkable match, stripe for stripe, with the magnetic time scale developed on land established the theory of sea floor spreading and showed the site of mid-ocean ridges to be the start of two huge conveyor belts that moved the ocean floor aside. It also established the *rate* at which the ocean floor was moving, almost an inch (2.5 cm) a year. This can be determined because the time between reversals is known so that the width of crust formed in a given time interval can be measured. Furthermore, the direction of motion can be determined because it is perpendicular to the magnetic stripes.

4.2.5 Spreading Confirmed

Sea floor spreading was thus introduced as an exciting new hypothesis with the potential to explain the motion of continents. But how could this new hypothesis be tested? Confirmation came from the drilling ship "Glomar Challenger" (Fig. 4.14a) as it probed the ocean depths between Africa and South America and recovered samples of the sea floor. Prior to this survey, the ocean floor was inaccessible. For centuries, scientists believed that the ocean basins were ancient features of the Earth's surface, and thought they must be floored by a rigid crust that had formed when the planet first cooled from a molten ball. The oceans were therefore expected to yield a crust as old as the Earth itself, with a continuous blanket of sediment above it. So it was argued that by drilling through the layers of sediment into the crust below, a core could be extracted that would read like an encyclopedia of geological time, each layer becoming older as the core got deeper.

From its 60 meter (140 foot) derrick, the Glomar Challenger could lower 6.5 kilometers (4 miles) of drill-pipe, which it kept in place by using a sea bed sonar beacon that could be detected by shipboard hydrophones. This information was fed through a computer that automatically activated the main propeller and side thrusters to hold the vessel in a stable position (Fig. 4.14b). But when core was recovered from the drill stem and analyzed, the results astounded the scientific world.

Contrary to the expectation of a complete record of sedimentation, the rocks of the ocean floor proved to be young and only a thin veneer of geologically recent sediment was found to cover them. The drilling did confirm that the Atlantic Ocean basin had a hard crustal floor, but it was not the ancient crust scientists had envisaged. The record of approximately 95 percent of geologic time was missing. So the sea floor was indeed young, as the hypothesis of sea floor spreading predicted. The sediment also contained fossils of microscopic sea organisms and these revealed a progressive increase in the age of the oldest sediment on the sea floor with increasing distance from the Mid-Atlantic Ridge. The fossils also confirmed the rate of sea floor spreading at nearly an inch (2.5 cm) per year. So the drilling confirmed that the crustal floor of the Atlantic Ocean was in motion, moving as if it were part of two huge conveyor belts that continually carry it away from the Mid-Atlantic Ridge in opposite directions.

It was not until the early 1970's that scientists gained a first-hand look at the site of sea floor spreading. As part of a joint oceanographic survey dubbed "Project FAMOUS," American and French submersibles first explored a tiny portion of the crest of the Mid-Atlantic Ridge in 1973, at a site southwest of the Azores (see Living on Earth: The Bizarre World of the Mid-Ocean Ridges, in Chapter 8). Three kilometers below the surface, at the bottom of a rift valley the size of the Grand Canyon, scientists found distinctive, bulbous flows known as **pillow lavas** characteristic of newly formed submarine basalt, and great fissures in the seabed, evidence of both the creation of new ocean floor and its separation along the axis of the Mid-Atlantic Ridge.

4.3 Subduction: Where Does the Old Ocean Crust Go?

As a result of sea floor spreading, new oceanic crust is created at the site of mid-ocean ridges and, once formed, is carried away in both directions from the ridge crest. Although the rate of spreading is slow, its effect over geologic time is staggering. Spreading at a rate of 2.5 cm (one inch) per year in either direction, Atlantic-sized ocean basins over 5000 kilometers (3100 miles) across could be created in a mere 100 million years. But what happens to all this crust once it has been created? Without a means of destroying it at the rate at which it is produced, the Earth would expand and swell up like a balloon. So where was the old ocean crust going? The answer to this question lay beneath another dramatic feature of the ocean floors, the deep ocean trenches, and with it came our modern understanding of plate tectonics, which united the principles of continental drift and sea floor spreading into one coherent theory.

(a)

Bow
Thrusters
Hydrophones
Stern
Thrusters

Sonar Signal
Sonar Signal

Drill String

This ship is kept on station by a sonar beacon dropped to the drill site. Detected by shipboard hydrophones, the signals are fed to a computer that corrects for drift by controlling the ship's propellor and side thrusters.

Sonar
Beacon

Up-and-down motion caused by waves is absorbed by telescoping "bumper subs." Rotating drill string turns the bit for extracting core samples up to 30 feet long.

Core sample

(b)

Figure 4.14

(a) The research ship *Glomar Challenger*. (b) This computer-assisted ocean drilling vessel uses a sonar beacon on the sea bed to maintain her position while sediment cores over a kilometer long are drilled and retrieved from the floor of the ocean several kilometers below its keel.

Every year, the tiny equatorial island of Ascension in the middle of the Atlantic Ocean witnesses a remarkable mystery of evolution. This barren volcanic peak on the Mid-Atlantic Ridge is the nesting ground of the green turtle (Fig. 4.15a). Each year, large numbers of green turtles, some weighing as much as 180 kilograms (400 pounds), crawl ashore on the island's few small beaches to lay their eggs in the sand. What makes this event so extraordinary is the enormous distance that each turtle must swim in order to reach this remote destination. From their home along the coast of Brazil the turtles embark on a hazardous, 2-month voyage that will take them across half the width of the Atlantic Ocean, a journey of some 2000 kilometers (1250 miles). They make this journey without food and against a strong equatorial current. Why do they attempt such an arduous passage and why do they travel so far to hatch their young?

It is easy to see why turtles might choose an island for their nesting site because there are far fewer predators on islands. Although hundreds of miles of sandy beaches lie along the coast of Brazil, they are within reach of a variety of carnivores that would devour the eggs as well as the nesting adults. But what made the turtles choose such a distant island and how did they know that it was there? Continental drift and sea floor spreading provide an elegant explanation of just how such a situation might have slowly come about over millions of years.

Green turtles first evolved about 200 million years ago when the supercontinent of Pangea was breaking up. At this time, the ocean separating Africa from South America was a narrow sea about as wide as a broad valley. Any island on the Mid-Atlantic Ridge would therefore have been both visible and readily accessible from the South American coastline. As the Atlantic Ocean slowly widened, the distance from South America to any island located on the Mid-Atlantic Ridge would have grown progressively longer (Fig. 4.15b). Unusually active volcanism at this location on the ridge would have ensured the continued existence of an island where Ascension Island now stands. Guided by instinct once the island could no longer be seen, each generation of turtles would have traveled a little further. As they did so, natural selection would have favored the sturdiest swimmers, slowly giving rise to the powerful turtles of today that are strong enough to swim half way across an ocean to an island that only their instinct tells them is there.

If this is the explanation, then the very genetic code of the green turtle may have been programmed by continental drift. Shaped by the drifting continents, their remarkable behavior would provide living testimony to the process of sea floor spreading, which every year carries South America a little further from the Mid-Atlantic Ridge.

Evidence for the location of the site of ocean floor destruction came from mapping the location of earthquakes. These are not distributed randomly on the Earth's surface, rather they define narrow zones (Fig. 4.16). Numerous shallow earthquakes trace the winding course of the world's mid-ocean ridges as they circle the globe on the ocean floor. Others occur beneath the world's great mountain belts. But most earthquakes define narrow zones that lie in close proximity to another important feature of the ocean floor, the **deep ocean trenches.** Discovered by oceanographic surveys before the advent of sea floor spreading, these trenches are narrow, curved depressions in the ocean floor that can reach depths of over 11 kilometers (7 miles) and extend for thousands of kilometers. They occur in many regions but are especially characteristic of the margins of the Pacific Ocean.

Ironically, it was the worldwide network of seismic stations set up to monitor underground nuclear testing during the Cold War that confirmed trenches as the sites of ocean floor consumption. These stations showed that earthquakes bordering the Pacific always occurred on one side of a trench and not the other, and that the depth to the source of the earthquakes always increased the farther one moved away from the trench (Fig. 4.17). The point in the Earth's interior at which an earthquake is generated is known as the **focus** of that earthquake. Thus, the deep ocean trenches marked the start of inclined zones of earthquake foci that plunged into the Earth's interior at angles that averaged 45 degrees. Originally termed **Benioff zones** after the American geophysicist Hugo Benioff who pioneered their study before the advent of sea floor spreading, the progressive increase in depth to the foci of these earthquakes was now seen to reflect the violent descent of oceanic crust back into the Earth's heated interior.

Figure 4.16 (right)
The world's earthquake zones defined by the distribution of earthquakes over a period of nine years between 1977 and 1986. Darker areas of the ocean floor outline the deep ocean trenches.

(a)

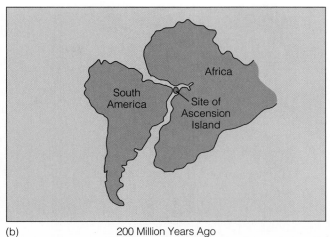

(b)   200 Million Years Ago

Today

Figure 4.15

(a) To lay their eggs, green turtles swim from their home along the coast of Brazil to the lonely island of Ascension in the mid-Atlantic, and then swim laboriously back again. (b) They may have acquired this habit before the breakup of Pangea, each generation traveling further than their ancestors as the Atlantic Ocean widened, until today's marathon journey of some 2000 kilometers (1250 miles) was established.

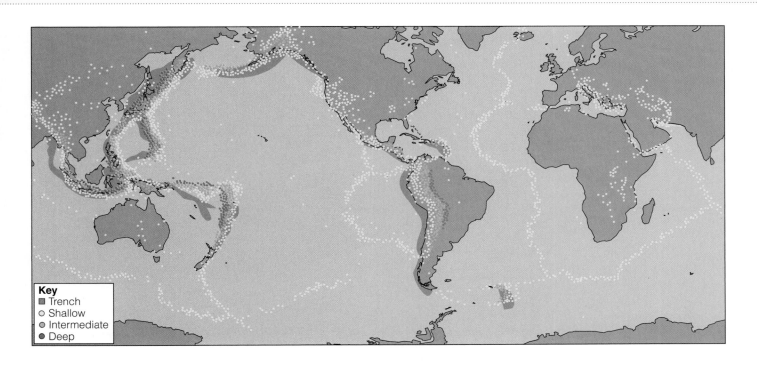

Key
■ Trench
○ Shallow
◔ Intermediate
● Deep

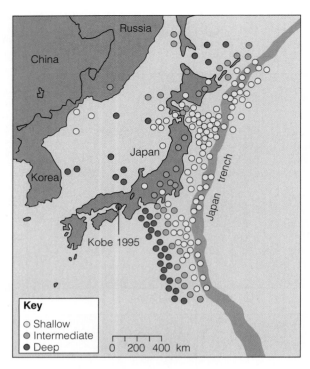

Figure 4.17
Distribution of shallow, intermediate, and deep earthquakes for part of the Pacific "Ring of Fire" in the vicinity of the Japan trench. Note that the earthquakes only occur on one side of the trench, the side on which the plate subducts.

The process by which ocean floor is consumed is called **subduction** (Fig. 4.18), and the plunging Benioff zone of earthquake foci with which it is associated is produced because earthquakes are generated along the upper boundary of the descending slab of oceanic crust as it is jolted downward. The process of subduction also generates magmas. As the downgoing slab reaches greater depths and is heated, water is first driven off and the slab eventually starts to melt. Water has a great capacity to carry heat, and as this water vapor moves upward it promotes melting in the rocks above. The resulting magmas rise buoyantly toward the surface to fuel the volcanoes that tower over the deep ocean trenches. Indeed, the pronounced line of volcanoes bordering the trenches that encircle the Pacific is often called the Pacific "Ring of Fire" (Fig. 4.19).

The distribution of volcanoes around the rim of the Pacific reveals a series of belts, like that of the Aleutian Islands, whose shape is distinctly curved (Fig. 4.19) but parallel to that of the ocean trenches. Indeed, the line of volcanoes above a subduction zone is known as a **volcanic arc.** But why should they be curved in this way? The answer lies in the curvature of the Earth. Because the Earth is spherical, the ocean floor has a curved surface. The trench produced by the subduction of ocean floor is consequently curved, just like a knife cut on the surface of an apple, or an indentation on the surface of a ping-pong ball. The curvature of the line of volcanoes above a subduction zone is simply inherited from the curved shape of the downgoing slab as it cuts into the crust.

Figure 4.18
The subduction of oceanic crust and its relationship to the Benioff zone defined by earthquake depth (note exaggerated vertical scale of crust and volcanic arc).

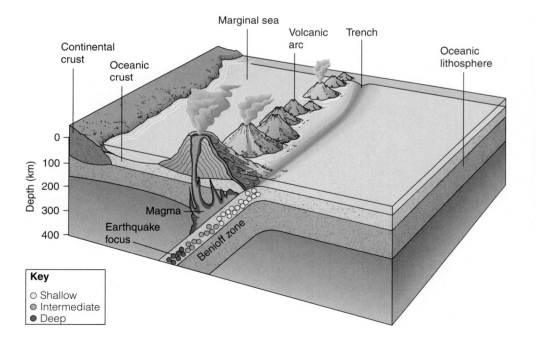

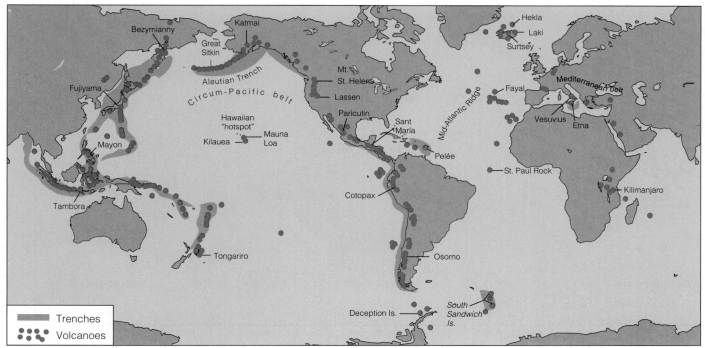

Figure 4.19
The global distribution of recent volcanoes showing the almost continuous belt of volcanism around the Pacific rim, the "Ring of Fire." Segments of the belt are curved because of the spherical shape of the Earth.

4.4 Plate Tectonics

With the recognition of sea floor spreading at mid-ocean ridges and subduction at deep ocean trenches came the revolution in Earth sciences we call **plate tectonics,** named from the realization that the Earth's rigid outer shell was broken into huge moving plates and from the Greek *tecton*, which means "to build." Nearly 50 years after Wegener's provocative ideas were first published, plate tectonics provided the missing mechanism for continental drift, and in so doing, transformed geology as profoundly as evolution transformed biology and relativity transformed physics.

According to plate tectonics, the world's earthquake zones outline large slabs or **plates** thousands of kilometers across but only 50 to 150 kilometers (30 to 100 miles) thick (Fig. 4.20). They are approximately as thick in proportion to the Earth as an eggshell is to an egg. These plates are in constant motion, their incessant jostling, one against the other, causing frequent earthquakes. Plate boundaries therefore lie within the Earth's most seismically active regions. So it is no surprise that the earthquake zones of Figure 4.16 match the plate boundaries of Figure 4.20.

About a dozen large plates and numerous smaller ones define the **lithosphere,** or outer rocky layer of the Earth, and "float" on a weak and partially molten portion of the Earth's interior known as the **asthenosphere.** We will learn more about these two layers when we examine the Earth's internal structure in Chapter 7. Each plate moves relative to other plates in response to the circulation of the Earth's heated interior. As we will learn in Chapter 5 when we examine plate tectonics in detail, the effects are most dramatic where two plates meet. It is at these **plate boundaries,** where plates either move away from each other, move toward each other, or slide past each other, that the effects of plate tectonics are most clearly seen.

According to plate tectonics, the Earth's lithosphere is broken into jagged fragments like a giant eggshell. As these fragments or plates move away from mid-ocean ridges and toward deep ocean trenches, they jostle each other for position, interacting in three ways as they do so (Fig. 4.20).

At boundaries where plates move apart, continents may be torn in two with new oceans forming between them. New oceanic lithosphere is created by sea floor spreading at mid-ocean ridges as the lavas and the rocks beneath them cool and thicken to form a rigid layer. Where plates slide past each other, great crustal fractures are formed such as the San

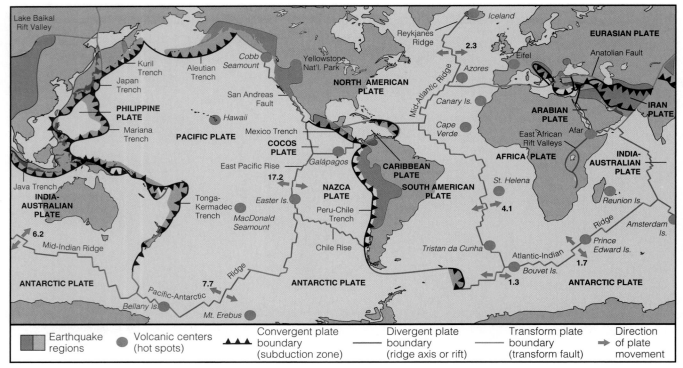

Figure 4.20
The interlocking network of lithospheric plates that make up the Earth's outer shell. The arrows show the present direction of sea floor spreading and the rate of plate motion in centimeters per year. Note that individual plates may include both continental and oceanic crust and that there are three types of plate boundaries: convergent boundaries where plates move toward each other, divergent boundaries where they move away from each other, and transform boundaries where they slide past each other.

Andreas Fault in California. Crust is neither created nor destroyed along these great faults but destructive earthquakes are common as the two plates grind against each other. Where plates converge, old oceanic lithosphere is consumed beneath either more buoyant (less dense) continental crust or younger ocean floor when one plate bends downward and is subducted beneath the other. Where the subducting plate bends before sinking, a great ocean trench up to 11 km (7 mi) deep is formed, and as it angles down into the Earth and melts, it fuels volcanoes in the plate above. Where the overriding plate is oceanic, the result is a curved line of volcanic islands like those of the Aleutians. But where the upper plate is continental, the collision causes uplift and the result is a volcanic mountain range like that of the Andes. If both converging plates carry continents, however, the collision is dramatic since neither plate can subduct completely because continental lithosphere is too buoyant (more about this in Chapter 6). Volcanoes are not common in these continent–continent collision zones because the subduction which creates new magma grinds to a halt. Instead, one continent overrides the other, the impact forcing the crust skyward to build great mountain ranges like those of the Himalayas.

With the advent of plate tectonics in the mid-twentieth century, scientists gained the first comprehensive theory about the forces that shape the planet. With this revolution in scientific thought, the concept of continental drift finally gained widespread acceptance, the oceans became quite young features of the Earth's surface, and the planet became an active one with a complex and dynamic history. In this new view of the Earth, mountains, oceans and continents lost their permanence and became just ephemeral expressions of a continual cycle of creation and destruction, just as James Hutton had envisaged almost two centuries earlier.

The supercontinent Pangea had indeed existed as hypothesized by Wegener, and the southern continents were once united to form Gondwanaland. Europe had separated from North America, South America had separated from Africa, and Australia had separated from Antarctica. But the continents did not have to plough their way through the solid rock of the sea floor as Wegener had thought. Instead, the ocean floors were themselves moving, passively carrying the continents with them like great pieces of luggage on huge conveyor belts.

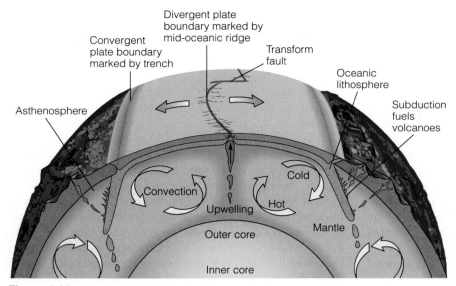

Figure 4.21

The Earth's internal heat drives its lithospheric plates. The lithosphere consists of the crust and the rigid upper part of the mantle and floats on the semimolten asthenosphere beneath. The plates are thought to move as a result of convection in the underlying mantle in which warm material from the interior rises toward the surface, cools and, on cooling, descends back into the interior. Each of the plates are in motion with respect to one another and their earthquake-prone boundaries are defined according to this relative motion. At divergent boundaries, marked by mid-ocean ridge spreading centers, the plates move apart and upwelling magma creates new ocean floor. Along transform faults, such as those that offset mid-ocean ridges, the plates slide past each other and snag, jostling the crust as they do so. At convergent boundaries, marked by deep ocean trenches, the plates move together and one plate plunges under the other in the process of subduction. The inclined boundary between the opposing plates defines the Benioff zone of earthquake foci. Descent of the plunging plate generates magma that fuels the volcanoes on the leading edge of the overriding plate.

4.5 Chapter Summary

For centuries scholars have pondered the origin of the forces that shape our planet. But it was not until the second half of the twentieth century that scientists finally arrived at the comprehensive theory of the Earth we call plate tectonics. The development of this theory is a fascinating chapter in our quest to understand the planet we live on and provides an instructive lesson in the way in which that understanding has been gained. The theory of plate tectonics holds that the Earth's rigid outer shell or lithosphere is broken into large moving slabs or plates that ride upon a yielding layer of the Earth's interior known as the asthenosphere. The interaction of these plates accounts for many of the major geologic processes that fashion our planet, causing entire continents and oceans to be rearranged as the shifting plates reshape the Earth's surface.

This new view of the Earth, in which even the continents and oceans have no permanence but instead are created and destroyed as a consequence of plate motion, has revolutionized geology as profoundly as evolution and relativity transformed biology and physics. But as recently as the early twentieth century these were radical ideas. When Alfred Wegener proposed that the continents moved in the early 1900s, his hypothesis of continental drift was met with fierce opposition from a scientific community who considered the Earth's major surface features to be fixed and permanent. Wegener amassed a wealth of evidence in support of his hypothesis including the jigsaw fit of coastlines, the continuity of geologic structures and the presence of identical rocks and fossils on continents now separated by oceans, and the geologic record of unaccountable shifts in the Earth's major climatic belts. But in believing that it was only the continents that moved, Wegener was unable to provide a mechanism that would permit continental drift to take place. Because no force would permit entire continents to plough their way across the ocean floors, his ideas were dismissed by the scientists of his day.

The scientific method had correctly found a weakness in Wegener's hypothesis and, in a theme that will recur many times in our examination of Earth science, the remaining circumstantial evidence upon which his hypothesis was built failed to convince the scientific community. But his ideas lingered on, and while he would not live to see it, his vision of moving continents was to be revived by later discoveries on the floor of the ocean.

In the mid-twentieth century, studies of the Earth's magnetic field and exploration of the world's ocean floors would reveal a planet more dynamic that even Wegener had envisioned. Rocks such as basalt lavas can record the orientation of the Earth's magnetic field at the time they crystallize. But when the orientation of the Earth's magnetic field was determined from ancient basalt lavas, in a study known as paleomagnetism, the results suggested that either the continents or the Earth's magnetic poles had moved. Polar wander curves that trace the pole's apparent movement through time, also differed from one continent to another. But the Earth's magnetic poles are geographically fixed by the planet's axis of rotation and so cannot move to the extent required to explain the data. Instead, these scientists were faced with the inescapable conclusion that it was the continents that had moved and that each continent had moved independently.

In an even more surprising discovery, studies of paleomagnetism revealed that the Earth's magnetic field periodically reverses itself, interchanging the north and south magnetic poles. Although the cause of these magnetic reversals is not fully understood, their pattern was to reveal the mechanism needed for continental drift to occur. In the North Atlantic, curious bands of magnetization arranged symmetrically about the Mid-Atlantic Ridge had been identified in the basaltic rocks on the sea bed. The precise match of these bands with the pattern of magnetic reversals established the theory of sea floor spreading. According to this theory, the ocean floor continuously spreads away from the world's mid-ocean ridges as if part of a system of huge conveyor belts. The molten basalt that oozes out at the ridge crest to form new ocean crust as the sea floor moves aside, becomes imprinted with the Earth's prevailing magnetic field as it cools, and so produces symmetrical bands of magnetization as it is carried away.

Our modern understanding of plate tectonics was born with the confirmation of sea floor spreading from samples recovered from the sea bed by the ocean drilling vessel "Glomar Challenger", and the recognition that the Benioff zone of earthquakes beneath the deep ocean trenches marked the sites of sea floor consumption or subduction. According to this theory, the Earth's rocky outer layer or lithosphere is fractured like a giant eggshell into huge moving plates outlined by the world's earthquake zones. Driven by convective circulation in the Earth's heated interior, these plates move from mid-ocean ridges toward subduction zones on the asthenosphere, a partially molten interior layer, and passively carry the continents with them as they do so. Along mid-ocean ridges new oceanic lithosphere is created. Below the deep ocean trenches, old oceanic lithosphere angles down into the Earth's interior, generating the plunging Benioff zones of earthquake foci and fueling volcanoes in the plate above as it is consumed. Where approaching plates carry buoyant continents, mountain-building collision ensues that ultimately brings their relative motion to a stop. Where plates slide past one another along great crustal fractures, lithosphere is neither created nor destroyed but the faults are prone to snag and slip as the plates jostle each other for position. It is at plate boundaries, therefore, that the effects of plate tectonics are most apparent because it is here that so many of the Earth's most important dynamic processes are shaped.

By providing continental drift, sea floor spreading, and subduction with an underlying mechanism, plate tectonics has shown the Earth to be a truly dynamic planet. This realization has revolutionized our understanding of the planet we live on by providing the first comprehensive explanation of the world's largest geologic features.

Key Terms and Concepts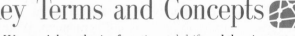

Look for the highlighted items on the web at:
WWW.BROOKSCOLE.COM

- Wegener's hypothesis of continental drift and the circumstantial evidence used to support the existence of the supercontinent Pangea, including the jigsaw fit of continents, the continuity across oceans of ancient mountain belts, major faults, and rock formations, as well as the paradox of ancient climates and puzzling distribution of fossils
- The rejection of continental drift by most scientists of the day because of Wegener's inability to propose a credible mechanism
- The Earth's magnetic field, magnetic inclination, and paleomagnetism
- Apparent polar wander curves suggest movement of the continents relative to the poles, revitalizing the notion of drifting continents
- Normal and reversed magnetic polarity, magnetic epochs, and magnetic events
- Paleomagnetic reversals, when used to interpret magnetic anomalies on the ocean floor, suggest conveyor-like movement of the crust of the oceans by the process of sea floor spreading

- Sea floor spreading is confirmed by the deep sea drilling of the Glomar Challenger
- Recognition of deep ocean trenches and Benioff zones of earthquakes as the sites of ocean floor destruction by the process of subduction. Associated magmatism accounts for the formation of volcanic arcs like those of the Pacific Ring of Fire
- Continental drift, sea floor spreading, and subduction unite to form one coherent theory of plate tectonics
- Rigid lithospheric plates are outlined by the world's earthquake zones and move on the partially molten asthenosphere in response to circulation of the Earth's heated interior
- The effects of plate tectonics are most obvious at plate boundaries where plates either move apart to create oceans, or slide past each other along great crustal fractures, or collide to form subduction zones and mountain ranges

Review Questions

1. Why was the jigsaw fit of continents so easily dismissed by scientists of Wegener's day?
2. Which mountain belts in western Europe did Wegener believe to be continuations of the Appalachian mountains of eastern North America?
3. How did Wegener use the distribution of Late Carboniferous glacial deposits in arguing for the existence of continental drift?
4. How did continental drift account for the distribution of *Lystrosaurus* fossils?
5. What mechanisms did Wegener propose to account for continental drift?

6. What is magnetic inclination and how can it be used to determine latitude?
7. How can the pattern of magnetic anomalies on the ocean floor be used to establish the rate of sea floor spreading?
8. What did the drilling vessel "Glomar Challenger" discover that astounded the scientific world?
9. What are Benioff zones and how are they produced?
10. Why are subduction zones associated with curved lines of volcanoes?

Study Questions

 Research the highlighted question using InfoTrac College Edition.

1. Wegener did not live to see his hypothesis of continental drift vindicated. Why was this and what does it tell us about science and the scientific method?
2. How is the polar wander curve of a continent established, and why does the curve show only apparent polar wander?
3. Figure 4.12 shows the times of reversals in the Earth's magnetic field over the past four and a half million years. What is the evidence on which the existence of these reversals is based?

4. How does sea floor spreading account for the pattern of magnetic anomalies on the ocean floors?
5. Account for the striking correlation between the distribution of earthquake epicenters (Fig. 4.16) and the shape of the Earth's lithospheric plates (Fig. 4.20).

Plate Tectonics: Plates and Plate Boundaries

We now know that the Earth's surface is broken into a mosaic of shifting plates whose movement over the course of geologic time has dramatically changed the geography of our planet. This dynamic view of the Earth forms the basis of "plate tectonics", a theory that has revolutionized our understanding of the forces that shape our planet.

If one examines the global distribution of earthquakes a definite pattern is revealed in their activity. Most earthquakes occur in relatively narrow zones. On a map, these zones link in a pattern that resembles a cracked eggshell. This analogy is appropriate. Only the planet's outermost rocky layer, or lithosphere, is cracked in this fashion, and this layer is about as thick in proportion to the Earth as an eggshell is to an egg.

The cracks represent tears in the Earth's lithosphere where, like a pressure cooker with a broken lid, most of its bottled-up internal energy is unleashed. The cracks surround coherent pieces of lithosphere we call plates. Many of these plates, and the earthquake-prone regions that identify their boundaries, can be readily identified on satellite photographs of the Earth. But a slow-running satellite video camera with an inexhaustible tape would reveal an additional, more dynamic dimension.

Unlike a cracked eggshell, the Earth's lithospheric plates are in motion with respect to each other. There is enormous resistance to this motion along the margins of plates, but all must finally succumb to the incessant driving forces from the Earth's interior. When failure ultimately occurs, an enormous amount of pent-up energy, easily dwarfing the world's entire nuclear arsenal, is abruptly released in what we recognize as an earthquake. Earthquakes are related to the tearing and grinding that are concentrated at the margins of plates because, like deep wounds that never quite heal, the boundaries between plates are repeatedly re-broken as they jostle for position.

The details of plate motion and the mechanism behind it are still hotly debated, but there is a general consensus that it is caused by the circulation of the Earth's heated interior. However, at speeds of only centimeters per year, the movement of plates is no faster than the rate at which your fingernails grow. As a result, plate tectonics is a very slow process. But over millions of years, even this very slow process can create and destroy entire ocean basins, and carry continents across the face of the Earth.

In this chapter we will develop these themes, and show how plate tectonics explains the origin and development of almost all the major surface features of our planet. We will first look at the structure and buoyancy of plates, and then examine each of the three types of plate boundary; divergent, convergent, and transform. We will then turn to the enigmatic features known as "hot spots" that track the movement of plates. Finally, we examine the driving forces that may be responsible for plate movement.

We will find that the process of plate tectonics gives us our air, our water and the nutrients required for the creation and sustenance of life. It is nature's vast manufacturing plant, and the underlying cause of almost every feature we think of as unique about our planet. To emphasize this point, in subsequent chapters we will show that Earth is the only planet in the Solar System capable of maintaining plate tectonics and, as a result, it is the only planet capable of sustaining life as we know it.

5.1 Moving Plates

Plate tectonics revolutionized the Earth sciences, transforming geology as profoundly as evolution transformed biology and relativity transformed physics. The theory not only vindicated Wegener's vision of moving continents, but also provided a mechanism that made such movement possible. The recognition of sea floor spreading at mid-ocean ridges where crust is created, and subduction below deep-ocean trenches where crust is destroyed, provides a framework for understanding the origin and distribution of many of the Earth's most dramatic features.

As we learned in Chapter 4, the term plate tectonics is derived from the realization that the Earth's rigid outer shell is broken into huge moving slabs or **plates** and from the Greek *tecton*, meaning "to build." According to plate tectonics, the Earth's rocky outer layer is broken into large jagged fragments, much like a broken eggshell. Each rigid plate may be thousands of kilometers across, but is on average only about 100 km (60 mi) thick, about as thick in proportion to the Earth as an eggshell is to an egg.

Like a fragment of eggshell, a plate is not flat but conforms to the spherical shape of the Earth. Unlike a broken eggshell, however, each plate moves relative to other plates. The most intense geologic activity associated with plate movement tends to occurs at their margins. At these **plate boundaries** plates either separate, move together, or slide past each other. Irrespective of which type of motion occurs, the constant jostling of their edges generates earthquakes. So earthquakes are not randomly distributed, but rather define narrow zones that coincide with the world's plate boundaries (Fig. 5.1). As a result, the Earth's tectonic plates are clearly outlined by maps that show the distribution of earthquakes on the Earth's surface.

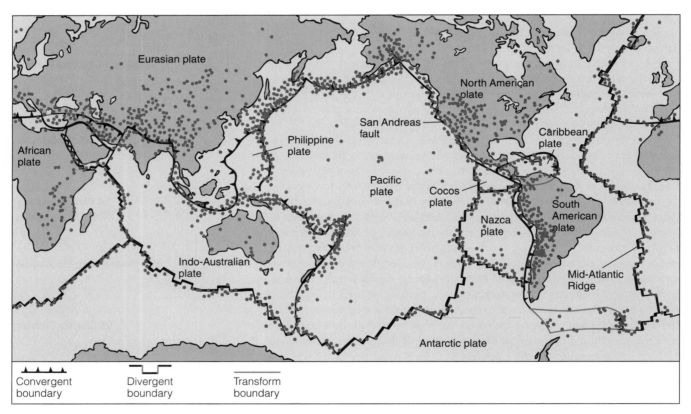

Figure 5.1
The distribution of earthquakes (dots) and their close coincidence with the boundaries of the Earth's major lithospheric plates. The lithosphere consists of the crust and the rigid upper part of the mantle and is about 100 km (60 mi) thick. The plates move with respect to each other and their boundaries are defined by this relative motion. Divergent boundaries, where plates move apart, are marked by mid-ocean ridge spreading centers and zones of continental rifting. Convergent boundaries, where plates move together, are marked by deep ocean trenches along which one plate plunges under the other in the process of subduction, or where continents are in collision, by great mountain ranges. At transform boundaries plates slide past each other along great crustal faults. Most volcanic eruptions take place at plate boundaries. Eruptions are quiet where plates separate but explosive where they collide.

So far, approximately a dozen large plates and numerous smaller ones have been identified (Fig. 4.20). Some, such as the Pacific plate, are almost entirely covered by ocean water and so are largely oceanic. Others, like the Eurasian plate, are largely continental. However, most plates carry both continent and oceans so that as they move, the continents travel with the oceans, riding passively like baggage on an airport conveyor belt, rather than ploughing through the ocean floor as Wegener had envisioned. This shift in focus from moving continents to moving plates removed the largest obstacle facing Wegener's hypothesis of continental drift and ushered in our modern view of a dynamic Earth.

Scientists generally believe that a theory is on the right track if it is both feasible and explains a host of previously unexplained and apparently unrelated natural phenomena. As we examine plate tectonics, we will find that this is especially true of plate tectonic theory. Today this theory forms the basis of our understanding of many such phenomena, including the origin of oceans and continents, the formation of the Earth's major surface features, and the distribution of earthquakes, volcanoes, and mountain belts. It further accounts for all of the phenomena used by Wegener in support of continental drift, that we encountered in Chapter 4.

The way in which plates move, and the ways in which they interact with each other, are fundamental to our understanding of many of the major geologic processes affecting the Earth's surface. As we examine plate boundaries and the processes associated with them, the theory of plate tectonics will further reveal itself to be a unifying principle or **paradigm** that has provided science with a powerful new way of looking at the Earth's dynamic systems. But what exactly is a plate and how are plates supported in a manner that permits them to move? To answer these questions we must first examine the internal structure of plates and the principle of isostasy which accounts for the buoyancy that enables plates to "float".

5.2   The Structure of Plates

If the Earth was rigid to the core, plate tectonics would not be possible. Movement of the Earth's tectonic plates is possible because they effectively "float" on a weak, or soft, layer in the Earth's interior. But what makes the underlying weak layer soft and how do we know it is there? To answer these questions we must examine the evidence for a weak layer and some of the variations in physical and chemical properties that occur within the Earth's outermost rock layers.

5.2.1 The Low-velocity Zone

Because the weak layer that allows plates to move is hidden from direct observation, its presence must be detected indirectly. The study of energy patterns released from earthquakes (more on this in Chapter 7) provides this evidence. When an earthquake occurs, energy is released in all directions and can be detected at seismic recording stations in many parts of the

world. Although we are most familiar with the destructive waves that propagate across the Earth's surface, earthquake waves also propagate downwards into the Earth's interior. The speed at which the energy propagates depends on the type of material it encounters along its path, just as an athlete may run the same distance at different speeds on wet and dry race-tracks. In particular, the velocity of earthquake waves varies with the density and rigidity of the rocks through which they pass. The energy travels most rapidly through rigid material because it can respond faster to vibrations. However, at depths of about 100 km to 250 km (62 to 155 mi) below the Earth's surface, earthquake waves are observed to lose their speed and slow down. This means that the rocks at this depth must be softer or partially melted and so absorb some of the wave's energy. Just as a stick of butter becomes softer after it has been sitting on the counter for a while than it is when fresh from the refrigerator, yet remains solid because it retains a shape of its own, so rocks at this level of the Earth's interior become softer yet remain solid as they get hotter. This is especially true for materials like butter that are close to, but have not yet reached, their melting points. Because of its effect on earthquake waves this interval is called the **low velocity zone** (Fig. 5.2), about which we will learn more in Chapter 7. The story behind the soft region that permits plate motion, however, does not end with the low velocity zone.

5.2.2 Asthenosphere and Lithosphere

In reality, the soft zone extends to somewhat greater depths than those that significantly influence earthquake waves because it is the presence of molten liquid that has the greatest effect on earthquake wave velocity. Although small amounts of melt are thought to be present only in the low velocity zone, rocks for some distance below this interval remain close enough to their melting points to be soft.

This broader region of soft rocks, which includes the low velocity zone, is known as the **asthenosphere** (Fig. 5.2) from

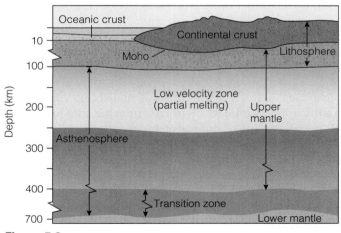

Figure 5.2

Subdivision of the Earth's outer layers into crust and mantle based on rock type, and lithosphere and asthenosphere based on rigidity.

the Greek for "weak zone". The asthenosphere permits movement of the rigid tectonic plates above by providing a zone of detachment that effectively separates them from the mantle material below. The plates are therefore free to move sideways if acted upon by a large enough force.

For the most part, the tectonic plates riding atop the asthenosphere behave in a more typically rock-like fashion because temperatures above the asthenosphere, in the upper 100 km (60 mi) of the Earth, are well below those required for melting. As a result, this outer shell of the Earth behaves in a rigid fashion, much like butter from a freezer, and is known as the **lithosphere** (Fig. 5.2) from the Greek *lithos* for "rock." Because the Earth's tectonic plates are made up of these lithospheric rocks, they too are rigid, and move as coherent pieces on the pliable asthenosphere beneath. In the same way, a frozen slab of butter placed on top of a soft one, will move with the soft slab as it changes shape.

The distinction between the asthenosphere and lithosphere is based on the contrasting physical property of the rocks in these regions rather than the composition of the rocks themselves. In fact, the rocks at the base of the lithosphere are very similar in composition to those in the asthenosphere, but because of the difference in temperature, they act in a rigid fashion in the cooler lithosphere but become pliable at the higher temperatures of the asthenosphere. The base of the lithosphere then is positioned at a depth where temperatures first become high enough to make the rocks soft. Because this may occur over a broad zone, the boundary between the lithosphere and asthenosphere is gradual rather than sharp, and so is sometimes hard to detect. This is not the case, however, for another important boundary that occurs within the lithosphere, that between the crust and the mantle.

5.2.3 Layers of the Lithosphere

The lithosphere can be internally subdivided into layers based on rock-types of contrasting chemical composition (Fig. 5.2). The top of the lithosphere comprises familiar rocks like granite and basalt that are common on the Earth's surface and make up the planet's outermost rocky layer known as the **crust**. There are basically two types of crust: continental and oceanic. **Continental crust** is made up of rocks that are broadly of granitic composition, whereas **oceanic crust** is overwhelmingly basaltic. Beneath the crust, a denser, compositionally distinct ultramafic rock called peridotite (see Chapter 2) predominates. Rocks of this composition make up much of the Earth's **upper mantle** which extends from the crust to the base of the asthenosphere. Immediately beneath the crust, where they are cool and rigid, these upper mantle rocks are considered to form part of the lithosphere. At greater depths, where increased temperatures cause them to become pliable, similar upper mantle rocks form much of the asthenosphere. The cool, rigid lithosphere therefore includes both the Earth's crust and part of the upper mantle, whereas the soft asthenosphere lies entirely within the upper mantle, as shown on Figure 5.2.

The rocks in the upper mantle remain soft to depths of up to about 700 km (440 mi), which marks the base of the asthenosphere. Below this, in the **lower mantle**, they become rigid again because the enormous pressures at this depth prevent them from becoming pliable, even while the temperature continues to rise. The broad zone of very dense iron- and magnesium-silicate-rich rocks that make up the lower mantle extends from the asthenosphere to the base of the mantle at a depth of about 2900 km (1800 mi).

The crust and upper mantle are made up of quite different rocks and so have different chemical compositions. The average densities of continental and oceanic crust, for example, are 2.7 and 2.9 respectively, compared to a density in the upper mantle of 3.3. The boundary between the crust and upper mantle, unlike that between the lithosphere and asthenosphere, is sharp and so is readily detected by earthquake waves which move abruptly faster upon entering the denser peridotitic rocks of the upper mantle. Named after its Yugoslavian discoverer, this sharp boundary is known as the Mohorovicic Discontinuity, or **Moho** for short.

Detailed analysis of the energy released by earthquakes also provides important clues to the hidden internal composition and thickness of the Earth's crust because the path and speed of this energy depends on the nature of the crustal rocks through which it is traveling. The results of such studies show that the crust varies greatly in both composition and thickness (Fig. 5.2). The basaltic crust of the ocean floors is quite thin, typically averaging only 5 km (3 mi) in thickness. The granitic crust of the continents, on the other hand, is of greater but highly variable thickness, averaging about 35 km (22 mi) but increasing to as much as 80 km (50 mi) below high mountain ranges like the Himalayas.

In summary, the Earth's lithospheric plates are made up of rigid crustal and upper mantle rocks, and are free to move because they rest on a weak zone within the upper mantle known as the asthenosphere. The upper mantle portion of the lithosphere is made up of rocks similar in composition to those of the asthenosphere, whereas the crustal component is of variable composition and highly variable thickness. This last point raises some important questions. Why do the crustal components of plates show such variability? Why, for example, do the continental portions of plates mostly stand above sea level, whereas their oceanic portions are almost entirely submarine? What does the marked difference in their elevation mean and what does this tell us about the buoyancy of plates? Indeed, how do the plates "float" on the asthenosphere? The answers to these questions lead us to another dynamic property of the Earth's interior, the phenomenon known as isostasy.

5.3 Buoyancy and the Principle of Isostasy

According to earthquake studies, the crust is thin beneath the oceans and thickest below high mountain belts. The crust is consequently thin where its surface is at a low elevation, and thickest where it is high. This means that the base of the crust

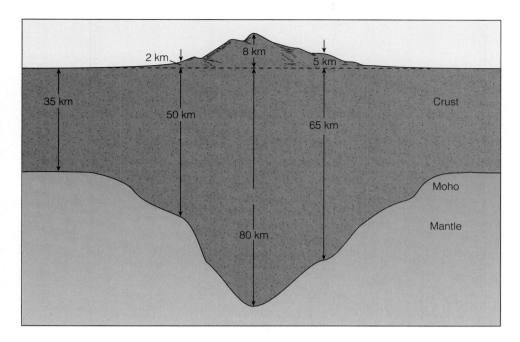

Figure 5.3
Earth's surface topography is mirrored in an exaggerated fashion by the shape of the Moho. This is because mountains are supported by light crustal roots that float on the denser mantle like an iceberg floats on seawater.

(the Moho) has a shape that mirrors in an exaggerated fashion the overlying shape of the Earth's surface topography (Fig. 5.3). In short, the higher the crustal elevation, the deeper the crustal "root". But why should this be so?

The reason behind this curious relationship is one of buoyancy. The level at which any object floats depends upon its buoyancy. The greater the object's buoyancy, the higher the level at which it will float. An empty cargo ship, for example, is more buoyant than a laden one and so rides higher in the water (Fig. 5.4). For the same reason, lighter continental crust rides higher than denser oceanic crust. The relationship between crustal thickness and elevation therefore shows that the crust of the Earth is in a state of buoyant equilibrium and is effectively "floating". The principle that describes this condition is called **isostasy** from the Greek for "equal standing". But to fully understand how it works, we must examine buoyancy more closely.

The buoyancy of an object actually depends upon two factors, the object's thickness and its density. The relationship of buoyancy to thickness is best illustrated by examining the floating behavior of objects of the same density but different size. Take, for example, floating icebergs. All icebergs float such that only 10 percent of their volume is exposed above sea level (Fig. 5.5). This is because the buoyancy of an iceberg depends on its density contrast with respect to ocean water, and the density of ice (0.9) is 10 percent less than that of water (1.0). The well-known expression "the tip of the iceberg" therefore comes from the fact that 90 percent of any iceberg is hidden from view. As the submerged portion of an iceberg is always nine times larger than its exposed portion, larger icebergs not only ride higher, but also extend to greater depths than smaller ones. For example, an iceberg 10 m (33 ft) thick will float such that only 1 m (3 ft) stands above sea level, whereas a 100 m (330 ft) thick iceberg will stand 10 m (33 ft) above sea level. In other words, for an iceberg to stand high in the water it must have a deep, submerged portion to support it.

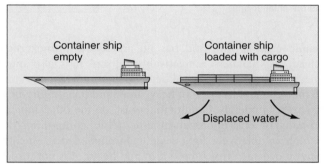

Figure 5.4
Floating container ships illustrate the principle of buoyancy. An empty container ship rides higher in the water than one laden with cargo because it is more buoyant. The laden container ship will sink until it displaces a volume of water equal in weight to that of the ship and its cargo.

Continents behave in much the same way in that they "float" on the pliable asthenosphere below. They stand highest where there are mountains and so must be thickest below major mountain belts. Hence, mountain belts are supported by buoyant "roots" that project deep into the mantle (Fig. 5.3). For continents, the depth to which this root projects below the average thickness of continental crust is about four and a half times more than the mountain's elevation. So for Mt. Everest to stand 8 km (5 mi) high, it must be supported by a crustal root that extends the average thickness of continental crust (35 km or 22 mi) by a further 36 km (22.2 mi). The thickness of the crust beneath the Himalayas actually approaches 80 km (50 mi).

This floating model also explains the contrast in thickness and surface elevation between oceanic and continental crust. The contrast occurs because both continental and oceanic crust effectively "float" on the denser mantle beneath. But

Figure 5.5
Floating icebergs demonstrate the concept of isostasy. According to the principle of isostasy, the depth to which a floating object sinks depends upon its thickness and its density. All icebergs, being of the same density, sink in water such that the same proportion of their volume (90%) is submerged; the thicker the iceberg, the greater this volume will be. If ice above the water level melts, the iceberg will rise to maintain the same proportion of ice above and below the water line. Thus the larger iceberg rides higher but also extends to greater depth than the smaller one.

oceanic crust is thin and has no anomalously thick roots. Instead, its thickness is relatively uniform. It is also much denser than continental crust because its composition is largely basaltic so that it is rich in heavy iron- and magnesium-bearing (mafic) minerals. Continental crust, on the other hand, is composed of rocks of broadly granitic (felsic) composition that are rich in relatively light-weight alumina-bearing silicates. Because oceanic crust is both thin and heavy, it is not surprising that it occurs in those areas where elevations are lowest. As water collects at the lowest elevations, these areas have also become the world's ocean basins.

Viewed from space, the contrasting elevation of the continents and oceans is one of the most striking features of Planet Earth. That we live on a split-level planet is most clearly displayed on a graph known as the **hypsometric curve** (Fig. 5.6). This shows the proportion of the Earth's surface that stands at or above a given elevation or depth. From the curve it is clear that much of the Earth's surface stands at one of two elevations; either between 1000 m (3300 ft) and sea level, or between 4000 m and 5000 m (13,000 to 16,400 ft) below sea level. Representing 20.8 percent and 22.6 percent of the Earth's surface, these two elevation ranges mark the **continental platforms** (0 to 1000 m above sea level) and the oceanic **abyssal plains** (4000 to 5000 m below sea level) respectively, and demonstrate the remarkable control that crustal density and thickness exert on the planet's surface elevation.

Thus the Earth's crust is in a state of buoyant equilibrium, the continents stand highest where they are thickest, and the oceans have formed in regions floored by denser basaltic crust. It is this balance or "equal standing" between crustal elevation, crustal thickness and crustal density, that is described by the principle of isostasy. Not only does isostasy explain why mountains stand high, but it also accounts for the subdivision of the

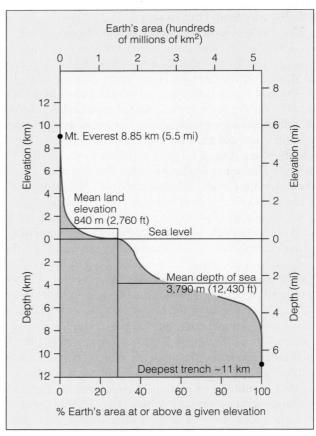

Figure 5.6
The hypsometric curve. A graph showing the proportion of the Earth's surface at or above a given elevation or depth. Note that less than a third of the Earth's surface is above sea level and more than half lies below a depth of 3000 m (10,000 ft).

Earth's surface into continents and ocean basins. According to the principle of isostasy, the edge of a continent occurs where the crust thins as it changes its composition from a broadly granitic one to the basaltic composition that makes up the ocean floor. This boundary between continental and oceanic crust does not necessarily mark the edge of the physical ocean. This is because ocean water is not confined by the boundary, but instead moves freely on top of the lithosphere. In fact, ocean waters tend to flood the feather edges of continents so that continents are bordered by areas covered by shallow water known as the **continental shelves** (see Chapter 8).

According to the principle of isostasy, the Earth's major topographic variations are a function of crustal thickness and the density of crustal rocks. If either of these is changed, the crust responds by establishing a new "floating" balance or **isostatic equilibrium**. During the last Ice Age, for example, the crust beneath the advancing continental ice sheets was first depressed under the weight of the ice, only to rebound again when the ice sheet retreated (Fig. 5.7).

In a similar fashion, as the summits of mountains are lowered by erosion, so the crust rises in response to the reduced load, and the crustal root supporting the mountains shrinks (Fig. 5.8). This response is known as **isostatic rebound.** The net result is a

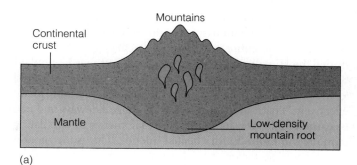

(a)

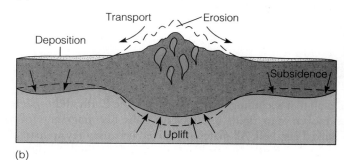

(b)

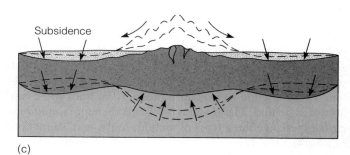

(c)

Figure 5.8

Sequence illustrating how erosion and isostatic readjustment combine to thin the crust in mountainous regions. As the mountains are eroded, isostatic rebound causes uplift of the mountain root. Deposition of sediment, on the other hand, causes crustal subsidence.

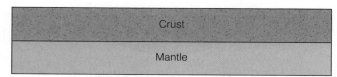

(a) Crust and mantle before continental glaciation

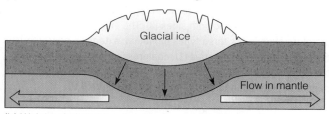

(b) Weight of glacial ice depresses crust into the mantle

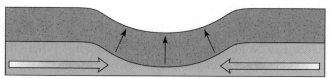

(c) Ice melts and isostatic rebound begins as weight is removed from crust

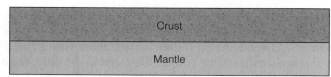

(d) Complete isostatic rebound restores crust to its original position

Figure 5.7

Isostatic subsidence of the crust under the weight of continental ice sheets is followed by isostatic rebound when the ice sheets retreat.

loss of elevation due to erosion which is offset by the rise towards the surface of the guts of the mountain chain in such a fashion that isostatic equilibrium is maintained. It is this process of erosion compensated by uplift by which progressively deeper levels of the crust are brought to the surface, that makes the rock cycle (see Chapter 2) possible. Hence, ancient mountain belts that have undergone long periods of erosion often expose metamorphic rocks that once resided deep within the root of the mountains. The process of erosion (which we examined in Chapter 2) therefore causes readjustments to isostatic equilibrium that slowly reduce the mountain belt to normal crustal thicknesses. When this is achieved, once deeply buried portions of the mountain belt will be exposed at the surface.

In summary, the principle of isostasy describes the state of buoyant equilibrium that exists in the Earth's crust. It accounts for the contrasting elevations of the continents and ocean floors, and the striking relationship between the crustal thickness and surface elevation of the continents. It also demonstrates that the Earth's crustal fragments and the plates in which they are embedded effectively "float" on a pliable mantle below. Crustal erosion is consequently compensated by isostatic rebound. This rebound not only drives the rock cycle but has also eliminated entire mountain belts that may have once rivaled the Himalayas. In the process, the crust may be uplifted by as much as 50 km (30 mi). Yet this distance is tiny compared to the horizontal distances moved by plates. Because plates effectively "float," they are also free to move sideways and can be carried in this way for thousands of kilometers by circulating currents within the Earth's heated interior. This is the motion that drives plate tectonics, and it is to this movement and the plate boundaries along which it occurs that we now turn.

5.4 Plate Boundaries

Recall that according to plate tectonics, the Earth's lithosphere is broken into rigid, interlocking slabs like fragments of a cracked eggshell. All plates are curved because they are part of a spherical Earth. But unlike the cracked eggshell, the lithospheric fragments we call plates float on a dense, pliable asthenosphere, and are therefore free to move relative to each other like rafts floating on a stream. As a consequence, all plates must interact with neighboring plates along their mutual boundaries. In the simplified case of Figure 5.9, movement of Plate A to the left requires it to slide along its top and bottom margins, while an overlap is produced in front and a gap is created behind. The margins of Plate A therefore *simultaneously* experience three types of motion. At the rear, its motion relative to Plate B is one of separation and **extension.** At the front, it is one of convergence and **compression.** Along its top and bottom margins, it is one of translation or **shear.**

Actual plate boundaries behave in much the same way as Plates A and B in Figure 5.9. As they move away from mid-ocean ridges and towards deep ocean trenches, they jostle each other for position, interacting in the same three ways as Plates A and B: separating, colliding, or sliding past each other. The three types of plate boundaries that result from these interactions are termed divergent, convergent, and transform boundaries depending on their sense of movement (Fig. 5.10). As reasoned by Vine and Matthews (see Chapter 4), **divergent plate boundaries,** where two plates move apart, are marked by spreading centered on a mid-ocean ridge and create new oceanic lithosphere. Because these boundaries form new crust, they are often called **constructive** margins. **Convergent plate boundaries,** where two plates are in collision, are marked by subduction zones along which one plate is forced beneath the other and consumed. These boundaries are also termed **destructive** plate margins. **Transform plate boundaries,** where two plates slide past each other in translation, are marked by great crustal fractures. Transform boundaries neither create nor destroy lithosphere but form links between one plate boundary and another. They are also known as **conservative** margins. In the following sections we will examine the nature of these three types of plate boundaries and the surface features they create in more detail.

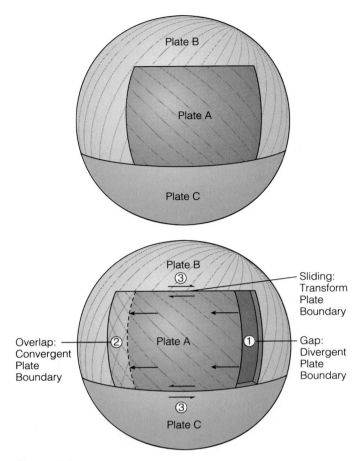

Figure 5.9

Plate boundaries at work. As Plate A moves to the left, a gap is formed behind it (1), an overlap with Plate B is formed in front (2), and sliding occurs along both sides (3). The margins of Plate A therefore experience three types of motion: extension characteristic of divergent plate boundaries (1), compression characteristic of convergent plate boundaries (2), and shear characteristic of transform plate boundaries (3), as shown in Figure 5.10.

5.4.1 Divergent Boundaries— Creating Oceans

At divergent plate boundaries, where plates move apart, continents may be torn in two to form new oceans. Where this occurs new oceanic lithosphere is created by sea floor spreading at mid-ocean ridges as the lavas and the rocks beneath them cool and thicken to form a rigid layer (Fig. 5.11). The birth of oceans starts on dry land with the breakup of a continent. When two plates start to separate, basaltic magma rises from the mantle below, and the overlying lithosphere bulges. As more magma is inserted into the continental crust like wedges into wood, the crust stretches and thins. Rising batches of magma create a line of separation, and in the early stages of rifting, continental blocks settle into faults in much the same way that the keystone of an arch would settle if the arch was pulled apart (inset, Fig. 5.11b). This settling or subsidence

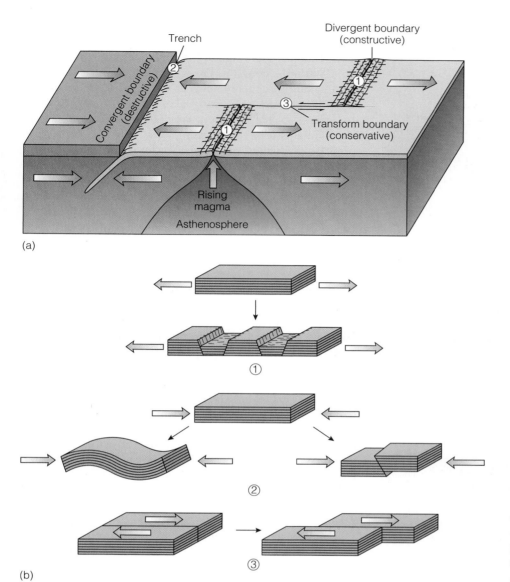

Trench

Divergent boundary
(constructive)

Convergent boundary
(destructive)

②

①

③

Transform boundary
(conservative)

①

Rising
magma

Asthenosphere

(a)

①

②

③

(b)

Figure 5.10

(a) Three types of plate boundaries are distinguished on the basis of their relative motion: (1) divergent, (2) convergent, and (3) transform. (b) The three types of motion match those of the margins of Plate A in Figure 5.9: (1) extension at divergent boundaries causes rifting, (2) compression at convergent boundaries produces buckling and shortening, and (3) translation at transform boundaries causes shear.

forms a steep-walled **rift valley.** This new valley commonly becomes a pathway for major streams. As a result stream and lake deposits accumulate on the floor of the valley in addition to flows of basaltic lava. When the rift has torn completely through the continent and reaches a coastline, seawater floods in and the first shallow-marine sediments are deposited on the rift floor. As the continental plates separate further, more water flows in and the upwelling magma starts to build new basaltic ocean floor. A small but progressively widening ocean develops between the now drifting continents. The thinned margins of the continents slowly subside as they move away from the heat source at the zone of divergence, and are progressively flooded and covered by unfaulted sedimentary rocks to form continental shelves.

In this way, continental rifting creates two plates from one and produces an ocean centered along the original line of separation. Once the two continents have separated and oceanic crust has developed between them, the divergent plate bound-

ary between the two plates is a **mid-ocean ridge,** a submarine mountain chain buoyed by underlying mantle heat. At the highest point of the ridge, basaltic ocean crust is created by sea-floor spreading. As the plates move apart, fractures open in the rift valley at the ridge crest and are continually filled by magma derived from the hot, partially molten mantle below. The near-vertical sheets of basalt that are produced by the filling of fractures are called dikes (see Chapter 2). In this way, new oceanic crust is continually inserted at the summit of the ridge, and all previously formed ocean floor is progressively wedged further and further apart. This implies that the crust is newest at the spreading ridge and gets progressively older with increasing distance from the ridge crest. The process is like inserting a series of books into the center of a bookshelf such that the newer books push the older ones apart. As older ocean floor is carried away from the ridge, it cools, contracts, and subsides as it does so, and a full ocean thousands of kilometers across is opened as a result.

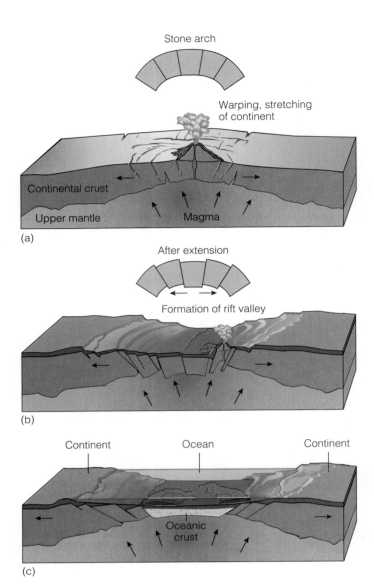

Stone arch

Warping, stretching
of continent

Continental crust

Upper mantle   Magma

(a)

After extension

Formation of rift valley

(b)

Continent   Ocean   Continent

Oceanic
crust

(c)

Figure 5.11

Formation of a divergent plate boundary with the breakup of a continent.
Insets show similarity of the process to the movement of the keystone on
stretching a stone arch. (a) As the lithosphere is uplifted and stretched,
cracks develop in the continental crust and molten basalt begins to rise
from the hot mantle below. (b) Settling of continental blocks along faults
creates a rift valley, on the floor of which basalt lava flows erupt and
non-marine sedimentary rocks are deposited. Often the fault blocks tilt
as they subside. The present-day East African Rift Valley resembles this
phase. (c) As the rift continues to open, the two new continents become
separated by a growing ocean, and new oceanic crust forms between
them. The stretched margins of both continents subside as they move
away from the source of heat and become blanketed by shallow-marine
sedimentary rocks. The present-day Red Sea resembles this phase.

Over the past 200 million years, just such a process created
the Atlantic Ocean. The present-day Atlantic can therefore be
visualized as a giant expanding bathtub, bordered by the Americas
to the west, and by Europe and Africa to the east, with a continuous ridge running down its center that is anchored to the base of
the bathtub and rises to within a few inches of its water surface.

A similar process in its early stages can be seen operating
in East Africa (Fig. 5.12), where the Great Rift Valley cuts
across Ethiopia southward to Mozambique, a distance of 4500
km (2800 mi). Along this divergent plate boundary a slab of
continental lithosphere known as the Somali plate, may one
day split from Africa. As rifting takes place, magmas rise to fill
the fissures so that volcanoes are commonplace, the largest,
Mt. Kilimanjaro in Kenya, rising over 5750 m (19,000 ft) to
form Africa's highest mountain. To the north, where the rift
system is flooded by the Red Sea and the Gulf of Aden, separation has reached the next stage with the birth of a new
ocean between Africa and the Arabian Peninsula. All three
arms of the rift system meet in Ethiopia's "Afar Triangle" to
form a Y-shaped plate junction known as a **triple point.**

The Earth's system of divergent plate boundaries form a
network of **mid-ocean ridges** that girdle the globe like a seam on
a baseball (see Fig. 5.1). The rates of divergence on this system
vary significantly, ranging from a rapid 17 cm/year (6.8 in/year)
on the fast-spreading East Pacific Rise off South America, to a
barely perceptible 0.1 cm/year in Africa's Rift Valley. Mid-ocean
ridges are found in all of the world's major oceans and form an
interconnected system of undersea mountains that rarely break
the surface, but which rise two to three km (6500 to 10,000 ft)
above the surrounding sea floor and stretch continuously for
65,000 km (40,000 mi). Typically their summits are one to five
km (0.6 to 3 mi) below the ocean surface.

The mid-ocean ridges constitute the greatest mountain
range on Earth. They are mountains because they rise above
the abyssal plains of the surrounding ocean floor, although
their summits rarely stand above sea level so they are mostly
hidden from view by a cover of ocean water. Only in Iceland
does a significant portion of a mid-ocean ridge (the Mid-
Atlantic Ridge) rise above the surface of an ocean.

Mid-ocean ridges, however, are quite unlike the more
familiar mountain ranges on land, which owe their origin to
compression and thickening of continental crust along zones
of plate convergence (discussed in more detail in the next section). Mid-ocean ridges are, by contrast, the products of extension and separation. The rift valley at their summits, where
basaltic lavas erupt to create new oceanic crust as the sea floor
spreads, is evidence of this extension. They are therefore high,
not because they are being compressed, but because they are
swollen with heat, being thermally buoyed by the hot mantle
beneath. As sea-floor spreading carries the oceanic crust away
from a ridge, the cooling seabed subsides and ultimately
becomes part of the flat abyssal plain of the deep ocean. So
mid-ocean ridges are broad areas of thermal uplift. In consequence, they tend to be far wider than most continental mountain belts, and as they are thermally buoyed, they do not
require a crustal root to support their weight. Unlike continental mountains, therefore, they are not in isostatic equilibrium like an iceberg, but rather are in thermal equilibrium,
being only as high as they are hot.

Rift valleys continue to develop at the crests of mid-ocean
ridges for the same reason that they form where continents are
tearing apart. The influence of mantle heat not only uplifts the
mid-ocean ridge, it also stretches the oceanic crust, causing it to

Some ophiolites, like the Troodos Complex of Cyprus, preserve a complete cross section of sea floor that demonstrates how the oceanic crust is constructed (Fig. 5.16). At the top of the sequence are deep-ocean sediments laid down in the ancient sea. Next come bulbous **pillow lavas,** which formed when basalt magma was extruded onto the seabed. Then there is a layer of basalt that solidified in conduits leading from the magma chamber to the surface to form dikes. The repeated opening of fractures by sea floor spreading produces a complex of **sheeted dikes.** Below these lie slowly cooled, coarse-grained basaltic rocks called gabbros, that once filled the magma chamber of a mid-oceanic ridge. At the base of the sequence, dark, heavy, magnesium- and iron-rich mantle rocks called peridotites were once part of the bottom layer of the lithosphere. Some of these peridotites are thought to have formed on the floor of the magma chamber that fed the overlying dikes and lavas. Here heavy crystals settling out of the magma chamber as it cooled would have slowly accumulated to produce concentrated layers of early crystallizing magnesium- and iron-rich minerals (see Chapter 2). Some of the other peridotites, however, are truly upper mantle rocks left behind after partial melting had extracted the basaltic magma to form the overlying oceanic crust.

So, divergent plate boundaries are those along which plates separate. They are responsible for the rifting of continents, the opening of oceans, the development of mid-ocean ridges, and the creation of new oceanic crust. They are constructive plate margins, producing new ocean floor that spreads away from the high-standing mid-ocean ridges, cooling and subsiding as it does so. But where does all this oceanic crust go? To answer this question we must examine another type of plate boundary, that associated with plate collision. It is to these convergent plate boundaries where oceanic crust meets it fate, that we now turn.

5.4.2 Convergent Boundaries—Recycling Crust and Building Continents

Assuming the size of the Earth to be constant, the creation of new lithosphere must be balanced by the destruction of old lithosphere. The contrast in the maximum ages of continental crust (up to 4 billion years) and oceanic crust (less than 200 million years) suggest that the oceanic lithosphere is preferentially destroyed. As can be seen in Figure 5.9, the destruction of lithosphere is accomplished at convergent plate boundaries, where two plates overlap. What happens at such boundaries depends on whether the overlapping plates are continental or oceanic. Because of this, three types of convergent plate boundary are possible: ocean-ocean, ocean-continent, and continent-continent. The two types of plate boundaries that consume oceanic lithosphere, ocean-ocean, and ocean-continent, have quite different characteristics and we will examine each in turn. The third type of convergent plate boundary, that of continent-continent collision, is examined in Chapter 6. But first, we must see how the overlap between two plates is actually achieved. To do this, we turn to the process responsible, that of subduction.

Subduction

Along convergent plate boundaries, two plates come together and one angles down beneath the other in a process known as **subduction** (Fig. 5.17). The inclined zone along which this happens is called a **subduction zone.** In general, it is the denser plate that is subducted beneath the lighter, more buoyant one. Because of this, it is usually oceanic lithosphere that is consumed. The broadly granitic rocks of the continental crust are less dense and, hence, far more buoyant than the dense basalts of the ocean floors. So if one of the colliding plates is continental, the denser oceanic lithosphere angles downwards beneath it. Such is the case today along the western margin of South America where the oceanic Nazca plate plunges beneath the continental South American plate. Because of their contrast in density, oceanic lithosphere created at mid-ocean ridges tends to be destroyed and recycled, whereas buoyant continental crust is preserved.

If two oceanic plates converge, it is again the denser plate that is preferentially consumed. However, in this case, the outcome is often determined by the relative age of the converging plates. Because oceanic crust cools and becomes denser as it ages, ocean-ocean convergence tends to preferentially destroy the older oceanic lithosphere. In this way, old oceanic lithosphere is preferentially recycled and the age of the ocean floor is kept young. Many examples of this style of subduction occur today in the western Pacific Ocean.

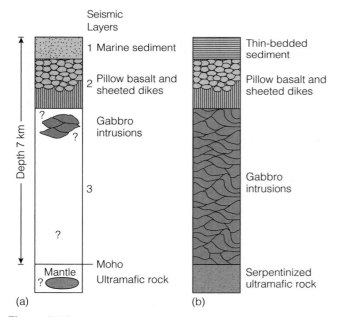

Figure 5.16

Comparison of oceanic crust and an ophiolite sequence. (a) Structure of oceanic crust determined from seismic studies and drilling. (b) Typical ophiolite sequence, like that of the Troodos Complex in Cyprus, found in mountain belts on land. Thicknesses are approximate because the sequence is usually faulted.

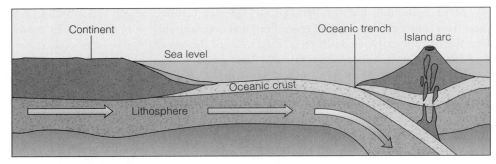

(a) Ocean-ocean convergence

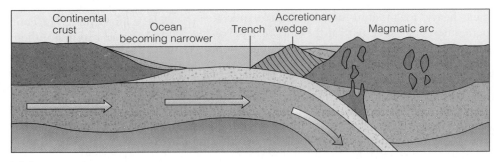

(b) Ocean-continent convergence

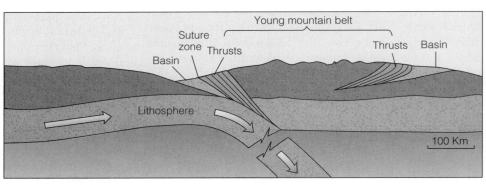

(c) Continent-continent convergence

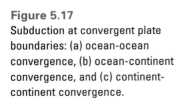

Figure 5.17
Subduction at convergent plate boundaries: (a) ocean-ocean convergence, (b) ocean-continent convergence, and (c) continent-continent convergence.

But what if both plates carry buoyant continents? Such continent-continent convergence is ultimately inevitable if the subduction of oceanic lithosphere consumes the ocean floor between two continents. When the two continents eventually meet, the one on the subducting plate will be dragged down into the subduction zone. But continental lithosphere is too buoyant to be subducted completely. As a result, the process of subduction eventually comes to a halt following the collision of continents. But the effects of continental collision are dramatic, as we shall learn in Chapter 6, when we turn our attention to the process of mountain building.

Prior to the terminal development of collisional mountains, however, other majors features are characteristic of the subduction process. All subduction zones, for example, are marked by earthquakes, deep ocean trenches, and volcanoes. But whereas some create curved chains of volcanic islands, others produce explosive volcanoes on land. These features and the reasons behind their development are discussed in the following section.

Characteristics of Subduction Zones

The downgoing oceanic slab in a subduction zone is cold and rigid, and so may penetrate deep into the mantle before it is consumed. Because earthquakes occur in cold brittle rock, those associated with subduction may therefore occur at much greater depths than the earthquakes associated with spreading ridges. (Recall that at a spreading ridge, hot magma sits relatively close to the surface, rendering the crust less rigid at relatively shallow depths). In fact, earthquakes within a subducting slab may come from depths as great as 600 km (375 mi), in contrast to the very shallow (often less than 5 km) depths typical of earthquakes at mid-ocean ridges. Subduction zone earthquakes also have a variety of sources.

Some subduction zone earthquakes are generated where the cold, and therefore brittle, oceanic plate bends before it subducts. Others occur if the downgoing slab breaks up before it is consumed in the mantle. But most are produced by slippage

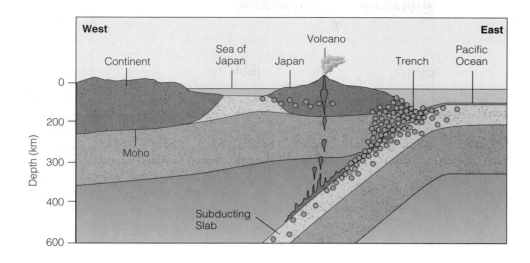

Figure 5.18
Japanese earthquakes in cross-section. Note correspondence of subducting slab with inclined Benioff zone of earthquake foci.

along the subduction zone itself, where the two plates rub together. Because rocks conduct heat very slowly, it takes a long time for the subducting plate to heat up and become soft. Until this occurs, the downgoing slab maintains its rigidity and so is capable of breaking along brittle, earthquake-producing fractures within it unlike the far hotter and more pliable asthenosphere that surrounds it. In fact, it is only in subduction zones that earthquakes can originate from depths of more than 100 km (62 mi). Subduction zone earthquakes, like those of the Japanese volcanic arc (Fig. 5.18), are consequently confined to the descending lithosphere, so their points of origin or "foci" become deeper in the direction of subduction. As we learned in Chapter 4, the existence of these inclined zones of earthquake foci was known before the advent of plate tectonics, and they are often referred to as **Benioff zones** after the American geophysicist Hugo Benioff, who pioneered their study.

The site of subduction is also marked by a deep ocean trench where the oceanic plate bends before sinking. Ocean trenches are produced by frictional drag between the colliding plates which causes the subducting plate to pull the edge of the overriding plate down with it, creating some the deepest points on the ocean floor. At more than 11,000 m (36,100 ft) below sea level and some 5,500 m (18,000 ft) below the adjacent seabed, the floor of the Mariana Trench in the western Pacific is the deepest point on the Earth's surface.

Subduction is also marked by volcanoes, which tower above the trench and are fueled by melting and the ascent of magmas released in or above the downgoing slab as it plunges into the Earth's interior. Where the overriding plate is oceanic, the volcanoes form a curved line of islands that run parallel to the adjacent trench (Fig. 5.19), like those of the Aleutian Islands in the northern Pacific. These curved chains of volcanic islands are called **island arcs.** The curvature of island arcs reflect the spherical shape of the Earth, and therefore the curvature of the surface of the subducting slab. Where the overriding plate is continental, however, magma rising from the subduction zone causes uplift of the continent's leading edge to produce a volcanic mountain range parallel to the coast, like

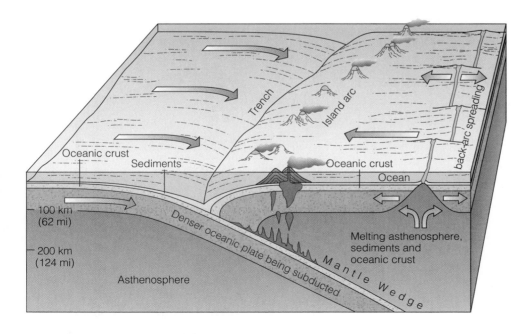

Figure 5.19
Ocean-ocean collision produces a curved line of volcanoes known as an island arc while the oceanward retreat of the trench stretches the ocean floor behind the arc, causing back-arc spreading. Magmas produced by melting of parts of the downgoing ocean floor and overlying the mantle wedge are predominantly basaltic.

that of the Andes Mountains in South America (Fig. 5.20). We will return to examine the origin of these mountains and their relationship to subduction in Chapter 6.

Where the subducting oceanic lithosphere is very old, the subduction zone is commonly steep because the downgoing slab is too dense to be supported buoyantly. As a result, the slab collapses into the asthenosphere, rolling back oceanward as it does so. The action of this "roll back" may pull the overriding plate behind the arc, causing its oceanic crust to crack. As molten rock in the mantle rises to fill the crack, a small spreading zone is created behind the island arc, like a miniature version of a mid-ocean ridge (see Figure 5.19). This process is called **back-arc spreading** and the basin it opens behind the arc is a **back-arc basin** floored by oceanic crust. Much of the western Pacific Ocean has been affected by back-arc spreading, which accounts for the seas that occur between the islands of the western Pacific and the coastline of Asia. The floors of the Sea of Japan and the Philippine Sea, for example, were formed by such a process.

Types of Subduction

As lithospheric plates may be oceanic or continental, subduction zones can be of two types: ocean-ocean and ocean-continent. Ocean trenches and volcanic islands like the Aleutians mark the sites of present-day ocean-ocean subduction, whereas ocean trenches and volcanic mountains like the Andes mark the sites of ocean-continent collision. These differences produce marked variations in the characteristics of subduction zones as we will now discover by examining the two types in more detail.

Ocean-Ocean Convergence. When two oceanic plates converge, the denser plate is generally subducted. Typically, the denser plate is also the older one, because the density of oceanic crust increases as the crust gets colder with age. In fact, old oceanic lithosphere will ultimately become denser than the

underlying asthenosphere so that its eventual subduction is inevitable. The main resistance to the subduction of old oceanic lithosphere is its own rigidity. But this resistance is eventually overcome as the plate's density increases. Like a tablecloth sliding off a table, once initiated, the subduction of old oceanic lithosphere can become very rapid indeed, often resulting in "roll back" of the subduction zone and the development of a back-arc basin.

The product of subduction (Fig. 5.19) is the formation of a deep ocean trench, produced as the subducting plate pulls the edge of the overriding plate down with it, and a volcanic island arc like that of the Aleutians. The curvature of island arcs is parallel to that of the neighboring deep ocean trenches, implying a direct relationship to the subduction process. The curvature of both the arc and trench is caused by the spherical shape of the Earth. As the subduction zone plunges into the Earth's interior, it forces the trench to adopt a curved shape, just as a dent in a ping-pong ball adopts a circular outline. As a result, both curvatures are always convex toward the subducting plate.

The volcanoes that make up the island arc are the product of subduction and occur on the overriding plate. As the downgoing oceanic slab angles into the Earth's interior it is heated and, when the temperatures become high enough, its ocean-floor basalts start to melt. The slab also releases water and other gases formerly trapped in minerals and fractures in the subducting ocean floor. These fluids rise into the wedge-shaped area of mantle (or **mantle wedge**) above the subduction zone where they promote melting of the mantle rocks. Those magmas derived from the mantle wedge, like all melts from the mantle, are basaltic in composition. Those produced by melting of the subducting slab itself are somewhat more silica-rich and are termed "basaltic-andesites" because they are transitional in composition between basalts and andesites (see Chapter 2). Both types of magma rise through the overriding oceanic plate and

Figure 5.20
Ocean-continent collision illustrated by the convergent plate boundary bordering the western margin of present-day South America. Magmas are produced from three sources. Melting of parts of the downgoing ocean floor and melting of the overlying mantle wedge produces basaltic liquids. Melting of continental crust by the rising basaltic magmas produces viscous felsic magma which may cool to form granite or may mix with the basaltic liquids to produce intermediate andesite magmas.

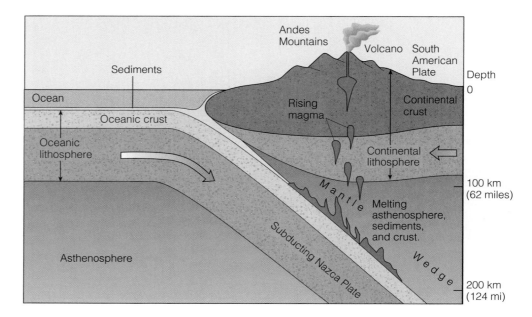

may reach the surface where eruptions build an arcuate line of largely basaltic volcanoes parallel to the trench (Fig. 5.19).

The gap between the trench and the arcuate line of volcanic islands (the "arc-trench gap") reflects the angle of subduction. Magmas can not be produced until the downgoing slab becomes hot enough to melt, a situation that usually occurs only after the slab has reached depths of 100 km (62 mi) or more. The volcanoes, which form above the point at which melting first occurs, are consequently offset from the trench (Fig. 5.19). Because steeply dipping subduction zones attain this depth closer to the trench than shallow-dipping ones, magmas rising from steep subduction zones generate arcs that are closer to the trench. In either case, however, the relative position of the arc and trench can be used to determine the direction or **polarity** of subduction, because subduction that starts at a trench will always be directed toward and beneath the neighboring volcanic arc.

In summary, Benioff zones of earthquakes, deep ocean trenches, volcanic island arcs, and back-arc basins are all features typical of ocean-ocean convergence. Volcanism is largely basaltic and the distance between the trench and arc reflects the subducting plate's angle of descent. Some of these features are also typical of ocean-continent convergence but, as we shall now see, others differ markedly.

Ocean-Continent Convergence. Unlike ocean-ocean collision, subduction beneath continents results in volcanic eruptions that are often highly explosive and, as we shall learn in Chapter 6, produces mountains. Ocean-continent convergence also shrinks oceans, as dense basaltic ocean floor is subducted beneath more buoyant continental lithosphere (Fig. 5.20). This kind of subduction generates some of the Earth's most violent internal actions.

Seventy-five percent of the Earth's active land volcanoes lie alongside the offshore trenches that reveal the lines of ocean-continent collision surrounding the Pacific Ocean. The edge of the Pacific is under attack, its floor shrinking as it is consumed beneath as many as six different plates along the subduction zones that rim this ocean. The volcanoes fueled by the subduction of its ocean floor form the 48,000 km- (30,000 mi-) long **Pacific Ring of Fire** (Fig. 4.19). As is the case for convergence between oceanic plates, the magmas produced above these subduction zones are mainly basaltic, and are formed by partial melting of either the downgoing oceanic slab or the overlying mantle wedge where melting is triggered by the addition of water and other gases from the subducted oceanic crust. Where the overriding plate is continental, however, these fluid-charged magmas must punch their way through the continental crust. The ascending fluid-charged magma is an efficient transporter of heat, and promotes partial melting of the overlying continental crust to produce granitic melts. These coexisting but compositionally contrasting magmas may then mix. Unlike basalt magmas which are very fluid and permit their dissolved gases to escape easily, continental crustal melts are more viscous or sticky because they are richer in silicon and poorer in iron and magnesium. They are generally andesitic or rhyolitic in composition (see Chapter 2), and are charged with liquefied gases under tremendous pressure. As the pods of melt rise to fill near-surface magma chambers, these gases expand, fracturing the overlying rock and giving vent to explosive volcanic eruptions. As we learned in Chapter 2, magma chambers can be likened to pressure cookers with lids of brittle crust. As the pressure within the chamber builds, cracks appear in the crustal lid through which the overpressured magma surges upwards and escapes violently at the surface in the form of a pyroclastic eruption.

The devastating pyroclastic eruption of Mt. Pinatubo in the Philippines in June 1991 (Fig. 5.21), exemplifies the often explosive volcanism associated with ocean-continent collision. This eruption, the largest in over 80 years, blasted 5.5 cubic km (1.3 cubic mi) of ash and some 20 million tons of sulfur dioxide into the atmosphere in a matter of hours. By blocking out sunlight, this blanket of gas and dust cooled the planet by a degree or so for several years (Fig. 1.6).

The eruption of Mount St. Helens in Washington State in May 1980 (Fig. 5.22), was significantly smaller than that of Mt. Pinatubo, ejecting a mere 1.0 cubic km (0.2 cubic mi) of material into the atmosphere. Nevertheless, it tore almost 400 m (1300 ft) from the summit of the volcano, and devastated an area of 200 square km (75 square mi).

Figure 5.21
The major eruption of Mt. Pinatubo on June 15, 1991, as seen from Clarke Air Force Base in the Philippines.

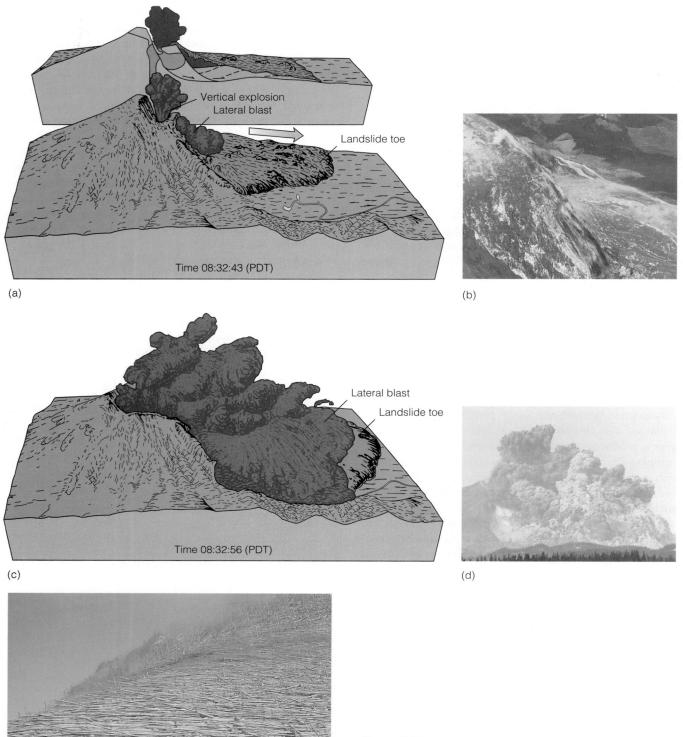

(a)

Vertical explosion
Lateral blast

Landslide toe

Time 08:32:43 (PDT)

(b)

Lateral blast

Landslide toe

Time 08:32:56 (PDT)

(c)

(d)

(e)

Figure 5.22

The climactic eruption of Mount St. Helens in the Cascades on May 18, 1980, which blew out the side of the mountain following a massive landslide. (a), (b) Following an earthquake generated by rising magma, the volcano's unstable north face collapsed in a massive landslide. (c), (d) The landslide, in turn, released the pressure on the magma within the volcano, which exploded in a violent sideways or lateral blast of pyroclastic ash. (e) Resembling flattened straw, an entire forest of trees is felled by the blast.

Both of these devastating eruptions are examples of subduction zone volcanism and stand in sharp contrast to the generally quiescent volcanism of mid-ocean ridges. Mid-ocean ridges do not form such serious "pressure cookers" because they lack a thick crustal lid and the basalt magmas they produce are both less viscous and less gas-charged than the magmas of subduction zones. Mid-ocean ridges are also highly fractured areas that facilitate the ascent of magma rather than allowing the build up of pressure. Mid-ocean ridge volcanism is exemplified by the lava fountains and fast-flowing lava flows witnessed on the island of Iceland, which sits astride the Mid-Atlantic Ridge.

As we shall learn in Chapter 6, the explosive volcanism characteristic of ocean-continent convergence represents only a small fraction of the magmatism associated with subduction. Most magmas never reach the surface, but instead cool to form vast bodies of plutonic rock within the overriding continental plate. This influx of hot buoyant magma together with the compression that often accompanies plate convergence, uplifts the leading edge of the continental plate into a range of mountains parallel to the offshore trench. As a result, the volcanoes associated with ocean-continent collisions occur within coastal mountain ranges like those of the Andes or Cascades.

Volcanic mountains rather than island arcs are therefore typical of ocean-continent convergence. At the same time, volcanism tends to be explosive and is generally of a more silica-rich andesitic or rhyolitic composition (see Chapter 2) rather than basaltic. Each of these characteristics reflect the presence of a continental plate above the subduction zone.

Controls on Subduction Style: Marianas versus Andean

Subduction at continental margins does not always cause crustal compression. The Andes, for example, preserve a more complex history. Some 150 million years ago, back-arc spreading occurred in the southern Andes, opening up a back-arc basin that later closed. This shows that the behavior of a subduction zone can change with time. Indeed, we sometimes talk of a continuum between two end-member types of subduction: Marianas and Andean (Fig. 5.23). One end-member (Marianas-type) is related to the more-or-less passive subduction of old, dense oceanic crust, which typically occurs in oceanic settings. The other (Andean-type) is related to the forceful subduction of oceanic crust beneath continental plate margins.

During **Marianas-type subduction,** named after the Marianas Islands of the western Pacific, a very old and heavy oceanic plate willingly dives beneath younger, more buoyant ocean floor so that the process of subduction is fairly passive. The two plates consequently meet with less force so that major earthquakes are few, roll-back of the sinking plate is common so that back-arc basins form, and the angle of subduction is steep. In fact, part of the heavy downgoing slab may even become detached and sink into the mantle. (Fig. 5.23a).

This situation clearly contrasts with the situation in the modern Andes where there is no back-arc spreading and mountains have developed instead. Indeed, during **Andean-type subduction,** named after the modern Andes, young buoyant ocean floor is "forced" to subduct. The plates consequently meet with great force so that earthquakes are frequent and strong (releasing a hundred times more energy than Marianas-type systems), the angle of subduction is shallow, and compression rather than roll-back occurs, creating a mountain belt parallel to the oceanic trench (Fig. 5.23).

Although the Andes today are the type example of Andean-type subduction, evidence in the Andes for back-arc spreading (a feature of Marianas-type subduction) some 150 million years ago, demonstrates that this was not always the case. This shows that the two types of subduction are not mutually exclusive but may change from one to the other.

Modern Japan, for example, is now part of an Andean-type system as its recent history of destructive earthquakes testifies. Prior to 20 million years ago, however, Japan was part of a Marianas-type system with active back-arc spreading forming the floor of the Japan Sea. Indeed, Japan was once attached to the Russian coast of Asia, but was spalled off by back-arc extension and drifted oceanward as the Japan Sea opened.

One cause for this change in the character of subduction with time is a progressive decrease in the age of the subducting ocean floor. When subduction first starts, ancient oceanic crust is consumed which collapses into the mantle and rolls back, giving rise to Marianas-type subduction. As subduction continues, however, the subducting ocean floor becomes younger and more buoyant. It must therefore be forced to subduct, causing compression, which closes the back-arc basin and gives rise to Andean-type subduction. A progressive decrease in the dip of the subduction zone may also accompany this change because Marianas-type subduction is steep whereas that of Andean-type systems is typically shallow. Where this occurs, the site of magma generation also changes with time, adding to the complexity of the subduction zone.

Subduction and the Growth of Continents

A progressive evolution from Marianas- to Andean-type subduction may be expected to accompany the closure of an ocean, because the age of the subducting ocean floor should decrease as the mid-ocean ridge approaches the subduction zone. As a result, ocean-continent subduction along the margins of a closing ocean is likely to become increasingly Andean-type with time. The magmas generated by this subduction will be added to the continent above the subduction zone. Similarly, island arcs, initially generated by Marianas-type subduction within the ocean basin, will continue to grow as the ocean shrinks, but must ultimately collide with a continental margin as the ocean closes. In this way, subduction must ultimately contribute to the growth of continents.

In fact, subduction of oceanic crust is nature's vast recycling program. As dense oceanic crust is destroyed by subduction, its destruction triggers melting and the ascent of more buoyant magmas which cool to form new continental crust in islands arcs and volcanic mountain belts. Thus, the granitic crust of continents is produced above subduction zones as a

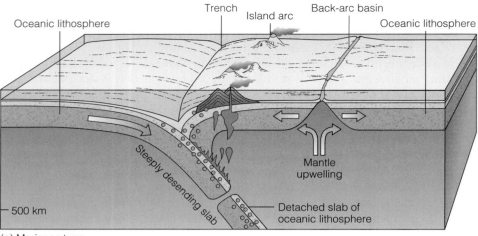

Oceanic lithosphere

Trench Island arc Back-arc basin
Oceanic lithosphere

Steeply desending slab

Mantle upwelling

— 500 km

Detached slab of
oceanic lithosphere

(a) Marianas type

 Earthquake foci

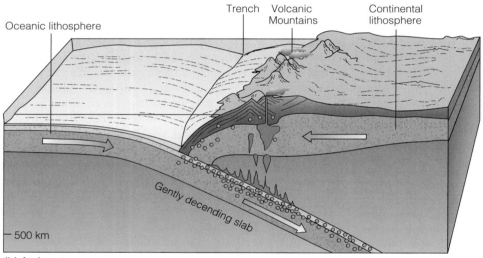

Trench Volcanic
Mountains

Continental
lithosphere

Oceanic lithosphere

Gently decending slab

— 500 km

(b) Andean type

Figure 5.23

Contrasting styles of subduction. (a) Marianas-type subduction produced by the collapse of old
oceanic lithosphere is characterized by back-arc spreading, a steeply-dipping subduction zone,
and few major earthquakes. (b) Andean-type subduction produced by the forceful consumption
of young oceanic lithosphere is characterized by a shallow-dipping subduction zone and frequent
severe earthquakes. The combination of compression and magmatism forms mountains.

consequence of the destruction of ocean floor. Volcanic island
arcs are where continents first start to form, slowly growing from
"microcontinents" to full-size continents as subduction proceeds with time. Because of the buoyancy and elevation of their
crust, island arcs are generally not subducted, but tend instead
to coalesce into larger bodies through "microcontinental" collisions, or become welded to the leading edge of larger continents when they are swept into ocean-continent collision
zones. Because ocean-continent subduction zones where these
processes take place can only form at continental margins, sub-
duction-zone magmas and colliding volcanic island arcs are

typically added to the edge of continents. As a result, continents
tend to grow sideways with time such that they often have a
nucleus of ancient rocks surrounded by progressively younger
ones (see Figure 6.25). We will return to the dramatic story of
collision and its role in the growth of continents when we
examine mountain building in more detail in the next chapter

The previous sections serve to emphasize the wide range of
processes associated with the subduction of oceanic lithosphere
at convergent plate boundaries. Subduction may occur
beneath oceanic crust, in which case the product is a volcanic
island arc, or it may be consumed below continental crust in

which case volcanic mountains are formed. Subduction may be of Marianas- or Andean-type, or of a variety intermediate between these two. Subduction causes magmatism which may be largely plutonic or largely volcanic, and volcanism may be basaltic and quiescent, or intermediate to felsic and highly explosive. Subduction also leads to collision and, as we shall learn in Chapter 6, this may involve "microcontinents" and affect quite small regions, or may be climactic and involve entire continents. No other type of plate boundary exhibits such diversity and, as we shall now see, the complexity of convergent margins contrasts with the comparative simplicity of the third and last type of plate boundary, that of transform faults.

5.4.3 Transform Boundaries— Fracturing the Land

Where two plates struggle to slide past each other, the boundary between them is a fracture in the crust known as a **transform fault**. Plate motion here is one of horizontal translation and crust is neither created nor destroyed. The plate boundary is consequently a "conservative" one, the transform fault simply linking two other kinds of active plate boundaries. At either end of a transform fault, therefore, the movement on the fault abruptly ends and is taken up by (or is "transformed" into) movement of another kind, such as spreading, subduction, or transform motion on another fault (Fig. 5.24). However, the fault itself is rarely smooth. If the rocks along a transform fault lock while the pressure or "stress" from plate motion continues, "strain" builds up in the rocks on either side of the fault so

that they slowly bend or deform, storing up elastic energy like a spring ever drawn tighter. Finally, when the spring snaps, rocks on either side of the fault jerk violently past each other and earthquake shock waves are sent out in all directions. Such **stick-slip** motion is typical of transform faults and accounts for the earthquakes with which they are often associated. As the fault movement occurs close to the surface, usually at depths of less than 20 km (12 mi), earthquakes associated with transform plate boundaries are often more damaging than deeper ones of similar magnitude.

Transform plate boundaries are of two principle types. The vast majority of transform faults link offset mid-ocean ridges, or provide links between other plate boundaries in an entirely oceanic setting. Such transform faults occur on the sea floor and are termed **oceanic transforms**. Far less common are transform faults that separate two continental plates. These are termed **continental transforms** and have a far greater impact on society because they occur on land.

Oceanic Transforms

Transform faults are extremely common in oceanic settings where they most frequently offset mid-ocean ridges perpendicular to their ridge crests, to produce the rectilinear pattern of ridges and transforms characteristic of spreading centers (Fig. 5.25). The pattern is rectilinear because divergent plate boundaries are usually perpendicular to the spreading direction whereas conservative or transform plate boundaries are roughly parallel to plate movement. Because of this relationship, transform faults tend to intersect mid-ocean ridges at high angles.

At first glance, it would appear as if the mid-ocean ridges were once continuous and had been offset by horizontal movements along the transform faults. But the direction of offset along oceanic transforms is precisely the opposite to that which would be needed to produce the ridge offsets. In Figure 5.26, for example, the ridge is offset to the right although the transform fault linking the two ridge segments moves the far side of the fault to the left. This is because the direction of movement on either side of the fault is determined by the relative plate motion produced by sea-floor spreading at the ridge crests. In fact, it is only between the offset ridge crests that the sea floor is moving in opposite directions, so it is only this segment of the transform fault that is active. Seismic activity on oceanic transforms, which is generally both frequent and shallow, is similarly restricted to the segment of the fault between two offset ridge crests. Beyond the offset ridge crests, the sea floor moves in the same direction on either side of the transform, so the fracture is merely a seismically inactive scar.

The origin of the transform offsets in mid-ocean ridges is not fully understood, but it appears to be the result of physical constraints that inhibit divergence along a curved plate boundary (Fig. 5.27). For example, if two oceanic plates begin to diverge along a curved boundary, the original curves are forced to readjust into a series of right-angle segments. The offsets may also be inherited from the initial continental rifting stage of ocean opening. Where pre-existing lines of weakness in the

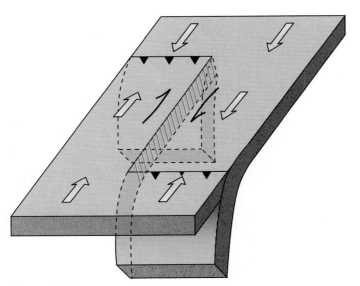

Figure 5.24
Transform plate boundary (lined) linking two subduction zones. Note that motion on the transform fault abruptly ends at either end of the fault and is taken up by, or "transformed" into, convergent plate motion. The transform fault consequently transfers the convergent motion from one subduction zone to the other.

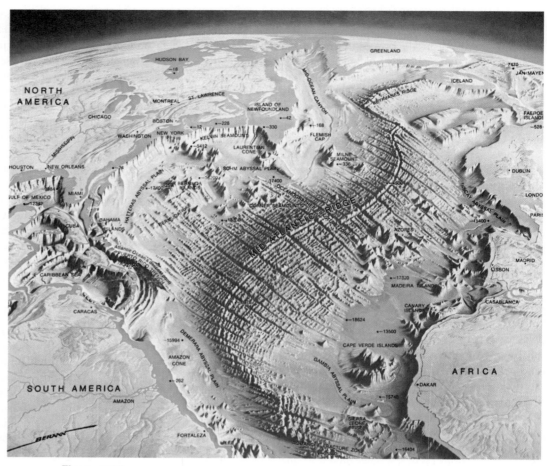

Figure 5.25
Oceanic transform faults offset the Mid-Atlantic Ridge at high angles along a portion of the Atlantic Ocean. Depths are in feet.

Figure 5.26
Transform fault and fracture zones along a mid-ocean ridge.

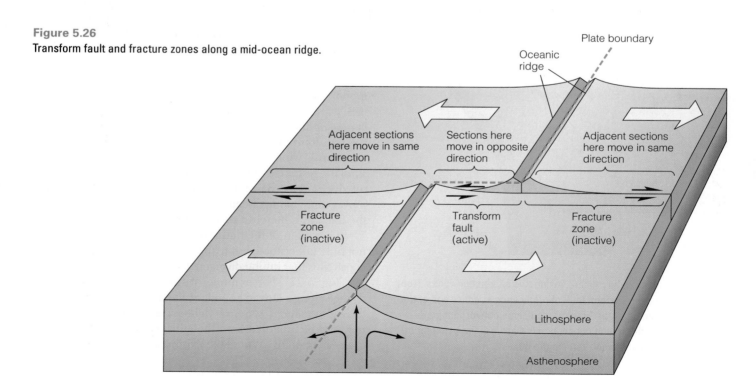

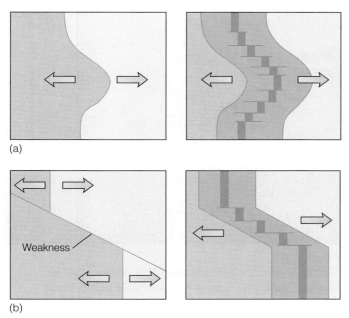

Figure 5.27
Origin of transform offsets in mid-ocean ridges. (a) From a curved plate boundary between oceanic plates. (b) From a pre-existing line of weakness between continental plates.

continent produced rifts at an angle to the direction of spreading, physical constraints will again cause the developing divergent plate boundary to readjust into a series of ridge segments offset by transform faults.

Other oceanic transform faults link mid-ocean ridges to subduction zones or link one subduction zone to another (Fig. 5.24). In so doing, transform faults provide the means by which oceanic crust is transported from the site of its creation to its destruction. In the northeastern Pacific, for example, it is the Mendocino transform fault that permits oceanic crust of the Juan de Fuca plate to be transported to the Cascade subduction zone beneath the Pacific Northwest, whereas the Queen Charlotte transform fault off the coast of British Columbia transports ocean floor from the Juan de Fuca Ridge toward the Aleutian Trench beneath Alaska (Fig. 5.28).

Continental Transforms

Only a few transform faults intersect continents. However, they tend to be longer and more continuous than their oceanic counterparts and do not display the simple rectilinear geometry of oceanic transforms. As they cut through the differing materials of the continental crust, continental transform faults continually exploit any weakness they encounter and so may bend and alter their path. As we shall learn in Chapter 6, these bends produce local areas of compression or extension that cause either uplift or small "pull-apart" basins to open along the fault zone.

The best known continental transform faults are the Dead Sea Transform of the Levant, where the two plates (African and Arabian) move in the same direction but at differing speeds, and the San Andreas Fault of California, where the adjacent plates (North American and Pacific) move in opposite

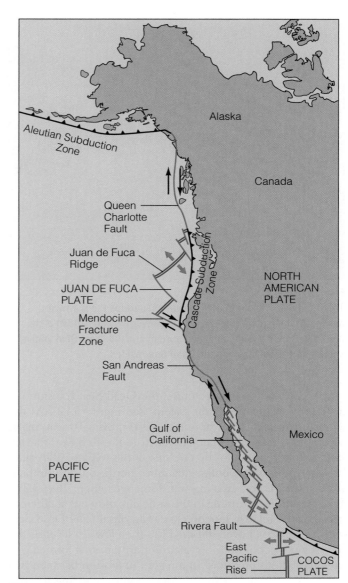

Figure 5.28
Plate boundaries along the west coast of North America, showing the Queen Charlotte, Mendocino, and San Andreas transform faults, the Juan de Fuca and Gulf of California ridges, and the Aleutian and Cascade trenches.

directions. These transform faults serve as links between other plate boundaries. The Dead Sea Transform (Fig. 5.29) links the spreading ridge in the Red Sea to part of the convergent zone of the Taurus-Zagros Mountains in Turkey (Fig. 5.12), whereas the San Andreas fault system links the spreading center in the Gulf of California to the subduction zone beneath the Cascade Mountains (Fig. 5.28). Continental transform faults are said to display either **right-lateral** motion or **left-lateral** motion depending on whether the *far* side of the fault (as viewed by an observer on the ground) moves to the right or left. As we shall see, the Dead Sea Transform is left-lateral, whereas the San Andreas Fault is right-lateral. In either case, however, the resulting stick-slip motion threatens the surrounding region with earthquakes.

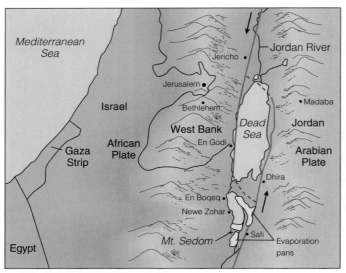

Figure 5.29
The Dead Sea Transform of the Levant separates the northward moving Arabian plate (right) from the slower moving African plate (left). Overlapping faults between Jordan and the West Bank outline a pull-apart basin in which the Dead Sea resides.

According to plate tectonics, the Dead Sea Transform is an extension of Africa's rift system (see Figure 5.12). As the Arabian plate split away from the slower moving African plate, separating along a line perpendicular to the Red Sea, translational motion began to slip one side of the Jordan Valley sideways past the other. Both sides move north, but the Arabian plate moves faster, creating transform movement (Fig 5.29). Like two athletes running in the same direction but at different speeds, their relative motion creates an offset. The Dead Sea, with its surface at over 400 m (1320 ft) below sea level, is the lowest land-locked body of water on Earth, having formed in a pull-apart basin produced in a region of tension between two overlapping strands of the transform fault separating the two plates. Relative motion between the Arabian and African plates began 15 million years ago and the total offset between them has now reached almost 110 km (70 mi).

The San Andreas Fault of California forms a 1600 km (1000 mi) long scar most clearly visible in the Carrizo Plain (Fig. 5.30), northwest of Los Angeles, where offset streams testify to active transform movement. Half of the over 20 m (66 ft) offsets on these streams occurred during the 1857 Fort Tejon earthquake. In April 1906, up to 6.5 m (21 ft) of abrupt movement on the San Francisco segment of the San Andreas Fault produced one of the worst earthquakes in U.S. history. At magnitude 8.3 on the Richter scale, the San Francisco earthquake rippled the ground in waves 1 m (3.3 ft) high and 20 m (66 ft) from crest to crest, tilting and destroying building at least 30 km (19 mi) from the fault. However, most of the damage, and most of the 700 or so lives that were lost, occurred as a result of fires ignited by overturned kerosene lamps and wood stoves. Within a few days, the city of San Francisco sustained over half a billion dollars in damage (equivalent to nearly 10 billion in today's dollars) and almost 3000 acres of the city center lay in ruins.

Figure 5.30
The San Andreas Fault on the Carizzo Plain, northwest of Los Angeles, separates the North American (left) and Pacific (right) plates.

The San Andreas Fault separates the North American and Pacific plates and, on average, moves at 3–4 cm (1–1.5 in) per year. Volcanic rocks of the Miocene Epoch (see Chapter 3) at Pinnacles National Monument, southwest of Hollister (Fig. 5.31), continue on the other side of the fault near Fort Tejon; an offset of 300 km (186 mi) in the 24 million years since they erupted. In another 30 million years, Los Angeles on the Pacific plate will have moved north of San Francisco on the North American plate.

However, segments of the fault are not moving at uniform rates. Some are locked whereas others move slowly but continuously by the process known as **fault creep.** Those with rates of fault creep of over 2 cm (1 in) per year are unlikely to experience severe earthquakes because stresses are relieved as fast as they build up. However, other segments, like the one that runs through San Francisco, are locked and therefore are building up strain that must one day be released in a major earthquake.

Because locks on the fault last 125 years on average, the segment in greatest danger is that north and east of Los Angeles where such a jolt is now overdue since the Fort Tejon earthquake occurred well over 125 years ago. Although many earthquake epicenters lie along the fault itself, numerous earthquakes on the California-Nevada border (Fig. 5.31) show that not all the plate motion is absorbed by the San Andreas system. The plate boundary is really a broad fragmented zone rather than a single major fault.

Although both the Dead Sea and San Andreas transform faults are active plate boundaries, ancient inactive transform faults may still scar mountain belts that developed along ancient plate boundaries. Loch Ness on Scotland's Great Glen Fault is home not only to the fabled Loch Ness Monster but lies over part of an ancient transform fault that, in vivid testament to Wegener's theory of continental drift, was torn in two with the opening of the Atlantic Ocean (Fig. 5.32) . Some 310 million years ago, when Europe and North America were part of the same continent, a transform fault cut across the future islands of Newfoundland, Ireland and Great Britain. When

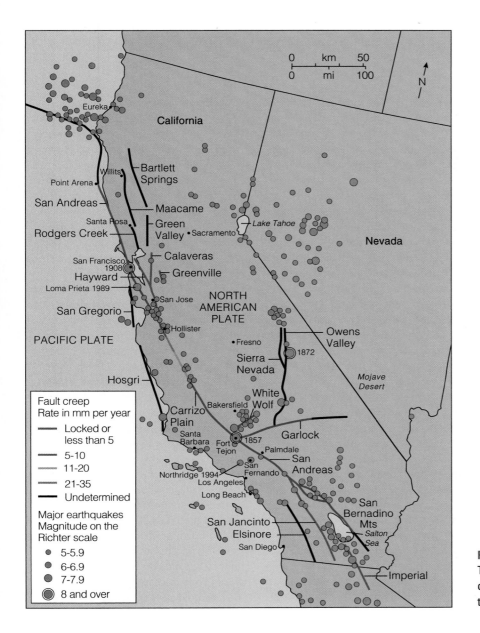

Figure 5.31

The San Andreas Fault of California showing rates of fault creep and the distribution of major (magnitude greater than 5.0) earthquake epicenters.

the Atlantic Ocean opened about 200 million years ago, the old fault broke apart so that the continuation of the Great Glen Fault now lies more than 4800 km (3000 mi) away in the Cabot Fault of Newfoundland.

Similarly, the Minas fault system of the Canadian Appalachians slices through the center of Nova Scotia (Fig. 5.34) and represents a right-lateral transform fault that was active

during the destruction of the Iapetus Ocean and the assembly of Pangea in the Paleozoic.

These ancient scars never quite heal and remain as fundamental zones of weakness in the continental crust. In the future, when the Atlantic Ocean starts to close and subduction begins along its margins, these wounds will be among the first to reopen. In fact, modern stresses along the eastern seaboard

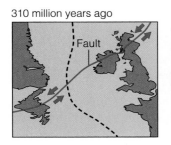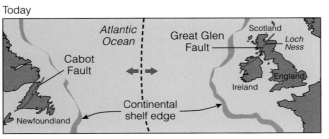

Figure 5.32

The Great Glen Fault of Scotland and the Cabot Fault of Newfoundland represent an ancient transform plate boundary active 310 million years ago but severed when the Atlantic Ocean opened 200 million years ago.

In 1868, after a major earthquake shook the San Francisco region, the San Francisco Daily Examiner proclaimed that such quakes were too rare to be worried about. More than 100 years later, plate tectonics has provided the conceptual framework to understand earthquakes, and we now realize that such earthquakes are inevitable. Today the threat of serious earthquakes is an accepted fact of life that more than 30 million Californians live with, and billions of dollars are currently being spent on renovating or "retrofitting" existing buildings, bridges, and roads to try to out-muscle the next one.

Unfortunately, this heightened knowledge has been acquired by tragic experience. In 1906, more than 700 people perished when San Francisco was reduced to a pile of ashes by raging fires that followed the earthquake that is estimated to have had a magnitude of 7.8 (8.3 on the now disused Richter Scale) when a segment of the fault lurched 6.5 m (21 ft) northwards in just a few seconds. Modern estimates of an earthquake's magnitude are based on the total energy released by the quake, which is calculated, in part, by multiplying the area of the faults rupture surface by the distance the Earth moves along the rupture. In 1971 an earthquake with a magnitude of 6.6 centered on the San Fernando Valley claimed 65 lives. The two most recent major earthquakes, centered in Loma Prieta, about 110 km (70 mi) south of San Francisco (October, 1989), and Northridge, less than 40 km (25 mi) northwest of Los Angeles (January, 1994), demonstrate how the complexity of earthquake patterns relate to the variations in the stress levels along the fault itself, and how this renders earthquake prediction more difficult than had previously been anticipated.

The Loma Prieta earthquake, with an intensity of 7.1, killed 63 people and caused extensive damage, estimated at six billion dollars in the San Francisco area. It was related to a rupture along the San Andreas Fault, some 10 km (6 mi) below the surface. A plot of seismic activity over the previous 20 years along the San Andreas Fault from San Francisco in the north to Parkfield in the south (Fig. 5.33) revealed that Loma Prieta was located above one of three major "seismic gaps", the others being in the Parkfield and San Francisco-Portola Mountains region. Seismic gaps are marked by few earthquakes, which means that these areas are locked segments of the fault and therefore are storing, rather than releasing the energy associated with the fault motion. Loma Prieta snapped because the stored stress finally exceeded a critical threshold.

Figure 5.33 shows that the earthquake and its aftershocks were preferentially concentrated in the Loma Prieta seismic gap. An implication of this analysis is that the other seismic gaps in the San Francisco and Parkfield areas remain at high risk. It is estimated that an earthquake between magnitudes 7.5 and 8 would be required to alleviate this built-up strain.

Whereas the Loma Prieta earthquake, tragic as it was, was considered typical of continental transform faults, the Northridge earthquake awakened our senses to their enigmatic character. The fault which ruptured 18 km (11.1 mi) beneath Northridge that January morning was a "blind" fault, or one that is not exposed at the surface, and therefore the pattern of stress buildup which led to its rupture was undetectable. The earthquake, with a magnitude of 6.7, claimed 61 lives and resulted in damage estimated at 20 billion dollars, a figure second only to Hurricane Andrew in 1992.

The geologic record clearly indicates that blind faults, such as the one that ruptured beneath Northridge, are a common feature where the curvature of the fault jams together blocks of crust on opposing sides of the fault, and so causes compression. However, because the break between the two pieces is not exposed at the surface, these faults are notoriously difficult to monitor. The only signs of their existence are the mountain ranges that represent the squeezed out portions of the crust which rise above them (see Chapter 6).

Can Earthquakes be Predicted?

When it comes to predicting earthquakes, the underlying problem is that in a matter of seconds, devastating earthquakes can release the stored energy along a fault that may have taken millions of years to accumulate. So while earthquake prone areas can be readily identified from seismic risk maps, predicting exactly when an earthquake might occur is a very difficult task. Typically, seismologists can estimate a probability that a fault will have a serious earthquake within a certain time-frame. For example, it is estimated that the Hayward Fault, which is part of the San Andreas fault system, has a 30% probability of generating an earthquake with magnitude greater than 7.0 in the next 30 years. Although this would encourage local residents to take precautions such as storing emergency supplies and learning earthquake drills, its terms are too broad to be useful in averting an instantaneous catastrophe.

The main thrust of earthquake predictions is to identify "precursors", or measurable features whose patterns change immediately prior to the main quake. As with volcanic predictions, Global Positioning Systems using a network of satellites and receiving stations can measure movements of the crust on a cm scale (less than 1/2 in). Thus, accelerated or retarded motions can be measured along the length of the fault and unusual activity, such as microseismic events, can be identified.

Similarly, rocks undergo an increase in volume immediately before rupturing, as a result of the creation of microcracks, and the consequent influx of water to fill these cracks. These tendencies can be measured with sensitive geophysical instruments. For example, since water has a high electrical resistance, its influx can result in increases in subsurface resistivity. Similarly, movement of water into the microcracks results in changes in local well-water levels. In some instances, agitated animal behavior has been noted prior to earthquakes, suggesting that they are more sensitive to microseismic

SAN FRANCISCO
BAY AREA EARTHQUAKES

Figure 5.33

Epicenters on the San Andreas fault system in the San Francisco Bay area. Main shock of the 1989 Loma Prieta earthquake shown by large white symbol; others record some of the 6000 aftershocks. Yellow symbols show distribution of earthquake epicenters (magnitude greater than 2.0) for a twenty-year period prior to 1989. Note how the white symbols fill a seismic gap revealed by the absence of yellow symbols in the vicinity of Loma Prieta.

precursors than humans. All agree that much more research is needed, and given the potential for an unimaginable disaster, the stakes are very high indeed.

In California, building codes require that a structure should be able to withstand a 25 second main shock. The main support structures of a building (its foundation, floors, walls, and roof) are all tied together to enhance resistance to horizontal and vertical shaking. The nature of the subsurface is also critical. For example, the 1985 earthquake that struck Mexico City and caused 9500 fatalities had its epicenter more than 200 km (125 mi) away. So why was this earthquake so devastating? The reason is that all materials have a natural resonance. That is, they are capable of vibrating at a fixed frequency. Like the glass shattered by the high notes of an opera singer, resonance is energy transferred between bodies with the same natural frequency. In Mexico City, the buildings and the subsurface soil had the same natural frequency as the shock wave from the earthquake. Just as a child's swing has to be pushed at the right time for the efficient transfer of energy, the earthquakes shock waves were very efficiently transferred to the buildings with devastating consequences.

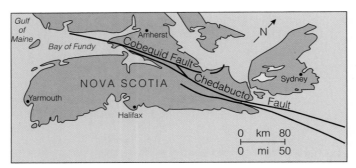

Figure 5.34
Like the Great Glen Fault, the Minas fault system of Nova Scotia, made up of the Cobequid and Chedabucto faults, represents an ancient transform plate boundary active during the assembly of Pangea in the Late Paleozoic.

5.4.4 Hotspots

Although plate tectonic theory explains many of the world's major crustal features, like mid-ocean ridges and mountain ranges, others are not fully accounted for. **Hotspots,** the existence of which was first proposed by the eminent Canadian geophysicist, J. Tuzo Wilson, are one of the more puzzling features of plate tectonic theory and their origin is uncertain. As their name suggests, hotspots are relatively small areas of higher than average heat flow associated with volcanoes. Although hotspots are essentially fixed point-sources of heat, over time they often create a line of extinct volcanic features that ends in an area of active volcanism. In the oceanic realm, hotspots form volcanic islands which slowly subside on becoming extinct. As a result, the line of extinct volcanic features first develop fringing reefs, and then become atolls. Finally, they become submerged seamounts as they get older (Fig. 5.35).

Although some hotspots lie on plate boundaries, most occur in the interior region of plates. Therefore, unlike most volcanic activity, hotspots cannot be directly tied to processes occurring at plate boundaries. Instead, they are widely attributed to giant plumes of heat welling up from the deep mantle. These **mantle plumes,** the existence of which was first proposed by Jason Morgan at Princeton University, push against the lithosphere, doming it like a blister. So, in contrast to plate boundary volcanism, the source of hotspot volcanism apparently lies deep in the mantle beneath the realm of plate tectonics.

Wherever hotspots burns through plates, volcanoes erupt. But the volcano is built on a moving plate whereas the plume beneath it is fixed. Just as a sewing machine generates a line of stitches as the cloth moves past the needle, so a plume repeatedly punctures the crust as the plate moves past. In this way, a line of volcanoes is produced as the moving plate carries each hotspot-built mountain away while a new volcano is built over the plume. Where the plate is oceanic, the line of volcanoes forms a volcanic island chain, which eventually becomes a

of North America, although far smaller than those of active plate margins, generate hundreds of microseismic events each year, many of which are sited on the ancient transform faults of the Appalachian mountains.

In summary, transform plate boundaries are major fractures in the lithosphere along which one plate slides by another. Most transform faults link plate boundaries on the ocean floor so that the earthquakes produced by their jarring motion pose little threat to human populations. But some intersect continents and generate earthquakes that pose a serious threat to society. The fractures themselves continue to be lines of continental weakness long after all motion on the fault has ceased.

Although earthquakes and volcanoes are two of the hallmarks of plate boundaries, active volcanism also occurs in certain isolated areas that lie far from any plate boundary. Termed "hotspots", these puzzling features are not readily accounted for by plate tectonic theory, yet they are thought to play a fundamental role in the breakup of continents and their existence has been used to document plate motions. But how can such features be both related and unrelated to plate tectonics? In the next section we explore this curious relationship by examining these hotspots and their possible origin.

Figure 5.35
Origin of linear chains of volcanic islands and seamounts produced by movement of an oceanic plate over a stationary hotspot. Note progression from an active volcanic island, to an inactive volcanic island with a fringing reef, to an atoll, and finally a seamount, with increasing age and distance from the hotspot. Age of the islands increases to the left.

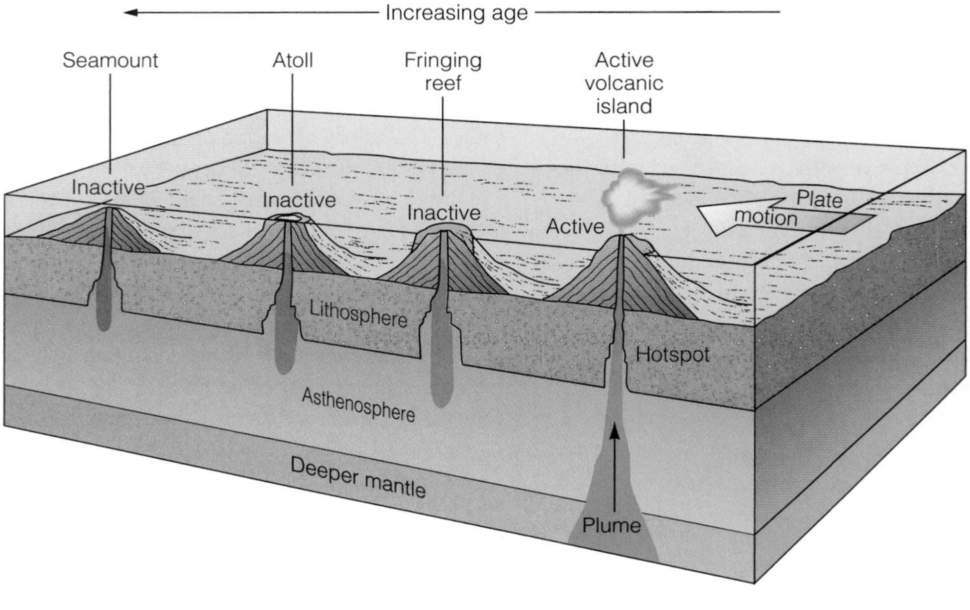

seamount chain as the ocean floor cools and subsides and the volcano is eroded by the sea (Fig. 5.35).

Although this model explains the pattern of volcanism, the origin of hotspots remains a mystery because they often lie far from plate boundaries where most volcanoes are located. Given that they remain in relatively fixed positions over very long periods of time, many geologists think that hotspots must arise from a very deep source. The most commonly cited explanation is one of deep-seated convection in which columns of heat rise from a thermally unstable layer thought to exist at the core-mantle boundary. At this boundary, the solid lower mantle is in contact with the liquid outer core. What initially creates these rising columns is very speculative and some have even suggested that they might be deep expressions of extraterrestrial impacts. Other geologists, however, have proposed much shallower sources immediately beneath the plates, created by processes that include the development of crustal rifts, friction between the plate and the mantle, and the local accumulation of mantle heat produced by the thermal blanketing effect of continental crust. Thermal blanketing occurs because continent crust does not dissipate heat as effectively as oceanic crust. So large continents are thought to trap mantle heat. The number of hotspots is also uncertain with estimates ranging from 20 to 120 depending on such criteria as the level of activity, lava chemistry, and the degree of doming. Figure 5.36, which excludes the polar regions, shows the distribution of some of the better known hotspots.

Two types of hotspots, oceanic and continental, are distinguished on the basis of the type of crust on which they occur. While the origin of the two types may be the same, there are major differences in the nature of their volcanism.

Oceanic Hotspots

Best known of all the world's hotspots is Hawaii. The Hawaiian-Emperor seamount chain (Fig. 5.37) marks the passage of the Pacific plate over a plume for the past 70 million years. The oldest seamounts, now eroded by time, stand at the line's northern end where the Pacific plate dives beneath the Aleutian Trench. Although Hawaii may be the best-known, other hotspots pepper the Pacific. Bora Bora, and the Society Islands, for example, are the product of a younger hotspot than Hawaii but record the typical sequence from **volcanic island** to **fringing reef** to **atoll** to **seamount**, produced as each volcano cools, subsides, and is beveled by erosion as it moves off the hotspot (see Figure 5.35). Easter Island and Darwin's Galapagos Islands in the eastern Pacific have formed over other plumes. In the Atlantic Ocean, the Azores and Canary Islands formed above plumes, although the latter has been inactive for 150 years. Likewise, the island of Reunion lies above a plume far from rifts and subduction zones in the Indian Ocean.

The progressive increase in the age of volcanic activity to the northwest of Hawaii is evident in the increasingly eroded

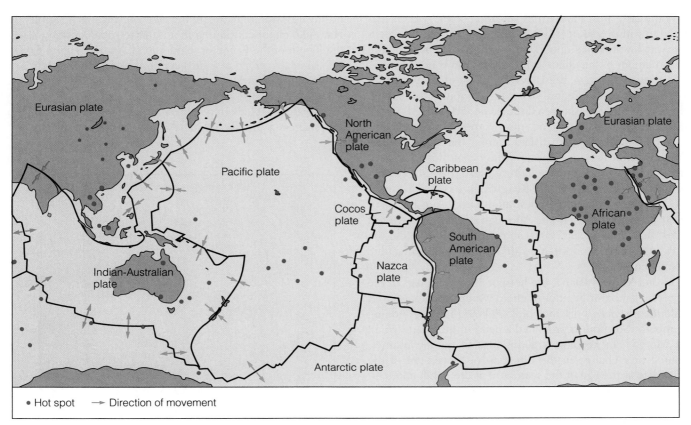

Figure 5.36
Global distribution of major hotspots on the Earth's surface thought to overlie rising mantle plumes.

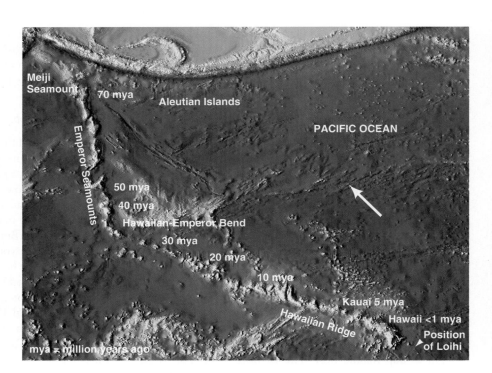

Figure 5.37

Movement of the Pacific plate over the Hawaiian hotspot has carried away volcano after volcano to form a hotspot trail composed of the Emperor Seamounts and Hawaiian Islands. A bend between the two reveals that the Pacific seafloor once followed a more northerly course. The oldest seamount, formed over the plume 70 million years ago, now stands ready to be subducted into the Aleutian Trench at the line's northern end.

and, hence, older appearance and lower elevation of the volcanoes in this direction. This apparent increase in age is also borne out by radiometric dating (Fig. 5.38). Present-day volcanic activity is observed on Mauna Loa and Kilauea which lie over the Hawaiian plume on the southeast side of Hawaii. Further southeast, the submarine volcano Loihi, which has reached to within 960 m (3170 ft) of the sea's surface, is inching its way towards islandhood, and will one day become the newest member of the chain as the islands move northwest.

Northwest of Kauai, the Hawaiian island chain continues as a dotted line of atolls, sand islets, and seamounts that stretch all the way to Midway Island, a distance of 3500 km (2170 mi). These unimpressive little islands now worn flat by the sea during their long journey are in fact among the world's tallest mountains, built flow by flow from the ocean floor 6 km (or nearly 4 mi) down to form a chain of more than fifty extinct volcanoes. On Hawaii, Mauna Kea, which has lain dormant for 3600 years, is the world's tallest mountain, rising 10,000 m (33,000 ft) from the sea-floor (Although higher, Mt. Everest rises less than 9000 m (29,700 ft) above sea level). It is also the world's largest single mountain, with a volume of 40,000 cubic km (9600 cubic mi).

Beyond Midway Island, the hotspot track continues in the form of the Emperor Seamount chain which reaches as far as the Aleutian trench, a distance of 2500 km (1550 mi). Adding the Emperor Seamounts produced a bend in the chain which Jason Morgan, in keeping with his idea of mantle plumes, attributed to a change in the direction of movement of the Pacific plate at the age of the seamount at the bend, about 40 million years ago. Because hotspots remain more or less stationary, seamount chains provide evidence of the direction and speed of the plate on which they sit, while past plate configurations can be reconstructed by sliding the plate back

towards the present hotspot along the seamount track. Hence the bend between the Hawaiian and Emperor chains suggested to Morgan that the presently northwest-moving Pacific plate had been moving directly north prior to 40 million years ago. This idea has since been borne out by dating, the progressive northwestward increase in the age of volcanism along the Hawaiian chain continuing in an uninterrupted fashion along the length of the Emperor seamount chain (see Figure 5.37).

Morgan's proposal is further confirmed by the existence of similar kinks of identical age in nearby hotspot trails (Fig. 5.39). This relationship is similar to the effect achieved by inserting several sticks of chalk into a sheet of cardboard, and moving the sheet across a blackboard. The chalk sticks trace out parallel

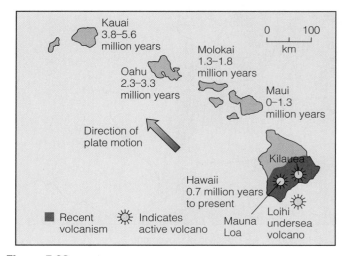

Figure 5.38

The Hawaiian Islands and the age of their volcanic rocks.

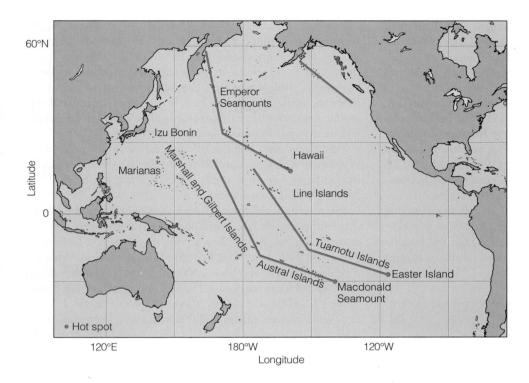

Figure 5.39
Kinked hotspot trajectories in the Pacific. The kink in each island chain occurred about 40 million years ago and is related to a change in the direction of motion of the Pacific plate.

lines that kink simultaneously when the direction of the sheet's movement is changed. In a neat geochronological fit, the kink in each hotspot trail occurs in the vicinity of a 40 million year old seamount. The timing of this change in movement coincides with a global re-organization of plate motion thought to have been brought about by the collision of India with Asia.

Continental Hotspots

While oceanic hotspots provide dramatic evidence of the movement of the Earth's lithospheric plates, continental hotspots may play an active role in determining the position of certain plate boundaries. Many scientists believe that hotspots and their under-

lying plumes drive the wedge that splits continents apart (Fig. 5.40), first doming and then cracking the continental crust into characteristic Y-shaped rifts which meet at a **triple point** like that of Ethiopia's Afar Triangle at the southern end of the Red Sea where the East African Rift meets the Gulf of Aden (Fig. 5.12). If this is the case, then continental hotspots play a fundamental role in the initiation and geometry of continental breakup and the opening of ocean basins. They may also be responsible for the development of many oceanic hotspots into which they must evolve if their activity persists after continental breakup.

Typically two of the rifts that meet at a triple point widen into oceans while the third may either widen as well or becomes inactive to form a special type of **failed rift** (see Dig

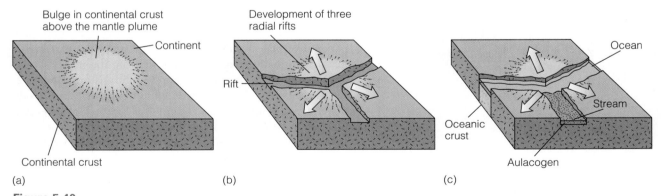

Figure 5.40
Continental breakup caused by a mantle plume. (a) A dome forms over a mantle plume rising beneath a continent. (b) Three radial rifts develop due to outward radial flow from the top of the mantle plume. (c) The continent separates into two pieces along two of the three rifts, with new ocean floor forming between the diverging continents. The third rift becomes an inactive failed rift (aulacogen) filled with continental sediment.

In: Failed Rifts, page 114) called an **aulacogen** (from the Greek *aulax*, a furrow) that will later channel major streams into the new ocean and slowly become filled with sediment. An excellent example of an aulacogen occurs at the elbow of Africa's west coast where the Benue Trough brings the River Niger to the Atlantic Ocean (Fig. 5.41). Following continental breakup, the hotspot beneath such a triple point will ultimately lie near a mid-ocean ridge, accounting for the large number of hotspots located at or close to the Mid-Atlantic Ridge and the mid-ocean ridges of the southern Indian Ocean (Fig. 5.36).

In contrast to hotspots like Hawaii, continental hotspots must burn their way through many kilometers of granitic continental crust. As a result, their record of activity suggests long volcanic silences and short cataclysmic explosions of pyroclastic material. The reason for this is that mantle-derived basaltic magma is sufficiently hot to melt the granitic continental crust during its ascent. Melting of continental crust produces magma with the same composition as granite. This gas-rich magma is less dense but more viscous than basalt, so that although it rises more rapidly, it retains its gases allowing pressures to build explosively. Bulging beyond capacity, the ground above the magma chamber finally cracks, releasing the pressure and triggering a catastrophic eruption. Only after this does the deeper basaltic magma flood out as lava flows (Fig. 5.42).

Such is the history of the Yellowstone hotspot in Wyoming, which is manifest at the Earth's surface today by hot springs, geysers, and other hydrothermal activity. Although we do not associate Yellowstone National Park with volcanic eruptions, huge explosions have rocked the region three times in the last 2 million years and are likely to do so again. Each has occurred on a fairly regular schedule of once every 700,000 years. The first and largest occurred about 2 million years ago, and blew 2500 cubic km (600 cubic mi) of ash skyward in an explosion 15,000 times greater than that of Mount St. Helens in 1980. Magma exploded again 1.3 million years ago and again 600,000 years ago, both explosions strewing 1000 cubic km (240 cubic mi) of ash across the western United States from Montana to Louisiana. In each case, collapse of the ground following the violent escape of huge volumes of ash and lava produced massive crater-like structures. Large, steep-walled and broadly circular structures produced by the collapse of volcanoes following an eruption are called **calderas.**

The size of the Yellowstone caldera (45 x 75 km or 28 x 47 mi) is so large that it was not discovered until detailed mapping was carried out in the 1950's and 60's. Forests and younger volcanic flows now obscure much of its outline so that it is very difficult to see from ground level. Beneath the caldera, earthquake waves have detected a column of heated rock extending more than 400 km (250 mi) down into the Earth's interior, supporting the presence of a mantle plume below the national park. As the North American continent moves southwest at about 4 cm (1.5 in) per year, a hotspot trace was formed along what is now the Snake River Plain, comprising flow upon flow of basaltic lava (Fig. 5.43).

So hotspots can be both related and unrelated to plate tectonics. Because they are stationary, their origin must be independent of plate tectonic processes, the most popular theory for their formation suggesting that they are the manifestation of mantle plumes that rise in narrow columns from the core-mantle boundary. Yet in providing the wedge that splits continents apart, hotspots may control the location of divergent plate boundaries. In so doing, they tend to become associated with mid-ocean ridges, and so show a relationship with plate tectonic processes.

Because they are stationary, hotspots also provide dramatic evidence of plate tectonics, recording not only the direction but also the speed of the plates through which they burn. But they do not tell us why the plates are moving. So what does cause plate movement? What forces are involved and to what do they owe their origin? To answer these questions, we now turn to the potential driving mechanisms for plate tectonics.

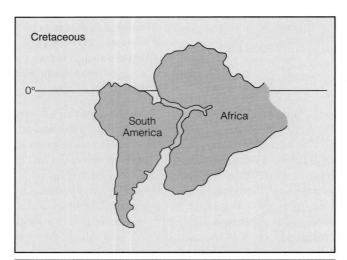

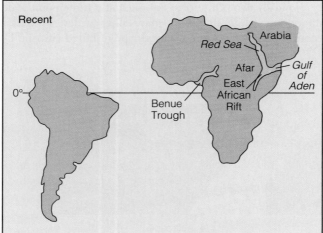

Figure 5.41
The Benue Trough is thought to be an aulacogen that formed as a result of continental rifting between Africa and South America during the Cretaceous period, about 140 million years ago. The East African Rift Valley may also be an aulacogen that formed when Africa and Arabia began separating 25 million years ago.

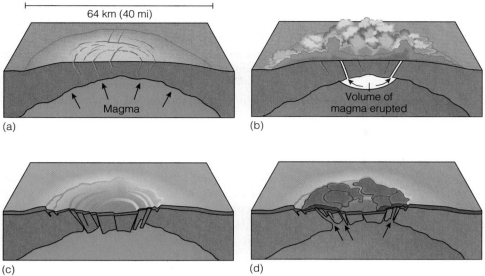

Figure 5.42

Stages in the development of the Yellowstone caldera 600,000 years ago. (a) Swelling magma chamber arches the overlying rocks into a broad dome, producing a series of arcuate fractures at the dome crest which propagate downwards towards the magma chamber. (b) Ring-shaped fractures tap the magma chamber, the uppermost part of which comprises rhyolite with large quantities of dissolved gases. With the sudden release of pressure, tremendous amounts of hot gases and molten rock are erupted, most of which move outward across the landscape as vast ash flows, rapidly covering thousands of square kilometers. (c) Crust above the magma chamber collapses several thousand feet along the ring-shaped fractures to form a caldera. (d) Later rise of basaltic magma through the ring-shaped fractures produces lava flows which spread across the caldera floor.

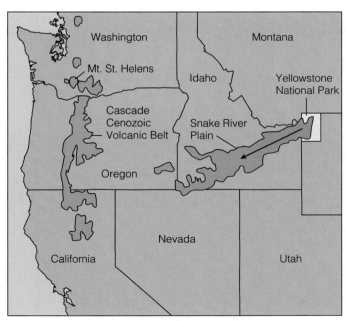

Figure 5.43

The track of the Yellowstone hotspot (arrow) across southern Idaho is revealed by the flood basalts of the Snake River Plain.

5.5 Plate-driving Mechanisms

Although plate tectonic theories account for many of the features of our lithosphere, considerable uncertainty exists about the specific forces responsible for plate motion. In a sense, the very same philosophical problems exist today as existed when Wegener proposed the idea that continents ploughed their way through the oceans. But our constraints are far better, and even with this uncertainty, the strength of plate tectonics lies in its ability to explain and relate a wide variety of geologic phenomena that were previously poorly understood; far more than even Wegener could have imagined. And while there may be many forces influencing plate motion, the main driving mechanism is clear. It is the flow of heat from the interior of the planet to its surface.

Basically, plate movement occurs as a result of the convective circulation of the Earth's heated interior and is powered by the decay of naturally occurring radioactive elements in the mantle. This convective circulation locally causes plates to move apart or spread, to come together or collide, or to slide past each other in translation. However, how convection actually makes the plates move is still poorly understood. Are they pulled from in front, pushed from behind, dragged from below, or is it some combination of these?

In the Pacific Ocean, the plates may be self-propelled. The Marianas-type subduction typical of the western Pacific, occurs because of the gravitational instability of old, dense oceanic crust. As the leading edge of the plate subducts, its weight pulls the rest of the plate towards the trench, just as a tablecloth will slide off a table when enough hangs over the edge. This force is called **slab pull** (Fig. 5.44) and occurs because the subducting lithosphere is denser and heavier than the asthenosphere through which it is sinking. But plate motion also occurs in the Atlantic where there is very little subduction. Yet it is clear that the opening of the Atlantic Ocean has resulted in the westward drift of North and South America, and the eastward drift of Europe and Africa. Here the plates may be partially propelled by sliding off the Mid-Atlantic Ridge. This gravitational force is called **ridge push,** the elevated position of the ridge allowing the plate to simply slide downhill like a toboggan. Deep-seated convective circulation may also drive the plates, the hot mantle rising beneath ridges or hotspots and spreading out laterally below the lithosphere, propelling the plates as it circulates. In this case the plates move like rafts in a moving stream. The force by which this motion of the asthenosphere may carry the plates along (or slow them down) is called **mantle drag.**

The most important of these forces appears to be slab pull because the fastest moving plates, like the Cocos and Nazca plates of the eastern Pacific (Fig. 5.1), are those with the greatest lengths of subducting edges in relation to their size pulling the plate along. On the other hand, because non-subducting plates like that of the North American plate also move, ridge push may also be important. The role of mantle drag is uncertain. If it was a major driving force, those plates with the largest surface areas over which the force could operate, should be the fastest moving. The

fact that they are not suggests that it is a drag force which opposes slab pull and ridge push, and slows the plates rather than carries them. The presence of continental crust also slows plates down. The deep roots of the continental lithosphere may exert a drag as they plough through the asthenosphere, much like a sea anchor slows the movement of a boat.

Clearly, there is much to be learned about the driving mechanisms behind plate motion. At a more fundamental level, however, we know it is some form of thermal convection in the mantle that is ultimately responsible for their movement. Thermal convection is a pattern of circulation in a fluid in which hot material rises and cools while cold material sinks and is heated. In the mantle, upwelling of hot mantle material to hotspots or ridges is coupled by rigid plate motion to the downwelling of cold oceanic lithosphere at subduction zones. In this sense, plate motion is simply a near-surface manifestation of mantle circulation.

5.6 Chapter Summary

With the recognition of sea-floor spreading at mid-ocean ridges and subduction below the deep ocean trenches came the revolution in Earth Sciences called plate tectonics. The theory of plate tectonics holds that the Earth's rigid outer shell or lithosphere is divided into huge, moving slabs or plates that ride upon a pliable layer of the Earth's interior known as the asthenosphere. The most intense geologic activity associated with plate movement occurs at their margins, where plates either separate, move together, or slide past each other. Hence, plate boundaries are most clearly outlined by the distribution of earthquakes generated by their jostling.

Plates incorporate both the crust and rigid portion of the upper mantle, and may carry both the granitic crust of continents

Figure 5.44
The forces that cause plates to move are thought to be created by gravity-driven slab-pull or ridge-push mechanisms. During slab-pull, the weight of the subducting slab pulls the rest of the plate down the subduction zone behind it. During ridge-push, the plate slides down-slope towards the trench from the ridge crest under the influence of gravity. Mantle drag occurs between the upper mantle and the asthenosphere and may also contribute to lithospheric plate movement.

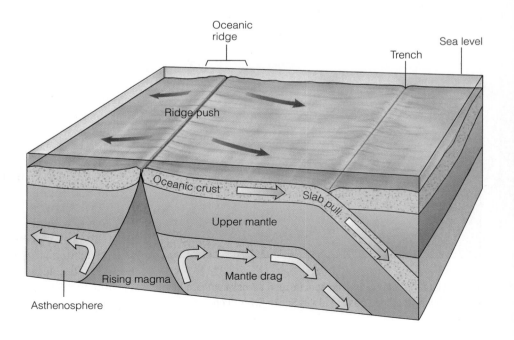

and the basaltic crust of the ocean floors. Because these effectively "float" on the denser mantle beneath, they are kept in a state of buoyant equilibrium called isostasy. Like any buoyant object, the level at which they float depends upon both their density and their thickness. Thus the thicker, lighter continents ride higher than the thinner, denser crust of the ocean floors so that continents usually lie above sea level while the ocean floors lie below as water fills the lower areas.

Understanding the ways in which plates move and interact has provided a comprehensive explanation for many of the Earth's major geologic processes and features, including the origin and distribution of earthquakes, volcanoes, and mountain belts.

Three types of plate boundaries result from the mutual interactions of moving plates. Divergent plate boundaries occur where adjacent plates move away from each other, and are marked by basalt volcanism, sea-floor spreading centered on a mid-ocean ridge, and the creation of new oceanic lithosphere. Convergent plate boundaries occur where two plates collide and are marked by deep ocean trenches, subduction zones and the development of island arcs or volcanic mountains. Convergent plate boundaries mark the sites of plate destruction and may culminate in continental collision. Transform plate boundaries link divergent or convergent boundaries and occur where two plates slide past one another. They are marked by earthquake-prone crustal fractures or faults.

At divergent boundaries, the birth of an ocean starts with the breakup of a continent and the building of new ocean floor between the walls of the rift valley that develops as two continental plates separate. Basalt dikes injected into fractures in the rift valley floor slowly wedge the continent apart; a process that continues at a mid-ocean ridge once separation is achieved, until a full-sized ocean is formed. Rates of divergence range from 0.1 to 17 cm (0.02 to almost 7 in) per year. Rifts that fail to open become filled with sediment to form buried zones of weakness in continental interiors called failed rifts. Continental rifts that successfully open evolve to become mid-ocean ridges, which encircle the globe in all major oceans to form a continuous undersea mountain chain over 65,000 km (40,000 mi) long. Unlike continental mountain ranges, which owe their origin to plate convergence, mid-ocean ridges are high only because they are thermally uplifted by the hot mantle beneath. So as the ocean floor spreads away from a ridge, it cools and subsides, ultimately forming the abyssal plain of the deep ocean about 4–5 km below sea level.

Rare fragments of ocean floor preserved on land are known as ophiolites, and reveal the internal structure of oceanic lithosphere. The layers of ophiolites mirror layers of the ocean floor. Beneath deep ocean sediments come basalt pillow lavas produced by submarine eruption. Below these is a complex of sheeted dikes that fed the lavas. And below these is a gabbro-filled magma chamber that fed the dikes, and which is floored in turn by upper-mantle peridotites.

At convergent boundaries along subduction zones, dense oceanic crust angles into the Earth's interior and is consumed. Its descent from the trench is marked by an inclined Benioff zone of earthquakes and, as temperatures mount, by the release of magmas. The magmas either rise to form volcanic island arcs where the overriding plate is oceanic, or uplift the leading edge of continental plates to produce a volcanic mountain range. The curvature of these volcanic arcs is caused by the spherical shape of the Earth, and variations in the style of subduction often reflects the age and buoyancy of the crust being subducted. In Marianas-type subduction, the crust is old and dense, and subduction is fairly passive so that the angle of the descending plate is steep and major earthquakes are few. With collapse of the subducting slab and rollback of the trench, spreading may occur behind the arc, producing a back-arc basin that resembles a miniature ocean. Such back-arc basins are typical of the western Pacific. Alternatively, in Andean-type subduction, young, buoyant crust is forced to descend. Andean-type subduction is associated with strong earthquakes, a shallow angle of subduction, and coastal volcanic mountains like the Andes.

The net effect of sea-floor spreading and subduction, the two main components of plate tectonics, is to effectively recycle the ocean floor. In so doing, the continents grow laterally with time as continental crust is produced from the magmas generated by the destruction of oceanic lithosphere. Continental crust is less dense than oceanic, and young ocean floor is more buoyant than old. Because the least buoyant plate is subducted, the ocean floor is kept young through recycling whereas the continents are not. As a result, the oldest ocean floor is less than 200 million years old, whereas the oldest continental rocks are almost 4 billion years old.

Along transform plate boundaries crust is neither created nor destroyed but their stick-slip motion produces frequent shallow earthquakes. Transform faults are most common in the oceans where they offset mid-ocean ridges in a rectilinear pattern, or link ridges to subduction zones or subduction zones to each other. Only a few, like the San Andreas Fault of California, intersect continents. After all motion along the fault has ceased, ancient continental transforms remain as zones of crustal weakness that are prone to occasional earthquakes. This inherent weakness also makes them susceptible to being reactivated again in the future.

Hotspots are small, isolated areas of volcanism that are not readily accounted for by plate tectonic theory since most, like that of Hawaii, occur in the interior of plates. They are essentially fixed sources of heat and so produce linear trails of volcanic features that become extinct as plate motion carries the features away from the hotspot. In the oceans, this produces basaltic seamount chains as the extinct volcanoes stop erupting and are beveled by the sea. Changes in the direction of plate motion produce kinked lines of seamounts, like those of the western Pacific. Hotspots are most frequently attributed to mantle plumes rising from a thermally unstable layer at the core-mantle boundary. However, shallow sources resulting from crustal rifting, plate friction, or thermal blanketing have also been proposed.

On continents, hotspots may drive the wedge that splits continents apart, and so play a role in the opening of ocean basins. Following breakup, such hotspots will come to lie near a mid-ocean ridge. As they must burn their way through continental crust, continental hotspots (like that of Yellowstone), are associated with both basalt lavas and rhyolite volcanism that can be highly explosive.

Plate movement is ultimately the result of convective circulation in the Earth's mantle, which is powered by heat generated from the decay of radioactive elements. What actually governs their movement, however, is less clear. The fast-moving plates of the Pacific appear to be pulled from in front by the weight of the downgoing slab, a process known as slab pull.

The slower-moving Atlantic floor, however, lacks encircling subduction zones, and appears to be pushed from behind by sliding off the Mid-Atlantic Ridge, a process called ridge push. Alternatively, plates may be dragged from below, either by the motion of the asthenosphere or by the circulation associated with hotspots. However this process, called mantle drag, may also be a resisting force, serving to oppose slab pull and ridge push and slowing the plates rather than carrying them.

Key Terms and Concepts

Look for the highlighted items on the web at:
WWW.BROOKSCOLE.COM

- The paradigm of plate tectonics holds that the Earth's rigid lithosphere is broken into plates that move relative to each other along plate boundaries

- Plate motion accounts for the origin and distribution of many of the Earth's most dramatic features, including earthquakes, volcanoes, mountain ranges, and oceans

- Plates move on the soft asthenosphere which contains a seismic low velocity zone

- Earth's crust is separated from the mantle by a sharp compositional boundary called the Moho

- Earth's surface topography is a function of crustal thickness and density because the crust is in a state of buoyant equilibrium known as isostasy

- The hypsometric curve reveals the Earth to be a split-level planet with two elevation ranges marking the continental platforms and abyssal plains

- Plate boundaries are of three types: divergent (constructive), convergent (destructive), and transform (conservative)

- Divergent plate boundaries, marked by continental rift valleys or mid-oceanic ridges, create new oceanic lithosphere

- Ophiolites containing pillow basalts, sheeted dike complexes, and gabbros are preserved fragments of oceanic lithosphere and reveal its internal structure

- Subduction zones at convergent plate boundaries are marked by deep ocean trenches and inclined Benioff zones of earthquake foci. Ocean-ocean convergence creates island arcs whereas ocean-continent convergence produces volcanic mountains

- Subduction zone magmatism includes mafic magmas produced by melting of the downgoing slab and partial melting of the overlying mantle wedge and, if the overriding plate is continental, intermediate-felsic magmas produced by melting at the base of the continent

- Marianas-type subduction of old oceanic lithosphere produces a steep subduction zone, few earthquakes, an island arc and back-arc spreading. Andean-type subduction of young ocean floor produces a shallow subduction zone, numerous earthquakes, and a trench-parallel belt of compressional volcanic mountains

- Transform faults, along which plates slide past each other in stick-slip fashion, are of two types, oceanic and continental. They show either right-lateral or left-lateral motion

- Hotspots are thought to be the product of mantle plumes rising from the core-mantle boundary

- Active volcanic islands formed as the lithosphere moves over an oceanic hotspot become inactive and develop fringing reefs as they move off the hotspot. They are progressively eroded, first to atolls and then to lines of submantle seamounts

- Continental hotspots form triple points that govern the line of continental breakup or become inactive and produce failed rifts called aulacogens. Collapse following explosive volcanism leads to the formation of calderas

- Forces potentially responsible for the movement of plates include slab pull, ridge push, and mantle drag

Review Questions

1. What is the principal difference between the theory of Plate Tectonics and Wegener's concept of Continental Drift?

2. What causes the rocks of the asthenosphere and the low velocity zone to be soft?

3. How is it possible for mantle rocks to occur in both the asthenosphere and the lithosphere?

4. Why does the Moho beneath the continents mirror in an exaggerated fashion their surface topography?

5. Describe the process of continental breakup from crustal rifting to the birth of a new ocean.

6. What happens to continental rifts that fail to open into ocean basins?

7. How does the structure of ophiolites relate to processes occurring at mid-ocean ridges?

8. Why are deep ocean trenches and Benioff zones of earthquakes associated with the subduction of oceanic lithosphere?

9. Why are volcanic island arcs curved?

10. Why are the volcanic eruptions of ocean-continent convergence so much more violent than those of mid-ocean ridges?

11. Why is the age of the oldest continental crust so much older than that of the oldest ocean floor?

12. The relative motion of oceanic transform faults is opposite to that suggested by the offset they produce at the crest of mid-ocean ridges. Why is this so?

13. What are kinked hotspot trails and how are they produced?

14. In what way might hotspots drive the wedge than splits continents apart?

15. What three driving forces may be responsible for plate motion and how has their relative importance been established?

Study Questions

1. From the hypsometric curve (Fig. 5.6) it is evident that there is a difference in elevation between continental and oceanic crust. Why is this and how does it support models of isostasy based on both crustal thickness and density?

2. How does the "bookshelf" model for a mid-oceanic spreading ridge (Fig. 5.13) account for the age distribution and magnetic anomaly patterns of the ocean floor (see Fig. 4.13)?

3. What characteristics distinguish Marianas-type subduction from Andean-type subduction, and what controls the change from one type to the other?

4. Transform faults are so called because they permit one type of plate motion to be "transformed" into another. What is meant by this and how is it achieved?

5. Lines of volcanic islands such as those of the Hawaiian chain are not readily accounted for by plate tectonic theory, yet can be used to deduce the direction of plate motion. Explain this paradox.

CHAPTER 6

Mountain Building

In this chapter you will learn the relationship between the origin of mountains and plate tectonic activity. The world's greatest mountain chains are actually hidden from view; they rise 2–3 km (6500–10,000 ft) above the floors of the oceans to form the Earth's interconnected network of mid-ocean ridges. We will find that these mountains are generated by the ascent of buoyant magma associated with sea-floor spreading. Underwater mountains in the form of seamounts are also formed in the wake of oceanic hotspots. Because subduction preferentially destroys oceanic crust, these mountains are rarely preserved in the geologic record.

The mountains with which we are most familiar occur at convergent margins. As we shall see, they form in a variety of ways. Some mountains are a direct product of subduction. Where two oceanic plates converge, the product of subduction is simply a chain of volcanic island arcs such as those of the western Pacific Ocean. Where subduction of an oceanic plate occurs beneath one capped by continental crust, mountains are produced like those of the modern Andes.

Other mountain belts are the result of collision. Ongoing subduction may eventually transport small or large blocks of crust into the subduction zone where they collide with and adhere to the continental margin of the overriding plate. We will find that the western margin of North America has suffered repeated collisions of this kind over the past 200 million years that have added an average of 600 km to the width of continental North America. Continental collision occurs when oceanic crust between intervening continents is completely subducted. One of the most dramatic examples of continental collision is the Himalayas, which started to form when India rammed into the underbelly of Asia approximately 40 million years ago. As we shall see, this collision continues to the present day, and its effects extend well beyond the Himalayas to impact most of southeastern Asia.

Plate tectonic theories are now routinely applied to mountain belts up to 2.5 billion years old. These studies are hindered, however, by the limited preservation of ancient oceanic crust which makes the identification of ancient oceans very difficult. Despite these difficulties, we will find that the role of plate tectonics in the evolution of many ancient mountain belts is well understood.

We will also see how rocks deform in response to the stresses associated with mountain building. We will find that, at low pressures and temperatures, rocks behave in a brittle manner, and break into fractures or faults. At higher temperatures and pressures, on the other hand, rocks tend to bend into folds.

Finally, we examine the basis of the theory of mountain building, by revisiting the production and interpretation of geologic maps. These maps are fundamental to our understanding of mountain building because they provide the basic geologic information needed to interpret tectonic events.

6.1   Introduction

Mountains provide some of the most breathtaking scenery on Earth and are testament to the awesome powers of nature (Fig. 6.1). Their beauty has provided inspiration for artists, poets, and writers, as well as recreation for skiers, hikers, and naturalists. Our desire to understand the world around us and the natural forces responsible for such beauty, helped to pave the way for the recognition of plate tectonics as a fundamental geologic process. Indeed, the explanation of the origin of mountains is one of the most impressive successes of the theory of plate tectonics.

Yet as recently as the late 1950's, before the concept of drifting continents was widely accepted, our understanding of the origin of mountain belts was very incomplete (Fig. 6.2). At that time, the prevailing view held that the evolution of mountains began with the development of long, relatively narrow troughs called **geosynclines**, in which thick accumulations of sediments were deposited. Over millions of years, these sedimentary piles and the rocks beneath them were deformed, metamorphosed, and intruded by igneous rocks. These processes then resulted in uplift and the generation of mountain belts. There were, however, many fundamental problems with this model, including an explanation for the origin of the troughs, the source of the sediments, and the mechanisms responsible for the uplift. Although many geoscientists realized that this theory was unsatisfactory, it was widely held to be the simplest explanation of the available data.

By the late 1960's, the principles of continental drift, sea-floor spreading, and plate tectonics had provided a new conceptual framework for understanding the origin of mountains. We now know that mountains may form along each type of plate boundary, albeit by different processes. Although hidden from view by a cover of ocean water, mountains on the ocean floor are formed both at divergent plate margins by the process of sea-floor spreading, and within oceanic plates by hotspots that pierce the oceanic crust to form linear island chains. On land, mountains form along convergent plate margins, either as a direct result of subduction, or by the collision of relatively small crustal blocks known as terranes, or by the collision between two continental landmasses. In some areas, mountains may also form along transform plate margins. This may seem counterintuitive as crust is neither created nor destroyed at such margins. Nevertheless, they possess localized areas of compression in which the crust is squeezed and uplifted to form mountains. They also possess localized areas of tension in which the crust is pulled apart to form basins. We begin our discussion with the unseen mountains of the ocean floor.

Figure 6.1
The majestic Tetons rise above the Snake River in northwestern Wyoming.

6.2   Mountains at Divergent Margins

In Chapter 5, we learned that mid-ocean ridges form at divergent (constructive) plate margins, where new oceanic crust is created by upwelling from the hot asthenosphere below (Fig. 6.3). In tectonically stable areas of the oceans, the ocean floor pre-

Figure 6.2
Geosynclinal model for the origin of mountain belts in the Appalachians. Continental and shallow water sediments were thought to be deposited in a shallow trough known as a miogeosyncline. Deep-water sediments and associated volcanics were thought to be deposited in a larger trough called a eugeosyncline. The origin of these geosynclines and the mechanism by which mountains originated from them were major problems with this model.

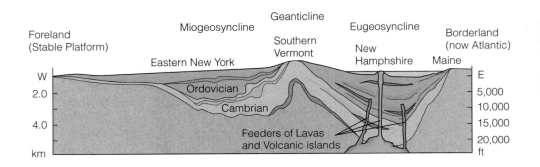

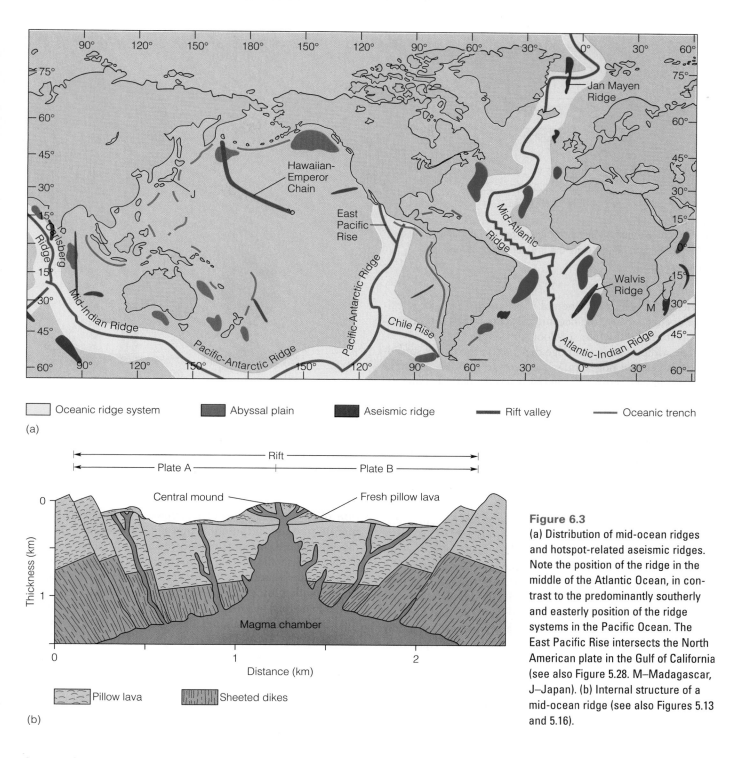

(a)

| Oceanic ridge system | Abyssal plain | Aseismic ridge | —— Rift valley | —— Oceanic trench |

Figure 6.3
(a) Distribution of mid-ocean ridges and hotspot-related aseismic ridges. Note the position of the ridge in the middle of the Atlantic Ocean, in contrast to the predominantly southerly and easterly position of the ridge systems in the Pacific Ocean. The East Pacific Rise intersects the North American plate in the Gulf of California (see also Figure 5.28. M–Madagascar, J–Japan). (b) Internal structure of a mid-ocean ridge (see also Figures 5.13 and 5.16).

(b)

dominantly lies about 4.0 km (13,000 ft) below sea level. Mid-ocean ridges, however, typically stand 2 to 3 km (6500 to 10,000 ft) above the surrounding ocean floor. Thus, they are rarely exposed above sea level. Nevertheless, their impressive dimensions place them among the greatest mountain chains on Earth. Together, they form an interconnected network of ridges 65,000 km (40,300 mi) in length and between 1000 and 4000 km (620–2480 mi) in width (Fig. 6.3a).

The origin of these mountains is quite different from that of continental mountains. As we learned in Chapter 5, conti-

nental mountains have thick roots, as dictated by the principles of isostasy. Beneath mid-ocean ridges, however, the crust is very thin (Fig. 6.3b). Unlike their continental counterparts, the height of a mid-ocean ridge cannot be related to buoyancy of the crust. Instead, its height is related to the buoyancy provided by the upwelling of hot magma from the mantle asthenosphere which underlies the ridge and fuels the sea-floor spreading process. Like continental mountains, the elevation of mid-ocean ridges also decreases with time. But because these ridges lie below sea level, they are not subject to the degree of erosion

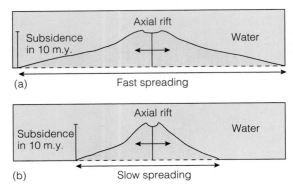

Figure 6.4
The shape of mid-ocean ridges resulting from (a) fast- and (b) slow-spreading ridges. Note the greater volume of the fast-spreading ridges.

experienced by their continental counterparts. Instead, subsidence occurs because the oceanic crust cools as it is transported away from the ridge by sea-floor spreading.

The width of an ocean ridge depends on the rate of spreading. Fast-spreading ridges are wider because the new, hot (and therefore elevated) crust is carried away more rapidly. In contrast, slow-spreading ridges are narrow because the crust subsides close to the ridge crest. As a consequence, fast-spreading ridges occupy a much larger volume than slow-spreading ridges (Fig. 6.4).

Where hotspots occur along spreading ridges, vigorous plumes of ascending hot material may provide sufficient buoyancy that the mid-ocean ridge becomes elevated above sea level. Iceland on the Mid-Atlantic Ridge and Easter Island on the East Pacific Rise are examples of this process. The additional buoyancy enables direct study of mid-ocean ridge processes by bringing the ridge above sea level.

As impressive as these mountains are today, they are unlikely to be preserved in the geologic record. Because their elevation is due to the heat provided by the magma ascending from the underlying mantle, they inevitably subside when sea-floor spreading ceases or when the location of spreading changes position. Oceanic crust is also preferentially subducted, and only very rarely are fragments preserved as ophiolites (see Chapter 5, section 5.4.1).

6.3 Mountains within Oceanic Plates

Another type of underwater mountain belt forms in the interior of plates where hotspots pierce the oceanic crust. There are probably more than 40 hotspots beneath the modern ocean floors (the exact number is uncertain). As we learned in Chapter 5, these hotspots form long linear chains of seamounts that stretch thousands of kilometers across ocean basins (Fig. 6.3a). Although there is some controversy about their origin, hotspots appear to represent the surface expres-

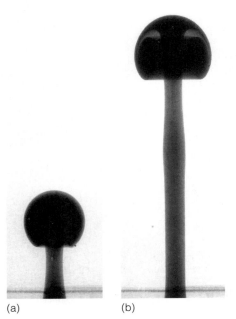

(a)   (b)

Figure 6.5
Most geoscientists believe that mantle plumes ascend from the core–mantle boundary, about 2900 km (1800 mi) below the surface. The ascent of a mantle plume is modeled using a buoyant solution of glucose syrup. (a) The plume establishes and maintains a relatively narrow central feeder pipe known as a plume tail, and

sion of focused columns of hot upwelling mantle known as **mantle plumes.** Because this upwelling comes from beneath the lithosphere, hotspots remain relatively fixed. Recent evidence suggests the mantle plumes may ascend from near the core–mantle boundary. Sophisticated modeling suggests that these mantle plumes typically have a relatively narrow central feeder zone that, at a depth of 40 to 60 km (25–40 mi) beneath the surface, balloons to some 400–1000 km (250 to 620 mi) in diameter (Fig. 6.5). The mantle plumes provide sufficient buoyancy to uplift the lithosphere and create mountains. In most situations the uplift is 1–2 km (3280–6500 ft). For vigorous mantle plumes, such as Hawaii, the uplift is greater than 6 km (20,000 ft) which is sufficient to produce an island. To the southeast of Hawaii, Loihi, the next member in the Hawaiian Island chain, stands 5 km (16,000 ft) above the sea floor and is currently within 960 m (3,150 ft) of the ocean surface.

As we learned in Chapter 5, each island that forms over the mantle plume is transported away from the hotspot by sea-floor spreading, cooling and subsiding as it does so. Like mid-ocean ridges, the elevation of these island chains reflects the buoyancy provided by the underlying mantle heat source. As the crust cools with distance from the hotspot, each member of the island chain subsides, eventually subsiding below sea level to form a chain of seamounts. As only the active volcanic island lying over the mantle plume is prone to earthquakes, such seamount chains are also referred to as **aseismic ridges.**

(c) (d)

(b–c) balloons below the lithosphere to eventually form (d) a plume head some 400–1000 km (250–600 mi) in diameter. The plume causes volcanism and uplift of the lithospheric plate. As the plume material cools, it may adhere beneath or "underplate" the lithospheric plate.

6.4 Mountains Produced by Subduction

As we also learned in Chapter 5, where two plates converge, the denser plate is almost always subducted beneath the lighter plate. As the denser plate sinks, it is both heated and deprived of its gases, stimulating the generation of magma either at the subduction zone or within the overlying mantle wedge. The resulting mountains have many different expressions depending on the nature of the upper plate and that of the subducting slab. Examples of subduction-related mountain building occur along the borders of the Pacific Ocean, in the so-called "Ring of Fire" (Fig. 6.6). The nature of this mountain building varies in style from one part of the Pacific margin to another. In the western Pacific Ocean, there are several examples of **ocean–ocean convergence** where one oceanic plate is subducted beneath another. In the eastern Pacific Ocean, on the other hand, **ocean–continent convergence,** in which oceanic crust is subducted beneath continental crust, occurs along much of the length of the west coasts of North and South America. The generation of mountains in both cases results primarily from the subduction process.

6.4.1 Ocean-Ocean Convergence

When two oceanic plates converge, the older plate will be preferentially subducted because older oceanic crust is denser than younger oceanic crust. In Chapter 5, we referred to the subduction of old oceanic crust as Marianas-type. This style of subduction results in the generation of a volcanic island arc (Fig. 6.7). When the subducting slab reaches a depth of about 100 km (62 mi) its gaseous components are driven off, instigating melting in the mantle wedge above the subduction zone. Because there is no continental crust with which the ascending magmas can interact, the magmas that produce the island arc are largely mantle-derived and so are predominantly basaltic in composition. Without continental crust to slow their ascent, a significant portion of the ascending magmas reach the surface to form volcanoes.

Complex tectonic activity also occurs in the vicinity of the trench. As subduction proceeds, elevated portions of the ocean floor are scraped off the subducted slab and mix with sediments derived from the erosion of the island arc. This chaotic mixture is plastered onto the leading edge of the upper plate to form a wedge-shaped package of metamorphosed and highly deformed rocks, known as an **accretionary wedge** (Fig. 6.7).

The island arc is produced on the upper plate behind the trench because the subducting slab descends beneath it.

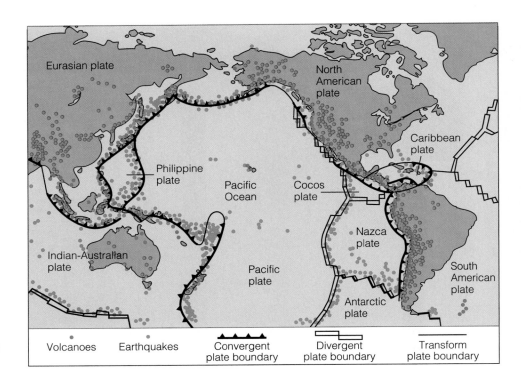

Figure 6.6
Subduction along the Pacific margin and the almost continuous belt of volcanism around the Pacific Rim known as the Pacific "Ring of Fire."

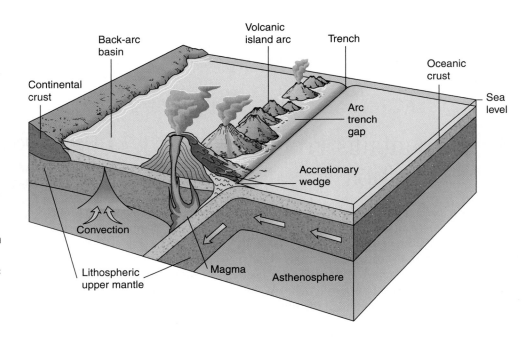

Figure 6.7
Ocean–ocean convergence where an older, denser oceanic plate is subducted beneath a younger plate. This style of subduction zone is typical of the "Marianas-type" in the western Pacific Ocean. The older oceanic plate angles down the subduction zone where it instigates melting in the overlying mantle. The ascent of magma forms a volcanic island arc. Spreading behind the arc may open up a back arc basin. The region between the trench and the volcanic arc is known as the arc-trench gap.

Between the trench and the volcanic arc is a region known as the **arc-trench gap,** the width of which is dependent on the dip of the subduction zone. In steeply dipping subduction zones, the arc-trench gap is narrow because magma ascends to the surface close to the trench. In more shallowly dipping subduction zones, magma ascent occurs much further from the trench and the arc-trench gap may be several hundred kilometers wide.

6.4.2 Ocean-Continent Convergence

As we learned in Chapter 5, one of the best known examples of ocean–continent convergence occurs where the oceanic Nazca plate is being subducted beneath the continental edge of the South American plate (Fig. 6.8). The end result of the process is the generation of the volcanically active Andes Mountains which extend the length of the western margin of South America.

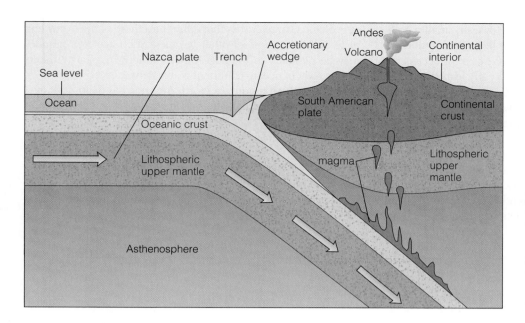

Figure 6.8
The oceanic Nazca plate is subducted beneath the South American plate. The descent of the slab instigates melting in the overlying mantle and in the continental crust. The ascent of these melts combined with compression associated with convergence forms the Andes Mountains.

Several of the processes associated with ocean–ocean convergence, such as the generation of magma, the development of an accretionary wedge, and the formation on an arc-trench gap, also accompany ocean–continent convergence. Deep-water sediments and sedimentary rocks may be scraped off the subducting plate at the trench to become added to the continent's leading edge. Similarly, a section of oceanic crust may break away and escape subduction by becoming incorporated into the mountains of the overriding plate as an ophiolite.

In contrast to ocean–ocean convergence, however, the basaltic magma ascending from the subduction zone and the mantle above interacts with the thick continental crust as it ascends to the surface. The heat associated with the rising basaltic magma is sufficient to promote melting near the base of the continental crust. Magma derived from continental crust is more silica-rich than that derived from the mantle. As a result, Andean-type subduction zones generate a much wider range of magma compositions than those of Marianas-type subduction zones in which basaltic compositions predominate.

Much of the magma generated by subduction beneath continents never reaches the surface because its ascent is impeded by a thick barrier of continental crust. Instead, the magma becomes lodged within the continental crust of the overriding plate where it cools to form great bodies of granitic rock. As we learned in Chapter 2, these igneous rocks are classified as "plutonic" because they crystallize beneath the surface. They only become exposed at the Earth's surface if the overlying crust is eroded away. Remember that slow-cooled plutonic rocks are readily distinguished from volcanic rocks by the relative size of their crystals, which may be several centimeters across. These plutonic rocks, which are the stuff of continents, are generally granitic or dioritic in composition (see classification in Chapter 2). In this way, new continental crust is produced by the destruction of old ocean floor. Thus, the process of subduction creates new continental lithosphere, just as sea-floor spreading produces new oceanic crust.

Of the magmas that do reach the surface to fuel volcanoes, the predominant composition of Andean-type volcanic rocks is intermediate to felsic. In fact, the intermediate volcanic rock "andesite" (see Chapter 2) is named after the Andes mountains.

The presence of abundant hot magma within the continental crust, coupled with the compression that accompanies plate convergence, causes the leading edge of the continental plate to crumple and rise into a mountain chain like the Andes. The Andes parallel the offshore Peru–Chile trench, demonstrating a relationship between subduction and mountain building. Although subduction-related mountains resemble collisional mountains like the Himalayas in their appearance, they are of quite different origin. Unlike the Himalayas, for example, the Andes Mountains are primarily volcanic mountains, as they lie above a subduction zone. Mountains produced when continents collide are not generally volcanic because subduction soon comes to a halt following continent–continent collision. We will discuss the origin of collisional mountains in the next section.

6.5 Mountains Produced by Collision

Mountains produced by collision also occur at convergent margins. Collision may occur between a continental margin and small crustal fragments known as terranes, or between two continents.

6.5.1 Collision of Terranes

As we have seen, the process of subduction may ultimately lead to ocean closure and a collision between the continents that once faced each other across the ocean. However, smaller collisions usually precede this climactic event because the subducting ocean floor is far from featureless. Instead, most oceans are dotted with islands and the floors of oceans carry features such as mid-oceanic and aseismic ridges (Fig. 6.3a), oceanic plateaux like those of the southern Pacific, and small continental blocks or **microcontinents** like Madagascar and Japan. Like baggage riding passively atop a carousel at an airport, when features such as these encounter a subduction zone, they tend to be scraped off the descending plate and are added or "accreted" to the leading edge of the overriding plate as they collide with it (Fig. 6.9). Each such encounter produces a block of land known as an **accreted terrane**, or more simply, a terrane, which will have a specific set of geologic characteristics that differ from those of the overriding plate and from those of other terranes. Such accreted terranes may be plastered to the edge of a continent in great numbers along zones of ocean–continent convergence.

The western margin of North America has been built by a succession of such collisions (Fig. 6.10). It is thought that these terranes were once dispersed within the eastern Pacific ocean (in a similar fashion to those of today's western Pacific) and that they subsequently collided with the North American margin. The process appears to have commenced with the breakup of Pangea, following which North America moved predominantly westward. As it did so, subduction of the eastern Pacific occurred beneath its western margin.

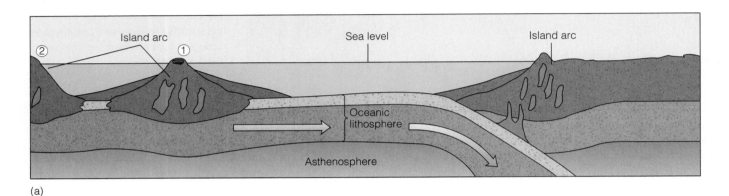

(a)

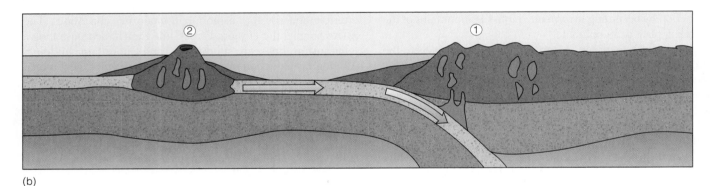

(b)

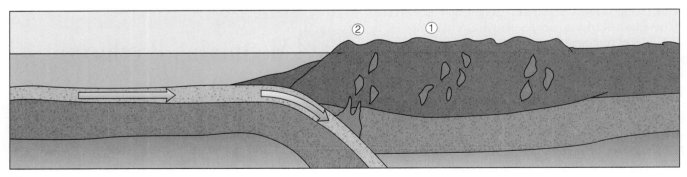

(c)

Figure 6.9
Model for the accretion of island arc complexes to a continental margin. (a) Two island arc complexes (1 and 2) sit outboard from a continental margin. (b) The nearest arc accretes when the intervening ocean floor is subducted. (c) Eventually, the second island arc complex accretes as well. The net effect is the growth of the continental margin oceanward.

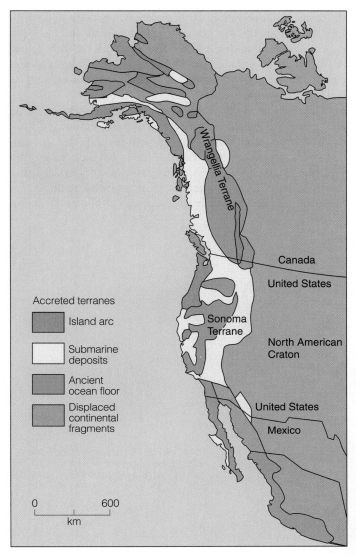

Figure 6.10
Terranes accreted to the western North American margin over the past 200 million years consist of remnants of island arcs, submarine sedimentary deposits, ophiolite suites uplifted from the ancient ocean floor, and continental fragments displaced by faults. Wrangellia, for example, is an island arc that collided with North America some 100 million years ago and has since been dispersed along the continental margin by faults like the present-day San Andreas.

An estimate of how much oceanic crust was consumed by the westward motion of North America can be obtained by looking at the position of the spreading ridges in the Pacific Ocean (Fig. 6.3). Sea-floor spreading is normally a symmetric process that creates new oceanic lithosphere in equal amounts on the two diverging plates. As a result, spreading ridges, like that of the Atlantic Ocean, tend to be located in the mid-ocean. In the Pacific Ocean, however, the spreading ridge occupies a highly asymmetric position near the ocean's eastern margin. This implies that much of the oceanic crust that once lay to the east of the ridge has been consumed. In fact, the North American plate has actually collided with the ridge along the

Californian coastline. During the consumption of this oceanic crust, terranes that once lay embedded within it collided with the North American continental margin (Fig. 6.10).

More than 200 such terranes have been identified along the North American continental margin between Mexico and Alaska. They have been found inland as far as Colorado and Utah. Over the past 200 million years, the piecemeal accretion of these terranes has added an average of 600 km (360 mi) of real estate along the length of the western margin of North America. This addition amounts to about 20% of continental North America (Fig. 6.10).

Terranes have two important stages to their history; the first reflects the setting in which they originated, and the second reflects their history of collision with a continental margin. Evidence that these terranes originated within the Pacific Ocean is recorded in their geologic histories. Perhaps the best example of this can be found in the rocks exposed around San Francisco Bay (Fig. 6.11). To the north of the Golden Gate Bridge, the sedimentary rocks of the Marin Headlands are typical of ocean floor sediments and contain microscopic oceanic

Figure 6.11
A view from San Francisco of accreted terranes in the San Francisco Bay area. In the background, the Marin Headlands are typical of ocean floor sediments that were deposited in the Pacific Ocean more than 100 million years ago. The geology of Alcatraz Island (center) is characteristic of near-shore deposits. Angel Island (out of view) contains rocks that probably formed in a subduction zone.

organisms. This region is believed to be part of a terrane that existed for 100 million years in the Pacific Ocean prior to its accretion to the North American margin. The geology of nearby Alcatraz Island, however, is quite different. Its fossils and sedimentary rocks are characteristic of near-shore deposits. To the south of Alcatraz, Angel Island contains deformed and metamorphosed rocks that may have been formed in a subduction zone or during the collision between two plates.

These three terranes formed in very different environments at great distances from each other and at varying distances from the continental margin. Their present side-by-side location is attributed to their accretion to the continental margin of western North America. Similar analysis of other terranes has revealed the scale and importance of this accretionary process to the evolution of continents. For example, paleomagnetic and fossil evidence indicates that much of British Columbia, the Yukon, and Alaska are made up of terranes that originated much nearer the equator. Other parts of Alaska may have drifted across the entire width of the Pacific Ocean (Fig. 6.12).

Most of the Rocky Mountains also comprise often far-traveled terranes that have been accreted to North America during the past 200 million years. Parts of Nevada originated as a volcanic island arc off the Pacific coast. In fact, when the dinosaurs roamed the Earth, some 150 million years ago, the Rocky Mountains as we know them did not exist at all. At that time, the Pacific coast lay many hundreds of kilometers to the

Figure 6.12
How Mesozoic crustal fragments from the southern Pacific may have become accreted to the North American continental margin and transported to Alaska during the Cenozoic.

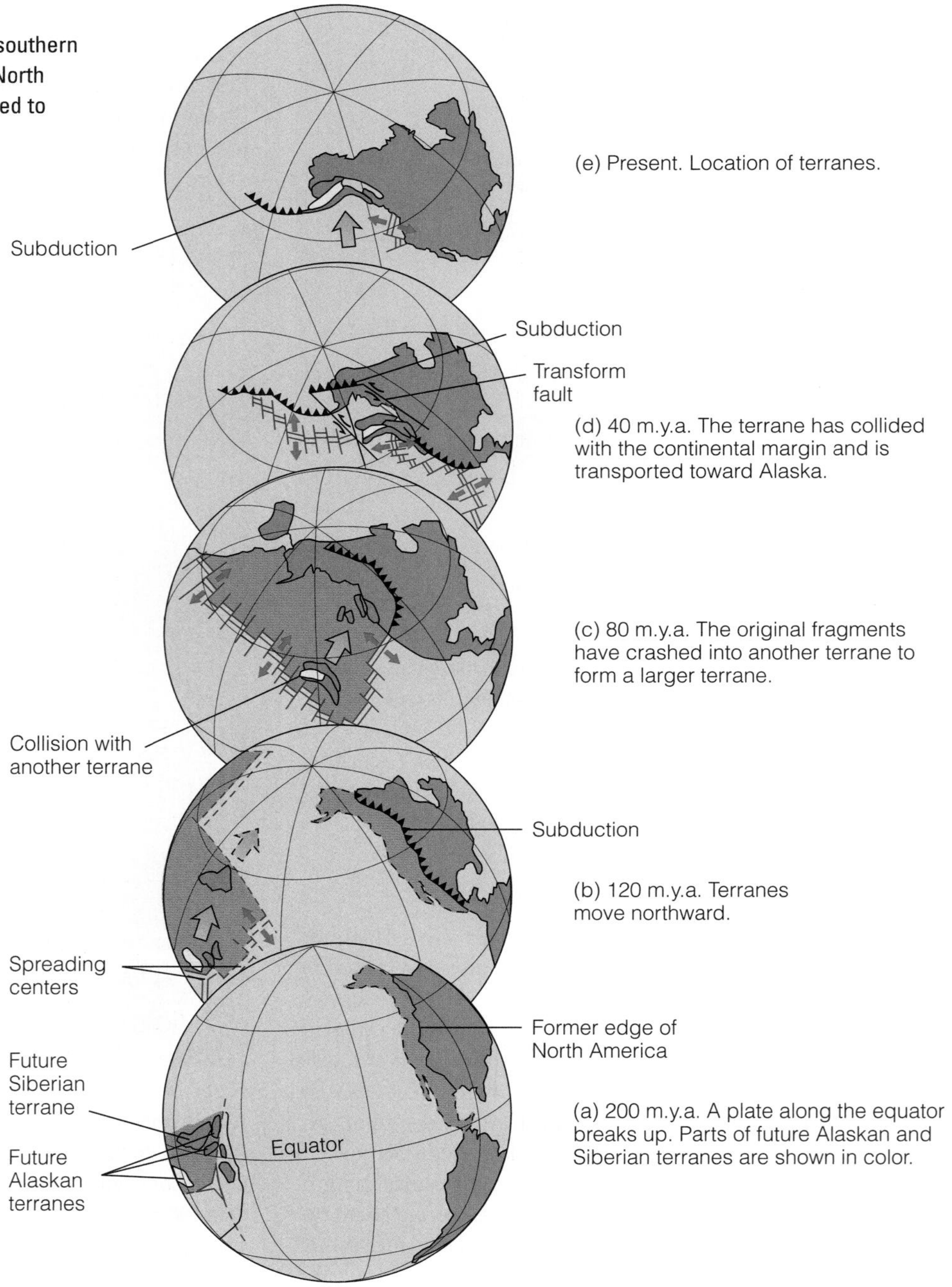

Subduction

(e) Present. Location of terranes.

Subduction

Transform fault

(d) 40 m.y.a. The terrane has collided with the continental margin and is transported toward Alaska.

Collision with another terrane

(c) 80 m.y.a. The original fragments have crashed into another terrane to form a larger terrane.

Subduction

(b) 120 m.y.a. Terranes move northward.

Spreading centers

Former edge of North America

Future Siberian terrane

Future Alaskan terranes

Equator

(a) 200 m.y.a. A plate along the equator breaks up. Parts of future Alaskan and Siberian terranes are shown in color.

east of its present position, and was characterized by a broad continental shelf like that of today's Atlantic seaboard.

Many microcontinental collisions are oblique side-swipes rather than head-on collisions with the continental margin. As a result, many of the terranes of western North America were subsequently broken up by faults and have since been dispersed by fault movement along the length of the continental margin (Fig. 6.12d, e). 100 million years ago, for example, an island arc somewhere in the Pacific Ocean called **Wrangellia** (after the Wrangell Mountains in Alaska) drifted north and collided with North America (Fig. 6.13) and broke up. Some blocks of Wrangellia were thrust onto the continent while others continued northward on faults similar to the present day San Andreas. By 40 million years ago, fragments of Wrangellia had been dispersed along the continental margin from Alaska to Idaho (Fig. 6.10) Western North America has experienced at least fifty such collisions. Some of these terranes seemed to have formed at or near their present positions, but others, such as Wrangellia, are clearly far traveled (Fig. 6.13). Because their original relationship with respect to North America is usually uncertain, fragments added to the continent in this way are often referred to as **exotic** or **suspect terranes.**

It is now clear that many of the world's mountain belts include complex mosaics of accreted terranes. Some of these small crustal blocks may have been returned to their original locations. That is, they may have been spalled off the continent, perhaps as a result of back-arc spreading (see Chapter 5), and then returned to the continent by closure of the back-arc basin (Fig. 6.14). Other crustal blocks, however, are clearly exotic and have traveled great distances. The wandering and accretion of these terranes is quite a recently recognized geologic process but clearly plays an important role in continental growth as a prelude to continent–continent collision.

6.5.2 Continent-Continent Collision

The most dramatic plate collisions take place when the two colliding plates carry continents. Continents ride passively on plates like baggage on a conveyor belt. But subduction destroys the oceanic crust between them so that collisions between continents are inevitable. Because continental lithosphere is too buoyant to be subducted completely, plate convergence following continent–continent collision must ultimately grind to a halt. But as one continent overrides the other in the process of stopping, the ensuing collision and crustal thickening forces the crust skyward to build enormous mountain ranges in which huge faults develop as the rocks in the impact zone are squeezed upward and sideways. Cessation of subduction and the presence of overthickened continental crust stops the creation and ascent of magma to the surface so that volcanoes are not common in continental collision zones. But such collisions result in the formation of some of the most spectacular and geologically complex regions on Earth.

If the rate of subduction in an ocean basin exceeds the rate at which new crust is generated at its mid-ocean ridge, the process of subduction may ultimately lead to the complete closure of the ocean. Given that the Earth has a constant radius, any discrepancy between the rates of subduction and spreading in one ocean basin must be compensated for by enhanced subduction or spreading in another ocean basin. Plate reconstructions for the last 150 million years (see Figure 6.15), for example, show that the destruction of the ancient Tethys Sea, which once separated Africa and India from southern Europe and Asia, was accompanied by the opening of the Atlantic Ocean.

The higher rate of subduction over spreading in a closing ocean has implications for the evolution of the ocean's subduction zones. Initially, the destruction of old oceanic crust is likely to produce Marianas-type subduction zones with back-arc basins. However, as the ocean basin shrinks, forced subduction of younger oceanic crust must gradually become dominant. Hence, Andean-type subduction is likely to become increasingly prevalent producing trench-parallel volcanic mountains on the overriding continental plate. Subduction terminates when the continents, riding like baggage atop a conveyor belt of oceanic crust, inevitably collide.

Like two colliding automobiles, the impact causes widespread crumpling, distortion, and dislocation of rock strata as the ocean finally closes. The opposing continental margins meet head on and meld together in massive mountain ranges now far from the sea. Leftover slices of ocean floor may also be pushed up to mark the boundary or **suture** between the formerly separate continents. The approaching continents also have irregularly shaped margins that can profoundly influence the style of the resulting mountain-building activity. Protruding parts of the margins or "promontories," for example, tend to bear the brunt of the collision, while recesses or

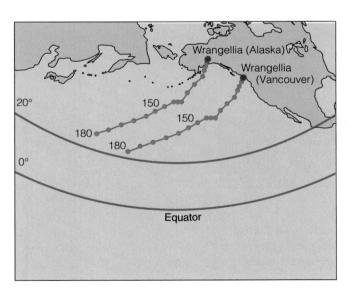

Figure 6.13
The migration of Wrangellia from a position within the northern Pacific Ocean 180 million years ago, until its accretion to the continental margin some 100 million years ago. Originally coincident, displacement of the Alaska path relative to that of Vancouver reflects dispersal of Wrangellia parallel to the continental margin following its accretion to North America.

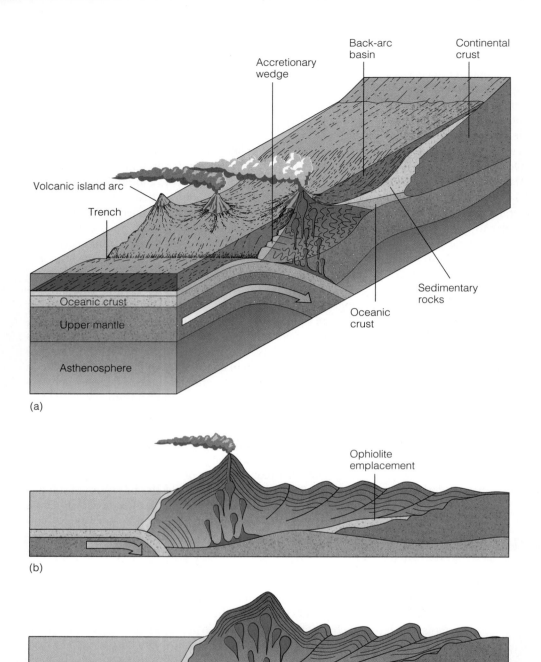

Figure 6.14

Accretion of a terrane by consumption of a back-arc basin. (a) The volcanic island arc is initially separated from the continental margin by a back-arc basin. (b) and (c) Continued subduction results in deformation of the back-arc basin and collision of the island arc with the continental margin.

"embayments" in the margins are likely to experience less of an impact.

Detailed studies of continental collision zones show these to be extremely complex regions that can cause events to take place far from the former plate edges. Although the overall sense is one of compression and convergence, important regions of translational movement and even divergence may occur.

Nowhere is this better illustrated than in eastern Asia, where the building of the Himalayas has created a collisional logjam of enormous scale between the continents of India and Asia. (Fig. 6.16). Approximately 60 million years ago, after the breakup of Pangea had split India from Africa's eastern flank (Fig. 6.15), India drifted rapidly northward at 10 cm (4 in.) a year, until about 40 million years ago, when it collided with

(a) 150 m.y.a.

(b) 60 m.y.a.

(c) Today

Figure 6.15
Continental reconstructions for (a) 150 m.y.a., (b) 60 m.y.a., and (c) Today showing the separation of India from Africa and the destruction of the Tethys Sea as India moved rapidly northward and rammed into Asia. During this time, the Atlantic Ocean progressively widened.

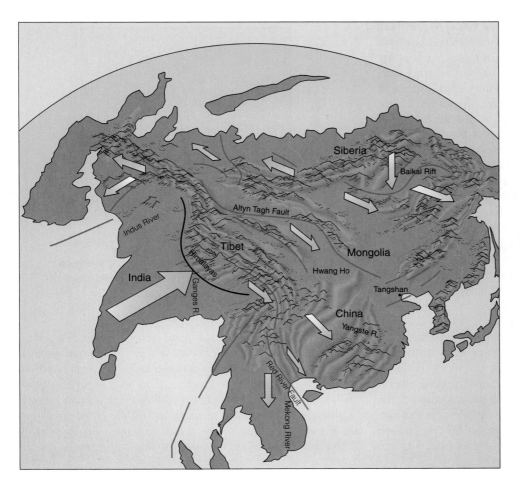

Figure 6.16
Collisional logjam of eastern Asia that started 40 million years ago when India first made contact with Tibet. India has now penetrated 2000 km (1240 mi) into the Eurasian plate pushing up the Himalayas and the high Tibetan Plateau. Pinned against stable Siberia, China and Indochina are being squeezed toward the Pacific Ocean. Arrows on their sides show motion along major faults such as the Altyn Tagh and Red River; flat arrows show movement of crustal blocks. As they jostle for position, one block far from the collision zone is pushed apart, opening up the Baikal Rift that harbors Siberia's Lake Baikal, which is over a mile deep.

the underbelly of the Eurasian plate to form the Himalayan Mountains and the Tibetan Plateau. A similar process also generated the European Alps where the northward migration of the African plate has rammed into Europe.

The crescent-shaped Himalayas stretch 2500 km (1550 mi) and contain thirty of the world's highest peaks, including the 8796 m (29,028 ft) Mt. Everest. Behind the mountains stretches the Tibetan Plateau where the average elevation is higher than most of the highest mountains of the continental United States. Beyond this stretches China whose earthquake-prone faults have developed thousands of kilometers from the point of collision as one continent pushes the other out of the way.

The building of the Himalayas is still in progress. Since the collision began, India has penetrated a further 2000 km (1240 mi) into the Eurasian plate, pushing up the Himalayas, and thickening the crust beneath the Tibetan Plateau. It is also breaking up China and Indochina which are being squeezed sideways to give India room to move north, a process known as "escape." The detailed explanation for the region's complex geology is controversial, but the effect has been modeled experimentally by pushing a rigid "indenter" (India) into a layer of plasticine (Fig. 6.17). The pattern of faulting so produced best mimics that of southern Asia when the layer is free to move on one side (the Pacific) but not on the other (Siberia). According to the model, parts of China and central Asia (as far west as Iran and Armenia on the Caspian Sea) have been moved aside along major strike-slip faults, some of which are longer than the San Andreas Fault. As the fault blocks jostle each other, some far from the collision zone have been wedged apart, resulting in local extension. Such extension opened up the rift valley that today harbors Siberia's Lake Baikal (see Fig. 6.16). Formed by the forces of collision and not by breakup forces like those of the East African Rift Valley, Lake Baikal is the world's deepest lake, with a depth of over 1600 m (5300 ft). It traps twenty percent of the world's fresh water, more than in all of the Great Lakes put together!

The Himalayan Mountains contain a complex mixture of rocks caught between the converging continents. Having originated in widely different environments and locations, they are now welded together by collision. The force of collision uplifted portions of oceanic crust from the subducting plate and squeezed them into the suture joining the two continents to form a band of ophiolites. Fossiliferous sedimentary rocks that were deposited in a tropical Tethys Sea, now lie under a cover of snow and ice near the roof of the Himalayas. Parts of the crust of southern Asia have also been stripped of their sedimentary cover so that granites are exposed.

Today, the *lowest* elevation in the Tibetan Plateau is 3660 m (12,000 ft) and more than 17,000 glaciers cover its surface. In fact, Tibet is so high that it tends to move downward under its own weight like cream cheese, opening small, north-south rift valleys as it does so. As India continues to push northward, the Tibetan Plateau is squeezed against the crustal blocks that surround it. Stable Eurasia and the Siberian Shield constrain its movement to the north and west, so it moves east, cascading into

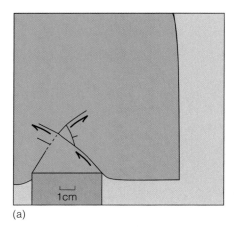

(a)

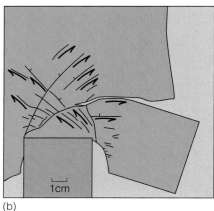

(b)

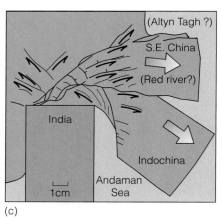

(c)

Figure 6.17

Indenter model explaining Asian tectonics showing progressive collision (a-c) of a rigid block (representing India) and plasticine (representing Asia). Note the squeezing out.

southeast Asia and China as a major tectonic logjam (Fig. 6.16). Hence, China is a tectonically active region despite its distance from the site of the collision between India and Asia. Nowhere was this more tragically apparent than in Tangshan, east of Beijing, in 1976. The devastating earthquake that struck this city is thought to have killed as many as 700,000 people and may have owed its origin to the forces that push up the mighty Himalayas thousands of kilometers to the southwest.

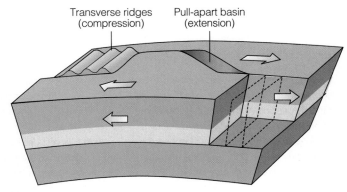

6.6 Mountains Produced along Transform Faults

As we learned in Chapter 5, transform plate boundaries are conservative, that is, they neither create nor destroy the crust. Instead, the plates on either side slide past each other. We know from studies along the San Andreas Fault, that this sliding is not a simple process. The San Andreas Fault is but one of a large number of faults along which motion occurs between the Pacific and North American plates (Fig. 6.18, see also Fig. 5.31). This motion is not smooth; for example, in some places the fault sticks, and stresses build up that eventually culminate in earthquake activity. In other areas, the faults bend as they exploit weaknesses in the continental crust.

Over millions of years, the complex pattern of faults may result in locally intense stresses that can form mountains. When transform faults bend, the rocks on either side are either extended or compressed depending on the direction of the bend and the direction of movement of the fault (Fig. 6.19). As a simple demonstration of this, cut a piece of paper in two along a line with a kink in it like a Z, and then slide the two pieces of paper along the cut line. Depending on which way the paper is moved, the kink will either generate a gap between the pieces or a zone of local compression where the edges stick and the paper folds. These two reactions mirror what are called divergent and convergent bends along a transform fault.

Figure 6.19

Model showing how local zones of compression and extension may occur along a transform fault leading to the development of a pull-apart basin and transverse ridges at bends in a transform fault.

At **divergent** (separating) **bends** in a transform fault, the crust may be stretched until it breaks and collapses to create a long, narrow depression or **pull-apart basin** that may collect sediment or become filled with water. The basin and its contents may resemble that of a young continental rift. However, the localized nature of the basin provides a clear distinction because pull-apart basins are restricted to divergent bends whereas rifts are continuous linear troughs. Examples of pull-apart basins include the Dead Sea between Jordan and the West Bank (see Fig. 5.29), and the Salton Sea of southern California.

In contrast, at **convergent bends** where the transform fault curves in the other direction, plate motion jams the blocks of crust together creating folded mountains or **transverse ridges** across the line of the fault. California's Transverse Ranges, north of Los Angeles, occur at a convergent bend in the San Andreas Fault (Fig. 6.18). In fact, it was movement on a buried fault within this convergent bend that was responsible for the devastating Northridge earthquake in January, 1994 (see Chapter 5, Living on Earth: Predicting Earthquakes—Some Lessons from California, pages 128–129).

Both pull-apart basins and transverse ridges vary in size but are often heavily faulted areas with minor volcanism and hot springs. As the San Andreas fault system illustrates, both may also develop simultaneously on the same continental transform depending on the direction of the bends in the fault trace.

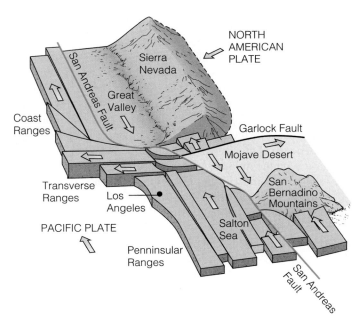

Figure 6.18

San Andreas fault system in the region of the "Big Bend." Because of its orientation, the bend is a region of local compression that has resulted in the formation of the Transverse Ranges.

6.7 Mountains in the Past

By the mid-1960's, the relationship between the origin of modern mountain belts and plate tectonics had gained wide acceptance. But how far back this process could be extrapolated into the geologic past was a matter of considerable debate. The principle difficulty in extrapolating the process backward is that oceanic

The mountains of the Himalayas are the predominant influence on the environment in southern Asia, one of the world's most densely populated regions. Because the mountains are high enough to influence atmospheric circulation, they have a profound effect on the prevailing climate. The climate of India, as we will learn in Chapter 12, is dominated by the monsoon, which every summer brings rains from the Indian Ocean. As the moisture-laden clouds track northeast over India, they are forced up by the Himalayan foothills. As the air cools, the last of the monsoon's moisture is precipitated as either snow or rain. Compaction of the snow into ice results in the thousands of glaciers that carve the mountains into rugged peaks. The rain and glacial melt-water feed the headwaters of some of the world's greatest rivers (Fig. 6.20), including those of the Ganges, Indus, and Mekong. In addition, in the uplifted Tibetan Plateau to the north of the Himalayas, lie the headwaters of important easterly flowing rivers, such as the Yangzte and Hwang Ho (the Yellow River).

The Himalayas also exert an influence far to the north. The air that travels northward over the mountains is dry because it has been stripped of its moisture. As a result, some of the worlds largest deserts occur in western China.

Together, the action of glaciers and running water in the Himalayas ensure the continued cycle of uplift and erosion. The tremendous erosional power of the Ganges is indicated by the amount of sediment it carries to the Indian Ocean, as much as 1450 million metric tons annually. This is higher than any other river in the world. By comparison, the Mississippi River carries a mere 310 million metric tons annually to the Gulf of Mexico.

As rivers flow to the sea, their paths are influenced by ongoing tectonic activity. Rivers exploit weaknesses in the bedrock, and fault zones are particularly weak because the rocks along them are highly fractured. As a result, many of the faults thought to be accounted for by the indenter model (see Figure 6.17) have major rivers flowing down their lengths. As we will learn in Chapter 10,

some of the world's earliest civilizations began on the floodplains of these famous rivers, which provide both a water supply and a fertile soil for agriculture.

In densely populated Asia today, many communities still live along these river valleys, relying on the river water for their survival. Unfortunately, this means that many communities are also prone to earthquakes, because the fault zones that the rivers follow are active ones. These rivers also overflow their banks each year, depositing fine-grained silt on the adjacent floodplain (see also Chapter 10). The silt provides a vital replenishment of nutrients to the soil in these agriculturally intensive regions. However, the flooding also brings tragedy. The Hwang Ho River is the deadliest in the world. In 1887, its flooding caused 900,000 fatalities, and in 1931, 4 million people died when flooding was followed by starvation and disease.

Indeed, it is in an effort to control the flooding of the Yangtze River (and harness its enormous energy for hydroelectric power)

---

crust is destroyed by subduction and rarely survives for more than 200 million years. As a result, direct evidence of sea-floor spreading in the ancient geologic record could not be found.

Today, most geologists believe that the essential elements of plate tectonic processes were in place by the beginning of the Proterozoic, some 2.5 billion years ago, and some believe that a modified form of plate tectonics may have commenced at the beginning of the Archean, about 4.0 billion years ago (see Chapter 13). However, it took a provocative paper by the Canadian geophysicist, J. Tuzo Wilson, entitled "Did the Atlantic Ocean close and then re-open?" to provide the key to unlocking the mysteries of these ancient plate motions.

6.7.1 The Wilson Cycle

In 1966, J. Tuzo Wilson proposed that the Appalachian–Caledonian mountain belt of eastern North America and western Europe (see Chapter 3) was formed by the destruction of an ocean that predated the Atlantic. In his model (Figs. 6.21a, b), the evolution of the Appalachian–Caledonian mountain belt began in the Late Precambrian and Cambrian (between 600 and 500 million years ago) with the deposition of thick sequences of shallow marine rocks at the margins of this former ocean. The similarity between these rocks and those of modern continental

shelves suggested that they had been deposited in a similar environment. But two distinct fossil assemblages occurred in these rocks. One assemblage called the "Pacific Realm" occurred along the northwestern (North American) side of the mountain belt, while the other, called the "Atlantic Realm," occurred along the southeastern (European) side. The distinctiveness of these fossil assemblages suggested that the shallow marine sedimentary rocks in which they occurred must have been separated by an ocean at the time of their deposition. This ocean could not have been the modern Atlantic, because the Atlantic Ocean only started to form with the breakup of Pangea approximately 200 million years ago. Instead, it must have been an older ocean that had been destroyed by the time Pangea formed.

The abundance of Ordovician volcanic rocks (490 to 440 million years old) with chemical compositions that resembled those of modern island arcs suggested that this former ocean began to subduct in the Ordovician. Extensive deformation, about 400 to 360 million years ago was attributed to continent–continent collision. This collision brought the two fossil assemblages together and resulted in the formation of the Appalachian–Caledonide mountain belt. Subsequent field work in Newfoundland has identified relics of this ancient ocean. Although the vast majority of its ocean floor was destroyed by subduction, slivers of oceanic crust were trapped by the colliding continents and have been preserved as ophiolites (see Chapter 5).

Figure 6.20

Radar of the Himalayan mountains, their drainage systems, and the Ganges River.

that the Chinese are now constructing the controversial Three Gorges Dam. When completed this will be the most powerful dam ever built, and China's most ambitious project since the Great Wall was constructed 2000 years ago.

Despite the monsoon's erosional on-slaught, the Himalayas remain the world's highest mountains because, with an uplift rate of 5 mm a year (about one inch every 5 years), they are still rising faster than they are being eroded. However, as erosion strips off the tops of the mountains ever deeper segments of the mountain belt are exposed at the surface as the crust rises to re-establish isostatic equilibrium (see Figure 5.8). In this way the Himalayas are a vivid example of isostasy and the rock cycle. As the guts of the mountain belt are exposed, igneous and metamorphic rocks formed deep within the crust are weathered and eroded, and the products of erosion are transported downstream to be deposited as new sediment in vast submarine fans at the mouths of the Indus and Ganges rivers.

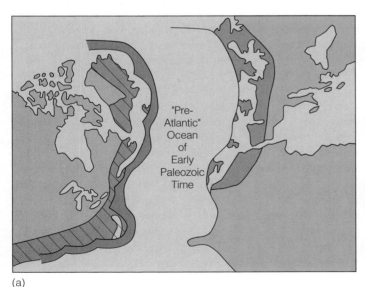

(a)

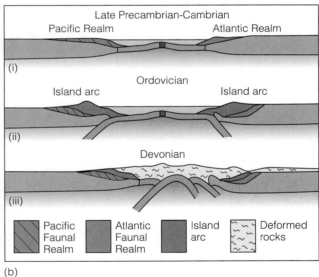

(b)

Figure 6.21

(a) Tuzo Wilson's model for the Appalachian–Caledonide mountain belt was based on the recognition of different faunal realms known as Pacific and Atlantic, and the inference that an ocean lay between them. The width of this ocean was unknown. He also interpreted a belt of Ordovician volcanic sequences as being remnants of an ancient island arc sequence formed by subduction. (b) The classic Wilson cycle applied to the Appalachian–Caledonide mountain chain. (i) Late Precambrian to Cambrian continental rifting led to the formation of an ancient ocean flanked by continental shelf deposits. (ii) Subduction in this ocean commenced in the Ordovician, dismembered the continental shelves, and resulted in the formation of island arcs. (iii) Continued subduction led to closure of this ocean which occurred some time in the Devonian.

Partial Melting and the Making of Continental Crust: Nature's Ultimate Recycling Program

As we learned in Chapter 5, new crust is created at mid-oceanic ridges, is transported away from the ridge by sea-floor spreading, and is ultimately consumed at a subduction zone. Collisions between buoyant continental crust and dense sea floor always result in the preferential subduction of oceanic crust. In fact, continents may grow laterally by scraping more buoyant crustal fragments off the descending oceanic plates. Hence, the interior regions of continents, like the continental shield of Canada, contain ancient rocks and are typically surrounded by belts of progressively younger rock (Fig. 6.22).

But why should mid-ocean ridges produce basaltic crust while subduction generates granitic rocks? The answer lies in the way magma is formed. When most substances begin to melt, there is always solid residue left behind because the process of melting is incomplete. We refer to this process as **partial melting.** Some chemical components are preferentially incorporated into the melt, whereas others are selectively retained in the residue. As a result, the composition of the melt is different from that of the residue, and the compositions of both melt and residue differ from that of the original material. So the composition of a liquid produced by partial melting is never the same as the solid starting material. In general, lightweight components tend to concentrate in the melt which, in geologic settings, will rise because they are buoyant and so are separated from the relatively dense residue.

This general principle also holds for the melting processes that occur in the mantle. As the ocean floors spread, partially molten mantle material wells up from beneath the mid-ocean ridges to form new ocean floor. This mantle is composed of the rock type known as peridotite which consists of a variety of minerals dominated by the magnesium-rich silicates, olivine, and pyroxene (see Chapter 2). Mixed in with these, however, are small amounts of lighter silicate minerals such as feldspar, which is an important component of basalt and granite. Partial melting of the mantle in the asthenosphere beneath the mid-ocean ridges selectively incorporates these lighter minerals much like distillation selectively concentrates lighter alcohol. The liquid produced by partially melting mantle rocks is consequently less dense than peridotite, and cools to produce the dark, moderately dense rock called basalt. Thus, the magma that rises buoyantly to fill the gap produced as two plates separate has the composition of a basalt because this is the composition of the liquid produced by partial melting in the mantle. The residue left behind in the mantle is referred to as **depleted mantle** because it is mantle material from which basalt has already been extracted.

At a subduction zone, in contrast, it is the basaltic ocean floor that is carried down into the hot asthenosphere. In contrast to its occurrence at mid-ocean ridges where it is the product of partial melting, basaltic crust at subduction zones becomes the *source* of partial melts. By the time basaltic crust reaches the subduction zone, it will have been altered as a result of its chemical interaction with the superheated waters of submarine vents and its long exposure to seawater. As a result, it will contain many minerals that can store water in their crystal structures. It will also be covered with a thick veneer of sediment, most of which will be scraped off at the trench although some will be subducted. The highly fractured oceanic crust also contains huge quantities of water which will be carried into the subduction zone. As a result of this varied intake, the melting processes associated with subduction are far more complex than those occurring at mid-ocean ridges.

Heating of the downgoing slab as it subducts recycles the basaltic oceanic crust, its veneer of subducted sediment, and its content of ocean water. This recycling is in fact a second fractionation since the starting material this time is predominantly basalt, the product of the first fractionation at mid-ocean ridges. Partial melting of the altered basaltic crust and any subducted sediment further concentrates the lighter minerals so that the liquids produced are less dense and more silica-rich. The water and other gases that are driven off the descending slab as it heats up, rise into the mantle rocks above the subduction zone where they promote partial melting in the mantle wedge. Because melting occurs within the wedge of mantle material, basalt liquids are once again produced, although they differ subtly in composition from those generated at mid-ocean ridges because they are produced at different depths. In ocean–continent collision zones, these basalt liquids may, in turn, trigger partial melting of the continental crust as they rise toward the surface. This is possible because the melting temperature of basalt is above 1100°C, whereas that of granite can be as low as 750°C. Just as hot water poured on butter causes the butter to melt, so the passage of basaltic liquids through continental crust may cause sufficient local

From the idea that oceans may close and then re-open came the concept of cycles of tectonic activity. In what came to be known as the "Wilson cycle," continental rifting following continental breakup leads to the development of an Atlantic-type continental shelf on which thick sequences of sedimentary rocks are deposited. Eventually, subduction commences leading to the development of volcanic arcs. When the rate of subduction exceeds the rate of sea-floor spreading, the ocean begins to close, culminating in continental collision. Many of the types of mountain building activity we have examined in this chapter are contained within this cycle. As the ocean begins to close, for example, mountains would be produced by subduction. These mountains would resemble those of volcanic island arcs or Andean-type continental margins. As subduction proceeds and the ocean gets narrower, terranes would become accreted to the continental margin. Eventually, when the oceanic crust is consumed, mountains are produced by continental

Figure 6.22

Map showing the major age provinces of North America. Note how the continental core of ancient rocks (cratons) tends to be surrounded by progressively younger mountain belts, suggesting that the continent has grown through a series of collisions. (b.y.a., billions of years ago)

heating to allow the granite to partially melt, further concentrating light minerals as a result. In this case, the ancient continental crust is recycled to produce magma, which ultimately cools and crystallizes to form new continental crust.

Each time partial melting occurs, the lighter minerals are concentrated in the magma so that, with every episode of subduction, the continents become progressively richer in silicon and poorer in iron and

magnesium. If this siliceous magma reaches the surface, the resulting volcanic rock is a rhyolite (recall from Chapter 2, that rhyolite is a felsic volcanic rock rich in light-weight elements). If the magma is trapped beneath the surface, it cools slowly to form the felsic plutonic rock, granite.

The process of partial melting not only explains the compositional difference between oceanic crust and the mantle from which it is derived, but it also accounts for

the composition of continental crust and its difference from that of the ocean floor. Partial melting selectively concentrates the lighter minerals so that, with every episode of subduction, the continents have become progressively more granitic in composition. Because these compositional differences have a strong influence on buoyancy, partial melting is also responsible for the fact that plate tectonics preferentially recycles oceanic crust while preserving continental crust.

collision. Thus the development of a mountain chain such as the Appalachian–Caledonide belt may involve several episodes of tectonic activity, beginning with subduction followed by terrane accretion, and culminating in continental collision.

The Wilson cycle is one of ocean opening and closing. But what determines when oceans open and close? The answer to this question may lie in an even larger tectonic cycle known as the supercontinent cycle.

6.7.2 The Supercontinent Cycle

Increased understanding of the processes involved in the Wilson cycle has led to an expansion of Wilson's ideas into what has become known as the Supercontinent cycle (Fig. 6.23). As we shall see, this approach places mountain building activity in a global context, and in so doing, demonstrates that continental breakup in one part of the world necessitates mountain building activity in another.

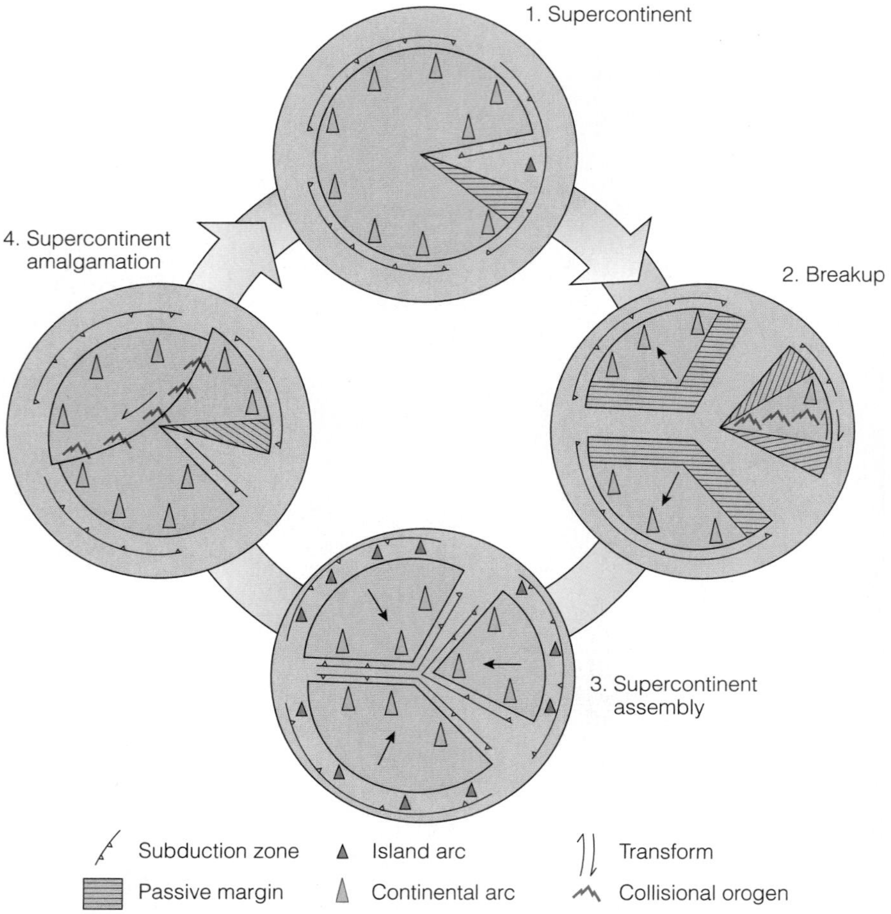

Figure 6.23
The Supercontinent cycle. The supercontinent (stage 1) is flanked by subduction zones which generate continental arcs. The supercontinent traps mantle heat beneath it resulting in break-up (stage 2). Breakup results in formation of new oceans and mid-oceanic ridges. The oceans are flanked by continental shelves, known as passive margins, that are characterized by thick sequences of sedimentary rocks. Eventually, the aging oceans become gravitationally unstable and begin to subduct (stage 3) and a new supercontinent begins to reassemble. This results in the destruction of the continental shelves and the formation of island arcs. Ultimately (stage 4), the oceans are destroyed by subduction and the new supercontinent amalgamates. This supercontinent traps mantle heat once again and the cycle starts anew.

1. Supercontinent

2. Breakup

3. Supercontinent assembly

4. Supercontinent amalgamation

Subduction zone ▲ Island arc Transform

Passive margin ▲ Continental arc Collisional orogen

Precise radiometric dating has shown that intense episodes of both collisional mountain building and continental rifting appear to occur at approximately 500 million year intervals (Fig. 6.24). For mountain building, the most recent episode occurred 300 million years ago with the amalgamation of the supercontinent Pangea. This was preceded by other periods of intense mountain building about 650 to 600 million, 1.1, 1.6, 2.1, and 2.6 billion years ago. These periods are likewise thought to represent episodes of supercontinent amalgamation. Similarly, the most recent period of continental rifting commenced about 200 million years ago with the breakup of Pangea. This episode was preceded by intense periods of rifting about 550 million, 1.0, 1.5, and 2.0 billion years ago, each of which is thought to represent a period of supercontinent breakup and dispersal. So the supercontinent that formed about 650 million years ago began to breakup 550 million years ago, and reassembled to form Pangea about 300 million years ago.

To understand the driving mechanism of this cycle, we must first examine why Pangea or any other supercontinent would break up. Several theories that might account for the breakup of supercontinents have been proposed. The most popular centers on the insulating properties of supercontinents. Because supercontinents are poor conductors of heat compared to thinner ocean crust, they act like an insulating blanket that blocks the escape of heat from the Earth's interior. Just as a book on an electric blanket gets hot because it prevents heat

from escaping, so a supercontinent causes heat to build up in the mantle beneath it. As the heat accumulates, magmas are generated in the mantle while the supercontinent arches upwards and cracks. As the magma is injected into the cracks, the supercontinent is progressively pried apart. Eventually, continental breakup takes place and new oceanic crust is created as the continental fragments drift apart (Fig. 6.23, stage 2).

If this model is correct, supercontinents have in-built obsolescence; their very formation sowing the seeds of their own destruction. As the vast volumes of magma injected into the supercontinent are derived from the mantle, they are predominantly basaltic in composition. These volcanic outpourings can be so voluminous that they are commonly known as **flood basalts.** As the supercontinent breaks up, oceans are generated between the dispersing continental fragments. These so-called "interior oceans" are fringed by continental shelves just like the present-day Atlantic Ocean. As the interior oceans expand, however, the "exterior ocean" surrounding the supercontinent must shrink, just as the modern Pacific Ocean has shrunk over the past 200 million years. In this way, subduction and the accretion of terranes along the western margin of North America may well be related to the opening of the Atlantic Ocean and the breakup of Pangea.

But the continents cannot continue to drift apart indefinitely. As we learned in Chapter 5, oceanic lithosphere becomes colder and denser as it ages. When it is 80 million years old, it has the same density as the underlying asthenosphere. Older ocean-

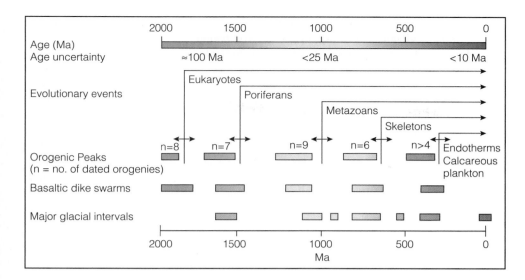

Figure 6.24
Summary of periodic events over the past 2 billion years. Orogenic peaks represent episodes of mountain building and supercontinent amalgamation. Basaltic dike swarms represent episodes of crustal extension indicating continental rifting. Note that these rifting events lag orogenic peaks. Evolutionary events occur during periods of thick shelf sediment accumulation that is indicative of the formation of new oceans. Major glacial intervals (ice ages) tend to coincide with supercontinents.

ic lithosphere is therefore gravitationally unstable and has a tendency to subduct. Forces that act against subduction, such as the rigidity of the lithosphere (which prevents it from bending) and frictional resistance to the descent of the oceanic slab, prolong the lifespan of the oceanic lithosphere. But eventually this resistance is overcome and subduction is inevitable (Fig. 6.23, stage 3). The geologic record suggests that this occurs before the oceanic lithosphere attains an age of about 200 million years. The margins of the central Atlantic Ocean are floored by oceanic crust as much as 180 million years old. This part of the ocean is therefore approaching its maximum age and should soon begin to subduct. When that happens, the Atlantic will begin to shrink, and another supercontinent will form in about 200 million years when the opposing continental margins collide.

When a supercontinent amalgamates (Fig. 6.23, stage 4) mountain belts related to the collision of individual continents will come to lie in the interior of the supercontinent. As the interior oceans close and subduction within them ceases, consumption of oceanic lithosphere is likely to become concentrated around the periphery of the supercontinent where volcanic arcs and Andean-type mountain belts will form. The relationship between these "interior" and "peripheral" mountain belts is most clearly seen in the positions of those associated with Pangea (Fig. 6.25). When the supercontinent breaks up once again, the peripheral mountain building activity is likely to continue along the leading edges of the dispersing continental fragments, particularly where these are oriented roughly perpendicular to the direction of breakup.

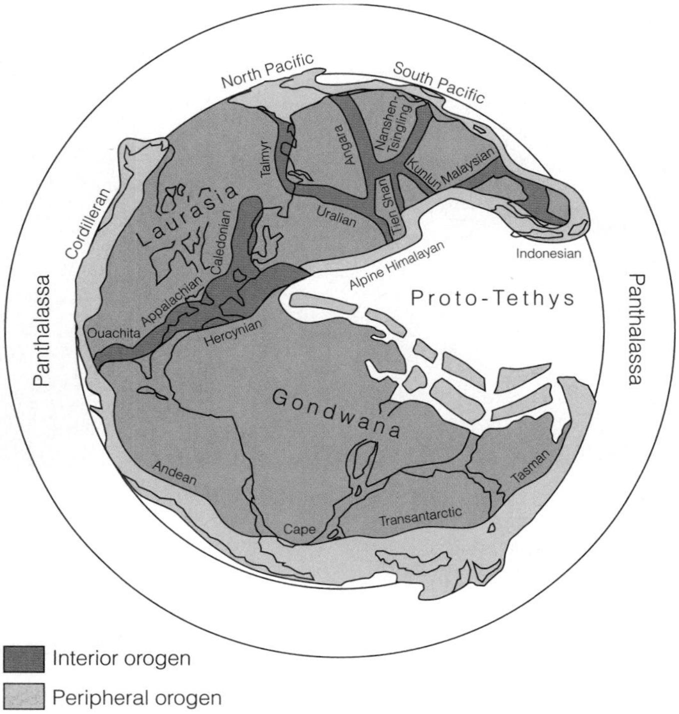

Figure 6.25
Late Paleozoic reconstruction of the supercontinent Pangea showing the distribution of interior orogenic belts associated with continent–continent collision and peripheral orogenic belts around the edge of the supercontinent that are characterized by volcanic arcs and terrane accretion.

Interior orogen

Peripheral orogen

The generation and destruction of the interior oceans are similar to that envisaged in the Wilson cycle. The Supercontinent cycle, however, draws attention to the relationship between the generation of interior oceans, and processes taking place in the exterior ocean. The potential relationship between the breakup of Pangea and mountain-building activity in western North America is an example of this. In addition, the model offers a mechanism that may explain why supercontinents occur, and may account for the apparently episodic nature of mountain-building activity.

6.8 Mountain Building and Deformation of Rocks

Mountain building subjects rocks to stresses that differ substantially from those under which they formed. Because of this, deformation and metamorphism of rocks are manifestations of mountain-building activity (Fig. 6.26). As we learned in Chapter 2, metamorphism refers to changes in the mineral content and texture of rocks, whereas deformation results in changes in their shape and volume. The development of metamorphic foliations (see page 46), for example, occurs as the result of applied stresses and so is directly related to the deformation that accompanied metamorphism.

The response of rocks to applied stresses depends on the temperature, pressure, and strain rate (Fig. 6.27). Low temperatures and pressures, as well as rapid strain rates, favor brittle behavior whereas high temperatures and pressures, along with slow strain rates, favor bending and smearing. The effect of strain rate on rocks is similar to the behavior of butter taken from the refrigerator. If a stick of butter is subjected to a high strain rate by abruptly bending it, it will break in a brittle fashion. However, if the stick is subjected to a low strain rate by flexing it slowly, it will behave in a ductile fashion and bend or smear.

Applied stresses can be of several types (Fig. 6.28). They may be compressional such that a rock is deformed by inward-acting stresses or they may be extensional and outward acting. The stresses associated with mountain building are usually compressional. Extensional stresses occur where the Earth's crust is being ripped apart, as is the case in the East African Rift Valley we examined in Chapter 5. Shear is another form of applied stress that results when forces act parallel to each other but in opposite directions. We have encountered this form of stress on faults like the San Andreas, where one crustal block slips sideways past another.

6.8.1 Faults

Because of these controls on rock behavior, rocks near the Earth's surface, where temperatures are relatively low and strain rates are usually high, typically behave in a brittle fashion and break during deformation. This results in the formation of fractures. Once a fracture forms, however, a plane of weakness is produced along which motion may occur. Fractures on which significant movement has taken place are called **faults** (Fig. 6.28). In the rock record, faults are identified by the severing and displacement of once continuous rock

Figure 6.26
Deformation of stratified rocks in western Newfoundland during the formation of the Appalachian mountain belt. (Photo John Waldron)

Figure 6.27
Rocks may respond to stress either in (a) a ductile fashion and bend to form a fold, or (b) in a brittle fashion and break to form a fault.

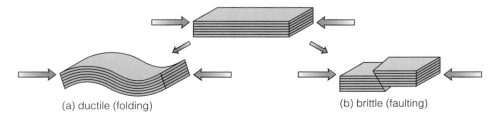

(a) ductile (folding)   (b) brittle (faulting)

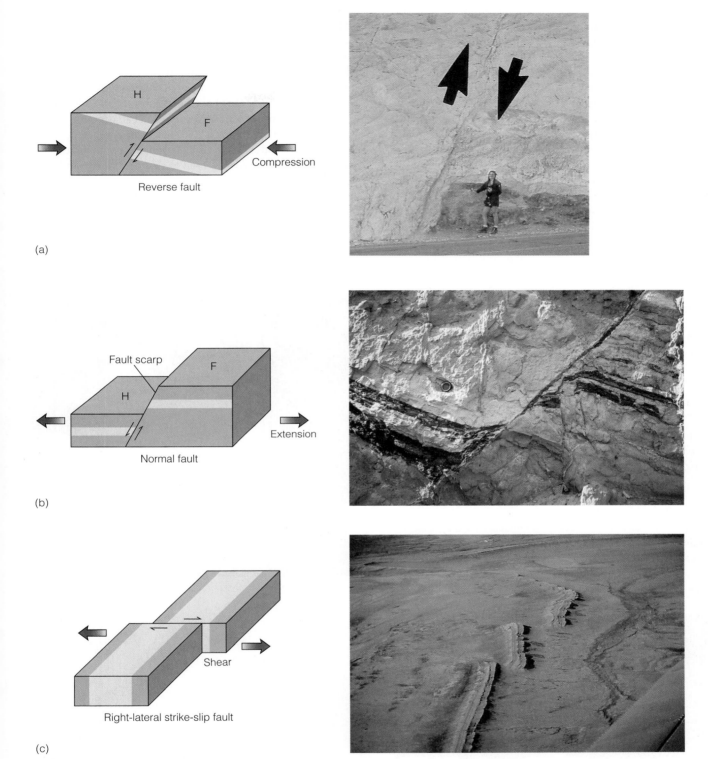

Figure 6.28

The relationship between stress and faulting. (a) Inward-acting stresses are compressional. They are typical of mountain building and produce reverse faulting and thrusts where the hanging wall is heaved upward relative to the footwall. (b) Outward-acting stresses are extensional and are typical of rifts. Dark brown layers of siltstone are truncated and offset by this inclined fault. The rocks on the hanging wall (left) were transported downward relative to the rocks in the footwall (right) (Photo Brendan Murphy). (c) Shear occurs when stresses act parallel to each other but in opposite directions. In this style of deformation, one crustal block slips sideways past another. For inclined faults (a) and (b), H denotes hanging wall, F denotes footwall.

layers. In modern settings, displacements along faults are a direct cause of earthquakes. Hence, faults recognized in the rock record provide evidence of ancient seismic activity.

The type of faulting that occurs depends on the applied stresses. Many faults are inclined relative to the horizontal. If so, one block occurs above the fault plane, and is known as the "hanging wall," so named by miners who hung their lanterns on this block. The other block, on which the miners stood, occurs below the fault plane and is known as the "footwall."

In compressional environments, the hanging wall is heaved upward relative to the footwall. If the fault is relatively steep, it is known as a **reverse fault,** if it is shallow it is termed a **thrust fault** (Fig. 6.28a).

In extensional environments, the hanging wall moves downward relative to the footwall, a situation known as a **normal fault** (Fig. 6.28b). This type of faulting is typical of extensional environments such as mid-oceanic ridges and continental rifts (see Chapter 5). Note that a gap may occur between beds in the vicinity of the fault.

Faults like the San Andreas Fault, where one block of crust moves sideways past another, are known as **strike-slip faults** because the movement, or **slip** of the fault is parallel to its surface trace or strike (see Figure 6.28c). Transform faults are a special type of strike-slip fault that occur along a plate boundary.

6.8.2 Folds

At deeper levels in the Earth's crust, where temperatures and pressures are elevated and strain rates are slow, rocks deform by bending and smearing, rather than by breaking. Like the pages

of a bent paperback book, rock layers at depth deform into buckles known as **folds** in which the layers are alternately arched upward to form **anticlines** and downward to form **synclines** (Fig. 6.29a). Just as the sequence of pages in the paperback is not changed when it is bent, folded rock layers generally maintain their original sequence during deformation even though their geometry may have been substantially altered.

Folds are the result of compression, and can occur on a wide variety of scales, ranging from millimeters to kilometers across (Figs. 6.29b, c). Folds also have highly variable orientations. Two important features of folds are used to describe their orientation (Fig. 6.30). First, the fold is divided into two by an imaginary surface known as the **axial plane,** the orientation of which is perpendicular to the direction of compression. The orientation of this plane can be measured. An "upright" fold, for example, has a near vertical axial plane (see Figure 6.30b), whereas an "inclined"

Figure 6.29
Photographs of folds at various scales. (a) Anticlines and synclines, Minas Shore, Nova Scotia, Canada. (Photo Brendan Murphy) (b) Complicated folding patterns in metamorphic rocks, Nova Scotia (note dime for scale). (Photo Brendan Murphy) (c) Large scale anticline in the Canadian Cordillera. (Photo Glen Stockmal)

(b)

(a)

(c)

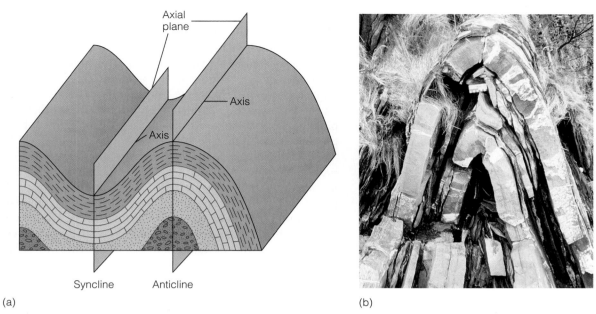

(a)

(b)

Figure 6.30
(a) The location of the axial plane and axis of folds. (b) A fold with a vertical axial plane and a horizontal fold axis. (Photo John Waldron)

fold has one that is tilted (see Figure 6.26). Folds with approximately horizontal axial planes are termed "recumbent." The axial plane cuts across each of the folded layers along an imaginary line known as the **fold axis**. The orientation of this line can also be measured, such that folds can be described as "horizontal" or "plunging" depending on the slope of their fold axis.

When metamorphism accompanies folding, the metamorphic minerals tend to flatten, rotate, or recrystallize perpendicular to the direction of compression. Well-cleaved minerals such as muscovite, biotite, or chlorite impart these characteristics on the metamorphic rock, resulting in the development of a splitting foliation or cleavage (see Chapter 2) parallel to the axial plane (Fig. 6.31).

(a)

(b)

Figure 6.31
The distinction between bedding and foliation. (a) Bedding is essentially horizontal and defined by differences in composition and color, whereas the foliation is vertical, and reflects the preferred orientation of platey minerals. (b) Bedding is folded whereas the foliation has a consistently steep orientation perpendicular to the direction of compression. (Photos John Waldron)

6.8.3 Unconformities Revisited

In Chapter 3, we learned how James Hutton deduced that the beds of strata beneath an unconformity on the River Jed (Fig. 3.4) must have been tilted prior to erosion and the deposition of the upper layers. This tilting is a manifestation of the stresses applied to the lower layers following their deposition. As Hutton correctly deduced, a significant time gap must have existed between the deposition of the upper and lower strata, during which time portions of the lower strata were removed by erosion. Therefore, the unconformity indicates that a segment of the geologic record is missing at that locality. On the River Jed, the relationship is an **angular unconformity** because an angular discordance exists between the upper and lower strata. We are now in a position to place Hutton's ingenious insights into a plate tectonic context. For example, a spectacular angular unconformity is exposed along the shoreline of Nova Scotia, Canada (Fig. 6.32). The lower layers consist of steeply dipping Carboniferous rocks (about 350 million years old), and the upper layers consist of shallowly dipping Triassic rocks (about 220 million years old). The Carboniferous rocks were deformed during continental collision associated with the amalgamation of Pangea. This collision generated mountains which were then eroded. The Triassic rocks were deposited as Pangea broke up after the collisional event, and contain fragments derived from the erosion of these mountains.

Besides angular unconformities, two other types of unconformity can occur in nature. A **nonconformity,** occurs when a plutonic igneous rock or a metamorphic rock is directly overlain by younger sedimentary strata (Fig. 6.33a). Because both plutonic and metamorphic rocks form beneath the Earth's surface, they must have been exhumed by uplift and erosion prior to the deposition of the younger strata. This history is similar to that which is occurring in the modern Himalayas where uplift and erosion expose the internal guts of the mountain chain. The overlying strata are significantly younger than the plutonic rocks so that, once again, a significant portion of the geologic record is missing.

The third type of unconformity, a **disconformity,** is a more subtle one (Fig. 6.33b). The basic definition of unconformities still applies, namely that the rock units above and below the unconformity surface have significantly different ages. However, if the lower units were not tilted, but merely uplifted and eroded prior to the deposition of the upper units, they would remain essentially horizontal, and therefore, parallel to the younger strata deposited on top of them. Because the older and younger strata are parallel, identification of a disconformity is more difficult than either an angular unconformity or a nonconformity. Identification is best achieved by precisely dating the two sequences using fossil assemblages. Dating will determine if a significant portion of the geologic record is missing between the deposition of the two sets of strata. Because the lower units are not folded, it is unlikely that they were directly involved in collisional mountain building. In fact, disconformities tend to occur in regions that are distant from those most intensely affected by mountain building. They are also developed when continental shelves become exposed by falling sea level. As we shall see in Chapter 8, sea level has fluctuated throughout geologic time, primarily in response to tectonic and glacial events.

Figure 6.32

Photograph of an angular unconformity in the cliffs of the Minas Basin, Nova Scotia, Canada. The steeply dipping beds are Early Carboniferous continental clastic sedimentary rocks and were deposited before Appalachian continental collision during the formation of Pangea deformed them. The gently dipping Triassic beds above are also continental clastic rocks, but were deposited after the formation of Pangea and were derived from the erosion of the Appalachian mountains as Pangea began to break up. (Photo Brendan Murphy)

(a)

(b)

Figure 6.33

(a) Nonconformity surface in Scotland separating sedimentary rocks (above) from metamorphic rocks (below). Because the metamorphic rocks were formed well below the Earth's surface and the sedimentary rocks were deposited on the surface, they could not have been deposited in sequence and a significant gap must exist between their respective ages. (b) Disconformity surface separating Early Ordovician (pale) from Late Ordovician (dark) rocks. (Photos John Waldron)

6.9 Mountain Building and Geologic Maps

The geologic map, which shows the distribution of the various rock types exposed at the Earth's surface, is the basis for unraveling the complex evolution of both modern and ancient mountain belts. These maps are the cornerstone of tectonic interpretations and show the distribution and orientation of all rock units at the surface of the Earth and the nature of the geologic contacts between them.

The contacts between strata can be either conformable, which indicates the strata were deposited in sequence with no significant break between them, or unconformable, which suggests a considerable portion of the geologic record is missing, or faulted, which indicates that the original relationship between the strata has been severed by displacement along the fault surface.

The identification of the nature of geologic contacts is crucial to the recognition of the tectonic events responsible for them, and illustrates the importance of being able to describe the various orientations of strata. In our discussion of geologic maps in Chapter 3 (Dig In: Geologic Maps, pp. 64–67), we learned that two parameters, known as strike and dip, are used to define the

orientation of surfaces such as bedding planes. The **strike** of an individual bed is the compass direction of an imaginary horizontal line on the bedding plane measured relative to North. The **dip** is a measurement of the maximum slope of the plane.

Using this approach, the presence of folds, faults and unconformities can be deduced from outcrops (Fig. 6.34) and maps (Fig. 6.35). Where rock strata have been bent into folds, for example, a characteristic outcrop pattern is often produced in which individual beds zigzag across the map. The noses of anticlines point in the direction of their plunge, whereas those of the synclines point in the opposite direction (Fig. 6.35). In addition, the strike and dip of the beds change around the fold; note especially that the strike direction is parallel to the contact between the rock units.

In Dig In: Geologic Maps, pp. 64–67, we compared the construction of a geologic map to that of completing a jigsaw puzzle. Outcrops are pieces of the jigsaw puzzle. Many rocks are hidden beneath a cover of soil and vegetation, and these represent pieces that are missing from the puzzle. Nevertheless, a geologic map is required to show the nature of these buried rocks, as if the landscape was stripped of its soil and vegetation. Now imagine how you would deduce the presence of folds if the rocks in the area were poorly exposed.

In Figure 6.36, the outcrops are color-coded according to rock type; blue is used for limestone, grey for shale, and yellow

Figure 6.34
Identification of a fold in outcrop. Because the water line is horizontal, it indicates the direction of strike of the bed for each outcrop. The dip is the line of maximum inclination along the plane and can also be obtained at each outcrop (see also Figure 3.14). Note that the bedding changes its orientation from left to right and in so doing defines the outline of a fold. The layers arch upward and so the fold is an anticline. (Photo Damian Nance)

for sandstone. The symbol **V** is used for volcanic rock. Although only 40% of the area has exposed bedrock, there is sufficient information to produce a well-constrained geologic map by "filling in the gaps" between the outcrops. Several outcrops expose the boundary, or contact, between two adjacent rock units. These outcrops are particularly important because they provide vital control on the accuracy of the map. These contacts can be traced parallel to their strike across the unexposed terrain. If this approach is followed on both sides of the stream, it is obvious that the rock units converge toward the north. In fact, the strike and dip of rock units change systematically around the fold. Note that in the central portion of the fold, known as the "hinge", the orientation of the beds is significantly different from that on the sides ("limbs") of the fold, that occur to the east and west. Even in this location, however, the strike remains parallel to the contact between adjacent rock units.

In Figure 6.37, many of the layers are folded in a similar fashion to Figure 6.36. In the southern part of the map, however, it is clear that the conglomerate (brown) outcrops do not follow this pattern. Instead, the layers dip gently to the south, and their outcrop pattern "cuts across" the fold. Using the principle of cross-cutting relationships (Chapter 3), this suggests that the conglomerates were deposited after folding had taken place, indicating the presence of an angular unconformity between the conglomerate beds and the folded strata.

In Figure 6.38, the folded pattern in the north of the map cannot be simply extrapolated in the direction of their strike to the south. Although the same rock units occur, they are displaced relative to those in the north. This is an example of a fault which has truncated and displaced the rock units along the fault plane (see also Fig. 6.28).

A typical geologic map has many of the above elements. In Figure 6.39, for example, the limestone and sandstone layers are folded in the central and eastern part of the map. The granite truncates these layers. This is typical of bodies such as granite, which intrude across rock strata as they ascend toward the surface. This relationship indicates that the granite is younger than the limestone, and sandstone, and younger than the deformational event that folded them. The conglomerate, which has a very shallow dip, is not folded, and cuts across the granite, the limestone, and the sandstone. Where the conglomerate overlies the limestone and sandstone (A–B and C–D, Fig. 6.39), the contact is an angular unconformity. According to the principle of cross-cutting relationships, this indicates that the conglomerates are younger than the limestone and sandstone. Where it overlies the granite (B–C), the contact is a nonconformity.

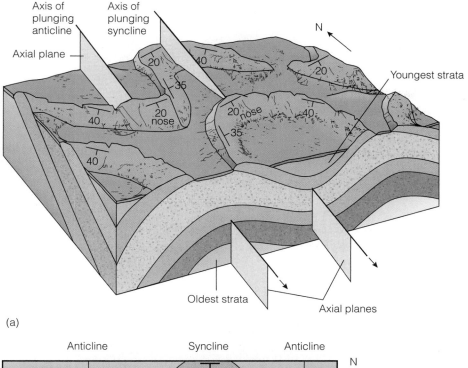

Axis of plunging anticline

Axis of plunging syncline

Axial plane

20
40
35
20
20
nose
40
20
nose
40
35
40
40

N

Youngest strata

Oldest strata

Axial planes

(a)

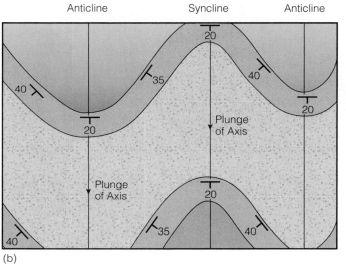

Anticline   Syncline   Anticline

N

20

35

40

40

20

20

Plunge
of Axis

20

Plunge
of Axis

40

35

40

(b)

Figure 6.35
(a) Block diagram of an anticline and syncline. Note that the closure or "nose" of the anticline points in the direction of plunge. The nose of the syncline points in the opposite direction. (b) Zigzag outcrop pattern of folds portrayed on a map, with appropriate dips and strikes.

To the west, a normal fault separates the conglomerate from metamorphic rocks. Because the fault cuts the conglomerate, and nothing cuts the fault, the fault must be the youngest feature in the area. The relationship of the metamorphic rocks to the sedimentary rocks to the east is unclear. The normal fault could have caused considerable displacement, so that the location of the metamorphic rocks relative to the sedimentary rocks prior to the faulting is uncertain. Identifying ambiguous relationships is one of the key roles played by geologic maps, and in this case has revealed a problem that needs further attention. Perhaps the age of metamorphism could be determined by radiometric dating (see Chapter 3), or the composition of the clasts in the conglomerate could be examined to determine whether they contain fragments of the metamorphic rocks. Such clasts would show that the metamorphic rocks were older,

and indeed were exposed at the surface at the time the conglomerate was deposited.

When a region has been adequately mapped, specimens from the outcrops are selected for laboratory investigation which might include geochronological, paleontological, geochemical, and paleomagnetic analysis.

In summary, geologic maps identify key features, such as stratigraphy, igneous intrusions, folds, faults, and unconformities, from which the evolution of the mapped area can be deduced. These maps show the distribution of rock units at the surface of the Earth, and must necessarily infer this distribution where the rocks are hidden beneath a cover of soil and vegetation. Orientational data taken from the outcrop (such as the strike and dip of bedding) and the interpretation of exposed geologic contacts are crucial to the completion of the map.

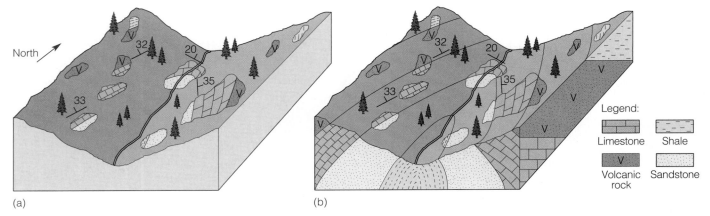

Figure 6.36
Construction of a geologic map. (a) Outcrops plotted with appropriate strike and dip data for beds. Patterns indicate rock type. (b) Interpretation of the data showing the distribution of contacts. Note these contacts are parallel to the strike of adjacent rocks.

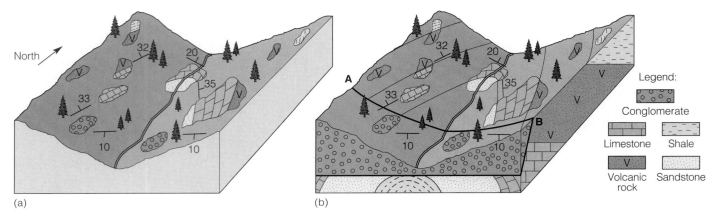

Figure 6.37
Similar map to Figure 6.36, except for the presence of a conglomerate in the southern part of the map area. The map shows the conglomerate cutting across the folded strata to the north indicating the presence of an angular unconformity (A-B).

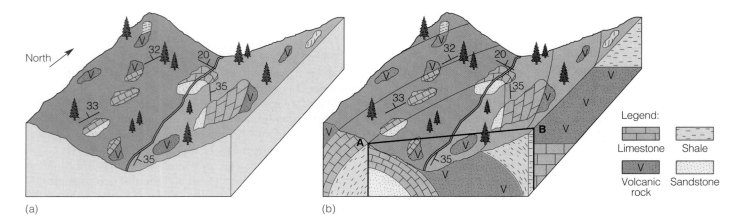

Figure 6.38
Similar map to Figure 6.36 except in the southern part of the map area where the folded pattern in the north cannot be simply extrapolated. The map interprets the rocks in the southern part of the map to have been displaced to the west by a fault (A-B).

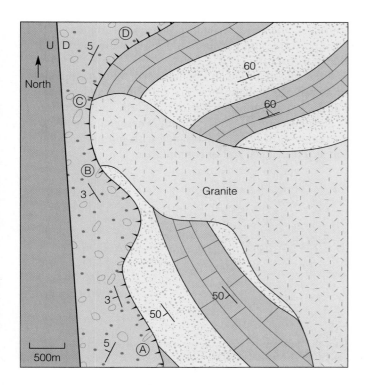

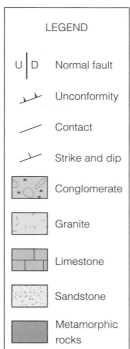

LEGEND

U | D   Normal fault

   Unconformity

   Contact

   Strike and dip

   Conglomerate

   Granite

   Limestone

   Sandstone

   Metamorphic rocks

Figure 6.39

Typical geologic map. The normal fault in the west of the map area cuts the conglomerate, and is the youngest event in the area. The conglomerate was deposited unconformably upon the granite (B–C), sandstone and limestone (A–B and C–D). The contact along B–C is a nonconformity, along A–B and C–D it is an angular unconformity. The granite intrudes the limestone–sandstone sequence and is therefore younger than both the limestone and sandstone, but is older than the conglomerate. The relationship between these rocks and the metamorphic rocks to the west of the normal fault is unknown because there are no rock types in common on either side of the fault. (U and D denote upthrown and downthrown sides of fault, respectively.)

6.10   Chapter Summary

Mountains form along each type of plate boundary and in the interior of plates. At divergent boundaries, mid-oceanic ridges stand 2–3 km (6500–10,000 ft) above the sea floor. Their height is related to the buoyancy provided by upwelling of magma from the asthenosphere. In the interior of oceanic plates, focused upwellings of hot mantle material known as plumes can create mountains that stand up to 6 km (20,000 ft) above the sea floor.

At convergent margins, mountains can be produced by subduction, by the accretion of terranes, and by continental collision. Subduction-related mountains can be produced by convergence between two oceanic plates (in which case the oldest plate is subducted) or by convergence between an oceanic and continental plate. In the latter case, the dense oceanic plate is subducted beneath the continental plate, as is the case in the modern Andes.

The process of subduction ultimately leads to ocean closure and continental collision. However, smaller collisions are first produced as mid-oceanic ridges, island arcs, microcontinents, and other high-standing features of the ocean floor encounter the trench and are scraped off the descending plate and accreted or attached to the overriding plate. These encounters produce a complex mosaic of accreted terranes each with its own set of geologic characteristics, and are now recognized as playing an important role in the early evolution of mountain belts. The west coast of North America has been built up by the accretion of more than 200 terranes which have widened the North American continent by an average of 600 km (360 mi).

The most dramatic plate collisions occur when subduction destroys the oceanic crust between two continents, forcing one continent to override the other along a suture zone. The buoyancy of continental lithosphere ultimately brings subduction and its associated volcanic activity to a halt. But the ensuing collision and crustal thickening produce massive landlocked mountain ranges and cause great crustal blocks to move upward and sideways along huge faults that can develop far from the former plate edges. The collision between India and Asia is a modern example of continental collision, and models show that regions thousands of kilometers away can be affected. Mountains along transform faults are produced by local compression along bends in the fault zone.

A critical problem in evaluating the role of plate tectonics in ancient mountain belts is that oceanic crust more than 200 million years old is very poorly preserved. However, J. Tuzo Wilson's suggestion that the Appalachian–Caledonide mountain belt was formed by continental collision after the destruction of an ancient ocean, opened the door to the application of plate tectonics to ancient mountain belts. The idea that oceans closed and then reopened spawned the concept of the Wilson cycle. The cycle begins with a rifting event which leads to breakup and dispersal of continental fragments and the generation of a new ocean. Eventually, subduction begins in this ocean, and the ocean begins to close, culminating in continental collision.

The Supercontinent cycle is an expansion of Wilson's ideas and reflects an increased understanding of plate tectonic processes. According to this cycle, intense episodes of mountain building that occurred about 2.6 billion, 2.1 billion, 1.6 billion, 1.1 billion, 650 million, and 300 million years ago reflect times of supercontinent amalgamation, whereas widespread evidence of rifting about 2.0 billion, 1.5 billion, 1.0 billion, 550 million, and 200 million years ago represent times of supercontinent fragmentation. Fragmentation of a supercontinent is attributed to magmatism associated with the buildup of heat beneath the insulating continental crust. Supercontinent amalgamation occurs because the oceanic crust that forms between the dispersing continental fragments becomes gravitationally unstable as it ages and must eventually subduct. As it does so, the adjacent continental margin may undergo subduction-related orogeny, followed by the accretion of terranes, and finally continental collision.

Mountain building is accompanied by deformation and metamorphism. Deformation near the surface of the Earth generally occurs at low pressure, low temperatures, and high strain rates, and generates faults. At deeper crustal levels, higher temperatures and pressures and lower strain rates result in folding or smearing.

Geologic maps are an essential element of tectonic interpretations. These maps show the distribution and orientation of all rock units at the surface of the Earth and the nature of the geologic contacts between them. These maps facilitate the interpretation of features associated with mountain building, such as folds, faults, and unconformities.

Key Terms and Concepts

Look for the highlighted items on the web at:
WWW.BROOKSCOLE.COM

- Mountains form at mid-oceanic ridges crested by upwelling of magma from the asthenosphere

- Mountains also form in the interior of oceanic plates at hotspots, which are situated above focused upwellings of hot mantle, known as mantle plumes

- At convergent margins, mountains can be produced by subduction, by the accretion of terranes, and by continental collision

- Mountains formed above subduction zones are related to compression and to the generation and ascent of magma

- Where two oceanic plates converge, the oldest plate is preferentially subducted and mountains are formed as basaltic magma rises to the surface to form curved lines of volcanic islands called island arcs

- Where oceanic and continental plates converge, the presence of abundant hot magma in continental crust coupled with compression causes volcanic mountain belts to form parallel to the coast

- Material scraped off a subducting plate forms an accretionary wedge

- Elevated areas on the ocean floors, such as island arcs, seamounts, ocean ridges, and microcontinents are accreted to the continental margin as terranes

- The effects of continent–continent convergence extend far beyond the suture where collision occurs, and are often preceded by the accretion of terranes

- Movement on curved transform faults creates pull-apart basins at divergent bends and transverse ridges at convergent bends

- The Wilson cycle of tectonic activity commences with rifting and the development of an ocean followed by subduction and the formation of volcanic arcs, and finally continent–continent collision

- Intense episodes of mountain building and continental rifting are explained by a Supercontinent cycle with a duration of about 500 million years. Supercontinent amalgamation is followed by breakup and the generation of new oceans, and then by subduction which ultimately leads to reassembly of a supercontinent

- Mountain building is accompanied by deformation and metamorphism

- Tectonic events may be deciphered by geologic maps, which show the distribution of rock units near the Earth's surface

- Geologic maps allow the recognition of features associated within mountain building, such as folds, faults, and unconformities

- Folded layers are alternatively arched upward and downward to form anticlines and synclines

- Fold orientation is described by an axial plane which is an imaginary surface perpendicular to the direction of compression and a fold axis, an imaginary line where the axial plane cuts across each of the folded layers

- Unconformities form where there is a significant time gap in the geologic record

- An angular unconformity occurs where there is an angular discordance between the upper and lower strata, a nonconformity occurs where a plutonic or metamorphic rock is directly overlain by sedimentary strata, a disconformity occurs where rock strata above and below the unconformity are parallel

Review Questions

1. What were the main problems with the geosynclinal model for mountain building?

2. What are hotspots and mantle plumes?

3. How has the accretion of suspect terranes contributed to the growth of western North America?

4. What are the implications of the indenter model for the pattern of faulting associated with the collision of India and Asia?

5. Why are bends in continental transform faults said to be either divergent or convergent, and what features are associated with each?

6. What is the fundamental problem for interpreting the role of plate tectonics in ancient mountain belts?

7. Summarize the reasons why supercontinents eventually fragment and break up.

8. Under what conditions will rocks respond to stresses in a brittle manner?

9. Why are disconformities difficult to recognize in the geologic record?

10. What is the difference between the axial plane and the fold axis of a fold?

Study Questions

 Research the highlighted question using InfoTrac College Edition.

1. What are the fundamental differences between the genesis of mountains in oceanic crust and the genesis of those in continental crust?

2. What aspects of plate tectonic theories account for deficiencies in the geosynclinal model for mountain building?

3. What is the evidence that mantle plumes exist beneath hotspots?

4. Explain how the accretion of terranes along the western margin of North America may be related to the fragmentation of Pangea.

5. In the indenter model for continental collision, Asia was modeled with plasticine, whereas India was modeled with a rigid block. Why not the other way around?

6. What is the relationship between the formation of oceans and the genesis of mountain belts?

7. In what way does the Supercontinent cycle build on the concepts of the Wilson cycle?

8. Explain why the San Andreas Fault is an example of shear stresses acting along a plate boundary.

9. By cutting a piece of paper along a curved line, explain how mountains and basins can form along continental transform faults.

10. At an unconformable contact, part of the geologic record is missing. How does this happen, and what is its potential geologic significance?

CHAPTER 7

The Earth's Interior

Our knowledge of Earth materials and the concepts of geologic time and the theory of plate tectonics sets the stage for an examination of the solid Earth. We now know that the Earth is 4.6 billion years old and that during its long history, the face of the planet has changed many times as a result of plate tectonic recycling. But what is the Earth made of and how do we know? And what processes go on in the Earth's interior that make plate tectonics possible?

The only rocks available for study are those at the Earth's surface and those recovered by drilling or from mines. Our direct knowledge of the Earth's interior is consequently confined to just a few tens of kilometers of the Earth's surface. With its radius of almost 6400 km (4000 mi), this represents only a minute fraction of the Earth itself, and it is most unlikely that the rocks exposed at the surface are in any way representative of those of the Earth's interior. So what is the inside of the Earth made of, and how have we determined this?

To learn about the interior of the Earth we must use indirect means, employing phenomena such as earthquakes, gravity, magnetism, and heat flow to apply the laws of physics to processes in the Earth's interior. Earthquake waves, which can be used in a similar fashion to X-rays, have provided the most detailed information about the Earth's internal structure. They have revealed the Earth to be a layered planet with a massive metallic core some 3500 km (2150 mi) in radius at the center, around which is a 2900 km (1800 mi) thick rocky layer or mantle with a thin skin or crust rarely more than a few tens of kilometers wide. They further show that each of these layers can be subdivided. Oceanic crust is shown to be distinct from continental crust, the mantle can be separated into upper and lower portions, and a liquid outer core apparently envelopes a solid inner one.

Besides showing differences in composition, the Earth's interior is also layered according to density, from the center of the core to the outer reaches of the atmosphere. As a result of a wide variety of geologic processes, heavy elements have sunk to become concentrated in the Earth's core, whereas light elements have risen toward the surface layers where plate tectonics occurs. The more gaseous elements have escaped the Earth's interior altogether and are now concentrated in the hydrosphere and atmosphere.

From studies of the Earth's gravity, we have deduced that the density of the Earth increases dramatically with depth, and that while the mantle is made of rocky material, the core is predominantly iron. From the pattern of circulation in the liquid outer core and the pattern of heat flow in the mantle, we are beginning to understand the way the Earth's interior circulates and, in consequence, what powers the Earth's magnetic field and what drives plate tectonics.

7.1 Introduction

We learned in Chapter 2 that the crust of the Earth is made up of igneous, sedimentary, and metamorphic rocks, upon which lies a surface layer of sediment, soil, and water, and the plants and animals that make up the biosphere. This crust, together with the rigid upper mantle, comprises the lithosphere, the rigid outer shell of the Earth. We also learned in Chapters 4 and 5 that this lithosphere, which floats on a pliable weak layer called the asthenosphere, is divided into plates and that relative motion between the plates results in the phenomenon we call plate tectonics.

But what of the Earth beneath the lithosphere? As it is hidden from direct observation, we must use indirect means to investigate its composition and structure. The task is a challenging one. Although deep boreholes have been drilled into the crust, none have penetrated to depths greater than 12 km (7.5 mi). This is less than 0.2% of the 6370 km (3960 mi) distance to the Earth's center. Of course, the rock cycle may bring igneous and metamorphic rocks to the Earth's surface that formed at deeper levels of the crust. Some volcanic rocks contain material that formed in the mantle. As a melt forces its way upward, it can sometimes rip off samples from the surrounding regions and transport them toward the surface. In addition, igneous intrusions that carry high-pressure diamonds are thought to be related to the rapid ascent of magma from deep in the mantle, beneath both the crust and the lithospheric plates and the asthenosphere. All of these samples have been the subject of numerous laboratory and theoretical investigations and provide vital information on the pressure, temperature, and composition of at least part of the mantle. But even so, these examples are few and far between and only a minute portion of the Earth's interior has been made available for direct study by this process.

Despite these difficulties, there is surprising agreement among geoscientists on the structure and composition of the Earth's interior. In addition, there are exciting new revelations that may finally provide a direct link between the Earth's internal processes and plate tectonics, something that has remained elusive since the time of Wegener. In this chapter, we review our current understanding of the Earth's hidden interior by examining the evidence provided by the Earth's internal heat, its density, and the path of earthquake waves through its interior. We will see how this evidence has allowed us to recognize and subdivide the Earth's three main internal layers, the crust, mantle, and core. We will also learn that it is now becoming possible to produce three-dimensional pictures of the Earth's interior in much the same way that a CAT scan provides a three-dimensional view of the human body.

7.2 Clues from Heat and Density

The first important clue to the nature of the Earth beneath the crust comes from understanding the Earth's internal heat. It is a popular misconception that we are standing on solid ground that is anchored to the interior of the Earth. We are not. The continents we are standing on are actually floating like rafts in a crowded stream, moving at rates between 1 and 17 cm (1 and 4 in.) per year. Over millions of years, this motion has created entire ocean basins when moving continents diverge from each other, only to close them again when continents converge and collide. These motions are powered by heat energy that has been escaping from the interior of the Earth for over 4.5 billion years. Volcanoes are the most obvious and dramatic manifestation of the escape of this internal heat. However, heat flow from the surface of the Earth to the atmosphere can be measured everywhere on our planet, telling us that our planet is cooling (see Figure 1.2). But where is this heat coming from and how is it being transported from deep within the Earth to the planet's surface? To answer these questions we must turn our attention to the structure and composition of the Earth's interior.

Processes in the interior of our planet are hidden from view and so can only be examined indirectly. So how different is the Earth's interior from its surface? Laboratory experiments tell us that most of the minerals present in surface rocks would not be stable at depths any greater than about 40 km (25 mi) below the surface. The average density of continental rocks is about 2.7, that is, an average continental rock weighs about 2.7 times as much as the same volume of water. However, the average density of the Earth is 5.52. Calculations show that this difference in density cannot be due only to increasing pressure with depth, implying that the interior of the Earth must be vastly different from its exterior. In the absence of direct observations, Earth scientists use geophysical methods, or the application of physical laws, to probe the Earth's interior. Surprisingly, these methods give us a good first approximation of the composition of the Earth from core to surface.

We know for instance, that the Earth is an internally layered planet, being stratified according to density from its upper atmosphere to its inner core. But how have we determined this? To answer these questions we must first examine the nature of earthquake shock waves because it is these waves that have revealed the layered nature of our planet by providing us with an X-ray picture of the Earth's interior.

7.3 Earthquake Waves

As we learned in Chapters 3 and 6, where two rock masses struggle to slide past each other, the boundary between them is a fracture in the crust known as a **fault.** If the rocks along the fault lock while the pressure or **stress** continues, distortion, or **strain** builds up in the rocks on either side of the fault so that they slowly bend or deform, storing up elastic energy like a spring ever drawn tighter (Fig. 7.1). Finally, when the spring snaps, rocks on either side of the fault rupture and jerk violently past each other, sending out earthquake shock waves, known as **seismic waves,** in all directions for periods that can last for several minutes. This is the "stick-slip" motion of faults we encountered in Chapters 5 and 6, and forms an explanation for the origin of earthquakes known as the **elastic rebound theory.**

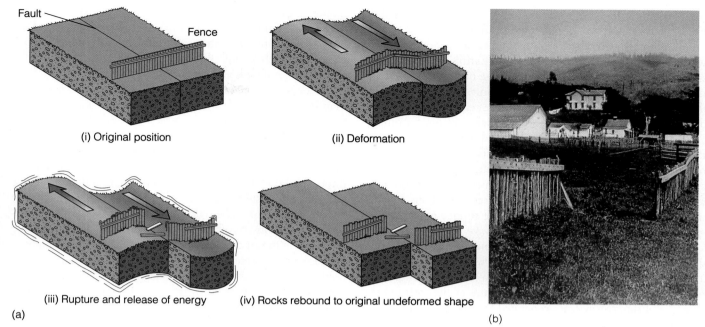

Figure 7.1

Elastic rebound theory for the cause of earthquakes. (a) Stresses acting on a rock cause the rock to deform (strain) by bending. This stores energy as the strain builds over a long period of time (i, ii). Finally, the strength of the rock is exceeded and it abruptly ruptures (iii), releasing energy in the form of earthquake waves as the rocks on either side of the fault rebound to their undeformed shape by movement along the fault (iv). (b) Offset of a fence in Marin County, California, measured 2.5 m (8 ft) following the 1906 San Francisco earthquake. Horizontal (strike-slip) offset is shown by the arrows in (ii) and (iii); rocks can also move vertically (dip-slip) along the fault plane.

The vibrations of an earthquake begin at the **focus** where a locked fault suddenly lets go (Fig. 7.2). From here, seismic waves spread outward in all directions in much the same way that ripples radiate when a pebble is thrown into a pond. The waves soon ripple through the Earth's interior, literally filling the Earth just as sound waves fill a ringing bell. The resulting pattern of interfering vibrations is an extremely complicated one, and to simplify matters, seismic waves are often described in terms of **rays** rather than waves, in the same way that the illumination from a light source is described in terms of light rays.

Several varieties of shock waves are produced by earthquakes in the same way that a branch, when snapped with your hands, produces vibrations that can be heard as well as felt. Thus the sound of the branch snapping is not responsible for the jolt you feel in your wrists, although both are produced as the branch breaks. In a similar fashion, seismic waves are of different types and can be grouped into three varieties.

The seismic waves with which we are most familiar are those that cause most of the damage associated with earthquakes. These waves are of several types but all travel along the Earth's surface and so are grouped together as **surface waves.** Surface waves produce low-frequency vibrations that either roll the ground like ocean waves or whip it sideways. Compared to other seismic waves, they travel quite slowly and take longer to diminish. As a result, they can sway buildings at great distances from the earthquake **epicenter,** which is the point on the Earth's surface directly above the earthquake focus (Fig. 7.2). These surface waves are, however, of limited

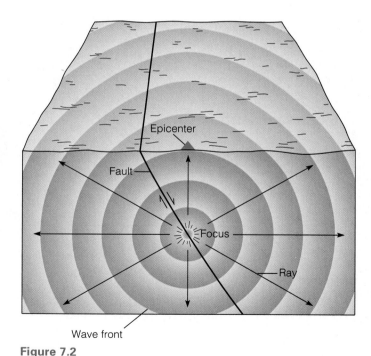

Figure 7.2

Seismic wave fronts radiate outward from the focus of an earthquake where a locked fault lets go. They first reach the Earth's surface at the epicenter lying directly above the focus. Note that the direction of motion of the earthquake waves can also be shown as rays perpendicular to the wave fronts.

use in probing the Earth because they only travel along the Earth's surface and do not penetrate its interior.

In contrast, two other important types of seismic waves, P waves and S waves, are directed into the Earth's interior, and it is from these that much of our knowledge of the Earth's internal structure is derived. Just as a car handles better on asphalt than gravel, so the speed at which these seismic waves travel depends upon the material through which they are traveling. Because of this, important information about the materials of the Earth's interior, especially their density and rigidity, can be obtained from variations in the travel times of seismic waves.

Primary or **P waves** are so called because, traveling through crustal rocks at speeds up to 7 km per second (about 16,000 mi/hour), they are the first to arrive at the surface or at seismographic stations equipped with seismic wave detection devices known as **seismometers**. Seismometers contain a weighted pendulum attached to a pen that produces a wiggly line (known as a **seismogram**) on a revolving paper-covered drum whenever movement of the ground shakes the frame of the instrument (Fig. 7.3). Primary waves are generated by sudden compression of the ground at the site of an earthquake, like a push on the end of a stretched spring (Fig. 7.4a). Hence, they are compressional waves (like sound waves), alternately squeezing and expanding the material in their path in the direction in which they travel. Because P waves travel like sound waves, they set the air in motion as they leave the Earth's surface, producing the locomotivelike roar associated with earthquakes. The exact speed of a P wave depends on the nature of the material through which it passes. It travels rapidly through compact dense solids, but is slowed by hot liquids and gases.

Secondary or **S waves** are so called because they are slower than P waves so that their characteristic signals are recorded by seismometers *after* those of P waves (Fig. 7.3c). S waves, which travel through the crust at speeds up to 4.5 km per second (about 10,000 mi/hour), are generated by the shearing or sliding motion at an earthquake site. As a result, they can only travel through solid materials since only solids have a definite shape. They are also called shear waves and behave like oscillations in a rope (Fig. 7.4b). When a rope is shaken, a series of up-and-down oscillations travel down its length. But simply shaking one's arm in the air or under water produces no such oscillations because air and water are fluid mediums and have no shape of their own. As up-and-down or side-to-side oscillations, shear waves cause the material in their path to heave sideways, perpendicular to the direction in which the wave is traveling.

To compare, P waves extend and compress the materials they encounter, causing them to expand and shrink, whereas S waves cause these materials to change their shape but not their volume (Fig. 7.5). Solids transmit both P and S waves. However, P waves can travel through liquids whereas S waves

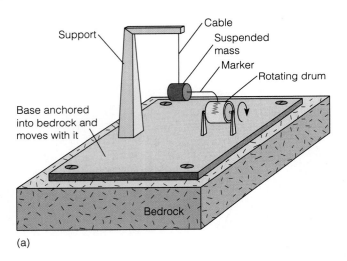

(a)

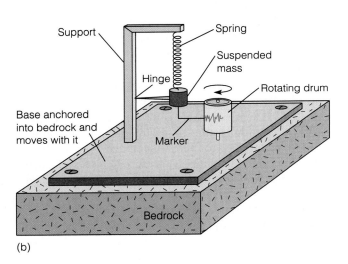

(b)

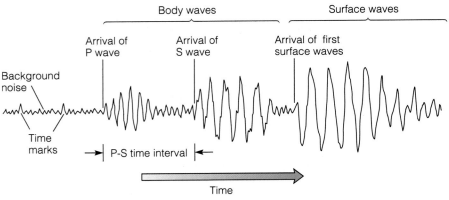

(c)

Figure 7.3

(a) A seismometer (seismograph) for recording horizontal ground motion. The mass is suspended from the column by a wire and swings like a pendulum when the ground moves horizontally. (b) A vertical-motion seismometer operates on the same principle. In both instruments, a pen attached to the mass records the motion on a moving strip of paper. (c) The wavy line the pen traces out is a seismogram that shows the first arrivals of P and S waves, followed by surface waves.

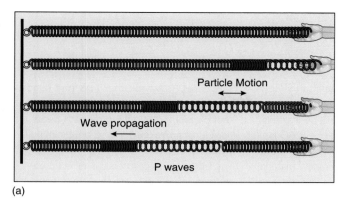

(a)

Particle Motion

Wave propagation

P waves

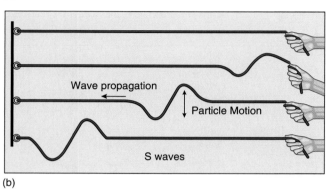

Wave propagation

Particle Motion

S waves

(b)

Figure 7.4
Particle motion in seismic waves. (a) A P wave is illustrated by a sudden push on the end of a stretched spring. The particles vibrate parallel to the direction of wave propagation. (b) An S wave is illustrated by shaking a loop along a rope. The particles vibrate perpendicular to the direction of wave propagation.

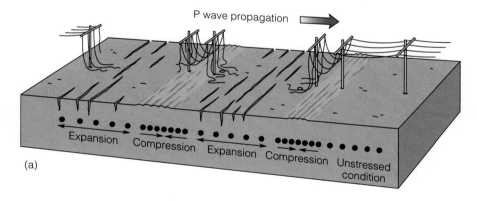

P wave propagation

Expansion   Compression   Expansion   Compression   Unstressed condition

(a)

Figure 7.5
Motion of the ground during the passage of (a) P waves, which cause the ground to compress and expand (or dilate), and (b) S waves, which cause it to move in all directions perpendicular to the wave path (only the horizontal motion is shown here).

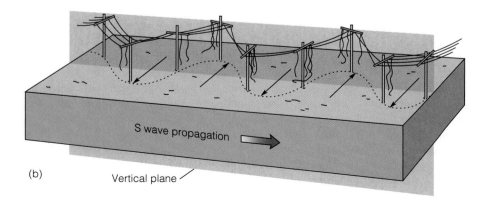

S wave propagation

(b)   Vertical plane

Wave Refraction

The change of direction when waves pass from one material to another is known as **refraction,** and is summarized in Figure 7.6. When wave fronts, with a characteristic speed for the material in which they are traveling (represented on Figure 7.6 by their spacing), encounter a different material, the ray path will generally meet the boundary at an angle. Because of this, one part of a wave front will cross into the new material before the other. When this occurs, one part of the wave front will be traveling through the new material (segment AC in Layer 2), while the other remains in the old (segment AB in Layer 1). If the new material (Layer 2) is less rigid, the wave velocity for segment AC is less than that for segment AB and the spacing between wave fronts decreases. As a result, the wave front bends so that the ray path in the new medium is not parallel to the original ray path, but is kinked or **refracted** and so travels in a different direction.

In order to describe this refraction, a reference line is selected that is perpendicular to the boundary between the layers. In the case presented here, the ray path is refracted

Figure 7.6

Refraction of seismic waves. A series of wave fronts emanating from an earthquake travel at the same speed (represented by their separation) in Layer 1 but at a slower speed (represented by their smaller separation) in the less rigid Layer 2. Because the ray path meets the boundary between the layers at an angle (for example, at A) one part of a wave front reaches Layer 2 before the other, and so starts to move at the slower speed sooner. Segment AB (in Layer 1) of the wave front therefore travels further than segment AC (in Layer 2) in the same time interval so that the wave front bends. The ray path perpendicular to the wave fronts is also bent or refracted on entering Layer 2, and the spacing between wave fronts becomes narrower.

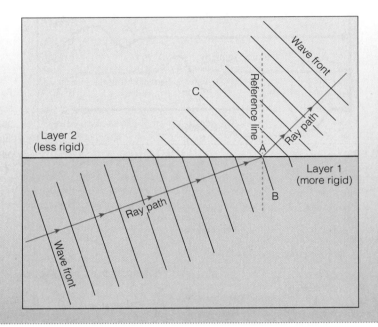

cannot. This is because liquids behave **elastically** under compression. Like a compressed spring, an **elastic** material is one that will rebound when released from the deforming force. Liquids exhibit elastic behavior when compressed because they expand again when the compression is removed. Liquids can therefore transmit P waves. Unlike solids, however, liquids have no shape of their own and simply adopt the shape of their container. Because they have no resistance to changes in shape, liquids cannot transmit S waves.

Both P waves and S waves travel through the Earth's interior and so are grouped together as **body waves.** But even when traveling through solid materials, the speeds of P and S waves vary because different materials react differently to compression and shear. For example, a diving board and trampoline respond quite differently when jumped on, although both are solid. Materials respond fastest to P and S waves when they are dense and rigid. So the speed of body waves depends on the material through which they are traveling, increasing with increased density and rigidity.

7.4 Journey to the Center of the Earth

Our knowledge of the Earth's interior is derived from several indirect sources, the most important of which is **seismology,** the study of earthquake waves. Because the speed with which these shock waves travel depends upon the nature of the material through which they pass, the pattern created by seismic waves as they travel through the Earth's interior provides insight into the nature of the materials they encounter at depth. As we will now see, it was from this source that geoscientists first learned that the Earth had a crust, mantle, and core, and that its outer core was liquid. But to understand how this was achieved, we must first examine more closely the way in which seismic waves travel through the Earth.

Because the Earth has a spherical shape, seismic waves radiating from a source close to the Earth's surface will probe

toward the reference line as it travels into the less rigid material. The opposite occurs when the wave travels from a less rigid material into a more rigid one. The ray path refracts *away from* the reference line.

The effect can be illustrated by a line of marchers walking arm-in-arm along a desert highway (Fig. 7.7). The marchers maintain a constant speed when walking on the road, but abruptly slow down if they walk in the sand beside it. If the marchers head off the road at an angle, some will reach the sand while the others are still marching on the road. As a result, their lines will pivot and their path will bend because some are marching faster than others.

When seismic waves travel across a boundary between materials with sharply contrasting rigidity, wave refraction is pronounced. Even within essentially the same material, the degree of rigidity increases with density, and density increases with depth. (Denser materials are more rigid because their atoms are more closely packed and their average bond lengths are consequently shorter). Because densities increase with depth, the area below the wave is a region of faster velocity than the area above. As a result, seismic ray paths progressively bend up toward the surface as they propagate downward. The result is a concave, rather than linear path, for the seismic ray as shown in Figure 7.8.

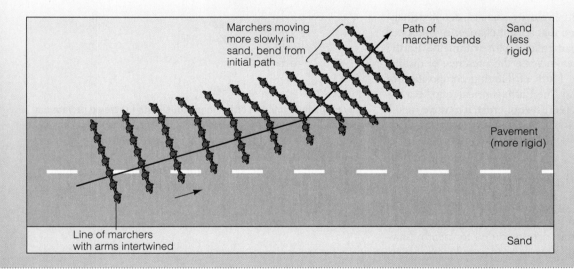

Figure 7.7
The path followed by ranks of marchers walking arm-in-arm on a desert highway is bent as they head off the road at an angle because those on the highway are marching faster than those on the sand. Seismic waves are bent in a similar fashion when crossing boundaries between materials in which they travel at different speeds.

different regions of the Earth's interior depending upon their line of travel or **ray path**. As the depth of penetration increases with the length of the ray path, seismic waves from sources located at great distances from a recording station probe further into the Earth's interior than those from earthquakes closer to the station (Fig. 7.8). Regardless of the route taken, however, much of the seismic energy released by an earthquake eventually returns to the Earth's surface where it can be analyzed at many recording stations. As with any journey, the time it takes to travel depends on the distance involved and the conditions encountered along the route. Thus the time it takes for seismic waves to reach a recording station depends on the distance between the station and the focus of the earthquake, and the properties of the rocks encountered along the ray path. Because earthquake waves are sent out in all directions, the energy from any one earthquake can be picked up by thousands of recording stations worldwide so that the travel times along thousands of different ray paths can be examined. Any variations in travel time that cannot be accounted for by differences in the distance trav-

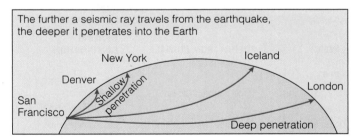

Figure 7.8
Seismic rays traveling from an earthquake in San Francisco to receivers in Denver, New York, Iceland, and London, penetrate more deeply into the Earth's interior the further they travel from the earthquake focus and travel in a concave path.

eled, correspond to changes in rock properties along the ray path. Over the last 30 years, information from each of the hundreds of moderate earthquakes that occur annually in various parts of the world, has been compiled and processed by powerful computers.

The results have provided a surprisingly thorough understanding of the composition of the Earth's interior (Table 7-1).

If the Earth was perfectly homogeneous, the ray paths of seismic waves would spread out from the focus of an earthquake in all directions at constant speeds and along straight travel paths (Fig. 7.9a). In reality, the speed of earthquake waves increases with depth because the density of the materials they encounter increases as the pressure rises in the Earth's deep interior. Because of this density increase, seismic waves move faster below than they do above. This causes seismic waves to adopt curved ray paths, bending up toward the surface as they travel downward into regions of faster velocity (see Figure 7.9b, c and Dig In: Wave Refraction, on page 180).

The effect is analogous to that of a car that drifts from the highway onto a soft shoulder. Since the curbside wheels lose traction before those on the driver's side, the car slows on one side while the faster-moving driver's side wheels cause the car to veer into the shoulder.

In addition to these gradual velocity changes, detailed analysis of certain ray paths showed that abrupt changes in the speed of seismic waves occur at particular depths within the Earth's interior. These discontinuities indicate the presence of distinct layers of material within the Earth of differing composition or physical properties. In this way, the Earth is revealed to be a layered planet with a very thin outer layer or **crust**, a massive metallic center or **core**, and a thick intervening rocky layer or **mantle** (Fig. 7.10). But how were these discontinuities discovered?

7.4.1 A Layered Planet Revealed

Evidence for the existence of an outer crustal layer was first reported in 1909 by Andrija Mohorovicic, a Yugoslavian seismologist. By analyzing seismic data recorded for nearby shal-

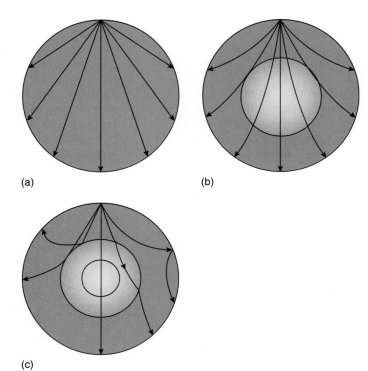

(a)

(b)

(c)

Figure 7.9
Ray paths for a uniform or homogeneous Earth (a) would radiate out in straight lines from an earthquake focus. For an inhomogeneous Earth (b) in which density, or rigidity, increases with depth, ray paths would be smooth curves that bend upward toward the surface. For a layered Earth (c) some ray paths are reflected by the boundaries between layers while others are bent, or refracted.

Table 7-1
General Characteristics of the Earth's Internal Layers

LAYER	DEPTH FROM SURFACE	GENERAL COMPOSITION	AVERAGE TEMPERATURE	AVERAGE DENSITY	PROPORTION OF THE EARTH'S TOTAL MASS	PROPORTION OF THE EARTH'S TOTAL VOLUME
Crust			500°C (900°F)		0.4%	<1%
Continental crust	From surface to 35 km (22 mi)	Granite/diorite (aluminum silicates)		2.7 g/cm³		
Oceanic crust	From sea level to 11 km (7 mi)	Basalt (magnesium silicates)		2.9 g/cm³		
Mantle	From 11 km to 2900 km (from 7 mi to 1800 mi)	Iron and magnesium silicates	2500°C (4500°F)	4.5 g/cm³	68.1%	83%
Core					31.5%	16%
Outer core	From 2900 km to 5100 km (from 1800 mi to 3200 mi)	Liquid iron and sulfur	4600°C (8500°F)	11.8 g/cm³		
Inner core	From 5100 km to 6370 km (from 3200 mi to 3960 mi)	Solid iron and nickel	5000°C (9000°F)	16.0 g/cm³		
Whole earth	6370 km (3960 mi)			5.5 g/cm³		

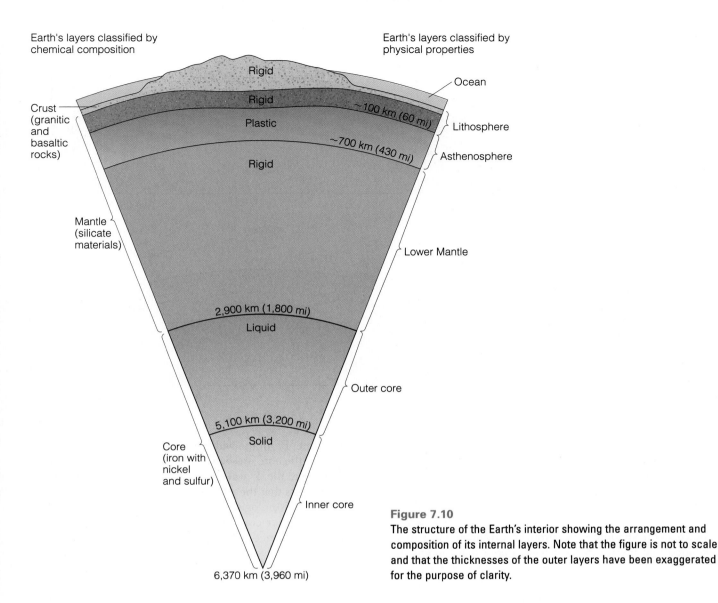

Earth's layers classified by chemical composition

Earth's layers classified by physical properties

Crust (granitic and basaltic rocks)

Mantle (silicate materials)

Core (iron with nickel and sulfur)

Rigid

Rigid

Plastic

Rigid

Liquid

Solid

Ocean

~100 km (60 mi)

~700 km (430 mi)

2,900 km (1,800 mi)

5,100 km (3,200 mi)

6,370 km (3,960 mi)

Lithosphere

Asthenosphere

Lower Mantle

Outer core

Inner core

Figure 7.10

The structure of the Earth's interior showing the arrangement and composition of its internal layers. Note that the figure is not to scale and that the thicknesses of the outer layers have been exaggerated for the purpose of clarity.

low earthquakes, he was able to show that P waves arriving at recording stations more than 200 km (125 mi) from the epicenter of an earthquake traveled significantly faster than those arriving at stations lying closer to the epicenter. He was therefore able to conclude that those P waves which followed the deeper ray paths to more distant stations must travel at greater speed and therefore through rocks of higher density than those encountered along the shallower ray paths to nearby stations (Fig. 7.11). Thus the first P waves to arrive at recording stations less than 200 km (125 mi) from the epicenter had traveled the shortest route directly through the crust. Beyond 200 km, however, the first P waves to arrive were those that had traveled the longer route at higher speed through the denser rocks at depth. The abrupt jump in speed observed at the more distant recording stations showed that the boundary separating the shallow crustal rocks from the denser rocks below was a sharp one, and that it occurred at a depth of 30 to 40 km (20–25 mi). This

boundary, which separates rocks of the continental crust from those the underlying mantle, is formally named the **Mohorovicic discontinuity** in honor of its discoverer, but is better known simply as the **Moho.**

Shortly after the discovery of the Moho, the second major boundary of the Earth's interior, between the mantle and the core, was discovered by the German seismologist Beno Gutenberg. This time however, the boundary was not detected by an abrupt change in the arrival times of seismic waves, but in their almost complete failure to arrive at all. For any earthquake of sufficient magnitude that it can be detected at recording stations many thousands of kilometers from the epicenter, there is always a belt encircling the globe in which neither P waves nor S waves are detected (Fig. 7.12). At about 11,500 km (7200 mi) or 103 degrees of latitude from the epicenter of the earthquake both waves die out completely. S waves cannot be detected beyond this distance (Fig 7.12a)

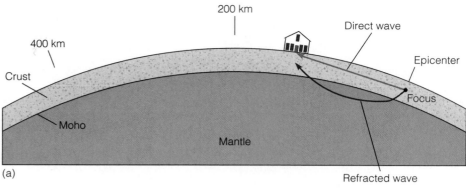

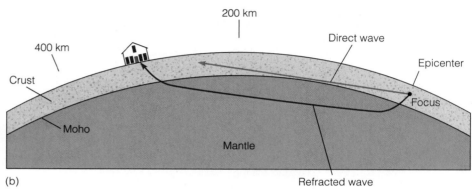

Figure 7.11
P wave ray paths from the focus of a shallow earthquake to two recording stations reveal the existence of the Mohorovicic Discontinuity or Moho. (a) Within 200 km of the epicenter, the nearer station receives the direct waves first. (b) Beyond 200 km, however, the first P wave to arrive traveled further than the direct wave because of its higher speed through the denser rocks of the mantle.

although, at distances greater than 16,000 km (10,000 mi) or 143 degrees of latitude from the epicenter, P waves reappear with much delayed arrival times (Fig. 7.12b).

So, for any earthquake that can be detected far from its source, a ring-like shadow is produced that is free from direct P and S waves. The ring extends from 103 to 143 degrees and is known as the **shadow zone.** From its existence, Gutenberg concluded that the Earth must have a core that hinders the transmission of seismic waves. From the failure of S waves to reappear, at least a portion of this core was deemed to be liquid. The major delay observed in the arrival of P waves beyond the shadow zone also supported the presence of a liquid layer below the mantle, which would slow and refract the P waves as they enter the core.

But P waves, which will travel through liquids, were also obstructed within the shadow zone (Fig. 7.12b). From their absence, Gutenberg concluded that the core must act on P waves in much the same way as a lens focuses light by changing the path of light rays. In fact, because the liquid core is less rigid than the lowermost mantle, the P waves are bent inward as they enter the core and exit the core. The net result is to direct them toward the area of the globe opposite the earthquake epicenter instead of the surrounding region of shadow (Fig. 7.12a). The ray paths of P waves bend downward into the core for the same reason that they bend upward toward the surface in the mantle—the less rigid material causes them to bend toward it. In this case, seismic velocities are higher at the base of the mantle than in the outermost core. The net result is that P waves are bent inward and away from the 103° to 143°

shadow zone. From the width of the P wave shadow zone, the diameter of the core was determined to be 6940 km (4320 mi).

However, the P wave shadow zone is not totally dark, which suggests that the core is not fluid throughout. In 1936, the Danish seismologist, Inge Lehmann, proposed the existence of an **inner core**, which she suspected to be solid, to explain the presence of faint P waves within the shadow zone that could only have been reflected from a discontinuity within the core. The faster speed of P waves within the inner core (consistent with its solid nature) causes their ray paths to be bent into the shadow zone because, this time, they are being slowed from above and so bend outward. The diameter of this inner core is now known to be 2540 km (1580 mi), but its exact size was not determined until underground nuclear tests performed in the early 1960's produced simulated earthquakes, the time and location of which were precisely known. The liquid portion of the core is therefore the **outer core**, the presence of which entirely prevents S waves from penetrating the core–mantle boundary, thereby accounting for their absence both within the P wave shadow zone and beyond it (Fig. 7.12b).

From studies such as these we have learned that the Earth is a layered planet with a crust, a mantle and a core, the outer part of which is liquid. We have also learned that this layering is one of density and that it actually extends beyond the crust to the hydrosphere and atmosphere. Over the past 30 years, however, the study of seismology has become more sophisticated and the seismic data base has improved. As a result, the gross subdivision of the Earth's interior into crust, mantle, and core has been greatly refined. More recent studies of seismic

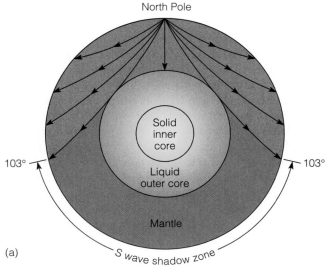

(a)

Figure 7.12
Pattern of P and S wave ray paths from an earthquake epicenter at the North Pole. (a) S waves are blocked by the Earth's liquid outer core so that no ray paths reach the surface in the S wave shadow zone south of 103 degrees latitude. (b) P waves are refracted so that no ray paths reach the surface in the P wave shadow zone between 103 and 143 degrees latitude.

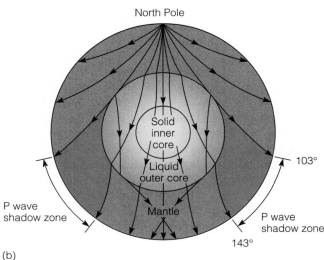

(b)

wave velocities have shown that, like the core, the Earth's crust and mantle can also be subdivided. Coupled with other studies that have shed light on the composition and density of the materials that comprise the Earth, we have come to our modern understanding of the structure and composition of the planet's interior (Fig. 7.13). But to learn how we have reached this modern understanding, we must first examine each of the Earth's major subdivisions in more detail.

7.4.2 The Crust

The crust of the Earth is not only the thinnest of the Earth's many layers but it also shows the greatest degree of variation in its component materials due to the dynamic processes that occur at or near the Earth's surface. On the continents, for example, the plants and animals of the biosphere sit atop a thin veneer of soil, sediment, and sedimentary rock that, in turn, overlies a crust composed of metamorphic and igneous rocks of broadly granitic to granodioritic composition. Like all granitic rocks, the crust of the continents is thought to be enriched with the elements potassium, sodium, and silicon, and to have a density of about 2.7 g/cm^3. But given that the average thickness of the continental crust is 30–40 km (20–25 mi), and the deepest borehole has yet

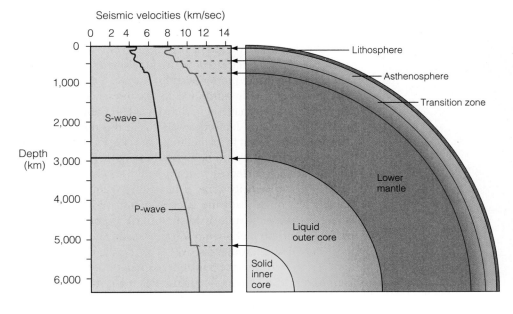

Figure 7.13
Variations of P and S wave velocities with depth and their relationship to the Earth's internal layers. Abrupt changes in the velocities of these waves define the boundaries between layers. A velocity decrease at a depth of about 100 km (60 mi) marks the boundary between the rigid lithosphere and the less rigid asthenosphere. Several velocity increases occur in the upper mantle to depths of about 700 km (450 mi). A large velocity decrease marks the core–mantle boundary at a depth of 2900 km (1800 mi). Beyond this S waves will not penetrate because the outer core is liquid. P wave velocities increase slightly in the solid inner core at a depth of 5100 km (3200 mi).

to penetrate this crust to a depth of more than 12 km (7 mi), little is known of the composition of continental crust at depth.

In 1925, a seismic discontinuity was detected within the continental crust. This boundary separates the **upper crust** from a slightly denser **lower crust**, where the velocity of seismic waves was found to be somewhat greater. Based on the available evidence, the lower crust is thought to be broadly dioritic in composition and highly metamorphosed due to elevated temperatures and pressures.

Even faster seismic wave velocities were found to occur in the crust beneath the oceans, demonstrating a compositional distinction between **oceanic crust** and **continental crust**. Despite being removed from direct observation by several kilometers of seawater, the broadly basaltic composition predicted for the oceanic crust on the basis of its seismic velocities, was later confirmed by deep-sea drilling. Compared to the continents, the crust of the ocean floors is thought to be enriched with the elements calcium, magnesium, and iron, and to have a density of about 2.9.

As we learned in Chapter 5, oceanic crust is also much thinner than continental crust (Fig. 7.14). The distance of the Moho (the base of the crust) below the ocean floor occurs at a remarkably constant depth of 6 to 7 km (about 4 mi). In contrast, the Moho below the continental crust lies as a depth that is rarely less than 20 km (12 mi) and may exceed 80 km (50 mi) beneath major mountain ranges. In both cases, however, the Moho delineates an abrupt compositional boundary separating lower crustal rocks of dioritic to basaltic composition from the ultramafic rocks of the uppermost mantle below.

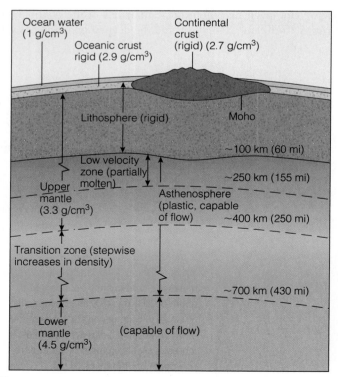

Figure 7.14
Cross section of the Earth's crust and part of the mantle showing layering within the lithosphere and density variations with depth.

7.4.3 The Mantle

Extending from the base of the crust to the core–mantle boundary at a depth of 2900 km (1800 mi), the mantle is the largest of the Earth's internal subdivisions (Table 10.1). The mantle gets its name because it surrounds or "mantles" the core. Because mantle samples are only very rarely brought to the surface, most of our knowledge about its composition is based on variations in the velocity of seismic waves and on experimental studies of mineral behavior at very high pressures and temperatures. From these data, the mantle is generally assumed to be largely solid, chemically the same throughout (chemically homogeneous), and made up of ultramafic rocks full of iron- and magnesium-rich silicate minerals. From the occasional samples brought to the surface, the uppermost mantle is thought to be made of peridotite, a rock composed mainly of the minerals olivine and pyroxene with a density of 3.3 g/cm³. Experimental studies, however, show that these minerals could not survive the pressures that exist lower down in the mantle, so that the mineralogy of the rocks there must be different even if the overall composition of the rock is the same.

As we learned in Chapter 5, the upper mantle is cool enough to be rigid, and together with the overlying crust, makes up the outer rocky layer of the Earth, or **lithosphere** (Fig. 7.14), that constitutes the Earth's tectonic plates. In contrast, from the base of the lithosphere (at a depth of 100 km or 62 mi), to a depth of as much as 700 km (430 mi), the mantle is hot and soft, and so can be penetrated by the rigid slabs at subduction zones. It is this weak layer, called the **asthenosphere**, that makes movement of the overlying lithospheric plates possible and is responsible for isostasy (see Chapter 5). Although the asthenosphere includes a small amount of melt, it is not the presence of this melt that makes the layer weak. Rather the layer is thought to be inherently soft because the mantle throughout this depth interval is sufficiently close to its melting point to be yielding (Fig. 7.15). The asthenosphere is also thought to be a globally continuous layer, unlike a discontinuous region within it, the low-velocity zone.

The melt in the asthenosphere creates a region known as the **low-velocity zone**. This zone occurs within the upper mantle between the depths of 100 and 250 km (62–155 mi). It is associated with a marked decrease in seismic wave velocities. The effects are more pronounced for S waves than for P waves and this is generally attributed to the presence of a small amount of molten material in the zone. The presence of such melt is likely at this depth because it is in this region that the mantle comes closest to its melting point (Fig. 7.15), and as little as 1% of liquid would be needed to produce the observed effects. Although detected beneath both the oceans and continents, the low-velocity zone is not entirely continuous. It is notably absent beneath the ancient continental shield areas where the upper mantle is likely to have experienced many past cycles of subduction and so is less readily melted.

Other discontinuities discovered within the Earth's mantle are thought to record the transformation of upper mantle minerals to ones with mineral structures that are stable at higher pressures. A series of stepwise increases in seismic wave velocities indicating such transformations occur between the depths of 400 and 700 km (250–430 mi). This interval, which is known

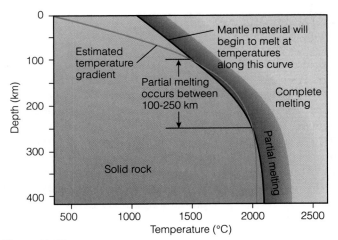

Figure 7.15
Melting in the mantle. The presence of melt results in a zone of low seismic velocity.

as the **transition zone**, separates the **upper mantle** from the **lower mantle**. Unlike the abrupt compositional boundary at the Moho, however, the discontinuities of the transition zone are not sharp, and occur instead over a significant depth interval. Hence they are generally attributed to changes in the crystal structure of the minerals present rather than representing changes in chemistry of the mantle rocks. Under the intense pressures of the Earth's interior, minerals are forced to adopt new structures in which their component atoms are more tightly packed. These **phase changes**, as they are known, increase the density of the rocks concerned (and therefore, the velocity of seismic waves passing through them), but do not affect their composition. The transition zone is thought to be a region in the mantle where several different minerals, such as olivine and pyroxene, undergo such phase changes.

Below the transition zone, in the lower mantle, evidence of such phase changes is absent. Instead of stepwise increases, seismic velocities increase smoothly (see Figure 7.13) in response to the steady increase in density with depth. This trend abruptly ends at the base of the mantle where a profound discontinuity marks the boundary of the outer core.

7.4.4 The Outer Core

With a diameter of 6940 km (4320 mi), the Earth's massive metallic center or core is a little larger than the planet Mars with its diameter of 6796 km (4216 mi). The core–mantle boundary at a depth of 2900 km (1800 mi) marks a profound change in both the composition and physical state of the material of the Earth's interior. Across this boundary, P waves are dramatically slowed, S waves are halted entirely, and estimates of density increase by a factor of two to values of nearly 12. The complete loss of S waves in the outer core shows it to be liquid while the dramatic rise in density precludes a composition in any way similar to that of the mantle. Even under the enormous pressures that exist in the core, silicate minerals such as those of the mantle could not have their atoms packed tightly enough to account

for such a high density. So what is the Earth's outer core made of, and how has this been determined? Once again, the Earth scientist becomes a detective, piecing together clues from a wide variety of sources. The total mass of the Earth is 5.98×10^{24} kg (that is, 5.98 multiplied by 10, twenty-four times). This has been determined from the gravitational attraction between the planets, from the time it takes the Earth to revolve around the Sun (the length of an Earth year), and from the time it takes the Earth to rotate about its own axis (the length of an Earth day). The volume of the Earth is 1.09×10^{12} cubic km. This figure can be determined because the shape of the Earth is known. If the Earth's mass is divided by its volume, an average density of 5.5 is obtained. This means that an average piece of the Earth weighs five and a half times as much as the same volume of water.

Density is a measure of how tightly packed the internal atomic structure of a substance is, and is one of the chief factors controlling the speed with which earthquake waves travel through the Earth's interior. From the speed of earthquake waves, we can tell that most of the rocks within the Earth's mantle and crust have densities well below 5.5 (Table 7-1). This implies that the material of the Earth's outer core must have a density well in excess of 5.5. In fact, calculations show that core densities between 10 and 13 are needed in order to achieve an average Earth density of this value. For the outer core, the only liquids of high enough density are molten metals, and the only abundant metal in the Solar System with the appropriate density is iron.

Other lines of evidence that support an iron-rich core relate to the Earth's place in the Solar System, a topic we will explore more fully in Chapter 15. The most important of these comes from the composition of meteorites. As members of the Solar System, meteorites are considered to be representative of the primitive material from which the Earth originally accreted. Those known as **stony meteorites** are made of rocky material that closely resembles the peridotite of the Earth's mantle. Other meteorites, however, are metallic and mainly composed of iron and nickel. These **iron meteorites** are thought to resemble the material of the Earth's core. An iron-rich core is also supported by the relative scarcity of this metal at the Earth's surface when compared to its abundance in the Solar System. Although it is a common element, iron is far less abundant in the Earth's crustal rocks than it is, for example, in meteorites. This suggests that the metal must be far more common in the Earth's interior if the average composition of the planet is to be brought in line with that of the Solar System. Finally, iron is a good electrical conductor. The motion of an iron-rich molten outer core would help generate the electric current loops that are thought to power, in a dynamo-like fashion, the Earth's magnetic field (see Dig In on page 190).

However, at the pressures that exist in the outer core, the density of iron alone would be slightly higher than that indicated by the seismic data. For this reason, the outer core is thought to be a molten mixture of either iron and sulfur or, possibly, iron and oxygen. Both sulfur and oxygen are common light elements that alloy with iron and facilitate melting. Their addition to the mixture would consequently lower both the density of the outer core and the melting point of the material of which it is comprised. A liquid comprising 88 percent iron and 12 percent sulfur, for example, would not only have the

The mantle makes up over 80% of the Earth's volume and almost 70% of its mass. This vast portion of the Earth's interior lies only a few kilometers below the ocean floors, but it is everywhere hidden from us by the Earth's crust and cannot yet be sampled directly. However, fragments of oceanic lithosphere that include uppermost mantle rocks are occasionally preserved in ophiolites. In addition, samples, or inclusions, of deeper mantle material are sometimes brought to the surface by volcanic eruptions. These rare samples of mantle rock provide important information on the composition of the mantle and the conditions that exist within it. Inclusions of mantle rocks can be quite common in certain basalt lavas of oceanic island and continental volcanoes, but the most remarkable of these mantle samples are the diamond-bearing inclusions found in diamond pipes.

Diamond pipes are the central conduits of former volcanoes and contain a fragmented ultramafic rock called **kimberlite** in which diamonds are occasionally dispersed. This bluish peridotitic rock is named after the deposit at Kimberley, South Africa, which is famous for its diamond mines. Shaped like carrots and often accompanied by sills and dikes (Fig. 7.16), diamond pipes extend to depths of up to 1000 m (3300 ft). But they are thought to have formed as a result of the explosive decompression of highly gas-charged magma rising rapidly from great depths. Diamond pipes occur in areas of ancient stable crust on virtually every continent, but the best known are those of southern Africa, Siberia, and western Australia. Well-known North American kimberlites occur in the Canadian Shield of Ontario and Quebec, along the Colorado–Wyoming border, in Arkansas and upstate New York, and in the southern Appalachians. In addition, exciting new discoveries of diamond-bearing kimberlites have been recently made in the Northwest Territories of Canada.

Not all kimberlites contain diamonds, and diamonds are very rare even in those that are diamond-bearing. On average, kimberlites contain only one part diamond to every 20 million parts of rock, so that several tons of ore must usually be mined to produce a single diamond. As a result, diamond miners rarely see the fabulous gemstones they seek. But the presence of diamond in these rocks has a very important bearing on the origin of kimberlites.

Diamond is the high-pressure form of crystalline carbon (see Figure 2.3) and can be produced artificially only at pressures above 55 kilobars and temperatures in excess of 1500°C (2700°F). This implies that diamond-bearing kimberlites must come from depths of at least 150 km (93 mi) below the Earth's surface. An actual depth of about 180 km (112 mi) is widely assumed for the origin of kimberlite magmas, although depths of more than twice this value have been suggested on the basis of certain experimental data.

Formed under these immense pressures, diamond is the hardest known natural material, and this quality, together with its highly attractive appearance when cut, makes it the most sought after of all gemstones.

Unfortunately, the unique conditions of its formation invalidate a fashionable sentiment.

---

correct density but would also produce the same P-wave velocities as those in the outer core. However, other elements, including nickel, silicon, and, possibly, hydrogen, are likely to be present in small amounts.

7.4.5 The Inner Core

At a depth of 5100 km (3200 mi), a small increase occurs in P-wave velocities (Fig. 7.13) that is attributed to the presence of a solid inner core, the diameter of which, at 2540 km (1580 mi), is similar to that of the planet Pluto. The bending of ray paths across the inner core boundary due to the velocity increase, also accounts for the presence of faint P waves within the shadow zone. However, since temperatures within the Earth's core are thought to rise from about 4000°C at the core-mantle boundary to values of about 5000°C at the inner core boundary, the existence of a solid inner core raises an obvious question. How can the inner core be solid when the outer core, which is not as hot, is molten? The answer is that the composition of the inner core is probably pure iron or a pure iron–nickel alloy. It therefore lacks the small quantities of lighter elements like sulfur that are thought to be present in the outer core, and which act to lower the melting point of the outer core material.

In fact, it is likely that the Earth's solid inner core has evolved over time, and that early in Earth's history the entire core was molten. But as the Earth cooled, the core has slowly separated such that the heavier iron-rich components sank toward the inner core and solidified, while the lighter components became concentrated in the outer core where their presence ensured that the core material remains above its melting point. In this way the core developed an internal structure in which a circulating liquid shell surrounds a solid interior.

As the Earth continues to cool, the inner core may continue to grow at the expense of the outer core. Indeed, billions of years from now, the core may conceivably become entirely solid, as it is believed to be the case on Mars. If this was to occur, the Earth might lose its magnetic properties, perhaps losing at the same time, the heat engine that drives plate tectonics.

7.5 Circulation in the Earth's Interior

Although the bulk of the Earth's interior is solid, much of it is thought to be slowly circulating. Some form of circulation is needed in the liquid outer core in order to generate the Earth's magnetic field, and circulation within all or part of the solid

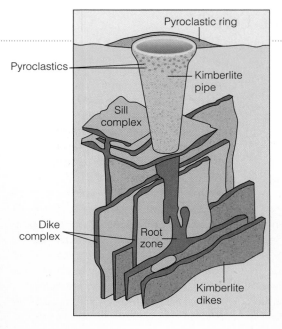

Pyroclastic ring

Pyroclastics

Kimberlite pipe

Sill complex

Dike complex

Root zone

Kimberlite dikes

Figure 7.16

Magmatic system of a kimberlite (diamond) pipe and associated dikes and sills from the explosive (pyroclastic) volcanic rocks at the surface to the feeder dikes of the root zone at a depth of 1 kilometer or more.

Another characteristic of diamonds that is not generally known is their antiquity. Whereas the kimberlites of diamond pipes range in age from Precambrian to Cretaceous (see Figure 3.17), the diamonds they contain are far older. Those from the mines of Kimberley, for example, are 3.3 billion years old. Indeed, the occurrence of diamond-bearing kimberlites in only the oldest rocks of the continents suggests that their formation may have been peculiar to the early Earth. If so, diamonds may hold clues to the nature of the Earth's early mantle. Perhaps it was only when the Earth was young, and then only beneath the thickest portions of the early continents, that the unusual conditions necessary for diamond formation were met. At some later date when kimberlite magmas rose violently from the mantle, they ripped off fragments of the lithosphere as they sped their way to the surface. But if diamonds are ancient, then only beneath the oldest parts of the continents would these fragments occasionally be diamond-bearing. Kimberlites that broke to the surface in other areas would be devoid of diamonds, as seems to be the case.

Contrary to popular opinion, a diamond is *not* forever! Because diamonds are formed deep within the mantle under very different physical conditions from those that exist at the Earth's surface, they are actually unstable. Indeed, if subjected to crustal pressures and temperatures, diamond will eventually break down to the far less romantic mineral, graphite, that we use as the "lead" in pencils. The occurrence of diamonds at the Earth's surface is therefore fortuitous. They exist here only because their rise from the mantle was too fast to allow them to break down, and they remain unchanged only because their conversion to graphite is extremely slow. But eventually, all diamonds will succumb to this degrading alteration.

mantle is believed to be ultimately responsible for the motion of the Earth's lithospheric plates.

Circulation of the Earth's interior is the principle means by which the planet loses its internal heat. Heat can travel through stationary materials by conduction (the process that makes a metal poker too hot to hold when it is held in a fire). However, circulation of the Earth's interior is thought to be driven largely by **thermal convection**. This process, whereby warm, less dense material rises while cool, denser material sinks, is far more efficient than conduction as a means of transporting heat because the material itself moves. In this way, hot material is transferred from the interior toward the exterior of the planet, and the heat produced in the interior travels to the surface where it can be radiated away into space. Because the process also encourages the segregation of material according to their densities, it has also played a major role in developing the Earth's internal layering.

Much of the Earth's internal heat comes from the breakdown of naturally occurring radioactive elements. Although radioactive elements are not very common in the mantle, the mantle contains about 83 percent of the Earth's volume (Table 7-1) so that even with very small amounts, a significant quantity of heat is generated by the radioactive decay of these elements. Chief among these are the elements uranium (isotopes ^{238}U and ^{235}U), thorium (^{232}Th), and potassium (^{40}K). The decay of these three

elements produces more than enough heat to cause the mantle to convect thermally.

Although radioactive elements may occur in the core, they are unlikely to be present in sufficient quantities to allow the outer core to circulate by way of thermal convection. Instead, the favored mechanism for the circulation needed in the outer core in order to power the dynamo that generates the Earth's magnetic field, is one of **compositional convection**. This form of convection relies, like thermal convection, on light material rising while heavy material sinks. In compositional convection, however, the contrast in density is due to differences in composition rather than differences in temperature. In the case of the outer core, circulation is thought to be driven by the buoyancy of the less dense material left when iron solidifies and attaches itself to the surface of the inner core. This links convection in the outer core to the growth of the inner core.

But while convective circulation can be visualized in the liquid outer core, how can the mantle convect if it is solid? The answer is that materials do not have to be liquid in order to move. The slow downhill movement of valley glaciers provides an everyday example of such flow. Similarly, given enough time, a large ball of silly putty will slowly flatten itself into a disk. In both cases, movement is slow and occurs while the material (ice or putty) is solid. This kind of movement is known as solid-state flow or **creep**. The ability of materials to behave in this fashion is

The Earth's Magnetism and Its Dynamic Core

Einstein stated that the origin of the Earth's magnetic field was one of the five greatest unsolved problems in physics. A satisfactory explanation remains elusive to this day. The Earth's magnetic field is a region of magnetic force that surrounds and penetrates the planet. Within this region, invisible lines of magnetic force form a pattern of closed loops that emerge from the planet in the southern hemisphere and, after looping far out into space, return to Earth in the northern hemisphere. Thus the surface expression of the magnetic force, used for navigation and pathfinding, represents only a tiny fraction of the entire field.

The pattern of magnetic lines of force around the Earth is similar to that displayed when iron filings are shaken onto a piece of paper placed over a simple bar magnet (Fig. 7.17a). Thus the Earth's magnetic field, like the bar magnet, can be said to have a north and south pole. Within the Earth's magnetic field, the north–south lines of mag-

netic force control the orientation of any smaller magnetized object providing it is free to move. As a result, a compass needle, which is a small magnet pivoted about its center of gravity, points toward the north and south magnetic poles. Because the Earth's magnetic poles lie close to the geographic North Pole and South Pole about which the planet rotates, the compass has proved to be a powerful tool in the exploration and navigation of our lands and seas.

The north magnetic pole is actually offset from the geographic north pole by an angle of 11.50 degrees and wobbles slowly around it (Fig. 7.17b). Today, for example, the north magnetic pole lies on Bathurst Island in the Canadian Arctic whereas the south magnetic pole lies in the Indian Ocean between Australia and Antarctica. Over the centuries, however, both magnetic poles have moved in as more-or-less circular fashion around their respective geographic poles. The intensity of the magnetic field has also varied with time,

waxing and waning over the millennia. As we learned in Chapter 4, the field can also point north or south and, over millions of years, its polarity has reversed itself many times.

Exactly how the Earth's magnetic field is produced is not known with certainty. In 1600, William Gilbert attributed the field to the presence of permanently magnetized materials deep within the Earth's interior. But this would not account for the field's variations with time and is precluded by the very high temperatures of the Earth's interior. Instead, the field is now thought to be generated by electrical currents within the Earth's outer core. Approximately 2200 km (1400 mi) thick and 2900 km (1800 mi) below the Earth's surface, the outer core is a highly conductive metallic layer rich in iron with small amounts of impurities. At the 4600°C (8300°F) temperatures that prevail in the outer core, the iron and sulfur form a molten liquid that slowly circulates as it is heated. This convection is affected by the Earth's rotation. The two forces of

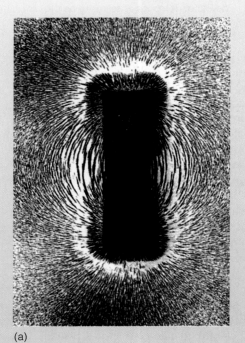

(a)

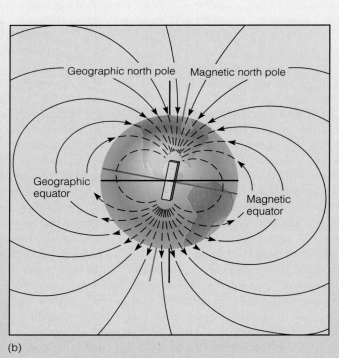

(b)

Figure 7.17

(a) Magnetic field pattern of a simple bar magnet revealed by iron filings which align themselves along the lines of magnetic force. (b) Magnetic field pattern of the Earth with lines of force that closely resemble those produced by the bar magnet.

convection and rotation combine to create columns of spiraling liquid that move parallel to the Earth's rotational axis (Fig. 7.18a) As the columns interact with the Earth's magnetic field, the mechanical energy of the spirals generates electricity in a fashion similar to an electromagnet (Fig. 7.18b). The electric current, in turn, produces a second magnetic field that reinforces the Earth's magnetic field. Thus the Earth's outer core behaves like a self-sustaining dynamo, generating electric currents which enhance the magnetic field that produces them. For as long as circulation continues in its outer core, the Earth's magnetic field will be sustained.

An exciting new discovery is that the inner core is rotating eastward faster than the rest of the planet. This would tend to accentuate the mechanical energy of the spirals and hence the dynamo effect. The reason for this disparity in rotation is uncertain. As a whole, the Earth's rotation is slowing down due to the gravitational attraction of the Moon and the Sun. As we will learn in Chapter 9, this gravitational attraction is responsible for tidal movements in the oceans which create frictional forces that apply a brake to the rotating planet. It could be that this break is less effective on the inner core because it is separated from the mantle by a fluid outer core. Alternatively, the faster rotation of the inner core may be the result of unusual magnetic flow patterns identified in computer simulations toward the base of the outer core.

The direction of the Earth's magnetic field at the surface is imposed by motion of the liquid outer core, which is itself influenced by the Earth's rotation. As a result, the Earth's magnetic poles are closely aligned with its rotational axis and have been throughout Earth history. Nevertheless, the Earth's magnetic field is known to reverse its polarity; that is, the north magnetic pole becomes the south magnetic pole and vice versa, so that a compass needle would reverse the direction in which it points. The cause of these reversals is not fully understood although they are not unique to our planet. Reversals in the Sun's magnetic field are even more frequent, occurring at intervals of about 22 years.

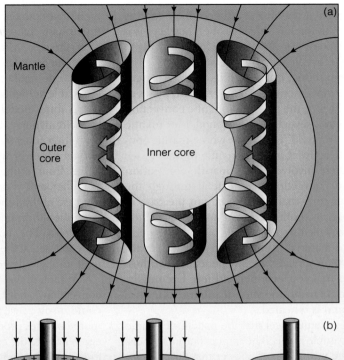

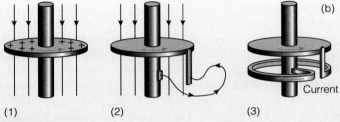

Figure 7.18

(a) The Earth's magnetic field may be generated by a dynamo in the Earth's outer core set up by the spiralling flow of electrically conducting metallic liquid which circulates under the influence of the Earth's rotation. Although the details of how the dynamo works are not known, the resulting magnetic field is similar to that produced by a bar magnet aligned north–south at the Earth's center; the curved lines of magnetic force are parallel to the Earth's surface at the equator and become more steeply inclined as the poles are approached. (b) The process is believed to be similar to that of a self-exciting dynamo. In (1), a disc is made to rotate in a magnetic field so that an electric potential is generated between the rim and axle, but no current flows because the circuit is not complete. In (2) the external circuit allows an electric current to flow. In (3), this current flows in a loop below the disc and so produces a vertical magnetic field, such that the original field can be removed.

Reversals in the Earth's magnetic field presumably reflect some chaotic aspect of the process that produces the field, perhaps due to turbulence or shifts in the pattern of flow in the outer core. Such changes in flow pattern are thought to cause the intensity of the Earth's magnetic field to fluctuate with time and might even result in its weakening to a magnitude of zero. We know, for example, that the Earth's magnetic field was much stronger 2000 years ago and that, at its present rate of decay, it should reach zero intensity within the next two millennia. Following such an event the Earth's magnetic field would slowly regenerate. But it might not do so with the same polarity and if it was to build with the opposite polarity, a magnetic reversal would be produced.

greatly increased at elevated temperatures, and especially at temperatures near the melting point of the material involved. As we learned when examining the asthenosphere, materials become soft and yielding as they approach their melting points, and under such conditions are capable of flow. Although the mantle is closest to its melting point within the asthenosphere (and is thought to just exceed it in the low velocity zone), nowhere is it very far from its melting point (Fig. 7.19). Therefore it is likely to be capable of slow convective flow in the solid state.

If the mantle is convecting, it is tempting to link its circulatory motion with the movement of the lithospheric plates at the Earth's surface. However, the pattern of mantle convection is not well understood and several different models of mantle convection have been proposed (Fig. 7.20): whole-mantle convection, layered mantle convection, and mantle plumes.

The simplest of these involves **whole-mantle convection** in which several large convection cells extend through the entire mantle from the core–mantle boundary to the base of the lithosphere, and include both the upper and lower mantle layers (Fig. 7.20a). To link this motion to plate movement, warm, buoyant mantle material is considered to rise beneath the mid-ocean ridges, then spreads laterally beneath the lithosphere, which it carries in a conveyor-belt fashion. Eventually, the material cools and, beneath the sites of subduction, sinks back into the mantle and is reheated.

However, some geoscientists feel that the pattern of mantle flow is unlikely to be as simple as this. They argue that a major discontinuity separating the upper from the lower mantle may profoundly influence the pattern of mantle circulation. In the model for **layered mantle convection** (Fig. 7.20b), more vigorous circulation in the upper mantle (where the mantle rocks are closest to their melting points) is thought to occur independently of the slower convection of the more rigid lower mantle. The two convection systems are therefore decoupled across the transition zone

between the upper and lower mantle. Layered convection would have to be the case if that boundary between the upper and lower mantle is a compositional one because whole-mantle convection would mix the two layers and so remove any compositional boundary between them. Whole-mantle convection is only possible if that boundary is produced by changes in the crystal structure (phase changes) of the minerals in the mantle, as is believed to be the case. Minerals crossing such a phase boundary do not move the boundary itself, but simply change their internal structure to one stable at lower pressure (in areas of upwelling) or to one stable at higher pressure (in areas of downflow).

A third model for mantle convection (Fig. 7.20c) suggests that circulation takes the form of focused columns of mantle upwelling and a much more diffuse system of return flow. These narrow zones of upwelling or **mantle plumes** are believed to come from a thermally unstable layer at the core–mantle boundary and rise through the entire mantle until they reach the base of the lithosphere. Here they spread out sideways, carrying the lithospheric plates away from the sites of upwelling as they do so. The more diffuse return flow is thought, once again, to be focused at the sites of subduction where cold, dense oceanic lithosphere descends into the mantle.

Which of these three models most closely describes mantle convection is not known. Indeed, in reality, mantle convection could involve several mechanisms, and is almost certain to be a more complex process than that described by any of these models. Whatever the case, the answer may soon be revealed, thanks to the development of a powerful new technique for studying the Earth's interior. This technique, known as seismic tomography, is capable of providing detailed three-dimensional maps of the Earth's mantle. As we shall now learn, the technique has already revealed tantalizing clues on the pattern of temperature variation within the mantle, and shows the potential for revealing further details on the true nature of mantle convection.

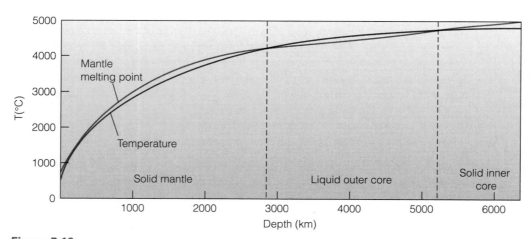

Figure 7.19
Temperatures within the Earth as revealed by seismic wave velocities compared to the melting point of the materials of the Earth's interior as determined from their composition and the pressure conditions that exist at depth. Note that the material of the outer core is above its melting point whereas that of the inner core is not.

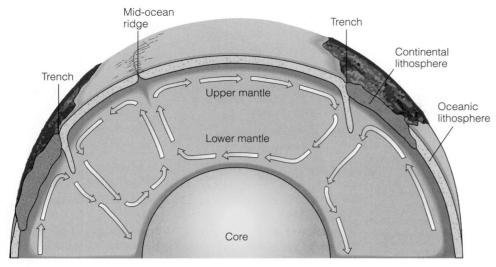

(a) Whole-mantle convection

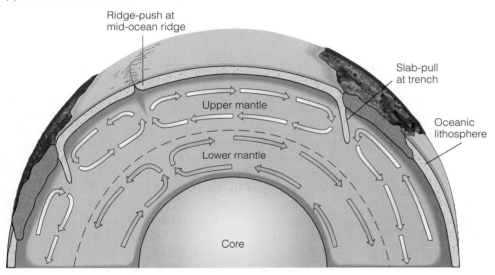

(b) Layered mantle convection

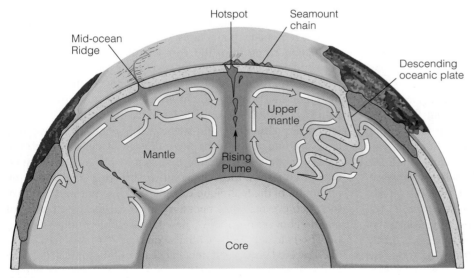

(c) Mantle plume convection

Figure 7.20
Proposed models for mantle convection:
(a) whole-mantle convection in which convection cells involving the entire mantle carry the lithosphere like a conveyor belt, (b) two-layer convection in which faster circulation of the upper mantle, driven by slab pull at subduction zones and ridge push at mid-ocean ridges, is decoupled from the slower circulation of the lower mantle, and (c) hot mantle plumes in which upwelling is confined to narrow plumes while downwelling occurs where subducting oceanic crust plunges deep into the mantle.

7.5.1 Seismic Tomography: 3-D Maps of the Mantle

Seismic tomography is a new technique that promises to improve our knowledge of the structure of the mantle in a dramatic fashion. Although still in its infancy, it is already indicating a pattern of convective flow in the mantle that is far more complex than that suggested by any previous model. The principle of seismic tomography is similar to that of the CAT scan (computerized axial tomography) used in the medical profession. A CAT scan is an X-ray procedure in which large numbers of X-ray images of a patient are taken from various angles by rotating the X-ray beam about the patient's body. Because the behavior of X-rays depends on the internal structure of the patient, rotation of the X-ray beam allows the structure to be imaged in the third dimension. Computers are then used to combine the X-ray data so that a three-dimensional image of the patient's internal organs can be created.

Seismic tomography uses the same approach to create three-dimensional images of the Earth's interior by combining seismic data from large numbers of earthquakes. As earthquakes occur in many places on the Earth's surface, information can be synthesized from many sources, mimicking the CAT scan effect of the rotating X-ray beam. Because of its relatively uniform composition and structure, variations of seismic velocity in the mantle are probably due to temperature differences that are, in turn, related to mantle convection. Hotter areas of the mantle, because they are less rigid, have slower seismic velocities than cooler areas at the same depth. Thus areas of unusually low seismic velocities are thought to indicate regions of hotter mantle, whereas areas of unusually high velocities are thought to depict cool regions. Seismic tomography's sophisticated computer analysis of many hundreds of slices through the mantle have made it possible for geoscientists to draw maps of seismic wave velocity and, therefore, mantle temperature, for various depths in the mantle.

The principle of seismic tomography is illustrated in Figure 7.21. In this diagram, the dark area represents a region of the mantle in which seismic velocities are anomalously slow. This, in turn, implies that the region is anomalously hot. The travel times for those rays (DD′, EE′, FF′, GG′) that pass through the shaded area will therefore be longer than the travel times of those rays that do not pass through this region. Given enough ray paths, it becomes possible to map out the anomalous region in three dimensions, and from their travel times, determine whether the region is anomalously hot or cold.

Although this technique works well in principle, there are some problems. For example, there are not enough seismic stations and too few earthquakes to produce the number of ray paths needed to map out the detailed structure of the mantle. In addition, little is known of the lower mantle because the resolution of the technique, that is, its ability to see detail, decreases with depth. Nevertheless, tomographic imaging of the upper mantle suggests a pattern of circulation that is far more complex than one of simple convection cells. Take, for example, the images illustrated in Figure 7.22 which show the distribution of unusually hot and cold areas in the upper mantle in both map view and cross section.

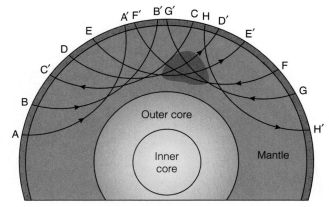

Figure 7.21

Principle of seismic tomography. The dark region has a lower wave velocity (because it is hotter than the surrounding region). If seismic travel times are measured for rays that travel equal distances AA′, BB′, etc., those for rays that travel through the shaded region will take longer, from which it is possible to deduce the approximate location and shape of the anomalously hot region.

When portrayed in map view (Fig. 7.22a) tomographic results show a strong relationship between the velocity structure of the mantle and the present distribution of lithospheric plates to a depth of 100 km (62 mi). Thus hot areas of low seismic velocity (in red) generally coincide with spreading ridges whereas cold areas of high velocity (in blue) occur beneath the oldest parts of the continents. This implies that, at least in the upper 100 km of the mantle, the pattern of convection is relatively straightforward.

At a depth of 300 km (186 mi), however, the pattern in plan view is dramatically different, and the correlation with present-day plates is significantly diminished. The mantle beneath some spreading ridges, like the one to the south of Australia, appear as hot at 100 km, but is cold at 300 km. For others, like that in the Red Sea, the reverse is true, that is, they are hotter at depth. The pattern changes again at a depth of 500 km (311 mi). According to geoscientists favoring a layered convection model, such changes argue against whole-mantle convection, because, if divergent plate boundaries were the direct expression of circulation rising from the core–mantle boundary, unusually hot conditions beneath the ridges should exist to great depths.

In contrast to the ridges, however, most of the continents maintain their velocity structure with depth, appearing cold in plan view at 100, 300, and 500 km. This is more consistent with large-scale convection and suggests that continents have very deep and old roots. At its present stage of development, the images produced by seismic tomography are not yet detailed enough to resolve these apparently conflicting results. But the resolution of the technique is being steadily improved and may soon achieve the level necessary to reveal the true nature of mantle convention. In fact, new tomographic results (see Dig In: The Fate of Subducted Slabs, on page 196) may have identified severed slabs of subducted oceanic lithosphere at the base of the mantle, which would firmly favor some form of whole-mantle convection.

Displayed in cross section, however tomographic imaging of the upper mantle reveals a complex pattern of low and high

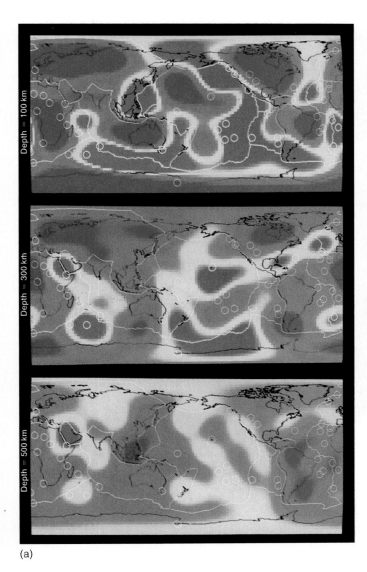

(a)

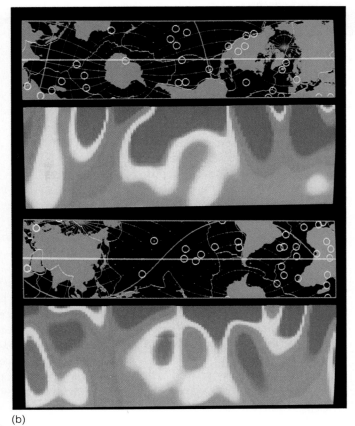

(b)

Figure 7.22

Tomographic scans of seismic wave velocities in the upper mantle (a) in map view at depths of 100, 300, and 500 km, and (b) in cross section (indicated by line) to a depth of 670 km. Regions in red have low seismic velocities indicating they are hot. Regions in blue have high seismic velocities indicating they are cold.

velocity regions that are difficult to relate to any simple model of thermal convection. In Figure 17.22b, the pattern of hot and cold regions are shown to the base of the upper mantle at a depth of 670 km (415 mi) directly beneath two lines that girdle the Earth, one from pole to pole and the other on a diagonal course across the Pacific Ocean. Once again, a correlation with present plates can be seen at shallow depths, with slow (hot) regions under mid-ocean ridges and other volcanically active areas, and fast (cold) regions below subduction zones and old continental interiors. However, only the high velocity structure beneath the continents extends to any great depth. In fact, the velocity of seismic waves appears to remain higher beneath the continents than it does beneath the oceans to depths as great as 500 km (more than 300 mi). Whether this observation reflects differences in the temperature or composition of the mantle beneath continents is uncertain, but it may have important implications on the nature of the plates themselves. If, as seismic tomography suggests, the continents have rigid roots that extend to depths of 500 km or more, then these roots cannot equate directly to the continental plates, because the thickness of these plates, based on P and S wave velocity data for the

lithosphere–asthenosphere boundary, is thought to be less than a quarter of this value.

Seismic tomography is therefore raising important new questions about the nature of plates and their relationship to the lithosphere, the answers to which are certain to further our understanding of plate tectonic processes. The technique also promises to reveal many new details of the Earth's interior as the quality of the pictures it provides improves. We may therefore be on the verge of resolving the nature of mantle convection and, in doing so, we may finally solve Wegener's problem with the mechanism of continental drift by coming to understand how the movement of the plates is linked to that of the mantle.

The principle function of mantle convection is to bring the heat generated in the Earth's interior to the surface, where it is eventually radiated into space. Plate tectonics is part of this process; sea-floor spreading bringing hot new crust to the surface where it cools, while subduction returns old cold crust to the Earth's interior. At the same time, circulation of the mantle ensures the continual addition of its lighter components to the crust, and so contributes to the development of an Earth

The Fate of Subducted Slabs

Exciting new results on the fate of subducted slabs have swung the debate over mantle circulation firmly in favor of a form of whole-mantle convection. Astonishingly, tomographic results have identified anomalously cold bodies, thought to represent the remnants of severed subducted slabs, that have penetrated the lower mantle as far as the core–mantle boundary (Fig. 7.23). Because they penetrate the transition between the upper and lower mantle, there is apparently no impenetrable barrier to mantle flow across this zone, removing the main obstacle to the idea of whole-mantle convection. In fact, it appears that the core–mantle boundary may be the graveyard of subducted slabs, potentially establishing a direct link between plate tectonics and the deep interior of the Earth.

Even so, many problems remain to be solved. First, the descent of such slabs demands a compensating return flow, like the water that rushes up when a diver enters the water. The pattern of this flow, however, remains ambiguous. The main candidate is that of mantle plumes, which many geoscientists believe to ascend from the core–mantle boundary. In addition, velocity patterns beneath some subduction zones suggest that not all descending plates penetrate the transition zone between the upper and lower mantle, but instead appear to be deflected by it. Perhaps only the oldest, coldest, and therefore densest oceanic plates are heavy enough to push their way through the transition zone, whereas younger, more buoyant subducting plates are not, and so are deflected by it.

Some of the ambiguity inherent in seismic tomography is clearly related to its pre-

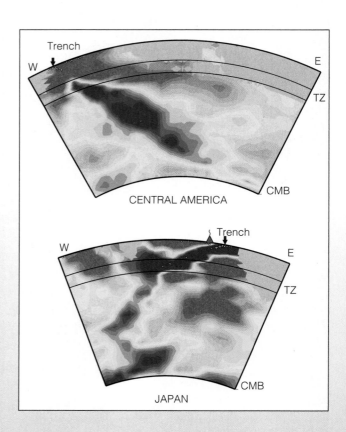

Figure 7.23
Vertical slices through the Earth's mantle beneath Central America and Japan, showing the distribution of colder (seismically fast) material (in blue) and hotter (seismically slow) material (in red) to the core–mantle boundary (CMB) based on seismic tomography. The distribution of colder material suggests that the subducting slabs beneath both areas have penetrated the transition zone (TZ) and extend to the base of the mantle.

sent level of resolution, and will doubtless become clearer in the future as the resolution of the technique improves with the gathering of more data. For example, there is currently no obvious evidence for deep mantle plumes, although their tails may simply be too narrow to detect at this level of resolution. Also, because mantle convection is such a slow process, the pattern of heat distribution and therefore, the velocity structure of the mantle, is likely to depend on past plate configurations as well as the present one.

Thus the mantle may have a "better memory" for ancient tectonic events than the crust, which is involved in endless recycling and renewal. Much better correlations may therefore emerge when tomographic results of higher resolution are compared to the pattern of plate motions integrated over very long periods of time. Although the results of such studies are still some years away, seismic tomography clearly promises dramatic improvements in our understanding of mantle processes in the not-too-distant future.

---

that is layered according to its density. The lightest components have escaped the Earth's interior altogether and extend this density layering beyond the crust to the hydrosphere and atmosphere. To continue our examination of the planet we must therefore extend our inquiry to the two reservoirs in which these lightest components are now concentrated. We do this in the next section of this book by first turning our attention to the hydrosphere and to the water of the oceans and continents that nurtures the life of the planet's biosphere.

7.6 Chapter Summary

The rocks of the Earth's surface represent only a minute fraction of the Earth itself, and provide few clues on the nature of the Earth's interior. Instead, the Earth's interior must be studied indirectly by applying geophysical methods to some of the Earth's internal processes, such as earthquakes, gravity, magnetism, and heat flow.

Earthquake waves have revealed that the Earth is a layered planet. These seismic waves are sent out in all directions from an earthquake focus when a locked fault abruptly lets go, and can be used like X-rays to probe the planet's interior. Like shock waves, they include several forms of vibrations. Body waves, which travel through the Earth's interior, are of two types. Primary or P waves travel like sound waves, vibrating the material through which they pass in the direction of their propagation. P waves are compressional waves and can travel through any medium, although they travel fastest through compact solids. They are the first to be detected at recording stations on devices known as seismometers. Secondary or S waves move like the oscillations in a rope, vibrating the material through which they pass at right angles to the direction of their propagation. S waves are shear waves and can only travel through solids. Surface waves, like ocean waves, are confined to the Earth's surface. Although often the most damaging of seismic waves, they provide little information on the Earth's interior.

Because their speed depends on the material through which they pass, the pattern of body waves as they travel through the Earth's interior provides clues to its internal structure. In this way, the Earth has been shown to be layered with a thin outer crust, an intervening dense rocky mantle, and a metallic core. An abrupt discontinuity known as the Moho separates the crust from the mantle, and a shadow zone, in which no seismic waves are recorded, testifies to the presence of a core. The layering of the Earth is therefore one of density and extends beyond the crust to the hydrosphere and atmosphere.

On average, the Earth's crust above the Moho varies in thickness from 7 km (4 mi) beneath the oceans to 30 to 40 km (20–25 mi) beneath the continents. Seismic wave velocities also distinguish oceanic crust of broadly basaltic composition, from continental crust, which is separated by a further discontinuity into a generally granitic upper crust and a more dioritic lower crust.

The mantle is the largest of the Earth's internal subdivisions, extending from the base of the crust to the core–mantle boundary at a depth of 2900 km (1800 mi). It is thought to be solid, chemically homogeneous and made up of iron- and magnesium-rich silicate minerals. The transition zone between the upper mantle and the lower mantle is defined by a series of discontinuities between 400 and 700 km (between 250 and 430 mi) depth. These discontinuities are thought to be the result of changes in mineral structure (phase changes) rather than composition. A decrease in seismic velocities between 100 and 250 km (between 60 and 155 mi) depth marks the low velocity zone, and is attributed to the presence of a small amount of melt. Above the low velocity zone, the upper mantle is cool enough to be rigid and, together with the crust, forms the lithosphere. This rigid layer makes up the Earth's tectonic plates, and rests on a yielding layer of the mantle known as the asthenosphere. It is the asthenosphere, which may extend to the base of the transition zone, that enables the plates above to move.

With a diameter of 6940 km (4320 mi), the core of the Earth is just a little larger than Mars. A discontinuity within it at a depth of 5100 km (3200 mi), separates a liquid outer core from a solid inner core. Seismic velocities, coupled with evidence from the Earth's gravity and magnetic field, and from the composition of meteorites, suggest that the outer core is a mixture of either iron and sulfur, or possibly iron and oxygen, whereas the inner core is pure iron or an iron–nickel alloy.

Much of the Earth's interior is thought to be slowly circulating. Movement in the liquid outer core must power the dynamo that generates the Earth's magnetic field, and is thought to be driven by the sinking of iron toward the inner core as it solidifies. Thermal convection of the mantle, which is ultimately responsible for plate motion, is powered by the decay of radioactive elements, and takes place in the solid state by a process known as creep. However, the pattern of mantle convection is not well understood, and models have been proposed that variously advocate whole-mantle convection, in which the entire mantle convects, layered convection, with separate convection systems above and below the transition zone, and mantle plumes, in which focused areas of upwelling emanate from the core–mantle boundary.

A new technique that promises dramatic improvements in our understanding of mantle processes is that of seismic tomography. This utilizes data from large numbers of earthquakes from many sources to produce a three-dimensional image of the planet's interior, in much the same way that a CAT scan uses X-rays to image a patient's internal organs. Although not yet of high enough resolution to map out the fine structure of the mantle, tomographic imaging of the mantle suggests an extremely complex pattern of circulation, only the uppermost 100 km (62 mi) of which shows a strong correlation with the present configuration of lithospheric plates. The velocity structure beneath continents may extend 500 km (more than 300 mi) below the surface, suggesting they have very deep roots. Present ambiguities in the method preclude definitive conclusions about the nature of mantle convection and the real promise of the technique is still some years away. On the one hand, dramatically different patterns of hot and cold mantle at depth beneath the oceans argue against a simple form of deep mantle convection. On the other, most recent images identifying the penetration of severed subducted slabs to the lower mantle, and perhaps the core–mantle boundary, argue for a direct link between lower mantle processes and plate tectonics, suggesting that the whole mantle may indeed convect.

Key Terms and Concepts

Look for the highlighted items on the web at:
WWW.BROOKSCOLE.COM

- The nature of the Earth's interior has been determined indirectly from earthquakes, gravity, magnetism, and heat flow studies
- According to the elastic rebound theory, earthquakes occur when stress builds up on locked faults, straining the rocks on either side until they rupture
- Seismic waves generated at the focus of an earthquake spread out in all directions much like light rays from a light source, first reaching the Earth's surface at the earthquake epicenter
- Seismic vibrations detected by seismometers and recorded as seismograms, are of three types: primary or P waves, secondary or S waves and surface waves. P and S waves are body waves and are used to provide an X-ray of the Earth's interior
- The Earth is layered according to its density with a thin crust, a dense rocky mantle and a metallic core. The Moho marks the boundary between the crust and mantle, while the shadow zone shows that the outer core is liquid
- Oceanic crust is largely basaltic whereas continental crust has a generally granitic upper portion and a more dioritic lower part

- A transition zone in which mineral phase changes occur separates the upper mantle from the lower mantle
- The lithosphere, which contains both the crust and part of the upper mantle, is rigid and overlies a hotter and more pliable asthenosphere that contains the low velocity zone in which partial melting occurs
- Circulation of the mantle by creep is thought to be driven by thermal convection, whereas that of the liquid outer core is probably powered by compositional convection
- Three models have been proposed to describe the pattern of mantle circulation; whole-mantle convection, layered mantle convection, and mantle plumes
- Seismic tomography, which uses data from many earthquakes to CAT-scan the Earth's interior, is a promising new technique that should greatly improve our understanding of mantle circulation

Review Questions

1. How can density be used to show that the Earth's interior is vastly different from its surface?
2. What is the elastic rebound theory and how is it used to account for the occurrence of earthquakes?
3. How do P waves differ from S waves?
4. Why do seismic waves adopt curved ray paths as they travel through the Earth's interior?
5. How were P wave velocities used in the discovery of the Moho?
6. What is the significance of the P and S wave shadow zone?
7. What is the relationship between the low velocity zone and the asthenosphere?

8. What is the composition of the outer core and how has this been determined?
9. How can the inner core be solid when the outer core, which is not as hot, be molten?
10. What is the difference between thermal convection and compositional convection, and in what areas of the Earth's interior is each likely to be found?
11. How is it possible for the mantle to convect if the material of which it is comprised is solid?
12. In what way is seismic tomography analogous to a CAT scan?

Study Questions

Research the highlighted question using InfoTrac College Edition.

1. What are seismic waves and how are they distinguished?

2. How is the boundary between the crust and mantle defined, and how does it differ from the boundary between the lithosphere and asthenosphere?

3. What do the variations of P and S wave velocities with depth (Fig. 7.13) tell us about the major internal divisions of the Earth?

4. Explain the origin of the P wave shadow zone.

5. Explain the origin of the S wave shadow zone.

6. If the earthquake occurred at the equator instead of the north pole, where would the shadow zones occur?

7. Why is the transition between the crust and the mantle a sharp one, whereas that between the upper and lower mantle is gradational?

8. What do seismic tomographic scans of the upper mantle (Fig. 7.22) reveal about mantle convection and the validity of mantle convection models (Fig. 7.20)?

Solid Earth

Solid Earth

The main processes affecting the solid Earth are the release of heat from Earth's core to the surface and the movement of plates on the surface. Plate tectonics is key to understanding the evolution of solid Earth and Earth's other reservoirs.

Astronomy

Earth has developed into a layered planet according to density based on the laws of gravity. All terrestrial planets are layered in a similar manner, and all matter in the universe follows the law of gravitational attraction. Heat from the cooling Earth and other cooling planets radiates into the universe.

Atmosphere

Volcanic eruptions near plate boundaries and at hotspots contribute gases to the Earth's atmosphere. Some of these gases are held above the Earth's surface by gravity, the remainder leak into space. The atmosphere radiates the heat from the cooling Earth into the universe while holding some of it to Earth.

Hydrosphere

Ocean basins are created by divergent plate movements. Submarine vents at mid-oceanic ridges change the composition of ocean water. Water in solid and liquid forms erodes the land and contributes to the rock cycle.

Biosphere

Moving plates and changing continental land connections create different environments for Earth's living creatures and plants. Plant and animal life modify continental crust and contribute to the rock cycle. Fossils were important evidence in the development of the theory of plate tectonics.

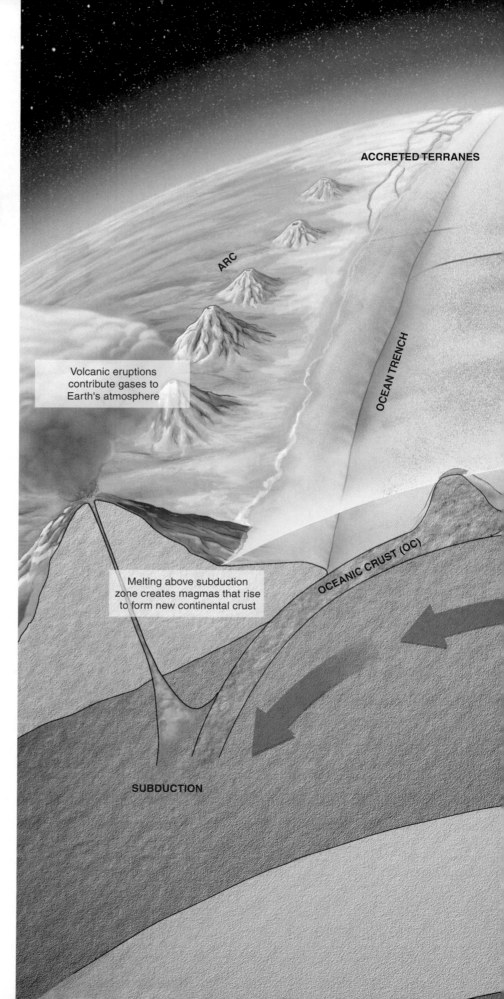

ACCRETED TERRANES

ARC

OCEAN TRENCH

OCEANIC CRUST (OC)

Volcanic eruptions contribute gases to Earth's atmosphere

Melting above subduction zone creates magmas that rise to form new continental crust

SUBDUCTION

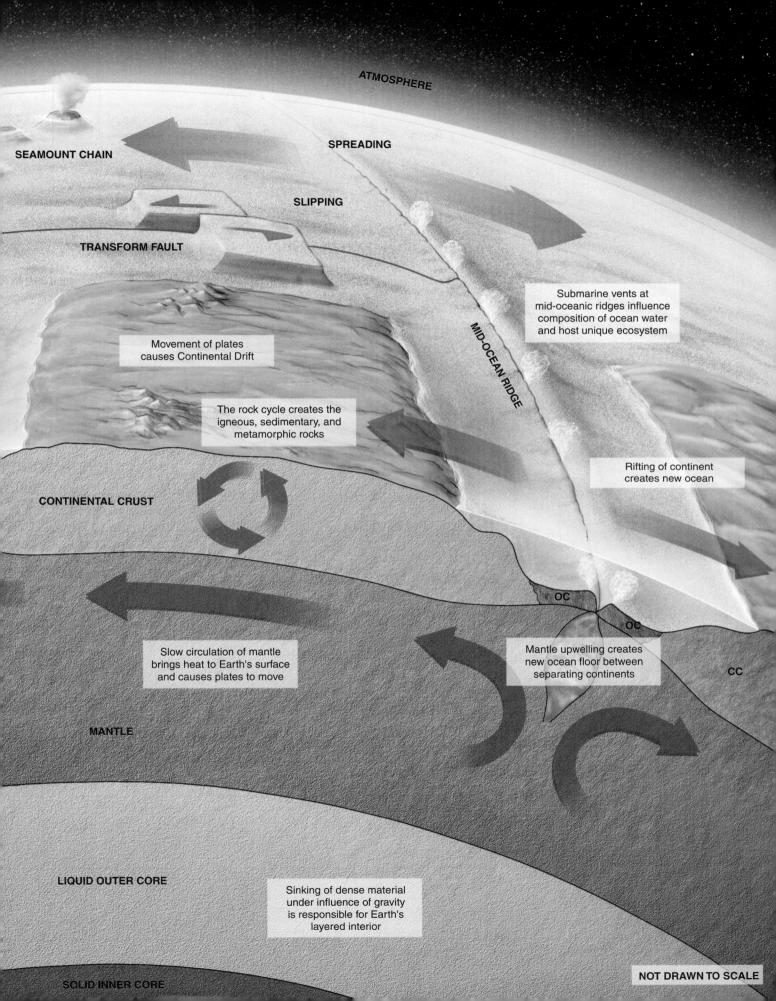

PART 2

The Hydrosphere:
The Origin and Motion
of Water on Our Planet

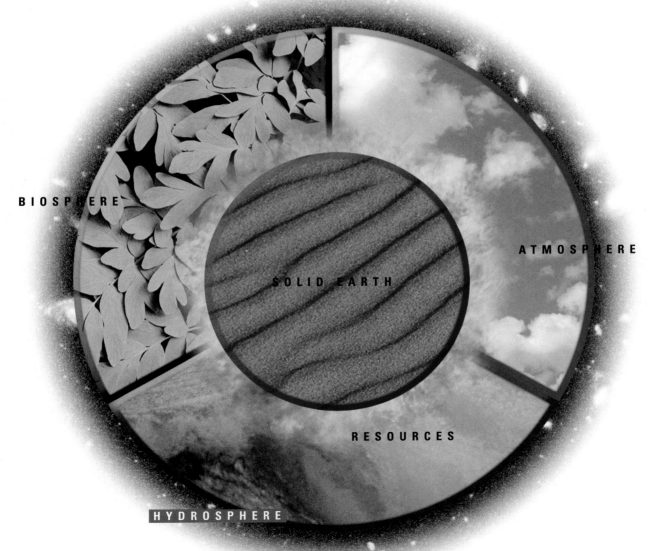

BIOSPHERE

ATMOSPHERE

SOLID EARTH

RESOURCES

HYDROSPHERE

STARS & PLANETS

CHAPTER

8

Ocean Water

From our background in plate tectonics, we know that individual oceans such as the Atlantic are only temporary and survive at the whim of plate motion. In this chapter we explore the mysteries of these ocean basins, their origin, characteristics, and the composition of their water. Our purpose is twofold. First, to form a bridge from plate tectonics to the generation and composition of ocean basins, and second, to set the stage for the material in the next chapter—ocean dynamics and the movement and circulation of ocean waters.

To understand the factors which govern the chemical composition of the ocean basins, we first discuss the uniqueness of the water molecule, which accounts for many of water's unusual, but vital, properties. These properties, together with the position of the Earth within the Solar System, explain the abundance of liquid water on the Earth's surface in contrast with its virtual absence on the other terrestrial planets. The influence of the Sun is also important in providing a driving mechanism for the hydrologic cycle, which describes the evaporation of water from the oceans, and its return flow to the oceans by way of continental drainage.

You will learn that "sea level" isn't really level at all; it is an average taken of a surface distorted by winds and gravity. In addition, we will find that this average "level" has fluctuated throughout geologic time and is very sensitive to climate changes and plate tectonic processes which either expand or contract the ocean basins.

Plate tectonics also helps explain the important and variable characteristics of continental margins, contrasting those that are tectonically active because they lie along plate margins, with those that are tectonically passive because they do not. We will find that the style of tectonic activity along the continental margin influences the shape of the adjacent ocean basin.

Our knowledge of plate tectonics also constrains theories involving the origin of water on our planet and the composition of sea water. For example, the geologic record indicates that oceans formed early in the Earth's history, probably as early as 4 billion years ago. However, without some crude form of plate tectonics, it is unlikely that the oceans could have existed. As we have learned, the contrasting compositions and densities of continental and oceanic crust causes oceanic crust to be relatively depressed, allowing water to collect in the ocean basins.

The salty composition of ocean water is related to water's unusual properties, and a delicate balance that exists between continental erosion and the activity of submarine hydrothermal systems near mid-ocean ridges. Since water is a strong solvent, it is rarely pure. Fresh water draining from continents strips minerals and soils of loosely held elements, including sodium. At the same time, emanations from hydrothermal vents on the deep ocean floor, introduce elements such as chlorine. Thus, ocean water chemistry, which is dominated by these two elements, is fundamentally linked to plate tectonic activity.

By the end of this chapter, the fundamental link between plate tectonics and ocean basin development and evolution will be evident.

8.1 Introduction

8.1.1 Water on Planet Earth

The influence of water on the history of the human race is more profound than that of any other commodity. In some parts of the world, water is taken for granted, but in others, securing access to it is a prerequisite to survival. The vital importance of water was recognized by many early civilizations and most of the world's major cities were originally settled on rivers or where rivers met the sea.

Water is an essential ingredient of all living things and without it the Earth would be as lifeless as the other planets in the Solar System. The abundance of water, which covers some 71% of the Earth's surface, is the first thing that an extra-terrestrial visitor would notice about our planet. The Earth is the only planet in the Solar System with liquid water on its surface. Venus, often referred to as our "sister" planet, has a surface temperature of almost 470°C (880°F), which is far too hot for water to exist in its liquid state. Mars, located further from the Sun than the Earth, has an average surface temperature between −20°C (−4°F) and −120°C (−184°F), which is too cold for liquid water. Mars, however, may contain water ice on its surface in the form of permafrost and polar ice caps. Liquid water may also exist beneath its surface, trapped within relatively warm rocks.

The average surface temperature on Earth is 13°C (56°F) which is ideal for maintaining liquid water on its surface. Like Goldilocks's porridge, surface temperatures on our nearest planetary neighbors, Venus and Mars, are either too hot or too cold, whereas the Earth, which is located between them, is "just right"!

Water occurs on the Earth in all of its natural states: liquid, solid, and gas. It forms a very thin veneer of liquid on our planet's surface. Even in oceans, the average depth of water (3.8 km or 12,500 ft) is very small compared to the average radius of the Earth (6370 km or 3950 mi). At present, the vast majority of water occurs in our oceans (about 97.2%). About 2% is locked in glaciers, icecaps, and snow, and about 0.65% occurs as groundwater and surface water on continental landmasses (Fig. 8.1). Water vapor (the gaseous state of water) and water droplets also occur in the lower 10 km (6 mi) of our atmosphere. Although conspicuous as clouds, however, on average these constitute a minute percentage (less than 0.001%) of our planet's water.

8.1.2 Mysteries of the Ocean Deep

Our current knowledge on the motion of water in the Earth's surface reservoirs owes much to relatively recent exploration of the deep ocean. Our ancestors did not have the advantage of modern technology, and in the absence of a clear understanding of the origin and motion of water on our planet, myths and legends persisted for centuries. Not until the Renaissance period, which began in the 15th century, were these superstitions seriously challenged. This period was one of the most dramatic episodes in the history of the human race. Based in Europe, it was a time of great intellectual ferment that spawned major innovations and experimentation in art, culture, and science. Traditional views and myths were set aside on exciting voyages of discovery.

The voyages of Christopher Columbus and other sea-faring pioneers such as Amerigo Vespucci (after whom America was named) and Vasco de Gama between the 15th and 17th centuries transformed the popular view of planet Earth. Although most educated people of the time had accepted that the Earth was round, rather than flat, they also believed that the Earth was much smaller than it actually is. Columbus set sail convinced he would reach the far east islands of Asia after only 5000 km (3000 mi) of sailing. These islands were actually more than twice that distance away. He died not knowing that he had actually discovered North America.

These voyages complemented the scientific experiments, observations and theories of Copernicus, Galileo, Newton, and others so that by the end of the Renaissance period, the currently held view of a spherical Earth as one of several planets orbiting the Sun had been established.

Although early maps showed that a considerable portion of the Earth's surface was covered by ocean water, very little was known about what lay beneath its surface. As we have learned, this ignorance was demonstrated as recently as the early 20th century by the popular rejection of Alfred Wegener's hypothesis of continental drift. Even Wegener assumed that for continental drift to occur, continents would have to plough across the ocean floor (see Chapter 4). Geophysicists argued that this mechanism defied the laws of physics. They pointed out that continents are made up of rocks like granite which are softer and less dense than many basalts of the ocean floor and so could not possibly plough across them.

In was not until later in the 20th century that the nature of the ocean floor was finally revealed. The period between 1910 and 1960 was one of observation, measurement, and data collection, which laid the groundwork for the concept of sea floor spreading and modern oceanography. Important advances were made using echo sounders (Fig. 8.2) which are devices that bounce a pulse of energy off the seabed and relate the time lapse between the transmission and reception of the pulse to ocean depth. The longer the time lapse to the return of the signal, the deeper the ocean bottom. Echo sounders allowed scientists to study variations in the depth to the ocean floor, and so identify elevated areas such as mid-ocean ridges and deep linear troughs that we now know to be the sites of subduction.

In addition, the development of radiometric dating (Chapter 3) and deep sea drilling technologies enabled the precise determination of the absolute ages of sediments and ocean crust recovered from drill cores and dredges. This showed that the ocean floor was less than 200 million years old and anomalously young compared to the continents.

At the same time, the Earth was conducting its own natural experiments. On November 18, 1929, at 3:31 P.M. Greenwich Mean Time, a powerful earthquake struck the Grand Banks of Newfoundland. The aftermath of this earthquake severed a number of submarine cables that had been laid across the Atlantic seabed between North America and Europe to aid communications between these two continents (Fig. 8.3a).

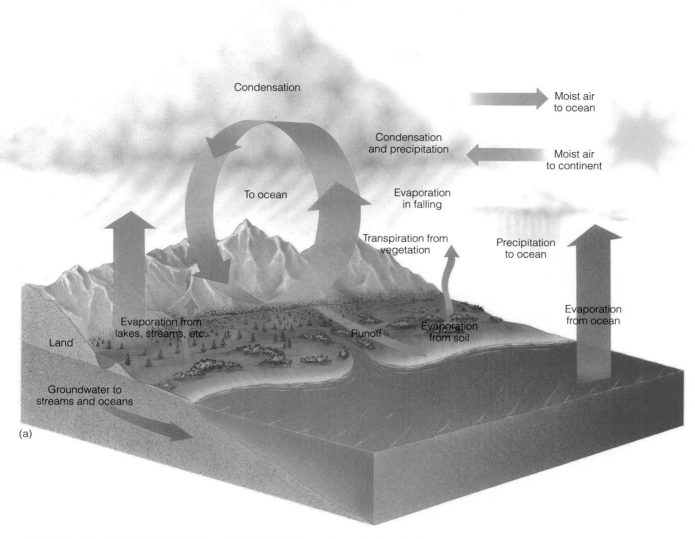

(a)

Water on Earth		
Location	Volume (km^3)	Percentage of Total
Oceans	1,327,500,000	97.20%
Icecaps and glaciers	29,315,000	2.15%
Groundwater	8,442,580	0.63%
Freshwater and saline lakes, inland seas	230,325	0.016%
Atmosphere at sea level	12,982	0.001%
Average in stream channels	1,255	0.0001%

(b)

Figure 8.1
The hydrologic cycle and the relative abundance of water at or near the surface of the Earth.
Water evaporates and rises from the ocean and forms clouds. Clouds precipitate either over land
or sea. Much of the precipitation falling on land drains back to the ocean by way of streams,
completing the cycle as it does so.

The Western Union Cable near the epicenter of the earthquake broke almost immediately. Over the next 24 hours, a series of cables were broken in sequence oceanward from the epicenter. This baffling turn of events was finally explained in the early 1950's when it was proposed that the earthquake generated an underwater avalanche that transported sediment downslope in what is known as a turbidity current from the continental margin toward the deep ocean floor (Fig. 8.3b).

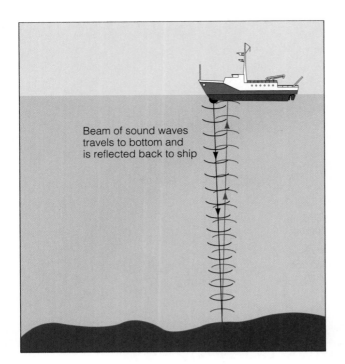

Figure 8.2
Echo sounders beam sound waves to the ocean bottom. The time that elapses before the reflected signal is received back at the ship allows the distance to the ocean bottom to be determined and provides information on ocean basin topography.

Turbidity currents are made up of dense sediment-fluid mixtures which travel just above the seabed. The velocity of this particular turbidity current calculated from the timing and sequence of breakages, was estimated to be close to 100 km (62 mi) per hour. This new interpretation radically changed the prevailing view of ocean sedimentary processes from one of steady slow sedimentation to one that was dynamic and profoundly influenced by tectonic activity.

Our modern understanding of the oceans, their water, the sediments that drape their underlying crust, and the life the oceans sustain, has greatly increased over the past 40 years. However, there are still many important problems and challenges to be overcome. Many of these arise from the special properties of the water molecule, which dictates its behavior in the oceans.

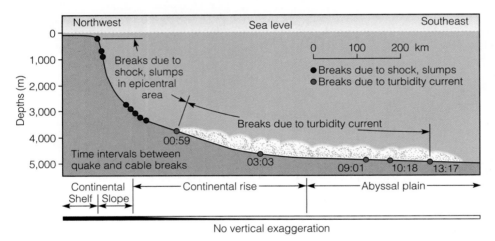

(a)

(b)

Figure 8.3
(a) A vertically exaggerated profile of the Grand Banks of Newfoundland where a 1929 earthquake triggered a turbidity current that transported sediment toward the abyssal plain. The rate of transport was determined by the sequence of submarine cable breaks. (b) Photograph of a turbidity current in motion. This shot was taken along the submarine slope off the island of Jamaica.

8.2 The Water Molecule

8.2.1 Water: A Unique Substance

The importance of water to the Earth's reservoirs and to the human race is a function of the special properties of the water molecule. As we shall learn, if water behaved like most molecules, the Earth would almost certainly be a lifeless planet.

The following are some of the unusual properties of water.

1. **Water occurs as a gas (water vapor), a liquid, and a solid (ice) on the Earth's surface.** Few other substances can be found naturally in all three states.

The water molecule is composed of two hydrogen atoms that are bonded by a single oxygen atom. Consequently, it is a very light molecule with a molecular weight of 18, the oxygen contributing 16 units of mass while each hydrogen contributes one. In comparison, other light molecules, some with masses two to three times that of water, such as nitrogen and carbon dioxide (with molecular weights of 28 and 44, respectively), occur only as gases in the Earth's atmosphere. In contrast, most water is present in liquid form, which helps maintain its presence on the Earth's surface. The difference between water in a liquid, solid, or vapor state lies in the arrangement and degree of bonding of the water molecules. Ice is a crystalline solid, made up of relatively loosely packed water molecules, whereas the molecules in liquid water are more tightly packed. This causes water to be denser in the liquid state than it is as a solid. Water vapor, on the other hand, is less dense than either liquid or solid water, having the least number of bonds and a less structured arrangement of water molecules.

The reason that most water on Earth occurs in the liquid state is related to its relatively high boiling temperature (100°C) compared to other light molecules. In other words, a greater amount of energy must be supplied to liquid water in order to break the bonds necessary to form water vapor, than is the case for most light molecules.

2. **Whether in a lake, the sea, or your favorite cold drink, ice floats on liquid water.** Although this is a commonplace observation, it is a very unusual property. For the vast majority of chemical compounds, solids are denser than liquids of the same composition, and so would sink under the same circumstances. This is because molecules of solid material are, for the majority of substances, more closely packed than liquids. Fortunately for us, the converse is true of water. The molecules

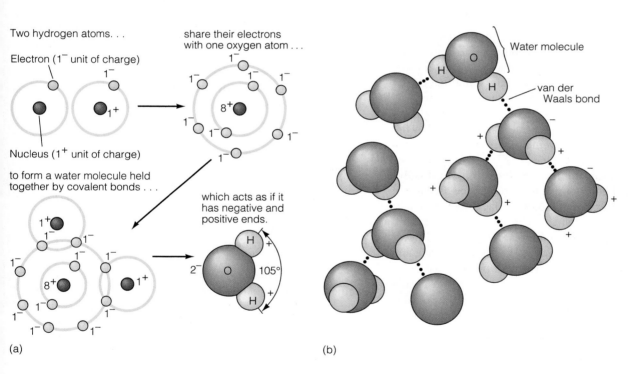

(a)

(b)

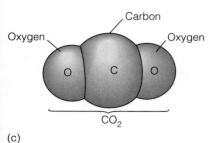

(c)

Figure 8.4
(a) Bonding and polarity in a single water molecule. (b) The water molecule forms from the bonding between hydrogen and oxygen. The molecule is highly polarized, which results in attraction of neighboring molecules. These van der Waals bonds are responsible for the cohesion of the water molecule and many of its unusual properties. (c) The molecular structure of carbon dioxide is vastly different from water. The lack of strong interaction between molecules allows carbon dioxide to be a gas at room temperature.

in ice, although structured, are more loosely packed than in liquid water. As a result, ice is less dense than water. If sea ice sank to the ocean bottom, the entire circulation system of the oceans would be altered. Given that life on land evolved from the sea, this would have fundamental implications for the origin and evolution of life on Earth.

Water's change in density from liquid to solid also produces a volume change. When water freezes, the solid form expands to a volume that is 9% greater than the same mass of water. So, within confined areas, the freezing of water produces a dramatic force. For example, freeze-thaw weather cycles play havoc with roads as water collects in potholes, freezes, and expands, and enlarges the pothole. This process is repeated with each freeze-thaw cycle creating ever larger potholes. As we will learn in Chapter 15, from a geologic perspective, the volume change is responsible for frost wedging of surface rocks, which pries rocks apart when water in fractures freezes. Cracking and subsequent weathering of rock in this fashion forms an effective means of erosion.

3. **The maximum density of fresh water occurs at 4°C and not at the freezing point (0°C).** As the temperature drops from 4°C to the freezing point, water becomes increasingly lighter. This property has direct consequences on bodies of fresh water, like lakes, where low surface temperatures (of around 4°C) produce a surface water layer that is denser than the water beneath it. This generates a convection system as the dense surface water sinks and the bottom waters rise to replace it. The end result is a circulation of the lake water that causes mixing of nutrients within it. In contrast to fresh water, the maximum density of seawater occurs below its freezing point (around −1.9°C). Consequently, seawater is not affected by the same churning process. The movement of seawater will be discussed in depth in the next chapter.

4. **Water is a strong solvent and has the ability to dissolve most other chemical compounds to some degree.** Because it is a strong solvent, water is rarely pure. In fact, the vast majority of naturally occurring elements can be identified in both ocean and fresh water. This is both good news and bad news. The good news is that this property allows for the storage and migration of the nutrients of life. The bad news is that water supplies are susceptible to pollution, because many toxic substances can also dissolve in water. Seawater also has the ability to dissolve gases, the majority being derived from the atmosphere. Water vapor, nitrogen, oxygen, and carbon dioxide are the principle gases absorbed in the ocean.

5. **Visible light can penetrate water to depths of up to 200 m (660 ft) before it is absorbed.** The ocean surface is typically transparent, allowing the passage of light through its uppermost layers. This property is essential for photosynthesis in the oceans, which, in turn, nurtures life at the base of the food chain. It is probable that the first light-loving organisms on Earth, some 3.86 billion years ago, formed in pools of shallow water penetrated by sunlight.

6. **Water has the capacity to store large quantities of heat with relatively little increase in temperature.** Ocean water temperatures vary within a small range from −1.9°C (28°F) to 30°C (86°F). This contrasts with continental landmasses, which absorb heat during the day and emit or lose heat at night. Daily temperature variations on land vary by as much as 100°C (180°F). Since the temperature of ocean water is more uniform than that of the continents, oceans have a strong moderating influence on our climate, particularly in maritime regions. Ocean water also buffers, or moderates, some of the effects of global warming by storing excess heat.

8.2.2 Bonding and the Water Molecule

Most of the above properties are the direct result of the bonding within and between water molecules, and the unusual way in which atoms of hydrogen and oxygen combine. Hydrogen atoms have a strong tendency to donate negatively charged electrons and, consequently, are positively charged. The larger oxygen atom has a strong attraction for these electrons, and is therefore negatively charged. The water molecule, like any molecule, is organized so that it is as stable as possible. This is achieved by maximizing the forces of attraction, and minimizing forces of repulsion. As we can see from Figure 8.4, the properties of hydrogen and oxygen atoms cause the water molecule to be strongly polarized; that is, one end of the molecule is positively charged, while the other is negatively charged.

The negative end of water molecules strongly attracts the positive end of adjacent water molecules, and vice versa, resulting in the formation of van der Waals bonds between adjacent molecules (see Chapter 2). The breaking of these bonds requires energy, which explains the ability of liquid water to absorb large quantities of heat, the high boiling temperature of water compared to other light molecules, and the significant amount of energy (known as the latent heat) required at this temperature before each water molecule can become independent and evaporate.

In contrast, carbon dioxide has a completely different molecular structure (Fig. 8.4), with a much less polarized distribution of positive and negative charges and, therefore, far less attraction for its neighbors. Because carbon dioxide has little interaction between its molecules, it is gaseous at room temperature.

The strong polarity of water gives it a cohesion or compactness that does not exist in carbon dioxide, so that although carbon dioxide has a greater molecular weight, water is denser and exists as a liquid at room temperature. When water freezes into snowflakes, its molecules rearrange into a very open six-sided (hexagonal) structure which is about 9% larger by volume than the equivalent mass of liquid water. Therefore, ice is less dense than, and consequently floats on, liquid water.

8.3 The Hydrologic Cycle

Although water's special properties help maintain its presence on the surface of our planet, the interaction between water and the Sun's radiant energy governs its motion between the continents, the oceans, and the atmosphere. This interaction is called the **hydrologic cycle**.

As we discussed in Chapter 1, radiant energy from the Sun promotes the evaporation of ocean water into the atmosphere. This moist warm air is swept onto land where it rises, cools, and condenses. Since the cooler air can no longer retain the moisture as a dissolved gas, water droplets form and coalesce until they surpass a threshold size. Then precipitation occurs as either rain or snow, (depending on the surrounding air temperature) and the moisture falls to the ground. Much of this moisture is ultimately returned to the ocean.

It is important to realize that water also occurs in the Earth's biosphere and in the solid Earth. The influence of these two reservoirs on the hydrologic cycle is more subtle than that of the Sun and the atmosphere, but it is crucial for life on Earth. Water is an essential ingredient of living tissue. Plants and animals continually absorb and expel water during the process of living. In plants, the process is known as **transpiration.** In the solid Earth, water migrates through the pores and cavities of soil, sediments, and rocks. For the most part, this water is derived from precipitation and much of it ultimately finds its way into streams, and drains back to the sea (Fig. 8.1).

Other forms of water in the solid Earth are more subtle, but their influence is nonetheless important. From our previous discussions of plate tectonics (Chapter 5), we know water can be absorbed and expelled from the crystal structure of minerals depending on the geologic environment. In this sense, they show some similarities to living organisms which also absorb and expel water. For example, when basalt magma is extruded at mid-ocean ridges, it absorbs some of the ocean water, which changes the mineral content of the rock. In this environment, new minerals grow that absorb water into their crystal structures. As we learned in Chapter 2, the chemical formulae of these minerals betray the presence of stored water in the symbol "OH." An example of this is illustrated in the for-

mula for chrysotile, a form of asbestos, which may be formed when oceanic crust absorbs water:

$$Mg_3Si_2O_5(OH)_4$$

When the basaltic oceanic crust is subducted (Fig. 8.5) these hydrous or "water-bearing" minerals become unstable, breaking down due to the increased temperature and pressure. As a result, the water locked within them is released. As the fluid rises, it promotes melting in the rocks it encounters and may find its way into magma chambers above the subduction zone, only to be released into the atmosphere by volcanic activity.

So the biosphere, solid Earth, and atmosphere all recycle water, constantly moving it from one reservoir to another. In addition, the relative proportion of water in the different reservoirs has fluctuated throughout geologic time. For example, during an ice age, like the one which ended about 11,000 years ago, a significant proportion of water that evaporated from the oceans did not return to the sea, but instead was "stored" in ice sheets. As a result, sea level was much lower than it is today. At the end of an ice age, when the glaciers melt, the water is returned to the ocean and sea level rises.

8.4 Sea Level

The motion of water from one reservoir to another fundamentally affects the amount of water in the ocean basins, and hence sea level. Sea level, in turn, controls the position of our coastlines, and therefore has had a profound influence on the development of major urban centers.

A fundamental property of all liquids is that, unlike solids, they have no shape of their own. So they adopt the shape of the container they are in, with an upper surface that is horizontal. As water drains from the continents (Fig. 8.1), it fills the vast

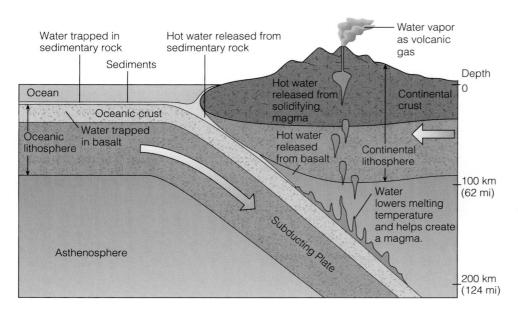

Figure 8.5
Schematic diagram of subduction emphasizing the location and behavior of water. Water trapped in basaltic oceanic crust or in marine sediments can be recycled by subduction processes. Seawater is released in the subduction zone promoting melting in the overlying lithosphere. Much of this water may eventually be vented to the atmosphere by way of volcanoes. Some of this vented water comes directly from the Earth's interior and is referred to as juvenile water.

containers we call the ocean basins. We call the average upper surface of ocean water **sea level**, and all elevations on land (positive) and depths at sea (negative) are measured relative to it.

Unlike a small container, however, a casual glance at any ocean reveals that ocean water does not have a smooth "level" surface because other forces act upon it. The most obvious of these forces is wind, which distorts the surface into waves. On a daily basis, tidal forces (see Chapter 9) related to the gravitational pull between the Earth, Moon, and Sun also affect water level and are "averaged" when sea level is calculated.

Gravity also distorts sea level in more subtle ways. Gravitational attraction between two bodies varies according to the mass of each body and the distance that separates them. Bodies with large masses attract adjacent bodies toward them. For the most part, the thickness of oceanic crust beneath the ocean water is constant (see Chapter 5), and therefore the gravitational attraction between ocean water and the underlying ocean crust is also relatively constant. However, the vast underwater mountain chain that forms the mid-ocean ridges constitutes an excess mass with a stronger attraction for ocean water. As a consequence, the sea surface mounds above mid-ocean ridges, producing a bulge in the oceans that has been detected by satellite (Fig. 8.6).

The effects of wind, tides, and gravity have distorted the sea's surface throughout geologic time. Other processes result in variations in sea level because they affect the volume of water in the ocean basin. An ocean basin can be likened to a bathtub partially filled with water. Sea level in such a basin remains constant, only if (1) the amount of water draining into the basin equals the amount taken out (by evaporation in the case of oceans), and (2) the shape of the basin remains the same. The onset of an ice age changes the balance between evaporation and drainage, whereas plate tectonics affects the shape of the ocean basin.

8.4.1 Sea Level and Ice Ages

If the amount of water draining into an ocean basin is less than that evaporating from it, sea level will drop. During an ice age,

evaporated water that becomes trapped in polar air masses falls as snow, rather than rain. If the snow does not melt in the spring, it will eventually become compressed into glacial ice as it is buried beneath successive layers of snow. Since this water does not drain back to the ocean basin, sea level drops. Of course, the opposite happens when glaciers melt. The water returns to the oceans and sea level rises.

During the last Ice Age, which peaked about 18,000 years ago, glacial ice up to 3 km (2 mi) deep covered many of the continents in the northern hemisphere. At that time, sea level is estimated to have been 100 m (330 ft) lower than it is today. Thus, much of the present continental shelves, that are today flooded by less than this depth of water, would have been exposed. The North American Atlantic coastline, for example, was about 200 km (120 mi) east of its present location (Fig. 8.7). Britain and France were joined, eliminating the need for a channel tunnel or "chunnel" beneath the English Channel. Similarly, a land bridge across the Bering Strait linked Alaska and Siberia, and many of the islands of Indonesia were connected to southeast Asia. As anthropologists will vouch, this period of relatively low sea level had a profound influence on the migration of animals, including humans. For example, humans migrated across the Bering Strait from Asia to North America during this glacial episode.

At the end of the last Ice Age, some 11,000 years ago, most of the continental glaciers melted and, as a consequence, sea level rose as the ice slowly retreated toward the poles. So if the remaining glaciers start to melt as a consequence of global warming, the sea will continue its invasion of coastal regions (Fig. 8.7). Low-lying coastal states such as Florida, Louisiana and the Carolinas will be most threatened, and cities such as New York, Orlando, Houston, and Memphis will be submerged.

The realization that sea level would change as the consequence of ice ages has spurred scientists to search for evidence of such events in the more ancient geologic record. In the mid 1970's, Peter Vail and a team of researchers from Exxon Corporation conducted a detailed analysis of drill core from sedimentary rocks deposited on the world's continental shelves. They proposed that similar changes in sea level revealed in these cores over the past 250 million years were similarly related to the waxing and waning of ice sheets. According to their hypothesis, as the sea retreated during an ice age, portions of the continental shelf became exposed. As a result, they were eroded, creating a gap in the geologic record known as an unconformity (see Chapter 3). Since all continental shelves appeared to possess unconformities of exactly the same age, the process responsible must have been global in scale. Vail and his team concluded that the only known process that was rapid enough to cause these unconformities was fluctuations in the sizes of major ice sheets.

As we shall learn in Chapter 14, a single cycle of glacial advance and retreat during an ice age typically lasts on the order of 100,000 years. However, there are cycles of much

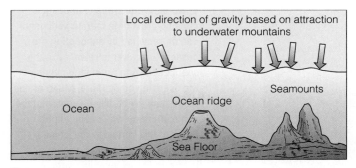

Figure 8.6
Distortion of sea surface above an ocean ridge due to the gravitational attraction between seawater and the oceanic mountains.

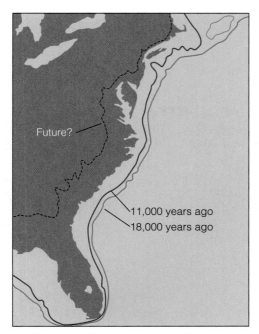

Figure 8.7
The position of the eastern North American coastline 18,000 years ago and its future position if the present-day ice caps melt. The position of the coastline 18,000 years ago, was located up to 200 km (125 mi) to the east, exposing the continental shelves. If all polar ice caps were to melt, sea level would rise by about 100 m (330 ft), submerging much of the eastern seaboard of North America.

longer duration that also affect sea level. These are the cycles associated with plate tectonics.

8.4.2 Sea Level and Plate Tectonics

Plate tectonic activity slowly, but inexorably, changes the shape of ocean basins. Just as the same volume of water will fill bathtubs of different size to different "levels" depending on the shape of the bathtub, so sea level changes with the tectonic evolution of both continental and oceanic crust.

Consider, for example, the effect on sea level of the supercontinent cycle we examined in Chapter 6, illustrated by the assembly and breakup of Pangea. Before breakup, the insulating properties of continental crust result in the trapping of mantle heat beneath the supercontinent (Fig. 8.8; Stage 1). This causes the supercontinent to become elevated and, as a consequence, sea level drops. This situation changes rapidly when the supercontinent rifts, and the continental fragments drift apart.

As the new oceans form, ocean ridges develop between each of the continental fragments. In addition, as the continental fragments drift apart, their edges or margins subside. The result is a dramatic rise in sea level (Fig. 8.8; Stage 2). There are three reasons for this. First, the development of new

mid-ocean ridges produces vast new underwater mountain ranges similar to the one at the center of the modern Atlantic Ocean. These mountains displace the water of the ocean. Just as placing a large object in a bathtub displaces the water upward, so the new submarine mountains cause sea level to rise upward and outward across the continental margins. The result is the generation of shallow seas floored by continental crust.

Second, as drifting proceeds, the dispersed continental fragments cool and subside as they move away from the hot regions of mantle upwelling at the ocean ridge crest. Just as lowering the sides of a bathtub filled to the brim causes the water to spill out, the effect once again is to allow sea water to flood over the continental margin.

Third, most geologists speculate that continental breakup results in global warming because greenhouse gases are introduced into the atmosphere by the volcanoes associated with the rifting process. Global warming would cause melting of any continental glaciers, resulting in a rise in sea level. Unfortunately, it is difficult to obtain reliable estimates of the volume of ice that may have melted when Pangea broke up because the size of continental ice sheets during any stage of the supercontinent cycle is unknown.

As sea floor spreading becomes established, however, the edges of the newly formed ocean crust get older. They also get further away from the mid-ocean ridge and become cooler as they do so. As a result, they become denser so that the ocean crust starts to sag, ultimately forming ocean trenches as subduction begins. Like a depression forming in the bottom of a bathtub, this sagging of the ocean floor causes sea level to slowly fall so that the continental shelves become emergent.

Subduction, however, begins the process of ocean destruction. As the width of the ocean shrinks, the oldest and heaviest ocean crust is preferentially destroyed. As a consequence, the ocean floor becomes progressively younger and more buoyant. As the floor of the ocean rises, so too does sea level (Fig. 8.8; Stage 3). If the ocean closes and a new supercontinent forms, however, sea level will begin to drop once more, as the continental crust is once again uplifted by trapped mantle heat and the cycle begins anew (Fig. 8.8; Stage 4).

The geologic record shows remarkable correspondence to this long-term pattern of sea level change. For example, following the breakup of a supercontinent in latest Precambrian time (about 600 million years ago), extensive shallow marine continental shelf deposits of Early Cambrian age (545 million years old) were formed at the edges of many of the dispersing continental fragments, consistent with rising sea level. Similarly, following the breakup of Pangea in the Triassic, continental shelf deposits of Jurassic and Cretaceous age (between 190 and 66 million years old) formed along the periphery of the Atlantic Ocean in response to rising sea level and the flooding of its continental edges. In both cases of supercontinent breakup, the newly flooded shelves formed an ideal habitat for flourishing ecosystems that are well preserved in the fossil record.

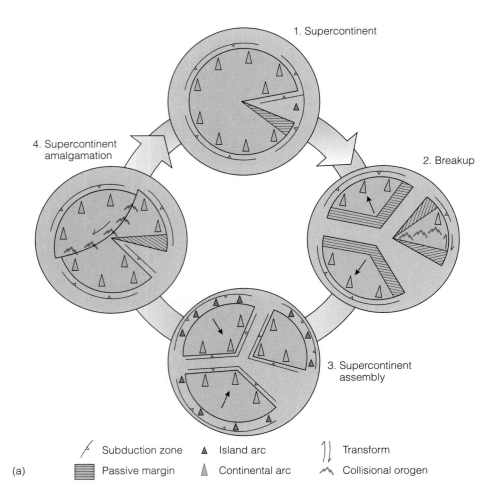

(a)

- ⟋ Subduction zone
- ▬ Passive margin
- ▲ Island arc
- ▲ Continental arc
- ⟩⟩ Transform
- ⟿ Collisional orogen

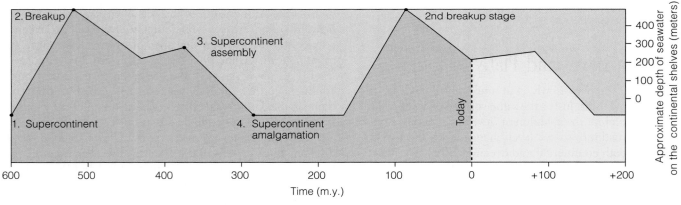

(b)

Figure 8.8

(a) Supercontinent cycle (see Chapter 6). (b) The supercontinent experiences uplift before continental breakup is initiated resulting in a low sea level (1). Supercontinent breakup results in formation of new oceans. The development of mid-ocean ridge mountains, increased volcanic activity, and the progressive subsidence of the crust of the dispersing continents contribute to a sharp rise in sea level (2). As the new oceans continue to widen, the average age of the ocean floor gets older, and sea level falls. Subduction of these oceans causes sea level to rise once more (3) only to fall again as amalgamation (4) traps mantle heat beneath the new supercontinent.

8.5 Ocean Basins and Plate Tectonics

In addition to their influence on sea level, plate motions govern the distribution of oceans and continents, and the nature of the contact between them. For example, although modern oceans make up 71% of the Earth's surface, a higher proportion of water is located in the southern hemisphere where less than 35% of the continental landmasses are found. The present distribution of the continents and oceans reflects sea floor spreading since the breakup of Pangea (Fig. 8.9).

As we learned in Chapters 4 and 5, the geology of modern ocean basins is dominated by *three* components. In addition to *water*, they contain *sediments* that are derived from the erosion of adjacent landmasses and the decay of organisms that lived within the ocean, and *crust* which underlies both the water and sediments. The crust itself consists of two components of contrasting composition and density. Most is oceanic crust which forms at mid-ocean ridges and extends from these ridges to the continental margins. The rest is submerged continental crust which underlies the continental shelves and represents the flooded edges of continents. The continental shelves are essentially an extension of continental crust beneath the ocean water.

Plate tectonics also influences important features such as the level of seismic activity along continental margins, the depth (bathymetry) of the oceans, and the development of continental shelves, which provide a habitat for flourishing ecosystems.

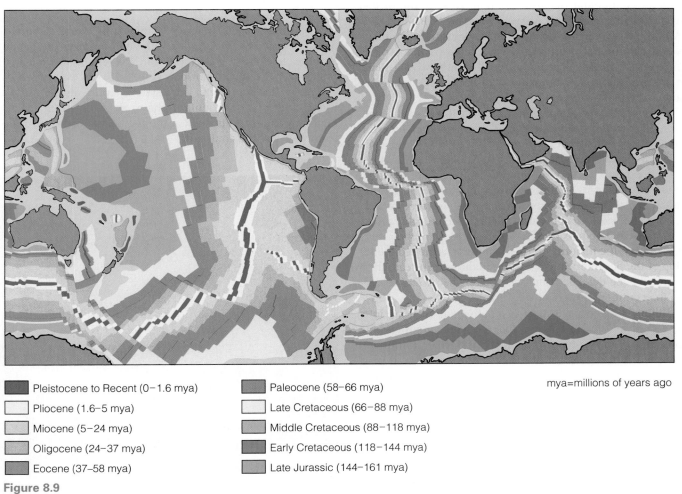

- ■ Pleistocene to Recent (0–1.6 mya)
- □ Pliocene (1.6–5 mya)
- Miocene (5–24 mya)
- Oligocene (24–37 mya)
- Eocene (37–58 mya)
- Paleocene (58–66 mya)
- □ Late Cretaceous (66–88 mya)
- Middle Cretaceous (88–118 mya)
- Early Cretaceous (118–144 mya)
- Late Jurassic (144–161 mya)

mya=millions of years ago

Figure 8.9
The influence of sea floor spreading on the world's ocean basins. The youngest crust is adjacent to modern spreading ridges, and its age increases sequentially with distance from those ridges. Note the symmetry of the Atlantic Ocean reflecting the presence of the spreading ridge in the middle of the ocean. This contrasts with the asymmetry of the Pacific Ocean where the spreading ridge is located closer to its eastern margin and intersects the coastline in California.

8.5.1 Active and Passive Continental Margins

As we have learned, a plate boundary is defined by tectonically active processes such as subduction, rifting, or transform faulting. If we look at a plate tectonic map of the modern Earth (see Figure 4.20), the margins of the oldest ocean basin, the Pacific Ocean, can be seen to coincide, in many cases, with plate boundaries represented by either major transform faults (such as the San Andreas Fault) or subduction zones (such as those adjacent to Japan and the Aleutian Islands). For this reason, ocean margins of this type are known as **active** or **Pacific-type** margins. The active continental margins of the Pacific Ocean are therefore prone to earthquakes and volcanic activity, and the circum-Pacific is often referred to as the "Ring of Fire."

Other continental margins, exemplified by those of the Atlantic Ocean are relatively passive, and so are termed **passive** or **Atlantic-type** margins. Along these margins, the boundary between continental crust and oceanic crust is *not* a plate boundary but, instead, occurs within the interior of a plate (see Figure 4.20). So, in contrast to Pacific-type margins, Atlantic-type margins are not seismically active.

However, the present contrast between the Atlantic and Pacific Oceans will not last forever. Given that oceanic crust generally does not survive more than 200 million years, subduction of the Atlantic Ocean crust is ultimately inevitable.

In the future, when subduction starts in the Atlantic Ocean, portions of its continental margins will become plate boundaries along which oceanic crust will be destroyed. When this happens, these margins will become tectonically active like those of the modern Pacific Ocean. Because of their relative buoyancy, the majority of the less dense marine sediments on top of the ocean crust will not be subducted. Instead, they will probably be detached and preserved by becoming accreted to the continental margin (see Chapter 5).

Although the classification of continental margins into "active" and "passive" is a useful one, the synonymous classification into "Pacific-type" and "Atlantic-type" (although entrenched in the geologic literature) is less useful and may invite confusion, especially when dealing with the geologic past or the future. In the future, for example, the Atlantic Ocean may resemble the modern Pacific, being fringed by subduction zones and transform faults. When this happens, the Atlantic Ocean will have become bordered by "Pacific-type" margins.

8.5.2 Ocean Bathymetry: A View of the Planet's Surface without Water

For centuries, mariners have pondered the depths of the oceans. Until recently, however, this information has been hidden by sea water. We now know that plate tectonics profoundly influences the bathymetry (depth) of the ocean, forming hidden mountain ranges at mid-ocean ridges and deep trenches at subduction zones (Fig. 8.10).

The strategic importance of ocean bathymetry was recognized during World War II, and methods to investigate ocean depths have been refined ever since. Echo sounders (Fig. 8.2), which provide images of the submarine topography, identified the submerged mountain ranges of mid-oceanic ridges and seamount chains, as well as the deep trenches at the sites of subduction. More recent technological advances in satellite and remote sensing imagery are similar in principle but allow precise depth determinations over wide areas. Today we can study underwater bathymetry as if the planet was completely devoid of water.

We now have detailed information on the bathymetry of most of our oceans. For our purposes, the most important result of this research is the fact that ocean depth and its variations are consistent with plate tectonic theories concerning the ocean crust. Hot crustal features are elevated, while cold features are depressed.

The most obvious elevated topographic features on the ocean floor are the mountains of the mid-ocean ridges where new oceanic crust is formed, and the chains of volcanic islands associated with hotspots (Fig. 8.10). As we learned in Chapter 5, the mid-ocean ridge system represents the world's widest and longest mountain chain. By comparison, continental mountain belts, such as the Himalayas, are small-scale features with little continuity.

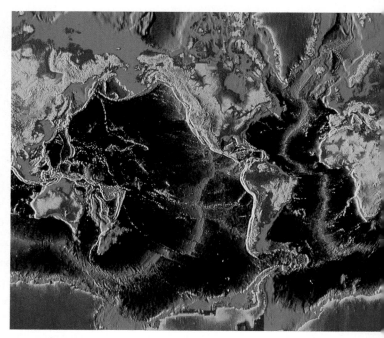

Figure 8.10
Ocean basin bathymetry showing the varied topography of the ocean floor. The ocean spreading ridges stand out as a great submarine mountain chain, the subduction zones as deep trenches, and continental shelves as shallow regions along passive continental margins.

Ocean trenches are the deepest parts of the oceans. They are associated with subduction zones and the destruction of oceanic crust. The bottom of the Mariana's Trench in the southwestern Pacific Ocean, for example, is 11,000 m (about 7.5 mi) below sea level.

On a global scale, the contrast between oceanic and continental elevations is most apparent in the diagram known as the hypsometric curve (Fig. 8.11; see also Chapter 5). This diagram plots the percentage of the Earth's crust that occurs *at or above* a given elevation. In it we can see that only about 30% of the Earth's surface has an elevation higher than sea level, about 60% has an elevation higher than the abyssal plains (typically about 4 km below sea level), and 99% of the Earth's surface stands higher than the deep ocean trenches (typically between 6 and 11 km below sea level).

It is also evident from this diagram that the dominant elevation of the ocean crust is about 3.8 km (2.4 mi) below sea level. This elevation occurs in tectonically stable areas, and represents the tracts of oceanic crust we call **abyssal plain**, which lie between mid-ocean ridges and continental margins.

The dominant elevation of continental crust (about 0.8 km or 0.5 mi above sea level) also occurs in tectonically stable regions. This contrast between the dominant elevations of stable continental and oceanic crust characterizes the Earth as a "split-level" planet, and is consistent with the concepts of isostasy and plate tectonics outlined in Chapter 5. The Earth's split-level surface plan reflects the density contrast between the oceanic and continental lithospheres, and the fact that both "float" on the yielding asthenosphere, which is located between 5 and 100 km below the Earth's surface. It is this density contrast that allows water to drain from the relatively elevated continents and collect in the relatively depressed ocean basins.

8.5.3 Features of Continental Margins

From bathymetric studies, we now have a thorough understanding of the features of continental margins. They are characterized by continental shelves, which simply represent the submerged extension of continental crust beneath the oceans (Fig. 8.10 and 8.12). The width of the continental shelf depends on sea level and whether the continental margin is passive or active. As we have learned, continental shelves are prone to submergence or exhumation depending on the rise and fall of sea level.

Passive continental margins tend to be wider than active margins because the continental crust extends unhindered beneath the sea. These shelves are typically more than 200 km (120 mi) wide (Fig. 8.12). The Siberian shelf, for example, extends more than 1000 km (620 mi) beneath the Arctic sea (Fig. 8.10). Along active margins, on the other hand, the continental crust is typically terminated by either a transform fault or a subduction zone and so the continental shelf is relatively narrow.

On average, continental shelves have a very gentle incline (1.7 m per km or 9 ft per mi). However, they are not smooth planar features. Like exposed continental crust they contain areas of uplifted blocks and down-dropped troughs which reflect the jostling of the crust as it subsides. In addition, they are dissected by **submarine canyons** (Fig. 8.13) which are narrow, deep, V-shaped valleys that represent the extension of rivers and their estuaries onto the continental shelf (see also sedimentary rocks, Chapter 2). These canyons are the preferred route for the transport of sediment to the deep ocean floor by turbidity currents.

As the hypsometric curve (Fig. 8.11) indicates, at some distance from the coast, continental crust eventually gives way to denser oceanic crust which rests at a greater depth. Thus, the continental shelf terminates and gives way to the abyssal plain of the deep ocean floor. A transition zone lies between these two regions and consists of the **continental slope** (which may have a gradient of up to 25°), and the **continental rise,** which occurs at the base of the continental slope.

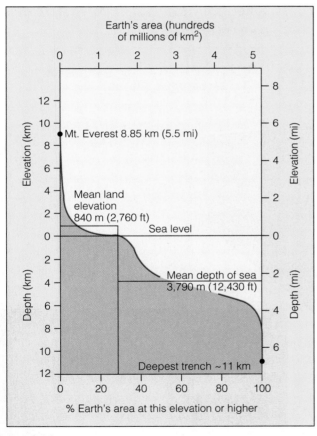

Figure 8.11
Hypsometric curve showing the distribution of the Earth's surface crust above–12 km below sea level. Note that the bottom axis refers to a cumulative percentage of Earth's surface area. For example, 30% of the Earth's crust is above sea level, while 60% of the Earth's crust has an elevation higher than the average level of the abyssal plain.

Figure 8.12
Typical features of a passive continental margin (a) with vertical exaggeration illustrating the shelf, slope and rise. The deep ocean floor is known as the abyssal plain. (b) The same continental margin with minimal vertical exaggeration.

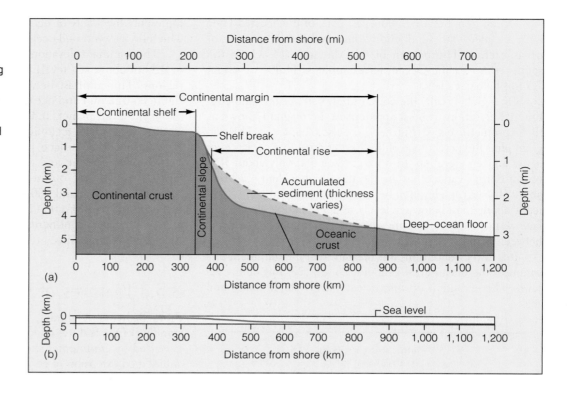

Figure 8.13
(a) A submarine canyon dissects the continental shelf transporting sediment from the canyon heads to the deep ocean floor. (b) Photograph of sediment cascading down submarine canyon.

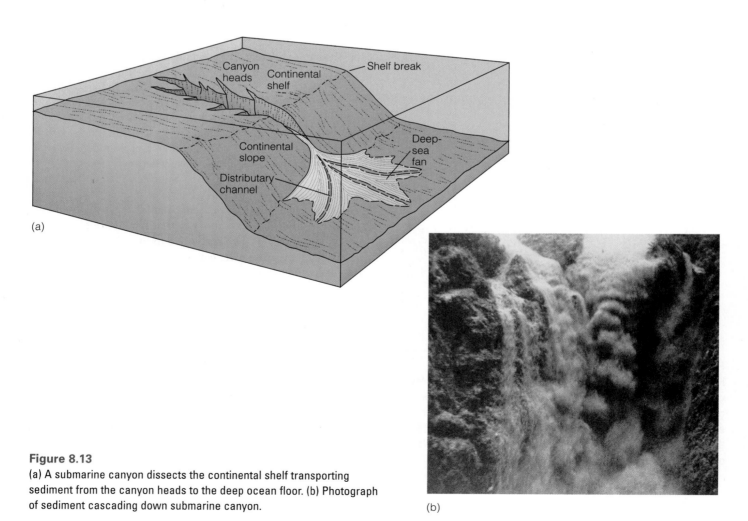

8.5.4 Deposition of Sediments and Sedimentary Rocks

In the Atlantic Ocean, vast volumes of sediments are deposited on continental shelves, especially where major river systems meet the sea. For example, the Mississippi River alone deposits 1 million tons of sediment into the Gulf of Mexico each day (Fig. 8.14). These sediments are deposited as a thick pile of layers that fill localized troughs and are dammed behind uplifted blocks. The immense weight of these thick sedimentary piles can cause the crust to further settle, which, in turn, allows even more sediment to be deposited. The total thickness of shelf accumulation can be as much as 15 km (8 miles).

In warm regions, where the sediment supply is not overwhelming, limestone-producing ecosystems flourish. For example, reefs (such as the Great Barrier Reef of Australia) and their associated ecosystems grow in shallow, relatively clear, sediment-free water, generally less than 50 m (165 ft) deep, at temperatures exceeding 20°C (68°F). The presence of these environments in the geologic record is preserved in limestone beds which predominantly consist of shells and the skeletal remains of animals.

The lack of tectonic activity and the very gentle slope of continental shelves makes these sediment layers very stable. As we have learned from the 1929 earthquake off the coast of Newfoundland, however, occasional seismic activity may trigger submarine avalanches. When this occurs, the sediment becomes unstable and hurtles down submarine canyons on the continental slope in turbidity currents. These sediments are typically deposited in submarine fans on the continental rise at the mouth of the canyons.

Our understanding of the geology of modern continental shelves is important in determining the early history of the Earth. In contrast to oceanic crust, which is preferentially subducted and so is lost from the ancient geological record, the continental margins are commonly preserved, and thus provide a record of ancient oceans. For example, thick piles of limestones and marine sandstones that resemble the sediments deposited on modern continental shelves date back 3.86 billion years. By applying the principle of actualism (Chapter 3), these ancient rocks are thought to have been deposited in a similar geologic environment along the margins of ancient oceans. These sedimentary piles are preserved in the geologic record because they were deposited on top of continental crust, rather than on oceanic crust (see Figure 8.12). In addition, sequences resembling the deep-sea fans of modern continental rises are also preserved. These sediments are less dense than the basaltic rocks of the underlying oceanic crust. As a result, they are commonly scraped off the oceanic plate when it is finally subducted and become attached to the adjacent continental margin. The preservation of these sedimentary sequences is very important because they provide irrefutable evidence that ocean basins have existed for the past 3.86 billion years, although the ocean basins that underlay them have since been destroyed.

Since we know that modern continental shelves harbor abundant life, the search for primitive life is concentrated in the most ancient representatives of shelf environments. Some of the oldest known fossils are found in 3.5 billion-year-old shelf limestones in western Australia. Our understanding of the geology of modern continental shelves can therefore be applied to the primitive Earth, the nature of its oceans, and the primitive life within those oceans.

Figure 8.14
Aerial view of the Mississippi delta.

8.6 Composition of Ocean Water

The composition of ocean water is influenced by its unusual properties. Despite the fact that water is a strong solvent, sea water is generally about 97.5% pure. Water's unique properties as a solvent stem from its electrically polarized molecular structure. This not only attracts adjacent water molecules, but also attracts (to varying degrees) every other element in the periodic table. In its simplest sense, negative ions such as chloride (the chlorine ion, Cl^-) and sulfate (SO_4^{2-}) are attracted to the positive end of the water molecule, while positively charged ions such as sodium (Na^+), magnesium (Mg^{2+}), and calcium (Ca^{2+}) are attracted to the negative end (Fig. 8.15). The major constituents of seawater (other than H_2O) are shown in Figure 8.16.

Despite the fact that streams and submarine vents bring major influxes of chemistry into the oceans, most ocean water from around the world approximates its average chemical composition. Experiments and calculations show that it takes one to two thousand years to circulate and mix chemical components

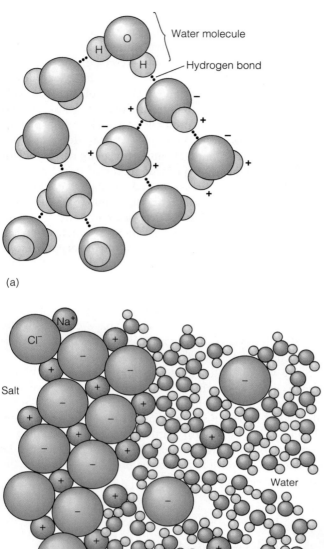

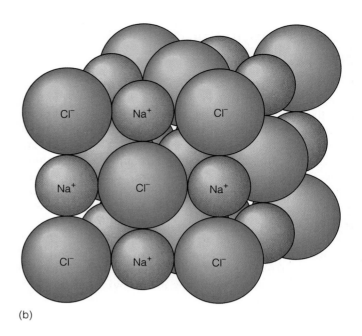

Figure 8.15
Solubility of common ions in water. (a) The structure of the water molecule. (b) The ionic structure of salt, Na^+Cl^-. (c) The solubility of salt. Note the positive sodium ions are attracted to the negative end of the water molecule, while the negative chloride ions are attracted to the positive end.

in the present ocean basins. Since the vast majority of ocean water has resided in the oceans for billions of years, it has had ample time to mix.

The salty taste of sea water is due to concentrations of sodium and chlorine, essential constituents of table salt. However, salt is also a technical term used to describe all the dissolved solids or ions in sea water. Determining the amount of salt in sea water, or its *salinity*, provides scientists with an important tool for tracking ocean water circulation and characterizing water masses at depth. The **salinity index** for water includes the total sum of the concentrations of all dissolved ions in a specified vol-

ume of water. Because this number is small, it is expressed in parts per thousand (per mil, which has the symbol ‰) instead of parts per hundred (percent, which has the symbol %). The salinity of normal ocean water ranges from 33 to 37 per mil (33 to 37‰). However, under special circumstances, it can be as high as 40‰. This occurs in relatively small, isolated bodies of water such as the Red Sea, that are undergoing significant evaporation but have limited influx of fresh water. As a result, salts become concentrated as evaporation proceeds. Other areas, such as the Baltic Sea, consist of cold waters, which undergo little evaporation and are diluted by an influx of abundant fresh

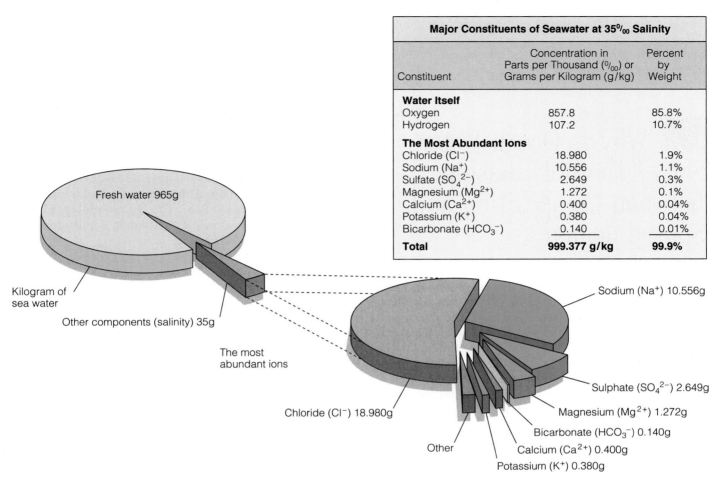

Major Constituents of Seawater at 35⁰/₀₀ Salinity		
Constituent	Concentration in Parts per Thousand (⁰/₀₀) or Grams per Kilogram (g/kg)	Percent by Weight
Water Itself		
Oxygen	857.8	85.8%
Hydrogen	107.2	10.7%
The Most Abundant Ions		
Chloride (Cl^-)	18.980	1.9%
Sodium (Na^+)	10.556	1.1%
Sulfate (SO_4^{2-})	2.649	0.3%
Magnesium (Mg^{2+})	1.272	0.1%
Calcium (Ca^{2+})	0.400	0.04%
Potassium (K^+)	0.380	0.04%
Bicarbonate (HCO_3^-)	0.140	0.01%
Total	**999.377 g/kg**	**99.9%**

Fresh water 965g

Kilogram of sea water

Other components (salinity) 35g

The most abundant ions

Sodium (Na^+) 10.556g

Chloride (Cl^-) 18.980g

Sulphate (SO_4^{2-}) 2.649g

Magnesium (Mg^{2+}) 1.272g

Bicarbonate (HCO_3^-) 0.140g

Other

Calcium (Ca^{2+}) 0.400g

Potassium (K^+) 0.380g

Figure 8.16
Relative abundance of the major constituents of a kilogram of sea water. Numbers adjacent to each component refers to the concentration in grams per kilogram.

stream water. In this situation, the salinity can be as low as 10‰. Low salinity marine waters are termed "brackish."

Ocean water also contains a significant proportion of dissolved gases, most notably carbon dioxide, nitrogen, and oxygen. Although these gases, like the ions, are dissolved in ocean water, the integrity of their molecular structure stays intact. As a consequence, they are not as tightly held as the ions, and are constantly interchanging with the atmosphere. In relatively still waters, the gases tend to escape into the atmosphere and the water becomes stagnant in the same way that a glass of soda pop will go flat if it is left standing. In open seas, this interchange is more dynamic because the moving water and blowing wind have more time to interact. It is also more obvious, producing the foam that forms in waves and breakers close to the shore.

Ocean water has an especially strong affinity for carbon dioxide. The gas dissolves and combines with hydrogen and other ions, and plays an important role in oceanic life processes. In fact, ocean water holds as much as 62 times more carbon

dioxide than is present in the atmosphere. As a consequence, ocean water can regulate the carbon dioxide content in the atmosphere. If the amount of carbon dioxide in the atmosphere increases (due perhaps to fossil fuel emissions), the oceans will absorb much of this increase. Alternatively, if atmospheric carbon dioxide levels fall, ocean waters tend to emit more carbon dioxide into the atmosphere. By regulating the carbon dioxide content in the atmosphere, oceans mediate the "greenhouse effect" and so exert a considerable influence on average global atmospheric temperatures.

Nitrogen accounts for nearly 50% of all dissolved gases in ocean water. Although marine organisms require nitrogen to build proteins, the nitrogen gas is not in a form that can be utilized for this process. Instead, nitrogen compounds required to support life are generally recycled among organisms.

Oxygen accounts for about 36% of dissolved gases. It is primarily produced by photosynthetic marine plants and is utilized by marine animals that extract the oxygen with their gills.

Much of the oxygen produced by these plants contributes to the oxygen content in the atmosphere. This process has had a profound effect on the evolution of the atmosphere throughout much of geologic time.

8.7 Why Is Ocean Water Salty? The Relationship between Composition and Source Region

The hydrologic cycle shows that the drainage of water from the continents to the oceans plays a major role in the transfer of chemical components to the oceans. Water on continental landmasses interacts with rocks and soil. The chemistry of stream water is, therefore, heavily dependent on the average chemistry of continental landmasses, and the ability of the elements involved to dissolve in water. Stream water, although fresh, often contains minor quantities of those elements that do dissolve. Elements such as sodium, potassium, and calcium are stripped from their crystals with relative ease and the relatively high concentrations of these elements in ocean water is, therefore, mainly the result of continental drainage. Other elements, such as silicon and aluminum, form an essential framework in many crystals and are difficult to dissolve. As a result, they are largely absent in stream water, and so are not enriched in ocean water.

Despite the fact that this relationship was recognized as early as the 17th century, it was not until the advent of plate tectonics that the full extent of the geologic influence on ocean water composition was understood. There are some serious discrepancies between the chemistry of seawater and that of the streams that drain into the sea. The most obvious discrepancy relates to the high concentrations of chloride ions (Cl^-) in seawater. Chlorine is relatively rare in rocks of the continental crust (Fig. 8.17), but is an essential ingredient of the salty oceans. With the recognition of sea floor spreading, scientists hypothesized that the source of chlorine and other elements with anomalously high concentrations in ocean water might be related to submarine volcanic activity. Evidence cited in favor of this hypothesis at that time was the discovery of gas bubbles, rich in chlorine and other anomalous elements, in samples of ultramafic rocks. As we discussed in Chapter 5, ultramafic rocks lie beneath the submarine basalts of the ocean floor and are thought to be the parent from which basaltic lavas are formed. The scientists therefore hypothesized that chlorine might be released from ultramafic rocks when they melted. The chlorine would then be transported toward the ocean floor and released into sea water through volcanic vents. The obvious test of this hypothesis was to investigate submarine volcanic activity and search for anomalous concentrations of chlorine-rich water in their vicinity.

In the late 1970s, a submersible called Alvin descended with three scientists to investigate an ocean ridge near the Galapagos Islands off the coast of Peru. They described for the first time the phenomenal spectacle of deep-sea hot springs in which jets of superheated water reaching temperatures as high as 350°C (662°F) were continuously erupting from cracks in the sea floor (see Living on Earth: The Bizarre World of Mid-Ocean Ridges, pages 224–225). These jets emerged from mineralized chimneys standing as high as 13 m (40 ft) above the sea floor. The hot springs proved to be rich in chlorine and metals, confirming a submarine volcanic source for chlorine in sea water.

These hot springs are part of a complex circulation pattern of water that is driven by the presence of magma (with a temperature of up to 1000°C, 1800°F) in relatively shallow crustal reservoirs at mid-ocean ridges. As the sea floor spreads apart, magma erupts onto the seabed to form new oceanic crust (see Chapter 4). Near-freezing sea water percolates down through the hot crust and may be heated to temperatures exceeding 400°C (750°F) before jetting upward through fractures in the crust to form hot springs or vents on the seabed. At these hot temperatures, the interaction of the water with ultramafic rock very efficiently dissolves and transports chlorine and a variety of metals to the sea floor (Fig. 8.18).

In summary, source regions exert an important control on seawater chemistry. Chemical contributions come from both continental and oceanic crust, revealing the delicate balance of natural systems. Ocean water is salty because it contains chlorine and dissolved ions such as sodium. But the chlorine is derived from oceanic crust, whereas the sodium is derived from continental crust. Despite their independent origins, however, the average concentration of salt in ocean water (33 to 37 ‰)

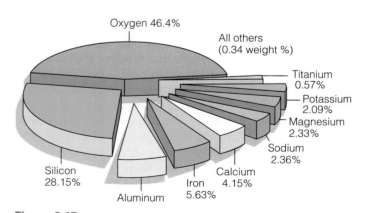

Figure 8.17
The average composition of continental crust. In comparison to Figure 8.16, note the abundance of Si^{4+} and Al^{3+}, and the lack of Cl^-.

(a)

(b)

Figure 8.18
(a) Worm tubes develop in regions where hot springs rich in metals and chlorine are vented.
(b) Photograph of a hot spring known as a "black smoker" taken during a dive by the submersible Alvin. The metals precipitate to form chimneys when they come in contact with cold sea water.

is ideal for present-day ocean life. Indeed, if the salt content rose any higher than 60‰, it would disrupt the cell walls of many marine organisms.

8.8 The Origin of Ocean Waters

Although the first part of this chapter addresses why we have water on our planet and the role plate tectonics plays in its distribution, it has not addressed a key question. Where did the water originally come from? Many students immediately point to the hydrologic cycle, but this is actually a chicken-and-egg type of argument. How does one get rain without evaporation from ocean water? How does one get ocean water without rainfall and the streams that bring this water to the ocean? Thus the hydrologic cycle, although a very important concept, does not address the question of how water *originated* on our planet. Many natural cycles have this problem. The cycle operates today because all parts are in place, but the origin of the cycle is more problematic. In this section we will find that there may be many sources, ranging from the Earth's interior to the burning of tiny comets that continually enter our atmosphere.

Fortunately, the geologic record provides us with some constraints. The existence of sedimentary rocks more than 3.86 billion years old with characteristics similar to those of modern marine sediments, suggests that ocean water has existed on the Earth throughout most of geologic history. Evidence of 4 billion-year-old continental rocks also suggests that some form of continents existed early in Earth history.

So the origin of ocean waters may well be related to the early history of the planet. It is difficult to be absolutely sure of the exact sequence of events. In this situation, scientists commonly resort to educated speculation. Because there is general agreement on the early evolution of the Earth, scientists look at the probable behavior of water in that context. There are several hypotheses concerning the origin of water. They are not mutually exclusive, and it is conceivable that there were many sources.

One hypothesis suggests that water formed as a consequence of the formation of the planet as it condensed from a hot vaporized cloud. When the Earth first formed, its interior temperatures were much higher than today. This intense heat was related to the greater abundance of radioactive elements (see Chapter 3) and the sinking of metals to the Earth's core. Under these conditions, water trapped in the interior of the planet would have boiled and risen toward the Earth's surface where it would have been vented from volcanoes and fractures as water vapor. This vapor would have formed a shroud of thick clouds

As we have learned, the mid-ocean ridge system is the largest geologic feature on Earth. It is populated by unique communities of animals, and is a natural laboratory for studying the creative forces of the planet at work. Yet humans had set foot on the Moon before scientists first explored the mid-oceanic ridge, just a little more than a mile below the surface of the oceans. When exploration of mid-ocean ridges finally got underway, the results were startling and provided insights into the nature of volcanic activity, the venting of superheated fluids, and the bizarre life-forms that depend on this activity. The surveys also provided fundamental information about the composition of the rocks of the ocean floor and its sediment cover, the nature of its unique and delicate ecosystem, and confirmed scientists suspicions about the source of chlorine in ocean water.

The first exploratory dives were carried out in the early 1970s as part of a project dubbed "FAMOUS," for French-American Mid-Ocean Undersea Study. Using submersibles such as *Alvin* and the French *Cyana,* scientists explored the ocean floor at a dive site chosen some 640 km (400 mi) southwest of the Azores in an area typical of the Mid-Atlantic Ridge (Fig. 8.19a).

The site selected occurs along the divergent boundary between the African and North American plates, which are separating at about 2 cm (0.8 in)/year. This separation is marked by a rift at the ridge crest which measures 3 to 30 km (2 to 20 mi) across and 1.5 km (1 mi) deep. These dimensions are similar in scale to those of the Grand Canyon.

Scientists observed that the ocean floor at this depth is almost entirely volcanic, comprising basalt which drapes the valley slopes in bulbous flows called pillow lavas (see Chapter 5). These bolster-like forms are characteristic of lava extruded underwater, the freezing of its surface producing pillow-like bulges when it comes into contact with seawater. Since the interior of the pillow is still molten, the pressure eventually bursts open the pillows and stretches them into tubes or bulbous extrusions. As anticipated, the scientists were able to confirm that these eruptions and lava flows differ markedly from those on land. Instead of hot molten streams of lava, these pillow basalts displayed small red cracks that gleamed and then winked out as they chilled against the near-freezing seawater. Unlike continental eruptions, their rapid solidification and the enormous pressures at such great depths prevented the escape of significant quantities of gases.

More dramatic discoveries accompanied a 1979 dive to the Galapagos Rift, northeast of the Galapagos Islands in the eastern Pacific. Although it was known that the mid-ocean ridge system was elevated because the rocks were hot, measurements taken at the FAMOUS dive site found the rocks to be cooler than expected. Why would this be the case? Geologists at the time speculated that cold ocean water was penetrating the ocean crust through fractures, absorbing heat from the rocks, and returning to release that heat at the sea floor through some form of hot spring. It was not until the Galapagos Rift was explored, however, that such submarine hot springs were discovered. Analysis of the **hydrothermal fluid** emanating from the hot springs (hydrothermal literally means "hot water"), indicated that they were rich in chlorine, and confirmed speculations for a submarine source of chlorine in ocean water.

The hot springs were surrounded by extraordinary life forms that make up one of the Earth's few complex ecosystems *not* based on photosynthesis (Fig. 8.19b, c). How could such a system sustain life? Instead of sunlight, these oases of life are nourished by minerals and heat. Bacteria flourish in the warmth by metabolizing hydrogen sulfide in the vent water. Unlike life that relies on the Sun's energy, the vent bacteria use energy liberated when the hydrogen sulfide from the vent combines with oxygen in the sea water. They then use this energy to convert inorganic carbon dioxide in the sea water into organic compounds, a process known as **chemosynthesis.**

The vent bacteria form the base of a food chain involving animals that either live on the bacteria, live in symbiosis with them, or live on the animals that eat the bacteria. Giant, foot-long albino clams and mussels live in the vent's warmth by filtering and digesting bacteria from the warm seawater which contains as many as a billion bacteria per liter (Fig. 8.19d). Growing far faster than related shallow-water species, the giant clams must develop and multiply before the hot spring dies out. In an adaption to the low levels of oxygen in the vent water, they also have red flesh rich in the same hemoglobin that colors our own blood red. Red-plumed tube worms more than 3.6 m (12 ft) in length also thrive near the vents where they live in symbiosis with the bacteria. The bacteria pack the tube worm's sacklike bodies and use oxygen filtered through the tube worm's plumes to metabolize hydrogen sulfide into food for their hosts. Higher in the food chain lie albino anemones, crabs, and vent fish that eat the clams, mussels, and tube worms.

The complex life of these sea floor oases is therefore based, not on photosynthesis which depends on sunlight, but on chemosynthesis, which is powered by chemical reactions. Subsequent explorations found similar marine communities at many other locations suggesting that this is a common feature of mid-ocean ridges. Since the process is independent of the Sun and occurs in an environment shielded from ultraviolet radiation by ocean water, it may have a significant bearing on the origin of life. Perhaps it was at the Earth's first mid-ocean ridges rather than in shallow seas, that life first developed from lifeless molecules.

But what is the origin of the hot springs? At these submarine vents, sea water seeps down to magma beneath the ridge and is superheated to temperatures as high as 350°C (662°F). This hot caustic brine becomes saturated with minerals leached from the surrounding basalt. As the superheated water spews back onto the ocean floor, it builds chimneys which form as the

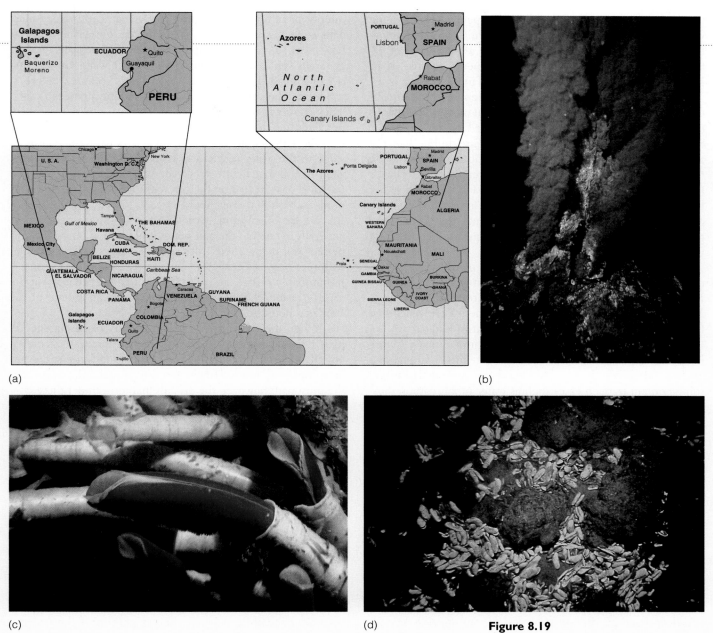

(a)

(b)

(c)

(d)

Figure 8.19

(a) Location of the Azores and Galapagos islands. (b) Photograph of a "black smoker." (c) Giant tube worms thrive around hot springs and "chimneys" in the Galapagos Rift. (d) Giant clams found near the heated vents of ridges.

minerals dissolved in the vent water crystallize on meeting the near-freezing seawater (Fig. 8.19b). The crystalline precipitate forms a sooty soup of particles that belch from the chimneys, called "black smokers" and "white smokers" depending on the mineral mix. The minerals in the soup include iron and manganese oxides, zinc and copper sulfides, and a little silver. On precipitation, they form such commercially important ore minerals as chalcopyrite (copper-iron sulfide) and sphalerite (zinc sulfide). But how exploitable these ore deposits are so far from land, and who they belong to, is uncertain. However, it is clear that many major ore deposits now on land may have originally formed in this way. As techniques to explore these resources become more advanced, exploitation of the sea floor with its abundant deposits and delicate ecosystems is likely to become one of the major environmental issues of the twenty-first century.

that encircled the Earth. Gradually, as temperatures cooled, the vapor would have condensed to form torrential rains that initially collected in pools and lakes. Eventually, as oceanic crust settled to a lower elevation than the continents because of its higher density, water would have collected in the oceans.

This hypothesis is supported by evidence that abundant volcanic activity occurred on the terrestrial planets and moons early in the history of the Solar System (see Chapter 15). Since the interiors of the other terrestrial planets are similar to that of the Earth, it is probable that similar processes occurred on these planets. However, the lack of water on these planets today suggests that the water vapor was lost to space either because surface temperatures remained too high (as in the case on Venus), or because their small size and, hence, low gravity, prevented them from holding a significant atmosphere (as in the case on Mars).

Although this hypothesis is perfectly plausible, it is difficult to confirm or refute because uncontaminated remnants of material from the early Solar System are difficult to find. Meteorites, which penetrate our atmosphere from space and fall to the Earth, may well be such remnants. Radioactive isotope dating consistently reveals that they formed nearly 4.6 billion years ago, at the same time as the birth of the Solar System. Although these meteorites have complex histories, some contain trapped gases such as water vapor, carbon dioxide, and nitrogen. They therefore contain the appropriate chemistry necessary to form the chemical constituents of our primitive atmosphere and oceans.

Another theory for the origin of water suggests that it is related to the entry of celestial bodies into our atmosphere. For example, icy comets continually burn up in the Earth's upper atmosphere and, when they do so, may contribute water to the Earth's surface. Recent observations of comets doing just that in our atmosphere have substantiated this as a viable process, although the extent to which it has contributed to the total amount of water on our planet is uncertain.

There is also general agreement that the initial supply of water to the Earth's surface has been continually supplemented by water added by plate tectonic processes through volcanism and deep fractures. Because volcanism is an ongoing process, some researchers have suggested that the total volume of water on our planet may well be increasing. However, there is disagreement on how much water has been added in this way. Water vapor is one of the main gases vented into the atmosphere by volcanic eruptions. In addition, hydrothermal systems also pump water vapor into the atmosphere and oceans. Over the 4.6 billion years of Earth's history, this activity could have resulted in the addition of significant amounts of water to the Earth's surface reservoirs from its interior. But how much of this water vapor is **juvenile** (that is, *directly* derived from the Earth's interior), and how much is simply recycled near-surface water? The subject is one of considerable debate. For example, as we have learned, volcanic eruptions at destructive plate margins generally vent recycled rather than juvenile water. Subducting oceanic crust is commonly water-rich, because, before it subducts, it becomes hydrated by interaction with ocean water. This recycled ocean water is released as the oceanic slab descends into the Earth's interior, and finds its way into magma sites above the subduction zone (Fig. 8.20). In a similar fashion, some of the water released at mid-ocean ridge hydrothermal vents may also be mainly recycled. As seawater percolates down fractures, it is heated by the hot rocks it encounters, and so is pumped back to the sea floor.

On the other hand, geoscientists who study igneous rocks generally agree that the components of water are presently dispersed well below the Earth's surface, occurring in particular within or between crystals in the upper mantle. This situation has probably existed since the mantle first formed early in the Earth's history. Is the mantle, then, the main source of the Earth's watery surface?

When magmas form in the mantle as a consequence of plate tectonic processes, experiments show that water is preferentially incorporated into the melt. As the melts rise to the surface, they cool and solidify by forming crystals (Fig. 8.20). For the most part, however, the crystals that form under these conditions contain no water and, consequently, water becomes increasingly concentrated in the remaining melt as crystallization proceeds. Eruptions of the magma chamber to form volcanoes would eject this water vapor into the atmosphere. Alternatively, since every liquid has a limited tolerance, or *saturation level*, for foreign components, the melt's saturation level for water may eventually be exceeded as the concentration of water in the melt increases. If this happens, the excess water may bind together and form small reservoirs of hot water deep below the surface. This water may then be literally pumped out under pressure through volcanic vents and associated fractures, which act like gigantic cracks in the Earth's plumbing system.

The origin of water on planet Earth is still a matter of important scientific debate: what proportion of the water is derived from the Earth's interior compared to that derived from the melting of its extra-terrestrial visitors? Has the amount of water in the Earth's surface reservoirs increased with time, and if so, by how much, and by what process? Or, did the vast majority of water originate early in Earth history and has it since been recycled by plate tectonics and the hydrologic cycle? The models proposed are not mutually exclusive, however. Water may have had many sources. But the relative contribution from each model is likely to remain an important scientific question for some time to come.

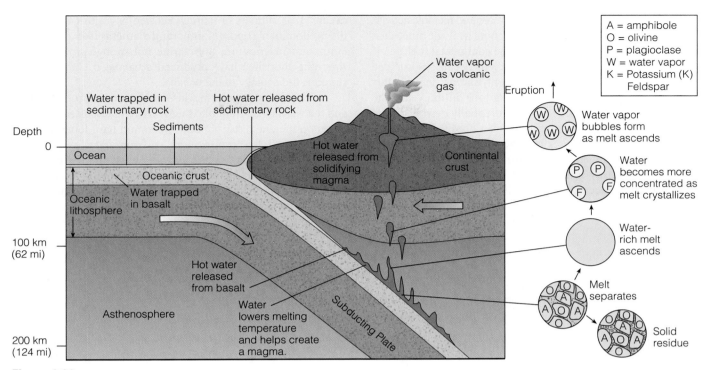

Figure 8.20

Figure showing how water is released from a crystallizing magma and ultimately escapes to the atmosphere and hydrosphere. The plate tectonic environment is the same as that depicted in Figure 8.5 in which water trapped in basaltic oceanic crust or in marine sediments is recycled by subduction processes. The water is released from the subduction zone and promotes melting in the overlying lithosphere. The inset schematically illustrates melting involving A (amphibole), which contains water within its crystal structure, and (O) olivine, which does not. As the magma ascends to the surface, it crystallizes potassium feldspar (K) and plagioclase (P) that do not contain water, further concentrating the water in the magma. As it continues to ascend, increasing concentration and decreasing pressure cause the separation of a water-rich fluid from the magma. Much of this water may eventually be vented to the atmosphere by way of volcanoes.

 8.9 Chapter Summary

Water is our prime resource and distinguishes the Earth from all other planets in the Solar System. It has highly unusual properties largely because of the electrically polarized structure of the water molecule and the attraction of individual molecules to their nearest neighbors. Water has a high boiling point relative to compounds of similar molecular weight, and therefore is present as a liquid at the Earth's surface. Since the liquid form of water has a more densely packed structure than ice, ice floats on water.

Water is a strong solvent, which means that it is rarely pure. It can be penetrated by visible light, a property that is essential to photosynthesis in the oceans. Its ability to store large quantities of heat is an important moderating influence on climate.

The presence of water on the Earth's surface is not only the result of these unusual properties. It is also related to the Earth's position in the "comfort zone" of our Solar System between Venus and Mars. In addition, processes that occurred during the formation of the Earth led to the exhumation of water that was dispersed within the planet's interior.

The contrast in the densities of crustal rocks (continental crust 2.7; oceanic crust 2.9) accounts for the trapping of water in what we call ocean basins since the heavier ocean crust subsides further into the underlying mantle. This influence provides support for the theory that all crustal plates effectively "float" on a yielding layer called the asthenosphere.

Sea level fluctuates according to climate and tectonic activity. It has risen and fallen dramatically throughout geologic time, alternately causing flooding or exposure of the continental shelves.

Evidence for the existence of ancient oceans is provided by the preservation on land of ancient marine sedimentary sequences. The oldest of these is close to 4 billion years old. Although plate tectonic processes preferentially destroy oceanic crust by subduction, sedimentary rocks deposited on ancient continental shelves and rises may be preserved as a result of their accretion to the continent during plate collisions.

Ocean chemistry is influenced by both continental and oceanic crustal processes. These influences combine to give oceans their saltiness or salinity. Erosion of continents provides the predominant supply of important elements like sodium, calcium, and magnesium. Submarine hot springs provides chlorine. Together they have produced a balanced chemistry that supports the ecology of the oceans.

In this chapter, we have addressed the origin of ocean basins and the waters that drain into them. We have discussed the influences on the chemistry of ocean water and have found them to be the result of a combination of continental and oceanic processes. In the next chapter, we turn our attention to ocean circulation, and to the waves and currents that play a fundamental role in distributing this chemistry throughout the oceans.

Key Terms and Concepts

Look for the highlighted items on the web at:
WWW.BROOKSCOLE.COM

- Earth is the only planet with abundant water in the Solar System
- The unique properties of water, its presence as a solid, liquid, and gas on the Earth's surface, include its unusual molecular structure, density, transparency, and its ability to store heat
- The hydrologic cycle describes the dynamic motion of water from one reservoir to another
- Sea level is affected by factors such as wind, gravity, ice ages, and tectonics
- Supercontinent breakup and amalgamation also affect sea level
- Continental margins may be either tectonically active or passive
- Three components of ocean basins are water, sediment, and crust
- Bathymetry displays the relationship between ocean depth and plate tectonics

- The hypsometric curve displays the relationship between density and elevation of the Earth's crust
- Continental shelf sediments are commonly preserved in the geologic record
- Earth as a "split-level" planet is consistent with concepts of isostasy and plate tectonics
- The composition of seawater is influenced by the absorption and release of atmospheric gases, continental drainage (which supplies Na^+), and submarine vents (which supplies Cl^-)
- Origin of ocean water may be related to the early history of the Earth, the burn-up of celestial bodies in the atmosphere, and/or ongoing plate tectonics and hydrothermal systems

Review Questions

1. Why is the Earth the only planet in the Solar System with abundant liquid water on its surface?
2. Why does sea level drop during an ice age?
3. Why do continental shelves preferentially occur in passive, rather than active tectonic environments?
4. Sketch the essential elements of the hydrologic cycle.
5. Why do we think that the hydrologic cycle may have existed since the early part of Earth's history?

6. Why are the deposits of ancient continental shelves generally preserved?
7. What are the three processes that may have contributed to the origin of water on our planet.
8. Why does water have such unusual chemical and physical properties?
9. Why is ocean water salty?

Study Questions

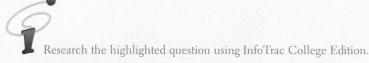

Research the highlighted question using InfoTrac College Edition.

1. In earlier chapters we showed that the rate of sea floor spreading can vary with time. How do you think these variations may affect global sea level?

2. Explain the contrasting compositions of average continental crust and average seawater.

3. Explain how the hydrologic cycle works. How do you think that this cycle would be affected by depletion of the ozone layer?

4. The hypsometric curve (Fig. 8.11) shows the proportion of the Earth's surface that stands at or above a given elevation relative to sea level. Briefly explain why the Earth's continental crust stands at a higher elevation than its oceanic crust. How does this explanation support the theories of isostasy and plate tectonics?

5. The sodium and chlorine concentrations of ocean water appear to be controlled by separate processes, namely the erosion of continents and degassing from submarine vents. Does this mean that the salinity of ocean water is fortuitous?

6. Why does sea level rise during supercontinent breakup?

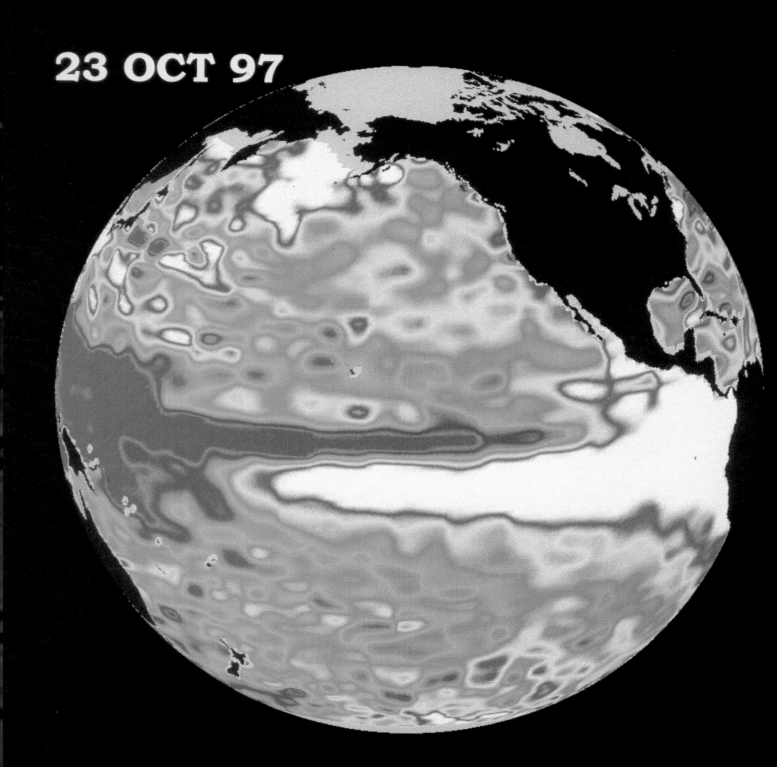

23 OCT 97

CHAPTER 9

Circulation of Ocean Water

In this chapter you will learn the rules that govern the dynamics of the oceans. Understanding these rules is fundamental to the proper management of the coastal regions, where much of the world's population resides, and the delicate marine ecology upon which we depend.

We will first focus on the energy exchange between the atmosphere and oceans, and explain how winds that blow across the ocean surface generate ocean waves. We will learn why the characteristics of these waves vary from the broad swells of the open ocean to the breakers near coastal regions, a property enjoyed by surfing enthusiasts.

The Earth's oceans do not form a uniformly layered watery envelope around the planet. Rather, the ocean surface is distorted, tugged and pulled by gravity which results in both bulges and depressions. Gravitational interaction between the Earth and the Moon generates high and low tides on a daily basis. Although the Sun's influence on tides is less than that of the Moon's, we will discover that it can amplify or dampen the daily tidal ranges resulting in the phenomena of spring and neap tides. The remainder of the chapter will focus on ocean circulation. We will discover that temperature, composition, and density contrasts between equatorial and polar ocean waters drive global-scale circulation. Waters starting from the icy northern Atlantic have been traced into the deep waters of the southern Atlantic, while those from the Antarctic have been traced into the deep waters of the northern Atlantic and Pacific oceans. This illustrates the vast scale of ocean circulation, and the ingenious methods used in its detection.

Although equatorial to polar contrasts drive circulation, we will find that most of the actual movement of water is concentrated in ocean currents, which are narrow channels of relatively fast-moving water within the oceans. These currents often separate large masses of water having contrasting temperatures and compositions. Their paths are greatly influenced by the Coriolis effect, which is related to the spin of the Earth, and are guided by the positions of the continents.

We will find that the relationship between wind and water is more subtle than is immediately apparent. In fact, because of the spin of the Earth, the net movement of ocean water is approximately perpendicular to the wind direction. This has important consequences for the behavior of currents in coastal regions, and the marine ecology which depends upon them.

The most dramatic example of our urgent need to understand ocean water circulation is demonstrated by the phenomenon known as El Niño. A periodic anomalous warming of part of the Pacific Ocean, El Niño apparently disrupts traditional global-scale weather patterns. This causes crop failures, torrential rains in deserts, droughts in regions that depend on rain, record numbers of hurricanes and typhoons, and substantial damage to the Earth's ecosystems.

9.1 Introduction

The ceaseless motion of our ocean water is, perhaps, the most dramatic evidence of the dynamic nature of our planet. Along our coastlines, for example, we can see the interaction between each of the Earth's surface reservoirs where the atmosphere and hydrosphere meet the solid Earth. Ocean waves, driven by the winds in the atmosphere, can batter coastlines with pressures on the order of 2500 kg per square meter (6000 lb per square foot) and carve out breathtaking scenery (Fig. 9.1). This relentless pounding erodes our coastline by disintegrating the rocks in the wave's path and sweeping the resulting sediment out to sea. Here it is later deposited to form the sandy beaches and continental shelves that provide life with such an hospitable environment in which to flourish.

In this chapter, we will discuss the science of oceanography (the study of oceans) and the circulation of ocean water. We will find that this motion is ultimately driven by the temperature and density contrasts between equatorial and polar waters. These contrasts produce ocean currents whose paths are profoundly affected by the Earth's rotation and the positions of the major continental landmasses. First, we will focus on the general description and principles of ocean waves, followed by an application of these principles to coastal regions. We will explain the origin of tides, most readily recognized by the daily rise and fall of ocean waters. Next, we discuss the circulation of ocean waters between the poles and the equator that is driven by temperature contrasts, but whose path is influenced by the spin of the Earth. The combination of these two effects guides our ocean currents. Finally, we apply our knowledge to one of the most enigmatic and dramatic features of all ocean circulation, the phenomenon known as El Niño.

9.2 Waves

To the casual observer, waves are the most obvious feature of ocean water dynamics. They occur when energy from the wind is absorbed by the water's surface. This relationship is obvious to anyone standing on the shoreline. On a windy day, waves are generally higher than they are on a calm day. As one travels out to sea, these same winds generate more gentle swells, in contrast to the breakers that form near shore.

Winds over ocean waters are generated by variations in atmospheric pressure across the width of an ocean (see Chapter 11). The effects of the wind depend on its strength, its duration, and the area over which it blows. For example, gentle winds of only 1 km per hour (0.6 mi per hour) can cause small ripples to form on a relatively smooth ocean surface that then move in the direction of the wind. As the wind's velocity increases, the height of the waves also increases. However, if the gentle wind blows over the ocean for a long period of time, the advancing ripples increase the surface area on which the winds can operate. This results in increased frictional resistance and more efficient transfer of energy from the wind to the water, causing the waves to become larger. This is because once a wave is initiated, it loses very little energy as it travels across an ocean. In fact, waves breaking on a shoreline could have been initiated anywhere within the ocean basin. If consistent, gentle winds can create large waves in this way, it follows that major storms in the ocean are capable of generating very large waves that might not break until they reach coastlines thousands of kilometers away.

The terminology used to describe ocean waves is similar to that used to describe any wave. The high and low points on a wave surface are known, respectively, as **crests** and **troughs**. The vertical distance between these two points, that is, their differ-

Figure 9.1
Waves batter the cliffs of Ka Lae, Big Island, Hawaii.

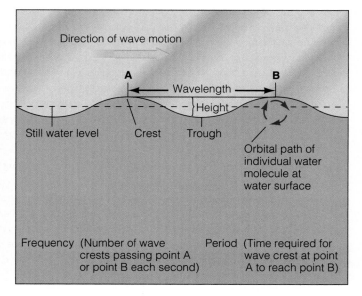

Figure 9.2
Wave terminology and the motion of water in a wave. The orbital motion of water decreases with depth and essentially ceases at a depth equal to one half the wavelength (called the wave base).

ence in elevation, is called the **wave height**, whereas the **amplitude** is the vertical distance the wave moves above or below sea level. Hence, the amplitude is half the wave height. The **wavelength** is the horizontal distance from any point on the wave to the equivalent point on the adjacent wave. It is generally measured from crest to crest or from trough to trough (Fig. 9.2).

As we have learned, waves are the result of interaction between the atmosphere and the surface waters of the hydrosphere. The wind's energy produces a disturbance in the water that moves across the water's surface. The rate at which this disturbance moves over the surface is called the **wave velocity**, and is dependent on a variety of factors. The most important of these are the wavelength and water depth. In deep water, subtle variations in water depth have very little effect on wave velocity.

However, in shallow water, slight changes in water depth become very important and wave velocity decreases dramatically with decreasing water depth.

The motion of waves is a form of energy that is rapidly lost along coastlines when a wave breaks on the shore. When waves are viewed at sea, it appears that the water simply rises and falls, and does not move with the wave. This is consistent with the definition of a wave as the means of transferring energy through a medium with no net movement of that medium. A familiar example can illustrate this. Imagine a seagull floating on the ocean's surface. The actual movement of the water is approximated by the motion of the gull which appears to bob up and down rather than moving forward with the water (Fig. 9.3a). In actuality, floating objects follow a circular path as the wave goes by. The seagull, for example, rises and is pushed forward as a wave crest goes past, only to fall backward into the trough that follows. The water behaves in a similar manner. It does not move horizontally with the waves, but instead moves up and down in a more or less circular path. Consequently, there is no net movement of the water (Fig. 9.3b). This circular path is often called the *orbital motion of water particles*. Over a period of time, the water, like the gull, does move forward, but this motion is more dependent on the frictional drag between the water and the wind.

Figure 9.3
Orbital motion of water particles. (a) A floating seagull travels in a circular path (see dots) as the wave migrates past. This demonstrates that the water itself does not move with the wave. The dots track the changing location of the center of the seagull. (b) Orbital motion of water particles as a wave migrates by. The orbital motion progressively decreases with depth and ceases at the wave base which lies at a depth approximately equal to half the wavelength.

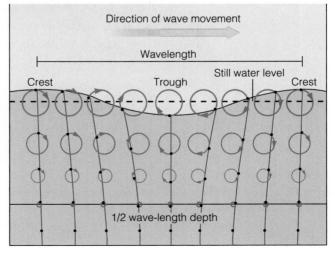

(a)

(b)

Perhaps the most dramatic form of sea waves are generated by submarine earthquakes. An earthquake releases a vast amount of energy. If it occurs beneath the sea, a portion of that energy may be transferred to the ocean water above, generating powerful seismic sea waves that can migrate across an entire ocean basin. When they reach the coastline, the waves unleash their energy and cause widespread devastation. These waves are commonly called "tidal waves," but this is a misnomer since they have nothing to do with the Earth's tides. They are better known as **seismic sea waves**, or **tsunamis**, the Japanese word for them (meaning "harbor wave").

Unlike a normal wave which concentrates its energy at the surface of the sea, a seismic sea wave carries its energy from the ocean bottom (where the earthquake occurs) to the surface (Fig. 9.4). Ships at sea would scarcely notice a seismic ocean wave passing, for in the deep ocean a tsunami raises only a broad, gentle swell little more than a meter high. But the shallower sea floor close to land compresses the wave and focuses its power into a series of major waves that may reach heights of over 15 m (50 ft).

Since the Pacific Ocean is predominantly rimmed by subduction zones (see Chapters 5 and 6), earthquakes (and hence tsunamis) are far more common in the Pacific Ocean than they are in the Atlantic Ocean. Following an earthquake in the Aleutians in 1946, a killer tsunami sped across the Pacific like a ripple across a pond but at speeds of about 750 km (472 mi) per hour (Fig. 9.5). Slamming into the Hawaiian Islands just 5 hours later, the waves killed at least 150 people.

In 1960, an earthquake off the coast of Chile not only resulted in 4000 deaths on the mainland, it also initiated a seismic sea wave that moved at nearly 700 km (440 mi) per hour. Its progress across the Pacific Ocean, shown in 5 hour intervals, may be seen in Figure 9.6. Nearly 24 hours after the earthquake, the seismic sea wave reached Japan where it caused 180 additional fatalities.

Today, an oceanwide tsunami warning system, based in Hawaii, detects tsunami-generating earthquakes and predicts their time of arrival. In this way, those areas at risk can be alerted before the tsunami arrive and the potential loss of life is greatly reduced.

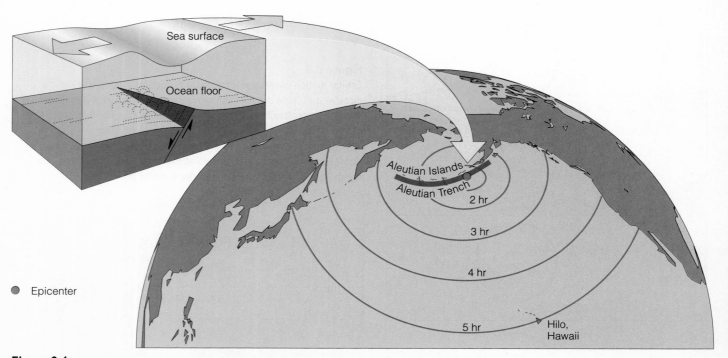

Figure 9.4
Development of the tsunami on April 1, 1946 following an earthquake in the ocean floor related to its subduction in the Aleutian trench. The displacement lifted the water above, generating a wave that moved outward at a speed of 212 meters per second (472 miles per hour). Five hours later, the tsunami had struck the Hawaiian Islands and the north coast of Japan.

Figure 9.5
Destruction associated with the April 1, 1946 tsunami in the Aleutian Islands. A lighthouse (left) whose foundation was 15 m (50 ft) above sea level was completely destroyed (right).

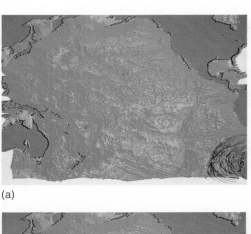
(a)

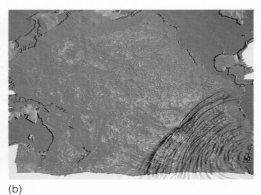
(b)

(c)

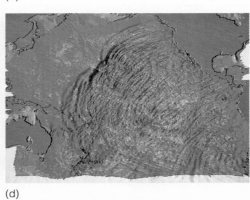
(d)

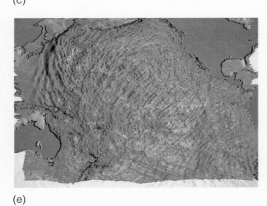
(e)

Figure 9.6
Computer simulation of the 1960 Chilean tsunami generated by an earthquake in the subduction zone below the Peru-Chile Trench. The tsunami migrated across the entire Pacific basin to reach Japan 22.5 hours later. (a) 2.5 hours after earthquake, (b) after 7.5 hours, (c) after 12.5 hours, (d) after 17.5 hours, (e) after 22.5 hours.

Below the water surface, we can get an idea of the relationship between waves and water motion by adding varying weights to a set of corks such that they suspend at different depths. The motion is seen to change with depth. In coastal regions, where the water depth is relatively shallow, the circular pattern near the surface becomes oval-shaped at depth as the "up and down" motion progressively diminishes. In deeper water, the depth has far less influence and the corks travel in circular paths of decreasing diameter until, at about a meter below sea level, there is virtually no motion at all. This is because deeper waters are more protected from surface winds.

Thus, waves generally reflect the interaction of surface water and wind, but not the net movement of water. We shall now focus on coastal regions where wave velocity becomes very sensitive to changes in water depth.

9.3 Coastal Regions

Waves generated in mid-ocean will migrate virtually unhindered until they hit the coastline. In certain circumstances, for example during hurricanes, the large waves generated at the storm center can travel thousands of miles to batter a coastline that may otherwise be experiencing tranquil air conditions. This creates the rather spectacular sight of very high breakers on an otherwise very calm day.

As waves approach a gently inclined shoreline, there is little loss of energy until the waves break on the shore since the water itself does not actually move forward with the wave. This suggests that the energy lost to friction with the seabed is negligible. However, wavelengths are observed to decrease, while wave amplitudes increase. These observations are attributed to the decreasing water depth and the restricting influence of shoaling on wave motion (Fig. 9.7) which also causes the wave velocity to decrease sharply. Consequently, as waves interact with coastlines, their wavelengths diminish and their velocity is reduced. However, the same number of wave crests must pass by a given point in a certain time interval. In order for this to happen, the wavelength must decrease as the wave slows down. As a result, the wave energy is more concentrated, and its amplitude increases. A critical point of wave stability occurs when the wave attains a height equal to about one seventh of the wavelength. At this stage, the crests become unstable and start to lean forward and spill over into the troughs to form **breakers**. The area between the outermost breaker and the beach is known as the **surf zone**. (Fig. 9.7).

Waves release their energy either by crashing on the beach or by pounding the cliffs that stand like immovable objects before the waves. At this stage, frictional interaction between the ocean water and seabed becomes very important and, as we have seen, plays a major role in the sedimentary processes of coastal regions.

Although the cliffs may seem like an impenetrable barrier, waves slowly, but inevitably, wear them down by erosion. The

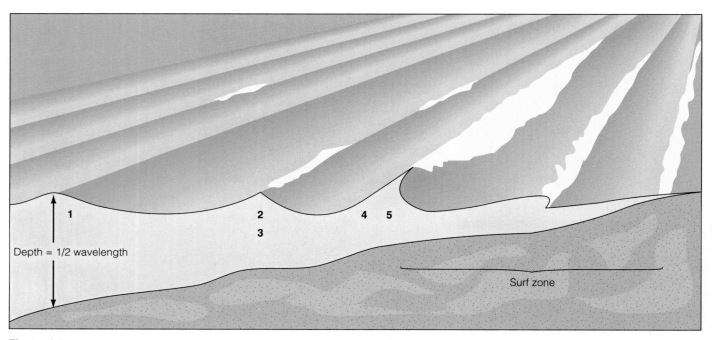

Figure 9.7
Ocean breakers along the shore. (1) Although the swell is influenced by the shallowing bottom as the wave migrates to shore, there is no frictional interaction. (2) Wave crests rise and become peaked as the wave's energy is concentrated into shallower water. (3) The wavelength shortens. (4),(5). The waves attain a height equal to about one seventh of its wavelength and begin to form breakers. This occurs when the ratio of wave height to water depth approaches 3:4.

resistance of cliffs depends on the material of which they are made. For example, the unconsolidated sediments on Cape Cod in Massachusetts are no match for the ocean waves and erosion is forcing these sediments to retreat at a hasty rate of about 1 meter per year. On the other hand, hard crystalline rocks, such as granite, offer much sterner resistance and therefore retreat much more slowly. Many spectacular cliffs composed of hard crystalline rock occur around the coastlines of the Atlantic and Pacific oceans. An irregular indented shoreline is often related to a variable geology, with headlands composed of hard crystalline rocks, while bays are formed by the more effective erosion and retreat of less resistant rocks (Fig. 9.8).

Waves rarely approach a shore head-on; they generally approach at an angle. However, as the water retreats, the backwash is always down the slope of the beach. Thus particles follow a zigzag pattern along the beach as they are carried in suspension, first by advancing waves and then by retreating down slope (Fig. 9.9).

An incoming wave that approaches the shore obliquely to its movement has two components. One, perpendicular to the shoreline, gives a measure of the erosive power of the crashing wave. The other, parallel to the shore, is produced within the surf and is known as a **longshore current** (Fig. 9.9). This current causes the sand within the surf zone to drift slowly along the shore.

Figure 9.8
Resistant bedrock in the background forms a headland and weaker bedrock in the foreground forms a bay and a sandy beach. Note also the ocean breakers that define the surf zone (see Figure 9.7).

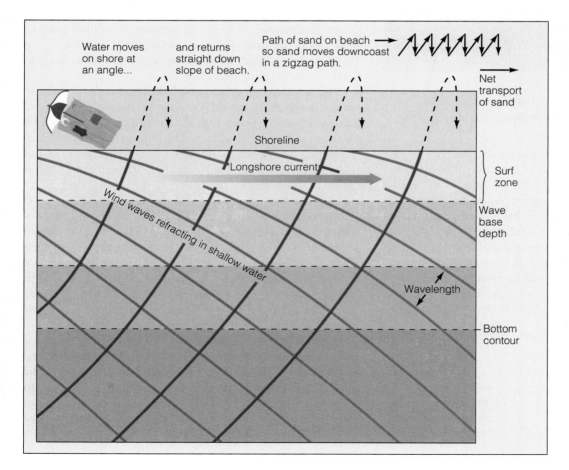

Figure 9.9
When waves are oblique to the shoreline, one end of the wave encounters the shallow water (upper left) before the other (lower right). As a result the wave bends or refracts. Although the wave hits the shoreline obliquely, the return flow is directly down the slope of the beach. This combination means that individual sand grains carried in suspension by the incoming water and return flow follow a zigzag path down the beach. The component of wave motion that is parallel to the shoreline is called the longshore current.

Figure 9.10
Groins act as an artificial barrier to the migration of sand along a beach. Note accumulation of sand on the top side of each groin.

The net effect of these two components is that beaches are on the move, the sand moving in the direction of the longshore current. On many beaches along both the Atlantic and Pacific coasts, a million tons or more of sand are removed each year. The migration of sand may be temporarily halted by the construction of **groins**, which are walls built at right angles to the longshore current that act as traps for the migrating sand (Fig. 9.10).

One of the predicted effects of global warming is that the increased energy of waves will result in more powerful coastal erosion. If so, the natural barrier of rocks and sediment that protects many of our coastal cities will be slowly whittled away.

9.4 Tides

The daily rise and fall of ocean waters have been recognized for millennia; the relationship of this cycle to the rising and setting of the Moon has been the source of many myths and legends.

This tendency was further observed to be amplified when the Moon was full or new, so that high tides were higher and low tides lower than usual. In contrast, during quarter moon phases, the differences between high and low tides were observed to be more subdued (Fig. 9.11 and Fig. 9.12).

The relationship between tides and the position of the Moon was first explained scientifically by Sir Isaac Newton who linked the tides to his theories of gravitational attraction. Gravity is the fundamental force of attraction between any two objects. It is proportional to the masses of the two objects and their distance apart. The greater the masses, the greater the attraction. However, the attraction diminishes very rapidly as the distance between the two objects increases. As a consequence, the Moon exerts a greater influence on the Earth's tides than does the Sun. This is because the Moon, although it has far less mass than the Sun, is much closer to the Earth. The Moon lies at an average distance of only 384,400 km (240,250 mi) from the Earth, compared to the average distance of about 150 million km (96 million mi) to the Sun. As a result, the Sun's influence on tides is less than half (46%) of that of the Moon's.

Since the Moon is held in orbit by the Earth, there must be a balance between their mutual gravitational attraction and the motion of the Moon itself. This balance is similar to that of a ball rotating at the end of a string about a central object. In order for the ball to maintain its orbit, the string must remain taut, that is, the ball must have a certain momentum. If the balance is disturbed, the orbit will be destroyed. For example, if the motion of the ball is reduced, it collapses toward the center. Alternatively, if the string breaks, the ball flies away. In both cases, the orbit is lost. Gravitational attraction is the "invisible string" that balances the Moon's motion and prevents it from escaping into space.

These gravitational forces are exactly balanced at the center of mass in the Earth–Moon system. However, since gravitational attraction varies with distance, the forces cannot be balanced at the Earth's surface. The portion of the Earth's surface facing the Moon is closer to the Moon than the center of mass, and therefore the gravitational pull from the Moon is higher on that side. The side facing away from the Moon is further away, so that the gravitational pull is lower.

Since the Earth's surface is solid, we do not notice a distortion of the surface itself. However, ocean water is not pinned to the solid Earth. It is free to move in response to these gravitational influences so that its varying behavior is far more obvious (Fig. 9.13). The water on the near side of the Earth feels the pull toward the Moon more strongly than the solid Earth does, creating a bulge in the ocean surface that points toward the Moon. This is known as a **direct high tide**.

On the far side of the Earth, the thin veneer of ocean water is further away from the Moon than the center of mass and is therefore pulled toward the Moon less strongly than the solid Earth. As a consequence, the solid Earth shifts beneath the ocean water toward the Moon, literally leaving the ocean water behind. This too creates a bulge in the ocean surface and is known as an **indirect high tide**.

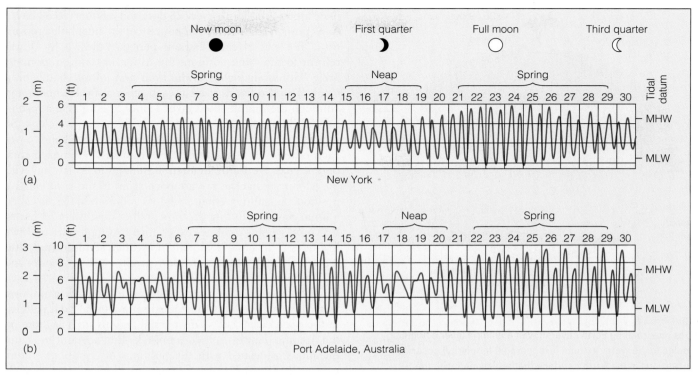

Figure 9.11

Tidal records for a typical month in (a) New York and (b) Port Adelaide, Australia. The biggest difference in high and low tide levels occurs at full moon and new moon; the smallest difference occurs during the quarter moon phases. MHW, mean high tide; MLW, mean low tide.

Figure 9.12

Photos of high and low tides in the Bay of Fundy, Nova Scotia, Canada.

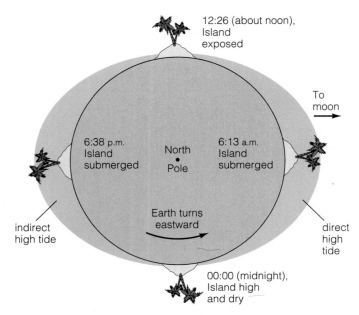

12:26 (about noon),
Island
exposed

To moon →

6:38 p.m.
Island
submerged

North
• Pole

6:13 a.m.
Island
submerged

Earth turns
eastward

indirect
high tide

direct
high
tide

00:00 (midnight),
Island high
and dry

Figure 9.13
Tides on a rotating Earth covered with a uniform layer of water. A bulge in the water surface (exaggerated for clarity) occurs on the portions of the Earth's surface that are aligned with the Moon. Note the island experiences two high tides and two low tides in a 24-hour period.

once around the Earth every 28 days and 8 hours (28.33 days). As a consequence, the *actual position* of the tidal bulge moves. So at the end of each 24-hour period, although the Earth returns to the same position, the Moon has moved along its orbit. This is analogous to the hour and minute hands on a clock. The two hands overlap each other at twelve noon, and although the minute hand takes 1 hour to complete a revolution, as it does so, the hour hand advances to one o'clock. So it takes the clock another 6 minutes before the two hands cover each other once again (Fig. 9.14). Similarly, after each rotation, it takes the Earth another 50 minutes to "catch up" with the Moon, so that the two are once again in the same relative position. This time period, 24 hours and 50 minutes, is called a **lunar day**. It is during this time, rather than every 24 hours, that most coastal localities experience two high tides and two low tides. This basic tidal pattern is called **semi-diurnal**, during which each change of tide typically occurs 6 hours and 12.5 minutes apart (Fig. 9.15a).

This model, which applies to an Earth with a uniform layer of water, explains the typical semi-diurnal tidal patterns in the open ocean and many coastal regions such as those on the Atlantic. However, it does not take into account the complications associated with the positions of continental landmasses, particularly where they partially enclose relatively small, confined bodies of water such as the Gulf of Mexico. In

An analogy of these two high tides can be found in two ways of emptying a bottle filled with water. The direct and most obvious method is by gently inverting the container, so that gravity pulls the water from the bottle. This is analogous to the direct high tide where the water is pulled by gravity from the solid Earth. The indirect way is to rapidly shift the bottle downward, so that the water is left behind and spills out. This is analogous to the indirect high tide, where the Earth is pulled away by gravity from the water. Both techniques separate the solid bottle from the water it encloses, just like the solid Earth is shifted relative to its watery envelope.

Thus, at any given time, two bulges of ocean water occur on the Earth's surface, one where the surface is closest to the Moon, the other where it is farthest away. Furthermore, because the bulges form at the expense of other regions, depressions are produced in those areas furthest removed from them. These are therefore regions of low tides. Since the solid Earth rotates once every 24 hours about its axis, it moves *beneath* these tidal bulges and depressions, giving the appearance of migrating tides. As the Earth completes one rotation, every location will ideally experience two high tides, one direct, and the other indirect. Between the two high tides, two low tides occur (Fig. 9.13).

If the Moon was stationary, we would see high and low tides at identical times each day, because the position of the bulges would not change and the solid Earth would merely rotate beneath them. But, the Moon is not stationary. It orbits

Watch with hands overlapping
at 12 o'clock noon

Same watch 1 hour later–
hands do not overlap

Same watch at the next time the
hands overlap: 1:06 AM/PM

Figure 9.14
An analog watch at 12:00, 1:00, and 1:06 showing overlap of hour and minute hands at 12:00 and 1:06, but not at 1:00.

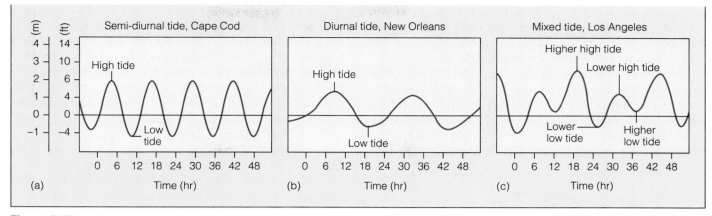

Figure 9.15
Tidal curves for (a) semi-diurnal, (b) diurnal, and (c) mixed tidal patterns.

these situations, the sea behaves like water sloshing back and forth in a container. This motion may modify the classical semi-diurnal pattern of tides. For example, most areas in the Gulf of Mexico see only one high tide and one low tide every lunar day, a pattern that is known as **diurnal** (Fig. 9.15b). Much of the Pacific coastline is also characterized by a diurnal tidal pattern or one that is mixed (Fig. 9.16). A **mixed** tidal pattern is one in which two high tides and two low tides occur each lunar day, but one high tide is significantly higher than the other, and one low tide is significantly lower (Fig. 9.15c).

Briefly, then, the gravitational attraction between the Earth and Moon, together with the location of the continents, explains

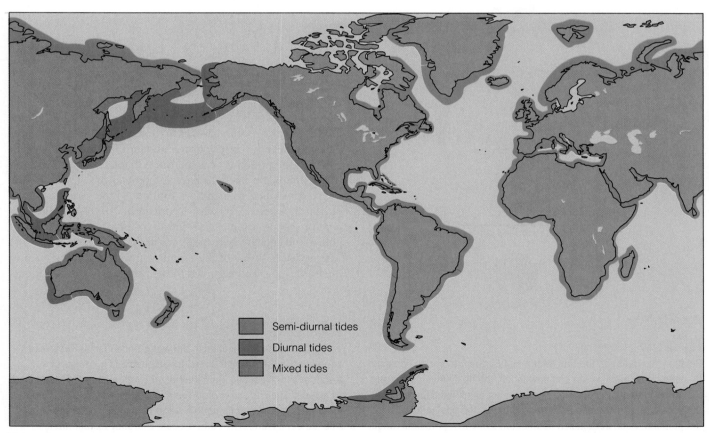

Figure 9.16
Worldwide distribution of semi-diurnal, diurnal, and mixed tidal patterns.

the origin and timing of tidal patterns. But in any one locality, why do tidal ranges (that is, the difference in water height between high tide and low tide) vary during the month (Fig. 9.11)? And why do these variations match the phases of the Moon?

Variations in tidal range are also explained by gravitational attraction, but this time it is due to the additional influence of the Sun (Fig. 9.17). Although the influence of the Sun on tides is less than that of the Moon, it is nonetheless significant. Depending on its position in the sky relative to the Moon, the Sun can either amplify or dampen daily tidal

ranges. **Spring tides** occur when the Sun and Moon are most closely aligned with the Earth. During this time, their respective gravitational pulls work in concert, and the tidal range is amplified (Fig. 9.17a). **Neap tides** (Fig. 9.17b) occur when the Sun and Moon are in maximum misalignment with the Earth. In these positions, the respective gravitational pulls work against each other, and so dampen the tidal range. But when do these alignments and misalignments occur?

As with all objects in the Solar System, half of the Moon's surface is always illuminated by the Sun. What we see as the "phases of the Moon" reflects the fact that only a portion of the illuminated half is generally visible from Earth (Fig. 9.17). However, when the Moon is on the far side of the Earth relative to the Sun, that is, in the opposite part of the sky, we see its entire illuminated half which we refer to as a "full moon". When the Moon is aligned between the Sun and the Earth, that is, in the same portion of the sky, we do not see the Moon at all because its illuminated half is facing away from us. We refer to this phase as the "new moon". In both these situations, the Sun, the Earth, and the Moon are most closely arranged in a line, and the additional gravitational attraction produces spring tides.

Quarter moons represent phases half way between these two alignments, when the Sun and Moon are most misaligned (that is, at right angles) with respect to Earth so that their respective gravitational pulls work against each other. The result is a dampening of the tidal range which produces neap tides.

9.5 A General Overview of Ocean Water Circulation

Waves and tides have a daily impact on coastal communities. However, in the ocean basin as a whole, global-scale circulation patterns affect the distribution of modern marine ecologies. By applying the principle of actualism (Chapter 3), ocean circulation patterns are likely to have influenced this distribution in the past and may, in fact, have influenced the evolution of life on Earth. Ocean circulation is driven by contrasts in the temperature, composition, and density of ocean waters from the pole to the equator, both at the surface and at depth (Fig. 9.18). The actual pathway of the circulating waters is influenced by the Earth's spin and is guided by the positions of the continents.

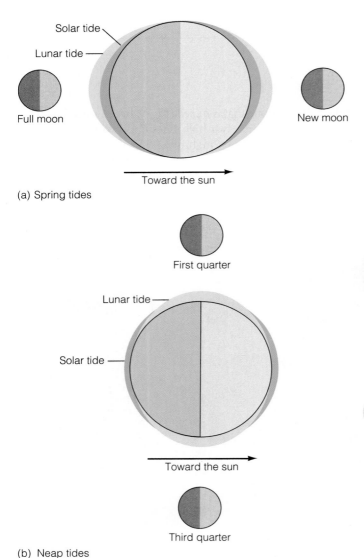

(a) Spring tides

(b) Neap tides

Figure 9.17
Relationship between the Sun, Moon and Earth at (a) spring tides and (b) neap tides. At full moon and new moon, the tidal influences of the Sun and Moon act in concert, resulting in an amplified tidal range. At the first-quarter moon and third-quarter moon, the tidal influences of the Sun and Moon act against each other, dampening the tidal range.

Figure 9.18 (right)
(a) Surface water circulation in the world's oceans. (b) Deep water circulation in the Atlantic Ocean from Greenland in the north to Antarctica in the south. The direction of water movement is indicated by arrows. The surface layer is too thin to show at this scale. Note that the denser Antarctic bottom water is found beneath the North Atlantic deep water. The vertical scale is greatly exaggerated.

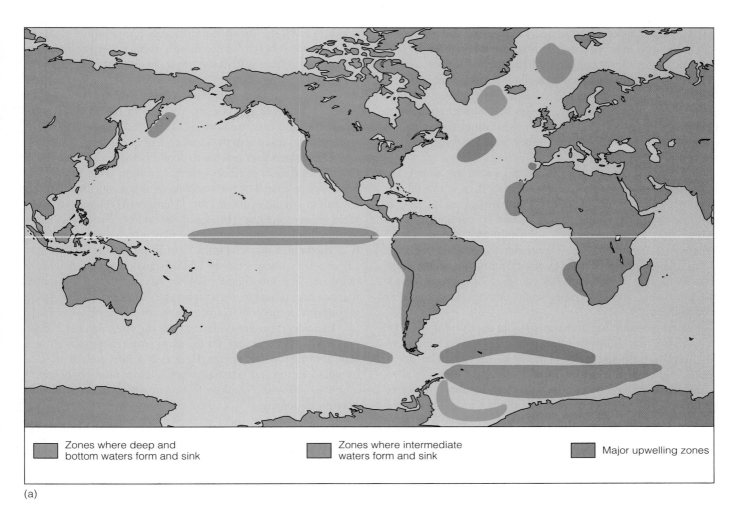

▨ Zones where deep and bottom waters form and sink	▨ Zones where intermediate waters form and sink	▨ Major upwelling zones

(a)

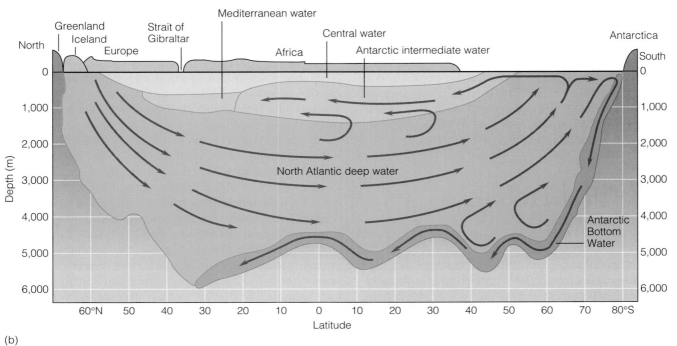

(b)

The Earth's polar waters are colder than the equatorial waters because they receive less direct solar radiation. In fact, water near the poles is cold enough to produce sea ice. Sea ice preferentially excludes salt so that when it forms, salt becomes concentrated in the adjacent seawater. The coldness, coupled with the increase in salinity, makes the adjacent sea water denser which causes it to sink. Thus the distinctive qualities of polar waters: cold temperature, high salinity, and high density are related to one another.

Of the two polar regions, Antarctic waters are denser and more saline than Arctic waters because they surround the Antarctic ice cap, which is made up primarily of sea ice. In contrast, the Arctic ice sheets are surrounded by continental landmasses that contribute significant freshwater runoff. Dense bottom Arctic water does not migrate into the Atlantic basin due to topographic barriers. However, sea ice formation occurs in the subpolar regions of the North Atlantic (Fig. 9.18a), near Iceland and Greenland, and as a consequence, dense surface waters are formed there.

Extensive sampling of ocean water at depths of up to 5000 m (16,400 ft) has shown that these cold, dense polar waters sink and migrate from the poles beneath all of the major ocean basins. Although the migration of these deep waters is slow, (averaging 5 km or 3 mi per day), they can be traced underwater because they are compositionally distinct from surface and near-surface water masses (Fig. 9.18b). The Antarctic Bottom Water has been traced as far north as the Aleutian Islands in the Pacific Ocean (55°–60° N) and to 45° N in the Atlantic Ocean. Similarly, sinking of subpolar waters in the northern hemisphere results in the North Atlantic Deep Water mass that travels southward toward the equator and beyond to Antarctica where mixing with other water bodies results in the loss of its identity.

In the tropics, less dense surface waters have an average temperature of about 25°C (77°F). These waters are warmed by the absorption of radiant energy from the Sun and are swept by wind-driven surface ocean currents toward the poles. This in turn allows the deeper, colder water at the tropics to rise toward the surface (Fig. 9.18a). The combined migration of deep polar waters toward the equator and surface equatorial waters toward the poles results in the highly efficient global-scale circulation of ocean water, heat, and nutrients.

9.6 Vertical Profiles of Ocean Water

In general, the global-scale, convection-driven system of circulation described above results in a vertical profile of ocean water which typically consists of three layers that differ in temperature, density, and salinity.

The *surface layer* is in direct contact with the atmosphere and is warmed by radiant energy from the Sun. As a consequence, its temperature varies with latitude and is affected by seasonal changes. As an example, the temperature of ocean surface waters shown for the month of August typically ranges from about 28° to 30°C in the tropics to close to freezing in the polar regions (Fig. 9.19).

The surface layer is also moved horizontally by wind and wave action. This allows deeper waters to rise to the surface where they too become warmed. The process results in significant vertical mixing in these areas (Fig. 9.18).

The surface layer extends to depths between 100 and 500 m (330–1640 ft) and overlies a sharp transition zone termed the *intermediate layer*. The intermediate layer is a zone that typically extends to a depth of about 1 km (0.6 mi) and is characterized by rapid changes in temperature, salinity, and/or density (Fig. 9.20). Within this layer, the temperature may drop by between 5° and 8°C, the salinity may vary from 33 to 35‰, and the average density rises from about 1.024 to 1.028. Depending on which factor dominates, the intermediate layer is varyingly referred to as the thermocline (changes in temperature), halocline (changes in salinity), or pycnocline (changes in density) (Fig. 9.20).

Beneath the intermediate layer, the *deep layer* is a cold, dense zone of water that, for the most part, reaches the surface only at or near polar regions. Hence, its waters are not affected by solar heating and it has an average temperature between 3° and 4°C (38°–40° F), close to that at which water attains its maximum density.

Although ocean water is efficiently circulated between the equatorial and polar regions, the ocean profile is relatively stable because the less dense waters occur at the surface and overlie denser waters at depth. As the ocean water circulates, it changes to match the characteristics of its local environment. For example, as the deep dense water rises to the surface in equatorial regions, it becomes warmer, less saline, and less dense, and so attains the characteristics of surface waters. Thus, vertical profiles in temperature, salinity, and density remain relatively unaffected by ocean circulation.

By analogy, a series of snapshots of a baggage-less carousel at an airport would look identical even though we know the carousel is moving. Similarly, successive recordings of ocean water profiles look almost identical, despite the fact that the ocean water is migrating.

It is important, however, to realize that modern circulation is driven by temperature contrasts between the hot equatorial and cold, glaciated, polar regions. Today, the presence of polar ice exerts a considerable influence on polar temperatures, helping to keep them cold. This maintains the global temperature contrast. In the geologic record, however, evidence for polar ice is sparse. Modern ocean water circulation may therefore be quite different from that which existed throughout most of the geologic record. For example, during the middle Cretaceous (about 100 million years ago), average atmospheric temperatures were significantly higher than they are today and there was no polar ice. In fact, most of continental North America was submerged beneath a thin veneer of seawater. The effect of these warm conditions on oceanic circulation and the marine ecology of the period is as yet unknown.

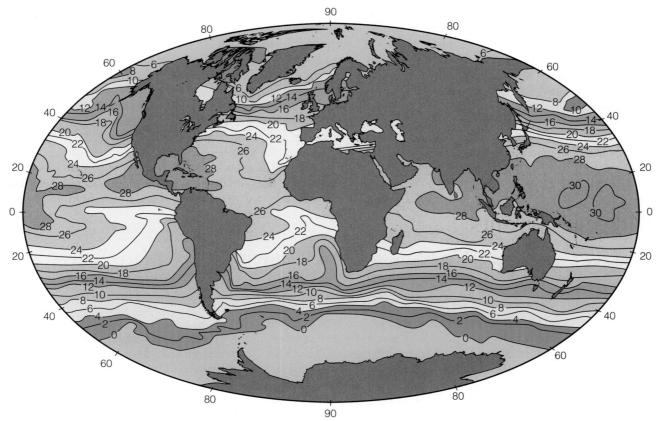

Figure 9.19

Typical temperature variations in ocean surface water for the month of August. The warmest temperatures are found in the tropical oceans (30°C) and decrease poleward to values close to freezing temperatures.

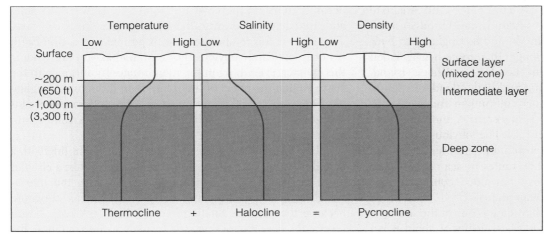

Figure 9.20

Variation in the physical and chemical properties of sea water with depth. The intermediate layer is a zone of rapid change between 200 m and 1000 m (3300 ft) depth variously called the (a) thermocline, (b) halocline, and (c) pycnocline depending, respectively, on whether temperature, salinity or density are being emphasized. Note that these are not independent properties, for example the density of sea water depends on both temperature and salinity.

Yet an understanding of the effect of polar ice on oceanic circulation is fundamental. If current fears of global warming are realized and the ice caps melt, future generations may have a Cretaceous-type world to contend with. We have to wonder what would happen to ocean circulation should this occur. Would there be a mass extinction of marine life and, if so, are some of the extinctions documented in the geologic record related to similar phenomena? What would be the repercussions of such an extinction on the food chain and, consequently, the presence of human life on Earth?

9.7 Ocean Currents

Unlike waves, which only cause significant movement of surface water near coastlines, ocean currents describe the *actual* movement of surface ocean waters. These currents move in definite and predictable directions, and with much swifter velocities than those of surrounding waters. This circulation system is an effective means of distributing heat and nutrients in the oceans, and is driven by the variations in the physical properties of ocean water from the poles to the equator. Ocean currents are analogous to streams on land which are the focus for the swift transportation of fresh water back to the sea. Currents, such as the Gulf Stream, are relatively narrow channels of swift-moving surface ocean water and commonly separate slower-moving, cool polar waters from those of the warmer tropics. Although they may not be as obviously defined as their continental counterparts, the ocean too has preferred pathways for the motion of water.

The existence of ocean currents was recognized by mariners in the early 16th century, long before the scientific community learned of their presence from Benjamin Franklin in 1786. Franklin proposed the existence of the Gulf Stream to explain discrepancies in the trans-Atlantic travel times of ships. Mail-carrying ships sailing from Falmouth, England, to New York, commonly took 2 weeks longer than more heavily laden merchant ships traveling from London to Rhode Island. Captains piloting the ships bound for Rhode Island were familiar with the northeasterly-flowing Gulf Stream and avoided it, whereas those bound for New York were not, and were often slowed down by sailing against it. The obvious economic implications of this discovery spurred further observations and measurements on all major trade routes by sea captains and their crews. Satellite technology has since confirmed and refined many of these early measurements.

We now know that the Gulf Stream is one of the strongest currents, and moves with an average velocity of about 4 km (2.5 mi) per hour. It is generally 50–80 km (30 to 50 mi) in width, and carries a volume of between 40 and 90 million cubic meters of water per second.

There are 38 major surface currents in today's oceans. Many of them, including the Gulf Stream, are part of the large-scale circular motion of surface waters in major ocean basins

called **gyres** (Fig. 9.21). For example, the North Atlantic Gyre consists of the Gulf, North Atlantic, Canary, and North Equatorial currents that together describe the circular motion of the surface waters across the entire width of the North Atlantic Ocean. Other gyres occur in the South Atlantic, North Pacific, South Pacific, and South Indian oceans. These gyres are as wide as the oceans in which they occur, and are centered on a subtropical zone of high atmospheric pressure. They describe a clockwise path in the northern hemisphere and a counter-clockwise one in the southern hemisphere.

But why do ocean currents occur, and why do they form gyres? Why does the water in a gyre move in a circular path, and why does a gyre's sense of rotation vary from one hemisphere to another?

9.7.1 Movement of Surface Waters and the Coriolis Effect

The answers to these questions were provided by the 19th century French mathematician, Gaspard de Coriolis. As we have learned, the net tendency of surface waters is to move away from the equator and toward the poles. Along this journey other factors affect the details of the pathway, including the location and shape of the continental margins, and the direction and velocity of the prevailing winds. However, Gaspard de Coriolis demonstrated that the most important factor is an apparent force due to the Earth's rotation, which is now known as the **Coriolis effect**. The combination of these factors modify the pathways of surface ocean waters into currents and gyres.

In essence, the effect is connected to one of Newton's laws of motion, that bodies will move in a straight line unless acted upon by another force. Indeed, if the Earth did not rotate, ocean currents would move in straight lines between the poles and the equator. However, the Earth rotates on its own axis. The effect of this rotation on objects pinned to the solid Earth is not immediately obvious. For example, as you are reading this chapter, everything in the room rotates in unison with the Earth so that the effect of the rotation on the room is negligible. However, as de Coriolis noted, the atmosphere and oceans are not pinned to the solid Earth; instead the solid Earth rotates *beneath* them.

To appreciate this effect, try drawing a straight line with a pencil from the edge of a slowly rotating object (like a circular piece of paper) to its center. Even if you draw the line as straight as you possibly can, the line will have a curvature because the paper is rotating beneath it (Fig. 9.22a). If you reverse the direction of spin, and try to draw the same line, the line will again be curved, but this time in the opposite sense (Fig. 9.22b). This is because the spinning object is moving in the opposite direction *beneath* the pencil. A similar effect would be produced if you were to try to run straight across a rotating merry-go-round. If there was wet paint on your feet, and you examined your footsteps, you would find that they

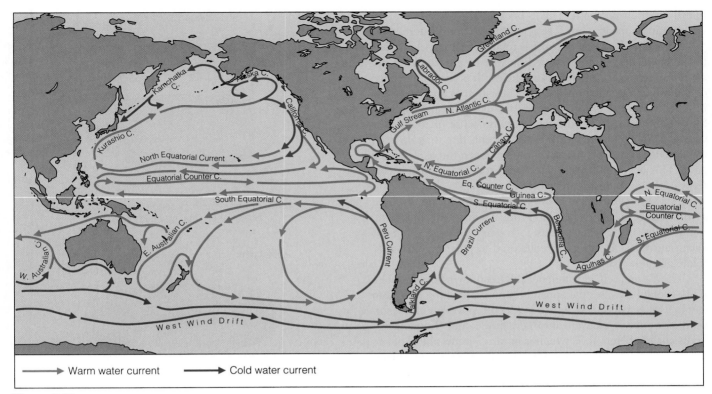

Figure 9.21
The world's major ocean currents trace a clockwise path in the northern hemisphere and a counter-clockwise path in the southern hemisphere. They also tend to span the width of the ocean in which they reside, implying that their paths are guided by the continental margins.

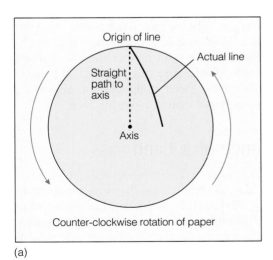

(a)

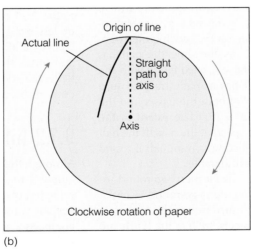

(b)

Figure 9.22
Figure showing attempts at drawing a straight line on a rotating object (a) counter-clockwise spin (b) clockwise spin. Although the hand guides the pencil in a straight line, the result is a curved path because of the motion of the paper beneath.

traced out a curved path. This is because the merry-go-round was rotating beneath your feet as you were running across it.

If the circular piece of paper is rotating counter-clockwise, the sense of deflection of your line will be clockwise. If the center or axis is taken to represent the north pole, and the edge of the paper the equator, then the situation is similar to the influence of the Coriolic effect in the northern hemisphere. Since the Earth has a counter-clockwise sense of rotation as viewed from above the North Pole, currents that migrate from the equatorial regions toward the poles trace out a clockwise path. This accounts for the clockwise rotation of ocean currents in the northern hemisphere (Fig. 9.21).

The reverse situation occurs in the southern hemisphere as illustrated in Figure 9.22b. From a viewpoint above the South Pole, the Earth appears to rotate clockwise. Thus as ocean currents migrate from equatorial to polar regions, they rotate counter-clockwise.

The Coriolis effect will deflect all free-moving objects. The Earth's rotation causes the apparent movement of the Sun from east to west across our sky which is why we move our clocks forward as we travel eastward into new time zones (London is 5 hours ahead of New York, which is in turn 3 hours ahead of Los Angeles). As a result of this rotation, all free-moving objects (such as a missile in the sky, seawater or the air we breathe) appear to veer clockwise in the northern hemisphere and counter-clockwise in the southern hemisphere. For example, as the Earth rotates beneath a descending spacecraft, the landing site moves as the spacecraft travels toward it. Even if the spacecraft were to continue to travel in a straight line, to an observer on the Earth's surface it would appear to progressively veer because the observer is anchored to a surface that is rotating.

Most athletes are familiar with moving reference frames such as those in Figure 9.22. In team sports, for example, a pass is rarely made directly at a teammate because the teammate is moving. Rather the pass is made to the position where the receiver will be when the ball arrives. Now imagine the athlete is dealing with a rotating reference frame such as a merry-go-round. A pass directed at a teammate on the moving merry-go-round will miss its target because, by the time the ball has arrived, the intended receiver will have rotated along with the merry-go-round. Although the ball has actually traveled in a straight line, from the point-of-view of both the passer and the receiver, it will appear to veer (Fig. 9.23). The extent of the apparent deflection depends on the speed of the pass. If the ball was thrown hard, the apparent deflection is minimal; if tossed slowly, the apparent deflection is considerable.

In familiar day-to-day situations like a merry-go-round we are aware of the rotation because the merry-go-round is a relatively small object and we can see it turn full circle. However, the effects of the rotation of large objects, such as the Earth, are not immediately obvious. The apparent deflection of an athlete's pass because of the Earth's rotation is negligible because the pass is over a very short distance compared with the size of the rotating body (the Earth). However, currents travel across the width of an entire ocean and are therefore significantly deflected (Fig. 9.24a). As the water of the Gulf Stream is direct-

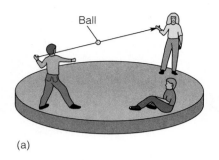

Stationary platform

Ball

(a)

Apparent path as seen by observer on rotating platform

Actual path

Observer

(b)

Figure 9.23
(a) A ball thrown in a straight line to a receiver on a stationary platform. (b) A ball thrown in a straight line to a receiver on a moving merry-go-round will miss its target. This is because the merry-go-round (and therefore the receiver) rotates while the ball is in flight. The perception of the observer will be that the ball followed a curved path passing behind the receiver.

ed out of the Gulf of Mexico toward western Europe, the solid Earth rotates beneath it like the merry-go-round rotates beneath the receiver. Thus, the current is continually deflected to the right even though (like the ball) it is traveling in a more or less straight line. This continuous deflection traces out a clockwise loop, like a car that turns right at every intersection, and forms the North Atlantic gyre (Fig. 9.24b). The rate of deflection depends on the speed of the current, fast-moving currents being less deflected than slow-moving ones.

9.7.2 Influence of a Landmass

Although the velocity of ocean currents is typically about 10 km (about 6 mi) per day, in certain areas they may locally attain velocities of about 160 km (100 mi) per day. Enhanced velocities are particularly common where currents run adjacent to landmasses, as is the case for the Gulf Stream off eastern North America. The reason for this is that the edge of a landmass creates a relatively stable path parallel to the coastline for a strong smooth current to follow. For example, off the coast of North Carolina, the Gulf Stream moves at about 9 km/hr (5.5 mi/hr). However, off the coast of Newfoundland, the coastline turns north, while the prevailing winds continue to drive the current

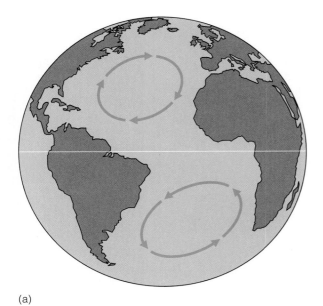

(a)

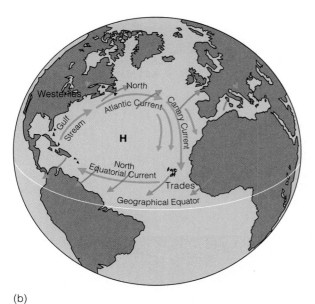

(b)

Figure 9.24
(a) Oceanic circulation influenced by the Sun's heat, atmospheric winds, gravity, and the Coriolis effect. (b) The North Atlantic gyre is made up of four interconnected current systems which together impart a clockwise rotation. Short superimposed arrows show the prevailing winds that influence major oceanic circulation. H signifies a region of high atmospheric pressure in the center of the gyre.

northeastward. Hence, the Gulf Stream loses the guiding influence of the coastline and, therefore, its stability, and slows down to less than 3 km/hr (2 mi/hr) and starts to meander across the Atlantic Ocean. This is analogous to a stream that is fast-flowing and follows a relatively straight path as it is guided down a mountain side, but begins to meander as it hits a relatively flat floodplain.

As the Gulf Stream loses the guiding influence of the coastline and meanders, it forms complex loops of water circulation, called **mesoscale eddies**. These loops are smaller than gyres and are also impermanent, unstable features. If conditions such as the strength of the prevailing winds change, the loops commonly become cut off to form **rings** (Fig. 9.25).

Since the Gulf Stream divides cold northerly waters, from relatively warm tropical waters, the rings are cored by either relatively warm or relatively cold water. Warm rings get dispatched northward into relatively cool waters, whereas cold rings get dispatched southward. Their existence is clearly shown by infrared satellite imagery which is highly sensitive to temperature variations and can readily detect both the number and the dimensions of these rings (Fig. 9.26). They are generally about 200 km (about 120 mi) in diameter, and the rotating currents within them move at about 25 km (15 mi) per day. They typically complete one rotation every 2 to 5 days and may last for up to 2 years.

The generation and migration of these rings is a highly efficient mechanism for mixing polar and tropical surface waters and their ecologies. As long as the ring current exists, the isolated ecology within them is in a cocoon, and temporarily spared the effects of the surrounding waters. As the ring current inevitably slows, the ecosystem dies. The phenomena is of considerable interest not only to ecologists and oceanographers but also to population biologists who study their influence on the migration of fish stocks.

9.7.3 Influence of Wind and Atmospheric Circulation

The frictional force of wind on surface waters exerts a major control on the motion of ocean water. This is readily apparent on a stormy day as wind-driven waters pound the coastline with increased ferocity. However, the true extent of this relationship is revealed by the very close correspondence between the directions of ocean currents and those of the prevailing winds (Fig. 9.24). This relationship also occurs on a global scale. For example, in the North Atlantic and South Pacific, winds associated with zones of high atmospheric pressure closely match the centers of the gyres, and, likewise move clockwise in the northern hemisphere and counter-clockwise in the southern hemisphere.

Wind affects surface waters to a depth of about 100 m (330 ft). Since the wind is only in contact with the surface, the velocity of these surface waters decreases rapidly with depth. So submarines can often avoid stormy waters by diving below 100 meters depth.

Within the surface waters, the effects of the wind are transmitted progressively downward as each thin water layer drags the layer immediately below it. As a result, the velocity progressively decreases due to increased frictional resistance.

This diminishing velocity means that the Coriolis effect varies with water depth within the surface waters. Since the extent of the deflection due to the Coriolis effect is greatest at

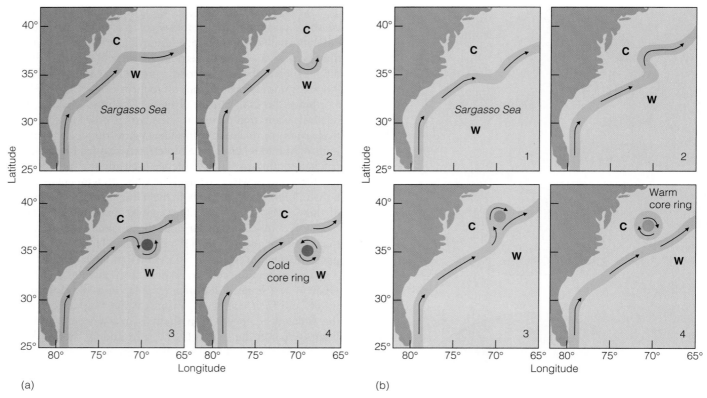

(a)   (b)

Figure 9.25

Schematic development of mesoscale eddies and ring currents in the North Atlantic Ocean. As the Gulf Stream tracks northeastward, it loses the guiding influence of the coastline and begins to develop "meanders" or eddies. When an eddy gets cut off, the Gulf Stream temporarily straightens and, depending on the local geometry, either produces (a) a cold core ring trapped in warm southern waters (1–4) or (b) a warm core ring trapped in cold northern waters (1–4). In this way ocean waters of contrasting properties can mix. (C = cold water, W = warm water)

Figure 9.26

Satellite mosaic infrared image taken in June 1984 of mesoscale eddies and ring currents that developed in the Gulf Stream south of Nova Scotia, Canada. The brightest color (red) depicts water with a surface temperature of 27°C (80°F). With decreasing temperature, colors change to orange, yellow, green, blue, and purple which represents the coldest water.

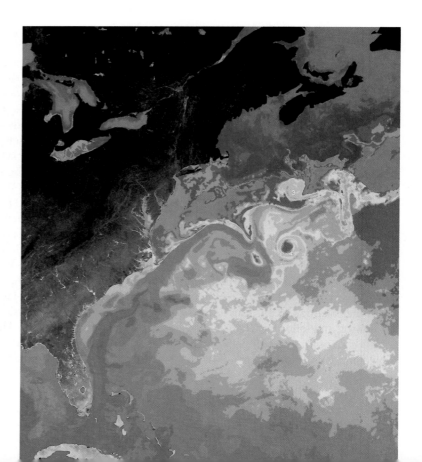

low velocities (remember the merry-go-round), the deflection becomes progressively greater with depth (Fig. 9.27). Movement of the uppermost layer of water (here referred to as layer 1) moves approximately 45° to the wind direction and drags the layer immediately beneath it (layer 2). However, layer 2 moves a little more slowly than layer 1 due to frictional resistance. Layer 2 is consequently deflected more than layer 1 and, in the northern hemisphere, is deflected clockwise or to the right. The motion of layer 2 similarly drags layer 3 immediately beneath it. But layer 3 moves more slowly than layer 2, and, as a result, experiences a greater deflection to the right than the one above it. Below 100 m (330 ft), the effect is negligible.

The effect generates what is commonly known as the **Ekman spiral** of flow in the upper 100 m (330 ft) or so of near-surface waters. The direction of each arrow in Figure 9.27 represents the direction of water motion, and the length is proportional to its velocity. Thus, the velocity of each layer is represented as a vector, and the average flow of the ocean water is represented by the sum of the vectors (given by the average direction and length of the arrows). This is not parallel to the wind. Instead, the sum of the vectors yields a direction that is commonly perpendicular to that of the prevailing wind. Because of the Coriolis effect, this mass movement is to the right of the prevailing wind direction in the northern hemisphere, and to the left in the southern hemisphere.

Consider, for example, the relationship between a persistent wind direction and the net movement of water on a west-facing coastline in the northern hemisphere (Fig. 9.28). If the persistent wind direction is from south to north, the net flow of water is to the east toward the coastline. At the coastline, **downwelling** occurs as the water is forced downward by the landmass (Fig. 9.28a). In contrast, if the persistent wind direction is from north to south, the net flow of water is to the west, away from the coastline. The stripping away of these surface waters allows **upwelling** of deep waters that are rich in nutrients such as nitrates and phosphate. (Fig. 9.28b)

These nutrients support high populations of microscopic marine organisms known as **plankton**, which in turn support other marine organisms. Such areas of upwelling represent only 1% of the ocean surface, yet they support more than 50% of the fish population. Notable areas include the west coasts of South America and Africa in the Southern Hemisphere. As our discussion of the Ekman spiral explains, the net flow of water should be at approximately 90° and, this time, counter-

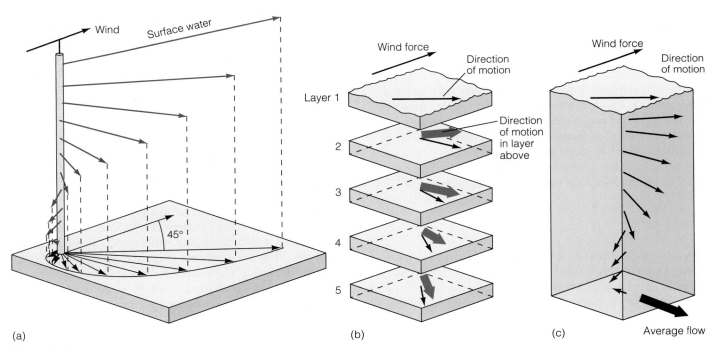

(a)   (b)   (c)   Average flow

Figure 9.27
The origin of the Ekman spiral which results in the net flow of ocean water moving almost perpendicular to the prevailing wind. Length of the arrows is proportional to the speed of the current in each layer. (a) Spiral model showing the motion of successively deeper layers of water in relation to the surface wind. (b) A water layer model where the uppermost layer moves approximately 45° to the wind. The layer beneath (layer 2) moves more slowly (narrow arrow) and is deflected to the right in the northern hemisphere because of the Coriolis effect. Similarly layer 3 is deflected to the right of layer 2, and so on. (c) The flow of water changes in a spiral fashion from the top downward. The theoretical net flow of water is the average of these vectors and nearly 90° to the wind direction.

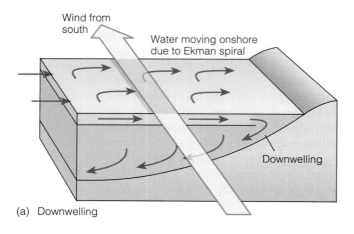

(a) Downwelling

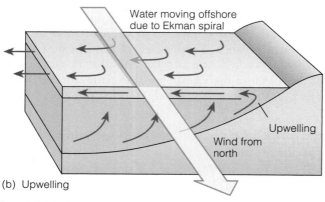

(b) Upwelling

Figure 9.28
Downwelling and upwelling currents at a shoreline in the northern hemisphere. (a) A wind from the south drives the surface waters to the right (eastward) onto the shore. This blanket of warm water blocks the ascent of cold waters and results in downwelling. (b) A wind from the north drives the average flow to the right (westward) resulting in the motion of near surface waters away from the coastline. This allows upwelling of cold nutritious waters from below.

clockwise with respect to the direction of the prevailing winds. Along both coastlines, winds from the south combine with the Coriolis effect to produce an east to west flow of surface waters away from the coastline. This allows the upwelling of deeper nutritious waters, so it is no coincidence that these regions are among the richest fishing waters in the world.

9.8 Putting It All Together: El Niño

We have learned that the relationship between wind, current direction, and coastline is a very important one and can profoundly influence the stability of ecosystems. In this section, we

apply what we have learned to the phenomenon known as El Niño, perhaps the most notorious example of this relationship. Over the last 20 years, scientists have realized that anomalous behavior in the eastern Pacific Ocean near the coastlines of Peru and Ecuador can have a dramatic effect on world climate. Toward the end of each year, warm currents raise water temperatures by 1°–2 °C (2 °–4 ° F) in this region. Local fishermen named these currents "El Niño" (Spanish for "the Christ child") because they noticed their occurrence around Christmas time. The fishermen also noted that the occurrence of these warm currents coincided with a sharp decline in fish catches. Traditionally, they take a break from fishing around Christmas time. In most years, the break lasts a few months, by which time the waters become cool again, fish stocks are replenished, and they are able to resume fishing. However, every 2–7 years, a warming of more than 6°C (11°F) occurs and the break in the fishing season extends until May or June. In 1982, the warming was so profound that the fishing harvest completely failed due to a decline in anchovies and the migration of sardine stocks southward to escape the warmer waters. The delicate balance of oceanic ecosystems is such that the decline in fish stocks resulted in a decline in the seabirds that feed on the fish, and a decrease in guano, a fertilizer excreted by many of these seabirds.

Scientists now reserve the term **El Niño** for these exceptionally warm periods which are separated by years of more normal cooler circulation. Although the 1982–1983 El Niño was the first to gain the attention of the public, it is clear that there have been at least ten such periods in the last 40 years.

The images shown in Figures 9.29a and 9.30a show a comparison of water temperature in the eastern Pacific for the month of May in a normal year (1988) and in an abnormal El Niño year (1992). In a normal year, cold nutritious waters occur along the South American coastline (displayed by the dark green colors) and a narrow tongue of cold waters typically extends across much of the eastern Pacific Ocean surface at the equator. In an El Niño year, however, both the coastal and equatorial waters are relatively warm.

How does an El Niño occur? In a normal year, two important wind directions influence western South America (Fig. 9.29b). In the central and southern parts of the continent, the prevailing winds are from the southeast. Near the equator, east-to-west winds are dominant. The southeasterly winds, known as the trade winds, combined with Ekman transport in the southern hemisphere, cause a net movement of ocean water to the west, that is, away from the coastline. This allows the upwelling of relatively cold deeper waters that provide nutrients for the abundant plankton (Fig. 9.31). Because they lie at the base of the food chain, these microorganisms provide the nutrients to sustain the fish populations.

El Niños occur when, for some unknown reason, the southeasterly winds diminish. As a consequence, the winds no longer drive the ocean water westward and the blanket of warm water stretches to the South American coast, thereby inhibiting the ascent of cold nutritious waters to the surface (Fig. 9.30b).

We now know that El Niño's are an integral part of a global-scale shift in weather patterns. In 1982–1983, over 250 cm (100 in) of rain fell in the Atacarra desert of Peru, and estuaries

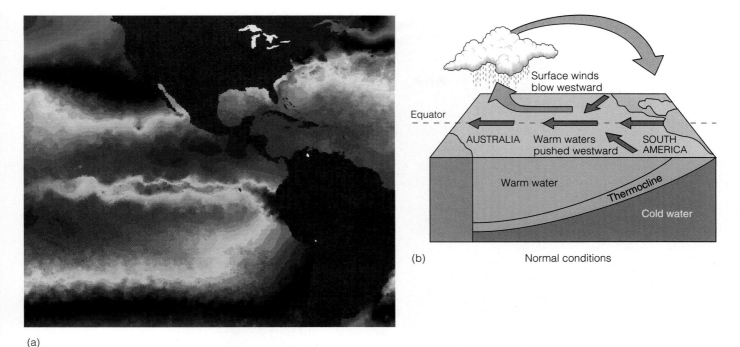

(a)

Figure 9.29

Normal year. (a) A satellite image taken in May 1988 showing normal upwelling of cold water along the South American coastline (displayed by the dark green colors) and a narrow tongue of cold equatorial surface water extending well across the eastern Pacific Ocean. (b) Winds blowing north-westward have driven warm waters away from the South American coastline resulting in a relatively shallow thermocline and upwelling of cold coastal water. Compare with Figure 9.30.

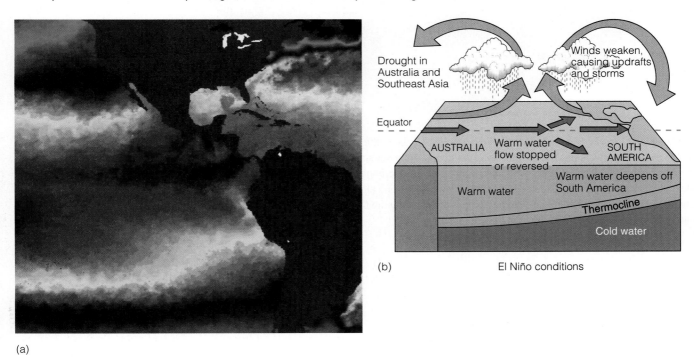

(a)

Figure 9.30

An El Niño year. (a) A satellite image taken in May 1992 showing a blanket of warm water (red and brown) that stretches from the South American coast across the equator. (b) The diminishing southeasterly winds can no longer drive the ocean water westward thereby inhibiting the ascent of cold nutritious waters to the surface. Note the deeper position of the thermocline compared to Figure 9.29.

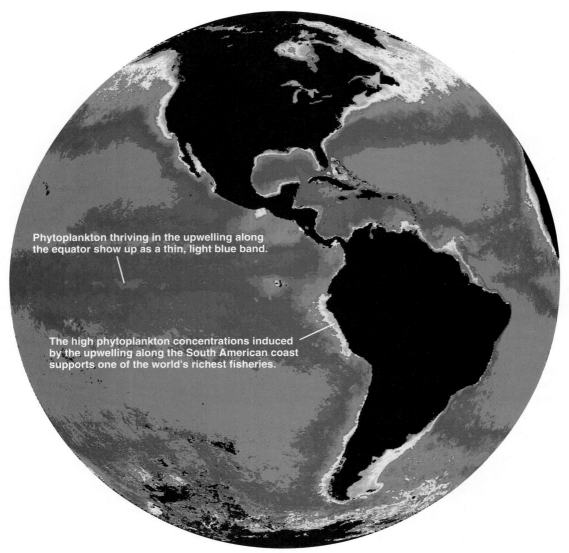

Phytoplankton thriving in the upwelling along the equator show up as a thin, light blue band.

The high phytoplankton concentrations induced by the upwelling along the South American coast supports one of the world's richest fisheries.

Figure 9.31
A satellite image showing average phytoplankton productivity along the coast of Peru and Ecuador over a period of several years. The images show variations in the amount of chlorophyll, a pigment that traps sunlight and initiates photosynthesis. Light colors (red, orange, green) highlight the highly productive coastal regions. Blue colors represent areas of upwelling along the equatorial regions as surface waters are either deflected to the north or the south by the Coriolis effect. Purple colors are areas of low productivity and occur in the centers of gyres.

were flooded. Typhoons were deflected from their normal paths and devastated the islands of Tahiti and Hawaii. Monsoon rains fell over the central Pacific Ocean instead of southern Asia, leading to droughts in countries such as India which depend on the monsoons for their water supply. Winter storms battered the Californian coastline. The changing winds over the Pacific Ocean caused sea level to rise by more than 30 cm (12 in) in the eastern Pacific, and to fall in the western Pacific, exposing the upper layers of the fragile coral reefs of Australia and damaging their ecosystems.

In January of 1995, an El Niño triggered record temperatures in North America, 6°C higher than normal. The effects of this were highly variable across the continent. The Californian

coastline was pounded by 10 m (33 ft) waves and by 2 weeks of torrential rains that resulted in 9 fatalities and damage estimated at $1.3 billion. In the Rocky Mountains, the precipitation produced a massive snow cover, to the delight of skiers, but in eastern North America, high temperatures produced a virtually snowless winter. Other areas also had anomalous weather. For example, there was widespread drought in Central America, the Caribbean, and Central Australia.

How could an El Niño be responsible for these widespread events? Two key observations resulted in a major breakthrough in our understanding of the El Niño effect. The first is that the anomalously warm waters reported by fishermen during an El Niño actually extend along the equator westward across the

Pacific Ocean. The second is that during an El Niño, trade winds are generally lighter and rainfall heavier than normal in this equatorial region. As the trade winds in South America die down, the normal westerly motion of water slows, allowing warm Pacific waters to spread eastward. These warm tropical waters are moisture-laden and generate abundant tropical thunderstorms (Fig. 9.30b). This profoundly affects both oceanic and atmospheric circulation in the Pacific Ocean, re-orienting the southern jet stream, so that it points directly at California. This jet stream is the preferred pathway for weather systems in the central Pacific Ocean and, as a result, system after system pounded the Californian coastline for 2 weeks (more on jet streams in Chapter 12).

As meteorologists and oceanographers turned their attention to the origin of El Niño, they soon realized that an important clue was provided by a key piece of research undertaken in the 1920's by Sir Gilbert Walker, a British scientist. His examination of climate records revealed that regions of high atmospheric pressures see-sawed from one side of the Pacific Ocean to the other, a phenomenon known as the **Southern Oscillation** (Fig 9.32). In most years, the atmospheric pressure is highest in the eastern Pacific. Since winds travel from high pressure toward regions of low pressure, the surface winds flow across the Pacific Ocean from the southeast toward the west (Fig. 9.32a). It is the winds emanating from this high pressure system that help to drive warm equatorial waters

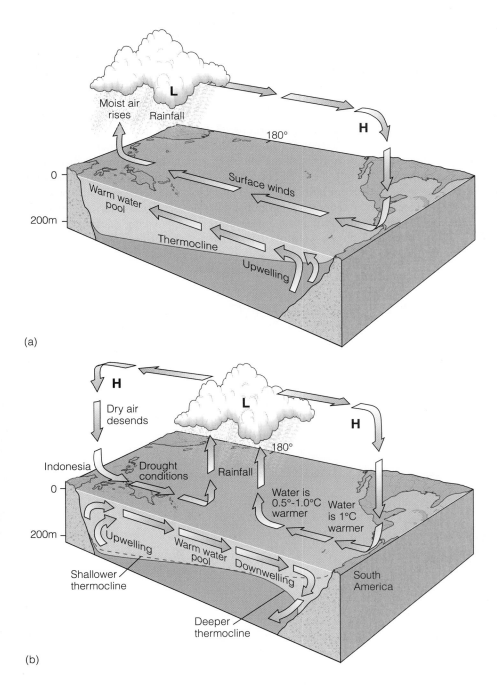

(a)

(b)

Figure 9.32
The Southern Oscillation. (a) In most years, atmospheric pressure is highest in the eastern Pacific Ocean and the resulting westward winds allow upwelling along the Peruvian coast and drive moisture westward. (b) In El Niño years, a region of high atmospheric pressure also occurs in the western Pacific Ocean. H, regions of high atmospheric pressure; L, regions of low atmospheric pressure.

westward, thereby facilitating upwelling of nutrients along the Peruvian coast. However, in years when the atmospheric pressure is also high in the western Pacific, these winds are severely diminished, or are even reversed. In this situation, warm waters blanket the Peruvian coastline, inhibiting the upwelling (Fig. 9.32b).

But what of other ocean waters? There is increasing evidence that the El Niños of the eastern Pacific Ocean may be just one of several such phenomena. Similar spreading of warm water masses have also been found in the Indian and Southern Atlantic oceans (Fig. 9.33). Most recently, the European climate has been shown to be predominantly influenced by phenomena similar to that of El Niño. A typical European winter is dominated by the circulation of winds between low pressure centers over Iceland and high pressure centers in the Azores off the coast of Portugal. As winds swirl around these pressure centers, they blow over the Gulf Stream. If the pressure difference between the two centers is high, then the winds are strong, and more heat is pumped into the European continent, leading to a relatively mild winter. When the difference is low, the winds slacken and an anomalously cold winter results. As we learn more about the relationship between winds and ocean circulation, we have found that processes similar in principle to those of an El Niño may be very common indeed. If so, then such processes may indeed be a phenomenon that influences global-scale weather patterns.

Until recently, scientists believed that El Niños were a periodic disturbance in the eastern Southern Pacific Ocean, occurring in cycles of 2 to 7 years. However, in the last 20 years they seem to have occurred more often. Many scientists blame global warming, where warm atmospheric conditions prevent the complete dissipation of the El Niño, allowing it to leave the seeds from which to build up again the following year. Other scientists say simply that climate is "nonlinear," that is, it is far too complex to predict or even expect any simple cyclicity in its behavior.

What is clear however, is that El Niño's have global-scale implications to life, crops, and property. The advent of satellite technology gives us some advance warning and ensures that they no longer take us by complete surprise. At the time of writing (late 1997) the most intense El Niño of this century is underway. The tell-tale signs of this event began as early as April 1997 (Fig. 9.34), a time when the ocean waters of the Eastern Pacific normally decrease rather than increase in temperature. Figure 9.34, however, clearly shows the extent of warm water (in red/dark red) across much of the Pacific Ocean. By September, surface water temperatures were already 5°C (9°F) warmer than normal. In contrast, during the previous most intense El Niño of 1982, the anomalous warmth of the equatorial waters only became obvious in September.

Regions that are susceptible to anomalous temperature and precipitation during an El Niño are shown in Figure 9.35. As significant portions of the world are affected, does an El Niño trigger a global-scale shift in the traditional climate belts, or is it one of the effects of such a shift? At present this is difficult to tell. El Niño probably ranks second only to our seasons as an influence on global-scale climate. Its origin and potential relationship to global climate change is one of the most intriguing and important problems facing ocean and atmospheric scientists.

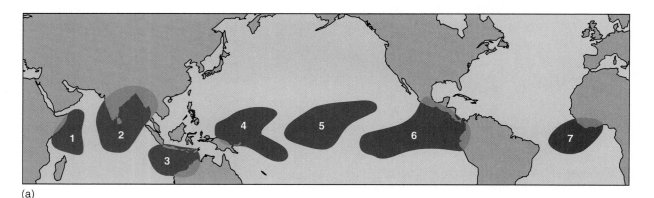

(a)

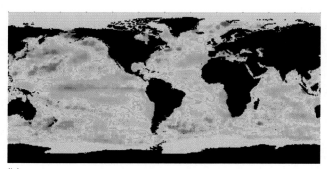

(b)

Figure 9.33

(a) Map showing El Niño conditions (1 to 7) in the equatorial regions of the Indian, Pacific, and Southern Atlantic oceans. (b) Regions of warm equatorial water indentified by red and orange colors.

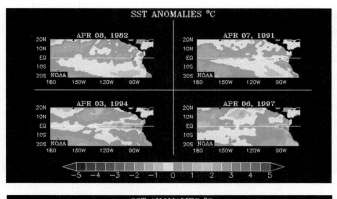

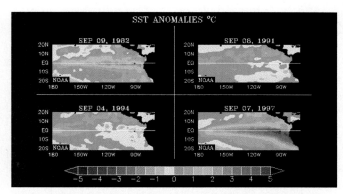

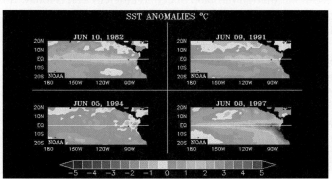

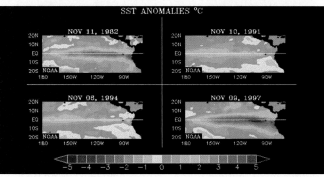

Figure 9.34

Comparative ocean surface temperatures in April, June, September, and November in El Niño years (1982 and 1997) and normal years (1991 and 1994).

WARM EPISODE RELATIONSHIPS   DECEMBER - FEBRUARY

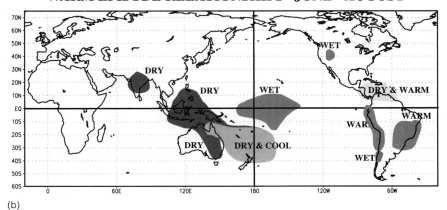

(a)

Figure 9.35

Climate models indicate which regions are most likely to experience anomalous weather during an El Niño from (a) December–February and (b) June–August.

WARM EPISODE RELATIONSHIPS   JUNE - AUGUST

(b)

Chapter Summary

The oceans are dynamic and in ceaseless motion. Although this is most obvious along our coastlines in the form of waves, tides, and coastal erosion, the most fundamental circulation occurs in the wide ocean. Waves are disturbances generated by wind that move across the ocean surface and can travel unhindered across an entire ocean basin. But they do not represent the actual movement of the water. In coastal regions, the slowing of wave velocities is compensated by increasing wave heights until a stability threshold is exceeded and breakers occur.

The shoreline's resistance to coastal erosion depends on the type of bedrock. A shoreline with weak bedrock or unconsolidated sediments commonly recedes at about 1 meter per year. The sediment on beaches commonly migrates because of the obliquity of the incoming waves, which generates a longshore current.

Tides are caused by the gravitational attraction between the Earth and the Moon. This attraction produces bulges and depressions in the ocean water surface that the solid Earth rotates beneath. High tides occur when the solid Earth rotates beneath a bulge, while low tides occur when it rotates beneath a depression. Changes in tidal ranges between spring tides and neap tides match the phases of the Moon and are, respectively, related to the alignment and misalignment of the Earth, Moon, and Sun. When all three are aligned, their respective influences act together producing spring tides. When the three are at maximum misalignment they act against each other and produce neap tides. These simple tidal patterns assume an Earth with a uniform layer of water, and are substantially modified by the influence of continental landmasses or in confined bodies of water.

Ocean water circulation describes the actual movement of ocean water which is driven by contrasts in temperature, density, and salinity from the poles to the equator. Since sea ice preferentially excludes salt, salt becomes concentrated in polar waters. Polar sea water is therefore dense and sinks as it migrates toward the equator. Its characteristic composition allows its motion to be tracked. For example, sea water generated in the Antarctic Ocean has been traced at depth into the northern Atlantic and Pacific oceans. This widespread deep ocean circulation, combined with the motion of surface waters from the hot equatorial to the cold polar regions, ensures efficient circulation of the oceans and the broad distribution of nutrients.

The presence of polar ice is fundamental to the efficiency of this process, because it provides the temperature, composition, and density contrasts that drives it. Vertical and horizontal compositional changes are also influenced by contrasts in polar and equatorial behavior of ocean water. Since the presence of polar ice is unusual in the geologic past, it is likely that ocean circulation has rarely been as efficient as it is today. It is possible that some of the mass extinctions in the marine fossil record have been brought about by demise of polar ice sheets.

Ocean water migrates in gyres or currents, which are narrow zones of concentrated high velocity flow. As water migrates in these gyres, it moves across the surface of a rotating Earth. As a consequence, it follows a curved path due to the Coriolis effect. This apparent force gives a clockwise sense of rotation to gyres in the northern hemisphere, and a counter-clockwise rotation in the southern hemisphere.

The continents exert a considerable influence on the width and path of the gyres, many of which span the entire width of the ocean in which they reside. For example, the Gulf Stream is guided by the eastern coastline of North America until it reaches open water, where it begins to meander and form mesoscale eddies. These eddies are zones of important mixing between polar and tropical waters and may affect the ecological evolution of the ocean and fish populations.

Wind also plays an important role in ocean circulation, but because of the Ekman spiral, the net flow of ocean water tends to be almost at right angles to the prevailing wind. If the wind drives warm surface waters away from the shoreline, this facilitates upwelling of cold nutritious deep ocean waters.

Should these winds temporarily cease or wane, the effects can be global in scale. This is almost certainly the case for El Niño, where the waning of southeasterly trade winds off the coast of South America results in the spread of warm tropical waters across the entire Pacific Ocean. This fundamentally affects the position of the jet streams and, consequently, impacts global-scale weather systems.

Key Terms and Concepts

Look for the highlighted items on the web at:
WWW.BROOKSCOLE.COM

- Motion of waves reflects the action of wind on the ocean surface
- Water depth influences particle motion, wave length, wave height, and the generation of breakers
- Coastline shape is influenced by strong versus weak bedrock
- Longshore currents result in the migration of beach sands
- Semi-diurnal tides are explained by an Earth with a uniform layer of water affected by the gravitational pull of the moon
- Diurnal and mixed tidal patterns are due to the complicating effects of continental positions and confined bodies of water

- Relationship of Earth-Moon-Sun alignment and misalignment explain tidal ranges and spring and neap tides
- The size of an ocean basin influences the scale of oceanic circulation
- Ocean circulation is driven by contrasts in the temperature, density, and salinity of ocean waters from the pole to the equator, both at the surface and at depth
- Water around sea ice is dense, cold and sinks; equatorial waters migrate to the poles. Polar waters can be traced along the ocean bottom where they typically move at 5 km (3 mi) per day. Warm

surface waters extend to depths between 100 and 500 m (330–1600 ft)

- The paucity of ice ages in the geologic record suggests modern oceanic circulation may be atypical of the past
- Ocean currents describe the actual movement of surface waters. As ocean currents move water from the equator to the poles, they are acted on by the Earth's spin, which induces the Coriolis effect and produces large-scale circular motion in ocean basins, called gyres. The gyres are clockwise in the northern hemisphere, and counter-clockwise in the southern hemisphere
- The actual pathway of ocean water circulation is influenced by the Earth's spin and is guided by the positions of the continents
- When the Gulf Stream loses the guiding influence of the North American coastline it develops mesoscale eddies, and warm and cold rings. Warm tropical and cold polar waters mix, influencing ocean ecology
- There is a close correspondence between ocean currents and prevailing winds
- Surface water movements due to the wind are transmitted progressively downward, each water layer drags the layer immedi-

ately below and the velocity progressively decreases due to increased frictional resistance. Since water at depth moves slower than surface water, deflection due to the Coriolis effect is greater at depth, generating the Ekman spiral which extends down to 100 m (330 ft)

- Bulk mass movement of water is the sum of the vectors created by the Ekman spiral, and is perpendicular to the direction of the prevailing wind
- Net flow of water away from coastlines allows upwelling of deep nutritious waters. Net flow of water toward coast results in downwelling
- El Niño is the most notorious example of the relationship between wind, current direction, and the stability of ecosystems. In an El Niño, anomalously warm moisture-laden waters extend across the Pacific Ocean when southeasterly winds off continental South America diminish
- Development of El Niño is monitored by satellite technology. The effects of El Niño are accompanied by a shift in the jet stream, which causes a global-scale shift in weather patterns

Review Questions

1. Sketch a typical wave and label it with the appropriate terminology.
2. Explain why it is incorrect to use the term "tidal wave" to describe a tsunami.
3. Explain why breakers occur.
4. Why do some coastlines retreat faster than others?
5. Explain the difference between direct and indirect high tides.

6. Explain the origin of semi-diurnal tides, and why they don't occur along all coastlines.
7. What are the fundamental factors driving ocean circulation?
8. What factors influence the actual path that ocean currents take?
9. Why is the net motion of surface ocean water currents generally perpendicular to the wind direction?

Study Questions

Research the highlighted question using InfoTrac College Edition.

1. If spring tides are generated twice every month by alignment between the Earth, Moon and Sun, then why do you think that lunar and solar eclipses are so rare?
2. Explain the relationship between tidal ranges and the gravitational pull of the Moon and Sun during a typical month.
3. Explain why major surface ocean currents have a clockwise sense of rotation in the northern hemisphere, and a counter-clockwise sense of rotation in the southern hemisphere. Why do the major surface ocean currents vary in size (for example, compare the Gulf Stream and Peru currents)?
4. Explain the relationship between wind direction and the presence of downwelling or upwelling currents at a coast line. In the southern hemisphere, a wind blows to the north along the east coast of a continent. Would this produce upwelling or downwelling, and why?

5. Sketch a westerly facing coastline in the southern hemisphere. Explain what conditions would favor upwelling over downwelling currents along this coastline, and why?
6. Explain what happens to the circulation of ocean waters in the southern Pacific Ocean during an El Niño. Why does an El Niño have such a devastating effect on the fishing industry?
7. What events trigger an El Niño?
8. What happens to mesoscale eddies as they move away from the Gulf Stream, and what are the implications to weather patterns and ecosystems?
9. Why may an El Niño have global-scale effects?

10

Fresh Water

In this chapter you will learn about our most precious resource—fresh water. We depend on this resource for our quality of life, yet we take it for granted. We are most familiar with fresh surface waters such as those that occur in streams, lakes, and swamps. These surface waters are an essential element of the hydrologic cycle (see Chapter 1) and the continental drainage systems that transport water toward the ocean.

However, as we shall see, an appreciable quantity of water is also stored in the ground, and is known as groundwater. In fact, groundwater is our main water resource for domestic purposes and is mined from the ground for these uses, just like any other commodity. Groundwater resources also support agricultural and industrial activities which foster economic prosperity and so tend to take place in areas that are highly populated. This places further strain on the resource in terms of both quality and quantity. Because we consume groundwater faster than nature can replenish it, it is a nonrenewable resource. The quantity and quality of groundwater will therefore diminish unless it is carefully managed.

In order to understand why we have fresh water, we first discuss the hydrologic cycle and what happens to precipitation after it falls on the ground. Some becomes surface runoff and flows back to the ocean in streams. But precipitation is not evenly distributed. Instead, some parts of the world are characterized by deserts while others have abundant rainfall. This unequal distribution is the key to the supply of fresh water and we will explain why this is so. We will also discuss global-scale precipitation patterns and the ways in which these can be substantially modified in coastal and mountainous regions.

Groundwater is the result of precipitation percolating through the ground to form aquifers of water-bearing sediment or rock. At a certain depth below the surface, water fills all fractures in the sediment or rock and all gaps between their constituent grains. As we shall see, locating the upper surface of this water-saturated zone, known as the water table, is fundamental to the successful exploitation of this resource.

Surface waters form large drainage basins, predominantly consisting of streams and lakes. Large rivers like the Mississippi and Amazon have vast drainage basins and carry up to 10 billion tons of sediment to the oceans each year. Mature streams occupy wide valleys that are prone to flooding. We will learn about flood prevention and the dilemma it poses.

We need water to survive. But in order to exploit this resource we first must find it. We will show how sophisticated exploration methods, such as electrical pictures of the rocks and sediments underground, help identify aquifers and provide targets for the drilling of water wells.

For many of our needs, water must be of high quality. But in order to monitor water quality, we must first understand what processes influence its chemical composition. As we learned in Chapter 8, water is a very strong solvent. Thus it is very susceptible to contamination both by natural processes and human activities. We will find that the process of weathering provides a fundamental control on water quality by causing chemical alteration and the physical disintegration of minerals and rocks below the surface. However, our actions too can modify the composition of water and limit its use. We will discuss how our actions have compromised our groundwater supply and the unforeseen and sometimes horrific consequences of diverting the flow of surface waters.

We emphasize the importance of proper management and long-range planning in ensuring the most appropriate exploitation of water resources. Signs that overexploitation has occurred are clear. They include diminishing well pressures, declining local streamflow, land subsidence, and the intrusion of saltwater into coastal aquifers.

We hope this chapter convinces you that quality fresh water is a priceless resource, and one that we can no longer take for granted. Carefully monitored and managed, we have enough for our needs. But when it comes to looking after our fresh water supply, we may well be our own worst enemy.

10.1 Introduction

Fresh water, whether in streams, lakes, or underground, has influenced the evolution of the human race more than any other substance. Early civilizations developed along important river valleys such as those of the Nile and Tigris-Euphrates in the middle East, the Indus in Pakistan, and the Yellow River in China. The rivers provided water for the crops that nourished these cradles of civilization. Archaeologists call these "hydraulic civilizations" because of their almost total dependence on water. The ancient Greeks even understood the basic elements of the hydrologic cycle. Nearly 2400 years ago, Aristotle (384 to 322 B.C.) observed that rainwater penetrating the ground percolated back to the surface to form springs, and that most precipitation on land ultimately found its way into streams that drained to the sea. The need for a reliable supply of water prompted these early civilizations to interfere with the natural hydrologic cycle. One of the first dams, for example, was built nearly 5000 years ago near Cairo by the ancient Egyptians. The Greeks too built dams prior to their conquest by the Romans. The Romans mastered this technology, building dams in many parts of their empire and aqueducts which channeled the ponded water to their cities (Fig. 10.1). They

also built public baths and toilets and used drains and sewers to carry the waste water away. The Chinese were likewise innovators in water technology. For example, in 610 A.D., after more than 300 years of construction, they completed the Grand Canal which connected the Yellow and Yangtze rivers.

The "hydraulic civilizations" flourished because of their recognition of the importance of water. Nearly 1500 years later, little has changed. We still intercept fresh water on its way back to the ocean and use it for a wide range of purposes. There is a general consensus among scientists that fresh, uncontaminated water will be the most important resource of the 21st Century. For people who live on parched lands, this is nothing new. Fresh water has always been their most important resource, the delicate balance of its supply and demand playing a critical role in their day-to-day lives.

Until very recently, the availability of fresh water in the industrialized nations of North America and western Europe has been taken for granted. Indeed, in most areas, supply in the form of rainfall greatly exceeded the demand. This oversupply could be re-routed to the few regions that experienced arid conditions like those in the heavily populated southwestern United States. The luxury of having access to water meant that efforts could be concentrated on industrialization, in contrast to the world's desert regions where such access is a day-to-day challenge.

However, we are receiving the early warning signals of severe problems to come. Population pressure, inefficiency, waste, and contamination threaten our continued access to the most precious of resources. The availability and conservation of fresh uncontaminated water, and the location and ownership of the supply, are likely to become key issues in the very near future.

The reason for this is simple. Because of our metabolism, we can survive without food for weeks, but we can only live for a few days without water. Our bodies require 4 liters (about 1 gallon) of water per day. Thus the 300 million people in North America alone require 1.2 billion liters (about 310 million gallons) of drinking water per day, and this is just the tip of the iceberg. We

Figure 10.1
Built in the first century A.D., this remarkably well-preserved Roman aqueduct supplied drinking water to the city of Segovia in central Spain for almost 2000 years. Constructed entirely without mortar, its 118 arches span 760 m (2500 ft) and stand 28 m (92 ft) above the valley floor. (Photo Damian Nance)

also use water for domestic purposes (in cooking, bathing, and cleaning), for agriculture (in irrigation and for livestock), and for industry (in manufacturing). In North America, this demand is a staggering 1.8 trillion liters (about 470 billion gallons) per day. As we shall see, most of this water is extracted from the ground as groundwater. But our usage has increased to the extent that we are now consuming water at a much greater rate than nature is replenishing it through rainfall. In this sense, fresh water has become a non-renewable resource, even though there are ways to recycle and re-use fresh water, as there are with all resources. We mine groundwater just like we mine any other resource, and just like oil and gas, groundwater can run out.

As you will learn in this chapter, ingenious modern methods to find new resources in the ground are buying us time to change our consumption habits. But this will not be enough to spare us from the severe problems that lie ahead. Economists often argue that when a society goes into economic debt, it is spending tomorrow's earnings and passing the debt burden on to its children or grandchildren. Similarly, if a society consumes water faster than it is being replenished, it passes on a "water debt" to succeeding generations by depleting a resource that took millions of years to form. Sooner or later we will have to recognize fresh water for what it is, our most precious resource.

In this chapter, we first outline the behavior of groundwater stored in the sediments and rocks beneath the Earth's surface, and then turn to the behavior of waters that flow at the Earth's surface. We then examine the ways in which we exploit water by interfering with its normal path as described by the hydrologic cycle.

You will notice that the chapter contains more terminology than usual (Fig. 10.2). This is not a change in approach. The topic is one of the few areas in Earth Science where terminology has pervaded everyday English language. Indeed, many of the

terms outlined below may be found in a standard dictionary and may be useful to you if you ever deal with a real estate agent or a building contractor.

10.2 The Hydrologic Cycle and the Water Budget

Fresh water represents less than 3% of the water on the Earth. Its abundance is dwarfed by the immense volume of saltwater in our oceans (Fig. 10.3). When water evaporates from the oceans, it leaves most of its impurities behind. Thus evaporated water is water in its purest natural form. It contains almost no dissolved salts (only a few milligrams per liter) and very little in the way of liquid or gaseous impurities. This is what we call "fresh" water, and it is the precipitation of evaporated water that forms our fresh water resources on land (Fig. 10.4). About 425,000 cubic km (101,000 cubic mi) of water evaporate annually from the oceans and an additional 71,000 cubic km (17,000 cubic mi) evaporate on land. On average, this moisture spends about 11 days in the atmosphere before it is returned to the Earth's surface as rain or snow. During this time, the moisture interacts with solid particles in the air and consequently contains small but measurable concentrations of sodium, potassium, calcium, magnesium and chlorine. It also interacts with atmospheric gases such as sulfur dioxide (to form dilute sulfuric acid) and carbon dioxide (to form dilute carbonic acid) and so tends to be slightly acidic when it falls again as rain or snow.

Because annual changes in sea level are negligible, the annual amount of evaporation must balance annual precipitation. We know, of course, that sea level can change significantly (see Chapter 8). But these changes occur over millennia and so are negligible over the course of a single year. Therefore, each year there must be 496,000 cubic km (118,000 cubic mi) of precipitation. Because the Earth's surface is dominated by oceans, most precipitation (385,000 cubic km or 92,000 cubic mi) falls directly back into the sea. Of the 111,000 cubic km (26,000 cubic mi) of precipitation that falls on land, 71,000 cubic km (17,000 cubic mi) is evaporated. Of the remaining 40,000 cubic km (9000 cubic mi), the vast majority either infiltrates below the surface to become groundwater, or is returned to the sea by continental drainage. A very small amount may contribute to the growth of ice sheets and glaciers.

Each year, the amount of infiltrating groundwater must be balanced by the amount of groundwater lost to the surface. Again, if this were not the case, sea level would change from one year to the next. We will explain how this balance is achieved later in the chapter. But a total of 40,000 cubic km (9000 cubic mi) of water must eventually drain from the land to the sea each year. However, as we shall see, it takes many different routes to do so.

Currently, about three-quarters of the Earth's fresh water occurs in ice sheets and mountain glaciers (Fig. 10.3 and Table 10-1). Although precipitation in polar regions is low, evaporation is negligible. Indeed, as a result, ice sheets represent hundreds of thousands of years of snow accumulation and store vast quantities of fresh water. This abundance has

Figure 10.2
Everyday terminology related to fresh-water geology. (Photo Brendan Murphy)

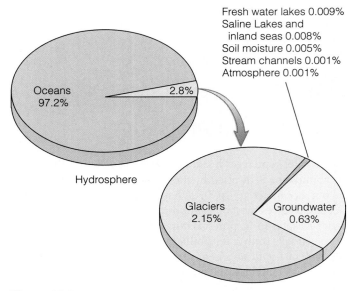

Fresh water lakes 0.009%
Saline Lakes and
  inland seas 0.008%
Soil moisture 0.005%
Stream channels 0.001%
Atmosphere 0.001%

Oceans
97.2%

2.8%

Hydrosphere

Glaciers
2.15%

Groundwater
0.63%

Figure 10.3
Inventories of water in the oceans, the atmosphere, and on land.

Table 10-1

Water on Earth in km³

LOCATION	VOLUME	PERCENT OF TOTAL	PERCENT FRESH WATER
Oceans	1,327,500,000	97.20	—
Glaciers	29,315,000	2.15	72
Groundwater	8,442,580	.63	22
Freshwater and saline lakes and inland seas	230,325	.016	6
Atmosphere at sea level	12,982	.001	trace
Average in stream channels	1,255	.0001	trace

probably fluctuated significantly throughout geologic time, as ice ages have come and gone. As we shall see in Chapter 14, the distribution of ice sheets and glaciers has changed dramatically over the past 2 million years. At one time, ice sheets covered more than half of continental North America, Europe, and Asia. Their many cycles of advance and retreat have had a major influence on the sculpting of our landscape. In this chapter, however, we concentrate on water in the ground (groundwater) and on the Earth's surface (surface water) which together form a vital resource in the world's most populated regions.

Groundwater constitutes about 22% of fresh water (Table 10-1) and is stored in rocks and soils below the ground, in what is termed the "subsurface." **Surface water** in lakes, streams, swamps, and springs comprises about 6% of the fresh water supply. A tiny fraction occurs in the atmosphere and biosphere.

The hydrologic cycle shown in Figure 10.4 gives an overview of the global "water budget" by showing the amount of water involved in evaporation, precipitation and continental drainage. On a global scale, however, precipitation is very unequally distributed, so that the abundance of fresh water varies from one region to the next. In order to understand why this is so, we must examine global precipitation patterns.

10.3 Global Precipitation Patterns

The distribution pattern of average annual precipitation is shown in Figure 10.5a. Net precipitation, or the amount of precipitation minus the amount of water that evaporates again, is shown in Figure 10.5b. These figures show that sub-equatorial regions (those within 10° latitude of the equator) and middle latitude regions (those between 45° and 65°) have a surplus of water, whereas subtropical regions (between 15° and 30°) have a net deficit.

Figure 10.4
Simplified hydrologic cycle. Numbers in brackets show the amount of water stored in ice sheets, groundwater, surface water (streams and lakes), atmosphere, and oceans. Other numbers show the amount of moisture in transport, including evaporation, vapor transport, precipitation, and surface runoff (continental drainage). All numbers represent the volume of water in millions of cubic kilometers (see Table 10.1).

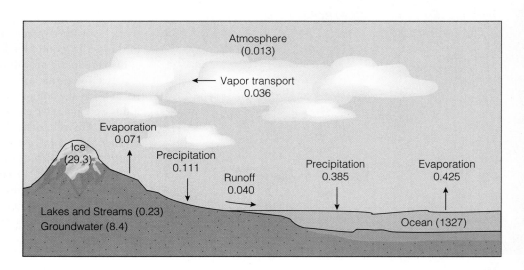

Atmosphere
(0.013)

Vapor transport
0.036

Evaporation
0.071

Ice
(29.3)

Precipitation
0.111

Runoff
0.040

Precipitation
0.385

Evaporation
0.425

Lakes and Streams (0.23)
Groundwater (8.4)

Ocean (1327)

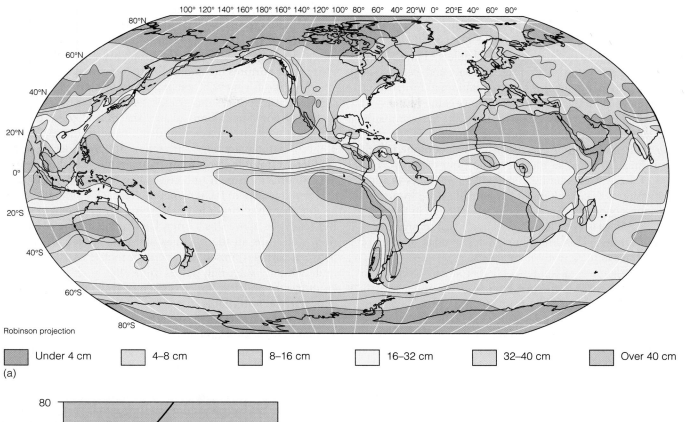

(a)

| Under 4 cm | 4–8 cm | 8–16 cm | 16–32 cm | 32–40 cm | Over 40 cm |

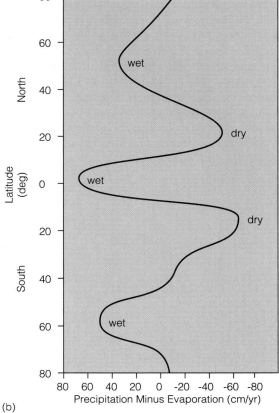

(b)

Figure 10.5

(a) Global annual average precipitation varies significantly. Areas close to the equator and at mid-latitudes have abundant precipitation, whereas subtropical and polar regions have low precipitation. (b) Net annual precipitation (precipitation minus evaporation) as a function of latitude. Positive values represent net precipitation whereas negative values represent net evaporation. Precipitation is high at the equator, generally low between 15°–30° latitude where deserts are located (see Chapter 12), high at mid-latitudes (45°–65°), and low in polar regions. This pattern is especially evident in the southern hemisphere because it is less modified by the presence of landmasses than in the northern hemisphere. Used by permission of the Estate of Eric Mose.

The surplus of fresh water in equatorial regions is caused by the large supply of moisture from adjacent warm oceans and the strong updrafts of hot equatorial air over the land. (This phenomenon is described in more detail in Chapters 11 and 12.) As the warm moist equatorial air rises, it cools and condenses, resulting in torrential rainfall. The precipitation, in turn, causes the warm air migrating toward the poles to become very dry. As a result, evaporation greatly exceeds rainfall in the warm subtropical latitudes leading to the formation of deserts. The mid-latitude fresh water surplus is created by an atmospheric belt where warm subtropical air meets cool air from the poles. The meeting of these two contrasting bodies of air produces weather fronts and storms that result in a lot of rainfall. As we learned in the previous section, there is very little precipitation in the polar regions. But there is even less evaporation because of the very cold temperatures. This imbalance yields a small surplus of water that leads to the growth of ice sheets.

There are some local influences that produce exceptions to this global trend, such as the proximity to ocean water. At any given latitude, coastal areas, like those on the eastern seaboard of North America, have much more precipitation than continental interiors. High mountains can also influence rainfall. The Sierra Nevada and Rocky Mountains in western North America, for example, form a north-south barrier to rain-bearing clouds. This barrier creates deserts behind the mountains in the region of so-called "rain shadow" (see Chapter 11). So two areas at the same latitude can have very different amounts of annual precipitation if they lie on opposite sides of a high mountain range.

10.4 Groundwater

Water stored beneath the Earth's surface in sediments and rocks as **groundwater** is the largest reservoir of accessible fresh water. In North America, it supplies between 40 to 50% of the current demand. Although groundwater occupies a small proportion of the hydrologic cycle in terms of its relative volume, it nevertheless forms the vital link between the evaporation of water from the ocean and its return to the sea by way of continental drainage (Fig. 10.6). This temporary storage of water in the ground nurtures life and facilitates our survival.

Beneath its veneer of soil and vegetation, the Earth's land surface consists of unconsolidated sediment and rock. Sediment contains billions of tiny spaces and rocks contain abundant fractures. At a certain depth beneath the surface, all of these openings are filled with water. But how does the water get there? Groundwater is formed as rain migrates directly into the subsurface through unconsolidated soil, sand, and gravel and into fractures in the rock below. Bedrock eventually forms an impenetrable barrier to this downward movement of water. But in a zone above this barrier, known as the **zone of saturation,** water fills all the fractures and pore spaces (the spaces between particles) in soil, sediment, or rock (Fig. 10.7). These underground water-rich layers form the resource we call groundwater and the upper surface of this saturated zone is known as the **water table.**

If we stand outside in the pouring rain for long enough, we often say we are "saturated," meaning that we couldn't possibly get any wetter. The subsurface below the water table is also saturated, because water occupies all the spaces available to it, so that the sediments and the rocks cannot get any wetter. The level of the water table may be determined by simply digging a hole. When the water table is reached, water will start filling the hole. When making sand castles on a beach, for example, water floods in as you dig below the water table because the sand is saturated with water.

Above the zone of saturation, the sediment is undersaturated and the pore spaces consist of a mixture of air and water. This region is known as the **zone of aeration.** The water table, therefore, is the surface separating these two zones. In a very thin zone immediately above the water table, a "capillary fringe" occurs as groundwater rises upward because of surface tension (which binds together the surface molecules of the water) and

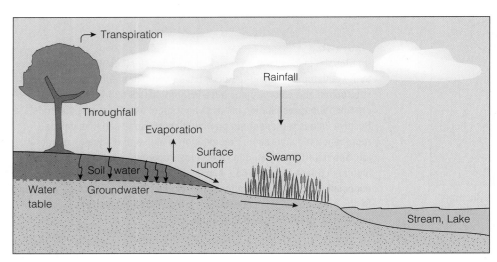

Figure 10.6
Pathways of water near the land surface.

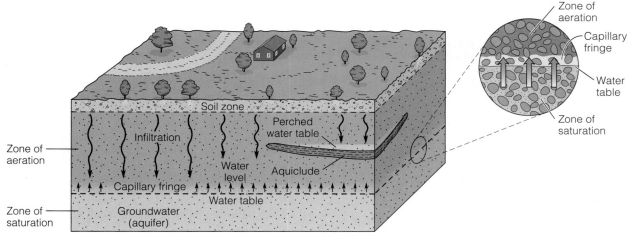

Figure 10.7
Typical distribution of groundwater in the subsurface and the formation of the water table. The downward migration of water is eventually stopped by impenetrable bedrock. Above this barrier, a zone of saturation occurs in which water saturates any fractured bedrock or unconsolidated sediments by filling all the available pore spaces. Above this zone, pores in the zone of aeration are a mixture of air and water. Inset shows the capillary fringe where water rises a few centimeters upward into the zone of aeration because of surface tension.

the tendency of water to wet solid surfaces (Fig. 10.7). The water climbs upward as bonds form between the water and the solid surface of the soil, sediment, or rock.

If you plan to build a house or set up an industrial plant, the supply and quality of water in the rocks and sediments of the ground beneath your property will affect the security of your investment. Groundwater supply is a delicate balance between **recharge** provided by precipitation, and **discharge** of the groundwater at springs, where the water table intersects the land surface. Changes in the balance between recharge and discharge are indicated by changes in the level of the water table. As we shall see, groundwater travels slowly, moving at a rate of meters per year. Given the slow speed at which groundwater moves, lowered water tables caused by periods of extended drought or overuse may take years to return to their former levels. On the other hand, too much recharge in the form of rain may cause flooding. Once the subsurface becomes saturated, it cannot hold any more water so that any subsequent precipitation results in rapid surface runoff (excess discharge), swelling the nearby streams.

10.4.1 Porosity and Permeability

The balance between recharge and discharge is affected by the nature of the subsurface sediments and rocks. Two factors, porosity and permeability, reflect how much water a given subsurface can hold, and how easily groundwater can move through it. As water is absorbed into unconsolidated sediments, it can only occupy the spaces or pores between adjacent mineral grains. So the amount of water an unconsolidated sediment can hold depends on the volume of space between the grains. For example, where sediments are composed of rounded grains of rough-

ly equal size (a condition known as "well-sorted," see Chapter 2), the grains are stacked together like billiard balls and large spaces exist between them (Fig. 10.8a). These sediments are said to have a high **porosity**. On the other hand, sediments with grains of different sizes have severely reduced porosity because small grains may fill the pore spaces between the larger ones (Fig. 10.8b). Fine clays have a very high percentage of minute pores (high porosity), and can therefore hold large volumes of water when saturated. If clays are compacted to form the rock shale, however, the porosity is greatly reduced (Table 10-2).

If porosity is a measure of the relative volume of pore spaces in a sedimentary material, **permeability** is a measure of how well these pore spaces allow a fluid (in this case water) to flow through the material. So permeability is the crucial factor in determining

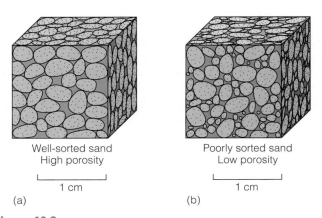

Figure 10.8
Schematic diagram showing sediments with (a) high porosity and (b) low porosity. Well-rounded and well-sorted sand grains have high porosity, whereas poorly sorted sands have low porosity.

Table 10-2

Typical Porosity Values for Different Materials

MATERIAL	PERCENTAGE POROSITY
Unconsolidated sediment	
Soil	55
Gravel	20–20
Sand	25–50
Silt	35–50
Clay	50–70
Rocks	
Sandstone	5–30
Shale	0–10
Solution activity in limestone, dolostone	10–30
Fractured basalt	5–40
Fractured granite	10

(U.S. Geological Survey, Water Supply Paper 2220 (1983) and others.)

the groundwater potential of a region because groundwater flow is essential if it is to form a resource. Certainly many porous materials such as well-sorted sandstone are also permeable, but there are also important exceptions. For good permeability, the pores need to be a sufficient size and they need to be interconnected. Clays, for example, are highly porous, but they are not permeable because their pore sizes are too small. In fact, the pore spaces in clay are so minute that water molecules loosely bond to adjacent clay particles which stops their flow. On the other hand, limestones commonly contain large pores, but can also be impermeable because the pores may not be connected.

There are also examples of permeable rocks with low porosity. Igneous rocks such as basalts or granites generally have very low porosity, because their grains are interlocking. Yet they are often systematically fractured (Fig. 10.9), which greatly increases their permeability if the fractures are interconnected (see Table 10-2).

Most groundwater occurs within 1 km (0.6 mi) of the surface. But in any region the extent of infiltration depends on a wide variety of factors. Typically infiltration is poor in regions where impermeable sediments or bedrock is exposed at the surface, where the topography is steep so that most precipitation becomes surface runoff, and where the vegetation is dense and so impedes the fall of water which may evaporate before it can reach the ground. Regions with high rainfall may therefore have a rather limited supply of groundwater if they possess some or all of these characteristics.

Recharge, on the other hand, is most efficient in regions with gentle slopes and light vegetation that are underlain by unconsolidated, porous and permeable sedimentary material. Water can also infiltrate fractured bedrock. In special circumstances, water can penetrate to considerable depths along deep crustal faults related to modern or ancient seismic activity. In general, however, the increasing pressures at depth tends to close fractures so that the depth of infiltration into the bedrock is usually limited.

10.4.2 Aquifers

A rock body or layer of unconsolidated sediment that can both store and transmit water below the Earth's surface is known as an **aquifer**, from the Latin for "to bear water." An aquifer is therefore both porous and permeable, and is the most productive source of groundwater.

In general, wells must be drilled into an aquifer when there are no surface supplies of fresh water. The wells must penetrate below the water table into the zone of saturation, so that changes in the level of the water table do not cause the well to dry up (Fig. 10.10). Ideally, they should be deep enough to provide water even in times of drought. This is rarely the case, however, and long dry summers can cause havoc with well water supply.

There are two types of aquifers: unconfined and confined. The difference between them is a function of local geology and topography. **Unconfined aquifers** are those whose upper surface coincides with the water table (Fig. 10.7). **Confined aquifers** occur where the water-bearing rocks or sediments are sandwiched between impermeable layers. This situation is common where well-sorted sandstones are interbedded between impermeable clay layers.

Unconfined Aquifers

Unconfined aquifers store vast quantities of water, and are common in relatively flat regions. The reason is that many flat regions have had long histories of meandering stream systems. These stream systems deposit well-sorted sands and gravels that are ideal for both storing and transmitting water (we will discuss the origin of meandering streams later in this chapter). As a meandering stream develops, its channels migrate back and

Figure 10.9

Systematically fractured bedrock of high permeability but low porosity on the south coast of Newfoundland, Canada.

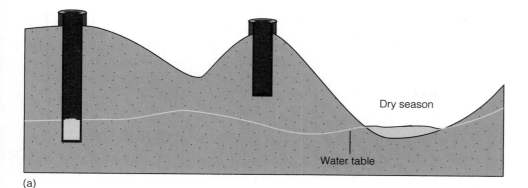

(a)

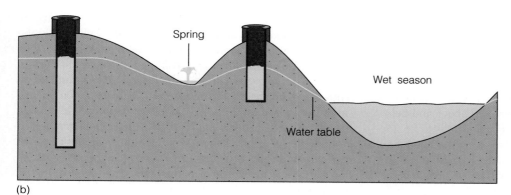

(b)

Figure 10.10

Wells drilled into an aquifer must penetrate below the water table. Note the lower level of the water table during the dry season. The water table feeds wells and springs. Therefore, changes in the level of the water table from the wet to the dry season affect the availability of water. The change shown here, for example, would cause the shallow well to dry up during the dry season.

forth across the floodplain. The sand and gravel deposits therefore encompass the entire floodplain and may be very extensive. One such aquifer, known as the High Plains (or Ogallala) aquifer, extends from southern Wyoming, Nebraska, and South Dakota in the north to New Mexico and Texas in the south (Fig. 10.11). Consisting of sands, gravels, and clays, it provides water for much of the agricultural heartland of the central United States.

The sands and gravels slope gently to the east away from the Rocky Mountains. The aquifer is recharged by precipitation falling in the mountains which then flows eastward beneath the surface, within these permeable sediments. Rainfall in the region itself also makes a significant contribution, typically ranging from 35 to 55 cm/year (14 to 22 in./year). However, in times of drought this contribution is severely reduced.

The depth to the water table, which forms the upper surface of the High Plains aquifer, varies locally from 15 m (50 ft) to 66 m (220 ft). The thickness of the aquifer itself is also highly variable, from about one m (3.3 ft) to 200 m (660 ft). The exploitation of this vast groundwater resource was provoked by the disastrous "Dust Bowl" droughts of the 1930's and late 1940's. Today it is mined by more than 170,000 wells that

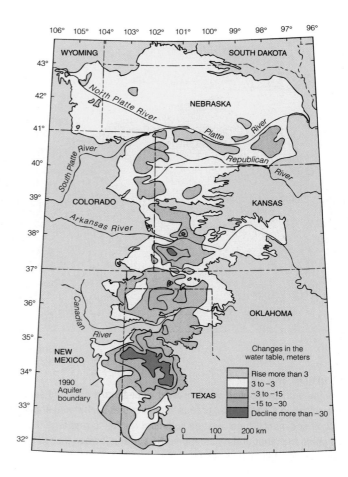

Figure 10.11

High Plains aquifer of the mid-continental United States showing changes in the level of the water table between 1980 and 1990. Although in some regions the water table rose during that time period, in most regions it dropped significantly. The Rocky Mountains are located to the west of the map area.

together extract 30% of all groundwater used for irrigation in the United States. As a consequence, the thickness of the aquifer has changed. These changes are most easily seen as variations in the level of the water table (Fig. 10.11). Between 1980 and 1990, for example, most areas show moderate to severe decline. Only a few show moderate to significant rise. Such studies of the High Plains aquifer and the changing distribution of the water resource demonstrate the importance of monitoring in the long-range management of water resources.

Confined Aquifers

One of the most important sources of groundwater has historically come from **artesian wells** (see Living on Earth: The Historical Importance of Aquifers on page 272), named after the French village of Artois (called "Artesium" by the Romans), in northern France. Water rises naturally up an artesian well and, in some cases, flows freely at the surface (Fig. 10.12). Artesian wells form in confined aquifers when water flows preferentially in a permeable, porous sediment, that is confined above and below by low-permeability rock or sediment. For the wells to be artesian, there must be sufficient rainfall to keep the aquifer recharged. If these conditions are met, wells located at a lower elevation than that of the recharge area will be artesian. Whether they flow freely or have to be pumped, however, depends on the elevation of the wellhead relative to a feature known as the poten-

tiometric surface. The artesian well at Artois has provided free-flowing water for centuries. In some instances, water rises naturally through fractures in the rock to form **artesian springs.** If artesian wells and springs are to be maintained, the water pressure must be carefully monitored, a reduction in well pressure being a sensitive indicator of overproduction.

10.4.3 The Flow of Groundwater

Because groundwater is such an important resource for both domestic and industrial purposes, a knowledge of the distribution of water and the level of the water table in the subsurface is essential for both city and rural communities. But as well diggers, building contractors, and hydrogeologists are well aware, the level of the water table in a region can be highly variable. The Earth's surface is irregular, and rocks and soils beneath the surface have varying capacities to absorb and store groundwater. Thus the depth below the surface at which saturation occurs, varies from place to place. In regions with appreciable rainfall, the water table crudely mirrors the surface topography. So groundwater generally flows "downhill" under the influence of gravity toward valleys. Figure 10.13 shows the regional distribution of groundwater typical of an unconfined aquifer. Eventually, the level of the water table intersects the Earth's surface and becomes surface water that discharges into streams, lakes, and

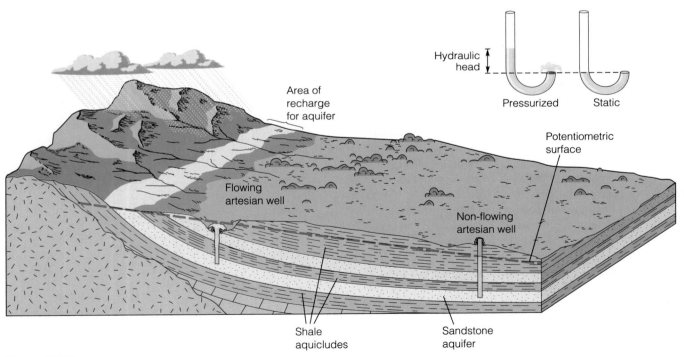

Figure 10.12

Flowing and non-flowing artesian wells drilled into a confined aquifer. Water flows preferentially downslope in a permeable, porous sediment that is confined above and below by low-permeability rock or sediment. There must be sufficient rainfall to recharge the aquifer. Wells located at a lower elevation than that of the potentiometric surface (below dashed line) will flow freely. Wells above the elevation will be non-flowing. Inset shows an analogous situation in which a "hydraulic head" in a hose causes water to flow out at the lower end.

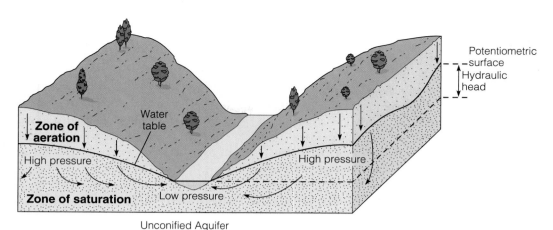

Figure 10.13
Groundwater flow moves downward under the influence of gravity through the zone of aeration into the zone of saturation. Due to its hydraulic head, most water then migrates from high-pressure to low-pressure regions in the zone of saturation and may become surface water and drain into lakes, streams, or swamps, where the water table intersects the land surface.

swamps. Variations in the elevation of the water table create adjacent zones of high and low pressure, the difference in which is known as the **hydraulic head.** It is this difference that results in groundwater flow. This natural flow establishes an important link between groundwater and surface water.

The flow of groundwater varies depending on the characteristics of the rocks, sediment, and soils through which it is passing. Because of frictional resistance, groundwater flow tends to be very slow, usually at rates of just a few meters per year. The highest flow rates measured in North America are less than 300 m (1000 ft) per year and occur only in exceptionally permeable material. Most of this groundwater flow is eventually gravity-fed into adjacent streams where it is transported to the sea. The overall groundwater supply is maintained by recharge.

In the 19th Century, French engineer Henry Darcy reasoned that the groundwater flow rate between two wells was related to the distance between the wells and the difference in the elevation of their water tables. Flow is generally faster in regions where significant variations in water table levels occur over short distances. In any aquifer, water rises in a well up to a level known as the **potentiometric surface.** In unconfined aquifers, this surface coincides with the level of the water table (Fig. 10.13). In a confined aquifer, the surface is gently inclined (dashed lines, Fig. 10.12). A natural pressure, or hydraulic head, exists between the surface and any water outlet at a lower elevation (see Fig. 10.12). It is this difference in elevation that produces the "water pressure" we get when a tap is turned on. This is why the water pressure in the showers of university dormitories is much greater on the ground floor than it is on the top floor. Similarly, within a municipal water system, houses at the bottom of valleys have higher water pressure than those on the tops of hills. Taps or water outlets at higher elevations than the potentiometric surface must have water pumped to them. Many municipal water systems create an artificial hydraulic head by storing the town water in elevated reservoirs or water towers so that water flows freely between the reservoir and the dwellings or commercial buildings of the town.

Where the water table intersects the surface, groundwater seeps out to form **springs.** As a result, springs are especially common in valleys within mountainous regions (Figs. 10.10b and 10.15). Before the properties of the water table were understood, the origin of springs had no obvious explanation and they

became the source of many myths and legends. Today we know that they are evidence of subsurface groundwater.

Springs also occur where water is forced up to the surface along fractures. In this situation, the spring is like a faucet in a municipal water system. The water flows due to the hydraulic head that allows water to ascend toward the surface along fractures provided they lie below the potentiometric surface.

Figure 10.15
A mountain spring in Co. Wicklow, Ireland. (Photo Cindy Murphy)

The London Basin (Fig. 10.14) contains a famous example of a confined aquifer. This aquifer is nearly 100 km (62 mi) across and underlies much of southeastern England, including the city of London. The aquifer is in a layer of chalk, which is both permeable and porous. The chalk is sandwiched between impermeable clay layers. In the Chiltern Hills to the north and west, and the North Downs to the south and east, the upper clay layer has been eroded so as to expose the chalk. Rainfall in these areas sinks into the chalk and migrates to the center of the basin.

The supply of water from this aquifer has been of great historical significance in the development of London as a major world city. The needs of the city, however, have taken their toll. The demand for water has exceeded the aquifer's replenishment, even in a region famous for its rain. When the fountains in Trafalgar Square were first constructed in the mid-19th Century, water tapped from the chalk aquifer gushed freely at the surface. Today, the water level in the chalk has fallen by over 130 m (425 ft). Because the chalk is no longer saturated, the aquifer is no longer artesian and water must now be pumped from deep within it if it is to reach the surface.

Fresh water used to percolate from the chalk into London's Thames River. However, depletion of the aquifer has caused the flow to reverse itself. Today the Thames water seeps into the aquifer and contaminates it.

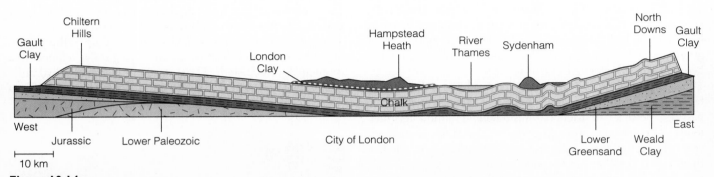

Figure 10.14

A cross-section, about 100 km (62 mi) across, showing the confined chalk aquifer beneath the London Basin. Precipitation in the Chiltern Hills and North Downs is absorbed by the layer of chalk and flows as groundwater beneath the London Basin. The aquifer is confined by impermeable clay layers.

An oasis, then, is merely a spring that occurs in desert regions, marking the local emergence of artesian water at the surface. In the Palm Springs region of southern California, for example, an oasis is located along the edge of the Mojave Desert in a basin bounded on both sides by high mountains. A series of faults, which includes the San Andreas, carve their way through the basin. Precipitation falling on the nearby mountains drains into the subsurface beneath the basin where it forms a precious groundwater resource beneath the desert. The pronounced difference in the elevation of the water table beneath the mountains and its level below the desert lowlands forms a very strong hydraulic head. Because the desert surface is well below the potentiometric surface, the pressure difference forces the water up fractures and faults beneath the desert to the surface. The resulting oasis (Fig. 10.16) provides vital nourishment to the parched vegetation of the Mojave Desert.

Occasionally, the temperature of spring water is anomalously warm, as is the case in parts of the Palm Springs region. Where this occurs, the water is a potential source of geothermal energy (*geo* means earth, *thermal* means heat). In the following section we will learn how groundwater becomes heated and examine several of the spectacular features that occur as a result.

10.4.4 Heated Groundwater: A Source of Geothermal Energy?

At depths below about 20 m (66 ft), water is generally unaffected by weather and climate at the surface, and its temperature is influenced instead by the heat inside the Earth. As miners who work deep underground will verify, temperature increases with depth, a fact that reflects the flow of heat from the interior of the Earth to its surface. On average, temperature increases about 2.5°C (4.5°F) for every 100 m (330 ft) depth, although this rise can be considerably greater in volcanically active regions. As we discussed in Chapter 8, water has a tremendous capacity for absorbing this heat.

Occasionally, it is possible for groundwater to penetrate the crust far below the regional water table. As water seeps down major faults, for example, it becomes warmed by the rocks around it (Fig. 10.17a, b). If this water is then able to rise rapidly to the surface, it may retain some of this heat so that its temperature is significantly higher than that of surface water. As a result, water derived from deep artesian wells is routinely warmer than ambient surface waters. This heat is a source of

(a)   (b)

Figure 10.16

(a) An oasis (center) in the desert (foreground) near Palm Springs. Groundwater migrates underground from the mountainous areas (background) to beneath the valley. It then seeps up along fractures and faults associated with the San Andreas fault zone. The oasis occupies a narrow zone above a fracture. (b) Lush vegetation in the oasis. (Photos Cindy Murphy)

geothermal energy and if the water reaches the surface, a **hot spring** will form. It is not surprising that hot springs are especially abundant in tectonically active areas adjacent to deeply penetrating faults and fractures.

In volcanically active regions, warm rocks adjacent to the magma chambers will heat groundwater, converting it to steam which then escapes to the surface through fissures to form steam vents known as **fumaroles** (Fig. 10.17c). The area most famous for these is the Valley of Ten Thousand Smokes in Alaska.

In some situations, superheated groundwater may persist as liquid rather than turning to steam because of the confining pressure of the rocks. This situation resembles a pressure cooker in which the uppermost crust is the lid. Like the steam that escapes through the valve on the lid of a pressure cooker, the superheated water will endeavor to escape through any flaw it can exploit in the overlying crust. Because most crust is scarred by faults which reflect either current or ancient tectonic activity, such flaws are abundant. Eventually the superheated water finds an avenue and forces its way upward. As it does so, its pressure is reduced, allowing the water to turn into steam. This conversion results in rapid expansion, as the steam occupies more space than the water did. The expansion creates a violent subsurface explosion that expels the water toward the surface where it forms a boiling hot spring of water and steam (Fig. 10.17). After each eruption, the vacated space in the subsurface is filled with more groundwater and the process repeats itself in a cyclic fashion. The result is a **geyser** (from the Icelandic word, *geysir*, for "gusher").

The most famous geyser is Old Faithful in Yellowstone National Park, Wyoming, which produces a natural spectacle that attracts thousands of tourists each year (Fig. 10.18) . This region has the highest heat flow in continental North America

(see Chapter 1), which is testament to its recent and prolonged volcanic history. As we learned in Chapter 5, this history is believed to reflect the presence of a hotspot directly beneath the Yellowstone region. Fortunately, its volcanic eruptions are far less frequent than those of its geysers. On average, the Yellowstone hotspot produces major volcanic eruptions every 700,000 years. The time between eruptions of Old Faithful, on the other hand, varies from 35 to 80 minutes but are typically about 66 minutes apart. The height of the geyser is usually around 45 m (150 ft).

Groundwater heated in this fashion is a potential source of geothermal energy (see Chapter 18). In Iceland, where the mid-Atlantic Ridge stands above sea level, more than 50 wells tap the geothermal energy produced from groundwater heated by local magmatism. These wells provide nearly 50% of the island's energy needs. Geothermal energy stations near Vesuvius, the famous active volcano in Italy, provide energy equivalent to that produced by four coal-fired power plants.

In the southwestern United States, anomalously hot rocks occur within a few kilometers of the surface because of modern tectonic activity, and provide an important regional source of geothermal energy. But it is northern California, with its high regional heat flow, that boasts the largest area of geothermal development in the world. It is estimated that by the early 21st Century, this region will supply 25% of California's energy needs.

10.4.5 The Dissolving Power of Groundwater

As we have learned, moisture evaporating from oceans is water in its purest natural form. However, in the atmosphere, this

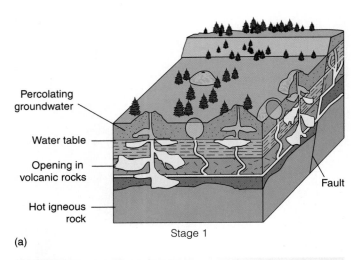

Percolating groundwater

Water table

Opening in volcanic rocks

Hot igneous rock

Fault

(a)   Stage 1

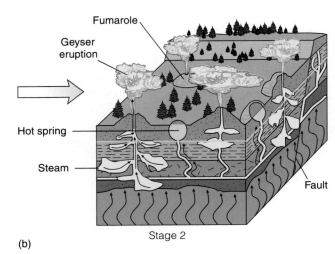

Fumarole

Geyser eruption

Hot spring

Steam

Fault

(b)   Stage 2

(c)

Figure 10.17

(a) Heating and pressurization of groundwater. (a) Water penetrating into the deep subsurface (by way of faults or fractures) is warmed by the rocks beneath, especially in volcanically active areas where hot rocks lie close to the surface. Deeper water is trapped in caverns and is under pressure. (b) When the water is heated to boiling, a sudden drop in pressure causes the water to turn into steam and ascend to the surface to form geysers. (c) Steam may be continuously emitted from the region around a volcanic vent, Hawaiian Islands.

moisture interacts with atmospheric gases and pollutants. As a result, rainwater is mildly acidic and when it infiltrates the subsurface, its property as a solvent enables it to dissolve material from rocks and unconsolidated sediments. This is a form of **chemical weathering** (Chapter 2) whereby rocks and minerals in the subsurface are broken down by their interaction with groundwater. The extent of chemical weathering is influenced by climate and is much greater in humid tropical regions, where there is abundant rainfall, than in polar regions, where permafrost hinders groundwater infiltration.

The chemistry of groundwater is also strongly influenced by the composition of the subsurface materials it encounters and the ease with which they dissolve. Minerals have different susceptibilities to weathering and do not dissolve at the same rate (Table 10-3). Some minerals, like quartz and muscovite have strong bonds and are therefore highly resistant to chemical weathering. Other silicates, like plagioclase, have relatively weak bonds and readily break down into clay minerals

by releasing elements such as sodium and calcium into solution. Carbonate minerals, such as calcite and dolomite, which are typically rich in calcium, magnesium, and iron, are highly soluble and can be almost entirely removed in solution. Groundwater analyses tend to be enriched in these soluble elements. Because most groundwater eventually finds its way to the sea, the process of chemical weathering has contributed to the salinity of the oceans throughout most of geologic time, as we learned in Chapter 8.

Local bedrock also has a profound influence on water chemistry. Much of the central United States and eastern Canada, for example, are underlain by a bedrock of limestone or gypsum, both rich in calcium. Water that contains large quantities of bedrock-derived dissolved ions (such as calcium, magnesium, and iron) is known as **hard water.** The presence of dissolved ions can pose problems. In many hot water domestic appliances and industrial machines, for example, the dissolved calcium will form a scum residue as the water cools. In

Figure 10.18
Old Faithful geyser in Yellowstone National Park, Wyoming.

regions underlain by other highly soluble materials, such as salt, the water is rarely suitable for drinking.

Water's properties as a solvent are greatly enhanced at elevated temperatures. As a result, many hot springs have anomalously high mineral concentrations. In many cultures, bathing in water with an elevated mineral content is thought to be healthy, and regions where such waters occur commonly have health resorts known as "spas."

Caves and Karst Topography

Karsts are features that are produced when groundwater dissolves significant amounts of bedrock. Gypsum, salt, and limestone are especially soluble, and groundwater passing through them dissolves their minerals and carries them off in solution. Because limestone is the most abundant of these rock types, the formation of limestone caves is perhaps the best known illustration of a karst feature (Fig. 10.19). On entering a limestone cave, one is struck by the bewildering labyrinth of caverns and passageways, and by remarkable columns known as stalactites and stalagmites.

But how do these features form? As water infiltrates limestone through its many fractures, it slowly dissolves the rock to make ever wider channels and passageways. Wider passageways increase the amount of water that can infiltrate. Once the groundwater reaches the water table, it flows sideways expanding the openings still further as the underground streams continue to dissolve away the limestone. If the water table then drops, these passageways are left stranded above it and form an intricate system of interconnected caves and caverns. However, drops of groundwater continue to seep into the caves through fractures in the limestone roof, and form small deposits of calcite (from the dissolved calcium) as the water evaporates from the ceilings. With each drop, a thin layer of calcite is deposited over the preceding one. The layering creates a column of calcite projecting downward from the roof known as a **stalactite** (Fig. 10.20a). During this entire process, an opening is maintained in the center of the stalactite. As a result, the stalactite has a hollow tube in its center, like a soda straw. As drops fall onto the floor of the cave, layers of calcite form stubby mounds that grow upward from the ground to form **stalagmites.**

The surface expression of a karst is known as **karst topography.** The dissolving of bedrock beneath the Earth's surface can result in circular surface depressions called **sinkholes** (Fig. 10.20b) that form either directly by the dissolving process itself, or indirectly by collapse of the roof of caves under the weight of the overlying rocks. Because gypsum and salt are much more soluble than limestone, sinkholes in bedrock made up of these minerals can develop in a matter of days. Should streams encounter a sinkhole, they may disappear into it, eventually returning to the surface elsewhere as springs. Karst landscapes primarily occur in humid tropical regions where there is abundant rainfall and lush vegetation. They may also occur in humid temperate

Table 10-3
Resistance to weathering of common rock-forming minerals

Mineral		
Halite		Non-silicates
Gypsum-anhydrite		
Pyrite		
Calcite		
Dolomite		
Volcanic glass		
Olivine	Increasing stability	Silicates
Ca-plagioclase		
Pyroxenes		
Ca-Na plagioclase		
Amphiboles		
Na-plagioclase		
Biotite		
K-feldspar		
Muscovite		
Quartz		
Kaolinite (clay material)		
Hematite		Oxides

Note: Minerals are listed in order of increasing resistance to weathering.

(Exact positions for some minerals can change due to effects of grain size, climate, etc.) Source: Berner and Berner, 1995.

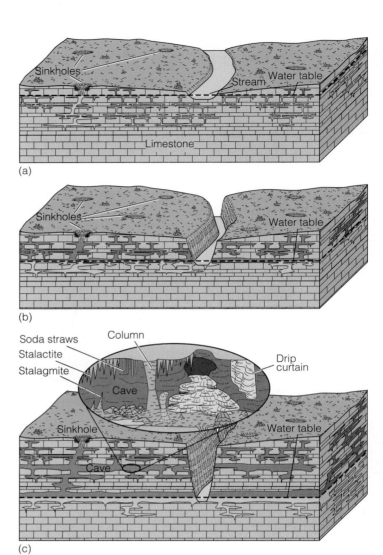

Figure 10.19
Stages in the formation of limestone caves. (a) Groundwater migrating though the zone of aeration dissolves carbonate rock. (b) Groundwater moves along the water table to form interconnected subsurface channels as dissolved rock is transported to streams on the surface. (c) As the stream continues to erode and the water table drops, caves are formed in the abandoned channels.

Figure 10.20
(a) Stalactites project downward from the cave roof and stalagmites grow upward from the cave floor at Lehman Caves in Nevada. (b) A circular sinkhole formed by the dissolving of gypsum bedrock in Levy County, Florida.

climates at mid-latitudes. The abundance of limestone bedrock in the United States, for example, has resulted in the development of karst topography on up to 15% of its surface area.

In summary, groundwater is our largest reservoir of accessible water and is mined below the water table in the zone of saturation. The water table level is influenced by the amount of precipitation and the ability of the subsurface to store water (as dictated by its porosity). Aquifers occur in porous and permeable bedrock, and may be confined or unconfined.

Aquifers are recharged by rainfall which may occur in surrounding mountainous regions. The groundwater then flows downhill under the influence of gravity, and although its motion is very slow, it can travel thousands of miles from its source.

Groundwater may seep out to become surface water where the water table intersects the land surface to form springs. Springs also occur where groundwater is forced upward to the surface along fractures or faults. Where this occurs in the desert, an oasis forms. Water forced up in this way is generally anomalously warm and may be the source of geothermal energy. This trait is accentuated in volcanically active regions. Groundwater can dissolve soluble rocks such as limestone, salt, and gypsum to form features such as caves and sinkholes.

It is important to realize that groundwater and surface waters are linked. Because changes in sea level are insignificant on an annual basis, the hydrologic cycle must be balanced. Therefore the amount of precipitation penetrating the subsurface to become groundwater must be balanced by the amount leaving the subsurface to become surface water. However, because groundwater moves very slowly in the subsurface, the groundwater forming today may take hundreds or even thousands of years to become surface water. Similarly, groundwater gushing from springs today may represent precipitation that occurred thousands of years ago.

Surface water may also become groundwater. During flooding, for example, water may penetrate the subsurface beneath the floodplain of a stream to become groundwater. In the next section, we will learn about the origin and behavior of surface waters, their drainage basins, and the factors that influence continental drainage.

10.5 Surface Water

Each year, about 40,000 cubic km (9000 cubic mi) of fresh water drains back to the sea. This drainage is not uniform, but instead is concentrated in narrow channels or streams, and in lakes, swamps, and springs (Fig. 10.6). The term "river" is used for streams with considerable volume and well-defined channels.

Some precipitation never becomes groundwater; instead it simply drains downhill as surface runoff. Runoff occurs for many reasons but is particularly common in areas of impermeable bedrock or sediment, after a major storm when the subsurface is already saturated, and at the end of a long drought if the surface becomes a hard, dry crust.

Eventually the force of gravity guides both the groundwater and the runoff toward stream valleys, and the majority of running water drains back to the sea by stream flow. This drainage has had an enormous impact on human development, and it is no coincidence that the world's most populated inland cities have developed along stream valleys where fresh water supports industrial, agricultural, and household activities. In the next section we will see that streams do not occur in isolation; instead they form vast, complicated, interconnected networks we call drainage basins.

10.5.1 Drainage Basins

As we travel from the headwaters to the mouth of a stream, the stream is fed by a complex set of side streams called **tributaries.** For example, the Mississippi River is fed by countless streams, the largest of which are the Arkansas, Ohio, and Missouri rivers (Fig. 10.21). The main stream is fed by all its tributaries and all of their tributaries. Each time a tributary enters the main stream, the amount of water in the main stream channel increases. Thus the headwaters of the Mississippi River contain only a tiny fraction of the water that issues from its mouth in the Gulf of Mexico. The total area that contributes water to a stream system is known as its **drainage basin.** The Mississippi River drainage basin covers 3,222,000 square km (1,244,000 square mi) and is the largest in North America.

Drainage basins may be grouped together into **drainage outlets** depending on which body of water the streams ultimately discharge into (Fig. 10.22). In North America, the mountain belts on the east and west coasts are the dominant influence on drainage patterns. A number of drainage basins originate along the eastern flank of the Appalachian Mountains and discharge into the Atlantic Ocean. Similarly, drainage basins on the western side of the Cordillera discharge into the Pacific Ocean. In the continental interior, drainage is guided between these two mountain ranges. For example, the headwaters of the Missouri and Arkansas rivers drain from the eastern slopes of the Rocky Mountains toward the Gulf of Mexico. The headwaters of the Ohio River, to the east, drain toward the Gulf of Mexico from the western slopes of the Appalachians. To the north, in Canada, major drainage outlets occur in Hudson Bay

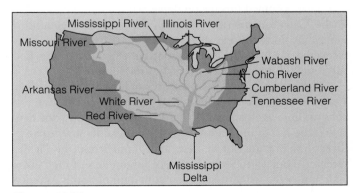

Figure 10.21

The Mississippi River drainage basin includes the total area from which precipitation reaches the Mississippi River. The basin is 3.2 million square km (1.25 million square mi) in area.

Figure 10.22
Drainage basins are separated by areas of higher topography. For example, the "Continental Divide" of the Rocky Mountains separates the Mississippi River drainage basin from streams that flow westward to the Pacific Ocean.

and the Arctic Ocean. This drainage pattern is actually dictated by the last Ice Age. The weight of the enormous ice sheets that once occupied this region depressed the continental crust in the Canadian interior. Although the crust has rebounded somewhat since the ice sheets retreated, it has still not fully recovered. As a result, many important streams like the Red River and Churchill River, drain toward Hudson Bay.

The history of the Amazon River of South America offers a most impressive demonstration of the influence of mountain belts on drainage patterns (Fig. 10.23). 100 million years ago, before the rise of the Andes Mountains, the ancestral Amazon River flowed from east to west, and drained into the Pacific Ocean. About 15 million years ago, subduction of the Nazca plate under the western continental margin of South America pushed up the Andes, and the resulting topographic barrier blocked this drainage and created a large inland lake. Eventually, streams draining into the Atlantic Ocean eroded back until they tapped the water of this lake, and the modern drainage basin of the Amazon River was born.

10.5.2 Stream Gradient and Base Level

Running surface water flows many orders of magnitude faster than groundwater, varying from one km (0.6 mi) per hour to as much as 35 km (22 mi) per hour. Remember that groundwater typically moves at rates of only a few meters per year.

In order for streams to flow, they must have a gradient, that is, the elevation of the stream valley must decrease from the headwaters to the sea. One of the most important variables influencing a stream's flow is its **average gradient,** that is, the vertical drop between the headwaters of the stream and the sea, divided by the distance the stream travels. For example, if a stream's headwaters occur at an elevation of 1000 m (3300 ft) above sea level, and it flows 500 km (380 mi) to the sea, the average gradient is 2 m per km or 10 ft per mile (Fig. 10.24a). However, the gradient of a major stream is not uniform. Instead, it changes along the length of the stream, slackening

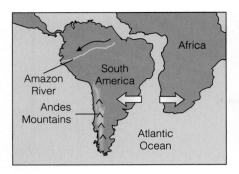 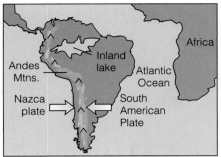 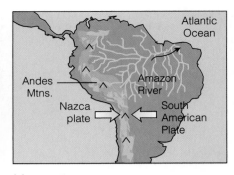

(a) 100 m.y.a. (b) 15 m.y.a. (c) present

Figure 10.23

Influence of mountain belts on drainage patterns. The ancestral Amazon River flowed westward into the Pacific before the rise of the Andes Mountains, about 15 million years ago. When this occurred, an inland lake formed which was eventually tapped by headward-eroding streams that drained toward the Atlantic.

from its elevated mountain source to the relatively gentle plains of the coastal region. In the headwater region, the gradient can be as much as 10–20 m/km (50–100 ft/mi), whereas the gradient in coastal regions may only be 2–3 cm/km (1–2 in./mi).

Because of this changing gradient, streams are relatively straight at their headwaters, but tend to snake or **meander** as they approach the coastal regions (Fig. 10.24b and 10.25). The Missouri and Amazon rivers display excellent examples of this pattern.

Because a gradient must be maintained for water to flow, a stream can never erode below the body of water into which it discharges. The elevation of this water body defines what is known as the **base level** of the stream. Because most streams flow into the sea, sea level is the **ultimate base level**. However, **local base levels** may occur along the path of a stream. For example, where lakes occur in a valley, they form a barrier to the upstream segment of the stream (Fig. 10.26a, c). Because this segment cannot erode below the level of the lake, the lake serves as a local base level. The nature of the underlying bedrock may also affect the local base level. Streams preferentially erode weaker rocks, such that a resistant rock may also form a local base level for the upstream segment of the stream (Fig. 10.26b, d). In certain circumstances, this can lead to the generation of **rapids** or **waterfalls** (Fig. 10.27).

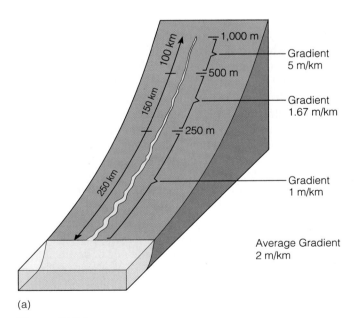

(a)

(b)

Figure 10.24

(a) Average gradient of a stream is the vertical drop between the headwaters and the sea, divided by the distance the stream travels. (b) Drainage pattern from a Hawaiian volcano. Note the relatively straight path near its source (left), and the tendency to meander near the coast. (Photo John Buckland-Nicks)

(a)   (b)

Figure 10.25
The meandering streams of the Mississippi River. (a) Near New Orleans. (b) South of Baton Rouge.
The wide floodplain is especially evident in (a). (Photos Joe Martinez)

10.5.3 Stages in Stream Development

Because continents are being continuously worn down by erosion, the average gradient of a stream will decrease with time. Local regions of steeper gradient (like those of rapids and waterfalls) are eventually eroded and the stream develops a smooth gradient (see Figure 10.24). In this scenario, there is little erosion or deposition. In reality, however, a stream's environment is a dynamic one, with the stream constantly responding to local changes along its length. Suppose, for example, that a collapse of a stream bank or a landslide dumps a body of sediment into the stream channel (Fig. 10.28). The front of the sediment body locally has a steep gradient which is preferentially removed by erosion. As a result, the sediment load is transported downstream and the smooth gradient is restored.

A stream is also affected by changes in base level, and especially changes in sea level. As we discussed in Chapter 8, these changes may occur over thousands or millions of years and are related to a variety of processes including the waxing and waning of ice sheets and tectonic uplift or subsidence. Because of this, the ultimate base level of a drainage system may rise or fall, and the average gradient of a stream may change drastically. If sea level rises, for example, stream valleys are invaded by the sea, leading to the formation of drowned valleys. If sea level falls, the energy of streams increases, resulting in a new cycle of erosion, transport, and deposition as the streams endeavor to attain a new smooth profile.

10.5.4 Moving Water: An Agent of Weathering and Erosion

The role of moving water in chiseling our landscape has been known for centuries. In Homer's Iliad, Scamander the river god cries out to "fill your channels with water from the springs, replenish your mountain streams, raise a great surge, and send it down, seething with logs and boulders." Leonardo da Vinci (1452–1519) noted that the banks of the River Arno contain "a mass of stones from various localities." Scamander and Da Vinci were actually describing the essential elements of erosion, transport, and deposition that accompany stream flow.

Our landscape is sculpted through the complimentary processes of **weathering** and **erosion.** As we learned in our examination of the Rock Cycle (Chapter 2), rocks undergo mechanical and chemical weathering when they are exposed to air and moisture (Section 10.4.5). Water also infiltrates the subsurface, where it dissolves the unconsolidated sediments and rocks it encounters, attacking both the edges of the mineral grains and the weak planes such as cleavage within them. Erosion results in the removal of rock and sediment, and the creation of a stream channel. The rock fragments or **clasts** produced by erosion ultimately are carried in streams.

Streams may contain both a dissolved component (which depends on the extent of chemical weathering), and a solid **sediment load** plucked from the rocks and sediments over which the stream flows. Depending on the size of the clast, the sediment load may be carried in suspension by the stream, or moved by traction in a hopping motion along the stream bed. Deposition of the load occurs in slow-moving portions of the stream or at the stream mouth, where the rate of flow is slowed on entering standing water. Up to 75% of a stream's sediment load is deposited along its course. The remainder settles where the stream enters the sea. About 9 to 10 billion tons of sediment are transported to the oceans each year. The Mississippi River alone transports 500 million tons of sediment. The influence of mountain belts on sediment supply is demonstrated by the Ganges River, which erodes the mighty Himalayas and contributes the world's greatest annual sediment load of 1500

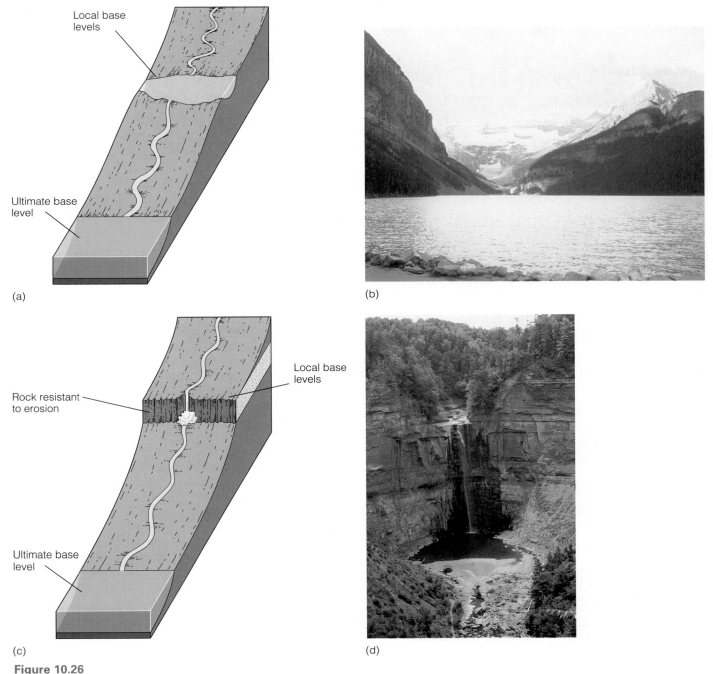

(a)

(b)

(c)

(d)

Figure 10.26

Examples of local base levels below which the upstream segment of a stream cannot erode.
(a) Where the stream flows into a lake, the lake forms a local base level. (b) The local base level at
Lake Louise, Canadian Cordillera. The stream running through the valley (background) cannot
erode below the base level represented by the level of water in the lake. (Photo Joe Martinez) (c)
Where the stream flows across resistant rock layers, the local base level is at the top of the
waterfall. (d) The local base level at Taughanock Falls, New York, where a resistant sandstone at
the top of the falls overlies a weaker layer of shale. (Photo Brendan Murphy)

million tons. This load is nearly 15% of the total amount of
sediment transported to the world's oceans.

The erosion, transport, and deposition of sediment by
streams is controlled by the size of the sediment particle and
the speed of the stream (Fig. 10.29). As you might expect, large

gravel-sized clasts can only be moved by fast-flowing streams,
whereas fine-grained clays are readily transported even in the
slowest of streams. The varying mobility of clasts in streams
leads to **sorting**, where particles deposited at a particular locality tend to be similar in size. Sorting is a fundamental charac-

Figure 10.27
Niagara Falls on the border between New York State and Ontario, Canada. Above the falls, a local base level is formed by the resistant rock.

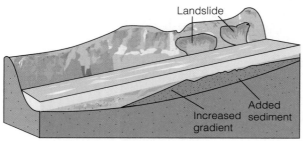

(a) New load of sediment added

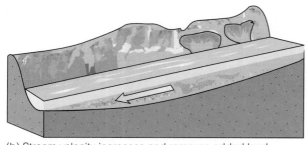

(b) Stream velocity increases and removes added load

Figure 10.28
When sediment is dumped in a stream, the gradient of the stream increases, causing the water to flow faster. This results in removal of the sediment such that the average gradient of the stream is restored.

teristic of stream deposition and, as we learned earlier, may lead to the development of aquifers.

10.5.5 Stream Flow and Discharge

No two streams are identical, but they do have common traits. Because streams derive their energy from gravity, many tend to

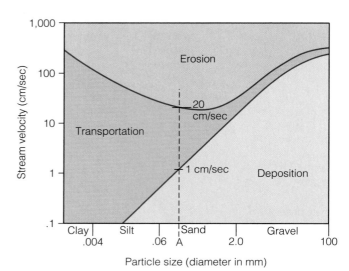

Figure 10.29
The relationship between the velocity of the stream and particle size controls erosion, transport, and deposition of sediment. To pick up and remove a particle of size A, for example, the stream velocity must be in excess of 20 cm/sec (8 in./sec). This particle will be transported as long as the stream velocity is greater than 1 cm/sec. When the stream flow falls below 1 cm/sec, the particle is deposited.

form in mountainous regions where the gradient is steepest. This energy means that the stream has the capacity to do work, that is, it can erode, transport, and deposit material.

A stream forms when water carves a clearly defined route, or **channel**, by removing the rocks and soil in its path. The scouring action at the bottom of the channel and the undermining of the banks leads to the development of a stream valley with a V-shaped profile (Fig. 10.30). The valley then acts as a guide to the flow of water.

The stream velocity is greatest where frictional resistance is least. Contrary to our visual impression of fast-moving mountain streams and lazy, meandering rivers, the average velocity of streams actually *increases* from its headwaters to its mouth. In the headwaters, nearly 95% of a stream's energy is dissipated by friction because the valleys are narrow and the stream bed is highly irregular. The *average* velocity is therefore quite low (less than half of that at its mouth), although at any one time *some* of the water may be moving very rapidly. Further downstream where the channel widens and less energy is lost to friction, the average velocity is higher. In straight segments of the stream, the velocity of the water is highest in the center of the channel, where the frictional resistance is least, and decreases toward the banks (Fig. 10.31a). Where the stream curves, frictional resistance is highest on the inner part of the bend and the region of maximum velocity shifts toward the outer part (Fig. 10.31b).

Downstream, more water enters the channel from tributaries, and the stream channel tends to widen and deepen as a result. The gradient also tends to slacken downstream (see Figure 10.24), which results in the development of loops or **meanders** in the stream channel (see Figure 10.25b) When a meander forms, the depth and velocity of the stream vary

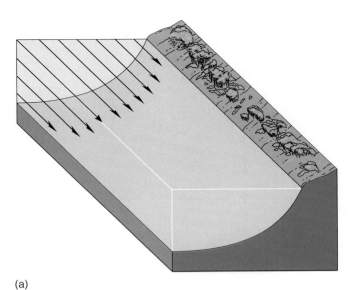

(a)

Figure 10.30
Erosion by streams cuts linear valleys into the landscape that are V-shaped in profile.

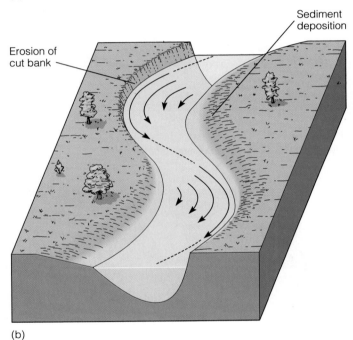

Sediment deposition

Erosion of cut bank

(b)

Figure 10.31
Stream velocity in (a) straight segments and (b) meanders. Length of arrows is proportional to stream velocity.

markedly from one side of the channel to the other. As the stream heads into the bend, it erodes the outer part of the meander creating a deep channel. In contrast, flow along the inner side of the meander is relatively slow, which keeps the stream shallow (Fig. 10.31b).

This difference in the velocity of a meandering stream from one bank to the other means that the position of a channel is constantly changing as erosion on one side is approximately balanced by deposition on the other. Stream deposits consequently migrate laterally and can become very extensive. Over time, stream migration leads to a widening of the valley into one with a flat floor known as a **floodplain** (Fig. 10.32). As we discussed earlier, this process of lateral migration is very important in the formation of the unconfined aquifers.

The sinuous configuration of meanders is such that only thin necks of land separate adjacent loops. With continued erosion of the outer banks, the intervening neck of land may be eroded away, giving the stream a more direct route to the sea (Fig. 10.33). The abandoned meander, now isolated from the main channel, forms a crescent-shaped lake called an **oxbow**.

10.5.6 Floods

Flooding occurs when a stream's channel can no longer contain the water flowing through it. This can happen for a variety of reasons. Most floods commence with a period of prolonged or heavy rainfall. In most regions, the ground can absorb a portion of this rainfall. Should heavy rain continue, however, the ground becomes saturated and surface runoff is fed directly into the channel. This results in a temporary

imbalance in the water budget of the stream. Water input along the length of the stream and its tributaries exceeds the output at its mouth. Eventually the water level rises, flows over its banks and across the flat floodplain that surrounds a stream (Fig. 10.34a,b).

When this happens, the stream immediately slows down, and the sediment held in suspension is deposited on the floodplain. The finer-grained material tends to spread across the floodplain, whereas the coarser-grained sediment is deposited adjacent to the riverbank. With each successive flood, the coarser material builds up to form a protective barrier to the flood (Fig. 10.35), known as a **natural levee** (after the French

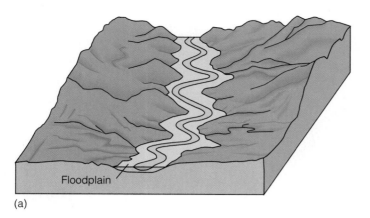

(a)

(b)

Figure 10.32
Idealized stages in the development of a floodplain.

lever, which means "to raise"). The natural levee is generally the highest point on the floodplain.

Many flood-prone rivers have artificial levees constructed so that the volume of water a stream can hold is increased. Some of this water is diverted into spillways which are cut through the levees. However, to be effective, levees must be built along the entire length of the stream. If they are only built along the lower part of the river, the floods upstream will invade the floodplain behind them. If they are only built along the upper part of the river, the concentrated fury of the water will be unleashed onto the floodplain downstream.

In the summer of 1993, 60 cm (24 in) of rain fell in a four-day period in southern Minnesota which brought about North America's worst flood since 1878, when records were first kept. Most of the flooding was concentrated along an 800 km (500 mi) stretch of the upper Mississippi Valley, from St. Louis, Missouri, to St. Paul, Minnesota. Here the stream channels are relatively shallow and narrow, and so could not hold the enormous volume of water. The flood claimed dozens of lives, displaced more than 50,000 people, and caused more than $10 billion in damage.

Ice jams are also a common cause of flooding. When frozen streams thaw in the spring, ice jams may form a natural dam which blocks the normal course of water flow and causes flooding up to the height of the dam in the valley upstream. The 1996 flood of the Red River which affected North Dakota and Manitoba, Canada, testifies to the importance of accurate

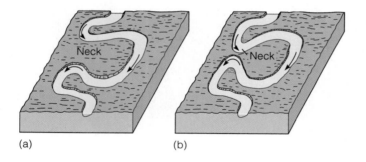

(a)   (b)

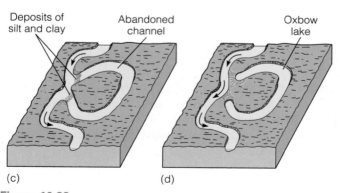

(c)   (d)

(e)

Figure 10.33
Formation of oxbow lakes. (a) Erosion occurs on the outside banks of a meandering stream. (b) This results in the formation of a narrow neck. (c) Eventually the stream cuts off the meander in order to form a straighter channel. (d) The abandoned channel becomes an oxbow lake. (e) Aerial photo of an oxbow lake, Tallahatchie River, Mississippi. The old scars (right) represent the migration of meanders.

(a)

(b)

Figure 10.34
Floods occur when there is too much water for the stream channel to carry. As a result, streams overflow their banks, wreaking havoc on communities built on the floodplain.

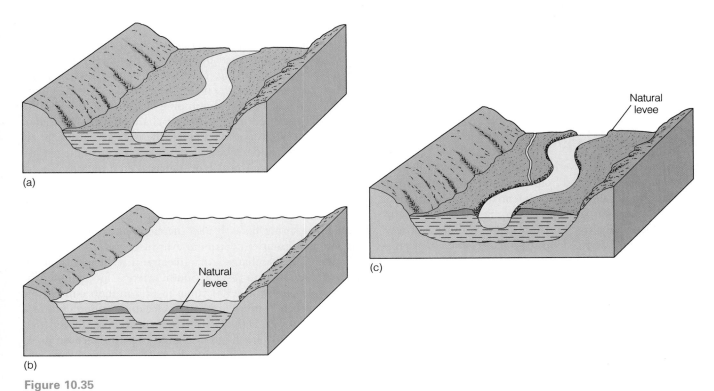
(a)

(b)

(c)

Figure 10.35
Formation of natural levees along stream banks. (a) Stream under normal conditions. (b) Stream begins to flood, and, as water spreads across the floodplain, it deposits coarser material near the banks. (c) When the flood recedes, a natural levee is exposed.

flood forecasting. In parts of North Dakota, dikes were built to withstand a forecasted flood of 15 m (50 ft). Unfortunately, the river crested at 16.5 m (54 ft) so that the flood water spilled over the dikes. In Manitoba, up to 1400 m³ (51,000 cubic ft) of water per second was diverted into artificial channels around the city of Winnipeg, so that the damage was not as severe as it was in North Dakota. These channels offered the river a straighter route and, by increasing the stream gradient, the water moved more rapidly away from the zone of potential flooding. In the case of Winnipeg, the channelways worked; however, it is a dangerous strategy. The faster-flowing river has greater power to erode. Thus the channelways may offer Winnipeg only temporary protection from the ravages of the Red River.

Flood-control dams are usually built upstream from densely populated areas. They are generally made of earth or concrete, and serve to block and divert water from its natural path. However, when they fail, they too can result in catastrophic flooding. One of the most infamous floods in North American history occurred in Johnstown, Pennsylvania, in 1889 when an earthen dam gave way on the Conemaugh River, and 2200 inhabitants of the town perished.

Many scientists believe that floods are an inevitable consequence of surface runoff, and that it is only a matter of time until levees are breached. In their view, our strategies may help contain ordinary floods, but not *extraordinary* ones. A better strategy might be to bow to the forces of nature by trying to control human use of flood-prone lands. This view was strengthened by the failure of levees during the massive flooding along the Missouri and Mississippi rivers in 1993.

In the aftermath of these floods there were two diverging points of view among the affected communities. Some suggested that the river should be allowed to reclaim its natural course and that development in flood-prone regions should be restricted. Others suggested that higher flood walls should be built to give them better "protection."

10.5.7 Coastal Regions

Streams carry both a dissolved and a solid load toward the sea. The dissolved load is derived from the chemical weathering of bedrock and from water-sediment interaction during sediment transport. Both loads are discharged into the sea at the mouths of streams. Depending on the nature of the bedrock and sediment in the stream valley, each stream has its own characteristic chemistry and so pours a distinctive chemical cocktail into the sea. As we learned in Chapter 8, the abundance of elements such as sodium, calcium, and magnesium in ocean waters is related to their abundance in streams and to the efficiency of continental erosion. However, most of the world's ocean approximates its average composition because the chemistry introduced by steams is very efficiently dispersed and mixed by ocean currents.

As a stream meets the sea, its velocity is dramatically reduced. As a result, the stream dumps its sediment load because

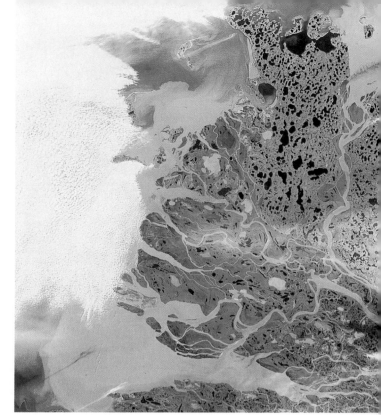

Figure 10.36
A delta forms where a stream empties into the ocean. The stream's flow stops and its sediment load is dumped and forms a triangular shaped sedimentary deposit (see also Figure 8.14).

the size of the sedimentary clasts that it can carry is related to its velocity. A fan-shaped deposit may therefore form at the mouth of the stream (Fig. 10.36). This deposit is known as a **delta** because its shape resembles the symbol (Δ) for the Greek letter delta. The continual dumping of sediment impedes the stream flow and restricts it to narrow channels within the delta. As the delta builds out into the sea, the pattern of its channels commonly resembles a bird's foot. In fact, many of the world's major streams have "bird's-foot" deltas at their mouths.

The **discharge** of a stream is the volume of water flowing past a reference point in a given unit of time. The discharge at any one point is not constant and can change after a heavy rainfall or spring thaw. It also increases progressively downstream as tributaries contribute ever more water to the main channel. Thus discharge is at a maximum at the mouth of the stream. The Amazon River in South America, for example, discharges a colossal 212,400 cubic m (7.5 million cubic ft) of water per second into the Atlantic Ocean, a volume that represents about 20% of the Earth's total stream flow (Table 10-4). The Amazon River's huge discharge is partly because of its large drainage basin, and partly because of the abundant supply of water from the wet continental interior. The Mississippi River, by comparison, discharges only a fraction of this amount (17,300 cubic m per second) into the Gulf of Mexico.

Table 10-4
World's Largest Rivers Ranked by Discharge

			DRAINAGE AREA		AVERAGE DISCHARGE	
RANK	RIVER	COUNTRY	SQUARE KILOMETERS	SQUARE MILES	CUBIC METERS PER SECOND	CUBIC FEET PER SECOND
1	Amazon	Brazil	5,778,000	2,231,000	212,400	7,500,000
2	Congo	Zaire	4,014,500	1,550,000	39,650	1,400,000
3	Yangtze	China	1,942,500	750,000	21,800	770,000
4	Brahmaputra	Bangladesh	935,000	361,000	19,800	700,000
5	Ganges	India	1,059,300	409,000	18,700	660,000
6	Yenisei	Russia	2,590,000	1,000,000	17,400	614,000
7	Mississippi	United States	3,222,000	1,244,000	17,300	611,000
8	Orinoco	Venezuela	880,600	340,000	17,000	600,000
9	Lena	Russia	2,424,000	936,000	15,500	547,000
10	Parana	Argentina	2,305,000	890,000	14,900	526,000

10.6 Human Exploitation of Fresh Water

All life requires water to survive and most animals attempt to manipulate the environment to suit their own purposes. However, none interfere with the natural hydrologic cycle to the extent that humans do. Modern society throughout most of the developed world tacitly assumes the limitless supply of water. But, because we consume water faster than nature can replenish it, water is actually a non-renewable resource. Every type of interference, small or large, has implications for the sustainability of water as a resource.

In order to make responsible decisions on water management, we need to understand the ramifications of our actions. In this section, we show how we locate and mine groundwater, the consequences of its overuse, and the problems of contamination. We then look at our use of surface waters, and how the natural flow of streams is re-routed for our benefit. The most obvious examples of this manipulation is the damming of surface waters to form reservoirs and the diverting of streams for irrigation purposes.

It is important to realize at the outset that contamination of fresh water isn't always due to human activity, but can also be related to natural phenomena. For example, groundwater and surface waters in the vicinity of ore deposits often contain high concentrations of the metals in the deposit as a result of water-ore interactions. In fact, mineral exploration companies routinely analyze water in order to detect elevated concentrations of metals which might indicate proximity to metal deposits. Other toxic elements such as arsenic may not be concentrated enough to form an ore body but may nevertheless be abundant in the vicinity of ore deposits, such as gold.

These toxic elements may find their way into the local water supply. In fact, there are documented examples of low-level arsenic poisoning in populations living near active or abandoned gold mines.

10.6.1 Exploration for Groundwater: Imaging the Subsurface

Before groundwater can be exploited, it first needs to be located. Some of you may be familiar with the practice of dowsing, in which a "water witch" walks back and forth across an area holding a Y-shaped stick until the bottom part of the Y jerks downward. According to the water witch, underground water may be found directly below this location. Although the technique has been heavily criticized by scientists for more than 200 years, there are more than 25,000 practicing water witches in North America today. Scientists cannot find a valid reason for the behavior of the Y-shaped stick, and statistics show that the success rate of the method is no better than chance.

It is true that water witches often locate water. However, in regions of adequate rainfall it would be hard to find an area that did not have some water below the surface. This implies that drilling would find at least some water almost anywhere. It is also probable that an experienced water witch knows that wells dug on hill tops are less likely to be successful than those dug in valleys, and that the vegetation may change where water lies near the surface.

Most scientific exploration focuses on finding the level of the water table. The first step is to examine topographic maps which show the distribution of streams, lakes, hills, and valleys. The location of streams and lakes determines where the water table intersects the surface, so their positions provide important clues for solving the puzzle of groundwater distribution.

Aquifers are not confined to regions of high precipitation. Because underground water can travel hundreds of kilometers from its source, water supplies can often be found beneath the surface of very dry regions. In addition, global-scale factors also come into play, sometimes providing water potential in the most surprising regions. During the last Ice Age (approximately 10,000 to 70,000 years ago) many modern deserts had more hospitable climates with significantly higher rainfall. Beneath the Sahara Desert, for example, beds of sandstone (known as the Nubian sandstone) were recharged by rain in the highlands of Libya and are now estimated to contain more than 18,000 cubic km (4300 cubic mi) of water. The average flow rate is estimated to be 50 m (165 ft) per year. The water currently being mined from this aquifer in Egypt is 1500 km (900 mi) from the recharge area, and so has taken about 30,000 years to reach there. This water actually fell as precipitation during the last Ice Age.

The Nubian sandstone underlies most of the Sahara Desert at a depth of about 100 m (330 ft). Locally, the sandstone is folded upward into anticlines (see Chapter 6) and reaches the surface (Fig. 10.37). Where this happens, the water seeps out at the surface to form an **oasis,** a local spring of artesian water in the desert. In some instances, the sandstone layer is offset by a fault. Like a leak in a plumbing system, water escapes from the sandstone layer and rises up the fault to form an oasis at the surface.

The Nubian aquifer is a major source of water in western Egypt and Libya. But as large as this resource is, it is no longer being renewed by modern rainfall. Its water is therefore a non-renewable resource, and care must be taken to ensure that it is both well managed and used efficiently.

Figure 10.37

A cross-section beneath the Sahara Desert from the mountains of Libya in the west to the Nile River in the east. The Nubian aquifer is within a sandstone that extends beneath the Sahara Desert. Rain falling in the Libyan mountains seeps into the ground and flows within the sandstone. The oases occur either in the crests of anticlinal folds or where faults which cut the sandstone extend to the surface.

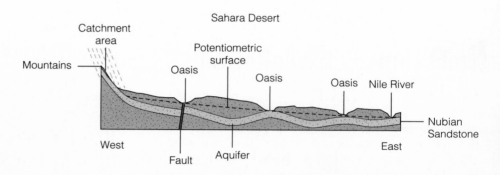

But how do we use these clues to find water in the subsurface between their locations?

Hydrogeologists are concerned with the quantity and quality of water resources, and the depth at which water may be found. They are also aware that groundwater can move hundreds of kilometers from its source and can even be found beneath deserts (see Living on Earth: Water in the Sahara Desert).

As a first step, they determine the sedimentary sequence or stratigraphy of the region (see Chapter 3), so that permeable and porous sediments can be identified. It is also important to identify impermeable sediments such as clay so that the nature of the potential aquifer (confined or unconfined) may be evaluated. The region's stratigraphy may be exposed in road cuts, on hillsides, and along the banks of streams where erosion has cut down into older sediment layers. By putting together this information, the stratigraphy of the region may be compiled using the same methods of stratigraphic correlation discussed in Chapter 3.

In many regions, a series of exploratory boreholes are drilled in order to check the stratigraphy and accurately determine the level of the water table. Sensitive instruments are then lowered down these boreholes in order to measure critical physical properties (such as the electrical characteristics of rocks and water), and identify the telltale signs of water saturation.

As we cannot see the subsurface directly, the most commonly used techniques are those that try to create a picture, or an image, of the subsurface. Much like an X-ray image gives information on bone structure and body tissue, an electrical image of the subsurface can readily identify potential aquifers. Pure fresh water is an extremely poor conductor of electricity and is therefore highly resistant to the passage of an electrical current. This property is analogous to that of bone, which blocks the passage of X-rays and so enables its detection on an X-ray image. The characteristically resistive property of fresh water makes its presence obvious in electrical images. These images are also sensitive to the differing electrical properties of sands and clay, and so can detect variations in porosity and permeability. They can also provide information on pore-water composition so that water quality may be determined. For example, poor quality water that may contain a high concentration of salts is much more conductive (less resistive) than fresh water.

Borehole exploration, although useful, provides only limited information since it is dependent on the number of wells drilled, and the confidence with which the hydrogeologist can determine the level of the water table between well sites. Another method, known as **magnetotellurics** has had better success in pinpointing both the depth and extent of aquifers by providing a more complete electrical image of the subsurface. The magnetotelluric method utilizes natural electrical currents in the outer atmosphere (recall our discussions on the Earth's magnetism in Chapter 7) which penetrate the Earth's surface and interact with Earth materials. The currents induce electrical fields in Earth materials which conduct electricity to varying degrees, depending on the nature of the material. Sensitive equipment can measure this interaction (Fig. 10.38a) and use it to produce a regional electrical image that provides a picture of the subsurface.

Magnetotellurics has been used with considerable success in California, which because of climate and population pressure, has a particularly acute water problem. Figure 10.38b shows an example from Merced County, California, where the subsurface consists of sandstones and impermeable clay layers.

The subsurface image produced by the magnetotelluric survey identifies potential water-bearing zones as regions of relatively low electrical conductivity (or high resistivity) (on the left in Fig. 10.38b). The image shows that the water-rich sandstone occurs in a lozenge-shaped lense. Because of this shape, hydrogeologists interpret the sandstone to have been deposited in former stream channels. This particular image aided the selection of drill targets that later confirmed the predictions of the image. A test well drilled at station 75 obtained excellent fresh water from the high resistivity regions at depths between 100 m and 150 m (330–500 ft).

Given the incomplete nature of borehole surveys, it is unlikely that they could have detected these water-rich lenses.

Many hydrogeological surveys now use magnetotellurics to constrain favorable target zones, and then follow the survey with borehole drilling directed at those targets.

In summary, to locate groundwater resources, interpretation of surface maps and sediment stratigraphy must be combined with subsurface information derived from boreholes. Electrical images of the subsurface are especially useful in identifying zones of relatively fresh water, and in estimating its quality and quantity. These methods take advantage of the highly resistive electrical properties of relatively pure groundwater.

10.6.2 Exploitation of Groundwater

When exploration locates a usable aquifer, a series of wells are sunk below the water table and the water is pumped out. The situation is similar to drinking water from a glass with a straw. As the water is sucked up, a cone-shaped depression forms in the water surface around the straw. This depression only recovers when you stop sucking on the straw, by which time the water level in the glass is lower. In a similar way, a cone-shaped depression forms around a well that only levels out again when pumping of the well stops. But if the withdrawal exceeds the recharge, this level will be at a lower elevation.

Municipalities almost always withdraw more water from the ground than nature is able to replenish. As a result, a **cone of depression** forms around each well (Fig. 10.39), that reflects the diminishing water supply. Eventually, the water table is lowered below the well and the well goes dry. Because groundwater flow is so slow, the cone of depression may persist for some time after the well is abandoned.

Other complications can also occur as a result of extracting groundwater. When groundwater is extracted from unconsolidated sediment, there is a loss of volume in the subsurface, which may result in land settling or **subsidence.** Between 1925 and 1975, large amounts of groundwater were pumped out of the

(a)

Figure 10.38

Use of magnetotellurics in groundwater exploration. (a) Equipment used in magnetotelluric surveys. (b) An electrical image of the subsurface in Merced County California. The freshest water occurs in highly resistive zones shown on the left of the image (courtesy Electromagnetic Instruments, Inc.).

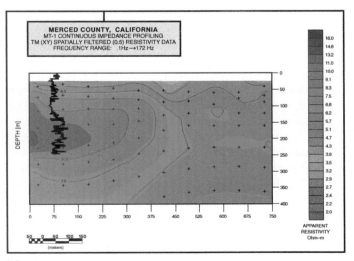

(b)

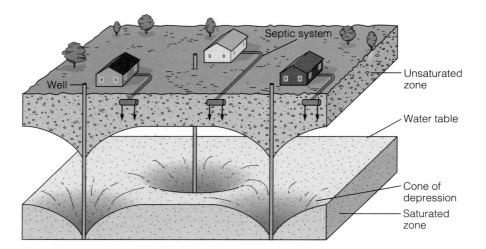

Figure 10.39
If water is withdrawn from the ground faster than it is replenished, cones of depression form in the water table directly beneath wells. In areas where the wells are closely spaced, these cones of depression may overlap.

ground beneath the agriculturally intensive San Joaquin Valley of California. As a result, the land subsided by approximately 9 m (29 ft) (Fig. 10.40a). In heavily populated regions, subsidence can severely damage a community's buildings, roads, water lines, and sewage schemes. The famous Leaning Tower of Pisa in Italy owes some of its tilt to subsidence associated with groundwater withdrawal from an aquifer (Fig. 10.40b). Mexico City, with the high demand for water posed by its population and climate, has subsided by more than 7 m (23 ft) in recent times.

In coastal communities, subsidence associated with the extraction of groundwater may lead to the invasion of seawater. Venice, in northern Italy, for example, has subsided 3 m (10 ft) in the past 1500 years. It is therefore not surprising that Venice has waterways for streets (Fig. 10.40c). The city therefore provides a clear example of the potential problems of subsidence caused by the overtaxation of an aquifer. Houston and New Orleans along the Gulf of Mexico are beset by similar problems.

As fresh water is depleted and the water table is lowered, water is drawn in from adjacent regions. This may lead to groundwater contamination, either from marine water held in permeable sediments in coastal communities, or from contaminated groundwater drawn in from adjacent geological formations. An invasion by marine water occurred in the New York aquifer in the 1960s as a direct result of its overproduction. Because salt water is denser than fresh water, it will normally sink below any fresh water in an aquifer. However, as fresh water is withdrawn and cones of depression form, the water pressure is reduced, which allows the saline water to migrate upward. This phenomena is known as a **cone of ascension** (Fig. 10.41).

Electrical imaging of the subsurface can readily detect contamination by marine water because fresh water is resistive and salt water is conductive. A study of marine contamination in the San Joaquin Valley shows how these images reveal the extent of contamination (Fig. 10.42). This region is a particularly important agricultural belt, producing close to 2 billion dollars worth of produce annually. Previous studies indicated that seawater invasion had migrated more than 11 km inland since 1980, but its influence on the aquifer was unclear.

The electrical image in Figure 10.42 clearly defines a shallow (less than 50 m or 165 ft) conductive portion of the aquifer

(a)

Figure 10.40
The effect of groundwater withdrawal. (a) Subsidence in the San Joaquin Valley. (b) The Leaning Tower of Pisa. (c) Invasion of sea water, Venice, Italy.

(b)

(c)

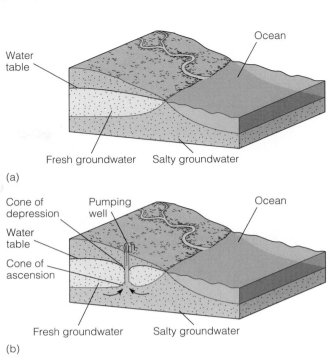

(a)

(b)

(c)

Figure 10.41
Invasion of the aquifer by marine water in coastal regions. (a) Fresh water is less dense than salt water and lies above it in the aquifer. (b) As groundwater is withdrawn, cones of depression form (see Figure 10.39) and a cone of ascension may form that results in the saline water rising to contaminate the fresh water aquifer. (c) The aquifer can be artificially recharged by pumping fresh water down a recharge well.

which identifies an area that has been extensively contaminated by salt water. Beneath this layer, however, there are important zones of high resistivity (blue colors) between the 300 to 600 m and 750 to 1000 m intervals along the survey. These are interpreted as zones of good quality water. At depths between 100 and 200 m (330–660 ft), the quality of the groundwater is highly variable. Beneath the 100 to 400 m intervals, for example, the survey has identified regions with conductive water, that are thought to represent cones of ascension (compare with Figure 10.41b), that are characteristic of the overproduction of an aquifer and the ascent of salty water.

To counteract this type of problem, some ingenious feats of engineering have been accomplished. For example, the aquifer beneath Chicago is artificially replenished by pumping in fresh water from Lake Superior. In the same way, groundwater used

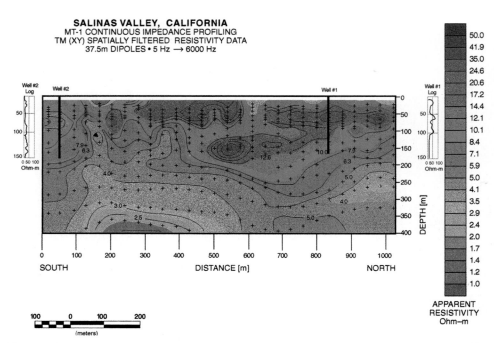

Figure 10.42

An electrical image 1 kilometer in length of the subsurface in Salinas, California. The freshest water occurs in highly resistive zones (blue) between 300 and 600 m and 700 to 1000 m intervals along the survey. Note the cones of ascension recognized by regions of highly conductive salt water between 100 to 400 m (courtesy Electromagnetic Instruments, Inc.).

in nonpolluting industrial processes can be recycled by pumping it back into the ground. Nonhazardous biodegradable liquids are sprayed on the land surface, and purify as they percolate down toward the water table. These examples of groundwater recycling may become more common as we continue to mine our most precious resource.

The study of the San Joaquin Valley is a clear example of the dilemma of groundwater exploitation. A vast water supply is needed to support the valley's agricultural produce, yet overproduction of the aquifer threatens the very industry that it is being used to support. More recently, aqueducts have been constructed that transport surface waters from northern California and the Sierra Nevada Mountains to the otherwise arid agricultural belts in southwestern California. This solution, however, is not without its problems.

10.6.3 Contamination of Groundwater

Unfortunately, the unique qualities that make water a very special substance, also render it susceptible to contamination. Its properties as an effective solvent mean that many chemicals, benign, and toxic, dissolve in water. Any soluble material, whether it occurs in the atmosphere, on the surface, or beneath the ground, can potentially contaminate the groundwater. Some examples of this contamination are shown in Figure 10.43. Rainwater scavenges soluble pollutants in the atmosphere, including sulfur (generating "acid rain") and toxic metals such as mercury or lead. This contaminated water may eventually

infiltrate the ground to contaminate the aquifer. Perhaps the most obvious example of contamination occurs at landfill or municipal garbage sites, in which horrific cocktails of toxic substances are created from an array of discarded industrial and household goods. As water percolates through these sites, elements leached from the material they contain may contaminate nearby aquifers (Fig. 10.43a). In order to protect against contamination, many landfills are lined with impermeable claylike materials. However, leaks in the system inevitably occur, especially in old abandoned landfills that may have been built to less stringent standards. The identification and cleanup of such sites are controversial subjects that rightly involve the entire community.

Hydrogeologic studies that establish the regional groundwater flow patterns are essential in the selection of new landfill sites. They are also essential for monitoring both abandoned and existing sites in order to determine their susceptibility to contamination, and for identifying any leakage at the earliest opportunity. In general, new sites are built in regions with thick deposits of impermeable clay well above the water table. Ideally, this blocks the passage of toxic substances into any underlying aquifer. Regions with permeable materials above the water table, such as sands and gravels, or rocks that typically contain fractures such as basalts or limestones, should be avoided.

Contamination from landfill sites, as serious as it is, pales in comparison to that from agricultural sources (Fig. 10.43b). The percolation of rain through fields laced with insecticides and fertilizers, for example, can contaminate an entire aquifer. Nitrate from fertilizers is the most hazardous contaminant in

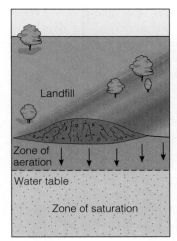

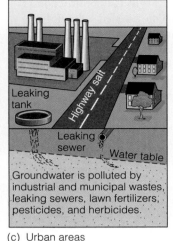
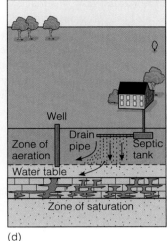

| (a) Landfill | (b) Rural areas | (c) Urban areas | (d) |

Figure 10.43

Contamination of groundwater in (a) landfill sites where pollutants can be carried into the zone of saturation, (b) rural areas in the vicinity of septic tanks and in general by use of fertilizers, pesticides, and herbicides, and (c) urban areas due to leaking gasoline tanks or sewage pipes. The identification of the source of contamination is possible by matching the chemistry of the pollutant in the groundwater with its source. (d) Problems associated with the placement of septic tanks.

groundwater and has been linked to methemoglobenemia ("blue baby" sickness) as well as stomach cancer.

Industrial waste and sewage are other potential contaminants (Fig. 10.43c). In many areas, sewage is stored in septic tanks that are housed in permeable soil. The tank slowly releases the sewage into the soil where the organic toxins are decomposed by their reaction with oxygen and microorganisms, and metal contaminants adhere to clay minerals. Problems with septic tanks arise if they are improperly designed, develop leaks, or are improperly positioned. In addition, if the contamination continues for an extended period of time, the capacity of the soil to filter the toxins may eventually be exceeded.

In Figure 10.43d, for example, the septic tank is located too close to the water table. The contaminated water leaks into the groundwater and may reach neighboring wells. A better placement would be to locate the septic tank well above the water table, preferably in a low to moderately permeable sandstone, so that water escaping from the tank is cleansed before it reaches the water table.

In ideal circumstances, the escaping water can be purified within a few tens of meters of the septic tank. The key is to locate the septic tank in sediment with a low enough permeability that the passage of sewage water through it is sufficiently slow for it to have time to react with the sediment. Those of us who live in rural communities may be familiar with the standard application of this principle in the "percolation test", which generally must be passed before a building permit is issued. The purpose of the test is to determine whether a septic tank and drain field can be constructed for a house without harming the local environment. The test simply involves digging a hole and filling it with water. If the water drains away too quickly, it is likely that contaminants would pass through and could contaminate the water table. The sediment is con-

sequently deemed to be too permeable and the test is failed. The test is also failed if the water fails to drain away because the sediment is too impermeable. A successful outcome requires the water to drain away slowly.

Because aquifers behave in the same way, they too have the ability to clean themselves. In principal, the self-cleansing potential of most aquifers should be enough to allow sufficient time to identify and eliminate the source of the pollution. In practice, however, the changing chemistry of these low-permeability materials is rarely monitored, and the contamination is often identified only after the cleansing capacity of the aquifer has been exceeded.

The most obvious long-term solution to the cleansing of aquifers is the elimination of the sources of pollution. Given that groundwater can travel considerable distances, the culprit may sometimes be difficult to identify. However, sophisticated laboratory techniques applied to groundwater samples now analyze a wide range of chemical elements and a variety of isotopes, including those of hydrogen, carbon, and sulfur. These isotopes, in particular, often serve as a fingerprint that can be used to identify the pollution source.

But one country's environmental policing can have political and economic consequences for others, as well as themselves. For example, the new global economy has given corporations the option of moving to countries with less stringent environmental regulations, rather than assuming the significant costs of making factories less polluting in their own countries.

10.6.4 Exploitation of Surface Water

As we have discussed, the exploitation of surface waters is as ancient as civilization itself. Perhaps our most visible interference with the natural flow of surface waters is in the construction of dams, and the reservoirs of water that are ponded behind them. Dams are constructed for many purposes, including flood control (see Section 10.5.6), hydroelectric power (see Chapter 18), irrigation, municipal water supplies, and recreation. Although their construction can bring many initial benefits, they are not without significant long-term disadvantages (Fig. 10.44).

Large dams flood vast areas and destroy natural wildlife habitats. They also alter the natural pattern of stream sedi-mentation, with surprising results. Essentially, this pattern changes because the reservoir formed behind the dam is an artificial lake, the water level of which acts as a local base level (Fig. 10.45, see also Figure 10.26a). The average gradient behind the dam is therefore decreased, so that the stream's capacity to erode is diminished. In addition, the flow of the stream is halted when it reaches the reservoir, causing its sediment load to be deposited. As a result, the water released below the dam is essentially sediment-free. The stream consequently

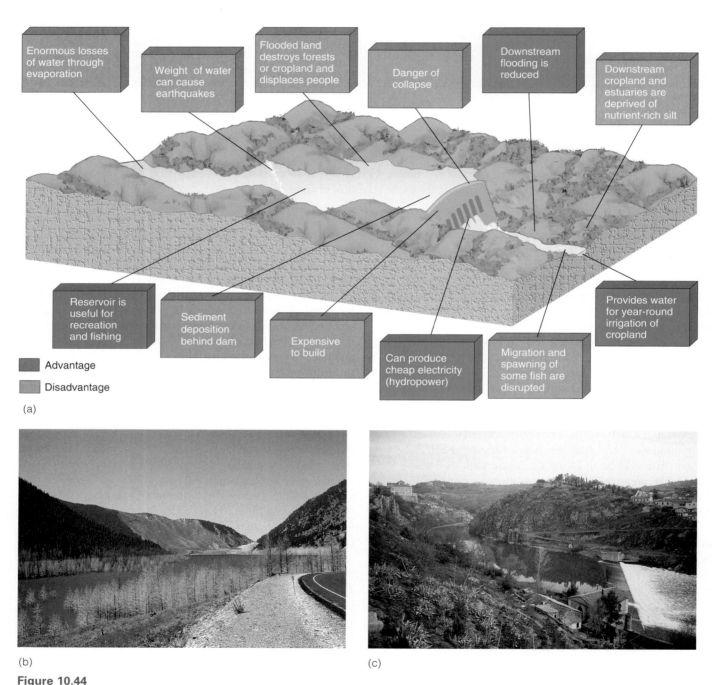

Figure 10.44

(a) Advantages and disadvantages of large dams and reservoirs. (b) An example of a dam formed by natural processes. A landslide (region of no vegetation in the background) dammed the stream (foreground). (Photo Joe Martinez) (c) Artificial dam in Rio Tajo, Spain. (Photo Damian Nance)

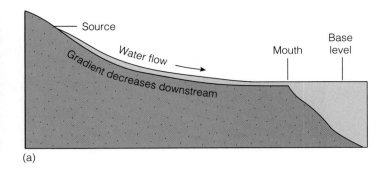

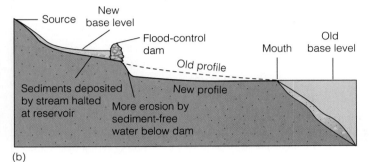

(a)

(b)

Figure 10.45

Profile of a stream (a) before and (b) after dam construction.

Figure 10.46

The James Bay Project in northern Quebec. If completed, the project would alter the flow of 19 rivers to produce hydroelectric power. Phase I has been completed but Phase II has been postponed indefinitely.

flows faster and its energy is used primarily to erode the underlying sediment and bedrock. The construction of a dam therefore alters a stream's natural profile such that its sediment load is prematurely deposited in the reservoir and its ability to erode downstream is enhanced.

This change in the natural cycle of a stream can have many unforeseen consequences. Left to their own devices, streams deposit about 75% of their sediment load on their floodplains, where it provides essential nutrients for the soil. Before the construction of the Aswan High Dam in Egypt, for example, annual flooding of the Nile River deposited between 6 to 15 cm (2.5 to 6 in) of new soil per century. This natural fertilizer replenished the nutrients extracted from the soil by intense agricultural activity. Since construction of the dam was completed in 1970, the transport of sediment downstream has been blocked, with the result that artificial fertilizers must now be used. In addition, the downstream portion of the Nile now flows faster and the river has undermined bridges and smaller dams. The amount of sediment reaching the Mediterranean Sea has also been drastically reduced, and the Nile Delta is now subsiding faster than sediment can build it up. This sediment also provided nutrients for the marine ecology so that the fishing industry that once thrived in the region has been decimated, with the loss of 30,000 jobs.

One of the world's most ambitious surface water projects started in the mid 1970's in northern Quebec, Canada (Fig. 10.46). The 50-year plan is to construct 600 dams which would alter or reverse the flow of 19 major rivers at a cost of $60 billion. When finished, the project, devised to provide cheap hydroelectric power to Quebec and the northeastern United States, would flood an area of 176,000 square km (68,000

square mi), which is about the size of Germany. After 20 years, Phase 1 has already been completed at a cost of $16 billion. In 1994, Phase II was postponed because of objections by the indigenous Cree people who claimed that their ancestral hunting grounds would be destroyed, and because environmental concerns led the New York State government to cancel contracts with Hydro Quebec (see also Chapter 18).

In arid climates, water is commonly diverted for irrigation. However, the intense heat results in rapid evaporation and the deposition of a salt residue which is transported downward into the soil. If the water table is well below the root zone of the crops, the salt can be flushed away with more water. However, excessive irrigation in flat low-lying regions produces a shallow water table, so that the salt cannot easily drain away and the roots of the crops are damaged.

Nowhere is the demand for water more intense than in the densely populated, parched regions of southern California. An intricate array of dams and aqueducts pump water from the Sierra Nevada and the relatively water-rich north (Fig. 10.48). Agriculture uses more than 80% of this water, and two crops alone, alfalfa and cotton, require as much water as the domestic needs of the entire state.

As we have seen, groundwater and surface water are interconnected, so that contamination of one can have adverse affects on the other. For example, contaminated groundwater may leak into springs, streams and lakes where the water table intersects the surface. The draining of the Aral Sea (see Living on Earth on page 296) is an example of the environmental catastrophe that can happen when surface water diverted into

THE DEATH OF THE ARAL SEA: AN ENVIRONMENTAL NIGHTMARE

Thirty years ago, the Aral Sea, located on the border between Uzbekistan and Kazakhstan (Fig. 10.47), was the world's fourth largest lake with an annual harvest of about 45,000 tons of fish that supported 60,000 people. It covered 69,000 square km (26,500 square mi) and was larger than all but the largest of the Great Lakes. However, since that time, the Amu Dar'ya and Syr Dar'ya rivers that feed the lake have been diverted to irrigate cotton, vegetable, fruit, and rice crops. The result has been an environmental catastrophe. The Aral Sea has lost more than 50% of its original surface area and 75% of its water volume. The rivers feeding it have been reduced to mere trickles. In most places, the coastline has retreated between 50 and 150 km (30 to 90 mi) and the average depth of the remaining water has decreased from 16 m (52 ft) to 9 m (30 ft). Its fishing communities are now stranded and the fishing industry is dead, the fish being unable to survive in the increasingly salty water. Unchecked, the lake will be reduced to three stagnant pools of water surrounded by parched desert within 20 years.

The region's climate has also been affected. Because the lake is no longer a moderating influence, the region has hotter summers and colder winters. The exposed sand is a salt desert—a toxic mixture of salt, pesticides, and defoliants. Dust storms strip these chemicals from the soil, and the wind carries them for distances of up to 300 km (186 mi). As a result, wide regions of Uzbekistan and Kazakhstan have become contaminated. Now that there is less rain, crop yields are much

Figure 10.47
The Aral Sea was once the fourth largest fresh water lake in the world. In 1960, the Syr Dar'ya and Amu Day'ra rivers that flowed into it were diverted for irrigation purposes. The size of the Aral Sea has been shrinking ever since, with horrific environmental consequences.

lower. In order to combat this, fields are laced with herbicides, insecticides, and fertilizers, which have percolated downward and have contaminated the groundwater.

The chemicals, dust storms, and contaminated groundwater have had catastrophic effects on the local population. Because of high concentrations of heavy metals and other toxic elements in the drinking water, there are critical mineral deficiencies in almost all of the population of the region. 99% of the population in northern Uzbekistan, for example, are anemic because their bodies are unable to absorb nutrients such as calcium

and iron. Children eat chalk to satisfy their craving for calcium, and have a high risk of developing brain damage. The rates of cancer, tuberculosis, typhoid, and liver and kidney diseases have also increased rapidly.

In 1992, the countries that border the Aral Sea signed an agreement on the management of its waters. Environmentalists say that the Aral Sea could be restored eventually if the irrigation canals were shut down, but that it would take decades and cost billions of dollars. Meanwhile, the economy of the region is still heavily dependent on revenue from agriculture and no end to this environmental nightmare is in sight.

Figure 10.48
A system of dams and aqueducts transfers water to the populated regions of California and Arizona.

canals results in contamination of both groundwater and surface water.

It is clear that we have much to learn about the consequences of our interference with the natural cycle of fresh water. We depend on water for our survival, yet our methods of exploitation are often poorly conceived and the results are unforeseen. Perhaps the best legacy we can leave our children is a sustainable planet. But if we wish to achieve this goal, we will need to learn more about water and its complex behavior.

10.7 Chapter Summary

Although it represents only 22% of the total fresh water budget, groundwater is likely to become the most important resource of the 21st Century. Its supply is threatened by population pressure, inefficiency, waste, and contamination. Surface waters have been historically important to the evolution of human society, and continue to be a major source of fresh water.

In North America, 1.8 trillion liters of fresh water are used daily, primarily for domestic, industrial, and agricultural purposes. This consumption rate is much higher than nature can replenish through rainfall. Groundwater is consequently a non-renewable resource, and we mine it just like any other resource. But its property as a solvent means that it is easily contaminated by natural sources and human-related activities such as industrial development and agriculture.

According to the hydrologic cycle, fresh water is derived from the evaporation of ocean water. When evaporation occurs, impurities such as salt are left behind, rendering the water vapor relatively pure, or fresh. The evaporated water spends an average of 11 days in the atmosphere. During this time it reacts with atmospheric gases so that precipitation tends to be mildly acidic. Although precipitation occurs preferentially in subequatorial regions (within 10° of the equator) and at mid-latitudes, these trends are substantially modified in coastal regions and mountainous areas.

The supply of groundwater is a delicate balance between discharge and recharge. Groundwater is discharged into surface waters such as streams, lakes, swamps, and springs. Recharge occurs when precipitation enters the subsurface. Favorable conditions, such as permeable surface sediments and relatively flat terrain, facilitate the downward passage of water until it is eventually blocked by rock or impermeable sediment. Water accumulates above this barrier and forms a zone of saturation where the pore spaces of the sediment are filled with water. The upper surface of this zone is known as the water table. The depth to the water table varies considerably and in areas of adequate rainfall tends to mirror the topography. If the subsurface sediment can store and transmit water, it is known as an aquifer, and can form an important groundwater resource for a region.

Aquifers are not merely confined to regions of high precipitation. The regional flow of groundwater in the subsurface is such that it can travel hundreds of kilometers from its source, so that important supplies can be found beneath desert regions. Aquifers such as those in Libya and Egypt formed during the cooler climate conditions of the last Ice Age (10,000 to 70,000 years ago) and provide vital groundwater resources beneath these arid lands.

In tectonically active areas, groundwater has the capacity to absorb significant quantities of heat, and may be a source of geothermal energy. The composition of groundwater is influenced by chemical weathering, which in turn is related to mineral stability. In general, minerals with strong bonds in their crystal structure (such as quartz) are relatively resistant whereas those containing weak bonds (such as plagioclase) are more reactive.

Groundwater becomes surface water where the water table intersects the Earth's surface. Accessible surface water consists of streams, lakes, swamps, and springs. Streams occur in vast drainage basins whose distribution in North America is primarily controlled by the position of the Appalachian and Cordilleran mountains. As a stream runs from its headwaters to its mouth, its gradient decreases, its average velocity increases, and its relatively straight path gives way to meanders.

Under normal conditions, a stream can never erode below sea level, although it will respond to changes in sea level. It carves out a V-shaped valley through the complementary processes of weathering and erosion. The position of a stream constantly changes and it eventually creates a wide valley which may be prone to flooding. Up to 75% of a stream's sediment load is deposited along its course and nearly 10 billion tons of sediment are transported to the oceans each year. At the ocean, the sediment is deposited in deltas and affects the composition of ocean water.

Our dependence on water to survive provides the impetus for groundwater exploration. Because fresh groundwater is a poor conductor of electricity, fresh water aquifers can be readily recognized in subsurface electrical images. Exploitation of groundwater requires careful management and planning. Where water is withdrawn faster than nature can replenish it, cones of depression form and land subsidence may occur. In coastal regions, where seawater may lie beneath the fresh water aquifer, excessive withdrawal may also lead to ascending plumes of saline water which invade and contaminate the aquifer. Groundwater is also easily contaminated by a variety of phenomena including acid rain, the application of fertilizers, and poorly located septic tanks and landfill sites.

Our exploitation of surface water may yield some initial benefits, but it often has unforeseen long-term disadvantages. Damming and re-routing of streams alters the stream profile and the distribution of sediment which replenishes the soil. Like groundwater, we have much to learn about the consequences of our interference with the natural flow of surface water.

Key Terms and Concepts

- Fresh water sources include glaciers, groundwater (water in the ground), and surface water (streams, lakes, springs, and swamps)

- Fresh water is a non-renewable resource because it is consumed faster than it is replenished by rainfall

- Fresh water is primarily derived by evaporation of ocean water and subsequent precipitation on land

- Fresh water spends an average of 11 days in the atmosphere during which time it becomes mildly acidic due to interaction with atmospheric gases

- Groundwater supply depends on the balance between discharge and recharge. Groundwater is discharged into streams, lakes, springs, or swamps. It is recharged when precipitation infiltrates the subsurface

- The downward passage of water is eventually blocked by impermeable rock, and water accumulates above this barrier to form the zone of saturation. The water table is the upper surface of the zone of saturation. The zone of aeration occurs above it

- Groundwater becomes a resource, or an aquifer, if the zone of saturation can store and transport water. In order to do so, an aquifer must be both porous (with a large volume of pore spaces between the sediment grains) and permeable (with pore spaces that are interconnected and large enough to allow water to flow)

- Unconfined aquifers are aquifers whose upper surface coincides with the water table. Confined aquifers occur where the water-bearing rocks or sediments are sandwiched between impermeable layers

- Groundwater generally flows "downhill" toward valleys at a few meters per year under the influence of gravity

- Because groundwater flows beneath the subsurface, it can be found beneath desert regions. Because it flows slowly, groundwater forming today may take thousands of years to become surface water

- Groundwater becomes surface water where the water table intersects the surface. The intersection of the water table with the surface commonly generates springs. Springs also occur where water is forced up to the surface along fractures. An oasis is a spring that occurs in desert regions

- Deep penetrating groundwater may absorb the Earth's internal heat and so becomes a source of geothermal energy and hot springs. In volcanically active regions, groundwater is converted to steam which escapes to the surface to form fumaroles. Cyclic emissions of superheated groundwater are known as geysers

- Because water is a strong solvent, it is the primary agent of chemical weathering and its composition is influenced by the extent of chemical weathering and the composition of the subsurface materials. Water that contains large quantities of dissolved ions is known as hard water

- Karsts are features that are produced when groundwater dissolves bedrock. Gypsum and limestone are very soluble, leading to the formation of caves, associated columns known as stalactites and stalagmites, and sinkholes

- Each year, about 40,000 cubic km of fresh water drains back to the sea

- Continental drainage is concentrated in streams, lakes, swamps and springs

- Surface runoff occurs where precipitation encounters impermeable or saturated bedrock or sediment at the surface

- Streams occur in connected networks known as a drainage basin, which may be grouped together into drainage outlets depending on which body of water the streams ultimately discharge into

- The average gradient of the stream is the vertical drop from the stream's headwaters to the sea divided by the distance traveled. Assuming constant sea level, the average gradient of streams decreases with time

- Because of the decreasing gradient along their lengths, streams tend to be straighter at their headwaters and meander as they approach the coastal regions

- Streams preferentially erode weaker rocks relative to hard, resistant, crystalline rocks, and this can lead to the formation of waterfalls

- Rising sea level may lead to drowned valleys. Falling sea level results in a new cycle of erosion and deposition

- Landscape is sculpted through the complimentary processes of weathering and erosion

- Streams may contain both a dissolved component and a solid sediment load which can be moved in suspension or by traction (a hopping motion along the stream bed), depending on the size of the clast and the velocity of the stream

- The average velocity of streams *increases* from its headwaters to its mouth

- Streams form V-shaped valleys. Their paths migrate with time, producing floodplains

- Flooding occurs when a stream's channel can no longer contain the water flowing through it and water input exceeds output at the mouth. This is generally the result of excess precipitation or ice jams

- Successive floods deposit coarser material along stream banks which builds up to form a natural levee

- The composition of stream water profoundly affects the composition of ocean water

- When a large stream discharges into the sea, a fan-shaped deposit is formed, known as a delta

- Exploration for groundwater focuses on finding the level of the water table and imaging the subsurface

- Fresh water is a poor conductor of electricity and so its location in the subsurface can be constrained by electrical images

- When water is withdrawn from an aquifer faster than nature can replenish it, cones of depression form around each well, and land settling or subsidence may occur

- Saline water sinks below fresh water in the aquifer, but may rise in a cone of ascension as fresh water is withdrawn from the aquifer

- Because water is an effective solvent, it is susceptible to contamination, especially adjacent to landfill sites. Entire aquifers can be contaminated by the percolation of rain through fields laced with insecticides and fertilizers

- Aquifers have the capacity to cleanse themselves

- The construction of dams to form reservoirs and aqueducts alters the natural flow of surface waters. These constructions affect a stream's sediment load, and the supply of nutrients to the stream's floodplain

Review Questions

1. Why is groundwater classified as "fresh water"?
2. Why is groundwater a non-renewable resource?
3. What forms of fresh water are not considered groundwater?
4. How does the water table form?
5. Why is the balance between discharge and recharge so important?
6. How does an aquifer form?
7. What is the relationship between chemical weathering, mineral composition, and the composition of groundwater.
8. Why does the average velocity of a stream increase from its headwaters to its mouth?
9. How does the composition of sediment affect ocean water composition?
10. How do natural levees form?
11. Explain how deltas obtain the shape of a "bird's foot".
12. Why does saltwater invade aquifers in coastal communities?
13. Why is water so easily contaminated?

Study Questions

Research the highlighted question using InfoTrac College Edition.

1. Four features that characterize overproduction of aquifers have been mentioned in this chapter. What are they, and explain how aquifers are sensitive to overproduction?
2. What measures would you recommend to decrease groundwater consumption?
3. Why is the water table level so variable within the same aquifer?
4. What type of material is porous but not permeable?
5. What type of material is permeable but not porous?
6. How are cones of depression formed, and why does the water table remain depressed beneath a well, rather than spread out and become uniform?
7. Would global warming serve to increase or decrease our groundwater supply? Explain.
8. What do you think is the most appropriate method of flood control?
9. Why do electrical images of the subsurface provide information on groundwater location?
10. What is the most appropriate location for (a) landfill sites (b) septic tanks? Explain.
11. How may re-routing of surface waters affect the composition of groundwater?

Hydrosphere

Hydrosphere

The hydrologic cycle moves water from the ocean to the atmosphere to the land and finally back to the ocean. This cycle brings water to all of Earth's reservoirs and makes life possible. Ocean circulation distributes heat and nutrients.

Astronomy

Heat from the Sun promotes the evaporation of water and the temperature differences in the oceans. The gravitational pull of both the Moon and the Sun on water as the Earth spins causes the tides. The burning of comets in the Earth's atmosphere may be the source of some of the Earth's water.

Atmosphere

Water vapor is stored in the atmosphere and falls as precipitation. Water in the atmosphere stores heat, and this heat affects atmospheric circulation and climate.

Biosphere

Fresh water is crucial for all life on Earth. Plants and animals cycle water through their bodies and add water vapor to the atmosphere. Marine life uses ocean water as a source of nutrients, for shelter, and as a home.

Solid Earth

Water in both liquid and solid forms is stored within the solid Earth. Volcanic processes release water vapor into the air. Water also plays a large role in carving the landscape. Continents affect movement of ocean currents. The solid Earth acts as a filter as groundwater moves through the crust.

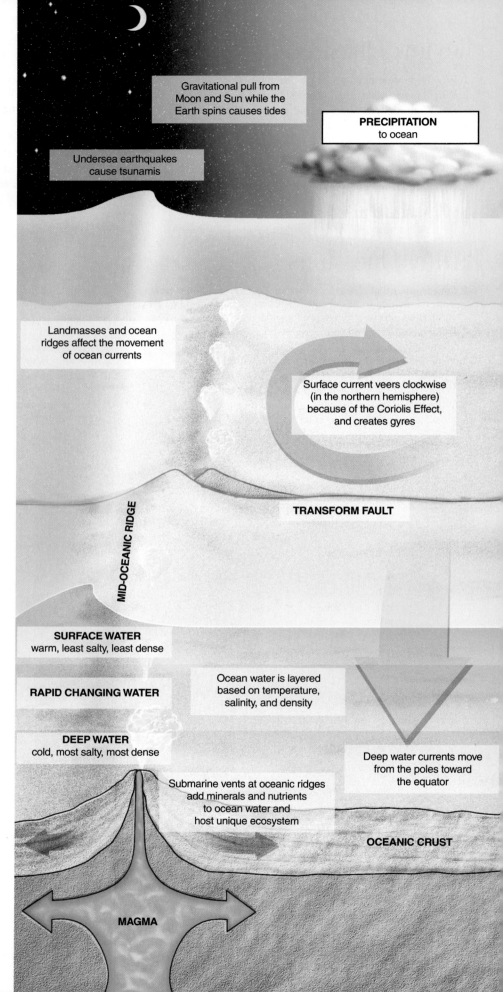

Gravitational pull from Moon and Sun while the Earth spins causes tides

PRECIPITATION
to ocean

Undersea earthquakes cause tsunamis

Landmasses and ocean ridges affect the movement of ocean currents

Surface current veers clockwise (in the northern hemisphere) because of the Coriolis Effect, and creates gyres

TRANSFORM FAULT

MID-OCEANIC RIDGE

SURFACE WATER
warm, least salty, least dense

Ocean water is layered based on temperature, salinity, and density

RAPID CHANGING WATER

DEEP WATER
cold, most salty, most dense

Deep water currents move from the poles toward the equator

Submarine vents at oceanic ridges add minerals and nutrients to ocean water and host unique ecosystem

OCEANIC CRUST

MAGMA

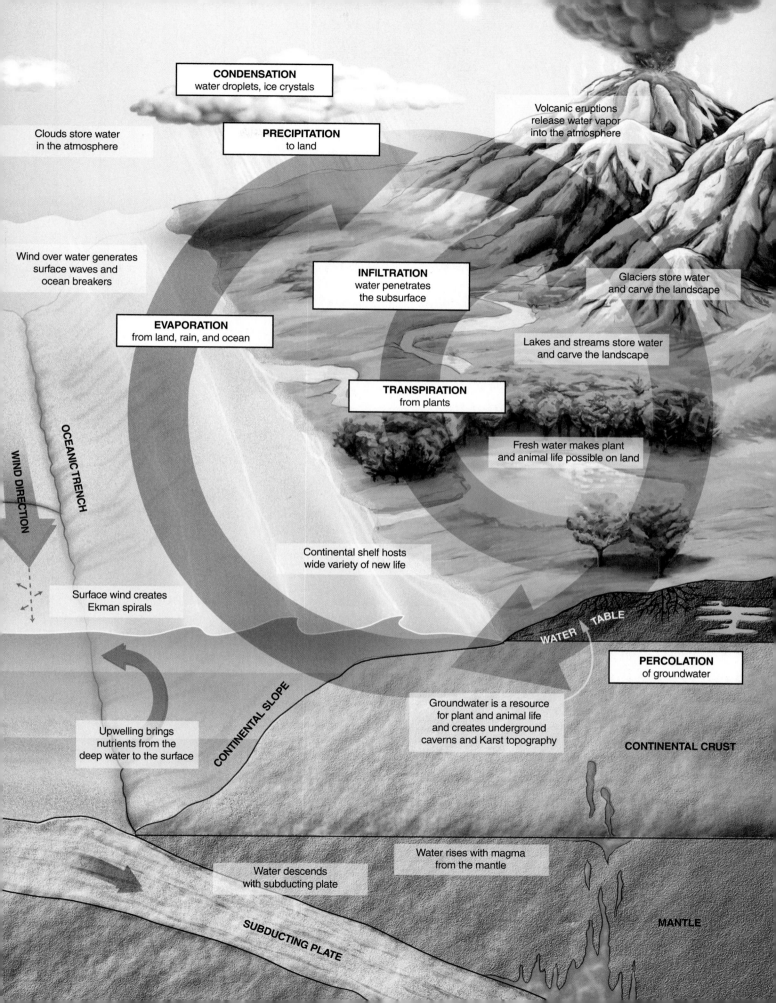

CONDENSATION
water droplets, ice crystals

Clouds store water
in the atmosphere

PRECIPITATION
to land

Volcanic eruptions
release water vapor
into the atmosphere

Wind over water generates
surface waves and
ocean breakers

INFILTRATION
water penetrates
the subsurface

Glaciers store water
and carve the landscape

EVAPORATION
from land, rain, and ocean

Lakes and streams store water
and carve the landscape

TRANSPIRATION
from plants

Fresh water makes plant
and animal life possible on land

WIND DIRECTION

OCEANIC TRENCH

Surface wind creates
Ekman spirals

Continental shelf hosts
wide variety of new life

WATER TABLE

PERCOLATION
of groundwater

Upwelling brings
nutrients from the
deep water to the surface

CONTINENTAL SLOPE

Groundwater is a resource
for plant and animal life
and creates underground
caverns and Karst topography

CONTINENTAL CRUST

Water rises with magma
from the mantle

Water descends
with subducting plate

MANTLE

SUBDUCTING PLATE

PART 3

Atmosphere:
The Air We Breathe

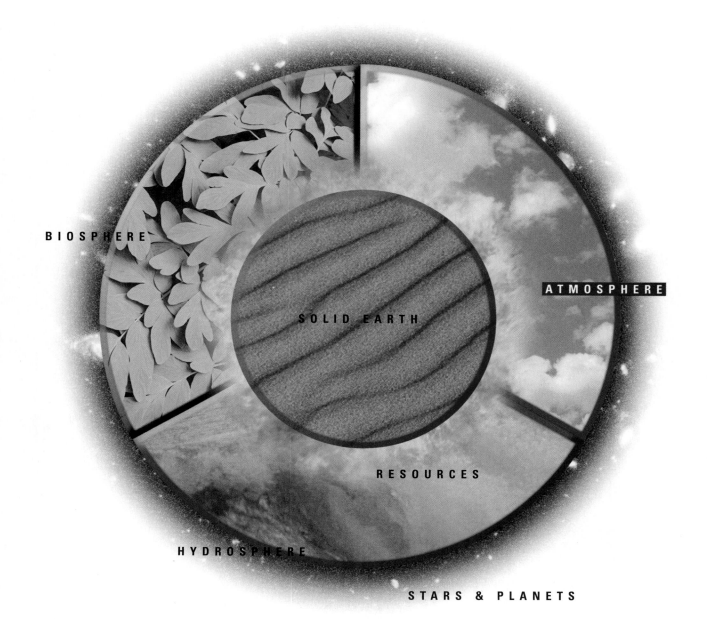

BIOSPHERE

SOLID EARTH

ATMOSPHERE

RESOURCES

HYDROSPHERE

STARS & PLANETS

11

The Atmosphere

In previous chapters, we saw that the Earth is layered according to density, from its dense iron-rich core, through its silicate-rich mantle and crust, to its watery hydrosphere and finally to its gaseous envelope, the atmosphere. In this chapter you will gain an understanding of the origin of our atmosphere as gaseous emanations from volcanoes that are held at the Earth's surface by the forces of gravity. Over geologic time, these noxious gases have been refined by processes in the hydrosphere and biosphere to produce the air we breathe.

We will learn that the atmosphere itself is also layered according to its density and the decreasing pressures that exist above the Earth's surface. This surface absorbs solar energy, and so acts like a radiator. As a result, temperatures decrease upward from the Earth's surface for about 12 km (7.5 mi). This lowest layer of the atmosphere, known as the troposphere, seals in the Earth's moisture. This is because, as warm air rises it cools and condenses and so eventually returns the moisture to the Earth's surface as precipitation.

Above this layer is the stratosphere, which includes the ozone layer. As we will see, the presence of ozone in the atmosphere not only shields the biosphere from the harmful effects of ultraviolet radiation from the Sun but also profoundly influences the atmosphere's thermal structure.

This chapter covers the major features of the atmosphere, including such characteristics as atmospheric pressure, temperature, and relative humidity that combine to influence weather and climate. Geographical features such as latitude, mountains, and proximity of oceans are also examined as these have an important moderating influence. By the end of this chapter, you should have a grasp of the important concepts needed to understand the following chapter on atmospheric circulation. We begin by considering the atmosphere's chemical composition.

11.1 Introduction: The Composition of the Atmosphere

The Earth's atmosphere not only sustains life, but also forms a shield that filters out harmful radiation from the Sun and protects the Earth from the constant showers of small extraterrestrial meteors. One breath of air contains 10^{22} molecules (that is, 10 multiplied by 10 twenty-two times). This figure is similar in magnitude to the number of stars in the universe. There are about 10^{44} molecules in the atmosphere, and it is probable that most of the molecules you inhale in your next breath were previously in someone else's lungs. Thus, in a very real way, the atmosphere is something we all share.

The atmosphere consists of gases and trace amounts of suspended particles from volcanoes, nuclear explosions, and industrial pollutants. Although atmospheric gases can be detected up to 700 km (434 mi) above the Earth's surface, 99% of these gases typically occur within the first 32 km (20 mi).

The composition of the Earth's atmosphere is unique in the Solar System. In contrast to its nearest planetary neighbors, Venus and Mars (the atmospheres of which are made up mainly of carbon dioxide—see Table 11-1), the Earth's atmosphere is dominated by nitrogen (78%) and oxygen (21%). The gas argon accounts for most of the remaining 1%. Carbon dioxide accounts for a mere 0.03%. These are average amounts however, as the concentration of some gases (most notably water vapor) can be highly variable. During wet weather, water vapor can locally account for as much as 4% of the atmosphere and is concentrated in minute droplets of moisture in clouds, fog or steam. In dry weather, water vapor may be virtually absent from the atmosphere.

The venting of gases to create an atmosphere is an end product of the igneous processes in magma chambers that we discussed in Chapter 2. As magma chambers cool, the crystals that form first (olivine, pyroxene, and plagioclase) preferentially exclude volatile components such as water, carbon dioxide, and nitrogen from their crystal structures. As a result, the concentrations of the volatile components in the magma increases as the magma cools. Just as they do in a carbonated beverage, the volatiles rise to the top and are trapped at the roof. Like a giant pressure cooker with a weak lid, the pressure exerted by the volatile gases steadily builds at the roof until the gases finally explode their way through the overlying rocks and vent into the atmosphere (Figs. 11.1 and 11.2).

The eruption of Mount Pinatubo in the Philippines in 1991 (Fig. 11.1, see also Chapter 1) provides an excellent example of the relationship between volcanic activity and the composition of the atmosphere. This eruption caused a global change in the composition of the Earth's atmosphere as 20 million tons of sulfur dioxide were vented into it. Within three weeks of the eruption, a shroud of sulfur dioxide-enriched atmosphere had encircled the Earth and become part of the atmosphere's global circulation pattern.

If one volcano can have such a measurable effect, imagine the influence on the composition of the atmosphere of 4.6 billion years of volcanic activity. Direct evidence from more ancient volcanic activity has been largely obliterated by the constant recycling of our planet's surface as a result of plate tectonics. But if,

Table 11–1

Comparison Between the Composition of the Atmospheres of Venus, Earth, and Mars

VENUS		EARTH[a]		MARS[b]	
GAS	PERCENT VOLUME	GAS	PERCENT VOLUME	GAS	PERCENT VOLUME
CO_2	96.5	N_2 (nitrogen)	78.1	CO_2	95
N_2	3.5	O_2 (oxygen)	20.9	N_2	2.7
SO_2	0.015	H_2O (water vapor)	0.05 to 4 (variable)	Ar	1.6
H_2O	0.01	Ar (argon)	0.9	CO	0.6
Ar	0.007	CO_2 (carbon dioxide)	0.03	O_2	0.15
CO	0.002	Ne (neon)	0.0018	H_2O	0.03
He	0.001	He (helium)	0.0005	Kr	Trace
O_2	≤0.002	CH_4 (methane)	0.0002	Xe	Trace
Ne	0.0007	Kr (krypton)	0.0001		
H_2S	0.0003	H_2 (hydrogen)	0.00005		
C_2H_6	0.0002	N_2O (nitrous oxide)	0.00005		
HCl	0.00004	Xe (xenon)	0.000009		

[a] Compositions are for near-surface conditions, with terrestrial data other than H_2O tabulated for dry conditions. CO_2 on Earth is probably increasing by 2% to 3% of the listed amount in each decade, because we are burning so much fossil fuel. This activity may be modifying Earth's climate.

[b] Amounts of gases vary slightly with season and time of day. H_2O is especially variable. Some CO_2 condenses out of the atmosphere into the winter polar cap; changing cap sizes cause small changes in the total Martian atmospheric pressure.

Figure 11.1
Photograph showing the eruption of Mount Pinatubo in the Phillipines in 1991. See also Chapter 1.

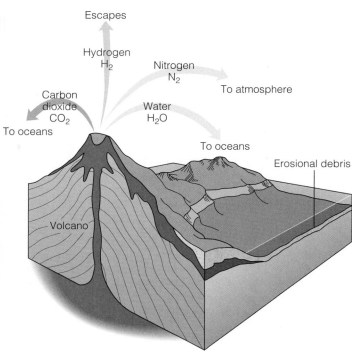

Figure 11.2
Outgassing from volcanic eruptions to the atmosphere. Nitrogen accumulates in the atmosphere at the expense of water and carbon dioxide which are concentrated in the hydrosphere. Hydrogen escapes into space.

as is generally believed, the Earth has been cooling down since it first formed, then volcanic activity is likely to have been very abundant early in Earth history such that the atmosphere would have originated soon after the Earth formed.

Volcanic activity releases the trapped volatile components of the inner Earth, and gravitational attraction prevents most of them from escaping into space. The larger the planet, the greater the gravitational hold on its atmosphere. As we shall learn in Chapter 15, the relatively small size of Mars provides an insufficient gravitational pull on its atmospheric gases. As a result, they slowly leaked into space. Because volcanic activity on Mars ceased about 1 billion years ago, the leaking of the Martian atmosphere into space is no longer compensated by volcanic activity. This explains why its atmospheric pressures are so much lower than those on Earth.

Although the Earth's volcanic activity and gravitational attraction may account for the existence of its atmosphere, they do not explain its composition. There is an obvious mismatch between the composition of the Earth's atmosphere

(which is overwhelmingly dominated by nitrogen and oxygen) and the gases vented from modern volcanoes. Samples taken from active volcanoes show that water vapor typically comprises between 50 and 80% of the gases emitted. Smaller and highly variable amounts of carbon dioxide, nitrogen, sulfur dioxide, and hydrogen sulfide and trace amounts of hydrogen, carbon monoxide, and chlorine, are also vented. This composition is probably typical of the gases released from ancient volcanoes, because the magma was formed by melting similar source materials and would therefore have concentrated similar gases.

The compositional mismatch between gases vented from volcanoes and the composition of the atmosphere occurs because of interactions between the Earth's surface reservoirs. Processes from within the Earth may give rise to the atmospheric gases, but interactions between the hydrosphere and biosphere have fundamentally changed the relative abundance of these gases to give an atmosphere that is dominated by nitrogen and oxygen. But how does this happen? Water vapor emitted from a volcano becomes part of the hydrologic cycle. The vast majority of this water is extracted from the atmosphere as rain or snow and eventually ends up in the oceans. Carbon dioxide is highly soluble in water, and so is extracted in rainfall and absorbed by the oceans. As we discussed in Chapter 8, ocean water has a very strong affinity for carbon dioxide. Carbon dioxide is also extracted from the atmosphere by its interaction with the biosphere. In the presence of sunlight, photosynthetic organisms in the oceans (such as phytoplankton) and vegetation on land (like that of the tropical rain

There is a general misconception that nature is entirely benevolent and that our environmental problems are all of our own making. This is not completely true. Radon gas is a natural constituent of our atmosphere, although it is present only in minute amounts. In the open air, radon is harmless. However, indoors, radon concentrations can build up to between 20 and 15,000 times that amount. At these concentrations, radon gas is deadly. It is estimated that inhalation of radon gas causes between 5000 and 25,000 cases of lung cancer every year in North America alone, a figure comparable with the number of deaths from automobile or home accidents. Radon gas inhalation is second only to smoking as the leading cause of lung cancer.

Radon is but one link in the chain of reactions that converts radioactive uranium-238 into stable lead (Fig. 11.3). As uranium decays, it sets off a ripple effect like falling dominoes. Although radon is formed during this decay, it too is unstable and decays to polonium which itself decays and so on, until a stable variety of lead (^{206}Pb) is ultimately produced. The very short half-life of radon decay (only 3.8 days) is a measure of its profound instability. Radon atoms decay by ejecting fundamental particles, known as alpha particles (see Chapter 3), from their nuclei. When radon gas is inhaled, the alpha particles damage biological tissue and cause lung cancer.

The danger with radon is not only its inherent instability, it is also extremely mobile. Since radon is a gas, it can readily escape from uranium-rich rocks and soils. It seeps into houses and buildings through any openings such as gaps between floors and

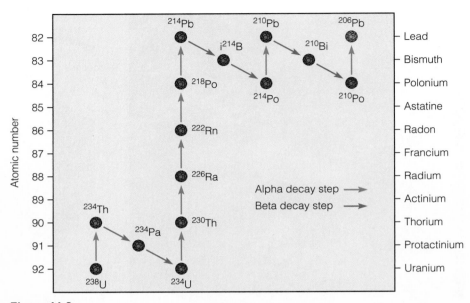

Figure 11.3
Radioactive uranium with an atomic mass of 238 (^{238}U), converts to lead-206 (^{206}Pb) in a number of steps. During these steps, radon-222 (^{222}Rn) is produced by radioactive decay of radium-226 (^{226}Ra). Radon then decays to polonium-218 (^{218}Po) with a half-life of 3.8 days. The alpha particle ejected from the radon nucleus destroys biological tissue, and is a leading cause of lung cancer.

walls, cracks in walls and foundations, sump holes, and holes for plumbing (Fig. 11.4). When radon seeps into a house or building, it circulates and accumulates as indoor air, especially in well-insulated, poorly-ventilated houses. Since radon gas is heavy (its atomic mass is 222, see Figure 11.3), its buildup is most concentrated in basements. If domestic water comes from a spring or well, it too may bring radon gas into the home. Well water may absorb radon from underground rocks and soils, only to release it, for example, during a shower. In the northern U.S. and Canada, accumulation of radon in domestic

houses is most likely to occur in the wintertime. However, the problem can be greatly alleviated if basement and indoor air is continually exchanged with the air outside.

Buildings with uranium-rich building stone such as granite are also at risk. Modern buildings have air circulation systems that generally compound the problem.

Radon introduced into a human environment in these circumstances is often inhaled by humans and lodges in their respiratory system. Its extremely rapid radioactive decay can cause random and deadly chemical changes as it passes through the human body.

forests) not only extract carbon dioxide, but also introduce oxygen into the atmosphere.

But what about nitrogen, the most abundant atmospheric gas? Nitrogen is relatively unreactive so that its concentration has built up throughout geologic time simply because it has not been extracted from the atmosphere in any significant way. Indeed, in agricultural areas, the lack of nitrogen in soils is typically compensated by the application of fertilizers. Some forms of bacteria can extract nitrogen from the atmosphere, but this amount is relatively small.

The atmosphere is continually bombarded by radiant energy from the Sun. As we will see, this interaction is primarily responsible for air circulation on the planet, for the Earth's climate belts and for our day-to-day weather. However, the interaction is also responsible for subtle but important chemical changes in our atmosphere, the most notable of which is the production of ozone gas.

Ozone is continuously being produced in the Earth's atmosphere by the interaction of the Sun's radiant energy with oxygen molecules (Fig. 11.5). At any given instant, a small pro-

Because radon is radioactive, its presence can be readily detected at a minimal cost. However, we are not all equally at risk. Some homes and buildings contain high concentrations of the gas; others have very low concentrations. There are two main factors that govern these variations—the nature of a building's ventilation and the local geologic environment. A poorly-ventilated house built on uranium-rich rocks and soils is most at risk. However, once the possible radon threat has been identified, it is relatively easy and inexpensive to minimize exposure by adjusting ventilation systems.

Since radon is a product of uranium decay, radon emissions are highest near uranium-rich rocks and soils. Uranium occurs in very minor amounts in most rocks. Its average abundance in continental rocks is about 2.8 parts per million or 0.00028%. However, it is often concentrated by a factor of 10 to 100 in rocks such as granite, black shale, coal, and limestone. Granite is an igneous rock, the product of the cooling and solidifying of a magma well below the Earth's surface. As the molten liquid cools, crystals form and the chemical elements separate and concentrate depending on their chemical and physical properties (see Chapter 2). These properties determine whether they can fit into minerals such as quartz and feldspar which are the principal components of granite. Uranium is an element that finds it difficult to fit into these common minerals and, as a result, slowly concentrates in the remaining molten liquid as these minerals form. When the remaining liquid finally crystallizes, the uranium in the last crystals can be very concentrated indeed. After the solid granite forms, millions of years of erosion may strip off the Earth's upper layers and eventually expose the solidified granite. If some of its uranium-rich minerals are exposed at the surface, or if soil forms on top of the granite, radon emissions may be significant.

Sedimentary rocks such as black shale, coal, and limestone also concentrate uranium. These rocks form from the consolidated remains of ancient life. Organic compounds typically concentrate uranium with the result that these sediments are now uranium-rich rocks capable of generating radon. Soils and glacial deposits that overlie uranium-bearing rocks may also be uranium-rich. Radon can also escape from uranium-rich rocks that are deeply buried. Ancient faults and fractures that cut these rocks are often scars that have not healed, and may concentrate and facilitate the escape of radon to the surface.

By implication, regions rich in uranium deposits are most at risk. As you know, there is a heated debate about the exploitation of uranium as a resource. In some instances, however, the removal of uranium-rich rocks and soils may actually be an act of remediation, and may benefit the local environment by removing the source of radon gas.

The potential hazard of radon demonstrates that not all natural processes are beneficial to humans—some may be lethal. It also demonstrates the importance of linking scientists of different fields. In this case, it is unlikely that we will identify areas of potential risk unless the medical and geoscience communities collaborate with their data and interpretations. There is a need for cross-referencing the regional distribution of lung cancer rates with the composition of the rocks and soils of a region.

There is cause for hope, however, in evidence that once humans are removed from areas of excess radon, their bodies recover. If this is correct, identification of radon gas problems and remediation may not only halt future damage, it may also reverse the damage already sustained.

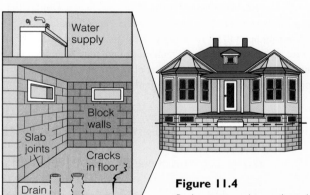

Figure 11.4

Radon enters a house through cracks in the foundation, drains, sumps, joints, or between blocks in walls.

portion of atmospheric oxygen molecules, which consist of two oxygen atoms (O) bonded together, are split apart by radiant energy. When these bonds are broken, two unstable (and therefore highly reactive) oxygen atoms (O) are produced. These unstable atoms collide with surviving oxygen molecules to form ozone (O_3), a gas consisting of three bonded oxygen atoms. In shorthand, the reaction is:

$$O_2 + energy \rightarrow 2\ O$$
$$O_2 + O \rightarrow O_3$$

In summary, volcanic activity is a prerequisite to the formation of an atmosphere. To retain these gases however, a planet must be sufficiently large that its gravitational attraction will prevent the gases from leaking into space. Mars is too small to retain an atmosphere, but the Earth, like Venus, is sufficiently large to do so. However, unlike Venus, the Earth's systems interact. Its volcanic gases have been transformed from a poisonous atmosphere to the life-sustaining air we breathe by their interaction with water and life. Radiant energy from the Sun facilitates the circulation of this air, and also

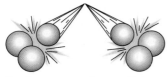

High-energy ultraviolet radiation strikes an oxygen molecule (O_2)...

...and causes it to split into two free oxygen atoms (O).

The free oxygen atoms collide with molecules of oxygen...

...to form ozone molecules (O_3).

Figure 11.5
Ozone in the atmosphere is produced when reactive oxygen atoms (O) that have been split from oxygen molecules (O_2) by ultraviolet radiation from the sun, combine with other oxygen molecules to form ozone (O_3).

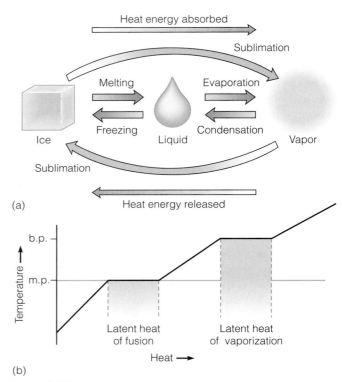

Figure 11.6
(a) Heat energy is absorbed and released in reactions between ice, liquid water, and water vapor. (b) As ice melts to form water (melting point, m.p) the applied heat is used to break the molecular structure of the ice (latent heat of fusion) and the reaction occurs without a rise in temperature. When all the ice is converted to liquid water, the temperature rises once again as more heat is applied until water reaches boiling point (b.p). Then the applied heat is used to break the attraction between the water molecules until they become independent and evaporate (latent heat of vaporization) The temperature remains constant until evaporation of the liquid is complete.

causes subtle but important changes in its composition, including the production of ozone.

11.2 Interaction between Water and Air

As we learned in Chapter 8, water exists on the Earth in all three physical states: solid, liquid, and gas (water vapor) (Fig. 11.6a). Water evaporated from the oceans spends an average of 11 days in the atmosphere, before it is returned to the surface as precipitation. While water vapor may occupy only a small fraction of the atmosphere, its interplay with ocean water is fundamental in atmospheric processes. In addition, the special properties of water, especially its ability to absorb heat, influences atmospheric circulation patterns, climate, and our day-to-day weather.

When a substance is heated, its temperature usually rises. Because water has a tremendous capacity to absorb heat, its temperature rises relatively slowly. One measure of heat is the **calorie,** which is defined as the amount of heat required to raise one gram of water by 1 degree Celsius. Twenty calories of heat will therefore change the temperature of 1 gram of liquid water by 20 degrees Celsius, or 20 grams of water by 1 degree Celsius. But for liquid water to evaporate, enough heat must be absorbed by the water molecules on the ocean surface to allow them to overcome the relatively strong attraction that each water molecule has for its nearest neighbors. As we learned in Chapter 8, the absorption of this heat does not result in a temperature change because the energy is used to overcome this attraction. It is therefore known as **latent heat,** where latent means "concealed" or "veiled" (Fig. 11.6b). It takes 540 calories of heat to overcome the attractive force between water molecules and convert 1 gram of liquid water into water vapor. On the other hand, this tremendous amount of energy is released when water vapor cools down. This explains why scalding by steam is so much more painful than scalds from boiling water.

When a threshold amount of energy is absorbed, the water molecules behave independently and evaporate to become water vapor that enters the atmosphere. As they do so, they rise and cool, eventually converting back to a liquid state, in the form of clouds or fog. This process is known as **condensation** during which the tremendous energy absorbed during evaporation is released. As we will discover in Chapter 12, this energy release plays a critical role in the development of violent weather systems.

The interaction between the atmosphere and the oceans also affects the distribution of other chemicals. Volatile gases that are dissolved in ocean water, such as carbon dioxide, nitrogen, and oxygen, are only loosely held and so accompany the water vapor as it evaporates into the atmosphere. Of the 425,000 cubic km (93,000 cubic mi) of ocean water that evaporates each year, about 25% is precipitated onto the continents (Fig. 11.7). This water, in combination with gases in the atmosphere, reacts with rocks and minerals, extracting soluble elements such as calcium and sodium which it carries back to the oceans.

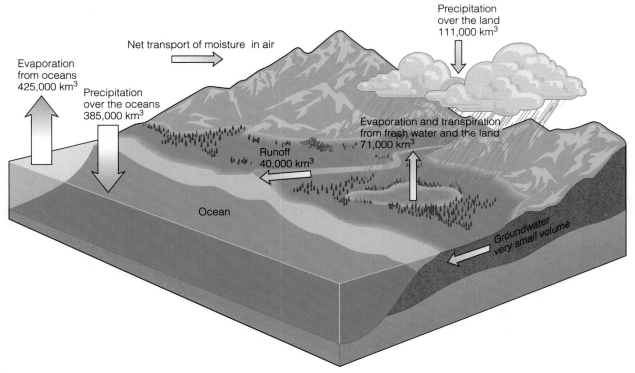

Figure 11.7
The global water budget showing the volume of water involved annually in each part of the hydrologic cycle.

In summary, the interaction between air and water is governed by the role of water and its ability to absorb heat. Evaporation from ocean water occurs when a threshold of energy is reached and water molecules behave independently. Volatile gases accompany water vapor into the atmosphere, and when precipitation occurs, soluble elements from rocks and minerals are extracted and transported back to the oceans.

11.3 The Layers of the Atmosphere

As an airplane takes off and rises toward its cruising altitude of about 11 km (36,000 ft), it is obvious to any window-seat passenger that our atmosphere is not uniform, but layered. As the plane flies above the clouds into clear blue skies, the outside temperature plummets to a frigid −50°C (−58°F). The air pressure outside drops so low that the cabin has to be pressurized to protect the passengers and safety devices, such as oxygen masks, exist should the cabin pressure fall. Indeed we experience a pressure change whenever an aircraft rapidly changes its altitude, and our ears may "pop" to adjust to it.

The layering of the atmosphere is associated with systematic changes in temperature and pressure that occur with increasing distance from the Earth's surface. These variations are shown in Figure 11.8a. Because each layer

encircles the Earth, they, like the atmosphere, are given the suffix *sphere*. These layers are the troposphere, stratosphere, mesosphere, and thermosphere. The boundaries between these layers are given the suffix *pause*. The tropopause, for example, is the boundary layer between the troposphere and the stratosphere.

11.3.1 The Troposphere

The lowest layer, the **troposphere**, contains about 80% of the mass of the atmosphere and extends from ground level to an altitude that varies from 8 to 18 km. The troposphere is thickest near the equator and thinnest near the poles (Figure 11.8b). A sharp rise in the thickness of the troposphere occurs at mid-latitudes and, as we shall see in Chapter 12, this exerts a fundamental control on global-scale weather patterns.

The term troposphere is derived from the Greek *tropos*, meaning "to mix" or "turn." This is an appropriate term, as the troposphere is the main layer of atmospheric circulation. Most of the Earth's weather, for example, is governed by air circulation in the lowermost portion of the troposphere. Aircraft typically fly in the upper portions of this layer where they are well above the cloud cover and so escape the unstable weather conditions below. Frequent flyers know that air turbulence is most commonly felt during takeoff and landing when the aircraft is at relatively low altitudes.

The Earth's surface acts like a radiator in a room. Just as it is colder the further one gets from a radiator, so the temperature

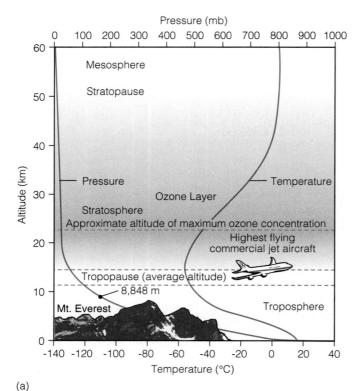

(a)

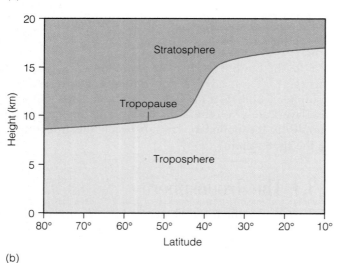

(b)

Figure 11.8
(a) Layers of the atmosphere in relation to the average profile of air temperature above the Earth's surface. Pressure decreases drastically with increasing elevation above the surface. (b) Variations in the thickness of the troposphere with latitude. A jet stream, known as the polar jet, is typically located between 40° and 45° latitude where there is a sharp change in the altitude of the tropopause.

of the troposphere decreases with increasing distance from the Earth's surface. If we sit on rocks exposed to sunlight on a hot day, the rocks are often warm because they have absorbed some of the solar energy. In fact, at the Earth's surface, the air temperature is greatly influenced by the absorption of solar energy by rocks, vegetation, and water. This

energy radiates back as heat into the atmosphere causing the air to be warmest at the Earth's surface. Air cools as it rises because the reduced atmospheric pressure allows it to expand. Cooling occurs because the air uses energy to push aside the surrounding air as it expands. The trend of decreasing air temperature with altitude is especially obvious in mountainous regions, where, on wet days, rain at the bottom of valleys often gives way to snow before we reach the mountain tops. As skiing enthusiasts will vouch, snow and glaciers persist on high mountain peaks because the air is colder at higher altitudes (Fig. 11.9). In fact, the air temperature decreases by an average of 6.5°C per km (or 3.6°F per 1000 ft) above the Earth's surface and temperatures as low as −60°C (−76°F) occur at the top of the troposphere.

This profile of decreasing temperature with altitude is fundamental to our weather patterns, and ensures the maintenance of water on our planet. As warm moist air rises, it cools, and the moisture it contains condenses to form minute water droplets, or clouds. Recall from Chapter 6, for example, the effect of the Himalayas on the moisture-laden air of the monsoons. The condensing water droplets grow bigger until they reach a critical threshold in size, whereupon they fall back to Earth as precipitation. But it is only the thermal profile in the lower atmosphere that allows this to occur. Without a temperature profile that causes moist air to be cooled as it rises, the evaporating water might escape the planet altogether.

As warm air rises, it is replaced by denser, cooler air, which then becomes heated by the warmth of the Earth's surface. Thus, the troposphere is in constant motion, and this motion is responsible for our changing patterns of weather.

11.3.2 The Tropopause

The **tropopause**, or boundary between the troposphere and the overlying stratosphere, is a surface of equal pressure (or **isobar**) extending from the equator to the poles. At this height, the atmospheric pressure is one thousandth of the pressure at the Earth's surface. As shown in Figure 11.8b, the height of this boundary decreases from 18 km (11 mi) near the equator to about 8 km (5 mi) near the poles. The decrease is not gradual, but instead occurs in a narrow zone at mid-latitudes where cold dense polar air meets warm tropical air in the troposphere below. This sharp change has important implications for the formation of jet streams. **Jet streams** are narrow zones of swift (120–240 km/hr, or 70–150 mi/hr) high-altitude winds that profoundly influence global-scale weather patterns. We will learn more about jet streams in Chapter 12.

11.3.3 The Stratosphere

The next layer above the troposphere is called the **stratosphere** which stretches from the tropopause to an altitude of about 50 km (30 mi). The stratosphere, a term derived from the Greek *stratos*, meaning "to smooth out," is above the height of most clouds. For reasons we will learn shortly, the temperature of the air actually increases with altitude in the stratosphere,

(a)

(b)

Figure 11.9
(a) The mountain peaks and glaciers of Mont Blanc, Swiss Alps, remain snow-capped even in summer because the temperature of the troposphere decreases with altitude. (b) For the same reason, high altitude ski slopes can remain open long after all snow in the lowlands has melted.

(Fig. 11.8) so that the layer is very stable with limited turbulence, except at times of solar storms.

The composition of the stratosphere is dominated by nitrogen (about 79%) and oxygen (20%). The stratosphere contains 90% of all atmospheric ozone, including the layer commonly referred to as the **ozone layer.** This term is a little misleading because ozone forms only a very tiny proportion of the stratosphere (0.001%), accounting for only one in every 100,000 molecules of gas. In fact, if all the ozone in the stratosphere was compressed together, it would form a layer of pure ozone only about 3 mm (about one-tenth of an inch) thick.

Despite its low abundance, ozone plays a vital role in the maintenance of life on Earth, and exerts a powerful influence on the thermal profile of the atmosphere. Ozone helps to prevent harmful ultraviolet rays from reaching the Earth's surface by absorbing this radiation. This absorption heats the atmosphere at the level of the ozone layer causing the temperature to rise dramatically in the stratosphere to an average value close to 0°C. The debate on the consequences of potential ozone layer depletion generally focuses on increased ultraviolet ray penetration and its effects at the Earth's surface. However, it is also clear that ozone layer depletion could affect the whole thermal structure of our atmosphere and so influence the Earth's ability to retain water.

11.3.4 The Mesosphere

The **mesosphere** lies above the stratosphere and is the layer where incoming meteors first begin to burn. The boundary between the mesosphere and stratosphere, or **stratopause**, is the point at which the temperature once again begins to diminish with increasing altitude. This change is attributed to the decreasing influence of ozone. The mesosphere therefore does not absorb any significant portion of solar radiation. In fact, the temperature along the upper boundary of the mesosphere may be as low as −100°C (−148°F).

11.3.5 The Thermosphere

Above the upper boundary of the mesosphere, at about 100 km (62 mi) above the Earth's surface, lies the outer atmosphere or **thermosphere.** Air in the thermosphere is very thin and highly responsive to incoming solar radiation. It is especially affected by the varying velocity of the solar wind which bombards the outermost atmosphere. As we shall learn in Chapter 16, the **solar wind** is a stream of ionized gases blown away from the Sun at supersonic velocities. The velocity of the solar wind is profoundly influenced by the abundance of sunspots, which are cold, dark areas near the surface of the Sun. As we will also learn in Chapter 16, the abundance of sunspots varies in an eleven-year period known as the sunspot cycle. During periods of peak activity in this cycle, the velocity of the solar wind is greatly enhanced, and the temperature in the thermosphere can rise as high as 1225°C (2237°F) as a result of the solar bombardment. During periods of low sunspot activity, the velocity of the solar wind is significantly reduced, and temperatures in the outer atmosphere fall to as low as 225°C (437°F). These dramatic temperature changes occur simply because there are very few molecules present in the outer atmosphere to absorb and distribute the heat. As a result, temperatures progressively rise from the base of the thermosphere to the outermost portions of the atmosphere, some 200 km above the surface.

The thermosphere is sometimes called the **ionosphere**, especially if the electrical properties of the layer are being emphasized. Strictly speaking, the ionosphere is defined as the

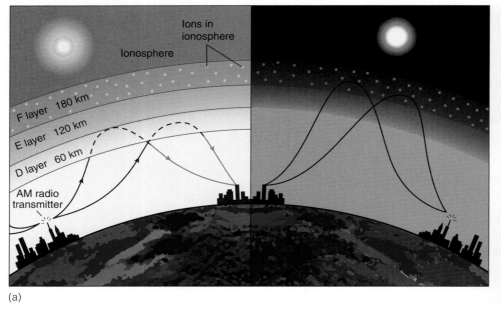

(a)

(b)

Figure 11.10

(a) Behavior of the ionosphere. During the day, the lower ionosphere (known as the D layer) strongly absorbs and weakens AM radio waves. At night, the upper ionosphere (F layer) reflects AM radio waves allowing them to be picked up by distant receivers. (b) Electrical disturbances in the Earth's magnetic field flow downward and interact with the Earth's atmosphere causing the aurora.

electrified region in the upper atmosphere. In fact, this region begins in the mesosphere about 60 km (37 mi) above the surface, and extends upward through all of the thermosphere. During the daytime, direct radiation from the Sun increases ionization in the upper atmosphere, whereas at night the disturbance tends to dissipate.

The ionosphere was discovered in the early 20th century shortly after the radio was invented. At that time, it was thought that radio signals could not travel great distances along the Earth's curved surface because radio waves were absorbed by the ground. However, in 1901, Marconi sent a message from England to Newfoundland using shortwave radio. Within a year, it was suggested that radio waves could travel around the world by being reflected off the ionosphere. This hypothesis, confirmed by later experiments, was a giant step in the development of our modern global-scale communications (Fig. 11.10a).

Electrical disturbances within the ionosphere are visible in the form of **auroras** (Fig. 11.10b), the dramatic curtains of light sometimes visible in the night sky at high latitudes. Also known as the "northern lights," they are the source of many myths and legends. The native tribes of northern Canada believed them to represent the torches of dueling gods. In southern Europe, people believed that the sky was on fire and that the Earth was coming to an end. These spectacular electrical disturbances are thought to be related to solar flares and the power of the solar wind. They will be discussed further in Chapter 16.

In summary, the Earth's atmosphere has a thermal profile that is unique in the Solar System. The Earth's surface absorbs and radiates solar energy back into the atmosphere. As a consequence, temperature decreases with altitude in the troposphere. This thermal structure helps to confine water and water vapor to regions near to the planet's surface.

The dramatic change in temperature from the troposphere to the stratosphere is attributed to the ability of ozone to absorb radiation. Our current concern about ozone depletion and the penetration of ultraviolet radiation, is not just limited to our exposure to this radiation and its relationship to skin cancer. Enhanced penetration of ultraviolet radiation may well affect the thermal profile of our lower atmosphere, thereby profoundly affecting the hydrologic cycle.

11.4 Weather and Climate

Weather is a description of atmospheric conditions at any one time in a given place. This description includes an evaluation of atmospheric pressure, air temperature, and relative humidity. Most recently, another item has been added to the description, the UV index, reflecting our increasing concern about ultraviolet (UV) radiation.

A weather forecast for a given region typically predicts what these atmospheric conditions are expected to be over the next 24 to 48 hours. This forecast is usually derived from atmospheric data in surrounding regions and by keeping track of the motion of air masses. The advent of satellite and computer technology has made major advances in the science of weather forecasting.

Climate is a generalized description of the totality of weather variations, usually on an annual basis. In North America, we can predict the seasonal changes in temperature, pressure, and relative humidity associated with spring, summer, fall, and winter. Depending on where you live, seasonal changes are manifest in different ways and are used to define climate belts (Fig. 11.11a). For example, if you live away from the coast near the Canada-United States border, you will be used to having hot summers and cold winters. This pattern of seasonal change is referred to as a continental climate. If you live in the far north, you will be familiar with the bitter cold, especially in the winter, and the climate is referred to as a polar climate.

In the western part of North America, the climate is profoundly influenced by the mountains. The mountains themselves may be snow-capped for a considerable portion of the year, and have a characteristic climate of their own described as a highland climate. The mountains block the eastward passage of warm moist air from the Pacific and so produce a dry region or **rain shadow** on the eastern side of the mountain range. In the mountainous regions of the southwestern United States, for example, this Pacific air is forced upward by the Coast Range and Sierra Nevada mountains, so that precipitation occurs preferentially on the westward side of the mountains (Fig. 11.11b). As a consequence, the air loses its moisture before it reaches central Nevada on the eastern side of the mountains. Regions east of the mountains consequently tend to be arid or semi-arid desert (Fig. 11.11c). The western coastal regions experience the moderating influence of the adjacent Pacific Ocean. The warm climates of the southern coastal regions of the United States are known as mediterranean climates, because of their similarity to those of southern Europe. The Canadian coastal regions, because they are further north, have slightly cooler climates called temperate or subtemperate.

11.4.1 Atmospheric Pressure and Wind

The Earth's atmosphere has a mass of five thousand million tons, the vast majority of which lies within 18 km (11 mi) of the Earth's surface. The force of gravity maintains this mass in an envelope that surrounds the entire Earth. The downward force of gravity gives the atmosphere a pressure, or a force per unit area. Gravity is thus a global force that contributes to **atmospheric pressure**. However, local conditions leading to rising and falling air masses also contribute to the total atmospheric pressure and will be discussed below.

Atmospheric pressure can be demonstrated by a very simple experiment. A tube filled with mercury is immersed in a beaker also containing mercury. If the closed end of the tube is held in an upright position, a column of mercury is held in the tube at a certain height above the beaker. If this procedure is repeated, and each of the tubes are positioned side-by-side, the height of the column in each tube is identical (Fig. 11.12), irrespective of the diameter of the tube.

The height of the mercury in each tube is directly related to the pressure of the surrounding atmosphere. Indeed, an adaptation of this procedure is still a very popular way of measuring atmospheric pressure. The pressure, or downward force exerted by the atmosphere on the surface of the mercury in the beaker, is sufficient to support the column of mercury in the tube. The higher the pressure, the higher the column of mer-

cury supported. At sea level, the average height of the mercury column is 76 cm (29.9 in.).

The pressure supporting the column reflects the weight of all the air in the atmosphere that is vertically above the surface of mercury in the beaker. Because of this, air pressure decreases dramatically with height above the Earth's surface (see Figure 11.8a). For example, at an altitude of 11 km (close to the cruising altitude of a jet airplane), the air pressure is only 25% of what it is at the Earth's surface. This is simply because there is less air above this altitude than there is at ground level.

Because of turbulence in the troposphere, air currents rise and fall. Warm, less dense air rises from the Earth's surface where it cools, and cool air descends toward the Earth's surface where it is warmed. Rising warm air creates a local area of low pressure, because the overlying column of air is anomalously light. Descending cold air creates a local area of high pressure because the overlying column of air is anomalously heavy. Air masses tend to rise or descend in columns or plumes, creating low pressure and high pressure centers (Fig. 11.13). Low pressure centers are also known as **cyclones,** and high pressure centers are known as **anticyclones.** Because these air masses are in constant motion, atmospheric pressure in any one location can vary from day-to-day or even from hour-to-hour. Monitoring and predicting these pressure changes is fundamental to weather forecasting. As we know from weather charts, our atmosphere develops centers of high and low pressure on any given day, so that the height of a column of mercury in the tube may vary as time goes by or from place to place.

Winds are a direct result of variations in atmospheric pressure. Pressure differences cause the air to move which, like all fluids, flows from high pressure regions to regions of low pressure. For example, the rush of wind we experience when we enter an air conditioned building is caused by the difference between the controlled air pressure of the building and the uncontrolled air pressure of the atmosphere outside.

Weather forecasters can predict wind velocity by studying the location and strength of high and low pressure air masses in a region. If variations in air pressure are small, we experience a relatively calm day. However, if the high and low pressure centers are close to each other and the pressure difference between them is high, strong winds would be predicted. This relationship is similar to a slope on a hill. The slope is steep where summit and valley are close together and gentle where they are far apart. Like a ball rolling down a hill, the wind is strongest when the pressure gradient (or "slope") is greatest.

On weather maps, variations in atmospheric pressure are represented by lines called **isobars** (Fig. 11.13). An isobar is a line connecting locations of equal atmospheric pressure. It is analogous to a contour line on a topographic map which connects points of equal elevation on the Earth's surface. The pattern of isobars depicts pressure variations in the atmosphere, just as contours depict the shape of the Earth's surface. Thus, closely spaced isobars indicate strong winds, just as closely spaced contours indicate steep slopes. For example, on a weather map of North America (Fig. 11.14), the pattern of the isobars suggests that the winds should be strongest in the southern mid-continent (between 90 to 100 km/hour or 55 to

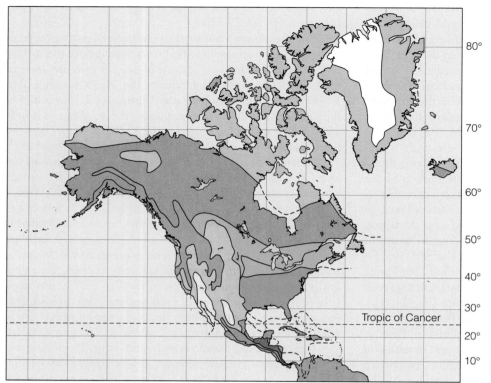
(a)

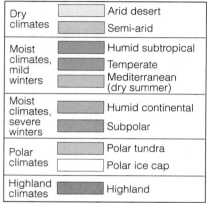

Dry climates		Arid desert
		Semi-arid
Moist climates, mild winters		Humid subtropical
		Temperate
		Mediterranean (dry summer)
Moist climates, severe winters		Humid continental
		Subpolar
Polar climates		Polar tundra
		Polar ice cap
Highland climates		Highland

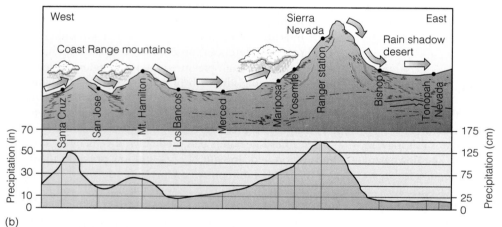
(b)

(c)

Figure 11.11
Controls of climate in North America.
(a) In eastern North America, climate belts run subparallel to latitude. In western North America, the mountains have a profound influence. (b) Effect of coastal mountain ranges on annual average precipitation from the Pacific Ocean to western Nevada. (c) Moist air rising over the Sierra Nevada Mountains cools to form clouds that precipitate their moisture (background), creating a cloud-free rain shadow (foreground) in the valley. (Photo John Buckland-Nicks)

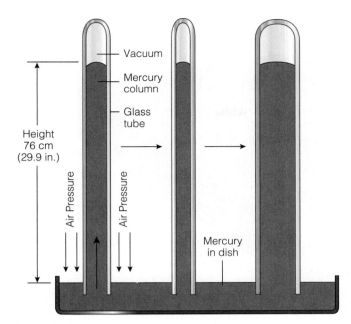

Figure 11.12
Mercury barometer. The height of the column of mercury in a glass tube is a measure of atmospheric pressure and remains the same regardless of the diameter of the tube.

60 mi/hr) where the isobars are relatively closely spaced, and weakest across much of southern Canada and the northern United States (24 to 40 km/hr or 15 to 25 mi/hr).

Seasonal variations in weather may also be influenced by long-term cycles in rising or falling atmospheric pressures. Indeed, these cycles may help define climate belts. Generally, as the interior of a continent heats up in the summer, air rises and creates a low pressure center. As it cools in the winter, the air becomes heavier, and a high pressure center develops. Because air moves from high pressure centers to low pressure centers, cold winter air from continental interiors spreads out and exerts a dominant influence on winter climate.

Winter winds tend to cool warm objects down to the temperature of the surrounding air. The stronger the wind, the faster the rate of cooling. The temperature of the human body averages 36°C (97°F), and in the winter season this temperature is generally considerably higher than that of the air. Our bodies therefore lose heat. The rate at which our bodies lose heat is controlled by a thin insulating layer of air immediately adjacent to our skin that is held stationary by friction. On a relatively calm winter's day, this insulation can reduce our rate of heat loss so that we feel relatively comfortable. If there is a sudden gust of wind however, we feel chilled because the wind substantially reduces the thickness of the insulating layer of air, and our bodies lose heat more rapidly. So the rate at which our bodies lose heat is partly dependent on the strength of the wind. We call this the **windchill factor**, which is a measure of the rate of cooling of exposed parts of the skin to the temperature of the surrounding air. For example, the combination of a wind speed of 40 km/hr (25 mi/hr) and an air temperature of −7°C (20°F) will cool your body as rapidly as an air temperature of −26°C, (−15°F) with no wind (Table 11-2). In order to convey this important information to the public, a weather forecaster will report an air temperature of −7°C (20°F) and a windchill factor of −26°C (−15°F).

11.4.2 Heat and Temperature

Temperature is generally thought to be a measurement of heat and a mercury thermometer is the most familiar method of measuring it. As heat is applied, the mercury inside the thermometer expands and rises up the tube. As the mercury is heated, its atoms or molecules move with greater velocity, and are said to possess a greater kinetic energy (the energy of motion). **Heat** is the *total* kinetic energy of all atoms or molecules in a substance. **Temperature** is a measure of the *average* kinetic energy within a body. As heat is applied to a substance, its atoms move with greater average velocity, and so its temperature rises.

Because heat is the total amount of kinetic energy in a substance, its quantity depends on the volume of the substance. Temperature, on the other hand, is an average value, and so is not volume-dependent. For example, if a cup of water and a bathtub full of water have the same temperature, the bathtub of water will have more heat. As a consequence, the water in the bathtub will take longer to cool to the surrounding temperature than the water in the cup. Because of their vast volume, ocean waters consequently store far more heat than smaller bodies of water with similar temperatures.

In solids, atoms and molecules remain relatively coherent when heated. They respond to increasing heat by vibrating more violently around relatively fixed positions. In gases, such as those in our atmosphere, the motion of atoms and molecules is much less well defined, and their movement when heated is much more chaotic. This property influences their behavior when they are subjected to solar radiation.

The atmosphere plays a fundamental role in balancing the Earth's heat budget. For every 100 units of radiant energy from the Sun, 26 are reflected back into space by dust and clouds and 4 are reflected by the Earth's surface (Fig. 11.15). A total of 70 units are absorbed, 19 by ozone and clouds in the atmosphere, and the remaining 51 by the Earth's surface. To balance this budget, 70 units of energy must be radiated back into space. If less were radiated, the planet would warm up. If more, the planet would cool down. Of the 70 units absorbed, 64 are radiated back into space from the atmosphere and 6 from the Earth's surface. This budget bears on our concern about the greenhouse gas content of the atmosphere. An increase in the concentration of greenhouse gases such as carbon dioxide may affect this delicate balance by blocking the escape of radiated heat into space. The importance of greenhouse gases to atmospheric temperatures will be discussed in detail in Chapter 20.

As we have learned, when a body absorbs heat, its temperature generally rises. However, different bodies have varying abilities to absorb heat. On a sunny day, for example, the grass is generally cooler than any adjacent rocks or the asphalt on a nearby road. Because the grass consists mostly of water, it can absorb more of the heat and so its temperature does not rise as

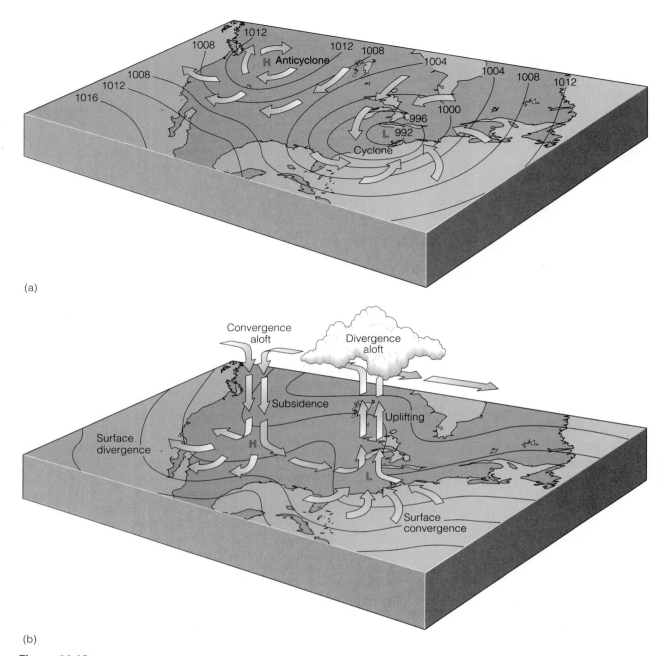

(a)

(b)

Figure 11.13
(a) Weather chart showing bulls-eye pattern of low pressure (cyclone) and high pressure (anticyclone) centers. The lines represent isobars. Each isobar is shown as a line connecting air of equal pressure. The numbers reflect the atmospheric pressure appropriate for each isobar. Note the values are highest surrounding the high pressure center and are lowest surrounding the low pressure center. (b) Pattern of air flow on the surface and aloft associated with high and low pressure centers.

sharply as the rocks or asphalt, even though each has received approximately the same amount of radiant energy from the Sun. This is because different materials absorb and radiate heat in their own characteristic manner depending on their particular atomic or molecular structures. For any given material, the relationship between heat, and the rise in temperature associated with it, is a characteristic property of that material known as its **heat capacity.** Heat capacity is defined as the quantity of heat required to raise the temperature of a substance by 1°C.

Air has a very low heat capacity. It responds very quickly to changing heat supply, and as we well know, its temperature can fluctuate dramatically. Water has a tremendous capacity to store heat, and therefore responds very slowly and can absorb a lot of heat from the Sun before its temperature rises. Ocean waters

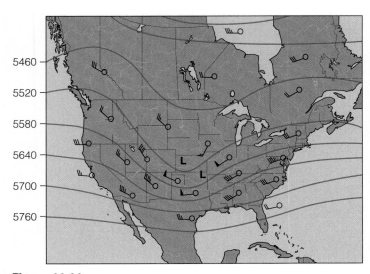

	Statute miles/hour	Knots
⊚	Calm	Calm
—	1-2	1-2
⌐	3-8	3-7
⌐	9-14	8-12
⌐	15-20	13-17
⌐	21-25	18-12
⌐	26-31	23-27
⌐	32-37	28-32
⌐	38-43	33-37
⌐	44-49	38-42
⌐	50-54	43-47
⌐	55-60	48-52
⌐	61-66	53-57
⌐	67-71	58-62
⌐	72-77	63-67
⌐	78-83	68-72
⌐	84-89	73-77
⌐	119-123	103-107

Figure 11.14

Weather map with isobars, and wind velocities. Note wind velocities are highest where the isobars are closely spaced because there is a strong gradient in atmospheric pressure.

Table 11–2

Windchill Equivalent Temperature (°F). A 25-mi/hr wind combined with an air temperature of 20°F produces a windchill equivalent temperature of –15°F

	AIR TEMPERATURE (F°)																
	35	30	25	20	15	10	5	0	–3	–10	–15	–20	–25	–30	–35	–40	–45
5	32	27	22	16	11	6	0	–5	–10	–15	–21	–26	–31	–36	–42	–47	–52
10	22	16	10	3	–3	–9	–15	–22	–27	–34	–40	–46	–52	–58	–64	–71	–77
15	16	9	2	–5	–11	–18	–25	–31	–38	–45	–51	–58	–65	–72	–78	–85	–92
20	12	4	–3	–10	–17	–24	–31	–39	–46	–53	–60	–67	–74	–81	–88	–95	–103
25	8	1	–7	–15	–22	–29	–36	–44	–51	–59	–66	–74	–81	–88	–96	–103	–110
30	6	–2	–10	–18	–25	–33	–41	–49	–56	–64	–71	–79	–86	–93	–101	–109	–116
35	4	–4	–12	–20	–27	–35	–43	–52	–58	–67	–74	–82	–89	–97	–105	–113	–120
40	3	–5	–13	–21	–29	–37	–45	–53	–60	–69	–76	–84	–92	–100	–107	–115	–125
45	2	–6	–14	–22	–30	–38	–46	–54	–62	–70	–78	–85	–93	–102	–109	–117	–125

(Wind speed (miles/hr) labels the rows at left.)

along the coast of eastern North America consequently stay cool well into the summer, their slow rise in temperature lagging well behind the rise in air temperature over the land. In Atlantic Canada, for example, the ocean water never really warms up in the summertime although air temperatures can reach 30 to 35°C (86 to 95°F). Thus coastal regions tend to have much more moderate summers than those of inland regions because the breezes that originate over ocean water remain cool.

For the same reasons, the air temperature in the wintertime drops much more suddenly than the temperature of ocean water. As a result, coastal regions have warmer winters than inland regions. The varying response of air and water to seasonal changes in heat supply explains the contrasting temperatures between the air and the surface waters of oceans and lakes. This contrast is most pronounced in coastal regions and is one of the main causes of coastal fog.

11.4.3 Relative Humidity

Humidity is a term we use often to describe an aspect of our weather without knowing exactly what it is. For most of us, humidity is a kind of comfort index. We know that humid days are uncomfortable, and that it is somehow related to the moisture content of the air. This moisture content varies from about 4% to 0.004%. Clouds and fog (which is essentially a ground-hugging cloud) are the most obvious manifestations of the presence of water in our atmosphere. However, small amounts of moisture are always dispersed throughout the air. When the air temperature drops, its presence is betrayed by the occurrence of frost on cold clear nights, ice on windshields, or dew and fog in the mornings.

Technically, **relative humidity** is the relationship between the temperature of the air and the amount of water vapor the

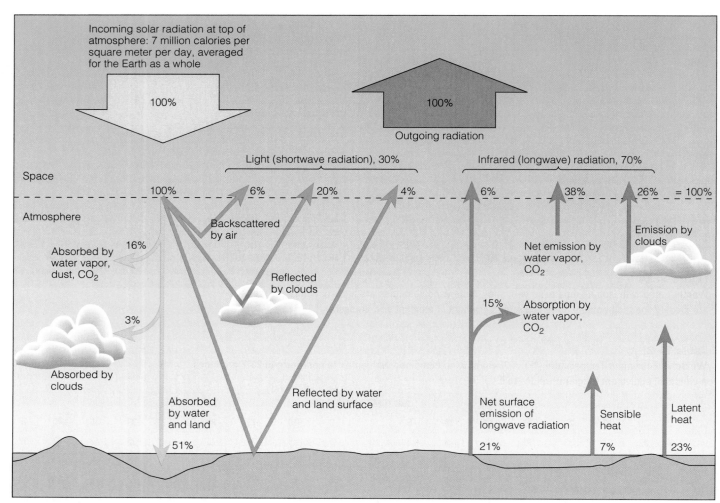

Figure 11.15

Heat budget from the Sun. On average, 30% of solar energy is reflected by the upper atmosphere and the Earth's surface, and is returned to space (see shortwave radiation). The remaining 70% is absorbed: 19% by the atmosphere and clouds and 51% at the Earth's surface (left). To balance the Earth's heat budget, the absorbed heat must be radiated back into space (see longwave radiation).

air contains. At any given temperature, there is a maximum amount of moisture that the air can hold. When we are told, for example, that the relative humidity is 50%, it means that the air contains half the maximum amount of water it can hold at that temperature. On hot and humid summer days, the moisture content is close to the maximum level. Because water has a tremendous ability to store heat, this makes the air very oppressive, and it is difficult for us to feel cool.

The concept of relative humidity is very similar to dissolving sugar in tea. If the tea is hot, a considerable amount of sugar can be dissolved in it. As the tea cools, its ability to hold the sugar in solution diminishes even though the concentration of the sugar in the tea is the same. Upon further cooling, the sugar reaches saturation level and sugar crystals are deposited at the bottom of the cup. These crystals will re-dissolve if the tea is reheated.

Similarly, the air's ability to hold water varies as the air temperature changes. Because cool air can hold much less

moisture than warm air, the relative humidity of rising air increases as it cools, even if the concentration of water vapor in the air remains constant. If the air continues to cool, the relative humidity may rise to saturation level. When this occurs, the total amount of moisture in the air is the maximum amount the air can hold. The water vapor is then no longer invisible and forms the visible droplets we call clouds. If cooling continues, the droplets become more abundant, coalescing and growing larger until they are too heavy to be held suspended in the air. As a result, precipitation occurs as either rain or snow, depending on the air temperature.

Typically, surface air temperatures are cooler during the night than they are during the day. Because the cool night air can hold much less moisture than warm daytime air, its relative humidity is higher (Fig. 11.16). As the ground continues to cool, it may become colder than the air immediately above it. When this occurs, the ground resembles a refrigerator rather than a radiator. A layer of cold dense air develops immediately above the ground, which because of its density, moves toward low lying areas. What happens next depends on how low the surface temperature drops. If it remains cool, but above freezing, the cold ground-hugging air condenses to form fog. In the morning, the fog burns off as its upper surface becomes exposed to sunlight. If the night-time temperature drops suddenly below the freezing point, the moisture may crystallize, just as the sugar would in cold tea. This can result in frost and black ice on the roads. A common misconception is that icy roads only result when a wet surface freezes, or when a cover of snow is compressed. But this is not always necessary. In fact, a sudden drop in air temperature alone can be extremely dangerous, particularly in the critical early morning period when an initial thaw may generate a thin film of water on top of the ice.

Many of the important characteristics of our atmosphere are the result of an important interplay between atmospheric pressure, temperature, and relative humidity (Fig. 11.17). As an air current rises or descends, it may do so without a significant exchange of heat with the surrounding air. As we have learned, air pressure decreases with height above the surface. As an air current rises, it therefore encounters surrounding air with a lower pressure. Consequently, it expands as its molecules push aside the molecules of the surrounding air. Because it uses up some of its own energy to expand, however, the average speed of its molecules is reduced, and therefore its temperature drops. The drop in temperature causes a rise in the relative humidity and clouds to form. This explains why low pressure systems, which are characterized by ascending air currents are typically associated with cloudy weather.

Alternatively, as an air current descends, its pressure increases and its molecules become compressed by the surrounding air. This increases the average speed of the molecules, which causes a rise in temperature and a decrease in relative humidity. As a result, high pressure systems, which are characterized by descending air currents, are typically associated with clear skies.

Thus, where little or no heat is exchanged with the surrounding air mass, expansion of air results in cooling, increasing relative humidity and cloudy skies, whereas compression of air results in warming, decreasing relative humidity and clear, sunny weather.

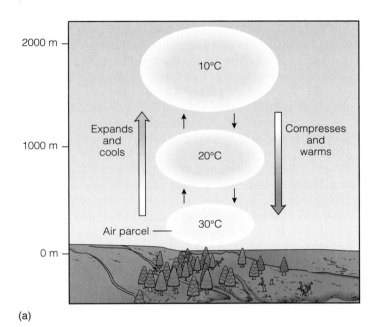

(a)

(b)

Figure 11.17
(a) Ascending air parcel cools as it expands. This increases the relative humidity, leading to the formation of clouds. Descending air parcel warms as it compresses. This decreases the relative humidity resulting in clear skies. (b) Warm moist air evaporating from the ocean expands and cools as it rises. Its relative humidity increases, leading to the formation of clouds.

Figure 11.16
Assuming stable weather conditions, where the air is cool in the mornings, the relative humidity is high. As the air warms up toward the afternoon, the relative humidity decreases because the warmed air has an increased capacity to retain water.

11.5 Chapter Summary

Our Earth is layered according to density from its heavy iron-rich core to its gaseous atmosphere. The atmosphere extends to the outermost reaches of our planet, forming the boundary between Earth and the space beyond. The atmosphere absorbs solar radiation, which drives its circulation and profoundly influences our weather and climate. At the same time, the atmosphere screens out some of the harmful effects of the Sun's radiation.

The atmosphere consists of gases vented from 4.6 billion years of volcanic activity. Its composition has been substantially modified by photosynthesis, the hydrologic cycle, and the production of a protective ozone layer through its interaction with the Sun's radiant energy. Gravity prevents atmospheric gases from dissipating into space.

Like the solid Earth, the atmosphere is also layered, with most of the moisture and weather being confined to its lowermost layer, or troposphere, which varies from 8 to 18 km (5 to 11 mi) in thickness. The troposphere seals in the Earth's moisture, because warm air cools and condenses as it rises in the troposphere, and so moisture eventually returns to the Earth's surface as precipitation. Because the Earth's surface absorbs solar radiation, it acts as a radiator. The air temperature within the troposphere consequently decreases with altitude. This trend is reversed in the ozone-bearing stratosphere above, because ozone absorbs the ultraviolet portion of solar radiation which heats this level of the atmosphere. The temperature decreases once again in the overlying mesosphere as the influence of ozone wanes. Temperature increases once again in the outermost thermosphere due to the effect of solar radiation on the very sparse atmosphere.

Day-to-day atmospheric conditions are referred to as weather, whereas long-term variations in weather, usually described on an annual basis, are defined as climate. Both are related to variations in atmospheric pressure, temperature, and relative humidity. Atmospheric pressure results primarily from the downward pull of gravity on the atmosphere. Winds are the result of variations in atmospheric pressure, and the strength of the wind is related to the pressure gradient between regions of high and low pressure.

Heat is a measure of the total amount of kinetic energy in a body, whereas temperature is a measure of the body's average kinetic energy. As the air heats up, its gases move with greater velocity and the temperature rises. Because air responds much more rapidly to changing heat supply than water, oceans have a moderating influence on the climate of coastal regions.

Relative humidity describes the relationship between air and the amount of moisture it contains. As air cools, its relative humidity rises, leading to the formation of clouds, fog, or precipitation.

Key Terms and Concepts

Look for the highlighted items on the web at:
WWW.BROOKSCOLE.COM

- The origin of the Earth's atmosphere is related to volcanic activity. Its gases are held in by gravity, and their composition has been refined by interaction with the biosphere and hydrosphere

- The layers of the atmosphere comprise the troposphere, near the Earth's surface, overlain by the stratosphere, mesosphere, and thermosphere

- Temperature and pressure variations in the atmosphere are influenced by the Earth's surface (which acts as a radiator) and by the absorption of ultraviolet energy by ozone in the stratosphere

- Most of the Earth's weather is confined to the troposphere

- Variation in air pressure, temperature, and relative humidity have an important effect on day-to-day weather

- Weather is a description of atmospheric conditions at any one time. Climate is a generalized description of weather variations on an annual basis

- Isobars are lines that connect locations of equal atmospheric pressure. Winds depend on pressure gradients, and are strong where isobars are relatively closely spaced

- Windchill factor is a measure of the rate at which a hot body cools to the surrounding temperature

- Heat is the total amount of kinetic energy in a substance, and temperature is the average kinetic energy of the substance

- Different materials have differing capacities to absorb heat and therefore have different heat capacities. Air has a much lower heat capacity than water

- Relative humidity is a measure of the maximum amount of moisture the air can hold at a given temperature

- Air holds more water vapor at warmer temperatures than colder temperatures

- Rising air expands and cools. As it does so, its relative humidity rises, and eventually saturation is met, giving rise to water droplets that may coalesce to form clouds and perhaps precipitation

- Descending air compresses and warms, and its relative humidity decreases, leading to warm sunny weather

Review Questions

1. Why is nitrogen the most abundant gas in the Earth's atmosphere?
2. What dictates the upper boundary of the tropopause?
3. Why does atmospheric pressure decrease with altitude?
4. Why is the troposphere known as the layer of mixing?
5. What are *rain shadows*, and what controls their location?
6. Why does the temperature profile for air decrease in the troposphere?

7. What is the distinction between heat and temperature?
8. What is the relationship between pressure, temperature, relative humidity, and the formation of clouds?
9. Explain the presence of early morning dew or fog.
10. Why are low pressure centers associated with cloudy weather?

Study Questions

Research the highlighted question using InfoTrac College Edition.

1. Explain the change in the temperature profile of the atmosphere that occurs at the tropopause. What is its significance?
2. What do you think would happen to the temperature profile of the atmosphere upon depletion of the ozone layer and what would be the environmental consequences?
3. What important role does the ozone layer play in the temperature structure of the atmosphere and why?
4. Explain what is meant by heat capacity. In view of the expense of home heating, is it less expensive to heat a humid room or a dry room if both have the same cool temperature? What effect would a de-humidifier have on the heating bill?
5. In what way is the heat capacity of water important in moderating climate near coastal regions in northern North America?

6. What effect does the output from Earth's internal heat flow have on the thermal profile of the atmosphere? Explain.
7. Why does air cool as it rises?
8. The formation of clouds can occur in a rising air mass without the addition of water or external input or output of heat. Explain.
9. Why is radio wave reception better at night, and why do auroras affect this reception?
10. By looking out of the window, can you deduce whether the weather is being influenced by a high or a low pressure system? Explain.

12

Circulation of the Atmosphere

Like the oceans, the atmosphere is in constant motion. Atmospheric circulation is driven by temperature contrasts between the Earth's polar and equatorial regions, and its path is influenced by the Coriolis effect caused by the Earth's spin. This circulation produces rising and descending air masses. In this chapter, we will see how atmospheric circulation influences our weather and climate, and how it generates weather frontal systems.

We start the chapter studying the effect of the tilt of the Earth's axis, and find that it explains the changing seasons. We then study heat transport from equatorial to polar regions. Equatorial regions receive more energy from the Sun than they radiate back from their surface, whereas the reverse is true of polar regions. This imbalance creates a strong equatorial to polar flow of heat. Complications in this flow pattern result in a persistent global circulation pattern that has a prime influence on weather and climate. We will find that this pattern can explain the distribution of important features such as the world's major deserts and the marine wind patterns upon which our ancestors depended for trade.

Like ocean currents, atmospheric circulation is also profoundly influenced by the Coriolis effect (due to the Earth's spin). Because of turbulence in the troposphere, air masses tend to descend and ascend in columns, or plumes, generating high and low pressure centers, respectively. We shall find that because of the Coriolis effect, in the northern hemisphere winds circulate around high pressure centers in a clockwise direction and around low pressure centers in a counterclockwise direction.

Interaction between contrasting air masses may result in the development of frontal systems. We will see that different cloud patterns may reveal the nature of this interaction. The ongoing tussle between air masses is influenced by the presence of swift high-altitude winds, known as jet streams, which have a daily influence on weather patterns.

We will learn how, in certain situations, the forces of nature may periodically become amplified to create major storms. These storms may be local, as in the case of most thunderstorms and tornadoes, or regional, as in the case of hurricanes and monsoons. By the end of this chapter, you should have a firmer understanding of the delicate, yet dynamic character of the atmosphere and its complex interaction with the hydrosphere and solid Earth.

12.1 Atmospheric Circulation

Atmospheric circulation is global in scale, and any gases introduced into the atmosphere, either naturally from volcanoes or by pollution, are distributed globally. This is demonstrated by the remarkably uniform composition of the atmosphere in various parts of the world. Whenever volcanic eruptions occur, the noxious gases emanated are rapidly dispersed in the atmosphere. When Mount Pinatubo erupted in 1991, for example, it took a mere three weeks before its telltale signature could be detected worldwide in the analyses of atmospheric chemistry (see Chapter 1). Similarly, radioactive particles from the 1986 nuclear accident at Chernobyl in the former Soviet Union were detected worldwide within days of the explosion.

Several important factors combine to influence the pattern of atmospheric circulation. First, the tilt of the Earth's axis in relation to the Sun profoundly affects the amount of radiant energy received at different points on the Earth's surface. This tilt controls the length of day time versus night, and is the main cause of our changing seasons. Second, the motion of hot air from equatorial to polar regions and cold air in the opposite direction provides an explanation for the world's climate belts. Third, as air masses migrate, they come under the influence of the Earth's spin, resulting in an apparent deflection in the path of air masses known as the *Coriolis effect*. Named after the French mathematician who first described it, the Coriolis effect was discussed in detail in Chapter 8, in our examination of ocean water circulation. Like ocean water, the air is not attached to the Earth beneath it, and as the solid Earth spins beneath its watery and atmospheric envelopes, the paths of both waters currents and air masses are deflected by this spin.

12.1.1 The Effect of the Earth's Tilt

The Earth is not an upright planet, it leans a little on its side and is tilted relative to the plane of its orbit around the Sun. The amount of tilt, however, has varied throughout geologic time from about 21.5° to 24.5°. It is presently inclined at 23.5° (Fig. 12.1) and is decreasing its tilt by a small amount annually.

What causes this tilt is unclear. One possibility may relate to the fact that the Earth is not a perfect sphere. But the most widely accepted hypothesis is that the tilt was caused by a massive impact with a Mars-sized body soon after the Earth formed (we will learn in Chapter 15 that this collision may have also resulted in the formation of the Moon). Most of the other planets in the Solar System are also tilted, but to varying degrees.

The tilt of the Earth's axis means that the Sun does not shine directly overhead at the equator for more than two fleeting moments per year. Because the tilt is 23.5°, on one day each year the Sun shines directly overhead at 23.5°N latitude. This day is June 21st, and the latitude is called the Tropic of Cancer. On this day, all localities in the northern hemisphere have maximum daylight (summer), while all localities in the southern hemisphere have their longest nights (winter). On December 21st, the situation is reversed and the Sun shines directly overhead at 23.5°S latitude. This latitude is called the

Figure 12.1 (top right)

The tilt of Earth's axis relative to the Sun is 23.5°, so the terminator (AB), which is the boundary between night and day, is inclined at an angle of 23.5° to the Earth's axis. In the summer, the northern hemisphere is tilted toward the Sun and therefore has more direct and longer hours of sunlight. However, the sunlight that reaches the Earth's surface in the far northern latitudes passes through a thicker layer of atmosphere (which absorbs, scatters, and reflects solar radiation) and so is less intense than the sunlight shining on equatorial latitudes. On June 21st (shown here), the Sun shines directly on latitude 23.5°N, which defines the northern limit of the tropics (the Tropic of Cancer). In this orientation, areas south of the Antarctic Circle receive no sunlight. The Antarctic Circle has a latitude 66.5°S (90°–23.5°). Conversely, areas north of the Arctic Circle (at latitude 66.5°N) receive 24 hours of sunlight.

Tropic of Capricorn. In the six months between these two cardinal dates, the Sun shines most directly on positions between these latitudes, gradually shifting over a six month period from one tropic to the other. From June 21st until December 21st, the length of daylight in the northern hemisphere progressively decreases to a minimum, while it increases to a maximum in the southern hemisphere (Fig. 12.2).

So in any year, there are only two occasions when the Sun shines directly on the equator. These two occasions occur on March 20th and September 22nd, the days that are known as the equinoxes. On these two days, all areas on Earth have twelve hours of daylight and twelve hours of darkness.

As you know, the Earth is hottest in the tropics and coldest near the poles. This is because the Sun's rays strike the Earth most directly in the tropical region and least directly at the poles. In turn, this influences the amount of solar radiation that reaches the surface. Solar radiation reaching the Earth's surface in higher latitudes is less intense than it is in lower latitude equatorial regions because the radiation travels through a thicker layer of atmospheric gases which absorb, reflect, and scatter it (See Figure 12.1).

Day and night exist on Earth because only half the planet can face the Sun at any one time. The half that faces the Sun is always illuminated, whereas the opposing half is always in darkness. When it is daytime we are on the illuminated side, and when it is nighttime, we are on the dark side. In Figure 12.1, the boundary line between day and night, or the "terminator," is shown by the line A–B. Because of the Earth's tilt, this line does

Figure 12.2 (bottom right)

The Earth revolves around the Sun with the northern hemisphere pointing toward to Sun during the northern hemisphere summer and away from it during the northern hemisphere winter. Because the Earth's axis is tilted at 23.5°, the Sun shines directly on latitude 23.5°N (the Tropic of Cancer) on June 21st. This date marks the summer solstice in the northern hemisphere when the day is longest. On December 21st (the winter solstice), the Sun shines directly on latitude 23.5°S (the Tropic of Capricorn) and the northern hemisphere day is shortest. On the equinoxes of March 20th and September 22nd, day and night are the same length (12 hours) everywhere on Earth. The diagrams are not to scale.

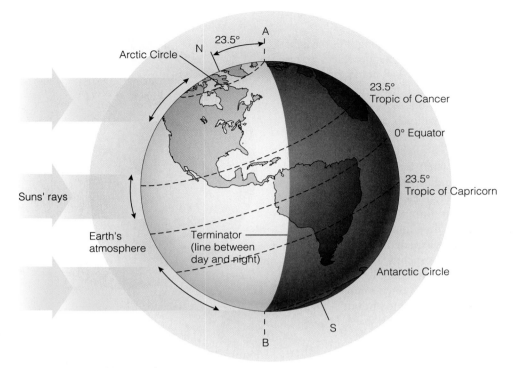

Summer in the Northern Hemisphere (June 21st)

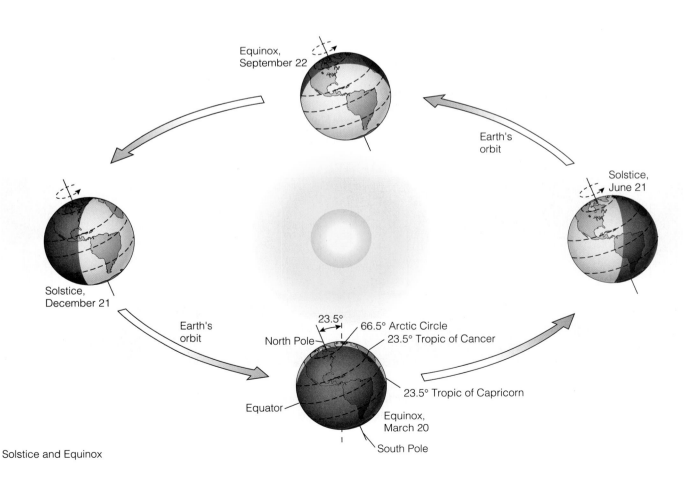

Solstice and Equinox

not coincide with the Earth's north-south axis of rotation, but instead makes an angle of 23.5° with this axis.

The Earth's tilt causes variations in the length of daylight from one place to another. To see how this happens, let's examine Figure 12.1 in more detail. The line A–B marks the boundary between day and night. As the Earth spins, the areas crossing A–B experience sunrise on one side of the Earth and sunset on the other. The Earth's rotational axis (N–S) is drawn through the north and south poles. As you would expect, the angle between A–B and N–S is 23.5°, the tilt of the Earth's axis.

Notice that on June 21st, when the north pole is tilted toward the Sun, the area surrounding the north pole between lines A–B and N–S receives sunlight all day. Rotate any point on the Earth's surface within that illuminated area around the Earth's axis and it remains on the illuminated side of the Earth. These far northern latitudes experience 24 hours of daylight for months at a time, which explains why this region is called "The Land of the Midnight Sun" (Fig. 12.3). The Arctic Circle marks the southern limit of this phenomena. It occurs at 66.5° north latitude, that is 90° of latitude minus 23.5°, the angle of tilt. Because of the tilt, the Sun's rays shine directly on 23.5°N, which is defined as the Tropic of Cancer.

Although solar radiation may not be as intense in northerly latitudes as it is further south during the northern hemisphere summer, there is more of it. The amount of daylight on June 21st decreases southward. Northern Canadian cities such as Edmonton and Yellowknife, for example, have more daylight than Seattle or New York, and these two cities have more daylight than Los Angeles or Orlando. Cities on the equator have 12 hours of daylight, while the south pole has 24 hours of darkness.

In the southern hemisphere, the equivalent situation is observed on December 21st (Fig. 12.4). On this day, the Sun's rays shine directly on 23.5°S, which is defined as the Tropic of Capricorn. In far southerly latitudes beyond the Antarctic Circle at 66.5° south, the midnight Sun shines and the amount of daylight decreases northward. On this day, the

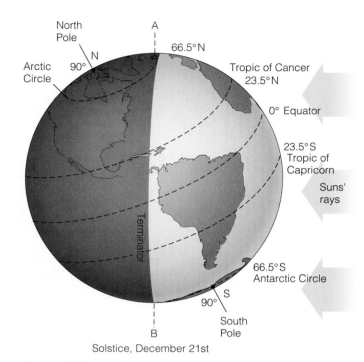

Figure 12.4
Winter in the northern hemisphere (December 21st).

north pole has 24 hours of darkness. At the equator, day and night are always 12 hours long, regardless of the time of year.

To illustrate further the effect of the tilt of the Earth's axis, consider a hypothetical planet that spins at the same rate as the Earth (once every 24 hours) but has no tilt, so that the terminator (A–B) and axis of rotation (N–S) coincide (Fig. 12.5). In this case, all locations would have 12 hours of daylight and 12 hours of darkness. In addition, upon rotation neither hemisphere would preferentially point towards the Sun, so the planet would have no seasons.

Figure 12.3
Land of the Midnight Sun. These 24 exposures of the Sun were taken in northern Greenland (latitude 78°N) in July between 7 PM (left) and 6 PM (right). Because of the northerly latitude, the Sun did not set.

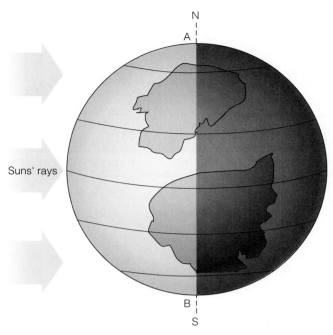

Figure 12.5
A planet with no tilt would have an axis (N–S) that coincides with the terminator (A–B). As a result, all locations on its surface would receive 12 hours of daylight and 12 hours of darkness, and the planet would have no seasons.

12.1.2 Heat Transport

Temperature contrasts play an important role in both local and global-scale atmospheric circulation. Maritime regions, such as Atlantic Canada or the Pacific Northwest, show the effects of local temperature contrasts most clearly. As we described in Chapter 11, the air temperature above the land on a warm summer day is higher than the air temperature above the oceans. Because of this, the hotter continental air rises, and the cooler sea air comes in to take its place. This creates an onshore breeze that helps to cool off a hot summer's afternoon (Fig. 12.6a). Unfortunately, the warm air may create clouds as it rises, so the day could be cloudy if the breeze does not blow the clouds away. During the night, the reverse situation may occur, especially in the early summer. The sea temperature barely falls at night because water retains its heat. But in the absence of sunlight, the air above the land may cool off dramatically, and its temperature may reach a value below that of the sea air. In this situation, the warmer sea air rises and is replaced by the cooler land air, reversing the direction of the breeze from onshore to off-shore (Fig. 12.6b). At twilight and at dawn, the air temperatures above the land and sea are approximately equal, and the wind dies down. Except for the mosquitoes who take advantage of these calm conditions, twilight is a beautifully tranquil time because the strength of the wind is considerably reduced. At maritime airports, which are notoriously tur-

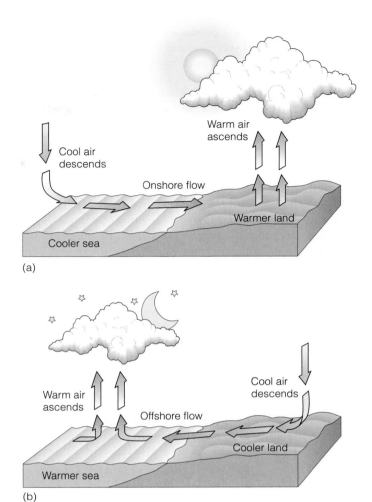

Figure 12.6
The typical flow of air in a coastal region during relatively stable weather conditions. (a) In the afternoon, the land is warmer than the sea, and warm air rising from the land is replaced by a cool moist onshore breeze. (b) At night, as the land cools off and the now warmer air above the ocean preferentially rises, it is replaced by cooler drier air from the land generating an offshore breeze.

bulent for much of the day because of the contrasting air temperatures, twilight is often the most comfortable time for aircraft landings.

These same principles can be applied on a global scale where atmospheric circulation is driven by the temperature contrast between the warm equatorial region and the cold polar regions. Although, on average, the amount of incoming solar energy is balanced by outgoing energy, this balance does not occur at each latitude (Fig. 12.7a). High-latitude polar regions lose more energy than they receive, whereas low-latitude equatorial and tropical regions gain more energy than they return to space. Heat transport between these regions prevents both of them from becoming progressively more extreme (Fig. 12.7b).

Figure 12.7

(a) Variation with latitude of the average annual solar radiation absorbed (red line) and emitted (blue line) by the Earth and the atmosphere. (b) The net surplus at low latitudes combined with the net deficit at high latitudes generates a strong equatorial to polar flow of heat. (c) Convection currents in a room when there is a hot wall (with a radiator) and a cold wall (with an exterior window).

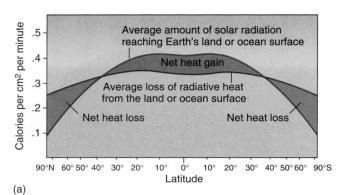

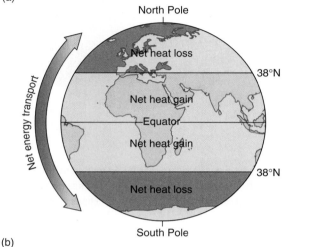

(a)

(b)

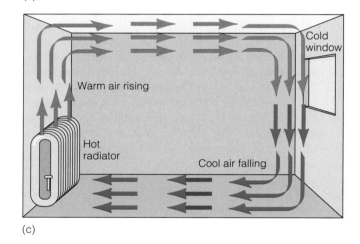

(c)

Warm air rising at the equator sets up a circulation system not unlike that within a heated room. Most rooms have a warm interior wall and a cold exterior wall, analogous to the equator and poles. Warm air rises to the ceiling by the warm wall and cold dense air takes its place near the floor. By the cold wall, cool air sinks and warm air takes its place near the ceiling. This is the essence of a convection system, which is fundamental to the circulation of gases and liquids (Fig. 12.7c).

One obvious difference between the room analogy and global air circulation is the scale of the convection. This scale is not just a function of the size of the system. The length, breadth, and height of a typical room are at least of broadly similar dimensions such that only a single circulation system, known as a **convection cell** develops. However, the distance from the equator to the poles is about 10,000 km (6,200 mi), whereas the height of the troposphere within which circulation occurs is only 8 to 18 km (5 to 11 mi) (Fig. 12.8a). This, together with the Earth's spin, complicates the atmospheric circulation system, and results in the development of several convection cells.

Typically warm air rises from the equator and heads towards the poles. Because atmospheric pressure is just the weight of the column of air above a unit area, the ascent of warm air in equatorial regions creates a band of low pressure systems. As the air moves towards the poles, it cools. It also converges because it is moving on a sphere. To witness this effect, simply place your hands on top of a globe near the equator. In this position, your hands represent the Earth's atmosphere at the equator. But notice as your hands move poleward, the shape of the globe starts to bring them together (Fig. 12.8b). In the same way, the air converges as it moves poleward, causing the air pressure to increase. At about 30 degrees north and south of the equator the increased pressure causes the air to descend, generating a persistent band of high pressure centers (known as the subtropical high) at these latitudes (Fig. 12.8c).

The downwelling air in the subtropical high region is dry because most of its moisture has already fallen. Air rising from the equator cools as it rises and its water vapor condenses such that rain falls in equatorial regions nearly every day. By the time the convection cells arrive at 30°N and 30°S, however, the air is dry.

So it should come as no surprise that the world's greatest deserts occur in the subtropical high pressure zones. The Sonora and Mojave deserts of the southwestern United States, the Sahara Desert, and the Gobi Desert of western China all occur at about 30° latitude in the northern hemisphere. Similarly, the Atacama Desert of Chile, the Kalahari Desert of southern Africa and the great Australian Desert all lie at about 30° latitude in the southern hemisphere. There is a direct causal relationship between the downwelling of this warm dry air and the occurrence of deserts. The desert regions of the southwestern United States are among the hottest and driest places on Earth because they are not only located below downwelling dry air, but also lie in the rain shadow of the Rocky Mountains, as we learned in Chapter 11. For example, Death Valley, California, averages only 4.5 cm (1.8 in) of precipitation annually and has endured some of the highest temperatures ever recorded (up to 57°C or 134°F).

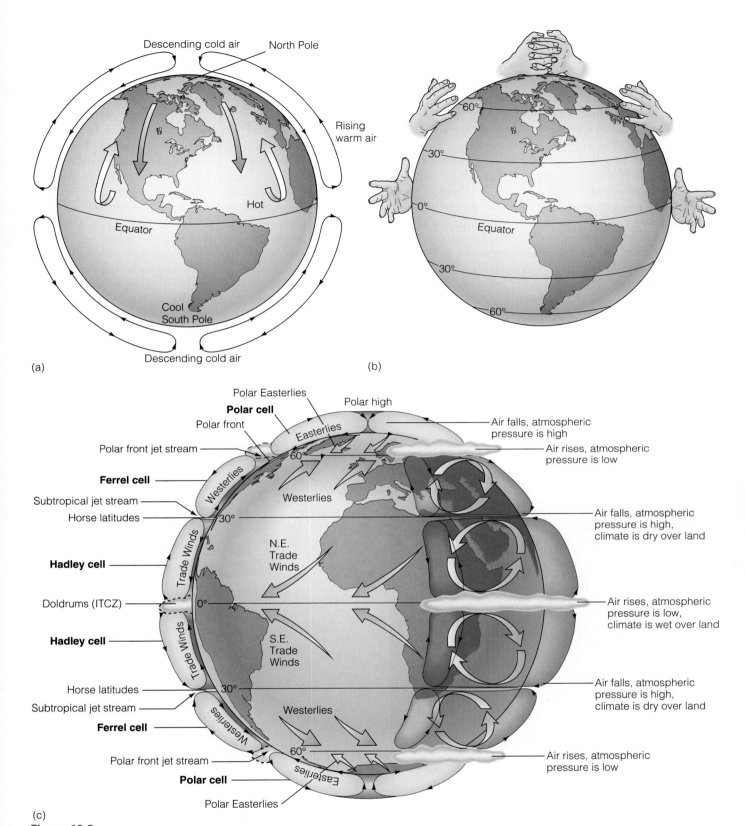

(a)

(b)

(c)

Figure 12.8

(a) Schematic global circulation model if uneven solar heating was the only variable (see Figure 12.7). Hot air rises at the equator and cold air descends at the poles. Complications arise from the spin of the Earth, the thinness of the atmosphere, and the convergence of air as it migrates toward the poles. (b) Model explaining convergence of air as it migrates towards the poles. The hands on the equator represent the atmosphere, and the bunching of hands as they move towards the poles represent the converging air. (c) Instead of one giant convection system, heat transfer occurs in steps generating three cells (Hadley, Ferrel, Polar) between the equator and each pole. This results in persistent global air circulation patterns. Rising air generates bands of low pressure (such as the "doldrums" in the intertropical convergence zone, or ITCZ) and descending air produces bands of high pressure (such as the subtropical highs of the horse latitudes, which generate deserts when they occur over land).

The descending air masses are remarkably similar to the venting systems in air-conditioned buildings (Fig. 12.9). Air is commonly forced out of the building through vents near the bottom of the walls. Nothing grows beside these vents, not even grass. These areas are local mini-deserts, caused by the blow-drying of the ground in the vicinity of the vents. Deserts are formed in a similar manner; the downwelling air literally blow-drying the land.

The convection cells between the equator and the 30° latitudes are called **Hadley cells,** after George Hadley, a British mathematician who discovered them in the 18th Century. As we shall see, these convection cells are only part of a global air circulation system in which warm air eventually migrates towards the poles. In fact, heat transfer between the equator and the poles actually occurs in three stages (see Figure 12.8c). In addition to Hadley cells, **Ferrel cells** occur between 30° and 60° latitude, and **Polar cells** occur between 60° latitude and the poles.

12.1.3 Coriolis Effect on Atmospheric Circulation

As we discussed in Chapter 9, most objects are anchored to the Earth's crust, and rotate with the Earth as it turns. However, water and air are free to move across the surface of the solid Earth as it spins beneath them. This creates an apparent force known as the **Coriolis effect.** In the merry-go-round analogy we used in Chapter 9, a ball thrown in a straight line from the moving merry-go-round to a catcher standing beside it will be seen differently by the thrower and the catcher. The catcher will observe the straight line path of the ball, but to the thrower on the merry-go-round, the path will *appear* to curve because the merry-go-round spins beneath the ball while the ball is in flight (Fig. 12.10). If the merry-go-round spins counterclockwise (the same direction that the Earth spins when viewed from above the north pole), the thrower will see the ball apparently veering to the right (or clockwise). Alternatively, if the merry-go-round spins clockwise (the same direction as the Earth spins when viewed from above the south pole), the thrower will see the ball apparently veering to the left (or counterclockwise).

Air masses behave in a similar manner to the ball. Figure 12.11a shows the counterclockwise rotation of the Earth viewed from above the north pole and the clockwise sense of rotation this gives to air masses. In Figure 12.11b the effect is mimicked by the flight path of a jet moving from A toward B, which from space is seen to be straight. As the jet moves, however, the Earth rotates beneath it so the flight path apparently veers to the right (Fig. 12.11c). In the southern hemisphere, (Fig. 12.12), where the Earth viewed from above the south pole has a clockwise rotation, an air mass or flight path moving in a straight line

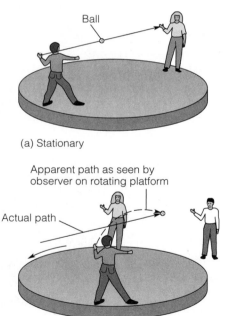

Figure 12.10
The Coriolis effect in a game of catch. (a) The merry-go-round is not rotating and a ball thrown in a straight line reaches its target. (b) The merry-go-round rotates and a ball thrown in exactly the same manner as in (a) now appears to the thrower to curve. Note that the catcher not on the merry-go-round sees the true, straight line path of the ball.

Figure 12.9
Local "desert-like" conditions produced near air vent. Constant "blow-drying" of the adjacent land results in arid conditions. (Photo Brendan Murphy)

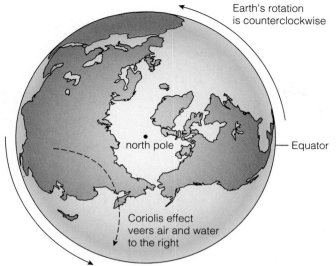

Earth's rotation
is counterclockwise

north pole

Equator

Coriolis effect
veers air and water
to the right

(a)

Figure 12.11
Coriolis effect in the northern hemisphere.
(a) Earth's rotation is counterclockwise as
viewed from above the north pole resulting
in a clockwise sense of rotation of air
masses. The effect is illustrated in the flight
path of a jet which is in a straight line as
observed from space (b), but *appears* to
verge to the right to an observer on the
surface (c) because of the Earth's
counterclockwise rotation.

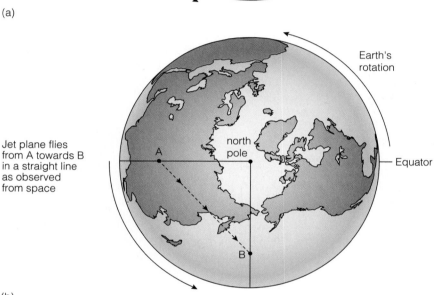

Earth's
rotation

Jet plane flies
from A towards B
in a straight line
as observed
from space

A

north
pole

Equator

B

(b)

Person at A looking toward B
sees flight path to the right

Earth's
rotation

A′

north
pole

Equator

Apparent
path

A

B

Actual path

B′

Due to Earth's rotation,
the jet's straight flight path
actually brings it to B′. But
a person at B will have
moved counterclockwise
and so observes an apparent
deflection in the flight path.

(c)

Figure 12.12
Coriolis effect in the southern hemisphere. (a) Earth's rotation is clockwise as viewed from beneath the south pole resulting in a counterclockwise sense of rotation of air masses. The effect is illustrated by the flight path of a jet which is in a straight line as observed from space (b), but *appears* to verge to the left to an observer on the surface (c) because of the Earth's clockwise rotation.

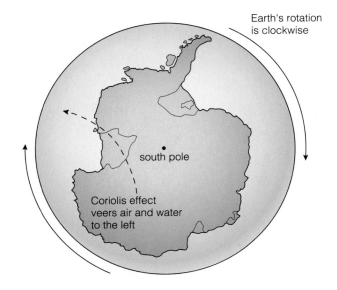

Earth's rotation is clockwise

south pole

Coriolis effect veers air and water to the left

(a)

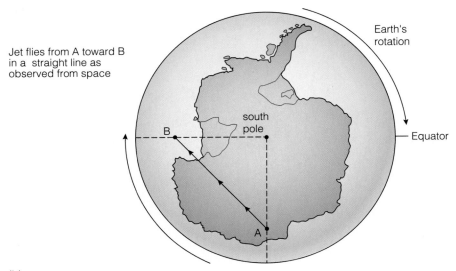

Jet flies from A toward B in a straight line as observed from space

Earth's rotation

Equator

B

south pole

A

(b)

Due to Earth's rotation, the jet's straight line flight path actually brings it to B′. But a person at B will have moved clockwise and so observes an apparent deflection in the flight path.

Earth's rotation

Equator

B

B′

south pole

Apparent path

A

A

Actual path

Person at A looking toward B sees flight path to the left

(c)

apparently veers to the left, and therefore has a counterclockwise sense of rotation.

Because we move in unison with the Earth, we rarely notice the influence of the Coriolis effect. The effect is also minimal for small-scale winds such as onshore breezes. Where winds blow over vast distances, however, such as those associated with regional high and low pressure centers, the influence of the Coriolis effect is appreciable and affects the path of weather systems.

12.1.4 Circulation in High and Low Pressure Regions

In the previous chapter we learned that winds are the result of variations in atmospheric pressure, and that the strength of the wind is related to the pressure gradient between regions of high and low pressure. Because of turbulence in the troposphere, air tends to ascend and descend in columns or **plumes** (Fig. 12.13a). These plumes mean that zones of high and low pressure occur in specific centralized regions, rather than continuous linear belts. These regions are represented on weather charts by a bulls-eye pattern of isobars (Fig. 12.13b). For example, areas of high pressure have maximum pressure values at the center, which decrease outward. High pressure centers pump air outward in all directions toward low pressure regions, so the wind direction is initially at right angles to the isobars. But as the air moves away from the high pressure center, it is acted upon by the Coriolis effect which, in the northern hemisphere, deflects it to the right so that it becomes virtually parallel with the isobars. In the northern hemisphere, wind patterns around high pressure areas consequently move

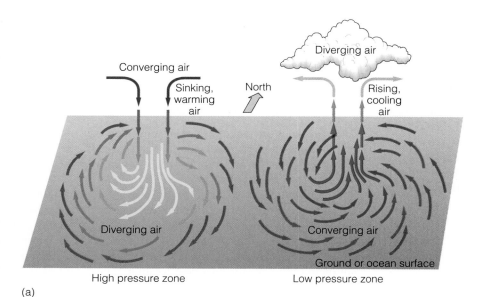

Figure 12.13
(a) Winds and air masses associated with high and low pressure centers in the northern hemisphere. Descending air plumes create high pressure centers, whereas ascending plumes create low pressure centers. (b) Typical daily pattern of high and low pressure centers on a map of North America. Arrows representing wind directions have a clockwise distribution around high pressure centers, and a counterclockwise distribution around low pressure centers. Contours represent isobars or lines of equal atmospheric pressure.

in a clockwise direction parallel to the isobars. This pattern is often called an **anticyclone** (Fig. 12.14a).

Conversely, low pressure regions in the northern hemisphere have counterclockwise air circulation. At first, this might seem odd, given the clockwise rotation due to the Coriolis effect. But examine Figure 12.14b which shows a low pressure center surrounded by regions of high pressure. The low pressure region *receives* air from each of the high pressure regions that surround it. As each mass of air approaches the low pressure center, it is rotated to the right. Viewed from the low pressure center, however, each mass of air appears to be deflected to the left. The net effect is a counterclockwise sense of air motion around the low pressure center to form a system commonly called a **cyclone**.

The Coriolis effect on high and low pressure centers in the southern hemisphere produces the reverse direction of flow from that in the northern hemisphere (Fig. 12.15). Air circulation around high pressure regions is counterclockwise while circulation around low pressure regions is clockwise.

The airflow that is typically associated with cyclones and anticyclones is shown in Figure 12.16. As warm air rises at the low pressure center, the air cools and its relative humidity rises (see Chapter 11). This results in cloudy weather typically associated with low pressure centers. This rising air is replaced by a convergence of surface air. As air descends at high pressure centers, the air warms and its relative humidity decreases. This results in relatively clear skies. High pressure centers have a net outflow of surface air which is replaced by converging air from above. This simple combination of the cyclonic and anticyclonic airflow profoundly influences the weather pattern of much of North America.

Northern Hemisphere

(a) Anticyclone

(b) Cyclone

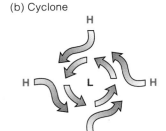

Air moves away from high pressure center and is rotated clockwise

Clockwise movement of incoming air, causes low pressure center to rotate counterclockwise

Figure 12.14
Air circulation around high and low pressure centers in the northern hemisphere. (a) The high pressure center (also known as an anticyclone) pumps out air in all directions which veers to the right because of the Coriolis effect, resulting in a clockwise circulation. (b) A low pressure center (also know as a cyclone) attracts air from all directions. As the air moves towards the low pressure center, it is again deflected to the right. Cumulatively, this results in counterclockwise flow (see Figure 12.13b).

Southern Hemisphere

(a) Anticyclone

(b) Cyclone

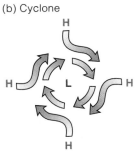

Air moves away from high pressure center in all directions and is rotated counterclockwise

Counterclockwise movement of incoming air causes low pressure center to rotate clockwise

Figure 12.15
Air circulation around high and low pressure centers in the southern hemisphere produces counterclockwise air circulation around high pressure centers and clockwise air circulation around low pressure centers because the Coriolis effect causes air currents to veer to the left.

12.1.5 Fronts and Clouds

The pattern of interaction between warm and cool air masses may not always be as regular as that shown in Figure 12.16. Contrasting air masses may occur in very close proximity to one another. When this happens, the interaction between them results in the development of **fronts**. This term was introduced by meteorologists in the early 20th Century who likened the tussles between air masses to the military fronts of the First World War. These contrasting air masses may have different atmospheric pressures, temperatures and humidity. So an advancing front means changing weather. The extent of the change depends on how sharp these contrasts are.

Clouds are the most obvious manifestation of colliding air masses and, as we know, they form when warm air rises and begins to cool down. As the air cools, the relative humidity increases resulting in the condensation of water vapor. Certain cloud formations are associated with weather fronts, and can be used to monitor their motion. Although hundreds of cloud formations have been identified, there are three main types, cirrus, stratus, and cumulus (Fig. 12.17). **Cirrus** (derived from the Latin "to curl") is a wispy cloud formation that occurs some 10 to 14 km above the Earth's surface and travels rapidly at speeds of up to 160 km (100 mi) per hour. Because of their high altitude, cirrus clouds are composed entirely of ice crystals.

Low-altitude clouds are entirely composed of water droplets. **Stratus** (derived from the Latin for "layer") is a relatively low altitude (2 to 6 km), flat cloud formation that occurs in relatively stable moist air. These clouds often bring steady drizzle or showers.

Cumulus (derived from the Latin for "heap" or "stack") is a puffy cloud formation in which clouds are piled on top of one another, and signifies unstable rising air currents.

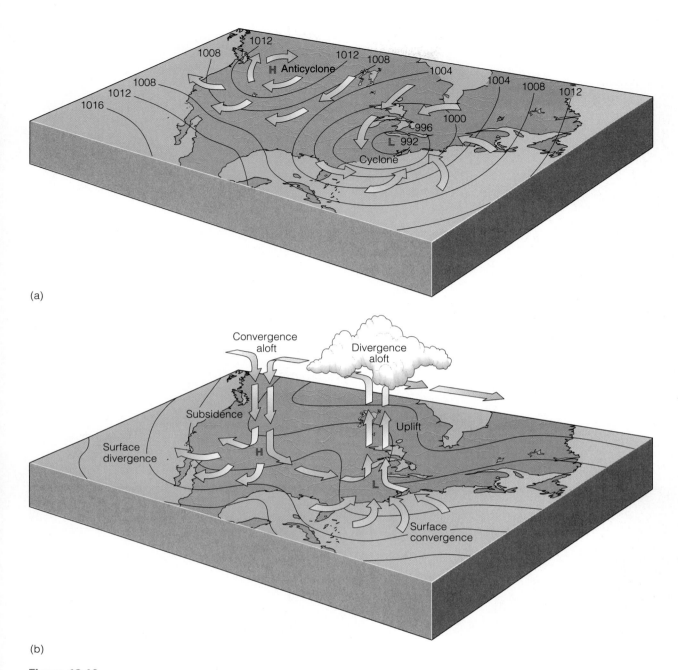

Figure 12.16

(a) Typical airflow pattern associated with surface low pressure (cyclone) and surface high pressure (anticyclone) centers. Note the clockwise flow of air around the high pressure center and the counterclockwise flow of air around the low pressure center. (b) Rising warm air above the low pressure center cools as it rises to create cloudy conditions. The rising air is replaced by converging surface air. Air associated with the high pressure region warms as it descends, resulting in relatively clear skies and is accompanied by an outflow of surface air. The combination of cyclone and anticyclone circulation may affect the airflow continent-wide.

Cumulus clouds are often small and confined to within 2 km of the Earth's surface. If further instability develops, however, they may tower upwards and develop into higher altitude **cumulonimbus** clouds (at 9 km) or even higher altitude **anvil** clouds (at 14 km), which are also known as thunderheads. Cumulonimbus and anvil clouds form from the rapid uplift of air associated with sharply contrasting temperatures and pressures in the colliding air masses. As a result, they are filled with energy, and in certain circumstances can give rise to violent thunderstorms.

Fronts are classified according to the relative temperature of the advancing air mass. The two most common examples

(a)

(b)

(c)

(d)

(e)

Figure 12.17
Distribution and appearance of various cloud formations. (a) General distribution of cloud type with altitude. (b) Cirrus. (c) Stratus. (d) Cumulus. (e) Cumulonimbus.

are cold fronts and warm fronts. The type of front is determined by the temperature of the faster moving air. In a **cold front,** cold air overtakes the warm air. In a **warm front,** the warm air overruns the cold air. In both cases, precipitation of some sort is very likely as the warmer, lighter air is forced upwards where it condenses into clouds. However, the extent of precipitation may be very different.

At a cold front (Fig. 12.18), the advancing dense air wedges beneath the relatively light warm air. The effect is like opening a window on a cold winter's day. You can instantly feel the cold dense air around your ankles as it rushes in through the window and descends as a wedge beneath the warm indoor air.

Cold fronts behave in a similar fashion. The warm air is forced to rise rapidly upwards. As we learned in Chapter 11, condensation of rising air releases an enormous amount of heat, known as latent heat. In the summer time, when air is typically very warm, the release of this heat energy amplifies the vertical updraft and may give rise to the typical thunderhead clouds associated with violent summer thunderstorms.

In contrast, a warm front is associated with advancing lighter air which rides over the top of the cold dense air. The rise of the lighter, warmer air is a gentle one and, as a result, cloudiness and precipitation are spread over a much wider area. To an observer on the ground, the first sign of an approaching warm front is that of high wispy cirrus clouds in the distance. These are typically about 800 km (500 mi) ahead of the front, (Fig. 12.19) and given their average speed of 160

km per hour, they usually precede it by about 5 hours. As the front gets closer, the cirrus clouds pass overhead and are replaced by progressively lower, thicker stratus clouds immediately prior to precipitation.

Another important type of front is known as an **occluded front.** This may occur when a faster moving cold front overtakes a warm front. The development of an occluded front is shown in Figure 12.20. To the south of line BB′ (Fig. 12.20a), the cold front and the warm front are separated by a body of warm air (section AA′, Fig. 12.20b). At BB′, the cold front has caught up to the warm front (Fig. 12.20c) and north of this line (section CC′, Fig. 12.20d) overtakes it as an occluded front. In this case, the two bodies of cold air associated with each front have collided and have forced the warm air between them to rise (section CC′). The net result is the uplift of the warm air so that it is no longer in contact with the ground. Although this uplift causes clouds to form, the frontal system has no contact with the ground and may pass overhead without causing a significant difference in surface temperatures.

A weather map for North America depicting a typical daily tussle between contrasting air masses is shown in Figure 12.21a, while the development of the low pressure center is shown in Figure 12.21b. Cold arctic air penetrating into the continental interior meets warm, humid air from the Gulf of Mexico along a frontal system (Stage 1). The temperature and density contrasts between them does not allow simple mixing and the air masses initially move parallel to each other as a **stationary front.**

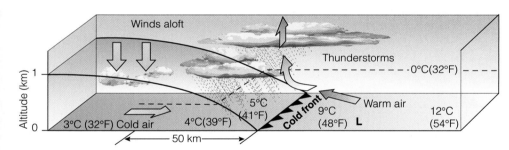

Figure 12.18
Schematic representation of weather across a cold front, where cold air wedges beneath warm air, resulting in the rapid ascent of the warm air and the development of thunderstorms.

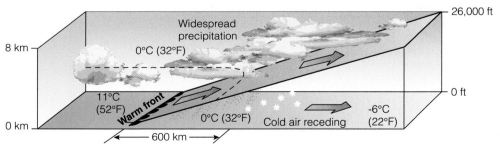

Figure 12.19
Schematic representation of weather across a warm front where warm air rides over cold air, generating widespread cloudiness and precipitation.

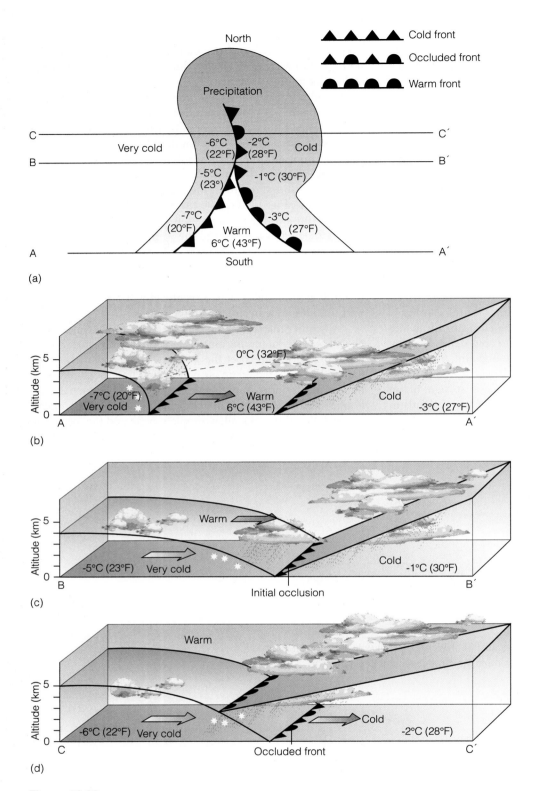

Figure 12.20

A schematic representation of the development of a cold occluded front. (a) To the south, the cold front and warm front are separated by a body of warm air. To the north, the faster moving cold front catches up with the warm front and forces the warm air off the ground, resulting in the development of an occluded front. (b)(c)(d) show three sections across the frontal system. In (b) section AA' a cold front and a warm front are separated by a body of warm air. In (c) section BB' the cold front has caught up to the warm front. In (d) section CC' the cold front has overtaken the warm front, resulting in an occluded front.

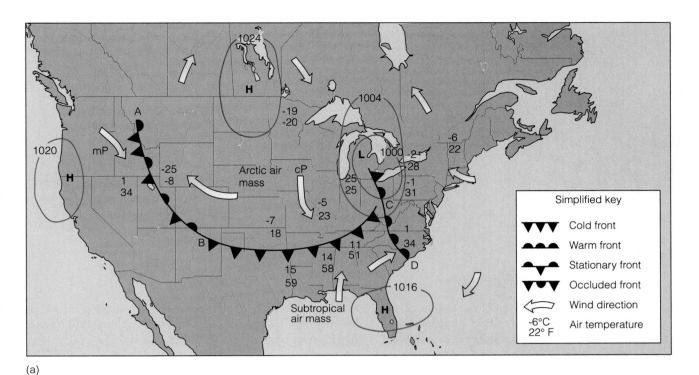

(a)

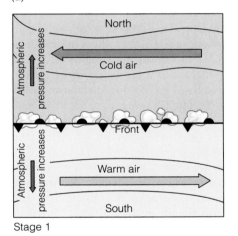

Stage 1

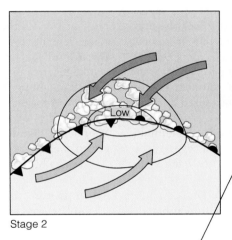

Stage 2

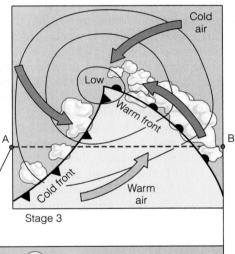

Stage 3

(b)

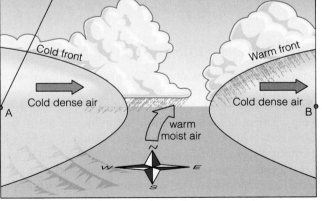

Figure 12.21

(a) Typical relationship between cold and warm air masses on a simplified weather map of North America. (b) Model for developing the frontal systems (arrows depict airflow) in which cold arctic air in the continental interior meets warm humid air from the Gulf of Mexico along a frontal system (Stage 1). The frontal system bends (Stages 2–3) and a tongue of arctic air penetrates southward, while a counterflow of warm moist air migrates northward (Stages 4–5). Eventually, the low pressure system separates from the front (Stage 6).

(continued)

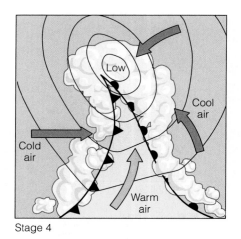
Stage 4

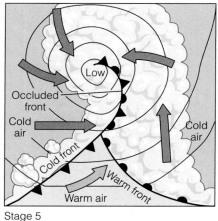
Stage 5

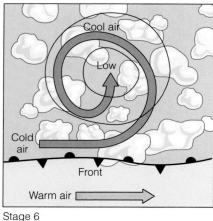
Stage 6

Figure 12.21b, continued

Eventually, however, local areas of high and low pressure develop that bend the front (Stages 2–3). This allows a tongue of arctic air to penetrate southward, and, as a result, a counterflow of warm humid air moves northward (Stages 4–5). Eventually, a second stationary front develops (Stage 6).

Note that the overall direction of airflow (arrows, Fig. 12.21a and b) is consistent with what we have learned. Clockwise flow occurs around high pressure systems and counterclockwise flow occurs around low pressure systems. The dry arctic air penetrates southward beneath the warm humid air creating a cold front that will dominate the weather for that day in the southern United States. The weather in this region will be characterized by thunderstorms as depicted in Figure 12.18.

To the north of Florida, the warm air overrides the northerly moving cold air, producing a warm front characterized by widespread precipitation, as shown in Figure 12.19. To the south of the Great Lakes, the warm air has been wedged upwards, creating an occluded front like that shown in Figure 12.20.

In summary, the ongoing turbulence in the troposphere is characterized by the daily struggle between contrasting air masses. This tussle forms frontal systems that profoundly influence our weather.

12.1.6 Jet Streams and Fronts

The daily wrestle between air masses is most pronounced where contrasting air masses meet and the struggle is influenced by swift high-altitude winds known as **jet streams** that flow across the top of these boundaries. The most influential jet streams occur where cold polar air and warm mid-latitude air fight for supremacy. A sharp boundary between polar and mid-latitude air masses occurs in both the northern and the southern hemisphere. This tussle defines the **polar front** and the jet stream associated with it is known as the **polar front jet** (see Figure 12.8c). Another example of contrasting air masses occurs at about 30° latitude north and south where tropical air meets mid-latitude

air. This location defines the position of the **subtropical jet**. As discussed in Chapter 11, jet streams form an important part of the global air circulation system (Fig. 12.22).

Jet streams were discovered during the Second World War by jet aircraft pilots who were surprised to find very strong winds when flying at high altitude. Jet streams are wind tunnels, that is, narrow high speed currents of wind, similar in some respects to ocean currents such as the Gulf Stream. Because of the rotation of the Earth, jet streams flow eastwards. Jet streams are therefore like tail winds for eastbound aircraft, enabling them to fly at greater speeds relative to the ground. For westbound flights, however, they are like head winds, and pilots try to avoid them. This explains why eastbound jets take less time to cross North America than the return, westbound journey.

Typically, jet streams are located where there is rapid variation in the altitude of the tropopause (see Chapter 11). Although the tropopause (the boundary between the troposphere and stratosphere) has an average altitude of about 12 km (7.5 mi), its height can vary significantly from about 8 to 18 km (5 to 11 mi). Jet streams are located where the change in altitude is especially rapid. At their centers, jet streams can travel at speeds between 120 and 240 km (70 to 150 mi) per hour.

In North America, the polar front jet profoundly influences our day-to-day weather and our seasons. The speed of this jet stream is typically about 60 km (35 mi) per hour in the summer, and about 150 km (85 mi) per hour in the winter. It separates cool polar air (to its north) from warm mid-latitude air and affects their interactions.

Every evening, weather forecasters draw our attention to the ongoing tussle between air masses at the polar front by pointing out the position and speed of this jet stream. When the speed of the jet stream increases, air flows out of it, much like a raging river overflows its banks in a storm. The outflow of air causes a low pressure region (or cyclone) to occur beneath the jet stream as the lower air rushes in and is pulled up by the overflow. When the speed of the jet stream decreases, air flows into

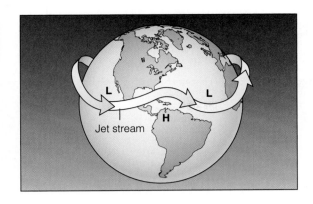

(a)

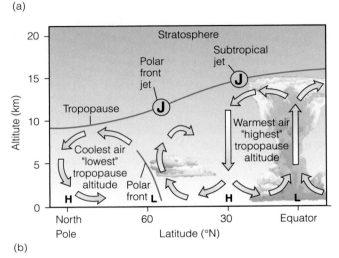

(b)

Figure 12.22

Location of major jet streams in the northern hemisphere (see also Figure 12.8c). (a) A jet stream, such as the subtropical jet shown here, is a fast-flowing current of air that moves in a predominantly west-to-east direction. (b) The polar front jet forms in a narrow zone of rapid change in elevation of the tropopause and is the boundary between cold northerly air and warm southerly air. The subtropical jet similarly occurs at a change in elevation of the tropopause.

the stream from the surrounding upper regions. This inflow of upper air causes a high pressure region to form below as the air descends into the lower regions.

Because of the interaction between cold and warm air masses that are pulled into and out of the jet stream, many weather fronts, both cold and warm, are spawned near the jet stream. In addition, the frontal systems become pulled by the jet stream and tend to follow its path. As a result, the weather in a region is very changeable when the jet stream is positioned overhead.

The variation in the position of the polar jet stream during the year is a major indicator of climate (Fig. 12.23). In the winter, cold, dense Arctic air wins the tussle with the mid-latitude air and can plunge far into the southern United States. In the summer, warm air pushes its way northward fueled by increased radiant energy from the Sun.

The motion of these air masses changes the position of the jet stream because they change the placement of the boundary between the warm and cool air masses. In the North American summer, the jet stream is generally positioned at northerly latitudes well within Canada. The predominant climatic influence is therefore provided by the warm air from the south. The two air masses become stable, and the jet stream develops a linear east-west orientation. In the fall, as the air cools and becomes more dense in northern latitudes, high pressure regions develop and begin to pump cold Arctic air southward distorting the jet stream into a contorted snake-like path. So, fall weather can be highly changeable as the tug-of-war between competing air masses alternately drives the jet stream northward and southward across the continent. For example, Arctic air plunging into the continent interior can cause the jet stream to bend southward which may result in a counterflow of warm moist ocean air northward along the Atlantic coastline. As a result, temperatures in New England and Atlantic Canada can exceed those in the central United States for a short period in the late Fall.

By winter, the Arctic air has won the battle, and the jet stream moves south into another relatively stable east-west path. Cold Arctic air consequently settles across much of North America. The first hint of spring comes with distortion of the jet stream once again, this time because of warm air pushing northward. This distortion also results in very changeable weather until summer settles in. The juxtaposition of high and low pressure systems spawned by the changing velocity of the jet stream can produce high winds and storms that are common to many regions in springtime.

The factors that influence the formation and location of the jet stream have a daily influence on weather patterns. However, the forces of nature may periodically conspire to amplify some of these effects. To understand why this happens, we must turn our attention to the origin of storms.

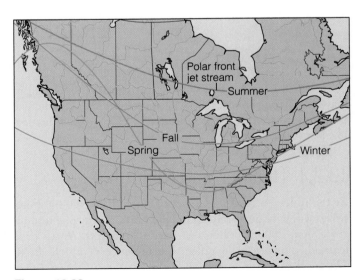

Figure 12.23

The typical position of the polar front jet stream in continental North America during summer, fall, winter, and spring.

THE ICE AGE OF 1998: BLAME IT ON EL NIÑO

In the earliest days of 1998, a freak ice storm engulfed the northeastern United States and eastern Canada (Fig. 12.24). Freezing rain is a common winter phenomenon in this part of the world, but generally only lasts for a few hours before it changes to either rain or snow. In early January, however, freezing rain fell for several days causing a buildup of ice which eventually toppled trees and power lines and brought the region to a standstill. Many scientists blamed El Niño which, as we learned in Chapter 9, occurs when warm water builds up off the west coast of South America.

But how does El Niño affect the climate of North America? It does so by altering the position of the polar front jet stream, which has a profound effect on weather patterns. In a normal winter, this powerful current normally carries warm Pacific air along a line across the central United States (Fig. 12.24).

However, due to the effects of El Niño, the jet stream was much further south and was carrying warm moist air from the Gulf of Mexico. As it traveled eastward, its path was diverted northward by a high pressure system off the Atlantic Coast (Fig. 12.25). Moving north across the northeastern United States it allowed warm, moist air

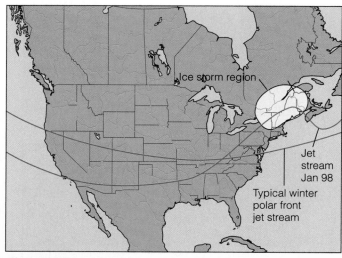

Figure 12.24
The normal location of the polar front jet stream and its location during the 1998 ice storm of the northeastern United States and eastern Canada.

from the Gulf of Mexico to push northward. This air mass encountered cold, dense arctic air. Unable to move the cold air, it instead rode over the top of it, forming a warm front. As the warm air rose, it cooled and precipitation began as rain. But as the rain fell through the layer of cold air below, it chilled and turned into freezing rain. The jet stream brought a relentless supply of warm moist air to the region which ensured that

the freezing rain lasted for days. Under the weight of ice, power lines snapped and pylons collapsed, depriving much of the region of electrical power.

But worse was to follow. While the freezing rain lasted, temperatures remained just below freezing. Cold, to be sure, but not dangerously so. When the storm abated, however, the temperature plunged as low as −30°C (−22°F) as the influence of the cold

12.2 Storms

Storms are the result of collisions between greatly contrasting air masses. Storms can occur as local, short-lived phenomena, like thunderstorms, or as events affecting a wide geographic area, like hurricanes.

12.2.1 Thunder and Lightning

By the time you read this sentence, more than 500 lightning strikes will have occurred around the Earth. At any one instant, more than 2000 thunderstorms are in progress. As we have learned, thunderstorms reflect the presence of a cold front. Thunderheads can stretch vertically from a height of 1.6 to 14 km (1 to 8 mi), and so have highly contrasting temperatures (up to 40°C, 72°F) from bottom to top.

For lightning to occur, separate regions of positive and negative charge must exist within the thunderhead. The mechanism responsible for this is uncertain. But since this unstable air is in

constant rapid motion, particles within it collide violently generating positive and negative electrical charges as they do so. The positive charges consist of hydrogen ions (H^+) and so are relatively light and rise towards the top of the thunderhead, whereas the negative charges (OH^-) are heavier and sink toward its base (Fig. 12.27). The contrast in electrical charge may build up to a threshold, and at that point it shorts out within the cloud. The result is a broad flash sometimes called **sheet lightning**, which is produced within the cloud itself and does not reach the ground.

But the negative charges at the base of the cloud also repel negative charges from the ground below, thereby inducing a positive charge on the surface. Eventually, the contrasting charges between the cloud and the ground may reach a threshold at which the negative charges overcome the air resistance and flow down from a "leader" at the base of the cloud to meet the positive charges flowing up from the Earth's surface (Fig. 12.28). The positive charges move upward through any conducting object, including buildings, trees, and even humans. The opposite charges usually meet about 30 m (100 ft) above the ground, generating a jagged bright

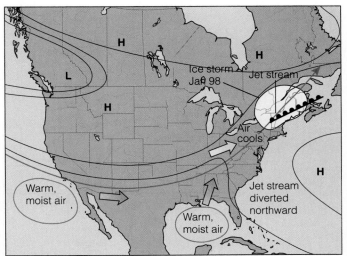

Figure 12.25
During the 1998 ice storm, a region of high pressure in the Atlantic Ocean diverted the path of the moist air from the Gulf of Mexico northward toward eastern Canada.

Figure 12.26
Damage to the power lines and trees in southern Quebec caused by the ice storm of January 1998.

arctic air returned. In a region stripped of its power, the consequences were devastating. There were at least 24 fatalities. Some even died from asphyxiation as a result of using outdoor barbeques to heat their homes. Two million homes were without power for more than two weeks. Thousands of people were evacuated from their homes and spent weeks in community shelters. At one stage during the storm, an evacuation of the city of Montreal was contemplated. Preliminary estimates of the cost of the damage exceed 2 billion dollars and it may take two years to complete the repairs (Fig. 12.26).

The ice storm of 1998 shows how interrelated our Earth systems are, and the importance of the polar front jet stream in influencing our day-to-day weather. It is also clear that El Niño can exert a global-scale influence by altering the position of jet streams.

lightning bolt known as **forked lightning** (Fig. 12.29). When this happens, the positive charges run up into the cloud in a series of pulses. This is called a "return stroke" and travels so quickly that our eyes cannot see the motion and it appears as a continuous flash of light. The power involved may be as much as 100,000 to 400,000 amps for every millionth of a second.

Forked lightning can be very dangerous, especially in flat outdoor areas like golf courses. Because forked lightning is often associated with torrential rain, golfers commonly seek shelter beneath trees. As is evident from Figure 12.28, however, this is the very worst place to shelter because tall objects form the path of least resistance for grounding the electrical charges. If you cannot make it back to your car, the safest position is to squat on the balls of your feet. This will minimize the surface area of your contact with the ground and will protect your heart in the event that you are struck.

In addition, water circulating in a tree conducts electricity, so when struck by lightning, it may explode as the sap is instantly vaporized. Thus forked lightning may also start forest fires. Furthermore, the old adage "lightning never strikes the same place twice" has no scientific validity. Many tall buildings have been repeatedly hit by lightning simply because they are tall. Because of this, many are protected by lightning rods, which were first devised by Benjamin Franklin. An iron rod is positioned above the highest point of a building, from which a cable leads to the ground. Because iron is a much better conductor than the building materials, the rod takes the direct hit and the electrical current is grounded, thereby protecting the building (Fig. 12.30).

A lightning stroke can heat the air through which it travels to an incredible 30,000°C (50,000°F). This extreme heat generates **thunder**, which is caused by the rapid expansion of the air and the formation of a compression wave that can be heard as it moves through the relatively still air. Because lightning occurs in various parts of the thunderhead, there is rarely a single clap of thunder. Instead, a rumbling sound is produced that seems to roll across the sky.

Because the speed of light is so much faster than the speed of sound, we see the lightning stroke before we hear the thunder. Light from the stroke, traveling at 300,000 km (186,000 mi) per second, is seen in virtually the same instant as it occurs. Sound,

Figure 12.27
Typical distribution of negative and positive charges in a mature thunderhead.

Lighter positive charges rise

Heavier negative charges sink

Negative ground charges are repelled away by negative cloud charges producing positive charge at surface

Figure 12.28
Model for forked lightning. (a) The negative charge concentrated at the bottom of the cloud becomes large enough to overcome the air's resistance and develops a "leader" pointing towards the ground. (b) An upward flow of positive charges from the ground concentrates at elevated points. (c) The downward flow of negative charges meets the upward flow of positive charges and a strong electric current known as a return stroke carries the positive charges into the cloud.

Stepped leader

Return stroke

(a)   (b)   (c)

Figure 12.29
Forked lightning strikes a mountainside.

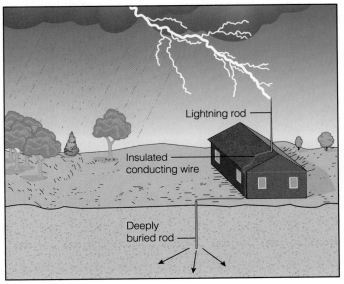

Figure 12.30
Principle of the lightning conductor. A lightning rod extending above the building attracts the downward flow of electrons from thunderclouds. After the lightning strikes the rod, the charge is directed along an insulated conducting wire and into the ground where it is dissipated.

on the other hand, moves in air at about 340 m (1100 ft) per second. How far away lightning has struck can therefore be estimated by measuring the time lapse between the lightning strike and the thunder. For example, a count of 5 seconds between the two indicates that the lightning struck 1.7 km (1.1 mi) away. If you hear the thunder in the same instant you see the lightning, the storm is directly overhead.

12.2.2 Tornadoes

Tornadoes are frightening storms because of their concentrated violence and unpredictable behavior. The word tornado is derived from the Spanish word for "thunder" and the storms are so named because of the thunder-like roar associated with

them. Several hundred tornadoes strike North America every year. They claim hundreds of lives and cause damage to buildings and crops costing tens of millions of dollars.

Tornado watches begin when weather forecasters detect a layer of warm moist air beneath cool dense air (Fig. 12.31). This typically happens in the central plains of the United States and southern Canada during the summer, when cool air descending from the Rocky Mountains is blown on top of warm summer air. The presence of cool dense air above relatively light warm air is very unstable, and the warm air rises rapidly and violently to create a narrow zone of very intense low pressure and a series of funnel-shaped clouds similar to thunderheads. Indeed thunderstorms often precede tornadoes. The release of latent heat from the ascending warm moist air accentuates the updraft and causes the low pressure to intensify. As a result, air rushes in rapidly at ground level from all directions. This creates a vortex with swirling winds of up to 350 km (220 mi) per hour (Fig. 12.32). In the southern United States, these cold and warm air masses typically converge in a region known as "Tornado Alley", which experiences the world's greatest number of twisters (Fig. 12.33).

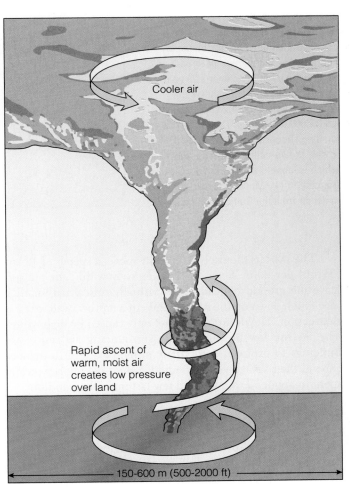

Figure 12.31
Schematic development of a tornado. Warm moist air at the surface rises and meets the cold, dry air. Air rushes in at ground level from all directions and the spin of the tornado begins.

Figure 12.32
Tornadoes begin where a layer of unstable warm, moist air occurs beneath cool, dense air. The warm air ascends rapidly to create a zone of intense low pressure.

Figure 12.33
A map showing Tornado Alley, which experiences the world's highest concentration of twisters. The upper figure shows the number of tornadoes reported in each state over a 25 year period. The lower figure shows the number of twisters per 25,600 square km (10,000 square mi). The darker shading shows the regions most at risk, the so-called "Tornado Alley" which is located where cool dry air from the Rocky Mountains meets warm moist air from the Gulf of Mexico.

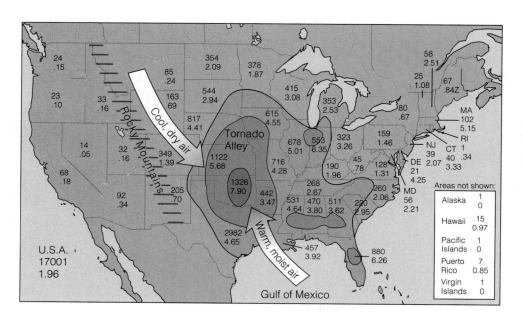

The twisters themselves typically move at speeds of 40 km (25 mi) per hour. They can spin entirely within the air and leave the ground beneath them virtually unscathed, or they may touch down on the surface and cut a narrow swath of total destruction. The destruction is not only caused by high winds. The intense low pressure of the surrounding air can cause buildings to explode, particularly airtight trailers. Tornadoes expend their savage violence rapidly; their energy is typically spent in about 30 minutes and few last more than an hour.

12.2.3 Hurricanes

Hurricanes are also characterized by regions of intense low pressure and high winds. But, in contrast to tornadoes, hurricanes develop over tropical ocean waters and affect a much wider region (Fig. 12.34). They can grow as large as 2000 km (1250 mi) in diameter, and can last for days or even weeks. They form in all of the world's major oceans, with the excep-

tion of the South Atlantic, and are also known as "tropical cyclones" or, in Asia, as "typhoons" (after the Chinese "tai-fung," meaning "wind which strikes").

Most of the hurricanes that affect the eastern seaboard of the United States commence as low pressure systems over the tropical waters of the Atlantic Ocean or over northwest Africa, about 5 to 20 degrees north of the equator. As they are swept westward towards the coast of the Americas by the trade winds, they intensify as each system is fed by the warm Atlantic waters.

Hurricanes draw their energy and moisture from warm tropical oceans. It is estimated that the surface temperature of the ocean water must exceed 27°C (80°F) if the chain of events leading to the development of a hurricane is to be initiated. So weather forecasters carefully monitor ocean water temperatures in order to predict hurricane development. One of the many concerns of global warming is that these temperatures may be attained over wider areas of ocean water and for

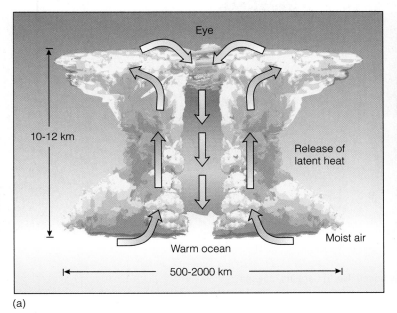

(a)

(b)

Figure 12.34
(a) Typical pattern of wind flow associated with intense low pressure centers which, over tropical ocean waters, can cause hurricanes to develop (note vertical exaggeration). (b) Hurricane Gilbert on September 15th, 1988 affecting the entire region around the Gulf of Mexico. The counterclockwise sense of rotation and the eye of the hurricane are clearly visible.

longer periods of time. As a result more hurricanes of greater power may be generated.

As a low pressure region develops because of rising warm air, more air is drawn in from the surrounding regions. As with all cyclones, the cooling moist air releases stored latent heat as it rises, and this additional heat further intensifies the updraft (Fig. 12.34a). When the temperature of the ocean water is above a critical 27°C (80°F), however, a chain of events is initiated whereby the low pressure region gorges itself on the surrounding warm moist air, drawing in more and more moisture and energy. The effect becomes a vicious circle in a literal sense; as more moisture is drawn in, a more violent updraft is created, which intensifies the low pressure and draws in even more moisture and energy. Winds associated with this motion can attain speeds in excess of 200 km (125 mi) per hour.

Because hurricanes are low pressure regions, the Coriolis effect in the northern hemisphere produces a counterclockwise rotation of air around the hurricane's center (Fig. 12.34b). This central region is a column of air known as the "eye" of the hurricane.

Paradoxically, hurricane winds are least intense at the center of the eye. In fact, in many instances, the eye has no cloud cover at all. This is because the principal updrafts occur in the circulating region immediately surrounding the eye. Dry air from as high as 10 km (6 mi) may be sucked down into the eye producing clear weather. As a result, a temporary respite in the hurricane occurs when the path of the hurricane is directly overhead. Those unfamiliar with the hurricane's geometry may emerge from shelters at this time, believing the hurricane

to be over. In fact, it is only half over. As the eye passes overhead, the trailing edge of the hurricane strikes with winds from the opposite direction that are often more ferocious than those at the leading edge.

In September 1926, a temporary lull in a hurricane affecting the Miami coastline fatally attracted adventurous bathers to swim in the high surf. However, as the eye passed by, the hurricane's ferocity was once more unleashed and hundreds were swept to their deaths.

When hurricanes reach the coast, they are cut off from their energy source (the warm ocean water), and as a consequence, their energy rapidly dissipates. As the winds diminish below 119 km (74 mi) per hour, hurricanes are downgraded to tropical storms. Nevertheless, as they expend their energy, hurricanes can wreak havoc on coastal and near coastal communities (Fig. 12.35). Those hurricanes that remain at sea are ultimately blown northward, and lose energy as they encounter cooler ocean water. For these reasons, hurricanes rarely survive to affect the northeastern United States or Atlantic Canada.

Hurricane Andrew, which struck southern Florida in August of 1992, was the costliest natural disaster in United States history, claiming 53 lives and causing damage estimated at $20 billion (Fig. 12.36). This is costlier than even the Northridge earthquake of 1994 (see Chapter 5). The hurricane began as a disturbance in northwest Africa, and fed by the warm Atlantic waters as it swept westward, attained the status of a hurricane on August 22nd (Fig. 12.36a). Normally, hurricanes that follow this path veer northward to strike the

(a)

(b)

Figure 12.35
(a) Before, and (b) after the destruction of the beachfront at Folley beach, North Carolina, caused by Hurricane Hugo in 1989.

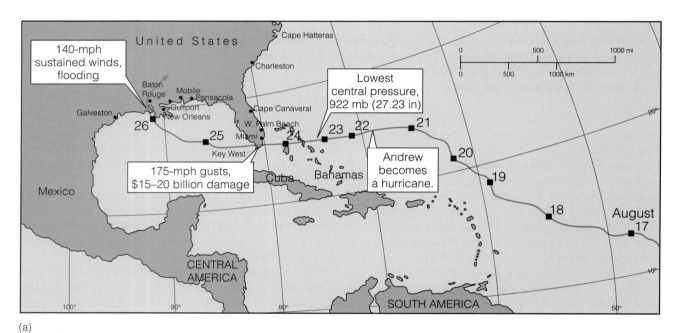

(a)

Figure 12.36
(a) The path of Hurricane Andrew from August 17th to the 26th, 1992, recorded at 10 PM CDT each day. The storm began as a low pressure center in North Africa that intensified over the warm tropical waters of the Atlantic Ocean and became a hurricane on August 22nd when its winds exceeded 119 km (74 mi) per hour. (b) Satellite image of Hurricane Andrew as the eye crossed the southern tip of Florida on August 24th, 1992.

(b)

Carolinas. However, intense high pressure centered in Canada altered this normal course, and the hurricane was deflected across the southern tip of Florida (Fig. 12.36b). Because this region of land is flat and narrow, the hurricane continued westward where it drew renewed energy from the warm waters of the Gulf of Mexico. Tracking northwestward, it made landfall a second time, this time striking the coast near Baton Rouge with winds up to 140 km (90 mi) per hour.

Each year in the United States, hurricanes result in scores of fatalities and billions of dollars in damage (Fig. 12.37). The hurricane season of 1995 was the worst on record, increasing fears that global warming may be causing hurricanes to become more intense and more numerous.

Other regions in the world are equally vulnerable, if not more so. The tropical cyclones spawned in the Indian Ocean cause widespread flooding in the low-lying coastal regions of Bangladesh. In April and May of 1991, a tropical cyclone, as ferocious as any hurricane on record in the United States, caused more than 100,000 fatalities. In 1970, more than 500,000 people lost their lives to flooding when the Ganges River broke its banks as a result of a tropical cyclone.

In summary, storm systems develop when contrasting air masses interact. The nature of the interaction can vary from local (such as tornadoes) to regional (such as hurricanes) in scale. Storms are a consequence of the amplification of the inherent instability in the troposphere. Although storm systems may periodically affect our day-to-day weather, they usually dissipate rapidly. But the effects of other patterns of circulation can be much more far-reaching. It is to these long-lasting air circulation patterns and their effects on climate that we now turn.

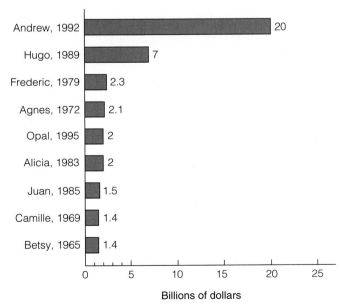

Figure 12.37
The most expensive hurricanes in U.S. history. Losses estimated in billions of dollars at the time of their occurrence.

12.3 Persistent Global Atmospheric Circulation Patterns

Every region on Earth is characterized by a certain type of climate that describes the region's annual cycle of weather patterns. Although prediction of the day-to-day weather may be fraught with uncertainty, these irregularities generally average out on an annual basis. We know what to expect from our four seasons. This predictability is due to persistent global air circulation patterns which are shown for July and January in Figure 12.38.

As we learned earlier, air circulates between the equator and poles in a series of convection cells called Hadley, Ferrel, and Polar cells (see Figure 12.8c). Hadley cells are characterized by zones of low pressure in equatorial regions where moist warm air rises, and zones of high pressure at 30°N and 30°S latitudes where this warm air is downdrafted. This combination controls atmospheric convection between these latitudes. Because of the Coriolis effect, air rotates clockwise around high pressure centers in the northern hemisphere. This movement generates a system of persistent winds that blow from the southwest north of 30°N and from the northeast in the tropics and subtropical regions.

The northeasterly winds were discovered by mariners who utilized them for trade. They were consequently dubbed the "trade winds." These trade winds mainly blow toward the southwest, which explains why Columbus reached the West Indies rather than landing on continental North America when he set sail for the New World. Sailors took maximum advantage of these trade winds when traveling from Europe to America by using a southerly route. They returned by a more northerly route to take advantage of the "westerlies," which mainly blow from the west toward the northeast.

At the equator, air converges from both the northern and southern Hadley cells in a region called the **intertropical convergence zone.** The band of low pressure in this zone shows little variation in pressure. Winds are consequently light, and mariners dubbed this region the "doldrums." The monotony of the weather in this region has given rise to the expression "down in the doldrums."

Similarly, high pressures with limited pressure variation produce very gentle winds at latitudes close to 30° north (in the vicinity of the Bermuda high) and 30° south. According to nautical legend, ships sailing to the Americas often languished in these regions due to the lack of winds. When food supplies diminished, horses were either thrown overboard to lighten the cargo or were eaten. Consequently, these regions became known as the **horse latitudes.**

North of the horse latitudes in the northern hemisphere (and south of them in the southern hemisphere) the predominant influence of the simple convection system of the Hadley Cell wanes. Although the downwelling air from the Hadley Cell stimulates convection in the cooler air to the north (the Ferrel Cell, see Figure 12.8c), its effects are weaker and are blended with two other important phenomena: the Coriolis

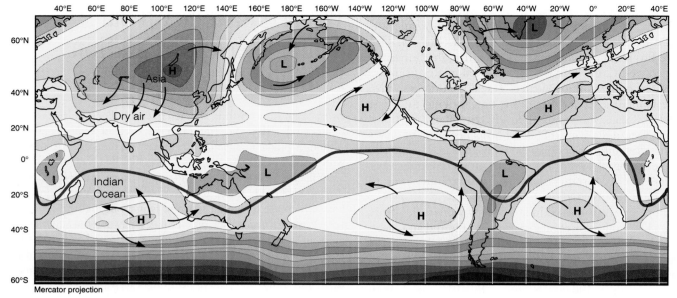

(a) January (Northern Hemisphere Winter)

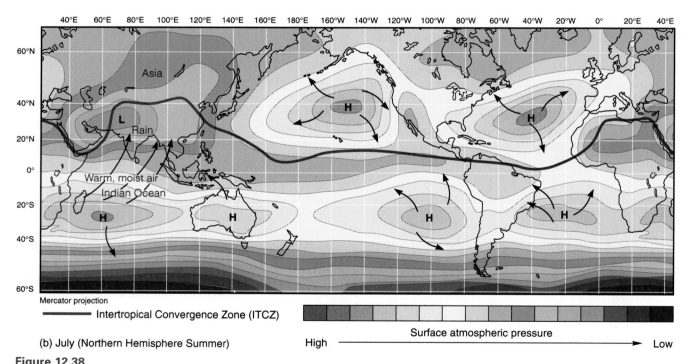

Intertropical Convergence Zone (ITCZ)

(b) July (Northern Hemisphere Summer)

Surface atmospheric pressure
High ——————————————————→ Low

Figure 12.38
Average sea level atmospheric pressure and surface wind-flow patterns for (a) January and
(b) July. The heavy line shows the average position of the intertropical convergence zone (ITCZ).
Note that the high pressure systems dominating the Asian continent in January give way to low
pressure systems in July. This is important in the explanation of monsoon climates.

effect and the polar front jet stream. The Coriolis effect is stronger at higher latitudes, and the wanderings of the jet stream separating cold polar and mild mid-latitude air masses generates powerful frontal systems in the region known as the polar front. As we have seen, this front has a profound effect on climate and the seasons. As a consequence, the weather at mid-latitudes is very changeable.

12.3.1 Monsoons

Perhaps the most important phenomenon related to persistent circulation patterns is the monsoon season that affects much of southeastern Asia. The word **monsoon** is derived from the Arabic word "mausim" for seasons. Areas which have a monsoon climate show dramatic seasonal changes in weather patterns.

Global Circulation and Past Continental Configurations

In this chapter, we have discussed the relationship between atmospheric circulation, weather, and climate in the modern world. We know from our discussions on plate tectonics (Chapters 5 and 6) that continental plates are adrift, and that over geologic time, their positions on the globe may have changed significantly.

In a general sense, there is nothing inherent in the principles of global air circulation that restricts their use to the present configuration of continents. Although modern and ancient climates may differ in detail, many of the principles outlined in this chapter may be applied to past continental configurations and can provide important clues about ancient climate. The explanations of circulation, including Hadley cells, polar jet streams, monsoons, and rain shadow regions, apply equally well to the past. The features shown in Figure 12.38 may therefore be superimposed on any continental configuration. In this way, we can investigate the history of ancient climates recorded in rocks.

Certain sedimentary rocks are indicative of climate. In Chapter 4, we learned how Wegener used certain rock features to promote his hypothesis of Continental Drift. For example, the distribution of ancient glacial deposits suggest that these regions once resided in high latitudes. Evaporite rocks such as salt and gypsum, on the other hand, indicate arid climate conditions in which evaporation exceeded rainfall, a scenario most typical of warm shallow seas with restricted circulation. A modern example is the Red Sea,

which occurs in just such an environment. Coral limestones are indicative of warm shallow seas with more open circulation. Dune sandstones and thick, red sedimentary sequences are representative of desert regions. On the other hand, coal forms from the burial of continental vegetation in both warm and cool climates but is most characteristic of equatorial and tropical regions.

Some differences in detail exist, however, between the modern and ancient world. For example, modern global circulation is driven by the temperature contrast between the warm equatorial and cold polar regions. This contrast is accentuated by the presence of polar ice caps. But the geologic record suggests that extensive polar ice is a relatively rare occurrence, so that the temperature contrast between the equator and the poles may not always have been so dramatic. This might have affected the size of Hadley cells, as well as the number of frontal systems developed by the polar front jet streams. Even so, it is clear that convection must have occurred, and that the size of the convection cells would have dictated the location of climate belts such as those associated with ancient deserts.

Climate models for the distant past can only identify long-term climatic changes. At present, they do not have the resolution to tackle short-term climatic changes, such as the advance and retreat of continental ice sheets over the last 200,000 years. Nevertheless, during some periods of geologic history, the principles we have determined from the modern world appear to

apply, at least in a general way, because the predicted climate matches the distribution of climate-sensitive sedimentary rocks.

Using computer models, Chris Scotese and his colleagues have produced paleogeographic and paleoclimatic maps for the last 550 million years. For example, reconstructions for the Late Paleozoic show Pangea stretching from pole to pole (Fig. 12.39). High pressure systems positioned at the 30°N and 30°S latitudes predict arid conditions on land and the formation of evaporites in shallow seas. These systems also suggest that the west coast of North America would have been dominated by northerly winds. As we learned in Chapter 9, this condition would have driven warm surface waters westward, facilitating marine upwelling and the development of vibrant ecological systems. Just such conditions are recorded in the marine sedimentary rocks of western North America at that time. Evidence of glaciation, noted by Wegener, occurred in South America, Africa, India, and Australia when these continents occupied much higher southerly latitudes than they do at present.

Mountains would also have had an important effect on climate and seasonality at this time. As Pangea was formed by continental collision, it would have had many collisional mountain belts in its interior. In addition, it would have had Andean-like mountains around its periphery, produced by the subduction of oceanic crust. All of these mountains would have been major barriers

(continued)

These patterns occur most notably in eastern India, in many parts of the Indochina peninsula, and in the Philippines, where a long dry season gives way to torrential rains. Some regions get nearly 10 m (400 in) of rain every summer.

Although these rains bring widespread flooding, they are probably the most benevolent floods on Earth. Highly innovative irrigation schemes utilize the water for agriculture, and indeed, should the monsoon rains fail as they have in the past, widespread starvation can result. Archaeologists speculate that this type of failure may have led to the sudden disappearance of the Harrapan people who resided along the Indus Valley in Pakistan until about 5000 years ago.

The origin of the monsoon rains lies in global-scale circulation patterns (Fig. 12.38). In mid-winter, cold air generates a

stable high pressure region in the center of the Asian continent. Like all high pressure centers (see Figure 12.16b), it pumps cold dry air outward, resulting in a dry season. In mid-summer, however, the continental interior heats up, and the presence of warm light air creates a low pressure region in central Asia which draws in air from its surroundings. At the same time, a summer high pressure region exists in the Indian Ocean because of the configuration of the Hadley cells. The low pressure system in Asia draws in warm moist tropical air from a vast tract of the Indian Ocean. This moisture-laden air strikes the southern Asian coastline where it initiates torrential rainfall.

The overall importance of the monsoon rains to the densely populated regions of southern Asia cannot be over-emphasized. The failure of these rains in any given year is catastrophic

Global Circulation and Past Continental Configurations, continued

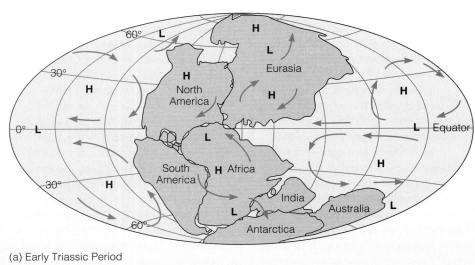

(a) Early Triassic Period

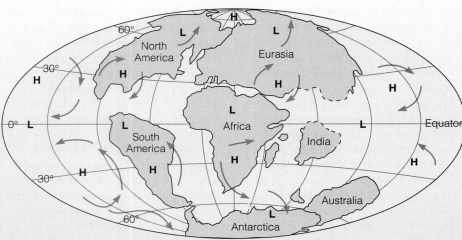

(b) Cretaceous Period

Figure 12.39
Persistent global air circulation patterns associated with Pangea. Note the presence of low pressure systems along the equator, and high pressure systems at 30° latitude, similar to modern patterns.

to the flow of moisture in the air, and would have led to the generation of numerous rain shadow deserts.

With a large area of land straddling the equator, Pangea would also have had very strong responses to seasonal changes in radiation. A monsoonal climate similar to that of modern southern Asia is therefore likely to have developed. In the northern hemisphere of Pangea, for example, cooling during wintertime would have resulted in the development of high pressure regions over the continent. This would be a major source of dry continental air. In the summertime, however, low pressure systems would have developed as the interior of the continent heated up. These would have drawn in warm moist air from the high pressure systems that would have formed over the ocean. The effects would have been accentuated by the presence of mountains at the continental margin, which would have caused most of this moisture to precipitate.

The equatorial regions of Pangea would also have been very humid, promoting the growth of vegetation and eventually the formation of coal. The very important coal belt, stretching from North America to eastern Europe, reflects the equatorial and near-coastal position of this part of Pangea at that time.

For the reconstruction of Pangea, therefore, the relationship between atmospheric circulation, weather, and climate derived from the modern world, predicts global patterns of circulation and climate that closely match the clues to Pangean climate preserved in the rock record.

because it means a failure of the crops. As we discussed in Chapter 8, failure of monsoon rains has been linked to the development of El Niño. Meteorologists monitor the development and strength of atmospheric pressure systems and water temperatures in the Indian and southern Pacific oceans in order to predict the timing and intensity of the monsoon rains. Pressure contrasts between continental Asian air and the air above the Indian Ocean are essential if the monsoon rains are to be drawn northward. Every year, these systems are carefully monitored so that advance warnings may be issued to governments and aid agencies should it appear that the monsoon rains will fail.

12.4 Chapter Summary

Weather and climate are profoundly influenced by the Earth's tilt and by variations in atmospheric pressure, temperature and relative humidity. Atmospheric circulation is a form of heat transport in which warm air is driven towards the poles to meet cold polar air moving in the opposite direction. Persistent global circulation patterns, including Hadley cells, account for the formation of deserts and the intense seasonality exhibited by the monsoon climates of southern Asia.

As air masses move, their path is determined by the Coriolis effect which influences all fluids that move long distances over the Earth's surface. High pressure centers are regions of relatively dense, descending air. They act as air pumps that blow cold air away from their centers. In the northern hemisphere, the Coriolis effect imparts a clockwise sense of rotation as the air moves because of the eastward rotation of the Earth.

Conversely, low pressure centers are regions of relatively warm ascending air. As the air rises, it cools and the moisture within it condenses. This releases the latent heat stored within it, amplifying the ascent. Low pressure regions draw in the surrounding air, the Coriolis effect imparting a counterclockwise sense of rotation on its movement in the northern hemisphere.

Jet streams occur where there are sudden changes in the altitude of the tropopause and generally mark boundaries between large contrasting air masses. The most important of these is the polar front jet stream which generates fronts, cyclones, and anticyclones in the higher latitudes.

Storms occur when highly contrasting air masses collide. Thunderheads develop when cold fronts wedge beneath warm air and intensify their uplift. Tornadoes occur over land when cool air derived from mountainous regions finds itself on top of warm, light air. This highly unstable situation causes rapid uplift, which draws in air from the surrounding regions. Hurricanes typically originate over tropical ocean waters when the ocean temperature attains a critical value. When this occurs, low pressure regions initiate and intensify as they feed on a plentiful supply of warm moist air. A monsoon climate is one of contrasting seasons related to changes in atmospheric circulation in the continental interior of Asia. The rainy season occurs during the summer, when the ascent of warm air in the Asian interior generates a low pressure system which draws in warm moist air from the Indian Ocean.

Key Terms and Concepts

Look for the highlighted items on the web at:
WWW.BROOKSCOLE.COM

- Atmospheric circulation is global in scale and remarkably uniform in various parts of the world. Circulation is influenced by the Earth's tilt, heat transfer between the equatorial and polar regions, and the Earth's spin which results in the Coriolis effect

- Contrasting air temperatures over the land and over the ocean influences the weather and climate of coastal regions

- Global-scale circulation is driven by contrasting heat between equatorial and polar regions. The resulting heat transport occurs in a system of convecting cells (Hadley, Ferrel, and Polar) that profoundly influence climate belts, including the location of deserts where the warm, dry air of the Hadley cells descend at latitudes 30°N and 30°S

- Because of turbulence in the troposphere, air tends to ascend or descend in plumes. Ascending air creates low pressure centers (or cyclones); descending air creates high pressure centers (or anticyclones)

- High pressure centers pump out air whereas low pressure centers receive air from the high pressure centers that surround them

- The Coriolis effect in the northern hemisphere results in clockwise circulation of air around high pressure centers and counterclockwise circulation of air around low pressure centers

- Fronts develop when contrasting air masses collide and warmer, lighter air is forced upwards to form clouds. At a cold front, dense cold air wedges beneath warm light air. At a warm front, warm air rides over the top of cold dense air. An occluded front occurs where a faster moving cold front overtakes a warm front

- Jet streams occur where there are rapid variations in the altitude of the tropopause. Jet streams are swift high-altitude winds that flow eastward at velocities of up to 240 km (150 mi) per hour. Where the jet stream bends, air may either flow out to produce low pressure centers, or flow in to form high pressure centers

- Because jet streams separate contrasting air masses, fronts develop below them. As a result, the location of the jet streams profoundly influences weather and climate

- Storms result from the collision between greatly contrasting air masses. 2000 thunderstorms occur at any one time. Sheet lightning occurs where the electrical charge within a thundercloud exceeds a threshold. Forked lightning occurs when opposite charges at the base of the cloud and on the ground exceed a threshold

- Tornadoes are spawned when a layer of warm moist air occurs beneath cool dense air. The warm air rises rapidly and violently to create a narrow region of intense low pressure, and a vortex of swirling winds

- Hurricanes are intense low pressure systems that occur over wide regions. They typically develop as a result of a chain of events over warm tropical waters when warm moist air rises rapidly and creates an intense low pressure that feeds on the surrounding warm air

- Monsoons are related to persistent global circulation patterns and the seasonal pressure changes in the interior of the Asian continent. In the summer, a low pressure system in the center of the continent draws in warm moist air from the Indian and southwestern Pacific oceans resulting in torrential rainfall in southern Asia. Periodic failure of the monsoon rains has been linked to El Niño

Review Questions

1. How do regions of high pressure and low pressure originate and develop?

2. Explain the counterclockwise rotation of low pressure systems in the northern hemisphere.

3. Why does the jet stream have such an important influence on continental weather systems?

4. What is the relationship between the jet stream and the elevation of the tropopause?

5. What atmospheric conditions allow the formation of tornadoes to be predicted?

6. Explain the relationship between low pressure systems and the development of hurricanes.

7. Why are deserts located at 30°N and 30°S latitudes?

8. Explain the principles of persistent global circulation patterns.

9. What is the origin of the "doldrums"?

10. Explain the contrasting seasons associated with monsoon climates.

Study Questions

Research the highlighted questions using InfoTrac College Edition.

1. Explain the duration of sunlight (see Figure 12.4) during the southern hemisphere summer (December).

2. Figure 12.7c shows an example of convection in a room. How may this example be applied on a global scale to explain the worldwide distribution of deserts?

3. Figure 12.40 shows a global reconstruction 180 million years ago. Show on the diagram where you would expect deserts to form at that time and give a brief explanation of your reasoning.

4. Figure 12.13a shows wind directions around high and low pressure regions in the northern hemisphere. Using similar principles, explain the wind directions related to high and low pressures in the southern hemisphere.

5. What is the origin and implication of the Hadley cells? How do these cells influence climate of both continents and oceans? What is their relationship to the "doldrums"?

6. What is the origin of the Ferrel cells? How are they related to the development of the Hadley cells?

7. What is the relationship between global air circulation (Fig. 12.38), the jet stream (Fig. 12.22), and the generation of cold and warm fronts (Figs. 12.18 and 12.19)?

8. As hurricanes develop, why do their low pressure systems continue to intensify?

9. Using Figure 12.38, explain why monsoon rains typically occur within a limited time period each year. In what way would an El Niño affect this typical pattern?

10. If you were in charge of an aid agency in Southern Asia, how would you recommend monitoring the development of the monsoon rains?

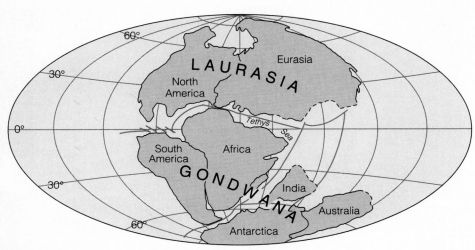

180 m.y.a

Figure 12.40
Figure to accompany Study Question 12.3: Global reconstruction for the Jurassic Period (180 million years ago).

Atmosphere

Solid Earth

Volcanic eruptions first created and continue to add to the atmosphere. The Earth's mass has enough gravity to keep the vented gases from being released into space. Earth's surface absorbs and reflects solar energy, affecting weather patterns. Mountains affect the flow of air and influence the location of deserts, storms, and tornadoes. Wind and precipitation erode the Earth's surface.

Hydrosphere

The oceans absorb oxygen and carbon dioxide and help to modulate air temperature. Warm moist air over oceans feeds hurricanes. The hydrologic cycle adds water vapor to the atmosphere. Water vapor forms clouds and falls to Earth as precipitation. Air circulation affects ocean currents.

Atmosphere

Atmosphere is layered according to temperature and density. Solar radiation heats the layers of the atmosphere. From these basic influences, complex weather and climate patterns develop. The ozone layer absorbs ultraviolet radiation.

Biosphere

Plant and animal life modify the atmosphere's composition by photosynthesis and respiration. The protection of the ozone layer makes life possible. Human activity adds pollution and greenhouse gases to the atmosphere.

Astronomy

The Earth's tilt makes seasons possible. The solar wind and incoming solar radiation affect global temperatures and weather patterns. The Coriolis effect, due to the Earth's spin, causes counterclockwise movement of low pressure systems (in the northern hemisphere). Disturbances in the ionosphere are seen as auroras.

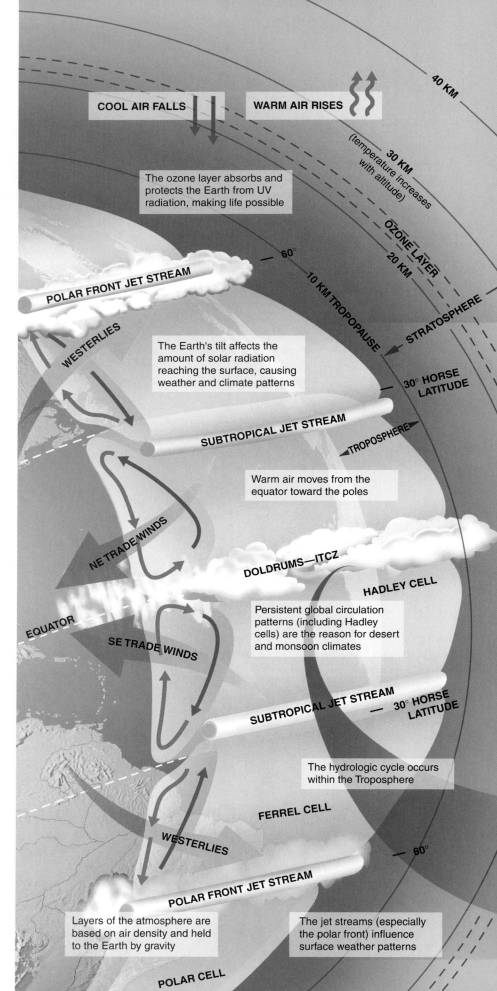

COOL AIR FALLS

WARM AIR RISES

The ozone layer absorbs and protects the Earth from UV radiation, making life possible

40 KM

30 KM

(temperature increases with altitude)

OZONE LAYER

20 KM

10 KM TROPOPAUSE

STRATOSPHERE

60°

POLAR FRONT JET STREAM

WESTERLIES

The Earth's tilt affects the amount of solar radiation reaching the surface, causing weather and climate patterns

30° HORSE LATITUDE

SUBTROPICAL JET STREAM

TROPOSPHERE

Warm air moves from the equator toward the poles

NE TRADE WINDS

DOLDRUMS—ITCZ

HADLEY CELL

Persistent global circulation patterns (including Hadley cells) are the reason for desert and monsoon climates

EQUATOR

SE TRADE WINDS

SUBTROPICAL JET STREAM

30° HORSE LATITUDE

The hydrologic cycle occurs within the Troposphere

FERREL CELL

WESTERLIES

60°

POLAR FRONT JET STREAM

Layers of the atmosphere are based on air density and held to the Earth by gravity

The jet streams (especially the polar front) influence surface weather patterns

POLAR CELL

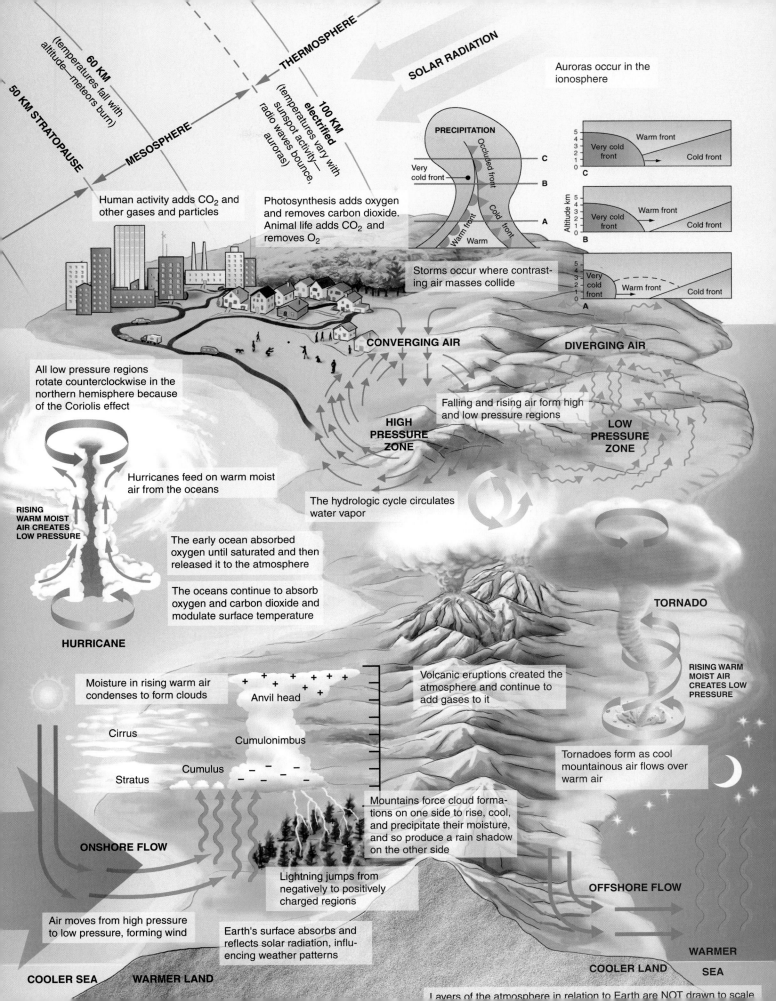

THERMOSPHERE

SOLAR RADIATION

100 KM
electrified
(temperatures vary with
sunspot activity—
radio waves bounce,
auroras)

60 KM
(temperatures fall with
altitude—meteors burn)

MESOSPHERE

50 KM STRATOPAUSE

Auroras occur in the
ionosphere

Human activity adds CO_2 and
other gases and particles

Photosynthesis adds oxygen
and removes carbon dioxide.
Animal life adds CO_2 and
removes O_2

PRECIPITATION

Occluded front

Very
cold front

Warm front

Cold front

Warm

C

B

A

5
4
3
2
1
0
Very cold
front
Warm front
Cold front
C

Altitude km
5
4
3
2
1
0
Very cold
front
Warm front
Cold front
B

5
4
3
2
1
0
Very
cold
front
Warm front
Cold front
A

Storms occur where contrasting air masses collide

CONVERGING AIR

DIVERGING AIR

All low pressure regions
rotate counterclockwise in the
northern hemisphere because
of the Coriolis effect

Falling and rising air form high
and low pressure regions

HIGH
PRESSURE
ZONE

LOW
PRESSURE
ZONE

Hurricanes feed on warm moist
air from the oceans

RISING
WARM MOIST
AIR CREATES
LOW PRESSURE

The hydrologic cycle circulates
water vapor

The early ocean absorbed
oxygen until saturated and then
released it to the atmosphere

The oceans continue to absorb
oxygen and carbon dioxide and
modulate surface temperature

HURRICANE

TORNADO

RISING WARM
MOIST AIR
CREATES LOW
PRESSURE

Volcanic eruptions created the
atmosphere and continue to
add gases to it

Moisture in rising warm air
condenses to form clouds

+ + + + + +
Anvil head
+

Cirrus

Cumulonimbus

Cumulus

Stratus

– – –
– – –
+

Tornadoes form as cool
mountainous air flows over
warm air

Mountains force cloud formations on one side to rise, cool,
and precipitate their moisture,
and so produce a rain shadow
on the other side

ONSHORE FLOW

Lightning jumps from
negatively to positively
charged regions

OFFSHORE FLOW

Air moves from high pressure
to low pressure, forming wind

Earth's surface absorbs and
reflects solar radiation, influencing weather patterns

WARMER

COOLER SEA   WARMER LAND

COOLER LAND   SEA

Layers of the atmosphere in relation to Earth are NOT drawn to scale

PART 4

The Evolution of the Earth Systems and the Biosphere

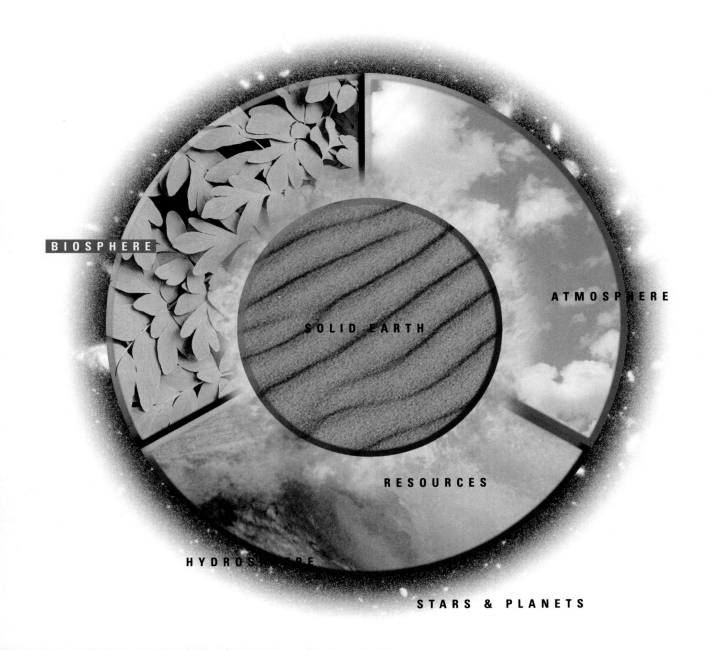

BIOSPHERE

SOLID EARTH

ATMOSPHERE

RESOURCES

HYDROSPHERE

STARS & PLANETS

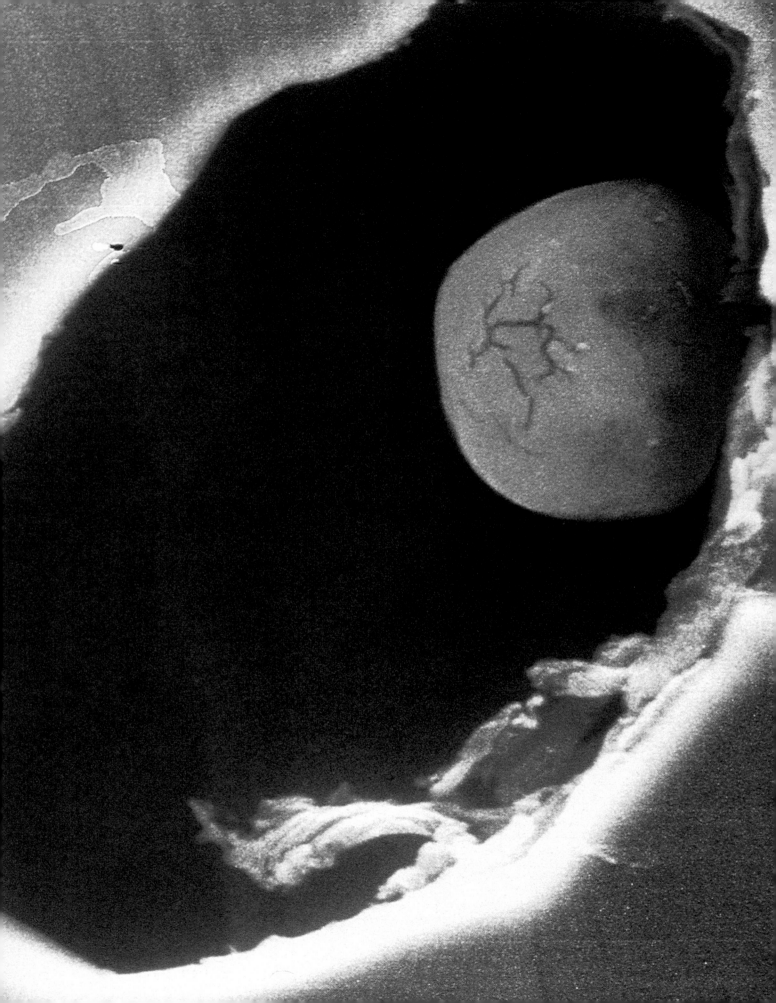

CHAPTER 13

Evolution of the Earth Systems I: The Precambrian

hapters 13 and 14 outline the major evolutionary events in the Earth's geologic history that transformed the planet from its tumultuous and hostile beginnings almost 4.6 billion years ago to the fertile life-sustaining planet we live on today. Together, they describe the origin and evolution of the biosphere and its interaction with the solid Earth, the hydrosphere and the atmosphere. The geologic record, although incomplete, provides the basis for our understanding and reconstruction of these interactions.

This chapter deals with the Precambrian time period which accounts for 88% of geologic time, beginning with the birth of Planet Earth nearly 4.6 billion years ago, and ending with the appearance of shelly fossils at the start of the Paleozoic Era, 545 million years ago. We will find that the very early history of the Earth was dominated by processes that divided the Earth into layers according to density. These layers are the core, the mantle, a primitive crust, the hydrosphere and the atmosphere.

The Precambrian is divided into three eons, the Hadean (4.6 to 4.0 billion years ago), followed by the Archean (4.0 to 2.5 billion years ago) and the Proterozoic (2.5 billion to 545 million years ago). The Hadean Eon marks the interval between the formation of the Earth and the Earth's oldest rocks. At the time of writing, the oldest known rocks occur in the Canadian Shield and are 3.96 billion years old. The Hadean Eon, for which we have no direct record, therefore represents about 13% of geologic time. The beginning of the Archean Eon marks the start of the rock record, almost four billion years ago, and broadly coincides with the first appearance of life, sometime before 3.86 billion years ago. The Archean rock record suggests that some crude form of plate tectonics probably began before the end of the eon, 2.5 billion years ago. During much of the Archean and subsequent Proterozoic eons, continents grew from small nuclei to large stable regions called cratons. Although controversial, most geoscientists now believe that some primitive form of plate tectonics influenced this growth.

We will outline the three popular theories for the origin of life. Most scientists believe that life began in intertidal pools on continental shelves. In these environments, cellular life could have reproduced and would have been protected from deadly solar radiation. Recent evidence suggests that a deep sea origin may also be possible adjacent to volcanic vents. A third theory, supported by recent studies of the Hale-Bopp comet, proposes that the young Earth may have been seeded with life through its collision with comets.

Irrespective of its mode of origin, it is generally agreed that early life probably consisted of single-celled organisms. We will see that organic compounds organized themselves in such a way that reproduction became possible. However, since reproduction was asexual rather than sexual, evolution was slow. In addition, the Earth's early microscopic organisms almost certainly evolved in bodies of water which were in contact with an oxygen-deficient atmosphere. These microorganisms were consequently anaerobic, meaning they did not require oxygen. Yet life rapidly became capable of photosynthesis, which produced oxygen. It is likely that this oxygen was absorbed by the ocean waters before it reached the atmosphere. Eventually, between 2.0 and 1.8 billion years ago, the ability of the oceans to absorb the oxygen produced by the Earth's early life was exceeded and oxygen became an atmospheric gas. As oxygen is a poison to anaerobic life, this "oxygen crisis" must have triggered a mass extinction. About 1.8 billion years ago, the first aerobic organisms appeared, and by about 1 billion years ago, the oxygen content in the atmosphere had crossed the 2% threshold needed to form ozone. This led to the first establishment of the ozone layer and a protective shield from ultraviolet radiation.

Towards the end of the Proterozoic, between 900 and 545 million years ago, the geologic record is better preserved and scientists can attempt realistic plate reconstructions. A major period of mountain-building activity produced global-scale mountain belts and culminated towards the end of the Proterozoic with the formation of a supercontinent. During this time, increasing atmospheric oxygen and the presence of an ozone shield opened up a range of ecological niches, or different areas that could support life. About 600 million years ago, a major evolutionary event saw the first appearance of a diverse assemblage of soft-bodied animals known as the Ediacaran fauna. These fossils probably represent the ancestors of the profuse shelly animals or fauna of the Early Cambrian Period which marks the beginning of the Paleozoic Era.

You will learn that the Earth's evolution has involved persistent interactions between the Earth systems. It will also become apparent that the same interactions will guide the future evolution of our planet. Therefore, the lessons to be learned from Earth history are fundamentally important both to our understanding of the pace and rhythm of global change throughout geologic time and in providing the backdrop for the study of modern environmental change induced by human activity. In the modern world, an accelerated rate of change, or an interaction hitherto undocumented, may well provide evidence of human-induced (or anthropogenic) change. We will explore these ideas further in the last section of the book.

13.1 Introduction

Today's Earth is fundamentally different from the primitive hostile planet that formed nearly 4.6 billion years ago (Fig. 13.1). But how did the primordial Earth evolve from its tumultuous beginnings to its relatively benign condition today? In this chapter, we embark on a journey through time and explore how the fundamental interactions of the Earth systems have guided the evolution of the planet. For the most part, this evolution has

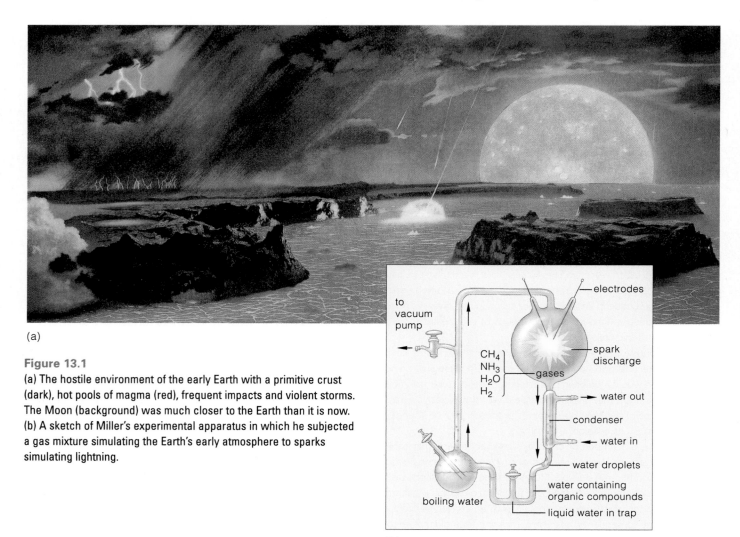

(a)

Figure 13.1
(a) The hostile environment of the early Earth with a primitive crust (dark), hot pools of magma (red), frequent impacts and violent storms. The Moon (background) was much closer to the Earth than it is now.
(b) A sketch of Miller's experimental apparatus in which he subjected a gas mixture simulating the Earth's early atmosphere to sparks simulating lightning.

(b)

been slow and progressive. However, at other times it appears that catastrophic events may have triggered sudden change. This is best documented in the biosphere, where evolution is punctuated by several mass extinction events, each of which annihilated most of life on Earth, leaving only the most opportunistic species to survive and evolve.

The first 4 billion years of the Earth's history, from its birth 4.6 billion years ago to the emergence of the first shelly fossils 545 million years ago, is called the **Precambrian**. This interval, which represents 88% of geologic time, is subdivided into three eons. The **Hadean Eon** represents the interval between the formation of the Earth and the age of its oldest rocks. There is therefore no direct record of this era. To date, the oldest known rocks are gneisses in the Northwest Territories of the Canadian Shield which record an age of 3.96 billion years. We can therefore derive no information from the geologic record for the first 600 million years of Earth history. The **Archean Eon** (meaning "ancient") stretches from 4.0 to 2.5 billion years ago and was followed by the **Proterozoic Eon** ("earlier life") which extends nearly 2 billion years until 545 million years ago.

As we will learn in Chapter 14, the history of the **Phanerozoic Eon** ("visible life"), which represents the last 545 million years, is more completely understood than either the Archean or Proterozoic. Because the Phanerozoic is the youngest eon, its geology has been better preserved. In addition, the fossil record of the Phanerozoic is far more complete, giving a much better control on the age of the rock strata (recall relative age dating from Chapter 3).

13.2 Principles and Processes in the Earth's Evolution

The observation that the Earth systems interact is not a modern concept. Ancient civilizations described the interaction between Earth, wind and water that, together with fire, were believed to be the basic elements from which all things were made. Although modern science has determined that atoms are the fundamental building blocks of matter, descriptions of the four "elements" by ancient civilizations suggest that they recognized the importance of the Earth systems and their interactions.

In more recent times, the revolutionary and controversial writings of Charles Darwin showed how the evolution of life is fundamentally affected by its environment. Although some major controversies remain concerning the origin of life, the basic elements of life's early evolution are now clear. The slow evolution from single-celled organisms to soft-bodied multicellular life dominated the Archean and Proterozoic. During the Phanerozoic Eon (see Chapter 14), life evolved into shelly and skeletal organisms which are more readily preserved than their soft-bodied predecessors, explaining the superior fossil record of this eon.

But what of the evolution of the other Earth systems, the atmosphere, hydrosphere, and the solid Earth? Surely the evolution of these systems must have had a profound influence on the biosphere, which depends on them for its survival. Here we return to one of the founding principles of geology, that of Actualism which states that "the present is the key to the past."

13.2.1 Principle of Actualism Revisited

At first glance, it may seem that the principle of Actualism provides the key to unlocking the mysteries of the Earth's early history. According to this principle we can interpret ancient history by looking at modern analogues. For example, our understanding of sedimentary depositional processes on the modern sea floor allows us to recognize and interpret such features in ancient rocks. This approach yields a wealth of information about the ancient lithosphere (including both its oceanic and continental crust), the evolution of which was probably dominated throughout much of geologic time by plate tectonic processes similar to those active today. Using this approach, we can find evidence for the existence of oceans during the Archean in the preservation of Archean deposits similar to those of modern continental margins. The deposits show that continental margins, and therefore oceans, had developed by this time. Likewise, because modern volcanoes occur mostly in tectonically active settings (such as continental rifts, mid-ocean ridges, subduction zones, and hotspots), many ancient volcanic rocks can be interpreted to have formed in a similar way. Modern volcanism also contributes to the composition of our atmosphere, so we can assume a similar relationship in the past. Using the same approach, we will see that ancient regions of highly deformed and metamorphosed rocks can be viewed as ancient analogues of more recent mountain belts such as the Alps and the Himalayas.

But while the principle of Actualism is a very useful guide, how far back in time can we extend it? Despite some similarities between the Archean, Proterozoic, and Phanerozoic eons, there are also some fundamental differences. Each eon exhibits some unique characteristics, which suggests that the principle of Actualism may have its limitations.

The most obvious example is in the application of plate tectonic theory. There is little doubt that plate tectonics is the dominant process shaping the evolution of the Earth's systems in the modern world. But how far back in time does plate tectonics as we know it extend? Certainly not as far as the Earth's beginnings. Modern plate tectonic interactions are dominated by the *horizontal* motion of plates. However, the formation of the layered Earth, from its core with the densest elements, to the atmosphere with the lightest ones, occurred early in its history. This separation of elements according to their density indicates the importance of *vertical* movement due to gravity in the Earth's early history (see Table 13-1). Indeed, many scientists believe that it was gravitational collapse that formed the Earth and caused its rapid segregation into layers of inwardly increasing density. Because the effects of plate tectonics are largely confined to

Table 13-1

Average Density of the Earth's Layers	g/cm³
Atmosphere (at sea level)	0.0129
Hydrosphere	1.0
Lithosphere	
Continental crust	2.7
Oceanic crust	2.9
Mantle	5.0
Core	13.0

the upper 500 km of the Earth, its role in forming the fundamental divisions of the Earth's interior must have been a minor one at best. And so while plate tectonics may be the dominant influence on the modern evolution of the Earth's upper layers, sorting by gravity may have had a more important role in the Hadean and Early Archean.

But if plate tectonic processes were less important during the Earth's early history, when and how did they evolve to their modern dominance? Some geoscientists believe that a transition occurred between the Archean and the Proterozoic, during which vertically-dominated processes of the youthful Earth led to the production of thin microplates, which then moved horizontally, amalgamating to form progressively larger plates. In order to understand the evolution of Earth's systems, it is therefore clear that we need an appreciation of Earth history.

13.3 Earth Systems in the Hadean and Archean

13.3.1 Hadean to Middle Archean (4.6 to 3.0 Billion Years Ago)

One of the key questions we seek to explore in this chapter is how the Earth's systems evolved from their fragile origins some 4.6 billion years ago to the complex interdependence they show today. Models of planetary accretion indicate that the early history of the Earth involved the rapid development of the core, some models suggesting that this might have happened within 20 million years of the Earth's formation. This rapid evolution is related to the fact that the planet was probably much hotter at that time than it is today. Several processes that primarily occurred early in the Earth's history supplied this heat. The sinking of iron to form the Earth's core, for example, would have

released heat in the form of gravitational energy. There would also have been abundant meteorite impacts on the Earth's surface because of the turbulent nature of the early Solar System, and because of the lack of adequate protection by the primitive atmosphere. Upon impact, these meteorites would have released vast amounts of energy as heat. Radioactive isotopes also give off heat when they decay to more stable daughter products (see Chapter 3) and would have been more abundant in the interior of the early Earth. The decay of the unstable isotope of Aluminum (^{26}Al) to a stable isotope of Magnesium (^{26}Mg) is a particularly important contributor to the heat budget of the primitive Earth because ^{26}Al was both abundant and decayed rapidly, with a half-life of only 720,000 years.

None of these processes produce as much heat today. Virtually no heat is supplied today by gravitational collapse because the Earth's layers have already formed. Similarly, negligible heat is provided by meteorite impacts because meteor showers are now far less frequent and the Earth's atmospheric shield now burns up most extraterrestrial objects before they impact on its surface. Although the production of heat by radioactive decay remains an important source of heat, it is estimated to be an order of magnitude less than what it was when the Earth was young. This is because the significant heat contribution to the early Earth was from very unstable radioisotopes with short half lives of less than 1 million years. Today, 4.6 billion years after the Earth's formation, the overwhelming majority of these unstable elements have already broken down to their stable daughters, and so cannot yield additional heat (Fig. 13.2).

Because the production of heat within the Earth's interior was considerably higher when the Earth was young, melting in

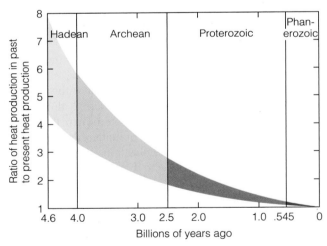

Figure 13.2

Heat production from radioactive decay through time, comparing the ratio of radiogenic heat produced in the past to that of the present time. This heat production has decreased exponentially because little remains today of the highly unstable elements with short half-lives. The width of the colored band reflects the degree of uncertainty.

the mantle, and therefore volcanic activity, would have been far more pronounced. The ascent of magma and the cooling of lava at the Earth's surface would have soon led to the formation of a delicate "proto-crust." Field investigations and laboratory analyses show that the compositions of these Early Archean lavas were generally similar to those of today. This indicates that the mantle source of these ancient lavas was of basically similar composition to that of the modern mantle.

But what about our early atmosphere? Given the dominance of hydrogen and helium in the universe, our early atmosphere probably contained significant quantities of these gases. However, the Earth's gravity is not strong enough to retain these gases so they would have leaked out into space. But if the Earth lost its early atmosphere, it was soon replaced by another one derived from the Earth's interior. The mantle is not only the source of magma, it is also the source of the gases released to the atmosphere by volcanoes. Given a similar mantle composition, volcanic gases emitted from Hadean volcanoes are likely to have been similar to those emitted today. If so, these volcanoes would have introduced abundant water vapor, and lesser amounts of carbon dioxide, carbon monoxide, sulfur dioxide, sulfur, chlorine, nitrogen, argon, and hydrogen into the primitive atmosphere. In the modern world, this atmospheric composition has become profoundly modified by the biosphere's production of oxygen during photosynthesis (the synthesis of organic compounds from water and carbon dioxide using sunlight). However, in the lifeless Hadean, the Earth's first atmosphere is likely to have been rich in these volcanic gases, and devoid of oxygen. It was therefore quite unlike the atmosphere of today.

And when and how did our hydrosphere form? To answer this we need to know when water first existed on the Earth's surface. The evidence for this comes from clastic sedimentary rocks which provide a record of the flow of water because they are formed by the cycle of erosion, transport, and deposition. The oldest definitive sedimentary rocks typical of ocean deposits and so requiring the existence of ocean water, are more than 3.8 billion years old. These rocks are typical of those formed at modern continental margins, and were likely formed in a similar environment. Such rocks are preserved in the geologic record because they are too light to subduct. So, as the adjacent oceanic crust is subducted, they tend to become attached to the overriding continental crust and are preserved (see Chapters 5 and 6). The presence of these ancient sedimentary sequences implies that oceans must have already existed by this time. But was there water on our planet at an even earlier time?

Very early in the Earth's evolution, the water vapor introduced into the atmosphere by volcanic activity would have reached a saturation level, and it would have rained. But did this rainfall accumulate in oceans, and if so, was there a hydrologic cycle? The answer is probably not. For a hydrologic cycle to develop, topography is needed, so that water can accumulate in topographically depressed areas as it does today in modern ocean basins. As we learned in Chapter 8, the topography in the modern world plays a key role in this process, the contrasting densities of continental crust (2.7 g/cm³) and oceanic crust (2.9 g/cm³) ensuring that the lighter continents float higher on the asthenosphere than the denser ocean floors. As a result, water drains from the continents and accumulates in ocean basins.

Therefore, in order to develop a hydrosphere similar to that of today, relatively buoyant continental crust is needed to provide the necessary topography for water to accumulate. So when did continental crust first form? Because evidence of the Earth's early history is poorly preserved in the rock record, this is a controversial subject. The oldest dated continental rocks that were originally igneous come from the Slave Province in the Northwest Territories of Canada, and contain zircon crystals that are 3.96 billion years old. However, the oldest zircon crystals come from metamorphosed sedimentary rocks in Western Australia. These crystals vary in age, but the oldest is 4.27 billion years old. The continent from which these zircons were derived must have been at least this old because the sedimentary rocks (and the zircons they contain) were derived by erosion of continental crust. Although this is indirect evidence, recall from our discussions of geochronology (Chapter 3) that zircon ages typically date the *age of crystallization of the zircon crystal*. So the zircon crystals in the rocks of Western Australia must have originally formed in an igneous rock, such as granite, 4.27 billion years ago before being removed by erosion and incorporated in the sedimentary rock (Fig. 13.3).

So the zircon ages do not yield the depositional age of the sedimentary rock but, instead, give the age of a granitic body whose location is unknown. It may seem foolhardy to base models of continental growth and the origin of the hydrosphere on a few tiny zircon crystals that originally crystallized in a rock that we cannot find! Yet, this is firm evidence that some exposed continental crust existed almost 4.3 billion years ago.

We may therefore have had low-lying areas in which water could accumulate to form primitive oceans (or "proto-oceans") approximately 4.3 billion years ago. But what changes would this have caused in the early Earth? Such proto-oceans would have initiated the hydrologic cycle of evaporation and precipitation. Precipitation would then have extracted water vapor from the atmosphere. Because water has the ability to dissolve substantial quantities of carbon dioxide (modern ocean water, for example, contains 62 times more carbon dioxide than the atmosphere), it is likely that the accumulation of ocean water also extracted significant amounts of carbon dioxide from the ancient atmosphere. This reduction of atmospheric carbon dioxide and water vapor would have weakened the greenhouse effect, resulting in cooling of the Earth's surface.

In summary, conditions in the Hadean and Early Archean would have resembled a planet on "fast-forward" compared to the relatively pedestrian pace of change that occurs in the modern world. This rapid evolution reflects a much hotter planet, and, as you will now learn, paved the way for the origin of life on Earth.

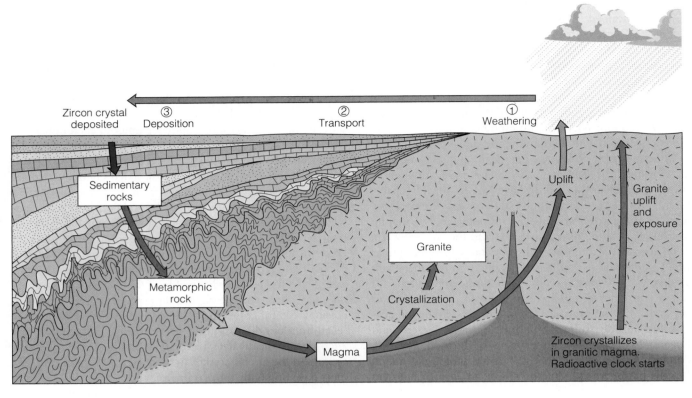

Figure 13.3
How zircon crystals end up in a sedimentary rock. The zircon first crystallizes in an igneous melt. Its radioactive clock starts ticking at that time and is virtually unaffected by subsequent geologic events. Eventually, because of uplift, the igneous rock is exposed at the surface where it is eroded. Zircon crystals plucked from the rock become incorporated in sediment. However, the age of the zircon crystal is unaffected by sedimentary processes and reflects its time of crystallization from the igneous melt.

13.3.2 Origin and Early Evolution of Life

To understand how life originated, we must first have a precise definition of life. This is not an easy task, but two characteristics are essential; the ability to reproduce and the ability to self-regulate (that is the ability to use materials in the environment to sustain orderly growth). Viruses, for example, have the ability to reproduce but cannot regulate themselves and therefore are not considered to be living things.

In the modern world, all organisms that have the essential characteristics of life are composed of discrete units called **cells.** Cells are the smallest unit that have the properties of life and all organisms are composed of one or more of them. Within these cells, the ability of life to reproduce and self-regulate is based on the interaction between compounds known as **nucleic acids** and **proteins.** Nucleic acids, are composed of sugar, a phosphate group and a nitrogen compound all of which are bonded together in complex patterns. One form of nucleic acid, known as **DNA** (deoxyribonucleic acid), carries the genetic information of all organisms except bacteria (Fig. 13.4). Its primary role is to contribute that information to the production of **enzymes,** which are forms of protein that help speed up and regulate chemical reactions in the cell. Another form of nucleic acid, known as **RNA** (ribonucleic acid), carries out these encoded instructions. The reactions in the cell facilitate the replication of genetic information and the survival of the genetic code with reproduction. Thus, the DNA blueprint is passed from generation to generation, carrying instructions for growth and functional development.

As we shall see, the oldest record of life occurs in the Early Archean. It is unclear whether DNA or RNA existed at that time. However, if they did not exist, very similar molecules must have performed these vital roles, since life clearly survived.

The evolution of the Earth's layers and the development of continents and oceans may have provided the environment for the genesis of life. In 1953, Stanley Miller, then a graduate student at the University of Chicago, investigated the relationship between the hostile environment of the primitive Earth and the origin of life. In his experiments, he subjected a gas mixture simulating the early atmosphere (consisting of hydrogen, water vapor, methane, and ammonia) to electrical discharge (or sparking) that simulated lightning (see Fig. 13.1b). He left the experiment running overnight. When he came back the next morning, the mixture in the flask had become cloudy and yellow.

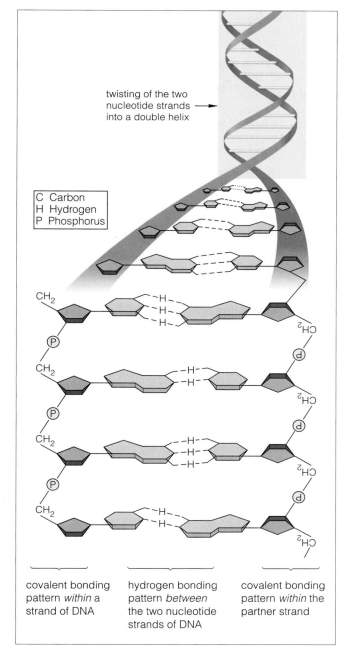

twisting of the two nucleotide strands into a double helix

C Carbon
H Hydrogen
P Phosphorus

CH₂
CH₂
CH₂
CH₂

covalent bonding pattern *within* a strand of DNA

hydrogen bonding pattern *between* the two nucleotide strands of DNA

covalent bonding pattern *within* the partner strand

Figure 13.4
Bonding patterns and twisted ("double-helix") structure of the DNA molecule. The various strands of the molecule are connected by hydrogen bonds.

Analysis revealed that amino acids had formed. This result is significant because amino acids, when bound together into long chains, form proteins, the basic constituents of all life.

However, many scientists now believe that the Earth's early atmosphere was significantly different in composition from that used by Miller in his experiments. They believe that the atmosphere was richer in carbon dioxide and nitrogen and did not have significant quantities of methane or ammonia. When this mixture is subjected to sparking, only a very tiny amount of organic molecules are produced. As a result, scientists are now actively considering other theories for the origin of life.

Some researchers believe, for example, that the genesis of life may have been stimulated by the high velocity impacts of extraterrestrial objects such as comets or meteorites. These impacts were probably very common in the turbulent early history of the Solar System (see Chapters 15 and 16). Certain meteorites and comets contain carbon and hydrocarbons. The burning of these bodies in the primitive atmosphere may have yielded amino acids in the manner described in Miller's experiments. If so, the essential elements of life may have formed in the primitive atmosphere and would have been ultimately absorbed into the biosphere. This view was strengthened in 1997 by studies of the Hale-Bopp comet which revealed the presence of water, ammonia, hydrogen cyanide, and formaldehyde, a mixture that can react to produce amino acids (see Living on Earth: Extraterrestrial Life on Earth?, page 372).

Another hypothesis attributes the origin of life on Earth to submarine volcanic activity. Such activity would have been abundant in the Archean, such that sea floor hydrothermal vents similar to the black smokers observed on the modern ocean floor are likely to have been commonplace. As we learned in Chapter 8, modern hydrothermal vents are associated with a special type of bacteria that sustain a unique ecosystem. The superheated fluids provide the energy that enables bacteria to convert inorganic molecules that are abundant close to the vent into organic compounds. The conversion process that the bacteria use is known as **chemosynthesis**, because the organisms use the energy from chemical reactions (in contrast to photosynthesis which uses light energy) to synthesize organic matter. Therefore chemosynthesis can take place in the absence of light. If life originated on the ocean floor, the deep ocean water would also have protected it from deadly ultraviolet radiation.

Support for a volcanic origin for life has also been supplied by studies like those currently underway in the geothermal waters of Yellowstone National Park. Here, Anna-Louise Reysenbauch and her colleagues have discovered primitive species of bacteria that can only grow in waters with temperatures close to 90°C (190°F). These bacteria are close relatives of the most primitive forms of Archean life and consume sulfur compounds, iron, and other toxic substances.

Irrespective how life came about, it must have interacted with other Earth systems for its survival in order to extract and utilize energy and nutrients from the environment. This is because energy is needed to assemble the nutrients into a cell-like structure.

In addition, all life forms need a supportive physical environment and a template upon which to develop. In the Archean, this role may have been performed by clay minerals. Clay minerals, the essential ingredient of shales (see Chapter 2), were abundant in the Archean oceans and are well represented in Archean sedimentary sequences. The clay mineral structure is very long and thin, like a wafer. The clay layers bond easily with amino acids. In addition, clay layers separate into individual wafers when they interact with water. This would have provided amino acids with the opportunity to invade the clay structure and may have afforded life the necessary protection in an early hostile environment. If so, then these environments would have had a selective advantage over other environments in competing for the available amino acids.

Unfortunately, the geologic record of ancient life is very sparse and not without controversy. As a result, it cannot distinguish between the various theories on the origin of life. However, carbon spheres have recently been described from 3.86 billion year old rocks in Greenland (Fig. 13.5a). If these spheres are organic, as some scientists believe, and so contain carbon that was assembled in cells, then they would represent the earliest record of life. However, other scientists debate the organic nature of these spheres, and favor instead an inorganic origin. The earliest widely accepted examples of life come from 3.5 billion year old rocks in Australia, and from 3.0 billion year old rocks in South Africa. Both rocks contain examples of what are known as **stromatolites,** representatives of which still exist today (Fig. 13.5b and 13.5c).

Stromatolites are sedimentary rocks produced in shallow marine environments by the binding of sediment with blue-green algae, or cyanobacteria. These primitive microorganisms form algal mats on which sediment adheres because the algae are gelatinous and sticky. Building up layer-by-layer, the algae form cabbage-shaped colonies or reefs of finely laminated stromatolites. The organisms themselves were simple single-celled structures known as **prokaryotes.** Prokaryotes are the most primitive of single-celled organisms, lacking complex internal structures such as a cell nucleus. Reproduction was asexual, that is, the offspring arose from a single parent and inherited the genes of that parent.

Because modern varieties of prokaryotes strongly resemble their primitive ancestors and are capable of photosynthesis (which produces oxygen), it is assumed that their primitive

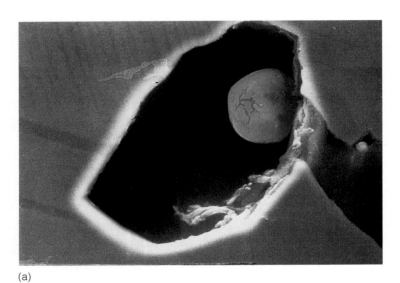

(a)

Figure 13.5
(a) This tiny sphere of carbon may be the oldest evidence of life on Earth. The carbon sphere (magnified 6000 times) is located in a cavity that was etched from a 3.86 billion year old rock found off the coast of Greenland of blue-green algae or cyanobacteria. (b) Formation of stromatolites from algal mats. (i) Algal mat in daylight. (ii) Sediment is trapped by the algal mat during the daytime. (iii) The sediment is bound together during darkness. (iv) Various types of stromatolites including mats, columns, and columns linked by mats.
(c) Recent stromatolites exposed at low tide in Shark Bay, Australia.

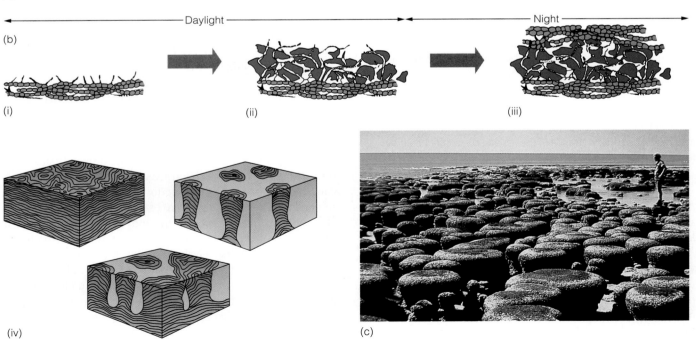

(b) Daylight — Night

(i) (ii) (iii)

(iv) (c)

ancestors had the same traits. Given the lack of oxygen in the primitive atmosphere, however, they cannot have required oxygen for survival, but, instead must have flourished in oxygen-free, or *anaerobic* environments.

So here we have a paradox. Life capable of producing oxygen by photosynthesis actually lived in an oxygen-free environment. Indeed, oxygen is poisonous to anaerobic organisms. So how could oxygen-intolerant organisms have survived the oxygen that they themselves produced? This apparent contradiction was possible because of the peculiar conditions that existed in the water of the Archean oceans. Unlike today's oceans, these waters were enriched in dissolved iron which would have been introduced by the abundant submarine volcanic activity. As you know, if you leave a lump of iron in the open air, it will rapidly rust because the iron oxidizes by absorbing oxygen out of the atmosphere. Therefore iron has the ability to absorb oxygen. As the primitive ocean waters con-tained no oxygen gas, the iron would have existed in a reduced (Fe^{2+}) rather than oxidized (Fe^{3+}) state. The oxygen produced by photosynthesis is thought to have been consumed by oxidizing the iron in the ocean water. Reduced iron (Fe^{2+}) dissolves easily in ocean water, but upon oxidation (to Fe^{3+}), it loses this ability and forms precipitates of iron oxide minerals. The rock record contains abundant evidence to support this explanation. Many regions dominated by Archean rocks contain rock layers known as **Banded Iron Formations**, whose origin is attributed to the oxidizing of ocean waters by photosynthetic blue green algae, and the settling of iron oxide minerals to the bottom of the seabed. Because the oxygen produced by photosynthesis was used up in this process, the oxygen-intolerant organisms were protected from its harmful effects. In addition, the atmosphere would have remained free of oxygen because it would not have left the oceans in which it was produced (Fig. 13.6).

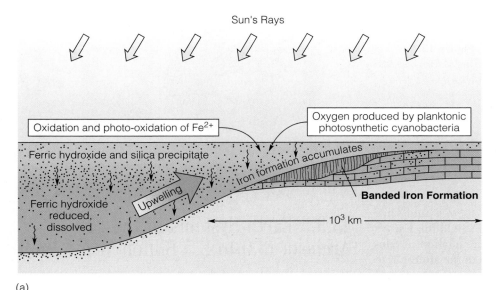

(a)

Figure 13.6

(a) Model for the formation of Banded Iron Formation. Oxygen produced by photosynthetic organisms in the oceans oxidizes iron in ocean water from Fe^{2+} (ferrous) to Fe^{3+} (ferric). Fe^{2+} is soluble, whereas Fe^{3+} is insoluble and forms compounds that precipitate and sink to the ocean floor to form Banded Iron Formation. (b) Banded Iron Formation, containing red and grey layers, both rich in iron-bearing minerals.

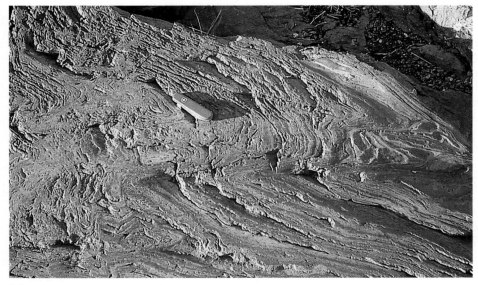

(b)

In our experience, there have been very few discoveries that have set the scientific world abuzz like the recent suggestions that life may have existed on Mars, and indeed may still exist there. The studies, if validated, also have implications for the origin of life on Earth.

Two independent research investigations by teams of American, Canadian, and British scientists have provided new and controversial evidence that supports such a claim. These scientists admit that their evidence is circumstantial, but in the true spirit of scientific investigation they have drawn attention to the possibility so that other scientists may find ways to test their hypotheses. By either confirming or refuting their findings, science moves forward.

The material studied was not recovered from Mars at all; it was recovered from the ice sheet of Antarctica (Fig. 13.7). It has been accepted for several years that gases trapped in some meteorites are identical to the Martian atmosphere and therefore provide

strong evidence for a Martian origin. If so, these meteorites must have been blasted off the Martian surface by an asteroid collision, which expelled vast chunks of rock into space. Eventually some of these chunks became caught in the Earth's gravitational field and fell to its surface as meteorites. Meteorites are relatively easy to find in the unspoiled terrain of Antarctica. They can be

recovered either from the surface itself, or at the front of glaciers where the heaving of ice along fractures drags up buried meteorites.

These particular meteorite samples are believed to come from Mars because they possess the chemical and isotopic signature characteristic of Martian crust, and the composition of gases trapped within them is identical to that of the Martian atmosphere.

Figure 13.7
This Martian meteorite recovered from the ice sheet of Antarctica contains miniscule rods, similar in shape to fossilized bacteria.

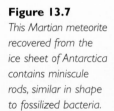

With no oxygen in the atmosphere, there would have been no ozone, and so no ozone layer to shield the Earth's surface from the Sun's harmful ultraviolet radiation. A concentration of at least 2% oxygen is needed in the atmosphere to form ozone, and this level probably was not reached until about 2 billion years ago. Ultraviolet radiation breaks down amino acids, and so attacks an essential raw material for life's development. So while the production of amino acids may have been a *necessary* precursor to life, it would not have been *sufficient*. Life would have also needed some protection from solar radiation. Yet photosynthetic life would also have needed exposure to sunlight in order to use the Sun's energy to synthesize organic matter from water and carbon dioxide. These apparently conflicting needs may have been met in intertidal pools in which the water would have offered some protection from the searing radiation of the Archean world.

In addition to witnessing the development of life, the geologic record demonstrates that the Earth's surface reservoirs, the solid Earth, biosphere, hydrosphere, and atmosphere, had become well established by the Late Archean. As we shall see, however, the Late Archean (3.0 to 2.5 billion years ago) was a major period in the development of continents, and sowed the seeds for the development of the large continental masses that we see today.

13.3.3 Earth Systems in the Late Archean (3.0 to 2.5 Billion Years Ago)

The geologic record of the Late Archean documents a very slow pace of change in the evolution of the biosphere and atmosphere from 3.0 to 2.5 billion years ago. The slow pace in the evolution of life may be due to the limitation of prokaryotes to asexual, rather than sexual, reproduction. In asexual reproduction, the only way for a favorable mutation to survive is for the organism that possessed it to outdivide the others. Thus a favorable mutation might allow one individual to reproduce countless replicas of itself.

The Late Archean atmosphere must still have lacked free oxygen because Late Archean Banded Iron Formations are abundant. Furthermore, many Late Archean clastic sedimentary rocks contain grains of minerals such as pyrite, an iron sulfide (FeS_2), and uraninite, a uranium oxide (UO_2), which readily break down in the presence of oxygen. Their abundance and stability implies an oxygen-free atmosphere.

However, there is plenty of evidence from the rock record of major developments in the solid Earth during this time period. In contrast to the fragmentary rock record of the Early and Middle Archean, the abundance of granitic igneous rocks with ages in the 3.0 to 2.5 billion year range

The first sample analyzed by the American and Canadian team was dated using radiometric techniques (see Chapter 2). It was found to be 4.5 billion years old, and to have been fractured 4 billion years ago. The carbonate material believed to be organic gave an age of 3.6 billion years and is thought to have formed by solutions percolating along the fractures. At this time, Mars was probably much warmer and wetter than it is today, so that conditions may have been favorable for life. Age dating also shows that this sample exploded off the Martian surface 16 million years ago, and landed in Antarctica 13,000 years ago. The British team re-analyzed this rock and confirmed these results. They also analyzed a second sample which was only 175 million years old and was blasted off the Martian surface a mere 600,000 years ago.

Both teams reported the presence within the meteorites of minuscule rods similar in shape to fossilized bacteria. The British team also found the telltale signature of "microbially-produced methane" similar to that produced by common bacteria. The isotopic character of this material matches that of the most primitive life forms on Earth.

In addition to the obvious implications for Martian life, the first sample raises the possibility that life on Earth may have been "seeded" by Mars. If life did exist on Mars 3.6 billion years ago, it is similar in age to the earliest indications of life on Earth. Given the abundance of asteroidal impacts on planetary surfaces in the early history of the Solar System, chunks of Martian crust could easily have made the journey to Earth (see Chapter 15). If so, they may well have provided the raw materials for life to commence on Earth.

The second younger meteorite sample is also significant because it suggests the presence of Martian life in the relatively recent past, sometime between 175 million years ago (the age of the rock) and 600,000 years ago (the time of asteroidal impact). If so, life on Mars must have been relatively robust to have survived the enormous changes in the Martian climate. Furthermore, as the climatic conditions of Mars have not changed substantially over the last 200 million years, the sample suggests that life may still exist there today.

How do we evaluate such data and observations? Let us look at the assumptions. We must first believe that the geochemical and isotopic signatures of the carbon compounds are indeed diagnostic of life, and that there is no rival "inorganic" explanation. Second, we must believe the analytical evidence that the meteorites are derived from Mars. In this case, the impressive similarity between the trapped gases and the Martian atmosphere is compelling. Third, we must believe that the organic molecules survived the catastrophic impact that blasted them off the Martian surface, and that the sample was uncontaminated by its time spent on Earth.

Clearly much remains to be done. At the time of writing, NASA is planning to land a robot on Mars by the year 2005 in order to bring back samples from the Martian surface. In this way, the hypothesis will be put through its most rigorous test.

---

suggest that this was a period of intense tectonic activity and rapid continental growth.

The product of this rapid growth was the formation of continental nuclei, known as **cratons** by 2.5 billion years ago. These relatively small microcontinental plates have remained largely stable since that time, and form the core of modern continental landmasses (Fig. 13.8a).

Late Archean cratons are made up of two main types of rock: granite-gneiss complexes and greenstone belts. The **granite-gneiss complexes** contain typical continental igneous and metamorphic rocks. Their presence in the cratons firmly establishes that abundant continental crust existed in the Late Archean. **Greenstone belts** are typically made up of basalt and ultramafic volcanic rock layers at the base, followed by a thick sequence of basalts, andesites, and rhyolites, and capped by thick marine sedimentary rocks (Fig. 13.8b). These sedimentary sequences often contain Banded Iron Formations and grains of pyrite and uraninite which imply continued oxygen-poor conditions. The oldest greenstone belt is about 3.8 billion years old. However, they become far more abundant in the rock record between 3.0 and 2.5 billion years ago.

The origin of greenstone belts is controversial. The ultramafic volcanic rocks of greenstone belts are highly unusual in the geologic record. They are similar in composition to the Earth's mantle, leading geologists to believe that they formed by whole-sale melting of mantle rocks. This requires very high temperatures (1350–1400°C), and provides further evidence that heat generation in the mantle was much higher in the Archean and Early Proterozoic (which also has greenstone belts) than it has been since that time. The basalt lavas are commonly pillowed (Fig. 13.8c), indicating a submarine origin like those of modern mid-ocean ridges. However, the thickness and composition of the sedimentary cap to the greenstone belt succession implies proximity to a continental source of sedimentary material. Indeed, some of these sedimentary sequences contain stromatolite reefs, suggesting deposition in a shallow marine environment like that of a continental platform or shelf. So greenstone belts must have formed adjacent to continental landmasses and could *not* have formed in a mid-ocean ridge environment which would have been far from land. Proximity to a continental landmass is also supported by the presence of abundant granitic rocks that intrude the greenstone belts (Fig. 13.9). As we learned in Chapter 5, granitic rocks are exceedingly rare in mid-oceanic environments because most granites are formed by melting of continental crust.

Two tectonic models for the origin of greenstone belts have been proposed (Fig. 13.10), both of which involve rifting to form a basin. One model involves continental rifting and the formation of greenstone belts between two continental blocks. The other suggests that greenstone belts may have

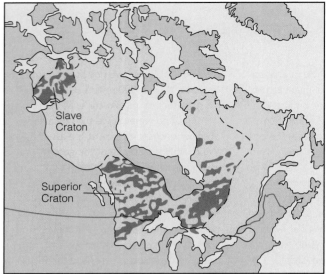

(a)

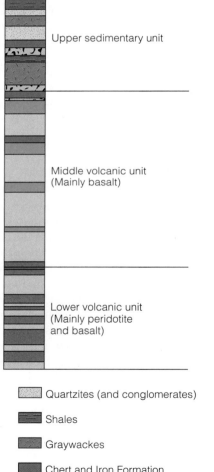

Upper sedimentary unit

Middle volcanic unit
(Mainly basalt)

Lower volcanic unit
(Mainly peridotite
and basalt)

Quartzites (and conglomerates)

Shales

Graywackes

Chert and Iron Formation

Rhyolitic volcanics or intrusives

Andesites and dacites

Basalts

Ultramafic flows and sills

(b)

(c)

Figure 13.8
(a) Distribution of late Archean greenstone belts (green) and granite gneiss complexes (brown). (b) Typical stratigraphy of greenstone belts consisting of lower mafic to ultramafic layers overlain by layers dominated first by basalt, and then by sedimentary rocks. (c) Basalt pillow lavas occur among the volcanic rocks of greenstone belts, providing evidence for submarine volcanism (see Chapter 2).

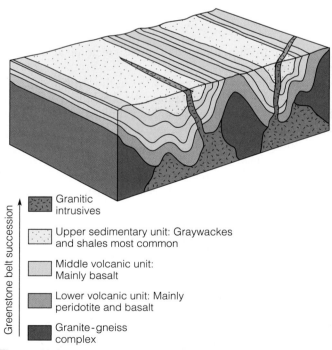

Greenstone belt succession →

Granitic
intrusives

Upper sedimentary unit: Graywackes
and shales most common

Middle volcanic unit:
Mainly basalt

Lower volcanic unit: Mainly
peridotite and basalt

Granite-gneiss
complex

Figure 13.9
Deformed greenstone belts with granitic intrusive rocks. Overall, the layers of the greenstone belts are bent downward into a synclinal structure (see Chapter 6). Granites were probably derived from melting of continental crust, implying that greenstone belts are more likely to have formed close to a continental setting rather than in a mid-ocean environment.

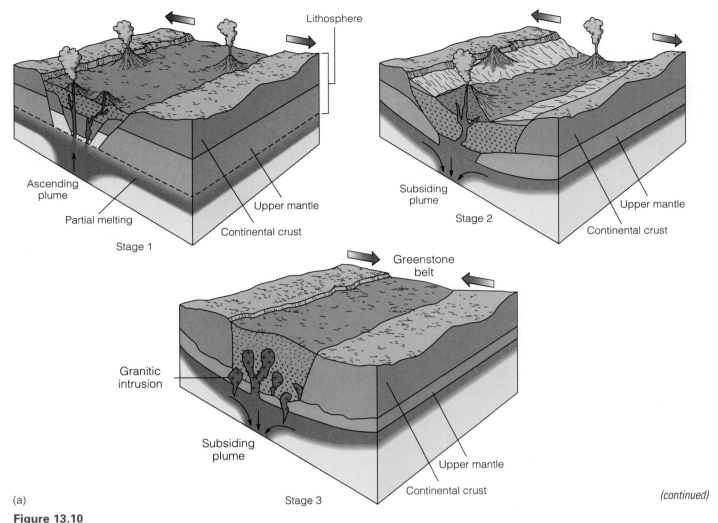

Figure 13.10

Two models for the formation of greenstone belts (a) Continental rifting model. In Stage 1, an ascending plume causes rifting and volcanism. A basin forms as the plume cools and the crust subsides (Stage 2). Closure of the rift causes deformation into a synclinal structure (see Figure 13.9), melting of continental crust, and intrusion of granite (Stage 3). (b) Back-arc model in which rifting behind a volcanic arc forms a back-arc basin (Stage 1). Basaltic lava and sediments are derived from the continent and fill the basin (Stage 2). Closure of the basin causes deformation, the formation of a large synclinal structure and generates the granites which ascend to intrude the greenstone belt (Stage 3).

(continued)

formed by rifting behind a volcanic arc in a "back-arc" setting (see Chapters 5 and 6) like that of the modern Sea of Japan behind the Japanese island arc. In the latter model, associations with continental crust such as stromatolite reefs would have occurred only near the basin flanks.

The continental rifting model may be analogous to modern plate tectonic settings such as the East African rift. But such an environment in the Late Archean could have developed from the ascent and descent of columns of heat in the mantle in the absence of horizontal plate motions. The back-arc model implies that greenstone belts were formed by the movement of plates. If this model is correct, then some form of plate tectonics was operative by Late Archean time.

With the present state of knowledge, it is difficult to select one model in favor of the other. Indeed, it may be a mistake to do so. In our modern world, volcanic rocks occur in many different tectonic environments, and greenstone belts may similarly reflect a variety of Archean environments. However, the application of these models to any one region can be tested by detailed field mapping aimed at establishing the relative ages and characteristics of the rocks, and complemented by geochronological and geochemical studies aimed at establishing their absolute ages and compositions. As described below, such studies tend to favor a variant of the back-arc model.

An excellent example of Archean geology is provided by the Superior craton in Canada (Fig. 13.11). This craton consists of a succession of granite-gneiss and greenstone belts that had become amalgamated into a stable craton by the end of the Archean. But how did this happen? Most geoscientists believe that the Superior craton formed from the amalgamation of smaller cratons, or

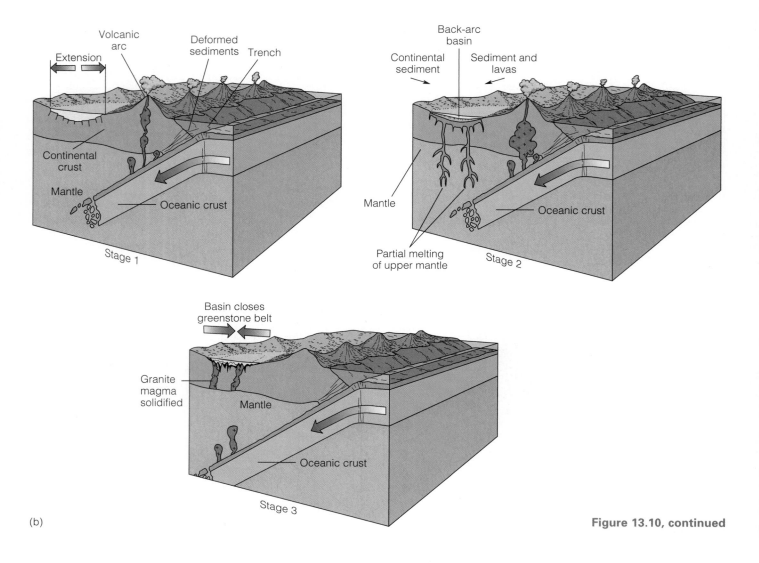

Volcanic arc
Extension
Deformed sediments
Trench
Continental crust
Mantle
Oceanic crust
Stage 1

Back-arc basin
Continental sediment
Sediment and lavas
Mantle
Oceanic crust
Partial melting of upper mantle
Stage 2

Basin closes greenstone belt
Granite magma solidified
Mantle
Oceanic crust
Stage 3

(b)

Figure 13.10, continued

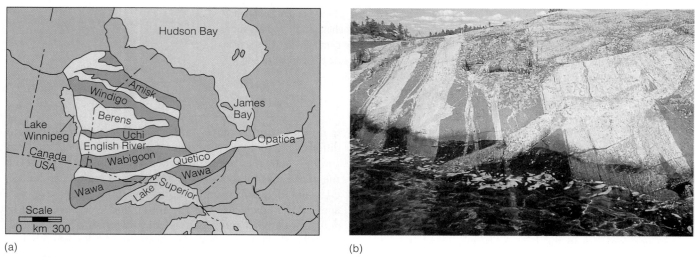

Hudson Bay
Amisk
Windigo
Berens
James Bay
Lake Winnipeg
Uchi
Opatica
English River
Canada USA
Wabigoon
Quetico
Wawa
Wawa
Lake Superior
Scale
0   km   300

(a)

(b)

Figure 13.11
(a) Map of the Superior craton, showing the distribution of granite-gneiss complexes (tan color) and greenstone belts (green). (b) The granite-gneiss complexes are dominated by metamorphosed continental rocks containing minerals that formed when pre-existing rocks were buried deep in the crust. The style of layering and folding is typical of high grade metamorphic rocks (see Chapter 2).

microplates, that collided as a result of plate tectonic processes (Fig. 13.12). In this model, a number of volcanic arcs existed which contained both granite-gneiss complexes (that were generated above the subduction zone) and greenstone belts (which represent back-arc magmatism). About 2.5 billion years ago, the volcanic arc complexes collided to form the Superior craton when subduction consumed the oceanic crust in between them. Metamorphosed sedimentary rocks resembling those of modern accretionary wedges (see Chapter 6) mark the sites of the former subduction zones.

This model, if correct, may be an important example of the type of primitive plate tectonic activity that occurred during the Late Archean and may set the stage for the progressive development of modern plate tectonic processes during the Proterozoic.

But how might plate tectonics have started? The answer may lie in the existence of continental crust, a significant amount of which had formed by the Late Archean. As we learned in Chapter 6 (see the Supercontinent Cycle), continental crust is a good insulator and traps the heat escaping from the mantle below (Fig. 13.13). By preventing the heat from escaping, the first continents may have caused convection cells to develop in the mantle that started to move the overlying plates horizontally. Subduction zones would then have developed above the descending limbs of these convection cells. Although the existence of some form of plate tectonics in the Late Archean seems likely, during the ensuing Proterozoic Eon there is no question that plate tectonics, more or less in its modern form, was active.

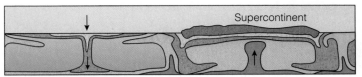

(a) mantle upwelling (red) beneath supercontinent

(b) breakup, generation of new ocean

(c) continental dispersal

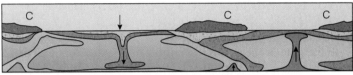

(d) Continents converge towards site of mantle downwelling (grey)

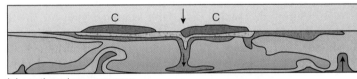

(e) continued convergence

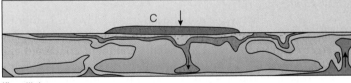

(f) collision

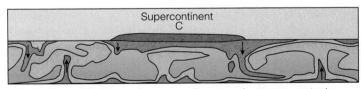

(g) mantle upwelling beneath supercontinent, cycle starts over again

Figure 13.13
Model for heat accumulation beneath a supercontinent. (a) Convection cells form as mantle upwelling occurs beneath an insulating continent and subduction zones occur above the sites of downwelling. (b and c) Continued upwelling causes the continent to rift apart resulting in a change in the heat pattern within the mantle. (d and e) As the continents drift apart, they are drawn toward sites of mantle downwelling. (f) Continents eventually collide to form a new supercontinent. (g) The cycle starts again as heat begins to build up beneath the supercontinent.

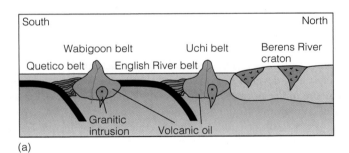

(a)

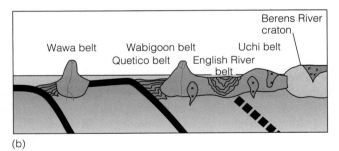

(b)

Figure 13.12
Plate tectonic model for the origin of the Superior craton from the amalgamation of smaller microplates (see Figure 13.11). This is a north-south section of the craton. (a) Initially, a number of volcanic arcs contained both granite-gneiss complexes and greenstone belts. (b) Eventually, the volcanic arcs collided when subduction consumed the oceanic crust between them, leaving only belts of deformed sedimentary rocks to mark the sites of the former subduction zones.

13.4 Earth Systems in the Proterozoic

The Proterozoic Eon spans nearly 2 billion years, or about 42% of geologic time. Its record is better preserved than that of the Archean, but it is not as well preserved as that of the Phanerozoic. The record is sufficient, however, to recognize important changes in each of the Earth's surface reservoirs. By the end of the Proterozoic, for example, diverse multicellular organisms capable of sexual reproduction had evolved, radically altering the biosphere. Oxygen had become a significant component of the atmosphere and plate tectonic processes were operating within the lithosphere in a similar fashion to those of today. Changes in the hydrosphere are more difficult to assess, but as we shall see, the proportion of continents and oceans towards the end of the Proterozoic was similar to that of today, suggesting similar mechanisms for the transport of heat and nutrients.

13.4.1 Early Proterozoic (2.5 to 1.6 Billion Years Ago)

If the Late Archean was responsible for the development of the cratonic nuclei of continents, then the Early Proterozoic saw these nuclei stitched together to form large continental landmasses. This was a profound step in the Earth systems evolution. During this period, the stable central regions or shield areas of many of the Earth's continents, known as **continental shields**, were formed. At the same time, greenstone belts became rare, suggesting that a transition had occurred from Archean microplate tectonics to larger scale Proterozoic plate tectonics, similar (but, as we shall see, not identical) to that of the modern world.

By 1.8 billion years ago, the landmass known as **Laurentia** (the predecessor of modern North America) was one of these larger continents (Fig. 13.14). This continent contained cratonic areas of Archean age, stitched together by Early Proterozoic (2.0 to 1.8 billion year old) mountains such as the Trans-Hudson belt. Geoscientists from the Geological Survey of Canada (see Dig In: Plate Tectonics in the Early Proterozoic?, page 380) believe that the ancient Trans-Hudson mountain belt was produced by the collision of the flanking Archean cratons (Superior, Wyoming, and Hearn). The model shown in Figure 13.15 is important because it represents the oldest example of a Wilson cycle (see Chapter 6) and suggests that plate tectonics operating in a similar manner to the present day may have been responsible for the development of Laurentia.

A Wilson cycle implies the initial development of a continental rift, accompanied by volcanism and the deposition of clastic sediments (Fig. 13.15a). Continued rifting would have led to the development of an ocean basin (Fig. 13.15b) followed by a continental shelf (Fig. 13.15c). One of the most exciting findings has been the discovery of the fragmentary remains of 2 billion year old ocean floor rocks. Collectively, these rocks constitute an ophiolite (see Chapter 5), and are thought to represent

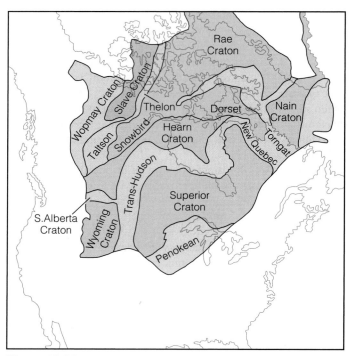

Figure 13.14

Map of Laurentia 1.8 billion years ago showing the various ancient mountain belts and cratons from which it was stitched together.

very rare portions of Early Proterozoic oceanic crust that were dismembered and heaved up onto the continental crust rather than being subducted. The significance of the discovery of ophiolites is that it suggests that ocean basins opened and closed in a manner similar to that of the modern Earth. When the development of ocean floor gave way to subduction (Fig. 13.15d), island arcs developed as the ocean began to close. Eventually, closure of the ocean resulted in continental collision, deformation, and abundant magmatic activity (Fig 13.15e).

As a result of collisions such as that recorded in the Trans-Hudson orogeny, continents grew in size and, as they did so, the length of their continental shelves increased. The Proterozoic record preserves thick and extensive sequences of sedimentary rocks. These rocks are similar to sediments that occur on modern continental shelves (such as those of the Atlantic Ocean), suggesting that they represent deposition in a similar environment.

The Early Proterozoic was not only an important period in the development of plate tectonics, but also witnessed profound developments in the evolution of the atmosphere. Prior to 2 billion years ago, the oxygen produced by photosynthetic organisms was absorbed by the oxidation of iron dissolved in ocean water, resulting in the production of Banded Iron Formations. Because of this absorption, ocean water was effectively controlling, or *buffering*, the oxygen composition of the atmosphere. More than 90% of all known Banded Iron Formations were formed between 2.5 and 2.0 billion years ago, and they have been virtually absent from the rock record since then. This suggests that, at around 2 billion years ago, the profusion of photosynthetic microorganisms living in the oceans

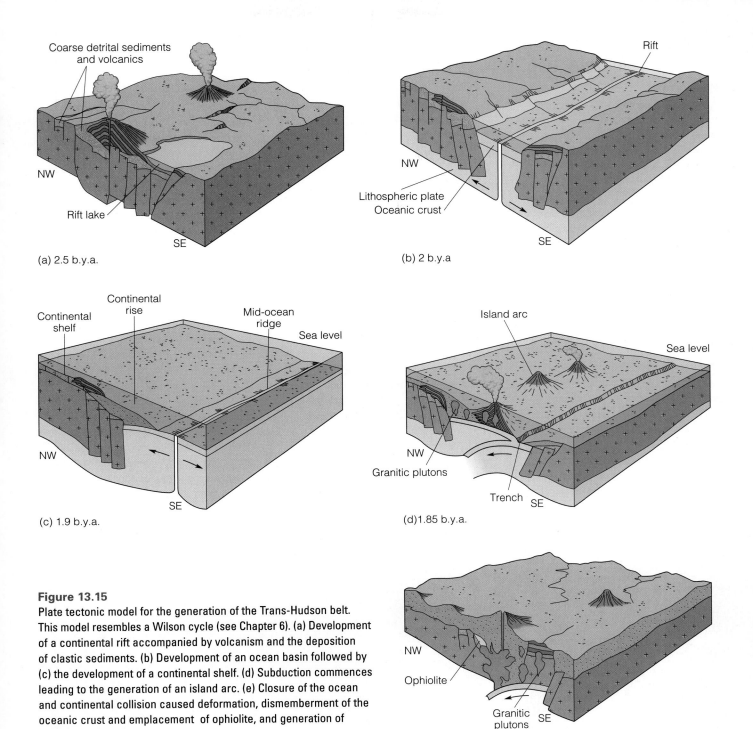

(a) 2.5 b.y.a.

Coarse detrital sediments and volcanics

Rift lake

NW

SE

(b) 2 b.y.a

Rift

Lithospheric plate
Oceanic crust

NW

SE

(c) 1.9 b.y.a.

Continental shelf

Continental rise

Mid-ocean ridge

Sea level

NW

SE

(d) 1.85 b.y.a.

Island arc

Sea level

Granitic plutons

Trench

NW

SE

(e) 1.8 b.y.a.

Ophiolite

Granitic plutons

NW

SE

Figure 13.15
Plate tectonic model for the generation of the Trans-Hudson belt. This model resembles a Wilson cycle (see Chapter 6). (a) Development of a continental rift accompanied by volcanism and the deposition of clastic sediments. (b) Development of an ocean basin followed by (c) the development of a continental shelf. (d) Subduction commences leading to the generation of an island arc. (e) Closure of the ocean and continental collision caused deformation, dismemberment of the oceanic crust and emplacement of ophiolite, and generation of intrusive rocks.

finally exceeded the oceans capacity to absorb the oxygen they produced. As a result, oxygen began to leak out of the oceans and so started to accumulate in the atmosphere. This accumulation would have eventually led to the development of a thin ozone layer because ozone is formed by the action of solar radiation on oxygen molecules. In turn, the development of an ozone layer would have facilitated for the first time the filtering of ultraviolet radiation.

The presence of oxygen in the atmosphere at this time is supported by evidence from the rock record. Continental sediments are in direct contact with the Earth's atmosphere when they are deposited and so are sensitive to atmospheric composition. Red continental sedimentary rocks, collectively known as "continental redbeds" make their first appearance in the rock record about 1.8 billion years ago. The red color of these sandstones and shales is attributed to the presence within them

DIG IN

Plate Tectonics in the Early Proterozoic?: A Field Trip to Northern Quebec

When did plate tectonics begin on Earth? There is probably no more fundamental a question in Earth Science. Studies of the very early history of the Earth suggest that plate tectonics was not a viable process in the Hadean. Instead, the formation of the core, mantle, crust, hydrosphere, and atmosphere, suggests that gravity played a dominant role in separating the Earth into layers according to density. It is also likely that the Earth's young surface was too hot for plate tectonics to operate, and any primitive crust that might have formed would have been destroyed by meteorite bombardment.

Many geoscientists believe the first signs of plate motions do not occur until the Late Archean, when stable cratonic areas were formed by the collision of microcontinents. However, it was not until the Early Proterozoic that there is definitive evidence for the existence of Wilson cycles (see Chapter 6). A Wilson cycle starts with continental rifting, followed in turn by the development of an Atlantic-type ocean basin, the destruction of this ocean by subduction, and, finally, continental collision when the intervening ocean basin is consumed.

In the early 1990's, a four year field project led by Marc St-Onge and Steve Lucas of the Geological Survey of Canada, probed the geology of the Trans-Hudson orogen, a candidate for an Early Proterozoic Wilson Cycle (Fig. 13.16). Each year, the project leaders coordinated the talents of up to twenty professional geologists, and graduate and undergraduate geology students (Fig. 13.16a) to map and interpret the distribution of the key rock units of the orogenic belt. They selected the northern region of Quebec as the study site because of the excellent exposure of bedrock. This remote area has no roads and is well north of the treeline. The climate is such that field work is restricted to a brief period between late June and early September. Samples collected during the field season were used in laboratory analyses in the fall and winter. During the field season, the field parties covered an average 20 km (12.5 mi) per day on foot, traversing across the streams and ridges of this rugged, but beautiful terrain. They used helicopters to transport the field parties to and from the base camp at the beginning and end of each day. After four

field seasons, a new geologic map covering thousands of square kilometers provided the basis for the interpretation of the tectonic history of the area.

The field parties mapped the distribution of the continental clastic sedimentary rocks that are the remnants of a rifting event. They found that these rocks were deposited on top of an older continental basement (Fig. 13.16b). They discovered the world's oldest ophiolite (Fig. 13.16c,d), proof that oceanic crust existed between adjacent blocks of continental crust 2 billion years ago. They also determined the distribution of volcanic complexes that represent ancient island arcs formed during subduction and closure of an ocean (Fig. 13.16e), and found evidence of complex deformation typical of continental collision (Fig. 13.16f). They studied the mineralogy of the highly deformed rocks in order to understand the degree of metamorphism. Many of the minerals they found could only be have been formed at deep levels in the continental crust, confirming their suspicions that the deformed rocks represent the exhumed interior of an ancient mountain belt.

(a)

(b)

of oxidized iron in minerals such as hematite (Fe_2O_3). This means that the rocks have been "rusted" and the only source of the oxygen available to accomplish this must have been in the atmosphere. So free oxygen must have been present in the atmosphere 1.8 billion years ago.

But how did life respond to these new atmospheric conditions? The answer is not very well! In fact, the introduction of oxygen into the atmosphere is likely to have triggered the first major mass extinction event in the geologic record. Oxygen is actually a toxic substance and organic matter can burn in the

(c)

(d)

(e)

(f)

(g)

Figure 13.16

Field trip to the dawn of plate tectonics; the Trans-Hudson belt of northern Quebec records an ancient Wilson Cycle. (a) The Geological Survey of Canada field party, 1994. (b) Complex geology of the ancient crystalline basement. (c) Volcanic rocks formed during rifting. The lavas formed "pillows" when they chilled against ocean water. (d) Sheeted dykes formed near the mid-oceanic ridge (see chapter 5). (c) and (d) are portions of the world's oldest ophiolite. (e) 1.82 billion year old granite formed during subduction contains large inclusion (center) or older rock. (f) Rocks in the background were thrust over rocks in the foreground during 1.8 billion year continental collision. (g) Intense deformation related to the continental collision.

Samples taken from each of these rock suites were investigated used the U-Pb dating technique (see Chapter 3) in order to establish their absolute ages. The rifted rocks were found to range in age from 2.04 to 1.92 billion years. The ophiolites were 2.0 billion years old, suggesting the existence of ocean floor between continental crustal blocks by that time. The volcanic arc rocks were found to be 1.86 to 1.83 billion years old, indicating that the ocean basin had begun to contract by that time. The continental collisional event was dated at 1.8 billion years old.

This detailed study clearly supports the existence of Wilson cycles in the Early Proterozoic. However, like many research projects, in answering one set of questions, the study generated another. The Wilson cycle in the Trans-Hudson orogen took only 200 million years to complete, from the time of initial rifting about 2 billion years ago, to the time of the culminating continental collision about 1.8 billion years ago. This is half the duration of the Wilson cycles typical of the last billion years (see Chapter 6). Were plate motions more rapid in the Early Proterozoic? If so, is plate tectonics slowing down, and will it cease in the future? These speculations may be the seeds of future field-based research projects of the type launched in northern Quebec.

presence of too much of it. Evolution has given most forms of modern life the capacity to produce enzymes that counteract this toxicity. But after some 2 billion years of living in the absence of free oxygen, the prokaryotes were ill-prepared for the abundance of this poison. They were faced with two choices in order to survive. They could either adapt to the changing conditions by producing the enzymes necessary to cope with oxygen toxicity, or they could retreat to areas that remained free of oxygen, like the restricted anaerobic environments of stagnant waters or muds. The geologic record suggests that some prokaryotes chose the latter, and survived the mass extinction by retreating into stagnant anaerobic ponds. In doing so, they gained additional time to adapt to the radically changing atmospheric composition.

Changes in the hydrosphere are difficult to assess because there are virtually no constraints on the position and size of the oceans. Nevertheless, it is clear that the Early Proterozoic witnessed major changes in atmospheric composition and the role of plate tectonics, both of which would have influenced the hydrosphere. By the end of the Early Proterozoic, 1.6 billion years ago, the biosphere was under stress and, as we shall see, was forced to adapt in order to survive. This adaptation would prove to be a major evolutionary step that paved the way for the proliferation of life on Earth.

13.4.2 Middle Proterozoic (1.6 to 0.9 Billion Years Ago)

The geologic record during the Middle Proterozoic documents the further development of large continental landmasses such as Laurentia (ancestral North America). As the oxygen content continued to increase in the atmosphere, the biosphere responded by evolving more complex single-celled organisms that were oxygen-tolerant. This was followed by the development of multicellular organisms that were aerobic and capable of both sexual and asexual reproduction.

The further development of Laurentia witnessed two important phases of orogenic activity. Between 1.8 and 1.6 billion years ago, a series of mountain belts (such as the Yavapai, and Mazatzal) developed along the southern flank of Laurentia. This was followed between 1.3 and 1.0 billion years ago by the development of the Grenville belt along Laurentia's eastern margin (see Dig In: The Growth of the North American Continent, page 385).

Sandwiched between these two mountain building events, from about 1.5 to 1.3 billion years ago, Laurentia appears to have undergone an important phase of continental rifting. This resulted in the widespread injection of magma into the crust in a pattern that seems to cut across the older Archean and Early Proterozoic belts (Fig. 13.17). The event produced abundant mafic dikes that intrude ancient host rocks, and the widespread intrusion of an unusual igneous rock called anorthosite. **Anorthosites** are coarse grained plutonic rocks overwhelmingly dominated by a single mineral, plagioclase (see Chapter 2). The abundance of mafic dikes suggests that the continents were being fractured by rifting, and that magma was injected into the fractures.

Like all igneous intrusions, those that formed during this period must have been fed from below by magmatic plumbing systems. The dikes and anorthosite complexes must therefore represent a relatively shallow expression of some deeper, more profound thermal disturbance that was probably caused by the upwelling of mantle heat. If the distribution of the exposed complexes is an indication of the shape of the thermal disturbance, then its position in the mantle must have been such that it cut across boundaries between the ancient cratonic nuclei. This type of magmatic activity is generally attributed to the combined effects of an anomalously large area of hot mantle known as a *superswell*, and the insulating properties of large regions of continental crust. As we discussed in Chapter 6, large insulating continents retard the escape of heat from the Earth's interior, causing it to accumulate beneath the continent (see Figure 13.13). This intense buildup of heat causes the continent to bulge upwards and fracture. Once the continent is fractured, the heat can start dissipating. As hot magma rises towards the surface some of it cools within the fractures to form dikes.

However, the origin of the anorthosites is unclear. Igneous melts are a blend of many chemical elements which, when they crystallize, usually produce a variety of minerals. But accumulations of predominantly one mineral can occur under certain conditions. We learned in Chapter 2 that if crystals are denser than the melt in which they reside (as is commonly the case), they will sink to the base of the magma chamber. Conversely, if they are less dense, they float to the top. In either case, accumulations of individual minerals may occur. There are other more complex mechanisms that can produce crystal accumulations of single minerals, any one of which may account for the formation of anorthosites. But each of these mechanisms is as effective today as it has been in the past. As a result, the mechanisms do not explain why anorthosites should have become abundant only during this narrow interval of Middle Proterozoic time. Nor do the mechanisms explain why anorthosites should preferentially occur in superswell environments. As yet, the relationship between the two remains uncertain.

Although some important features of the Middle Proterozoic history of North America are still poorly understood, the main developments are clear. The evolution of the crust during this time interval was dominated by two mountain building events, separated in time by an episode of rifting, and by the end of this period, the large continental landmass of Laurentia had been assembled (see Dig In: The Growth of the North American Continent, page 385). However, our database is too poor to attempt realistic plate reconstructions for this time period.

The main development in the biosphere during the Middle Proterozoic was the first appearance of more complex single-celled organisms known as **eukaryotes**. Eukaryotes differ from the more primitive prokaryotes in possessing a more organized cell structure that includes a cell nucleus (Fig. 13.18a). Although the emergence of eukaryotes is thought to have taken place 1.8 billion years ago or even earlier, the oldest fossil eukaryote cells occur in 1.4 billion year old rocks in California and Australia (Fig. 13.18b). Unlike prokaryotes, which are capable only of asexual reproduction, eukaryotes are capable of either asexual or sexual reproduction. Therefore, the emergence of the eukaryotes represents an important evolutionary step because sexual reproduction creates a diverse gene pool, and this was to become a dominant influence in subsequent evolution. Furthermore, the eukaryotes were aerobic, and so needed oxygen to survive. Life had therefore developed the enzymes necessary to counteract the toxicity of oxygen.

Throughout the Proterozoic, ongoing photosynthesis was gradually changing the composition of the atmosphere. The levels of carbon dioxide were decreasing while those of oxygen were increasing. It is estimated that toward the end of the Proterozoic, oxygen concentrations may have made up between 1% and 2%

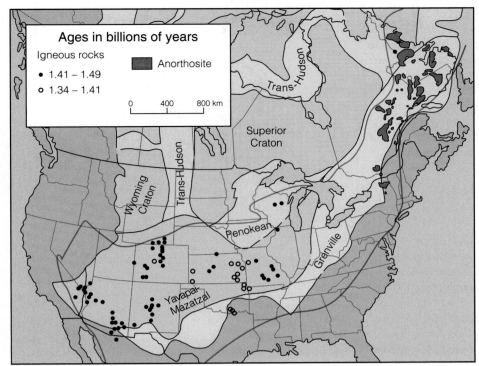

Figure 13.17

Distribution of North American igneous rocks between 1.5 and 1.3 billion years old in a broad band extending from Labrador in the northeast to California in the southwest. Their distribution may reflect the presence of an underlying mantle "superswell," which is thought to have generated them. This band of igneous rocks lies at an angle to the cratonic boundaries within Laurentia and may represent an important period of continental rifting.

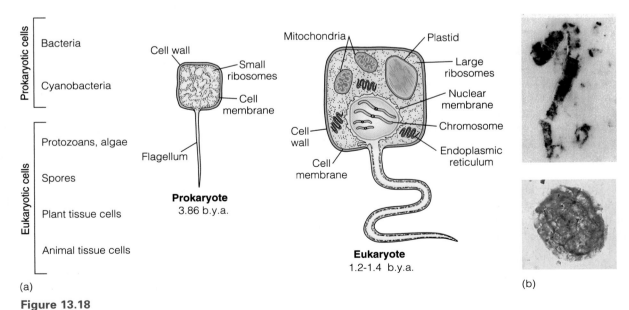

Figure 13.18

(a) A comparison between prokaryote and eukaryote cells. Note the better organization and more complex internal structure of the eukaryote cell. (b) Examples from Australia of 3.5 billion year old walled prokaryotic cells (above) and a 1.4 billion year old eukaryote (below).

of the atmosphere, compared to its modern level of 21%. This is close to the minimum concentration of oxygen required to produce ozone in the atmosphere. So a thin ozone layer is likely to have developed that would have provided organisms with some protection from the Sun's harmful ultraviolet radiation.

In summary, major changes in the biosphere during the Middle Proterozoic occurred in response to increasing atmospheric oxygen. Successful organisms became aerobic, and were capable of sexual reproduction. This paved the way for the more diverse biosphere we find in the Late Proterozoic.

13.4.3 Late Proterozoic (900 to 545 Million Years Ago)

By Late Proterozoic time, the geologic record is sufficiently well preserved that realistic plate tectonic reconstructions showing the possible positions of the major continental landmasses can be attempted (Fig. 13.19). However, this is a much more difficult task than it is for the last 200 million years, the period for which we have a sea floor record of magnetic reversals and preserved continental shelves. Almost all of the ocean floor that existed in the Late Proterozoic has long since been recycled back into the Earth's mantle by subduction. As a result, Late Proterozoic reconstructions are more speculative than those for the more recent geologic past, and are subject to ongoing refinement.

Late Proterozoic continental reconstructions center around Laurentia, Baltica (the predecessor of modern Europe), and Gondwana (a collage of the southern continents), largely because our database is best in these regions. Geologists have used several features of these ancient landmasses to piece together the continental jigsaw for that time period. Reconstructions of the distribution of the continents are based on matching patterns from one continental block to another. This is similar in principle to the technique Wegener used to reconstruct Pangea by showing, for example, the former continuity of the Appalachian-Caledonide mountain belt (see Chapter 4).

For many years, geologists working on the Late Proterozoic rocks of western North America had noted their similarity to the thick sedimentary successions of modern continental margins. Recent dating of igneous rocks in that area has suggested that the continental rifting needed to create such a margin occurred about 760 million years ago. The distribution of sedimentary rock types suggests that the margin faced an ocean to the west. Taken together, this requires a continental landmass to have separated from western North America some 760 million years ago. But which landmass was it? In 1991, Eldridge Moores and Ian Dalziel revived an earlier hypothesis by suggesting that the landmass was a continent made up of Australia and southwest Antarctica. The hypothesis was proposed on the basis of matching geology and the continuity of Proterozoic mountain belts.

Further evidence for the distribution of the continents during the Late Proterozoic comes from the proposed assembly of Gondwana. As we have learned, Laurentia contains mountain belts of varying ages (see Dig In: The Growth of the North American Continent). So too does Gondwana, among which are those known as the Pan-African mountain belts.

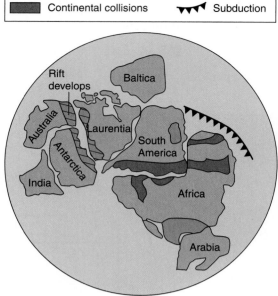

(a) 760 m.y.a.

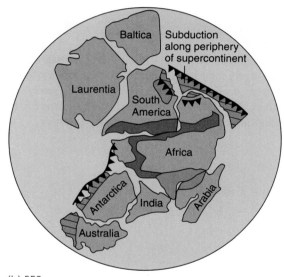

(b) 550 m.y.a.

Figure 13.19
These Late Proterozoic reconstructions suggest that breakup of a Late Proterozoic supercontinent at 760 m.y.a. may have led to the reassembly of a supercontinent of different geography by 550 m.y.a.

These widespread belts, the majority of which formed between 700 and 550 million years ago, have all the hallmarks of modern mountain belts produced by continental collision. This implies that the continental blocks on either side of the Pan-African belts were separated by oceans prior to their collision, and that their collision resulted in the amalgamation of Gondwana. Figure 13.19 shows two Late Proterozoic continental reconstructions based on the distribution of continental margins and the continuity of the mountain belts, each suggesting the existence of a supercontinent at that time.

Late Proterozoic life maintained a steady course of evolution. As atmospheric oxygen concentrations continued to rise,

The Growth of the North American Continent

The modern North American continent is an excellent illustration of the nature of continental growth by plate tectonic processes. Over the past 3 billion years (nearly 70% of geologic time), North America evolved from many small microplates to a large continental landmass. The diagrams illustrated in Figure 13.20 show how the continent grew from its ancient interior to its present exterior of relatively youthful mountain belts. These diagrams show when major episodes of deformation occurred, and how these

Figure 13.20

The growth of continental North America (Laurentia) in the Precambrian. (a) Archean cratons, like the Superior craton, were formed by the collision of microplates more than 2.5 billion years ago. These cratons were then amalgamated by 1.8 to 2.0 billion year old mountain building events, like the one that produced the Trans-Hudson belt. (b) Between 1.5 and 1.3 billion years, an important phase of rifting occurred. (c) This was succeeded 1.3 to 1.0 billion years ago by another mountain building event known as the Grenville. During the Phanerozoic (see Chapter 14), the growth of continental North America continued with the formation of the Appalachian Mountains along its western margin and the Cordillera-Rocky Mountains along its eastern margin.

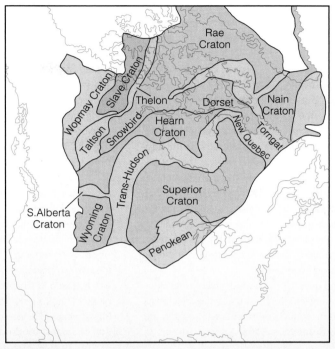

(a) 1.8-2 b.y.a.

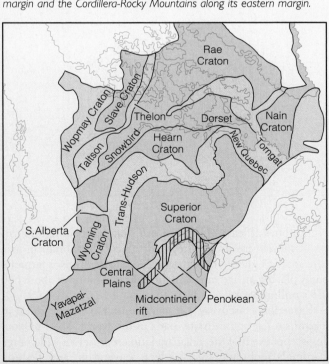

(b) 1.3-1.5 b.y.a.

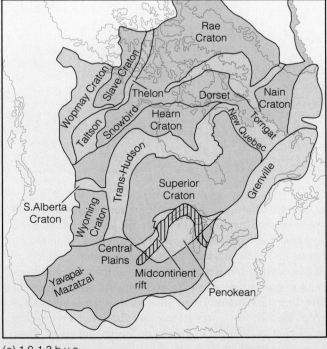

(c) 1.0-1.3 b.y.a.

☐ 1.0–1.3 billion	☐ 1.8–2.0 billion
☐ 1.6–1.8 billion	☐ >2.5 billion

(continued)

The Growth of the North American Continent, continued

regions of deformation have migrated outward to produce a stable continental nucleus.

In Figure 13.20a, several stable nuclei, or cratons, are identified. In these regions, all major tectonic activity had ceased by the end of the Archean about 2.5 billion years ago. These rigid cratonic areas can be thought of as "ball-bearings" surrounded by more malleable "putty." During later episodes of deformation, the rigidity of the cratons meant that tectonic stresses were deflected around them and so were concentrated in the adjacent weaker regions. As a result, these weaker regions became deformed and were ultimately welded onto the craton to build successively larger areas of stability. Thus, the cratons grew in size. The relative strength of cratonic areas and their ability to withstand subsequent stresses is conveyed in the term "Precambrian shield."

Figure 13.20a also identifies areas like the Trans-Hudson belt that were affected by a major episode of mountain building in the Early Proterozoic, some 1.8 to 2.0 billion years ago (see Dig In: Plate Tectonics in the Early Proterozoic?, page 380). By 1.8 billion years ago cratonic North America had grown considerably and by the end of the Proterozoic, two further belts had been added. The first was added in the southern United States some 1.8 to 1.6 billion years ago. Following a phase of rifting (Fig. 13.20b), the second (the Grenville belt) was added along the eastern margin of the craton between 1.3 and 1.0 billion years ago (Fig. 13.20c). By this time, cratonic North America had grown from small nuclei to about two-thirds of its present size. This ancestral version of North America is known as Laurentia.

During the Middle to Late Paleozoic, between 450 and 300 million years ago, the Appalachian mountain belt was formed along the eastern seaboard (see Chapter 14). As we discussed in Chapter 6, the west coast of North America grew steadily throughout the Mesozoic and Cenozoic through the accretion of suspect terranes.

It is important to realize that each major deformational event caused continental growth and that, in each case, the deformation and growth occurred along the edges of the continent. The continental interior of North America, on the other hand, was affected in only a relatively minor way.

The Wilson cycle (see Chapter 6) successfully explains why North America grew in this manner (Fig. 13.21). The Wilson cycle starts with the opening of an ocean and development of a continental margin (Fig. 13.21a). The ocean expands to a maximum width, and then begins to close by subduction (Fig. 13.21b). Closure is accompanied by terrane accretion and culminates in continental collision (Fig. 13.21c). These processes add new material to the continental margin, so that when a new ocean opens again, a young mountain belt is left behind at the edge of the continent. In this way, new mountain belts are progressively attached to the edge of the continent with each repetition of the cycle.

---

ozone concentrations increased while carbon dioxide levels fell as a result of photosynthesis. Higher ozone levels increased the atmospheric shield against ultraviolet radiation and opened up ecological niches in progressively shallower water. In this way hospitable environments were primed for the major evolutionary event that occurred about 700 million years ago. The event was the first appearance of multicellular animals known as **metazoa**. Best known among these are the latest Proterozoic examples that make up the **Ediacara fauna**. Fossils of these creatures are named after the site of their first discovery in 1947 by R. Sprigg, in the Ediacaran Hills of south-central Australia. At the time, they were thought to be Cambrian in age. But geologic mapping subsequently revealed that the fauna occurred in sandstones at least 150 m *beneath* strata containing the earliest Cambrian fossils. So they had to be Precambrian in age. More recent radiometric dating has confirmed this interpretation by placing an age range on the Ediacara fauna of about 600 to 545 million years.

Since their discovery, Ediacara fauna have been found in rocks of similar age in many parts of the world. Most paleontologists believe that the Ediacara fauna represent a wide range of animals including filter-feeders, scavengers, and floating organisms that lived in a near-shore environment. Although recent studies indicate that much of this fauna became extinct prior to the Phanerozoic, the discovery of these organisms kindled interest in more detailed analysis of Late Proterozoic rock

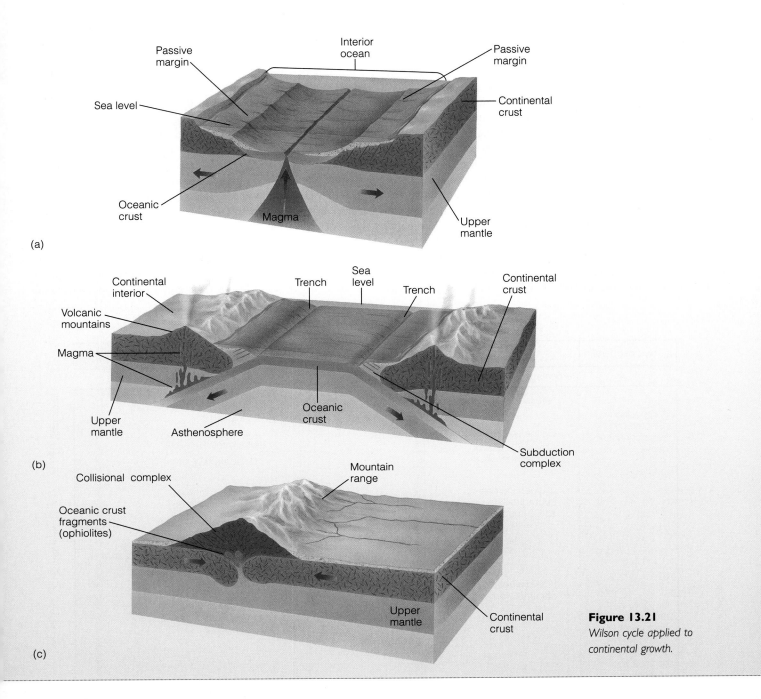

(a)

Passive margin
Interior ocean
Passive margin
Sea level
Continental crust
Oceanic crust
Magma
Upper mantle

(b)

Continental interior
Trench
Sea level
Trench
Continental crust
Volcanic mountains
Magma
Upper mantle
Asthenosphere
Oceanic crust
Subduction complex

(c)

Collisional complex
Mountain range
Oceanic crust fragments (ophiolites)
Upper mantle
Continental crust

Figure 13.21
Wilson cycle applied to continental growth.

sequences similar in age to the one in which the fossils were first found. This research revealed an impressive diversity of soft-bodied organisms. These creatures may be the ancestors of a wide range of organisms that developed the ability to secrete shells in the Early Paleozoic. As such, they may form an important evolutionary link to the more profuse fossil record of the Phanerozoic.

In summary, the Late Proterozoic was a period of widespread mountain building that led to the formation of a Late Proterozoic supercontinent. The reconstructions shown in Figure 13.19 show the changing geography of this supercontinent. As oxygen increased in the atmosphere, so too did ozone, strengthening the atmospheric shield against ultraviolet radiation. The development of an ozone layer set the stage for major evolutionary events, such as the first appearance of metazoan organisms about 700 million years ago.

13.5 Chapter Summary

Precambrian geology represents up to 88% of geologic time. In this chapter we have chronicled the key events and processes that governed the evolution of the Earth systems from the planet's hostile, tumultuous beginnings 4.6 billion years ago to the relatively benign environment it possessed when life suddenly exploded in the Late Proterozoic and earliest Phanerozoic. These events are summarized in Figure 13.22.

The planet became layered according to density very early in its history from its core to the top of its atmosphere. This attests to the importance of vertical, gravitational processes in the primitive Earth. The abundance of ancient meteorite impacts on other terrestrial planets (see Chapter 15) suggests that the Earth too was bombarded by impacts early in its history, and this may have aided in the formation of the Earth's internal layers. The early Earth was very active volcanically, and the gases emitted from these volcanoes would have controlled the composition of the planet's primitive atmosphere. Comparison with modern volcanic eruptions suggests that the dominant gases would have been water vapor and carbon dioxide with relatively minor nitrogen. When water vapor reached saturation level and the Earth's atmosphere began to cool, it started to rain. This rain had nowhere to collect until continental crust first

formed and generated the necessary topographic contrast between continents and oceans. The available data suggest this had happened at least 4.25 billion years ago. Because continental crust is less dense than oceanic crust, it floats at a higher elevation on the asthenosphere so that water started to flow from one to the other, eventually ponding in depressions we now call ocean basins. The hydrosphere and the hydrologic cycle commenced as water evaporated from the oceans and was swept onto land where it descended as rainfall and drained back to the oceans. The formation of the hydrosphere extracted huge quantities of water vapor and carbon dioxide (which is very soluble in water) from the atmosphere. Depletion of these greenhouse gases would have been responsible for significant global cooling.

The earliest signs of life are 3.86 billion years old. Life developed as single-celled organisms in an oxygen-free or anaerobic environment. Because the atmosphere contained no oxygen gas, it could not have contained ozone. Therefore

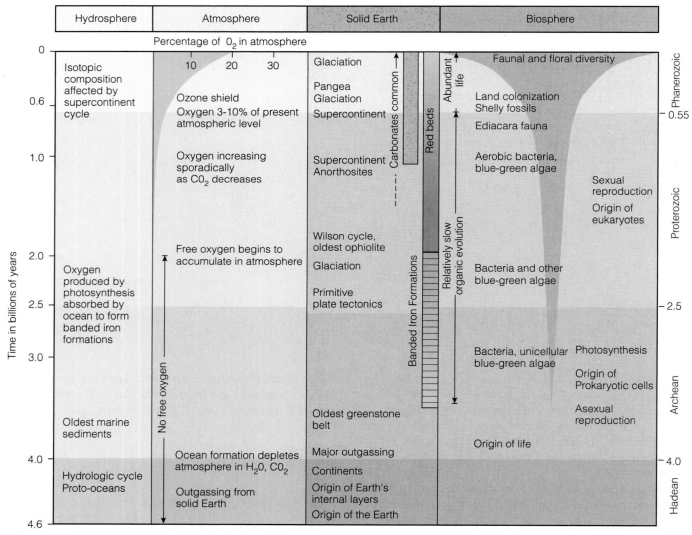

Figure 13.22
Major events in the evolution of the atmosphere, solid Earth, and biosphere during the Precambrian. The evolution of the hydrosphere is poorly documented.

the atmosphere could not offer any protection from the Sun's ultraviolet radiation. Life developed in oceanic environments that could offer partial or complete protection. The earliest preserved organisms are prokaryotes which are found in algal mats called stromatolites and probably resided in inter-tidal regions. These organisms were capable of photosynthesis, but the oxygen that was produced did not become part of the atmosphere. Instead, it was absorbed by the oxidation of iron dissolved in ocean water, producing layered iron-rich rocks known as Banded Iron Formations. Prokaryotes were only capable of asexual reproduction, and so evolved very slowly.

By the Late Archean, significant amounts of continental crust had been generated. As continental crust is a good insulator of heat escaping from the mantle, mantle-convection cells developed beneath the lithospheric shell. These cells were able to move overlying plates horizontally and a crude form of plate tectonics probably commenced.

Between 3.0 and 2.0 billion years ago, small microplates collided to form bigger plates and the stable cratonic shields that form the interior of many modern continents were formed. By the Early Proterozoic, there is strong evidence for the existence of Wilson cycles which involve the opening and closing of ocean basins due to plate tectonic activity.

About 2.0 billion years ago, the rate of photosynthesis exceeded the ocean's ability to absorb the oxygen it produced. So, for the first time, oxygen gas started to accumulate in the atmosphere, and its abundance slowly increased throughout the Proterozoic. Because oxygen is poisonous to anaerobic life, its presence triggered a mass extinction. The only surviving organisms were those that could adapt to the rapidly changing conditions. The first organisms that required the presence of oxygen were eukaryotes, which first appeared about 1.8 billion years ago. These organisms were capable of sexual reproduction, which helped establish a diverse, healthy gene pool. Toward the end of the Proterozoic, about 700 million years ago, multi-cellular animals or metazoa had developed, most notably the Ediacara fauna which first appeared some 600 million years ago. Their sparse preservation is attributed to the fact that they were soft-bodied.

Abundant Late Proterozoic mountain building activity provides strong evidence of the existence of a Late Proterozoic supercontinent, although its configuration is controversial.

Key Terms and Concepts

Look for the highlighted items on the web at:
WWW.BROOKSCOLE.COM

- Knowledge of the evolution of Earth systems is important to distinguishing between natural and anthropogenic (human-induced) change

- The Precambrian represents 88% of geologic time and is divided into three eons: the Hadean, Archean, and Proterozoic

- The rapid early evolution of the Earth resulted in the separation of the Earth into layers according to density, from its core of densest elements to its atmosphere of lightest ones

- The oldest rocks found to date are 3.96 billion years old

- The oldest zircon crystals are nearly 4.3 billion years old, and imply that continental crust existed by that time

- The oldest sedimentary rocks are typical of modern ocean deposits and are 3.86 billion years old, indicating that oceans and the hydrologic cycle existed by that time. These sediments also contain the first evidence for life

- The presence of light, elevated continental crust and dense, depressed oceanic crust provided the topography for water to drain and collect in ocean basins

- A primitive form of plate tectonics may have existed in the Late Archean

- All life is made up of discrete units called cells. Life has the ability to reproduce and self-regulate, which is based on the interaction between nucleic acids and proteins. Amino acids are the building blocks of proteins. Nucleic acids are composed of sugar, a phosphate group and a nitrogen compound

- One form of nucleic acid, deoxyribonucleic acid (or DNA), carries the genetic information of all organisms except bacteria. Another nucleic acid, ribonucleic acid (or RNA), carries out the instructions embedded in this genetic information

- DNA carries genetic information for the production of enzymes, which are a form of protein that helps speed up and regulate chemical reactions in a cell

- The origin of life is still debated; it may have occurred in intertidal pools where fragile organisms were partially protected from the Sun's poisonous ultraviolet rays, it may have occurred near deep sea volcanic vents, or it may have been seeded by extraterrestrial impacts of meteorites and comets

- Life had probably originated by 3.6 billion years ago. The oldest examples of life are found in stromatolites, which were formed by the binding of sediment with blue-green algae and were deposited in shallow marine environments

- Organisms in stromatolites consist of simple single-celled structures known as prokaryotes. Prokaryotes lived in an anaerobic environment, but were capable of photosynthesis. The oxygen produced was absorbed by dissolved iron in the oceans resulting in the formation of Banded Iron Formations

- Life may also been nurtured in deep sea environments near volcanic vents. Here sea water would have offered protection from solar radiation poisoning, and bacteria could have converted inorganic compounds into organic compounds by chemosynthesis

- Comets and meteorites also contain chemicals which can react to produce amino acids, the building blocks of proteins. Life may have been seeded by extraterrestrial collisions with the young Earth

- Because prokaryotes were only capable of asexual reproduction, the pace of evolution throughout the Archean was very slow

- The main development in the Late Archean was the production of stable continental nuclei, or cratons, which consist of granite gneiss complexes and narrow greenstone belts that are dominated by volcanic and sedimentary sequences. Greenstone belts may represent continental rift environments and/or back-arc settings like the modern Sea of Japan

- By the Early Proterozoic, cratonic nuclei became stitched together to form stable continental shields

- The history of the Trans-Hudson orogenic belt suggests that plate tectonics was operative 2 billion years ago

- Two billion years ago, oxygen produced in the oceans by photosynthesis finally leaked into the atmosphere resulting in a mass extinction of anaerobic prokaryotes. Life adapted in two ways; by producing enzymes to counteract this toxicity, or by retreating to restricted anaerobic environments

- By 1.8 billion years ago, Laurentia, (ancestral North America) had formed, probably by the plate tectonics processes depicted in the Wilson cycle

- Peculiar plagioclase-rich rocks called anorthosites, predominantly formed in a very limited time interval in the Middle Proterozoic. Their origin is unclear

- Eukaryotes, the first oxygen-tolerant or aerobic organisms, appeared in the Middle Proterozoic and were capable of sexual reproduction, which diversified the gene pool

- At a critical point in the Late Proterozoic, the oxygen content in the atmosphere was sufficient to form an ozone layer, which afforded life some protection from solar radiation Protection from solar radiation resulted in rapid evolution and expansion of ecological niches

- Multicellular animals or metazoa, like those of the Ediacara fauna that appeared in the Late Proterozoic, may represent the evolutionary link to the Cambrian shelly fauna of the earliest Phanerozoic

Review Questions

1. How did the primitive microcontinents form?

2. Why was the very early history of the Earth characterized by rapid evolution?

3. In which environment did life most likely originate and why?

4. What was the special role of Banded Iron Formations in the evolution of the atmosphere?

5. Outline the differences between prokaryotes, eukaryotes, and metazoan fossils.

6. Discuss the merits and importance of the models of greenstone belt formation.

7. What is the importance of the Trans-Hudson orogen to the formation of continental North America?

8. What is the key evidence in the rock record for the first presence of free oxygen in the atmosphere?

9. What is a "superswell"?

10. Why might the Ediacara fauna form an evolutionary link to the Phanerozoic Era?

Study Questions

 Research the highlighted question using InfoTrac College Edition.

1. Why do you think the vertical processes of planetary accretion, gravitational collapse, and layer formation eventually gave way to the horizontal processes of plate tectonics?

2. There is circumstantial evidence that some form of crude plate tectonics operated by the end of the Late Archean. What is this evidence and when does the first hard evidence of plate tectonic processes similar to those of today appear in the geologic record?

3. Outline the special role of oxygen in the evolution of Precambrian life.

4. Why is the resolution of the controversy of greenstone belt origin critical to our understanding of Archean and Proterozoic tectonic processes?

5. Explain why the discovery of a 2 billion year old ophiolite is strong evidence for the existence of plate tectonics at that time.

6. What are the problems associated with the theories of anorthosite genesis?

7. Why do you think the Late Proterozoic was a period of widespread orogenic activity?

8. Why is the transition from asexual to sexual reproduction an important evolutionary event?

CHAPTER 14

Evolution of the Earth Systems II: The Phanerozoic

In this chapter, you will learn how the Earth's Precambrian time interval gave way to a progressively more hospitable planet capable of sustaining the enormously diverse forms of life that inhabit the Earth today. The term Phanerozoic means "visible life," reflecting the first occurrence of shelly fossils at the beginning of the eon, 545 million years ago, and their importance thereafter. The Phanerozoic is divided into three eras, the Paleozoic, Mesozoic, and Cenozoic, which mean "ancient," "middle," and "recent" life, respectively, and reflect the evolution of life toward ever more modern forms.

The pathway from the early life of the Precambrian to the "visible life" of the Phanerozoic was one of slow evolutionary steps punctuated by catastrophe. Important innovations in the Phanerozoic, like the colonization of land by plants, were at first very tentative steps. The eventual success of these innovations occurred because they were appropriate responses to changing atmospheric conditions and the development of more hospitable environments. Along the way, however, we will find that there were many catastrophes. Those species that were opportunistic and adaptable were able to flourish in the wake of the catastrophes. These species have passed the torch of life along to the present day.

We will learn that the geologic history of the Phanerozoic is much better understood than that of the Precambrian. Our greater understanding of this eon is due to a combination of radiometric and fossil dating, and the fact that sedimentary layers containing these fossils are better preserved in the geologic record. In addition, we have an ocean floor record for the past 180 million years because it has yet to be subducted.

The existence of soft-bodied fossils in Late Precambrian sequences provides the evolutionary bridge to the Phanerozoic. Prior to their recognition, the beginning of the Phanerozoic was thought to represent the beginning of life, and its rapid worldwide introduction was of grave concern to evolutionary biologists who could not understand from these sophisticated life forms how life might have originated. However, as we shall see, the prevailing view now is that life was already flourishing by the end of the Precambrian and that the major evolutionary feature at the beginning of the Phanerozoic was the ability to develop shelly exteriors. We will learn that the plate tectonic context of this development in evolution was probably very important. The Late Precambrian and Early Cambrian was a period when the fragmentation of a Late Precambrian supercontinent resulted in the generation of continental shelves. This opened up new ecological niches for life to exploit.

The Paleozoic is the oldest era in the Phanerozoic. The era begins 545 million years ago with the breakup of a supercontinent and ends 245 million years ago with the assembly of a new supercontinent, called Pangea. Thus oceans created by continental rifting at the beginning of the Paleozoic had been consumed by subduction by the end of the era. During this era, life evolved in marine environments while the oxygen content in the atmosphere became more abundant and the ozone shield became more effective.

This change in atmospheric composition prompted the first tentative colonization of land by plants, some 470 to 440 million years ago. The 80 to 100 million year period that followed this colonization was an important time for plant evolution on land, setting the planet on a course towards its current atmospheric oxygen level of 21%.

The colonization of land produced vegetation and soils on a previously barren landscape allowing for the more effective retention of fresh water. By the Carboniferous Period, some 300 million years ago, vegetation had flourished to the point at which coal beds could develop.

The end of the Paleozoic was indeed the end of an era, recording what is probably the largest known mass extinction. The cause of this extinction is uncertain, but we shall find that plate tectonics may have played an important role because the destruction of oceans and the formation of Pangea resulted in the widespread loss of ecological niches.

The Mesozoic Era (245 to 66 million years ago) began with the spread of life forms that had survived the Late Paleozoic crisis. We shall learn that this truly was a case of the "survival of the fittest." Because their eggs could survive far from water, reptiles dominated over fish and amphibians. Opportunistic species flourished at the expense of their less adaptable neighbors. Because of the importance of reptiles and the dominance of a group of animals that evolved from them, the Mesozoic Era is widely known as the "Age of the Reptiles." The end of this era is marked by another mass extinction, the commonly accepted reason for which was an extraterrestrial impact. However, as we shall see, there is evidence that many life forms, including the dinosaurs, were already in decline and that the impact may well have been the "straw that broke the camel's back."

The Mesozoic Era was dominated by the breakup of Pangea and the development of the Atlantic Ocean. The resulting flooding of the continents was so extensive that about 100 million years ago only 15% of land was exposed above sea level. During this period, the data indicate that both atmospheric and ocean water temperatures were considerably warmer than they are today. This was caused by an intense greenhouse effect which resulted from high levels of atmospheric carbon dioxide produced by voluminous volcanism.

Since most of the oceanic crust and continental shelves of the Mesozoic are still intact, plate reconstructions for this era are much more reliable than those of the Paleozoic. As the Atlantic Ocean expanded, the Pacific Ocean contracted, and a succession of mountain building events commenced on the west coast of North America related to subduction and island arc-continent collisions.

The catastrophe that ended the Mesozoic Era presented life with another ecological fitness test. The surviving life forms adapted and flourished in the Cenozoic Era, during which many of the traits of modern life were developed. Many Cenozoic fossils have clearly recognizable modern descendants and represent evolutionary bridges to the modern biosphere.

During this era, expansion of the Atlantic Ocean continued at the expense of other major oceans, which were being consumed by subduction. Episodic subduction and orogeny continued around the margins of the Pacific Ocean, producing major collisional events in western North America that culminated in the birth of the San Andreas Fault system. In addition, collisional events in Europe and Asia led to the development of the Alpine-Himalayan mountain belt.

About 1.6 million years ago, the Ice Age began, marked by the cyclic advance and retreat of polar ice sheets. The history of the Ice Age is especially well documented in the northern hemisphere where it was responsible for shaping much of our modern landscape. Sea level changes occurring in response to these cycles intermittently provided land bridges that aided migration. During the same time period, modern humans evolved: an event that was to be a major evolutionary step, because the human species was the first to artificially interfere with the Earth systems.

As we progress through the Phanerozoic, we will develop an appreciation of the pace of natural change in the Earth systems. Furthermore, as we shall see in subsequent chapters, an understanding and appreciation of this global change provides the context for analyzing our modern environmental problems.

14.1 Introduction: Geologic Time Scale Revisited

The Phanerozoic Eon comprises less than 12% of geologic time. Yet, we understand this eon far better than the Precambrian. This greater understanding is reflected in the greater resolution, and hence finer subdivision, of the geologic time scale (Fig. 14.1). As we learned in Chapter 3, eons of geologic time are subdivided into eras and periods, much as this textbook is divided into parts, with each part containing a number of chapters. The **Phanerozoic Eon** (which means "visible life") started with the first appearance of shelly fossils in the geologic record 545 million years ago and continues to the present day. As implied by its name, the fossil record of the Phanerozoic is far more complete than that of the Archean or Proterozoic eons.

This superior fossil record enables the Phanerozoic Eon to be subdivided into three eras, the **Paleozoic** (meaning "ancient life") from 545 to 245 million years ago, the **Mesozoic** ("middle life") from 245 to 66 million years ago, and the **Cenozoic** ("recent life") from 66 million years ago to the present. Remember that the fossil record and the evolution of life were understood well before radioactivity was discovered. The relative time scale of eons, eras and periods was therefore in place long before radiometric dating established their absolute ages (see Chapter 3). Prior to this, only the sequence of events was understood, the age range they each represented was not.

It is important to realize that the geologic time scale is under constant scrutiny and refinement. The recent reinterpretation of the age of the Precambrian-Cambrian boundary is an excellent example of this. Until the early 1990s, an age of about 570 million years was widely accepted for this important time line. However, suspicions of the accuracy of this age were raised by the dating of volcanic rocks in Newfoundland, Canada, which lay at least 4 km (2.5 mi) *below* rocks that contained lowest Cambrian fossils. The volcanic rocks yielded an age of 565 million years and suggested a significantly younger depositional age for the rocks containing the lowest Cambrian fossils. These data focused attention on similar sequences worldwide, and as a result, further dates were obtained that placed the age of the boundary at about 545 million years.

Even in the 19th century, however, paleontologists understood that biological evolution in the Phanerozoic was punctuated by major extinction events that were global in scale. They used these events to subdivide geologic time so that each event can truly be said to define "the end of an era." At the end of the Mesozoic Era, for example, nearly 75% of all living things became extinct, including the dinosaurs. But paleontologists were not sure what caused these extinction events, nor did they know how long ago (in terms of millions of years) the events had occurred. Only with the advent of radioactive dating did the absolute age of these events become clear.

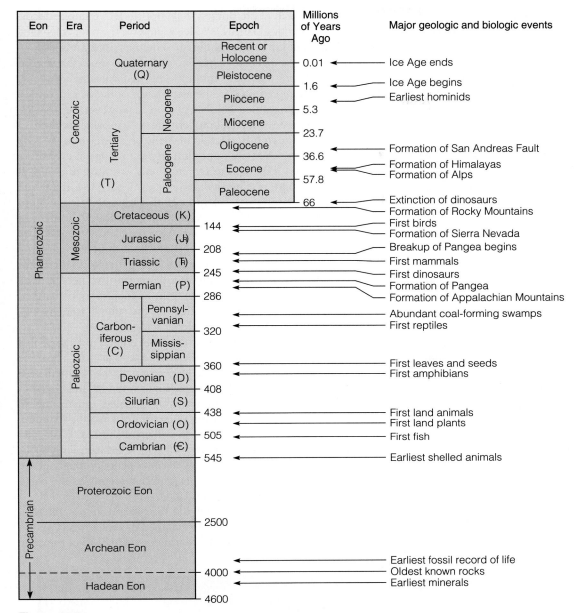

Eon	Era	Period	Epoch		Millions of Years Ago	Major geologic and biologic events
Phanerozoic	Cenozoic	Quaternary (Q)	Recent or Holocene		0.01	Ice Age ends
			Pleistocene		1.6	Ice Age begins
		Tertiary (T)	Pliocene	Neogene	5.3	Earliest hominids
			Miocene		23.7	
			Oligocene	Paleogene	36.6	Formation of San Andreas Fault
			Eocene		57.8	Formation of Himalayas / Formation of Alps
			Paleocene		66	Extinction of dinosaurs
	Mesozoic	Cretaceous (K)			144	Formation of Rocky Mountains / First birds
		Jurassic (J)			208	Formation of Sierra Nevada / Breakup of Pangea begins
		Triassic (Ŧ)			245	First mammals / First dinosaurs
	Paleozoic	Permian (P)			286	Formation of Pangea / Formation of Appalachian Mountains
		Carboniferous (C)	Pennsylvanian		320	Abundant coal-forming swamps / First reptiles
			Mississippian		360	First leaves and seeds
		Devonian (D)			408	First amphibians
		Silurian (S)			438	First land animals
		Ordovician (O)			505	First land plants / First fish
		Cambrian (Є)			545	Earliest shelled animals
Precambrian		Proterozoic Eon			2500	
		Archean Eon			4000	Earliest fossil record of life / Oldest known rocks
		Hadean Eon			4600	Earliest minerals

Figure 14.1

The geologic time scale. Note that the scale is not linear. The Phanerozoic Eon accounts for less than 12% of geologic time. Some of the important events in Phanerozoic history are shown on the right. The letters after each period are the standard symbols for that period.

14.2 Earth Systems in the Paleozoic Era (545 to 245 Million Years Ago)

The Paleozoic Era is divided into six periods (Fig. 14.1). Together, these record the fragmentation and dispersal of the Late Proterozoic supercontinent during the Early Paleozoic (Cambrian to Early Ordovician) and the coalescence and amal-gamation of the supercontinent Pangea during the Middle to Late Paleozoic (mid-Ordovician to Permian). The Early Paleozoic consequently witnessed the generation of new oceans fringed by continental shelves and a global-scale rise in sea level. This environment provided new ecological niches in which the marine biosphere could flourish. By the Middle Ordovician, many of these oceans began to subduct, leading to the genera-tion of volcanic island arcs. During the Late Ordovician and Silurian, island arc-continent collisions occurred in the narrow-ing oceans, culminating in the Devonian, Carboniferous, and Permian with ocean closure and continent-continent collisions.

14.2.1 Early Paleozoic: Cambrian to Early Ordovician

The Paleozoic Era is defined by the advent, evolution and subsequent extinction of a variety of life forms. The trilobite (so called because its hard outer shell has three lobes) is an excellent example of one of these life forms. Like the king crab, which it superficially resembles, the trilobite lived as a scavenger on the seabed in marine environments near shore. Trilobites first appeared at the beginning of the Paleozoic Era, and became extinct when the era came to an end 245 million years ago. Like many Paleozoic life forms, they have no modern relatives. Other invertebrate life forms that flourished in the Early Paleozoic such as brachiopods, echinoderms, and shelled mollusks do have modern relatives. Some of the more important marine invertebrates are summarized in Figure 14.2. Fish are the oldest vertebrates, first appearing in the Upper Cambrian, but were restricted to shallow marine environments until the Silurian Period.

As paleontologists compiled more and more information from various parts of the world, they recognized a simultaneous explosion in the fossil record which they used to define the start of the Paleozoic and the base of the Cambrian Period. Charles Darwin was particularly troubled by this apparent explosion of life, which he acknowledged was a weak link in his evolutionary hypothesis. As we learned in Chapter 13, the prevailing view of the time was that the Cambrian explosion represented the beginning of life itself.

We now know that this view was incorrect, and that evidence of life in the form of soft-bodied organisms extends back at least 3.5 billion years, and that ancestors of Cambrian species existed in the Late Proterozoic. It was the evolution of organisms with hard exteriors that accounts for their abrupt appearance in the rock record at the beginning of the Cambrian because the hard parts are readily preserved.

The fossils of the Burgess shale in British Columbia, discovered in 1909 by Charles D. Walcott, preserve the most celebrated examples of life in the Cambrian. The organisms are spectacularly preserved because of the fine carbon-rich muds that draped the flourishing marine community. All major invertebrate fossil groups are represented in the assemblage, some of which have since become extinct. Those that survived, however, became the basic stock from which all modern life on our planet evolved.

The sudden introduction of shelly organisms into the rock record is an extremely important step in biological evolution. But why did life begin to secrete shelly exteriors, and why at this point in time? The vital ingredients needed to form a shell include phosphorus, calcium, and carbonate. In modern oceans, these ingredients are the nutrients found in areas of marine upwelling, where cold, oxygenated waters ascend to nourish the marine community. But would this have been true in the Early Cambrian? Since the appearance of shelly fossils worldwide occurred virtually simultaneously, we need to look at the origin of these nutrients from a global perspective and their relationship to events affecting the solid Earth, hydrosphere, and atmosphere on a global scale.

From this viewpoint, the Cambrian Period is just a stage in a plate tectonic cycle that spans the entire Paleozoic Era. The cycle started at the beginning of the Paleozoic with the fragmentation and dispersal of a Late Proterozoic supercontinent and ended in the Late Paleozoic with the assembly of a new supercontinent (Pangea) (Fig. 14.3). Following rifting (Fig. 14.3a), the Late Proterozoic supercontinent started to drift apart (Fig. 14.3b). By the Early Cambrian, oceanic crust had formed between the dispersing continental blocks. The new oceans opened throughout the Cambrian, developing passive margins that resembled the continental shelves of the modern Atlantic.

Plate reconstructions for the Cambrian offer a potential explanation for the sudden appearance of shelly fossils. As we learned in Chapter 9, coastal marine upwellings from the deep

Cambrian			Ordovician	
Early	Middle	Late	Early	Middle
Arthropoda (Trilobites) →				
Brachiopoda →				
Echinodermata →				
Mollusca (Gastropoda) →				
Porifera →				
	Mollusca (Bivalvia) →			
	Mollusca (Cephalopoda) →			
		Protozoa →		
			Bryozoa →	
			Cnidaria (Tabulate corals) →	
				Cnidaria (Rugose corals) →

(a)

(b)

Figure 14.2

(a) Summary of important marine invertebrate life in the Early Paleozoic. (b) An Early Paleozoic trilobite.

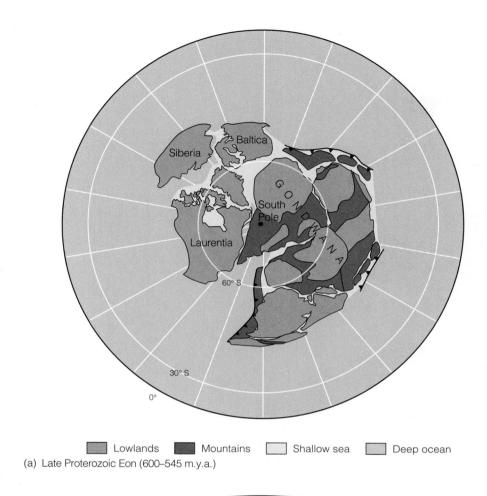

Lowlands   Mountains   Shallow sea   Deep ocean

(a) Late Proterozoic Eon (600–545 m.y.a.)

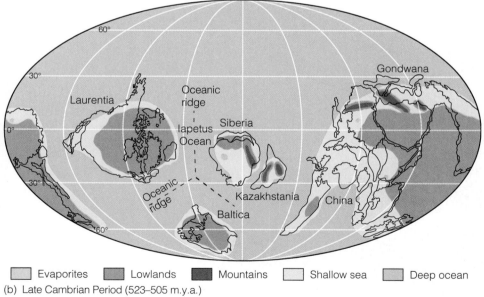

Evaporites   Lowlands   Mountains   Shallow sea   Deep ocean

(b) Late Cambrian Period (523–505 m.y.a.)

(continued)

Figure 14.3

Late Proterozoic-Paleozoic reconstructions and paleogeography for (a) Late Proterozoic, (b) Late Cambrian, (c) Late Ordovician, (d) Late Silurian, (e) Middle Devonian, (f) Early Carboniferous, (g) Late Carboniferous, and (h) Late Permian. Note that (a) shows a map projection centered on the south pole with the equator on the rim. On all other maps, the equator runs horizontally across the center of the projection. Together, the reconstructions show that the history of the Paleozoic was dominated by the breakup of a Late Proterozoic supercontinent and the formation of new oceans, followed by the destruction of these oceans and the amalgamation of Pangea.

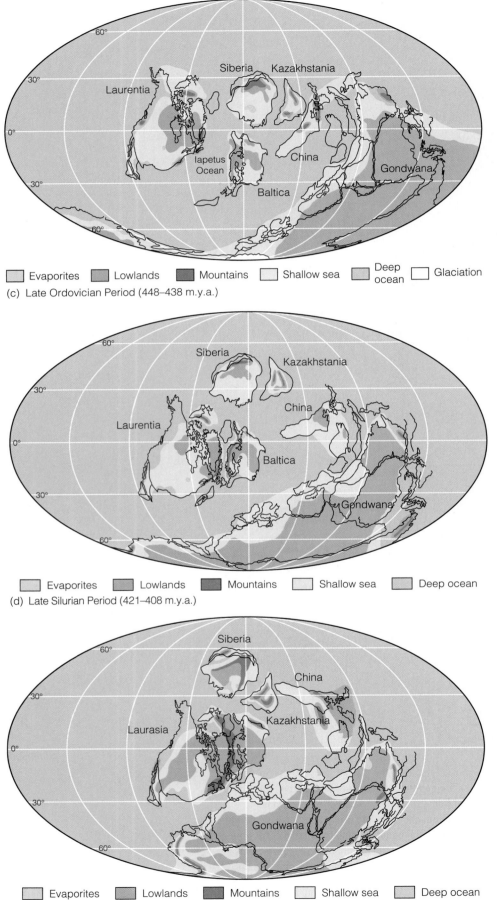

Evaporites Lowlands Mountains Shallow sea Deep ocean Glaciation

(c) Late Ordovician Period (448–438 m.y.a.)

Evaporites Lowlands Mountains Shallow sea Deep ocean

(d) Late Silurian Period (421–408 m.y.a.)

Evaporites Lowlands Mountains Shallow sea Deep ocean

(e) Middle Devonian Period (380–374 m.y.a.)

Figure 14.3, continued

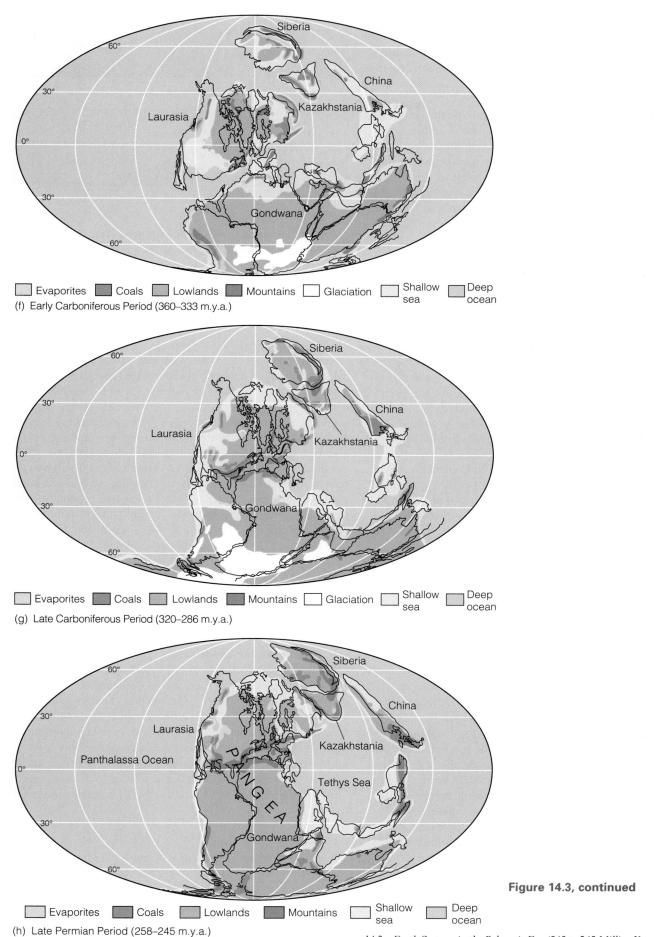

Evaporites　Coals　Lowlands　Mountains　Glaciation　Shallow sea　Deep ocean
(f) Early Carboniferous Period (360–333 m.y.a.)

Evaporites　Coals　Lowlands　Mountains　Glaciation　Shallow sea　Deep ocean
(g) Late Carboniferous Period (320–286 m.y.a.)

Figure 14.3, continued

Evaporites　Coals　Lowlands　Mountains　Shallow sea　Deep ocean
(h) Late Permian Period (258–245 m.y.a.)

ocean bring up nutrients that are vital to modern shallow marine ecosystems. In the Early Cambrian, the development of new oceans and new continental shelves would have provided an ideal environment for such upwellings. In addition, continental erosion may have provided the additional nutrients necessary for shell development.

Cambrian plate reconstructions suggest that the new continents of Laurentia, Baltica, and Siberia were drifting apart (Fig. 14.3b). We can therefore infer that spreading ridges existed between each of these continents. The reconstruction suggests a three-pronged ridge system with an angle of about 120° between each ridge axis. Such three-pronged systems are typical of continental breakup and would have been initiated at the continental rifting phase (see Chapter 5). These rifts later became mid-ocean ridges as the continents drifted apart and a new ocean opened between them. In this case, the new ocean occupied a similar position to the modern Atlantic and was named **Iapetus** after the father of Atlantis in Greek mythology.

As is typical of the drifting phase of supercontinent breakup, the Cambrian reconstruction shows extensive continental shelves fringing most continental landmasses. As modern oceans demonstrate, such shelves provide very hospitable environments for life to flourish. Evidence of this ancient margin of the Iapetus Ocean is well preserved today as a series of Cambrian continental shelf deposits exposed along the length of eastern North America from Newfoundland in Atlantic Canada to the Ouachita mountains in Oklahoma. Indeed, it is likely that the Iapetus Ocean of the Cambrian closely resembled the modern Atlantic Ocean.

Three complementary factors would have aided the development of continental shelves and so helped life to flourish during the Cambrian. First, the development of new mid-ocean ridges would have created vast underwater mountain ranges similar to the modern Mid-Atlantic Ridge. These new mountains would have displaced the ocean water, forcing it to rise upwards (Fig. 14.4). Just as climbing into a bathtub filled to the brim displaces the water over the sides, so the effect of these new submarine mountains would have been to displace ocean water upward and outward across the continental margins to create new continental shelves (see Chapter 8).

Second, the continents would have subsided as they drifted apart and moved away from the unusually hot regions of mantle upwelling at the mid-ocean ridges. Just as lowering the sides of a bathtub filled to the brim would also cause the water to spill over the edges, the effect of continental subsidence would allow seawater to encroach over the continental margin.

Third, most geologists speculate that the breakup of the Late Proterozoic supercontinent caused the melting of its continental glaciers. This melt-water would have also led to a rise in sea level. However, its overall contribution to the rise in Cambrian sea level is uncertain because we have no reliable estimate of the volume of ice that may have melted.

The widespread development of shallow continental margins in the Early Cambrian almost certainly contributed to the abundance of Early Cambrian fossils in the geologic record. These coastal marine environments harbor the vast majority of marine species. So, while many Cambrian species may have had

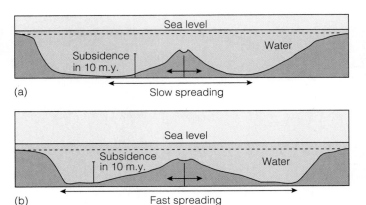

Figure 14.4

The relationship between mid-ocean ridge generation and sea level. A mid-ocean ridge displaces ocean water such that the water is displaced onto the continent, resulting in a greater area being covered by shallow seas. This reduces continental erosion because only rocks exposed above sea level can be eroded from the continents. (a) Slow spreading ridges displace less water than (b) fast spreading ridges.

Late Proterozoic ancestors, the availability of vast underpopulated habitat areas would have provided many new ecological niches in which life could flourish. Depending on the prevailing wind direction (see Chapter 9), some of these new coastal regions would have been zones of upwelling. Here, nutrients transported from the deep ocean to the continental shelves would have been absorbed into the biosphere. The calcium needed for shells would have had terrestrial sources. As we learned in Chapter 8, calcium is one of the most important constituents of continental crust. Continental erosion of Late Proterozoic mountain belts may have contributed significant amounts of calcium to the ocean, especially at continental shelves since these are built up from sediments derived from adjacent continental landmasses.

Direct evidence concerning the evolution of the atmosphere during the Early Paleozoic is difficult to obtain from the geologic record. It is generally assumed, however, that the proliferation of photosynthetic organisms would have produced a progressive increase in the quantity of oxygen in the atmosphere during this period.

The composition of the oceans appears to have been influenced by processes similar to those that affect modern ocean water; namely, continental erosion and emanations from mid-ocean ridges (see Chapter 8). Geoscientists track the impact of these two processes by using an isotopic tracer that is sensitive to their relative influences. Isotopic analyses of the element strontium (Sr) in marine sedimentary rocks are particularly suitable for this purpose. Strontium-87 (^{87}Sr) is produced by radioactive decay of rubidium-87 (^{87}Rb) with a half-life of 48.8 billion years.[1] On the other hand, strontium-86 (^{86}Sr) is a stable isotope of

---

[1] Recall from the notation introduced in Chapter 3 that strontium-87 (written shorthand ^{87}Sr) and rubidium-87 (^{87}Rb) have atomic weights of 87 atomic mass units, and strontium-86 (^{86}Sr) has an atomic mass of 86.

strontium and its quantity is constant. As a result, rubidium-rich rocks become progressively enriched in ^{87}Sr relative to ^{86}Sr as time goes by. They therefore have high $^{87}Sr/^{86}Sr$ ratios. Since rubidium is preferentially concentrated in rocks of the continental crust, continental erosion contributes a high $^{87}Sr/^{86}Sr$ ratio to the ocean water in contrast to mantle emanations, which contribute low $^{87}Sr/^{86}Sr$ ratios.

Sedimentary rocks which crystallize from seawater, such as evaporites and carbonates (see Chapter 2), preserve the $^{87}Sr/^{86}Sr$ ratio of the seawater in which they form. Figure 14.5 shows the wide variation in $^{87}Sr/^{86}Sr$ ratios obtained from such rocks at various times in the Phanerozoic. In the Cambrian, the $^{87}Sr/^{86}Sr$ ratio attains values above 0.709, the highest value of the Phanerozoic. This suggests an overwhelmingly continental influence on ocean water chemistry, consistent with the erosion of mountain belts formed during the amalgamation of the Late Proterozoic supercontinent over the previous 100–200 million years.

As we have discussed, however, the breakup and dispersal of the supercontinent in the Cambrian was accompanied by a large rise in sea level (see Figure 14.3b), which would have inhibited continental erosion. This, together with increased mantle emanations associated with the development of mid-ocean ridges caused the $^{87}Sr/^{86}Sr$ ratio of ocean water to drop steadily until the end of the Ordovician. Because they are derived from the mantle, mid-ocean ridge rocks and the submarine vents associated with them (see Chapter 8, Living on Earth: The Bizarre World of the Mid-Oceanic Ridges) have low $^{87}Sr/^{86}Sr$ ratios, and contribute that characteristic to ocean waters.

Subsequent rises and falls in the $^{87}Sr/^{86}Sr$ ratio of ocean water are related to individual continental collisions and the extent of mid-ocean ridge activity. The relatively high values achieved about 300 million years ago, for example, coincide with the uplift of the Appalachian mountains and the formation of Pangea (Fig. 14.3g). The use of $^{87}Sr/^{86}Sr$ ratios as a tracer shows that seawater chemistry was influenced throughout the Phanerozoic by continental erosion and mid-ocean ridge activity, just as it is today, and that the relative contributions of these two processes have changed with time in a way that matches global-scale plate tectonic events.

In summary, the breakup and dispersal of the Late Proterozoic supercontinent exerted a fundamental influence not only on the solid Earth, but also on the biosphere and hydrosphere, providing new ecological niches and essential nutrients for life, and influencing the chemistry of seawater. As we shall find, the influence of moving plates continued throughout the Middle and Late Paleozoic, as the dispersed continents started to approach and finally collide with one another with the amalgamation of Pangea.

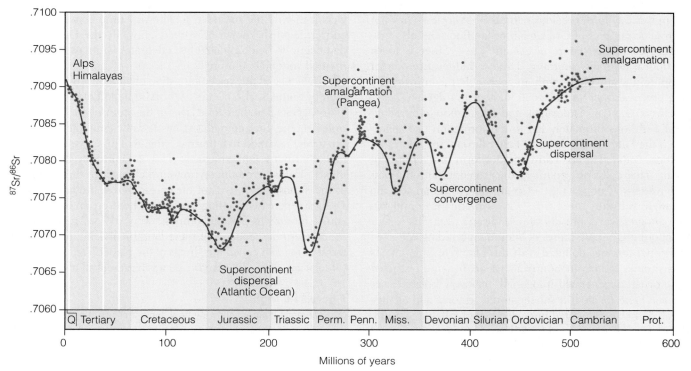

Figure 14.5
Variation in $^{87}Sr/^{86}Sr$ ratios with time throughout the Phanerozoic for marine carbonates. High $^{87}Sr/^{86}Sr$ ratio values reflect times of high input of continental sediments into the marine realm and occur after important episodes of mountain building. Low $^{87}Sr/^{86}Sr$ values reflect times of important mid-ocean ridge activity and reduced continent erosion.

14.2.2 Middle to Late Paleozoic: Middle Ordovician to Permian

By the Middle Ordovician, the distance between the continents had started to decrease, most notably between Baltica and Laurentia. Their convergence implies that subduction had commenced at one or both continental margins (Fig. 14.3c) at a rate exceeding that of sea-floor spreading. As a result, the Iapetus Ocean is likely to have resembled the modern Pacific Ocean during this time period, being fringed by subduction zones and volcanic island arcs.

Subduction along the eastern margin of Laurentia resulted in a number of distinct mountain building episodes that occurred prior to the final climactic collision that assembled Pangea at the end of the Paleozoic. First, the subduction process itself generated mountains, in the same way as it does in today's Andes (see Chapter 6). Second, the volcanic island arcs created by the subduction process were destined to collide with the adjacent continental margin prior to ocean closure.

In eastern Canada, the convergence had an additional twist. Small portions of the ocean floor between Laurentia and the approaching volcanic island arcs were thrust upward and preserved (a process known as **obduction**), rather than being destroyed by subduction (Fig. 14.6). Fragments of this ancient oceanic crust, known as ophiolites (see Chapter 5), are now exposed in southern Quebec and along the spectacularly rugged coast of western Newfoundland. Although obduction is a very rare process, the fragments of ancient oceanic crust preserved in this way form precious natural laboratories. They not only preserve samples of Ordovician ocean floor, but also provide clues about the nature of oceanic crust. Since modern ocean floor is normally submerged and so hidden from direct view, ophiolites provide us with an accessible means of studying the processes that take place at mid-ocean ridges.

Between the Late Ordovician and the Carboniferous, continental collisions ultimately led to the formation of Pangea. Between the Late Ordovician and Late Silurian, the continent of Baltica collided with Laurentia to form the continent Laurasia. This collision produced the Caledonide mountain belt of western Europe (Fig. 14.3c,d). The Silurian to Late Carboniferous history of eastern North America is dominated by the progressive approach of Gondwana (Fig. 14.3e,f,g). Throughout this period, pulses of mountain building occurred as individual terranes collided with the Laurasian margin. As Iapetus shrank, the circulation of its waters became restricted and the ocean degenerated into small, relatively isolated seas. These seas were located in subequatorial regions, and as they evaporated, they precipitated huge deposits of salt and gypsum.

The final event was the collision of Gondwana and Laurasia in the Late Carboniferous (Fig. 14.3g). This massive continental collision was responsible for building much of the Appalachian mountains during a mountain building event generally referred to as the "Alleghanian" orogeny (see Chapter 6). The collision involved folding on a grand scale (Fig. 14.7a) and the thrusting of rock sequences, one atop the other, over great distances. As a result, the continental crust was greatly thickened and Himalayan-style mountains were produced along the entire eastern seaboard of the United States. The overloaded crust behaved like a heavy block on a plastic substrate, generating a great trough known as a **foreland basin** in front of the advancing thrust sheets (Fig. 14.7b). The Appalachian Foreland Basin preserves the products of the erosion of the Appalachian mountains and extends across much of the mid-eastern United States.

In eastern Canada, a different type of basin formed because the region did not experience a "head-on" collision with Gondwana as did the eastern United States. Rather it experienced a series of "glancing blows" as the collision progressed. The contrast between this basin and that of the Appalachian Foreland Basin exemplifies the diverse tectonic expressions that occur in a continental collision zone as a result of the irregular shape of continental margins. A modern analogy can be seen in the collision of India with Asia during the formation of the Himalayan mountains (see Chapter 6). As India advanced on Asia, it dealt several glancing blows to the Indochina peninsula. The effect has been modeled by Molnar and Tapponier (Fig. 14.8) who showed that both squeezing and lateral escape of lithosphere occurs in the jaws of a continental collision, much as the contents of a sandwich tend to slip sideways when the sandwich is being eaten. This "tectonic escape" is performed by motion along major faults. In the northern Appalachians, this process produced the Maritimes Basin (see Figures 14.7b and 14.8) as spaces opened up along the moving faults.

In the biosphere, life went through some major evolutionary changes during the Middle to Late Paleozoic. By the Ordovician, marine life included for the first time corals, bryozoa, and jawless fish known as ostracoderms. By the Devonian, it had begun to be dominated by vertebrates (animals with a segmented vertebral column).

Evidence of the first colonization of land, first by plants and then by animals, also occurs in the Middle to Late Ordovician. This suggests that the ozone layer in the atmosphere had by this time become thick enough to provide sufficient protection from ultraviolet radiation in terrestrial habitats. This evolutionary step had enormous consequences. The colonization of land by plants must have produced profound changes in the chemistry of the atmosphere. As plant life spread across the land, photosynthesis increased, setting the atmosphere on an accelerated course towards its present composition of 21% oxygen. Indeed, the modern descendants of these land plants, the tropical rain forests, are often called the "lungs of the planet." Yet prior to the Ordovician, the Earth's landscape was barren of all but primitive

Figure 14.6 (right)
(a) Schematic diagram showing how ancient slices of oceanic crust, known as ophiolites, became preserved on Laurentia by obduction of back-arc ocean floor. (b) The "stratigraphy" of ophiolite suites preserved on land typically consists of peridotite and gabbros produced by processes occurring beneath mid-ocean ridges; (c) Sheeted dikes produced by the ascent of basaltic magma to the surface;
(d) Pillow lavas produced by extrusion and rapid chilling of lava on the ocean floor; and deep sea sediments deposited on the ridge.
(Photos Hank Williams)

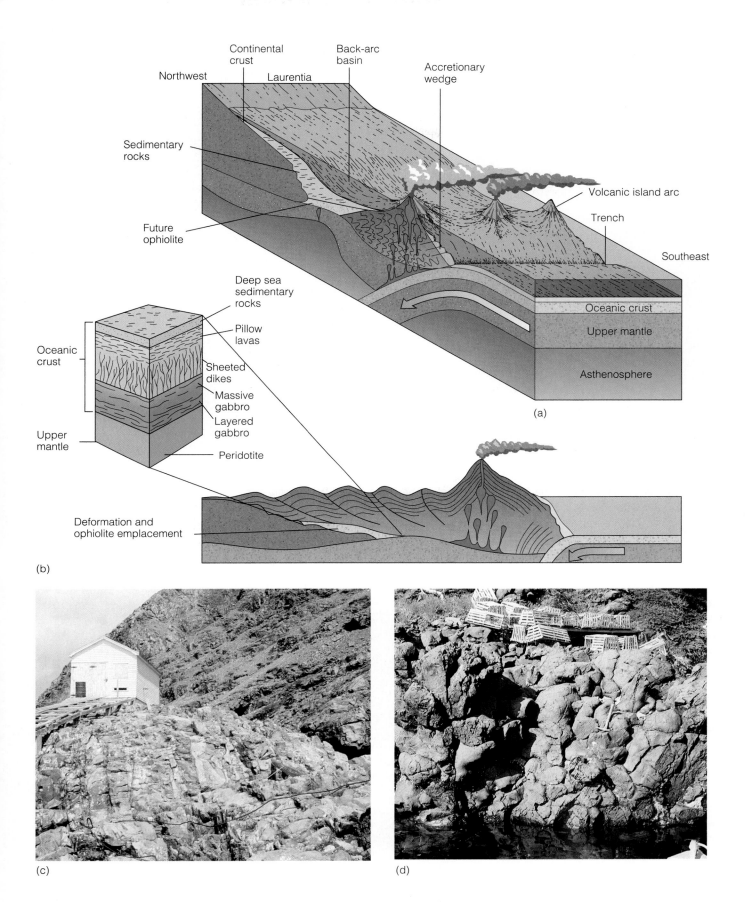

Northwest
Continental crust
Laurentia
Back-arc basin
Accretionary wedge

Sedimentary rocks

Future ophiolite

Volcanic island arc

Trench

Southeast

Oceanic crust

Upper mantle

Asthenosphere

(a)

Deep sea sedimentary rocks

Pillow lavas

Oceanic crust

Sheeted dikes

Massive gabbro

Layered gabbro

Upper mantle

Peridotite

Deformation and ophiolite emplacement

(b)

(c)

(d)

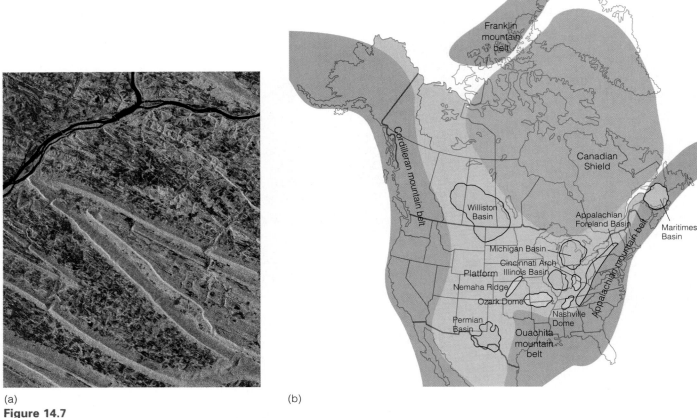

(a)   (b)

Figure 14.7
(a) Satellite image of folded strata in the Appalachian Mountains near Harrisburg, Pennsylvania. These rocks were deformed about 300 to 250 million years ago during the Alleghanian orogeny, as a result of continental collision between Laurasia (North America) and Gondwana during the formation of Pangea. (b) Relationship of the Appalachian mountain belt to the Appalachian Foreland Basin and the Maritimes Basin.

Figure 14.8
Molnar and Tapponier's collision model (Chapter 6) applied to the Appalachian mountain belt. Note that the deformation associated with progressive head-on collision (a to c) in the northeastern United States contrasts with the fault-related deformation in eastern Canada ("escaping" fault blocks moving parallel to the arrows).

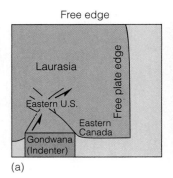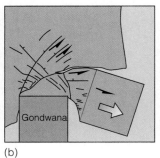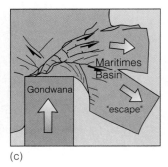

(a)   (b)   (c)

bacterial lichen. The lack of vegetation would have made the continents less able to retain water, so that rates of continental erosion were presumably much faster prior to life's colonization of land than they have been since.

Suspicions that colonization of land had occurred by the Ordovician were raised by experts who claimed that plants and spores in the following period, the Silurian, were too diverse and too evolved to be the first colonials. This argument is similar to that which used the profusion and diversity of Early Cambrian life to argue for an earlier, Late Proterozoic, ancestry. Following an intense search, Jane Grey and her co-workers at the University of Oregon found features that resembled the spores of primitive land plants in Middle to Late Ordovician sedimentary rocks in Libya.

The Devonian Period, which followed the Silurian, was a particularly important period for plant evolution, with the first appearance of seeds and leaves and, by the Late Devonian, the first trees and tree ferns in lowland areas (Fig. 14.9). By the

Figure 14.9
Plants like this modern horsetail (scouring rush) and ferns, resemble those that first evolved in the Late Devonian.

ment is then needed to compress the rotting organic matter into coal (see Chapter 18).

During the Carboniferous, the waxing and waning of the polar glaciers of Pangea caused global sea level to fluctuate, resulting in periodic marine incursions over the continental landmass of what is now North America and Western Europe. As these seas receded, they left swampy conditions where decaying vegetation could accumulate in oxygen-starved environments. Erosion of the elevated Appalachian-Caledonide mountain belt then rapidly buried these accumulations in sedimentary deposits.

The existence of coal beds during the Carboniferous is both a paleoclimatic and paleoenvironmental indicator, providing evidence of hot, humid, swampy conditions, consistent with the presumed subtropical position for Laurentia and Baltica (Europe) at that time (see Figures 14.3f,g and 14.10). The belt of Late Paleozoic coals extends from Alabama through Maritime Canada in North America, but continues into Europe, through Britain, France, Germany, and Poland. Indeed, the continuation of the North American coal belt into Europe was one of the lines of evidence that Wegener used to advocate continental drift (see Chapter 4).

If the land became colonized by plant life in the Ordovician, then animal life must have followed soon after. Burrows preserved in Late Ordovician continental sedimentary rocks (some 440 million years ago) are thought to be the handiwork of these first land animals, which were probably millipedes. Insects, snails, and spiders followed soon thereafter. Fresh water fish first appeared in the Silurian, and the first record of amphibians occurs in Late Devonian rocks of Greenland.

Amphibians became abundant worldwide in the Carboniferous and Permian, although many did not survive beyond the end of the Triassic. Their success was an early example of the food chain operating on continents, since they were attracted to

ensuing Carboniferous Period, continental plant life had become luxuriant and conditions in much of North America and Europe were ideal for the formation of coal deposits. Essential to the process of coal formation is the accumulation of vegetation in an oxygen-starved environment followed by rapid burial so that exposure of the organic matter to an oxidizing atmosphere is prevented. A thick pile of overlying sedi-

Figure 14.10
Reconstructed landscape of life and vegetation in a Late Carboniferous swamp.

The historical importance of "King Coal" cannot be overemphasized—Coal was the fuel of the industrial revolution, providing the energy necessary to power the heavy machinery on which industry depended. Because of their strategic importance and widespread occurrence in western Europe and North America, Carboniferous coal beds were the most intensely studied rocks of the 19ᵗʰ Century. Indeed, the name "Carboniferous" is derived from the abundant carbon in coal, and the Carboniferous Period was the first system of geologic time to be defined.

The strategic importance of coal also led to conflict, and many wars were fought to secure a guaranteed supply of the resource. Prior to the outbreak of the Second World War, for example, Hitler invaded the French coal fields of the Ruhr valley expressly for the purpose of securing fuel for the Nazi war machine.

The proliferation of industry and its dependence on fossil fuels led to the first significant assault by humans on Earth's global ecosystems. The consumption of coal to generate energy produced greenhouse gases that leaked into the atmosphere and may have contributed significantly to the problem of global warming.

It is interesting to note that the main centers of the Industrial Revolution (North America and Europe) owe their origin to the way in which the supercontinent Pangea split apart some 200 million years ago. By geologic accident, most of the coal deposits were largely left on the North American and European plates rather than in South America and Africa.

---

land by the food resource provided by land animals living in forested regions. Amphibians, in turn, provided the evolutionary link to reptiles, which became the dominant land animals of the Mesozoic Era. Reptiles owe their success to the development of the amniotic egg, one in which the embryo develops in a liquid-filled cavity. This development allowed reproduction to take place in a continental environment, rather than the aqueous setting required by amphibians. The oldest known reptiles occur in Late Carboniferous rocks in Nova Scotia, Canada (Fig. 14.11), where they are preserved inside fossil tree trunks. They are thought to have fallen inside hollow tree trunks and became buried by subsequent sedimentation.

In summary, the Middle to Late Paleozoic records the continental convergence and collisions that led to the formation of Pangea. During this period, the biosphere repeatedly adapted to changing environmental conditions. Photosynthesis led to the progressive enrichment of oxygen in the atmosphere, and the further development of an ozone layer. This, in turn, afforded the necessary protection for the colonization of land. By the end of the Paleozoic, continental plant life had proliferated to such an extent that coal beds, derived from compressed organic matter, were being deposited over wide areas.

The end of the Paleozoic Era is marked by the greatest mass extinction event of the Phanerozoic, one which can be truly said to mark the "end of an era." As we shall now see, however, the origin of this mass extinction has been one of the great controversies in paleontology.

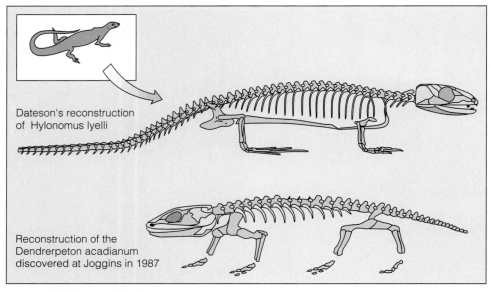

Dateson's reconstruction of Hylonomus lyelli

Reconstruction of the Dendrerpeton acadianum discovered at Joggins in 1987

Figure 14.11
Reconstruction of early reptiles found inside fossil tree trunks preserved in Late Carboniferous rocks at Joggins, Nova Scotia.

14.3 Late Paleozoic Extinctions: The End of an Era

Sudden changes in the Earth's biological population in narrow intervals of geologic time have occurred on several occasions in Earth history and are referred to as "mass extinctions" (Fig. 14.12). An understanding of the processes that can result in mass extinction, as well as those that allow life to flourish, is therefore important if we are to understand the controls on the Earth's ecosystems and the factors that might threaten the survival of different species.

Many hypotheses have been advanced to account for the various mass extinction events evident in the rock record. In all likelihood, each event was the product of a rare combination of circumstances that together resulted in the sudden and simultaneous demise of a wide variety of species. However, paleontologists generally agree that the largest mass extinction by far was the one that brought the Paleozoic Era to a close. This extinction event was not instantaneous. Instead, several relatively minor extinctions occurred during the last stages of the Permian, culminating in an abrupt event at the very end of the period. Together, these extinction events caused the permanent disappearance of about half of the planet's animal life. The marine realm was affected most acutely, with a loss of more than 90% of all species. Some marine invertebrate groups, like the trilobites and certain varieties of coral, were already in decline, having waned in diversity since the Devonian. For others, such as the blastoids, bryozoans, and many families of brachiopods, extinction came more abruptly. Vertebrate life was also profoundly affected, especially the amphibians, which lost some 70% of their species.

This extinction event, which is known as the "Permian crisis," has been documented for more than a century. But its cause is still uncertain. The rock record clearly indicates that the end of the Paleozoic was a time of major global climate change. Areas of luxuriant Carboniferous vegetation, for example, were succeeded by desert-like conditions in the Permian. So habitats clearly changed drastically, and ecologies would have certainly been stressed.

The end of the Paleozoic was also a period of major tectonic change associated with the amalgamation of Pangea. This event is also likely to have brought about environmental stress in a wide variety of ecological niches that may have ultimately triggered widespread extinction. Significant environmental changes, for example, would have resulted from (1) the loss of marine habitats as oceans closed and sea level fell, (2) changes in atmospheric composition associated with widespread volcanic activity and mountain building, and (3) the harsher climates associated with continental glaciation (see Figure 14.3g) and the development of rain shadows in areas adjacent to high mountain belts. Furthermore, with a single landmass and a single ocean, formerly separated species would have been thrown together and forced to compete. Faced with this threat to their survival, some species were successful while others became extinct. The contrasting fate of the amphibians and reptiles is an instructive example of the varying response to this environmental stress. Reptiles were far less affected by the changing environmental conditions and could out-compete the amphibians because, unlike the amphibians, their eggs had shells that could survive far from water. As a result, the reptiles thrived while amphibians went into decline.

As we shall now see, the dawn of the Mesozoic Era was a time of great opportunity for those species that survived the Permian mass extinction. These species took advantage of the new ecological niches available to them, and in so doing profoundly influenced the evolution of the Mesozoic biosphere.

14.4 Earth Systems in the Mesozoic Era (245 to 66 Million Years Ago): The Age of Reptiles

The era of the Mesozoic, meaning "middle life," spans the Triassic, Jurassic, and Cretaceous periods. The dawn of the era was first defined by paleontologists on the basis of biological innovation at the start of the Triassic. The end of the era came with a mass extinction at the close of the Cretaceous. Radiometric dating has since shown the era to be almost 180 million years in duration, spanning the interval from 245 to 66 million years ago. The geologic database for the Mesozoic is far supe-

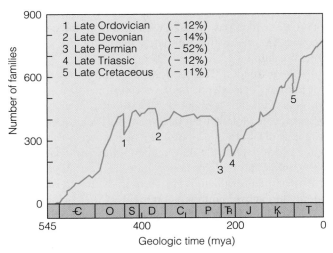

Figure 14.12
Diversity for marine vertebrate and invertebrate families during the Phanerozoic. Symbols for each geologic period are given in Figure 14.1. There have been at least five periods of mass extinctions during this interval (top left), three of them occurring in the Paleozoic. The mass extinction at the end of the Permian (P) was the most profound. About one-half of marine families (representing about 90% of all marine species) became extinct at this time.

rior to that of the Paleozoic. This is because Mesozoic rocks and their fossil record are better preserved, and, more importantly, because many Mesozoic continental shelves and much of its ocean floor remain intact. With this level of preservation, well-constrained plate reconstructions are possible.

14.4.1 Solid Earth

If the Paleozoic Era culminated in the amalgamation of Pangea, the Mesozoic Era documented its breakup and dispersal. On a global scale, the most profound development of the Mesozoic was the birth and progressive opening of many of the world's oceans, most notably the Atlantic Ocean, which continues to this day (Fig. 14.13).

As the Atlantic Ocean grew larger, however, the Pacific Ocean shrank as ocean floor was subducted along much of its perimeter. Thus, the breakup of Pangea set the stage for the major tectonic regimes that dominate the Earth today. The influence of this change in tectonics is well illustrated by its effect on North America (Fig. 14.14). During the Paleozoic, most mountain building in North America occurred along the east coast where it was related to subduction, collision, and the accretion of suspect terranes during closure of the Iapetus Ocean and the formation of Pangea.

With the breakup of Pangea in the Mesozoic, the focus of orogenic activity switched to the western margin of North America. The continental interior became submerged beneath widespread shallow seas floored by continental rather than oceanic crust. The invasion of these seas onto the continent was also related to the breakup of Pangea because the generation of new ocean ridges (such as the Mid-Atlantic Ridge), and the subsidence of continents as they drifted away from the original sites of rifting, combined to produce a massive rise in sea level.

The fragmentation of Pangea took place in several stages, beginning in the Late Triassic (see Figure 14.13). Prior to breaking up, the supercontinent had been in existence for about 100 million years. Eventually, perhaps as a result of the buildup of heat trapped beneath it, the supercontinent became unstable. Several unsuccessful attempts at rupture probably occurred prior to the actual breakup. One of the first symptoms of rupture can be seen in the abundance of igneous activity about 200 million years ago, along much of what is now the eastern seaboard of North America. The initial fragmentation of Pangea was accompanied by huge outpourings of basalt and widespread intrusions of mafic magma, as magma trapped in the upper mantle and lower crust exploited the developing fractures. In North America, one of the most celebrated examples of this activity is the Palisade's Sill[2] of New York State (Fig. 14.15). However, far more voluminous outpouring are preserved in India, southern Africa, and Brazil.

In maps showing the distribution of the continents at this time (see Figure 14.13a), we can see an embryonic Atlantic

Ocean basin opening between North America, Africa, and South America. Subsequent maps show the progressive growth of the Atlantic Ocean (see Figure 14.13b,c).

By the Jurassic Period, Africa and South America had split from North America. At this stage, the embryonic Atlantic was a relatively small, isolated, and subequatorial ocean basin. As we learned in Chapter 2, this environment is conducive to marine evaporation, and significant deposits of marine evaporites were deposited at that time. By the end of the Mesozoic, however, the final disintegration of the supercontinent had been achieved and the continental shelves that currently border the Atlantic Ocean had formed. Most notably, South America had separated from Africa, resulting in the demise of Gondwana, a continental entity that had existed since the Late Proterozoic. Since that time, the Atlantic Ocean has continued to widen by an average of about 5 cm (2 in) a year.

The reasons why Pangea broke up are not clear. One possibility, proposed by Don Anderson of the California Institute of Technology, is that the supercontinent blocked the escape of heat from the Earth's interior. Unlike ocean crust, which allows the Earth's internal heat to escape, continental crust acts as an insulator. So heat would have built up under the supercontinent just as it does beneath a book on an electric blanket. This would have caused the supercontinent to uplift and rupture, and might have ultimately split it apart (see also Section 6.7.2, The Supercontinent Cycle, in Chapter 6).

Another possibility, suggested by Andrew Hynes of McGill University, is that internal strains within the supercontinent tore it apart as it moved northward (see arrows in Figure 14.13a). This mechanism is supported by the location of certain rifts near the sites of Late Paleozoic mountain ranges. As we learned from the concept of isostasy (Chapter 5), high mountains have very deep roots, and these may have acted as anchors as the supercontinent moved northward. By slowing the supercontinent's motion, these roots may have focused stresses on the sites of Late Paleozoic mountains that ultimately led to rupture and continental breakup.

As we learned in Chapter 5, for the Earth to remain a constant size, old ocean floor must be consumed if a new ocean basin is to open. The formation of the Atlantic Ocean, for example, was accompanied by contraction of the Pacific Ocean as the continents of North and South America moved westward (see Figure 14.13c). Thus the development of the rift-related ("passive") continental margins that border the Atlantic Ocean was contemporaneous with an increase in subduction-related activity along the "active" continental margins of the Pacific Ocean.

During the Mesozoic, the western margin of North America was affected by several episodes of terrane accretion in which the floor of the Pacific behaved as a giant conveyor belt moving inexorably from spreading ridge to subduction zone. Like suitcases riding passively on top of the oceanic conveyor belt, each terrane became detached at the continental margin because its elevation clogged the subduction zone. As each was accreted to the continent, the locus of subduction shifted oceanward of the accreted terrane.

As we learned in Chapter 6, more than 200 individual terranes have been identified in western North America. Most were

---

[2] Remember from Chapter 2 that a sill is a slab-like igneous body intruded parallel to the bedding in the adjacent rocks.

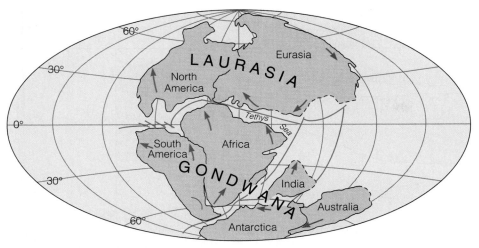

(a) Triassic Period (245–208 m.y.a)

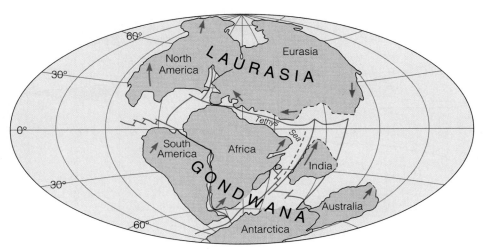

(b) Jurassic Period (208–144 m.y.a)

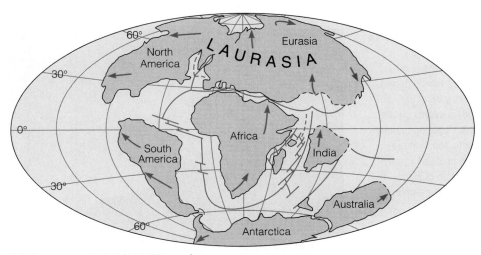

(c) Cretaceous Period (144–66 m.y.a)

Figure 14.13
Mesozoic continental reconstructions for the (a) Triassic, (b) Jurassic, and (c) Cretaceous periods. Laurasia was a continent consisting of present-day North America, Europe, and Asia. Gondwana consisted of present-day South America, Africa, Antarctica, India, and Australia. The breakup of Pangea resulted in the progressive opening of the Atlantic Ocean, and the westward migration of North America. Expansion of the Atlantic Ocean along spreading ridges (the red lines) led to contraction in other oceans, most notably the Pacific Ocean and the Tethys sea (between India, Africa, and Asia). The large arrows represent the average direction of plate motion during the time period represented.

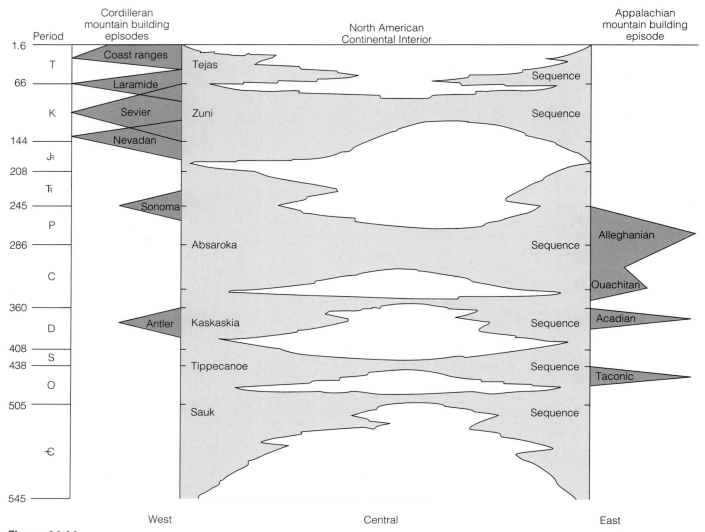

Period	Cordilleran mountain building episodes	North American Continental Interior	Appalachian mountain building episode

Period axis values: 1.6, T, 66, K, 144, J_R, 208, T_R, 245, P, 286, C, 360, D, 408, S, 438, O, 505, €, 545

Cordilleran episodes: Coast ranges, Laramide, Sevier, Nevadan, Sonoma, Antler

Continental Interior sequences: Tejas, Zuni, Absaroka, Kaskaskia, Tippecanoe, Sauk; Sequence (right side)

Appalachian episodes: Alleghanian, Ouachitan, Acadian, Taconic

West — Central — East

Figure 14.14

A summary of the geologic history along the western and eastern margins, and the continental interior of North America during the Phanerozoic. Blue regions represent the time and extent of marine deposition on the continent. The white region shows times where no sediment was deposited. The places where the white and blue regions make contact represent periods either of marine transgression (toward the continental interior) or regression (towards the continental margins). Note that the extent of marine transgression during the Cretaceous (K) indicates that shallow seas overlay all of the continental interior of North America. The major orogenic events on the west coast of North America are shown on the left. Those on the east coast of North America are shown on the right. Note, too, that the breakup of Pangea at the start of the Mesozoic commenced a period of prolonged orogenic activity along the western margin of North America, while the eastern margin remained tectonically passive.

Figure 14.15

Photograph of Palisade's sill, New York state. The intrusion of this sill is related to the earliest stages of separation of Africa from North America.

accreted during the Mesozoic, resulting in the outward growth of the North American continent by about 600 km (360 mi) and the formation of the Cordilleran mountain belt. The effects of these accretionary events were sometimes felt at considerable distances from the continental margin.

One such event is called the Sevier orogeny, and took place during the Cretaceous Period (see Figure 14.14). This was a particularly important episode of mountain building because it helped create an environment in which important reserves of oil and gas were formed. The event is also unusual in that it deformed rocks more than 800 km (500 mi) behind the magmatic arc (Fig. 14.16 a,b). The reasons for such widespread deformation are controversial. However, most geologists attribute the event's far-reaching effects to compression in a region that had been softened by rising magma from the subduction zone.

The deformation consisted of eastward-directed thrusts, which thickened and overloaded the continental crust of western North America. The crust consequently pressed down on the underlying asthenosphere forming a foreland basin, not unlike the one that developed in front of the Appalachians as a result of Alleghanian thrusting (see Figure 14.7b). The moat-like depression that formed in front of the Sevier thrust sheets

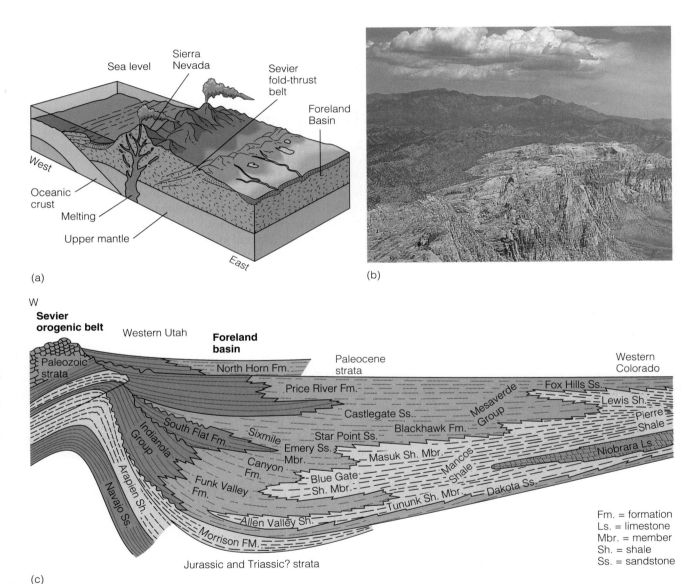

(a)

(b)

(c)

Figure 14.16

(a) Tectonic model for the Cretaceous Sevier orogeny showing subduction beneath the North American plate, and the Sevier fold-thrust belt associated with compression behind the magmatic arc. (b) Photograph (looking north) of a thrust within the Sevier fold-thrust belt marked by the boundary between the dark (left) and light (right) colored rock sequences. The thrust transported the dark colored rocks from west to east (left to right). (c) Schematic section immediately east of the fold-thrust belt showing sedimentary rocks in the foreland basin. Sediment accumulation was thickest in close proximity to the thrust belt. Note that this section is vertically exaggerated for clarity.

developed into a foreland basin east of the thrust belt. Along the western edge of this basin, thick sequences of coarse continental sediments were deposited (Fig. 14.16c). Further east, the supply of detritus from the mountains declined, and the sediments became progressively more fine-grained and even include some limestone layers. Still further east, the sediments were deposited in deltaic environments at the eastern edge of the basin.

The conditions that developed in the basin were ideal for the generation of organic-rich oil and natural gas source rocks, many of which were deposited in Mexico, the United States, and Canada at this time. As sea level rose and fell as a result of global tectonic factors (see Chapter 8), much of continental North America was episodically flooded by extensive shallow inland seas. These subequatorial seas (recorded in the Zuni sequence, see Figure 14.14) developed during periods of high sea level and were home to thriving marine communities. As sea level fell, however, these marine communities were rapidly buried by sediment derived from the erosion of the mountains to the west. Erosion and sedimentation were especially pronounced during the pulses of mountain building in the Cordillera. Rapid sedimentation and burial gave the organic-rich material little opportunity to decompose by reacting with oxygen. Instead, it broke down in a buried, oxygen-starved environment, leading to the formation of hydrocarbon deposits (see Chapter 18).

14.4.2 Biosphere

Life in the Mesozoic Era began with a period of biological innovation in the Triassic, and culminated with a renowned mass extinction at the end of the Cretaceous. The era has been known as the "Age of the Reptiles" since the studies of Swiss naturalist Louis Agassiz more than 150 years ago. As we have learned, reptiles evolved from amphibians in the Late Paleozoic, and out-competed the amphibians in the arid world of the Permian because their eggs had shells and could survive far from water. This allowed the embryo to grow in a protected, self-contained environment. During the Triassic, the reptiles evolved rapidly, giving rise to the dinosaurs and pterosaurs (the first vertebrate animals to fly), and came to dominate the land, the seas, and the sky (Fig. 14.17a).

The origin of birds is still hotly debated. Although many paleontologists believe that birds evolved in the Jurassic from carnivorous dinosaurs, "missing links" that could prove or refute the hypothesis have been elusive in the fossil record. Birds are poorly preserved because they have extremely fragile skeletons. On death, they commonly decompose when exposed to the atmosphere and are also subject to predation. Recently, however, fossil remains discovered in Madagascar show that early birds, about the size of ravens, had large and sharp claws that closely resembled those of the Velociraptor, the killer dinosaur made famous by the movie "Jurassic Park." This find suggests that birds did indeed evolve from dinosaurs during the Jurassic Period.

Without question, dinosaurs were the dominant animals of the Mesozoic Era (Fig. 14.17b). The term "dinosaur" was coined by the English zoologist, Sir Richard Owen, in 1842 and literally means "terrible lizard." In fact, the term is a misnomer because not all dinosaurs were terrible, and none of them were lizards! Dinosaurs included both carnivores (meateaters) and herbivores (plant-eaters), and occupied the most favorable environmental niches. Dinosaurs were also the beneficiaries of an expanding diversity of land plants, including grasses, seed ferns, trees, and the first flowers. Some Mesozoic trees resembled modern conifers while others were flowering plants and bore fruits, nuts, and seeds.

Mammals made their first appearance in the Triassic, but they were small and few and far between (Fig. 14.18). By the Cretaceous, however, both marsupial and placental mammals had evolved. Marsupials, like the modern kangaroo, lack a placenta so their young are born at an immature stage and must develop after birth in a specialized pouch. More advanced Mesozoic placental mammals included hedgehogs, moles, and shrews.

14.4.3 Hydrosphere

In the ocean basins, the fragmentation of Pangea provided new continental shelves and accompanying ecological niches in which marine life could develop. Chemical nutrients derived from the erosion of mountain chains and from coastal marine upwellings provided the opportunity for species that survived the "Permian crisis" to flourish. Rising sea levels associated with supercontinent dispersal in the Jurassic and Cretaceous caused widespread flooding of the continents so that warm shallow seas, ideal for the development of marine life, were commonplace. Subsidence of the dispersing continental fragments occurred as the mantle heat trapped beneath Pangea was dissipated, and the development of new mid-ocean ridges caused ocean water to be displaced upwards. The combination of these factors caused sea level to rise about 100 to 200 m (about 330 to 660 ft) higher than it is today. As a result, by the end of the Cretaceous Period only 15% of the Earth's surface was land, less than half that of today.

Mesozoic seas were rich in new fauna, including new invertebrates such as ammonites, belemnites, and oysters; new groups of corals; and echinoderms and foraminifera, all of which evolved from the survivors of the Permian crisis.

14.4.4 Atmosphere

In general, the climate during the Mesozoic was significantly warmer than it is today. This was especially so in the mid-Cretaceous, about 100 million years ago. The first indications in the rock record of an anomalously warm climate come with the widespread development of evaporite deposits and coral reefs at latitudes about 15° higher than those in which they occur today. Isotopic analyses of deep water sedimentary deposits suggest that water temperatures were about 15° to 20°C warmer than they are today. On land, coal deposits that formed under warm, wet, conditions are also found at higher latitudes (Fig. 14.19). Climate modeling suggests that average surface air temperatures were about 8°C (15°F) warmer than today, and that the temperature contrast between the equator and poles was much lower. This is attributed to an increased

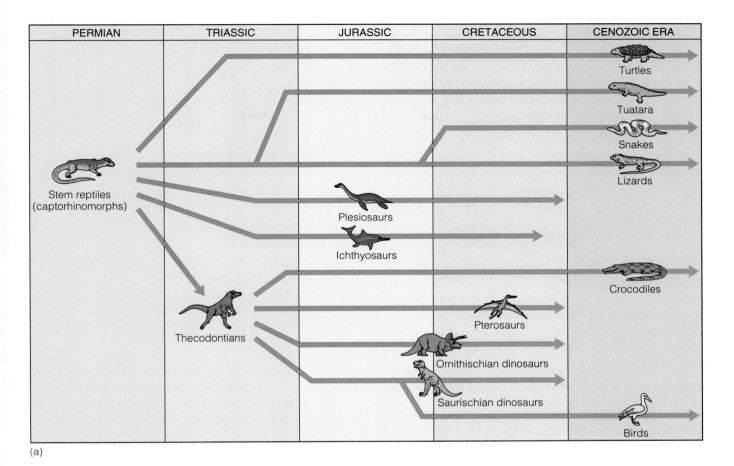

(a)

(b)

Figure 14.17

(a) Evolution of reptiles from the Permian Stem reptile during the Mesozoic. Note that many groups became extinct at the end of the Cretaceous. (b) Herd of large herbivorous dinosaurs under attack by smaller meat-eaters.

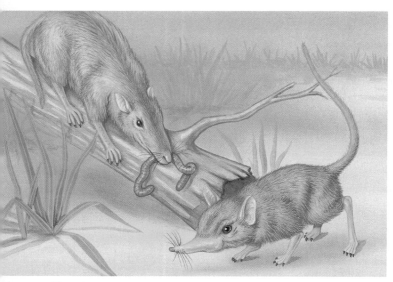

Figure 14.18
An artist's interpretation of some of the oldest known Mesozoic placental mammals.

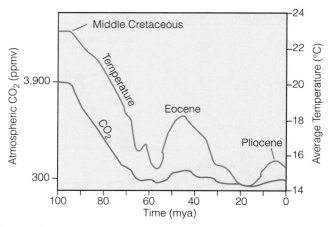

Figure 14.20
The close relationship between atmospheric carbon dioxide concentration (measured in parts per million by volume, ppmv) and average global temperature over the past 100 million years. Note the atmospheric carbon dioxide and average temperature values of the mid-Cretaceous are significantly higher than they are today.

greenhouse effect related to carbon dioxide concentrations in the atmosphere that were significantly higher than those of today (Fig. 14.20).

Why was the Cretaceous climate so warm? The geologic record indicates that, because of the fragmentation of Pangea, there was abundant oceanic and continental volcanism. The degassing of the mantle associated with this volcanism would have resulted in enhanced levels of greenhouse gases such as carbon dioxide in the atmosphere. In a less direct way, the lim-

ited extent of continental erosion was also an important contributing factor to global warming because erosion is the prime source of calcium in ocean waters. Calcium combines with the carbon dioxide in ocean water to form calcium carbonate, or limestone. The ocean water replenishes the carbon dioxide lost in this manner by extracting it from the atmosphere. In Chapter 8, we learned that oceans have a strong affinity for carbon dioxide and can moderate global warming by absorbing the gas from the atmosphere. However, the rise in sea level

Figure 14.19
Mid-Cretaceous landmasses (brown) and submerged continental shelves (light blue). The broad distribution of deposits such as evaporites (E), coal (C), and reef deposits (R) testify to the warm climate associated with this time period. Areas of submerged continental shelf reflect the high sea levels of that time. The dashed outlines show the edges of the continental shelves. Note that the area of exposed continent is significantly less than that of today, further reflecting high sea levels.

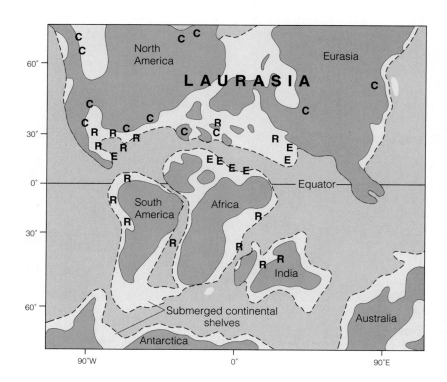

THE TECTONIC EVOLUTION OF WESTERN NORTH AMERICA: THE PAST IS THE KEY TO THE PRESENT

Western North America has been a geologic destination for over 200 million years. Indeed, this region is a "tectonic graveyard," the final resting place for hundreds of terranes that were transported across the Pacific Ocean and accreted to the continental margin. As a result, the continental margin of western North America has been tectonically active for hundreds of millions of years, and much of the modern instability of the region is rooted in this ancient geologic heritage.

Four important pulses of orogenic activity took place along the margin during the Mesozoic, each of which was related to subduction of Pacific Ocean floor and the accretion of Pacific suspect terranes (see Figure 14.14). These four events have combined to produce the mountain chain we call the Cordillera. The first was the Late Permian-Early Triassic Sonoma orogeny, followed by the Late Jurassic-Early Cretaceous Nevadan orogeny, the Middle to Late Cretaceous Sevier orogeny and the Late Cretaceous-Early Tertiary Laramide orogeny. Although differing in detail, each of these events records a pulse of deformation associated with ongoing subduction along the continental margin.

The Nevadan orogeny is an excellent example of the way in which the North American continent has grown as a result of Mesozoic plate tectonic activity. The orogeny was the result of subduction and collision of the continental margin of western North America and a microcontinent (Fig. 4.21). The accretion of this terrane expanded North America westward by about 200 km (125 mi).

Evidence of Late Triassic to Early Jurassic subduction is preserved in the rocks of the Sierra Nevada. These rocks comprise a wide area of Late Triassic to Early Jurassic granites intruded into Paleozoic sediments, and a zone made up of a chaotic mixture of rocks known as a "mélange" (Fig. 14.21a) that includes ophiolites and blueschists (see Chapter 2). Blueschists contain minerals

Figure 14.21
Schematic development of the Nevadan Orogeny from the Late Triassic to the Late Jurassic.

(such as the blue amphibole, glaucophane) that can only be produced under metamorphic conditions of high pressure and low temperature near the site of subduction (see also Chapter 6). The metamorphism is of Early Jurassic age. Taken together, the rocks of the Sierra Nevada record the former existence of a subduction zone of Late Triassic to Late Jurassic age in which chaotic rocks deposited in an ocean trench were metamorphosed as the oceanic crust began its angled descent into the subduction zone, and granitic rocks were formed above descending oceanic slab. Today, the granites form the mountains of the Sierra Nevada while the chaotic rocks make up a unit called the Franciscan complex.

In the Middle to Late Jurassic, the geologic picture gets more complicated with the appearance of volcanic rocks typical of an island arc complex (Fig. 14.21b). This magmatism cannot be related to the easterly subduction zone of the Sierra Nevada because it occurs on the western side of the trench. Instead, the arc is likely to have developed above a westerly dipping subduction zone. The two subduction systems led to the destruction of the intervening oceanic crust and the collision of the volcanic arc complex with the continental margin of western North America. This occurred in the Late Jurassic and marks the peak of the Nevadan orogeny (Fig. 14.21c).

Although it produced an important orogenic event, the collision was too small to affect the ongoing expansion of the Atlantic Ocean. As a result, subduction resumed along the reconfigured continental margin of western North America (Fig. 14.21c). Because of this, the Sevier and Laramide orogenies, while differing in detail, are also dominated by subduction-related processes.

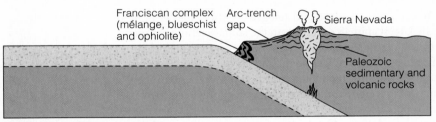

(a) Late Triassic–Early Jurassic

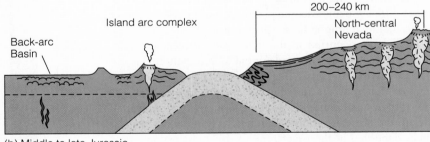

(b) Middle to late Jurassic

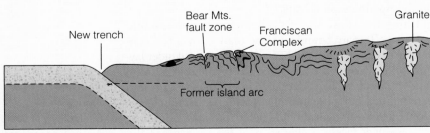

(c) Late Jurassic–Nevadan orogeny

associated with the breakup of Pangea greatly reduced this capability, because far less land was exposed above sea level. As a result, continental erosion was significantly reduced. During times of low continental erosion, the calcium content of the oceans is lower, so that less carbon dioxide is extracted from ocean water to form calcium carbonate. As a result, the extraction of carbon dioxide from the atmosphere by ocean water is also reduced, allowing the greenhouse gases introduced into the atmosphere by volcanic activity to build up.

14.5 The Life and Death of the Dinosaurs

By the start of the Jurassic Period, Pangea had already begun to slowly break apart. The presence of a landmass of supercontinent dimensions meant that climate patterns and barriers to the migration of land-dwelling organisms were very different from those of today. For example, climate patterns would not have shown the tremendous north to south variations that we are familiar with today, so that plants and animals would have occupied much wider regions.

These conditions were ideal for the dinosaurs, allowing them to hold sway as the dominant land animals for more than 140 million years. Once thought to be rather witless creatures, recent research has shown that dinosaurs were intelligent animals that hunted in packs and foraged for their young. Some were the largest animals ever to roam the land, weighing between 70 and 140 metric tons. Others were no bigger than a chicken.

Because of the long period of time during which they dominated life on Earth, their sudden extinction at the end of the Cretaceous (about 66 million years ago) has long been the subject of much investigation and discussion. But they were not alone in their demise. About 11% of marine families, and 75% of all known marine and terrestrial species also became extinct at this time. Since the microscopic plankton and shellfish of the oceans are included among these species, it is probable that a serious breakdown occurred in the food chain. Indeed, most paleontologists believe that no land animal weighing more than 25 kg (55 lb) survived the crisis. Because of this mass extinction, sediments deposited before and after the crisis show radical differences in their fossil content, and the evolutionary torch was passed from the dinosaurs to the mammals of the modern era.

Rocks deposited at the exact time of this mass extinction event have been subject to intense investigation. In 1979, a study in the Apennine mountains of Italy by Walter Alvarez, a geologist from the University of California at Berkeley, found the Cretaceous-Tertiary boundary to be marked by a two centimeter layer of clay sandwiched between younger and older layers of limestone. Both of the limestone layers contained marine fossils and so must have been deposited in an ancient sea before they were uplifted to form part of the Apennines. But the older layers contained fossils typical of life before the mass extinction while the younger layers contained only the survivors of the event. A chemical analysis of the intervening clay layer showed

a very unusual chemistry. Most notably, the clay's concentration of the rare element iridium was 30 times higher than that of ordinary clays, and 160 times higher than that of the adjacent limestone layers (Fig. 14.22). Iridium is an element that is very rare at the Earth's surface (having been concentrated into the Earth's core) but is very abundant in meteorites. So Alvarez and his father Luis interpreted the clay layer as fallout from an asteroid impact (Fig. 14.23). Since the fallout from such an event would have been a global phenomenon, the Alvarezes predicted that similar iridium-rich clays should occur in other areas that expose rocks of the same age. This indeed has proved to be the case, and most geoscientists now believe that a major extraterrestrial body (or bodies) struck the Earth 66 million years ago. The resulting dust cloud would have had the effect of a "nuclear winter," blocking out the Sun's heat and obscuring the sunlight on which photosynthetic organisms at the base of the food chain depend. Such a breakdown of the food chain would have triggered massive extinction of life.

Additional evidence supporting an impact hypothesis takes the form of shocked quartz crystals within the clay layer.

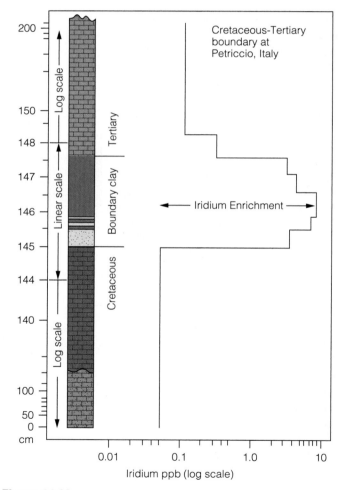

Figure 14.22
Stratigraphy of the Cretaceous-Tertiary boundary at Petriccio, Italy. A clay deposited at the time of the boundary is rich in iridium. This enrichment is attributed to the fallout from an extraterrestrial impact.

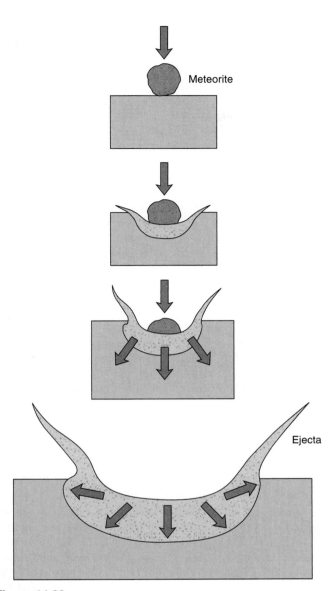

Figure 14.23

Schematic diagram showing the production of impact craters. The energy of the fast moving meteorite is converted into heat energy and compression. Debris (the grey zone) is ejected from the crater and is circulated within the atmosphere such that it is deposited over wide regions. By obscuring sunlight, on which photosynthetic organisms depend, the debris can cause a breakdown in the food chain. The debris also contains fragments which are characteristic of the high pressure of impacts. Reprinted from Donald Gault in J.K. Beatty, et al., in *The New Solar System*. Sky Publishing Co., p. 2, 1981 by permission of the publisher.

These damaged crystal structures can only be produced by intense shockwaves and are thought to be diagnostic of extraterrestrial impacts. Massive chaotic beds of clay and sandstone debris have also been found in 66 million-year-old sedimentary successions, and are thought to have been deposited by tsunami (giant ocean waves) caused by an impact at sea.

Furthermore, evidence of a "smoking gun" has now been found in the form of a crater about 180 km in diameter in the Yucatan Peninsula of Mexico. The crater is no longer visible at the surface because it has since been buried beneath a kilometer of younger sediment. In fact, it has only been detected using sensitive geophysical instruments. The size of the crater suggests that it was produced by a comet or asteroid some 10 to 15 km across. Calculations indicate that the impact would have penetrated the Earth's crust to a depth of 40 km, exploding with a force more than 10,000 times greater than the combined yield of the world's current nuclear arsenal. It is further calculated that close to 2000 cubic kilometers of debris would have been ejected into the atmosphere, creating a blanket of dust and smoke that would have lowered surface temperatures to below freezing for about 200 days.

It should be emphasized, however, that the impact theory for the end-Cretaceous mass extinction also has its critics. Some geoscientists attribute the event, and indeed the iridium anomaly, to an important period of volcanism about 66 million years ago that is spectacularly represented today in the Deccan Traps of India. In addition, paleontologists point out that the extinction of individual species did not all occur simultaneously, but instead took place in a succession of steps, suggesting a buildup of environmental stresses. There was also a marked decrease in average global surface temperatures of about 6°C between the middle and the end of the Cretaceous (see Figure 14.20) that may have caused such stresses. Indeed, it now seems clear that the extinction of the dinosaurs and their contemporaries took place at various times during the Late Cretaceous. Advocates of the impact hypothesis acknowledge this data, but claim that the impact was "the straw that broke the camel's back."

14.6 Earth Systems in the Cenozoic Era (66 Million Years Ago to Recent)

The Cenozoic Era (Cenozoic means "recent life") rose from the massive destruction of life that ended the Mesozoic Era. As we shall see, the Cenozoic Era represents a more or less orderly transition to our modern world. It is subdivided into two periods: the Tertiary, which began 66 million years ago and ended 1.6 million years ago with the beginning of the Ice Age; and the Quaternary, which brings Earth history to the present day and has been responsible for the development of much of our modern landscape.

Much of the tectonic activity of the Cenozoic has been described in Chapters 5 and 6 because our understanding of plate tectonic concepts is based largely on the Cenozoic rock record. In this section, however, we will recap some of the more important events. In the biosphere, modern plants and animals gradually became established in the wake of the mass extinction at the end of the Mesozoic and slowly evolved towards their present form. The evolution of our own species, *Homo sapiens*, is also a Cenozoic development. As we shall

Is the extinction of the dinosaurs the only example of extraterrestrial bombardment? The answer is no. The end-Cretaceous extinction event is only one of a number that have occurred throughout geologic time, and reports of iridium-rich clay layers like that at the Cretaceous-Tertiary boundary have been made in connection with other mass extinctions. Indeed, the timing of extinctions over the past 250 million years appears to show a cycle of 26 to 28 million years. In 1983, a group of scientists proposed that the periodicity may be linked to a regular cycle of extraterrestrial impacts. According to this hypothesis, the Earth comes under an intense barrage of comets or asteroids about every 26 million years. This barrage has been proposed to result from a disturbance in a cloud of comets (known as the "Oort cloud") that encircles the Sun far beyond the orbit of Pluto. It has been suggested that such a disturbance may have been caused by the oscillation of the Solar System above and below the median or galactic plane of the Milky Way. Because our galaxy is densest along this median plane, more frequent extraterrestrial bombardment might be expected as the Solar System passes through it.

Whatever the reason for the apparent cyclicity of mass extinctions, it is becoming increasingly clear that in order to fully understand the Earth, we must look beyond the planet. According to the hypothesis, however, we should not expect the next mass extinction for 10 to 15 million years.

---

see, while *Homo sapiens* is probably less than 200,000 years old, our ancestors, the primate family called *hominids*, have a fossil record dating back at least 4 million years. The evolution of *Homo sapiens*, however, has had an enormous impact on Earth systems, as we shall learn in Chapter 20, because ours is the first species to directly interfere with the fundamental rules by which the Earth systems have evolved over billions of years.

In the atmosphere, photosynthesis continued to be the most important modifying agent, progressively lowering carbon dioxide levels (see Figure 14.20) and increasing those of oxygen and nitrogen toward their modern values.

14.6.1 Solid Earth

On a global scale, the major events of the Cenozoic included the continued expansion of the Atlantic Ocean, the progressive convergence between Africa and Europe to form the Alps, and the collision between India and Asia to form the Himalayas (Fig. 14.24). Late Mesozoic to Early Cenozoic convergence first resulted in continental collision between India and Asia about 40 million years ago. The collision was more or less head-on and has been modeled as an indenter with lateral motion of fault slices away from the site of collision (Fig. 14.25; see also Chapter 6).

Comparison between the rock sequences of the two continents prior to the collisional event shows that abundant magmatic activity took place on the Asian continent, but not in India. The subduction that resulted in the collision must therefore have been directed under the Asian continent.

The warm Asian continental margin was relatively elevated when collision occurred, allowing the colder Indian plate to be pushed beneath Asia long after the ocean between them had disappeared. Indeed, convergence between India and Asia continues today and is responsible for tectonic instability in regions such as Tibet, inland China, and Siberia, that are far removed from the site of mountain building. As India thrusts beneath Asia, it has doubled the thickness of the continental crust, caus-

ing it to rise buoyantly. As a result, the Asian continental crust above the zone of underthrusting has been uplifted to produce the broad, elevated, seismically-prone Tibetan Plateau.

At the same time, the escape of fault slices of continental crust from the jaws of the collision has been a primary influence on the tectonic evolution of the Indochina peninsula. Many of the fault zones initiated by this process are still active and guide vast drainage systems that transport sediment from the Himalayas and the Tibetan Plateau to the Indian and Pacific oceans. The sediment is then deposited in deltas where the streams meet the oceans (see Chapter 10).

During the collision that formed the Alpine mountain chain, the Tethys Sea succumbed to subduction, leaving only the narrow, relatively isolated, sea we call the Mediterranean (Fig. 14.24). About 6 million years ago, the northward movement of Africa sealed off the Strait of Gibraltar, restricting the flow of ocean water between the Mediterranean Sea and the Atlantic Ocean. Even the modern Mediterranean evaporates faster than it is replenished by streams, and so must be resupplied with water from the Atlantic. Without this unhindered connection, the Mediterranean Sea virtually dried up 6 million years ago, and evaporite beds 3 km (2 mi) thick were deposited on the basin floor. This enormous deposit effectively lowered the salt content in the world's oceans, allowing seawater to freeze at higher temperatures. As a result, the polar ice sheets grew and sea level fell. About 500,000 years later, the barrier between the Mediterranean Sea and the Atlantic Ocean was breached, and open circulation was restored. In a gigantic waterfall, the Atlantic waters poured into the Mediterranean, filling the basin in about 100 years, and lowering global sea level by about 10 m (33 ft).

In contrast to the belts of continental collision in Europe and Asia, Early Cenozoic mountain building on the eastern Pacific margin is dominated by easterly-directed subduction. In South America, this resulted in the continued growth of the Andes Mountains (Figure 14.26a, see also Chapter 6). Along the western margin of North America, the Late Cretaceous–Early

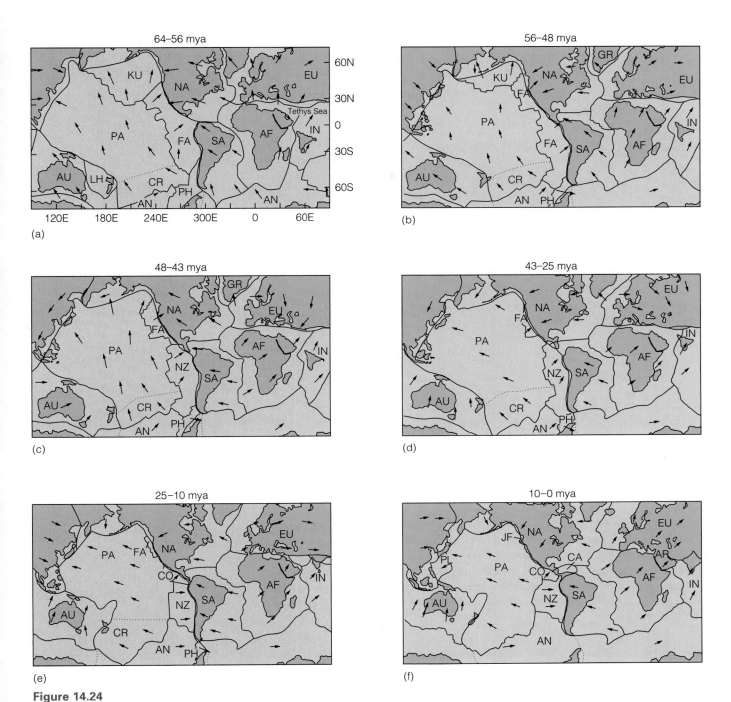

Figure 14.24
Plate reconstructions for the Cenozoic. Continued expansion of the Atlantic Ocean occurred at the expense of the western Pacific Ocean. Northward movement of the African and Indian plates resulted in collisions that formed the Alps and Himalayas, respectively. The arrows indicate the direction of plate motion. The abbreviations for plates are: AF, African; AN, Antarctican; AR, Arabian; AU, Australian; CA, Caribbean; CO, Cocos; CR, Chatham Rise; EU, Eurasian; FA, Farallon; GR, Greenland; IN, Indian; JF, Juan de Fuca; KU, Kula; LH, Lord Howe; NA, North American; NZ, Nazca; PA, Pacific; PL, Philippine; PH, Phoenix; SA, South American.

Tertiary Laramide orogeny was the predominant Cenozoic mountain building event. Although it too was a subduction-related event, the style of this orogeny differed somewhat from both the Andes and the Mesozoic orogenic events that preceded it. Much less magmatism was associated with the Laramide orogeny, and deformation was far more widespread than is typical of subduction-related deformation, extending into the continental interior by as much as 1000 km (620 mi). What could have changed to make this happen? Most geoscientists attribute the anomalous character of the Laramide event to what is

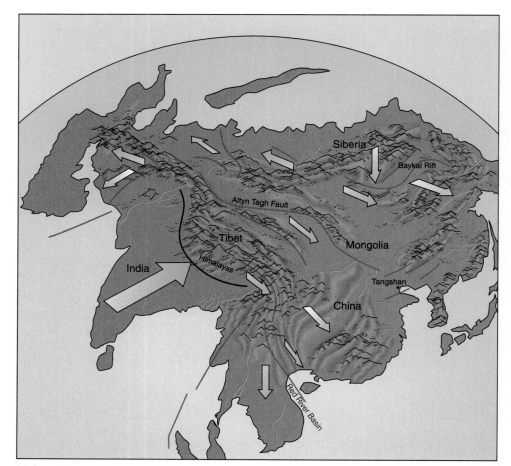

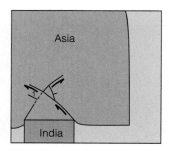

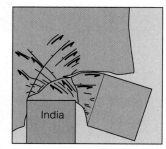

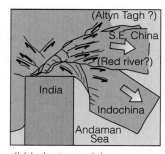

(a) Collision between India and Asia

(b) Indenter model

Figure 14.25

(a) The continent-continent collision between India and Asia that formed the Himalayas records the destruction of the Tethys Sea. The main frontal thrust system has formed where head-on collision occurs. Material is directed away from the collision zone along major faults (the small arrows). To the east, in the Indochina peninsula, the collision has produced predominantly strike-slip fault motion. (b) Indenter model for the Himalayas with India (representing the indenter) penetrating into the Asian continental interior causing present-day southern China and Indochina to "escape" along faults to the east.

known as "flat slab" subduction. In this extreme form of Andean-type subduction (see Chapter 5), the subducting slab becomes essentially subhorizontal and hugs the base of the lithosphere. As a result, the subduction zone does not penetrate to the depths necessary to initiate abundant melting, but instead extends far inland beneath the continental interior (Fig. 14.26b). Deformation associated with Laramide subduction consequently extended much farther from the continental margin than it had during earlier subduction-related orogenies.

Later in the Cenozoic, the contrasts between the Pacific and Atlantic oceans became more profound. This is most evident if the modern positions of their respective mid-ocean ridges are compared (see Figure 14.24f). The Atlantic Ocean is relatively symmetric with a centrally located Mid-Atlantic Ridge and well-developed continental shelves on its opposing margins. In the Pacific Ocean, on the other hand, the spreading ridge (known as the East Pacific Rise) lies on the eastern

side, and in areas like California, it actually intersects the coastline at a high angle (Fig. 14.27). This relationship is the key to understanding the origin of the San Andreas Fault system. As the Atlantic Ocean widened, the North American continent drifted westward, and the eastern half of the Pacific Ocean was consumed beneath it by subduction. As a result, the position of the East Pacific Rise became increasingly asymmetric. About 30 million years ago, North America collided with the East Pacific Rise (see Figure 14.24d), subduction stopped in that area, and the San Andreas fault system was born. The fault system then spread along the continental margin (Fig. 14.27a), and now forms the boundary between the North American and Pacific plates. Although motion along the San Andreas Fault is compatible with the northwesterly movement of the Pacific plate, development of the fault system terminated subduction along this segment of the continental margin.

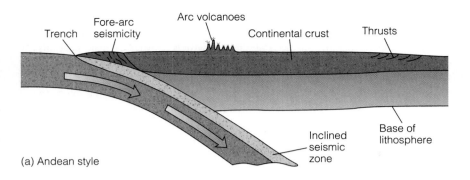

(a) Andean style

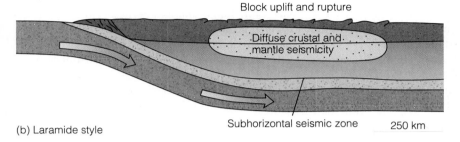

(b) Laramide style

Figure 14.26
Variations in the dip of the subduction zone associated with ocean-continent convergence. (a) Most of the Andean margin has an inclined subduction zone. (b) The subduction zone thought to be responsible for the Laramide orogeny is relatively shallow, becoming subhorizontal beneath the continental plate.

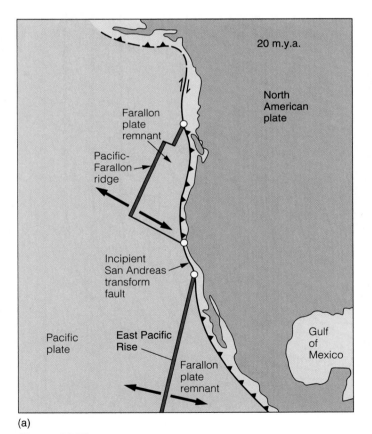

(a)

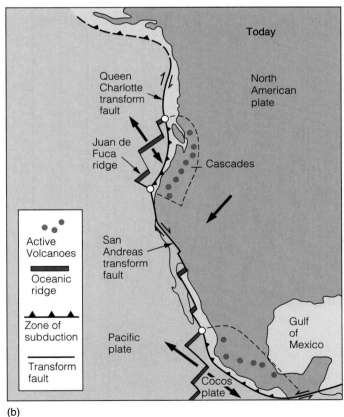

(b)

Figure 14.27
Schematic model depicting the evolution of the San Andreas Fault system over the past 20 million years. The fault system originated 30 million years ago when the Pacific-Farallon ridge system collided with the continental margin. This collision is related to the westward migration of North America since the breakup of Pangea, which has resulted in the expansion of the Atlantic Ocean and contraction of the Pacific Ocean (see Figure 14.24).

With the continued westward motion of North America, the San Andreas Fault spread to the northwest (Fig. 14.27b), eventually linking by way of an oceanic transform fault to the Juan de Fuca Ridge of the Pacific northwest. Note that subduction continues to the north of this oceanic transform fault, producing the modern volcanic activity of the Cascade Ranges.

14.6.2 Biosphere

As the name Cenozoic ("recent life") implies, many of the fossils in Cenozoic rocks are instantly recognizable and compare closely with their modern relatives. The Cenozoic Era is often referred to as the "Age of Mammals." Although mammals originated at about the same time as the dinosaurs in the Triassic Period, they were unable to compete with dinosaurs and remained small and relatively inconspicuous throughout the Mesozoic. The fossil record indicates that the largest Mesozoic mammal was probably about the size of a domestic cat. All that changed with the mass extinction at the end of the Mesozoic.

With dinosaurs out of the way, the mammals were set to take over and a major radiation event among placental mammals began in the Paleocene, which is the earliest part of the Tertiary Period (Fig. 14.28). But, as we shall see, if the mass extinction was a result of an extraterrestrial impact, then, the present dominance of mammals and, indeed, the existence of human life has involved more than an element of chance.

Many Mesozoic mammals were unable to compete successfully and became extinct. But, as we learned in the previous section, two groups of mammals, *marsupials*, (like the modern kangaroo) and *placental mammals* (such as hedgehogs, moles, and shrews) evolved in the Cretaceous. These mammals survived the mass extinction at the end of the Mesozoic and successfully exploited ecological niches vacated by the dinosaurs. Placental mammals develop inside the mother's uterus and are nurtured by the mother's body fluids, whereas marsupials develop in an external pouch. Because of their mode of reproduction, marsupials were the less successful of the two groups, and came to represent only 5% of Cenozoic life.

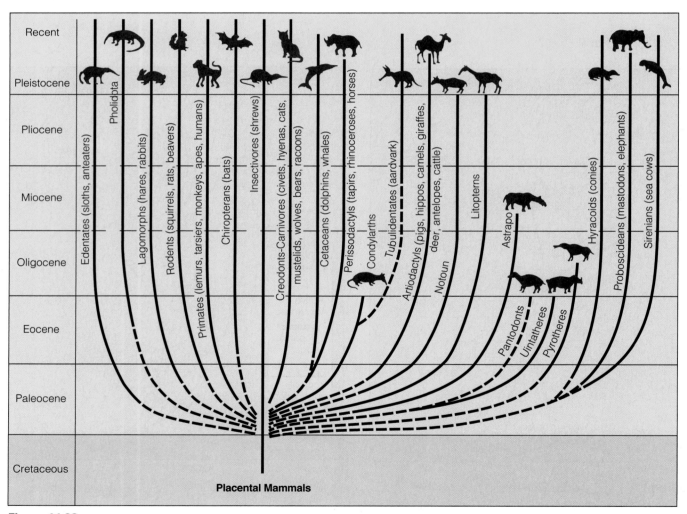

Figure 14.28
The mass extinction at the end of the Cretaceous provided the opportunity for radiation of the placental mammals, beginning in the Early Tertiary (Paleocene). The ancestry of modern placental mammals can be traced back to this radiation event.

The distribution of the Earth's tectonic plates also played an important role in the Cenozoic evolution of the biosphere. The continued breakup of Pangea caused the continents to become disconnected, in contrast to their juxtaposition throughout most of the Mesozoic. As a result, no one group of land animals gained worldwide dominance to the extent that dinosaurs had in the Mesozoic because they could not easily migrate from one continent to another. Marsupials became dominant on the con-

tinents of Australia and South America, which were isolated throughout most of the Cenozoic. At the same time, placental mammals gained dominance in North America, Europe, Asia, and Africa.

An important relationship between tectonic activity and evolution also occurs when formerly separated continents become connected, as demonstrated by the Cenozoic evolution of central America and the Caribbean (Fig. 14.29). In the Early Cenozoic

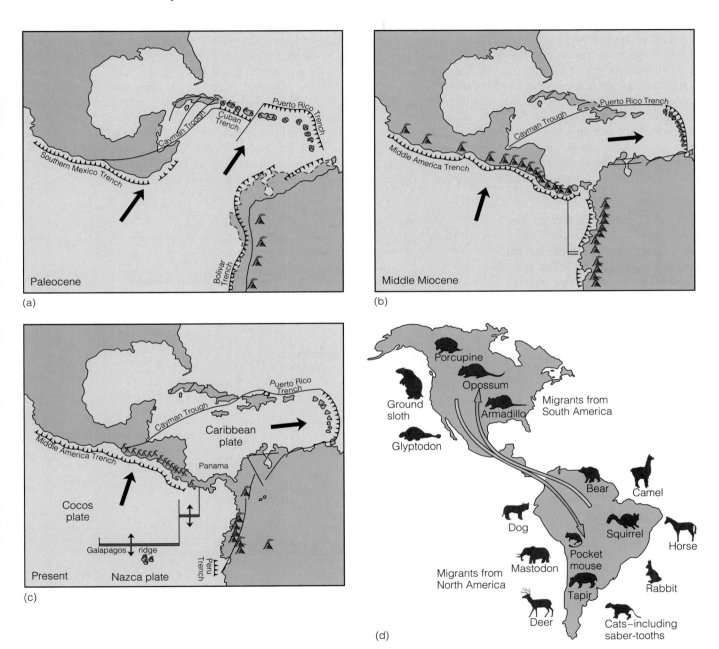

Figure 14.29
(a–c) Origin of the land bridge between North America and South America along the isthmus of Panama. The land bridge was formed 3 million years ago by a volcanic arc system (the red triangles) that developed along the isthmus. The black arrows show the direction of plate motion. (d) Migrations between North and South America of marsupial and placental mammals occurred when the isthmus of Panama joined North America and South America. Prior to that time, North America and South America were separated, and their respective fauna evolved in isolation.

(Paleocene), an essentially continuous subduction zone from North to South America was broken near the Caribbean, such that the Cuban and Puerto Rico trenches were well separated from those of southern Mexico and Peru. By the Middle to Late Tertiary (Middle Miocene), however, these trenches had merged, and subduction associated with the extended trench system resulted in the buildup of volcanic mountains. Three million years ago, this buildup formed the isthmus of Panama, creating a land bridge between North and South America. Prior to this event, North and South America had been isolated since the breakup of Pangea, and had become home to numerous examples of indigenous fauna that had evolved quite independently. The development of a land bridge facilitated the southward migration of placental mammals from North America to South America, resulting in widespread extinctions among the marsupials of South America. A few South American mammals, such as armadillos and opossums, successfully migrated in the opposite direction and are now found in North America.

The most significant Cenozoic development in the biosphere has been the dominance of our own species, *Homo sapiens*. Like all mammals, the characteristics of our species were developed throughout the Cenozoic. **Primates** are mammals characterized by a large brain, stereoscopic vision, and a grasping hand. **Hominids,** the primate family to which *Homo sapiens* belongs, are omnivores and are bipedal, that is, they have an upright posture and walk erect. They also have large, internally organized brains, show evidence of increasing dexterity, and have the ability to use sophisticated tools. At present, the oldest fossil remains of hominids belong to the genus *Australopithecines*. Discovered in Kenya in 1995 by Maeve Leakey and her colleagues, the remains are 4.1 million years old. Some paleontologists believe the genus *Australopithecines* is the ancestor of the genus *Homo* (Fig. 14.30a), and ultimately modern humans. According to this scheme, the oldest known species of *Australopithecines*, *A. anamensis*, gave way to *A. afarensis*, one branch of which evolved to become *Homo habilis*, the oldest species of the genus *Homo*. Other paleontologists believe that *Australopithecines* evolved independently from the human lineage (Fig. 14.30b), but that they shared a common ancestor that is, as yet, unidentified.

Figure 14.30
Two evolutionary schemes proposed for the ancestry of *Homo sapiens*. (a) Donald Johnson and Tim White propose that *Homo sapiens* evolved from a known ancestor *Australopithecines anamensis*. (b) Another scheme, proposed by Richard Leakey and his colleagues suggests that humans and *Australopithecines* evolved independently from a common ancestor that has yet to be identified.

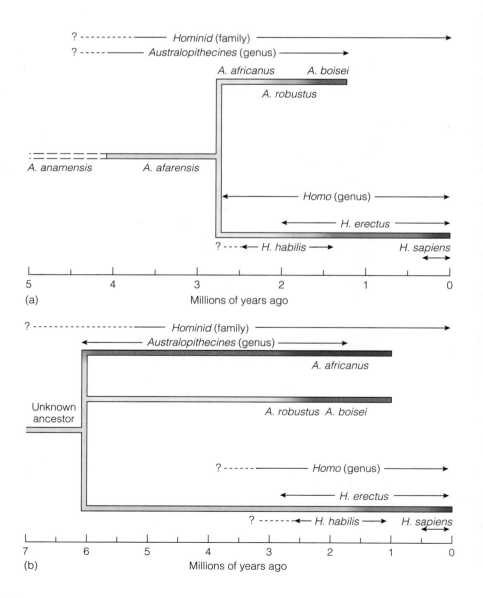

About 2.4 million years ago, *Homo habilis* arrived in Africa, and this species evolved first to *Homo erectus* and then to modern humans or *Homo sapiens*. One of the most important evolutionary trends in the hominids was that of increasing brain size. Prior to about 2.75 million years ago, hominids had a brain size of 380 to 450 cc (cubic cm). *Homo habilis*, which existed from 2.4 to 1.4 million years ago, had a brain size of about 700 cc. *Homo erectus*, which appeared about 1.8 million years ago and became extinct 100,000 to 200,000 years ago, had a brain size varying from 700 to 1300 cc. The average brain size of *Homo sapiens* is about 1350 cc.

There is a general consensus among anthropologists that *Homo erectus* migrated from Africa and spread throughout Eurasia between 1 and 2 million years ago, and that *Homo sapiens* evolved from *Homo erectus* between 200,000 and 100,000 years ago. How this evolution occurred, however, is the subject of heated controversy. One theory proposes that *Homo sapiens* evolved from the population that remained in Africa, and then spread throughout Eurasia, driving *Homo erectus* into extinction. A second theory suggests that *Homo sapiens* evolved from *Homo erectus* independently in a number of small, isolated communities in Eurasia.

14.6.3 Atmosphere

The climate in the Early Cenozoic was relatively warm (see Figures 14.20). Oxygen isotope data from deep sea sediments, for example, suggest that temperatures were about 3°C warmer than those of today. The ratio between the two isotopes of oxygen, oxygen-18 and oxygen-16 (expressed as $^{18}O/^{16}O$), is very sensitive to average seawater temperatures. Since ^{18}O is the heavier isotope, it is harder to evaporate than the lighter isotope, ^{16}O. However, at cold temperatures, the difference in evaporation of the heavy and light isotopes is especially pronounced. As a consequence, ^{18}O becomes enriched in colder ocean water relative to ^{16}O. Evidence of past climate is preserved in the shells of planktonic foraminifera, which record the isotopic ratio of the seawater in which they lived because their shells grow by absorbing water and nutrients from their surroundings. Thus, changes in the oxygen isotope ratios in foraminifera shells from one sedimentary layer to the next, reflect changes in ocean water temperature. The data indicate that a prolonged period of cooling occurred sometime between 58 and 35 million years ago (Fig. 14.31), leading to the formation of glaciers in the Antarctic. As we learned in

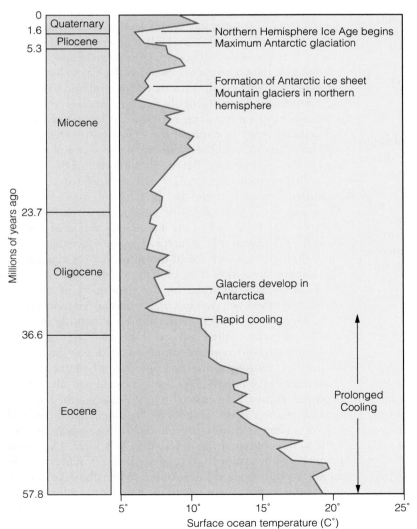

Figure 14.31

Surface ocean temperatures for the past 58 million years based on oxygen isotope data from foraminifera shells in sediment cores from the Pacific Ocean. Note the period of prolonged cooling preceding the formation of the Antarctic glaciers about 35 million years ago, the onset of mountain glaciation in the northern hemisphere about 10 million years ago, and the recent Ice Age in the northern hemisphere about 1.6 million years ago.

Chapter 8, cooling in the mid-Cenozoic was probably accompanied by significant changes in oceanic circulation patterns which are driven by the strong temperature contrast between warm equatorial and cold polar waters.

Compared to this dramatic cooling event, however, global temperature has fluctuated little in the past 35 million years (Fig. 14.31). Yet there is abundant evidence of the repeated advance and retreat of continental glaciers, particularly during the Quaternary Period of the last 1.6 million years, as we shall now discover.

14.7 The Ice Age: Sculpturing the Modern Landscape

The Ice Age, as the most recent episode of continental glaciation is commonly known, took place in the most recent geologic Period, the Quaternary. This period began 1.6 million years ago and continues to the present day. The Quaternary preserves evidence of many phases of ice sheet advance (known as a **glacial stage**) and retreat (known as an **interglacial stage**), during which much of our modern landscape was sculptured. In North America, the distribution of ice sheets in the Quaternary directly affected the landscape in northerly latitudes and in mountainous regions. However, the waxing and waning of ice sheets dramatically affected drainage patterns, which, as we learned in Chapter 10, also have a profound influence on our modern landscape.

The glaciers took advantage of weak rocks or ancient fault zones to scour out deep valleys. More resistant crystalline rocks, on the other hand, were often left as mountain peaks between adjacent glacial valleys. Following retreat of the ice, these valleys had a major influence on drainage patterns, and these, in turn, fostered the development of major population centers in recent times. Unconsolidated sediments deposited during interglacial stages store vast quantities of fresh water in the ground (see Chapter 10), and their varying ability to support agricultural activity has also had a profound influence on human development.

14.7.1 Glacial Features

In Chapter 3, we learned that modern geology owes its origin to such simple questions as "how old are the hills?" We now know that "the hills," and indeed most features of the landscape around us, were formed during the Quaternary Period, and, in terms of geologic time, are very young indeed. In contrast, the rocks that underlie the hills can be millions or even billions of years old.

This is not to say that landscapes did not exist in the past. Ever since continental vegetation first began to flourish toward the end of the Paleozoic Era, the processes governing the shape of the land have probably been similar to those of today. However, the relentless grinding of the rock cycle through geologic time has obliterated most features of these ancient landscapes. Only the more dramatic ancient landforms persist, and then only as pale shadows of their former selves. Some three hundred million years ago, for example, continental collision uplifted the Appalachian mountain chain to heights that would have rivaled the mighty Himalayas. The ravages of time have since reduced its lofty peaks to a series of relatively subdued hills and valleys. Some of the more rugged parts of the Appalachian chain in New England and Atlantic Canada have been further sculpted over the past million and a half years by the scouring action of ice sheets, which preferentially exploited their structural weaknesses.

Evidence of Quaternary glacial activity is widespread in much of North America because of the bulldozing effect of advancing ice, which pushes debris to the front and sides of a glacier, resulting in deposits called **moraines** (Fig. 14.32a). As a glacier grinds over bedrock, it also leaves scratch marks, or **striations** (Fig. 14.32b), which reveal the ice sheet's direction of motion when examined on a regional scale.

Glaciers do not really "retreat," rather, the ice at the front of the glacier simply melts away. In the same way, individual hairs of a receding hairline do not move backwards, but rather, the hair merely falls out! When glaciers recede, debris is deposited in a chaotic mixture of boulders, sand, and clay called **till** (see Figure 14.32a). Unlike other agents of erosion that tend to sort sediment according to its grain size, glaciers can produce deposits in which boulders the size of houses are separated, or "supported," by a matrix of sand and clay. This lack of sorting is diagnostic of tills and there are few other environments in which very large boulders can be transported significant distances. In contrast, the debris transported by ice may have traveled a sufficient distance that it is completely foreign to the local geology. Boulders transported in this fashion are called **erratics** (Fig. 14.32c) and their source can yield important information on the direction of ice movement. Many erratics on the plains of the American midwest, for example, can be traced to the Laurentian shield of eastern Canada. Till may also be sculptured by the movement of ice into egg-shaped mounds called **drumlins** that are elongate in the direction of ice flow (Fig. 14.32d).

Glacial lakes dammed by the deposition of till may record important information about annual climate conditions. Deposition in such lakes commonly consists of laminated sediments called **varves** (Fig. 14.33a). The fine laminations record seasonal variations in deposition over a single year. During spring and summer when the lake is thawed, drainage into the lake brings in sand and silt that are deposited in thin, light colored layers. But when the lake surface freezes over in the winter, only the fine-grained clay and organic matter that were previously held in suspension can settle to the bottom, giving a very thin dark layer. Thus each couplet of alternating light and dark layers represents an annual record of glacial deposition. When the lake surface thaws, any debris caught in the ice is suddenly released and plunges to the lake bed where it disrupts the varve layering.

(a)

(b)

(c)

(d)

Figure 14.32
Evidence of continental glaciation takes the form of (a) moraine made up of glacial debris (called till) deposited at the edge of glaciers, (b) striations or scratch marks on exposed bedrock that reveal the movement direction of the ice, (c) erratics (large, glacially transported boulders), and (d) drumlins (gradually sculptured moraines).

This style of deposition is also unique to glacial environments and the debris so released are called **dropstones**.

A record of the local vegetation may also be preserved in these lakes in the form of wind-blown pollen (Fig. 14.33b) that settles onto the lake water and is deposited with the sediment layers. Since the type of vegetation and its pollen are climate-sensitive, detailed studies of pollen distribution provide valuable information on paleoclimate.

The marine record of glaciation is more subtle because oceanic sediments lie well below sea level. As ice sheets melt, however, icebergs become detached from the main ice mass and may drift thousands of kilometers from the parent ice sheet. When they melt, they drop their load of rock debris, which plunges to the seabed to produce dropstones in the otherwise undisturbed sediment. Thus, the presence of dropstones in oceanic sediments provides evidence of contemporary glaciation, although the ice sheets themselves may be far removed from the site of deposition.

14.7.2 Extent of the Ice Sheets

One of the most impressive features of the Quaternary was the maximum extent of ice on the continental landmasses of North America and Europe (Fig. 14.34). The direction of ice flow has been determined from features such as striations, the distribution of erratics, and the orientation of drumlins. However, it is clear that ice sheets have waxed and waned several times during the Quaternary. Many areas contain several individual layers of till that are separated from one another by thin organic-rich soils. Each till was deposited during the retreat of an individual ice sheet while soils formed during the relatively warm conditions of the interglacial stages. Using the principle of Superposition (see Chapter 3), a sequence of glacial and interglacial events can be established. At least four distinct glacial episodes are recognized in North America, and six or seven episodes are recognized in Europe.

(a)

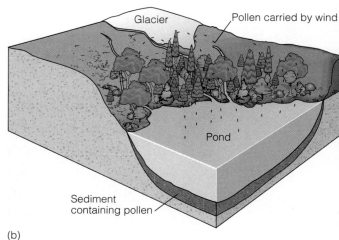

(b)

Figure 14.33
(a) Varves are thinly laminated sediments that represent seasonal changes in deposition over a single year. Dropstones (see large pale rock fragment) disrupt the varve layering and provide evidence of floating ice. (b) Pollen deposition in glacial lakes preserves a record of the local vegetation and can be used as a climate indicator.

At the height of the glaciation, more than 70 million cubic km (17 million cubic mi) of ice and snow covered the North American continent. This would have lowered sea level by about 100 m (330 ft), exposing the present-day continental shelves. As we learned in Chapter 8, should the remaining polar ice sheets melt, sea level would rise by another 70 m (230 ft), flooding much of the coastal regions of North America (Fig. 14.35).

But to determine the volume of ice and, hence, the effect of glaciation on sea level, we need to know how thick the ice

sheets were. Although some indication is provided by the glacial deposits, there are other, more subtle ways of deriving an estimate. When ice advances, it loads the continental crust beneath it. Because the continents ride on plates above a pliable asthenosphere, this loading depresses the continental crust, just as a boat floats lower in the water when it is laden. This conforms to the principle of isostasy, as outlined in Chapter 5.

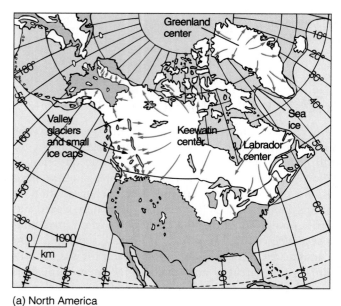

(a) North America

(b) Europe

Figure 14.34
Extent of Quaternary ice accumulation in (a) North America and (b) Europe. The arrows indicate the direction of ice flow determined from field investigations. Note how the arrows radiate from the centers of ice accumulation.

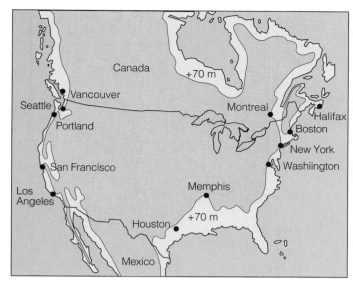

Figure 14.35
Map of North America showing the expected rise of sea level if the polar ice sheets were to melt. The +70m contour shows the position of the coastline in the event of such melting, and the threat to major coastal cities.

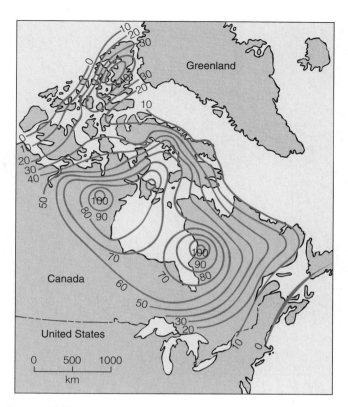

Figure 14.36
Isostatic rebound of the crust in eastern Canada following the last retreat of ice, some 10,000 years ago. The amount of uplift is given in meters over the past 6000 years. This amount is greatest where the ice sheet was thickest, and serves to identify centers of ice accumulation.

During peak glaciation, the crust beneath the ice sheets is estimated to have been depressed by as much as 300 m (1000 ft). When the ice melted and the load was removed, the crust started to rebound, in the same way that a boat bobs up when its cargo is off-loaded. Although the process is a slow one, the greater the load, the faster the rebound. Figure 14.36 shows the amount of isostatic rebound in eastern Canada over the past 6000 years from which the thickness of ice can be calculated. The greatest rebound occurs where the ice sheets were thickest, helping to identify former centers of accumulation. Isostatic rebound associated with the last glacial retreat continues to this day.

In eastern North America, however, the crust is flawed because it contains many ancient faults inherited from the formation of the Appalachian mountain belt. So the crust forms several fault-bounded rafts that rebound at different rates. As a result, the stresses near these ancient fault boundaries may build up to the point that their release generates mild earthquakes (generally less than magnitude 5 on the Richter scale). The distribution and style of these earthquakes also yields information on the extent of the former ice sheet. A record of recent seismic activity in eastern North America shows a profusion of mild earthquake activity (Fig. 14.37).

Glacial advances and retreats during the Quaternary profoundly affected sea level, the most dramatic manifestation of which is in the formation of coastal features such as **fjords** (also spelled fiords). Fjords are narrow submarine depressions that were once glaciated valleys and are presently submerged beneath the sea (Fig. 14.38). They are most common along the high latitude coasts of Canada, Greenland and Scandinavia. The depressions were formed by glaciers at a time when the valley bottom was above sea level. However, sea level rise associated with the melting of continental ice sheets over the past 10,000 years has resulted in the drowning of these valleys.

On the other hand, features known as **raised beaches** provide evidence of a drop in sea level (Fig. 14.39). These elevated regions are small plateaus that are covered with wave-washed sediment typical of modern beaches. Their present elevation, however, indicates that they have been raised from a past position at sea level. Together, the existence of fjords and raised beaches indicates that sea level fluctuated throughout the Quaternary.

As ice sheets waxed and waned during the Quaternary, there have been several glacial and interglacial stages. We are presently in an interglacial stage which started with the retreat of the most recent ice sheets some 10,000 years ago. Because of the special importance of this interglacial stage to our own development, we divide the Quaternary Period into two epochs; the **Pleistocene** Epoch (starting at the beginning of the Quaternary, 1.6 million years ago) and the present **Holocene** Epoch (starting with the most recent glacial retreat 10,000 years ago).

It should be emphasized, however, that there is nothing geologically distinctive about the Holocene. It is merely another interglacial interlude, little different from the others that preceded it. Implicit in this statement is that nature, if left to its own devices, will conspire to produce another glacial advance in the not-too-distant future. For the first time, however, the forces of nature may be pitted against those of humanity as fossil fuel emissions increase the concentration of greenhouse gases in the atmosphere, leading to global warming.

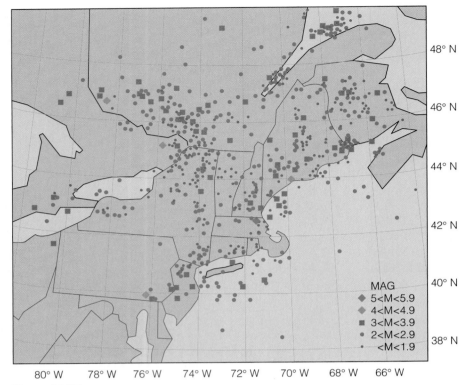

Figure 14.37
Earthquake activity in the northeastern United States and eastern Canada (1975–1985).

The legend on the map reads:

MAG
◆ 5<M<5.9
◆ 4<M<4.9
■ 3<M<3.9
● 2<M<2.9
· <M<1.9

Figure 14.38
A fjord forms when rising sea level causes a U-shaped glacial valley to be invaded by the sea, Norwegian coastline.

Figure 14.39
A raised beach on the coast of the Northumberland Strait, Nova Scotia, Canada. The old beach level is about 6 m (20 ft) higher than the present level.

Sculpturing the Landscape through Mass Wasting

Because of the effects of glaciation, continental drainage (see Chapter 10) and other surface processes, the Earth's surface is rarely flat, but instead consists of slopes inclined at varying angles. Some of these slopes are steep and become unstable, especially where weathering has weakened the rocks by breaking them apart. As a result, the rubble formed may slide downslope under the influence of gravity, a process known as **mass wasting**. Mass wasting is consequently distinct from erosional processes in that it does not require a transporting agent such as ice or water (Fig. 14.40).

Mass wasting works in concert with weathering to remove sediment from unstable slopes. Although this sediment may be dumped in valleys, and then transported to the sea by streams, the initial downslope motion is commonly a direct influence of gravity. **Landslides** are the most hazardous examples of mass wasting. Each year in North America, landslides result in more than 50 deaths and cause about two billion dollars in damages.

Factors that contribute to mass wasting include: the steepening of slopes, the amount of vegetation, the amount of pore water in the sediment, and seismic activity.

If you pour sugar onto a table, the growing mound of sugar that forms develops a stable slope. If you repeat the action, the second mound of sugar will develop an identical slope. In fact, many unconsolidated materials have a stable slope, which is usually between 25 and 40 degrees and is known as its **angle of repose.** If for some reason the angle increases, a condition known as oversteepening develops and the debris will slide downhill until the original slope is restored. Oversteepening commonly occurs when a stream undercuts its valley wall or when waves incessantly pound the base of a cliff. The rocks and sediments above become gravitationally unstable and so succumb to mass wasting (Fig. 14.40).

Plants have intricate root systems that bind the soil together and consequently make vegetated slopes more stable. However, when plants are removed, for example by defor-

estation, the soil is only loosely held and commonly starts to migrate downhill.

As discussed in Chapter 10, pore spaces in sediment may store appreciable amounts of water. As a result, the sediment becomes heavier. When the sediment is water-saturated, the sediment grains become lubricated and friction between adjacent grains is significantly reduced. The additional mass and the lubrication of individual grains dramatically reduces the stability of the sediment and drives it downslope.

One of the most tragic instances of mass wasting related to water saturation occurred in 1966 in the coal mining town of Aberfan, in South Wales (Fig. 14.41). For more than 50 years, the debris from underground coal mines had been dumped into large waste piles (called "tips") along the valley slope above the town. But on October 21, Tip 7 failed and the black sludge slid 800 meters (2600 feet) through cottages, across a canal, and over Pantglas Junior School. In just

(continued)

Figure 14.40
Example of mass wasting caused by failure of a hillslope. The resulting landslide temporarily dammed the stream below, which may have contributed to the instability of the hillslope by undercutting it.

Sculpturing the Landscape through Mass Wasting, continued

a few seconds, 144 people perished, including 116 young school children. The investigation into the tragedy showed that many of the tips, including Tip 7, were built on top of natural springs. The debris had become so saturated with water that the pressure of the pore water exceeded the friction between the grains. As a result, the sediment "liquefied," just like quicksand, and flowed like a liquid downhill.

In seismically active regions, mass wasting may be triggered by earthquakes. In fact, in many instances, seismically-triggered mass wasting causes more loss of life and property damage than the jolt from the earthquake itself. Areas that are most at risk are those that have oversteepened slopes, lack vegetation, or are water-saturated. In these instances, even a minor tremor may be enough to cause a landslide. Mount

Huascaran in western Peru has had two recent devastating landslides, both triggered by seismic activity. In 1962, about 4000 people died when a tremor caused an ice flow and avalanche, and in 1970, 25,000 people perished when an earthquake dislodged 50 million cubic km of mud, rock and water which then smashed through towns and villages in the valley below.

Figure 14.41
Tragedy in Aberfan, South Wales. Coal tips (in background) were positioned above the town. Tip number 7 was located above a natural spring which saturated the coal debris and turned it into sludge. Lubricated in this fashion, the debris moved downhill and buried part of the town.

14.8 Chapter Summary

The Phanerozoic Eon makes up less than 12% of all geologic time and is divided into three eras: the Paleozoic (545 to 245 million years ago), the Mesozoic (245 to 66 million years ago) and the Cenozoic (66 million years ago to today). The superior geologic record of these eras facilitates their further subdivision into a number of periods whose names and durations are given in Figure 14.1.

By the Early Paleozoic, the Late Proterozoic supercontinent started to break up and disperse, creating new continental shelves and opportunities for species to expand and flourish in new ecological niches. These earliest Paleozoic fossils were able to secrete hard shells, so their preservation in the geologic record is far superior to that of their Proterozoic ancestors. As the oxygen levels in the atmosphere increased to more than 2%, an ozone shield was created. Between about 470 and 440 million years ago, the biosphere spread from the oceans onto the continents. The first invasion of land was by plants, but animals followed as food became available. Flora and fauna diversified remarkably during the remainder of the Paleozoic, and oxygen values increased at an accelerated rate due to the spread of vegetation. But the Paleozoic fossil record documents both major radiation and mass extinction events. The origin of the various mass extinction events is controversial, although each appears to be related to environmental stresses associated with rapidly changing conditions.

By the end of the Paleozoic, some 245 million years ago, a new supercontinent, Pangea, had formed as the result of the collision of continental plates. The Appalachian orogenic event in eastern North America is one of many collisional mountain belts associated with Pangea's assembly. A huge mass extinction, known as the Permian crisis, occurred in the very latest stages of the Paleozoic. The origin of this crisis is uncertain, but it is speculated that a massive loss of ecological habitat caused by the formation of Pangea may have played a significant role.

The beginning of the Mesozoic is dominated by the break-up and dispersal of Pangea. Marine communities flourished in the new continental shelf habitats and the relatively benevolent climates that ensued. Mammals made their first appearance in the Triassic, and became well established by the Cretaceous. However, the Mesozoic Era was dominated by reptiles. Dinosaurs became the dominant land animal and were the beneficiaries of an expanding diversity of land plants.

Mesozoic plate reconstructions are well constrained since most of the ocean record of sea-floor spreading is preserved. Orogenic activity in North America shifted from the east to the west coast as the North American plate drifted westward. The growth of the Atlantic Ocean occurred at the expense of the Pacific Ocean, whose margins became dominated by subduction zones.

The warm Mesozoic climate was probably related to tectonic activity associated with the breakup of Pangea. Volcanism associated with continental breakup may have outgassed significant quantities of carbon dioxide into the atmosphere, while rising sea levels associated with continental dispersal may have inhibited continental erosion and the withdrawal of carbon dioxide from the atmosphere.

Most geoscientists believe that the mass extinction at the end of the Mesozoic Era, 66 million years ago, was caused by the impact of an asteroid or comet. This extinction wiped out the dinosaurs and the vast majority of both continental and marine species. However, global ecosystems are also widely believed to have been under environmental stress prior to the impact. If so, the extraterrestrial impact may simply have been the "straw that broke the camels back."

The Cenozoic Era brings us to the modern world. The positions and shapes of the continents are readily recognizable and most fossil forms have modern analogues. The Cenozoic is the "Age of the Mammals," who took advantage of the ecological niches vacated by the mass extinction of the dinosaurs at the end of the Mesozoic.

In North America, orogenic activity was concentrated on the west coast and reflects collisions of terranes with the continental margin and the development of the San Andreas fault system. This has resulted in the generation of mountain belts that fringe the coastline and the growth of the North American margin westward by an average of about 600 km (360 mi).

In the relatively recent past, two very significant events occurred. The first was the initiation of the modern Ice Age about 1.6 million years ago, during which ice sheets systematically advanced and retreated over the continental landmasses of the northern hemisphere. The Ice Age was largely responsible for sculpturing our modern landscape. The second was the evolution of modern humans or *Homo sapiens* some 200,000 to 100,000 years ago. As we shall see in subsequent chapters, humans were to become the first species to directly interfere with the fundamental rules of Earth systems, rules that have evolved over billions of years.

Key Terms and Concepts

Look for the highlighted items on the web at: WWW.BROOKSCOLE.COM

- The Phanerozoic Eon (visible life) started 545 million years ago and is subdivided into the Paleozoic, Mesozoic, and Cenozoic eras

- The better fossil record of the Phanerozoic allows higher resolution of the geologic time scale

- The Paleozoic Era (ancient life) spans the interval between the breakup of a Late Proterozoic supercontinent and the amalgamation of Pangea by the end of the era

- During the Paleozoic, new oceans such as Iapetus were formed, but by the end of the era these oceans had been consumed by subduction. Tiny remnants of these Paleozoic oceans are preserved as ophiolites

- An evolutionary bridge exists between Early Cambrian shelly fossils and soft-bodied Late Proterozoic fauna. Supercontinental breakup in the Early Paleozoic provided the environmental niches for biological innovation.

- The buildup of atmospheric oxygen and ozone facilitated the first colonization of land in the Ordovician Period, about 470 to 440 million years ago

- Late Paleozoic swampy conditions led to the generation of coal deposits

- Middle to Late Paleozoic subduction and the amalgamation of Pangea destroyed ecological niches and may have played a role in the mass extinction at the end of the Paleozoic Era

- The Mesozoic Era (middle life) is known as the Age of the Reptiles, the dawn of which was accompanied by the breakup of Pangea and by biological innovations

- The expansion of the Atlantic Ocean was accommodated by contraction in the Pacific Ocean, leading to a succession of orogenic events on the west coast of North America

- Continental vegetation and marine life flourished during the Mesozoic Era in unusually warm air and water temperatures that were probably related to ongoing volcanism

- Many species, including the dinosaurs, were in decline before the end of the Mesozoic Era. The mass extinction at the end of the Mesozoic Era was probably due to an extraterrestrial impact, but this may have been only the "straw that broke the camel's back."

- Most life forms of the Cenozoic Era (recent life) have modern analogues

- Major tectonic events during the Cenozoic Era include the formation of the Alpine-Himalayan mountain belt and the San Andreas fault system

- The modern Ice Age began 1.6 million years ago in the Quaternary. The cyclic advance and retreat of ice sheets (glacial and interglacial intervals) were accompanied by fall and rise (respectively) of sea level

- The generation of land bridges during times of low sea level facilitated migration of animal species

- Modern humans, *Homo sapiens*, evolved from *Homo habilis* over the past 2.4 million years

Review Questions

1. What is the importance of evolutionary links between the fauna preserved in Late Proterozoic and Cambrian fossiliferous rocks?
2. What type of tectonic setting favored the Cambrian evolution of shelly fauna? Explain.
3. When did the land become colonized by animal life, and what conditions facilitated the colonization?
4. Why would you never explore for coal in Cambrian rocks?
5. What is the potential relationship between the amalgamation of Pangea and the mass extinction at the end of the Paleozoic?
6. What important biological innovations dominated the Mesozoic Era?
7. If the dinosaurs were already in decline at the end of the Mesozoic, what is the hard evidence that their ultimate extinction was related to an extraterrestrial impact?
8. Why is the Cenozoic Era so called?
9. Why are ice ages so poorly preserved in the geologic record?
10. How might the advent of the Ice Age have aided in the migration of land-dwelling animals?

Study Questions

 Research the highlighted questions using InfoTrac College Edition.

1. Why does plate tectonics play such an important role in biological evolution?
2. Why are many Cambrian fossils shelly?
3. Why was there such a delay between the origin of marine life and the colonization of land?
4. How do you think a Precambrian landscape might compare with one of the Carboniferous? Explain.
5. Why did the supercontinent Pangea form?
6. Why are mass extinctions generally followed by major biological innovations?
7. What do you think led to the extinction of the dinosaurs?
8. What caused the climate of the Cretaceous Period to be so mild, and why did temperatures suddenly drop at the end of the period?
9. Ice ages are poorly preserved in the geologic record. Can you suggest why some periods in the geologic past preserve evidence of an ice age similar to that of the Quaternary Period?
10. How has your knowledge of the relentless evolution of the planet over 4.6 billion years, affected your views on the effects of human interference with the Earth systems?

Evolution

Solid Earth

The core, mantle, and crust formed early in the Earth's history as the Earth became layered according to density. Plates moved horizontally as plate tectonic cycles created and dispersed supercontinents. These cycles created shallow marine and land habitats for evolving life forms.

Hydrosphere

The oceans evolved as water vapor released in volcanic eruptions fell as rain into ocean basins. Oceans provided the first habitat safe from ultraviolet radiation, allowing life to begin and evolve. Ocean chemistry provided nutrients for organisms and helped moderate atmospheric chemistry, affecting global climate.

Atmosphere

The atmosphere evolved from the gases released in early volcanic eruptions. Oxygen released into the atmosphere 2 billion years ago from the oceans triggered the first mass extinction of anaerobic organisms. Eventually, an ozone layer developed enabling life to move onto land and continue evolving there. Climate fluctuations influenced life's evolution.

Biosphere

The evolution of life began with single-celled anaerobic asexually reproductive procaryotes, followed by single-celled eukaryotes, and then multicelled metazoa. From these simple beginnings, came the huge diversity of life we see today. Mass extinctions forced survival of the fittest. Humans evolved very recently and are the first animals to directly interfere with the Earth's natural cycles.

Astronomy

The original chemical elements responsible for life's beginnings may have come from an extraterrestrial impact. Early meteorite impacts on Earth's surface released enormous amounts of heat. An extraterrestrial impact is likely to have been the "last strand" in a series of events leading to the mass extinction at the end of the Mesozoic Period.

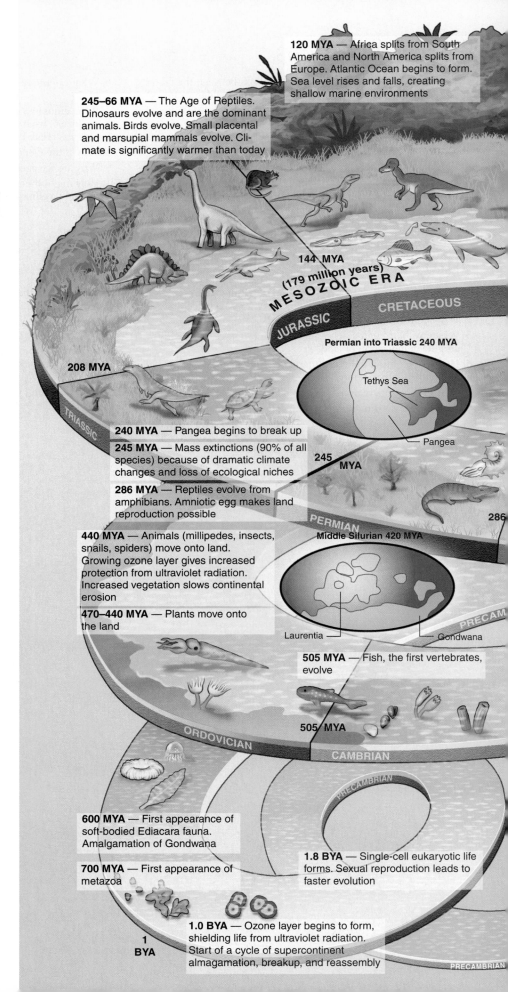

120 MYA — Africa splits from South America and North America splits from Europe. Atlantic Ocean begins to form. Sea level rises and falls, creating shallow marine environments

245–66 MYA — The Age of Reptiles. Dinosaurs evolve and are the dominant animals. Birds evolve. Small placental and marsupial mammals evolve. Climate is significantly warmer than today

144 MYA (179 million years)

MESOZOIC ERA

CRETACEOUS

JURASSIC

208 MYA

TRIASSIC

Permian into Triassic 240 MYA

Tethys Sea

Pangea

240 MYA — Pangea begins to break up

245 MYA — Mass extinctions (90% of all species) because of dramatic climate changes and loss of ecological niches

245 MYA

286 MYA — Reptiles evolve from amphibians. Amniotic egg makes land reproduction possible

PERMIAN

286

440 MYA — Animals (millipedes, insects, snails, spiders) move onto land. Growing ozone layer gives increased protection from ultraviolet radiation. Increased vegetation slows continental erosion

Middle Silurian 420 MYA

470–440 MYA — Plants move onto the land

PRECAM

Laurentia

Gondwana

505 MYA — Fish, the first vertebrates, evolve

505 MYA

ORDOVICIAN

CAMBRIAN

PRECAMBRIAN

600 MYA — First appearance of soft-bodied Ediacara fauna. Amalgamation of Gondwana

700 MYA — First appearance of metazoa

1.8 BYA — Single-cell eukaryotic life forms. Sexual reproduction leads to faster evolution

1 BYA

1.0 BYA — Ozone layer begins to form, shielding life from ultraviolet radiation. Start of a cycle of supercontinent almagamation, breakup, and reassembly

PRECAMBRIAN

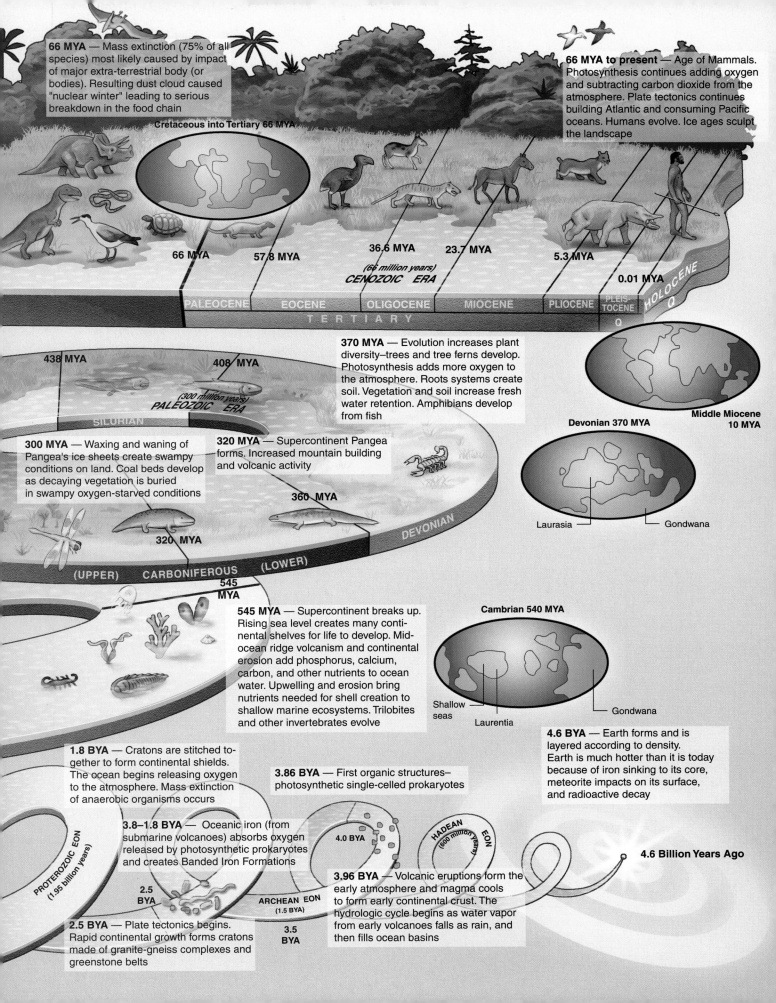

66 MYA — Mass extinction (75% of all species) most likely caused by impact of major extra-terrestrial body (or bodies). Resulting dust cloud caused "nuclear winter" leading to serious breakdown in the food chain

66 MYA to present — Age of Mammals. Photosynthesis continues adding oxygen and subtracting carbon dioxide from the atmosphere. Plate tectonics continues building Atlantic and consuming Pacific oceans. Humans evolve. Ice ages sculpt the landscape

Cretaceous into Tertiary 66 MYA

66 MYA   57.8 MYA   36.6 MYA   23.7 MYA   5.3 MYA

(66 million years)
CENOZOIC ERA

0.01 MYA

PALEOCENE   EOCENE   OLIGOCENE   MIOCENE   PLIOCENE   PLEIS-TOCENE   HOLOCENE   Q

T E R T I A R Y

370 MYA — Evolution increases plant diversity–trees and tree ferns develop. Photosynthesis adds more oxygen to the atmosphere. Roots systems create soil. Vegetation and soil increase fresh water retention. Amphibians develop from fish

Middle Miocene 10 MYA

Devonian 370 MYA

438 MYA   408 MYA

(300 million years)
PALEOZOIC ERA

SILURIAN

Laurasia   Gondwana

300 MYA — Waxing and waning of Pangea's ice sheets create swampy conditions on land. Coal beds develop as decaying vegetation is buried in swampy oxygen-starved conditions

320 MYA — Supercontinent Pangea forms. Increased mountain building and volcanic activity

360 MYA

320 MYA

DEVONIAN

(UPPER)   CARBONIFEROUS   (LOWER)

545 MYA

545 MYA — Supercontinent breaks up. Rising sea level creates many continental shelves for life to develop. Mid-ocean ridge volcanism and continental erosion add phosphorus, calcium, carbon, and other nutrients to ocean water. Upwelling and erosion bring nutrients needed for shell creation to shallow marine ecosystems. Trilobites and other invertebrates evolve

Cambrian 540 MYA

Shallow seas   Laurentia   Gondwana

4.6 BYA — Earth forms and is layered according to density. Earth is much hotter than it is today because of iron sinking to its core, meteorite impacts on its surface, and radioactive decay

1.8 BYA — Cratons are stitched together to form continental shields. The ocean begins releasing oxygen to the atmosphere. Mass extinction of anaerobic organisms occurs

3.86 BYA — First organic structures– photosynthetic single-celled prokaryotes

3.8–1.8 BYA — Oceanic iron (from submarine volcanoes) absorbs oxygen released by photosynthetic prokaryotes and creates Banded Iron Formations

4.0 BYA

HADEAN EON (600 million years)

4.6 Billion Years Ago

PROTEROZOIC EON (1.95 billion years)

2.5 BYA

ARCHEAN EON (1.5 BYA)

3.5 BYA

3.96 BYA — Volcanic eruptions form the early atmosphere and magma cools to form early continental crust. The hydrologic cycle begins as water vapor from early volcanoes falls as rain, and then fills ocean basins

2.5 BYA — Plate tectonics begins. Rapid continental growth forms cratons made of granite-gneiss complexes and greenstone belts

PART 5

Astronomy

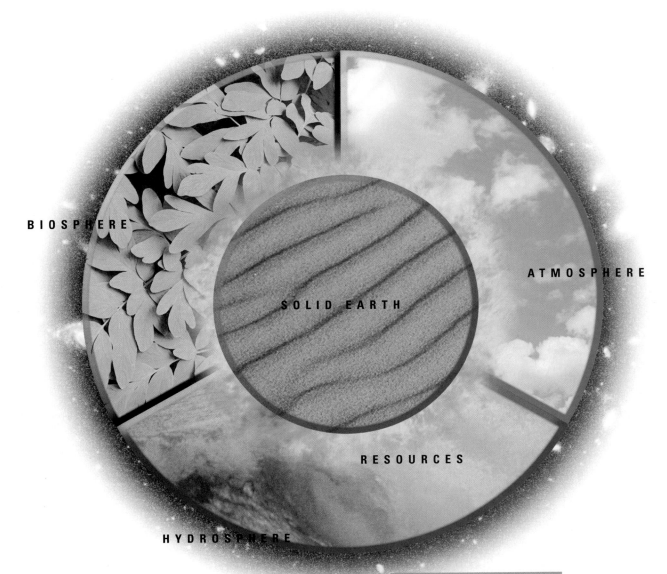

BIOSPHERE

ATMOSPHERE

SOLID EARTH

RESOURCES

HYDROSPHERE

STARS & PLANETS

The Planets

We know from the preceding chapters that plate tectonic activity on our own planet has given rise to the solid Earth and contributed to its watery and gaseous envelopes. As a result, the Earth is a planet that can sustain life. But are these processes unique to the Earth? What do we know of our planetary neighbors and the composition of their surfaces and their interiors? Do they have atmospheres and, if so, what factors control their composition? In this chapter, we will broaden our horizons by comparing the planets with each other and with the Earth. In doing so, we gain a different perspective on Earth processes that will help us better appreciate the uniqueness of our delicate, life-sustaining environment.

Our current knowledge of the planets is the product of centuries of scientific investigation and our discussion begins with the historical development of these concepts. We will see how the principal rules of planetary motion proposed in the 16th Century by Nicolaus Copernicus and later modified by Johannes Kepler still apply today. But as often happens with radical departures from scientific dogma, their ideas only became firmly established in the late 17th Century when Sir Isaac Newton established his laws of motion and gravitational attraction. Acting like a taut, invisible string, it is gravity that keeps the planets in orbit about the Sun.

After the laws of planetary motion were established, attention turned to the discovery of planets, and to speculations on their environment. Some were difficult to find. Pluto, for example, was not discovered until 1930. There were also many misconceptions about the composition and conditions on these planets. We will learn that eminent scientists like Sir Percival Lowell thought that the Martian landscape was scarred by the frantic work of Martian canal-builders seeking to transport water from the planet's poles to its parched equatorial regions.

With the coming of the modern space age, many of these myths were rapidly dispelled, but our planetary neighbors proved to be no less strange. Major surprises about the antiquity of the Moon's landscape were revealed by radioactive dating of samples brought back by the Apollo missions. The launching of space probes confirmed the unique status of Earth in the Solar System as the only planet that harbors abundant life. Of the terrestrial planets (Mercury, Venus, Earth, and Mars), Mercury was revealed to be geologically dead with a surface that is billions of years old. Venus, long regarded as our sister planet because of its similar size, proved to be a very different world. With its dense carbon dioxide-rich atmosphere, it is a nightmarish greenhouse with surface temperatures of about 470°C (880°F). Scientists suspect that it is still volcanically active, although no eruption has ever been seen.

The first indications from space probes suggested that Mars was a planet with a vanishingly thin atmosphere and a cold, ancient surface on which volcanic activity is thought to have virtually ceased about 1 billion years ago. However, recent evidence has highlighted the tantalizing possibility that primitive life-forms may once have existed in its warmer subsurface, and indeed may exist there still.

Many of the characteristics of the larger Jovian planets (Jupiter, Saturn, Uranus and Neptune) are still largely unknown. Their gaseous surfaces and distance from Earth make them relatively inaccessible. Space probes have increased our understanding of Jupiter's atmosphere, its color bands, and the Great Red Spot, a storm center that has existed for centuries. Jupiter's moons include one that is volcanically active (Io) and another (Ganymede) that may have liquid water beneath its icy surface. Saturn's spectacular rings have been shown to be dominated by chunks of ice. But we still know very little about the outermost planets, Uranus, Neptune, and Pluto. In fact, astronomers even debate whether Pluto should be considered a planet.

Recently, the revelations of the Hubble Space telescope, the possibility of extraterrestrial life, and the discovery of other Solar Systems have greatly intensified our efforts to understand our true position in the cosmos.

In this chapter, several themes will emerge. Each planet will be seen to have its own unique set of characteristics, while also sharing some common traits. Thus the terrestrial planets were all formed simultaneously from raw materials of similar composition to meteorites. However, their subsequent evolutions contrast with each other because of their difference in size and distance from the Sun. We will find that Earth had the unique ability to generate continental crust, and that the contrast between continental and oceanic crust was fundamental to the formation of oceans, the evolution of plate tectonics, and the development of life.

15.1 Introduction

Is there anybody out there? For centuries, artists, philosophers, and scientists alike have speculated on the possibility of intelligent life elsewhere in our universe. Indeed, few other subjects have so consistently inspired the intellectual curiosity of the human race. Recent evidence that life may have existed on Mars (see Chapter 13), and that one of Jupiter's moons may have water beneath its icy surface, has rekindled this interest.

Over the centuries, we have come to realize that the Earth belongs to a Solar System made up of nine planets that orbit a single star we call the Sun. Investigations have shown that each planet has its own unique characteristics, and that the Earth seems to be the only planet that can sustain life on its surface. This ability is related to a number of variables, including size, composition, active tectonics, and distance from the Sun. Together, these allow for a life-sustaining atmosphere and liquid water on the planet's surface.

Throughout our history, humans have devised ingenious ways of observing phenomena in our Solar System and the universe. Many of the key observations were made centuries before the advent of optical telescopes. Now, radio telescopes can listen to the signals of distant stars and hear the faint echoes from the farthest reaches of the universe. We have launched telescopes into space to search the heavens in ways never before possible. Humans have walked on the Moon and space probes have visited many of the planets in our Solar System. These sophisticated tools have vastly increased our knowledge of the Solar System and the universe beyond.

Researchers currently debate whether Mars could at one time have harbored primitive life and whether it may still do so beneath its frozen surface. Similarly, scientists speculate that Venus could have developed life early in its history and might do so again in the distant future. But these are hotly debated issues. Our quest is now focused beyond the edge of the Solar System, where celestial bodies do not orbit our Sun. Do other stars have planets of their own? Are there other "solar systems" and, if so, do any of their planets nurture environments that would facilitate the development of life?

In late 1995 and early 1996, Swiss astronomers documented the first planet orbiting another star, although its inferred surface temperature is much too great to support life. At the same time, Geoffrey Marcy at the San Francisco State University reported two planets orbiting stars in different constellations about 40 light years from Earth. One of these planets appears to have a watery surface and might even support life. It is clear that we are on the verge of a very exciting period in our understanding of the cosmos, and the Earth's position within it. Indeed, astronomers have now identified more planets outside the Solar System than exist within it.

In this chapter, we will learn about the planets of the Solar System, beginning with a look at historical views of the universe. We will then consider modern theories about the origin and evolution of the Solar System followed by a discussion of each of the planets and a comparison between them and the Earth. Finally, we will also examine the other bodies in our Solar System, such as asteroids, meteoroids, and comets and their role in our understanding of its development.

15.2 Development of Fundamental Concepts

15.2.1 An Earth-Centered or Sun-Centered System?

The efforts of astronomers to understand the significance of the Earth in the universe date to the earliest of recorded time, and stands as testament to our insatiable quest for knowledge. More than two thousand years ago, the ancient Greeks realized that the sky contained fundamentally different celestial objects. All but one star (known as Polaris, or the North Star) were observed to rotate in the night sky. However, their *relative* positions were constant, that is, they always stayed in the same positions relative to each other. Thus constellations (patterns of stars) such as the Big Dipper, changed their *position* in the night sky, but not their pattern, which could always be recognized. Other celestial bodies, however, apparently behaved in a more complex fashion. These bodies moved relative to each other and were called **planets**, after the Greek word for "wanderer."

The ancient Greeks understood that their observations could be explained equally well either by a rotating Earth or by a stationary Earth about which the heavens rotated. Because they could not prove that the Earth was rotating, they opted for the second explanation. This explanation sufficed for centuries and was formalized in the 2nd Century A.D. by Claudius Ptolemy (140 A.D.). Ptolemy placed the Earth at the center of the universe with all celestial objects revolving about it. This view of the cosmos became known as the **Ptolemaic** or **geocentric** (Earth-centered) system (*geo* is Greek for earth) (Fig. 15.1).

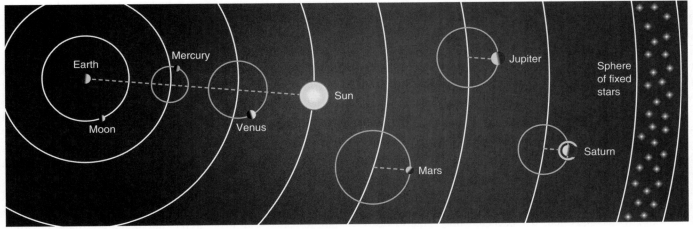

Figure 15.1
The geocentric (Earth-centered) Ptolemaic system in which the Sun, planets, and stars revolve around a fixed Earth. The small circular motions within the larger circles are known as epicycles and were thought to account for the changing relative motions of the planets as seen from Earth.

The Ptolemaic system survived virtually unchallenged until the 16th Century when more refined measurements revealed that the planets were not in the positions predicted by the geocentric model. This mismatch created a scenario that is quite common in science and generally results in two parallel approaches to the solution. The first is to try to modify, or fine-tune, the model in order to account for the discrepancy. The second is to discard the model altogether, and replace it with an alternative model that provides a better account of the observations.

Nicolaus Copernicus (1473–1543) decided that the discrepancies were too great to overcome merely by fine-tuning the geocentric model. So he discarded the Ptolemaic system and replaced it with one in which the Earth turns on its own axis and is but one of several planets that follow circular paths around the Sun. This revolutionary idea was proposed during the Renaissance, a period of great intellectual ferment in Europe when creativity and new ideas were encouraged.

Copernicus described many features of the Solar System that still form the basis of much of our current understanding. He proposed that the motion of the stars was due to the Earth's rotation about its axis, an axis that points towards the North Star. From the motion of the stars, he argued that the Earth rotated once a day. He also proposed that the Sun was the center of our Solar System and that the planets revolve around, or **orbit**, the Sun at different rates, accounting for their apparently complex motions in the sky. He was able to determine the order in which the planets revolved around the Sun, with closer planets moving faster than those farther out. This view of the Solar System (Fig. 15.2) is known as the **Copernican** or **heliocentric**

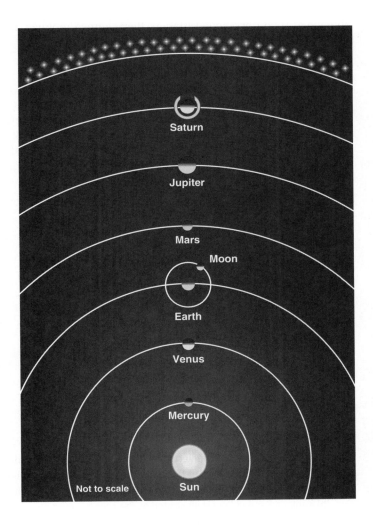

Figure 15.2
The heliocentric (Sun-centered) system as it was conceived by adherents to the Copernican system towards the end of the 17th Century. According to this scheme, the planets have perfectly circular orbits about the Sun. The outermost planets (Uranus, Neptune and Pluto) had not yet been discovered. The outer stars were considered to be quite distant from Earth, showing an understanding of the Solar System's position within the universe.

(Sun-centered) system (*helios* is the Greek word for the Sun). By assuming the Earth's axis had a tilt, he was able to provide an explanation for the four seasons. He further proposed that the Moon orbited the Earth, and that the stars must be "far away" because they remained in the same relative positions. These statements were the first to imply that our Solar System was but a tiny fraction of the universe.

Copernicus's model did not predict the positions of the planets any better than the geocentric model. However, the elegance of his arguments slowly won support, fueling a bitter debate between his supporters and those of the Ptolemaic system. At that time, the laws of gravity and planetary motion had not been established. Supporters of the Ptolemaic system correctly pointed out that the Earth must be rotating very quickly if, as Copernicus suggested, it completed one rotation every 24 hours (the Earth's equatorial speed is 1670 km/hr, or 1035 mi/hr). Surely, they argued, if this were the case, everything that was not pinned to the Earth's surface, including the oceans and atmosphere, would be hurled into space. Great winds would be created that would destroy everything in their path. The absence of such ongoing devastation, they claimed, implied a stationary Earth, consistent with the Ptolemaic system.

Supporters of the Copernican model responded by claiming that it was the nature of all earthly matter to move with the Earth. However, they could not demonstrate why this was so. We now know that it is the force of gravity, first described by Sir Isaac Newton more than 100 years later, that is responsible for maintaining the Earth's watery and gaseous envelopes. And as we have seen in previous chapters, the Earth's rotation does affect the movement of air and water.

Using detailed astronomical measurements, Johannes Kepler (1571–1630), found flaws in the Copernican system, which was based largely on theory rather than observation. Kepler deduced that the planets could not have perfectly circular orbits if the Sun was indeed the center of the Solar System. So he was faced with the same philosophical choice that had confronted Copernicus nearly a century earlier. In this case, however, he concluded that his observations could be reconciled if the paths of the planets around the Sun were elliptical, or oval, rather than circular. In doing so, he modified rather than discarded the Copernican system.

Kepler proposed several hypotheses that have been subsequently confirmed by so many observations that they have come to be known as **Kepler's Laws of Planetary Motion** (Fig. 15.3). These laws form the foundations of planetary science and allow us to predict the motion of planets today. However, it is important to realize that while Kepler attempted to explain the available data, he did not propose a mechanism that explained *why* the planets should obey these laws.

Kepler's First Law simply states:

The paths of the planets around the Sun are ellipses.

The fundamental geometric difference between a circle and an ellipse is that a circle has one focus, which we call the center of the circle, whereas an ellipse has two. A loop of string drawn taut by a single pin at one end and a pencil at the other, allows the pencil to describe a circle as it is moved around the pin. If the string is looped around two pins, however, the pencil will describe an oval or ellipse with a focal point at each pin (Fig. 15.3a). Kepler's calculations suggested that the Sun was positioned at one of these focal points. No object was known to reside at the other focal point.

However, Kepler also knew that a mere change in the geometry of planetary orbits did not satisfy all of his calculations.

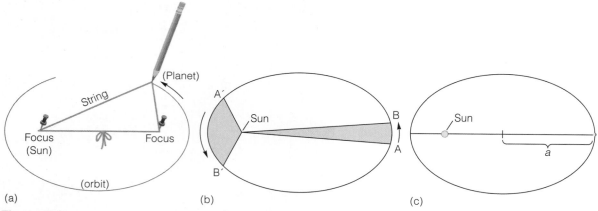

Figure 15.3
Kepler's Laws of planetary motion are based on elliptical orbits around the Sun. (a) Kepler's First Law, which describes the orbits of the planets, can be modeled by drawing an ellipse with two tacks and a loop of string. The tacks represent the foci of the ellipse, one of which represents the Sun. (b) Kepler's Second Law is illustrated by the area covered by the Earth's orbit at two different times of the year (represented by AB and A'B'). The Earth sweeps over equal areas (in blue) in equal times. Hence, the Earth takes as long to travel from A to B as it does to travel from A' to B'. The radius vector is the line that connects the Sun and the Earth and sweeps over the area travelled by the orbiting Earth. (c) Kepler's Third Law relates the time it takes a planet to orbit the Sun (the planet's orbital period) to the planet's average distance from the Sun (a), which equals half the length of the major axis of the ellipse (see Table 15-1).

In particular, the planets appeared to change their speeds during their path around the Sun. **Kepler's Second Law** states:

A planet moves about the Sun such that its radius vector sweeps out an equal area in equal times.

This statement sounds complicated, but it can be illustrated quite simply. Since the Earth has an elliptical orbit, its distance from the Sun changes systematically during the course of a year. The radius vector is simply an imaginary line between the Sun and the Earth. The changing position of the Earth relative to the Sun at various times of the year is shown in Figure 15.3b. If, for example, we compare the position of the Earth at the beginning and end of any given day, the radius vector always sweeps out the same area for each day (see blue shaded areas). One of the consequences of this law is that the Earth is actually traveling most slowly when it is farthest away from the Sun (in June), and most rapidly when it is closest (in December).

Kepler realized that while his second law explained the motion of any one planet, it did not explain why the planets take different amounts of time to complete a single revolution or orbit around the Sun. Astronomers already knew that planets farther from the Sun take longer to complete their orbits than those closer to it. For example, Mercury, the innermost planet, takes only 88 Earth days to complete an orbit, whereas the Earth takes a year (365.25 days), and Pluto, the outermost planet, takes 248 Earth years.

Clearly, a planet's average distance from the Sun and the time it takes the planet to make a single orbit around it are related, but how? Kepler devised a solution that underscored the order of our Solar System. **Kepler's Third Law** states that for any planet:

The ratio between the square of the time needed to make a revolution around the Sun, is proportional to the cube of its average distance from the Sun.

The third law is expressed by the formula

$$p^2 = a^3$$

where p represents the time taken for a planet to make one revolution, an interval known as the "orbital period," and a represents the planet's average distance from the Sun. It should be emphasized, however, that Kepler had no proof for his formula, which he constructed simply to match the observations. But...it worked! The application of this law is best demonstrated using examples.

By convention, the orbital period of a planet is measured relative to that of Earth (365.25 days). As we have learned, Mercury, the innermost planet, takes 88 Earth days to orbit the Sun. Its orbital period (p) is therefore 88/365.25 = 0.241 Earth years. The square of this value ($p^2 = 0.241 \times 0.241$) is 0.058. The distance of the planet from the Sun is also expressed relative to that of the Earth, which is defined as 1 astronomical unit (1 AU). Mercury is much closer to the Sun and orbits at an average distance of 0.387 AU. The cube of this value ($a^3 = 0.387 \times 0.387 \times 0.387$) is also 0.058.

Mars, which is farther from the Sun than the Earth is, has an orbital period of 687 Earth days, or 1.881 (687/365.25)

Earth years. The square of this value is 3.538. The average distance of Mars from the Sun is 1.523 AU. The cube of this value is 3.533. Although the agreement is not perfect, the two calculated values are once again very nearly equal. The same is true for the calculated values of p^2 and a^3 for each of the other planets (Table 15-1).

Through careful observation, ingenious insights, and calculations, Copernicus and Kepler established the geometric relationship between the planets and the Sun they orbit. However, the Copernican system was still criticized because no *mechanism* was proposed to account for the planets' systematic orbits around the Sun. Criticism is a recurring theme of many important scientific breakthroughs. As we discussed in Chapter 4, Alfred Wegener was to experience similar criticism 300 years later when he proposed the theory of Continental Drift without providing a mechanism that could account for the movement.

Nevertheless, the weight of observational data gradually converted scientists from a geocentric view to a heliocentric one. More importantly, a quantum leap in our ability to observe the heavens occurred early in the 17th Century with the invention of the optical telescope. Before the advent of the telescope, only five planets had been discovered. Named for the Roman gods Mercury, Venus, Mars, Jupiter, and Saturn, these planets, in addition to the Sun, the Earth, and the Moon, were thought to comprise the Solar System. But in 1609, Galileo Galilei (1564–1642), a contemporary of Kepler, made several fundamental discoveries with a newly built telescope. He discovered four of Jupiter's sixteen moons (now called the Galilean moons) and showed that they orbited Jupiter. His observations demonstrated for the first time that not all bodies revolved around the Earth. He also showed that the apparently changing illuminations or **phases** of Venus, as viewed from the Earth, were consistent with its proposed orbital motion around the Sun. These phases are similar in principle to the phases of the Moon. Galileo further discovered dark blemishes on the Sun's surface known as **sunspots.** The sunspots moved systematically across the Sun's surface,

Table 15-1

Verification of Kepler's Third Law for Each Planet*

PLANET	a (AU)	p (YEARS)	a^3	p^2
Mercury	0.387	0.241	0.058	0.058
Venus	0.723	0.615	0.378	0.378
Earth	1.000	1.000	1.000	1.000
Mars	1.523	1.881	3.533	3.538
Jupiter	5.203	11.86	140.85	140.66
Saturn	9.539	29.46	867.98	867.98
Uranus	19.18	84.01	7,055.79	7,057.68
Neptune	30.06	164.8	27,162.32	27,159.04
Pluto	39.44	248.4	61,349.46	61,702.56

*The minor disagreement between p^2 and a^3 seen here for the outermost planets do not indicate failures of Kepler's third law; instead they reflect inaccuracies in the measured values of p and a.

a is the distance from the Sun measured relative to Earth (known as an astronomical unit, AU) and p is the period for the planet (the time it takes to complete a revolution about the Sun, measured relative to Earth). In each case $p^2 = a^3$ as required by Kepler's Third Law.

which Galileo interpreted as showing that the Sun rotated about its own axis.

The mounting evidence that the Earth was not the center of the universe, but instead was part of a Sun-centered Solar System, ran contrary to the 17th Century religious dogma of western Europe. As an old man, Galileo was brought before the Inquisition in Rome. In 1633, at the age of 70, under the threat of torture, he was forced to renounce his scientific work. He died under house arrest in 1642.

15.2.2 Gravity and the Solar System

The last important piece of the puzzle of planetary orbits was provided by Sir Isaac Newton (1642–1727). In 1687 Newton published the book *Principia* in which he provided a theoretical basis for Kepler's work. Newton had previously developed calculus as well as the laws of motion and gravity. **Newton's First Law of Motion**, also known as the *principle of inertia*, states that:

Objects in motion remain in motion and objects at rest remain at rest, unless they are acted upon by an outside force.

The mathematics of this law indicates that the path of an unimpeded moving object should be a straight line. So, what forces cause planets to move along curved paths, changing both speed and direction? Newton realized that the maintenance of planets in their orbits must be a question of balanced forces. For example, when you swing a ball around on the end of a string, there are two counterbalancing forces controlling the path of its movement. First, there is the motion of the ball, which will move in a straight line unless another force acts upon it. Second, there is an inward force because the ball is connected to the string. The balance between these two forces causes the ball to move in a circular path. The orbit is lost as soon as the balance between these two forces is lost. If the string is cut, the ball flies off in what is initially a straight line path. If the motion of the ball decreases, then the orbit collapses. The Scottish sport of hammer throwing provides a good example of the two balancing forces (Fig. 15.4).

However, as the hammer thrower lets go, the hammer flies away. On the other hand, if the appropriate hammer speed is not maintained, the unfortunate thrower gets hammered!

Newton surmised that the Moon moves around the Earth as if held in orbit by an invisible string. The Moon does not fly off into space because there is a force called **gravity** that balances this motion and holds the Moon in place. Nor does the Moon fall to Earth, because of its motion. Newton successfully extended his theory to the Copernican system, with the Sun at the center of the Solar System and the planets held in orbit by the counterbalancing forces of gravity and orbital motion. Newton had finally answered the initial criticisms of the Copernican system. The oceans were not hurled into space by a rotating Earth because they too were held in place by gravity.

15.2.3 Discovery of the Outer Planets

After Newton's discoveries, important new advances were made as telescopes became more powerful and more data became available. In 1781, Uranus was discovered by the English astronomer William Herschel. Herschel first thought Uranus was a comet, but its status as a planet was confirmed by its motion relative to the stars and by detailed observations that showed it to orbit the Sun. Neptune was discovered in 1846 and Pluto in 1930.

Technology has developed so much over the past 60 years that it seems inconceivable that significant new planetary discoveries are left to be made. Yet, a belt of icy bodies, thought to represent the condensed remnants of the early Solar System has been recently located beyond the orbit of Pluto. This discovery will doubtless focus considerable attention on the outer reaches of the Solar System, and who knows what further discoveries lie ahead of us.

As scientists have learned more about the Solar System, with its central Sun and orbiting planets, they have speculated in ever more detail on how it came to exist. How did the planets form? Why do they move in only one direction? What is the origin of their satellite bodies? In the next section, we will examine theories about the origin of the Solar System, the Earth, and the Earth's satellite, the Moon.

Figure 15.4
A traditional Scottish sport, hammer-throwing, illustrates Newton's Laws of motion and gravity. (Photo Brendan Murphy)

15.3 The Origin and Early Evolution of the Solar System

The laws of Kepler and Newton provided the basis for hypotheses about the evolution of the Solar System. Two contrasting hypotheses (not unlike those first proposed for the origin and evolution of the Earth) were seriously considered; one was a catastrophic theory, the other was evolutionary.

The evolutionary theory, championed by Descartes in 1644, was expanded on some 100 years later by the German philosopher Immanuel Kant and the French mathematician

Pierre-Simon Laplace. Between them, they formulated the **nebula hypothesis** for the origin of the Solar System in which a rotating cloud of dust and gas is thought to have flattened into a disk as it contracted under the influence of gravity.

According to Newton's laws of motion and gravitational attraction, a spinning object will spin faster if its size decreases. Think of figure skating and the increased spin of ice skaters when their outstretched arms are drawn closer to their bodies (Fig. 15.5). The speed of rotation depends on the mass, the initial velocity, and the radius of the object. The combination of these factors is called the **angular momentum**. Because angular momentum must be conserved, a decrease in radius results in an increase in speed unless an outside force puts on the "brakes."

According to this model, the rates of rotation should increase toward the center of the Solar System. However, the Sun, which should rotate at the fastest rate because it lies at the center, takes a lazy 27 Earth days to complete one rotation.[1]

This problem provoked the two familiar and simultaneous approaches to any flawed theory. Some scientists discarded the evolutionary theory altogether, while others attempted to modify it. The catastrophic theory was an attempt to develop a new model. The theory proposed a "close encounter" between the Sun and another star, during which material was pulled away from the Sun. This material then condensed into the planetary bodies. However, the problem with the catastrophic theory was that the material pulled out should have been too hot to condense into planets and should instead have been lost into space.

At the same time, other scientists attempted to fine-tune the evolutionary theory by searching for an external force that might account for the anomalously slow rotation of the Sun. Such a force was found in the braking effect of the Sun's magnetism, which slows its rotation. With this modification, the evolutionary model is now favored and has become known as the **solar nebula theory** (Fig. 15.6), many essential elements of which are similar to those envisaged by Kant and Laplace more than 200 years ago.

What initiated the gravitational collapse or shrinking of the solar nebula in the first place, however, is unclear. Some scientists think it could have been caused by a nearby exploding star, an event called a **supernova**, which would have produced a powerful shockwave that rushed through space. If these scientists are correct, then ironically, after centuries of heated scientific debate, the catastrophic and evolutionary theories may be linked. The birth of the Solar System may have been triggered by the explosion of a nearby star (rather than the close encounter proposed by the catastrophic theory), and from that moment on, an evolutionary process involving gravitational collapse took over.

Figure 15.5

The principle of conservation of angular momentum as demonstrated by a spinning skater. The skater spins faster when the arms are pulled in.

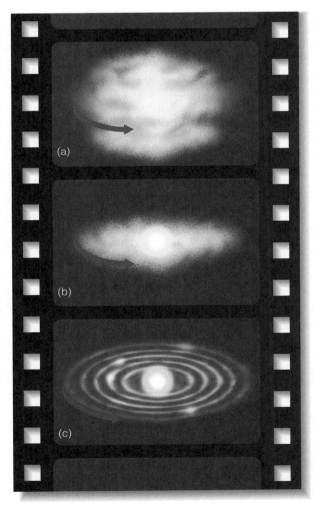

Figure 15.6

Solar nebula theory. (a) The solar nebula, a swirling cloud of gases and dust, flattens and rotates counterclockwise (b), forming (c) the Sun and the planets whose orbits lie in nearly the same plane.

---

[1] 27 Earth days is the Sun's average rate of rotation. In fact, the Sun's equator rotates faster (25 Earth days) than its poles (35 Earth days).

A supernova explosion produces a powerful shockwave that would compress the dust and gas of an adjacent nebula creating fluctuations in the density of nebula material. Rotation would then flatten the nebula into a disk (Fig. 15.6). The force of gravity would then take over in a chain of events that would induce further, irreversible collapse. As material began to collapse and spin, gravitational attraction between the particles would have been greatly enhanced. The greater the attraction, the greater the rate of collapse, and so on. In this way, some 90% of the solar nebula is thought to have condensed to form the Sun. This would have resulted in high enough temperatures at the Sun's core to initiate nuclear fusion, lighting the Solar System with the first sunlight.

Since the solar nebula began as a large interstellar dust cloud, it was presumably dominated by hydrogen with minor amounts of helium and traces of other elements. As the core of the primitive Sun became hot, intense radiation would have heated the gas and dust around it. In the inner part of the Solar System, the temperature was consequently high and most elements would have remained gaseous.

As this inner nebula cooled, however, those elements that become solid at the highest temperatures would have condensed first, to form small objects called **planetesimals** ranging in size from a few centimeters to tens of kilometers across. Computer models show that gravitational instability and turbulence in the swirling spiral disk associated with the early stages of the Solar System would have caused the planetesimals to cluster in clumps. Since the planetesimals were orbiting the Sun in the same direction, their relative velocities would have been minimal and their collisions relatively gentle. In this way, the planetesimals are thought to have stuck together, or **accreted**, in much the same way that a snowball is made by the accretion of snow particles. When the largest particles exceeded 100 km (62 mi) in diameter, other processes influenced their subsequent evolution, resulting in the formation of **protoplanets**: objects that were destined to become planets.

There are two popular models for the formation of planets by the process of accretion. If the planetesimals had similar compositions prior to accretion, then the internal layering of the innermost planets (core, mantle, crust) must have been the result of internal processes that took place *after* accretion (Fig. 15.7a). This is the **homogeneous** model of planetary accretion.

If, on the other hand, the nebula continued to cool as the planets formed, then the composition of the newly-formed planetesimals may have changed with time so that their accretion was inhomogeneous (Figure 15.7b). The wide variety of meteorite compositions suggests that the raw materials for the innermost planets were indeed quite varied, which would be consistent with the inhomogeneous model. According to this model, an iron-rich core accumulated first from the earliest most refractory

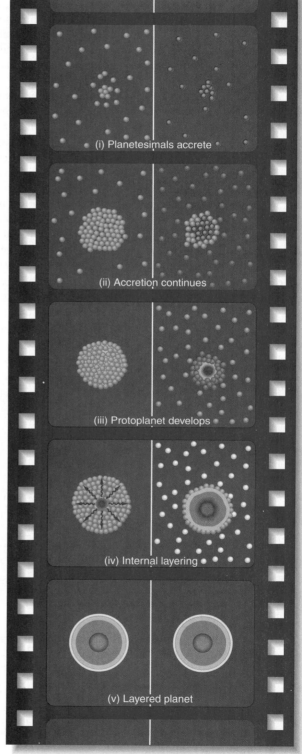

Figure 15.7

Two models shown here in sequence (i–v), have been proposed for the growth and early development of a planet by the accretion of planetesimals (small spherical bodies). (a) In the homogeneous model, the planet grows by the accretion of planetesimals with uniform composition. Layering (core, red; lower mantle, dark brown; upper mantle, pale brown) develops from internal processes (see text). (b) In the inhomogeneous model, planetesimals of different composition are generated as the solar nebula cooled. The first particles to accumulate are heavy metals (red) followed by silicates (brown). The planet's internal heat further aids differentiation into layers.

(a) Homogeneous model   (b) Inhomogeneous model

particles. The planets then grew by the accretion of lighter more silicate-rich planetesimals to form silicate mantles.

In either model, however, the formation of the innermost planets would have occurred so rapidly that the heat generated by the process of accretion would not have had time to dissipate. This heat of formation, combined with heat released by radioactive decay, may have caused the interior of the planets to melt, resulting in the sinking of heavy elements such as iron and nickel to form the planetary cores (as described in Chapter 7). Lighter elements, by contrast, would have risen to the surface of the planets to form primitive silicate crust. It is also conceivable that the heat released as the innermost planets accreted may have boiled off the lightest gaseous elements.

The innermost planets, Mercury, Venus, Earth, and Mars, have relatively high densities, suggesting that, like the Earth, they are essentially rocky spheres. Together they are termed **terrestrial planets** after *terra*, the Latin word for land.

In the outer part of the Solar System, temperatures would have been low enough for volatile components to condense around any small rocky core, forming a layer of hydrogen ice containing smaller quantities of helium and minor amounts of silicate minerals. These outer planets would also have grown rapidly. When they attained a mass about 10–15 times that of the Earth, they would have had sufficient gravitational attraction to capture gas directly from the solar nebula, until the gas in that region of the nebula was exhausted. Jupiter, located at 5.2 AU from the Sun, formed near the boundary of ice condensation in the solar nebula. As these outer planets grew and their internal pressures increased, their icy layers were converted from hydrogen ice into liquid metallic hydrogen. Table 15-2 summarizes the condensation sequence.

These outer planets have densities that are significantly lower than those of the terrestrial planets, in spite of their greater size. They are essentially gaseous bodies with poorly defined surfaces and, excluding Pluto, are termed **Jovian planets** after the largest of them, Jupiter (*Jove* is a Latin word for Jupiter).

Figure 15.8 shows the relative sizes of the planets compared to that of the Sun and Table 15-3 lists some of the current data for the planets of the Solar System. From this information, it is easy to see that the terrestrial planets and the Jovian planets differ in many important aspects.

The gravitational energy generated by the accretion of the planets, known as the **heat of formation**, guided their subsequent evolution. In fact, the loss of heat from planetary interiors is such a slow, inefficient process that even now, nearly 4.6 billion years later, the heat of formation is still an important component of the planets' internal heat engines. On Earth, for example, plate tectonics is partially driven by this heat, and Jupiter and Saturn still radiate more heat than they absorb from the Sun.

The predominantly counterclockwise motion of the planets and their moons (as viewed from above) is thought to reflect the original direction of rotation of the swirling nebula disk. Anomalies, such as the clockwise rotation of Venus and the varying angles of tilt of the planets, are attributed to the

Table 15-2

The Condensation Sequence, Showing That the Temperature of Formation of the Planets Decreases with Distance from the Sun

TEMPERATURE (K)	CONDENSATE	PLANET (ESTIMATED TEMPERATURE OF FORMATION; K*)
1500	Metal oxides	Mercury (1400)
1300	Metallic iron and nickel	
1200	Silicates	
1000	Feldspars	Venus (900)
680	Troilite (FeS)	Earth (600) Mars (450)
175	H_2O ice	Jovian (175)
150	Ammonia–water ice	
120	Methane–water ice	
65	Argon–neon ice	Pluto (65)

*K is in degrees Kelvin.

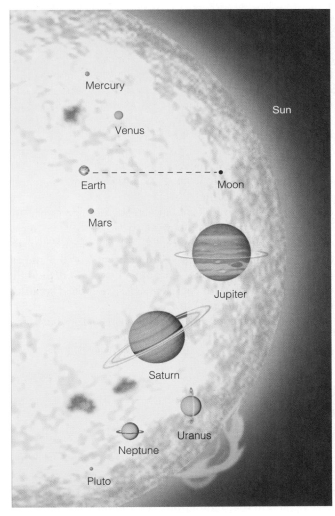

Figure 15.8

The relative sizes of the planets and the Sun. Note that the Jovian planets are bigger than the terrestrial planets, and all are dwarfed by the Sun.

Table 15-3

Comparison Between the Planets

PLANET	MEAN DISTANCE FROM SUN, EARTH = 1[a]	DIAMETER, THOUSANDS OF KM	MASS, EARTH = 1[b]	MEAN DENSITY, WATER = 1[c]	SURFACE GRAVITY, EARTH = 1[d]	ESCAPE SPEED, KM/S[e]	PERIOD OF ROTATION ON AXIS	PERIOD OF REVOLUTION AROUND SUN	ECCENTRICITY OF ORBIT[h]	INCLINATION OF ORBIT TO ECLIPTIC[i]	KNOWN SATELLITES
Mercury	0.39	4.9	0.055	5.4	0.38	4.3	59 days	88 days	0.21	7°00′	0
Venus	0.72	12.1	0.82	5.25	0.90	10.4	243 days[f]	225 days	0.01	3°34′	0
Earth	1.00	12.7	1.00	5.52	1.00	11.2	24 h	365 days	0.02	—	1
Mars	1.52	6.8	0.11	3.93	0.38	5.0	24.5 h	687 days	0.09	1°51′	2
Jupiter	5.20	140	318	1.33	2.6	60	10 h	11.9 yr	0.05	1°18′	16
Saturn	9.54	121	95	0.71	1.2	36	10 h	29.5 yr	0.06	2°29′	23
Uranus	19.1	52	15	1.27	1.1	22	16 h[g]	84 yr	0.05	0°46′	15
Neptune	30.1	49	17	1.70	1.2	24	16 h	165 yr	0.01	1°46′	8
Pluto	39.4	2.3	0.03	1.99	0.43	3.2	6 days	248 yr	0.25	17°12′	1

[a] The mean Earth-Sun distance is called the *astronomical unit,* where 1 AU = 1.496×10^8 km.

[b] The Earth's mass is 5.98×10^{24} kg.

[c] The density of water is 1 g/cm^3 = 10^3 kg/m^3.

[d] The acceleration of gravity at the Earth's surface is 9.8 m/s^2.

[e] Speed needed for permanent escape of gas from the planet's gravitational field.

[f] Venus rotates in the opposite direction from the other planets.

[g] The axis of rotation of Uranus is only 8° from the plane of its orbit.

[h] The difference between the minimum and maximum distances from the Sun divided by the average distance.

[i] The ecliptic is the plane of the Earth's orbit.

abundance of major impacts in the early Solar System. Evidence of such impacts and the turbulent early history of the Solar System is derived from the abundance of ancient craters on the terrestrial planets and the planetary moons.

These high-speed collisions are estimated to have occurred at up to 5400 km/hr (3350 mi/hr) and caused widespread cratering and volcanism on the terrestrial planets of the youthful Solar System. Indeed, as we will discuss in the next section, the Earth's Moon may have accreted from material splashed out of the Earth early in its history as a result of a collision with a Mars-sized, celestial object.

The more numerous moons of the Jovian planets have a variety of origins. The internal Galilean moons probably formed with Jupiter. They lie in Jupiter's equatorial plane and have nearly circular orbits. The outer moons, on the other hand, may have been independent celestial bodies "captured" by Jupiter's gravitational field.

In summary, the solar nebula theory best explains our current state of knowledge about the Solar System. As we learn more, however, we may refine the theory or even develop a new model.

15.4 The Origin of the Earth and Moon

As we have noted, the Earth and the other terrestrial planets are thought to have formed in a similar way. We will compare these planets to one another in a later section. In this section, we outline the origin of the Earth and its only natural satellite, the Moon. As we shall see, studies of the Moon have provided a wealth of information about the age and history of the Solar System.

15.4.1 The Earth

According to the solar nebula theory, the Earth is likely to have condensed from interplanetary dust, and to have become differentiated by gravity into a core, mantle, and crust very early in its evolution. At present, it is not possible to determine which model of planetary accretion (homogenous or inhomogenous, see Figure 15.7) best describes the origin of the Earth and its internal layers. It is conceivable that the most realistic scenario may incorporate elements of both models. As we learned in Chapter 13, the geologic record suggests that these layers formed very early in the Earth's history. In fact, model calculations suggest that the Earth's core could have formed in as little as 20 million years. The formation of the Earth's core by the gravitational collapse of iron caused further release of heat, and is thought to have amplified the density segregation of the Earth's internal layers in an event known as the **iron catastrophe**. As we discussed in Chapter 13, evidence on the Earth for the existence of 4.3 billion-year-old crust suggests that the internal layering of the Earth was achieved prior to that time.

It is certainly clear that the Earth's early evolution occurred at a hectic pace compared to the more pedestrian pace of more recent geologic time. In addition, the processes involved helped to form our hydrosphere and atmosphere, and in so doing, paved the way for life.

On Earth, the resulting layering comprised a dense iron- and nickel-rich core, a less dense silicate mantle, a buoyant crust, a watery hydrosphere, and a gaseous atmosphere. Initial outgassing occurred rapidly, producing a dense, primitive atmosphere rich in carbon dioxide, carbon monoxide, water vapor, methane, and ammonia. Plate tectonics had little or no role in this early differentiation. Instead, the thin fragile crust was constantly bombarded by comets and meteorites and may have been consumed as quickly as it formed.

15.4.2 The Moon

The Moon is much smaller than the Earth with a diameter of only 3476 km (2155 mi) compared with the Earth's 12,740 km (7900 mi). Yet it is our nearest neighbor, at an average distance of only 384,400 km (238,330 mi). Although the Moon looks much bigger when it is near the horizon than it does when it is high in the sky, this is merely an optical illusion. It is actually a spherical object with a fixed size.

Sir Isaac Newton showed that the Moon was the only celestial body that orbited the Earth. As we learned in the previous section, it is held in its orbit by the balance between its motion and its gravitational attraction to the Earth. In 1680, Giovanni Cassini discovered that the Moon rotates about its own axis at the same rate as the Earth. Because of this, the same side of the Moon always faces us. This is known as "synchronous rotation." Although the Moon rotates as it orbits the Earth, it does so only once per orbit. The illuminated surface we see at "full moon" is consequently the surface that always faces us (Fig. 15.9). Unless we venture into space, we will never be able to see the far side of the Moon directly.

In the late 18th Century, Joseph Legrange and Pierre Laplace speculated that the synchronous rotation of the Moon was not a coincidence. Instead, it could be explained by Newton's law of gravity if the Moon was slightly elongated (like an egg) rather than perfectly spherical. The elongation was thought to be a tidal influence, that is, it was thought to be caused by the gravitational pull of the Earth on the Moon. The Moon's shape, they speculated, should be deformed by this force and so should be elongate along a line pointing towards the Earth. The gravitational attraction between the Earth and the Moon would then act to keep the Moon's long axis pointing towards the Earth. This has since been confirmed by accurate measurements from space probes; one axis of the Moon is indeed about 2 to 3 km (1 to 2 mi) longer than the other and points towards the Earth.

Prior to 1969, when humans first landed on the Moon, the speculations of scientists concerning the composition and origin of the Moon were essentially untested. Viewed from the Earth, the light and dark areas on the lunar surface are the most obvious visible features (Fig. 15.10). With the advent of telescopes, the light-colored areas were (correctly) interpreted to be lunar highlands, standing above the ancient dark plains, or lowlands. However, early astronomers mistakenly thought that the dark areas resembled the Earth's ocean basins, and so called them **mare** (pronounced *ma-ray*), which is the Latin word for sea. At one time it was even thought that these basins might hold water, and that the Moon might therefore harbor life. However, even though the term "mare" (plural "maria") is firmly entrenched in the literature, we now know that these basins are waterless depressions.

The space race of the 1960s culminated with the lunar landings of the Apollo space program, and between 1969 and 1972, 12 humans walked on the Moon's surface. In addition, several kilograms of lunar rocks were brought back to Earth and have provided the key to unraveling the Moon's history. The Moon's surface is composed of igneous rocks with mineral compositions similar to those on Earth (Fig. 15.11). The brightness of the lunar highlands reflects the presence of feldspar, which is one of the most common minerals in the Earth's crust

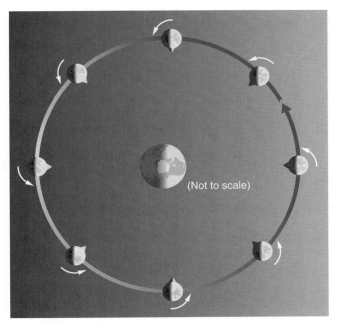

Figure 15.9
Synchronous rotation of the Moon. The Moon rotates counterclockwise on its own axis as it revolves about the Earth such that the same side always faces us. This is schematically represented by an exaggerated mountain on the Moon that changes its position as the Moon rotates so that it always faces the Earth. If the Moon did not rotate as it revolved, the position of the mountain in each instance would be identical and different parts of the Moon's surface would face the Earth.

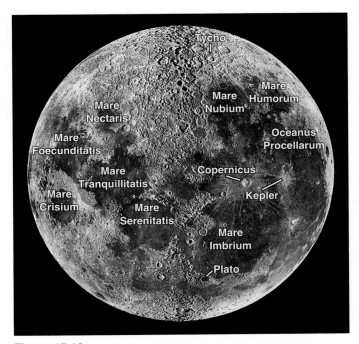

Figure 15.10
An inverted photo (as seen through a telescope) of the lunar surface that faces the Earth. Impact craters are especially evident in the middle portions of the photograph. Lunar maria are the dark smooth areas, whereas the lunar highlands are lighter and more heavily cratered.

Figure 15.11
Photograph of a sample of lunar basalt, displaying vesicles which represent the remnants of gaseous bubbles. This indicates that outgassing occurred on the Moon. However, the Moon was too small to retain these gases in an atmospheric envelope.

(see Chapter 2). But the most abundant rocks on the Moon are dark basalts, which form the floors of the lunar maria. These rocks are relatively rich in magnesium and iron and are very similar in composition to the basalts that underlie most of the Earth's oceans. Since basalts are known to be the product of volcanic activity, their presence on the Moon suggests the existence of similar source materials beneath the surface of both the Moon and the Earth.

But here the similarities end. The final Apollo mission, in December of 1972, specifically targeted a sampling area near the Taurus Mountains that was thought to be a candidate for recent volcanism. But when the age of these basalts was determined by radioactive dating, they proved to be in excess of 4 billion years old! Furthermore, investigations revealed that volcanic activity on the Moon virtually ceased 3 billion years ago. Rocks in the lunar maria were found to range in age from 3.0 to 3.8 billion years. Those of the lunar highlands prove to be even older, ranging up to 4.5 billion years in age.

The antiquity of these ages is significant. Although the age of the planets in our Solar System is thought to be close to 4.6 billion years, the 4.5 billion-year ages from the Moon provided hard *data* in support of this. The Solar System must be at least as old as any of the objects it contains. The dates from the Moon therefore provide a minimum estimate for the age of the Solar System.

Since the Earth is a geologically active planet, rocks with ages in excess of 3 billion years are very rare on its surface. We may therefore learn much about the early history of the Earth by studying ancient processes that occurred on the Moon. The lunar landscape has abundant craters and has obviously been affected by the impact of innumerable meteorites. With its vanishingly thin atmosphere, the Moon has little protection from such bombardment. In contrast, the Earth's atmosphere causes most celestial debris to burn up before they reach the surface in a phenomenon misleadingly referred to as "shooting stars." It has been calculated that impacts on the Moon would have had velocities close to 2.4 km/s, or 5360 mi/h (Fig. 15.12). Such impacts would have released sufficient heat energy to melt adjacent rocks and eject vast volumes of material to form a crater. It

(continued)

Figure 15.12
Schematic sequential representation (a–d) of the development of an impact crater on the Moon. The meteorite (a) strikes the lunar surface, (b) penetrates the surface, and (c) vaporizes. Shock waves fracture and eject the rock to form a circular crater. (d) Rebound can result in a raised central portion of the crater. (e) The lunar maria were formed by major impacts that broke the lunar crust and produced multi-ringed basins that were later flooded by basalt lava that rose up the fractures.

is estimated that the impact of a meteorite only 3 m (10 ft) in diameter could result in a crater 150 m (500 ft) wide. On the Moon, fallout from meteorite bombardment has produced a thin veneer of lunar "soil" 3 to 5 m (10 to 15 ft) thick.

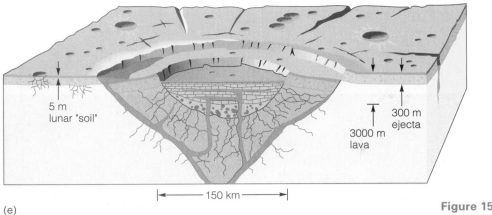

5 m
lunar "soil"

300 m
ejecta

3000 m
lava

|← 150 km →|

(e)

Figure 15.12, continued

Craters are typically more abundant in the older lunar highlands than they are in the younger maria (Fig. 15.13), suggesting that meteorite bombardment was much more common early in the Moon's history and had tapered off significantly by about 3.1 billion years ago. Given that these meteorites traveled through the entire Solar System, it seems certain that the Earth's early fragile crust was also affected by this activity. However, the evidence has since been lost on the Earth because of its restless geology.

It is widely believed that massive meteorite impacts were also responsible for the widespread volcanic activity that characterizes the Moon's early evolution. But what about the origin of the lunar maria? On the Earth, plate tectonic processes provide an excellent explanation for the origin of the topographically depressed regions of the ocean basins because dense oceanic crust rests on a plastic substrate we call the asthenosphere. But there is no evidence of plate tectonic processes on the Moon so we must search for a different mechanism.

The lower frequency of impact structures in the lunar maria implies that their basaltic floors formed after the main bombardments. However, most scientists believe that the lunar maria were also formed by impacts during a period of heavy bombardment between 4.6 and 3.9 billion years ago.

(a)

(b)

Figure 15.13
Photographs showing the contrast between (a) the heavily cratered lunar highlands which represent the oldest portions of the Moon's surface, and (b) the comparatively smooth lunar lowlands, or maria, which contain volcanic rocks and few craters.

PHASES AND ECLIPSES

The Moon and its phases are at the center of many myths and legends, and has captured the imagination of musicians and poets for centuries. In Chapter 9, we discussed the relationship between the phases of the Moon and the tides. Here we address the origin of the phases themselves. Although half of the Moon is always lit by the Sun, the part we see from Earth may be totally illuminated (full moon) or completely dark (new moon) as the relative positions of the Moon, the Earth, and the Sun change in the course of the Moon's 27.32-day revolution about the Earth.[1]

A cursory glance at the night sky reveals that the Sun is responsible for illuminating the Moon. The light that reaches the Earth from the Moon is simply reflected sunlight. Indeed, on clear nights, a full moon can provide appreciable reflected illumination.

Although half of the Moon is always illuminated, we can usually see only a portion of the illuminated half from Earth. This gives rise to what are called the **phases of the Moon** (Fig. 15.14a). Thus, the phase we call "full moon" occurs when the Moon's entire illuminated half faces towards us. This is achieved when the Moon is on the far side of the Earth relative to the Sun (Fig. 15.14b). The phase we call the "new moon," on the other hand, occurs when the illuminated half faces away from us. This is achieved when the Moon is between the Sun and the Earth. Crescent, quarter, and gibbous moons represent conditions in between these two extremes. As the Moon moves from new moon to full, the illuminated portion becomes larger, and it is said to be **waxing**. As the Moon moves from full moon to new, the Moon's illuminated portion becomes smaller, and it is said to be **waning**.

Remember, that the same side of the Moon always faces the Earth; it is just the shadow on the Moon that moves as the phases change. In a similar way, an observer on the Moon would see the Earth go through phases as the position of the Earth's shadow changed.

One troubling point that may have occurred to you is that at full moon, the

[1] The cycle of lunar phases takes 29.53 days or about 4 weeks. This is because the Earth orbits the Sun as the Moon orbits the Earth (see Tides, Chapter 9).

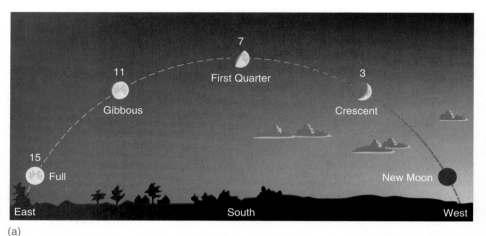

(a)

(b)

Figure 15.14

The phases of the Moon. (a) The changing celestial positions of the Moon at sunset waxes from a new moon to a full moon position in a 15-day period (as viewed in the northern hemisphere). (b) The relative positions of the Earth and Moon during the Moon's phases. Note that for each position of the Moon, the side facing the Sun is illuminated. During the full moon phase, that side faces Earth. During the new moon phase, the illuminated side faces away from Earth. At the first and third quarter phases, an observer on Earth sees only half the illuminated side. During the full moon phase, moonrise is at sunset. During the new moon phase, the setting of the Sun and Moon are simultaneous. Note the Moon's "peak" is for illustration only and reveals that the same side of the Moon continually points toward the Earth.

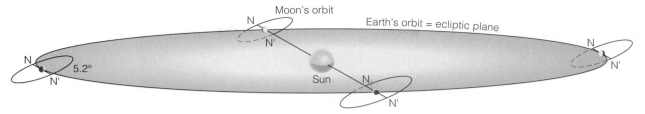

Figure 15.15

Schematic explanation of why eclipses are rare. The Moon's orbit is oblique (5.2°) to the Earth's orbit around the Sun, and there are only two periods each year (shown in front and rear) when perfect alignment between the Earth, Moon, and Sun might occur.

Earth lies between the Sun and the Moon in apparent alignment. So how do the Sun's rays reach the Moon at all? Would they not be blocked by the Earth? Similarly, at the new moon, the Moon is between the Sun and the Earth, once again in apparent alignment. Both situations should produce eclipses, and so we might expect one every 14.76 days. However, such eclipses are relatively rare. So why is this?

Like the Earth, the Moon orbits in a plane. If this plane were the same as the Earth's orbital plane, then eclipses would indeed be very common. However, the Moon's orbit is tilted with respect to that of the Earth by 5.2° (Fig. 15.15). For this reason, the Moon is rarely in strict alignment with the Sun and the Earth. Instead, the Moon generally passes either above or below the line joining the Sun and the Earth when in full moon or new moon positions, so that an eclipse does not occur. Only on those rare occasions when the Sun, the Earth, and the Moon are perfectly aligned do eclipses occur. If the Earth is between the Sun and the Moon, the shadow of the Earth falls across the illuminated side of the Moon producing a **lunar eclipse** (Fig. 15.16a,b). Ancient Greek astronomers realized that the circular shape of this shadow was strong evidence that the Earth was spherical.

A **solar eclipse** occurs during the new moon stage of the cycle when the Moon lies on a direct line between the Sun and the Earth (Fig. 15.16c,d). The shadow of the Moon falls on the Earth and the Moon appears to be almost exactly superimposed on the entire solar disk, resulting in a total eclipse of the Sun. This occurs because the "apparent" diameters of the Sun and the Moon are almost identical when viewed from the Earth. In reality, the diameter of the Sun is about 400 times greater than that of the Moon. However, the Sun is also approximately 400 times farther away during most of the Earth's orbit. As a result, the "apparent" diameters of the Sun and Moon as seen from the Earth are similar so that total solar eclipses occur whenever the Moon is perfectly aligned between the Sun and the Earth. Depending on their relative distances, however, the apparent diameter of the Moon can, at times, be less than that of the Sun, in which case a ring of sunlight remains visible even at the height of the eclipse. When this occurs, it is called an **annular eclipse**.

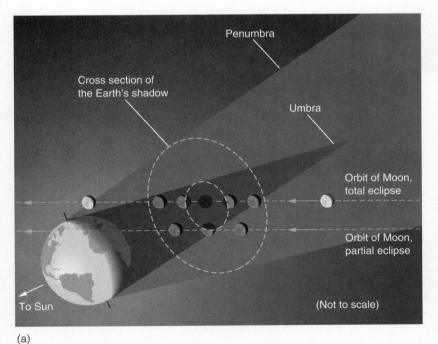

(a)

Figure 15.16

(a) During a total lunar eclipse, the Moon passes through the partially shaded region of the Earth's shadow (known as the penumbra) and into the totally shaded region (known as the umbra). (b) Multiple photo exposure of a lunar eclipse over a 5-hour period. (c) During a total solar eclipse, the Moon lies on a direct line between the Sun and the Earth. A total solar eclipse is seen only in the totally shaded region, or umbra. (d) Sequence of photographs showing the first half of a total solar eclipse.

(continued)

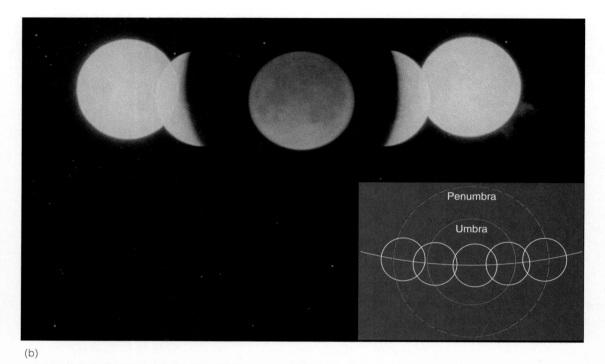

(b)

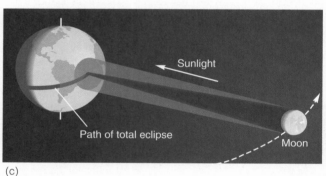

(c)

Figure 15.16, continued

This bombardment was later followed by widespread melting in the lunar subsurface which flooded the crater floors with successive flows of basalt from 3.8 to 3.2 billion years ago.

Until recently, there has been no evidence for the presence of water on the Moon. However, in 1998 data from NASA's Lunar Prospector mission revealed that ice crystals make up 1% of the lunar soil near the polar regions and also occur in some deep trenches where "the Sun doesn't shine."

The total supply of water is estimated to be between 10 and 330 million metric tons, and is thought to have been deposited by the impacts of comets.

The Moon's surface is characterized by the absence of oxygen. For example, rocks containing metallic iron are found on the Moon, whereas on Earth, metallic iron rapidly oxidizes (that is, it absorbs oxygen from the atmosphere) to form iron oxide, or rust.

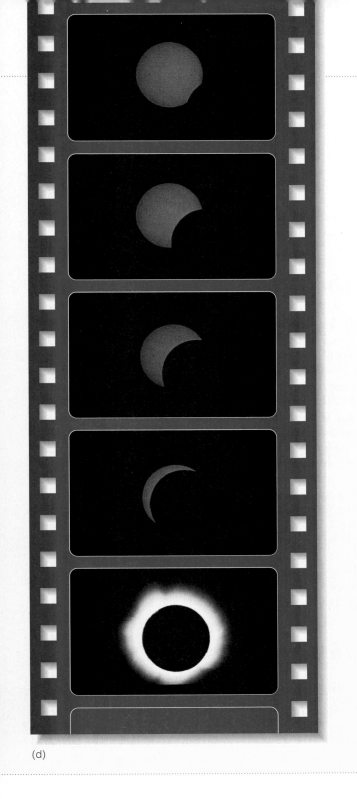

(d)

The Moon's ancient volcanic rocks, like those on Earth, originated from the melting of its interior. Since the rock produced in both cases is basalt, can we argue that the respective interiors of these bodies are similar? The answer is both yes and no.

A general comparison of their respective interiors can be obtained by comparing their average densities. The Moon's average density is 3.3 g/cm³, whereas that of the Earth is 5.5 g/cm³. Calculations show that this density difference cannot be attributed to the greater pressure that exists within the Earth because of its greater size. Therefore, the average composition of the two bodies must be fundamentally different. However, the average density of the Moon is similar to that of the Earth's mantle. Since the mantle is thought to be the source of basalts on Earth, an average lunar composition similar to that of the Earth's mantle would be consistent with the similarities between their respective volcanic rocks.

15.4.3 Origin of the Earth-Moon System

Despite nearly two thousand years of investigation and speculation, the origin of the Earth-Moon system is still uncertain. Three traditional hypotheses for lunar formation appear to have serious difficulties. The **condensation hypothesis**, which proposes that the Earth and Moon condensed simultaneously from the same raw materials, has been abandoned because of the Moon's much lower density. The **capture theory**, which proposes that the Moon formed somewhere else and was captured by the Earth's gravitational field, has also been discredited. To be captured in Earth's orbit, the approaching Moon would have to have had a very high velocity. But at such velocities, the stresses associated with capture by the Earth's gravitational field would almost certainly have led to the Moon's disintegration. The **fission hypothesis** proposes that the Moon split from a rapidly spinning young Earth shortly after its formation. However, the rate of spin required to split the Moon from the Earth is such that the Earth-Moon system should have a higher angular momentum than is currently observed.

A more recent model that draws its strength from the similarity of the Moon's average density with that of the Earth's mantle is now the most favored hypothesis (Fig. 15.17). Following the descent of dense liquid iron to the Earth's core, the impact of a very large object (perhaps the size of Mars) on the Earth's surface would have resulted in the wholesale expulsion of material of a similar composition to that of the Earth's mantle. The impact would have broken through the Earth's thin solid crust into the mantle, which at that time was predominantly liquid. The ejected material was probably a splash of hot liquid, which cooled and solidified in orbit around the Earth to form the Moon. In this scenario, the Moon's high angular momentum (about 10 km/s) is attributed to the impact.

Although this model is speculative, it at least fits the facts and constraints as we now know them. Moreover, if such collisions were common throughout the Solar System at this time, they would explain the variable tilts and periods of rotation of the planets shown in Table 15-3.

The features of the Moon indicate that its surface is essentially ancient, static, and lifeless. The absence of modern volcanic activity may account for its lack of an atmosphere because unlike the Earth, no new gases are being vented from active volcanoes. Because of its small size (and, hence, its low gravitational attraction for atmospheric gases), gases from ancient lunar volcanoes have long since been dissipated into space.

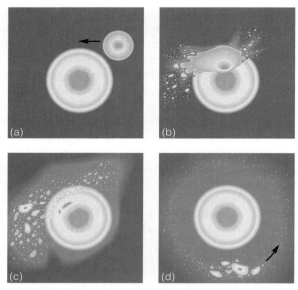

Figure 15.17
Origin of the Earth-Moon system from the impact of a Mars-sized body (a) on the Earth's surface (b). Note that the Earth's internal layers were developed at the time of impact (although they were probably not totally solid). Condensing debris from the impact of both bodies (c) formed an orbiting cloud around the Earth (d) from which the Moon began to accrete.

To summarize, the Moon preserves an ancient, lifeless surface in contrast to a dynamic Earth. But it is nevertheless our companion in space and the early histories of both bodies are undoubtedly closely related. Because of this, studies of the Moon can be used to fill important gaps in our knowledge of the Earth's early history.

In the next section, we will consider the terrestrial planets which appear to have a similar origin to that of the Earth. As we will see, the size and distance from the Sun has had a major influence on the evolution of these planets.

15.5 The Terrestrial Planets and Their Satellites

Figure 15.18 shows the arrangement of the planets in the Solar System as we now know it, from Mercury, the innermost planet, to Pluto, which is generally considered to be the outermost planet. Table 15-3 compares some of their important properties. As we journey through the Solar System examining each in turn, we will find that every planet is unique. But important common themes will emerge that will provide us with new insights into the processes that influenced the Earth's evolution. Do other planets have plate tectonics, for example? Do they have active volcanoes and atmospheres? Is there something about a planet's internal composition that governs its evolution? What makes each planet unique, and is Pluto really a planet?

A general idea of a planet's chemical composition is given by its mean density (Table 15-3). The innermost (terrestrial) planets, from Mercury to Mars, have relatively high densities suggesting that, like the Earth, they are essentially rocky spheres. In contrast, the outer (Jovian) planets (from Jupiter to Neptune) have significantly lower densities and are essentially gaseous bodies with poorly defined surfaces.

As we shall see, the planets also vary in other important characteristics primarily because of their size and distance from the Sun. Some planets and their moons, for example, preserve ancient surfaces that have been undisturbed for billions of years. Others have very young surfaces suggesting that they are tectonically active.

15.5.1 Size

Size is an important factor in models for planetary evolution. The innermost terrestrial planets are much smaller than the gaseous Jovian ones (Table 15-3; Fig. 15.8). For two terrestrial planets of similar composition, size gives an indication of the rate at which the planets cooled. Earth and Venus are the largest of the terrestrial planets and it may be no coincidence that these planets are still volcanically active. On the other hand, smaller Mercury and Mars cooled more rapidly and are essentially geologically dead.

The presence of atmospheres on Earth and Venus and their virtual absence on Mercury and Mars may also be due to factors that ultimately depend on size. Volcanic activity, which plays an important role in the generation of atmospheres, will cease first on smaller bodies. Furthermore, gases that may once have been expelled from ancient volcanoes are more loosely held by gravity on smaller planets and may escape into space. This is indicated on Table 15-3 by the lower escape velocity on Mercury and Mars relative to that of Earth and Venus. The **escape velocity** is the speed a gas needs in order to escape the gravitational attraction of a planet and dissipate into space.

15.5.2 Relative Distance from the Sun

Our Solar System is but a very tiny proportion of the universe; its size is dictated by the extent of the gravitational influence of the Sun on its nearest celestial neighbors. This gravitational attraction wanes rapidly with distance from the Sun. Pluto, the outermost planet, for example, has a very poorly constrained orbit, consistent with this diminishing influence.

A planet's distance from the Sun also influences the amount of solar radiation it receives. The distance of Mercury to the Sun (denoted 0.39 AU) is 39% of that of the Earth and, as a consequence, Mercury receives approximately 10 times more solar radiation than the Earth. A planet's distance from the Sun also influences its orbital period, that is, the amount of time (in Earth years) it takes a planet to complete one revolution around the Sun. The more distant the planet, the longer it takes to complete a single orbit.

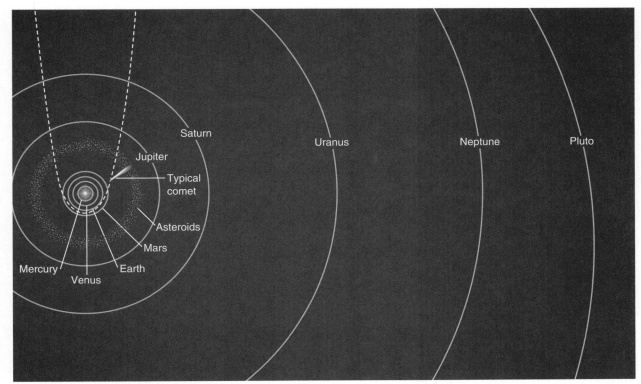

Figure 15.18
Arrangement of the planets in the Solar System as it is now known. The orbit (dashed line) of a
typical comet (with a "tail" pointing away from the Sun) and the Asteroid belt are also shown. Note
the skewed path of Pluto, which brings it closer to the Sun than Neptune during part of its orbit.

15.5.3 Mercury

The Romans named the planet Mercury, noted for its rapid movements, after Mercury, the messenger of the gods. Mercury is the second smallest planet in the Solar System (after Pluto) and, with a diameter of 4878 km (3024 mi), it is about 40% bigger than the Moon. Because of its small size and the intense solar radiation it receives, it is not surprising that Mercury has virtually no atmosphere. Only minute traces of sodium, hydrogen, helium, and other inert gases have been found. Any gases that do occur on the surface are only loosely held by Mercury's low gravity and readily acquire the necessary energy to reach their escape velocity from the intense solar radiation.

It takes Mercury only 88 Earth days to complete one orbit around the Sun. A Mercury "year" is short, due to its proximity to the Sun. However, it takes Mercury 59 Earth days to complete one rotation around its own axis. So a Mercury "day" (59 Earth days long) is approximately two-thirds the length of a Mercury "year" (59 divided by 88). The slowness of Mercury's rotation is probably due to the gravitational attraction of the Sun which acts as a brake.

The Mariner 10 spacecraft mission of 1974–75 showed that Mercury's surface more closely resembles the cratered surface of the Moon than the juvenile surface of the Earth (Fig. 15.19a). Indeed, the abundance of craters is such that scientists believe that Mercury's surface, like that of the Moon, must be similarly ancient. By implication, scientists have also concluded that

Mercury, although fractured (Fig. 15.19b) is an essentially dead planet with no recent volcanism. If ancient volcanic activity provided gases for a primitive atmosphere, these gas molecules have long since escaped into space. So Mercury has virtually no atmosphere and no chance of ever getting one.

The surface temperatures on Mercury vary from 430°C (800°F) at "midday" to −180°C (−290°F) at "midnight." Because Mercury rotates very slowly, the side that faces the Sun's intense radiation does so for a much longer period of time than it does on Earth and so becomes very hot. However, since Mercury has no atmosphere, there is no "greenhouse effect" to store this solar heat. Once the daylight side rotates into night, it loses heat rapidly and becomes very cold. In fact, in 1991, a team of scientists from the California Institute of Technology proposed that radar images of the planet's polar regions (an area shaded from solar radiation) were more typical of ice than rock. They suggested that vapor burned off impacting comets may have been trapped there as ice.

The mean density of Mercury is 5.4 g/cm³, which is comparable with that of the Earth (average density 5.5 g/cm³). However, calculations show that if the Earth were the same size as Mercury, its density would be only 4.2 g/cm³. This suggests that a larger proportion of Mercury consists of a metallic core compared to the proportion on Earth (Fig. 15.20). A weak magnetic field (about 1% of that of Earth) has also been detected, consistent with the presence of an iron-rich core. However, scientists debate whether this core is predominantly solid or liquid.

(a)

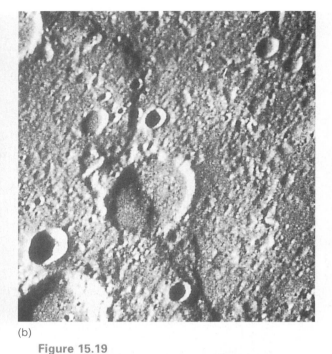

(b)

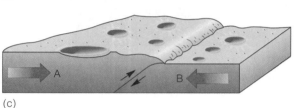

(c)

Figure 15.19

(a) Photomosaic of Mercury made by the Mariner 10 spacecraft. Craters are typically between 100–150 km (60–90 mi) across. (b) Photograph revealing the presence of faults on Mercury (upper left, to bottom center). These faults often transect the craters and produce cliffs. (c) A schematic diagram of the proposed geometry of these faults. They are thought to be related to contraction of the planet after early heating.

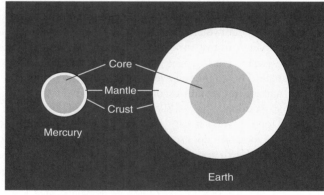

Figure 15.20

Internal layering of Mercury compared to the Earth. Note the thick core of Mercury relative to its mantle, as indicated by its relatively high density.

15.5.4 Venus

When Venus was named after the Roman goddess of love and beauty, scholars were unaware of the nightmarish conditions that exist at its surface. Instead, they saw it only as the brightest planet in the night sky. In the 17th Century, Galileo observed that Venus had "phases" much like those of the Moon. Like the Moon, half of the surface of Venus is illuminated by the Sun at any one time, but when viewed from the Earth different por-

tions of the illuminated disk are visible. Galileo correctly proposed that the phases were consistent with the Copernican system of the orbit of Venus around the Sun, rather than the Ptolemaic system of its orbit around a stationary Earth.

Until the advent of the space age, Venus was viewed as our planetary twin, perhaps even harboring life such as tree ferns and jungle creatures. This view was held because the two planets have approximately the same diameters (12,100 km or 7500 mi for Venus, 12,740 km or 7900 mi for Earth) and are neighbors in the Solar System. Observations also suggested that Venus had an atmosphere which supported the idea of life on its surface (Fig. 15.21). This atmosphere is most visible when Venus passes between the Earth and the Sun, and was first noted by the Russian astronomer Mikhail V. Lomonosov in 1761. However, its dense atmosphere camouflages a most inhospitable surface. Any theories that Venus may have harbored life were laid to rest when intense microwaves emanating from the planet revealed a surface temperature of about 470°C (880°F). Furthermore, the surface pressure below an atmosphere predominantly composed of carbon dioxide would be about 95 times that of the Earth. In fact, it is equivalent to the pressure that exists nearly one kilometer below the Earth's surface.

Paradoxically, before the modern space age we knew more about the interior of Venus than its surface. Although its mean density (5.25 g/cm³) is lower than that of the Earth (5.52 g/cm³), calculations show that if the Earth had the same diameter, it would also have the same density. This suggests that the mean

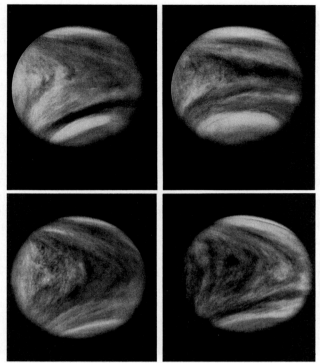

Figure 15.21

The swirling atmosphere of Venus as seen in photographs taken 5 hours apart. In order to see the contrasting features, the photographs were taken in ultraviolet light.

Figure 15.22

A mosaic image of the surface of Venus as revealed by radar, which can penetrate the thick cloud cover. The image shows both impact craters and volcanoes (near center).

composition of the two planets is similar, and that Venus has both a silicate-rich mantle and an iron-rich core. Unlike the Earth, however, Venus has no magnetic field.

Eventually, the deep cloud cover on Venus was penetrated using radar signals emitted from Earth and from space probes. Radar has short wavelengths and can be concentrated into narrow beams that can penetrate the cloud cover (Fig. 15.22). By bouncing radar off its surface, the time it takes for Venus to complete a single rotation about its own axis was determined. Prior to this, the thick pervasive cloud cover meant that this important information had eluded scientists. The results were astonishing. Venus proved to be the only terrestrial planet that is spinning backwards, rotating clockwise rather than counterclockwise as viewed from above. On Venus, therefore, the Sun rises in the "west" and sets in the "east," the opposite pattern to that on Earth. Venus also rotates rather slowly with a period of rotation (a Venus "day") that is 243 Earth days long. The reasons for the reverse spin and the slow period of rotation are uncertain but may account for the absence of a magnetic field on Venus.

A year on Venus is longer than a year on Mercury, but shorter than one on Earth. It takes 225 Earth days for Venus to complete one revolution around the Sun. This is consistent with Kepler's laws for increasing orbital periods with distance from the Sun. However, it implies that a Venus year (225 Earth days) is actually shorter than a Venus day (243 Earth days)!

Recent space probes that have orbited Venus and landed on its surface reveal that its atmosphere is almost entirely car-

bon dioxide (96.5%) with a minor amount of nitrogen (nearly 3.5%). There are trace amounts of many gases including sulfur dioxide and water vapor. Its abundant atmospheric carbon dioxide, immense atmospheric pressure, and closeness to the Sun make Venus a most inhospitable world. It is a celestial example of a "runaway greenhouse" with surface temperatures that are so high that no liquid water can exist on its surface. There is consequently little prospect of life on the planet. In fact, the first probe to land on its surface, the Soviet-built Venera 7, lasted only 23 minutes before being destroyed.

Atmospheric circulation on Venus is also very different from that on Earth. The cloud layers of Venus occur at a height of about 55 to 60 km (34 to 37 mi) above its surface, where the temperature is close to 0°C. This is more than 50 km (31 mi) higher than typical cloud elevations above the Earth. It does rain on Venus, but the drops of sulfuric acid that fall from the clouds do not reach the surface. Instead, the moisture heats rapidly as it falls, and rises again as it evaporates.

The reflection of radar signals from the surface of Venus have provided images of its topography (Fig. 15.22). Like an echo sounder measuring water depth in a lake or ocean (see Chapter 8), the time elapsed between the transmission of the radar signal and the reception of an "echo" gives an indication of surface elevation. The surface consists of elevated "mountainous" regions (15%), comparable in height with the Himalayas, less elevated plateaus (25%), and relatively low-lying plains (60%). In comparison to either Mercury or the Moon, Venus has relatively few impact craters, implying that its surface is relatively young and perhaps still volcanically active. Images from the Magellan probe in 1990 revealed volcanic vent structures, lava flows, and radial fracture patterns similar to those on Earth.

Although no volcanic activity has been detected directly, sensors have monitored subtle changes in the atmospheric concentration of some key gases, most notably sulfur dioxide, that are similar to those accompanying volcanism on the Earth. For example, in the three week period following the 1991 eruption of Mount Pinatubo in the Philippines, 20 million tons of sulfur dioxide gas were vented into the atmosphere (see Chapter 1).

If Venus is volcanically active, then could some form of plate tectonics operate on its surface? Venus does have some surface features that resemble Earth's oceanic trenches and associated subduction zones. However, long linear mountain belts like the Himalayas and Alps (which represent continent-continent collisions), or the Andes (which represent ocean-continent collisions) are absent. These differences may be due to the high surface temperatures on Venus, which would make its crust less rigid and more pliable than the crust of the Earth. This would inhibit Earth-like subduction, which is greatly facilitated by the presence of old, cold, and dense oceanic crust.

Venus may resemble a very primitive Earth, with thin, fragile crustal plates or wafers. If so, *processes* on Venus would provide insight into the Hadean or Early Archean geology of the Earth. In contrast to the Earth, which loses more than 70% of its heat at mid-ocean ridges, Venus expels most of its heat from large currents of hot magma that rise beneath the crust. It is speculated that cooler magma may sink into the interior and that the combination may cause minor horizontal movement of the crust. If so, these processes may give us a glimpse of the very primitive style that plate tectonics could have shown on Earth in the Early Archean (see Chapter 13).

As Venus cools, there is an intriguing possibility that it could follow a similar evolutionary path to that of the Earth. The planet may well develop some form of plate motion in the future as its cooling surface conditions approach those of the Earth. Since life on Earth initially evolved beneath an atmosphere similar to that of present day Venus, we cannot exclude the possibility that some form of life may eventually evolve there.

15.5.5 Mars

If Venus is a model for the primitive Earth, Mars may shed light on its fate. Because of its reddish-orange surface color, Mars was named by the Romans in honor of their god of war. Of all the planets, Mars is the one historically thought most likely to support life and the one whose surface has been most intensely studied by telescope. In 1895, Percival Lowell, drawing on the ideas of Father Angelo Secchi in Rome, interpreted streaky marks on the surface of Mars as Martian canals built for the purpose of transporting water from the cold polar areas to the hot equatorial regions. These interpretations inspired H.G. Wells to write "The War of The Worlds," and the famous Orson Wells radio production of the book which convinced many Americans that they were being invaded by Martians in 1938.

However, the space age put all fantasies of a race of Martians to rest. The first close-up pictures of Mars were obtained in 1965 from the Mariner 4 spacecraft, and both Soviet (1971–74) and American (1976, 1997) space probes have since landed on the Martian surface (Fig. 15.23). From these probes, scientists have concluded that there is little possibility of life on the Martian surface today. However, it remains possible that life might exist in warmer environments below the surface. There also remains a tantalizing possibility that primitive life forms may have existed either at or near the Martian surface in the distant past (see Chapter 13, Living on Earth: Extraterrestrial Life on Earth?). The red-orange color of Mars is due to the presence of oxidized iron, or rust on its surface. On Earth, rust is formed by absorption of oxygen from the atmosphere, the origin of which is attributed to photosynthesis. If a similar process of oxygen absorption was responsible for the coloration on Mars, then photosynthetic organisms may have once produced free oxygen in the Martian atmosphere.

Mars has two tiny, angular-shaped moons, Phobos (28 km long by 20 km wide, or 17 mi by 12 mi) and Deimos (16 by 12 km, or 10 by 7 mi), named after the sons of the Greek war god Ares. These small moons are thought to be asteroids that were captured into orbit around Mars as they traversed the Solar System.

Mars is more distant from the Sun (at about 1.5 AU) than is the Earth, and takes the equivalent of 687 Earth days to complete a single revolution about the Sun. But a Martian day is similar in length to an Earth day since Mars takes 24.5 hours to complete one rotation about its axis. At 3.93 g/cm³, the average density of Mars is the lowest of all the terrestrial planets, suggesting that its iron-rich core may be poorly developed. This would be compatible with the planet's lack of a magnetic field.

Figure 15.23

Photograph of the Martian surface from the Viking spacecraft. The dark streak across the middle of the planet is a deep canyon known as the *Valles Marineris*. Two examples of Martian volcanoes are visible along the left horizon. The red surface of Mars indicates the presence of iron oxides in its soil.

The iron-rich Martian surface is also atypical of the other terrestrial planets and it is speculated that, in contrast to the Earth, iron never sank to the core of Mars. This, in turn, would imply that Mars never experienced a meltdown like that postulated on the Earth during the "iron catastrophe" (see Section 15.4).

   The diameter of Mars is just over half that of the Earth, so it probably cooled significantly faster. In the southern hemisphere, much of its surface is densely cratered terrain similar to that of the Moon and Mercury. This suggests that the surfaces of these three bodies are of similar age. However, the northern hemisphere of Mars is less cratered, and suggests a period of renewed volcanism about one billion years ago. There is only limited evidence of more recent volcanic activity. These observations emphasize the importance of size in the evolution of a planet. It appears that Mars, which lies between the sizes of Mercury and Venus in size, became geologically extinct after smaller Mercury but before larger Venus.

As a consequence, Mars has only a very thin atmosphere with a surface atmospheric pressure only 0.7% of the pressure at sea level on Earth. The relative proportions of gases in the atmosphere are very similar to those of Venus (Table 15-4), being dominated by carbon dioxide (95%), nitrogen (3%), and argon (1.6%). Water vapor and oxygen are only present in trace amounts. However, the total amount of atmospheric gas is so small (as indicated by the very low atmospheric pressure it exerts) that it produces virtually no greenhouse effect. The absence of ozone in the atmosphere also implies that Mars has no protective screen for the Sun's ultraviolet radiation.

Surface temperatures on Mars are highly variable. The average is about −30°C (−22°F) consistent with the terrestrial planets' trend of decreasing average surface temperatures with

increasing distance from the Sun. Like the Earth, surface temperatures may vary considerably from pole to equator. At the equator, daytime temperatures may rise to about 25°C (77°F) but at night they might drop below −70°C (−94°F). At the poles, intensely cold temperatures of −120°C (−184°F) have been recorded by the Viking space probes. Despite the thin atmosphere, the contrast between polar and equatorial temperatures is sufficient to cause vigorous atmospheric circulation and is thought to be responsible for dust storms and the generation of Earth-like weather frontal systems.

Since Mars is tilted on its axis at a similar angle and in a similar direction to the Earth, it too has four seasons and a day of similar length. Like Earth, each season is about one quarter of its year. However, since Mars has an orbital period of 687 days, each season is just over 170 days in length, or close to six Earth months. Evidence of these seasons is confirmed by photographs taken above the Martian poles (Fig. 15.24). These photographs reveal the presence of ice caps which are believed to be composed of frozen carbon dioxide and water. The ice caps expand in the winter and contract in the summer, suggesting that they are formed by freezing of the Martian atmosphere.

Mars has a highly exaggerated topography, with very high mountains and deep valleys (see Figure 15.23). Despite their advanced age, Mars has the largest volcanoes known in the Solar System. The volcano Olympus Mons stands 24 km (15 mi) high, dwarfing the very highest mountains on Earth, which attain a mere 8.8 km (5.5 mi) in height, and the Earth's largest volcano, Mauna Kea in Hawaii (see Chapter 5) that rises a mere 10 km (6 mi) above the floor of the Pacific Ocean. But the Martian surface has none of the features we traditionally associate with plate tectonics, such as earthquake zones defining plate

Table 15-4

Comparison Between the Composition of the Atmospheres of Venus, Earth, and Mars

VENUS		EARTH[a]		MARS[b]	
GAS	PERCENT VOLUME	GAS	PERCENT VOLUME	GAS	PERCENT VOLUME
CO_2 (carbon dioxide)	96.5	N_2	78.1	CO_2	95
N_2 (nitrogen)	3.5	O_2	20.9	N_2	2.7
SO_2 (sulfur dioxide)	0.015	H_2O	0.05 to 2 (variable)	Ar	1.6
H_2O (water vapor)	0.01	Ar	0.9	CO	0.6
Ar (argon)	0.007	CO_2	0.03	O_2	0.15
CO (carbon monoxide)	0.002	Ne	0.0018	H_2O	0.03
He (helium)	0.001	He	0.0005	Kr	Trace
O_2 (oxygen)	≤0.002	CH_4 (methane)	0.0002	Xe	Trace
Ne (neon)	0.0007	Kr (krypton)	0.0001		
H_2S (hydrogen sulfide)	0.0003	H_2 (hydrogen)	0.00005		
C_2H_6 (ethane)	0.0002	N_2O (nitrous oxide)	0.00005		
HCl (hydrogen chloride)	0.00004	Xe (xenon)	0.000009		

[a]Compositions are for near-surface conditions, with terrestrial data other than H_2O tabulated for dry conditions. CO_2 on Earth is probably increasing by 2% to 3% of the listed amount in each decade, because we are burning so much fossil fuel. This activity may be modifying Earth's climate.

[b]Amounts of gases vary slightly with season and time of day. H_2O is especially variable. Some CO_2 condenses out of the atmosphere into the winter polar cap; changing cap sizes cause small changes in the total Martian atmospheric pressure.

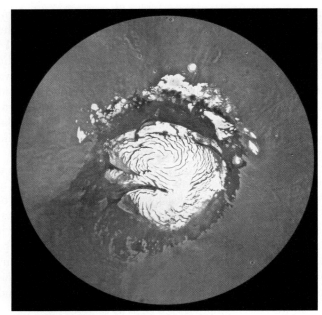

Figure 15.24
Computer-enhanced photomosaic of the Martian north polar ice cap.

boundaries. In the absence of plate tectonic processes, most planetary geologists believe the Martian volcanoes to be giant examples of hotspot type volcanism, possibly initiated by meteorite impacts. This raises the possibility that similar features on Earth, such as the Hawaiian chain, may have been initiated in a similar fashion. The exaggerated height of the volcanoes on Mars may reflect the greater extent of melting. Alternatively, since there are no moving tectonic plates on Mars, the products of hotspot volcanism may have accumulated in one place instead of producing the linear island chains characteristic of oceanic hotspots on Earth (Fig. 15.25). The height of the volcanoes implies that the Martian crust is thick enough to support the load. On Earth, elevation is influenced by isostasy because the lithospheric plates float buoyantly on the asthenosphere. It is therefore unlikely that Mars has an asthenosphere beneath its crust. The absence of an asthenosphere implies no detachment of the Martian crust from its deep interior and, hence, no possibility of plate motion.

The deep rifts on Mars are spectacular features. Many attain a depth of more than 4 km (2.5 mi), or nearly three times that of the Grand Canyon, and a length equivalent to that of the entire United States (see Figure 15.23). These rifts are the "canals" identified by Percival Lowell, but rather than being the work of desperate Martians, they are deep waterless natural canyons. However, they may not always have been waterless. In detail, the patterns of these canyons resemble drainage systems on Earth, but on a much grander scale. Some estimates suggest that the water that carved these canyons may have flowed at rates up to 10,000 times greater than that of the Mississippi River.

This interpretation has enormous significance because it suggests that water once flowed across the Martian surface. But if so, where did all the water go? There are three possibilities. The very low atmospheric pressure on Mars would have promoted rapid evaporation. The moisture may have therefore escaped the gravitational clutches of the planet, in which case it is lost forever. Alternatively, the water may exist as ice in the polar regions. The polar ice caps store vast quantities of liquid that if unleashed by melting would have tremendous powers of erosion. However, most of this ice is dry ice, consisting of carbon dioxide rather than water. A final possibility is that the water exists as "permafrost" in the Martian soils. On melting, it too could fuel catastrophic flooding and erosion (Fig. 15.26).

By counting the craters in these canyons, it is estimated that the huge floods that may have carved them occurred between 1 and 3 billion years ago. But if water flowed across the Martian surface in the past, how may this have happened? One possibility is that the release of water from the interior was triggered by meteorite impacts, like a needle pierces a water balloon. Another possibility is that the Martian climate was warmer in the past, as indicated by past periods of intense volcanic activity. Like the other terrestrial planets, ancient volcanic activity would certainly have vented greenhouse gases from the Martian interior to its atmosphere. During periods of enhanced volcanism, atmospheric pressures on Mars would have been higher so the greenhouse effect would have been greater and surface temperatures would have been higher. This would have facilitated the melting of polar ice and/or permafrost and may have triggered catastrophic flooding.

However, like the Earth, other more subtle influences may affect Martian surface temperatures. Over the past million years, for example, our climate has been driven in and out of ice ages by cyclic changes in Earth-Sun geometry produced by variations in the Earth's tilt, by the eccentricity of its orbit, and by its precession. We will examine these so-called Milankovitch cycles in Chapter 20. However, there is nothing inherent in these cycles that would restrict their occurrence to the Earth. So changes in Mars-Sun geometry may also have triggered the melting or growth of ice sheets on Mars.

Obviously, many fundamental questions about Mars remain to be resolved. If Mars was warmer in its past, then it may have experienced similar conditions to those in which life was initiated on Earth (see Chapter 13, Living on Earth: Extraterrestrial Life on Earth?). The presence of iron oxide-rich rocks at the surface implies an ancient atmosphere that must have been far richer in oxygen than the present one. However, "the window of opportunity" for life on the Martian surface was small. The fact that Mars cooled faster than the Earth means that any primitive life forms, if they had existed, would not have evolved to any significant degree. The very thin carbon dioxide-dominated atmosphere and the very low average temperature on Mars suggests that it is very unlikely that life could exist at its surface today. However, life may be possible in the warmer subsurface.

As the Earth continues to cool and its internal supply of heat wanes, it too will become unable to replenish its atmosphere or its hydrosphere. Hence Mars may illustrate the ultimate fate of our own planet.

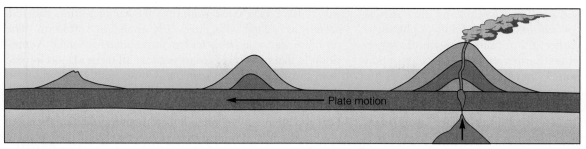

(a) Hotspots on Earth

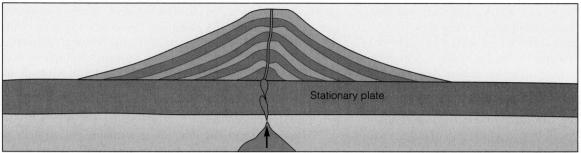

(b) Hotspots on Mars

Figure 15.25
Comparison between the hotspots on Earth and Mars. (a) The typical development of a volcanic island chain above a hotspot on Earth. As the Earth's lithospheric plates move, the mountains cool and subside as they are transported away from the hotspot. (b) A Martian volcano develops above a stationary hotspot on a crust that is not moving. As a result, Martian volcanoes can build to great size.

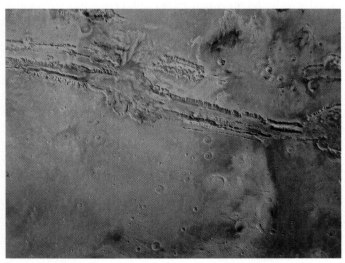

Figure 15.26
Photograph from the Viking space probe showing about 180 km (110 mi) of the Martian surface. Note evidence of erosion, suggesting the former presence of liquid water on the planet. Erosional features resemble Earth's continental drainage patterns (see Chapter 10). Water is trapped today in the polar ice caps and as permafrost in the Martian soil.

15.6 Comparison of the Terrestrial Planets

There are very important lessons to be learned about our own planet from the examination of our terrestrial planetary neighbors. The early history of the Solar System seems to have been a turbulent one, as evidenced by the preservation of abundant meteorite impacts on many of the terrestrial bodies. If we assume that the Solar System is at least as old as the material it contains, then radioactive age determinations on meteorites of up to 4.57 billion years provide us with a minimum age for its origin. The oldest age determinations on any large celestial body (4.5 billion years) come from the Moon.

15.6.1 Life

It now seems clear that the Earth is the only planet that can harbor life as we know it. Only Earth has the right combination of distance from the Sun, atmospheric composition, active tectonics, and a biosphere capable of photosynthesis. In the simplest sense, Earth is like Goldilocks' porridge. Venus is too hot, Mars is too cold, and the Earth is just right!

Soon after life gained a tentative foothold on the Earth, at least in geologic terms, it acquired the capability of photosynthesis. As mentioned earlier, the resulting release of oxygen had profound effects on the Earth's atmospheric composition and subsequent evolution. When this process took hold, the Earth's evolution was diverted onto a fundamentally different course from that of its celestial neighbors.

An inventory of carbon in the Earth's surface layers shows that 99.9% of it now resides in carbonate, a rock formed by biological and biochemical processes. The bulk of carbon dioxide vented from the Earth's volcanoes was consequently taken into the biosphere by photosynthesis and by other life processes, ultimately becoming a component in carbonate rock. The fact that remnants of the earliest life are preserved in such rocks provides evidence of this process. So life really *is* a highway, because it provides the route by which carbon dioxide travels from the atmosphere to its final destination in solid rock.

Since the earliest life existed in the absence of oxygen, it would have had no protective ozone shield because ozone in our atmosphere is produced by the interaction of oxygen with the Sun's radiant energy. Without its atmospheric shield, life is likely to have begun in places where water could offer some protection from ultraviolet radiation, either in intertidal pools or perhaps in deep ocean waters near primitive mid-ocean ridges (see Chapters 8 and 13).

The possibility that life once existed on the Martian surface cannot be ruled out. However, if life had evolved there, the planet's inability to replenish its atmosphere once volcanic activity ceased would have doomed whatever fragile life had gained a foothold. However, primitive life may still exist in specialized environments in the subsurface. Similarly, we cannot be sure that the evolution of Venus precludes the possibility of life on its surface either in the past or in the future. Earth's early life originated in a very similar carbon dioxide-rich environment and it is conceivable that Venus was more hospitable early in its history and may be so again as the planet cools.

15.6.2 Atmospheric Composition

The size of a planet is very important to the development of an atmosphere for two reasons. First, size greatly influences the rate of planetary cooling, and therefore the longevity of volcanic activity. Mercury was too small to maintain its volcanism, and Mars, which is larger than Mercury but smaller than the Earth, became essentially geologically dead about 1 billion years ago. Venus, which is of similar size to the Earth, may still be volcanically active. Active volcanism is needed to generate and replenish an atmosphere by venting gases from the planet's interior, and by bringing new chemical nutrients to the surface that life can use as it evolves. Second, size increases the gravitational attraction a planet has for its atmospheric gases. Mercury and Mars, for example, were too small to retain the gases vented by their ancient volcanism, so that gases eventually leaked into space.

The modern composition of the Earth's atmosphere is unique. However, this may not always have been the case. Once the last remnants of the solar nebula were expelled following ignition of the Sun's nuclear furnace, the terrestrial planets would have developed primitive atmospheres from the outgassing associated with their early abundant volcanism. The primitive Earth probably had an atmosphere of similar composition to that of Venus or Mars because all three planets formed from the same celestial material. Consequently, volcanism on these planets should have produced very similar gaseous products. The initial release of trapped vapors by internal planetary heat and intense meteorite bombardment should have therefore produced similar primitive atmospheres on each of these planets.

The presence of nitrogen in meteorites shows that the terrestrial planets were capable of releasing nitrogen to their atmospheres during volcanism. Indeed, nitrogen occurs among the gases vented from Earth's volcanoes. However, comparison with the atmospheres of the other terrestrial planets suggests that the predominance of nitrogen in our modern atmosphere has occurred largely by default. Its proportion on Earth has built up slowly over geologic time (to 78% today) because, unlike carbon dioxide and water, there is no efficient process that extracts nitrogen from the atmosphere.

The Earth's modern atmosphere is unique among the terrestrial planets because it contains oxygen (21%). When evolving organisms developed photosynthesis and so began to release oxygen, the Earth started down a very different path from that of its nearest neighbors. Increasing oxygen led to new aerobic life forms and to the development of an ozone layer that shielded the Earth's surface from harmful ultraviolet radiation.

15.6.3 Climate

Distance from the Sun is an important factor in determining climate on the terrestrial planets, because the greater the distance, the lower the intensity of solar radiation. But well-established theories on the origin and evolution of stars suggests that the Sun is 25 to 30% brighter today than it was when the Solar System originated, so that the intensity of solar luminosity has increased progressively with time. This introduces a dilemma known as the "faint-young-Sun paradox." As pointed out by the late Carl Sagan and George Mullen of Cornell University, if the Earth's atmosphere had always had the same composition as it does today, and the intensity of solar radiation (or the distance from the Sun) was the only factor determining its climate, then the Earth should have been ice-covered until about 2 billion years ago because the Sun was so much dimmer in the past. This conflicts with the geologic record of water-deposited sedimentary rocks that extends back almost 4 billion years.

Many scientists believe that the greenhouse gas content of the Earth's early atmosphere (notably carbon dioxide) was much higher than it is today and that it has decreased with time to balance the increase in solar luminosity. As a result, liquid water has been maintained on the Earth's surface. This may not be just chance. The Earth may have a built-in thermostat that controls our mean surface temperature by regulating the greenhouse gas content of its atmosphere, cooling the planet when it gets too hot and warming the planet when it gets too cold. The ability of the Earth to regulate its own environment is a key element of the Gaia hypothesis which will be discussed further in Chapter 19.

Mercury has virtually no atmosphere, so its surface temperature is almost exclusively controlled by solar radiation. On Venus, the temperature has become incredibly hot because of the high atmospheric pressures exerted by greenhouse gases such as carbon dioxide. Furthermore, since Venus does not harbor life, it does not have the ability to extract carbon dioxide from its atmosphere and so lower its surface temperature. However, in the early days of the Solar System, when the Sun was much dimmer than it is today, it is possible that surface temperatures were significantly lower.

As the Sun grew stronger, the high concentration of carbon dioxide combined with the intense pressure of its atmosphere led to a runaway greenhouse effect on Venus. Solar radiation penetrating the atmosphere to warm its surface would have been trapped, and surface temperatures would have risen dramatically. So if surface water ever existed on a young Venus, it eventually would have evaporated as the greenhouse effect intensified. The water vapor would have risen into the upper atmosphere where it would have escaped into space.

The atmosphere of Mars is also dominated by carbon dioxide (95%) but its atmospheric pressure is only about 0.7% of that on Earth. These very low atmospheric pressures indicate that the planet is no longer volcanically active and therefore has lost the ability to vent carbon dioxide into its atmosphere and so raise its surface temperature. Yet the networks of ancient channels indicate that water may have existed on the surface of Mars between 1 and 3 billion years ago. Photographs of the Martian surface show that the pattern of these ancient channels resembles that of stream channels on Earth. During this period, Mars would have had a surface atmospheric pressure more than 600 times its modern value. When volcanism on Mars ceased, precipitation and reaction with surface rocks in a manner similar to that on Earth would have reduced carbon dioxide in its atmosphere to modern levels in only 10 million years. Volcanism may have ceased on Mars because its small size (its mass is only 1/10 that of the Earth) allowed its interior to cool quickly. The small size of Mars also allows its atmosphere to leak into space, so that the modern Martian atmosphere is too thin to moderate its climate with a greenhouse effect.

15.6.4 Tectonics

It is clear from the preceding discussion that the Earth's evolutionary path rapidly diverged from that of Mars because of its larger size. Why it diverged from that of Venus is less obvious. Many of the processes ascribed to the early Earth apply equally to Venus. Admittedly, Venus is closer to the Sun and so may have had a hotter atmosphere. But the Earth's early atmosphere was also dominated by greenhouse gases. Nevertheless, the geologic record suggests that there was a fundamental parting of their evolutionary paths early in their respective histories.

Perhaps the key to this divergence is the contrasting tectonics on the two planets, as a result of which the Earth developed continental crust whereas Venus did not. The earliest records of life on Earth are found on ancient continental platforms, which provide relatively hospitable environments. As we learned in Chapter 13, buoyant continental crust was first generated early in Earth's history (certainly by 4.0 billion years ago) by a process known as *differentiation*, which results in the separation of light continental crust from the relatively dense interior. Initially this crust would have been rather thin and fragile, and much of it may have been destroyed by meteorite bombardment. Paradoxically, continued meteorite bombardment may have ultimately enhanced crustal differentiation, in the same way that agitating a relatively flat carbonated drink renews the migration of gas to the surface. Meteorite bombardment, therefore, may have destroyed the fragile crustal surface, but at the same time may have sped up the exhumation of light continental crust from the deep interior.

The geologic record indicates that differentiation of continental crust from the Earth's interior ultimately overcame the destructive effects of meteorite bombardment. At some stage, a critical thickness of continental crust must have been attained that could survive this bombardment. When this happened, the Earth's evolutionary path irrevocably diverged from that of Venus. Lighter continental crust is more buoyant than denser oceanic crust, so that, in contrast to Venus, the Earth developed elevated regions and vast ocean basins where water could collect. In short, exhumation of the continental crust began the Earth's hydrologic cycle, with drainage of water from elevated continental crust to depressed ocean basins.

Continental crust is also a much better insulator than oceanic crust and, like a book on an electric blanket, blocks the escape of heat from the Earth's interior to a far greater degree than oceanic crust. In fact 70% of the Earth's heat loss today occurs along mid-ocean ridges. This contrast indicates a profound asymmetry in the distribution of heat loss from the Earth's surface, and may have provided a basis for the initiation of a primitive form of plate tectonics. Plate tectonics, in turn, provided the environment for the origin of life. It is conceivable, therefore, that for life to evolve, a critical thickness of continental crust must first be attained.

In summary, it is clear that the Earth's early evolution occurred at a hectic pace when compared to the more pedestrian changes of the recent geologic past. The Earth became layered according to density very early in its history, from a dense iron-nickel rich core, through a silicate mantle, to a buoyant crust, and thence to watery and gaseous envelopes. The development of this density layering was primarily an internally driven process in which plate tectonics played little or no role. Although plate tectonics similar to the modern era may not have developed until the end of the Archean (about 2.5 billion years ago), a primitive form of plate tectonics may have commenced in the Early Archean upon the exhumation of continental crust and the development of asymmetric heat loss at the Earth's surface.

It is generally believed that plate tectonics would have been a more rapid process in its early primitive form than it is today. We must therefore ask ourselves whether there is a long-term trend of decreasing rates of plate motion as the Earth cools. If so, does the Earth have a finite life span, and given that plate tectonics is fundamental to the replenishment of the hydrosphere, atmosphere, and biosphere, will our planet end up as geologically dead as Mars?

15.7 The Jovian Planets and Their Satellites

The Jovian planets occupy the outer reaches of the Solar System (see Figure 15.18). The nearest of them, Jupiter, is more than five times farther from the Sun than is the Earth. The farthest, Neptune, is on average more than 30 times this distance. The greater distances from the Sun are reflected in progressively longer periods of revolution (Table 15-3). Jupiter, for example, takes 11.9 Earth years to cover its 5 billion km (3 billion mi) orbit around the Sun. Pluto, the most distant planet, has characteristics that do not fit its classification as a Jovian planet. In fact, astronomers debate whether Pluto should be considered a planet at all. Its orbit is highly irregular because of the waning gravitational attraction of the Sun at the edge of the Solar System. In fact, Pluto is sometimes closer to the Sun than Neptune is.

Several features of the Jovian planets yield important information about the origin of the Solar System. Calculations of their mean densities (ranging from 1.70 to 0.71 g/cm^3) reveal these bodies to be very different in composition from the inner terrestrial planets. The Jovian planets are thought to be largely composed of hydrogen and helium, the lightest elements in the periodic table. The internal structures of both Jupiter and Saturn consist of liquid molecular hydrogen (liquid H_2), which with increasing depth (and pressure) gives way to metallic hydrogen (H), and a relatively small core of rock and ice (Fig. 15.27a,b). Less is known of Neptune and Uranus,

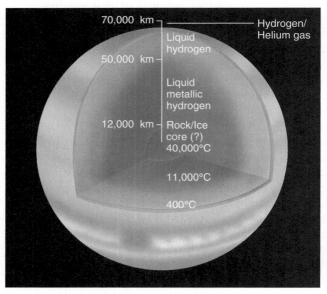

(a) Jupiter

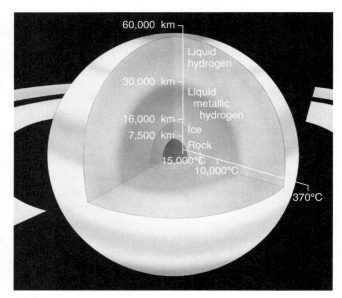

(b) Saturn

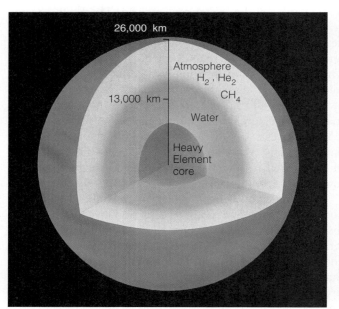

(c) Uranus

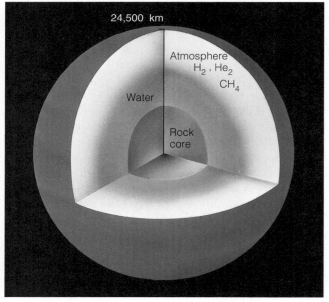

(d) Neptune

but it is generally assumed that they have similar internal structures to those of Jupiter and Saturn (Fig. 15.27c,d).

The fact that these planets are dominated by lighter elements in the outer reaches of the Solar System is central to models for its origin. As we learned in an earlier section, many astronomers believe that when the raw materials of the Solar System cooled, the Jovian planets grew rapidly enough to capture gas directly out of the solar nebula.

The enormous size and mass of the Jovian planets relative to the terrestrial planets implies a greater gravitational attraction for their atmospheres and hence greater escape velocities (Table 15-3). In addition, they tend to have a larger number of moons. Given the gaseous nature of the surfaces of Jovian planets, there is no way that these moons could have been jettisoned from the planets by a gigantic impact as is thought to be the case for the Earth-Moon system. In fact, the 1995 collision of the Shoemaker-Levy comet with Jupiter demonstrated that impacting bodies merely sink into the gaseous interior (see Living on Earth: A Comet Collides, page 471). Instead, the Jovian moons are likely to have different origins. Some are bodies of rock and ice that were captured and held in orbit by the strong gravitational attraction associated with these large planetary bodies. Others may have formed at the same time as their adjacent planet and were held in by gravitational attraction. As we shall see in the next sections, although the Jovian planets share similar characteristics, each has some unique features.

15.7.1 Jupiter

Jupiter, the largest planet, was named after the most important of the Roman gods. It contains 71% of all planetary mass in the Solar System. Its mass is 318 times that of the Earth, and its diameter is about 11 times larger (see Figure 15.8). It is remarkably similar in composition to the Sun and is made up mainly of hydrogen and helium with minor amounts of methane, ammonia and water (Table 15-5).

Although Jupiter has the composition of a star, it did not attain the size necessary to ignite nuclear reactions. A critical size is needed to ignite a star's nuclear furnace because the heat at the core of a star is largely generated by gravitational collapse. Because of its comparatively small size, the temperature at Jupiter's core is only 40,000°C (72,000°F), which is much too low to initiate a nuclear furnace. In fact, Jupiter would have to increase in size by a factor of 30 before this could happen.

Jupiter's atmosphere appears to be layered with an uppermost cloud layer dominated by ammonia and hydrogen sulfide, and a lowermost layer dominated by water vapor. The temperature of the highest clouds is estimated to be about −123°C (−189°F) based on the temperature at which ammonia is known to condense.

Figure 15.27 (left)
Inferred internal structures of the Jovian planets: (a) Jupiter, (b) Saturn, (c) Uranus, and (d) Neptune. Uranus and Neptune do not have liquid metallic hydrogen near their cores because their smaller masses provide insufficient interior pressures.

Table 15-5

Compositions of Jupiter and Saturn (by Mass)

MOLECULE	JUPITER (%)	SATURN (%)
H_2 (hydrogen)	78	88
He (helium)	19	11
H_2O (water vapor)	0.0001	—
CH_4 (methane)	0.2	0.6
NH_3 (ammonia)	0.5	0.2

Jupiter's most recognizable features are the color bands in the atmosphere that encircle the planet (Fig. 15.28a). These color bands are caused by variations in temperature and atmospheric composition (particularly elements such as carbon, sulfur, and phosphorous). These variations result in different degrees of reflectivity of sunlight. In general, red bands appear to represent cold descending layers, whereas blue bands are hot and rising (Fig. 15.29).

Jupiter's color bands are in constant and rapid motion. The movement of Jupiter's most famous feature, the Great Red Spot (Fig. 15.30), was discovered by Giovanni Cassini in 1665. From observations on the motion of the Great Red Spot, it is estimated that Jupiter has a 10-hour period of rotation.[2] This implies an incredible rotational speed at the equator of almost 45,000 km (28,000 mi) per hour! By way of comparison, Earth's equatorial speed of rotation is a mere 1670 km (1035 mi) per hour.

The Great Red Spot is the largest of numerous storm systems in Jupiter's atmosphere. It has a diameter close to 38,000 km (23,560 mi), which is three times the diameter of the Earth. This storm has lasted for at least 300 years and shows no signs of abating. The storms are driven by heat from the planet's interior, rather than by solar radiation. In fact, Jupiter, like Saturn, emits more heat from its surface than it absorbs from the Sun. This excess heat may be responsible for the enormous scale and longevity of Jupiter's storms.

Since Jupiter is a gaseous planet, the distinction between the planetary surface and its atmosphere is rather vague but is generally considered to mark a major compositional change. Jupiter's core is thought to be made up of a rock and ice mixture enclosed by a thick layer of liquid metallic hydrogen (see Figure 15.27a). Under extreme pressures, hydrogen's electrons are free to migrate (see Chapter 2, Metallic Bonds), which makes this region a good conductor of electricity.

15.7.2 Jupiter's Moons

Jupiter has 16 moons held in place by its strong gravitational attraction. Indeed, Galileo's observation that four moons orbited

---

[2] This 10-hour period of rotation is an average value. Like the Sun, Jupiter rotates faster at the equator than it does at the poles.

(a)

(b)

(c)

(d)

Figure 15.28
Photographs of the Jovian planets (a) Jupiter, (b) Saturn, (c) Uranus, (d) Neptune as obtained by the Voyager spacecraft. In (a), Jupiter's color bands and the Great Red Spot (lower left) are clearly visible. In (b), Saturn's rings and relatively poorly developed color bands can be seen. In (c), the surface of Uranus appears featureless. In (d), the Great Dark Spot of Neptune is visible in the center right portion of the image.

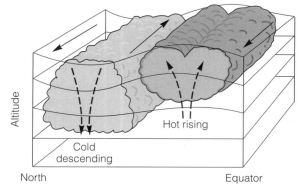

Figure 15.29
The color bands of Jupiter interpreted as ascending (blue) and descending (red) gases high in Jupiter's atmosphere.

Figure 15.30
The Great Red Spot of Jupiter, as photographed by the Galileo spacecraft in 1996, showing details of its internal turbulence and rotary flow.

Right on schedule, fragments of Comet Shoemaker-Levy 9 plunged toward Jupiter's surface in the summer of 1995 (Fig. 15.31). Traveling at 100 km (62 mi) per second, the energy released at impact from one fragment of the comet was an order of magnitude higher than the world's entire nuclear arsenal at the height of the Cold War.

The effects of impacts on Jupiter are very different from what they would be on Earth. Jupiter's larger mass means that it has a much stronger gravitational attraction and many astronomers believe that the Shoemaker-Levy comet got close enough to Jupiter to be snared by its gravity. After that, the comet fragmented, and its fate was sealed. On Earth, our atmosphere also offers a degree of protection from celestial objects. Many small comets and meteorites burn up in the Earth's upper atmosphere and only a barely detectable sprinkling of celestial ash reaches the surface. The small meteorite that hit the surface in St. Robert, Quebec in June 1994, was only a fraction of its original size when it entered the Earth's upper atmosphere.

Although the probability of impacts is far greater on Jupiter than it is on Earth, the effects of the impacts are less obvious. On Jupiter, there is no well-defined surface where the planet ends and its atmosphere begins. Because of this, prior to the impact, astronomers debated whether the effect of the impact of the Shoemaker-Levy comet on Jupiter would be any more dramatic than throwing a stone into a pool of water, although the speed at which the ephemeral effects would die away was unclear. In fact, huge plumes extending more than 3000 km above the surface were observed after the fragments "splashed down." Bright spots, the size of the Earth, were caused by fragments burning in the atmosphere. The visible evidence of impact lingered only for weeks, as dark spots gradually stretched and blended to form bands.

In contrast, the consequences of impacts on the Earth have altered the course of its evolutionary history. Although

Figure 15.31

Impact of the Comet Shoemaker-Levy 9 on the surface of Jupiter as seen by the Hubble Space Telescope. (a) Stages in the advancing comet as it approached Jupiter. (b) An infrared image showing the bright impact zones. (c) An ultraviolet image showing dark regions that developed after impact. (d) An infrared image showing a hotspot created by the impact.

most scientists subscribe to the principles of a slow progressive evolution of life on Earth, they are equally aware that catastrophic events such as extraterrestrial impacts have had a dramatic influence on the course of evolution, and in some instances may have reset the evolutionary clock. For example, as we learned in Chapter 14, the meteorite impact that is generally held to be responsible for the extinction of the dinosaurs 66 million years ago, paved the way for mammalian species to become dominant on the Earth. Today, the dominant member of this group is *Homo sapiens*. But whether a similar fate lies in store for us, or whether we deliver a pre-emptive strike on ourselves through neglecting our own environment, remains to be seen.

Jupiter was one of the final nails in the coffin of the geocentric system. These Galilean moons (Io, Europa, Ganymede, and Callisto) turned out to be the innermost of the sixteen satellites. Images from the Voyager space probe in the 1970s showed that the four Galilean moons are very different (Fig. 15.32). Information on the internal structure of these moons has been recently obtained by the Galilean spacecraft.

All four moons are fascinating. *Io* is comparable in size and density to our Moon, with a diameter of 3632 km (2252 mi) and a density of 3.55 g/cm³. It has a smooth, crater-free surface, which suggests that its surface is extremely young and geologically active. Its yellow-to-red coloration suggests an abundance of sulfur-rich compounds. Astonishingly, the Voyager images also captured an erupting volcano (Fig. 15.33), making Io the only other body in the Solar System on which active volcanism has so far been demonstrated. Although most scientists believe that Venus is also active, this belief has yet to be confirmed. It is thought that Io is literally turning itself inside out as a result of the opposing gravitational pull of Jupiter and its nearest satellite neighbor, Europa.

Europa is the next closest moon to Jupiter and is a mixture of rock and ice (Fig. 15.32). It has a density of about 3.0 g/cm³ and a diameter of 3126 km (1938 mi). Its icy surface was once thought to be very smooth, like a billiard ball. However, recent pictures taken by the Galileo spacecraft show that the surface is completely cracked, with ice flows and ice volcanoes that were probably produced by water erupting between the cracks. There are even speculations on the presence of a subterranean ocean and the possibility of life.

Figure 15.33
A computer enhanced image showing a volcano on Io (left) venting a plume of sulfur-rich ash.

Figure 15.32
Photographs of the Galilean moons of Jupiter (not to scale). Io (upper left), Europa (upper right), Ganymede (bottom left), and Callisto (bottom right). The Earth's Moon is in the center for comparison. Io has a sulfur-rich surface, Europa displays an icy crust, Callisto and Ganymede have icy surfaces with cores of rock.

Ganymede, with a diameter of 5262 km (3262 mi), is bigger than both Pluto and Mercury, and, with a density of 1.9 g/cm³, it is thought to be primarily composed of surface ice overlying a rocky core. Measurements taken by the Galileo spacecraft in 1996 indicate that Ganymede has a strong magnetic field, a clearly differentiated internal structure with a metallic core surrounded by a silicate mantle, and an ice shell up to 800 km (500 mi) in thickness. The magnetic field is thought to be generated in a similar manner to that on Earth with a core of molten iron that whirls around like a dynamo as the Moon spins and generates an electric current.

Ganymede's surface displays evidence of ice tectonics. Water appears to have oozed through fractures in the ice and crystallized on its surface. It is therefore speculated that, as on Europa, a layer of liquid water may exist below the icy surface.

Callisto, the outermost Galilean moon, is a little smaller than Mercury, with a diameter of 4800 km (3000 mi). It has a density of 1.8 g/cm³ and is therefore thought to be composed of a rock-ice mixture. Its crater density is similar to both the Moon and Mercury, suggesting that it preserves an ancient 4.5 billion-year-old surface, and may be the least active of all the Galilean moons.

The Galilean moons revolve in near circular orbits in Jupiter's equatorial plane. This observation, together with the decrease in the average density of these satellites with distance from Jupiter suggests that these moons may have formed along with Jupiter. Little is known about Jupiter's other moons. They are not distributed as regularly, and may have been captured by Jupiter's gravitational attraction.

15.7.3 Saturn

Saturn, the second largest planet in the Solar System, has a diameter of 120,660 km (75,400 mi). The Romans named the planet after their god Saturn, the father of Jupiter, for whom they also named Saturday, their traditional day of festivals. Saturn is the lightest of all the planets with a density of only 0.71 g/cm³, which is even less than that of ice (Table 15-3). Saturn is thought to have a similar internal structure and composition to that of Jupiter, being dominated by hydrogen and helium. However, its smaller size implies that its internal layers are under less pressure than the equivalent layers on Jupiter. As a result, it contains less liquid metallic hydrogen (see Figure 15.27b). Saturn also has color bands in its atmosphere, although the colors are more subdued than those of Jupiter (see Figure 15.28b).

Saturn is most famous for the spectacular flat rings that encircle the planet above its equator (Fig. 15.34). These extend for 170,000 km (105,400 mi) above its surface, yet vary in width, and can be as narrow as 2 km (1.2 mi). It has long been known that the rings were not solid, because distant stars, and indeed the planet's surface, could be seen through them. However, their reflectance suggested that they were not gaseous. Their composition remained a mystery until the early 1970s when observations suggested that they consisted of chunks of ice varying from 5 cm (2 in) to 100 m (330 ft) in size. This conclusion was later confirmed by images from the Voyager space probe.

At last count, 23 moons have been discovered around Saturn, the largest of which is the ice body, *Titan*. This moon has a surface temperature of −180°C (−290°F), and has an atmosphere rich in nitrogen.

Figure 15.34
Image of the rings of Saturn. The rings are made up of ice particles and are transparent, as evident where they cross in front of the planet's surface (left).

15.7.4 Uranus, Neptune, and Pluto

In comparison with Jupiter and Saturn, little is known about the remaining outer planets. As we have learned, they were discovered after the heliocentric (Copernican system) was well established. **Uranus** was named for the god of the heavens, the grandfather of Jupiter and the father of Saturn. Uranus is unique among the planets in that its axis of rotation is almost on its side, rather than upright like the other planets. As a consequence, it spins in the plane of its orbit, like a football, in contrast to the other planets, which spin perpendicular to their orbits like spinning tops. This unusual orientation is thought to be related to a massive celestial impact when the planet was forming. Uranus has nine narrow rings, and a clear outer atmosphere consisting of methane, hydrogen, and helium (see Figure 15.27c). Unlike Jupiter and Saturn, its surface is almost featureless.

Neptune was named for the god of the sea because of its blue color (see Figure 15.27d). This color is attributed to the abundance of methane in its atmosphere. Images from Voyager show that Neptune is a stormy planet with a large storm center called the "Great Dark Spot" that is similar to Jupiter's Great Red Spot (see Figure 15.28c).

Pluto, named for the god of the underworld, has a density of 1.99 (which is higher than any of the Jovian planets), and its composition is thought to consist of rock (80%) and ice (20%). Pluto was discovered in 1930 after a lengthy search was initiated when disturbances were observed in the positions of Uranus and Neptune which were attributed to the gravitational pull of another, then unknown, planet. Pluto is only 2300 km (1426 mi) in diameter, and so is smaller than our Moon. Its surface, which ranges in temperature from −233°C to −223°C (−387°F to −369°F), appears to be dominated by nitrogen ice (97%) and traces of methane ice. Recent observations that stars do not abruptly disappear when they pass behind Pluto, but appear to "wink out," suggests that Pluto has a gaseous atmosphere.

Astronomers still debate whether Pluto is a planet of similar status to the other eight. It does not easily conform with the major subdivision of the Solar System into inner terrestrial and outer Jovian planets. The strongest evidence for its planetary status came in 1978 when images of Pluto also revealed a satellite body, 1300 km (800 mi) across, named *Charon*. In 1990,

Figure 15.35
Image from the Hubble Space Telescope showing Pluto (lower left) and its moon Charon (upper right).

the Hubble Space Telescope took the first clear photo-images of both bodies (Fig. 15.35). Like the other planets, Pluto also orbits the Sun, albeit in a highly irregular manner. This irregular orbit is probably due to the waning gravitational attraction felt by Pluto because of its small size and its position at the outer edge of the Solar System. In fact, from January 1979 to March 1999 Pluto was closer to the Sun than Neptune.

Observations from the Hubble telescope indicate that Pluto may be a large representative of a vast number of icy bodies that orbit the Sun at the outer edge of the Solar System. These bodies are thought to represent remnants left over from the Solar System's formation.

15.8 Asteroids, Comets, and Meteoroids

The recent visit in early 1996 of the comet Hyakutake (pronounced hiyakoo-ta'-kee) and in 1997 of comet Hale-Bopp (Fig. 15.36a,b) has once again brought to our attention what an amazing celestial junkyard our Solar System is. When the Sun and planets formed, an amazing amount of debris was left over. These are preserved as an assortment of celestial bodies that have been virtually unmodified since that time, and therefore provide vital clues as to the age and origin of our Solar System.

An **asteroid** is a celestial body that varies from 1000 km (620 mi) to about 100 m (330 ft) in diameter. Most asteroids are located in a belt between Mars and Jupiter (between 2 and 3 AU from the Sun). They were first discovered in 1801 and more than 20,000 have now been identified. Many of these small bodies have impact craters on their surface (Fig. 15.37), indicating that they too were bombarded early in the history of the Solar System. Many of the smaller asteroids are thought to be the splintered fragments of planetoids that broke up as a result of such collisions. Asteroids are composed of rocky, metallic, and carbonaceous material.

Figure 15.37
Photograph taken by the Galileo spacecraft of an asteroid showing its pitted surface, interpreted to record collisions with small celestial objects. Note the irregular shape of the asteroid, a common feature of such bodies.

(a)

(b)

Figure 15.36
(a) Photograph of Comet Hyakutake taken in 1996, around the time of its closest approach to Earth.
(b) The Comet Hale-Bopp which reached its most favorable position for viewing in the Spring of 1997.

Comets are small bodies with a central nucleus ranging from 1 to 20 km (0.6 to 12.0 mi) across. They are composed of a mixture of dust, icy material, and carbon-rich soil, and can be thought of as dirty celestial snowballs. As the nucleus vaporizes, dust and gases are released to form the head, or coma, of the comet, and the comet tail. The coma may be up to 100,000 km (62,500 mi) in diameter and is the most visible part of the comet. The origin of comets remained a mystery until the 1950s, when Jan Oort, a Dutch astronomer, showed that the overwhelming majority of them came from the remotest parts of the Solar System, lying between 20,000 and 150,000 AU from the Sun. This region is now known as the *Oort cloud*. These comets may take a few hundred years to orbit the Sun. Another source of comets is believed to be located in an area beyond Pluto, named the *Kuiper belt*.

In contrast to other celestial bodies, comets have highly eccentric elliptical orbits around the Sun (Figure 15.38, see also Figure 15.18). These orbits may be eccentric, but they are

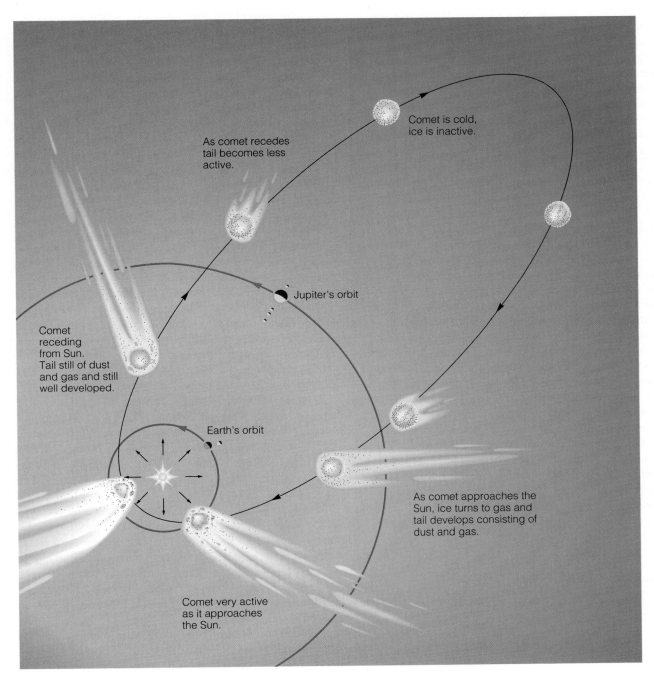

As comet recedes tail becomes less active.

Comet is cold, ice is inactive.

Comet receding from Sun. Tail still of dust and gas and still well developed.

Jupiter's orbit

Earth's orbit

As comet approaches the Sun, ice turns to gas and tail develops consisting of dust and gas.

Comet very active as it approaches the Sun.

Figure 15.38
Schematic representation (not to scale) of a comet's eccentric orbit compared to that of Earth and Jupiter. Note that the comet's tail, which develops as volatile ices become gaseous, always points away from the Sun and is produced by the interaction of the comet with the solar wind.

regular. In 1704, Edmond Halley deduced that the comets sighted in 1456, 1531, 1607, and 1682 had identical orbits so that the sightings were likely to record the return of the same comet every 75 years. His prediction of its return in 1758 was confirmed when a comet was sighted on Christmas day of that year.

Comets may also have long-term reappearances that may have had profound effects on the evolution of life on Earth. Although hotly debated, some studies of mass extinctions suggest that they occur every 26 to 28 million years. As we have learned, the most celebrated of these mass extinctions is attributed to an impact 66 million years ago that wiped out the dinosaurs and a large number of other species, and brought the Mesozoic Era to a close.

The origin of the long-term reappearances of comets is unclear. Recently, it has been proposed that an influx of Oort cloud comets might occur as a result of oscillation of the Solar System about a plane within our galaxy. Another suggestion is that the Sun might have a mysterious companion star, called *Nemesis*, with a highly elliptical orbit. If its orbit passed near the Sun, such a star might trigger instability in the Oort cloud, resulting in a storm of comets being propelled towards the inner Solar System. If some of these comets were to hit the Earth, mass extinctions might result. However, it should be emphasized that this hypothesis is very speculative; there is no independent evidence for the existence of Nemesis.

Comets are famous for their tails which always point directly away from the Sun (see Figures 15.18 and 15.38). The tails form as comets are blasted by the solar wind, a stream of ionized gases emitted from the surface of the Sun. In fact, it was by studying these tail patterns that the characteristics of the solar wind were first determined. As we learned in Chapter 13, analysis of comets reveals that they contain many kinds of carbon-, nitrogen-, and oxygen-bearing molecules. This implies that the primitive components of the Solar System contained the raw materials needed to initiate life (Table 15-6). Comet Hyukatake was the first comet in which the organic compounds methane (CH_4) and ethane (C_2H_6) were detected.

Meteoroids are smaller pieces of celestial debris, ranging up to a few meters in size. They can be either rocky or icy and so can be thought of as smaller versions of asteroids and comets. Once they enter the Earth's atmosphere they are referred to as either meteors or meteorites. A **meteor** is often mistakenly called "a shooting star." Of course, they are not stars at all. The light we see is not due to nuclear fusion (the essential process that lights the stars), but merely records the vaporization of meteors as they travel through the Earth's atmosphere. By definition, meteors burn up in the atmosphere, producing a bright trail as they do so, but do not strike the ground. Meteor showers may occur when clusters of meteors simultaneously enter the atmosphere. The most famous, the Perseid shower, occurs every year near the middle of August and many people enjoy watching the display of dozens of meteors per hour. These showers may be related to the dust scattered from comets. For example, the Perseid shower occurs when the Earth's orbit around the Sun takes us through the orbit of a comet known as 1862 III. Each year the disintegration of meteors adds 1000 tons of mass to the Earth as they disintegrate into dust in the upper atmosphere. The late Luis Alvarez, a Nobel laureate physicist from the University of California, has likened this to someone with a pepper shaker perpetually sprinkling meteoritic dust over the globe.

Meteoroids that do land on Earth are called **meteorites**. The locations of some craters thought to have been produced by meteorite impacts are shown in Figure 15.39. Meteorites collide with the Earth at speeds between 40,000 and 200,000 km/hr (25,000 and 125,000 mi/hr). It is thought that about 500 land each year, but only 1% of these are found. Of course, the chances of our lives being influenced by a meteorite impact are very remote, far less, for example, than our chances of winning a million dollar lottery. However, we have had some close calls. The asteroid Toutatis which is 5 km (3 mi) long and 2.5 km (1.5 mi) wide, has made several close encounters. In 1992, it came within 3.3 million kilometers (2 million mi) of Earth, and in Nov-ember 1996 it returned, this time missing Earth by a mere 5.3 million km (3.3 million mi). By astronomical standards, these are very small distances. In September of 2004, Toutatis will miss us by less than 1 million mi, or about four times the distance to the Moon. In 2028, we will have another "close call" with an as yet unnamed asteroid.

In North America, one of the best examples of a meteorite impact occurs near Flagstaff, Arizona (Fig. 15.40). This impact site is known as Barringer Crater or, more popularly, Meteor Crater. The impact occurred 50,000 years ago and produced a crater 1.2 km (0.75 mi) across and 200 m (660 ft) deep. It was formed by a meteorite thought to be little more than 90 m (295 ft) across.

Meteorites are quite variable in composition and probably have no single mode of origin. As discussed in Chapter 13, some meteorites appear to have originated on Mars, in a sort of interplanetary free trade. This type of meteorites is thought to have been blasted off the Martian surface by an impact and been caught by Earth's gravitational clutches. Other meteorites appear to be remnants of asteroids that formed early in the history of the Solar System.

Meteorites are classified according to their composition and texture. Some, known as **chondrites**, have peculiar spherical textures that have been simulated in the laboratory by rapid heating of meteoritic material followed by rapid cooling. This is consistent with melting and rapid crystallization associated with the collisional events at the birth of the Solar System. Many scientists believe the composition of these meteorites approximates

Table 15-6

Analysis of Compounds and Ions Detected in Comets

COMA	GAS TAIL
H, OH, O	CO, CO_2
C C_2, C_3, CH	H_2O, OH
CN, CO, CS, S	CH, CN, N_2
NH, NH_2, HCN, CH_3CN	C, Ca
Na, Fe, K, Ca, V	
Cr, Mn, Co, Ni, Cu	

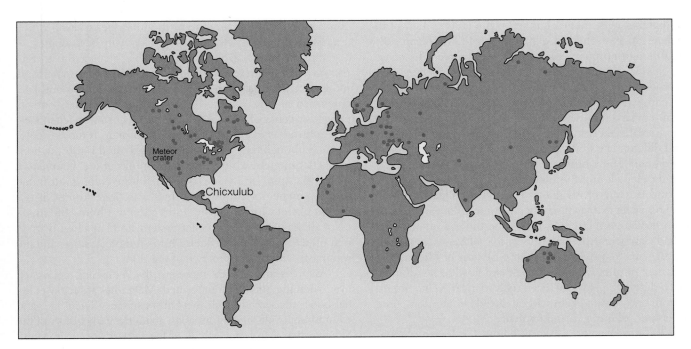

Figure 15.39

Locations of major craters (pink dots) located on the Earth's continents and interpreted to be the result of meteorite impacts. The location of Meteor Crater (see Figure. 15.40) and Chicxulub crater (thought to be related to the extinction of the Dinosaurs; see Chapter 14) are also identified.

Figure 15.40

Meteor Crater, near Flagstaff, Arizona (location shown in Figure 15.39). The crater is 1.2 km (0.75 mi) in diameter and 200 m (660 ft) deep. It formed 50,000 years ago when a meteorite 90 m (295 ft) in diameter blazed through the Earth's atmosphere. Although it splintered in flight, the majority of it impacted near Flagstaff to produce the crater seen here.

the average composition of the primitive Earth, before the iron and nickel-rich core formed. This interpretation is supported by hundreds of age dates on meteorite samples, all of which suggest that they formed between 4.5 and 4.6 billion years ago.

Other meteorites, known as **achondrites** are similar in composition to the Earth's mantle, suggesting that they were originally part of a larger body with internal layering similar to that of the Earth. **Iron meteorites** may represent the splintered cores of asteroids and, as such, are pieces of the primitive Solar System. This interpretation is supported by their ages, which are typically 4.57 billion years. Their composition is similar to the Earth's core, suggesting that collapse of iron to the core may have occurred in many celestial bodies.

15.9 Chapter Summary

Our sojourn through the Solar System has left us with a more complete understanding of the uniqueness of planet Earth. We are the only planet with active plate tectonics and life-sustaining watery and gaseous envelopes. The terrestrial inner planets all shared a similar origin and were made from similar meteorite-like materials, but each embarked on its own unique evolutionary course. Differences between the terrestrial planets are related to their varying sizes, the composition of their surfaces and subsurface rocks, and their distance from the Sun.

We have seen that smaller planets such as Mercury and Mars are essentially geologically inactive. Mercury became inactive nearly 4 billion years ago, and Mars has not had a major episode of volcanic eruptions for nearly a billion years. The degree of preservation of craters is a measure of the antiquity of these ancient surfaces. Because of their tectonic inactivity and small size, atmospheric gases are not replenished on these planets and are only loosely held. Over time, they have leaked into space. The comparatively low crater density on

Venus, together with subtle changes in the composition of its atmosphere, suggests that the planet might still be volcanically active, although eruptions have not been observed. Its dense carbon dioxide-rich atmosphere and very high surface temperatures render this a most inhospitable world.

By comparison with the other planets, a scenario for why the Earth embarked on its own unique course can be presented. The Earth is likely to have condensed from interplanetary dust and differentiated into a core, mantle, and crust very early in its evolution. Some primitive form of plate tectonics is likely to have developed, and the contrasting character of its continental and oceanic crusts allowed water to collect in ocean basins. This water extracted significant amounts of carbon dioxide from the atmosphere, by which time the evolution of the Earth had significantly departed from that of Venus and Mars, its nearest terrestrial neighbors. The evolution of life and the development of photosynthesis locked the Earth onto a fundamentally different path that eventually converted its atmosphere from one dominated by carbon dioxide to one composed almost entirely of nitrogen and oxygen.

We still have much to find out about the Jovian planets. Their low densities indicate that they are dominated by the lightest chemical elements (hydrogen and helium). Recent studies of Jupiter indicate that long-lasting storm systems (such as the Great Red Spot) are fed by energy from the planet's interior. Jupiter's color bands are related to subtle variations in the composition and temperature of its atmospheric gases, which are manifest in different degrees of reflectivity of sunlight.

Jupiter's moons have highly variable characteristics. Suspicions that modern volcanic activity might occur on one of these moons, Io, were first aroused by its low crater density, and later confirmed by observations. Pictures of another moon, Europa, provide evidence of a global ocean hidden beneath an icy surface.

Saturn has many similarities with Jupiter. It too has color bands (although their colors are subdued), and is most famous for its flat ringlets, which consist of chunks of ice. Uranus is a planet that spins on its side. Neptune's blue color is attributed to the presence of atmospheric methane. Like Jupiter, it is a violent planet with long-lived storm systems. Pluto is the smallest of all the planets in the Solar System, and has a highly irregular orbit that probably reflects Pluto's remote position and the waning gravitational influence of the Sun.

Asteroids and comets represent the remaining debris left over when the Solar System formed. Meteorites are rocky celestial bodes that collide with the Earth's surface. Their ages and chemistry provide vital information about the early history of the Solar System. Most, but not all, of the asteroids occur in a confined belt between the orbits of Mars and Jupiter. Comets are like dirty celestial snowballs. They are mixtures of dust and ice. Many come from the remotest parts of the Solar System and have highly eccentric orbits. Their famous tails always point directly away from the Sun, revealing the presence of the solar wind.

Key Terms and Concepts

- The Ptolemaic (geocentric) model placed the Earth at the center of the universe

- The Copernican (heliocentric) model proposed that planets orbited the Sun in perfectly circular orbits

- Kepler's Laws of Planetary Motion proposed that the paths of planets around the Sun are ellipses, that a planet moves about the Sun such that its radius vector sweeps out an equal area in equal times, and that the time needed for a planet to make a single revolution around the Sun is related to its average distance from the Sun

- In 1609, Galileo discovered the four Galilean moons of Jupiter (Io, Europa, Ganymede and Callisto) and showed that they orbited Jupiter

- Newton's Laws of Motion and Gravity provided the theoretical basis for understanding Kepler's laws

- The solar nebula theory is the most favored model for the origin of the Solar System in which a rotating cloud of dust and gas is thought to have flattened into a disk as it contracted under the influence of gravity

- Gravitational collapse and shrinking of the solar nebula may have been initiated by a supernova explosion

- As the inner nebula cooled, planetesimals accreted to form protoplanets, which were destined to become planets

- Depending on the composition of the planetesimals, accretion to form planets may have been homogeneous or inhomogeneous

- The innermost planets (Mercury, Venus, Earth, and Mars) are terrestrial (rocky) planets

- The outer planets (Jupiter, Saturn, Uranus, and Neptune) are Jovian (gaseous) planets

- Pluto does not have characteristics that fit either a terrestrial or Jovian classification

- The Earth probably condensed from interplanetary dust and was differentiated by gravity into a core, mantle, and crust very early in its evolution

- The gravitational collapse of iron to form the Earth's core may have occurred in an event known as the iron catastrophe

- The Moon rotates about its own axis at the same rate as the Earth

- The Moon's surface is heavily cratered and consists of bright lunar highlands and dark lunar maria
- Volcanic activity on the Moon ceased about 3 billion years ago
- Ice crystals, which make up 1% of lunar soil, are thought to have been deposited by cometary impacts
- The Moon may have been ejected from the Earth by the collision of a Mars-sized object with the Earth's surface
- The size of a terrestrial planet influences its rate of cooling, the longevity of its volcanic activity, and its atmospheric composition
- The age of the planetary surfaces can be estimated by the abundance of impact craters
- Mercury has a heavily cratered surface. It is a geologically dead planet with no recent volcanism and a vanishingly thin atmosphere
- Venus has a dense atmosphere dominated by carbon dioxide (96.5%) and a surface temperature of 470°C. Modern volcanic activity is suspected, but remains unproven
- Mars has exaggerated relief, polar ice caps, a very thin atmosphere, and very cold average surface temperatures. It may once have harbored life, and might still do in its subsurface. Volcanism on Mars virtually ceased about 1 billion years ago
- The uniqueness of planet Earth is related to its size, distance from the Sun, composition, active tectonics, generation and preservation of its atmosphere and hydrosphere, and life
- Jupiter is dominated by hydrogen and helium. Its color bands reflect subtle changes in the composition and temperature of its atmosphere. Its massive storms are driven by energy released from its interior
- Jupiter's innermost moons include Io (which has volcanism), Europa (which has a cracked icy surface that may overlie a subterranean ocean), Ganymede (which shows evidence of ice tectonics) and Callisto (with an ancient, heavily cratered surface)
- Saturn is similar in many ways to Jupiter. Its flat ringlets are chunks of ice
- Uranus, Neptune and Pluto are the least known and least understood planets
- Asteroids, comets, meteoroids, and meteorites are various forms of celestial debris

Review Questions

1. What were the key arguments in determining that the Earth is part of a Sun-centered system?
2. What important information do comets and meteorites yield about the early history of the Solar System? Explain.
3. Why do the planets have varying tilts?
4. Construct a model using a flashlight and two balls of different size that could be used to explain the phases of the Moon.
5. Why are Mercury and Mars considered geologically dead, and what are the implications of this for their respective atmospheres?
6. Why does Mars have the lowest density of the terrestrial planets?
7. Compare and contrast tectonics on Venus and Earth.
8. Why is there no well defined surface on Jupiter?
9. How was the nature of Saturn's rings determined?
10. Why is Pluto's orbit so eccentric?

Study Questions

 Research the highlighted question using InfoTrac College Edition.

1. The early history of the Earth is thought to have been profoundly influenced by meteorite impacts. However, there is no direct evidence for this on Earth. Where does the evidence come from and why has it been applied to the Earth?
2. How do the origins of the mountains on the Moon contrast with those of the mountains on the Earth?
3. Why do Mercury and the Moon have virtually no atmosphere?
4. Why is the Earth's atmosphere dominated by nitrogen (N_2) and oxygen (O_2) whereas the atmospheres on Venus and Mars are dominated by carbon dioxide (CO_2)?
5. The cores of the Earth and Venus are thought to be iron-rich. Yet the Earth is magnetic, but Venus is not. Why might this be so? (Hint: review the origins of the magnetic field on Earth; see Dig In: The Earth's Magnetism and its Dynamic Core, on page 190.)
6. Why is the Earth the only terrestrial planet in our Solar System with a water-rich surface?
7. What evidence is there that iron-rich meteorites are similar in chemical composition to that of the Earth's core?
8. Where do planetary atmospheres come from and when do they form?
9. What evidence points to the age of the Solar System as 4.6 billion years?
10. Do you think that there was once life on Mars? If so, what are your main arguments in favor of the idea? If not, what are your main objections?

16

The Sun and the Stars

Our everyday experiences repeatedly demonstrate the importance of the Sun's energy in stimulating the biosphere, in driving atmospheric and oceanic circulation, and in profoundly influencing our weather and climate. In this chapter we look at the Sun itself, and how the nuclear reactions in its core create the energy that emanates from its surface. We will find that this energy comes in a wide range of wavelengths, only a tiny portion of which is in the form of visible colors. Other portions of the spectrum include ultraviolet, infrared, and X-ray radiation.

We will discuss the ingenious detective work that unmasked the composition of the Sun, and how the element helium was actually detected in the Sun's spectrum before it was discovered on Earth.

We will find the Sun to be a great ball of gas held together by its own gravity. We will also find it to be a magnetic body that is layered from its core, which is powered by thermonuclear fusion, through a thick zone where energy travels by radiation to a zone dominated by convection, and a granular surface called the photosphere, which may contain dark depressed blemishes called sunspots. The appearance and disappearance of these sunspots appears to occur in cycles of just over 11 years.

The solar wind is a stream of ionized gases that are blown away from the Sun at high velocities and travel throughout the Solar System. The wind blasts our upper atmosphere and is the principal agent that carries solar disturbances to the Earth's surface. We will find that the speed and intensity of the solar wind have a direct bearing on global temperatures in the Earth's upper atmosphere and that the wind may be the instigator of auroras, the Earth's most spectacular natural light shows.

The intensity of the solar wind varies as a function of the sunspot cycle. We will discover that a 80–100 year period in the 17th Century when very few sunspots were observed overlaps a period of global-scale cooling known as the Little Ice Age.

Despite its importance to the Earth, we will discover that our Sun is not particularly special when we look beyond the Solar System. It is, in fact, a rather mediocre star. We will also examine recent evidence of other solar systems in which planets revolve around a central star. If verified, the existence of such solar systems has enormous consequences for our position in the universe.

We will also see that events in the cosmos can have subtle influences on the Earth. These influences can be detected in changes in chemistry recorded in the growth pattern of trees. From our discussion of the life cycle of stars, we will learn what the fate of our Sun will be when its nuclear fuel supply is exhausted. This discussion will also give us insights into the origin of the chemical elements in the periodic table. We will find that many are synthesized at the end of the life cycle of stars when they die in supernova explosions.

16.1 Introduction

We could not exist without the constant flow of energy we receive from the Sun. To many ancient civilizations, the Sun was regarded as a god. To the Egyptians, the Sun was the supreme god, named Ra. To the Greeks, the Sun was Helios, and to the Romans, Apollo. In about 500 B.C., a Greek philosopher, Anaxogaros, was exiled for declaring that the Sun was actually a great ball of fire.

Yet Anaxogaros was on the right track. Today we know that our Sun is a star: a body that can sustain nuclear reactions in its core. But the Sun is a rather mediocre star, compared to others in the heavens. It is of average size and is disparagingly referred to as a "Yellow Dwarf." The Sun was formed about 4.6 billion years ago at the center of a swirling cloud of gas which, when it succumbed to the force of gravity and collapsed, was of sufficient size to ignite.

Now, 4.6 billion years after its formation, the temperature at the center of the Sun is estimated to be about 15 million degrees Centigrade. It has a surface temperature of about 5700–5800°C (10,300–10,500°F). The Sun's diameter (1,392,000 km; 863,000 mi) is more than one hundred times the diameter of the Earth. It contains more than 99% of the mass of the Solar System, and it is this concentrated mass that provides the gravitational attraction that keeps the planets in orbit. In fact, the edge of the Solar System is defined as the position where the Sun's gravitational attraction is no longer the dominant force.

In this chapter, we will examine the Sun, its energy and composition, and its influence on the Earth. We will also consider the effects on Earth of events beyond the Solar System, and the attempts to detect them. We will look at stars and galaxies, and examine the life cycle of stars, a cycle which will give us insight into the origin and dramatic fate of the Sun. We will find that most of the chemical elements that form the basic building blocks of all Earth materials are created during the life cycle of stars. The chapter concludes with a look at the beginning of the universe itself.

16.2 The Sun's Energy and Composition

Radiant energy from the Sun drives many of the processes that take place near the Earth's surface. It is responsible for the circulation of the atmosphere and hydrosphere, and powers photosynthesis in the biosphere. In this section we will learn that the Sun's energy results from thermonuclear reactions at its core, and that the energy released has a wide variety of wavelengths which together make up the **electromagnetic spectrum**. We will then discuss the ingenious ways by which the composition of the Sun was determined, and look at variations in the character of the Sun from its core to its surface.

16.2.1 Thermonuclear Fusion

The Sun is a gaseous ball producing energy through nuclear reactions that take place within its core. The core is dominated by hydrogen nuclei (single protons) moving at such hurtling speeds that upon collision with each other, they fuse to create helium, a process known as **thermonuclear fusion**.

Under normal circumstances, positively charged particles like protons have a very strong tendency to repel one another. However, this tendency is overcome in the Sun's core by the extremely high temperatures (about 15 million °C, or 27 million °F) and consequently the high average velocity of the hydrogen nuclei (500 km/sec, over a million mi/hr). We have artificially duplicated this process on Earth in the hydrogen bomb. Thermonuclear fusion is the source of the hydrogen bomb's destructive energy. However, to attain the very high temperatures necessary to induce fusion, the hydrogen bomb must be detonated by the explosion of a companion uranium bomb.

The thermonuclear reaction can be illustrated by a number of steps (Fig. 16.1). The fundamental particles that collide are protons, each having a positive charge (step 1: Figure 16.1a). When the two protons fuse, one of the particles loses its positive charge and becomes a neutron. In addition, two subatomic particles are ejected. One of them is a **positron**, which is a particle with a positive charge but virtually no mass. It quickly collides with a negatively charged electron and is annihilated, producing a gamma ray upon its destruction. The other is a **neutrino**, which has no charge and almost no mass and travels at the speed of light (see Living on Earth: Neutrinos on page 484).

The particle fused by the collision in step 1 has one proton (and is therefore, by definition, hydrogen) and one neutron. This "heavy" hydrogen isotope is called **deuterium**. The deuterium nuclei so formed then collide with the predominant single protons (step 2: Figure 16.1b), producing an atom that contains three subatomic particles of similar mass. Because two of these particles are protons, the atom produced is helium and is designated helium-3 (^{3}He) because it has a mass of three units. Among all of the billions of collisions that take place in the Sun's core every second, collisions between two helium-3 atoms are inevitable (step 3: Figure 16.1c). The collision results in the production of a nucleus containing four of the six particles (two protons and two neutrons) and the ejection of the remaining two protons. The situation is somewhat analogous to a "break" in snooker or pool. When the carefully racked balls are hit by the cue ball, those at the back and sides are often ejected from the pack (Fig. 16.2).

There are two very important aspects of solar thermonuclear reaction. First, the Sun's supply of nuclear fuel is a nonrenewable resource. Six protons are consumed by the reaction and only two are released (see Figure 16.1). So there is a net consumption of four protons in the production of helium. As a result, the Sun, like all stars in the universe, has a finite life span because its fuel is not inexhaustible. Over the Sun's 4.6-billion-year history, the volume of helium in its core has risen from about 9% to 28%. Calculations show that the Sun is about halfway through its life, and will exhaust its fuel supply in another 5 billion years.

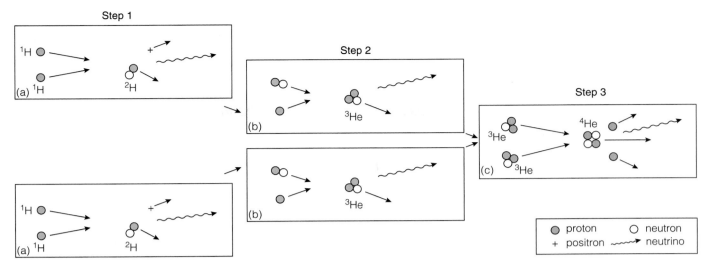

Figure 16.1

Thermonuclear fusion of hydrogen to helium takes place in several steps. The overall reaction results in the loss of mass, which is transformed into energy. (a) When the two protons fuse (^{1}H, step 1), one of them loses its positive charge and becomes a neutron. The particle formed is a heavy isotope of hydrogen called deuterium (^{2}H), which has a mass of two units. During this process, two subatomic particles are ejected: a positron (which is annihilated) and a neutrino, which has no charge, no mass, and travels at the speed of light. (b) When deuterium nuclei collide with single protons (step 2), the atom produced is helium-3 (^{3}He) which contains two protons (and so is helium) and one neutron, thus having a mass of three units. (c) Eventually collisions between two helium-3 atoms (step 3) produce a helium nucleus (^{4}He) containing four particles (two protons and two neutrons), ejecting the two remaining protons.

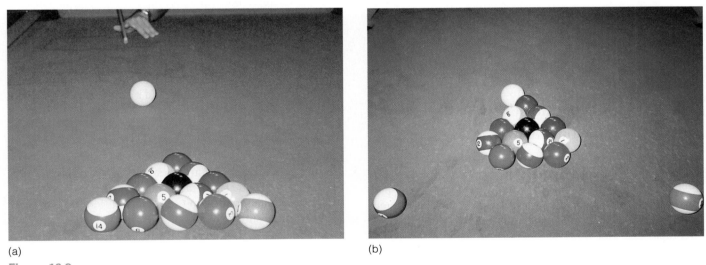

(a)   (b)

Figure 16.2

In a "break" in snooker or pool, the collision of the cue ball (white) with carefully racked balls can result in the ejection of balls from the back or sides of the pack. (Photos Brendan Murphy)

Second, mass is *not* conserved during the fusion process because the helium-4 atom produced (^{4}He) is 0.7% lighter than the four hydrogen atoms that are consumed. This loss of mass is converted into energy according to the relation eloquently expressed in Einstein's famous equation:

$$E = mc^2$$

where E is energy, m is mass, and c is the speed of light. Because the square of the speed of light is such a large number, the tiny amount of mass lost during this conversion produces enormous energy. It is this energy that radiates from the Sun's surface and permeates the Solar System.

The reaction in the Sun consumes 4 to 5 million tons of mass every second. Hence the amount of energy released from

A knowledge of the subatomic particles emitted by thermonuclear reactions is important to any understanding of the fusion processes occurring in the Sun's core. One such particle, the neutrino, is among the most pervasive forms of matter in the universe. It is estimated that between 100 and 1000 neutrinos occur in every centimeter of space and that a trillion neutrinos pass through each of us every second! Fortunately, they do not interact with our body tissue, and so have no effect on us. In fact, since neutrinos have almost no mass or charge, and travel at the speed of light, they pass unhindered through the entire Earth.

Neutrinos are not only produced in the Sun's core but also form during the collapse of stars in our galaxy and beyond. Thus their presence provides important insights into the processes that have shaped the universe.

Despite their abundance, they eluded detection for nearly 30 years after their existence was first theorized by Austrian physicist Wolfgang Pauli in the late 1920s. In 1933, Enrico Fermi named the particle "neutrino," which is Italian for "little neutral one." But they remained undetected until the early 1950s, when Clyde Cowan and Fred Reines of the Los Alamos National Observatory in New Mexico set out to find them because they considered this task "the hardest physical experiment" they could think of. Cowan and Reines reasoned that nuclear reactors should emit neutrinos, and they successfully detected their presence at the Savannah River reactor in South Carolina.

Since that time, ingenious methods have been used to study these particles. Neutrinos are difficult to detect near the Earth's surface because of interference with other forms of radiation. However, most forms of radiation cannot penetrate far into the Earth's crust. In deep mines, therefore, the undistorted signal of neutrinos can be detected as they pass through solutions sensitive to their presence. One such solution is carbon tetrachloride, which neutrinos interact with by turning chlorine atoms into radioactive argon.

Physicists have also devised experiments that use beams of high energy neutrinos to probe the internal structure of atoms. Neutrinos are also used in studying the "weak force." This is one of the fundamental forces involved in the nuclear reactions in the Sun and other stars and is responsible for radioactive decay of neutrons and other particles.

---

the Sun's core and eventually radiated into space is inconceivably vast. This energy travels as discrete packets of energy called **photons** which radiate through the Solar System at the speed of light (3.00×10^8 m/s, or 186,000 mi/s). Thus, the thermonuclear fusion in the Sun's core results in the emission of energy that permeates the entire Solar System. We call this energy **solar radiation.**

16.2.2 Electromagnetic Spectrum and the Sun's Radiation

As an object is heated, electrons orbiting the nucleus of its atoms become disturbed and emit electromagnetic radiation in the form of photons. Solar energy transmitted by moving photons is described in terms of its wavelengths, that is, the distance from one wave peak to the next. The Sun's radiant energy encompasses a wide range of different wavelengths that collectively make up the electromagnetic spectrum (Fig. 16.3).

The range of wavelengths emitted from a hot body depends on its temperature. The Sun shines with a continuous spectrum, spanning the interval from gamma rays with very short (10^{-12} m) wavelengths to radiowaves with very long ones (10^2 to 10^3 m). Our eyes have the ability to detect only a tiny portion of this spectrum. This is called the **visible spectrum**, which we recognize as "the colors of the rainbow." But we are sensitive to other forms of solar radiation. The invisible ultraviolet wavelength, for example, causes sunburn and has been linked to skin cancer. We can detect infrared radiation as heat. Indeed, when we talk of the greenhouse properties of atmospheric gases, we are aware that it is the ability of these gases to block the escape of reflected infrared radiation into space that is the cause of the "greenhouse effect." Shorter wavelengths, such as X-rays, are more penetrative than longer ones, and so we use artificially-generated X-rays to detect fractured bones through soft-bodied tissue and to scrutinize sealed luggage at airport security checkpoints.

Although the Sun emits a wide range of wavelengths, most of the energy it radiates is transmitted between the ultraviolet and infrared portions of the spectrum. This blend of wavelengths is known as "white light." However, within this range there is a peak wavelength of maximum intensity and it is from this that the temperature of the Sun's surface has been determined.

Atoms in hotter objects have more energy and therefore vibrate more rapidly than atoms in cooler bodies. Since more of their electrons are disturbed, they emit more radiation than cooler objects. In 1898, Wilheim Wien discovered an inverse relationship between the temperature of a body and the wavelength of the most intense radiation (λ_{max}) emitted from it (Fig. 16.4). The hotter the surface of a body, the *shorter* the wavelength of maximum intensity. As a result, the color transmitted from bodies hotter than the Sun is closer to the blue or ultraviolet end of the spectrum. Colder bodies emit energy with longer wavelengths and so appear red. Wien formulated an equation to explain this relationship which is known as **Wien's Law**:

$$\lambda_{max} = \frac{3,000,000}{T}$$

where T is the surface temperature of the body in degrees Kelvin[1] and λ_{max} is the wavelength of emitted radiation that has the maximum intensity.

---

[1] Degrees Kelvin can be converted to degrees Centigrade by subtracting 273. e.g. 283°K = 10°C.

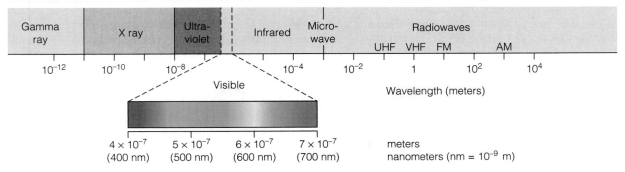

Figure 16.3

Solar energy encompasses a range of wavelengths, known as the electromagnetic spectrum, that vary from very short gamma rays (10^{-12} m) to very long wavelength radiowaves (10^2 to 10^4 m). Our eyes have the ability to detect only a tiny portion of this spectrum, which is known as the "visible spectrum." The solar radiation intensity varies according to wavelength, and peaks within the visible spectrum.

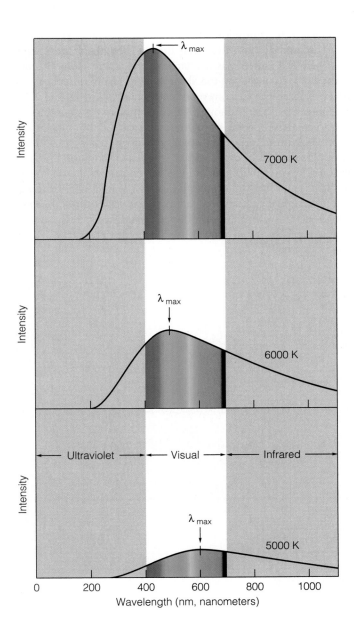

Figure 16.4

Application of Wien's Law showing the relationship between temperature, the intensity of radiation emitted from a hot body, and the wavelength of radiation that has the maximum intensity. The hotter the surface of a body, the *shorter* the wavelength of maximum intensity (the blue part of the spectrum, upper diagram). The Sun most closely corresponds with the middle picture, its wavelength of maximum intensity occurring within the blue-green region of the visible spectrum (510 nm).

From this relationship, the temperature at the surface of a radiating body can be estimated if the wavelength of maximum intensity (λ_{max}) of the emitted radiation is known. This peak wavelength can, in turn, be determined from the color of the radiating body. For example, an orange star emits radiation with maximum intensity corresponding to a wavelength of about 600 nm (see Figure 16.3). Substituting this value for λ in the equation, we obtain a temperature for the surface of the star of about 5000°K, or 4727°C (see Figure 16.4).

The Sun has a wavelength of maximum intensity of 5.1×10^{-7} m, which corresponds to a surface temperature of about 6000°K (5727°C). This wavelength also corresponds to the boundary between the green and blue portion of the visible spectrum. It is no coincidence that our eyes are most sensitive to this color, since our vision has evolved to match the peak wavelength of the spectrum. The Sun's visible color is actually a yellow-white, which is a blend of all the solar radiation.

The average amount of solar radiation reaching the Earth's outer atmosphere every second is 1.36 kilowatts per square meter (kw/m²), a number known as the **solar constant**. However, this number will not remain constant over the Sun's remaining life span. The energy output of a star is known to vary, especially in the early and later stages of its evolution.

For most of its life, however, the Sun will regulate its own energy output because it is entirely made up of gas. If, for example, the energy production in its core were to increase, its gases would expand, thereby cooling the core and decreasing the energy production. Conversely, if energy production in its core decreased, its gases would contract, and in so doing, would heat the core and increase energy production.

Only 50% of the solar energy reaching the Earth's outer atmosphere reaches the surface. The other 50% is absorbed by atmospheric gases, clouds, and particulates. The energy that does reach the Earth's surface is either absorbed or reflected by rocks, soils, vegetation, and water, but the reflection is partially blocked by greenhouse gases in the atmosphere (see Chapter 11).

16.2.3 The Composition of the Sun

The Sun's composition is dominated by hydrogen (92.1%) and helium (7.8%), the remainder being made up of heavier elements that are higher in the periodic table. But how has this composition been determined?

In 1814, Joseph von Fraunhofer studied the Sun's visible spectrum in detail. He found that the spectrum was not continuous, but instead was interrupted by dark lines (Fig. 16.5). He gave each black line a code letter to denote its position within the spectrum. He was able to deduce that the lines probably signified the *absence* of discrete wavelengths of light from an otherwise continuous solar spectrum. However, neither he, nor the scientists of his day, had any explanation for their occurrence. Since von Fraunhofer's observations were duplicated in many laboratories, the lines were assumed to represent some intrinsic property of the Sun's radiation, rather than being the result of local radiation interference on Earth.

Fifty years later, von Fraunhofer's meticulous observations bore fruit. Gustav Kirchoff observed a similar phenomenon of an interrupted spectrum in the light emitted when any single chemical element was burned. The ignited element emitted a color corresponding to a narrow portion of the visible spectrum. But whereas the same element gave identical results in repeated experiments, different elements emitted light of different wavelengths.

At first glance, the spectra of von Fraunhofer and those of Kirchoff look quite different (Fig. 16.6). Von Fraunhofer found a series of black lines superimposed on the visible spectrum; Kirchoff's spectrum, consisted of bright colored lines on a black background. However, the light emitted in Kirchoff's experiments gave bright **emission lines** that matched exactly the wavelength of the dark lines or **absorption lines** discovered by von Fraunhofer.

Combining these observations, Kirchoff deduced that the dark lines described by von Fraunhofer were related to the *absorption* of energy by the material of the Sun. Such absorption occurs when white light passes through a relatively cool low-density gas. Energy of specific wavelengths is absorbed by the gas atoms, giving rise to what is known as a dark-line or **absorption spectrum**. Conversely, the bright emission lines evident in his experiments were related to the *emission* of light from the ignited element. Emission occurs when a low density gas is excited and emits light. In this case, the pattern formed is known as the **emission spectrum**.

Kirchoff also reasoned that the wavelength of the energy absorbed or emitted was a characteristic of the element concerned. For example, on the absorption spectrum of von Fraunhofer, wavelength D corresponds to sodium, and H and K to calcium (Fig. 16.5). Although the Sun shines with a continuous spectrum, these specific wavelengths are absent from the absorption spectrum because they are absorbed by cooler gases in the solar atmosphere. In essence, then, the dark absorption lines discovered by von Fraunhofer are like fingerprints of the chemistry of the gases of the Sun's upper layers. Just as a flashlight shining through a grid casts a shadow pattern that is diagnostic of the grid, each element within the Sun's upper layers has its own characteristic black lines that allow its identity to be determined.

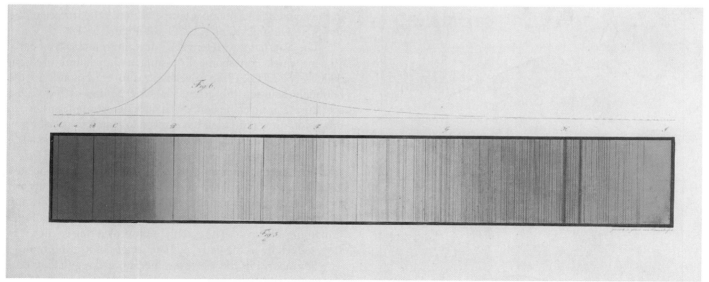

Figure 16.5
Fraunhofer lines are dark lines in the Sun's absorption spectrum, and signify the *absence* of discrete wavelengths of light from an otherwise continuous solar spectrum.

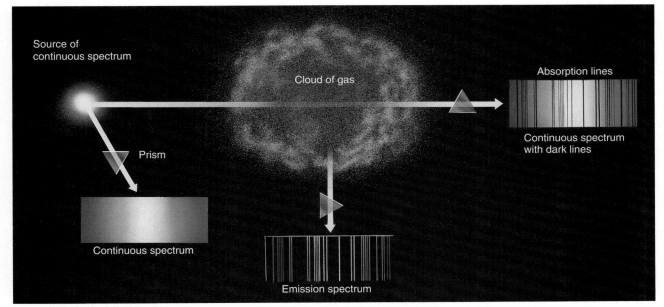

Figure 16.6

Comparison between absorption and emission spectra. Absorption spectra are dark lines signifying the absence of discrete wavelengths of light. Absorption of energy occurs when a cool gas is warmed and its electrons jump to a higher energy level. A telescope aimed at the light source through the gas would detect this absorption spectrum. Emission spectra occur when electrons in the gas drop to lower energy levels and emit light as they do so. A telescope aimed at the gas cloud but not at the light source would detect this emission spectrum. The bright emission lines match the wavelength of the dark absorption lines.

To fully understand the relationship between the absorption lines and the chemical composition of the Sun, we now examine what happens when an element is heated near the surface of the Sun. As we learned in Chapter 3, atoms have a nucleus composed of protons and neutrons with electrons orbiting around it. Protons are positively charged, electrons are negatively charged, and neutrons carry no charge. The chemical elements are distinguished from one another by the number of protons they contain. An element's charge is neutral if the number of protons is the same as the number of electrons. Electrons move around the nucleus in tightly constrained paths, known as orbitals, each of which can contain two electrons. Elements with larger numbers of electrons have several orbitals.

When an element absorbs radiant energy, some wavelengths contain just the right amount of energy, known as a **quantum**, to move an electron from its usual orbital to a higher one. The element still has the same number of electrons, but by absorbing the energy, the electron has jumped out of place into an **excited state** (Fig. 16.7a). Since each element has its own unique arrangement of electrons, each element absorbs its own characteristic portion, or quantum, of the supplied energy. Like an elevator in an apartment building that stops at every floor level but not between floors, an electron leaps to a specific higher energy level when the element absorbs the appropriate amount of energy. The jump is known as a quantum leap. As the element cools, the electron eventually returns to its normal, or **ground state**, releasing the quantum of energy as it does so.

If the Sun did not have an atmosphere, radiant energy (or photons) emitted from its surface would travel unhindered and a telescope aimed at the Sun would detect a continuous spectrum. However, as photons of energy are emitted from the Sun's surface, they interact with the gases of the solar atmosphere. Most of the photons pass through the gas unaffected, because only a small fraction of photons will have the wavelengths that the atmosphere can absorb. Those photons that do not pass through are absorbed by atoms within the atmosphere. Electrons within the affected atoms are excited for the briefest of moments, before returning to the ground state and emitting a new photon of energy. The original photons were travelling directly toward the telescope. However, new photons are emitted in random directions, and very few travel in the direction of the telescope (Fig. 16.7b). As a result, photons of light with these specific wavelengths are missing from the spectrum detected by the telescope.

The photons of energy that are absorbed are bounced around in the solar atmosphere, being absorbed and emitted repeatedly until they finally escape into space. If instead of aiming a telescope directly at the Sun, it is aimed to one side, we can detect the spectrum emitted from the gas atoms (Fig. 16.7c). The bright emission lines exactly match the wavelength of the dark lines in the absorption spectrum (Fig. 16.6). The emission lines also effectively behave as chemical fingerprints, allowing us to deduce the composition of the Sun's outermost layers. In addition, the intensity of a particular line reflects that element's abundance at the Sun's surface. In this way, the dominant

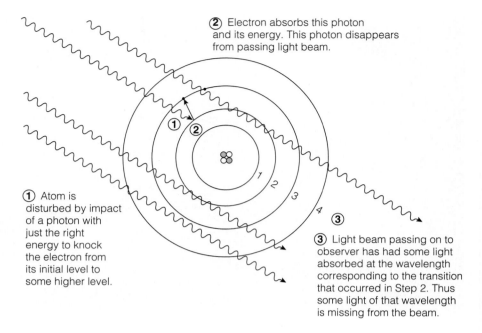

② Electron absorbs this photon and its energy. This photon disappears from passing light beam.

① Atom is disturbed by impact of a photon with just the right energy to knock the electron from its initial level to some higher level.

③ Light beam passing on to observer has had some light absorbed at the wavelength corresponding to the transition that occurred in Step 2. Thus some light of that wavelength is missing from the beam.

(a)

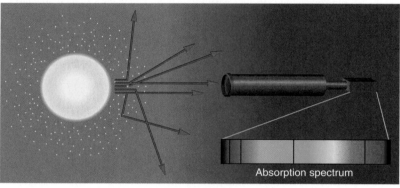

Absorption spectrum

(b)

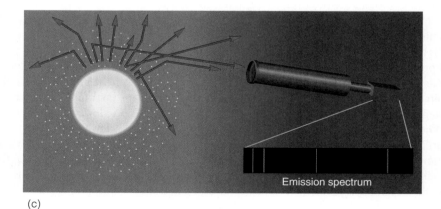

Emission spectrum

(c)

Figure 16.7

(a) As radiation passes through a gas cloud, an electron may be moved from its usual orbital in a gas atom to a higher one if it absorbs just the right amount of energy, known as a quantum. By absorbing the energy, the electron has jumped out of place into an excited state. The absorption of this wavelength by the atom is characteristic of the element and shows up as a dark line on an absorption spectrum. (b) The absorption spectrum of the Sun is produced when atoms in its atmosphere absorb photons with specific wavelengths. These photons are re-emitted in random directions, and consequently most do not reach a telescope directed at the Sun. As a result, these wavelengths appear as dark lines on an otherwise continuous spectrum. (c) When a telescope is directed away from the Sun, it can detect only the absorbed and re-emitted photons, producing an emission spectrum.

fingerprints identify the overwhelming presence of hydrogen and helium. Traces of oxygen, carbon, sodium, and iron have also been identified in a similar manner.

Interestingly, when the solar spectrum was first analyzed to deduce the chemical composition of the Sun, no match could be found for the absorption lines of one particular element. The unknown element was therefore given the name **helium**, after *helios*, the Greek word for the Sun. Helium had not yet been discovered on Earth. In fact, another 17 years would pass before its presence on our planet would be finally demonstrated.

As more information becomes available about the Sun's composition compared to that of other stars, we find its composition has an important anomaly. Relative to typical stars, the Sun has almost twice as much carbon and heavy elements (including iron and nickel). Astronomers believe that heavy elements like these are produced only late in the lives of stars, and then only in stars far more massive than the Sun. Upon the destruction of massive stars, the heavy elements are dispersed in a supernova explosion and so become incorporated into other solar systems. Their presence in the Sun therefore provides important evidence that the chemistry of our Solar System has been influenced by the death of older, more massive, stars.

16.2.4 From Center to Surface

The Sun, like almost all bodies in the Solar System, is layered from its core to its surface. But, unlike the terrestrial planets, these layers are not defined by changes in chemical composition. Instead, they are defined by the behavior of gaseous material; behavior that varies with changes in temperature, pressure, and density between the Sun's core and its surface.

The Sun's internal layering is based on the effects of the thermonuclear reactions that occur in its core. In a region known as the **radiation zone**, the energy produced by nuclear fusion in the core of the Sun travels towards the surface primarily by radiation (Fig. 16.8). Within 200,000 km (125,000 mi) of the surface, however, convection takes over as the dominant process of heat transfer as the gases cool. This defines the **convection zone** and occurs because the gases cool at different rates, and so produce temperature contrasts that stimulate convection (recall the controls on convection in the Earth's atmosphere, Chapter 11).

It may take a million years or more for the energy to get from the core to the surface of the Sun. Once emitted from the surface, however, it travels at the speed of light, taking a mere 8.3 minutes to travel the 150 million km (93 million mi) to the Earth.

A transition zone separates the convection zone and the overlying photosphere. The **photosphere** is the visible part of the Sun, best seen on a partly cloudy day. It is about 500 km (310 mi) thick. When this luminous disk is studied in greater detail, a granular appearance is revealed (Fig. 16.9a), which is attributed to the presence of convection cells. The granules are between 1000 and 5000 km (600 and 3000 mi) across. Their centers sit in the convection zone below, atop rising convection plumes that descend at the borders of the granule (Fig. 16.9b). Each granule maintains its position and shape for only a few minutes before it is destroyed and a new convection cell is created.

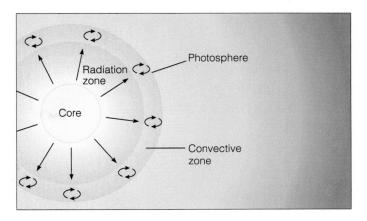

Figure 16.8

A cross-section through the Sun from its core to its surface. The core is defined by the extent of thermonuclear reaction. This energy radiates from the core through the radiation zone. Within 200,000 km (125,000 mi) of the surface, heat is transported towards the photosphere by convection.

As we have learned, the temperature of the photosphere is approximately 5700°C. However, the photosphere is broken by dark regions, known as **sunspots** (Fig. 16.10), which represent relatively depressed regions on the Sun's surface. Sunspots are, on average, about 1500°C cooler than adjacent areas. They are regions with strong magnetic fields (Fig. 16.9c) that are thought to be responsible for the regions' relatively cool temperatures because intense magnetism stops the gases from escaping, thereby inhibiting the efficient transfer of heat. The magnetic pattern in the vicinity of these sunspots resembles that of a bar magnet (Fig. 16.9d). Their dark color simply implies relatively cool temperatures in accordance with Wien's law, which states that color and temperature are related. Because they are cool, they are also depressed relative to the more elevated warm areas. Huge clouds of hot gas trapped in the magnetic field of the sunspots erupt just above the surface of the Sun. These eruptions last for minutes and are called **prominences** (Fig. 16.9e).

A highly irregular transition zone known as the **chromosphere** separates the photosphere from the overlying corona. Averaging some 5000 km (3300 mi) in thickness, the transition is dominated by **spicules**, which are like waves that rise to enormous heights (some higher than the Earth's diameter) and then fall and disappear (Fig. 16.9f). They typically last less than 15 minutes.

The base of the **chromosphere** has a temperature of about 10,000°C. Temperatures rise toward the upper part of the chromosphere, where they reach about 500,000°C. When viewed in isolation during a total eclipse of the Sun, the chromosphere appears pink in color. This color reflects the dominance of hydrogen in the Sun's composition. The uppermost part of the chromosphere represents the restless surface of the Sun that is commonly dubbed the "Solar Sea."

The chromosphere gives way to the **corona**, which is the spectacular solar crown we see during a total solar eclipse (Fig. 16.9g). Temperatures rise dramatically in the corona and can reach 2 million °C. These high temperatures are confirmed by X-ray telescopes,

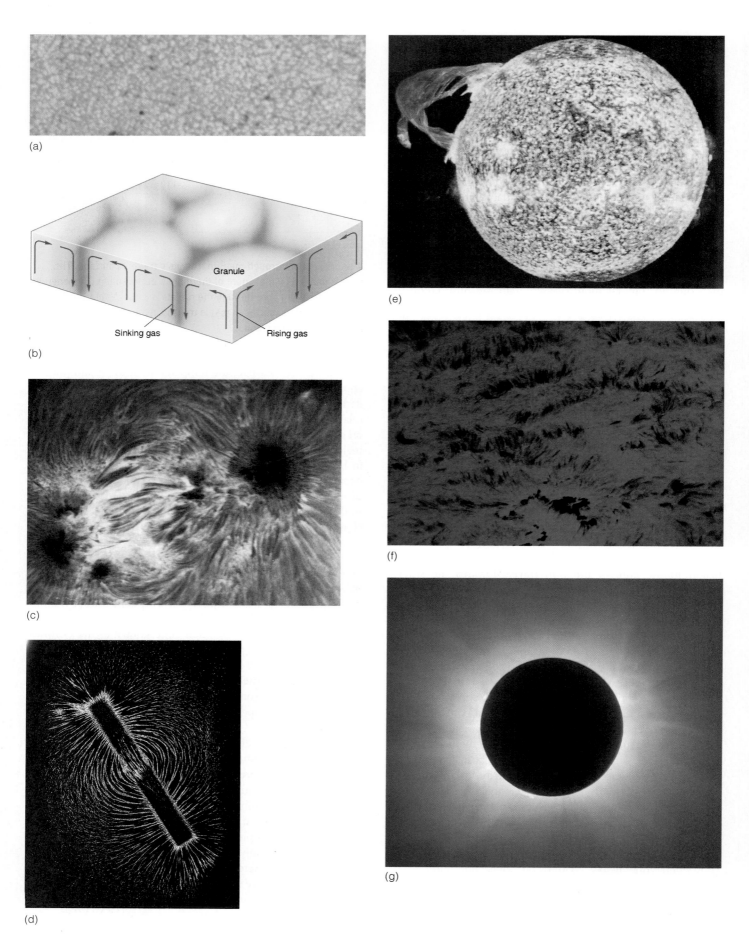

(a)

(b)

Granule

Sinking gas

Rising gas

(c)

(d)

(e)

(f)

(g)

Figure 16.9 (left)
Features of the Sun's surface. (a) Detailed visible-light photo of the solar photosphere shows granules between 1000 and 5000 km (600 to 3000 mi) across. (b) A model for the granules in the photosphere, showing their centers to occur above rising currents of relatively hot gas while their edges lie above descending currents of relatively cool gas. The pattern in the photosphere is influenced by the convection zone below. (c) A filtered image showing the solar activity in the chromosphere in the vicinity of sunspots. (d) The pattern of magnetism in the chromosphere in the vicinity of sunspots resembles that of iron filings around a bar magnet. (e) Prominences, or jets of hot gas in the vicinity of sunspots erupt from just above the Sun's surface. Note the granular nature of the photosphere. (f) Spicules, or clustered jets of solar gas. (g) View of the corona during a solar eclipse, several pink prominences are visible.

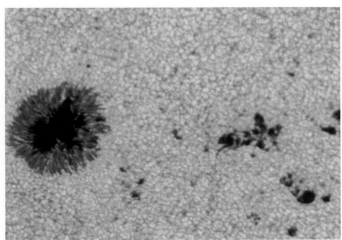

Figure 16.10

Sunspots are dark regions of cold, depressed, and anomalously magnetic areas on the Sun's surface. They were first described by Galileo.

which show that the corona emits energy largely in the very short X-ray wavelength portion of the spectrum, in accordance with Wien's Law. The corona's outer boundary is not defined because it merges with interplanetary space.

The Sun's internally layered structure was deduced by a variety of observations made in the 1970s and early 1980s. Careful observation of the granules showed that the whole Sun oscillates with a period of about 5 minutes. The origin of these oscillations was at first puzzling, but they were later shown to be caused by the reflection and refraction of sound waves traveling below the surface of the Sun. These sound waves are thought to be caused by turbulence in the convection zone. Regardless of their origin, however, the patterns of wave reflection and refraction were used to deduce the internal layering of the Sun in much the same way that seismic waves have been used to explore the Earth's interior (see Chapter 7).

16.3 From the Sun to the Earth

It is clear that the influence of the Sun extends far beyond its surface. Indeed, the radiation that emanates from its surface pervades the entire Solar System. There are, however, many subtle variations in solar activity that can have profound effects on the Earth. Among these are the effects of the solar wind and sunspots.

16.3.1 The Solar Wind

Within its body, the motion of the Sun's ionized gases sets up a magnetic field with lines of force that influence the entire Solar System. If this field were similar to that of the Earth's, it would resemble (on a gigantic scale) the "bar magnet" model we discussed when describing the Earth's magnetic field (see Chapters 4 and 7). However, the style of the Sun's magnetism is very different for several reasons. First, the Sun is much larger and is far more dynamic than the Earth, with temperatures many orders of magnitude higher. Second, the Sun rotates significantly faster at the equator (once in 25 days) than it does at the poles (once in every 35 days). The differential rotation distorts the magnetic field into a spiral. Finally, the magnetic field beyond the Sun's surface is further distorted by the solar wind.

The **solar wind** is produced by ionized gases blown away from the Sun's surface but is more concentrated in regions such as sunspots. The manifestation of this wind is most obvious when it interacts with comets (see Figure 15.35). Bombardment of the comet (a rock-ice mixture) by the solar wind produces a "tail" that always points away from the Sun, regardless of whether the comet is approaching the Sun or moving away from it. The tail represents the entrainment in the solar wind of the gases released from the comet.

The stream of charged particles that make up the solar wind blows past the Earth at 300 to 800 km/s (600,000 to 1,600,000 mi/hr) and distorts the Earth's magnetic field as it does so. This effect is most obvious during **solar flares**, which are major magnetic storm systems on the Sun's surface that release bursts of highly charged particles. These storms can last for weeks and greatly increase the density and speed of the solar wind. When it collides with the Earth's outer atmosphere (the ionosphere), the enhanced solar wind can cause a significant disruption to our communication systems, particularly for the shortwave signals which are reflected off the ionosphere.

Solar wind intensity may also have other, more long-lasting effects. As we shall learn in Chapter 20, some scientists propose that the activity of the solar wind helps to create or destroy ozone in the atmosphere. The solar wind also influences the region beyond the Earth known as the magnetosphere. The **magnetosphere** is a region encircling the Earth in which interference occurs between the Sun's magnetic field, the solar wind, and the Earth's magnetic field (Fig. 16.11). The geometry of the magnetosphere is governed by the bombardment of the Earth's magnetic field by particles in the solar wind. Like a large rock in a fast-moving stream, the Earth has a magnetic wake on the side facing away from the Sun where the magnetosphere is drawn out into a comet-like tail by the solar wind. But, on the side of the Earth facing the Sun, the magnetosphere parts the solar wind (just as a rock parts the water that moves past it) and is compressed and warmed as it does so.

Unlike the rock, however, the boundary of the magnetosphere is very porous, and charged particles may leak into it from the solar wind. These charged particles collect in regions known as the *Van Allen belts*, where they align themselves within the Earth's magnetic field just as iron filings line up adjacent to a bar magnet. In doing so, they produce electric currents which build until the charge reaches a threshold. Like an overloaded circuit system, the particles move down the Earth's magnetic lines of force towards the north or south magnetic poles. As they do so, they interact with atmospheric gases to produce fingers of light that stretch across the sky parallel to the Earth's magnetic field. This spectacular celestial light display is known as the **aurora** (Fig. 16.12); or more precisely, the *Aurora Borealis* or "Northern Lights" in the northern hemisphere and the *Aurora Australis* or "Southern Lights" in the southern hemisphere. Such auroras are more pronounced during solar flares when the solar wind is most intense. Auroras also produce some strong basic colors, which can be used to detect the presence of specific gases in the atmosphere. One of the dominant colors is green, which corresponds to a wavelength of 5.577×10^{-7} m (Fig. 16.3), and is typical of spectra emitted by monatomic oxygen (that is, O, rather than the more common O_2). Another is crimson red, which corresponds to a wavelength of 6.3×10^{-7} m and is consistent with the presence of nitrogen, the dominant gas in the atmosphere.

The auroras tell us that the solar wind penetrates the Earth's magnetic shield. Auroras also vary in frequency, telling us that the strength of the solar wind itself varies. This variation is partly correlated with the presence of sunspots on the surface of the Sun.

16.3.2 The Sunspot Cycle

Galileo's discovery of sunspots in 1610 dispelled the myth that the Sun was a homogenous body with a constant energy output. Since that time, meticulous records have been kept of their numbers and properties.[2] These numbers reveal a phenomenon known as the **sunspot cycle** (Figs. 16.13 and 16.14), where over an 11-year period, the number of sunspots declines and then increases to a peak (Fig. 16.15). Because of changes in the polarity of the Sun's magnetic field during this 11-year interval, a complete sunspot cycle is a little over 22 years in duration.

Recently, sunspot activity peaked in 1980 and 1991; the next peak is expected in 2002. Sunspot minima occurred in 1975, 1986, and 1997. Sunspots accelerate the particles that form the solar wind to speeds of up to 500–1000 km/s (1–2 million mi/h). The reason for this is not entirely clear; however, most scientists believe that the intense magnetic fields associated with sunspots literally hurl charged particles into space. Whatever the mechanism involved, peak sunspot activity greatly enhances the power of the solar wind bombarding our upper atmosphere. At the peak of a sunspot cycle, the upper atmosphere may attain a temperature of about 1225°C (2240°F). In contrast, the temperature during a sunspot minimum is only 225°C (440°F). The wide range in temperature reflects the sensitivity of a very sparse atmosphere to changes in solar wind velocities. But does this dramatic fluctuation in upper atmospheric temperatures affect the Earth's surface temperature and climate? At present there is no clear mechanism that would link the temperature in the upper atmosphere to that at the Earth's surface. However, the effect does not have to be great. The last major Ice Age, for example, was triggered by a mere 2°C drop in the average temperature at the Earth's surface.

At first glance, the relationship between sunspots, which are, after all, relatively *cold* areas on the surface of the Sun, and the relatively high temperatures their peak activity generates in the Earth's upper atmosphere, would appear to be counter-intuitive. Indeed, recent measurements demonstrate that large numbers of sunspots actually *reduce* solar output by as much as 0.1%. However, sunspots are associated with accelerated solar wind, and it is the solar wind that blasts our upper atmosphere. The potentially negative effects of reduced solar output at the

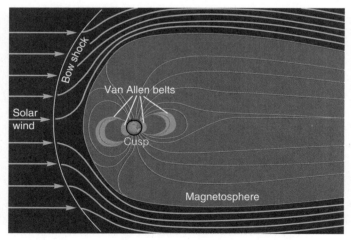

Figure 16.11
The magnetosphere is the region where the Earth's magnetic field (pink lines) and the Sun's magnetic field (blue lines) interact. The bow shock is the leading buttress of the Earth's magnetosphere against the solar wind, and the Van Allen belts are bands of charged particles that are trapped in the Earth's magnetic field. Note that the solar wind parallels the lines of the Sun's magnetic field.

[2] Observing the Sun through a telescope even for an instant can burn the retina of your eye. However, a telescopic image of the Sun is readily projected onto a screen, and from this the number and properties of sunspots can be examined.

(a)

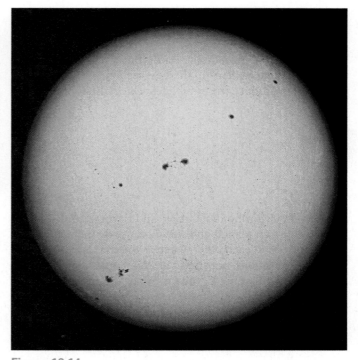

(b)

Figure 16.12
(a) The Aurora Borealis, as seen from the ground. The green color is indicative of the presence of oxygen atoms in the atmosphere.
(b) The Aurora Australis, as photographed from space by astronauts on the space shuttle Discovery. The colored lines are parallel to the Earth's magnetic lines of force.

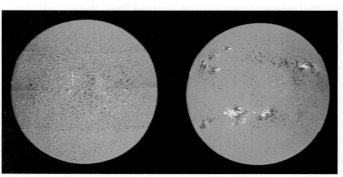

Figure 16.13
Two images of the Sun at minimum (left) and peak (right) times of the sunspot cycle. The image detects the intense magnetic fields above the sunspots, the dark and light areas corresponding to north and south polarities, respectively.

peak of the sunspot cycle are consequently overridden by the enhanced solar wind bombarding the Earth's ionosphere.

There is empirical and anecdotal evidence to support a connection between sunspots, solar wind, and the Earth's climate. In the 17th Century, immediately following their discovery, the records show that virtually no sunspots were observed (Fig. 16.15). This period of reduced solar activity is referred to

Figure 16.14
Observing the Sun through a telescope will immediately damage your eyesight. However, an image of the Sun can be projected onto a screen allowing the sunspots (dark areas) to be seen.

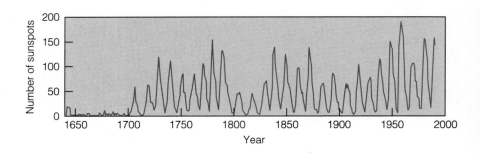

Figure 16.15
The sunspot cycle since 1640. Numbers are compiled from historical sources and show that there were few sunspots from 1645 to 1710. This period, known as the Maunder Minimum, overlaps a period of global cooling known as the "Little Ice Age."

as the **Maunder Minimum**, and may be linked to the phenomena known as the *Little Ice Age*—a period of cool weather in North America and Europe between about 1450 and 1850. During this period, glaciers in the Alps advanced to crush settlements in the valleys below. The normally ice-free canals of the Netherlands were frozen for up to three months each winter. French vineyards were severely damaged, and had to be harvested much later in the season. The astronomer William Herschel even identified a correlation between the sunspot cycle and the price of grain, by suggesting that reduced solar output resulted in a lower crop yield, which drove up the price.

Whatever caused the temporary hiatus in the sunspot cycle is as enigmatic as the origin of the cycle itself. However, its correlation with at least part of the Little Ice Age permits the speculative hypothesis that prolonged periods of decreased sunspot activity could result in prolonged cooling of Earth's uppermost atmosphere. And as we have discussed, even a minor connection between upper atmosphere and surface temperatures could result in a sufficient temperature drop to start an ice age.

The relationship among sunspots, the strength of the solar wind, and the temperatures of the upper atmosphere, provides strong evidence that changes in the short-term behavior of the Sun have a direct effect on the Earth. But what about the stars beyond the Solar System? In the next section, we will discover that they too produce measurable effects on the Earth.

16.4 Beyond the Solar System

The ancient civilizations of Greece and Egypt were able to distinguish between planets and stars because the planets moved rapidly relative to each other, whereas the stars appeared fixed so that constellations, or patterns of stars, could be recognized. It is speculated that the Egyptians believed that the spirit of a dead Pharaoh ascended to the heavens to form a star in the constellation Orion. Indeed, some archaeologists believe that the Egyptians built the pyramids so that each apex would point towards one of these stars.

Our modern understanding of the universe commenced when Sir Isaac Newton provided an understanding of the forces that govern planetary motion in the Solar System. It became clear that our Sun was but one among many stars, and

attention switched to the heavens beyond the edges of the Solar System. We now know that there is nothing particularly special about our Sun. It is a medium-sized star, and just one of 100 billion stars in our galaxy, the Milky Way. The Milky Way, in turn, is just one of billions of galaxies in the universe.

One of the major breakthroughs in our understanding of the universe came from Albert Einstein's theory of relativity, which explains the behavior of moving bodies in space and time. One application of this theory involves the relationship of time and the travel of light through space. When you look up at the night sky, you are looking into the past because the light you see emanating from celestial objects has taken a long time to reach us. The Sun is our closest star, and light emanating from its surface takes 8.3 minutes to reach us. What we see, therefore, is an image of the Sun as it looked 8.3 minutes ago. The next closest star, known as Proxima Centauri, is 4.3 light years away. A **light year** is the distance traveled by light in a single year (9.46 trillion km or 5.91 trillion mi). Hence, we see Proxima Centauri as it looked 4.3 years ago. In 1987, astronomers observed a catastrophic explosion, known as a supernova, that signaled the death of a remote star. But the explosion actually took place 170,000 years ago because the star was 170,000 light years away. The flash of light and neutrinos emitted from the explosion consequently took 170,000 years to get here (Fig. 16.16).

16.4.1 Distant Stellar Events and Their Effects on Earth

The observation by astronomers of supernova explosions demonstrates that the Earth is receptive to cosmic processes far beyond the Solar System. In fact, it is estimated that supernovae occur every 40 to 60 years in our galaxy, although their remoteness prevents us from seeing them. But can cosmic interactions cause changes in Earth processes? The answer to this question is probably yes, although the effects appear to be far more subtle than those produced by the Sun. The leaking of cosmic radiation into our atmosphere from *outside* the Solar System was first detected in the mid-1960s. But while we can detect this radiation, we do not know what its short-term or long-term effects on the planet might be. We know that our Sun and the Solar System revolve around the center of the Milky Way galaxy every 225 million years. But we do not know whether the influence of cosmic radiation from other stars differs subtly as we revolve.

(a)

(b)

Figure 16.16

(a) Before and (b) after the supernova explosion observed in 1987. The explosion actually took place 170,000 years earlier, but it took that amount of time for the flash of light and neutrinos emitted by the explosion to reach the Earth.

Yet conclusive evidence that cosmic rays penetrate our atmosphere comes from the study of radioactive isotopes (Fig. 16.17).

Our atmosphere is being continuously bombarded by radiation from both solar and cosmic rays. Cosmic radiation causes nuclear reactions to occur in a small percentage of the Earth's most abundant atmospheric gas, nitrogen. The most common isotope of nitrogen is nitrogen-14. By definition, the isotope has 7 protons and 7 neutrons which gives it an atomic mass of 14 units. The net effect of cosmic bombardment is to convert nitrogen-14 to carbon-14, which is the basis of radiocarbon dating (see Dig In: Radioactivity, on pages 70–71).

The reaction scheme (Fig. 16.17) first shows a cosmic neutron striking the nucleus of the nitrogen atom. Like the cue ball at the "break" in pool (see Figure 16.2), the atomic pack absorbs the incoming neutron (a process known as neutron capture) but also spits out a particle. In this case, the ejected particle is a proton. Thus, the original nitrogen atom now has only 6 protons, and so, by definition, has become a carbon atom. But it also has 8 neutrons rather than 7, so that its atomic mass stays the same. The product of the bombardment, then, is carbon-14, the radioactive isotope of carbon. With increasing cosmic bombardment, we should therefore expect an increase in the production of carbon-14 in the atmosphere. Since the properties of atomic elements are governed more by the number of protons they contain than by their mass, the carbon-14 produced by cosmic radiation is incorporated into living organisms of the biosphere, just like the common isotope carbon-12. This

Figure 16.17

Cosmic rays penetrate the atmosphere and result in the production of carbon-14 from nitrogen-14. When the Earth's magnetosphere is weak, intense cosmic bombardment occurs, producing abundant carbon-14. Carbon-14 is absorbed into living tissue while the organism lives but will eventually spontaneously decay back to nitrogen-14. So, a high concentration of nitrogen-14 in a dead organism means that it once contained abundant carbon-14, implying a weak magnetosphere.

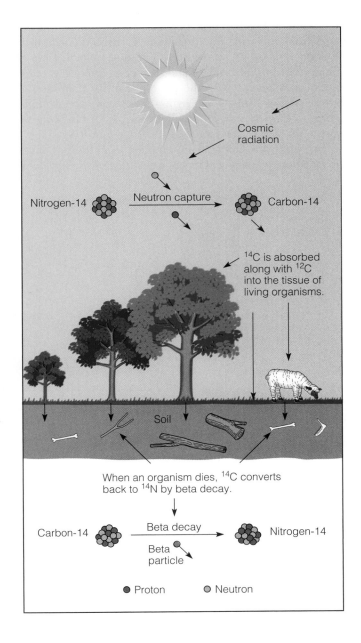

Cosmic radiation

Nitrogen-14   Neutron capture   Carbon-14

^{14}C is absorbed along with ^{12}C into the tissue of living organisms.

Soil

When an organism dies, ^{14}C converts back to ^{14}N by beta decay.

Carbon-14   Beta decay   Nitrogen-14

Beta particle

● Proton   ● Neutron

behavior of carbon, of course, contrasts with that of nitrogen, which resists incorporation into any of the other Earth systems and, as a result, is the most abundant gas in the atmosphere.

As long as an organism is alive and growing, carbon-14 will continue to be incorporated into its living cells. Thus, the abundance of carbon-14 relative to carbon-12 in the organism remains essentially constant. This occurs despite the fact that carbon-14 is radioactive and spontaneously decays back to nitrogen-14. During life, it is replaced as fast as it decays. As soon as the organism dies, however, the radioactive carbon-14 is no longer replaced and, over time, progressively decays to nitrogen-14. A high concentration of nitrogen-14 in a dead organism consequently implies that it contained a large amount of carbon-14 when it was alive, and so must have lived at a time when the atmosphere was subjected to a high rate of cosmic bombardment.

We can study the effects of cosmic radiation through time by examining the organic record of carbon-14 production in the biosphere. The best evidence for changing radiation levels comes from analyzing tree rings (Fig. 16.18) for variations in the concentration of nitrogen-14. The tree rings provide an annual record of nitrogen-14 and hence, an annual record of cosmic radiation levels in the atmosphere. Furthermore, all the tree rings within a given region should preserve a similar record, so they provide an excellent means for checking the consistency of the data obtained.

But while it is simple to examine the recent record of cosmic radiation by looking at recent tree ring growth, how do we extend such studies into the past? To do this we use stratigraphic principles similar to those of William Smith (see Chapter 3). Smith's method was to compare the sequence of individual rock layers in different areas by using the fossils they contained. In so doing, Smith was able to construct a "composite" stratigraphy in which the individual rock layers were arranged in a continuous sequence (see Figure 3.7). He did this by identifying key beds within individual sections that contained the same assemblage of fossils and so must have been deposited at the same time.

A similar approach permits the correlation of tree rings and has provided an excellent record of carbon-14 production over the last 7000 years. Tree rings are the result of climate variations. In semiarid climates, for example, trees show vigorous growth and produce wide rings in wet years, but in dry years, growth is inhibited and the rings are narrow. Hence, trees living in the same climate belt at the same time should show the same diagnostic sequence of tree rings. Figure 16.18 demonstrates the method of tree ring correlation, whereby tree ring patterns from living and dead trees are matched to provide a continuous sequence of growth rings that extends to progressively older trees.

When the tree rings are analyzed and carbon-14 levels in the atmosphere are deduced from nitrogen-14 values, it is clear that carbon-14 production has varied over the past 7000 years (Fig. 16.19). But how do we interpret this variation? To do so, it is important to identify and distinguish *first-order* and *second-order* trends. The first-order, or dominant, trend is

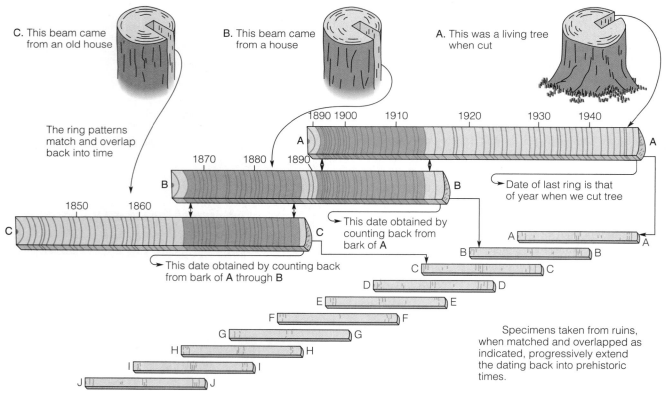

Figure 16.18
Correlation of tree-ring growth patterns produces a composite record that is continuous for the past 7000 years.

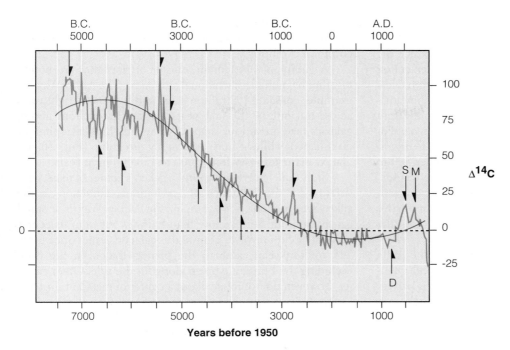

Figure 16.19

Variations in Carbon-14 ($\Delta^{14}C$) in tree rings over the past 7000 years (from 5000 B.C.). Carbon-14 production is high when the magnetosphere is weak. The smooth curve defines the first order trend and gives an average pattern of decreasing ^{14}C until about 1500 years ago. This pattern may represent variations in the strength of the Earth's magnetosphere which depends on both the strength of the Earth's magnetic field and the solar wind. The second order trends (see arrows) are shorter-term variations in the pattern and may link climate changes to variations in solar output. Deviation labeled M corresponds to the Maunder Minimum. Arrows labeled S and M together comprise the Little Ice Age. D signifies the Medieval Optimum. Reprinted by permission of NCAR.

obtained by smoothing out the curve to give an average pattern over the past 7000 years. The second-order trends are then deviations or "excursions" from this smoothed curve. We will examine the significance of each of these trends in turn.

Analysis of the first order trend shows that over the past 7000 years carbon-14 production has varied significantly on a long-term basis. Production reached a peak about 6500 years ago, then steadily declined until about 1500 years ago, and has been increasing gradually since then. Are there processes in stars that might affect long-term energy output? There may be, but if there are, they are very difficult to assess. We know that the life cycle of stars typically lasts millions to even billions of years, so that variations of this magnitude on a time scale of only 7000 years are most unlikely. Indeed we usually assume (rightly or wrongly) that cosmic bombardment is relatively constant. If this is the case, cosmic influences alone cannot account for the observed variations, so we must turn our attention to the Earth itself.

As we discussed earlier in this chapter, the magnetosphere forms a protective shield around the Earth that is produced by the *combined* effects of the Sun's magnetic field and that of the Earth. Hence, the production of carbon-14 in the Earth's atmosphere is *not* just a matter of cosmic bombardment, it is also a measure of the strength of this shield against the bombardment.

When the Earth's magnetosphere is weak, it is more readily penetrated by cosmic bombardment and more carbon-14 is produced. This weakness could arise from reductions in either the solar magnetic field or the Earth's magnetic field.

Recent measurements have shown that variations in the strength of both magnetic fields do occur. Although the reasons for these variations are unclear, changes in the intensity of either magnetic field affects the strength of the Earth's magnetosphere and hence the strength of the shield against cosmic bombardment. The shield is most effective when both magnetic fields are strongest.

Paleomagnetic studies indicate that the Earth's magnetic field was weak about 7000 to 6000 years ago (5000 B.C.), that it increased to a peak at about 500 A.D., and has been decreasing ever since. The first-order trend of the carbon-14 data closely matches these findings. The high production of carbon-14 about 7000 to 6000 years ago is consistent with an enhanced penetration of the magnetosphere by cosmic rays when the magnetic field was weak. As the intensity of the Earth's magnetic field increased to its maximum at about 500 A.D., carbon-14 production progressively decreased.

So the preferred interpretation for the first-order trends in carbon-14 production is one that attributes them to variations in the intensity of the Earth's magnetic field. These variations, in turn, affect how much cosmic radiation penetrates the magnetosphere and produces carbon-14.

But what about the second-order trends that produce the relatively small excursions from the average first-order curve? Since both positive and negative excursions occur within a few hundred years of each other, a much shorter-scale phenomenon is indicated. An important clue to its identity comes from more recent data, which suggest a correlation with climate. For example, there are two pronounced positive carbon-14 excursions (labeled S and M) that closely follow each other 300 to 500 years ago. Taken together, S and M span the Little Ice Age. The pronounced negative excursion, labeled D, coincides with what is known as the **Medieval Optimum**, a period of relative warmth that occurred about 800 to 1000 years ago, when the Vikings first settled in Greenland.

Reduced sunspot activity during at least part of the Little Ice Age implies a long period of diminished solar wind intensity. Since the strength of the magnetospheric shield is partially dependent on the strength of the solar wind, the shield may have been weakened at this time resulting in short-term increases in cosmic bombardment of the atmosphere and increased carbon-14 production.

So there is important evidence that while the effects of cosmic radiation may be subtle, cosmic rays do penetrate Earth's protective shield. At certain times the bombardment is more pronounced, depending on the strength of the Earth's protective shield, or magnetosphere, which is itself dependent on the strength of the Earth's magnetic field and the intensity of the solar wind. There may well be other more important consequences of this radiation, but, at present, scientists are only beginning to explore the topic.

16.5 Stars, Constellations, and Galaxies

Our knowledge of the universe beyond our galaxy is fraught with uncertainty. Estimates of the age of the universe range from 12 to 15 billion years. Given that the age of the Solar System is about 4.6 billion years, there is an appreciable time period for which it can provide little or no record. On the grounds of probability, however, most scientists believe that the Earth is unlikely to be unique because of the vast numbers of stars that could have orbiting bodies on which conditions duplicate those on Earth.

As viewed from Earth, many stars seem to occur in patterns called **constellations.** These patterns were noted by ancient civilizations, and elaborate mythologies are associated with them. Today, they are generally used to help an observer locate a particular star. Some, such as Polaris (the North Star), have been used for centuries to aid navigation. Because of its apparent position over the north pole, Polaris does not appear to move as the Earth rotates. The stars within

in a constellation need not be close to each other, and so, unlike galaxies, stars within constellations need not share a common evolution.

A **galaxy** is a systematic cluster of stars that are bound together by mutual gravitational attraction. The Milky Way, for example, contains 100 billion stars, and is 100,000 light years across but only 1000 light years thick. Viewed from above, galaxies take on various shapes. Many are elliptical, spiral, or disk-shaped while others are highly irregular (Fig. 16.20). Galaxy diameters vary from 5000 to 150,000 light years.

Stars tend to revolve around a galaxy's center of mass, consistent with the influence of gravitational attraction in the galaxy's formation. For example, the Milky Way is a spiral galaxy, and our Sun is more than halfway out from its center point. The Sun revolves around this galactic center once every 225 million years, and all the planets that orbit the Sun, including the Earth, are taken along for the ride. Other galaxies, however, do not rotate about a center of mass but, instead, appear to have a more random motion.

Galaxies also occur in clusters. The Milky Way lies in a cluster of about 20 galaxies. So far, millions of galaxies have been observed in the universe, and, as technology advances, we will almost certainly find more. The dimensions of the universe are so large as to be almost incomprehensible. Light waves travel 9.5 trillion km (about 5.9 trillion mi) per year, taking more than 12 billion years to travel across the known universe.

16.6 Life Cycle of Stars

A star is born when a large diffuse cloud of dust and gas shrinks into a smaller, denser cloud due to the force of gravity. Initially, the cloud might have a very low density (as low as 1 atom per cm^3) and temperatures as low as −253°C (−423°F). But as it shrinks, its density increases and the cloud heats up. Eventually, when the temperatures in its core become high enough to permit nuclear fusion, it ignites and shines. When the star ignites, the force of gravity is compensated by the energy given off by nuclear fusion and, as a result, contraction of the star ceases. As we have discussed, the temperatures in the center of the Sun is about 15 million degrees Celsius, which allows the thermonuclear fusion of hydrogen to form helium. Heavier elements occur in minute traces in the Sun, but their presence is believed to be the result of contributions from nearby supernova explosions while the solar nebula was forming. Most stars show a relationship between brightness (or luminosity) and surface temperature and plot on a trend known as the Main Sequence (Fig. 16.23). The Sun plots near the center of this trend. Larger stars are more luminous than the Sun because they have enormous surface areas. Stars that plot in the upper right are known as giants. Stars that plot in the lower left are dwarfs (about the size of the Earth).

The central core of a star is its nuclear furnace, but as you now know, a star's resources are not renewable. Eventually, all stars run out of energy and die. Since stars form by gravitational

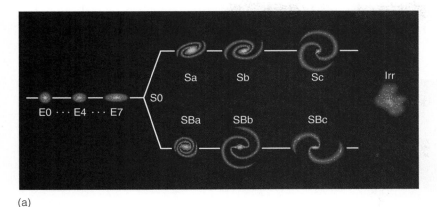

(a)

(b)

(c)

Figure 16.20

(a) "Tuning-fork" diagram showing a classification scheme for galaxies (E = elliptical, S = spiral, SB = barred spiral, Irr = irregular). This was once thought to represent an evolutionary sequence, from left to right, but is now used for descriptive purposes only. Examples of (b) spiral (Sb) and (c) elliptical (E0) galaxies are also shown.

collapse, larger stars have hotter cores so their thermonuclear fusion reactions are more rapid and the nuclear fusion process can form elements as heavy as iron (atomic number 26). As a result of more rapid thermonuclear reactions, stars use up their fuel in proportion to their mass, larger stars burning more quickly than smaller ones. In fact, the largest stars burn out in a matter of millions of years.

Throughout their lives, the size of stars is sustained by the thermonuclear reactions in their cores. The incredibly high energy produced by these reactions prevents the gravitational collapse of the core onto its own center. But as a star grows older and its supply of nuclear fuel begins to run low, it is no longer able to maintain its size. The fate of the star then depends on its mass. The collapse of massive stars (those with a mass more than four times that of the Sun) releases an enor-

mous amount of energy, resulting in rapid expansion as the star becomes a supergiant (Fig. 16.23), before dying in a cataclysmic explosion known as a **supernova**. Such explosions send out light and neutrinos, which travel across the universe and continuously bombard the Earth's atmosphere. Supernova explosions also release enormous amounts of energy. The very high pressures and temperatures that occur during the collapse of the star facilitate the synthesis of elements heavier than iron.

Mid-sized stars, like the Sun, do not share the same fate as larger stars, and do not experience such a cataclysmic ending. When the Sun's supply of hydrogen fuel begins to run out, for example, it will collapse until the radius of its core is reduced from about 170,000 km to 3500 km. As we now know, gravitational collapse releases enormous amounts of energy, which will make the Sun's core even hotter, and instigate a retaliation

The countless numbers of stars in the universe makes the likelihood of other solar systems very probable indeed. Thus, the existence of other solar systems, in which a collection of planets revolve around a central star, has long been hypothesized by astronomers. Until very recently, however, it had never been confirmed. Certainly, no scientist has come up with a rigorous analysis demonstrating why the relationship between the Sun and the Earth should be unique. But such intuitive arguments, although appealing, are not good enough for science. For decades, scientists have strived to demonstrate that other solar systems exist. This might be termed the "leprechaun" approach to science; if only one example could be found, the principle would be demonstrated and the door would open to countless possibilities!

There are several reasons why confirmation of orbiting bodies has proven difficult. Planets do not shine by themselves, they can only reflect the light of a nearby star and may be too dim to be directly observed. In addition, the brightness and size of a star makes it very difficult to detect the more subdued outlines of nearby orbiting bodies. If our Solar System is any indication, the planets will be much smaller than the adjacent star so that their size and finally, their distance from the Earth, may prevent their detection.

By the mid-1980s, however, important hints of other "solar systems" were discovered. In 1984, astronomers at an observatory in Chile found evidence of a disk-shaped cloud of debris surrounding a star in the constellation Pictor (Fig. 16.21). This debris was interpreted as a solar system in the process of formation.

In late 1995 and early 1996, the breakthrough came in separate observations from two different observatories. First, two Swiss astronomers reported the existence of a planet orbiting a star called 51 Pegasi. Other laboratories confirmed this observation

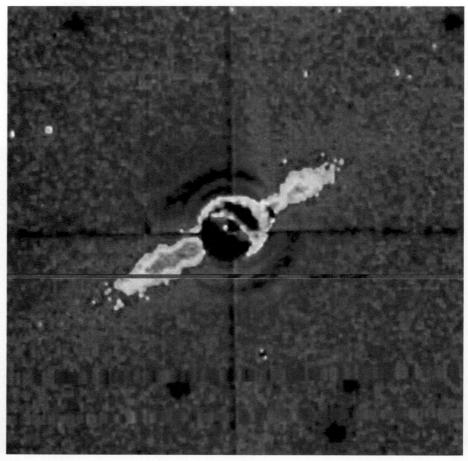

Figure 16.21

Beta Pictoris, a star in the constellation Pictor, is surrounded by a disk-shaped cloud of matter that may contain planet-sized objects, interpreted to be a solar system in the process of formation.

which demonstrated, for the first time, the existence of other "solar systems."

Second, Geoffrey Marcy at the San Francisco State University found evidence of two planets, one orbiting the star 70 Virginis in the constellation Virgo, and one orbiting 47 Ursae Major, a star within the Big Dipper (Fig. 16.22). Both discoveries have since been confirmed by other observatories and have caused considerable excitement among the scientific community. Indeed, we now know of more planets outside the Solar System than exist within it.

These observations have very profound consequences. If there are other "solar systems," then the possibilities for life elsewhere in the universe are vastly improved.

Since the planets themselves are invisible, their existence must be determined indirectly. When watching a hammer-throwing event, you might observe that the thrower wobbles as the hammer revolves as a consequence of gravitational interaction (see Figure 15.4). So if your view of the hammer was hindered by a person in the row in front of you, you could still deduce its existence by the behavior of the thrower. Although the relationship between a star and its planets is considerably more complicated, planets revolve around a star as if tied taut to it by an invisible string. As they do so, their gravitational pull results in rhythmic oscillations in the star which can be observed. So while a planet may be hidden from our view, the rhythmic oscillations of the star can be used to betray its presence.

Now that the "ice has been broken," most astronomers believe that there is nothing particularly special about our Solar System. Indeed there may be millions or even billions of similar systems in the universe. Evidence from the Hubble telescope suggests there may be countless examples of other planetary systems. If so, the odds that life exists elsewhere in the universe become very good indeed. If the relationship between the Earth and the Sun is anything to go by, then the inhabited planet would have to lie at an appropriate distance from its sun and would need to be volcanically active so as to have an atmosphere and surface water. But if there is life, what form would it take? Again, our own experience on Earth tells us that life evolves such that the fittest, most adaptable species survive. However, evolution is not as simple as that because it also contains an element of "Russian Roulette." For example, it is most unlikely that the human species would have become the dominant life form if the dinosaurs had not become extinct 66 million years ago. If theories that this extinction was the result of an extraterrestrial impact are correct, then our own existence may be largely the result of chance; we are potentially the fortuitous benefactors of a cataclysmic event.

So if life forms exist on the planets of other solar systems, it would be fruitless to speculate what form they would take. Nevertheless, this discovery has added fuel to a scientific and philosophical question that refuses to go away: Is there anybody out there?

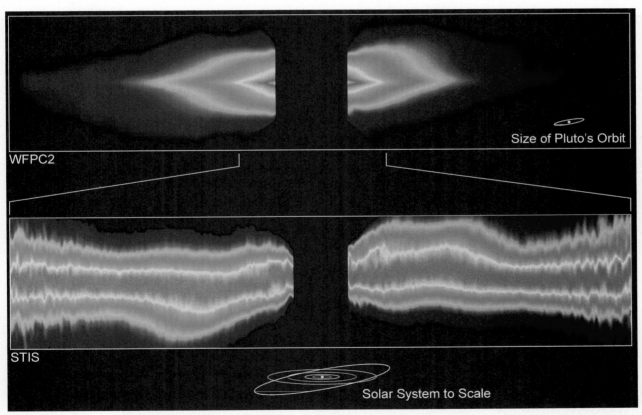

Figure 16.22
Photo from the Hubble space telescope showing the inner region of a 200-billion mile diameter dust disk around star Beta Pictoris. The warping in the disk indicates the possible presence of a planet. Visible light image (top) of disk, false-color (bottom) image processing brings out details of the disk structure.

so profound that the Sun will temporarily expand to 100 times its present diameter. When this occurs, the temperature at the surface of the Sun will drop to about 3000°C, and its color will dull from yellow to red (see Fig. 16.23). Hence at this stage in its development, the Sun would have become a **red giant**. Red giants are highly luminous because they have enormous surface areas. On the Earth, temperatures would rise sufficiently to melt the surface, while the atmosphere would be blown away and the oceans would vaporize. This then will be the final end of the Earth as it, and all the other planets of our Solar System, become engulfed by the expanding Sun.

Since, at this stage, the Sun will lack nuclear fuel, its existence as a red giant will be only temporary. Once the collapse of the Sun's material is complete, the red giant phase will be over, and the Sun will rapidly shrink again to form a tiny brilliant star known as a **white dwarf**. Reduced to a size comparable to that of the Earth, it may continue to emit residual energy for billions of years before it finally fades away to become a **dark dwarf** with negligible luminosity.

16.7 Origin of the Chemical Elements

The life cycle of stars provides insight into the origin of the chemical elements found in the universe. Naturally occurring chemical elements are numbered from 1 to 92, depending on the number of protons (or positive charges) in their atomic nuclei. The names we give these elements depend on this number because the number of protons determines an element's chemical properties.

But how do these elements form and, given that all protons are positively charged, how can their mutual repulsion be overcome so that as many as 92 protons can be packed into the nucleus of an atom? Because of the mutual repulsion of protons, it is not surprising that hydrogen, the simplest element, is by far the most abundant element in the universe, and is followed by the second simplest, helium. Together, these two

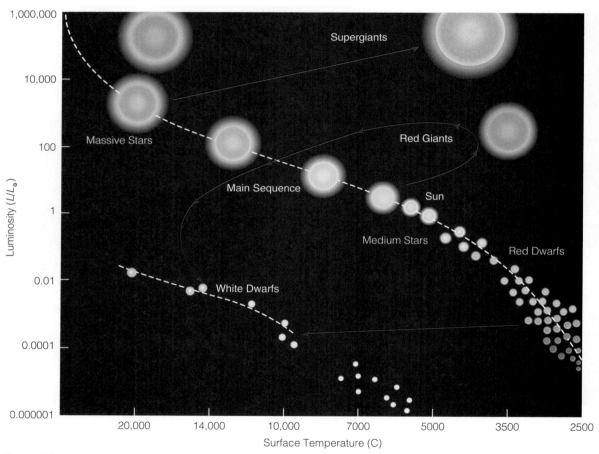

Figure 16.23

The main sequence or life cycle of stars. The brightness (or luminosity) of stars is compared to that of the Sun ($L/L_0 = 1$). Massive stars burn out rapidly and become supergiants before exploding as supernovae. When the hydrogen fuel of the Sun begins to run out, the Sun's core will collapse, resulting in its temporary expansion to form a red giant. After the red giant phase, the Sun will rapidly shrink again to form a white dwarf before finally fading to become a dark dwarf.

elements comprise 98% of the chemistry of the universe and are believed to have been the only elements present early in its history. The heavier elements have been synthesized progressively as the universe evolved.

But how were the elements heavier than helium formed? We know that nuclear fusion takes place in the core of stars, creating new material. But in mid-sized stars such as the Sun, these reactions overwhelmingly involve the production of helium from hydrogen. In larger stars, however, a succession of reactions may occur that can synthesize heavier elements. Helium atoms may fuse first to form carbon, then oxygen and, it is hypothesized, may ultimately form elements as heavy as iron, which has an atomic number of 26 (Fig. 16.24). This process depends greatly on a star's core temperature, with the production of heavier elements requiring ever higher temperatures. But because iron is the most tightly packed of all the elements, neither it nor the elements higher in the periodic table can release energy by thermonuclear fusion.

When larger stars die in supernova explosions, the unimaginably high pressures produced can overcome the mutual repulsion between protons to synthesize chemical elements heavier than iron. Thus many scientists believe that supernovae are responsible for producing all of the heavier, naturally occurring elements in the universe. Because these elements are only synthesized in such catastrophic events, they are relatively rare. Supernovae may also create a number of unstable elements by what can euphemistically be referred to as "shotgun marriages." These unstable elements are radioactive and, as we learned in Chapter 3, spontaneously break down into stable daughter products.

The explosion of supernovae also disperses the synthesized elements into space where they can become incorporated into nebulae to form the raw materials for future solar systems. Thus it is likely that the heavier chemical elements of the solar nebula and, indeed, of the Earth, were actually synthesized during the life cycles of unknown and now extinct stars.

16.8 Origin of the Universe: The Big Bang Theory

At one time, Albert Einstein believed that the universe was static; that it was neither expanding nor contracting. He later admitted that this was one of the most serious blunders in his career. Most astronomers are now convinced that the universe originated in an explosion of unimaginable proportions, an event known as the **Big Bang**. Most estimates of the age of this event place it between 12 and 15 billion years ago.

But what is the evidence for the Big Bang, and how is the age of the universe estimated? In 1929, Edwin Hubble observed that the galaxies were moving away from each other, implying that the universe is expanding. The rate at which the universe is expanding has since become known as the **Hubble constant**. To understand how he came to this conclusion, think of the sound from a whistle of a moving train (Fig. 16.25a). The pitch of the whistle sounds different depending on whether the train is moving towards you or away from you. The whistle itself has a constant pitch, which you can hear when the train is stationary. However, the train's motion distorts the sound waves. When it is approaching, the wavelengths are compressed and shortened, but when it is receding they are stretched and lengthened. As a result, the whistle sounds high-pitched if the train is approaching, but abruptly drops to a lower pitch as it passes by. This change in pitch is an example of the **Doppler effect**.

In a similar manner, the wavelengths of light emitted from galaxies as they move away from Earth are lengthened, and so are shifted towards the longer wavelength (red) end of the visible spectrum. This is known as the **red shift** (Figs. 16.25b and 16.26). Conversely, if a galaxy was moving towards us, the wavelength of its light would be shortened and shifted toward the blue end of the spectrum. But Hubble showed that all galaxies had spectral lines that were shifted towards the red end, and hence

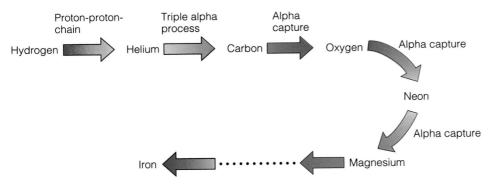

Figure 16.24

Thermonuclear fusion in stars of various masses, in which elements as heavy as iron (atomic number 26) are synthesized in a number of steps. Helium atoms fuse to form carbon, then oxygen and, depending on the star's core temperature, eventually elements as heavy as iron. An alpha particle is the equivalent of a helium nucleus which fuses with the nucleus of an atom during alpha capture to form a heavier element. Elements heavier than iron are synthesized in supernovae that mark the death of massive stars.

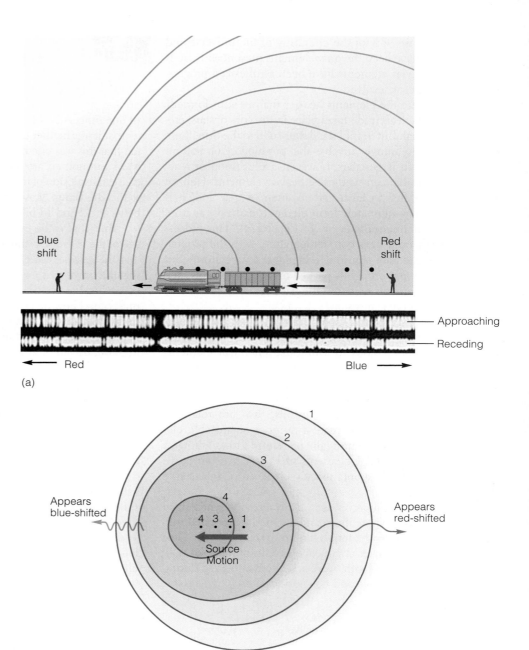

Figure 16.25

(a) The Doppler effect applied to sound waves. The whistle's sound waves reaching an observer on the left are compressed as the train approaches, so that a shorter wavelength (higher frequency) sound is heard. For the observer on the right, the sound waves are lengthened as the train moves away, so the sound has a lower frequency. (b) The Doppler effect applied to the motion of galaxies. As the source moves, the light it emits at stages 1–4 are shown as circles of wavefronts, centered on each of the four positions. The wavefronts are bunched up in the direction of motion, and the lightwaves are shifted to bluer wavelengths. The wavefronts are stretched out in the opposite direction and the lightwaves are shifted to redder wavelengths.

(a)

Blue shift

Red shift

Approaching

Receding

Red

Blue

Appears blue-shifted

Appears red-shifted

Source Motion

(b)

were moving away from the Earth. The amount of red shift gives a measurement of how fast the galaxies are moving away.

But if the galaxies are moving away from each other, they must have been closer together in the past. Hence, the receding galaxies have become a cornerstone of the Big Bang theory, in which the universe originated in a gigantic explosion and has been expanding ever since.

The expansion of the universe can be likened to the expansion of a loaf of raisin bread when it is baked. As the dough rises and the bread expands, the raisins become increasingly separated from each other (Fig. 16.27). Although the bread expands uniformly, neighboring raisins move apart more slowly compared to raisins that are far apart. Similarly, although the universe is expanding at a uniform rate (known as the Hubble

constant), neighboring galaxies move apart from one another more slowly than ones that are far apart.

It is tempting to envision that, at the time of the explosion, the universe originated from a single point. However, since at any one time the universe incorporates everything that exists, it has no edges and therefore no center. In the strictest sense, then, the universe at the time of the Big Bang should be thought of as an infinitely small, high energy region with an infinite density. Immediately after the Big Bang, (just 10 millionths of a second later), the universe was filled with high energy photons having a temperature of over 1 trillion (10^{12}) °K.

This region would have had so much heat that only radiation could exist. Space and time, in our conventional sense, would have had no meaning; they would have been infinitely

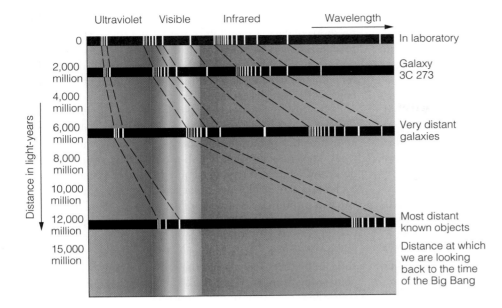

Figure 16.26
Simplified diagram of the electromagnetic spectrum showing the red shift of the galaxies (see dashed lines connecting spectra of galaxies). Light waves from receding objects are shifted toward longer wavelengths (i.e. toward the infrared portion of the spectrum). The amount of red shift is proportional to the objects' distance from Earth. The most distant known object is 12 billion years old.

warped around this region and nothing, not even empty space, would have existed beyond it.

According to Einstein's theories, space and time are linked. Without space, there is no time. So the question "what happened before the Big Bang?" cannot be asked because there was no "before." Time and space would have begun in the same instant as the explosion. If correct, these ideas have far-reaching scientific and philosophical implications.

If the Big Bang had been a perfect explosion, all the components of the universe would have been homogeneously distributed. There would be no stars, and therefore no planets. But the explosion did not distribute matter homogeneously. The inhomogeneity allowed regions of concentrated matter to become centers of high gravity that rapidly accumulated mass to become large groupings of stars, or galaxies. Estimates of the time lapse between the Big Bang and the formation of galaxies varies from 10^5 to 10^9 years.

The original hypothesis of the Big Bang has passed so many tests that it was elevated to the status of a theory. In fact, in 1965, two astrophysicists detected its "echo." While measuring the brightness of the sky, they inadvertently recorded a peculiar noise with their horn antennae. At first, they attrib-uted the noise to pigeons living inside the antennae. However, the noise persisted after the pigeons and their droppings were removed. Eventually, they concluded that the noise represented a weak cosmic microwave radiation that had been predicted as the remnant of the catastrophic explosion implicit in the Big Bang theory. Since then, scientists have debated whether the universe is "open" or "closed." That is, whether the universe will continue to expand indefinitely, or whether the pull of gravity will eventually overcome this expansion and reverse its direction, culminating in the "Big Crunch."

Looking back into the past is like putting a videotape on rewind. According to the Big Bang theory, the age of the universe is the time it would take to "rewind the tape" such that all the galaxies in the universe, and all the mass within them, would return to a single region of dense hot gas. Determination of the Hubble constant is therefore crucial to estimating the age of the universe.

The range in estimates for the age of the universe reflects the uncertainty in the distances between galaxies. Recent measurements and observations from the Hubble space telescope (named in honor of Edwin Hubble), have forced a thorough re-examination of the constant. Early results were embarrassing.

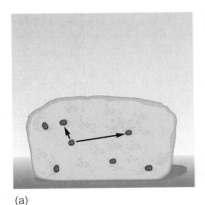

(a)

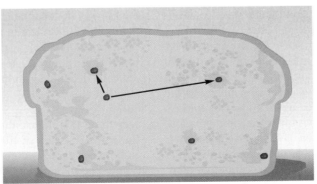

(b)

Figure 16.27
The uniform expansion of the universe can be represented by the expansion of raisin bread when it is baked. As the bread expands, all raisins move apart at a rate that is proportional to their separation. Neighboring raisins move apart from one another more slowly than raisins that are far apart.

They suggested that the Hubble constant was greater than previously believed. Just as a videotape that rewinds faster than expected will return a movie to its beginning more quickly, the higher Hubble constant implied that the universe might be as young as 7 billion years.

However, some galaxies are estimated to be 14 billion light years away, implying that they are at least 14 billion years old. Thus astronomers were confronted with a contradiction since the universe cannot possibly be younger than the stars within it. More recent determinations of the Hubble constant suggest that the universe is about 12–15 billion years old, and go a long way toward resolving the dilemma.

But many questions remain unanswered. Are our estimates of the distance to the galaxies correct? How can we better determine the value of the Hubble constant? Were there, in fact, many catastrophic explosions during the early history of the universe, rather than just one? If there were, perhaps they interfered with each other so that there is no Hubble *constant*. At the time of writing, the controversy has not been resolved. But exciting times lie ahead that may provide a deeper understanding of one of the most important scientific questions of all. . .*the origin of the universe.*

In summary, the universe appears to have originated in a cataclysmic explosion known as the Big Bang. The critical evidence for this theory is derived from the observation that the galaxies are moving away from each other and from the detection of cosmic radiation thought to represent the "echo" of the explosion. The age of the Big Bang is uncertain, but more recent estimates suggest it occurred 12–15 billion years ago.

16.9 Chapter Summary

The Sun is composed almost entirely of hydrogen (78%) and helium (21%), the two lightest elements in the periodic table. Trace amounts of other elements such as sodium also occur. The Sun's energy is created by thermonuclear reactions in its core, which forms helium from the fusion of hydrogen nuclei. However, this reaction does not conserve mass, and a mass loss of 0.7% is converted into energy according to Einstein's famous equation, $E=mc^2$. Some 4 to 5 million tons of mass are converted into energy every second. This energy is transmitted outward from the core and is emitted into space as solar radiation from the Sun's surface. It takes 1 million years for energy generated in the core to reach the surface. Once emitted, however, it travels at the speed of light and reaches Earth in a mere 8.3 minutes.

The radiation is emitted in a wide range of wavelengths, known as the electromagnetic spectrum, and only a portion of this radiation reaches the Earth's surface. Our eyes have the ability to detect a limited range of wavelengths known as the "visible spectrum," but the existence of invisible radiation on Earth can be readily confirmed. Invisible ultraviolet radiation, for example, causes sunburn whereas infrared radiation is sensed as heat and is directly related to the greenhouse effect.

The composition of the Sun has been determined by comparing the Sun's visible spectrum with absorption and emission spectra produced by elements ignited in a laboratory. The positions of dark (Fraunhofer) lines in the solar spectrum were recognized as being diagnostic of the Sun's composition. But no match could be found for the absorption lines of a particular element, which was given the name Helium. It was discovered on Earth 17 years later.

The Sun is layered from its thermonuclear core, through a radiation zone and a convection zone to its visible surface, known as the photosphere. The photosphere contains blemishes, known as sunspots which are relatively cool, magnetic, dark, and depressed regions. These blemishes come and go in cycles of just over 11 years. Solar activity around these sunspots (such as prominences and solar flares) are observed in the overlying chromosphere and even in the corona beyond.

The solar wind is the product of ionized gases that escape the Sun's surface. The wind permeates the entire Solar System and can be seen drawing out the tails of comets, which always point away from the Sun. It also bombards the Earth's outer atmosphere, causing upper atmospheric phenomena such as the auroras. The speed and intensity of the solar wind varies according to the sunspot cycle and may cause temperature fluctuations up to 1000°C in the upper atmosphere.

The 17th Century was a period when very few sunspots were observed, and overlaps with the period known as the Little Ice Age. It is possible that a prolonged period of diminished solar wind intensity allowed the upper atmosphere to cool sufficiently that surface temperatures were also affected.

There is mounting evidence that events outside our Solar System have also affected the evolution of the Earth. The most obvious of these is the synthesis of the Earth's chemical elements, which probably took place during the life cycle of stars and in distant supernova explosions caused by their death. Other effects are more subtle. They include important changes in the carbon-14 isotopic composition of tree rings from which an annual record of the penetration of cosmic radiation into our atmosphere has been obtained. Penetration is diminished when the Earth's magnetic field strengthens, or when the solar wind intensifies.

It appears that our Solar System is not unique, because there are stars elsewhere in the universe with planets revolving around them. Since the planets themselves are difficult to detect, their existence is best determined by the rhythmic oscillations of the central star. These oscillations are produced by the gravitational pull of the revolving planets.

The Big Bang theory suggests that the universe originated in a cataclysmic explosion. The cornerstone of the theory is that the universe is expanding at a constant rate. This rate is known as the Hubble constant. The detection of weak cosmic microwave radiation, thought to record the "echo" of the explosion, supports the theory. Determination of the age of the universe is critically dependent on the value of the Hubble constant, estimates of which vary. But some galaxies are estimated to be 14 billion light years away. Since the universe must be older than everything it contains, these observations provide a minimum age for the origin of the universe. Recent estimates of the Hubble constant are in reasonable agreement with these observations and suggest that the Big Bang occurred between 12 and 15 billion years ago.

Key Terms and Concepts

 Look for the highlighted items on the web at: WWW.BROOKSCOLE.COM

- The average amount of solar radiation falling on a square meter of the Earth every second is 1.36 kw, and is known as the solar constant

- Solar energy is produced by thermonuclear fusion, in which protons combine to form helium, converting mass into energy

- The Sun emits energy that can be characterized in terms of an electromagnetic spectrum that contains both visible and invisible radiation

- Fraunhofer lines represent the absorption of photons within the gas of the solar atmosphere and the lines are characteristic of the elements found within the Sun

- The composition of the Sun has been determined by its absorption and emission spectra

- The Sun is layered, having a central core which is surrounded in turn by a radiation zone, a convection zone, and a photosphere enveloped by a less dense chromosphere and corona

- The Sun's corona is its spectacular "crown" visible during solar eclipses

- Sunspots are small, cold, dark, depressed regions on the surface of the Sun and are the site of spectacular solar activity

- Solar wind consists of ionized gases emitted from the Sun's surface, as revealed by the tails of comets

- The Earth's magnetosphere is the region where its magnetic field interacts with that of the Sun

- The Van Allen belts are regions where charged particles collect in the magnetosphere

- Auroras are spectacular light shows produced by the interaction of charged particles with atmospheric gases at the north pole (Aurora Borealis) and south pole (Aurora Australis)

- The sunspot cycle describes progressive changes in the number of sunspots on the Sun's surface, from a maximum to a minimum number over an 11 year period. Because of changes in the polarity of the Sun's magnetic field, a complete sunspot cycle is 22 years in duration

- The sunspot cycle affects the intensity of the solar wind and the temperature of the Earth's upper atmosphere

- The Little Ice Age coincided with a prolonged disappearance of sunspots (part of which is named the Maunder Minimum), diminishing solar wind, and cooling of the Earth's upper atmosphere

- Stars occur in immense clusters called galaxies

- Cosmic radiation has a subtle effect on Earth, and its effect on the Earth's atmosphere is recorded in the carbon-14 isotope chemistry of tree rings, variations in which may be related to variations in the intensity of the Earth's magnetic field

- Other potential solar systems have been detected by rhythmic oscillations in stars, which reflect the presence of orbiting planets

- The life cycle of stars explains the presence of all naturally occurring chemical elements

- The Big Bang theory requires that all matter initially existed as a hot, high-density gas in an arbitrarily small region that exploded cataclysmically between 12 and 15 billion years ago. Supporting evidence includes the observed red shift of all galaxies, and the presence of background cosmic radiation

Review Questions

1. How can the Sun's spectrograph be used to identify the composition of the solar atmosphere?
2. How does the Sun's thermonuclear fusion process in the core yield energy and what is the significance of the energy?
3. Why do the tails of comets stream away from the Sun?
4. Explain the variations in carbon-14 concentrations in tree rings over the last 7000 years.
5. The Sun's surface temperature of 5700°C yields a wavelength in the visible spectrum which corresponds to a greenish-yellow color. Knowing that the average temperature on Earth is 15°C, where would you predict the wavelength of maximum intensity of Earth's radiation to lie? (Hint: apply Wien's Law.)
6. Why do the evolutionary histories of stars in the same constellation often have little in common with one another?
7. Explain the potential relationship between sunspots, solar wind, and the Earth's magnetosphere.
8. Explain the relationship between solar flares and auroras.
9. How did the naturally-occurring chemical elements form?
10. Why is the age of the universe unlikely to be only 7 billion years old?

Study Questions

 Research the highlighted question using InfoTrac College Edition.

1. Do you think our Solar System is unique? List the arguments for and against, in order of their relative importance.
2. How is matter converted into energy by thermonuclear fusion? If this is an essential aspect of all stars, why are the effects of solar rather than cosmic thermonuclear fusion more apparent on Earth?
3. Do all stars have the same electromagnetic spectrum? Explain.
4. How are absorption and emission spectra related to one another?
5. Why are sunspots dark?
6. What is the relationship between the auroras and the solar wind?
7. Why is the generation of carbon-14 an indication of enhanced cosmic ray penetration into our atmosphere? Are there other examples of first-order and second-order trends you have come across in this text? If yes, what are they?
8. Compare and contrast the internal layering of stars (like the Sun) and the internal layering of planets.
9. The Solar System will expire when the Sun dies. However, Mercury and Mars are already essentially geologically dead. What processes will dictate the ultimate fate of the Earth?
10. How can the Big Bang theory be tested?

Astronomy

Solid Earth

Volcanism keeps a terrestrial planet or moon geologically alive—creating its atmosphere and bringing elements and nutrients from its center to its surface. A large enough mass is required to hold an atmosphere in place.

Hydrosphere

H_2O occurs as ice in bodies too far from the Sun and evaporates from bodies too close to the Sun. Earth has a functioning hydrosphere because of the density difference between continental and oceanic crust. Mars has polar ice, and may have subsurface liquid water. Two of Jupiter's moons may contain subterranean oceans. The gravitational effect of the Sun and Moon combine to create tides on Earth.

Atmosphere

Terrestrial planets (Earth and Venus) have enough mass to gravitationally hold an atmosphere in place. The distinction between atmosphere and surface is blurred on the Jovian planets where large weather systems have been observed. Atmospheric composition affects surface temperatures.

Biosphere

Earth is the only planet we know that has an evolving biosphere. In other solar systems, planets with a similar size and distance from their sun may also contain life. Mars and Venus may once have harbored life, and as Venus cools, it might again.

Astronomy

The universe began 12 to 15 billion years ago with the Big Bang and has been expanding ever since. The life cycles of stars form every naturally occurring element. Solar systems form as solar *nebulae* flatten and condense due to gravity. Cosmic radiation/ neutrinos flow through and affect every part of the universe.

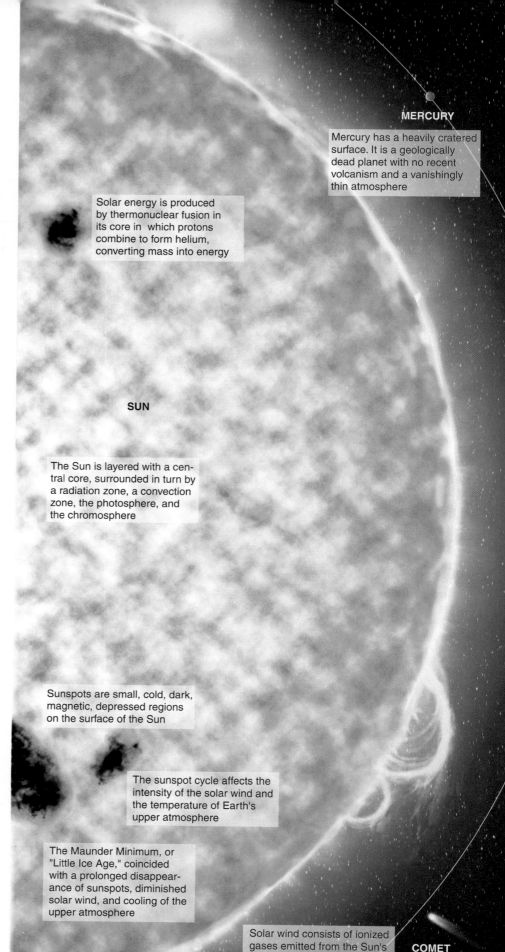

MERCURY

Mercury has a heavily cratered surface. It is a geologically dead planet with no recent volcanism and a vanishingly thin atmosphere

Solar energy is produced by thermonuclear fusion in its core in which protons combine to form helium, converting mass into energy

SUN

The Sun is layered with a central core, surrounded in turn by a radiation zone, a convection zone, the photosphere, and the chromosphere

Sunspots are small, cold, dark, magnetic, depressed regions on the surface of the Sun

The sunspot cycle affects the intensity of the solar wind and the temperature of Earth's upper atmosphere

The Maunder Minimum, or "Little Ice Age," coincided with a prolonged disappearance of sunspots, diminished solar wind, and cooling of the upper atmosphere

Solar wind consists of ionized gases emitted from the Sun's surface and revealed by the tails of comets

COMET

ercury, Venus, Earth and
ars) are terrestrial (rocky)
anets

Jupiter's innermost moons
include Io (which is the only
other body in the Solar
System on which active volcanism has been observed),
and Europa (which has a
cracked, icy surface that may
overlie a subterranean ocean)

have been dectected by rhythmic oscillations in stars which
reflect the presence of orbiting
planets

nus has a dense atmo-
here dominated by CO_2
.5%) and a surface temper-
ure of 460° C. Modern vol-
nic activity is suspected,
t remains unproven. Only
nus rotates clockwise

Cosmic radiation has a subtle
effect on the Earth, and its
effect on the Earth's atmosphere is recorded in the
carbon isotope chemistry of
tree rings

JUPITER

Jupiter is dominated by
hydrogen and helium. Its color
bands reflect subtle changes
in its atmosphere. Its massive
storms (e.g. Great Red Spot)
are driven by energy from
its interior

Saturn is similar in many ways
to Jupiter. Its flat ringlets are
chunks of ice

VENUS

The Moon may have been
ejected from the Earth by
the collision of a Mars-sized
object with the Earth's surface.
Volcanic activity on the Moon
ceased about 3 billion
years ago

SATURN

The outermost planets (Jupiter,
Saturn, Uranus, Neptune) are
Jovian (gaseous) planets
because they formed at a
distance from the Sun where
water remained solid and lighter
gases condensed

MOON

EARTH

The Moon rotates about its
own axis at the same rate as
the Earth

The uniqueness of planet
Earth is related to its size and
distance from the Sun, its composition and active tectonics,
and the generation and preservation of its atmosphere,
hydrosphere, and biosphere

ASTEROID BELT

Rock fragments

URANUS

The age of the planetary surfaces can be estimated by the
abundance of craters from
meteorite impacts

Uranus, Neptune and Pluto
are the least known and understood planets

MARS

Mars has exaggerated
mountains, polar ice caps,
a very thin atmosphere, and
very cold average surface
temperatures. It may once
have harbored life, and might
still in its subsurface. Volcanism on Mars virtually
ceased about 1 billion
years ago

The size of a terrestrial planet
influences its rate of cooling,
the longevity of its volcanic
activity, and its atmospheric
composition

NEPTUNE

Oort cloud—Icy bodies 50au
from the Sun—possible source
of comets and water/ice on
Earth, the Moon, and Mars

he Big Bang theory re-
uires that all matter
xisted in a small
egion that exploded
ataclysmically
etween 12 and 15
llion years ago. Our
olar System is around
billion years old

*Gravitational collapse
of the solar nebula to
form the Sun and
planets may have
been initiated by a
supernova explosion*

*The life cycle of stars
explains the origin of
all naturally occurring
chemical elements*

The classification of Pluto as a
planet is uncertain

PLUTO

Relative sizes of planets are drawn to scale; distances between orbits are NOT drawn to scale.

PART 6

Natural Resources

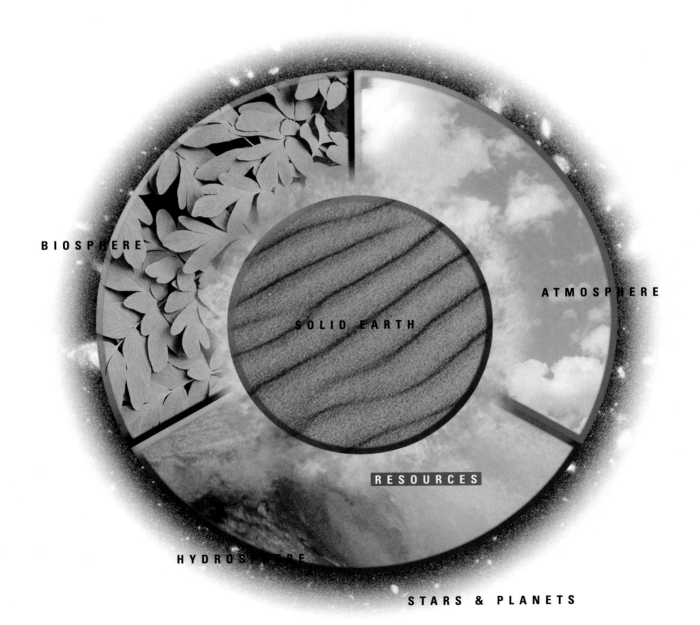

BIOSPHERE

ATMOSPHERE

SOLID EARTH

RESOURCES

HYDROSPHERE

STARS & PLANETS

CHAPTER 17

Mineral Resources

Ⓝowhere has the interaction of the Earth's reservoirs played a more vital role in the development of human society than it has in the history of our exploitation of the planet's rich mineral resources. Indeed, the very ages of human society are named for the mineral resources that their people first learned to utilize. Thus we refer to the earliest period of human history as the Stone Age because its people made tools from stone. Later, the ability to alloy copper and tin brought us into the Bronze Age, and our subsequent exploitation of iron marked the start of the Iron Age. Today, the Earth's mineral resources form the cornerstone of modern industrial society. Products made from mineral resources are so widespread in today's society that we often take them for granted, yet our lives are wholly dependent upon them, whether they are the metals of our machines, the industrial materials of buildings and roads, or the fertilizers we use to grow food.

Mineral resources take many forms but all are important inorganic materials that must be mined or quarried from the Earth. Most are ores of metals, such as lead, zinc, and aluminum. But many are nonmetallic minerals like gypsum and clay, and others are just sediment such as sand and gravel. In every case, however, the material must have an economic use, and must have been sufficiently concentrated in one place to make its exploitation profitable. For example, aluminum, the lightweight metal used in aircraft, beverage cans, and foil wrap, is one of the most common elements at the Earth's surface, being present in almost every rock type. Yet only under the most exceptional circumstances does aluminum become naturally concentrated in sufficient quantities to form a mineral deposit or orebody.

The widespread use we make of Earth's mineral resources and the relative scarcity and nonrenewable nature of even the most common Earth materials make mineral resources very valuable commodities. Indeed, the exploitation of its mineral resources is often vital to the economy of a nation favored with mineral wealth. Understanding the natural processes that lead to the selective concentration of certain minerals is therefore very important because it is essential to the success of mineral exploration. As we shall soon learn, the various mechanisms involved also provide a uniquely instructive demonstration of the interaction between the Earth's major reservoirs and the interdependence of the Earth's systems.

In this chapter, we will examine the different types of economic minerals, both metallic and nonmetallic (or industrial), their abundance and their human uses. We will also examine mineral deposits and learn where and how minerals become concentrated in sufficient quantities that they can be extracted economically. We will see how different natural processes can create mineral deposits and we will examine the exploration methods by which mineral deposits worth exploiting are discovered. Finally, we will discuss what the future might hold as the demand for mineral resources confronts their limited supply.

17.1 What Are Metals and Industrial Minerals?

A **mineral resource** is a naturally occurring accumulation of inorganic material that is taken from the Earth for the purpose of enhancing human life or human society. As such, it differs from a **mineral reserve** which is an *identified* resource that could be exploited for profit in today's marketplace. For the most part, mineral resources are the commodities of industry and include both **metallic minerals** and a variety of **nonmetallic (or industrial) minerals** that can be used as building materials, chemicals, or fertilizers. Materials such as coal, oil, and natural gas, while certainly valuable commodities, are not considered mineral resources because, unlike minerals, they are *organic* in origin. Instead, they are classified as **energy resources** and will be examined separately in Chapter 18. Fresh water, our most precious natural resource, was examined in Chapter 10.

The Earth's crust contains a wide variety of chemical elements, many of which are vital to society. Accumulations of these valuable inorganic resources are termed **mineral deposits**, and if they can be exploited economically, form **ore deposits**, or **orebodies**. Most of these elements are extremely rare, but some of the most widely used, such as iron and aluminum, are actually very common in crustal rocks. However, even the commonest elements rarely form ore deposits as most occur within silicate minerals of crustal rocks from which they are very difficult to extract. Only as a result of unusual geologic processes do these elements become sufficiently concentrated that they can be extracted from the crust economically.

As the product of geologic processes, the rates at which mineral resources form and become concentrated into economic deposits are extremely slow. They are therefore limited or finite resources that cannot be replaced by nature in a time frame remotely close to one of human duration. So unlike other resources, such as crops and timber, mineral resources are **nonrenewable** commodities. Their discovery and assessment are therefore of special concern to Earth scientists because the increasing global population and the spread of industry and technology have combined to ensure that the rate of consumption of mineral resources is increasing, whereas existing supplies and the rate at which new supplies are being discovered is diminishing. Consequently, the continued extraction of mineral resources must eventually lead to their depletion.

17.2 Distribution and Uses of Mineral Resources: Some Facts and Figures

Our dependence on mineral resources is vividly illustrated by an examination of the role they play in our everyday lives. From the fluoride toothpaste we reach for in the morning to the tungsten-filament light bulb we turn off at night, we constantly make use of mineral resources.

We use feldspar dust in toothpaste and soda ash in soap. We use talc in talcum powder and take dolomite in medicine. We use stainless steel cutlery and cook with cast-iron skillets. Our pans are copper-bottomed and we wrap our food in aluminum foil. There is quartz and lime in our glassware and lead in decorative crystal. We drink beverages from cans of aluminum and keep food in cans of tin-plate. We cook with baking soda and serve rock salt with our meals. We use powdered chert for scouring and put chlorine in our bleach. There is china clay in the porcelain of our toilets, sinks, and bathtubs. Rare earth elements coat the TV screen and mercury works our light switches. We wear gold and silver jewelry and trim our cars with chrome. We use platinum in the catalytic converter and fluorine in the air conditioner. Our pens are tipped with tungsten and our pencil "lead" is really graphite. We use limestone in cement and make concrete out of sand. There is gravel on the road and chalk in every classroom. We use gypsum for plastering and make bricks and tiles from clay. There is manganese in siding and titanium in paint. We use copper in our wiring and add zinc to make it brass. We use cobalt to form magnets and make paper using sulfur. We use potash, nitrate, and phosphorus if we fertilize our lawns, and we grace our public monuments with facades of polished stone. The list could go on and on.

The more important of these mineral resources are listed in Table 17-1. Many of these resources are metals, and these occur either quite commonly in the rocks of the Earth's crust (Table 17-2), and so are termed **abundant metals**; or they are rare (less than 0.1% by weight) and are termed **scarce metals**. Few, however, are mined as pure metals. Instead, most are found combined with elements like oxygen and sulfur as mineral oxides and sulfides, and are mined as **ores,** which is the name given to rocks that contain an economic mineral in sufficient quantity to make its exploitation profitable.

In contrast, most industrial minerals are very common in nature and so are classified by their use rather than their abundance. The greatest volumes of these minerals are used as building materials, but fertilizers and other raw chemicals form a second important category. However, some of the precious stones included in the third category of nonmetallic minerals, that of abrasives and gems, are very rare indeed.

17.3 Metallic Minerals

17.3.1 Abundant Metals

The abundant metals are among the most common elements in the Earth's crust because they form an important component of many silicate minerals. They can be extracted at profit from any number of already proven ore deposits called reserves, but only in the form of non-silicate minerals in which the metal is concentrated. They include iron, aluminum, magnesium, titanium, and manganese.

Table 17-1

Classification of Mineral Resources

METALLIC MINERALS		NONMETALLIC MINERALS		
Abundant Metals (greater than 0.1 percent of the crust)	**Scarce Metals** (less than 0.1 percent of the crust)	**Building Materials**	**Raw Chemicals**	**Abrasives and Gemstones**
Iron	Copper	Stone	Fertilizer	Diamond
Aluminum	Lead	Sand and Gravel	Nitrogen	Corundum
Magnesium	Zinc	Cement	Phosphorus	Garnet
Titanium	Chromium	Gypsum (Plaster)	Potash (Potassium)	Quartz
Manganese	Nickel		Halite	Emerald
	Gold		Sulfur	
	Silver			
	Platinum Group Elements			
	Platinum Palladium			
	Rhodium Iridium			
	Ruthenium Osmium			

Table 17-2

Relative Percentages by Weight of Chemical Elements in the Earth's Lithosphere

ELEMENT	% WEIGHT
Oxygen (O)	46.40
Silicon (Si)	28.15
Aluminum (Al)	8.23
Iron (Fe)	5.63
Calcium (Ca)	4.15
Sodium (Na)	2.36
Magnesium (Mg)	2.33
Potassium (K)	2.09
Titanium (Ti)	0.48
Manganese (Mn)	0.10
All others	0.08
Total	100.00

Iron (Fe)

Of all the Earth's mineral resources, iron is the most important. Current demand in the United States alone, for example, is approaching 120 million tons per year and is predicted to increase steadily into the new century. Iron is the dominant metal of industry (accounting for some 95% of all metals in the world) and is the basic ingredient of steel. When alloyed with other metals such as nickel, chromium, tungsten, cobalt, and manganese, iron is strong, relatively inexpensive, and resistant to corrosion. It frames our buildings, shapes our vehicles and is the metal of our machines.

Iron is the fourth most abundant element in the Earth's crust. Iron, however, is not found in a metallic or "native" state, but rather, is mined either in its partially oxidized form, magnetite (Fe_3O_4)—the mineral we met in our examination of paleomagnetism in Chapter 4—or in its fully oxidized form, hematite (Fe_2O_3), a mineral more familiar to us as rust. The hydrated iron oxide, limonite (FeO.OH), and the iron carbonate, siderite ($FeCO_3$), are also sources of iron ore. Although iron is found everywhere, economic deposits of iron ore are widely spaced. In North America (Fig. 17.1), the largest deposits are those in the Lake Superior region, the Vermillion Range, and the Labrador Trough of Canada, in which the ore is a Precambrian iron oxide of sedimentary origin known as Banded Iron Formation (Figure 17.2; see also Chapter 13).

Aluminum (Al)

Of all metals, aluminum shows the most rapid growth in its utilization by industry. In the United States, for example, demand for aluminum in the next decade is likely to approach 10 million tons per year, almost double that of the last decade. Aluminum is a lightweight, corrosion-resistant metal and when alloyed with copper, manganese, or magnesium, it can be as strong as steel. It is widely used in the construction of airplanes, it is the metal of choice for packaging and window frames, and it is increasingly replacing iron in the manufacture of vehicles.

Aluminum is the third most abundant element in the Earth's crust (Table 17-2) and is an important ingredient in many rock-forming minerals, especially feldspars and clays. However, it is difficult to extract aluminum from these minerals because it is tightly bonded within the silicate mineral structure. Instead, the principal source of aluminum is the ore *bauxite* (Fig. 17.3). This complex mixture of aluminum oxides and hydroxides is actually a soil formed by the weathering of aluminum-rich rocks in tropical climates. Most bauxite deposits are consequently found in tropical areas, the largest

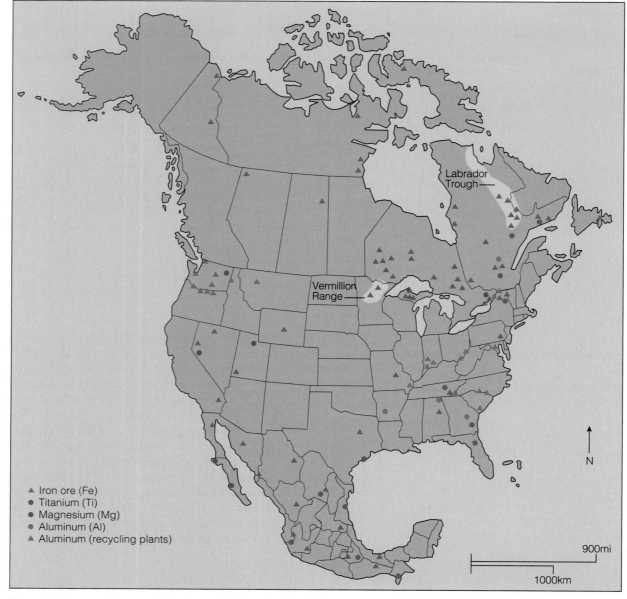

Figure 17.1
Distribution of the deposits of abundant metals (iron, aluminum, manganese, and titanium) in North America.

▲ Iron ore (Fe)
● Titanium (Ti)
● Magnesium (Mg)
● Aluminum (Al)
▲ Aluminum (recycling plants)

900mi

1000km

occurring in West Africa, Australia, Brazil, and Jamaica. Aluminum production, however, uses huge quantities of electricity, so that for processing, the ore is often transported to areas like Canada and Scandinavia where cheap hydroelectric power is available. The high production cost of aluminum is its chief disadvantage and is the reason we place such an emphasis on the recycling of aluminum products. Indeed, some 25 percent of the annual production of aluminum metal now comes from recycled material. The location of aluminum deposits in North America are shown in Figure 17.1.

Magnesium (Mg)

Magnesium is the lightest common metal and when alloyed with aluminum, is strong and resistant to corrosion. The U.S. demand for this metal, which is more expensive to produce than aluminum, is close to 600,000 tons per year. Most magnesium is alloyed with iron and steel for use in the aviation, transportation, and instrumentation industries. Although magnesium is a stable metal, it burns quite readily with a bright white light and, as a powder, is used in fireworks and flashbulbs.

Magnesium is the seventh most abundant element in the Earth's crust (Table 17-2) and the third most abundant element dissolved in sea water. Although it is an important ingredient of many rock-forming minerals, particularly olivine, and pyroxene (See Chapter 2), the magnesium is tightly bonded within the silicate structure. As a result, the principle ore mineral is the carbonate magnesite ($MgCO_3$), from which magnesium can be relatively easily extracted. Small deposits of this mineral are

Figure 17.2
Sample of Precambrian Banded Iron Formation from the Lake Superior region. Dark grey bands are hematite (Fe_2O_3), light red bands are jasper (hematitic chert).

Figure 17.3
Bauxite, the principal ore of aluminum, is produced by the tropical weathering of aluminum-rich rocks and commonly has a nodular appearance.

exploited for magnesium in Washington, Nevada, and Ontario (Fig. 17.1), but the most important reserves occur in Europe, Asia, and South America. In the United States, sea water and evaporite brines form additional sources of this metal.

Titanium (Ti)

Titanium is almost as light as aluminum but is a good deal stronger and resists corrosion even at high temperatures. It is therefore of particular value to the aerospace and defense industries where, for example, it is used in heat shields. However, most of the demand for titanium, which presently exceeds one million tons per year in the United States alone, is in the form of titanium oxide (TiO_2) as a white pigment for paint.

Titanium is the ninth most abundant element in the Earth's crust (Table 17-2) and is concentrated in several accessory rock-forming minerals such as ilmenite ($FeTiO_3$) and rutile (TiO_2). Indeed, most deposits of titanium ore occur where one of these two minerals, which are resistant to weathering, has become concentrated by sedimentary processes in beach sands. Notable deposits occur in Australia, India, Brazil, and Florida. The largest deposits of titanium ore, however, are those of Quebec and the Adirondacks (Fig. 17.1), where ilmenite has been concentrated by igneous processes in a rock type called anorthosite (see Chapter 13).

Manganese (Mn)

Manganese is seldom used on its own but it is an essential ingredient in iron and steel. In fact, every ton of cast iron and steel contains 6.5 kg (14 lbs) of manganese to protect it from oxidation and sulfur corrosion, and it is this demand that accounts for most of the approximately 700,000 tons of manganese used annually in the United States.

Manganese is the tenth most abundant element in the Earth's crust (Table 17-2) and accompanies iron and magnesium in a wide variety of rock-forming minerals. However, although it is widespread, manganese is rarely of sufficient concentration to be mined, and no significant ore deposits exist within Canada or the United States. The most common ore mineral is the oxide pyrolusite (MnO_2), although the carbonate rhodochrosite ($MnCO_3$) is common in some deposits. Manganese ore deposits are usually one of two types. Some, like the major deposits of South Africa, the Republic of Georgia, Australia, and Mexico, are the result of chemical processes in sedimentary rocks. Others, like those of Brazil and central Africa, form by enrichment as a result of weathering. Vast deposits of manganese nodules exist on today's ocean floors, but these are not yet an economically viable source of the metal.

17.3.2 Scarce Metals

Unlike the abundant metals that need only be concentrated a few times above their average abundances to form ore deposits, scarce metals must be concentrated by natural processes hundreds or even thousands of times above their average abundances if they are to be profitably exploited. As a result, the discovery and management of these rare, nonrenewable resources are significant challenges. Figure 17.4 shows the enrichment factors needed for the economic recovery of these metals from their ores. Scarce metals include molybdenum, tin, antimony, mercury, cadmium, and lithium, but the most important in terms of their significance to industry are copper, zinc, lead, chromium, nickel, and precious metals such as gold, silver, and the platinum group elements. Economic deposits of these scarce minerals are rare and widely scattered. Their distribution across North America is shown in Figure 17.5.

Copper (Cu)

Although the crustal abundance of copper is only 50 parts per million, it is the third most consumed metal behind iron and aluminum. Demand for copper in the United States alone, for

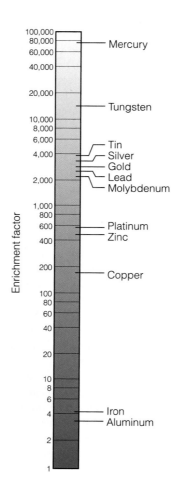

Figure 17.4
Enrichment factors needed for the economic recovery of metals from their ores. For a metal to be mined profitably, it must be concentrated in the ore deposit in an amount exceeding its average concentration in the Earth's crust. This enrichment factor is least for those metals, such as aluminum and iron, that are most abundant in the crust, and greatest for the metals that are scarcest, such as mercury and tungsten.

example, has exceeded 2 million tons per year for almost two decades and is presently approaching 3 million tons per year. Copper is a metal that is readily shaped or drawn out into fine wire, and its use dates back at least 8000 years. It is also an excellent electrical conductor, and it is this property that accounts for its primary use in electrical wiring. Copper is also used for plumbing and in coins, and is an important ingredient of brass (copper and zinc) and bronze (copper, tin, and zinc).

Copper occurs in numerous minerals, including the semi-precious stones malachite and azurite, and is one of the few non-precious metals to occur naturally in its metallic or "native" form. The most important ore minerals, however, are copper sulfides such as chalcopyrite ($CuFeS_2$) and chalcocite (Cu_2S). It is these sulfides, when disseminated in granitic bodies by igneous activity above ocean-continent subduction zones, that form the vast low-grade ore deposits (often less than 0.5% Cu), known as *porphyry copper* deposits, from which most copper is mined. Porphyry copper deposits form well defined belts parallel to subduction zones of Triassic or younger age, and are responsible for the huge open-pit copper mines of Montana, Utah (See Figure 17.14), and Arizona in the western United States, and those of Chile and Peru in the Andes. Copper sulfides may also be enriched to ore grade in shales by sedimentary processes. Deposits of this type are known as *kupferschiefer* after those of central Europe, and account for the mines of the Zambian-Zaire copper belt in Africa.

Zinc (Zn) and Lead (Pb)

These two metals are best described together because they commonly accompany each other in natural settings. Although their average crustal abundances are only 70 parts per million and 12.5 parts per million, respectively, zinc is the fourth most widely used metal after copper, and lead is the fifth. For the past decade, each metal has been consumed in the United States at a rate of about one and a half million tons per year. Zinc is used primarily for diecastings and as a corrosion-resistant coating on galvanized iron and steel. It is also alloyed with copper to make brass. The primary use of lead is in storage batteries, but it is also used in construction because of its resistance to corrosion, and in ammunition and radiation shielding because of its density. Its use in paint and as an antiknock additive in gasoline (once the second largest area of its consumption in North America) has been virtually eliminated because of the health hazard posed by its toxicity.

Although most of the world's production of zinc and lead come from just two minerals, the zinc sulfide sphalerite (ZnS) and the lead sulfide galena (PbS), economic deposits of these metals occur in a variety of settings. Concentrations of these minerals as a result of submarine volcanism in bodies of stratified volcanic, sedimentary, and metamorphic rocks are termed *massive sulfide* deposits, and owe their origin to black smokers on the floor of some ancient sea. The major lead-zinc deposits of eastern Australia, and the Kidd Creek, Sullivan, and Bathurst districts of Canada are of this type. The large lead-zinc deposits of the United States, however, are of the *Mississippi Valley* type. Named for their occurrence in the tri-state area of Oklahoma, Kansas, and Missouri, this type of deposit is hosted by carbonate rocks such as limestone and dolomite in rifted regions of continental interiors. Galena and sphalerite also fill mineralized fractures called **veins** that occur in association with mineralized granites.

Chromium (Cr)

The average abundance of chromium in the Earth's crust is only 100 parts per million. However, the metal is an essential component of stainless steel and nonferrous alloys, and is of critical strategic importance to the United States because the country has no chromium reserves. Consumption in the United States is currently about 500,000 tons per year.

Most crustal chromium is concentrated in the mineral chromite ($FeCr_2O_4$) which is the metal's only ore mineral. Chromium ores are restricted to iron- and magnesium-rich ultramafic igneous rocks where chromite has been concentrated by igneous processes. Because of this, the ores are found either in ophiolite bodies (which represent tectonically preserved remnants of ancient oceanic crust; see Chapter 5) or in large ultramafic igneous intrusions like the Bushveld complex of South Africa (Fig. 17.6). The main producers of chromium are South Africa and Russia.

Nickel (Ni)

Although the average crustal abundance of nickel is only 75 parts per million, it is vital to the manufacture of stainless steel and is

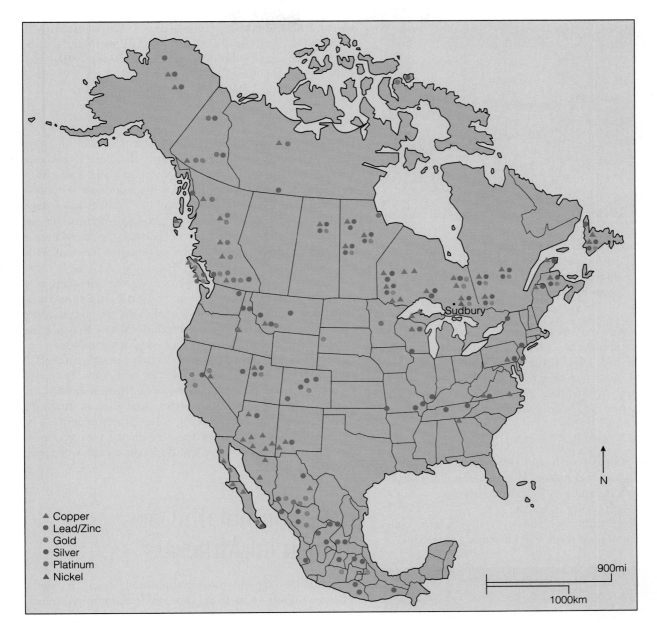

Figure 17.5
Distribution of the deposits of scarce metals (copper, lead and zinc, gold, silver, platinum, and nickel) in North America.

Legend:
- ▲ Copper
- ● Lead/Zinc
- ● Gold
- ● Silver
- ● Platinum
- ▲ Nickel

900mi
1000km

used widely in the chemical, petroleum, electrical, construction, and aerospace industries as an alloy. Current consumption in the United States is about 150,000 tons per year.

Nickel occurs in sulfide minerals and most is extracted from the mineral pentlandite [(FeNi)$_9$S$_8$)]. Like chromium ores, nickel-rich sulfides are often concentrated in rocks of mafic and ultramafic composition (see Chapter 2) as a result of their segregation by igneous processes. The huge deposit in Sudbury, Ontario, and others of similar size in Russia and Australia, are of this type. The largest reserves of nickel in the western hemisphere, however, are in Cuba where the metal has been concentrated as the result of intense tropical weathering of ultramafic rocks rich in trace amounts of nickel.

Gold (Au)

Gold is an extremely rare metal with a crustal abundance of only 3 parts per *billion*. In addition to its use as a precious metal in jewelry, gold is an excellent electrical conductor and performs critical functions in electrical equipment and in the aerospace industry. Gold also has a unique status among commodities as a long-term store of value and is the monetary base for international trade. Indeed, it is estimated that about a third of all the gold ever mined (some 32,000 tons) is stored in bank vaults and government treasuries like that at Fort Knox, Kentucky.

Gold is mined as a native metal in a variety of settings. It is found filling mineralized fractures (veins) in a variety of

Figure 17.6
Chromite layers (black) are concentrated towards the base of the plutonic Bushveld Complex in South Africa because the high density of the early-formed chromite crystals caused them to sink through the magma as the igneous body cooled.

igneous and metamorphic rocks (Fig. 17.7), and because it is heavy, it may accumulate as a **placer deposit** in the sand and gravel of stream beds. The huge placer-like Witwatersrand deposit in South Africa occurs in metamorphosed conglomerates of Archean age. World production of gold is dominated by South Africa, Russia, the United States, and Canada.

Silver (Ag)

Like gold, silver is an extremely rare metal with a crustal abundance of only 70 parts per billion. It is most familiar to us in the form of silverware and jewelry, but its major use is in the photographic and electronic industries.

Although silver occurs as a native metal, it is more common as a sulfide and most silver is produced as a by-product of

Figure 17.7
Gold on a sample of vein quartz from a mineralized fracture in igneous rock, Lizard Peninsula, U.K.

copper, lead and zinc smelting. Galena (PbS), particularly in vein deposits like those of northern Idaho, is the most common host to silver, which substitutes for some of the lead in the mineral's crystal structure. The largest reserves of silver occur in North America and Russia.

Platinum Group Elements

The extremely rare metals that make up the platinum group elements, namely platinum, palladium, rhodium, ruthenium, iridium, and osmium, have crustal concentrations of only a few parts per billion. Recall that the anomalous concentration of iridium in a thin clay layer at the Cretaceous-Tertiary boundary (Chapter 14) has been used as evidence for the impact of a dinosaur-killing asteroid because the abundance of iridium in asteroids and meteorites is far higher than that of the Earth's crust.

The principal use of the platinum group elements is as chemical catalysts in the catalytic converters of automobiles. As precious metals they are also used in jewelry, but they are more important to the chemical industry where they play an indispensable role in the production of pharmaceuticals.

The platinum group elements all occur as native metals in mantle-derived igneous rocks, but are commonly associated with sulfides and are by-products of the processing of nickel ores in ultramafic plutonic bodies like that of Sudbury, Ontario. They are also found in some placer deposits. The largest single occurrence of these elements, however, is in association with chromite in the Bushveld complex of South Africa (Fig. 17.6).

17.4 Nonmetallic or Industrial Minerals

In addition to metals, modern industrial society is wholly dependent upon a variety of nonmetallic or industrial minerals. Indeed, while these mineral resources are far less costly than metals because they are easier to exploit, their total value is much higher because they are utilized in far greater volumes. Most are used in the production of the concrete, brick, cement, and plasterboard that form the basis of the construction industry. Others, however, are used as raw chemicals in the manufacture of such products as fertilizers and industrial acids and alkalis. Yet others possess a rare natural hardness or beauty and are used as abrasives or gemstones.

17.4.1 Building Materials

In addition to lumber, the construction industry depends upon a variety of industrial minerals, and almost every rock type can be used for building. It is therefore not surprising that, of all natural resources, building materials are used in the largest volume. The consumption of construction materials in the United States alone, for example, amounts to more than $11 billion annually. Of these materials, the most important are stone, sand and gravel, limestone, and gypsum.

Stone

A variety of rocks, including granite, limestone, sandstone, marble, and slate (Fig. 17.8), are quarried simply for their use as stone. Some rocks, like marble, granite, and slate, are used as "dimension stone" to form the polished floors of our banks, and the facades and roofs of our public buildings and monuments. Most stone, however, is broken up, and some of this "crushed rock" is used to build roads and buildings. Road stone is usually chosen on the basis of whatever local rock is the most durable and can be crushed to the desired size. Igneous rocks, such as granites and basalts, are often used for this purpose. The crushed rock used most widely, however, is limestone. Pulverized limestone is the primary ingredient of cement, crushed limestone is used as a flux in steel making, and finely powdered limestone is used for agriculture or is burnt ("calcined") to produce the chemical "lime." The numbers involved are astounding. In the United States, for example, dimension stone valued at about $235 million is produced in some 230 quarries every year; 1.4 billion tons of crushed rock valued at almost $8 billion is produced annually from 3700 quarries, active in all but two states; and the yearly production of lime from limestone amounts to some 19 million tons valued at over $1 billion. World production of lime in 1997 was estimated to have been 124 million tons.

Sand and Gravel

As a mineral resource, unconsolidated sand and gravel rank second in importance after crushed stone in terms of production tonnage and, in the United States, is the only commodity produced in all 50 states. Sorted and deposited by flowing water (see Chapters 2 and 10), sand and gravel is a readily accessible resource and is a component of concrete aggregate, road base, asphalt, and fill. Production in the United States amounts to almost a billion tons per year.

Cement

Portland cement, which has the property of hardening when mixed with water, is the chief binding agent for concrete. Cement is produced by heating the right proportions of crushed limestone, dolomite, quartz sand, and oxides in a high-temperature kiln, and it must be added to an aggregate of sand and gravel or crushed rock to make concrete. In the United States, cement is consumed at a rate of almost 100 million tons per year. It is produced globally in amounts approaching 1.5 billion tons per year.

Gypsum

The evaporite mineral gypsum ($CaSO_4.2H_2O$), one of several minerals that form as the result of the evaporation of salt water, is the natural source of building plaster and plasterboard. When the mineral is heated in low-temperature kilns, much of its water of crystallization is driven off and plaster of Paris ($2CaSO_4.H_2O$) is formed. As those of us who have had the misfortune to break a bone can testify, this plaster rapidly sets on the addition of water by recrystallizing back to gypsum. Annually, some 100 million tons of gypsum are mined for the construction industry (Fig. 17.9).

17.4.2 Raw Chemicals

A number of nonmetallic minerals used simply as chemicals play a vital role in agriculture and industry. The most important of these are the various natural sources of potassium and phosphorus used in fertilizers, and the minerals sulfur and salt.

Figure 17.8
The now-abandoned Penrhyn slate quarry in Bethesda, North Wales, U.K., was once a major source of roofing slate. (Photo Damian Nance)

Figure 17.9
Gypsum deposits of the Carboniferous Windsor Group in Nova Scotia, Canada. (Photo Brendan Murphy)

Fertilizer

The intensive agriculture of modern society depletes the soil of many essential nutrients. To replace these nutrients, agriculture depends to a large degree on the use of inorganic fertilizers. The three nutrients most essential to plant growth are potassium, phosphorus, and nitrogen, and it is materials containing these elements, combined with agricultural lime and gypsum, that make fertilizer. Nitrogen (N) is extracted for its use in plant growth-enhancing nitrates (NO_3^-) from the atmosphere, but the other components of fertilizer must be mined.

Although potassium (K) is widely distributed in crustal rocks, economic potassium deposits are all found in marine evaporites. A variety of potassium-bearing minerals, marketed as "potash," are left behind by the evaporation of sea water. The most important of these is sylvite (KCl), almost all of which is used in the manufacture of fertilizer. The United States produces some 1.5 million tons of potash per year, mostly from southeastern New Mexico. The world production of potash, however, presently amounts to over 23 million tons annually, over a third of which comes from Saskatchewan in Canada (Fig. 17.10).

The principal source of phosphorus (P) is the mineral apatite, a calcium phosphate [$Ca_5(PO_4)_3(F, OH)$]. This occurs in minor amounts in many rocks but is concentrated in certain marine sedimentary deposits called **phosphorites** as a result of

oceanic upwelling along the continental shelves (see Chapter 9). Although two thirds of the world's reserves occur in Morocco and the Western Sahara, the United States is the leading producer of phosphate, supplying over 46 million tons annually from major marine phosphate deposits in Florida and North Carolina, and from the Permian Phosphoria Formation in Idaho (Fig. 17.10). Apatite is also mined from phosphate-rich igneous rocks at Cargill in northern Ontario.

Salt

In addition to its well-known use as a food additive (common table salt) and road de-icer, salt or halite (NaCl) plays a far more important role in the industrial production of the chloralkalis used to produce polyvinyl chloride (PVC), the chlorine and caustic soda used in pulp production, and the soda ash used in the manufacture of soaps and detergents.

The largest reserve of salt is in the ocean, and salt is produced from the solar evaporation of brine in many parts of the world. However, as we learned in Chapter 14, the evaporation of ancient seas has resulted in the formation of bedded evaporite deposits and salt domes that are preserved in the rock record. These deposits form the main source of salt. The United States has vast reserves of such deposits and is the leading producer of salt with a production of over 40 million tons per year. Salt deposits in the Gulf Coast region, the Permian Basin of west Texas and Oklahoma, the Salina Basin bordering the Great Lakes, and the Williston Basin of the Dakotas and Montana (Fig. 17.10), amount to a resource of more than 60 trillion tons. Canada, which produces some 12 million tons of salt annually from the northern Williston and Maritimes basins, is the third largest producer after China.

Sulfur

Sulfur (S) is such an important element to modern industrial society that the level of its consumption is used as an index of a nation's industrial development. Sulfuric acid (H_2SO_4), which accounts for 90 percent of sulfur consumption, is one of the most important industrial chemicals. Approximately two thirds of it is used in the production of fertilizer, but its powerful acidity also makes it vital to metal mining, paper, and plastics production, oil refining, and the chemical industry. In fact, sulfuric acid is used at one stage or another in the manufacture of most industrial products.

Sulfur is the only nonmetallic element to occur in its native state, most commonly in volcanic vents and hot springs (Fig. 17.11), and at the top of salt domes where it forms through the alteration of the evaporite mineral anhydrite ($CaSO_4$) by sulphate-reducing bacteria. From the latter source it is recovered by the Frasch process, which takes advantage of the low melting point of sulfur. Hot water is pumped into the sulfur-bearing formation, and the resulting mixture of water and molten sulfur is then pumped out. The main source of sulfur, however, is from petroleum as a by-product of the refining of crude oil and the purification of natural gas. The United States is the world's leading producer of sulfur with an annual production of almost 12 million tons, half of which is from Texas and Louisiana. Canada, with its annual production of

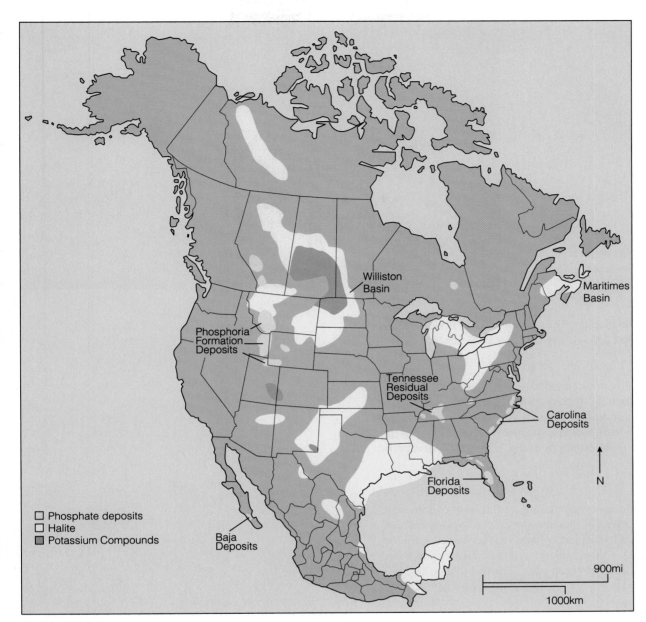

Figure 17.10
Phosphate and marine evaporite (salt and potassium) deposits of North America.

Legend:
☐ Phosphate deposits
☐ Halite
■ Potassium Compounds

Map labels: Williston Basin, Maritimes Basin, Phosphoria Formation Deposits, Tennessee Residual Deposits, Carolina Deposits, Florida Deposits, Baja Deposits

over 9 million tons, runs a close second. The annual world production of sulfur exceeds 50 million tons.

17.4.3 Abrasives and Gemstones

Because of their natural hardness, a variety of minerals are used as abrasives for cleaning, polishing, cutting, grinding, and drilling. Because of their natural beauty, some of the same minerals are also worn as gemstones.

Diamond

This, the hardest of all minerals, is a crystalline form of carbon (C) created at very high pressure and found only in intrusive pipes of **kimberlite** believed to have originated from deep with-

in the upper mantle (see Chapter 7, Living on Earth: Diamond Pipes: Windows to the Earth's Mantle). Those that for reasons of color or impurity are not suitable for gemstones are valuable as industrial diamonds for drilling and cutting. South Africa (Fig. 17.12) and Australia dominate world production of diamond stone but most industrial diamond powder is synthetic.

Corundum

This aluminum oxide or alumina (Al_2O_3) is the second hardest mineral, occurring in ancient metamorphic rocks in South Africa and India. It is used industrially as an abrasive, but also occurs in gem quality, forming ruby and sapphire when the natural color is either red or blue. Natural mixtures of corundum and magnetite form "emery," which is used in wear-resistant floors, polishing

Figure 17.11
Gathering sulfur from a vent in the crater of Kawah Ijen, a volcano in eastern Java.

cloth, and non-slip surfaces. Emery is mined near Peekskill, New York, but the largest deposits are in Turkey and Greece.

Garnet

Garnet, particularly the iron-aluminum variety almandine, is a common constituent of metamorphic rocks (Fig. 17.13) and is widely used as an industrial material for cutting, polishing, and sandblasting. Some 50,000 tons of garnet, almost a third of the world's annual supply, comes from the Adirondack Mountains of New York state.

Quartz

In addition to its use in glass making, casting, and construction, quartz sand and other siliceous abrasives are utilized in sandblasting and sandpaper. In addition, colored varieties of quartz, such as amethyst (purple), citrine (yellow), and rose quartz (pink), and other varieties of silica, like opal, agate, jasper, and onyx, are semi-precious and used in jewelry (Table 17-3; see also Figure 2.5).

Emerald

This is the only precious gem with a complex formula, being a brilliant green variety of the mineral beryl ($Be_3Al_2Si_6O_{18}$), the principal ore of the element beryllium. Emeralds occur in very coarse-grained granitic rocks called pegmatites, and most come from Columbia and Zambia. Blue-green beryl is aquamarine, one of the semiprecious gems whose characteristics are listed in Table 17-3.

Figure 17.12
The Kimberley Mine in South Africa produced over 14 million carats of diamond from a kimberlite pipe; the product of a violent, gas-rich volcanic explosion capable of bringing up fragments from deep within the Earth's interior (see Chapter 7, Living on Earth: Diamond Pipes: Windows to the Earth's Mantle). Now partially filled with water, the pit is 495 m (1625 ft) deep.

Figure 17.13
Large crystals of garnet in gneiss, Clarke Head, Nova Scotia. (Photo Brendan Murphy)

Table 17-3
Characteristics of the Semiprecious Gems

NAME	ORIGIN	HARDNESS	CHEMICAL CLASS
Amber	Organic	2–2 $\frac{1}{2}$	Organic
Apatite	Igneous	5	Phosphate
Beryl (aquamarine)	Igneous and metamorphic	8	Silicate
Chrysoberyl (alexandrite)	Igneous and metamorphic	8 $\frac{1}{2}$	Oxide
Feldspar (moonstone, amazonite, sunstone)	Igneous	6	Silicate
Garnet	Igneous and metamorphic	7–7 $\frac{1}{2}$	Silicate
Jadeite (jade)	Metamorphic	6 $\frac{1}{2}$	Silicate
Lazarite (lapis lazuli)	Metamorphic	5–5 $\frac{1}{2}$	Silicate
Malachite	Sedimentary	3 $\frac{1}{2}$–4	Carbonate
Nephrite (jade)	Metamorphic	5–6	Silicate
Opal (hydrous silica)	Sedimentary	5 $\frac{1}{2}$–6 $\frac{1}{2}$	Silicate
Pearl	Organic		Organic
Peridot (olivine)	Igneous	6	Silicate
Quartz (agate, jasper, smoky, rock crystal, amethyst, citrine, rose, tiger eye)	Igneous and sedimentary	7	Silicate
Spinel	Igneous	8	Oxide
Topaz	Igneous and metamorphic	8	Silicate
Tourmaline	Igneous and metamorphic	7 $\frac{1}{2}$	Silicate
Turquoise	Sedimentary	4–6	Phosphate
Zircon	Igneous	6 $\frac{1}{2}$–7	Silicate

Exploitation of the world's mineral resources has defined human development and lies at the very heart of modern industrial society. But the mining of our mineral resources has serious environmental consequences. The quest for mineral riches has damaged vast tracts of otherwise unspoiled land, and today threatens national parks and other ecologically sensitive areas. The waste produced by mining far exceeds that from all other sources, and even outstrips the natural erosion of the world's streams. The extraction and processing of ore consumes huge amounts of energy, and so is a major partner in the problem of global warming caused by the burning of fossil fuels. The refining of ore releases huge quantities of toxic gases into the atmosphere, and is a major contributor to acid rain. As a result, the environmental impact of our exploitation of mineral resources can be severe.

The amount of rock waste produced by mining activities is truly staggering, particularly when low-grade ores are recovered by surface or "open pit" methods. The Bingham Canyon copper mine in Utah, for example, has removed an entire mountain and excavated a pit over 3 km (2 mi) across and 1 km (3300 ft) deep (Fig. 17.14). In all,

this mine alone has removed 3.3 billion tons of material, an amount seven times greater than was excavated in the building of the Panama Canal. At the height of its production, the mining operation was moving 400,000 tons of material every day, three quarters of which was rock waste. At a grade of only 0.3% copper, the remaining 100,000 tons of ore that the mine processed daily, produced a further 97,000 tons of tailings. Worldwide, close to 30 billion tons of material a year, or almost twice the estimated amount of sediment carried by the world's streams, is moved by mining activities.

To move such vast quantities of material and process the extracted ore requires huge amounts of energy. Aluminum production and steelmaking, for example, are particularly energy intensive processes. Worldwide, the excavation and refining of ore uses up to one tenth of all the energy produced each year. Because most of this energy is supplied by the burning of fossil fuels, mining is a major contributor to the environmental problems caused by energy use, such as air pollution and the climate changes linked to increased atmospheric carbon dioxide levels.

Serious contamination problems are also associated with leaching of the waste material

produced by mining, and with the refining process known as "smelting" that extracts the metal from its ore. Although most of the rock waste is chemically inert, the finely ground tailings left behind after the ore has been concentrated can be highly reactive, producing acid mine drainage and contaminating surface waters with toxic metals. Up to 10,000 miles of streams are estimated to have been affected in this way in the United States alone. By laying waste to vast areas, mining can also cause the silting of lakes and streams as a result of increased erosion.

The smelting of ore can release huge quantities of pollutants into the atmosphere. The extraction of metals from metal sulfide ores, for example, is estimated to be responsible for the annual emission of six million tons of sulfur dioxide, the primary cause of acid rain. Smelting also emits arsenic, lead, and other toxic chemicals, and if left uncontrolled, can produce environmental disaster areas in which little of no vegetation will grow. Such an area around the old nickel smelters in Sudbury, Ontario, for example, once occupied over 100 square km (40 square mi).

The damage caused by the exploitation of mineral deposits is clearly very significant and it poses one of the major dilemmas facing

17.5 What Makes a Mineral Deposit?

A mineral deposit is an accumulation of a valuable inorganic resource, often in the form of an ore deposit or orebody that can be mined or quarried economically. The elements sought for in mineral deposits, however, are usually rare in the Earth's crust, or when relatively abundant, are rarely concentrated in one place. All mineral deposits therefore require the commodity they contain to have been concentrated by natural processes. The degree of concentration needed for economic extraction is determined by the cost of recovering the commodity. This, in turn, depends on the average concentration of the element in the Earth's crust. So what, then, is the composition of the crust?

There are 92 naturally occurring elements, and the Earth's crust contains a certain amount of each of them. Fully

99% of the Earth's continental crust, however, is made up of just nine elements (Fig. 17.15), and just two, oxygen and silicon, make up almost three quarters of this crust. It is not surprising, therefore, that most rock-forming minerals are silicates, and it is within these silicates that most of the common metallic elements occur.

The same chemical abundance statistics are true of the Earth's oceanic crust although, relative to granitic continental crust, it contains slightly more magnesium, calcium, and iron, and slightly less silicon and potassium, consistent with its basaltic composition. The mineral composition of ocean water, however, is quite different from that of the crust (see Chapter 8), the relative abundance of dissolved elements reflecting their solubility in sea water. Nevertheless, excluding the water itself, more than 95% of the ocean's composition is made up of only six dissolved ions, and sodium and chlorine alone account for more than 80% (Fig. 17.15). From these simple statistics, it is abundantly clear that massive concentration factors are required for economic mineral deposits to be produced for the vast majority of elements.

Figure 17.14

The Bingham Canyon copper mine in Utah is one of the largest in the world, having excavated over 3 billion tons of rock. At peak production 400,000 tons of material were moved daily to recover just 3000 tons of copper.

modern society. It would be easy to be critical of an industry responsible for such environmental damage. But mining is an inherently destructive industry that does not readily lend itself to environmental mitigation by pollution control. Put another way, mining is a very dirty business! Yet mining is an economic activity upon which civilization depends. The very prosperity of industrial nations like Canada and the United States, for example, and the high standard of living that their people enjoy, depends on their use of mineral products.

If the devastating environmental impact of mineral extraction is to be lessened, it cannot be done simply by demanding cleaner mines. Instead, more mineral-efficient economies are needed which emphasize the recycling, repairing, and re-manufacturing of mineral products. Also needed are substitutes for mineral products whose production is less damaging to the environment. A particularly successful example of such a substitute is the replacement of copper wire by fiber optics made from glass. Society

must also change its behavior. Although residents in the United States, for example, now recycle almost 50% of the aluminum they use, they still throw away 2.3 million tons of metal each year. As consumers, we must all learn to recycle metals more efficiently while, at the same time, lessening our dependence on mineral products. For in the end, it is only by reducing demand for new mineral-based materials that we can significantly reduce the environmental impact of their production.

17.5.1 Global Distribution of Mineral Deposits

Although crustal rocks contain large amounts of the abundant metals in their silicate minerals, these do not form an economic resource because the cost of extracting metals from stable rock-forming silicates is prohibitively high. Instead, most ore minerals are simpler compounds such as oxides, sulfides, and carbonates, from which the respective metals are more easily separated. Deposits of these minerals, however, occur only where natural processes have produced local concentrations of a metal that far exceeds its average composition in the crust. This is because the natural concentration of an element to the

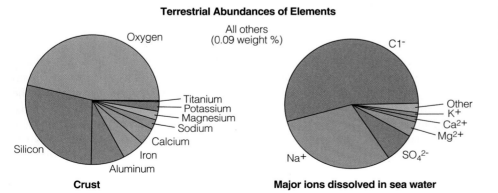

Figure 17.15
Relative abundance of elements in the Earth's crust and ocean waters.

point where it becomes economical to exploit only occurs under very exceptional circumstances. Mineral deposits of even the abundant metals are therefore rare, and deposits of elements with low crustal abundances are very rare indeed.

The various processes that naturally concentrate elements also tend to occur in very specific geologic settings. As a result, mineral deposits are far from evenly distributed worldwide, and the mineral resources of some countries are far greater than those of others. Even the size of a country is no guarantee of mineral wealth, as small countries blessed by a favorable geologic history may sometimes have major deposits of a particular mineral, while much larger countries have none. A mineral deposit must also be a resource that can be mined or quarried *economically*. The mineral wealth of a given country is therefore subject to technological developments and the vagaries of local politics and world economics. A mineral deposit that is presently uneconomic, for example, may become exploitable with improved extraction techniques, political subsidies, or a rise in ore prices on the world market. Conversely, political upheavals and environmental issues may render an existing economic deposit unworkable.

The changing economics of mineral deposits highlights the difference in the terms **reserve** and **resource**. A reserve is a deposit of known size and quality that *could* be worked profitably with today's technology and in today's political and economic climate. Mineral reserves are usually defined on the basis of drilling and subsequent assays of drill core. Because there is always some uncertainty in this process, they are further classified as *proven*, *probable*, and *possible*, depending on the level of confidence given to the estimates of a deposit's quality and quantity. Changes in technology, politics and economics can also affect these estimates and may move a deposit from one class to another or may render it uneconomic. Because of this, the world's reserves of a given mineral commodity may rise and fall significantly without actually changing the amount of the mineral known to be in the ground.

A resource, on the other hand, is a *potential* source of a mineral commodity either from a known but currently subeconomic deposit, or from a hypothetical one. Subeconomic resources may become reserves as economic conditions improve. Hypothetical resources are as yet undiscovered deposits, either in areas where such deposits are known to occur, or in areas where geologic criteria are indicative of their existence. A mineral resource is, therefore, a conjectural supply of a commodity. Consequently, whereas a reserve could be mined today, there is no guarantee that a resource will ever be mined.

The global distribution of metal reserves (Fig. 17.16) highlights their uneven allocation among the regions of the world. Whereas iron ore, for example, is distributed broadly in proportion to land area (with the notable exception of Africa), it is clear that the Americas have a disproportionate share of copper and molybdenum, that Asia holds most of the world's tin, that Africa dominates the chromium and manganese markets, and that tiny Cuba has a share of nickel that bears absolutely no relation to its size. This unequal distribution of metal reserves shows clearly that the processes by which metals are naturally concentrated are governed by factors other than geography. To understand the distribution, we must therefore turn to these processes and the specific geologic settings in which they occur.

17.5.2 Processes of Mineral Concentration: Chemical, Physical, and Biological

The process of mineral concentration can be illustrated with a simple example. Suppose you were to add a pinch of finely ground pepper to a cup full of salt and then stirred the two together until they were thoroughly mixed. Now suppose that the pepper represents a valuable mineral resource but only when it is concentrated. Clearly, the mixture would make a hopeless substitute for pure pepper! How would you go about extracting or separating the pepper from the salt so as to concentrate the pepper?

Several possibilities come to mind. You might pour the mixture into a glass of water and stir the water until all the salt had dissolved. You could then pour the salt water off, leaving the pepper behind. This is similar to the natural process by which aluminum becomes concentrated in bauxite by the selective removal of other elements during weathering. Furthermore, by leaving the pepper behind, the salt is concentrated in the water. So if the salt were the valuable commodity rather than the pepper, it could be recovered, like a natural evaporite deposit, by evaporating the salt water until the water was gone and the salt was left behind. This approach utilizes a difference in the *chemical properties* of salt and pepper, the fact that salt dissolves whereas pepper does not. This is the most obvious and one of the most effective methods of separation, but it is not the only one.

Finely ground pepper grains are also smaller than salt grains. With a sieve of the right mesh (one with holes in it larger than the pepper grounds but smaller than the grains of salt), you could shake the pepper through while the salt remained behind. This approach utilizes a difference in the *physical properties* of the salt and pepper, their grain size. However, it is unlikely to achieve as clean a separation as the first method because there is bound to be some variation in the size of the salt and pepper grains. Hence, the finest salt grains are likely to fall through the sieve while some of the coarser pepper grounds will not. Sedimentary processes take advantage of such physical differences to sort sediments according to their grain size (see Chapter 10).

Salt is also denser than pepper. So, if you were to slowly pour the mixture out of its container, blowing across the stream of grains as you did so, the lighter pepper grains would be blown further than the salt grains and could be collected separately. In a similar way, windblown sand in the desert tends to collect in sand dunes, leaving behind boulder fields known as "desert pavement." Alternatively, the salt and pepper mixture

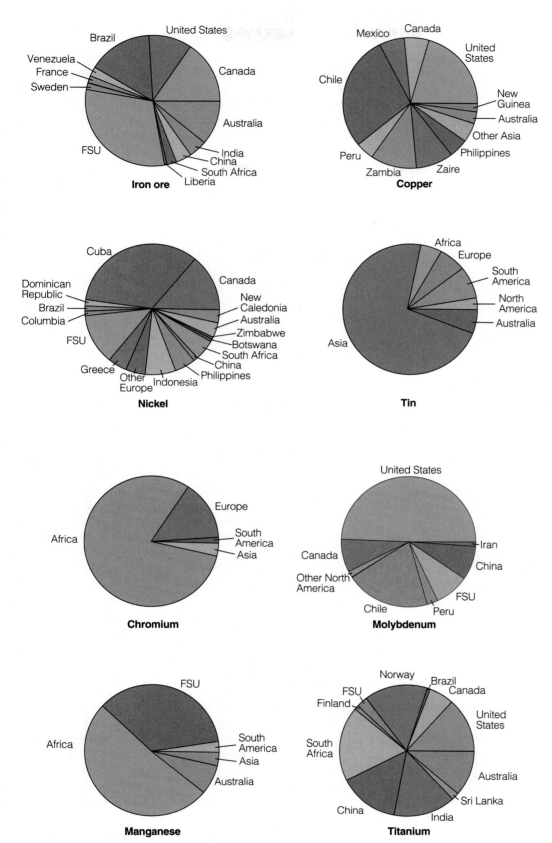

Figure 17.16
World distribution of metal reserves. FSU is the former Soviet Union.

could be spread out on a sheet of paper and gently shaken until the lighter pepper grains separate from the heavier salt. These approaches are similar to the ancient method of separating grain from chaff by winnowing and, like naturally occurring placer deposits, utilize another difference in the physical properties of salt and pepper, their density. However, the mechanical processes involved are fairly crude and are unlikely to produce a particularly good separation.

Another approach that utilizes the difference in the density of salt and pepper would be to pour the mixture into a tall glass of cooking oil. The heavier salt would immediately sink to the bottom of the glass but the lighter pepper would settle much more slowly. As soon as all the salt had settled, the oil could be poured off into a second container, in which the pepper could be allowed to settle out separately. The oil could then be poured off, leaving the pepper behind. Done carefully, this method could produce an excellent separation. However, it would leave you with the difficult "environmental" problem of having to separate the pepper from the oil.

Although simplistic, the processes by which minerals are naturally concentrated to form mineral deposits are similar to these examples, as indeed are the industrial methods we use to extract these minerals from their ores. Like our methods for separating salt and pepper, there are many different processes by which minerals can be concentrated. Different processes may apply to different minerals or to the same mineral under different conditions. But most, like the salt and pepper, are either **chemical** or **physical,** that is, they depend on chemical and physical differences between the ore mineral and its environment.

There is, however, a third method by which minerals can be concentrated that is not illustrated by any of those we used to separate the pepper from the salt. Minerals can also be concentrated by **biological** processes. To illustrate the biological concentration of an element, we can turn to the simple example of sea shells. Many marine invertebrates have shells made of calcium carbonate ($CaCO_3$), the raw materials for which are drawn from sea water. Yet the concentration of calcium in sea water is just over 400 parts per million (0.04%), whereas it comprises 40% of the calcium carbonate shell. In order to produce the shell, the marine organism must therefore concentrate calcium biologically by a factor of about 10,000.

The role of organisms in the formation of ore deposits is not nearly as obvious. Yet, in some cases, they are known to play an important part. Plant and animal activity is important in the formation of soils, for example, and certain tropical soils weather to form bauxite, the principal source of aluminum. Similarly, sulfide-fixing bacteria are common in hot mineral-charged waters like those that spout from the submarine vents of mid-oceanic ridges (see Chapter 8, Living on Earth: The Bizarre World of Mid-Oceanic Ridges) and may play an important role in the formation of some metal sulfides. The formation of sulfur from the alteration of the evaporite mineral anhydrite is likewise mediated by sulphate-reducing bacteria. Other bacteria synthesize iron oxide (Fe_3O_4) and may play an important role in the formation of certain sedimentary iron ores. Indeed, as we learned in Chapter 13, oxidation of the world's oceans by photosynthetic microorganisms in the Archean and early Proterozoic is thought to have played a key role in the development of Banded Iron Formation, the most important source of iron ore.

Like the salt and pepper, however, the principal mechanisms of ore formation are chemical and physical. Which mechanism acts to concentrate a particular mineral depends upon the ore in question and the geologic setting of the ore deposit. Ores that are produced by the segregation of molten rock, for example, are peculiar to igneous rocks, whereas those resulting from the circulation of hot, water-rich fluids occur in both igneous and metamorphic settings. Yet other ores, however, form as a result of deposition, evaporation, and weathering processes, and are confined to sedimentary environments.

17.6 Mineral Deposits in Igneous Rocks

The principal mechanisms of mineral concentration in igneous settings are those that result from mineral separation in a cooling magma, and those produced by the cooling of magmatically heated waters.

17.6.1 Magmatic Deposits

A body of magma cools through a wide temperature range from the time of its formation to the time it is fully crystalline (see Chapter 2). Like sugar in cooling tea, chemical components dissolved in the magma eventually reach saturation as they cool. As a result, they crystallize to form minerals. In general, minerals crystallize in a specific order. Those with the highest melting points crystallize first, followed by minerals with successively lower melting temperatures. Early-forming minerals are often silicates rich in magnesium and iron that are denser than the magma in which they crystallize and so settle to the bottom of the magma chamber. Silicate minerals richer in lightweight sodium and aluminum, on the other hand, are less dense than the magma and so float to the roof of the chamber. In either case, the remaining magma becomes depleted in the elements removed by the minerals and enriched in those left behind. The process is consequently one of chemical differentiation whereby the magma physically segregates into components of different composition as it cools and crystallizes. A similar form of segregation is involved in the separation of cream from milk, or milk solids from clarified butter.

Depending on their chemical affinity, non-silicate ore minerals may join the separating crystals or be left behind to become enriched in the remaining magma. In either case, however, they become concentrated. Economic minerals whose ore deposits are the product of **magmatic segregation** include chromium (see Figure 17.6), titanium, iron, platinum, and nickel.

As the crystallization of a magma proceeds, elements that do not have an affinity for the common rock-forming minerals become increasingly concentrated in the remaining magma. When the residual liquids finally crystallize, they form very coarse-grained bodies known as **pegmatites** (Fig. 17.17),

Figure 17.17
Pegmatite from northern Nova Scotia, Canada, showing large quartz (white) and feldspar (orange) crystals with radiating needles of the semiprecious mineral tourmaline (black). (Photo Brendan Murphy)

which can be extremely rich in these so-called *incompatible* elements, including water. The presence of water aids in the growth of large crystals because it reduces the viscosity or stickiness of the magma, making it more fluid so that the chemical components can move rapidly. Among the minerals sometimes present in pegmatites, are those that form the ores of beryllium, lithium, the rare earth elements, and gemstones such as emerald, all of which contain incompatible elements that have become concentrated in the last liquids to crystallize.

17.6.2 Hydrothermal Deposits

One of the most important methods of concentrating metals involves the circulation of hot magmatic water and waters heated by magmatic processes. Areas of mineralization produced by the action of heated waters are termed **hydrothermal** deposits from the Greek for "hot water." Just as sugar dissolves more readily in hot tea than in iced tea, hot water circulating at high pressure through fractured rocks has a greatly enhanced ability to dissolve minerals out of the material through which it passes. Once in solution, the water may carry these dissolved minerals considerable distances before falling temperatures and pressures, or changing chemical conditions drastically reduce their

solubility and cause them to be precipitated. For example, hot mineral-charged waters, or "brines," are particularly effective at carrying metals, which go into solution as complex chlorides and fluorides. Should the brine encounter gases such as oxygen or magmatic hydrogen sulfide (one oxidizing, the other reducing), the metals will react with the gases to form insoluble metal oxides and sulfides. Because they can no longer be held in solution, these metal-rich oxides and sulfides precipitate out onto the walls of the fractures to form mineral **veins** (Fig. 17.18), or permeate into the host rock and precipitate within them in a more dispersed fashion. Many important ore deposits are the product of hydrothermal activity, including those of tin, copper, lead, zinc, tungsten, silver, and gold.

We have already witnessed the effectiveness of hydrothermal activity in concentrating minerals in our examination of submarine vents at spreading centers (see Chapter 8, Living on Earth: The Bizarre World of Mid-Oceanic Ridges). Here, sea water seeps into the highly fractured rocks of the mid-ocean ridges and is progressively heated as it approaches the magma below. At temperatures hot enough to melt lead, the brine leaches metals from the surrounding basalt and carries them in solution back to the ocean floor. On meeting the ice-cold, oxygenated sea water, the dissolved metals quickly precipitate as a sooty cloud of oxide and sulfide particles. Belching from chimneys called **smokers** (see Figure 8.19d), the settling particles blanket the sea floor with deposits rich in iron, manganese, zinc, and copper. Although presently impractical to mine, these deposits may one day be exploited. Ophiolite complexes, which are slices of ocean floor preserved by tectonic upheaval, sometimes preserve ancient examples of these deposits. Mined for over 3000 years, the rich copper deposits of Cyprus in the eastern Mediterranean are believed to have been formed in this way.

Figure 17.18
Multiple veins of mineralized quartz in granite, Sierra Nevada, California.

17.8 Mineral Deposits in Sedimentary Rocks

The principal mechanisms of mineral concentration in sedimentary settings are those that result from relatively cold processes such as deposition, evaporation, solution, and weathering. These processes may work independently or in concert to produce rich deposits. The processes of erosion and weathering, for example, combine to reduce exposed rock to rubble or sedimentary detritus. The detritus so produced is then usually picked up and transported by the various agents of erosion, such as wind, water, and ice, and may ultimately reach the sea. In the process the detritus tends to become sorted according to its grain size. This is especially true if the transport mechanism is one of flowing water (see Chapter 10). As a result, well-sorted deposits of gravel, sand, and clay are produced in environments such as streams, beaches, and lakes, and may be of sufficient size to exploit as construction materials. Indeed, many industrial minerals owe their origin to the sorting process associated with sediment transport.

In addition, the transport of sedimentary detritus in flowing water also provides the opportunity for minerals to dissolve, so that they may later be concentrated by evaporation. The effects of evaporation are well known to those familiar with arid environments. In desert climates like those of the southwestern United States and the Middle East, temporary lakes fed by flash floods or winter snows slowly grow saltier as they evaporate, and finally form salt flats as they dry up altogether. Just such a process produced the Bonneville salt flats of northern Utah as a once vast glacial lake (Lake Bonneville) slowly shrank when the climate became arid following the last Ice Age (see Chapter 2, Living on Earth: Salt of the Earth). In much the same way, the constant evaporation of boiling water causes a scale of mineral salts to build up inside a kettle.

When an entire sea dries up, as the Mediterranean is believed to have done some 8 million years ago (see Chapter 14), vast deposits of mineral salts are left behind on the seabed. Similarly, thick deposits can form on the floor of any marine basin if it is subjected to very high levels of evaporation and is only weakly connected to an ocean. Protected by burial beneath younger marine sediments, these deposits of mineral salts become incorporated into the rock record as **evaporites.**

Evaporite deposits are important sources of a wide variety of nonmetallic minerals. The most important of these, common rock salt (halite), gypsum, and potash, are produced by the evaporation of sea water. Others, such as borax, used as a cleaning agent, are produced by the evaporation of alkaline lakes. Because of the low density of salt and the ease with which it flows under pressure, beds of rock salt may become mobilized upon burial (Fig. 17.19), rising up through the overlying strata in columns known as **salt domes.** These structures are particularly common in the Gulf Coast region of the southern United States and Mexico where they play a major role in the accumulation of oil and natural gas, as well as being important sources of common salt and sulfur (see Figures 18.5 and 18.7).

Figure 17.19
Underground at the Weeks Island salt mine in southern Louisiana. Contorted anhydrite-bearing layers (dark) reflect the upward movement of salt within the Weeks Island salt dome. (Photo Damian Nance)

In addition to providing raw material for deposition or evaporation elsewhere, weathering may produce mineral concentrations by leaving material behind. Such **residual** deposits of less easily weathered material can concentrate elements like iron, manganese, and nickel, and minerals such as bauxite and clay. As we have learned, bauxite (see Figure 17.3), produced by the chemical weathering of aluminum-rich rocks in tropical climates, is the principal source of aluminum. Chemical weathering of nickel-bearing rocks under similar climatic conditions is also responsible for the huge reserves of nickel in Cuba, the largest in the western hemisphere.

For minerals that do not dissolve, oxidize or react with water, the transport of sedimentary detritus in streams provides another opportunity for concentration, that of **mechanical enrichment.** Stable metals and ore minerals such as gold and cassiterite (the principal ore of tin) are denser than most sedimentary detritus, and so are not as easily moved by flowing water. As a result, where streams drain areas in which the bedrock is mineralized, eroded grains of metal and ore tend to be preferentially deposited wherever the velocity of the stream slackens. Concentrations of economic minerals formed in this way are known as **placer deposits,** and can be important sources of platinum, titanium, chromium, rare earths, diamonds, and other gemstones.

They are most important, however, as sources of gold and tin. In fact, placer deposits are the principal source of tin, and are also responsible for the huge Witwatersrand gold field in South Africa. It was also placer gold that led to the California Gold Rush when it was discovered at the base of the Sierra Nevada Mountains in 1849. Interestingly, the search for gold by panning uses exactly the same principle as that which produced the placer deposit, the swirling action of the water in the pan effectively separating the heavier gold from the unwanted sedimentary detritus.

Placer deposits are not restricted to streams, but are also found in the marine realm where they owe their origin to wave action and ocean currents. Gold-bearing placers, for example, occur off the coast of Alaska, chromite occurs off the coast of Oregon, and extensive titanium-mineral sands occur off the coasts of Florida, Georgia, and South Carolina. In each case, the ore minerals were derived by erosion of rock, transport by streams, and deposition in the sea.

Important marine sedimentary mineral deposits also occur where hot mineral-charged hydrothermal fluids are discharged directly into the sea. Chemical precipitation of the dissolved minerals brought about by oxidation and the sudden drop in temperature produce mineral-rich sedimentary strata known as **exhalites**. Significant deposits of iron, manganese, copper, and zinc have been formed in this way around submarine vents at spreading centers (see Chapter 8, Living on Earth: The Bizarre World of Mid-Oceanic Ridges). Presently uneconomic because of the cost of their recovery, these deposits on the deep ocean floor are likely to become one of the major environmental issues of the twenty first century as improving technology and rising metal prices make their exploitation both practical and financially attractive. Ancient exhalites preserved in exhumed oceanic sedimentary rocks are mined for their ore in many parts of the world.

17.8 Mineral Deposits in Metamorphic Rocks

Hydrothermal fluids are also responsible for mineralization in metamorphic environments. Under certain conditions, hydrothermal fluids emanating from a magma may chemically react with the rocks with which they come in contact. As a result, the fluids remove certain elements from the rock and replace them with others. When hot fluids, rich in silica (SiO_2), come into contact with limestones, for example, they react with the mineral calcite ($CaCO_3$, or calcium carbonate) by removing carbon dioxide and adding silicon, thereby replacing the calcium carbonate with calcium silicate ($CaSiO_3$). In the same way, the process of **replacement** may remove a rock-forming mineral and substitute an ore mineral in its place. The chemically

altered metamorphic rocks so produced are called **skarns**, and their formation is not uncommon in areas adjacent to granites that have been emplaced into limestones. Ore minerals concentrated by replacement include those of iron, copper, zinc, lead, tin, tungsten, molybdenum, graphite, gold, silver, manganese, and corundum.

17.9 Plate Tectonics and Mineral Deposits

The distribution of ore deposits is often related to the plate tectonic history of the areas in which they occur. This is because the various processes of mineralization usually occur in very specific geologic settings, and these settings are commonly associated with magmatic activity that tends to be focused at plate boundaries. Indeed, plate tectonics can be used to account for the occurrence of many types of mineral deposits, particularly those of metals. However, knowing that a particular type of mineral deposit is characteristic of a given plate tectonic setting does not guarantee the discovery of ore, because there is an enormous difference in scale between plate tectonic regimes which occupy wide regions, and mineral deposits, which for the most part, are very localized. Nevertheless, an understanding of the relationship between plate tectonics and mineralization can be used to guide exploration strategy, and so provides a powerful tool for mineral prospecting. Because many of the processes of mineral concentration are either directly or indirectly linked to magmatism, mineral deposits are most frequently associated with the processes of rifting and spreading at divergent plate boundaries and with the process of subduction at convergent plate boundaries, because it is in these plate tectonic settings that magmatism is most common.

17.9.1 Mineral Deposits at Divergent Plate Boundaries

A wide variety of mineral deposits occur at divergent plate boundaries where they are associated with the magmatism of sea-floor spreading. The most important of these are the product of the hydrothermal activity resulting from the circulation of sea water through the mid-ocean ridges. As we have learned from our examination of submarine vents (see Chapter 8, Living on Earth: The Bizarre World of Mid-Ocean Ridges) and hydrothermal deposits (see Section 17.6.2), the superheated brines produced as sea water infiltrates the hot, fractured seabed at spreading centers are highly acidic and dissolve a variety of metals out of the basalt crust with which they come in contact. Upon their return to the cold, oxidizing environment of the ocean floor, the metals precipitate as sulfides and oxides. While the exploitation of these minerals on the seabed is presently uneconomic, fragments of ancient ocean floor that have subsequently been emplaced onto the margins of continents during plate convergence are important resources. To date, deep-sea mining at an active spreading center has been contemplated only in the narrow Red Sea where metal-rich sediments along the axis of the ridge lie close to the shores of Saudi Arabia and the Sudan.

Ancient fragments of ocean floor are called ophiolites (see Figure 5.16), one of the best known of which is the Troodos Massif of Cyprus in the eastern Mediterranean. Formed by sea-floor spreading during the Cretaceous Period, about 120 million years ago, the Troodos ophiolite represents a fragment of the ancient Tethys Sea that once separated the southern continents of Gondwana from the northern continents of Laurasia (see Chapter 14). Emplaced over the African margin as the continent converged with southern Europe, the ophiolite has been mined since antiquity for massive sulfides of iron, copper, zinc, and cobalt formed by hydrothermal activity at an ancient mid-ocean ridge. In fact, the mines were so important in Roman and pre-Roman times that the very name of the island is derived from the Latin *cuprum* and the Greek *kupros*, meaning copper.

In addition to its hydrothermal ores, the Troodos ophiolite has been mined for its deposits of nickel sulfide and chromite. Unlike the hydrothermal ores, however, these deposits were produced at depth by the magmatic segregation of nickel and chromium in igneous bodies that cooled and crystallized beneath the ridge crest. Similar ore deposits are produced by magmatic concentration in continental rift settings where huge igneous bodies of basaltic composition cool and segregate at depth within the crust to form what are known as *layered igneous complexes.*

The best known of these layered igneous bodies is the vast Bushveld complex in South Africa, which contains the world's largest reserve of chromium, platinum group elements and vanadium-bearing magnetite, and is also mined for nickel. Here the ore minerals, which formed early in the crystallization process, have been concentrated toward the base of the intrusion because their high density caused them to sink. Although they often occur in layers that are only a few inches thick (see Figure 17.6), the layers extend over many thousands of square miles and so contain huge reserves of ore.

The origin of the 2.05 billion year old Bushveld complex is not fully understood, although some form of crustal rifting is certainly involved. Because the complex occupies an elliptical area and was intruded into a relatively isolated structural depression, it has been argued by some that it is the product of hot spot activity and by others that it is the result of meteorite impact. An impact origin for rifting has also been proposed for another important layered igneous complex, the elliptical Sudbury structure in Ontario, Canada. This structure, which formed 1.85 billion years ago, is a major source of nickel and copper sulfides, and its origin, like that of the Bushveld complex, is unresolved.

Continental rift environments are also associated with deposits of copper, zinc, and cobalt, like those of the Zambian copper belt, the Kupferschiefer in central Europe, and the copper country of Upper Michigan's Keweenaw Peninsula. In this case, the weathering of basalt lavas produced by the process of rifting released the metal content of the lavas, which was then transported and fixed in sedimentary rocks. Because these deposits occur in a specific part of the stratigraphy, they are known as *stratiform deposits.*

Rifting in stable cratonic regions is also associated with carbonate-hosted Mississippi Valley-type lead-zinc deposits. Advanced stages of rifting are associated with siltstone- or shale-hosted massive sulfide deposits rich in zinc, lead, and copper, such as those at Sullivan in British Columbia and Mount Isa in Australia, and sandstone-hosted deposits of uranium, like those of Saskatchewan in Canada.

17.9.2 Mineral Deposits at Convergent Plate Boundaries

The most important mineral deposits associated with convergent plate boundaries are those related to the subduction of oceanic crust at continental margins. As we have learned, subduction processes in this setting produce magmatic arcs in which linear belts of granitic magmas are emplaced into the continental crust of the overriding plate. Mineral concentrations resulting from hydrothermal activity are widespread in these granitic bodies and the rocks around them, and include porphyry copper deposits and a wide variety of vein deposits and skarns.

Because of the scale on which these processes occur, rocks with ore-grade values of copper may be widely distributed. This makes them attractive high-tonnage mining targets, the most important of which are the porphyry deposits. Formed by hydrothermal activity within fractured granitic bodies, these deposits often contain iron, molybdenum, gold, and silver in addition to copper. The global distribution of these deposits, most of which are between 15 and 150 million years old, clearly shows their relationship to modern or geologically quite recent subduction zones (Fig. 17.20). Most, such as those of South America and the Philippines, occur adjacent to active subduction zones. Others, such as those of western North America, mark the sites of recent subduction.

Porphyry copper deposits are low grade and usually only contain between 0.2% and 2% copper. But they are often large and so may contain as much as several million tons of copper in a single deposit. Because the extraction process is fairly simple, the deposits can be mined profitably despite their low grade. In many cases, however, the economic viability of such deposits is dependent upon a secondary process of mineral enrichment that is the result of weathering by percolating groundwater. By this process, which is known as *supergene enrichment*, copper is leached from near-surface parts of the deposit and precipitated at depth. As a result, lower levels of a deposit that might otherwise have been uneconomic can be sufficiently enriched in copper to form a low-grade orebody. However, huge volumes of ore must be processed and the open-pit mines created, such as Chuquicamata in Chile and Bingham Canyon in Utah (see Figure 17.14), are among the largest human-made excavations on the Earth's surface and were the only ones seen by the Apollo astronauts.

Skarn deposits formed by the action of magmatic fluids on limestones adjacent to the plutonic bodies are commonly rich in

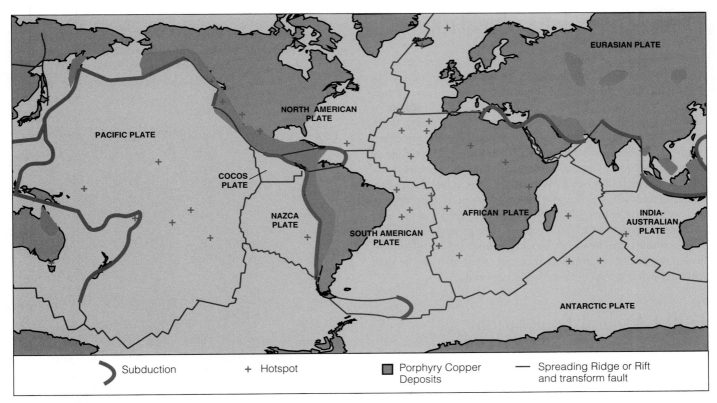

Figure 17.20
Porphyry copper deposits and their relationship to convergent plate boundaries.

iron, tungsten, tin, and copper. Hydrothermal fluids generated in response to arc magmatism also produce complex vein deposits that often contain copper, gold, and silver. Closer to the surface, hydrothermal fluids may be responsible for hot spring deposits rich in gold, mercury, antimony, and arsenic. Hydrothermal replacement and vein deposits on the inner (continentward) side of the arc often contain zinc, lead, and silver, and the plutonic rocks here may be rich in tin and tungsten.

Mineralization in convergent margin settings may also occur in the region behind the arc, where it is associated with crustal extension and the development of back-arc basins (see Chapter 5). Continental rift zones in areas of plate convergence such as the Basin and Range Province and Rio Grande rift in the United States, are associated with voluminous rhyolite volcanism and the formation of calderas (see Figure 5.42). Hydrothermal circulation fueled by this volcanism is responsible for the emplacement of large molybdenum deposits and smaller deposits of tungsten, tin, and uranium. The large molybdenum deposits associated with the Rio Grande rift, such as the Climax deposit of Colorado, are responsible for the disproportionately large reserves of this important steel-making element in the United States (Fig. 17.16).

Continental rifting behind the arc may lead to the opening of a back-arc basin floored by oceanic crust like the modern Japan Sea. Once established, these oceanic rift zones are characterized by hydrothermal mechanisms of mineral concentration at the spreading center that are similar to those at mid-ocean ridges. The most important of the resulting mineral deposits are the so-called *Kuroko-type* massive sulfides, named for those of northern Japan, which were formed by hydrothermal activity on the floor of the Japan Sea some 12 to 15 million years ago. Kuroko-type massive sulfide deposits are smaller but much higher grade than porphyry copper deposits and are rich in zinc, lead, and copper. In Canada, an ancient analogue for this environment is thought to account for the Paleozoic ore body at Bathurst in New Brunswick. In Late Archean greenstone belts, massive sulfide deposits (like those at Norando in Quebec, and Kidd Creek in Ontario) and a number of important gold deposits (like that at Homestake in South Dakota) are also thought to have formed in back-arc settings, lending support to the back-arc model for the formation of greenstone belts (see Chapter 13).

Within the arc, the magmas associated with subduction zones (see Chapter 5) are generated either by melting of the downgoing slab and the overlying mantle wedge and/or at continental subduction zones, by melting at the base of the continental crust. Although rising magmas from each of these sources power the hydrothermal circulation responsible for much of the mineralization at convergent plate boundaries, those coming from the subducting slab may supply many of the metals themselves. This is because the subducted ocean floor originally formed at a spreading center, and so may already be enriched in metals as a result of the hydrothermal processes occurring at the mid-ocean ridges.

17.10 Mineral Exploration

Mineral exploration has always involved the search for mineralization and environments that might be favorable for mineralization. In the days of the individual prospector with an eye for gold or ore, it was a skill learned from long experience on strenuous surveys. The prospector worked empirically, recognizing favorable patterns without necessarily understanding the processes involved. Modern exploration benefits from our knowledge of the mechanisms of mineral enrichment and the plate tectonic settings in which they occur. We can therefore select the geologic setting most likely to host the type of ore deposit being sought, and within this setting, identify the specific geologic environment most likely to be mineralized. Modern prospecting marries old discoveries and the wealth of accrued knowledge with sophisticated methods of exploration, including some that can be used without having to visit the area being explored.

17.10.1 Mineral Exploration on the Ground

The basic tool of exploration remains the geologic map (see Chapters 3 and 6). This is usually a simple topographic map that also shows the distribution of rock types that make up the bedrock immediately below the surface. The production of such a map is a very labor-intensive exercise, requiring an extensive field survey of the rocks exposed at the surface coupled with a review of any other information, such as that from drill hole data and soil surveys, that might reveal the nature of the bedrock in unexposed areas (see Chapter 13, Living on Earth: Plate Tectonics in the Early Proterozoic: A Field Trip to Northern Quebec). Once completed, the map enables the exploration geologist to identify areas where specific rock types occur, and locate features such as faults and fractures that may have provided pathways for mineralizing fluids. However, while this approach allows potential target areas to be selected, it does not guarantee the existence of an ore body and, even when present, the ore deposits themselves may not be exposed at the surface or may underlie areas that lack any surface exposure. To further narrow the target, therefore, more sophisticated geophysical and geochemical surveys are conducted.

Geophysical methods depend upon contrasts between the physical properties of ore bodies and those of their host rocks. Many ore bodies, for example, are made up of minerals of high density and so minutely increase the pull of the Earth's gravity on the land surface above them. Although not detectable to the human senses, minute changes in the local gravity field can be measured by sensitive instruments known as **gravimeters**. Because dense rocks in the subsurface cause gravity readings above them to rise above average, their presence is revealed by variations in the local gravity field known as *positive gravity anomalies* (see Chapter 5, Dig In: Failed Rifts). In a similar fashion, buried ore bodies that contain weakly magnetic minerals produce positive magnetic anomalies in the local magnetic field (Fig. 17.21), and these can be detected by sensitive instruments known as **magnetometers**.

Yet other deposits contain metallic minerals with electrical properties (recall our discussion of metallic bonds in Chapter 2). Many massive sulfide deposits, for example, have been detected by using their conductive properties. When stimulated by an electrical current, such deposits generate

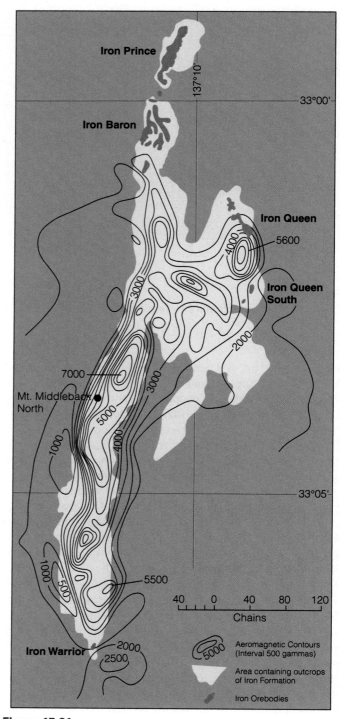

Figure 17.21

Contours of magnetic intensity over the Northern Middleback Range of South Australia, a major source of iron ore, show a striking correspondence with the area in which iron-bearing formations outcrop.

weak magnetic fields that can be detected by a variety of ground and airborne instrumentation.

The search for uranium deposits makes use of another physical property of minerals, that of radioactivity. Indeed, radioactivity can be used to detect concentrations of other minerals providing they contain, or occur in association with minerals that contain, uranium, thorium, and potassium, the three most common naturally occurring radioactive elements. Because these three elements occur to varying degrees in most rocks, surveys of natural radioactivity can also be used to distinguish different rock types and so can be effective tools for geologic mapping purposes.

Geochemical exploration techniques detect the presence of an ore body more directly, by using some aspect of its composition as a chemical fingerprint. Although the ore deposit itself may be buried, low levels of mineralization that occur around an ore deposit may be dispersed over wide areas. Chemical analyses of rock samples, soils, or streams that run through the region may therefore reveal telltale traces of the ore mineral being sought in the vicinity of the ore body. Trace amounts of the mineralization may even be taken up by plants, such that the deposit can be revealed by the chemical analysis of nearby plant materials. More commonly, the leaching of buried ore deposits by percolating groundwaters affects the local surface water chemistry, and this can be detected by chemically analyzing water samples from streams or lakes.

Although geophysical and geochemical exploration can focus the search for ore deposits, neither technique can actually *discover* an ore body. The only method available for this purpose is to drill a hole with a diamond drill bit. Of all exploration tools, therefore, the diamond drill is the most important because it is the drill hole and the samples recovered from drill core that provide the ultimate test in any exploration program. Only a drill hole will reveal what is actually present beneath the ground and, if an ore deposit is discovered, only a series of carefully placed drill holes will allow the ore body to be delineated.

17.10.2 Mineral Exploration from a Distance

The use of aerial photographs has long been an integral part of geologic mapping and mineral exploration. Indeed, where topographic base maps are unavailable or of poor quality, aerial photographs are an essential element of any exploration. They not only provide a means by which surface features can be located, but can also be combined in such a way as to allow the features to be perceived in three dimensions. With the advent of satellite technology, however, methods of mineral exploration have become available that do not require the area being examined to be visited at all. This form of "armchair" exploration or investigation from a distance is known as **remote sensing.** The dramatic discoveries we have made in our exploration of the Solar System through satellite images of the planets and their moons, is a familiar example of knowledge gained through remote sensing. Similarly, the spectacular photographs of the Earth's surface taken from the Space Shuttle are examples of the same technique applied to our own planet.

The application of remote sensing to exploration geology makes use of a variety of images of the Earth's surface taken by satellites (Fig. 17.22). Since the first Earth-resource satellite Landsat was launched in 1972, satellites have provided continuous surveillance of the Earth's surface at ever greater resolution. The images the satellites produce, however, are not the same as photographs. Cameras take photographs using visible light, that is, they are sensitive to all wavelengths of the visible spectrum. Satellites, on the other hand, use sensors known as *multispectral scanners* to make images from specific wavelengths, some of which may not be visible to the naked eye. The early Landsats, for example, utilized bands in the green, red, and invisible infrared portions of the spectrum. More recent observational satellites use a more advanced multispectral scanner called a *thematic mapper* that can scan many bands of the spectrum at much higher resolution.

As the satellite orbits at an altitude of about 700 km (435 mi), these wavelength bands are scanned along a strip of the Earth's surface 185 km (about 116 mi) wide directly below the scanner. The images received are then transmitted back to the Earth in digital format, where the data carried in the various bands are superimposed and the image is reprocessed in the form of a picture. The colors used for the various spectral bands that are scanned by the satellite are often arbitrarily chosen to highlight particular features. The color red, for example, is often used to denote the infrared light reflected from lush vegetation because it provides better contrast than green. Areas of diseased crops or regions of clear cut in a forest are consequently highlighted. Arbitrary colors must also be chosen for other portions of the spectrum which we cannot see. Satellite pictures, which superimpose each of the arbitrary colors assigned to the bands, are therefore referred to as *false color* images.

What the satellite actually detects, of course, is the reflection of sunlight off the Earth's surface. Objects bathed in sunlight receive a broad spectrum of solar radiation. However, as we have learned, the Earth's surface absorbs a portion of the solar spectrum so that only the other portions are reflected. The reflected portion of the visible spectrum, for example, gives an object its color. Similarly, the reflected portions of the invisible spectrum also vary between objects. As a result, an object often displays a characteristic pattern of reflected radiation that can be used to identify the material of which it is made. False color satellite images are well suited for this because they are readily manipulated and computer enhanced so that very specific differences can be highlighted. Features that might be of particular significance for mineral exploration, such as differences in vegetation, rock type or soil variety, can be emphasized.

Many ore deposits, for example, are associated with zones of alteration that have different reflectance characteristics from those of the adjacent host rock. In barren country lacking vegetation, these differences may actually show up on the satellite image as different colors. On the other hand, where dense vegetation is present, there may be subtle differences in the pattern or type of vegetation in the zone of alteration, which may also be highlighted on satellite images that show the appropriate combination of spectral bands. Similarly, changes in vegetation revealed in satellite images may distinguish different soil types and even trace element distributions in the soil that may be of

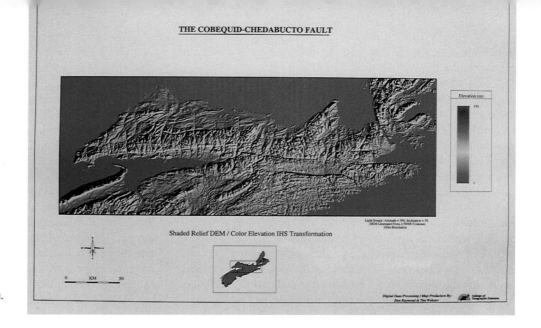

THE COBEQUID-CHEDABUCTO FAULT

Shaded Relief DEM / Color Elevation IHS Transformation

Figure 17.22
Satellite radar image of part of Nova Scotia, Canada, highlights linear features of the landscape or lineaments, many of which correspond to major faults of the Minas fault system. Compare with Figure 5.34.

relevance to mineral exploration. Satellite images can also be used to produce reconnaissance geologic maps, enabling the exploration geologists to target areas of potential economic interest within large regions that might otherwise be too remote or too inaccessible to explore, and focus their efforts and cut the costs of exploration in more accessible areas.

Remote sensing, however, is not restricted to multispectral scanning of reflected light, an obvious limitation of which is the requirement of clear weather. The spectacular images of the surface of Venus, for example, could not have been produced in this way because of the planet's thick cloud cover. Instead, these images were produced by radar. **Radar** is microwave radiation that can penetrate clouds. Radar images of the Earth's surface, such as those produced by SLAR (side-looking airborne radar) and SAR (synthetic aperture radar), are particularly sensitive to changes in the shape and roughness of the Earth's surface, and like aerial photographs, can be combined to give three-dimensional images (Fig. 17.22). Because the shape of surface features is often controlled by the structure and composition of the underlying bedrock, radar images are also useful for exploration purposes. By highlighting linear alignments of surface features known as **lineaments,** for example, they can reveal faults and fractures that may be the sites of mineralization, even when the primary features are not evident.

Remote sensing techniques have consequently provided the economic geologist with a whole range of new tools for mineral exploration. Unlike more conventional exploration methods on the ground, remote sensing can be applied worldwide, it is not limited by political frontiers, and it is often far less expensive. The resolution of satellite-based remote sensing techniques is also continually improving. The multispectral scanner on the first Landsat, for example, was unable to resolve objects much smaller than 200 m (660 ft) across. Today, the resolution of commercial satellite scanners is down to as little as 2 m (a hundred-fold improvement), and even higher resolutions are achieved by spy satellites. As with ground-based geophysical and geochemical exploration methods, however,

remote sensing techniques can only focus the search for ore deposits. They cannot determine whether a commercial ore body is present, nor can they delineate the body if one is found. For these purposes, the only methods of exploration that can provide answers remain those of basic geologic mapping followed by drilling.

17.11 What the Future Holds

As we have learned, mineral resources are nonrenewable and cannot be replaced. Yet the rate at which they are consumed is only likely to increase as global populations rise and less developed countries become increasingly industrial. Eventually, the demand for mineral resources must confront their limited supply. So what are the future patterns of supply and demand likely to be, and what can we do to meet our future needs?

17.11.1 Future Patterns of Supply and Demand

As any stock broker will tell you, attempting to determine the future availability of, and demand for, a particular mineral commodity is a very risky task. It is virtually impossible to predict new discoveries of a mineral or what effects a rapidly changing technology will have on its demand. One might have predicted, for example, that the "information age" with its dependence on electronic communications would have led to a spiraling increase in the demand for copper wiring. In reality, however, the opposite has occurred because the demand has been met technologically with the introduction of fiber-optic cables that contain no copper.

Because the global allocation of mineral resources is unevenly distributed, future demands for a given mineral commodity also vary dramatically from one nation to another.

Consider the metal chromium, for example, which is a vital component of stainless steel and automotive chrome plate. Globally, there are adequate reserves of chromium ore to meet world demand for the immediate future. However, this is of little comfort to the United States, which has virtually no chromium reserves. To support a vast manufacturing industry, the country is therefore wholly dependent on foreign sources for this strategically important metal, the largest reserves of which are in South Africa (see Figures 17.6 and 17.16). Future chromium supply to the United States is consequently of great concern despite the metal's plentiful reserves.

The bottom line, however, is that all minerals are nonrenewable resources because they are consumed far faster than nature can replenish them. Hence, the continued demand for mineral commodities must eventually result in shortages. Mineral exploration techniques are now widely believed to be sufficiently sophisticated and global in nature that most of the world's major high-grade mineral deposits on the continents have already been found. This means that future mining activity, even for relatively abundant mineral commodities such as the common metals, will eventually have to focus increasingly on the recovery of lower-grade ores on the continents or high-grade ores on the ocean floors, each at greater cost. Reserves of some of the scarce metals, on the other hand, are only sufficient for the next twenty-five years or so, at present rates of consumption. Future shortages in these metals are therefore likely, with lead, mercury, tin, and silver probably leading the way. These trends predict a steady rise in future metal prices as their ore reserves become increasingly scarce and expensive to extract.

Future trends for individual countries are likely to approximate models for the consumption of any limited resource. The history of metal consumption, for example, might follow the model illustrated in Figure 17.23. For any given country, the number of mines extracting a given metal initially increases as new deposits are found, and the amount of the metal produced increases in unison. As the deposits become depleted, however, the number of working mines declines, which leads to a decline in production. Continued demand must therefore be met by importing the metal from foreign sources at ever increasing levels.

Within this model, individual countries currently occupy a wide spectrum of positions. Some, such as Russia, have huge mineral reserves and expanding patterns of exploitation. Others, like the United States, have passed peak productivity and are becoming increasingly dependent on imported metals. Yet others, such as the United Kingdom and Japan, have almost exhausted their mineral resources and are now heavily dependent on foreign sources.

Just as there is inequity in the global distribution of mineral resources, so too is there inequity in their consumption. The developed industrial nations consume the lion's share, such that some 90% of the world's mineral production is consumed by only 5% of its population. These nations, however, have also produced the lion's share of mineral commodities. It was the rich mineral and energy resources of Europe, for example, that fueled the industrial revolution, and even today, it is the devel-

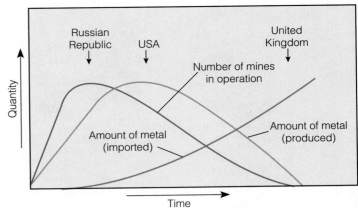

Figure 17.23
Model for the consumption of a limited resource.

oped nations of Russia, the United States, Canada, Australia, and South Africa that dominate world mineral production.

But this situation is rapidly changing. As the industrial nations exhaust their own mineral reserves, they are becoming increasingly dependent upon minerals imported from less-developed countries to supply the raw materials used in manufacturing. This is viewed by some as the exploitation of these countries by the developed nations. Others, however, claim that it is the capital gained from the exportation of their raw materials that is likely to fuel the future development of the less-developed nations.

17.11.2 Meeting Future Needs

In the future, increases in ore prices are likely to convert many presently subeconomic mineral resources into economic mineral reserves. However, other factors may also be important. For the exploitation of a mineral deposit to be profitable, the value of the product must outweigh the cost of its production. Production costs are those of extraction and processing, and future trends may see these rise or fall. Future technological developments resulting in increased recovery, for example, may reduce production costs by improving the techniques of extraction and processing. It is conceivable that old mine tailings could be profitably reworked in this fashion. As the richest ore deposits become exhausted, however, the mining of progressively lower-grade and less easily processed ores may cause these costs to increase. Environmental concerns and political situations may prevent the working of an otherwise profitable deposit. Given these uncertainties, it is both difficult and probably unwise to attempt to forecast what the future may hold for the mineral industry.

Mineral exploration, however, is certain to continue, and one area that is likely to receive considerable attention is that of the ocean floors. As we have learned, rich deposits of minerals associated with submarine vents at mid-ocean ridges exist on the seabed and some of these, like those of the Red Sea, are relatively accessible. However, these are not the only deposits known to occur on the seabed. On the Pacific Ocean floor, for example, huge deposits of manganese exist in the form of concretions known as **manganese nodules** (Fig. 17.24) Averaging

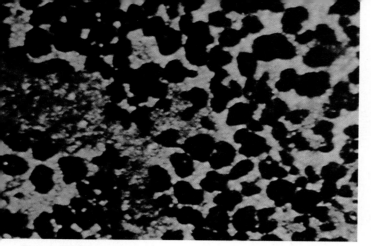

Figure 17.24
Manganese nodules measuring 5 to 10 cm (2 to 4 in) across on the deep ocean floor of the Pacific Ocean.

5 to 10 cm (2 to 4 in) across, these form as chemical precipitates in areas of slow deposition and contain iron, copper, nickel, and cobalt in addition to manganese.

Estimates of the size of this resource, at least for the metals manganese and copper, suggest it may be larger than those on land, and the technological difficulty of recovering the nodules by dredging or vacuuming the seabed is not insurmountable. However, the environmental impact of such an operation is entirely unknown and, because many such areas occur in international waters, there is presently no protocol for establishing national ownership and the right to mine, or for determining the responsibility for monitoring such mining activity. Some fundamental and politically difficult legal questions concerning mineral rights and policing responsibilities in international waters will therefore have to be addressed and agreed upon before any such mining operations can be undertaken.

There are also significant tracts of land that for political, geographic, or technological reasons have seen little in the way of mineral exploration and development. Chief among these are the polar regions. Technological developments have now made mining possible in the high Arctic. Canada, for example, has operating lead-zinc mines at Nanisivik on Baffin Island (73°N), and at Resolute on Cornwallis Island (75°N). However, the operating conditions are extreme, shipment by sea is limited to an ice-free period of only a few weeks at the height of the summer, and questions regarding the environmental impact of such mining in an ecologically sensitive area are still fiercely debated.

At the other pole, the Antarctic Treaty, which sets aside Antarctica for scientific research, has so far shielded the continent from mineral exploration and development. This protection, however, is not guaranteed indefinitely, and as future demands for mineral commodities increase, pressures to relax the conditions of the treaty are certain to increase.

This raises another issue in mineral exploitation that is likely to become an increasingly controversial topic in the future. As we have learned, mineral deposits occupy a minute fraction of the Earth's surface, and occur only where the vagaries of mineral-forming processes have placed them. Consequently, if society continues to depend on mineral resources for its well-being, future generations will be forced to carefully weigh the benefits that might be gained from mineral exploitation against those that might be realized from setting aside the same land for uses (such as parks) that would exclude mining activities.

In the long term, there is sure to be a debate about the necessity to extract minerals from ocean water, or even to expand the search for mineral resources beyond the Earth. The Moon, for example, is believed to be rich in iron and nickel, and the development of its mineral resources might be necessary if, in the future, a permanent lunar base is to be established and maintained.

In the immediate future, however, the continuing supply of mineral commodities is likely to be maintained by an increasing effort to conserve and recycle existing materials and, where possible, to develop cheaper and more abundant substitutes for scarce mineral resources. It is already worth our while to recycle aluminum because of the high energy costs associated with its production. As we have learned, for example, the United States recycles about 50% of its aluminum beverage cans. Further, it re-uses 15% to 20% of its old aluminum scrap. In the future, these figures are likely to increase with improvements in efficiency, and the recycling of other metals is likely to become as familiar to us as that of aluminum. We can also expect to see the scarcer metals increasingly replaced by other less costly substitutes in the same way that fiber-optic cables are today replacing the use of copper in electrical wiring.

17.12 Chapter Summary

Mineral resources are natural, inorganic materials taken from the Earth for the well-being of society. They include both metallic and nonmetallic minerals, and are nonrenewable because they are being used faster than nature can replenish them. Yet they are so much a part of our daily lives that we take them for granted.

Metallic mineral resources include both the abundant metals (iron, aluminum, magnesium, titanium, and manganese), which are quite common constituents of the Earth's crust, and the scarce metals, which comprise less than 0.1% of the crust by weight. Scarce metallic minerals include copper, zinc, lead, chromium, nickel, gold, silver, and the platinum group elements. Most metallic minerals are mined as ores, which are usually metal oxides or sulfides.

Nonmetallic or industrial minerals are generally quite common and are grouped according to their use. Most are used in the construction industry as building materials, such as stone, sand and gravel, cement, and gypsum (plaster). Others are used by industry as raw chemicals, such as potash and phosphorite (fertilizer), salt, and sulfur. Yet others, such as diamond, corundum, garnet, quartz, and emerald, possess a natural hardness and are used as abrasives and gemstones.

In order to form a mineral deposit, the local accumulation of these materials must be of sufficient quality and quantity to make their extraction profitable. Yet of the 92 naturally occurring elements, only nine comprise 99% of the Earth's crust and just six make up 95% of the dissolved ions in sea water. For the vast majority of elements, therefore, massive concentration factors are needed if a mineral resource is to be mined or quarried economically.

Mineral concentration by natural processes, however, occurs only under very exceptional conditions and in very specific geologic settings. Because of this, mineral deposits are very rare and they are far from evenly distributed worldwide. To be exploited, a mineral deposit must also be profitable. The economics of mineral deposits are therefore very volatile, being tied not only to the price of ore on the world market, but also to technological developments that might reduce the cost of extraction and processing, and to political conditions that can encourage or prevent a profitable deposit from being worked. For this reason, mineral reserves, which are known deposits that could be worked profitably today, are distinguished from mineral resources, which are potential mineral sources, either from presently subeconomic deposits or from predicted deposits that have yet to be found.

The processes by which minerals are concentrated vary with the ore in question and the geologic setting of the ore deposit. Most mechanisms, however, depend on differences between the physical and chemical properties of the ore mineral and those of its host. Density, grain size, chemical affinity, and solubility, for example, may all act to concentrate a given mineral. In some cases, enrichment is also mediated by biological processes, as is the case in the production of soils and the formation of sulfur by sulphate-reducing bacteria.

Mineral deposits in igneous rocks are usually the result of magmatic or hydrothermal processes. Magmatic deposits result from the gravity separation of early-formed minerals in a cooling magma because the minerals and the magma have different densities. This process is known as magmatic segregation and allows ore minerals to become enriched either as a result of their separation or because of their enhanced concentration in the magma left behind. Hydrothermal deposits result from the precipitation of minerals carried in solution by hot magmatic water or waters heated by magmatic processes. Mineralized fractures formed in this way are called veins. Hydrothermal activity is also responsible for mineralization at mid-ocean ridge submarine vents.

Mineral deposits in sedimentary rocks result from deposition, evaporation, and weathering. Mechanical enrichment during deposition reflects the natural tendency of flowing water to sort sediment according to its grain size and density. Concentrations of heavy minerals deposited in areas of relatively slow moving water are called placer deposits. Minerals precipitated by the evaporation of salt water are called evaporites and, in sufficient quantity, produce evaporite deposits. Weathering may leave behind concentrations of less easily weathered minerals known as residual deposits. Where precipitates from mineral-charged hydrothermal fluids become incorporated in sediment, they are termed exhalites.

Mineral deposits in metamorphic rocks are those produced by chemical reactions between circulating hydrothermal fluids and the rocks with which they come in contact. Mineral deposits are formed when the reactions remove one mineral and substitute an ore mineral in its place. The process is known as replacement and the chemically altered metamorphic rock so produced is a skarn.

Because mineralization is commonly associated with magmatic activity, the occurrence of mineral deposits is often related to the plate tectonic setting in which they occur. The relationship between plate tectonics and mineralization can be used as a reconnaissance exploration tool and is most apparent at convergent and divergent plate boundaries where magmatism is most common.

Mineralization at divergent plate boundaries is associated with the magmatism of rifting and sea-floor spreading. The circulation of sea water through mid-ocean ridges, witnessed by submarine vents called smokers, is responsible for widespread mineralization on the ocean floor. Mineral deposits may also result from magmatic segregation within the developing oceanic crust. In fact, the subduction of ocean floor previously enriched in metals is a likely contributor to the mineralization at convergent plate margins.

Most mineral deposits at convergent plate boundaries are associated with the emplacement of granitic bodies above the subduction zone. Mineral concentrations are the result of associated hydrothermal activity and replacement, and include porphyry copper deposits, vein deposits, and skarns. Where a magmatic arc is in extension, however, mineralization may also occur as the result of hydrothermal activity in continental rift zones developed behind the arc. And where rifting is successful, Kuroko-type massive sulfide deposits may form at the spreading ridges of back-arc basins.

Magmatic segregation of basaltic magmas is also characteristic of continental rift settings where it is responsible for the development of mineral-rich layered igneous bodies such as the Bushveld complex. Stratiform copper deposits, Mississippi Valley-type lead-zinc deposits, and siltstone- or shale-hosted massive sulfide deposits, are similarly typical of continental rift environments.

Once potential regions for mineralization have been identified on the basis of geologic mapping, the focused search for mineral deposits involves ground-based geophysical and geochemical surveys, and satellite-based remote sensing. Geophysical surveys seek to pinpoint ore bodies by highlighting contrasts in those physical properties, such as density, electrical properties, magnetism, and radioactivity, that distinguish the ore from its host rock. Geochemical surveys look for telltale anomalies in soil, groundwater, and surface water chemistry as indicators of an ore body's proximity.

Remote sensing techniques range from simple aerial photographs to sophisticated satellite imagery. Earth-resource satellites, such as Landsat, employ multispectral scanners and thematic mappers to image specific wavelengths of light reflected from the Earth's surface. The individual spectral bands can then be superimposed and computer enhanced in such a way that features of potential interest to mineral exploration, such as differences in vegetation, rock type, and soil variety, can be highlighted.

Other satellites use microwave radiation or radar to map the Earth's surface. Radar, which can penetrate cloud cover, is particularly sensitive to surface roughness, and can be used to highlight lineaments that may reveal hidden faults and mineralized fractures. As with ground-based surveys, however, remote sensing can only focus the search for ore deposits. Only drilling will allow the existence and size of an ore body to be tested.

Future patterns of supply and demand are almost impossible to predict for mineral commodities. Discoveries of new deposits, developments in technology and changes in political climate will all influence future trends. Demands will also vary dramatically from one nation to another, with the developed countries becoming increasingly dependent on the mineral reserves of less-developed nations.

Because minerals are nonrenewable resources, however, rising ore prices and shortages are likely to accompany future mining activity. As a result, many presently subeconomic deposits will become minable reserves, and mineral exploration will intensify in areas like the polar regions, which have so far received little attention. Exploration of the rich mineral deposits of the ocean floors is also likely.

In the long term, minerals may be extracted from the oceans themselves and the search for new resources may reach to the Moon. In the immediate future, however, the supply of mineral commodities is likely to be maintained by improved conservation and recycling, and by the development of cheaper and more abundant substitutes.

Key Terms and Concepts

 Look for the highlighted items on the web at:
WWW.BROOKSCOLE.COM

- Mineral resources are nonrenewable, natural accumulations of metallic and nonmetallic (industrial) minerals that are taken from the Earth for human use
- Mineral reserves are known mineral resources that could be profitably exploited in today's marketplace
- Ores are the materials from which minerals can be profitably extracted
- Metallic minerals form the ores of abundant metals (iron, aluminum, magnesium, titanium, and manganese), scarce metals (copper, zinc, lead, chromium, and nickel), and the precious metals (gold, silver, and the platinum group elements)
- Nonmetallic (industrial) minerals are those used as building materials (stone, sand and gravel, cement, and gypsum), raw chemicals (fertilizer, salt, and sulfur), and as abrasives and gemstones (diamond, corundum, garnet, quartz, and emerald)
- Mineral deposits are scarce and their global distribution is uneven
- Mineral concentration occurs by chemical, physical and biological methods

- Mineral deposits occur as magmatic and hydrothermal deposits in igneous rocks, as evaporites, residual deposits, placer deposits, and exhalites in sedimentary rocks, and as replacement deposits in metamorphic rocks
- At divergent plate boundaries mineral deposits are associated with sea floor spreading (massive sulfides and chromite) and continental rift environments (stratiform copper and Mississippi Valley-type lead-zinc deposits)
- Mineral deposits at convergent plate boundaries include porphyry copper deposits, vein deposits and skarns associated with arc magmatism, molybdenum deposits associated with extensional back-arc volcanism, and Kuroko-type massive sulfide deposits associated with back-arc spreading
- Mineral exploration involves geologic mapping, geophysical, and geochemical surveying, remote sensing and, finally, drilling
- Meeting future needs for increasingly limited mineral resources is likely to be met through improved extraction technology, further mineral exploration, the more widespread use of less expensive substitutes, and improved efforts at conservation and recycling

Review Questions

1. Using examples, distinguish between renewable and nonrenewable resources.
2. What is the most important abundant metal and how is it used?
3. What is bauxite and under what climatic conditions does it form?
4. What is the most important scarce metal and how is it used?
5. What two types of mineral deposit are the principal producers of lead and zinc?
6. What mineral commodities are used as construction materials?
7. Why is sulfur used as an index of a nation's industrial development?
8. What is the distinction between mineral reserves and mineral resources?
9. What are the three main types of processes by which minerals can be concentrated?
10. What are porphyry copper deposits and in what plate tectonic setting do they occur?
11. Give two examples of remote sensing techniques that are used for mineral exploration purposes.

Study Questions Research the highlighted question using InfoTrac College Edition.

1. Describe how magmatic processes may result in hydrothermal deposits.
2. Explain how processes at mid-oceanic ridge spreading centers generate the submarine vents we call smokers, and describe the mineralization with which they are associated.
3. How do sedimentary processes result in the concentration of minerals?
4. Why are metal reserves not equally distributed worldwide and how has this influenced the distribution of mineral wealth?
5. List as many examples as possible (other than those used to introduce Section 17.2), to illustrate the role mineral resources play in your daily life.
6. Given that aluminum is the most abundant metal and the third most abundant element in the Earth's crust, why do we make such an effort to recycle aluminum cans?
7. What are false color images and how are they used for mineral exploration purposes?
8. Show how models for the consumption of a limited resource can be used to predict future trends in the mining industry.
9. What is the difference between substitution and conservation of mineral commodities and how will this ensure a maintained supply of metals in the immediate future?
10. Why has there been relatively little mining in the polar regions? Is this predicted to change? If so, why?

CHAPTER 18

Energy Resources

Modern society cannot function without huge supplies of energy, the relatively cheap cost of which is responsible for the high standard of living enjoyed by the industrial nations. Energy is at the heart of industrial society, powering all sectors of the manufacturing, transportation, communication, and construction industries. It fuels our vehicles, and with this our freedom to move. It also lights our buildings at night and controls the temperature inside them. Indeed, without energy, many of us would literally freeze in the dark.

Although there are many sources of energy, we meet the vast majority of our energy needs by burning oil, natural gas, and coal. These energy sources are the fossil fuels. Like the mineral resources we examined in Chapter 17, fossil fuels are classified as nonrenewable commodities because we consume them faster than they can be replenished by natural processes. They cannot be recycled or reused, and cannot be replaced naturally on any human time scale because they form very slowly over long periods of geologic time. They also require unusual geologic conditions in which to form and only rarely accumulate in sufficient quantities to produce commercially exploitable deposits. An understanding of the natural processes that lead to their production is therefore essential to the business of fossil fuel exploration. Unlike mineral resources, however, fossil fuels are organic in origin, and represent energy stored for millions of years in the fossilized tissue of plants and animals.

The need for very specific geologic settings for the accumulation of fossil fuels is reflected in their uneven global distribution. Whereas it was local coal supplies that fueled the Industrial Revolution, for example, the developed nations of today are becoming increasingly dependent on foreign oil reserves, nearly 60% of which lie beneath the desert sands in the politically unsettled region around the Persian Gulf. Great inequity also exists in the consumption of fossil fuels, the industrial nations using far more per capita than less-developed countries.

Regardless of the differences in their distribution and consumption, however, the total supply of all fossil fuels is finite. Production of oil in the United States, for example, has already passed its peak, and the developed nations have experienced the chaos of shortages produced by political embargoes. Natural gas is present in larger quantities, and oil shales and tar sands that are presently sub-economic may be added to the oil reserves of the future. Coal is also present in large quantities but its future as a resource suffers from the environmental impact of its use. The future of fossil fuels is therefore likely to be one of diminishing production, such that alternative energy sources will have to be developed to take their place.

Many of these alternative energy sources are already in use. Nuclear power, driven by the fission of naturally occurring radioactive uranium, is another proven source of energy. Like the fossil fuels, however, it is nonrenewable, being dependent in this case on a mineral resource. Nuclear power is therefore of finite supply although, used in breeder reactors, present reserves of uranium could last for centuries. Heralded as the power source of the future, public concern over the safety of nuclear reactors, the security of nuclear fuel, and the unsolved problem of nuclear waste disposal, have greatly curtailed the development of the nuclear industry. The development of nuclear fusion, the same energy source that powers the Sun, has yet to be achieved economically. It would, however, be free from the problems associated with nuclear fission, and may one day provide energy from virtually limitless supplies of hydrogen.

Geothermal and solar energy, hydroelectric power, wind and tidal power, and energy obtained from biomass, have all been developed as alternative energy sources. Unlike fossil fuels and nuclear power, these are renewable energy sources because they do not depend on an exhaustible natural resource. However, only in certain areas can power be produced from such sources and, to date, the scale of these operations has been small. For the immediate future, therefore, we are likely to see little reduction in our dependency on fossil fuels.

18.1 Energy Supply and Demand

18.1.1 Energy Sources

For much of human history, wood was the dominant fuel and power was measured by the strength of individuals or the strength of their beasts of burden. Indeed, this is true today in the rural areas of many less-developed countries. Following the Industrial Revolution, however, the use of coal as a commercial fuel increased dramatically, and by the middle of the 19th Century, coal had replaced wood as the dominant source of energy. Oil, in turn, surpassed coal as the principal energy source in the middle of the 20th Century and has remained the dominant source of energy to this day (Fig. 18.1).

In 1997, for example, oil provided over 38% of the energy needs in the United States, with coal providing a little less than 23% and natural gas making up about 24% (Fig. 18.2). Nuclear power, which was heralded as the power source of the future when it was introduced in the 1950s, is beset with environmental and safety problems and provides only 7% of the nation's energy today. All other sources of energy together similarly provide less than 8% of the country's needs. Almost 77% of these energy needs were met by domestic production. But the remaining 23% was imported, mostly in the form of crude oil and crude oil products.

The figures for Canada and the rest of the world's energy consumption are similar to those of the United States. Oil provides 40% of our global energy needs, natural gas provides 25%, coal provides 23%, nuclear power provides 8% and hydroelectric power provides 4%. All other sources of energy combined provide a little over 0.1%.

Some 88% of the world's energy needs, therefore, are presently met by the three **fossil fuels,** oil, natural gas, and coal, so called because of their fossilized organic origin. A further 8% is met by nuclear fuel obtained from uranium ore. Almost all of our energy needs are consequently met by the consumption of **nonrenewable** energy sources, so called because each of these commodities is replaced naturally on a geologic time scale and cannot be recycled. Only a little over 4% of our global energy needs are met by **renewable** energy sources such as hydroelectric power, geothermal and solar energy, wind and tidal power, and energy obtained from biomass.

18.1.2 A Dichotomy in Global Energy Use

Despite a shared dependence on nonrenewable energy sources, there are striking differences in the pattern of global energy use that clearly distinguish the more developed nations from the less-developed ones. Many areas of less-developed countries, for example, are still heavily dependent on humanpower and the use of beasts of burden as sources of energy. More developed countries, on the other hand, particularly the heavily industrialized member nations of the Organization for Economic Cooperation and Development (OECD), expend huge quantities of generated energy powering their industries, transporting people and materials, and controlling the temperature of their buildings. As a result, the developed nations consume a disproportionate share of the total energy supply (Fig. 18.3).

Figure 18.1

Energy consumption in quadrillion (1,000,000,000,000,000) British thermal units (Btu) in the United States from 1850 through 1995, with projections to 2020. A Btu is the quantity of heat required to raise the temperature of one pound of water by 1°F. 1 million Btu approximately equals 1000 cubic feet of natural gas.

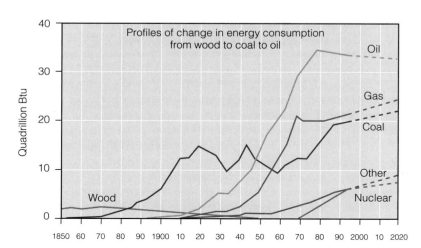

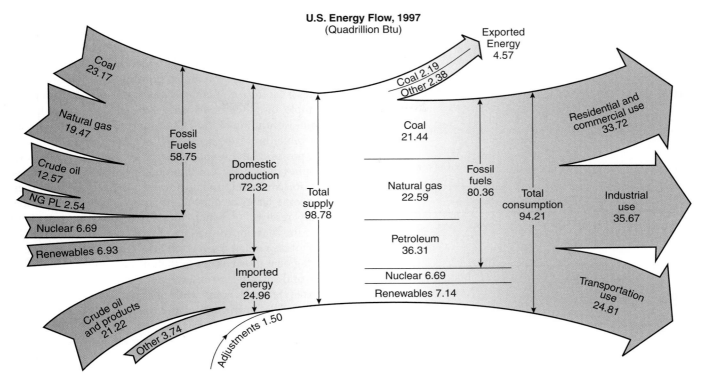

Figure 18.2
Energy flow in the United States for 1997, according to the U.S. Energy Information Administration. All figures for energy supply and consumption are in quadrillion Btu (British thermal units).

The United States, for example, consumes about 25% of the world's total energy, yet only accounts for 5% of the world's population. Of the total energy the country consumed in 1997, about 38% was used for industrial purposes, more than 26% was used for transportation, and the remaining 36% was used in residential and commercial buildings (see Figure 18.2). The less developed (non-OECD) countries, on the other hand, use less than half of the global energy supply, yet make up about 85% of the world's population.

Just take a look around you and it is easy to see why we in the developed nations are such gluttons for energy. From our cars to our computers, our air conditioners to our appliances, we are surrounded on every side by energy-consuming devices. Indeed, our very society and standard of living depend upon the availability of such devices and the huge supply of relatively cheap energy necessary to run them.

But while the developed nations clearly consume more than their fair share of energy, it is important to remember that a significant portion of the energy consumed produces the food, technology, and manufactured products upon which the whole world depends. Furthermore, the most rapid increase in energy consumption is also occurring, not in the developed nations, but in the non-OECD countries. In 1970, for example, the less-developed countries accounted for only about one-third of the world's total energy consumption. Today this figure is closer to one-half, and the total consumption of energy by these countries has more than doubled (Fig. 18.3). By the year 2010, the U.S. Energy Information Administration projects the total energy consumption by non-OECD countries to be about the same as that of OECD countries.

The OECD and non-OECD countries also differ markedly in the mix of energy sources they use. Coal, for example, is of far greater importance as an energy source in non-OECD countries, where it is of equal significance to oil (Fig. 18.4a).

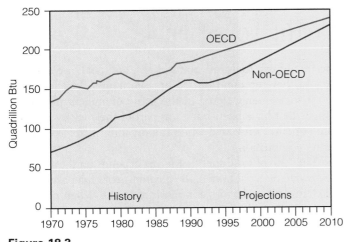

Figure 18.3
OEDC and non-OEDC energy consumption and projected energy consumption for the period between 1970 and 2010, according to the U.S. Energy Information Administration.

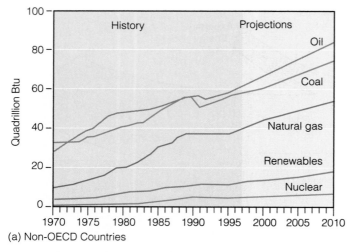

(a) Non-OECD Countries

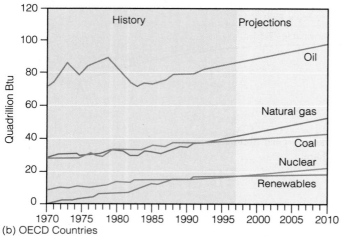

(b) OECD Countries

Figure 18.4
Energy consumption by fuel type from 1970, projected to 2010, according to the U.S. Energy Information Administration.
(a) In non-OECD countries, and (b) in countries of the OECD.

In countries of the OECD, on the other hand, the use of coal is of equal importance only to that of natural gas (Fig. 18.4b). The OECD countries also use about twice as much oil as any other energy source, while coal and natural gas compete for second place. The use of nuclear energy is also far more prevalent in the OECD than it is in non-OECD countries because of the cost associated with the construction of nuclear power plants. Consumption of renewable energy sources is similar for both groups of countries but the rate of growth in consumption is significantly greater in the non-OECD countries.

18.1.3 Energy Availability: A False Sense of Security

Despite our experience with energy shortages in the 1970s, brought about by the oil embargo imposed by OPEC (the Organization of Petroleum Exporting Countries), we still tend to think of energy as an unlimited resource. Although the oil glut of the 1980s and 1990s has done nothing to remedy this view, our almost complete dependence on nonrenewable energy sources shows that it is a serious falsehood.

Because fossil fuel accumulation only occurs in very specific geologic settings, the global distribution of these energy resources is far from equitable. Many of the OECD countries, for example, owe their status today as developed nations to the Industrial Revolution of the 18th Century. This revolution was only possible, however, because North America and western Europe possessed the supplies of coal necessary to fuel it, and the supplies of iron ore and limestone needed for the production of iron and steel. Today, however, the OECD countries have become far more dependent on oil and natural gas, the necessary reserves of which most of them do not possess. Hence, the developed nations have become increasingly dependent on foreign reserves of oil and gas, which have very different global distributions to that of coal. Indeed, over 70% of the world's natural gas reserves lie in the Middle East and the former Soviet Union, while almost 60% of the world's reserves of oil lie in and around the Persian Gulf, a region with a history of political instability.

Yet even with these vast reserves of oil and gas, the total supply of fossil fuels is finite. Like mineral deposits, nonrenewable energy supply is also subject to the distinction between energy **reserves**, which are known sources of energy that could be worked profitably in today's marketplace, and energy **resources**, which are potential sources of energy that are presently sub-economic or have yet to be found. As with mineral deposits, nonrenewable energy resources only form under very specific conditions and rarely accumulate in large enough quantities to make their exploitation profitable. If we are to recognize the implications of our dependence upon fossil fuels, an understanding of the natural processes that lead to their production is therefore essential, just as it is to the business of coal mining and petroleum exploration.

18.2 Energy in the Solid Earth: Fossil Fuels

The three fossil fuels: **oil, natural gas,** and **coal,** are so-called because they represent buried organic matter. In effect, they harbor solar energy stored in the tissue of ancient plants and animals that have been buried as part of the rock cycle. That stored energy is released when the organic matter is oxidized during combustion.

The type of fossil fuel that forms depends on the nature of the organic material buried, the environment of deposition, and the changes it undergoes as a result of burial. When organisms die, their remains may be trapped as they are buried by ongoing sedimentation. When the remains of oceanic microorganisms such as plankton are trapped in marine muds, for example, the rising temperatures associated with their progressive burial may cause their organic compounds to be transformed into oil and natural gas (mostly methane). Alternatively, if burial temperatures are insufficient for this transformation to occur, a wax-like material called *kerogen* is formed instead, and the rock is referred to as an oil shale. Land plants, on the other hand, are rich in woody tissue and so remain solid upon burial to form coal.

Despite their formation in rocks, the fossil fuels are not considered minerals because they are organic rather than inorganic in origin. However, they have been used as a natural resource for energy since the dawn of civilization. They first became important sources of energy with the Industrial Revolution of the 18th Century. Today, they have become the very lifeblood of industrial society.

18.2.1 Oil and Natural Gas

Crude oil from natural seeps has been used for purposes such as lubrication and caulking since ancient times. By the middle of the 19th Century shallow wells were being dug near Oil Springs, Ontario, to produce kerosene for lamps. The first commercial oil production began in Romania in 1856, and just three years later, the first oil well was drilled at Titusville, Pennsylvania. Here, on August 27, 1859, Edwin L. Drake struck oil at a depth of 21 m (70 ft) and started pumping as much as 35 barrels a day (1 barrel = 42 U.S. gallons or 159 liters). With this began our modern age of oil consumption, one that now demands a global rate of oil production of almost 65 million barrels a day.

Records of the use of natural gas go back at least 3000 years, when it was used by the Chinese to evaporate salt water for salt. Early in the 18th Century, gas made from coal was being used for street lighting, and by the early 19th Century commercial gas companies had been established in London and Baltimore. Today, the global production of natural gas has reached almost 80 trillion (80,000,000,000,000) cubic feet annually.

The term **petroleum** (from the Greek for "rock oil") applies to all naturally occurring organic substances that consist mainly of *hydrocarbons* (compounds of hydrogen and carbon), and so includes both oil and natural gas. Hydrocarbons, however, occur in a wide variety of combinations and can be solid (bitumen), semi-solid (asphalt, paraffin wax), liquid (gasoline, kerosene) or gaseous (methane, propane). Liquid petroleum is extracted from the ground as a complex mixture of hydrocarbons known as *crude oil*. During refining, the heavier, more complex molecules are first *cracked*, or broken into simpler ones, and these are then separated and removed by distillation to produce a wide variety of products. Many of these products, such as gasoline, kerosene, and fuel oil, are used as sources of energy for transportation and heating. Others, however, are used as lubricants, and yet others form vital raw materials for the chemical industry, where they are used in the production of such commodities as paints, fertilizers, insecticides, synthetic fibers, plastics, soaps, and pharmaceuticals.

Origin and Accumulation of Petroleum

In order for petroleum to form and accumulate in commercial quantities, four conditions must first be met. A *source rock* capable of producing oil or natural gas must be available for petroleum to

form, and must be buried to the depth needed to convert its organic matter to petroleum. A *reservoir rock* in which the petroleum, once formed, can be stored is also necessary, as is an impervious *seal* to provide a barrier to its upward movement. Finally, some form of *geologic structure* is required in order to trap the petroleum underground. Even when all these conditions are met, however, the petroleum must still be found. Because oil and gas are difficult to detect underground, most exploration techniques focus on identifying potential traps in sedimentary basins that are likely to have generated petroleum.

Petroleum is nearly always found in sedimentary rocks but requires rather unusual conditions in which to form. It is derived from organic matter (principally the remains of microscopic organisms such as plankton and bacteria) that has been incorporated in clays and buried to form organic-rich shale. Under normal circumstances, however, organic matter will decay long before it can be buried. In effect, decay reverses the process of photosynthesis by oxidizing the organic matter to water and carbon dioxide. For oil and natural gas to form, burial must take place in an oxygen deficient or *anaerobic* environment. Such environments develop where the burial of organic matter is rapid or occurs under stagnant conditions in which there is little water circulation to replace the oxygen used in the decay process. As we learned in Chapters 10 and 14, rapid burial is common in deltas (see Figure 10.36) and restricted water circulation can sometimes develop in lakes and marine environments. If these conditions are met, much of the organic matter may be preserved and the slow transformation from rotting organisms to oil and natural gas can begin. Rocks in which this has occurred are called **source rocks.**

The conversion of organic matter to petroleum is a complex process that occurs with burial. As the source rock becomes progressively buried to ever greater depths beneath a cover of younger sediments, it is subjected to ever increasing pressures and temperatures. At depths of 3 to 6 km (approximately 2 to 4 mi), a complex series of chemical reactions take place that effectively *refine* the heavy organic molecules into simpler hydrocarbons including methane, the main component of natural gas. The nature of the cracking process depends on the existing conditions of pressure and temperature, and the length of time over which it occurs. The process is also likely to involve varying degrees of bacterial action. As a result, the composition of the resulting crude oil varies from one source rock to another, accounting for the differences found in the oils from different deposits. At great depths and higher temperatures, crude oil breaks down and only natural gas is produced. The formation of oil is therefore restricted to a narrow range of temperature and pressure, which is appropriately termed the *oil window*.

Once formed, both oil and natural gas will start to migrate upwards because they are lighter than the sedimentary rocks in which they occur and are also lighter than the water contained within them. As we have learned, sedimentary rocks with a significant volume of void space are said to be *porous*, and those in which the void space is interconnected are said to be *permeable* (see Chapter 10). Movement of oil and natural gas is possible because most sedimentary rocks contain interconnected pores between their constituent grains and so are relatively permeable. Indeed, rocks such as sandstone and limestone are often sufficiently porous that they can hold petroleum in much the same way as a sponge holds water. Those that eventually store the oil in this way are called **reservoir rocks**.

Unless the upward migration of petroleum is impeded by some barrier, however, the oil and natural gas will migrate until they reach the surface, where they will dissipate and be lost. Barriers to petroleum migration are known as **seals** and are usually provided by a relatively impermeable mudstone or shale. By blocking its upward migration, the seal forces the petroleum to collect in a more porous reservoir rock below. Given sufficient time, however, most petroleum will eventually leak out by exploiting fractures and other imperfections in the seal. This may explain why oil and natural gas are typically found in relatively young rocks, usually those that are less than 100 million years old.

But in order for petroleum to accumulate, the seal must do more than simply impede its upward migration. It must also have a vault-like geometry that will prevent the fossil fuel from moving laterally, forcing it to collect in one area. This geometry is provided by an underground structure known as a **trap**, of which the two main types are structural and stratigraphic.

Structural traps occur as the result of either bending or fracturing of the sedimentary layers. The simplest is the anticlinal trap, in which the layers (including the seal) are folded into an arch or dome called an anticline (see Chapter 6) like an inverted bath tub (Fig. 18.5a). A seal that is arched in this way will cause any upward-migrating oil and natural gas to collect in any porous rocks below, the lighter gas eventually collecting above the heavier oil.

Anticlinal structures usually result from horizontal compression or fault movement, but they can also be produced by the upward movement of buried evaporite deposits such as salt. When thick beds of salt are buried, the low density of their constituent minerals coupled with the pressures exerted by the overlying strata, can cause the salt to punch its way upward through the overlying strata in huge, finger-like protrusions called **salt domes**. In the process, the overlying sedimentary layers are folded into anticlinal structures, and any punctured layer is drawn up against the wall of the salt dome. Any upward moving oil or natural gas will therefore tend to collect either below an impermeable layer in the anticline, or against the salt, which is also impermeable, and hence a very effective seal (Fig. 18.5b). Structural traps such as these are important sources of petroleum in the Gulf Coast region of the United States and Mexico, where more than 500 salt domes have been identified, both on land and below the shallow waters of the Gulf of Mexico.

Faults may also form impenetrable barriers to petroleum migration so that oil and natural gas can become trapped against them (Fig. 18.5c). During fault movement, an impermeable layer of clay may become smeared along the fracture surface so that the fault itself becomes an effective seal. Petroleum fields that develop in folded and faulted rocks may therefore contain many isolated reservoirs of oil and natural gas.

Stratigraphic traps are produced by changes in sedimentary lithology rather than by structural deformation. A permeable layer of sandstone, for example, may give way or *pinch out* laterally into an impervious shale, allowing petroleum to be trapped within the sandstone (Fig. 18.6a). Similarly, oil may be trapped in the porous limestone of an ancient coral reef that was later blanketed with impermeable mud. Indeed, many of the huge oil deposits of the Middle East have accumulated in traps of this sort. Alternatively, petroleum may be trapped below unconformities in tilted beds of permeable rock that are unconformably overlain and sealed against impervious mudstones or shales (Fig. 18.6b).

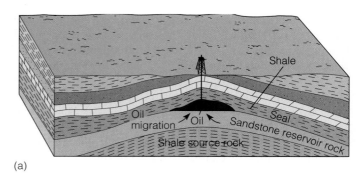

(a)

(b)

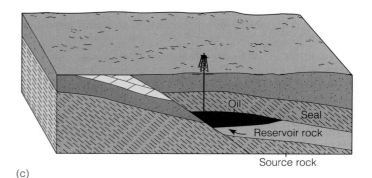

(c)

Figure 18.5
Types of structural petroleum traps (a) anticline, (b) salt dome, and (c) fault. Natural gas, if present, would be located at the top of the oil-bearing zone.

Occurrence and Plate Tectonic Setting of Petroleum

In providing a mechanism for the formation of sedimentary basins, plate tectonics can play an important role in the development and distribution of petroleum deposits. In fact, petroleum has been recovered from sedimentary basins associated

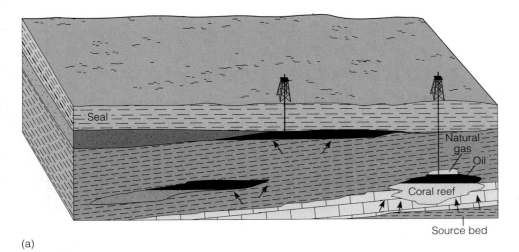

(a)

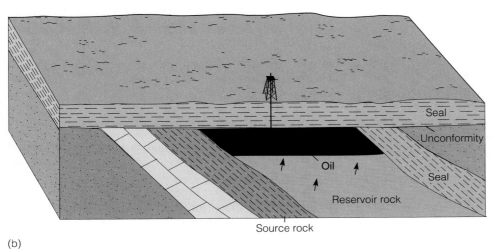

(b)

Figure 18.6
Types of stratigraphic petroleum traps (a) pinch out and reef, and (b) unconformity.

with all three types of plate boundaries. The continental rifts and narrow ocean basins associated with the early stages of plate divergence, for example, are often sites of organic-rich shale and evaporite deposition because marine conditions are restricted and circulation is poor. Later, as the ocean opens, sandstone deposition is common as sediments eroded from the continent are deposited on the developing continental shelf. The continental shelf environment may therefore contain all the elements necessary for the formation and storage of petroleum deposits. Once buried, the shale would provide a potential source rock while the sandstone could form a reservoir. The evaporite would provide an excellent seal and, when mobilized, could form traps in which petroleum can accumulate (see Figure 18.5b). Indeed, just such a combination is responsible for the development of oil and natural gas fields in the Gulf Coast region of the United States and Mexico.

Alternatively, where continental margin sediments are deformed by processes occurring at convergent plate boundaries, structural traps are produced when horizontal compression arches the sedimentary layering into anticlinal folds (see Figure 18.5a). Such is the case, for example, in some of the oil and natural gas fields in the mountain states of the western U.S. and the foothills of the Canadian Rockies.

Sedimentary basins with important petroleum potential also occur in association with transform fault boundaries, as is the case in California. Here, movement on the San Andreas Fault has led not only to the formation of sedimentary basins in which both source and reservoir rocks have accumulated, but has also been responsible for the deformation of the basin rocks and the resulting development of structural traps.

Exploration and Production of Petroleum

Once petroleum has accumulated in a trap, it must still be found if it is to be exploited as an energy resource. Direct detection of oil and natural gas in the subsurface is very difficult, so that most techniques used in petroleum exploration are designed to detect potential petroleum traps instead. Even when potential traps are found, however, most will not contain petroleum.

In addition to detailed geologic mapping, petroleum geologists use a wide variety of geophysical methods to help them in their search for oil and natural gas. Gravity surveys, for example, often reveal the presence of buried structures that, in sedimentary basins with the potential for petroleum production, might act as oil traps. Buried salt domes are readily identified by this method because of the low density of salt compared to the rocks around them.

The most widely used geophysical method, however, is one that uses seismic waves to "X-ray" the subsurface. In Chapter 7, we discussed a component of seismic energy that was bent or refracted as it penetrated the layers of the Earth's interior. A portion of this energy, however, is also reflected back to the surface and can be used for subsurface exploration. The technique is known as **seismic reflection** and uses either small explosions or equipment to vibrate the ground surface, generating shock waves which travel into the ground. These waves are then reflected by the different rock layers underground so that part of their energy returns to the surface, where its arrival is monitored. By computer analysis of the time lapse between the explosion and the arrival of this reflected energy, a detailed picture of the underground structure of the rock layers can be reconstructed (Fig. 18.7), allowing potential exploration targets to be identified.

Ultimately, however, the presence of petroleum within a potential trap must be tested by drilling. If petroleum is discovered, further test wells will be drilled to determine the size of the reservoir. At the same time, sensitive geophysical instruments are lowered into the drill holes so that various physical properties (density, conductivity, porosity, etc.) of the rock layers encountered can be measured. These properties, in turn, identify rock characteristics that can be used to improve the resolution of seismic surveys, and will allow the rock layers to be correlated from one drill hole to another. In this way, a three dimensional picture of the petroleum field can be constructed.

If petroleum is found in sufficient quantity, the field may go into production. Any oil trapped in a reservoir rock at depth is usually under very high pressure because buoyancy causes the water beneath the oil to push upward, while the pressure of the natural gas pushes downward from above (Fig. 18.8). In fact, the pressure is sometimes so high that when the seal is penetrated by a drill hole, the effect is like piercing an inflated balloon with a pin. The pressure is abruptly released and oil, gas, water, and even the drill pipe itself, may shoot into the air

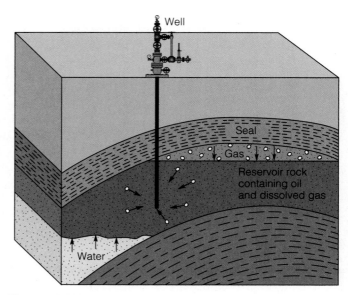

Figure 18.8
Oil in an anticline is driven upward by buoyant water pressure from below and meets with a downward pressure from the overlying natural gas.

as a *gusher*. Under controlled conditions, however, the oil and natural gas may flow naturally to the surface where they can be collected and treated.

The pressure in most successful oil wells, however, is not sufficient to bring the oil to the surface. Instead, they require pumping, and even then, no more than a quarter of the oil in the reservoir rock is actually recovered. To increase the amount of oil retrieved, *enhanced oil recovery* techniques are employed. These attempt to drive the oil out of one well by injecting pressurized water, gas, or steam into the reservoir rock through another well. Even with these methods, however, as much as 20% of the total oil is not recovered from the reservoir. The same problem exists for the recovery of natural gas. Thus, even the most successful exploration strategies leave much of the petroleum in the ground.

Petroleum Distribution and Reserves

Although sedimentary basins containing major petroleum fields occur on all continents, oil and natural gas remain rare commodities and their distribution is very erratic. More than three quarters of the world's petroleum, for example, comes from just a few percent of known petroleum deposits, and most comes from just a handful of huge fields called *supergiants*.

Although supergiant oil fields occur in Alaska and the Gulf of Mexico, the Russian Arctic and Europe's North Sea, almost 60% of the world's known reserves of oil lie in the Middle East. But while non-OPEC countries contain less than 25% of the world's proven oil reserves, they produce 60% of the world's oil. In fact, after Saudi Arabia, which produces more than 8 million barrels of oil a day, the main producers of oil are the United States (at almost 7 million barrels per day), and Russia and the North Sea countries, both of which produce

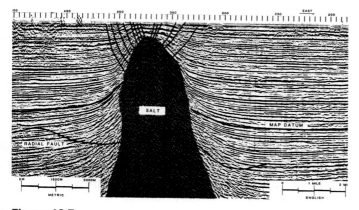

Figure 18.7
Interpretation of a seismic reflection profile of a Gulf Coast salt dome.

about 6 million barrels per day. In 1996, the world's proven oil reserve was estimated to be just over one trillion (1,000,000,000,000) barrels.

North America possesses less than 7% of this proven oil reserve, and almost two thirds of the North American reserves lie in Mexico (Fig. 18.9). Canadian oil represents less than half a percent of the world's proven reserve, and is being produced at a rate that will consume it in the very near future. Nearly all Canadian oil is produced in Alberta and Saskatchewan, whereas a large part of the oil reserve of the United States is contained within the North Slope oil field of Alaska. However, half of the 10 billion barrels of recoverable oil that the Alaskan field has been estimated to contain has already flowed through the Trans-Alaska Pipeline. At current production rates, U.S. oil reserves should last until the middle of the 21st Century.

Of the world's natural gas reserves, estimated at almost five quadrillion (5,000,000,000,000,000) cubic feet, 40% lie in the countries of the former Soviet Union, and a further 30% lie in the Middle East. North America possesses a mere 6% of this total reserve, just over half of which lies in the United States. As with oil, these reserves should be sufficient to last until the middle of the 21st Century at current production rates.

Significant areas of petroleum potential, however, remain to be explored. Many of the basins of China, for example, are

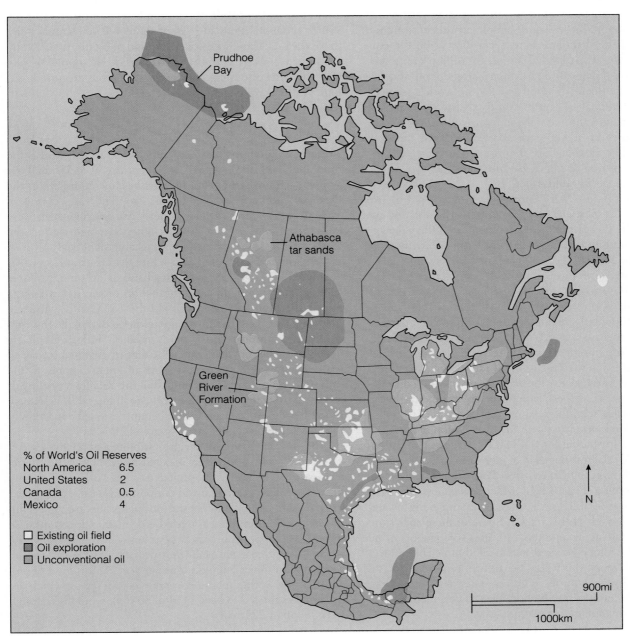

Figure 18.9
North American oil reserves (unconventional oil includes reserves in oil shales and tar sands).

only now being surveyed with modern geophysical equipment. Although the importance of offshore oil production has already been demonstrated in regions like the Gulf of Mexico and the North Sea, large areas of the world's continental shelf have yet to be fully explored. Production from much deeper prospective areas on the continental slope may also be opened up with new technology and improved procedures to counter the environmental hazards of offshore drilling and the transportation of oil and natural gas at sea. On land, improved methods of enhanced recovery may also increase the percentage of recoverable oil from existing fields.

Because of these factors, it is difficult to estimate how long our reserves of oil and natural gas will last. However, a simple indication of the longevity of these natural resources can be obtained by dividing the world's proven reserves by their global rate of consumption. In 1996, for example, world consumption of oil approached 65 million barrels per day, while the consumption of natural gas totaled almost 82 trillion cubic feet. Using current estimates of one trillion barrels of oil and five quadrillion cubic feet of natural gas for the world's reserves, global consumption at 1996 rates would exhaust the oil reserve in just 43 years and the natural gas reserve in only 61 years. Adding estimates of as yet undiscovered reserves, however, would more than double these figures, while increased efficiency and improved methods of enhanced recovery would increase their longevity still further.

The exploitation of "unconventional" petroleum resources, such as oil shales and tar sands, would also add significantly to the lifespan of these commodities. **Oil shales** are organic-rich mudstones that yield petroleum when they are heated (Figure 18.10a). Typically deposited in swampy lakes, stagnant lagoons, or marine basins with restricted circulation, they contain the wax-like substance *kerogen* and are considered high grade if, on treatment, they yield more than 25 gallons (about 100 liters) of oil per ton. Oil shales are used commercially in Russia and China, where they are typically developed by strip mining. However, the mining of oil shales is very problematic and the process of oil extraction and refining is expensive. As a result only the richest deposits are commercially viable and, even then, significant environmental problems are associated with their exploitation.

The problems associated with the exploitation of oil shales are particularly frustrating for the United States, since it contains the world's largest deposits of rich oil shale. The most important of these is the Green River Formation of Colorado, Utah, and Wyoming (Fig. 18.10b), which was deposited during the Eocene, some 50 million years ago, when huge fresh water lakes occupied much of this region. Some of these shales are capable of producing as much as 240 liters (65 gallons) of oil per ton, and the total oil reserve is estimated to be about 2 trillion barrels. Although this would be sufficient to meet U.S. oil demands for several centuries at current rates of consumption, the cost and difficulty of production have so far precluded the economic exploitation of this vast resource.

Tar sands are sandstones that contain an asphalt-like hydrocarbon called *bitumen* that is too thick and viscous to be pumped out of the ground. To be recovered, therefore, it must either be steam-heated prior to pumping, which encourages the tar to flow, or, where the deposit is shallow, it can be mined by open pit methods, treated, and refined.

The deposits are thought to originate in the same way as oil, but are later modified as a result of biodegradation and the escape of the more volatile components that leaves only the most viscous materials behind. Extensive deposits of tar sands occur in northern Canada, Russia, and Venezuela. The United States contains several small deposits, the largest of which is in Utah. The largest deposit in the world, however, is the Athabasca tar sands of northern Alberta (see Figure 18.9), which are estimated to contain 300 billion barrels of recoverable oil. With a production rate in 1996 of about 400,000 barrels a day, this is the only heavily exploited tar sand in North America and accounts for about 21% of Canada's crude oil supply. In general, the costs of mining and treating tar sands are simply too high for their exploitation to be economic.

18.2.2 Coal

The black, inflammable sedimentary rock we call **coal** is the most abundant of the fossil fuels, occurring in rocks of all the geologic periods since plant life first became abundant in the Devonian, some 390 million years ago. The vast deposits of eastern North America and western Europe, however, were formed between 285 to 300 million years ago during the Carboniferous, a period named for its abundant coals.

Although the use of coal as a fuel goes back at least as far as the Middle Ages, its widespread use as an energy source came about only with the introduction of the steam engine during the Industrial Revolution of the 18th Century. Indeed, it was the use of coal as an energy resource that made this fundamental transformation towards modern industrial society possible. By the middle of the 19th Century, coal had surpassed wood in providing the largest share of energy supply to the industrial nations, and it continued in this role for a century or more before being surpassed by oil (see Figure 18.1). Indeed, in the less-developed non-OEDC countries, coal continues to rival oil as the principle energy source to this day.

Because of its abundance, accessibility and relative ease of extraction, coal is a relatively inexpensive fossil fuel. It is therefore expected to be an important part of the world's energy supply in the future, particularly as diminishing petroleum reserves force the price of oil and natural gas higher. Following a sharp rise in coal consumption as a result of embargo-driven instabilities in the supply and cost of petroleum in the 1970s, world production of coal has stabilized over the past decade at an average of just over 5 billion tons per year. At this level of production, the world's total recoverable coal reserves, which are presently estimated to be 1,145 billion tons, would be enough to last another 225 years.

(a)

(b)

Figure 18.10
(a) Oil shale and the heavy oil that it produces when heated. (b) Distribution of oil shales in the Green River Formation of Colorado, Utah, and Wyoming.

In the map (b):

Idaho

Green River

Great Divide Basin

Wyoming

●Rock Springs

Basin

Washakie Basin

●Salt Lake City   Utah

River

Sand Wash Basin

Vernal
●

Colorado

Uinta Basin

White

River

Piceance Basin

●Rifle   River

Naval oil-shale reserve 2

Battlement Mesa

0   25   50 km

Green

Colorado

Grand Mesa

Area underlain by the Green River Formation in which the oil shale is unappraised or of low grade.

Area underlain by oil shale more than 3 m thick which yields 0.1 m³ or more oil per ton of shale.

Of the coal consumed, the bulk is used in the generation of electric power from steam turbines supplied from coal-fired boilers. However, coal is also converted into **coke,** a hard, porous and practically smokeless fuel that is produced when coal is baked and its volatile matter is driven off as *coal gas.* Coke plays a critical role in the smelting of iron and the manufacture of steel. In addition, coal provides the raw material for many organic chemicals as well as for nylon and many other plastics.

The use of coal as an energy source, however, is not without severe environmental problems. Strip mining for coal requires elaborate land restoration, while underground mining is not only a hazardous operation but can lead to surface subsidence and acid mine drainage. Methane gas released by underground coal mining operations is explosive; in addition, breathing methane gas can be fatal if sufficient amounts are present. Indeed, historically the risk of explosion or asphyxiation was so severe that miners resorted to carrying caged canaries into the mine with them. Being less tolerant to the effects of methane, the canaries would succumb before the miners: a clear warning that methane levels had become dangerously high. Furthermore, water used in the treatment of coal is too contaminated to be released to the environment, and coal dust can be highly combustible and, if inhaled, presents a serious health risk in the form of the miner's disease known as "black lung." Sulfur dioxide emissions associated with the burning of coal adversely affect ground level air quality and have been linked to the problem of acid rain. In addition, the burning of coal adds more carbon dioxide to the atmosphere than the combustion of any other fossil fuel. Recall that carbon dioxide is viewed as the most important of the greenhouse gases which are thought to be associated with global warming. Despite these adverse environmental impacts, however, the U.S. Energy Information Administration predicts a 50% increase in world coal consumption by the year 2015.

Origin and Accumulation of Coal

Coal is a sedimentary rock consisting almost entirely of organic carbon, and is formed from the burial and compaction of accumulated plant remains in fresh water and brackish swamps. So, unlike petroleum, which is largely the product of marine microorganisms, coal is formed from the decay of terrestrial plants. In order for preservation to occur, however, the decomposition of plant matter typical of terrestrial environments must be prevented. This is best achieved under oxygen-starved conditions such as those produced in the stagnant waters of fresh or brackish-water swamps.

Coal is the end product of the progressive burial of plant matter deposited under such oxygen-starved conditions. In order to form coal, the accumulated plant remains must first form **peat,** a water-saturated, organic-rich humus produced by bacterial decay that, when dried, is itself a low-grade fuel. Peat often forms in warm, humid swamplands that border marine basins. The Okeefenokee Swamp in Georgia and the Everglades of Florida, for example, contain large deposits of peat. In settings such as these, land subsidence allows occasional marine incursions to occur, which result in the deposition of sediments and the rapid burial of the peat deposit.

As the peat is buried beneath overlying sediment, it becomes progressively compacted and heated, which squeezes out the water and drives off the organic gases. As a result, the carbon content of the peat becomes concentrated, eventually transforming it into woody-looking *brown coal,* or **lignite.** Continued burial raises temperatures and pressures, and so concentrates carbon still further. This causes the lignite to undergo additional changes, first to **subbituminous coal** and then to **bituminous coal,** the two most common forms of coal in North America (Fig. 18.11). This change in structure and composition is a form of very low grade metamorphism. Further metamorphism of bituminous coal, usually accompanied by structural deformation in regions of plate convergence, produces the jet-black coal we call **anthracite.**

This progression from lignite to anthracite coincides with an increase in the **rank** of the coal, which is based on its carbon content and heat value on combustion. Coal quality or **grade,** on the other hand, is based on its purity. Typical contaminants include the sulfur and ash content, both of which have detrimental effects on the environment when the coal is burned. It is the combustion of high-sulfur coal, for example, that produces the toxic emissions of sulfur dioxide linked to acid rain. A coal's sulfur content, which is usually present as iron sulfide in the minerals pyrite and marcasite, is considered low when it comprises less than one percent of the coal, and high when the amount exceeds three percent. Pyrite and marcasite are natural by-products of coal formation and occur in highest concentrations in bituminous coal and lowest in subbituminous coal and anthracite.

Coal Distribution and Reserves

The formation and periodic burial of peat in shallow swamps on broad coastal plains ultimately produces a series of relatively thin coal seams interlayered with other sedimentary strata as part of a normal stratigraphic succession. Indeed, most coals deposits extend over large areas in a series of subhorizontal seams that range from several centimeters to a few meters in thickness. As a result, coal seams are quite easily located and there is usually little effort involved in coal exploration. In fact, it is likely that most of the world's coal deposits have already been identified.

Coal deposits are widely distributed, but are far more abundant in the northern hemisphere than they are in the south. Since the majority of coal was formed during the Late Carboniferous, this bias is largely a function of the world's geography at that time (see Figure 14.3a). During the Late Carboniferous, much of what is now North America and Europe (which host vast coal deposits) occupied positions near the equator in climatic regimes that were favorable for coal formation. Indeed, the former continuity of these deposits from North America to Europe played an important role in Wegener's argument for the existence of the supercontinent Pangea. At the same time, many of the southern continents of Gondwana lay in the south polar regions and were experiencing the very ice age that Wegener also used in support of his hypothesis of Continental Drift (see Figure 4.5). Although these conditions were unfavorable for coal formation in the southern continents, the burial of plant material near the Carboniferous equator was encouraged by the sea level

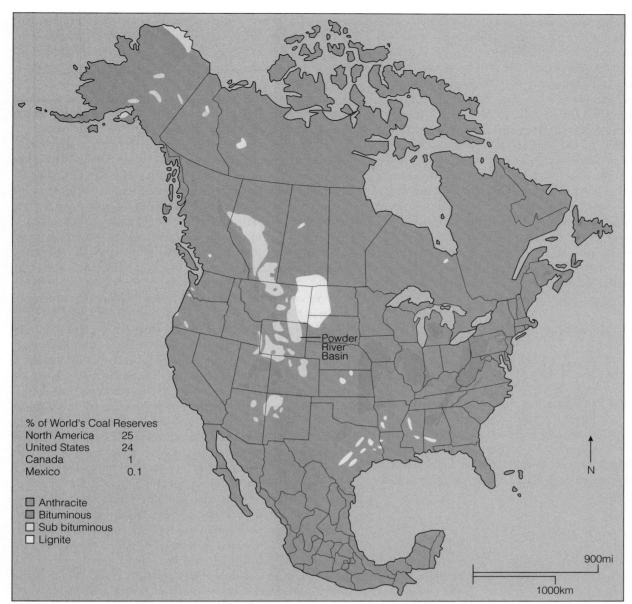

Figure 18.11
North American coal reserves, estimated in 1996.

% of World's Coal Reserves
North America   25
United States   24
Canada   1
Mexico   0.1

☐ Anthracite
☐ Bituminous
☐ Sub bituminous
☐ Lignite

Powder
River
Basin

N

900mi

1000km

changes associated with the advance and retreat of the polar ice sheet in the southern hemisphere.

The breakup of Pangea dismembered these once continuous belts of coal and glacial deposits, the precise location of the continental rifts determining the subsequent distribution of significant coal reserves. One could therefore argue that the fragmentation of Pangea, some 200 million years ago, has profoundly influenced the history of modern society.

Today, almost half of the world's total recoverable coal occurs in the former Soviet Union and the United States, each of which have recoverable reserves of about 270 billion tons (Fig. 18.12). They are followed by China and western Europe, each with reserves amounting to about half this figure. It is China and the United States, however, that are both the leading producers of coal and the leading consumers.

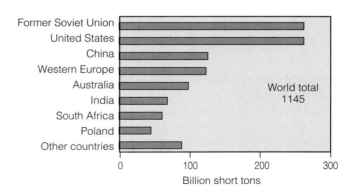

Figure 18.12
World coal reserves in 1997, according to the U.S. Energy Information Administration.

The United States, for example, produces a billion tons of coal annually, almost 90% of which is consumed in the production of electricity. Some of the largest reserves of high ranking bituminous coal and anthracite occur in the upper Midwest and Appalachia, and huge reserves of low grade lignite lie beneath the northern Plain States (Fig. 18.11). However, with the introduction of the federal Clean Air Act, which requires reductions in sulfur emissions from coal combustion, coal production has shifted to the western states, led by Wyoming, where the coals are largely of Cretaceous age and their sulfur content is low. At the same time, production of Carboniferous coals in Appalachia and the Midwest, led by West Virginia and Kentucky, has declined because of their high sulfur content. The need to either treat these high-sulfur coals to remove sulfur prior to combustion or scrub their emissions of sulfur dioxide when burned, in order to reduce air pollution, has increased the cost of these coals and made them a far less economic energy source for generating thermal electric power.

18.3 Energy in the Solid Earth: Alternative Sources

In addition to the fossil fuels, the solid Earth provides other sources of energy in the form of naturally occurring radioactive minerals, and shallow areas of hot rocks that can be used to provide geothermal energy. The former is strictly a mineral resource and, hence, is nonrenewable, but the latter is considered a renewable resource because it can be used without exhausting the energy source.

18.3.1 Nuclear Power

Nuclear power depends on the heat released when atoms of the naturally occurring radioactive element uranium (U) are split apart in a form of radioactive decay known as **nuclear fission.** In a nuclear reactor, the heat produced by the fissioning of uranium is used to make steam, which is then used to generate electricity. Nuclear power consequently depends on the exploitation of a mineral resource and the utilization of its radioactive properties under controlled conditions.

Uranium Deposits

Uranium, with an atomic number of 92, is the heaviest of the naturally occurring elements and has an average crustal abundance of about 2 parts per million. It is found in minor quantities in a wide variety of rocks, but is chiefly exploited as an oxide in the ore minerals *pitchblende* and *uraninite*. Economic deposits of these minerals are most common in sedimentary rocks such as sandstones and conglomerates, where their concentration is often the result of the chemical reduction of uranium dissolved in groundwater. Uranium exists in both oxidized (U^{6+}) and reduced (U^{4+}) forms, the former being highly soluble in groundwater, whereas the latter is highly insoluble. Uranium released into solution by the weathering of rocks under oxidizing conditions will therefore precipitate if reducing (oxygen-poor) conditions are encountered by percolating uranium-bearing groundwater. Hence, motion of groundwater from oxidizing to reducing environments can produce very high uranium concentrations.

Some of the richest uranium deposits, however, occur in early Precambrian rocks. As we learned in Chapter 13, during the early Precambrian atmospheric oxygen levels were extremely low; too low, in fact, to allow weathered uranium minerals to dissolve. As a result, uranium was able to accumulate with other heavy minerals in placer deposits (see Chapter 17), often in sandstones and conglomerates. Such deposits are mined for uranium in the Athabasca Basin of central Canada and in parts of Africa. In addition, uranium becomes highly concentrated in the latest stages of crystallization of magma. As a result, uranium forms important hydrothermal vein deposits in Australia, Canada, Africa, and France, and can also occur in late-stage igneous rocks such as pegmatites.

Almost half of the world's reserves of uranium, however, occur in North America (Fig. 18.13), and almost three quarters of these lie in the United States where they mostly take the form of sandstone ores produced by the reduction of uranium-bearing groundwater. New Mexico, Texas and Wyoming host almost three quarters of U.S. uranium reserves, which are estimated to be about 300 million pounds of recoverable uranium oxide (U_3O_8) at a cut-off grade based on an operating cost of $66/kg ($30/lb).

It is Canada that leads the world in uranium production, accounting for almost 30% of world total, followed by the former Soviet Union (24%) and Africa (18%). By contrast, production in the United States amounts to a mere 4% of the world total.

Exploration for uranium ore and the evaluation of potential uranium deposits makes use of the element's radioactivity, which can easily be detected at the surface and measured in boreholes using instruments such as geiger counters. The uranium ore is often mined by open pit or underground methods, but solution mining by in-situ leaching is also possible if the host rock is permeable. Following crushing and grinding, uranium is extracted from the ore by chemical leaching.

Nuclear Energy

Although all radioactive isotopes are unstable and spontaneously break down, some are so unstable that their atomic nuclei can be split by neutron bombardment, a process known as nuclear fission. When this occurs, the nucleus breaks into two fragments of roughly equal size and heat energy is released along with a number of neutrons (Fig. 18.14). If these neutrons then split the nuclei of neighboring atoms, a **chain reaction** is initiated as more and more neutrons are released to split still further nuclei. Uncontrolled chain reactions lead to nuclear explosions, but controlled reactions in which, on average, each fission event produces one of its own, provide the source of heat that is used to generate electricity in nuclear power plants. In nuclear reactors, controlled fission is achieved by inserting moderators that absorb some of the neutrons produced by each nuclear reaction. Failure to carefully monitor this process, however, can lead to a reactor meltdown like that which occurred at Chernobyl in the Ukraine in 1986.

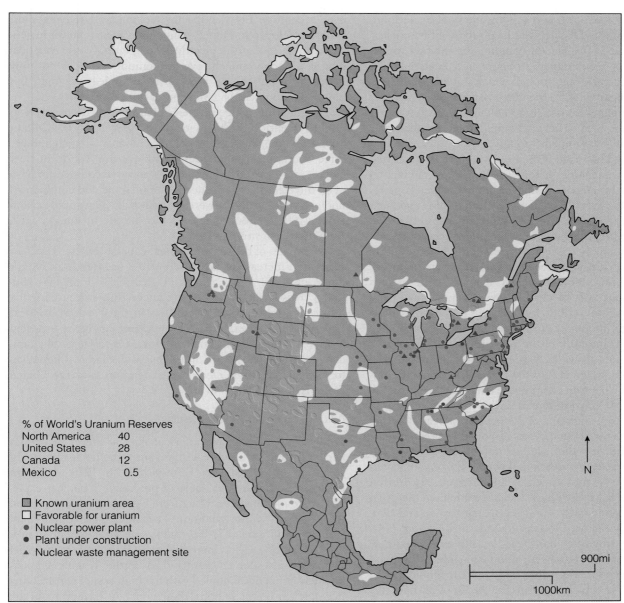

Figure 18.13
North American uranium reserves, and the distribution of North American nuclear power plants and nuclear waste management sites.

% of World's Uranium Reserves

North America	40
United States	28
Canada	12
Mexico	0.5

▪ Known uranium area
▫ Favorable for uranium
● Nuclear power plant
● Plant under construction
▲ Nuclear waste management site

900mi
1000km

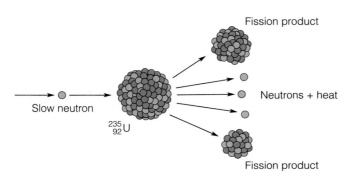

Figure 18.14
Nuclear fission. A neutron strikes the nucleus of a uranium-235 atom, producing fission fragments, free neutrons, and heat. The free neutrons can participate in additional fission reactions.

Fission product

Neutrons + heat

Slow neutron

$^{235}_{92}U$

Fission product

Almost all naturally occurring uranium consists of just two isotopes, ^{238}U and ^{235}U, only one of which (^{235}U) is fissionable. Uranium-235, however, only makes up 0.7% of natural uranium, the remaining 99.3% consisting of the heavier isotope ^{238}U. Some reactors, like those of the CANDU variety used in Canada, are designed to use natural uranium. In most commercial reactors, however, about 3.5% of the fissionable isotope is needed to produce a chain reaction. So natural uranium must be processed to concentrate or enrich this isotope before it can be used. The enriched uranium is then made into pellets and assembled into fuel rods that can be inserted into the reactor core where the chain reaction takes place.

Most commercial reactors in use in North America are **light-water reactors**, in which ordinary water is used both as the moderator that controls and sustains the nuclear reaction, and as the coolant that extracts the heat produced by nuclear fission (Fig. 18.15). The reactor core holding the fuel rods is housed within a containment structure and enclosed in a stainless steel reactor vessel through which water is circulated. Fission in the fuel rods starts the chain reaction by generating heat and releasing neutrons that promote further fission which generates further heat. The circulating water consequently becomes hot and is pumped through a heat exchanger that generates steam. The steam is then used to drive turbines that turn electrical generators. In this way, the water circulating through the reactor core is separated from that which supplies the turbines with steam. The chain reaction itself is regulated by control rods, which are made of material that absorbs excess neutrons and so prevents them from bombarding other nuclei. The chain reaction is consequently slowed when these are inserted into the core, and speeds up when they are removed.

Although conventional nuclear reactors such as this account for about 20% of electric power generation in the United States and only a little less than this in Canada, only a small fraction of the uranium in their fuel rods is fissionable ^{235}U. As a result, the bulk of the uranium, which is non-fissionable ^{238}U, goes to waste. Under neutron bombardment, however, ^{238}U is converted into the fissionable isotope of plutonium, ^{239}Pu, by neutron capture (see Chapter 3). Hence, the potential exists for converting part of the uranium waste into fissionable fuel by using some of the neutrons released during nuclear fission. This phenomena is exploited in **breeder reactors**, so-called because they produce more nuclear fuel than they use. In a breeder reactor, a fuel core of fissionable material (^{235}U or ^{239}Pu) is surrounded by a blanket of ^{238}U. Fission of the core material is used to generate electricity in the usual way. However, since each fission event releases several neutrons, yet is only required to produce one fission event of its own in order to sustain a chain reaction, the excess neutrons can be used to convert waste ^{238}U into fissionable ^{239}Pu fuel.

The most economical feature of breeder technology is its ability to use ^{238}U, which accounts for 99.3% of all naturally occurring uranium, as a fuel. As a result, the amount of uranium needed to power a breeder reactor is only about one hundredth of that used in a light-water reactor of equivalent capacity. Indeed, the amount of ^{238}U presently stored in waste stockpiles could be used to meet breeder reactor needs for well over a century. Breeder reactor technology has been most widely developed in France, which uses nuclear power to generate some 75% of its electrical energy. In North America, however, public concern over plutonium, which is not only highly toxic but can be used to make nuclear weapons, has halted development of a breeder reactor program.

Despite this concern, the United States led the world in nuclear electric power generation in 1996 (with 675 billion kilowatt-hours), followed by France and Japan. Together, these three countries generated about 60% of the world's nuclear electric power. Worldwide, nuclear power accounted for 23% of all the electricity generated in 1996. Some 430 commercial nuclear power stations were operating in 30 countries, and almost 100 additional nuclear units, mostly in the Far East, were under construction.

Estimates of the world's nuclear capacity by the year 2015 suggest an increase or decrease by 10% or more from today's figure of about 350 gigawatts (1 gigawatt = 1,000,000 kilowatts). This variation reflects the uncertainty surrounding the future of nuclear power. Countries in the Far East, such as China, South Korea, India, Pakistan, and Japan, are expected to increase their use of nuclear power. Owing to public concern, however, Western Europe may well choose alternatives to nuclear power for electricity generation, and the United States is almost certain to do so.

Nuclear Safety and Waste Disposal

Public confidence in the safety of nuclear power plants has been low in the United States ever since the nuclear accident at Three Mile Island in Pennsylvania in 1979. Although nuclear reactors are incapable of producing a nuclear explosion, core meltdown resulting from a runaway chain reaction (the so-called "China Syndrome") can destroy the containment structure and spread lethal radioactive contamination over wide areas. The 1986 Chernobyl disaster in the former Soviet Union was just such an event and has further heightened public safety concerns. Coupled with the high cost of constructing new nuclear power stations, these concerns have brought the nuclear industry in the United States to a virtual standstill.

However, many consider the safety record of the nuclear industry to be excellent and argue that coal-fired generating stations pose a far greater hazard to the environment than those using nuclear power. Uranium is also cheap in comparison to fossil fuels. One enriched uranium pellet, for example, has the energy of one ton of coal costing more than twice the price of the uranium pellet, or three barrels of oil costing almost seven times as much. Put another way, 1 ton of nuclear fuel produces as much energy as 20,000 tons of coal at less than half the price. But these figures do not consider the hidden costs of nuclear energy.

The construction of nuclear power stations, for example, has proven to be expensive, and concerns persist over nuclear safety, acts of terrorism, and the security of the plutonium. There is also the unresolved problem of nuclear waste disposal. Much of the waste material from nuclear power stations will remain dangerously radioactive for thousands of years. How best to dispose of this material without contaminating the

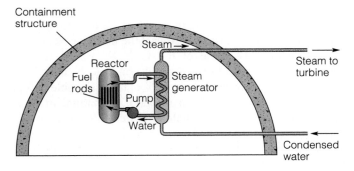

Figure 18.15
Main components of a light-water reactor.

environment is a highly charged issue and one that has greatly deterred the development of the nuclear industry. Critics argue that waste disposal and construction costs must be considered as part of the fuel cost, in which case the economic attractiveness of nuclear power is much reduced.

At present, most of the used nuclear fuel rods and other high-level radioactive wastes are stored temporarily in water-filled pools at individual reactor sites. It is intended that these materials will ultimately be buried in permanent underground disposal facilities, but at present no such facilities exist in the United States and only one, at Yucca Mountain in Nevada, is actively being evaluated. The same is true of other nations, although feasibility studies have been carried out in Canada, the United Kingdom, Germany, Sweden, and Japan. Most waste repository studies have focused on sites in salt deposits, which are virtually impervious to groundwater, and stable homogeneous igneous rocks like those at Yucca Mountain. However, salt is capable of flow and no igneous rock is free of fractures. It is therefore difficult to guarantee that any storage site, even in the most ideal of geologic settings, would remain fail-safe for the long period over which dangerous levels of radioactivity would persist.

18.3.2 Geothermal Energy

Another source of energy within the solid Earth is provided by natural heat from the Earth's interior. As we learned in Chapter 7, the Earth becomes progressively warmer with depth, as any deep miner can testify. This heat is a reflection of the planet's accretion, the formation of its internal layers, and the radioactivity in its interior. Some near-surface regions, however, are anomalously warm and provide a potential source of energy. The potential for such **geothermal energy** exists wherever water heated by shallow-level hot rocks or by deep circulation is close enough to the Earth's surface that it can be reached by borehole. In order to be economically viable, the water must be within 3 km (1.9 mi) of the surface. This is most common in areas of recent volcanism (at active plate boundaries or near hotspots), where molten bodies of magma lie close to the surface. Alternatively, some near-surface rocks, such as granites, may contain sufficient quantities of disseminated uranium minerals that they remain hot by virtue of radioactive decay. In this case, the energy source is the natural analogue of a nuclear power station.

In either case, the existence of hot rocks close to the surface allows them to be used to heat water, either to produce steam for generating electricity or to provide heat for buildings. This usually entails drilling into the hot rocks so that steam and hot water trapped in fractures can be brought to the surface to run a power plant (Fig. 18.16). Once used, the condensed steam and waste water are pumped underground again by way of an injection well, and reheated. Because the heat energy is replenished far faster than it is consumed, geothermal energy is classified as a renewable resource.

Evidence for subsurface geothermal activity often takes the form of hot springs, which have been used from the earliest of times. Much loved by the ancient Romans, hot springs and spas have also fostered the ancient Japanese art of bathing.

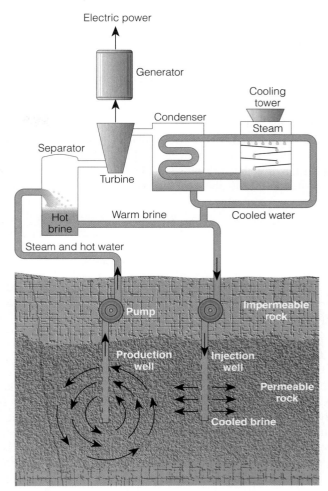

Figure 18.16

Geothermal energy is used to generate electricity from the steam produced when water is heated by hot rocks near the Earth's surface.

Geothermal energy was first used to generate electricity in Italy, in 1904, and has been used to heat homes in Iceland since 1928. In fact, geothermal power now provides some 75% of Iceland's population with inexpensive central heating. Today, the United States, Japan, New Zealand, Italy, Iceland, Mexico, and the Philippines utilize geothermal energy in regions close to active plate boundaries and hotspots where recent volcanism has occurred. The world's total geothermal output of about 6000 megawatts (1 megawatt = 1000 kilowatts), however, is a mere 0.4% of it's energy needs.

The United States first developed geothermal energy in 1960 around The Geysers, a huge geothermal resource near Napa Valley in northern California. Today, geothermal plants operate on 19 fields in California, Nevada, Utah, and Hawaii, all of which lie in areas of anomalously-high heat flow. The Geysers facility remains the largest and accounts for about 70% of the country's geothermal power. It provides close to 2000 megawatts of electricity, a capacity similar to that of two nuclear power stations. Nationwide, however, geothermal energy only provides about 0.5% of the country's electricity.

In fact, geothermal development in the United States has declined in recent years as the price of competing fuels (such as natural gas) has decreased; the demand for electricity has diminished; and federal funding for research and development has been reduced. The world's total geothermal resource is vast, and geothermal energy is relatively clean. But it is not without environmental consequences. Geothermal waters extract minerals from the rocks through which they pass and so are often toxic brines that are highly corrosive to machinery. Subsidence and small tremors due to the extraction and injection of geothermal waters have also been recorded at The Geysers. It has also been estimated that only about 1% of the energy of a geothermal reservoir is recoverable, so that even with improved technology, it is unlikely that geothermal power will ever satisfy more than a fraction of human energy needs. Nevertheless, geothermal energy in the United States is expected to supply about 1% of the country's energy needs by the early 21st Century.

18.4 Energy in the Hydrosphere

The energy of moving water is enormous, as anyone who has experienced whitewater rafting or a heavy coastal surf can testify. Indeed, water power has been used as a source of energy since the earliest of times. Prior to the Industrial Revolution, waterwheels were powering grist mills, sawmills, cotton mills, and a wide variety of mining operations throughout western Europe and North America. Tidal mills have also been used since Medieval times. More recently, water power has been applied to hydroelectric turbines, and efforts have been made to harness the energy in ocean waves and currents.

18.4.1 Tidal Power

The ebb and flow of the tides under the influence of the Moon's gravitational attraction (see Chapter 9), have been used as a source of power for more than a thousand years. Modern tidal power stations generate electricity by building a dam across a tidal basin. The basin is allowed to fill as the tide rises but is sealed at high tide, trapping the water as the tide begins to ebb. Once a sufficient head of water has developed, the basin is emptied through hydroelectric turbines.

The sites of tidal power plants are consequently restricted to coastal areas where the tidal range (the difference in height between high and low tide) is at least 5 m (16 ft), and a tidal basin of sufficient capacity exists. Worldwide, such sites number about a hundred and several small plants have been built in Canada, Norway, China, and the former Soviet Union. The first and largest tidal-power station was built across the estuary of the River Rance in northwest France, where the tidal range is 9 to 14 m (30 to 46 ft). Operating since 1967, the plant is capable of generating 240 megawatts of electricity from both the incoming and outgoing tides.

In the United States, tidal ranges of over 5 m (16 ft) occur only in Maine and Alaska. However, the world's greatest tidal range of about 16 m (52 ft), occurs in Canada's Bay of Fundy (Fig. 9.12), and a tidal power project that would take advantage of this has been under consideration by Canada and the United States for more than 50 years. The proposed project, known as the Passamaquoddy Site (Fig. 18.17), involves damming a series of bays at the mouth of the Bay of Fundy, and is anticipated to have a generating capacity of up to 345 megawatts. Feasibility studies have also been performed to explore the tidal power potential of the Severn Estuary in Britain.

Because of the limitations of tidal range and coastline shape, however, tidal power is unlikely to become anything more than a local energy option. Tidal power stations are also expensive to build, and while the energy source is pollution-free, the necessary interference with tidal flow is likely to be detrimental to the local marine life.

18.4.2 Wave Energy

One only need witness a coastal storm to recognize the enormous amounts of energy contained in waves. But harnessing this energy has proven difficult and, to date, no large wave-powered electrical generating facility has been built. Small wave-powered generators have been used to power buoys and lighthouses, and are usually equipped with funnels to focus the wave's energy. They are, however, susceptible to storm damage and corrosion by the salt water.

18.4.3 Hydroelectric Power

Hydroelectric power, which harnesses the potential energy of stream water as it flows downstream, is the world's primary source of renewable energy. In order to harness the energy effectively, however, the stream must first be dammed (see Chapter 10) so that water from the top of the dam can be channeled through turbines at its base (Fig. 18.18).

Hydroelectric power, which was first used about 100 years ago in Appleton, Wisconsin, generates as much electricity today as thirteen hundred medium-sized coal-burning power plants, and provides about 25% of the world's electricity needs. Indeed, Norway generates almost all of its electricity this way, while Canada generates more hydroelectricity than it needs, exporting the surplus to the United States. Canada, the United States, Brazil, Russia, and China are the five largest producers of hydroelectric power, their combined output accounting for more than 50% of the world total. In 1995, for example, Canada generated 330 billion kilowatt-hours of hydroelectricity, principally in the provinces of Quebec, British Columbia, and Newfoundland and Labrador where it accounts for more than 95% of the total electricity produced. At the same time, 324 billion kilowatt-hours of hydroelectric power was generated in the United States, principally in the states of the Pacific Northwest.

Hydroelectric power is commonly viewed as one of the simplest and most cost-effective forms of alternative energy. It is also easy to turn on and off, and so provides a useful source of

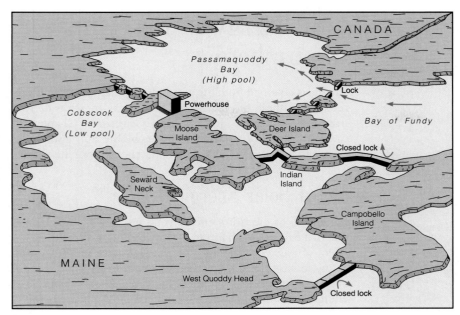

(a) High tide

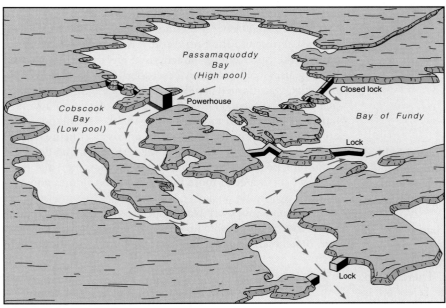

(b) Low tide

Figure 18.17
Area and operation of the proposed Passamaquoddy Site tidal power project, at the mouth of the Bay of Fundy.

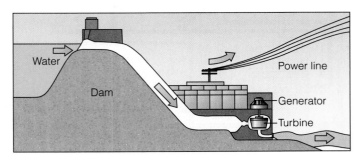

Figure 18.18
Basic components of a hydroelectric power station.

energy during times of peak demand. In fact, many hydroelectric installations have built water storage facilities for just that reason. These function by using excess power during off-peak periods to pump water to a high reservoir so that it can be channeled back down through electrical turbines at times of peak demand (Fig. 18.19).

Although hydroelectricity is a clean energy source, it is not without disadvantages. Large hydroelectric installations are very expensive to build and may be difficult to justify economically while oil prices are low. Hydroelectric reservoirs also drown vast tracts of land with significant ecological, social, and potential health consequences. In Egypt, for example, 80,000

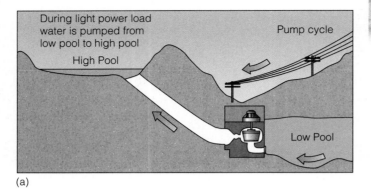

During light power load water is pumped from low pool to high pool

High Pool

Pump cycle

Low Pool

(a)

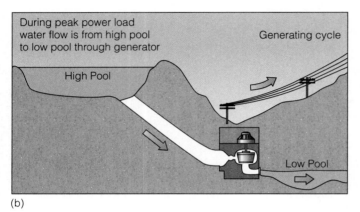

During peak power load water flow is from high pool to low pool through generator

High Pool

Generating cycle

Low Pool

(b)

Figure 18.19
Operation of a hydroelectric pump storage system.

people were displaced when Lake Nasser was created behind the Aswan High Dam. Wildlife habitats are similarly impacted in the flooded area, and the presence of a reservoir and dam may irreversibly alter the ecological balance both upstream and down (see Figure 10.44).

Hydroelectric reservoirs are also artificial basins and so are apt to become filled with sediment and silt up. Indeed, high sedimentation rates can limit the effective lifetime of a hydroelectric installation to as little as 20 years. There is also the potential threat of dam failure. Thus site selection is extremely important and should be chosen in a tectonically inactive region.

Despite the continuation of competitive fossil fuel prices, however, the U.S. Energy Information Administration expects a 50% increase in the worldwide use of hydroelectric power over the next twenty years, by which time it should account for some 10% of the world's energy consumption mix. But the potential for further development in the United States is likely to be limited because hydroelectric power is well established, and environmental and economic concerns are likely to preclude further dam building. Canada, on the other hand, has numerous plans to expand its hydroelectric resources despite decreasing demand for electricity. For the time being, at least, it has abandoned plans for large-scale installations, like the second and third phase of the James Bay Project in northern Quebec (see Figure 10.46), in the face of increased environmental concerns.

18.5 Energy in the Atmosphere

Energy is obtained from the atmosphere either directly or indirectly from the Sun's radiation. Direct solar energy is derived from the Sun's heat. However, as we learned in Chapter 11, uneven heating of atmospheric gases by the Sun also causes the air to move, allowing us to obtain energy from the wind.

18.5.1 Solar Energy

Direct use of the Sun's energy is an ancient technology and one that causes least harm to the environment. We witness the effect of passive solar radiation whenever we enter a warm, sunny room. Indeed, many houses have windows facing south and east to take full advantage of this "greenhouse effect." Solar energy is also abundantly available, easily modulated, and is readily accessible in remote areas. The process of converting solar energy into a usable form, however, is inefficient, expensive, and subject to changing weather conditions.

Current solar energy technologies are of two types; *solar thermal devices* use the Sun's heat, while *photovoltaic devices* convert the Sun's energy directly into electricity. Active solar heating uses collection devices called *solar panels*. Within these flat, glass-fronted panels, fluid circulated through pipes is heated by the Sun. The heat is then transferred to a storage tank and pumped to radiators (Fig. 18.20). Usually mounted on the roof, batteries of solar panels are used in many parts of the world to heat building space, water supplies, and domestic swimming pools. In the United States, for example, solar energy now heats water in more than 1.2 million homes and over 300,000 swimming pools.

Solar energy power plants operate on the solar thermal, rather than the photovoltaic, principle. A field of parabolic mirrors, or *heliostats*, designed to track the Sun, focus the sunlight

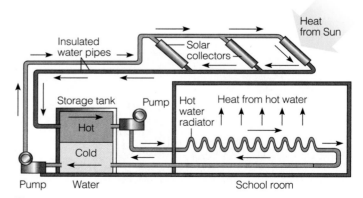

Figure 18.20
Operation of an active solar hot water heater using roof-mounted solar panels.

onto a central power tower filled with a fluid such as water, having the ability to store heat (Figure 18.21; see also Chapter 8). The heat generated by the focused sunlight can reach 1980°C (3600°F), turning the water into steam that drives a turbine to produce electricity. In the United States, which leads the world in solar electric power generation, there are now nine solar power plants in the Mojave Desert with a total generating capacity of 354 megawatts.

The basic photovoltaic device is the *solar cell* composed of semiconducting materials that produce electricity when sunlight is absorbed. To provide power, numerous interconnected cells are assembled in panels and either fixed on a roof or mounted to track the Sun. Solar cells are expensive, but they are used the world over to provide domestic electricity in remote regions. They are perhaps most familiar in North America as a means of powering highway call boxes. As yet, however, no large-scale photovoltaic systems have been commercialized.

18.5.2 Wind Power

The use of wind energy, like water power, is an ancient practice. Wind has been used to propel sailing ships for thousands of years and windmills have been milling grain and pumping water since the Middle Ages. Its use to generate electricity, however, is more recent, the first grid-connected wind generator having been used in Vermont during the Second World War. Today, most of the 3.4 gigawatts of wind power produced worldwide is in the United States and Europe.

The potential electrical generating capacity of the wind is enormous. In the United States, for example, it has been estimated to be twice the country's existing generating capacity, with the steady winds in the wide open states of Kansas, Texas, and North Dakota having the greatest potential power output. The wind is also a clean energy source. Like solar energy, however, winds are subject to changing weather conditions and are strongly influenced by the local terrain, so the usable portion of this resource is far smaller. In 1995, for example, only 3.2 billion kilowatt-hours of wind-generated electricity is estimated to have been produced in the United States, almost all of which was generated in California. Wind energy consumption is consequently small, however, it is growing more rapidly than that of any other renewable resource, having increased at an annual rate of about 10% throughout the 1990s. Nevertheless, its represents less than 0.5% of the country's total electrical generating capacity.

Commercially, wind energy is harnessed by *wind farms* comprising batteries of suitably sited wind turbines that can be used to generate electricity (Fig. 18.22). California's 3.2 billion kilowatt-hours of wind-generated electricity comes from about 16,000 such turbines. Almost all of Canada's production of wind-generated electricity comes from just 52 turbines on two such wind farms in Alberta. Sites are chosen on the basis of *wind power density*, a factor that takes into account the average wind speed, its variability, and the average air density. In general, sites with steady winds are preferable to those with more variable winds but the same average wind speeds, and lower ground altitudes are usually preferable to higher ones where

Figure 18.21
Solar One, a solar energy power plant in the Mohave Desert.

Figure 18.22
Wind farm near Palm Springs, California.

the air is thinner and greater wind speeds are required to generate the same energy. However, mountain passes, which tend to focus and accelerate air flow, form the sites of many Californian wind farms. Wind turbines are also mounted on towers to avoid the reduced air speeds due to frictional drag from buildings, vegetation, and topographic irregularities near the ground surface.

In the United States, it is the Great Plain states, the Midwest, the West, and the Northeast that have the most abundant wind resources. However, because of the intermittency of wind energy and the geographic restriction of suitable sites to harness it, wind power is likely to remain a local energy option.

18.6 Energy in the Biosphere

The oldest energy source of all comes from the burning of organic matter, its use dating back to the discovery of fire. Today, this resource is termed **biomass** and includes wood, municipal solid wastes, and biomass-derived liquid fuels such as ethanol. As a renewable energy source, biomass is second only to hydropower, providing about 3 quadrillion (3,000,000,000,000,000) British thermal units (Btu) of energy, or 45% of total renewable energy consumption.

18.6.1 Biomass

For most of human history, the principle source of energy was that produced from biomass. Indeed, it was not until the 20th Century that wood was displaced by other energy sources as the primary fuel supply, and even today, wood provides 80% of biomass energy consumption. In the United States, this contribution is equivalent to some 200,000 barrels of oil per day (about 1% of its daily oil consumption).

The most familiar form of energy obtained from wood is the energy of firewood, most of which is used for residential heating. However, fully two-thirds of this energy source is derived from wood waste, such as that produced by the pulp and paper industry. Energy derived from municipal solid waste, manufacturing waste, and landfill gas, is the second largest source of energy from biomass. Biomass ethanol derived from corn and used as a gasoline additive in "gasohol," is a distant third, although the production capacity for fuel ethanol in the United States reached 1.5 billion gallons in 1995.

Not surprisingly, the use of energy from biomass is very dependent upon proximity to a source of supply. In the United States, most energy obtained from wood is consumed in the forested South, ethanol use is greatest amid the cornfields of the Midwest, and consumption of energy generated from waste is highest in the population centers. Biomass also generates about 60 billion kilowatt-hours of industrial electricity in the United States, particularly in pulp and paper mills. The United States currently converts about 16% of its municipal solid waste into energy in some 116 waste-to-energy combustion facilities. However, it has been slower to develop its urban

waste as a fuel source than has Europe, some countries of which use more than a half of their municipal waste for energy production.

18.7 Energy Outlook for the Future

18.7.1 Future of Fossil Fuels

Because of their dependence on technological advances, economic conditions, and socio-political developments, future trends in the supply and demand for energy resources are very difficult to predict. However, one fact is clear. Despite an increase in the use of renewable resources, almost 90% of the world's energy consumption presently relies on fossil fuels, the total supply of which is finite. So how long will our present supplies of fossil fuels last, and what other energy options are available to take their place as the reserves of these fuels decline?

Production of oil in the United States, for example, has been declining since the early 1980s, yet the quantities of fuel the country consumes are now so great that even minor inequities between supply and demand can have serious repercussions. As a sobering example of this, the OPEC oil embargo of 1973 caused considerable disruption across western Europe and North America, despite the fact that total oil consumption in the United States, for example, was cut by less than 5%.

North American supplies of natural gas are present in greater quantities than those of oil, and large deposits of sub-economic oil shales in Alberta, Wyoming, and Colorado may one day be added to the oil reserves of the United States and Canada. Coal is also present in large quantities in both countries but the severe environmental problems associated with its use as a fuel are likely to limit its future as an energy resource. Eventually, the world's declining reserves of fossil fuels will necessitate the development of alternative energy sources. But what source of energy could replace fossil fuels by providing almost 90% of our global energy needs?

Nuclear energy certainly could, but would we want it to? Sufficient reserves of uranium ore exist in North America, for example, to fuel a nuclear power industry based on the breeder reactor that could provide all of the electricity needs of the United States and Canada for several hundred years. Nuclear power generation, however, is not without serious potential hazards and, following accidents like those at Three Mile Island in Pennsylvania and at Chernobyl in what is now the Ukraine, the nuclear power industry is beset by a lack of public confidence. There are also unresolved problems regarding radioactive waste disposal and the security of nuclear fuel in the face of terrorist activity, which constitute hidden costs to nuclear power. Finally, nuclear energy, like the fossil fuels, is obtained from a nonrenewable resource, and so would serve only to delay the eventual necessity to develop an alternative, renewable energy supply.

Unfortunately, most existing renewable sources of energy, such as geothermal energy, solar energy, hydroelectric power, wind power, and tidal power, are restricted to those areas on the Earth's surface where power can be produced by these means. As a result, they are typically small-scale operations that would be hopelessly inadequate as replacements for fossil fuels. Indeed, even hydroelectric power, which presently accounts for almost 95% of all power generated from renewable energy resources, provided only 7% of the world's primary energy production in 1995, and is not without serious environmental consequences (see Chapter 10).

A possible answer lies in the development of nuclear fusion, the same process that fuels the stars by fusing hydrogen into helium. As we discussed in Chapter 16, this fusion does not conserve mass; 0.7% of the hydrogen's mass is converted into energy. As an energy source, nuclear fusion would share few of the hazards of nuclear fission and could theoretically provide power from the virtually limitless supplies of hydrogen in seawater. However, the technology has yet to be developed and it is most unlikely that anything approaching a commercially feasible fusion reactor will be available until well into the 21st Century. In the meantime, we must make do with existing energy sources and so are likely to witness continued exploitation of our nonrenewable resources. Consequently, in their analysis of future trends in world energy consumption, the U.S. Energy Information Administration foresees no immediate reduction in our dependency on fossil fuels.

18.7.2 Future Trends in Supply and Demand

Having grown at an annual rate of 2.6% for two decades, world energy consumption totaled more than 362 quadrillion Btu in 1990. The annual growth rate of consumption is expected to slow to 1.6% over the next decade with the implementation of more efficient technologies, but even at this rate, the U.S. Energy Information Administration expects total energy consumption to have reached 470 quadrillion Btu by 2010. The bulk of the increased supply (86%) is expected to be met by fossil fuels, with almost 40% being met by oil alone. Because the cost of fossil fuels has remained competitive, the annual growth rate of nonrenewable energy consumption is also expected to slow to 2.3%, down from an average rate of 3.9% between 1970 and 1990. Little or no growth is expected in nuclear power because of public concerns regarding its safety. As a result, no reduction in our dependency on fossil fuels is expected, despite the projected decrease in the growth rate of global energy consumption.

Despite this projected decrease in growth rate, the global demand for energy will continue to rise as the world's population increases and the less-developed countries become increasingly industrialized. By 2010, for example, the world population is expected to be almost twice that of 1970, surpassing 7 billion. Similarly, economic activity in 2010, as measured by the world's gross domestic product, is projected to be more than triple that of 1970. Over the same period, energy consumption is expected to more than double. The majority of this growth, both in pop-ulation and economic activity, is expected to occur in the less-developed countries where present energy use is small compared to those of the OECD. But if these non-OECD countries were to start consuming as much energy per person as the OECD countries, estimates for their energy demand in 2010 would increase more than fivefold, bringing world energy consumption to a staggering 1560 quadrillion Btu.

18.8 Chapter Summary

Energy is the lifeblood of modern society and is consumed by the industrial nations at a staggering rate. The consumption of energy worldwide is already approaching 400 quadrillion British thermal units (Btu) per year, and can be expected to rise further as the world's population grows and the energy demands of the less-developed countries increase. Yet fully 96% of this energy is obtained from nonrenewable sources, the total supply of which is finite. An understanding of our energy resources and reserves, their availability and distribution, and the processes that lead to their production and use, is consequently of great importance.

Almost 90% of the world's energy needs are presently met by burning the three fossil fuels—oil, natural gas, and coal—each of which represents buried organic matter slowly transformed over geologic time. Oil and gas, known collectively as petroleum, are produced by the burial and heating of oceanic microorganisms under oxygen-starved conditions like those that can develop in nearshore environments with restricted water circulation. For petroleum to form and accumulate, however, four conditions must be met. First, a source rock capable of producing petroleum must be present, and must be buried to a sufficient depth. Second, a porous reservoir rock is needed to store the oil and natural gas. Third, an impervious seal is necessary to prevent their continued upward movement, and fourth, some form of geologic trap is required to hold the petroleum underground. This last condition is met by both structural traps produced by bending or fracturing of sedimentary layers, and sedimentary traps produced by changes in layer rock type.

Detection of potential petroleum traps is based mainly on seismic reflection, a geophysical method that employs penetrating shock waves to effectively X-ray the subsurface. Most traps, however, do not contain petroleum and once detected must still be tested through drilling.

Although oil and natural gas deposits are widespread, oil is a rare commodity and most is produced from just a handful of supergiant fields in the Persian Gulf, Alaska, the Gulf of Mexico, the Russian Arctic, and the North Sea. As of 1996, the world's proven oil reserve was estimated to be just over one trillion barrels, almost two thirds of which lies below the politically unsettled region of the Middle East. In 1996, global oil consumption was estimated to be about 65 million barrels a day, a rate that would exhaust this reserve in a little over 40 years. The world's reserve of natural gas, 70% of which lies in the countries of the former Soviet Union and the Middle East, is estimated at almost five quadrillion cubic feet. Global consumption of natural gas approached 82 trillion cubic feet in 1996, a rate that would exhaust this reserve in a little over 60 years.

Yet the longevity of the world's petroleum reserves is difficult to predict because areas of petroleum potential remain to be explored, and known areas may yield to improved technology and enhanced oil recovery techniques. Significant petroleum reserves also exist in oil shales and tar sands, although commercially viable methods of mining and treating these materials have yet to be developed on a large scale.

The most abundant fossil fuel is coal, which is produced by the burial and compaction of land plants. Coal is formed by the progressive burial of peat, which is transformed first to lignite (brown coal) with compaction, and then to subbituminous coal, bituminous coal, and anthracite with rising pressure and temperature. An increase in coal rank, which is based on carbon content, accompanies this progression. Coal grade, on the other hand, increases with lower sulfur and ash contents.

Coal deposits, which are mostly of Carboniferous age, are widely distributed and easy to find. Because of this, coal is a relatively inexpensive fuel and while its use is not without severe environmental consequences, it is likely to become an important part of the world's energy supply as diminishing petroleum reserves drive prices ever higher. Indeed, sufficient coal reserves exist to last over two hundred years at present levels of coal consumption. Almost half of these deposits occur in the United States and the former Soviet Union.

Nuclear power depends on the heat released by the breakdown or fission of the naturally occurring radioactive element, uranium. It is therefore dependent upon a mineral resource and, like the fossil fuels, is a nonrenewable energy source. Most uranium deposits occur in sandstones and conglomerates, either as precipitates from uranium-bearing groundwater or as Precambrian placer deposits, the largest reserves of which are in the United States and Canada.

In conventional nuclear reactors, the heat used to generate electricity is produced by controlling chain reactions in uranium enriched in its fissionable isotope (^{235}U). But the bulk of the uranium is not fissionable and so goes to waste. In breeder reactors, however, the chain reaction is also used to convert some of the non-fissionable uranium into fissionable plutonium fuel. Hence, breeder reactors require far less uranium fuel.

Nuclear power, led by the United States, France, and Japan, accounts for almost a quarter of the electricity generated worldwide but its future is uncertain. In the Far East, use of nuclear power is expected to increase. In the United States, however, the expense of nuclear power stations coupled with public concerns over reactor safety, the security of nuclear fuel, and the problems of nuclear waste, has brought the nuclear industry to a virtual standstill.

Geothermal energy is a renewable resource, the potential for which exists wherever hot rocks lie close enough to the surface to be reached by boreholes. Water heated by such rocks is used for heating and generating electricity, especially in areas of recent volcanism near active plate boundaries and hotspots. However, the recoverable resource is too small to satisfy more than a fraction of the world's energy needs. Although it is an important resource in Iceland, geothermal energy only provides about 0.5% of the electricity needs of the United States, the bulk of which is generated by The Geysers facility in northern California.

Most renewable energy resources, however, exist in the hydrosphere and atmosphere. The energy of moving water forms the basis of tidal power, wave energy and hydroelectricity. Tidal power stations, such as that on the River Rance in France, form local energy options in coastal areas with large tidal ranges and spacious tidal basins. Hydroelectricity, however, is the primary source of renewable energy and, led by Canada and the United States, generates about a quarter of the world's electricity. Hydroelectric power is also clean and cost effective, and its use is certain to increase. However, it requires that a dam and reservoir be constructed, which is expensive and also has significant ecological and social consequences.

Solar energy and wind power are renewable energy resources in the atmosphere. Solar energy is used in both solar thermal devices, such as solar panels, that use the Sun's heat, and photovoltaic devices, such as solar cells, that convert sunlight directly into electricity. Solar power plants, like those in California, use fields of parabolic mirrors to focus the Sun's heat onto a furnace. The use of wind energy, led by the United States and Europe, employs batteries of wind turbines to generate electricity at sites chosen for wind speed and constancy. Both energy sources, however, are expensive and intermittent, and neither provide more than a small fraction of our global energy needs.

Biomass, the energy resource that includes the burning of wood, municipal solid waste, and fuel ethanol, is second only to hydropower as a source of renewable energy. Most energy from biomass is derived from firewood and wood waste, but as much as 50% of the urban waste of some European countries is converted to energy in waste-to-energy combustion facilities.

Because of their dependence on technological and political uncertainties, future energy trends are difficult to predict. However, there is likely to be no immediate reduction in our dependence on fossil fuels, given the lack of public confidence in nuclear power and the inadequacy of renewable energy sources as replacements. A possible alternative lies in nuclear fusion, but this technology has yet to be developed.

Key Terms and Concepts

Look for the highlighted items on the web at: WWW.BROOKSCOLE.COM

- Fossil fuels (oil, natural gas, and coal) are nonrenewable energy sources of organic origin that provide almost 90% of the world's energy needs

- Four conditions: a source rock buried to an appropriate depth, a reservoir rock, a seal, and a trap, must be met for petroleum deposits to form

- Traps, which are subsurface structures that allow petroleum to accumulate, can be structural (anticline, salt dome, and fault) or stratigraphic (pinchout, reef, and unconformity)

- In providing a mechanism for the development of petroleum-producing sedimentary basins and the formation of stratigraphic and structural traps at plate boundaries, plate tectonics plays an important role in the distribution and development of petroleum deposits

- Seismic reflection, which uses shock waves to penetrate the subsurface, is the geophysical technique most widely used for petroleum exploration

- Unconventional sources of petroleum include oil shales, which yield petroleum when heated, and tar sands, which contain bitumen

- Coal is the end product of the progressive burial of plant material that yields peat, lignite (brown coal), subbituminous, and bituminous coal, and finally, anthracite

- Coal rank is based on carbon content whereas coal grade is based on purity

- Nuclear power, which provides about 8% of the world's energy needs, is driven by the fission of naturally occurring radioactive uranium, the fissionable isotope of which (^{235}U) must usually be enriched in order to produce a chain reaction

- In breeder reactors, fission of ^{235}U is used to convert waste ^{238}U into fissionable plutonium (^{239}Pu)

- Renewable energy sources, such as geothermal energy (which uses natural heat from the Earth's interior), solar energy, hydroelectric power, wind power, wave energy, and energy from biomass provide about 4% of the world's energy needs

Review Questions

1. What percentage of our global energy needs are met by each of the major energy sources?
2. What is the distinction between renewable and nonrenewable energy sources?
3. What are the fossil fuels and why are they so called?
4. What four conditions must be met for petroleum to accumulate in commercial quantities?
5. Why is most oil and natural gas found in rocks that are less than 100 million years old?
6. What are the three main types of structural and stratigraphic traps?

7. Why are coal deposits unlikely to be more than 400 million years old?
8. At present production rates, how long can the world's supply of coal be expected to last?
9. What is nuclear fission and under what conditions can it cause a chain reaction?
10. Why is the safe disposal of nuclear waste so difficult?
11. How is hydroelectric power generated?
12. What are three main sources of biomass energy?

Study Questions

 Research the highlighted question using InfoTrac College Edition.

1. How has humanity's use of fuel evolved through time?
2. How does the energy use of industrial nations differ from that of less-developed ones?
3. What controls whether buried organic matter is converted to petroleum or coal?
4. At current levels of consumption, what would be the longevity of the world's supply of oil and natural gas, and why is it difficult to be certain of this estimate?

5. What factors favored the formation of coal deposits in northeastern North America and western Europe during the Carboniferous Period?
6. What is the difference between conventional reactors, such as those of the light-water variety, and breeder reactors?
7. Why does the U.S. Energy Information Administration foresee no immediate reduction in our dependency on fossil fuels?

Earth Resources

Solid Earth

Plate tectonics is responsible for the formation of many ore deposits and the uplift and sedimentation that produces fossil fuels. Magmatism at active continental margins may yield geothermal energy. Evaporite and residual deposits may form along passive margins if the climate is suitable.

Hydrosphere

Circulation of the hydrosphere through the ocean floor produces exhalite deposits. Its movement creates tidal power, hydroelectric power, geothermal, and wave energy. It harbors microorganisms whose remains produce petroleum and the continental shelf sediments in which petroleum accumulates. Movement of fresh water results in erosion and mineral concentration in placer deposits.

Atmosphere

Atmospheric circulation is responsible for wind and wave energy and the hydrologic cycle on which the biosphere depends.

Biosphere

The biosphere mediates formation of certain ore deposits and produces raw materials for fossil fuels and biomass energy.

Astronomy

Solar radiation causes evaporation in salt basins, drives the atmospheric circulation, maintains the biosphere, and powers solar energy.

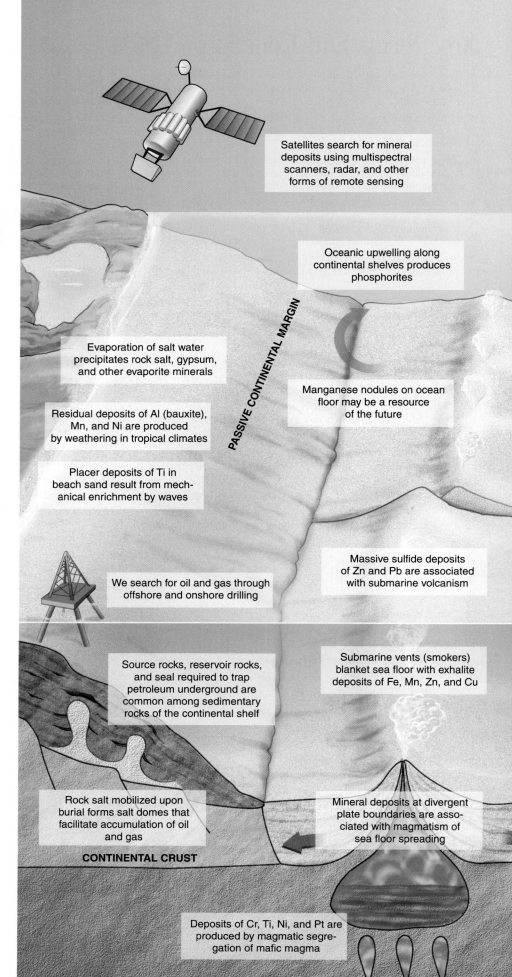

Satellites search for mineral deposits using multispectral scanners, radar, and other forms of remote sensing

Oceanic upwelling along continental shelves produces phosphorites

PASSIVE CONTINENTAL MARGIN

Evaporation of salt water precipitates rock salt, gypsum, and other evaporite minerals

Manganese nodules on ocean floor may be a resource of the future

Residual deposits of Al (bauxite), Mn, and Ni are produced by weathering in tropical climates

Placer deposits of Ti in beach sand result from mechanical enrichment by waves

Massive sulfide deposits of Zn and Pb are associated with submarine volcanism

We search for oil and gas through offshore and onshore drilling

Source rocks, reservoir rocks, and seal required to trap petroleum underground are common among sedimentary rocks of the continental shelf

Submarine vents (smokers) blanket sea floor with exhalite deposits of Fe, Mn, Zn, and Cu

Rock salt mobilized upon burial forms salt domes that facilitate accumulation of oil and gas

Mineral deposits at divergent plate boundaries are associated with magmatism of sea floor spreading

CONTINENTAL CRUST

Deposits of Cr, Ti, Ni, and Pt are produced by magmatic segregation of mafic magma

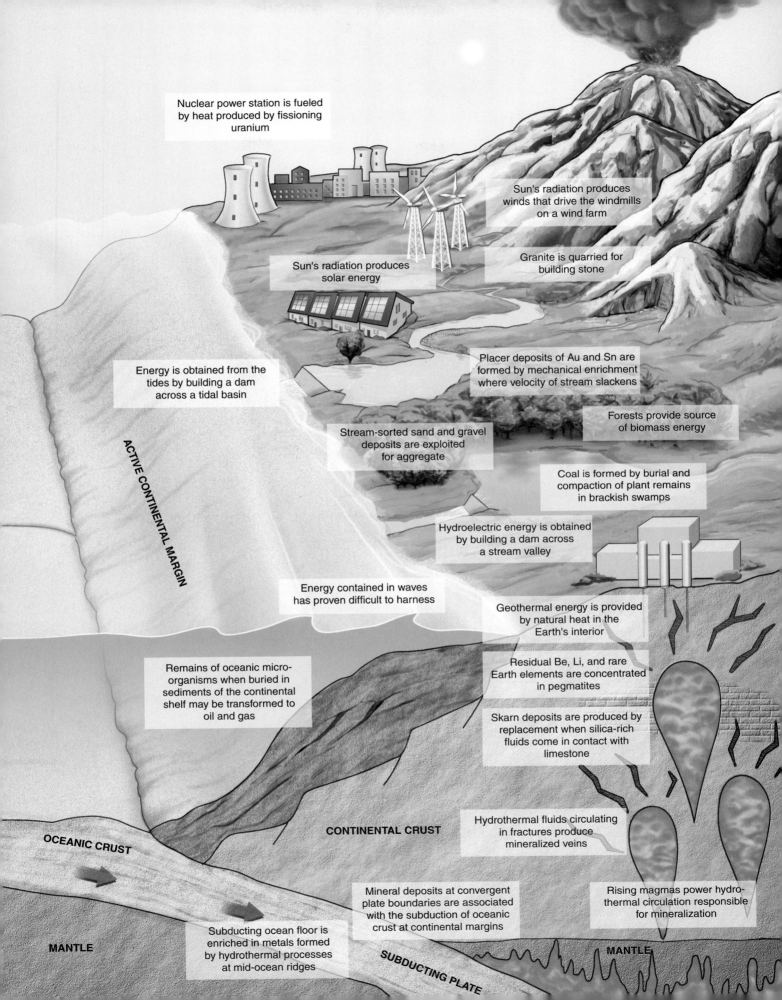

PART 7

Earth Science and the Environment

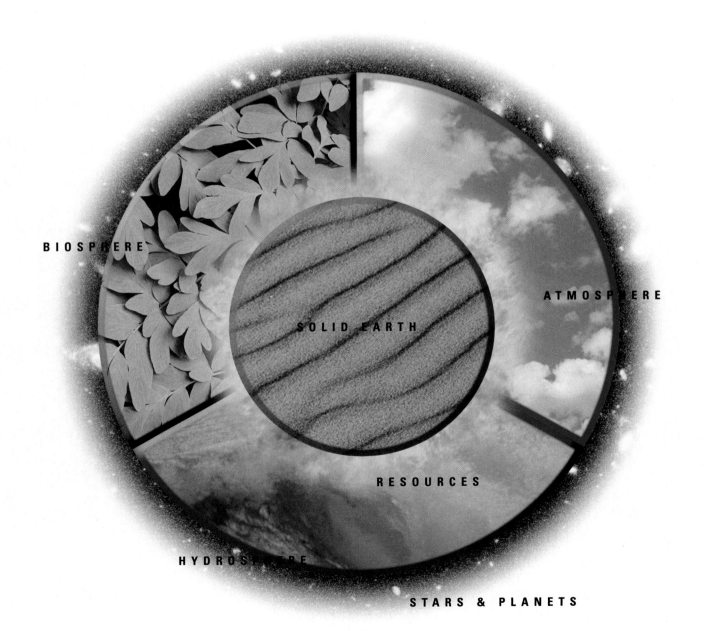

BIOSPHERE

ATMOSPHERE

SOLID EARTH

RESOURCES

HYDROSPHERE

STARS & PLANETS

19

Biogeochemistry: Interaction between the Earth's Reservoirs

In this chapter, we focus on the exciting interdisciplinary science of biogeochemistry. We will show that nature is inherently interdisciplinary and does not respect the artificial boundaries that we have constructed between the traditional sciences such as biology, chemistry, geology, and physics. We start with the best known example of biochemistry, the food chain, which describes the passage of energy and nutrients in the biological realm. We will see that the food chain has existed in various forms for hundreds of millions of years, and that the energy derived from the food chain can be "stored" in specialized geologic environments for later use. This stored energy is better known as fossil fuel (coal, oil, and natural gas).

We then turn to soil, the thin veneer of unconsolidated sediment that lies between bedrock and the atmosphere. By definition, soil must be able to support plant life. As we learned in Chapter 10, soil also contains air and water in pore spaces between sediment grains. Therefore, soil is a special environment where all the Earth's systems interact. We shall see that soil development depends on climate, drainage, the activity of microorganisms, and the nature of the underlying bedrock. Soil development also depends on the length of time in which these systems interact.

We then track the hypothetical path of a single carbon atom through the Earth's surface reservoirs, to show how chemical elements migrate from one geologic environment to another according to the laws of physics. As they do so, their migration is aided by radiant energy from the Sun and they may pass into and out of the biological realm. In short, the odyssey of a single carbon atom provides examples of many of the most important processes we have discussed in preceding chapters.

Finally, we look at the science of biogeochemistry in more detail, by examining the biogeochemical "cycles" of carbon, nitrogen, and phosphorus, which are the essential elements in DNA and RNA, the building blocks of life. We find that each element has its own migration path through the Earth surface reservoirs, but that interaction among these elements is needed to stimulate life. We speculate that this interaction takes place at times of major biological innovation in the geologic record.

At the beginning of this book, we suggested that if we understood nature's rhythms, we could detect the effects of human interference. We will find that the science of biogeochemistry offers very important insights in this regard. By laying out the natural pathways of chemical species as a kind of road map, we can readily identify where humans have forced artificial "shortcuts" or short circuits in the natural cycle.

19.1 Introduction

In the preceding chapters, we have repeatedly seen evidence both in the geologic record and in modern environments that interaction among the Earth's surface reservoirs has guided our planet's evolution for at least 4 billion years. We have seen how the movement of plates is the key to the formation of the solid Earth, and how associated volcanic activity has helped to create our hydrosphere and atmosphere (Chapters 8 and 11). We have learned how various connections between the hydrosphere and atmosphere control climate (Chapter 12) and the motion of water on our planet (as exemplified by the hydrologic cycle). We have also discovered that these reservoirs are stimulated from *above* by solar radiation (Chapters 8 and 16), and are fed from *below* by processes in the Earth's hidden interior (Chapter 6) that drive some of the planet's chemical constituents towards the surface. We also assume that this interaction will continue into the future. The study of the evolution of the Earth systems, therefore, provides the template by which to examine the natural pulse, or rhythm, of Planet Earth.

Concerns about the future of life on Earth have galvanized scientists into research on environmental issues. As you might imagine, this is a daunting task. Earth's systems are highly complex and scientists are still far from solving them. There are still too many unknowns and not enough equations! This implies that scientists must make some simplifying assumptions in order to obtain models that try to predict the future of our environment. Very often, the validity of these assumptions is hotly debated.

As an initial approach, it is important to first look at the broad perspective in order to identify the major questions that can only be solved by subsequent research. What are the major issues influencing our environment and the future of life on Earth? Living creatures need air and water to survive. They also need nutrients that come directly or indirectly from the solid Earth. Elements such as calcium and phosphorus are essential for the bones and shells of animals, and trace amounts of other elements are vital for the growth of vegetables and other plants. To facilitate this growth, these trace elements must cross the bridge from the inorganic solid Earth to the biosphere (Fig. 19.1). For example, plants absorb inorganic chemicals from the soil and produce organic matter by way of photosynthesis. Animals utilize this organic material, either by grazing on and digesting the plants directly (herbivores) or by eating other animals (carnivores). Decomposition of plants and animals eventually returns these chemicals to the soil (Fig. 19.2). Vegetation, then, is essentially a chemical extract from the soil beneath, and transforms chemicals from an inorganic form to an organic form. The chemistry of the soil is influenced by the decomposition of organic matter, by the disintegration of inorganic near-surface rocks, and by the absorption of chemicals from groundwater and the atmosphere.

The Sun also plays a key role in the evolution of life, not only by providing energy to promote the interaction and circulation of air and water, but also for photosynthesis. As volcanic activity continually replenishes our atmosphere with necessary gases, photosynthesis continually purifies their potentially toxic composition by absorbing carbon dioxide and producing oxygen. If volcanoes produce the raw materials,

Figure 19.1

Plate tectonics introduces inorganic chemicals into the atmosphere, hydrosphere and the solid Earth. These chemicals are then circulated and become trapped in soils where they are absorbed by plants (producers) and enter the organic realm. From there, they may be transferred along the food chain by consumers, either herbivores (plant-eaters) or carnivores (meat-eaters), until death and decomposition of the animal. This material is processed in the crust through burial and the cycle is completed.

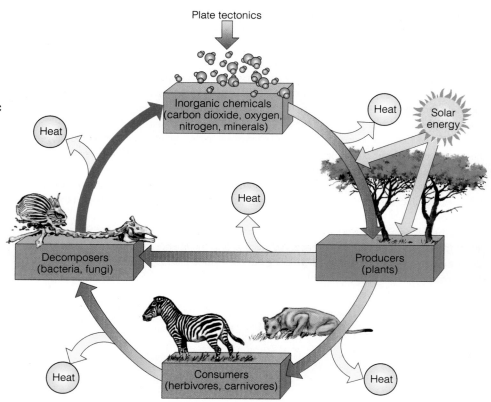

Energy and nutrient transport in the biosphere

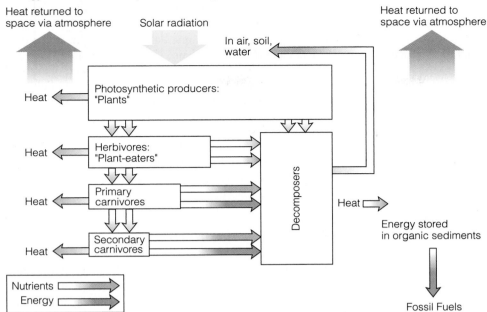

Heat returned to space via atmosphere

Solar radiation

Heat returned to space via atmosphere

In air, soil, water

Photosynthetic producers: "Plants"

Heat

Herbivores: "Plant-eaters"

Heat

Primary carnivores

Heat

Secondary carnivores

Heat

Decomposers

Heat

Nutrients

Energy

Energy stored in organic sediments

Fossil Fuels

Figure 19.2
The food chain. Solar radiation stimulates photosynthesis in plants. This energy is transmitted further down the food chain by herbivores (plant-eaters), primary carnivores (herbivore eaters), and secondary carnivores (carnivore eaters). However transmission is very inefficient, so that the further down the food chain, the less abundant the species. If decomposition of organic matter occurs in an oxygenated environment, the energy (in the form of heat) is returned to the atmosphere. If decomposition occurs in an oxygen-depleted environment, the energy may be stored in the solid Earth as fossil fuels.

photosynthesis is the factory that processes these materials to form a life-sustaining product.

Although the broad picture concerning the migration of chemical elements is clear, how do we go about systematically analyzing what seems to be a bewildering array of processes? As usual in science, we start from a foundation that we think gives us a good grasp of the basic processes. In this case, it is the food chain.

19.2 The Food Chain

Together with the hydrologic cycle, the **food chain** is probably the best known example of the interplay between the Earth's surface reservoirs and the Sun. Figure 19.2, which illustrates this interplay, is the first of a series of diagrams in this chapter that show the connections among the Earth's surface reservoirs. At first glance the connections may look almost impossible to decipher. But try to think of them as a type of road map that either energy or chemical elements use to travel. The roads can be either one-way or two-way streets. Sometimes the "traffic" can get a little heavy on some of the streets as chemical species travel from one destination to another. Therefore, the pathway needs to be efficient or the motion of energy or chemical species is impeded (a kind of chemical traffic-jam). The rectangular boxes in the diagram are the temporary rest stops along the journey, where energy and nutrients enter the biosphere as part of a living species.

The food chain is stimulated by radiant energy from the Sun and the availability of moisture. However, this is not a very efficient process. About 50% of the solar radiation that enters the Earth's atmosphere is either absorbed or reflected back into space and so never reaches the Earth's surface. For example, when a

cloud passes between us and the Sun, we can feel the temperature drop because the cloud either absorbs or reflects a portion of the Sun's energy. A further 30% of solar radiation is reflected off the Earth's surface. Altogether, less than 1% of the total radiant energy is utilized for photosynthesis. Yet this small proportion of radiation makes the food chain possible by stimulating photosynthesis in plants like those of modern tropical rain forests.

The base of the food chain therefore depends on photosynthesis and the presence of plants. In its simplest form, the photosynthetic reaction may be summarized as:

$$6CO_2 + 6H_2O + energy \rightarrow C_6H_{12}O_6 + 6O_2$$

carbon dioxide + water + energy → sugar + oxygen

As we learned in Chapter 13, the process of photosynthesis draws energy from the Sun's radiation to create carbohydrate sugar from carbon dioxide and water. In the process, oxygen is produced as a by-product. The transformed energy is stored in the plant's sugars and starts to be transmitted down the food chain when the plants are eaten by herbivores (plant-eaters). These, in turn, may be eaten by primary carnivores (herbivore eaters) or secondary carnivores (carnivore eaters) lying higher in the food chain (see Figure 19.2). There are, of course, many levels of carnivore activity. Humans, for example, are omnivores (eaters of plants and animals), which is one of the main reasons for their success in populating the planet.

The transmission of energy and nutrients down the food chain is not a particularly efficient process. The relative sizes of the boxes in Figure 19.2 give a rough indication of the relative abundance of each member of the food chain. In fact, producer plants only transmit about 10% of the stored energy to herbivores, and only 10% of that is passed on to primary carnivores. Similarly, only 10% of this energy is passed on to secondary carnivores. This is only 0.1% of the energy initially produced by photosynthesis and less than 0.001% of the solar radiation entering the atmosphere. The inefficiency of the food chain explains why there are fewer and fewer species as we follow its path from plants to secondary carnivores. So there are vast numbers of edible plants and numerous insects and rabbits, but fewer foxes and even fewer lions.

One of the reasons for this inefficient transmission of energy is that, like humans, plants and animals use some of this energy and also burn off calories in the process of living. In Figure 19.2 this loss of energy is depicted by the arrows labeled "Heat" leaving each of the boxes. This heat ultimately returns to space by way of the atmosphere.

However, upon the death of a plant or animal, the remainder of its unspent energy may be stored if the environment is favorable. Those members of the food chain that are not eaten eventually die and decompose. As we have seen (Chapter 18), the fate of decomposing organic material depends on the environment in which it occurs. If decomposition takes place in contact with the oxygen-rich atmosphere, the energy is lost to the atmosphere in the form of heat. In unconsolidated sediments, decomposition of plant roots, bacteria, fungi, and burrowing animals occurs immediately below the surface in an oxygen-deficient environment. The decomposition of this organic material results in a nutrient-rich soil called **humus**. The energy stored in this layer is a vital part of the food chain and is used by the agriculture industry to generate much of the food we eat.

Energy may also be preserved in oxygen-deficient environments such as those that exist in swamps and near the estuaries of major streams. Trees that die in swampy regions, like those of ancient tropical rain forests, may be rapidly buried and compressed, gradually forming peat and then coal deposits like those of the Carboniferous Period. Similarly, rapid burial of marine microorganisms in sediments of the continental shelves may initially preserve the material, but will gradually convert it to oil and natural gas as it becomes progressively heated and compacted with burial.

As we learned in Chapter 14, the efficiency of the food chain must have changed radically between the Ordovician and Carboniferous periods, some 470 to 290 million years ago. The fossil record demonstrates that the land became progressively colonized by plants and animals during this time, and by the end of the Carboniferous abundant coal deposits had formed from the decay of continental vegetation. The availability of fossil fuels as an energy source in these ancient rock formations is no quirk of nature. Fossil fuels are like energy banks, storing ancient portions of the Sun's radiant energy that have been transmitted down the food chain. The presence of fossil fuels in the geologic record suggests that energy has been transferred from one Earth reservoir to another for hundreds of millions of years, and is testament to the importance of the food chain throughout this portion of geologic history. In detail, since ecosystems vary from one habitat to another, so too does the operation of the food chain. Figures 19.3a and b show the essential differences between the major components of the food chain in a fresh water pond and those in an adjacent field.

The food chain is an important example of the significance of the interaction of the Earth systems. One aspect of this interaction, that of soil development, deserves special treatment. Along with air and water, soil is a vital resource that provides much of our food. In the next section, we show that soil development depends on the interaction of all the Earth's systems.

19.3 Soil Profiles

To an engineer, soil includes all unconsolidated sediment that overlies bedrock. To an Earth scientist, however, **soil** is defined as loose material at the Earth's surface that is capable of supporting plants and their root systems. For an Earth scientist, the ability of soil to sustain life is an essential aspect of its definition. This thin veneer of material that separates bedrock from the atmosphere not only consists of unconsolidated sediment, plant life, and a variety of microorganisms that depend on plant life, but also contains pore spaces that contain trapped air and water. Soils are therefore a venue in which all the Earth's systems come together and interact.

Most of the Earth's surface has a layer of unconsolidated sediment consisting of rock and mineral fragments produced by weathering of the underlying bedrock. However processes operating within this sediment may cause it to develop a layered structure capable of supporting life. The layered sequence as a whole is known as a **soil profile**, and it is this profile that dictates the use of the soil for agricultural purposes. The degree of layering depends on a number of factors, but the most important are climate, the length of time that soil-forming processes have been operative, and the action of microorganisms.

Soils are best developed in regions with warm climates and abundant rainfall. As we learned in Chapters 11 and 12, these favorable conditions occur in tropical regions and at mid-latitudes where microorganisms flourish (Fig. 19.4). The topography and the nature of the underlying bedrock also influence the development of soil layers and are important in explaining the variation of soil quality within the same climate belt.

A mature soil is one in which soil-forming processes have been operating for a considerable amount of time. A mature soil has a number of characteristic layers. The uppermost layer is

Figure 19.3

Major components of the food chain in (a) a fresh water pond and (b) a field adjacent to the pond. Producers (rooted plants and microscopic phytoplankton) absorb water and nutrients, and manufacture organic material by way of photosynthesis, absorbing carbon dioxide and releasing oxygen in the process. Consumers obtain energy by eating plants (primary consumers), by eating plant-eaters (secondary consumers), or by eating carnivores (tertiary consumers).

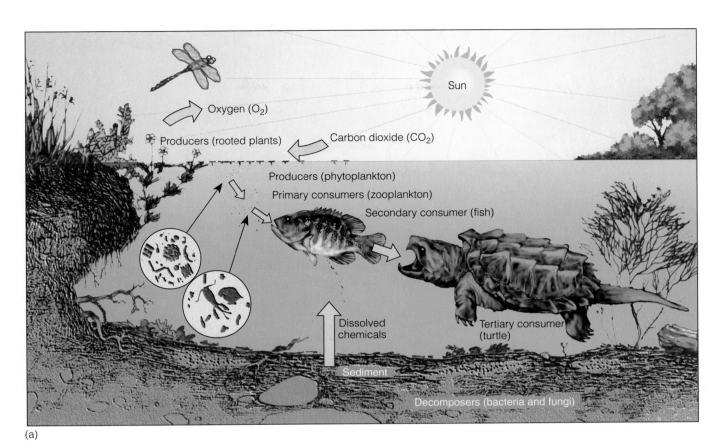

(a)

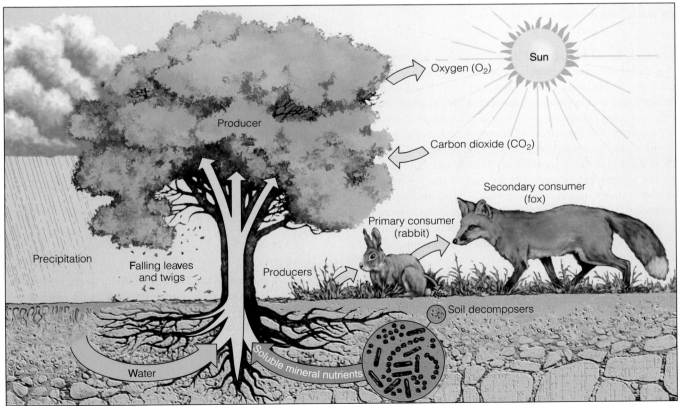

(b)

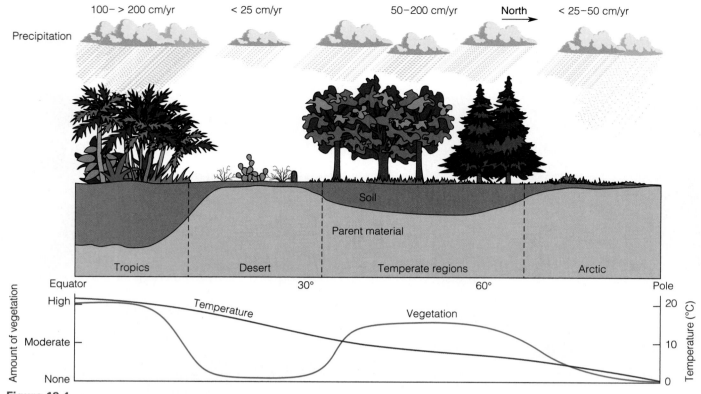

Figure 19.4

The typical variation of soil development as a function of climate. Soil is best developed in tropical and temperate regions where there is abundant precipitation. This facilitates the weathering of bedrock, the development of vegetation, and the action of microorganisms. Soil development is poor in desert regions due to lack of sufficient rainfall, and in polar regions because the cold climate restricts the development of vegetation.

referred to as the **O horizon**. This layer is only a few centimeters thick and is rich in decaying organic matter (Fig. 19.5). Beneath the O horizon is the **A horizon**, where plant roots, microorganisms, and burrowing animals live. Dead plant material is broken down by microorganisms, and the vigorous organic activity results in the formation of humus, which is also known as **topsoil**. Other than organic material, the A horizon consists mostly of clays and stable minerals such as quartz. The E layer constitutes a "zone of leaching" because percolating groundwater dissolves soluble material and generally transports the material downward. In humid climates, the soluble material may be completely removed from the soil. More generally, however, the material is deposited in a layer known as the **B horizon**, which consequently constitutes a "zone of accumulation."

There are some exceptions to the downward movement of groundwater. For example, in arid climates, groundwater is drawn upwards to the surface where it evaporates. In this case, minerals precipitate at the surface instead of accumulating in the B horizon. Beneath the B horizon is the **C horizon**, which consists mostly of broken up bedrock and forms a transition zone to the parent bedrock beneath.

Because of the processes that cause their development, the chemical composition of soils varies considerably from the top of the soil profile to the bottom. The chemistry of the uppermost layers (O and A horizons) is profoundly influenced by organic activ-

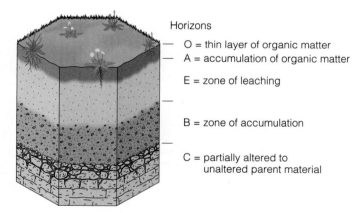

Figure 19.5

Typical profile in a mature soil. Soil is divided into layers, or horizons. The O horizon is rich in organic matter. The A horizon is also rich in organic matter but is leached of soluble material by groundwater. The leached material is deposited and accumulates in the B horizon. The C horizon consists of broken up bedrock.

ity, whereas the composition of the lowermost layer (the C horizon) is controlled by the composition of the underlying bedrock. The activity of microorganisms in the A horizon is important as they help absorb essential nutrients such as nitrogen and phosphorus from the atmosphere and from groundwater. Removal of these uppermost layers in areas subject to soil erosion can seriously affect the ability of the soil to sustain agricultural activity.

The food chain and soil development, although important, are but small portions of a much larger scheme that controls the migration of chemical elements in the Earth's surface reservoirs. The essential roles of carbon in the food chain, and as a component of greenhouse gases make it a very important element to track in its migration from one reservoir to another. To illustrate further the interaction between the Earth's four reservoirs, we will now follow the hypothetical path of a single carbon atom. We start with its introduction to the solid Earth by plate tectonic processes and its subsequent venting into the atmosphere. We then follow its absorption by the hydrosphere, its incorporation into a living organism, and its eventual return to the solid Earth as the organism dies and its remains become part of the fossil record.

19.4 The Odyssey of a Single Carbon Atom

Like the vast majority of the chemical elements at the Earth's surface, carbon is exhumed from the interior of the Earth by plate tectonic processes (Fig. 19.6). As we learned in Chapter 4, the Earth's oceanic lithosphere is created at mid-ocean ridges from magmatic activity beneath the ridge crest. So let us assume that our carbon atom is one of countless carbon atoms in a large magma chamber directly below a mid-ocean ridge (see Figure 19.6, 1). This chamber is like a hot water tank in a house, being continually tapped from above to create new lithosphere, and continually replenished from below with hot material from deeper magma chambers. Eventually, our carbon atom is pumped upwards as part of a surge of magma towards the crest of the mid-ocean ridge. Some neighboring carbon atoms combine with oxygen to form carbon dioxide (CO_2) and escape as gases. Others are pumped up fractures, and come out on the sea floor as constituents vented from black smokers (see Chapter 8, Living on Earth: The Bizarre World of the Mid-Ocean Ridges).

But, let us assume that our carbon atom doesn't escape. Instead it finds itself in a rapidly cooling magma and becomes trapped within solid rock adjacent to the mid-ocean ridge. Our atom therefore remains in the realm of the solid Earth, and, as part of the oceanic lithosphere, it must respond to the process of sea-floor spreading.

Spreading at a rate of a few centimeters per year, the ocean floor that contains our carbon atom slowly migrates away from the ridge. During this time, which may last hundreds of millions of years, the atom has a rather dull existence. All that changes, however, when the portion of oceanic crust in which our atom is trapped reaches a subduction zone (see Figure 19.6, 2). As the carbon atom is subducted, the temperature rises and it encounters steadily increasing pressures.

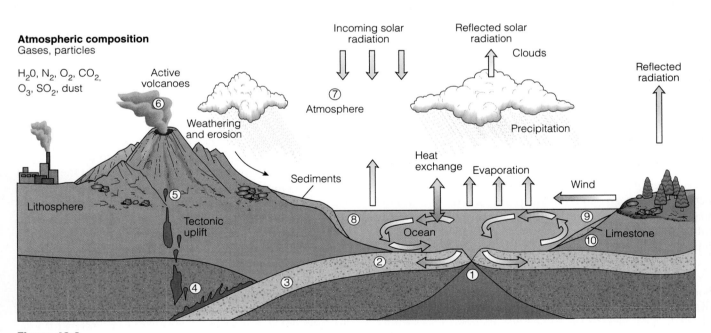

Figure 19.6

Odyssey of a single carbon atom. In this scheme, carbon is (1) introduced into the solid Earth at a mid-ocean ridge. It then (2) becomes part of the oceanic crust, and is (3) subducted. Eventually (4,5), it becomes part of a magmatic plumbing system that ultimately (6,7) expels the carbon into the atmosphere in the form of carbon dioxide gas. Ocean waters (8,9) then absorb the carbon dioxide where it becomes part of the biosphere. When the organism in which it resides dies, it sinks to the ocean bottom and becomes part of the fossil record (10).

After several million years of subduction, the atom's local environment has become so extreme that its ocean crust host is forced to adapt. Increasing pressure means that the rock progressively shrinks in volume and becomes more dense (see Figure 19.6, 3). The pressure can be relieved somewhat if new minerals form (a process called metamorphism; see Chapter 2). During metamorphism, loosely held gaseous components in the old minerals are expelled and are not incorporated in the new ones. Our expelled carbon atom finds some loose oxygen atoms to bond with to form carbon dioxide, and so escapes along with other volatile constituents which begin to rise upwards. Other less mobile chemical constituents like aluminum and titanium remain behind and must spend a longer time in the subduction zone.

As our carbon atom rises toward the Earth's surface in the form of carbon dioxide, it enters the overriding plate (see Figure 19.6, 4) . It will encounter many obstacles within this plate and will try to take advantage of any weaknesses, such as fractures, within it. However, since the process of subduction generates large zones of melting in the overriding plate, it is almost impossible for rising volatile constituents like carbon dioxide to completely avoid becoming incorporated into a magma.

Once incorporated, the volatile constituents are likely to be subjected to vigorous convection currents in the magma. During this process, our carbon atom may lose contact with the oxygens that were once its partners, and it may be forced into loose, temporary affiliations with all sorts of other chemical constituents. It will also encounter many other carbon atoms behaving in a similar fashion.

The magma, like the carbon atom, is less dense than the surrounding rocks and will exploit any opportunity to rise toward the surface. Seismic activity in near-surface rocks, for example, may open up new fractures that would allow the magma to rise, bringing our carbon atom as a passenger. As it rises, the carbon atom experiences a reduction in pressure and re-enters into a bonding arrangement with nearby oxygen, once again becoming part of a carbon dioxide molecule. As we shall see, this molecular partnership will prove to be a relatively long-lasting one.

As in any carbonated beverage, bubbles of gas are produced once the pressure is released. Just as a can of soda pop fizzes when its seal is broken, our carbon dioxide molecule becomes part of a gaseous bubble as pressures drop within the magma. Just below the surface, the rising magma finds its way into a smaller chamber (see Figure 19.6, 5) above which is a pipe, or conduit, that forms our molecule's eventual escape route to the surface.

As the magma sits in this near-surface chamber, our carbon dioxide bubble slowly rises to the top. If it were to reach the Earth's surface, it would escape passively into the atmosphere. Indeed, some of the carbon dioxide in the magma does just that. But before our carbon dioxide bubble reaches the top of the magma chamber, the volcano above erupts and our carbon dioxide molecule is jettisoned forcibly into the atmosphere (see Figure 19.6, 6).

Once the shock of the volcanic explosion has dissipated, our carbon dioxide molecule moves passively through the air in atmospheric circulation patterns (see Figure 19.6, 7). Although there are other carbon dioxide molecules in the air, they represent only a tiny minority (0.03%) of all atmospheric gases. However, their presence is important. Once in the atmosphere, carbon dioxide experiences the effects of solar radiation for the first time. As we learned in Chapter 16, this radiation arrives in a wide variety of wavelengths. For the most part, carbon dioxide is transparent to this radiation so that the dominant wavelengths pass easily through the molecule's structure en route to the Earth's surface. At the same time it experiences direct solar radiation, the carbon dioxide also experiences the reflection of this radiation off the Earth's surface. Some of this reflected radiation is essentially unchanged and so passes through the carbon dioxide again as it exits the atmosphere back into space. However, other portions of the radiation are affected by their interaction with the Earth's surface such that their dominant wavelength is now longer. The molecular structure of carbon dioxide is incompatible with this longer wavelength, and blocks its passage. As it takes part in this process, our molecule displays the properties of a greenhouse gas.

But, in all likelihood, our carbon dioxide molecule is destined to spend a relatively short time in the atmosphere because the ocean waters below have a strong affinity for this gas (Figure 19.6, 8). Oceans can hold as much as 62 times the amount of carbon dioxide held by the atmosphere. The average length of time a carbon dioxide molecule spends in the atmosphere (its "residence time") is about 200 years. By comparison, the residence time of carbon dioxide in the solid Earth is hundreds of millions of years. So within 200 years of entering the atmosphere, our molecule is likely to be dissolved in rain water, whereupon it falls to the oceans and is absorbed. It has now migrated from the atmosphere and has entered the hydrosphere. Carbon dioxide molecules extracted from the atmosphere in this manner can no longer play a part in the Earth's greenhouse effect. But if their extraction is not compensated by the introduction of new carbon dioxide molecules from other reservoirs, the result would be global cooling.

Dissolved in the ocean water, our original carbon atom is now rather loosely bound and becomes reacquainted with other dissolved elements. These include calcium and magnesium, which it last met in appreciable amounts in the magma chamber beneath the volcano. Whereas our carbon dioxide reached the ocean by way of the atmosphere, the calcium and magnesium traveled by a different route. These elements are less volatile than carbon dioxide and were expelled from the magma chamber in a lava flow, which cooled to form volcanic rock. However, rainfall near the volcano stripped the calcium and magnesium from the volcanic rock and transported them to the ocean by way of stream systems.

Once in the sea, our carbon atom is transported by ocean currents. But it doesn't move through ocean waters for long because life in the oceans has a tremendous affinity for carbon. Like sponges and corals, many of these life forms grow by continually absorbing and expelling ocean water. As they do so, they absorb the water's nutrients, among them calcium and carbon. After a few years, our carbon atom is transported to regions of upwelling in tropical waters, where life flourishes (see Figure 19.6, 9). Here, a microscopic foraminifera absorbs our carbon atom and uses it along with calcium, oxygen, and to a lesser extent, magnesium, to build its hard shell. The carbon and calcium combine with three oxygens to form calcium carbonate ($CaCO_3$). When, after a period of several years the organism dies, the calcium carbonate molecule (still held in the shell) sinks to the sea floor (see Figure 19.6, 10). As life continues to flourish near the ocean surface, more and more creatures die, and the shell that contains our carbon atom becomes progressively buried by younger layers of shell-rich material. Eventually, as the material around it becomes compressed, the shell becomes part of a hard rock known as limestone. Our carbon atom is now firmly trapped in solid rock and may stay there for hundreds of millions of years.

Indeed, the chances of our carbon atom ever escaping from its rocky tomb are remote. Table 19-1 shows the distribution of all carbon in the Earth's surface reservoirs. As the table demonstrates, the vast majority of this carbon (more than 99.99%) resides in limestone rock. Thus our carbon atom's long and tortuous path, which began in a magma chamber beneath a mid-ocean ridge and passed through all the Earth's surface reservoirs, has ended in limestone rock.

Throughout the 4.6 billion years of geologic time, countless carbon atoms have followed a similar route only to experience the same fate. The overwhelming proportion of the Earth's surface carbon that occurs in limestone tells us that limestone is a very efficient carbon trap. From this trap, our carbon atom has two unlikely, but possible escape routes. One will develop if the oceanic crust beneath the limestone begins to subduct. If the limestone is dragged down a subduction zone, the shell structure containing our carbon atom will ultimately become unstable and the carbon may once again be expelled (in the form of carbon dioxide) into the overriding plate. If this occurred, the carbon atom would follow its previous route through the solid Earth and the cycle would repeat.

Alternatively, tectonic upheavals may push the limestone above sea level. Although quite commonly found on the continents, the vast majority of limestones form in shallow marine environments and are preserved on land only by being tectonically uplifted. Because limestone is highly soluble (see Chapter 10), the uplifted limestone would begin to disintegrate once it was exposed to weathering, and our carbon atom might eventually be released into streams and drain back to the sea. Alternatively, if the local environment happened to be warm and swampy, the carbon might eventually be absorbed by plant life, which, on death, could be buried and compressed, and might ultimately form coal. If we were then to burn the lump of coal in which our carbon atom resided, it would be recombined with oxygen and returned to the atmosphere. Here it might reside for another 200 years or so before being absorbed by the ocean once again.

The odyssey of our carbon atom not only illustrates how the Earth systems connect and interact, but also serves as a model for the paths of other chemical elements. Like our carbon atom, the vast majority of chemical elements arrive at the surface from the Earth's interior, and are involved in several phases of igneous activity. And like our carbon atom, other elements may take a

Table 19-1

The Amount of Carbon (in Gigatons) in the Earth's Surface Reservoirs

Amount of carbon residing in different reservoirs Quantities in gigatons (10^9 metric tons)	
Atmosphere	710
Continents	
Biomass	590
Litter	60
Soil	1,670
Fossil fuels	5,000
Carbonate sediments and rocks	20,000,000
Oceans	
Biomass	4
Sea water	
Surface layers	680
Intermediate waters	8,200
Deep waters	26,000
Sediments	4,900

variety of paths through the atmosphere and hydrosphere, and may eventually be absorbed by life. Such complex pathways that begin with the processes of plate tectonics are probably taken by most of the chemical constituents we see around us, including the atoms and molecules of which each of us is made.

In the next section, we will learn that the interconnected nature of the Earth's surface systems has spawned a new discipline in science, that of biogeochemistry. As we shall see, this new and exciting field has a prominent role to play in the study of our environment.

19.5 Biogeochemistry: An Interdisciplinary Science

The odyssey of our carbon atom not only demonstrates the interconnected nature of the Earth's reservoirs, but also illustrates how a *chemical* species migrates through various *geologic* and *biological* environments, driven, in part, by the *physical* laws of gravity and motion. In view of this, many scientists now recognize that the compartmentalization of science into its traditional academic disciplines (biology, chemistry, geology and physics) hinders our understanding and our ability to resolve current environmental issues. On the other hand, **biogeochemistry** (which literally means "life-Earth-chemistry") is an interdisciplinary science ideally suited to the study of such issues.

We now know that almost all chemical elements are exhumed from the Earth's interior adjacent to mid-ocean ridges, although they are probably influenced by processes in a variety of plate tectonic settings. Many chemical elements are essential nutrients to life, while others inhibit its development. The choice of paths confronting each element is like a road map, and each may travel a different route at different speeds. However, their movement is not chaotic since there is a degree of order and predictability to their behavior. Elements with similar chemical properties tend to behave coherently and travel together.

However, for the purposes of illustrating the multidisciplinary science of biogeochemistry, we have selected three elements for special study: carbon, nitrogen, and phosphorus (Fig. 19.7). These elements are of particular importance because they are essential components of DNA and RNA, the chemical building blocks of life (see Chapter 13). They also have several features in common. Each is continually added to the Earth's surface reservoirs from below by exhumation from the Earth's interior. Much of the energy required to process, modify, and drive the movement cycles of these three elements comes from the radiant energy of the Sun. Carbon and nitrogen compounds are commonly volatile and, as a result, are generally exhumed directly into the atmosphere as volcanic gases. But while their means of entry into the atmosphere is very similar, their subsequent behavior is quite different. As we shall see, this difference in subsequent behavior explains why nitrogen is the Earth's most abundant atmospheric gas (78%), whereas carbon dioxide is present only in minute quantities (0.03%). Phosphorus, on the other hand, is a far less abundant element and commonly combines with oxygen and other elements such as calcium to form relatively stable minerals. As a result, it is not an important constituent of the atmosphere but instead displays a movement cycle that is concentrated in the solid Earth and hydrosphere.

19.6 The Carbon Cycle

In Chapter 8 we learned that carbon readily combines with other elements such as oxygen (to form carbon dioxide or carbon monoxide) and hydrogen (to form methane). But in situations where these other elements are in short supply, carbon occurs as a relatively pure element. In Chapter 18, we learned that rapid burial of organic matter in an oxygen-starved environment results in the relatively pure concentration of carbon we call coal.

The movement of carbon through the Earth's reservoirs (Figs. 19.6 and 19.7) is known as the carbon cycle. As presented in Figure 19.8, the cycle focuses on carbon alone, although as we learned in the previous section, carbon often becomes involved in "temporary arrangements of convenience" with other elements.

Like the path of the carbon atom whose odyssey we followed, the cycle illustrated in Figure 19.8a is like a road map that guides the journey of carbon atoms. It shows the input from volcanoes and solar energy, and is arranged with the continental and atmospheric portions of the cycle on the left-hand side and the oceanic portion on the right. The left-hand side also displays all the components of the food chain shown in Figure 19.2. The energy producers are plants which are capable of photosynthesis; they extract carbon dioxide from the atmosphere to produce oxygen and sugar, storing energy in the process. In Figure 19.8a, however, herbivores, and primary and secondary carnivores are grouped together as energy consumers.

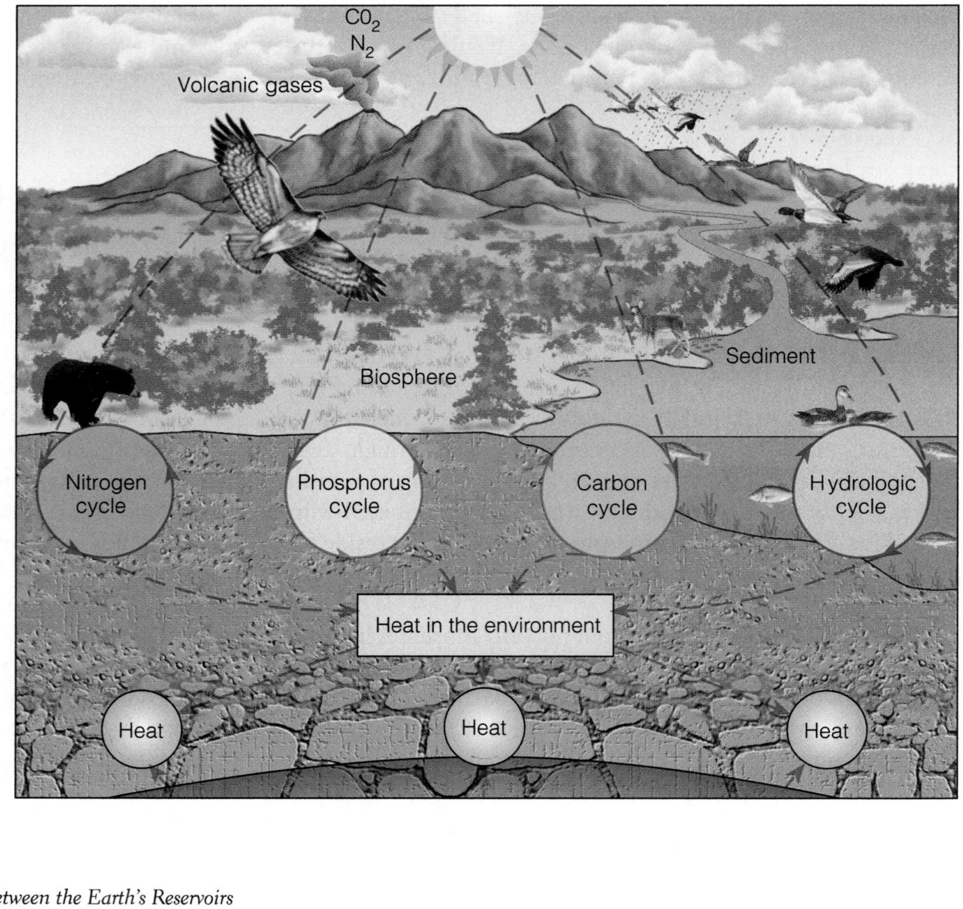

Figure 19.7
Schematic representation of the flow of energy (dashed lines) from the Sun through the biosphere and the cycling of carbon, nitrogen and phosphorus. The chemistry is predominantly provided by plate tectonic activity, and circulation is stimulated by solar energy and the hydrologic cycle.

THE CARBON CYCLE

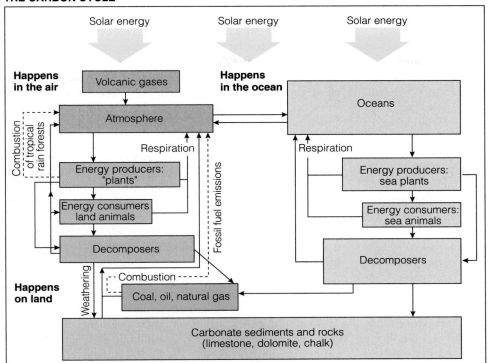

(a)

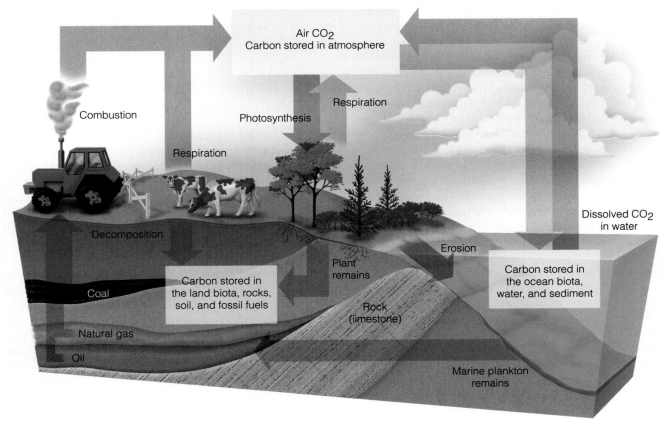

(b)

Figure 19.8

(a) The carbon biogeochemical cycle "map." The continental part of the cycle is on the left, the oceanic part is on the right. Human interference with the cycle is portrayed by dashed lines. (b). A simplified diagram of the carbon cycle.

On death, the organic material of both producers and consumers may decompose. As we learned earlier, recent evidence suggests that decomposition is aided by bacteria. In effect, the decomposition recycles the carbon because it is made available once again to the photosynthetic producers. Microorganisms in soil perform many short-term transformations of carbon from organic to inorganic and vice versa. If abundant oxygen is available, the decomposition may eventually convert the carbon into carbon dioxide, and as a result, many soils become enriched in this gas. Ultimately, the carbon dioxide escapes from the soil and returns to the atmosphere. In regions of extensive continental erosion, the soil may be eroded and transported to the sea, where the carbon is likely to become a constituent of carbonate sediments and eventually carbonate rocks. In oxygen-starved regions, burial of organic matter and the carbon it contains produces fossil fuel deposits (see Chapter 18). Such regions include swamps and deltas where high sedimentation rates ensure rapid burial.

The oceanic component of the diagram has a similar arrangement. Phytoplankton, which are microscopic plants, are the main photosynthetic producers, and all consumers, from zooplankton (floating marine animals) to mammals, are grouped together. Note that the atmosphere and oceans are linked by a "two-way street." Ocean water, with its strong affinity for carbon dioxide, extracts it from the atmosphere. Solar energy, on the other hand, evaporates ocean water into the atmosphere, which in turn results in precipitation and the erosion of continents, bringing elements to the ocean that will combine with carbon. This coupling between the atmosphere and hydrosphere "completes the circuit" and is fundamental to the exchange of energy between the Earth's surface reservoirs.

Marine life often flourishes in continental shelf environments, particularly in subtropical climates. In these environments, there is commonly an abundant supply of other chemical constituents such as calcium and magnesium for carbon to combine with. As a result, when the marine life dies and decomposes, the carbon typically resides in carbonate sediments which eventually becomes limestone (calcium carbonate).

But carbonate sediments can only form if the oceans are supplied with sufficient quantities of calcium and magnesium. As we learned in Chapter 8, these elements are not volatile and are transported to the sea following their extraction from rocks, minerals, and soils during continental erosion. If there is an insufficient supply of calcium, the carbon atoms will remain in the ocean water. The concentration of carbon dioxide in the ocean consequently builds up, and less is extracted from the atmosphere.

So carbon initially vented from volcanoes into the atmosphere in the form of carbon dioxide may travel highly convoluted but reasonably predictable routes through the Earth's surface reservoirs such that its movements (the carbon cycle) can be mapped. At any one time, there are countless carbon atoms moving like cars along these congested routes. In Figure 19.8a, each box represents a reservoir (like a "rest stop" on the highway), where carbon resides temporarily before it continues its travels. Many of these reservoirs have two exits for the carbon: one which recycles carbon dioxide back into the atmosphere, and another which sends the carbon further along the path of its ultimate destiny to be a component of rock.

In continental environments, this rock is commonly a fossil fuel such as coal. In oceanic environments, it is typically limestone. Life and death are important aspects of this highway system. Life uses the carbon dioxide introduced into the atmosphere. Death and decomposition begin the process of extracting carbon from the system and isolating it in rock. Calcium is critical to the cycle since the efficiency with which carbon dioxide is removed from the system depends on calcium's availability. Insufficient calcium can have a braking effect on the carbon cycle by decreasing the rate at which oceans absorb carbon dioxide. This results in a buildup of carbon dioxide in the atmosphere, which enhances the greenhouse effect and causes global warming.

The pathway of the carbon cycle is a biogeochemical version of the game "Chutes and Ladders" or "Snakes and Ladders," in which all of the Earth's surface reservoirs participate (Fig. 19.8b). The mobile carbon atom may land on a "chute" or "snake" that brings it back towards the start. However, it will eventually overcome all barriers, and come to reside at its final destination in either coal or carbonate sediments. It is no accident that carbonate rocks overwhelmingly dominate the modern distribution of carbon in the Earth's surface reservoirs (see Table 19-1). They are generally well preserved throughout the geologic record, so only a minor portion of the limestone produced has been subsequently destroyed by weathering. In essence, then, when our carbon atom finally enters carbonate rock, it has little chance of being recycled and is effectively removed from circulation.

19.6.1 Human Interference with the Carbon Cycle

Once the natural routes of the carbon cycle are laid out, unnatural distortion related to human interference in nature's biogeochemistry can be clearly recognized (Figure 19.8a, the dashed lines). Left to its own devices, nature has worked out a scheme whereby energy flow is distributed between the various reservoirs at a predetermined rate. For example, the sudden introduction of carbon dioxide into the atmosphere that accompanies a volcanic eruption is handled by the absorption of this gas by the Earth's other surface reservoirs.

When viewed in this context, the effects of some forms of human activity become obvious. Deforestation and burning of tropical rain forests (continental producers), for example, is a double-edged sword. It not only compromises the efficiency of photosynthesis (because there are fewer trees) and, therefore, the ability of the solid Earth to absorb carbon dioxide, but it also prematurely returns carbon dioxide to the atmosphere. The effect is analogous to an electrical short circuit. Energy and nutrients that would have naturally flowed down the food chain are prematurely diverted back to the atmosphere. In addition, soils are stripped of the root systems that bind them, and therefore are more easily eroded and transported to the ocean.

The burning of fossil fuels is another short circuit. Left to its own devices, nature has stored away this carbon, effectively removing it from circulation. However, the burning of fossil fuels feeds this stored carbon directly back to the atmosphere, a path nature never intended.

The situation in the oceans is equally precarious, although this is not as immediately obvious in either diagram of Figure 19.8. The flow of carbon dioxide from the hydrosphere, through the biosphere, and into the solid Earth is being threatened at two points along its route. First, depletion of the ozone layer (discussed in detail in Chapter 20) is allowing an increasing amount of the Sun's radiant ultraviolet energy to reach the Earth's surface. This radiation affects the life cycle of photosynthetic microorganisms (phytoplankton) in ocean water, inhibiting their ability to extract carbon dioxide from the atmosphere. As a result, the oceans are less likely to absorb the excess of carbon dioxide in the atmosphere. Second, the stability of modern coral reefs (which ultimately produce limestone) is being threatened in many regions by local pollution, compromising the passage of carbon dioxide from the hydrosphere into carbonate sediments.

It might be argued from Table 19-1 that the amount of carbon involved in these processes are insignificant compared to the amount in carbonate sediments and rocks, and therefore the effects are similarly insignificant. However, this is misleading. When carbon finally comes to reside in carbonate rocks or coal, it is essentially out of natural circulation. Hence, only the remaining amount of carbon is actually moving through the system. In the same way, it is the moving traffic that concerns us, not the parked cars.

Fortunately, we do not burn significant amounts of limestone. But we do burn vast amounts of fossil fuels, and relative to the amount of carbon in circulation, this forms a very significant source of atmospheric carbon dioxide. As we discussed in Chapter 1, the fact that humans measure time in days and months, whereas geologic processes may take millions of years, now takes on a more profound and ominous meaning. The burning of fossil fuels is being conducted on a human time scale, whereas the formation of these deposits took place on a geologic time scale and represents the culmination of geologic processes lasting many millions of years. We are therefore using up these resources far faster than the Earth's ability to respond. In fact, in a few hundred years, the burning of fossil fuels will have pumped back some 5000 gigatons of carbon into the atmosphere in the form of carbon dioxide. From this perspective, our actions are a gross violation of the Earth's natural contract for the transfer of energy and nutrients.

In summary, our study of the carbon cycle has given us the context in which to evaluate important environmental issues in a systematic way. Other chemical elements, such as nitrogen and phosphorus, are also important to the existence and stability of life, and in the following sections, we study the biogeochemistry of each of these elements in turn.

19.7 The Nitrogen Cycle

Nitrogen is the most abundant gas in our atmosphere and is used in the biosphere to make many organic compounds, including proteins, DNA, and RNA. An understanding of nitrogen's movement through Earth's various reservoirs is therefore fundamental to environmental analysis.

Nitrogen and carbon dioxide are both vented in significant quantities as gases from volcanoes. However, as we learned in Chapter 8, the two gases have widely differing concentrations in our atmosphere. The reason for this is that, with the exception of the atmosphere, the Earth's surface reservoirs have no great affinity for nitrogen gas. So the concentration of nitrogen is high in our atmosphere largely by default, since no other reservoir can absorb the gas in any significant quantity.

The affinity of the hydrosphere for nitrogen gas is sufficiently minor that analysis of the hydrosphere is omitted altogether from diagrams that display the nitrogen cycle (Figs. 19.9a and b). The main natural linkage between nitrogen in the atmosphere and the solid Earth is through biological activity. However, nitrogen cannot be used directly by multicellular life, although the availability of nitrogen can limit the development of ecosystems. Fortunately for the biosphere, lightning strikes and certain bacteria (Fig. 19.10) convert nitrogen into compounds that can be absorbed by soil. Blue-green algae (cyanobacteria), for example, mediate the extraction of nitrogen from the atmosphere into soil. Some plants (such as beans, peas, alfalfa, and clover) have "nitrogen-fixing" bacteria associated with their roots that likewise have the ability to draw nitrogen out of the atmosphere and create nitrogen compounds that can be absorbed by the soil. These compounds are soluble in water and when precipitation occurs, they are naturally added to the soil and are subsequently taken up by plants. In general, however, the natural affinity of the solid Earth for nitrogen is so poor that soils are impoverished in this element.

Indeed, some ("dentrifying") bacteria actually extract nitrogen from the soil. Whatever its source, nitrogen in soil is used to form proteins and amino acids. When plants are eaten by herbivores or omnivores, these important nutrients are passed down the food chain. Some nitrogen is then returned to the soils by natural animal waste and by decomposition of dead plant material.

Certain physical parameters may override simple chemical exchanges. For example, soils are loosely bound materials and so are readily eroded, particularly in active drainage basins. Where this occurs, soil erosion effectively impedes the transmission of nitrogen from the atmosphere to the base of the food chain.

In addition, the behavior of the nitrogen biogeochemical cycle has not always been the same. As we learned in Chapter 14, plant life did not flourish until the Devonian Period, just a little over 400 million years ago. Before that time, soils and nitrogen-fixing bacteria were scarce and the nitrogen biogeochemical cycle is likely to have been quite different from that of today.

THE NITROGEN CYCLE

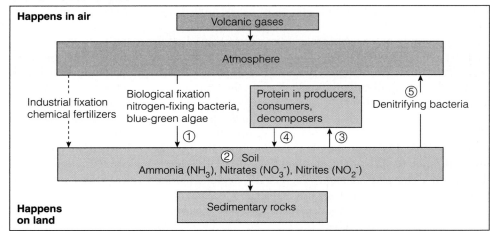

Figure 19.9

(a) The nitrogen biogeochemical cycle "map." Numbers 1 to 5 correspond with the same numbers on (b), a simplified diagram of the nitrogen cycle. Notice the hydrosphere is omitted.

(a)

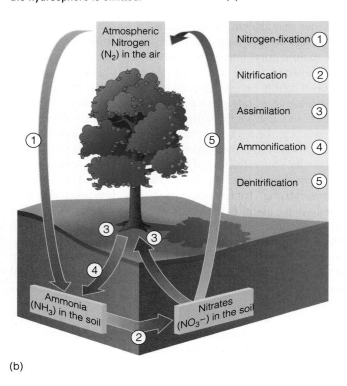

(b)

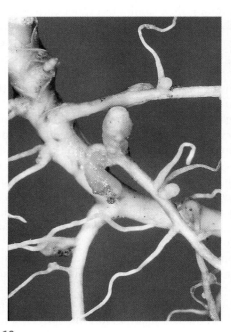

Figure 19.10

Nitrogen-fixing bacteria (such as *Rhizobium*) live on the root nodules of plants such as alfalfa, beans, clover and peas).

19.7.1 Human Interference with The Nitrogen Cycle

In most regions the nitrogen content of soils is insufficient to sustain agricultural activities. Indeed, humanity has long realized that in order to sustain agriculture, nitrogen needs to be repeatedly replenished in cropland soils. This is because plants deplete soils of nitrogen at a faster rate than nature can replace it. To counteract this tendency, some organic farmers grow plants such as alfalfa and beans so that the nitrogen-fixing bacteria associated with these plants can enrich the soils in nitrogen.

Other farmers artificially add chemical fertilizers, which are usually produced from atmospheric nitrogen by industrial fixation of nitrogen gas, or from mines that extract nitrogen-rich minerals from deposits of saltpeter (ammonium nitrate, NH_4NO_3). Fertilizers contain nitrogen in a form that dissolves in water and so adds ammonia (NH_3), nitrite (NO_2^-) and nitrate (NO_3^-) ions to the soil. The nitrogen compounds introduced by fertilizers are extracted from the topsoil by the crops which are harvested, necessitating repeated applications of fertilizer to maintain adequate nitrogen levels. Fertilization can be a very inefficient process, as a large proportion of the added nitrogen is easily washed out of the soil when it rains. Not surprisingly, ever-increasing application of fertilizers has resulted in the contamination of groundwater by nitrates (see Chapter 10). In many agricultural regions, a high percentage of rural wells show nitrate values in excess of 10 milligrams per liter, which is the government health standard. Scientists have found links between elevated levels of nitrate in drinking water and a variety of human diseases including stomach cancer, fetal abnormalities, and methaemoglobinaemia, or "blue baby syndrome." Excess nitrogen in streams and coastal systems can also lead to massive algal and bacterial growth, which absorbs oxygen and suffocates other aquatic life. This excess of plant nutrients in water is a condition known as **eutrophication**, and may make water unfit for human consumption.

In fertilized regions, the abundance of nitrogen in the soil results in higher concentrations of nitrogen and nitrous oxide (N_2O) gases being released from the soil into the atmosphere. Nitrous oxide is a strong greenhouse gas and has also been linked to ozone depletion. It is inert in the troposphere, and lasts for about 120 to 160 years in the stratosphere before it is broken down. All predictions indicate that global use of fertilizers and the areal extent of fertilizer application are both increasing, suggesting that these problems will become more acute in the future.

Our understanding of the environmental problems of soil erosion is less widely appreciated than those associated with fertilizers. Since soil is the main means by which nitrogen is extracted from the atmosphere, any erosion of the soil fundamentally affects the nitrogen cycle. In areas deforested for agricultural purposes, the soil is no longer stabilized by root systems. As a result, it becomes even more loosely bound and so is prone to rapid erosion. Furthermore, the absence of vegetation means that the soil is less protected from the wind and rain. In areas of deforestation, therefore, soil erosion can become a very major problem to agriculture.

In summary, nitrogen is the dominant gas of the atmosphere because it is not easily absorbed by the Earth's other reservoirs. However, nitrogen is an essential element in the biosphere, and its availability is a limiting factor in the development of ecosystems and agriculture. Whereas carbon and nitrogen are common elements in the Earth's surface reservoirs, phosphorous, the element we now turn our attention to, is present only in trace amounts. Yet it plays a vital role in biogeochemical cycling and is equally important in environmental studies.

19.8 The Phosphorus Cycle

Phosphorus (P) is an example of a chemical element that is essential to the biosphere, yet occurs in relatively minor quantities. It is found in DNA and RNA, and is essential in the making of proteins. It occurs in the cell membranes of all living things, and also in the hard parts of animals such as bones, shells, and teeth. In these materials, it usually occurs as a calcium phosphate mineral such as apatite. Unlike carbon and nitrogen, however, phosphorus forms few volatile compounds; therefore it is far more abundant in the solid Earth than it is in the atmosphere.

Phosphorus mainly occurs in compounds known as phosphates, in which phosphorus is bonded with oxygen.

Phosphorus is generally introduced into the Earth's surface reservoirs by igneous activity (Fig. 19.11). Although the phosphorus content in magma is generally less than 1%, it may become concentrated and eventually trapped in phosphorus-rich minerals such as apatite, as the magma cools to form a rock near the Earth's surface. At this stage, the phosphorus has had no direct contact with the atmosphere or hydrosphere. However, magmatic activity is commonly associated with mountain building, which in turn involves uplift and erosion. Ultimately, erosion will exhume the rocks containing phosphate minerals, exposing them to weathering, whereupon the phosphorous will either enrich the soil or wash into the sea. Like nitrogen, phosphorus is an essential nutrient for plant life. Some phosphate compounds are dissolved by groundwater, and their absorption by plants allows the element to enter the biosphere at the base of the food chain. However, many phosphorus compounds are only weakly soluble. As a result, the availability of phosphorus is an important limiting factor in the development of many ecosystems.

Phosphorus is also a vital nutrient to species higher up the food chain. Consumer species obtain this phosphorus by eating plants or by eating herbivorous animals. Animal waste or plant decomposition then returns most of this phosphorus to the soil.

As we have learned, the erosion of rocks and soils also facilitates the transport of phosphorus to the sea. Deep ocean waters have a particular affinity for phosphorus as a result of which the element is one of the important nutrients brought up by upwelling currents along continental margins (see Chapter 9). The availability of phosphorus results in increased organic productivity on the continental shelves, where marine life absorbs the phosphorus, particularly into its hard parts. For example, when we discussed the effect of El Niño on marine life (Chapter 9), we found that marine upwelling, if impeded for just a few months, produces a lack of nutrients that has drastic results.

On the death of animals and plants, the phosphorus they contain becomes concentrated in fossil-rich layers. In some situations, ocean currents extract or winnow other particles from these layers, so that a sedimentary rock is formed that is rich in phosphate and contains abundant fossil fragments.

When birds that prey on marine life ingest the phosphorus-rich material, much of it is later excreted in their droppings. Most simply falls to the sea and is recycled, but in populous nesting grounds, bird droppings may accumulate to form a phosphate-rich deposit called **guano**.

The importance of phosphorus to the shells of modern organisms has led paleontologists to speculate on the role of its abundance to the evolution of life. As we learned in Chapter 14, the explosion of life at the beginning of the Cambrian Period, some 545 million years ago, coincided with the evolution of marine organisms with hard body parts. This could have been sparked by an increase in the amount of phosphorus available to these ecosystems. Consequently, studies of the distribution of phosphorus in the geologic record and the mechanisms that control this distribution are important areas of current research.

THE PHOSPHORUS CYCLE

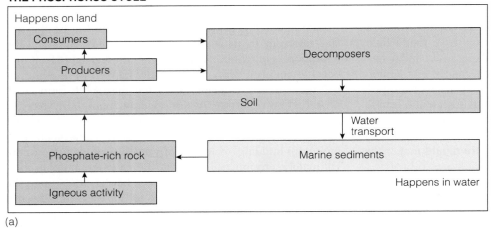

(a)

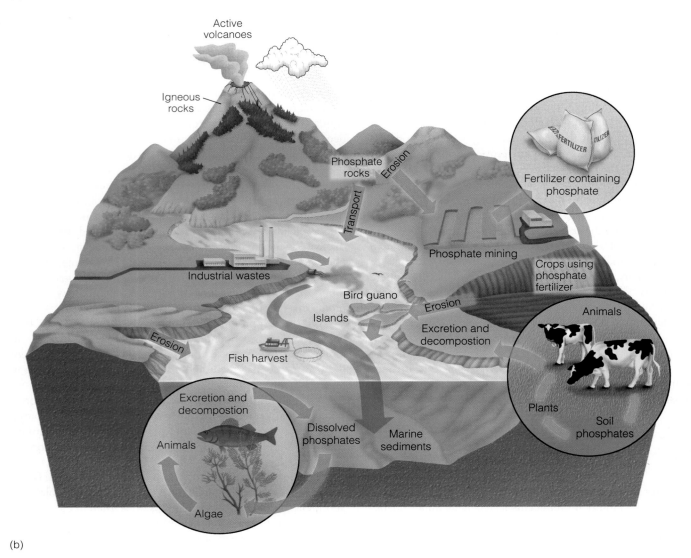

(b)

Figure 19.11

(a) The phosphorus biogeochemical "map." (b) A simplified diagram of the phosphorus cycle.

In contrast to our planetary neighbors, life on Earth has flourished. As we learned in Chapters 13 and 14, life on this planet has evolved from the simplest organisms to the complex and diverse species of today by adapting to changing environments. Life has survived mass extinctions brought about by environmental stresses and extraterrestrial impacts. Species that survived these extinctions were able to take advantage of the ecological niches left vacant as a result of them. Life on this planet is consequently robust and has survived some of the most unimaginable catastrophes, only to flourish again in the wake of the cataclysm.

Some scientists believe that the resilience of life is no accident. They believe that life does not merely adapt to its environment; instead it actually manipulates the environment to make it more hospitable. The ability of life on Earth to control its own environment is a key element of the **Gaia hypothesis.** Named after the Greek goddess of the Earth and proposed by Sir James Lovelock, an English scientist, and Lynn Margulis of Boston University, the Gaia hypothesis was the first significant attempt to draw attention to the importance of the inter-

action between the Earth's surface reservoirs in guiding the evolution of life. Proposed in the 1970s, the idea was contested in the 1980s by a rival hypothesis, known as **coevolution.** Championed by Stephen Schneider of Stanford University, coevolution proposed that climate and life "grew up together," in contrast to the claim of the Gaia hypothesis that life optimizes its own environment.

The Gaia Hypothesis

According to the Gaia hypothesis, the biosphere is a self-regulating system that controls its own chemical and physical environment. This implies that the Earth and the organisms that live on it are inextricably linked. In essence, the Gaia hypothesis is an expansion of the concept of homeostasis, by which the human body is able to maintain an optimum temperature range, despite variations in the temperature of its surroundings. In order to achieve homeostasis, the human body must be able to counteract the external changes imposed on it, so that the body's temperature is kept within its optimum range. This stabilizing effect is an example of a **negative feedback,** whereby any change invokes a

response that opposes it. According to Lovelock and Margulis, such negative feedbacks must have dominated the Earth's evolution in order for life to have survived and flourished. According to the Gaia hypothesis, the Earth's climate and chemical properties, now and throughout its history, have been optimal for life, because life has manipulated its environment in order to ensure that these optimal conditions are maintained.

Evidence cited in favor of the Gaia hypothesis includes the composition of the Earth's atmosphere, which is optimal for the proliferation of modern life. The high oxygen content in our atmosphere clearly distinguishes us from our planetary neighbors, the atmospheres of which are dominated by carbon dioxide (Table 19-2). This distinction cannot be attributed to outgassing during the accretion of these planets, because each is composed of similar materials and so should have acquired atmospheres of similar composition. In fact, without life, the Earth would have an atmosphere very similar to those of Venus and Mars. The high carbon dioxide content in a lifeless Earth's atmosphere

(continued)

19.8.1 Human Interference with the Phosphorus Cycle

The relative scarcity of phosphorus, coupled with its importance as a nutrient in soil, make it a valuable commodity. Phosphate is extensively mined; over $5 billion worth is extracted annually, mainly from sedimentary phosphatic deposits.

Humans absorb phosphorus in food and excrete it in their waste. To the extent that we are part of the biosphere, it could be argued that our influence on the phosphorus cycle is a natural phenomenon. However, human interference is related to deforestation, our agricultural methods, and the artificial ways in which we concentrate and treat human waste.

Deforestation affects the phosphorus cycle, because soil erosion as a result of deforestation removes the phosphorus held in soils. Erosion of clay-rich soils is of particular concern because they are commonly enriched in phosphorus.

In agricultural areas, phosphorus needs to be continually replenished by fertilizers and is added in the form of phosphate compounds, which are soluble. However, like nitrogen, excess

input of these compounds is wasteful, and rainfall may result in runoff of excess fertilizer from agricultural lands, a process that can contaminate groundwater and adjacent drainage basins.

Although phosphorus is a nutrient in small quantities, it is very toxic to the ecosystem when concentrated. Like nitrogen, excess phosphorus in streams and coastal systems can lead to eutrophication, in which massive blooms of algae and bacteria absorb oxygen, resulting in the suffocation of other aquatic life. Eutrophication can also be a problem in densely populated coastal areas where the influx of phosphate-rich industrial detergents can profoundly impact the adjacent marine ecology and nearby agricultural lands.

In the preceding sections, we have tried to demonstrate the importance of understanding the natural migration of chemical nutrients through the Earth's surface reservoirs. We presented three examples, carbon, nitrogen, and phosphorus, because of their importance to life. In principle, however, an understanding of the cycling of all naturally occurring elements is important. Only by understanding the natural cycles of these elements can we clearly recognize disruptions induced by human interference.

would have turned today's Earth into a nightmarish greenhouse with surface temperatures of about 290°C (554°F). Such a planet would have no free water on its surface and could not harbor life. But the Earth today has minimal carbon dioxide (0.03%) in its atmosphere, and instead contains significant amounts of oxygen (21%), resulting in a planet with an average surface temperature (15°C, 59°F), that is ideal for life.

As we learned in Chapters 13 and 14, the oxygen content in our atmosphere is primarily controlled by photosynthetic organisms, and has increased through time at the expense of carbon dioxide. Despite evidence of higher atmospheric carbon dioxide in the past, Lovelock and Margulis proposed that the existence of life for the past 3.8 billion years or more indicates that the Earth's average surface temperature has remained within tolerable levels throughout most of geologic time. The Sun was considerably dimmer in the early history of the Solar System, with an output of solar radiation that is estimated to have been about 30% less than that of today. A dimmer Sun would have reduced the greenhouse effect of the Earth's early CO_2-rich atmosphere. But as solar output increased, carbon dioxide concentrations in the atmosphere must have decreased in order to keep surface temperatures within a tolerable range. So how might this have been achieved?

According to Lovelock and Margulis, this example of self-regulation is not just a chance event. Lovelock showed how life might regulate the surface temperature of a planet like the Earth in a hypothetical "daisy world" (see Figure 19.12). The biosphere of Lovelock's "daisy world" consists of only three species: white daisies that grow better when it is warm, black daisies that grow better when it is cool, and cows that eat the daisies.

Now suppose the surface temperature of "daisy world" begins to rise. The white daisies, preferring warm conditions, would grow more rapidly and become more abundant than the black daisies. But since white objects reflect sunlight, the proliferation of white daisies would cause "daisy world" to cool down. At the same time, the cows would eat more white daisies than black ones, simply because they are more

common. The cooling temperatures now allow the black daisies to grow faster, until there are more black daisies than white. But black objects absorb more sunlight than white ones, so "daisy world" would begin to warm up. The cows would now eat more black daisies than white because they are more numerous. But as the temperature begins to rise, the white daisies grow faster once again. In this way, the daisies stabilize the surface temperatures of "daisy world" and do so even in the presence of cattle that eat them.

Lovelock's "daisy world" is a simple example of a "self-regulating negative feedback system" in which changing surface temperature is counteracted so as to stabilize the environment. This concept of a negative feedback may be applied to the understanding of the Earth's surface temperatures. In the case of the Earth, photosynthetic organisms were able to absorb atmospheric carbon dioxide as the Sun grew hotter, thereby reducing the greenhouse effect so as to keep the Earth's surface temperature within the optimum range for life. As we have learned, this process has been so effective that 99.99% of the carbon in the Earth's surface reservoirs is now stored in carbonate rock, which owes its origin largely to the activities of ancient life.

Other evidence cited in favor of the Gaia hypothesis includes the composition of the ocean waters, which is also conducive for life, being neither too alkaline nor too acidic. In fact, it is nearly neutral, which again is optimal for marine organisms. As we learned in Chapter 8, the salt content of the oceans is influenced by two parameters: continental erosion, which transports elements such as

sodium and calcium to the oceans, and hydrothermal activity at mid-ocean ridges, by which chlorine is introduced through igneous activity. Despite these two apparently unrelated processes, the average salt content of the oceans (about 3.4%) is also optimal for life. Too low a salt content would kill many marine organisms, whereas too high a salt content could disrupt their cell walls. The Gaia hypothesis claims that marine life maintains the salt content of the oceans at this optimal level.

The Gaia hypothesis also allows for *positive feedbacks* whereby any environmental change invokes a response that amplifies that change. The most commonly cited example of a positive feedback is one related to global warming. Any global warming resulting from increased atmospheric carbon dioxide levels causes more evaporation from the oceans, which raises the content of water vapor in the atmosphere. However, water vapor is an even stronger greenhouse gas than carbon dioxide, so its evaporation causes further global warming. Lovelock and Margulis argue that such positive feedbacks must be overridden by negative ones, or the Earth's environment would have evolved out of the narrow range of conditions that could harbor life.

The Gaia hypothesis is widely recognized as a significant advance in science because it was one of the first studies that focused attention on the interaction of the Earth's surface reservoirs. But one of its central arguments, that life optimizes its own environment, has come under criticism. In the next section, we present the rival hypothesis, coevolution, that suggests that Earth's systems evolved from their primitive origins to their modern state by mutually influencing each other.

Table 19-2
Comparison of Planetary Atmospheres

GAS	VENUS	MARS	EARTH WITHOUT LIFE	MODERN EARTH
Carbon dioxide	98%	95%	98%	0.03%
Nitrogen	1.9%	2.7%	1.9%	78%
Oxygen	Trace	0.13%	Trace	21%
Argon	0.1%	2%	0.1%	1%
Surface temperature (°C)	477	−53	290	15
Total pressure (bars)	90	0.0064	60	1.0

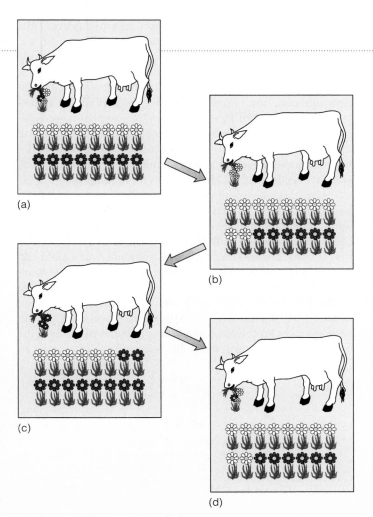

(a)

(b)

(c)

(d)

Figure 19.12

Lovelock's "daisy world" biosphere consisting of white daisies that grow better when it is warm, black daisies that grow better when it is cool, and cows who eat the daisies. In (a), the daisies are equal in abundance. But as the surface temperature increases (b), white daisies become more abundant. However, white daisies reflect solar energy so that their proliferation causes the planet to cool down. Meanwhile, cows eat more white daisies than black. But as the planet grows colder (c), black daisies grow faster and become more abundant. However, black daisies absorb more solar energy, causing the temperature to rise. At the same time cows eat more black daisies than white. As the planet warms up (d), white daisies proliferate once more. So the daisies stabilize surface conditions even in the presence of herbivores that consume them.

The Coevolution Hypothesis

The central theme of the coevolution hypothesis is that the Earth's surface reservoirs have evolved to their present state by mutual interaction rather than by biosphere control. As is common in science, the need for this new hypothesis was born out of criticisms of the Gaia hypothesis. Have the Earth's climate and chemical properties, for example, been optimal for life throughout its history? In Chapter 13, we learned that this has certainly not been the case. Throughout the Archean and the Early Proterozoic, for example, life was anaerobic and evolved in an oxygen-starved environment. During that time, life was capable of photosynthesis, but the oxygen it produced was absorbed by ocean waters. About 2.0 billion years ago, the ability of oceans to absorb oxygen was exceeded, and oxygen gas entered the atmosphere. This "oxygen crisis" triggered the first great mass extinction because free oxygen is toxic to anaerobic organisms.

It took another 800 million years or so for the evolutionary transition to aerobic life to occur. So, for nearly half of the Earth's history, life was producing its own poison. This is hardly a successful manipulation of its own environment.

Proponents of the coevolution hypothesis point out that the only reason life survived this situation was because it interacted with another of the Earth's surface reservoirs, in this case the hydrosphere, which was able to absorb life's toxic oxygen by-product for over 2 billion years.

Indeed, the geologic record shows that "optimization" of the conditions of life may well be in "the eye of the beholder." If the mass extinction at the end of the Permian was a result of environmental stress (see Chapter 14), then the environment at that time can scarcely be viewed as optimal for the 90% of marine species that were wiped out in the extinction event. These species obviously could not overcome the adverse

conditions and loss of ecological niches caused by the formation of Pangea. However, for the species that survived because they were able to out-compete their neighbors during the crisis, the event could be considered optimal because it eliminated their competitors. When Pangea broke up in the Mesozoic, these opportunistic species were able to flourish in the many new and vacant ecological niches that the breakup created.

In the case of both the Proterozoic oxygen crisis and the Permian mass extinction, interactions between the Earth's surface reservoirs led to mutual change, or coevolution. These changes were not necessarily ones that optimized conditions for life. Hence, most scientists now favor coevolution over the Gaia hypothesis. Nevertheless, they readily concede that the Gaia hypothesis was a brilliant and innovative insight into the interdependence of the Earth's surface reservoirs. So a hypothesis does not necessarily have to be correct to make an important contribution to the science. By stimulating and provoking the scientific community, new hypotheses may be formulated to challenge the older ones and bring us ever closer to a more rigorous understanding of Earth systems. In the words of Eldridge Moores, past-President of the Geological Society of America, "Outrageous but testable hypotheses advance the cause of science by galvanizing other workers into action to disprove them."

19.9 Chapter Summary

Nature does not respect the artificial boundaries between the traditional sciences and we must break down the boundaries if we are to fully understand natural processes. The interdisciplinary science of biogeochemistry is an example of how this is being done, and has provided important new insights into the evolution of the Earth and modern environmental problems.

The food chain, the best known example of biogeochemistry, has been operative since life began on Earth, although it must have changed dramatically when the land was first colonized by plants and animals. The food chain not only applies to the modern biosphere, but also explains why fossil fuels are an energy resource. Essentially, fossil fuels are like "energy banks" that store ancient portions of the Sun's radiant energy.

Soils, upon which the food chain depends, consist of components from each of the Earth's systems: sediment, organic matter, air, and water. Soil is a vital resource and is essential to viability of human life on Earth. Its development depends on climate, the amount of precipitation, organic activity, the nature of the underlying bedrock, and drainage.

Our study of the odyssey of the carbon atom revealed the complex paths that individual atoms and molecules take as they migrate from one surface reservoir to another. The carbon, nitrogen, and phosphorus cycles are particularly important because these three elements are the essential ingredients of DNA and RNA, the building blocks of life. The paths of these three cycles are like a highway map that enable us to distinguish between natural and human-induced influences. The latter are recognized as "shortcuts" in the natural cycle.

The geologic record demonstrates the importance of carbonate sediments in removing carbon dioxide from ocean water and isolating it in rock. But this can only occur if the ocean waters contain sufficient calcium and, to a lesser extent, magnesium, because the carbonate minerals that isolate carbon dioxide contain these two elements. If these elements are not present in sufficient quantity, carbon dioxide levels in ocean water will remain high and the extraction of carbon dioxide from the atmosphere is impeded. This has important consequences for the concentration of greenhouse gases in the atmosphere and, hence, for global warming.

Key Terms and Concepts

 Look for the highlighted items on the web at: WWW.BROOKSCOLE.COM

- Biogeochemical cycles are aspects of the new interdisciplinary science of biogeochemistry
- The food chain describes the migration of energy and nutrients in the modern and ancient biosphere
- Soils consist of components of each of the Earth's systems: sediment, organic matter, air, and water
- Soil development depends on climate, the amount of precipitation, organic activity, the nature of the underlying bedrock, and drainage
- The migration of a single carbon atom from one surface reservoir to another provides an example of the interdisciplinary approach to understanding our environment
- Soil profiles are layered and may show an O horizon (organic matter, A horizon (zone of leaching), B horizon (zone of accumulation), and C horizon (broken bedrock)
- Most atoms and molecules at the Earth's surface have been exhumed from the Earth's interior

- The carbon, nitrogen, and phosphorus cycles are particularly significant because of their importance to the biosphere
- The biogeochemical cycles of each element are laid out like a road map, facilitating the distinction between natural and human effects
- As a result of biogeochemical cycling, most carbon (>99.9%) has ended up in carbonate rock
- The burning of fossil fuels short-circuits the natural carbon cycle
- Nitrogen is concentrated in the atmosphere and is impoverished in the solid Earth
- Phosphorus is an essential ingredient of proteins and occurs in the hard parts of all living things
- Phosphorus is an important nutrient in areas of coastal marine upwelling and its availability may have been crucial to the major biological innovations of Earth history

Review Questions

1. In what way does biogeochemistry differ from more traditional sciences?
2. How long has the food chain been operative, and when in geologic history did it first begin to resemble the scheme outlined in Figure 19.1?
3. Explain why fossil fuels represent stored energy within the food chain.

4. What is the role of microorganisms in soils?
5. Of all the naturally occurring elements, why are carbon, nitrogen, and phosphorus so important?
6. Why are nitrogen fertilizers so important to modern agriculture?
7. Explain the relationship between phosphorus and modern marine ecology.

Study Questions

Research the highlighted question using InfoTrac College Edition.

1. What is the role of physical processes in biogeochemical cycles?

2. In what plate tectonic settings do you think the carbon, nitrogen, and phosphorus supply would facilitate biological innovation?

3. How do the plate tectonic settings identified in Question 2 match the geologic record of biological innovations outlined in chapters 13 and 14?

4. Lions are often referred to as being "at the top of the food chain." Do you think this is a fair statement or a misleading one? Explain.

5. What do you think now about the fossil fuel/global warming issue? Has the depiction of the carbon biogeochemical cycle clarified or changed your opinions? Explain.

6. Why is nitrogen the most abundant gas in the atmosphere?

7. On Figure 19.11a, (the biogeochemical cycle "map" for phosphorus), sketch the behavior of phosphate-rich guano (bird droppings).

8. Looking back through geologic time, when do you think guano deposits first became important?

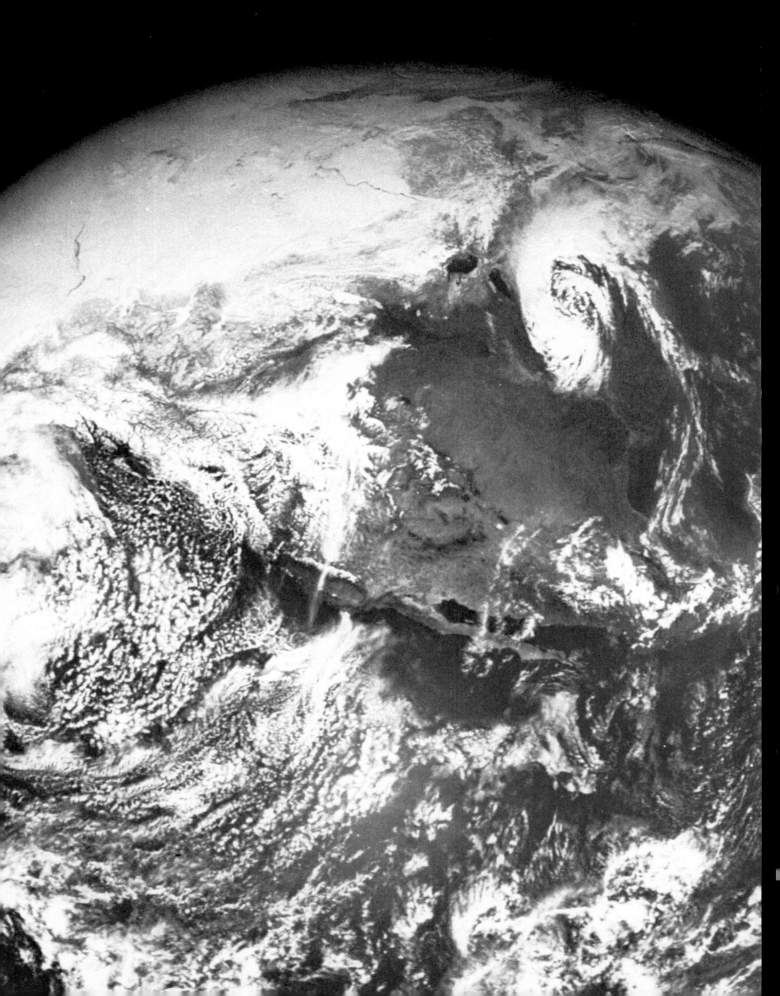

C H A P T E R

20

Our Fragile Atmosphere

We often assume that nature is benevolent and that environmental hazards are entirely of our own making. This assumption is not valid, as we are about to see. In this chapter we present two controversial issues connected with global change: ozone depletion and global warming. Understanding each of these issues will draw on what we've covered in previous chapters, and will require you to synthesize information about how the Earth's systems interact and affect each other. In each case, there is apparently conflicting evidence; some data suggesting that global change is merely a response to the Earth's natural rhythms, while other data suggest it is the result of the profound and irreversible influence of human activities.

Distinguishing between these two possibilities is fundamental to the planning of our own immediate future and that of future generations. As we mentioned in Chapter 1, the evolution of the Earth's systems over geologic time offers a unique perspective on these issues. By gaining an understanding of the Earth's natural rhythms over a range of geologic time, we can examine the data to see if there are distortions to those rhythms in the modern world. We are now in a position to make that evaluation.

Both of the issues we discuss are controversial and sometimes emotionally charged, but both have implications for our own survival and for life as we know it. Our need to understand the effect of human activity on the Earth is vital. This chapter cannot provide all the answers; science has not advanced to that stage. However, we will attempt to demonstrate the unique perspective that Earth science offers to environmental problems. We deliberately aim to be provocative in order to generate discussions among yourselves and with your instructor. We begin with the case of ozone depletion and conclude with the case of global warming.

Depletion of the ozone layer, which may result in increased incidence of skin cancer, is generally considered to be related to the introduction of chlorofluorocarbon compounds (CFCs) into the atmosphere. These compounds do not occur in nature. They are industrial products used in coolant systems. As we shall see, CFCs have a voracious appetite for ozone molecules. However, they are not the only culprits and, as with most environmental problems, alternative viewpoints exist. An alternative viewpoint in this case is that the ozone-depleting effects of CFCs are dwarfed by natural phenomena such as solar flares, sunspots, and volcanic eruptions. To reach our own conclusions, we will review the recent detective work that tries to identify the prime cause of ozone depletion in the Antarctic. In doing so, we will discover that ozone-depleting reactions form by-products that can only be attributed to CFCs.

To examine the case of global warming, we trace the history of the "greenhouse effect" as an example, peeling back the layers of time to look at progressively more and more of the geologic record. Atmospheric compositions of greenhouse gases and average global temperatures are well documented for the past 100 years. However, as we go further back in time, we cannot measure temperature directly, and must turn instead to proxy methods. Just as we use the expansion of mercury as a means of measuring temperature in a basic thermometer, we use oxygen isotope ratios as a "geothermometer" for measuring past temperatures because these, too, are sensitive to temperature changes. By looking at the oxygen isotope ratios preserved in polar ice, in marine sediments, and in ancient sedimentary rocks, we can deduce reasonably well-constrained patterns of global temperature change over the past 120,000 years, 2 million years, and about a billion years, respectively. These temperature changes show natural patterns of variation because they were developed long before the modern industrial era. We shall see that each time we delve deeper into the past, the temperatures inferred from the isotope data seem to show a correspondence with the geologic evidence for global greenhouses or icehouses. This gives us added confidence that the data are meaningful. We shall also see that these natural temperature fluctuations are far more dramatic than those which concern us today.

By examining the geologic record, we are also able to deduce Earth's major rhythms and pulses through time. Our conclusions may surprise you. But whether or not you agree with them, we hope that our approach provides you with a way to organize and assess the vast amount of apparently conflicting information concerning the environment.

20.1 Introduction

In Chapter 1, we stated that Earth science offers a unique perspective on environmental issues because it has the potential to distinguish natural from human-induced global change. By studying the origin and evolution of the Earth systems and the Earth's resources, we are now in a position to evaluate the effect of resource exploitation on the environment and on our future.

Until very recently, the Earth's evolutionary path was unaffected by our influence. We have seen that this evolution was a dynamic one, full of distinctive rhythms and beats that we have learned to listen to and have attempted to understand. We have learned that the Earth's major reservoirs interact in a manner that profoundly influences life. Can we now apply this knowledge in order to distinguish between natural change and that induced by humankind? Little new material is introduced in this chapter. However, like a deck of cards, opposing sides of the debate use the same evidence, or cards, to support their cause. The strength of the hand depends on the way in which the cards are assembled…and how they are played!

Many of us are concerned about excessive exposure to sunlight during the summer months, and the "UV index" is now a familiar part of the weather forecast. We are also aware that the ozone layer filters ultraviolet light from the Sun, and as we learned in chapters 13 and 14, the presence of ozone in the Earth's atmosphere has had a profound effect on evolution. So does our current concern with ozone depletion threaten life on Earth as we know it? Some scientists correlate this depletion of the ozone layer with our increased use of chemicals known as chlorofluorocarbons (also known as CFCs). Other scientists question the validity of this theory, pointing out that ozone levels have fluctuated during the past century and we may merely be in a temporary downward part of the cycle. These conflicting views have led to a heated debate. So is ozone depletion part of a natural cycle or is it induced by human activity? In the latter case, what is causing this depletion?

Climate change is also a key environmental issue that often focuses on the relationship between average global temperature and the greenhouse gas content of the atmosphere. As we learned in Chapter 11, greenhouse gases block the escape into space of solar radiation reflected off the Earth's surface. Some scientists believe that burning of fossil fuels has increased atmospheric greenhouse gas concentrations and that this increase is the prime cause of global warming. Other scientists argue that our modern climate is within the bounds of proven natural fluctuations and that climate change is a natural phenomenon. So are greenhouse gas concentrations in the atmosphere increasing, and if so, is this the result of the burning of fossil fuels?

In addition to the specific cases we address in this chapter, we must bear in mind that other problems will confront future generations. We cannot be certain what these will be. However, we hope our treatment provides an example of the kind of approach that may be used to provide solutions.

Much of what follows is provocative and controversial, and it is certainly not the last word. We invite you to criticize and debate the contents among yourselves and with your course instructor. Irrespective of whether you agree or disagree with what follows, we believe that one key question will continue to be asked. Can we distinguish between natural changes and those induced by human activities? The answer to this question is not straightforward. However, we hope to demonstrate that a knowledge and understanding of geologic processes and the geologic record provides a unique and essential insight into our modern environmental problems. We must understand the Earth if we are to live in harmony with it.

20.2 The Ozone Layer

Ozone is a gas that occurs in very minute quantities in our atmosphere. High up in our atmosphere, ozone acts as a vital natural filter to protect life on Earth from over-exposure to the Sun's harmful ultraviolet radiation. Ozone is particularly effective in absorbing a portion of ultraviolet radiation, thereby preventing these rays from reaching the Earth's surface (Fig. 20.1). It has recently become apparent that this protective function of the atmosphere is now being threatened by the depletion of ozone.

As we learned in Chapter 11, about 90% of the world's ozone occurs between 18 and 50 km (11 and 31 mi) above the Earth's surface in a layer known as the stratosphere. Maximum concentrations occur at an altitude of about 22 km (14 mi). This is above the height of most clouds. Supersonic aircraft, such as the Concorde, fly in the lower parts of the stratosphere at altitudes of about 20 km (12 mi). However, even in the stratosphere, ozone is only a very minor component of the atmosphere compared to nitrogen (about 79%) and oxygen (21%), since it only accounts for one in every 100,000 molecules of gas (0.001%). In fact, if all the ozone in the atmosphere was compressed to form a layer of pure ozone, the layer would only be about 3 millimeters (or one-tenth of an inch) thick. Despite these tiny amounts, the protection afforded by the ozone layer has probably been a major influence on the evolution of life on our planet for nearly 2 billion years.

In the last 20 years, we have begun to realize how important the ozone layer is to the existence of life, not only in filtering

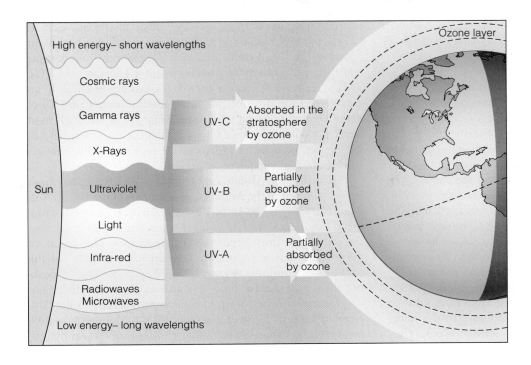

Figure 20.1

Absorption of ultraviolet radiation in the stratosphere by ozone. Note that in the range of wavelengths classified as ultraviolet, the shorter wavelengths (UV-C) are almost completely absorbed, whereas the longer wavelengths (UV-B and UV-A) are only partially absorbed.

harmful radiation, but also in influencing the circulation of water within the hydrologic cycle. The presence of ozone influences the entire thermal structure of our atmosphere (see Chapter 11). In the troposphere, which is confined to the lowermost 18 km of our atmosphere, the air temperature *decreases* from an average of about 15°C (59°F) at the surface to about −60°C (−76°F) at an altitude of 18 km. Thus, moisture evaporating from the Earth's surface cools as it rises, and so the hydrologic cycle is confined to the lower portions of the troposphere. The boundary between the troposphere and the overlying stratosphere is defined where the air temperature begins to *increase* with increasing altitude. Within the stratosphere, between altitudes of 18 km and 50 km, the average temperature progressively increases from −60°C to about 0°C (32°F) (Fig. 20.2). This increase is attributed to the ability of stratospheric ozone to absorb a component of ultraviolet radiation, thereby warming the air. The presence of ozone therefore influences the altitude of the boundary between troposphere and the stratosphere. Ozone depletion within the

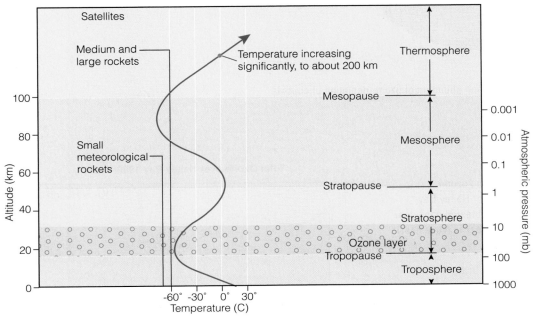

Figure 20.2

Thermal structure of the atmosphere emphasizing the importance of ozone. Note that the rise in temperature with increasing altitude in the stratosphere (which contains the ozone layer) contrasts with the profile in the troposphere (the lowermost portion of the atmosphere; see also Chapter 11).

stratosphere would affect the position of this boundary, and this may have important implications for the circulation patterns of water within the hydrologic cycle.

Hence, the stability of the thermal structure of the atmosphere is certainly affected by the stability of the ozone layer. What this structure might look like in the absence of the ozone layer is difficult to predict, but it would certainly affect the rate at which the Earth's atmosphere cools above its surface, the stability of the hydrologic cycle, and the ability of the planet to retain water evaporated from its oceans.

Just as we are realizing the importance of the ozone layer, we are also learning about the threats to its existence. In 1974, American scientists Molina and Rowland suggested that the integrity of the ozone layer may be damaged by chemicals known as chlorofluorocarbons (or CFCs), which contain bonded atoms of chlorine, fluorine, and carbon. CFCs are industrial chemicals that are extremely stable in the troposphere and rise from the Earth's surface without being absorbed or destroyed. In contrast, other chlorine compounds (such as those used in swimming pools and detergents) are absorbed by moisture in the lower atmosphere and are returned to the Earth's surface in rain. They do not reach the stratosphere and consequently pose little threat

to the ozone layer. More recently, other chemicals such as halons, methyl chloroform, and carbon tetrachloride have been added to the list of ozone-depleting compounds. Refrigerators and their foam insulation contain about 1 kilogram of CFCs, and car air conditioners use up to 5.0 kilograms of CFCs in an average lifetime. Together, these account for more than 85% of domestic and industrial CFC usage (Fig. 20.3). Although most new car models do not use CFCs in their cooling systems, it will take several years before older cars are replaced by these newer "ozone-friendly" models.

Since the early 1960s, ozone levels in the atmosphere have been measured in the Antarctic by members of the British Antarctic Survey (Fig. 20.4). In 1985, the survey reported that dramatic thinning of the ozone layer was occurring annually over the south pole from September to November. This region of thinning, which is commonly referred to as a "hole" in the ozone layer, continues to grow in size (Fig. 20.5). Strictly speaking, the region is not a "hole," but an area of extreme depletion.

Depletion levels of up to 60% were recorded in the Antarctic in the early 1990s, but depletion is not confined to remote polar regions. In the 1980s, average global ozone concentrations declined by 3%. In North America, where ozone depletion usu-

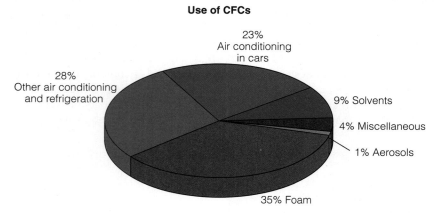

Use of CFCs

23%
Air conditioning
in cars

28%
Other air conditioning
and refrigeration

9% Solvents

4% Miscellaneous

1% Aerosols

35% Foam

Figure 20.3
Use of CFCs. CFCs are the most widely used and abundant of ozone-depleting chemicals (Source: Environment Canada).

Figure 20.4
Data from the British Antarctic Survey showing ozone depletion (measured in Dobson Units, DU) over Halley Bay between 1960 and 1990. The analyses were obtained each year during the Antarctic spring (i.e., from September to November).

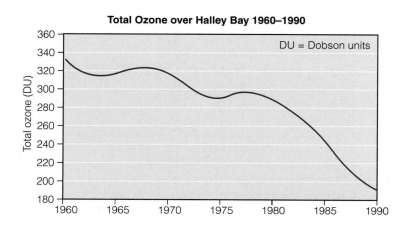

Total Ozone over Halley Bay 1960–1990

DU = Dobson units

Total ozone (DU)

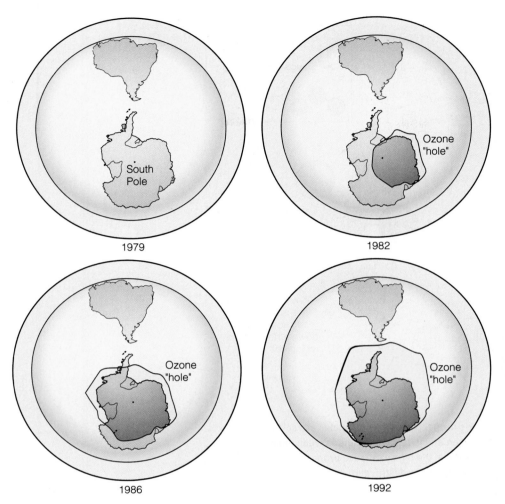

Figure 20.5
The "ozone hole" in the Antarctic region of the southern hemisphere in which thinning of the ozone layer occurs. The "hole" has grown enormously since 1979.

ally occurs between November and April, the largest depletion recorded so far was in 1993 when values averaged close to 15% below normal.

In 1987, 24 nations signed the Montreal Protocol, pledging to reduce production and consumption of CFCs by 1999. This target was amended in Copenhagen in 1992 to an agreement to eliminate production and consumption of CFCs by 1996. Some countries have timetables that target elimination of the production of ozone-depleting chemicals by the end of this century. But even keeping these timetables will not mean that the atmosphere will be immediately purged of CFCs, for two reasons. First, other countries may still produce these chemicals, and second, many of the chemicals can remain in the upper atmosphere as active ozone-depleting agents for up to 200 years.

The ozone depletion problem will therefore be with us well into the next century. But is there anything we can do about it? In order to approach the problem, we need to understand what ozone is, how it originally formed and how it is formed today, how it is distributed and how it protects us.

20.3 Formation of Ozone

Ozone is one of three important forms of oxygen. An ozone molecule consists of three atoms of oxygen that are bound together and is given the chemical symbol O_3. The vast majority of oxygen in our atmosphere occurs in molecules that contain two atoms, symbolized by O_2. More rarely, oxygen can exist as independent atoms, symbolized by O. These three different types of atmospheric oxygen continuously interact and interchange as they are bombarded by solar radiation. For example, ultraviolet radiation may strike a normal oxygen molecule and split it apart to form two single atomic units. These units are highly unstable and may attach themselves to neighboring oxygen molecules to form ozone (Fig. 20.6).

For solar radiation to create ozone by bombarding the atmosphere, a critical level of about 2% oxygen is needed in the atmosphere. At levels less than 2%, there aren't enough easily accessible neighboring oxygen molecules for single oxygen atoms to bond with. For about half of the Earth's 4.6 billion year

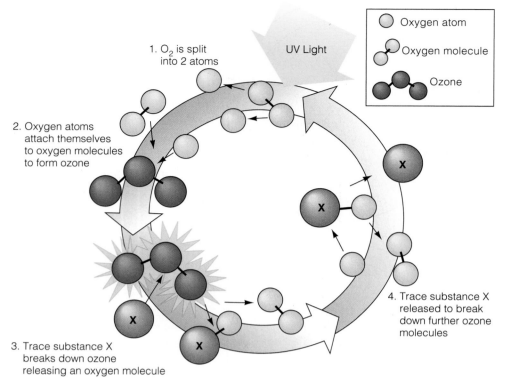

1. O₂ is split into 2 atoms

2. Oxygen atoms attach themselves to oxygen molecules to form ozone

3. Trace substance X breaks down ozone releasing an oxygen molecule

4. Trace substance X released to break down further ozone molecules

UV Light

Oxygen atom
Oxygen molecule
Ozone

Figure 20.6

Natural production and destruction of stratospheric ozone. (1) Oxygen molecules are split into atoms as they absorb ultaviolet light. (2) These oxygen atoms react with other oxygen molecules to form ozone. (3) Trace substances attack and break down the highly reactive ozone releasing an oxygen molecule. (4) The trace substance is released and can destroy further ozone molecules.

history, this critical 2% oxygen level had not been attained. The absence of oxygen implies that no ozone shield existed on the primitive Earth and, as a consequence, when life evolved it was probably restricted to the environments where ocean water could give partial screening from ultraviolet radiation. Despite the fact that oxygen-producing (photosynthetic) life has existed for more than 3.5 billion years, it is thought that iron-rich ocean waters absorbed this early oxygen just as iron metal left outdoors absorbs oxygen as it rusts (see Chapter 13). This absorption of oxygen converted soluble reduced iron (Fe^{2+}) to insoluble oxidized iron (Fe^{3+}). The result was the deposition of Banded Iron Formations. Just over 2 billion years ago, the ocean waters became unable to keep pace with oxygen production and, for the first time, free oxygen occurred in the atmosphere.

Ironically, the initial presence of oxygen in the atmosphere was a poison that almost wiped out the fragile life forms that were used to oxygen-depleted conditions. It probably took several hundred million years before organisms that could freely exist with oxygen flourished. Ozone began to form when the atmospheric oxygen level increased to about 2%. Photosynthetic algae probably increased the levels of oxygen (and therefore ozone) in the atmosphere quite slowly, until about 400 million years ago when plants first conquered the land (see Chapter 14). This evolutionary milestone was most likely made possible by the buildup of enough ozone in the

atmosphere to provide sufficient protection from ultraviolet radiation. Once continental vegetation appeared in the geologic record, increased photosynthesis probably accelerated the increase of oxygen and ozone in our atmosphere toward modern levels.

Our atmosphere today is ideal for the propagation of life on our planet. It interacts with radiation emitted from the Sun, absorbing, transmitting, and distributing heat and, as a result, is a fundamental influence on our climate.

All hot objects emit radiation. For example, we can feel radiated heat from a domestic heating appliance even though we cannot see the waves of heat traveling towards us from the heat source. Hot objects emit heat with a variety of wave patterns, and the range of their wavelengths depends on the temperature of the radiating object. As we discussed earlier, the Sun, with its intense heat, emits a very wide range of wave patterns, only a narrow portion of which can be detected by our eyes (see Figure 16.3, the electromagnetic spectrum). Like the variable sizes of ocean waves that break upon a beach, these different wave patterns vary in wavelength, that is, the distance measured between the crests of successive waves. Some have long wavelengths, such as radio waves, microwaves, and infrared rays. These waves generally pass through the atmosphere and reach the Earth's surface. Because of the ease of their passage through the atmosphere, these long waves are commonly used for communication purposes. Shorter waves such as ultraviolet radiation, X-rays, gamma rays, and cosmic rays tend to be absorbed in the upper atmosphere by a shield of gas particles. Ozone molecules are a particularly effective shield, filtering most of the ultraviolet radiation.

The formation of ozone requires the presence of oxygen in the atmosphere. But as we shall see in the next section, ozone is notoriously unstable and breaks down naturally. So at any one time, ozone levels in the atmosphere depend on the rate of ozone formation and the rate at which it is broken down.

20.4 Mechanisms and Implications of Ozone Depletion

Figure 20.6 shows a typical path of ozone creation and destruction. Ultraviolet light splits an oxygen molecule (O_2) into two oxygen atoms (O). As described above, these atoms are very unstable and each combines with an oxygen molecule to form an ozone molecule (O_3) composed of three oxygen atoms. But these new ozone molecules are also very unstable and can react with trace substances in the atmosphere. When they do so, the ozone molecule breaks down to yield, once again, a typical oxygen molecule and a single oxygen atom. The trace substance temporarily bonds with the single oxygen atom, but releases the atom when it encounters another single oxygen atom. These two oxygen atoms then combine to form a normal oxygen molecule. These latter two steps are very important because they mean that the trace substance is also liberated from the molecule and so is free to attack another ozone molecule and then another. These trace substances therefore have insatiable appetites for ozone molecules.

In the stratosphere, then, natural processes are constantly creating and destroying ozone. Ozone depletion results when ozone destruction occurs at a faster rate than ozone production. Ozone depletion in our atmosphere is now well documented, as are the statistics that associate increased health risks with exposure to excessive solar radiation. Since ozone moves throughout the stratosphere, the depletion measured in Antarctic ozone levels over the past 30 years is an indication of a much more widespread phenomenon. About 10% of the southern hemisphere is directly affected by this ozone loss, particularly New Zealand, Australia, and southern Argentina and Chile. In Canada, ozone levels have declined by about 4% since 1970, and it is projected that losses of up to 8% may occur by the year 2000. Some estimates suggest that a 10% loss in ozone would lead to a 25% increase in occurrences of non-melanoma skin cancer.

It is crucial to realize, however, that while ozone depletion is a documented fact, the relative contributions of natural versus human factors in this depletion are not. Although ozone is formed by the bombardment of solar radiation, we are unsure how variations in solar activity (reflected, for example,

in solar flares, sunspots, and the intensity of the solar wind) affect ozone levels. Indeed, many scientists still doubt that the depletion of ozone in our atmosphere is related to industrial activity. They point out that ozone depletion is a natural phenomenon related to volcanic activity. If this is the case, they argue, there is nothing that can be done about it.

The eruption of Mount Pinatubo in the Philippines in June 1991 provides an example of natural ozone destruction. For almost six months following this event, local ozone levels dropped by 20%. The eruption was powerful enough to inject mixtures of volcanic particles and gases (called aerosols) into the stratosphere. These aerosols acted as trace substances and facilitated ozone-depleting reactions. Images received from satellites showed that these aerosols spread around the globe within months of the eruption (see Figure 1.6b), causing widespread ozone depletion. So there is no doubt that volcanoes expel gases that attack the ozone layer. But is this the cause of the extensive depletion observed over the poles?

Recently, atmospheric scientists at NASA claim to have found the answer. The instability of ozone is such that a wide number of compounds (some introduced naturally, and others related to industrial activity) can cause its destruction. However, they claim to have detected the "fingerprints" of a massive presence of industrial chlorofluorocarbon gases (CFCs) in the Antarctic. The approach taken by these scientists is similar to that of a detective working on a criminal case. In most cases, there are no eyewitnesses. So the case is put together from forensic evidence that, it is claimed, provides proof beyond a reasonable doubt. These scientists claim to have found evidence that chlorofluorocarbons account for at least 75% of ozone depletion.

Their understanding of the chain of reactions triggered by ozone-CFC reactions have allowed them to detect these fingerprints. When CFCs reach the stratosphere (stage 1 of Figure 20.7) they are broken down by ultraviolet radiation. The reaction releases chlorine and fluorine. The fluorine combines with hydrogen to form hydrogen fluoride (hydrofluoric acid). But the chlorine, once released from the CFC molecule, reacts with ozone. The chlorine joins with the ozone molecule to form chlorine monoxide (ClO), leaving molecular oxygen (O_2) behind (stage 2, Figure 20.7).

Unfortunately, the damage doesn't stop there. The chlorine monoxide reacts with itself to make Cl_2O_2, which sunlight breaks apart, freeing the chlorine to attack additional ozone. It is calculated that a single chlorine atom may destroy 100,000 molecules of ozone in this way, so that CFC compounds that leak into the atmosphere today may destroy ozone for 20 to 50 years to come. In extremely cold atmospheric conditions, such as the Antarctic, another set of reactions also leads to ozone depletion. Chlorine monoxide (formed at stage 2) reacts with nitrogen dioxide to form chlorine nitrate (stage 3), which itself reacts with hydrogen chloride (hydrochloric acid) to form nitric acid and a chlorine molecule (stage 4). Ultraviolet radiation then splits the chlorine molecule to form two individual chlorine atoms (stage 5), which attack more ozone.

The details of the reactions in this diagram are not of fundamental importance here. What is most important to

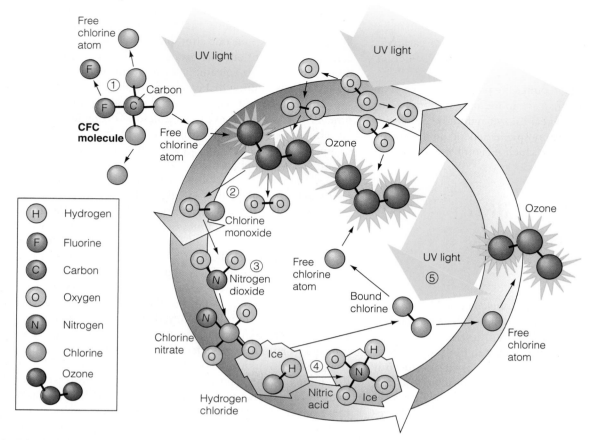

Figure 20.7

Destruction of ozone by CFCs. (1) Synthetic CFCs rise into the stratosphere. UV light reacts with CFCs and frees chlorine atoms. (2) Chlorine atoms attack the highly reactive ozone molecule, forming chlorine monoxide and an oxygen molecule (O_2). (3) In the cold polar conditions, chlorine monoxide reacts with nitrogen dioxide to form chlorine nitrate. (4) Chlorine nitrate reacts with hydrogen chloride in the stratospheric air. (5) This reaction frees the chlorine atom which can then attack another ozone molecule.

understand is that the main ozone-depleting agents are chemical contaminants like chlorine, and that one of the main agents introducing chlorine into the ozone layer are CFC molecules. The chlorine is released from the CFC and then goes on an ozone-gorging rampage. The rampage occurs because chlorine only forms temporary bonds with other substances and is ultimately released again as a free agent that can destroy more ozone. Try to follow this cycle in Figure 20.7, paying special attention to the fate of chlorine.

But how are the "fingerprints" of these reactions detected? If ozone depletion is related to the presence of CFCs in the ozone layer, then the relative concentrations of chlorine monoxide and hydrogen fluoride in the Antarctic atmosphere should be consistent with that predicted from the reactions involving CFC contamination. On the other hand, if ozone depletion is related to volcanic activity, there should be no obvious relationship. In fact, as far as we know, hydrogen fluoride cannot be produced by natural processes such as volcanism. According to atmospheric scientists, the results of the latest tests are conclu-

sive. They indicate that the atmosphere contains chlorine monoxide and hydrogen fluoride in the exact abundances that were predicted from the presence of CFCs. In the last 30 years, levels of chlorine in the atmosphere have risen dramatically from 1.5 to 3.5 parts per billion and are expected to reach 4.0 by the year 2000. This increase would account for the vast majority of the observed ozone depletion and is difficult to attribute to natural phenomena. This finding is equivalent to identifying fingerprints or obtaining DNA evidence at the scene of the crime.

800,000 metric tons of CFCs are released annually into the atmosphere. Ultimately, these compounds will diffuse through our atmosphere and into space. But as they rise through the stratosphere, they will repeatedly destroy ozone. The atmospheric lifetime and ozone depletion potential (ODP) of a variety of CFCs and related compounds are shown in Figure 20.8. The ozone depletion potential of each compound is graded relative to one of the more abundant CFC compounds (CFC-11), which is assigned a value of 1.0. Some compounds that contain bromine (Br) have a much higher ODP than the CFCs. At pre-

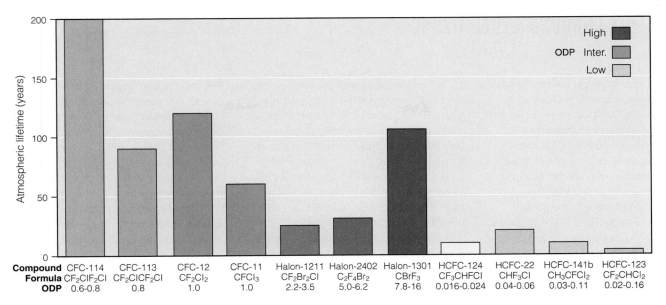

Figure 20.8

Comparison between various CFCs and related chemicals, in terms of their Ozone Depletion Potential (ODP) and duration in the atmosphere. Even if CFC emissions were to cease today, their effect on ozone would continue for many decades. ODP is measured relative to CFC-11 (the most abundant CFC compound), which is arbitrarily assigned a value of 1.0.

sent, however, they are not as abundant in the atmosphere. The implications of this diagram are clear: even if CFC emissions were to cease today, ozone depletion would continue.

There are some preliminary indications that the measures adopted by the Montreal Protocol have had a beneficial effect. Some recent data suggest that the ozone "hole" over Antarctica is beginning to repair itself. However, most calculations suggest that ozone levels will probably not return to normal until well into the 21st Century, even with the measures implemented by the protocol.

20.5 The Greenhouse Effect: A Matter of Balance

If we include water vapor, greenhouse gases account for less than 1% of the atmosphere. It is commonly assumed that the presence of such gases in our atmosphere is bad. This is not true. Greenhouse gases (such as water vapor, carbon dioxide, and methane) are essential naturally occurring parts of our atmosphere. They trap heat in the atmosphere and are responsible for global warmth, which greatly influences weather, climate, and the ecosystem. Without them the Earth would be, on average, 35°C (63°F) degrees colder: an uninhabitable, icy planet. Carbon dioxide is responsible for an estimated 60% of the global warming from greenhouse gases. Each methane molecule traps about 25 times as much heat as a carbon dioxide molecule, but because methane is much less abundant in the atmosphere, its total effect on global warming is much smaller.

Because of their ability to trap heat, many scientists believe that there is a cause-and-effect relationship between recent increases in the concentration of greenhouse gases in the atmosphere and the ongoing rise in average global temperature. The main question for these scientists is whether the increases are largely the result of industrial activity or whether they are due to natural phenomena. Since carbon dioxide accounts for most of the global warming from greenhouse gases, attention has been focused on the potential relationship between the abundance of carbon dioxide in the atmosphere and the burning of fossil fuels.

Current figures show that estimated resources of fossil fuels, such as coal, oil, and natural gas, can satisfy present consumption levels for several hundred more years. Policy-makers are therefore caught between pressures to search for alternate sources of energy (both because of the global warming issue and the finite nature of the resource) and the necessity to meet our immediate energy requirements. But what is the evidence that the increase in atmospheric carbon dioxide is related to fossil fuel consumption? And if this consumption leads to global warming, what are the effects of this warming? Are these effects sufficiently disturbing to warrant the dismantling of the very foundation of our modern technological era? These are important issues and their resolution is a global problem because air masses travel freely across political boundaries.

GROUND LEVEL OZONE

It is ironic that ozone gas, so important in our atmosphere, is a major pollutant at ground level (Fig. 20.9). Ozone gas is released into the environment by domestic use in bleaches and water purifiers. It has a noxious odor and can cause respiratory and other problems in humans and animals. It can also seriously damage crops and vegetation.

Ground level ozone is also a major component of smog (Fig. 20.10). It is called a secondary pollutant because it is not directly released into the atmosphere, but instead is produced when pollutants react with each other and sunlight. The newly created ozone gas is then carried away by prevailing winds. Carbon monoxide, on the other hand, is an example of a primary pollutant because it is released directly into the atmosphere from a car's exhaust. According to most government agencies, the "maximum acceptable level" of ground level ozone is 82 parts of ozone in a billion parts of air (or 82 parts per billion). Even this concentration, however, can cause breathing difficulties and air pollution.

Air pollution values can vary significantly within a given city. Official values for any city are averages of the highest concentrations from all recording stations in that city.

Readings are taken regularly, not only of ground level ozone, but also of the concentrations of airborne particles (which are small enough to be inhaled and can also cause breathing problems) and of primary pollutants such as sulfur dioxide, carbon monoxide, and nitrogen dioxide. Airborne particles are primarily introduced into the atmosphere by industrial and domestic burning. Statistics released by environmental agencies in North America show that urban air quality has improved in many cities over the past decade. In 1992, ground level ozone was below maximum acceptable levels 99.9%

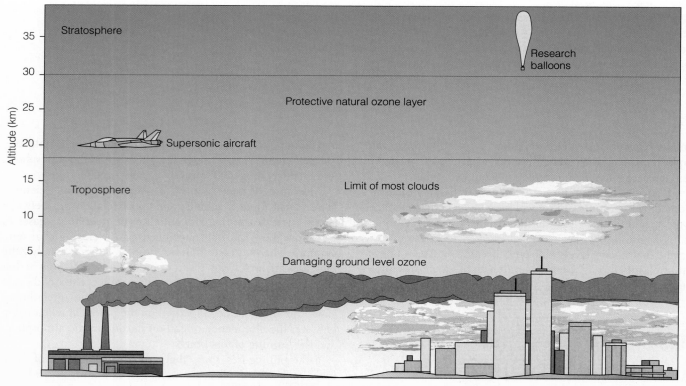

Figure 20.9

The distribution of ozone in the atmosphere. Ground level ozone is harmful to the environment, whereas the presence of ozone in the stratosphere is vital to the stability of the Earth's ecosystems. Smog-generating emissions such as those from a coal-burning power plant will produce ground level ozone and can be trapped near the Earth's surface by thermal inversions (see Figure 20.11).

Figure 20.10

Smog in an industrial city trapped by thermal inversion (see Figure 20.11).

of the time. However, the average concentration of ozone in many cities remain close to the maximum acceptable levels. Cities such as Los Angeles, New York, and Sydney, Australia, for example, have values close to twice the acceptable level.

Ground level ozone concentrations are particularly hard to control because they are strongly influenced by weather conditions. During the summer, sunny, hot, stagnant weather conditions favor ozone production. In winter, increased domestic burning introduces airborne particles that affect air quality. Local geography and local wind directions also play an important role in trapping pollutants. In general, the air is well mixed because air at the Earth's surface is heated by the Sun, and so rises and mixes with the cooler air above it. However, in many valleys, winds may bring in relatively dense, cool air that pushes the lighter warm air upwards. Nearby mountains may block the lateral escape of the warm air so that it becomes sandwiched between layers of cold air above and below, a condition called a **thermal inversion** (Fig. 20.11). Under such conditions, the warm layer of air forms a seal over the valley that traps the pollutants, including ground level ozone, below. In North America, the California coastal cities, the lower regions of the Fraser Valley of British Columbia, and the Gulf of Saint Lawrence are particularly prone to these inversions. When thermal inversions occur, the trapped airborne particulates react with sunlight to produce the familiar "brown cloud" problem.

Improving the quality of air is not easy. It not only mandates changes in our lifestyle, but also requires detailed knowledge of local geography, accurate weather forecasting and the ability to predict thermal inversions. However, significant improvements can be made. London, England, for example, was once famous for its smog but now has lower ground level ozone levels than many North American cities. The serious health problems that can arise from poor air quality should be sufficient incentive to undertake such improvements.

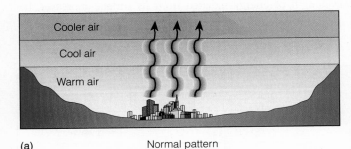

(a) Normal pattern

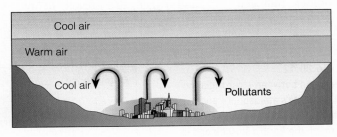

(b) Thermal inversion

Figure 20.11

(a) Normally the Sun warms the air near the Earth's surface. As a result, the air heats up, expands (thus becoming lighter), and rises as cold dense air from above descends to take its place. This mixing of air dilutes pollutants and keeps them from reaching dangerous concentrations. (b) Thermal inversions occur when a layer of cool, dense air occurs beneath warm, light air. The warm air acts as a lid trapping the pollutants near the surface.

20.6 Understanding the Greenhouse Effect

The term *greenhouse effect* conveys the impression that the interaction of sunlight on our atmosphere is similar to its effect on a greenhouse. On a sunny day, it is warmer inside a greenhouse than outside because the greenhouse is a heat trap. In our atmosphere, it is the greenhouse gases that trap heat, although, in truth, our atmosphere is a very poor greenhouse because these gases occur in very small quantities.

So how does this heat-trapping work? The Sun continuously bathes the Earth with waves of radiation that have a variety of wavelengths. In Chapter 16, and earlier in this chapter, we learned that this radiation ranges from very short waves like X-rays, to long waves like those used in radio and telecommunications. All hot bodies emit radiation from their surfaces, and the hotter the body the wider the range of wavelengths. The Sun, with a surface temperature of about 5700°C, emits solar radiation (known as the electromagnetic spectrum) with wavelengths that range from 10^{-10} m to 10^3 m (see Figure 16.3). Our eyes can only detect a very narrow range of these wavelengths (between 4×10^{-7} m and 7×10^{-7} m) known as the visible spectrum. Within this range, our eyes resolve the wavelengths into various colors which we see as the colors of the rainbow. Not all wavelengths of solar radiation are of equal intensity. The vast majority of the radiation has wavelengths between 10^{-8} and 10^{-6} m (that is, from infrared to ulraviolet) and incorporates all the colors of the visible spectrum. The dominant wavelength is 5.1×10^{-7} meters, which also falls within the visible portion of the spectrum. The combination of all wavelengths of the visible spectrum is what we call white light. Approximately 99% of our atmosphere (nitrogen, 78%, and oxygen, 21%) readily permits both the entry of solar radiation from space and the exit of the radiation that is reflected off the Earth's surface.

Although we cannot see it, we know that radiation of other wavelengths exists because we can detect its influence. For example, excessive exposure to invisible ultraviolet solar radiation can lead to sunburn. UV radiation's characteristic wavelength is 10^{-8} m, which is outside the range of our eyes sensitivity.

Solar radiation encounters numerous obstacles as it attempts to penetrate our atmosphere (see left side of Figure 20.12). Its direction is scattered upon entry. It is reflected by clouds and dust. It is absorbed by atmospheric gases such as water vapor and ozone, the latter preferentially absorbing ultraviolet radiation. In addition, radiation is simply reflected off the Earth's surface back through the atmosphere and out into space.

If atmospheric gases generally permit the passage of incoming solar energy, why do they block the escape of some of the reflected radiation? What happens to this energy when it interacts with the Earth's surface? We know from our own experience that the Earth's surface is not a pure reflector and that it absorbs some of this solar energy. For example, solar energy causes the rocks and soils of the land surface to heat up and moisture to evaporate from the oceans.

Like a lump of coal absorbing heat in a fire, the Earth's surface absorbs energy and radiates it back into the atmosphere. As with coal, we can feel the heat emitted from the Earth's surface even though we cannot see it because its wavelength is too long to be detected with the naked eye. The longer wavelength places this energy in the infrared rather than the visible portion of the electromagnetic spectrum. Humans also radiate energy with a similar wavelength. We call this body heat. Even though this energy is invisible, it can be picked up by detectors sensitive to infrared radiation and is sometimes used to locate missing persons.

Similarly, as the Earth's surface is warmed, the dominant wavelength of energy radiated from its surface increases, be-

Figure 20.12

Interaction between solar radiation and the Earth's atmosphere and surface. Much of this radiation does not reach the Earth's surface and is either absorbed in the atmosphere or reflected back into space. The wavelengths of the reflected radiation from the Earth's surface are increased, and because of this, the escape of some of this outgoing radiation is blocked by greenhouse gases.

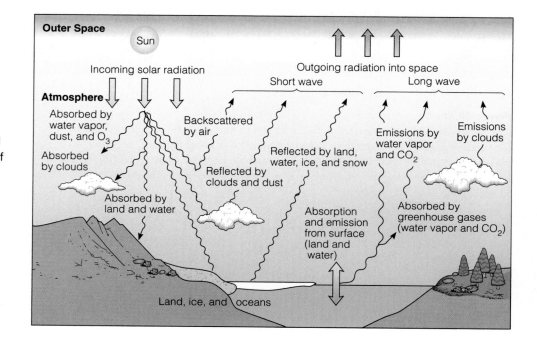

coming longer than that of the incoming radiation from the Sun, and shifting further into the infrared portion of the spectrum (see right side of Figure 20.12). Greenhouse gases block the escape of this reflected energy by absorbing these longer wavelengths. The higher the concentration of greenhouse gases in the atmosphere, the more heat is trapped. The trapped heat energy greatly influences global temperatures, atmospheric circulation, evaporation from oceans, motion of ocean currents, and weather patterns (see chapters 9 and 12).

Scientists generally agree that we need to understand what the desired range of greenhouse gas concentration in our atmosphere should be. It is a question of balance. The concentration must be enough to keep the planet tolerably warm, but not so much that the planet overheats. However, there is little agreement as to what this critical range might be, and even if agreement is reached, it is doubtful that we have the ability to artificially control it, or that the agreed-upon range would be the correct one.

Underpinning all of these arguments is the controversy over whether recent increases in greenhouse gas concentrations are related to modern industrialization or whether they are simply natural fluctuations. In order to address this fundamental issue, we must first understand the pattern and extent of any natural fluctuations, and to do so we must delve further and further back into geologic time. We start this process in the following section from a recent perspective by analyzing global temperatures over the past 40 years or so. As we go further back in geologic time, however, we will begin to recognize more extreme and longer lasting periods of global warming and global cooling. Only when we have understood the origin of these cycles will we be in a position to address the contentious relationship between fossil fuel emissions, atmospheric greenhouse gas content, and global warming.

20.7 A Recent Perspective

Since 1958, scientists have been acquiring accurate data on the composition of our atmosphere. The Mauna Loa observatory on the island of Hawaii is the site of some of this recent atmospheric data collection. The observatory is positioned near the top of an extinct volcano that was deemed suitable for monitoring the Earth's changing atmospheric compositions because of its altitude and remoteness. The site is free from the effects of local pollution and the air at the altitude of Mauna Loa (3500 m or 11,000 ft) is part of a pattern that circulates around the globe at approximately the same latitude. Changes in atmospheric concentrations are therefore indicative of global-scale change.

A graph of the carbon dioxide data collected at Mauna Loa over the last 40 years (Fig. 20.13) shows two trends. A first-order trend is derived by enveloping the data with a line joining the peaks and a line joining the troughs. This envelope shows a systematic increase in atmospheric carbon dioxide. Today's concentration, for example, is about 10% higher than it was about 40 years ago. Closer inspection of the envelope suggests that the increase is not linear but is accelerating, a relationship known as *exponential*. In other words, carbon dioxide levels are rising at an ever increasing rate.

Within the envelope there are smaller scale annual highs and lows that represent second-order trends. This effect is attributed to growing patterns within the biosphere in the northern hemisphere. During the spring and summer, plants are relatively active and absorb more carbon dioxide by photosynthesis than they produce by respiration. In the relatively dormant fall and winter months, on the other hand, respiration produces more carbon dioxide than is absorbed by photosynthesis. Despite these fluctuations on a yearly basis, however, it is clear that the annual effects of vegetation cannot negate the overall increase in carbon dioxide over the past 40 years.

During this same 40-year period, there has been a parallel increase in the rate at which carbon dioxide has been added to the atmosphere through the burning of fossil fuels, a trend that can be traced back to the middle of the last century (Fig. 20.14). Today, almost 6 billion metric tons of carbon dioxide are released into the atmosphere annually from fossil fuel emissions. The Earth's ecosystems have the ability to absorb about half this amount, but the remaining half causes carbon dioxide concentrations in the atmosphere to rise. The greenhouse capabilities of carbon dioxide, that is, its ability to seal in much of the Sun's radiant energy, suggests a connection between the increasing fossil fuel emissions, increasing atmospheric carbon dioxide levels, and global warming. Although these trends are suggestive, they do not, by themselves, prove a cause-and-effect relationship. So how can a potential connection be tested?

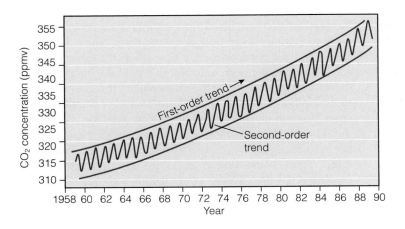

Figure 20.13

Data from the Mauna Loa Observatory in Hawaii showing the concentration of carbon dioxide in the air over the last 40 years, given in parts per million by volume (ppmv). 350 ppmv = 0.035%. The long-term first-order trend shows an overall increase in carbon dioxide, whereas short-term second-order trends reflect annual fluctuations due to varying productivity in the biosphere.

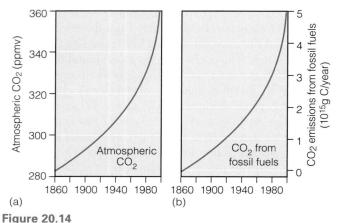

Figure 20.14

(a) Atmospheric concentrations of carbon dioxide and (b) carbon dioxide emissions from the burning of fossil fuels have increased in a similar manner over the last 140 years.

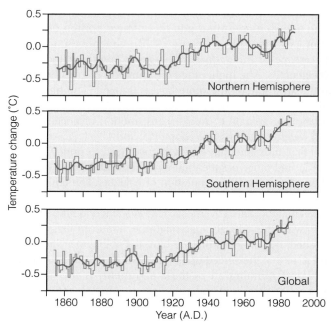

Figure 20.15

Relative change in the Earth's average surface temperature since 1860. Current models suggest that the pace of global warming will accelerate during the first half of the next century (Source: Climate Research Unit, 1991).

Prior to the 1950s, the information needed to calculate average air surface temperatures was sparse. Since that time, accurate temperature records have become available from many land stations. Most of these stations are located in the northern hemisphere. Each year, these records are manipulated in an identical fashion in order to yield an average global temperature. The accuracy of this method is open to debate, but its precision is high because each data point is derived using identical methods (see discussions on accuracy and precision in Chapter 3). As a result, the data are very sensitive to changes in relative temperature (see Fig. 20.15). The data indicate that the average global temperature is nearly 1°C higher than it was about 100 years ago, and that most of this increase took place since the 1920s. Computer models that match the recorded temperature changes since 1860 project a further 1°C increase in temperature over the next 25 years.

Although there is now almost universal agreement that the concentrations of greenhouse gases have increased in our atmosphere compared to their levels 40 years ago, and that average global temperatures have risen over the same time period, scientists continue to debate whether this correlation is coincidental, or whether a causal relationship is involved. Some scientists correctly point out that modern global temperatures are still comfortably within the bounds of natural temperature fluctuations as evidenced by the geologic record. There have been other periods in the past, millions of years before the present industrial era, when the Earth was a hot global greenhouse, and yet other times when it was a frozen global icehouse. For example, as we learned in Chapter 14, during part of the Cretaceous Period, about 100 million years ago, average global temperatures were about 8°C higher than they are today, and atmospheric carbon dioxide levels were probably an order of magnitude higher. In contrast, from about 11,000 to 70,000 years ago, ice sheets covered more than half of continental North America, advancing well beyond the Canadian-U.S. border. These natural swings are more dramatic and far more intense than those we have witnessed over the past 40 years. Although many of these scientists concede that human activity may have contributed to

recent trends, they maintain that nature may override these trends at a whim.

Other scientists acknowledge that a cause-and-effect relationship is uncertain. However, they feel that "we are conducting an experiment on ourselves" and by the time we find out the results, it may be too late to act. They believe that it is only prudent to decrease fossil fuel emissions until more research is done.

Recently, the debate has taken another shift. If there is indeed a causal relationship and the burning of fossil fuels continues to rise, one would predict that atmospheric carbon dioxide levels would likewise continue to rise. Until recently this was the case. In 1991, however, carbon dioxide concentrations in the atmosphere leveled off.

Although on the surface this would appear to be good news for environmental scientists, it is not. These scientists thought that they were beginning to understand the global budget of carbon dioxide. Now, just like economists who make their financial predictions based on certain assumptions, scientists must re-examine their budget calculations. Since fossil fuel emissions have not decreased, where did all the carbon dioxide go? If the ecosystem can only absorb half of the budget, why is the carbon dioxide not in the atmosphere? It is estimated that some 2 billion metric tons of carbon dioxide, or one-third of the annual carbon dioxide budget, is unaccounted for.

The explanations proposed, irrespective of their scientific fate, provide insights into the way science works. First, these recent figures were viewed as anomalous, and extraordinary explanations were invoked. Correlations with unusual events were proposed. For example, the "anomalous" trends began soon after the 1991 eruption of Mount Pinatubo in the Philippines, one of the largest volcanic eruptions this century. Investigations then focused on examining a potential cause-and-effect relationship. If carbon dioxide is usually taken out of the atmosphere by ecosystems, could the eruption have accelerated this process?

There is some evidence that this is indeed the case, but why it should be so is another matter. The evidence is derived from studying subtle variations in carbon isotopes in recent vegetation. There are two naturally occurring stable isotopes of the carbon atom, heavy and light. Carbon, as we learned in Chapter 2, has 6 protons. But it can have either 6 or 8 neutrons. The number of particles in the atomic nucleus is, therefore, either 12 (light carbon) or 14 (heavy carbon). During photosynthesis, plants prefer to absorb light carbon because it is the faster moving of the two isotopes. Accelerated photosynthesis means more pronounced absorption of light carbon into plants and, as a consequence, its depletion in the atmosphere. Hence, the atmosphere becomes enriched in heavy carbon.

An increase in the relative abundance of heavy carbon in the atmosphere following the 1991 eruption of Mount Pinatubo has been confirmed by measurements and is especially marked in the northern hemisphere. But how does this relate to volcanic activity? By pumping massive amounts of volcanic dust into the atmosphere, it is argued that the eruption of Pinatubo may have caused cooler, rainier climates. This, in turn, would increase growth in the forests of the northern hemisphere and thereby increase photosynthesis.

This hypothesis is hotly debated. However, the hypothesis did correctly predict that carbon dioxide levels in the atmosphere would resume their upward trend once the volcanic dust had settled. This is indeed what happened in 1994, and the upward trend has continued ever since. However, irrespective of the correct explanation for these data, it is clear that we are only beginning to understand the subtle linkages between geologic events and biological phenomena. As is common in scientific debates, the data has been used to support arguments on both sides of the global warming issue. Those who favor a link between the burning of fossil fuels and global warming claim that their view is supported by the resumption of the trend of the past 40 years once the volcanic dust had settled out of the atmosphere. However, other scientists claim that this resumption is an example of the overriding effect of nature.

It is clear that a better understanding is essential if the budget predictions of scientists are to become more reliable than those of economists! Falling back on the scientific method for insight, it is also clear that the hypotheses erected to explain the "anomalous" data are difficult to test simply because it is difficult to distinguish between natural and industrial influences. As a result, although there is an impressive correlation between fossil fuel emissions of carbon dioxide, increasing greenhouse gas concentrations in the atmosphere, and increasing average global temperatures, a cause-and-effect relationship is still debated. Our present climate is within the bounds of proven natural variability so that a large component of natural variation in modern global change cannot be dismissed. So how can we gain the better understanding that we need?

By delving into the past, we can look at pre-industrial natural variations in global temperatures. The geologic record offers a unique perspective because we can be sure that ancient temperature changes are entirely due to natural processes. We can then use the geologic data as a foundation to compare modern global change with global change in the geologic past. The geologic record of temperature variation is excellent over the last million years, and is reasonably good for up to a billion years. In the following sections, we will peel back the layers of time in order to discover more of the Earth's natural rhythms over ever longer periods. As we do so, we shall find that whenever we change the time-frame of our investigation we get a different signal or message. The apparently conflicting messages tell us that we have not heard enough of the Earth's rhythms to hum its song. Eventually, the issue will become startlingly clear, but not until it is viewed from a billion-year perspective.

20.8 The Recent Past: Variations in Solar Output

In the preceding section, we assumed that the changes in Earth's average surface temperatures reflected earthly processes. By implication then, we assumed that the amount of solar radiation entering the Earth's atmosphere is constant. In this section, we question that assumption and look at the role of variable solar output in influencing global temperatures.

The Sun is a star, and stars vary in their output of radiation. Since the greenhouse effect is in essence the result of trapped solar radiation near the Earth's surface, it has been argued that variations in solar activity and the energy output of the Sun may be important variables in influencing global warmth.

As we have learned (Chapter 16), the Sun has existed for at least 4.6 billion years. It has a diameter of about 1.4 million km (more than four times the distance from the Earth to the Moon) and is approximately 150 million km away from us. As a result, it takes solar radiation about 8.3 minutes to make the journey to the Earth.

The Sun consists of 92% hydrogen and 7–8% helium, the first two chemical elements in the periodic table. Energy radiated from the Sun is attributed to nuclear fusion reactions occurring within its core that convert hydrogen into helium. Less than 5% of the available hydrogen has been converted during the Sun's 4.6-billion-year life span. But are there variations in the output of solar radiation and, if so, is there a relationship between these variations and the Earth's surface temperatures?

As we learned in Chapter 16, the Sun's surface commonly contains small dark regions (up to 50,000 km across) called sunspots. These areas are dark because their temperature is about 1500°C cooler than the approximately 5700°C of the surrounding regions. They are cooler because their strong magnetism suppresses the escape of heat from the Sun's interior. The number of sunspots at any one time can be recorded with the aid of a long-range telescope by projecting the Sun's image onto a piece of paper and counting the number of dark spots.

The number of sunspots varies considerably and, over the course of a year, fluctuates within an average range from 200 to nearly zero (Fig. 20.16). Sunspot activity also shows a periodicity, known as the sunspot cycle. Peak years are 11.1 years

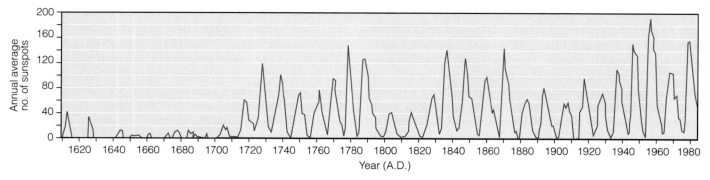

Figure 20.16
The annual mean number of sunspots since their discovery in 1610. From about 1720 to the present day, the number of sunspots shows an 11.1 year cycle. However, throughout much of the 17th Century, few sunspots were recorded. This time period is known as the Maunder Minimum and overlaps the Little Ice Age.

apart with recent peaks occurring in 1980 and 1991. Peak activity is next expected in the year 2002. Minimum activity also occurs at 11.1 year intervals, with recent minima occurring in 1975, 1986, and 1997. The reason for this periodicity is not well understood (and is not of direct concern to us here) but most scientists attribute it to the interaction between the Sun's magnetism and its complex, dynamic internal structure.

Sunspots accelerate the particles that form the solar wind to speeds of 500 km/s (over 1 million mi/h). This is important because the solar wind controls the abundance of charged particles that radiate away from the Sun. The effect of the solar wind is most visible in the tail of a comet, where it literally blasts the comet's surface and streams around it to form a tail of dust and gas that points directly away from the Sun.

But how does this activity affect the Earth, some 150 million km away? The interaction of the solar wind with the Earth's magnetic field causes the spectacular natural light display known as the aurora (called the *Aurora Borealis* in the northern hemisphere and the *Aurora Australis* in the southern hemisphere). However, as the solar wind blasts the Earth's upper atmosphere, it profoundly influences its average temperature. During peak sunspot activity this solar wind is much more powerful, and the temperature of the uppermost atmosphere rises dramatically from the 225°C typical of minimum sunspot activity to about 1225°C.

But does this remarkable variation in the temperature of the *uppermost* atmosphere affect the Earth's *surface* temperatures? There is some empirical (or statistical) evidence that this is the case. For the most part, the sunspot cycle over the past 300 years has been very regular and predictable. However, during the 17th Century relatively few sunspots were recorded and, for reasons that are not fully understood, the sunspot cycle appears to have been suppressed. Historic records document intervals of prolonged cold during this period, which forms part of an interval known as the *Little Ice Age*.

During this extended time of little sunspot activity, the solar wind was presumably less intense and the upper atmosphere relatively cool. However, no clear mechanism has been proposed to show how temperature variations in the upper atmosphere affect those at the Earth's surface. We also have lit-

tle idea how these variations might be affected by the greenhouse gas content in the lower atmosphere. Could a decrease in the intensity of the solar wind over a long period result in low temperature conditions in the upper atmosphere which may, in turn, induce periods of cooling at the Earth's surface? If so, the Little Ice Age could be related to decreased sunspot activity. The drop in average global surface temperature does not need to be substantial; average global temperatures during the Little Ice Age were probably less than 2°C colder than they are today.

So as we begin to peel back the layers of time, we find evidence that the output of solar energy varies and may be related to the development of sunspots and the strength of the solar wind. From a 300- to 400-year perspective, it seems that these phenomena may have had an important influence on our climate and could potentially cause global change.

20.9 Earth-Sun Geometry: Global Temperatures over the Past One Million Years

Variations in solar activity may not be the only influence that the Sun has on global temperatures on Earth. Many scientists believe that the relative positioning of the Earth to the Sun also has an effect. Astronomers have shown that these variations occur on the order of thousands of years, so to assess their potential effects, we are immediately confronted with problems. Anecdotal evidence, like that used in the previous section to correlate sunspot activity with the Little Ice Age, is lacking. And we do not have a time machine at our disposal with which to directly measure global temperature!

Since we cannot use direct temperature measurements, we must turn instead to "proxy" methods that determine ancient temperatures indirectly. To find a suitable temperature proxy, we must identify some measurable property that changes with temperature and is preserved in the geologic record. Furthermore, the property must also be one that measures global rather than local temperature changes. In a sense, an ordinary thermometer is an example of a modern proxy method of measuring local temperature. What we are actually measuring is not the temperature but the height of a fluid (usually mercury) in a narrow tube. Using the numbers on the thermometer, we convert this height into temperature. This method is successful because we know from repeated experiments how mercury will expand and rise up the thermometer tube as the temperature increases. This allows us to put a temperature scale on the tube so that we can then use the mercury level to distinguish, for example, between normal and feverish body temperatures.

The most obvious place to look for suitable proxies of global temperature change is in the oceans. Ocean water covers more than 70% of the Earth's surface and is very sensitive to changes in global temperature. This sensitivity is reflected in variations in surface ocean water temperature, and in the amount of moisture that evaporates into the atmosphere. But are there proxies in ocean water that are preserved in the geologic record?

As we have learned, most of the moisture evaporated from the oceans returns to the sea by way of the hydrologic cycle. Rain falling on land, for example, feeds into valleys, which drain to the sea. Winter snow may temporarily preserve this record of evaporation from the sea, but this snow melts in the spring. In polar regions, however, where the winter snow does not completely melt, the record of evaporation may be preserved (Fig. 20.17).

In polar regions, the snow that falls each year is compressed into a thin layer by the snowfall in subsequent years. With continued compaction, each year's snowfall is ultimately preserved as a thin layer of ice. Each ice layer preserves a record of the evaporation for that year, accumulating in sequential order like the pages in a book. Using the *principle of superposition* (Chapter 3), the topmost layers preserve a record of relatively recent evaporation from the oceans, whereas the lower layers preserve ancient records. In this way, the polar ice sheets have been found to preserve up to 200,000 years of evaporation history. But are there measurable features in this evaporation record that proxy ancient temperatures? The answer is yes.

As snow falls and is compressed, it traps bubbles of air. Ancient layers of ice therefore trap samples of the ancient atmosphere present at that time (Fig. 20.18). By measuring the concentrations of greenhouse gases in the bubbles of individual layers, we can get a year-by-year documentation of the past composition of the atmosphere.[1] If we assume that the present relationship between greenhouse gas concentrations and global temperatures is typical of the past (that is, if we apply the *principle of actualism*; see Chapter 3), we can use these concentrations to infer changes in global temperatures.

We know that the composition of our modern atmosphere is changing. But did it do so in the pre-industrial era? The answer is yes. In the last 10 years, scientists have obtained profiles of ancient temperature variations by drilling cores into the ice sheets of both polar regions. By carefully analyzing samples of the ancient atmosphere trapped in individual layers, a record of changing atmospheric greenhouse gas composition has been obtained. Gas bubbles analyzed in individual layers of these ice cores show systematic variations in the concentrations of carbon dioxide, methane, and other greenhouse gases from one layer to the next. Data from the Vostok station in Antarctica (Fig. 20.19), for example, show significant fluctuations in both carbon dioxide and methane. Elevated values (to the right in each graph) suggest warm interglacial periods, whereas low values are interpreted as periods of glaciation.

It is important to realize that these data do not actually measure the temperature of the ancient ice. Instead, we *estimate* changes in atmospheric temperature from the analysis of tiny samples of the ancient atmosphere trapped within the ice.

The exact temperature for each time period is not known with any certainty. However, the data do provide evidence of periods of relative warmth between about 150,000 and 120,000 years ago, a known interglacial period, and evidence of colder but fluctuating temperatures between 110,000 and

---

[1] Many scientists believe that the gases may have moved since their time of entrapment. However, this movement does not seem to be extensive enough to blur the warming and cooling trends identified.

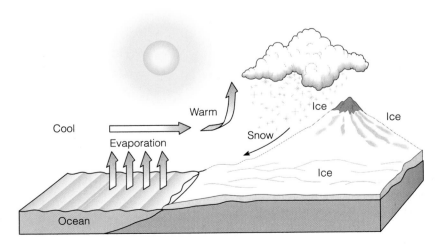

Figure 20.17
The hydrologic cycle adapted to emphasize the storage of evaporated water in polar ice sheets. Evaporated water that circulates over polar regions during the winter falls as snow and may not melt during the following summer. Eventually each winter's snow is compressed to form a layer of ice as the ice sheet grows.

Cool

Warm

Evaporation

Snow

Ice

Ice

Ice

Ocean

Figure 20.18
Polar ice crystals as seen under the microscope. The small dark circular areas are bubbles trapped within and between ice crystals. The bubbles preserve samples of the atmosphere at the time the ice formed. The gas may be extracted and its composition determined.

20.9.1 Supporting Data

First, ice cores from both the northern and southern polar ice sheets show very similar patterns. If they did not, scientists would be looking for local rather than global explanations for the patterns they show. Although there must be a local component, it does not appear to override the major global-scale patterns.

Second, as we have learned in Chapter 14, there is abundant geologic field evidence for episodes of widespread glaciation between 90,000 and 11,000 years ago followed by melting and the deposition of glacial sediments. This history is closely matched by fluctuations in the greenhouse gas concentrations preserved in the ice cores. Their concentrations were anomalously low between 90,000 and 11,000 years ago, but rapidly increased thereafter.

Third, the well-documented migration of plants and animals, including our own species, can be explained by the growth of ice sheets and the consequent drop in sea level which provided "land bridges" between the continents that aided migration (see Chapter 14).

Fourth, as we shall now see, other proxy measures of ocean water evaporation preserved both in the ice cores and in deep sea sediments, show similar patterns of global temperature change.

11,000 years ago, suggesting many major cycles of glacier formation and melting. Between 11,000 and 10,000 years ago, a dramatic period of global warming began, during which it is estimated that the average global temperature climbed an astonishing 7°C in only 40 years!

Modern ice sheets record variations in atmospheric composition over the past 200,000 years, implying that the relationship between greenhouse gas composition and temperature has existed over a considerable period of time, not merely during the industrial era. But, in the spirit of the scientific method, how do we test this hypothesis, and how do we know that data from narrow cores of ice are representative of global, rather than local phenomenon?

As we learned in Chapter 8, water which has evaporated from the oceans is different in composition from ocean water, the latter being saltwater (about 3.4% NaCl) whereas the former is freshwater. This differing composition indicates that the evaporation of ocean water cannot be a chemically passive process but, instead, is highly discriminatory. That is, evaporation selects some chemical components in preference to others. This subtle form of discrimination provides additional supporting evidence for the global temperature fluctuations deduced from the ice cores. The evidence is based on differences in oxygen isotopes. But what is this evidence and how can we best interpret it?

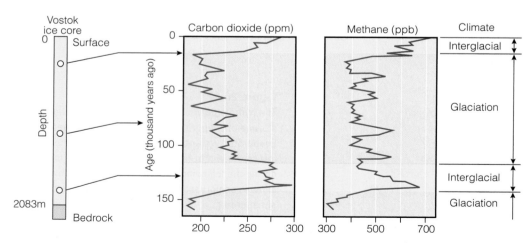

Figure 20.19
Layer-by-layer analysis of air bubbles in ice cores from the Vostok Station in Antarctica show changing concentration of greenhouse gases in the atmosphere over the past 150,000 years. The carbon dioxide and methane data show similar trends that match the geologic record (right) of known glacial and interglacial stages.

Although all water molecules are composed of two parts hydrogen and one part oxygen, both hydrogen and oxygen have more than one stable isotope. The main difference in the weight of water molecules is due to differences in the weights of the oxygen isotopes. As a result, molecules of water can vary in weight depending on whether or not they include the heavier oxygen isotope, that is, the one that has more neutrons in its nucleus.

By definition, all oxygen atoms contain eight protons in their nuclei. The vast majority (about 99.8%) also contain eight neutrons forming the isotope oxygen-16. However, a small proportion (0.2%) contain 10 neutrons, forming the heavier isotope oxygen-18.

No matter what the temperature, water molecules containing the heavier oxygen find it more difficult to evaporate than those containing light oxygen. However, this difficulty is more pronounced at low temperatures because, at high temperatures, more energy is available to help the heavier water molecules evaporate. In an "evaporation race" then, the lighter water molecule always wins, but if the conditions of the race change (for example, if the temperature rises), the winning margin is reduced. Hence, the "margin of victory" can be used to infer the race conditions which, in this case, are dictated by global temperature. Again, it is important to realize that these relationships, while preserved in glacial ice or sediments, are a function of ocean evaporation patterns and, therefore, global warmth.

When this method is applied to the layers of ice in the ice cores, an historic profile of global warmth is obtained from which we can identify warming and cooling trends. Figure 20.20 shows the layer-by-layer variations in the ratio of heavy oxygen (^{18}O) to light oxygen (^{16}O). This variation is shown on the x-axis with higher $^{18}O/^{16}O$ ratios indicating increasing temperature to the right. The units for the $^{18}O/^{16}O$ ratio are in parts per thousand (‰) as opposed to part per hundred, or percent (%). Once again, we cannot be certain of the absolute temperature recorded by the $^{18}O/^{16}O$ ratio in each layer, but we can be confident of the warming and cooling trends that the ratios identify.

Ice core drilled through the Greenland ice cap provides a record that goes back 120,000 years (Fig. 20.20). With some exceptions, the warm and cool periods predicted by oxygen isotopes are similar to those deduced from the analysis of the greenhouse gas content of the trapped atmospheric gases (see Figure 20.19). Like the trapped gas record, the oxygen isotope data also record a period of warmth about 120,000 years ago, and a cold period lasting from about 70,000 until 11,000 years ago, followed by a rapid warming trend. Although there are some discrepancies around 90,000 years ago (compare Figures 20.19 and 20.20), the general agreement between the two records gives added weight to the supposition that we are measuring global, rather than local, phenomena. Note that the global warming evident over the past 40 years or so cannot be identified in this data set because it is dwarfed by effects of older natural fluctuations that drove ice sheets back and forth across the continental landmasses.

A further test of the use of oxygen isotopes as a proxy for ancient temperature is provided by recent data from glacially-derived sediments in a Swiss lake. These sediments preserve about a 15,000-year record from which warm and cool periods

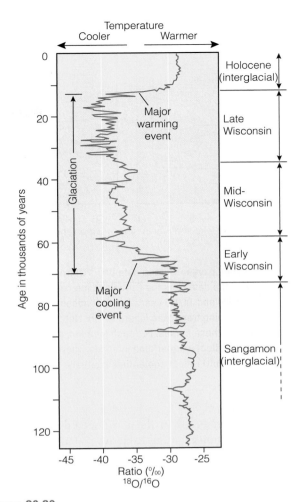

Figure 20.20

Layer-by-layer variation in oxygen isotopic ratios ($^{18}O/^{16}O$) over the last 120,000 years from the Greenland ice core. Higher $^{18}O/^{16}O$ ratios (on the right) reflect warmer time intervals whereas lower $^{18}O/^{16}O$ ratios (on the left) reflect cooler time intervals. The data are in broad agreement with ancient temperatures inferred from the greenhouse gas content of trapped air bubbles in the Vostok ice core. Adapted by permission of the publisher from W. Dansgaard, et al. *The Late Cenozoic Ages*, 37-56 ©1981 Yale University Press.

can be deduced. These periods closely match those found in the Greenland ice sheet and provide further evidence that the proxy records of temperature changes have global-scale significance (Fig. 20.21).

The ocean sediment record provides a different, yet related type of test. In a sense, the atmosphere and oceans can be thought of as complementary reservoirs, even though modern records of direct temperature measurements do not go back far enough to test this complementary relationship. However, there is another way to uncover it, again using oxygen isotope ratios. As temperatures drop, evaporation extracts moisture from the oceans with an increasing abundance of light oxygen isotopes. So during a period of glaciation, the oceans should become increasingly depleted in light oxygen. But how do we test this? If ice sheets preserve the evaporation record, what might preserve the record of what is left in the ocean?

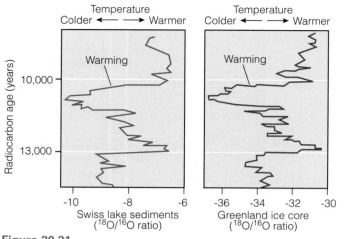

Figure 20.21

Comparison between oxygen isotope data ($^{18}O/^{16}O$) from Lake Agassiz in Switzerland and the data from the Greenland ice core for the period between 9,000 and 14,000 years ago that encompasses the end-Wisconsin warming trend (see Figure 20.20). Note the broad agreement between these curves. For each location, higher $^{18}O/^{16}O$ ratios (to the right) reflect warmer time intervals whereas lower $^{18}O/^{16}O$ ratios (to the left) reflect cooler time intervals.

Marine life preserved in the fossil record is thought to retain the information required to derive water temperatures. Evidence of past climate, for example, is recorded by floating microorganisms called *foraminifera*, which provide important information about the average temperature of surface ocean waters in the composition of their shells. Oxygen in the shell material of these marine organisms is believed to record the oxygen isotope ratio of the ocean water in which they lived because, while the animals are alive, their shells grow by exchanging water with their surroundings and therefore retain the oxygen isotope ratio of the ocean water. As we have learned, this ratio is very sensitive to the temperature of the sea water. When the foraminifera die, they cease to chemically interact with the seawater and sink to the bottom of the ocean, where they are buried by subsequent layers of sediment. As more foraminifera die and sink to the seabed, they accumulate in progressively higher sediment layers.

Changes in the oxygen isotope ratios of foraminifera from one layer to the next consequently reflect changes in ocean water temperature over time. When temperatures are cold, the lighter ^{16}O-enriched moisture is preferentially trapped in polar ice. As a result, the oceans become relatively enriched in ^{18}O compared to ^{16}O, and the high $^{18}O/^{16}O$ ratio recorded in foraminifera shells reflects these colder ocean water temperatures. As the temperature rises, the polar ice sheets melt, and ^{16}O-enriched melt water is liberated into ocean water, leading to a decrease in the $^{18}O/^{16}O$ ratio in foraminifera shells.

These changes in the foraminiferal $^{18}O/^{16}O$ ratios can be identified in sediment cores retrieved by ocean drilling surveys. Layer-by-layer analysis of the $^{18}O/^{16}O$ ratio reveal alter-

nating periods of warm and cold ocean surface water temperatures that match the alternation of glacial and interglacial stages (Fig. 20.22). Since ocean waters are well mixed (see Chapter 8), the $^{18}O/^{16}O$ ratios measured in fossil shells are thought to reflect global-scale conditions of the past.

With the information from the foraminifera, the proxy temperature records derived from ocean sediment and polar ice could be compared. Figures 20.20 to 20.22 are set up to show deduced relative temperatures increasing to the right. Note, however, that the oceanic oxygen isotope ratios on the axis in Figure 20.22 have the opposite sense to those of the ice core records in Figures 20.20 and 20.21, becoming more negative to the right. This is consistent with their complementary relationship; the relative enrichment of an oxygen isotope in one reservoir necessitating its relative depletion in the other.

The oceanic record obtained from foraminifera in the Pacific Ocean (Fig. 20.22) extends back one million years, but can be seen to confirm earlier findings. The periods of warming and cooling deduced in the upper part of the record (the last 120,000 years) compare very favorably with the ice core records shown in Figures 20.20 and 20.21, and with those deduced from the geologic record. Each indicates a major period of cooling about 80,000 years ago (Fig. 20.22) and a major warming trend starting about 11,000 years ago.

The above results give us confidence that each of the data sets reveal genuine periods of warming and cooling, allowing us to interpret warming and cooling trends over the past 1 million years. The temperature record over this time interval shows repeated oscillations between warm interglacial periods (like the present one), and cold glacial climates (like the Ice Age of 30,000 years ago). Oscillations of up to 10°C (18°F) are implied from geologic and biological evidence. From this perspective, our modern global temperature is at the high end of the natural scale, but its variation is not inconsistent with the natural pulse of high and low temperature cycles that has dominated the Earth's climate during the recent geologic past. Indeed, by extrapolating this pulse into the future, some scientists claim that another ice age is inevitable when temperatures begin to decrease. If this is correct, then the effects of modern global warming may ultimately be entirely overridden.

From the perspective of the above data, one could argue that our *current* period of global warming began as a natural phenomenon 11,000 years ago. This interpretation assumes a long-standing relationship between greenhouse gases and global temperature. If this assumption is correct, then in our global warming debate we must ask what causes these temperature fluctuations, and assess whether natural phenomena dwarf or override the effects of modern industrial pollution, or whether their combination has a multiplying effect.

It is difficult to evaluate the implications of these temperature variations on the global warming debate without understanding what causes the temperature oscillations in the first place. With the Pacific Ocean sediment core, we may finally be seeing enough of the record to identify some of the Earth's natural rhythms. According to this record, there have been 10 warm periods in the last 1 million years (see Figure 20.22),

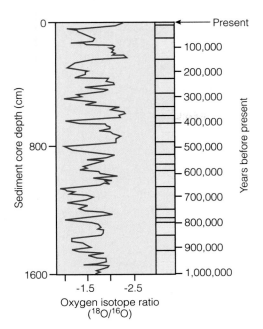

Figure 20.22
Oxygen isotope data ($^{18}O/^{16}O$ ratio) from foraminifera shells in sediments extending over the past 1 million years. The samples were obtained from drill core extracted from the Pacific Ocean. The graph is arranged so that warmer time intervals or lower (more negative) $^{18}O/^{16}O$ ratios are on the right whereas cooler time intervals, or higher (less negative) $^{18}O/^{16}O$ ratios are on the left. These match the alternation of interglacial (yellow) and glacial (blue) stages shown on the right. Note, however that the oceanic oxygen isotope ratios on the horizontal axis have the opposite sense to those of the ice core records in Figures 20.20 and 20.21, since an increase in $^{18}O/^{16}O$ ratios in one reservoir necessitates depletion in the other. Adapted by permission of the publisher from N.J. Shackelton and N. D. Opdyke, *Quarternary Research*, vol. 3, pp. 39-55, 1973, Academic Press, Inc.

suggesting a cyclicity of about 100,000 years. It is also clear from this record that, like a song, other rhythms are present as well. But why should these variations occur?

20.10 From the Sun to the Earth

In Chapter 14, we mentioned some of the possible causes of natural temperature changes with specific reference to the Ice Age. Building on the observations of John Croll, Milutin Milankovitch recognized in the early 20th Century that three forms of variations in the position of the Earth's axis and in its orbit around the Sun occur in predictable cycles and that these variations can cause the amount of solar energy reaching the Earth to vary by as much as 10% (Fig. 20.23).

First, according to Milankovitch's calculations, the Earth's orbit around the Sun varies in a systematic manner over periods lasting 100,000 years. At times the orbit is highly eccentric, at other times it is almost circular. This means that the distance from the Earth to the Sun, and hence the amount of the Sun's energy reaching the Earth, varies systematically in a cycle of this duration (Fig. 20.23a).

Second, Milankovitch's calculations showed that the tilt of the Earth's axis changes systematically from 21.5° to 24.5° in a motion that repeats itself every 41,000 years. A lower tilt means that the northern hemisphere gets less radiation in the summer and more in the winter (see Chapter 12). In other words, the contrast between summer and winter radiation becomes less pronounced (Fig. 20.23b).

Third, the wobble of the Earth's axis due to the varying gravitational pull of the Sun and the Moon on the Earth's equatorial bulge causes changes in the direction of the Earth's tilt. This is a process known as **precession**, during which the Earth's axis traces out the shape of a cone that, as Milankovitch showed, returns to an identical position every 26,000 years (Fig. 20.23c).

Milankovitch's calculations revealed the variations in the amount of solar radiation received by the Earth as a result of these three different aspects of Earth-Sun geometry, as well as the variations in the total combined signal of radiation produced by the summation of these effects. The result demonstrates a remarkable match between the combined signal of solar radiation received and the temperature variations derived from the oxygen isotope record (Fig. 20.24).

In the last million years, there have been ten periods of global warming, each separated by a period of glaciation (Fig. 20.22). Each cycle of warm or cold temperatures was therefore about 100,000 years in duration, a period similar in length to the cycle of the Earth's orbit around the Sun. Indeed, the obvious correlations and intuitive appeal of relating global temperatures to the effects of changing Earth-Sun geometry on the amount of solar energy reaching the Earth provides an elegant explanation of the temperature oscillations for the past few million years.

Many scientists now believe that while Milankovitch cycles may *initiate* global warming and cooling events, the cycles *alone* cannot account for the extent of the observed fluctuations in temperature. The calculated temperature fluctuations required to drive ice sheets back and forth across the continents (around 10°C or 18°F) are just too great to be explained by the 10% variation in solar energy received by the Earth during the cycles. Assuming these calculations are correct, the temperature swings associated with changes in Earth-Sun geometry must be made more dramatic by natural

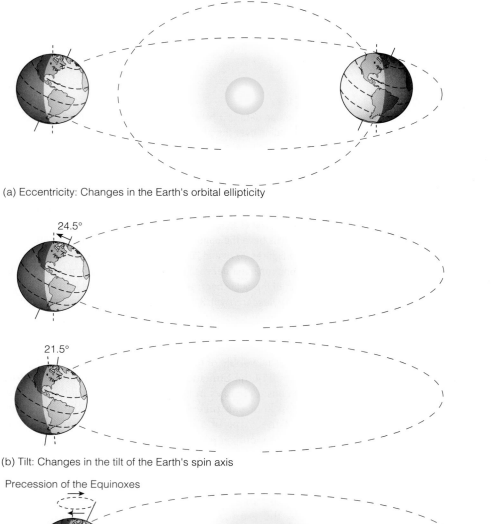

(a) Eccentricity: Changes in the Earth's orbital ellipticity

24.5°

21.5°

(b) Tilt: Changes in the tilt of the Earth's spin axis

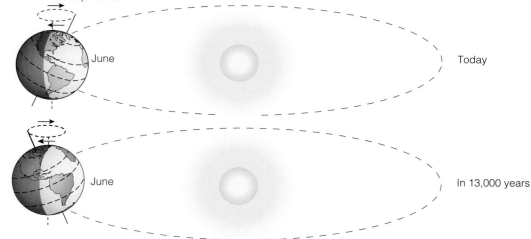

Precession of the Equinoxes

June

Today

June

In 13,000 years

(c) Precession: Changes in the direction of the Earth's spin axis without changes in the tilt angle

Figure 20.23

Milankovitch Cycles showing three types of cyclical changes in Earth-Sun geometry. (a) *Eccentricity:* The Earth has an elliptical orbit around the Sun. The shape of this orbit varies from being almost circular (low eccentricity) to more elliptical (high eccentricity). This change in shape occurs in a cycle lasting 100,000 years. (b) *Tilt:* The tilt of the Earth's axis varies from 21.5° to 24.5°. Presently the tilt is 23.5°. The change in tilt occurs in a cycle lasting 41,000 years. (c) *Precession:* The Earth wobbles on its axis like a spinning top. The axis makes one revolution every 26,000 years. In 13,000 years (lower diagram) the northern hemisphere will be tilted away from the Sun in June.

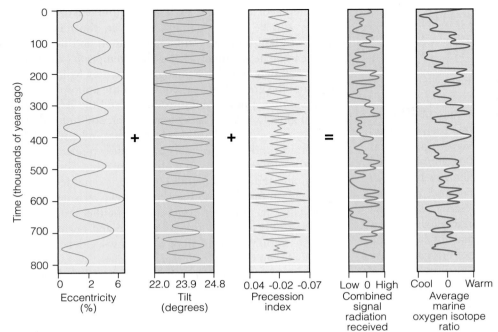

Figure 20.24
Curves showing variations in the amount of solar energy reaching the Earth as a result of variations in eccentricity, tilt, and precession as described by Milankovitch. The combined signal, which represents the sum of these effects, closely matches the marine oxygen isotopic data (right).

amplifiers in order for glacial and interglacial stages to occur. As we shall see, these amplifiers, which are known as **feedbacks**, act in such a manner as to increase the intensity of a warm or cold cycle once they commence.

20.11 Feedback Systems and the Gaia Hypothesis

British scientist James Lovelock popularized the importance of feedback mechanisms in the late 1970s. According to Lovelock, a **positive feedback** mechanism is one that causes the amplifier to enhance an existing trend (even though the effects may not be desirable). The effect is that of a *vicious circle* which continually reinforces itself because the amplification of the trend by the feedback mechanism causes enhancement that, in turn, leads to further amplification, and so on. For example, following the initial signs of a recession in our economy, there is often a general loss of confidence and the public stops buying. This amplifies the effects of the recession causing further loss of confidence. In enhancing the trend of the recession, the response of the public is a positive feedback.

A positive feedback in a climate cycle renders warm climates warmer or cold climates colder, bringing about runaway greenhouses or icehouses. For example, during periods of global cooling, ice sheets expand in polar regions. Ice sheets and the snow that accumulates on top of them have a high **albedo**: that is, they reflect, rather than absorb, a large portion of the Sun's heat. The glare and warmth associated with this reflection are familiar to skiers. Since an increased portion of the Sun's heat is now reflected back into space rather than absorbed, surface temperatures fall and the ice sheets expand further, covering more of the vegetation and rocks that once absorbed the Sun's heat. Hence, the surface temperature continues to drop. So the feedback creates larger and larger ice sheets which then amplify the global-scale cooling trend that produced them. Similarly, as the ice sheets retreat and the rocks beneath them are exposed once again, a global-scale warming trend ensues, promoting further retreat of the ice. These positive feedback effects may amplify the warming and cooling trends initiated by Milankovitch cycles.

Obviously, there must also be **negative feedback** mechanisms or the Earth's surface could never reverse a trend once it had been initiated. Negative feedbacks are self-regulators, like a thermostat, that counterbalance the effects of change to ensure that conditions remain within some reasonable constraints. If the door of a house is left open in the wintertime, the heat loss is counterbalanced by the heating system of the house. Likewise, if a house gets too warm, the heating system shuts down until the house loses heat to the outdoors. This counterbalancing is also known as a *virtuous circle*.

Since the geologic record demonstrates that neither a runaway icehouse nor a runaway greenhouse has ever occurred in the past, Lovelock believes that the Earth has been under the influence of self-regulating mechanisms throughout its 4.6 billion year history. His provocative ideas form part of a proposal known as the **Gaia hypothesis** after the Greek goddess of the Earth (see Chapter 19).

But how does the Gaia hypothesis work in practice? To better understand the Earth's self-regulation, we now examine the role of a key element such as calcium in the feedback cycle.

20.12 The Role of Calcium and Carbon Dioxide in the Feedback Cycle

The extraction of carbon dioxide from ocean water to form carbonate sediments such as limestone ($CaCO_3$) has been an important process throughout geologic history. However, in order for carbonate sediments to form, a vast supply of calcium is needed because the principle component of such sediments is calcium carbonate. So, if the calcium content of the oceans is increased, more carbon dioxide can be extracted from ocean water because more calcium carbonate can form. The ocean water would then attempt to replenish its carbon dioxide by extracting it from the atmosphere. This would reduce the amount of this greenhouse gas in the atmosphere and global cooling would ensue.

So the task of explaining natural fluctuations in surface temperatures now becomes one of explaining how calcium levels vary in the oceans. Continents, on average, contain much more calcium than ocean waters. As continents erode, they contribute their calcium to the oceans. So, the more material that is eroded from continents, the more calcium becomes enriched in ocean waters.

An ice age commences if moisture which evaporates from the oceans precipitates on land as snow and becomes compressed to form permanent ice, rather than flowing back to the oceans. This could be triggered by one of the Milankovitch cycles. As ice sheets grow in size they store an enormous quantity of frozen water so that sea level falls and continents become more emergent. During the last Ice Age, for example, sea level dropped by at least 100 m (330 ft). This fall in sea level exposed much of the continental shelves. With the land more emergent, streams become more powerful and, as a consequence, carry more sediment (and therefore calcium) to the ocean. Once in the ocean, the calcium can combine with carbon dioxide to form carbonate sediment. The ocean water, in turn, compensates for loss of carbon dioxide by extracting it from the atmosphere (recall from Chapter 9 that ocean water has a much stronger affinity for carbon dioxide than the atmosphere). This extraction reduces the atmosphere's greenhouse gas content and lowers global temperatures, causing the ice sheets to expand still further. So this is an example of a positive feedback, a vicious circle that makes a cold world even colder.

In this context, the role of carbon dioxide as a greenhouse gas is relegated to one of an amplifier of global change, rather than that of an instigator. Obviously, this cycle must be overridden or the world would never recover from an ice age. So how does this recovery happen? The Earth receives more solar radiation as it approaches the warm periods of Milankovitch cycles. As a result, the ice sheets begin to melt and sea level rises. The drowning of the continental shelves and coastal regions inhibits continental drainage so that the amount of calcium brought to the sea is reduced. As a result, the amount of carbon dioxide drawn down from the atmosphere into the oceans is also reduced so that global temperatures rise and the effects initiated by increased solar radiation are amplified. Once again this is a positive feedback effect.

Other feedback effects, such as the influence of the albedo, may also be very important. However, the relationship between carbon dioxide, continental erosion and the formation of carbonate sediments is one that attempts to explain the relationship between global change initiated by Milankovitch cycles and the greenhouse gas content of the atmosphere.

In summary, in our quest to understand the relationship between global warming, greenhouse gases, and industrialization, we have examined various factors that influence global temperatures and cause warming and cooling trends. We have detected cyclic oscillations in global temperature that may be initiated and terminated by the changing geometric relationships between the Sun and the Earth known as Milankovitch cycles. The effect of these cycles are then either amplified or dampened by a complex array of positive and negative feedback systems on the Earth, all of which involve delicate interactions between the Earth's surface reservoirs.

Taken at face value, this could lead us to a disturbing conclusion. The natural temperature oscillations documented over the past million years show more dramatic swings and have had far more powerful effects than those associated with the modern industrial era. If these cyclic temperature variations are projected into the future, they suggest that glaciers will once again advance across continental North America following today's interglacial climate. Does this mean that humanity is a relatively minor player when it comes to the environment? Many scientists feel that this is indeed the case and suggest that the Earth's cyclic mechanisms will override and repair any environmental damage caused by the modern industrial era.

However, we have still only viewed a tiny fraction of the Earth's 4.6 billion year history. Can we be sure, then, that we have heard enough of the Earth's song to identify all of the rhythms? The answer is emphatically no, as we shall see in startling fashion when we look further back into the geologic record. When we examine time periods of many millions of years, we find that plate tectonic processes play a major role in controlling global temperature.

20.13 Predictable Temperature Swings for at Least One Billion Years

Carbon dioxide is introduced into our atmosphere by volcanic activity. The vast majority of this carbon dioxide is preferentially taken up by the oceans because ocean waters can absorb 62 times more carbon dioxide than the atmosphere. But the oceans are not an infinite reservoir. Joly's ingenious method for calculating the age of the Earth on the basis of the sea's saltiness (Chapter 3) was in error because he assumed that all of the sodium carried to the sea by streams had remained there throughout the Earth's history. He was unaware that the formation of salt deposits had extracted sodium from the oceans. Similarly, the geologic record shows that carbon dioxide has been efficiently extracted from the oceans by organisms for at least 3 billion years to form carbonate sediments such as limestone. The oceans have responded by extracting carbon dioxide from the atmosphere, causing global temperatures to fall.

The biogeochemical highway linking volcanic activity, the atmosphere, and carbonate rock (Chapter 19) is therefore paved by life and death in the oceans. In this way, the solid Earth, atmosphere, biosphere, and hydrosphere are inextricably linked. The linkage has been so successful that 99.9% of the world's carbon has traveled down this biogeochemical highway and is now stored in carbonate sediments and sedimentary rocks. At the same time, the carbon dioxide content of the atmosphere has been reduced from about 98% early in Earth history to its present value of 0.03%. In Chapter 15, we learned that the composition of the Earth's primitive atmosphere was comparable with those of Mars and Venus. If the Earth's atmosphere had maintained this original level of carbon dioxide, surface temperatures now would be close to 300°C, which is far too hot to sustain life. The linkage between volcanic activity and the Earth's surface reservoirs is therefore crucial to the existence of life on this planet.

By providing the oceans with calcium, continental erosion indirectly affects the carbon dioxide content of the atmosphere, and hence the planet's average surface temperatures. Not surprisingly, the formation and erosion of mountain ranges have a particularly profound effect. In fact, a cyclic pattern of surface temperature variations that can be related to the formation and erosion of mountain belts is evident in the geologic record over the last billion years.

As we discussed in Chapter 5, most geoscientists believe that continents drift at the rate of a few centimeters a year as they float on top of a denser, yielding, layer beneath the Earth's lithosphere. The lithosphere is a thin, rigid layer that resembles a cracked eggshell. According to plate tectonics, the fractured segments join together to form a series of floating plates that are constantly interacting to produce volcanoes, earthquakes, and mountain belts. Like rafts floating on a stream, these plates collide and where their collisions involve continents, major mountain belts are generated. For example, the Himalayan mountains are the result of the collision of India with southern

Asia, while the Alps are related to the collision of Africa with southern Europe. Prior to collision, the continental plates approach each other as the crust of the intervening ocean is destroyed. This causes widespread volcanic activity, which pumps carbon dioxide into the atmosphere, resulting in global greenhouse warming. As the continents collide, however, several factors combine to turn the global greenhouse into an icehouse. First, the volcanic activity that was responsible for producing the greenhouse gas diminishes. Second, ice caps that develop and grow on the mountains produced by collision reflect solar radiation rather than absorb it. Third, erosion of these mountain belts rapidly contributes calcium and other nutrients to the oceans. The nutrients stimulate increased organic productivity and photosynthesis, while the calcium eventually combines with carbon dioxide in the ocean water to produce deposits of limestone. The ocean water then responds by extracting carbon dioxide from the atmosphere, thereby diminishing the greenhouse effect.

Conversely, when supercontinents break up and disperse, several factors conspire to raise the level of atmospheric greenhouse gases. First, volcanism increases so that atmospheric carbon dioxide levels are raised. Second, as the continents disperse, they subside because they are migrating away from the uplifted sites of active rifting and associated thermal upwelling. Once their passive margins subside below sea level, they can no longer be eroded, and so they no longer contribute calcium to the oceans. The Cretaceous Period, with its anomalously warm temperatures and submerged continents (only 15% of the Earth's surface was exposed), provides an example of the combined climatic effects of enhanced volcanism, subsidence, and diminished erosion rates.

An examination of the geologic record shows an impressive match between the history of ancient greenhouse and icehouse climates, and that of mountain building, limestone deposition, and spurts in organic productivity and biological diversity (Fig. 20.25). This correlation supports the current view of many geoscientists that plate tectonic processes such as mountain building and erosion dominate the natural controls

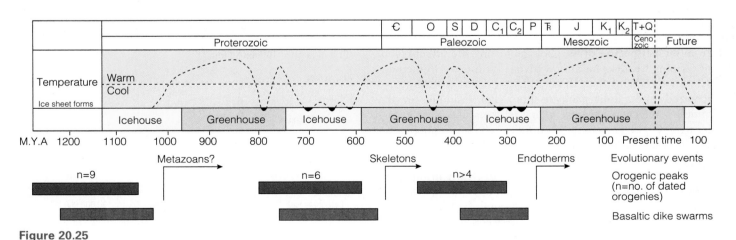

Figure 20.25

The relationship between global temperatures, "icehouse" and "greenhouse" climates, and the development of supercontinents based on episodes of mountain building (supercontinent amalgamation) and dike injection (continental rifting) for the past 1 billion years (see Figure 6.23). Note that orogenic peaks culminate in cold "icehouse" climates while basaltic dike swarms herald warm "greenhouse" conditions and major evolutionary events.

Table 20-1

Summary of the Natural Processes that Affect the Earth's Surface Temperature

NATURAL CYCLE	DURATION	DETECTION	EFFECT	RELIABLE RECORD
Sunspots	11.1 years 22.2 (sunspot cycle)	Telescope	Temperature changes in upper atmosphere; Little Ice Age	300 years
Earth-Sun cycles	<100,000 years (Milankovitch cycles)	Chemical analysis of evaporation record preserved in ice sheets, recent fossils	Cyclic retreat of ice sheets in the Quaternary	1.6 million years
Plate tectonics	200–400 million years (supercontinent cycles)	Isotopic analysis of the fossil record, inferences from climate models	Global scale, long-term change	1 billion years

of global warmth. As we discussed in Chapter 6, it has been long recognized that mountain building events are not randomly distributed in time but, instead, are concentrated into narrow portions of the geologic record. Figure 20.25 shows the three most recent events: a period about 1100 million years ago and a period 800 to 600 million years ago, both thought to be associated with the amalgamation of Late Proterozoic supercontinents, and a period 470 to 300 million years ago, associated with the amalgamation of Pangea. Each of these supercontinents coincided with the development of icehouse conditions.

In the preceding sections, we have peeled back the layers of time, starting with the relatively modern record and, by using temperature proxies, we subsequently examined an ever increasing portion of the geologic record. By organizing our discussion in this way, we have been able to examine a variety of influences that operate with different intensities and with cycles of different duration (Table 20-1). We are now in a position to put all these observations together. We now know how to recognize the Earth's beats and rhythms, and it is time to identify its song.

20.14 Putting It All Together: The Greenhouse Effect and Global Warmth

From the perspective of the geologic record, the present global warming trend seems almost trivial. Far more extreme temperature variations than those which concern us today have occurred throughout geologic time. For example, in the Middle Cretaceous, average global temperatures were a staggering 8°C (14°F) warmer than today, and atmospheric carbon dioxide levels were an order of magnitude higher. These extreme conditions lasted for millions of years, so that shorter-term Milankovitch or sunspot cycles were clearly insufficiently powerful to override them.

The major beat in the Earth's natural rhythms is, therefore, a tectonic one. The interplay between continental erosional rates, mountain building, the availability of continental shelves and biological activity exerts the primary control on global warmth. Plate tectonic processes, which involve the opening and closing of ocean basins and the fragmentation, dispersal, and amalgamation of continents, occur in cycles last-

ing hundreds of millions of years (see Section 6.7.2, The Supercontinent Cycle). The shorter Milankovitch and sunspot cycles are consequently relegated to the role of amplifiers.

Once this is established, the environmental effects related to human activity seem clear. Since the amount of carbon stored in fossil fuels is dwarfed by that in carbonate sediments (see Table 20-2), the effects of burning fossil fuels might appear to be negligible. This is misleading, however, because carbon residing in carbonate sediments is effectively out of natural circulation, and it is the amount of carbon in circulation that concerns us. Relative to the carbon in circulation, the burning of fossil fuels is a very significant source of atmospheric carbon dioxide. In fact, it is estimated that burning of fossil fuels will add some 5000 gigatons of carbon to the atmosphere over the next few hundred years, all of it from fossil fuels that nature took millions of years to form.

Our actions have consequently caused a short circuit in the carbon biogeochemical cycle since we are in violation of the Earth's natural contract for the transfer of energy and nutrients (see Chapter 19). Quite simply, we are consuming fossil fuels many, many orders of magnitude faster than nature can replenish them. Ocean waters cannot absorb carbon dioxide from the atmosphere at this accelerated rate. Therefore, burn-

Table 20-2

Amount of Carbon (in Gigatons) Residing in Different Surface Reservoirs*

(One gigaton is 10^9 metric tons)

Atmosphere	710
Biosphere	
Continental biomass	590
Oceanic biomass	4
Hydrosphere	
Ocean layers: Surface	680
Intermediate	8,200
Deep	26,000
Solid Earth	
Carbonate rock	20,000,000
Fossil fuels	5,000
Soil	1,670
Litter	60
Oceanic sediments	4,900

*Note the vast proportion resides in carbonate sediments. However, for the most part, this carbon is not "in circulation," but is stored.

Data from the U.S. Department of Energy, 1980.

ing fossil fuels directly increases carbon dioxide levels in the atmosphere. Given that atmospheric carbon dioxide concentrations have been correlated with global temperature changes throughout much of geologic time, it is unreasonable to argue against such a relationship in the modern world. Instead, from the perspective of the geologic record, there is a fundamental linkage between burning of fossil fuels, increases in the carbon dioxide content in the atmosphere, and global warming.

In fact it is the complete mismatch between the relatively slow rate at which carbon dioxide is naturally extracted from the atmosphere compared to the rapid rate at which it is being introduced into the atmosphere by human activity that is the strongest argument for a human-induced origin for modern global warming. But we can only arrive at this conclusion when we understand the long-term role played by plate tectonics in influencing greenhouse gas content in the atmosphere.

20.15 Conclusion

This chapter contains very little in the way of new principles or concepts. Instead, our discussion is essentially a synthesis of concepts discussed previously in different parts of this text. However, like a 52-card deck, it is the way the cards are assembled and played that counts and we have assembled these concepts in order to peel back the layers of time.

At the outset, we pointed out that our intention was to promote discussion among yourselves and with your instructor. You might disagree with some or even all of the preceding hypotheses. Certainly we could use more data and we must be prepared to continually re-examine and test our hypotheses as more data are obtained. However, we feel that our *approach* to the greenhouse effect demonstrates the unique perspective Earth science can offer to modern environmental problems. By peeling back the layers of time, we have been able to identify the Earth's natural rhythms. At each step, we have gained additional insights. We have identified the main plate tectonic beat that lasts for hundreds of millions of years and profoundly affects global surface temperatures. Other shorter beats, such as those related to the geometry of the Earth-Sun system and those related to solar variability, are less pronounced. We have also learned that only in identifying the main rhythms can any distortions related to the burning of fossil fuels be recognized. Like a flaw in a well-known song, we can now instantly recognize the distortion of the modern era.

There is no consensus among scientists as to the severity of our environmental problems. However, most think it is prudent to act to limit production of ozone-depleting chemicals and reduce our fossil fuel emissions. They are concerned that if we wait for the final piece of the jigsaw, it may be too late. Many of us can make a good guess at a jigsaw image with only 50% of the pieces in place because the patterns are sufficient for us to make a judgment call. Similarly, these scientists are willing to stand back from the scientific jigsaw, and make a judgment call based on what they can see.

Others, who believe in rigorous application of the scientific method have seen many hypotheses come and go. When the hypotheses were tested, they were found wanting and were cast aside. In like fashion, some of the issues discussed here may be replaced or superseded by others. But if current projections for world population in the next century are accurate, environmental issues will continue to be with us for as long as we utilize the resources on this planet.

20.16 Chapter Summary

One of the key problems in environmental analysis is the ability to distinguish between natural and human-induced global change. In this chapter, we have focused on key environmental issues such as ozone depletion and global warming, but the approach we describe can be applied to many global-scale problems.

Ozone gas is present in the atmosphere only in the minutest quantities and forms when oxygen is bombarded by ultraviolet radiation. Even in the so-called "ozone layer" it makes up less than 0.001% of atmospheric gases. The ozone layer occurs within the stratosphere where it absorbs ultraviolet radiation and protects the Earth's surface from the harmful effects of this radiation. As a result of this absorption, the stratosphere becomes warmer with distance from the Earth's surface, in contrast to the troposphere below.

Ozone depletion in the stratosphere is most pronounced in polar regions, where up to 60% depletion has occurred in the last 30 years. In 1974, it was proposed that ozone depletion was related to the presence of chlorofluorocarbons (CFCs) in the atmosphere. CFCs do not occur naturally and their presence in the atmosphere can be attributed to their use in refrigerators and air conditioners. CFCs are highly reactive with a voracious appetite for ozone, and NASA scientists have found the chemical fingerprint of CFCs in ozone-depleting reactions above Antarctica. The Montreal Protocol established limits on CFC production, but it is generally recognized that even if this agreement was rigidly adhered to, the ozone layer would not return to normal until well into the 21st century.

The presence of greenhouse gases in the atmosphere is vital to life on Earth, because without them our atmosphere would be 35°C (63°F) colder. The Earth's surface acts as a radiator and greenhouse gases block the return of infrared radiation into space. Modern global temperatures are well within the range of natural fluctuations. Over the last 40 years, however, the available data indicate that the concentration of greenhouse gases in the atmosphere and the average global temperature have been rising. But from this data alone, it is difficult to decide whether this is due to our increased consumption of fossil fuels or whether it merely reflects natural fluctuations.

If we look further back in time, we can identify different influences on global temperature that can help us resolve the question. Sunspots are dark, relatively cool, and intensely magnetic regions on the Sun's surface that accelerate the particles that form the solar wind. This, in turn, blasts the outer atmosphere of the Earth. Sunspot numbers shows a periodicity of about 11.1 years, with peak activity coinciding with anomalously warm temperatures in the Earth's upper atmosphere. However, in the

17th Century, a mysterious absence of sunspots coincided with part of the Little Ice Age, an anomalously cool period that lasted for most of that century. During this time, the solar wind was presumably less intense and the upper atmosphere relatively cool.

As we go further back in time, we cannot rely on direct temperature measurement and must resort to proxy methods. The evaporation record of oceans for the past 150,000 to 200,000 years is preserved as layers in the polar ice sheets, within which air bubbles trapped in pore spaces preserve remnants of the contemporary atmosphere. From the greenhouse gas content of these bubbles, the Earth's surface temperatures can be deduced. Oxygen isotopes ($^{18}O/^{16}O$) also act as proxy geothermometers. Water molecules containing heavy oxygen (^{18}O) are more difficult to evaporate than those molecules that contain light oxygen (^{16}O). This difficulty is more pronounced at lower temperatures, so that evaporated water has a relatively low $^{18}O/^{16}O$ ratio during times of anomalously low temperatures. This low ratio is then preserved as layers in polar ice sheets. There is an impressive correlation between the oxygen isotope data for the past 200,000 years and the geologic record of global climate change; relatively low $^{18}O/^{16}O$ ratios coincided with times of major continental glaciation during this period.

The atmosphere and oceans are complementary reservoirs, so while the atmosphere becomes increasingly enriched in light oxygen during periods of glaciation, the oceans become increasingly depleted. The shells of floating microorganisms, such as foraminifera, preserve the oxygen isotope ratio in ocean water because the shells grow by exchanging water with their surroundings. This exchange ceases when the organism dies and its skeleton sinks to become part of the sedimentary record preserving the isotopic information. A layer-by-layer analysis of the microorganisms in the sediment consequently reveals changes in surface water temperatures with time.

This record is well constrained for the last million years and suggests that temperature fluctuations of up to 10°C (18°F) have occurred under the influence of Milankovitch cycles caused by changes in Earth-Sun geometry. The record reveals 10 warm periods during this interval, suggesting a cyclicity of about 100,000 years. This cyclicity matches that of the eccentricity of the Earth's orbit around the Sun as determined by Milankovitch.

But while Milankovitch cycles probably initiate global warming and cooling events, they must be amplified or damp-ened by feedback systems, such as those described by the Gaia hypothesis, in order to produce climatic fluctuations evident in the proxy temperature record. A positive feedback mechanism amplifies an existing trend, whereas a negative one will dampen it. Negative feedback systems predominate on Earth, ensuring that both warming and cooling trends are ultimately reversed and that the environment remains equitable.

The roles of calcium and carbon dioxide provide important examples of feedback systems. Continental erosion, which provides calcium to the oceans, probably exerts an important influence on atmospheric carbon dioxide. As the calcium content of the oceans is increased, more carbon dioxide can be extracted from ocean water to form calcium carbonate (or limestone). The ocean then replenishes its carbon dioxide content by extracting it from the atmosphere. This reduces the greenhouse gas content in the atmosphere, causing global cooling.

The importance of the roles of calcium and carbon dioxide throughout geologic time is emphasized by an inventory of carbon's present-day distribution among the Earth's surface reservoirs. Almost all (99.9%) of the world's carbon is now stored in carbonate sediments and sedimentary rocks. So by providing calcium to the oceans, continental erosion has profoundly affected the carbon dioxide content of the atmosphere, and in doing so, has strongly influenced the Earth's average surface temperature. Continental erosion, which is tied to plate tectonic activity and the supercontinent cycle, is most pronounced during periods of mountain building. The extent of global change that this activity can bring about far exceeds that which can be caused by changing Earth-Sun geometry.

But supercontinental cycles operate over periods lasting hundreds of millions of years and fossil fuel deposits are formed in tectonic settings that are part of this cycle. Humans consequently consume fossil fuels at a rate that is many orders of magnitude faster than nature can replenish them, introducing greenhouse gases into the atmosphere at a rate that far exceeds nature's ability to absorb them. This implies that human activity has increased the greenhouse gas content of the atmosphere. Since this greenhouse gas content has correlated with average surface temperatures for at least the last billion years and probably throughout geologic time, it follows that human activity, fossil fuel emissions, and global warming are inextricably linked.

Key Terms and Concepts

Look for the highlighted items on the web at: WWW.BROOKSCOLE.COM

- One of the most difficult aspects of environmental analysis is to distinguish between natural and human-induced global change
- Ozone, which forms where oxygen molecules are bombarded by ultraviolet radiation, comprises less than 0.001% of the "ozone layer" in the stratosphere but has been depleted in polar regions by as much as 60% over the past 30 years
- Ozone absorbs harmful ultraviolet radiation. As a result, the stratosphere becomes warmer with distance from the Earth's surface, in contrast with the troposphere
- Ozone depletion has been linked to the presence of synthetic chlorofluorocarbons (CFCs) in the atmosphere, which have a voracious appetite for ozone and have left a chemical fingerprint of their involvement
- Ozone depletion can also occur by natural means, as demonstrated by the 1991 eruption of Mount Pinatubo, which temporarily caused ozone depletion of almost 20%
- Ground level ozone is a health hazard particularly where thermal inversions sandwich warm air between cold layers and so trap pollutants.
- Greenhouse gases block the return of infrared energy radiated from the Earth's surface, without which our average surface temperature would be 35°C (63°F) cooler.

- Global temperatures and the greenhouse gas content in the atmosphere have increased over the past 40 years, but are well within the range of natural fluctuations. Volcanic eruptions have caused temporary interruptions in these overall trends
- Sunspot cycles, which are typically 11.1 years in duration, affect temperatures in the upper atmosphere. The mysterious absence of sunspots in the 17th Century may be associated with a cooling event known as the Little Ice Age
- In order to decipher the geologic record, average surface temperatures are deduced indirectly by proxy methods, such as that which employs oxygen isotope ratios ($^{18}O/^{16}O$)
- Heavy oxygen (^{18}O) is more difficult to evaporate than light oxygen (^{16}O), and this difficulty is more pronounced during relatively cold periods. Layers of ice in the polar ice sheets preserve the evaporation record of the oceans, so their isotopic ratio is an indicator of average global temperature
- The atmosphere and oceans are complementary reservoirs, so while the atmosphere becomes increasingly enriched in light oxygen during a glacial period, the oceans become increasingly depleted. Organisms, such as foraminifera, whose shells grow by exchanging water with the oceans, preserve the ancient isotopic record of the oceans
- There is an impressive correlation between the geologic record of climate change and oxygen isotope data for both ice sheets and foraminifera. The record shows that the Earth's climate over

the past million years has been dominated by alternating warm and cool episodes with about a 100,000-year cyclicity
- Milankovitch cycles of Earth-Sun geometry, especially the 100,000-year cycle of eccentricity in the Earth's orbit around the Sun, form the most popular explanation for the initiation of cooling (glacial) and warming (interglacial) trends
- Cooling and warming trends are amplified or dampened by feedback systems, such as those popularized by James Lovelock's Gaia hypothesis. Negative feedback systems predominate on Earth ensuring that both types of trend can be reversed and the environment remains equitable
- Continental erosion, which provides calcium to the oceans, is probably an important control on atmospheric carbon dioxide. Calcium combines with carbon dioxide in the oceans to form calcium carbonate or limestone. The oceans respond by extracting carbon dioxide from the atmosphere, resulting in global cooling
- Continental erosion, and therefore calcium input into the oceans, is tied to the supercontinent cycle of mountain building, which lasts hundreds of millions of years. Fossil fuel deposits are formed in tectonic settings that are part of this supercontinent cycle
- We are consuming fossil fuels many orders of magnitude faster than nature can replenish them, and so are introducing fossil fuels into the atmosphere far faster than nature can remove them

Review Questions

1. When did ozone most likely appear in the Earth's atmosphere for the first time? What processes facilitated its appearance?
2. When was the relationship between ozone depletion and CFCs first proposed?
3. What effect do volcanic eruptions have on atmospheric ozone levels? Explain.
4. Ozone is one of three forms of oxygen. What are the other two?
5. Compare and contrast the origins of ground level ozone and stratospheric ozone.
6. Why is the greenhouse gas content of the atmosphere "a matter of balance?"
7. Why is Mauna Loa a suitable location for determining global atmospheric carbon dioxide concentrations?

8. Outline the relationship between sunspot activity and the Little Ice Age.
9. Why do oxygen isotopes behave like a proxy geothermometer?
10. Which of the three phenomena responsible for Milankovitch cycles appears to have the greatest influence on global temperature over the last million years? Explain.
11. What is the special role that calcium plays in influencing the greenhouse gas content of the atmosphere?
12. How is continental erosion related to the climate change?
13. How many orders of magnitude faster are we consuming fossil fuels than nature can replenish them?
14. Is the greenhouse effect fact or fiction?

Study Questions

Research the highlighted questions using InfoTrac College Edition.

1. If a relationship between CFCs and ozone depletion is confirmed, how can we be sure that it is related to human activity rather than being a natural phenomena?
2. Precisely what part of the CFC molecule attacks ozone?
3. What do you think the atmosphere's thermal profile looked like in the Precambrian?
4. Volcanic eruptions have been very common throughout Earth's history, yet they are thought to have had a negligible influence on atmospheric ozone concentration. Why is this?
5. Why should solar activity affect atmospheric ozone concentrations?
6. Debate the ozone depletion issue with your fellow students and your instructor. Include information that you have gathered from sources outside this course.
7. What critical data did NASA scientists obtain that strongly suggested a linkage between CFCs and ozone depletion?

8. Given that Milankovitch cycles range up to 100,000 years in duration, why is evidence of glaciation so sparse in the geologic record?
9. If incoming solar radiation can pass through greenhouse gases virtually unhindered, why do these gases block the return of this energy into space?
10. Why is the supply of calcium to the oceans an important factor to consider in the global warming debate?
11. Explain why plate tectonic cycles are the predominant influence on global surface temperatures throughout most of geologic time.
12. Debate among yourselves the problems of fossil fuel consumption and the greenhouse effect. Bring information from other sources to bear on the issue. The approach in this chapter was one of peeling back the layers of time. Can you suggest other approaches?
13. How does continental erosion influence global temperatures?

Systems and the Environment

Solid Earth

Most atoms and molecules at Earth's surface were exhumed from the Earth's interior. Biogeochemical cycles move elements through the solid Earth. Soils are made up of components of each of the reservoirs—sediment, organic matter, air, and water.

Hydrosphere

Biogeochemical cycles move elements through the hydrosphere. The ocean, by absorbing CO_2, moderates atmospheric levels of greenhouse gases.

Atmosphere

Biogeochemical cycles move elements through the atmosphere. The ozone layer protects life on Earth from harmful UV radiation. Greenhouse gases keep heat in the atmosphere.

Biosphere

Biogeochemical cycles move elements through the biosphere, making life possible. The food chain concentrates incoming solar energy into fossil fuels, which are a finite resource. Human interference affects both the depletion of the ozone layer and level of greenhouse gases.

Astronomy

Solar energy provides the energy for all of Earth's biogeochemical cycles. Variations in solar activity may affect ozone levels and global temperature.

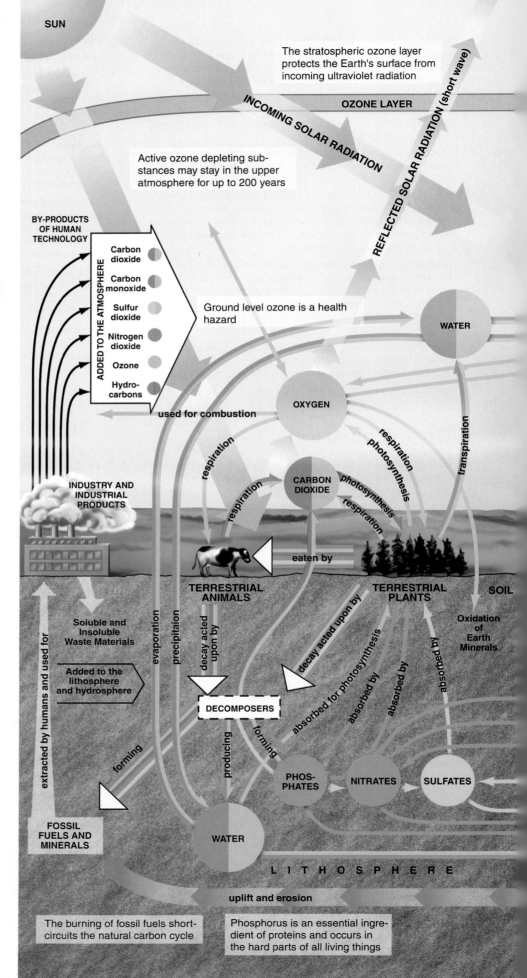

SUN

The stratospheric ozone layer protects the Earth's surface from incoming ultraviolet radiation

OZONE LAYER

INCOMING SOLAR RADIATION

REFLECTED SOLAR RADIATION (short wave)

Active ozone depleting substances may stay in the upper atmosphere for up to 200 years

BY-PRODUCTS OF HUMAN TECHNOLOGY

ADDED TO THE ATMOSPHERE

Carbon dioxide
Carbon monoxide
Sulfur dioxide
Nitrogen dioxide
Ozone
Hydrocarbons

Ground level ozone is a health hazard

used for combustion

WATER

OXYGEN

CARBON DIOXIDE

respiration
photosynthesis
transpiration
respiration
photosynthesis
respiration
respiration

INDUSTRY AND INDUSTRIAL PRODUCTS

eaten by

TERRESTRIAL ANIMALS

TERRESTRIAL PLANTS

SOIL

Soluble and Insoluble Waste Materials

Added to the lithosphere and hydrosphere

extracted by humans and used for

evaporation
precipitation

decay acted upon by

decay acted upon by

absorbed for photosynthesis
absorbed by
absorbed by
absorbed by

Oxidation of Earth Minerals

DECOMPOSERS

forming

producing

forming

PHOSPHATES

NITRATES

SULFATES

FOSSIL FUELS AND MINERALS

WATER

LITHOSPHERE

uplift and erosion

The burning of fossil fuels short-circuits the natural carbon cycle

Phosphorus is an essential ingredient of proteins and occurs in the hard parts of all living things

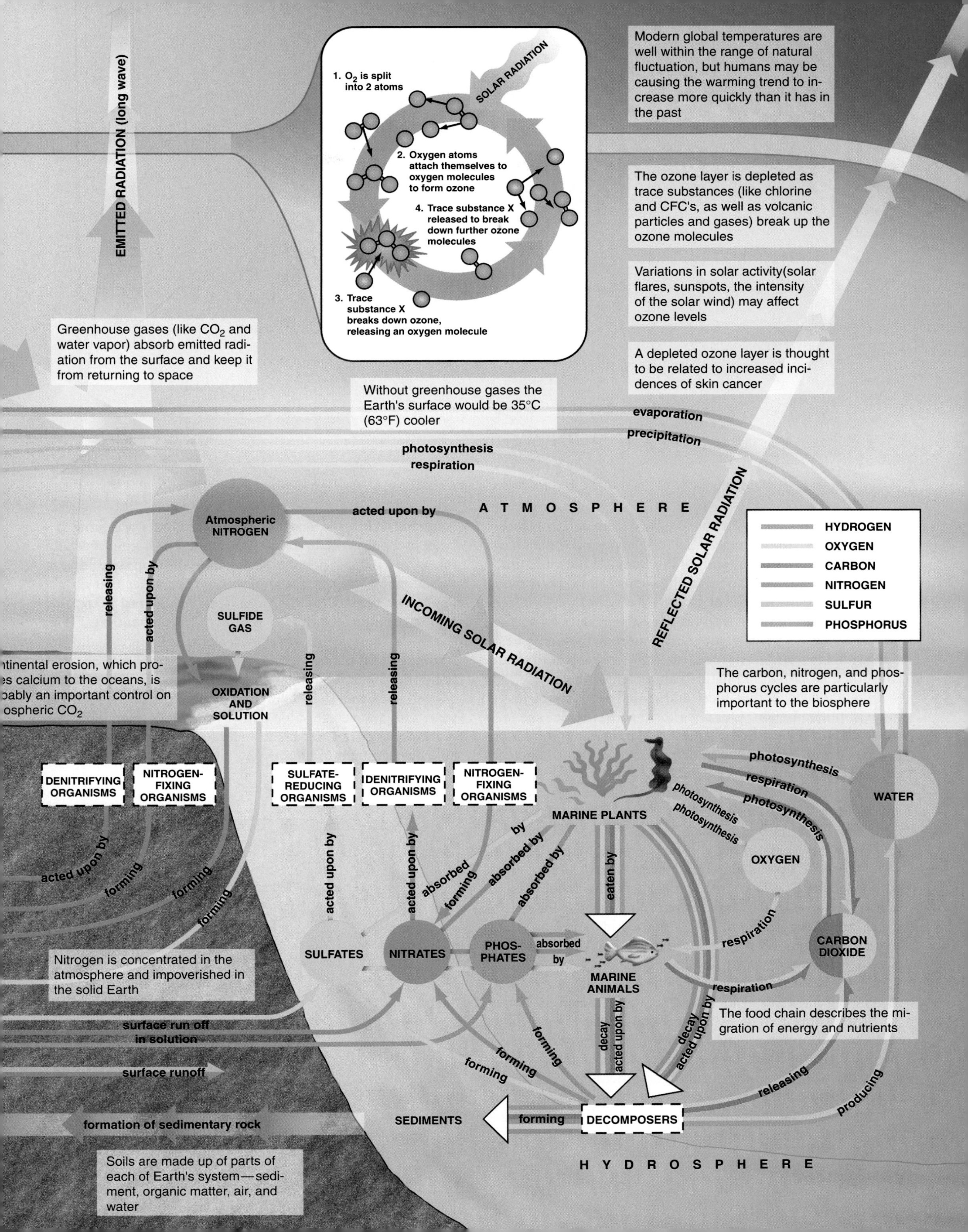

EMITTED RADIATION (long wave)

Modern global temperatures are well within the range of natural fluctuation, but humans may be causing the warming trend to increase more quickly than it has in the past

SOLAR RADIATION

1. O_2 is split into 2 atoms

2. Oxygen atoms attach themselves to oxygen molecules to form ozone

4. Trace substance X released to break down further ozone molecules

3. Trace substance X breaks down ozone, releasing an oxygen molecule

The ozone layer is depleted as trace substances (like chlorine and CFC's, as well as volcanic particles and gases) break up the ozone molecules

Variations in solar activity(solar flares, sunspots, the intensity of the solar wind) may affect ozone levels

A depleted ozone layer is thought to be related to increased incidences of skin cancer

Greenhouse gases (like CO_2 and water vapor) absorb emitted radiation from the surface and keep it from returning to space

Without greenhouse gases the Earth's surface would be 35°C (63°F) cooler

evaporation

precipitation

photosynthesis

respiration

A T M O S P H E R E

acted upon by

Atmospheric **NITROGEN**

releasing

acted upon by

SULFIDE GAS

REFLECTED SOLAR RADIATION

	HYDROGEN
	OXYGEN
	CARBON
	NITROGEN
	SULFUR
	PHOSPHORUS

INCOMING SOLAR RADIATION

releasing

releasing

ntinental erosion, which pro-
es calcium to the oceans, is
ably an important control on
ospheric CO_2

OXIDATION AND SOLUTION

The carbon, nitrogen, and phosphorus cycles are particularly important to the biosphere

DENITRIFYING ORGANISMS	NITROGEN-FIXING ORGANISMS

SULFATE-REDUCING ORGANISMS	DENITRIFYING ORGANISMS	NITROGEN-FIXING ORGANISMS

MARINE PLANTS

photosynthesis

respiration

photosynthesis

photosynthesis

photosynthesis

WATER

acted upon by

forming

forming

forming

acted upon by

acted upon by

absorbed by

forming

absorbed by

absorbed by

by

eaten by

OXYGEN

respiration

Nitrogen is concentrated in the atmosphere and impoverished in the solid Earth

SULFATES

NITRATES

PHOS-PHATES

absorbed by

MARINE ANIMALS

respiration

CARBON DIOXIDE

surface run off in solution

forming

forming

decay acted upon by

decay acted upon by

The food chain describes the migration of energy and nutrients

surface runoff

forming

releasing

producing

formation of sedimentary rock

SEDIMENTS

forming

DECOMPOSERS

Soils are made up of parts of each of Earth's system—sediment, organic matter, air, and water

H Y D R O S P H E R E

Afterword: The Fate of Planet Earth

The fate of the Earth is in our hands. We are certainly faced with severe environmental problems, many of which have been discussed in various parts of this book. At the same time, through basic and applied research, we now know more about the Earth systems than ever before. We know, for example, that our Earth is a place with only finite resources to support our existence. This is because most resources, including fresh water, minerals and fossil fuels, are nonrenewable and so are consumed far faster than nature can replenish them.

Pollution on both a local and global level is not just a modern phenomenon. Our ancestors had flawed sewage and water systems that caused plagues and famines, and pollution associated with Bronze Age smelting has recently been detected in the ice sheets of Greenland.

However, there are additional environmental pressures that are unique to modern society. Perhaps the most urgent pressure is the explosion in world population (Fig. A.1), which over the last 150 years has increased at a rapidly accelerating pace. In 1850, the world's population is estimated to have been about 1 billion people. By 1930, just 80 years later, it had doubled to 2 billion. However, it took only 45 years to double again, reaching 4 billion by 1975. In 1998, the world population had reached 6 billion. Although economists predict that population growth will slow down as developing nations become more developed, projections indicate that the Earth will have close to 10 billion inhabitants by the middle of the next century.

Recent history tells us that economic development and resource consumption are closely related. For example, a direct correlation exists between gross world product (GWP) and gross world energy consumption (measured in carbon dioxide emissions) for the last 40 years (Fig. A.2). As more nations become developed, this correlation will place increasing demands on our finite natural resources.

The realization that the Earth's resources are finite has been brought home to us by views of the Earth from space. In recent years, the concept of **sustainable development**, that is, the adoption of economic policies and activities that do not deplete or degrade natural resources, has gained increasing public attention. Until very recently, the economic value of the world's ecosystems was not adequately quantified, and as a result, the importance of ecosystems was given little weight in policy decisions. The value of the entire biosphere has recently been estimated to be between 16 and 54 trillion dollars per year, with an average value of $33 trillion per year. This estimate allows policy makers to include the benefits of sustainable development in their economic analyses.

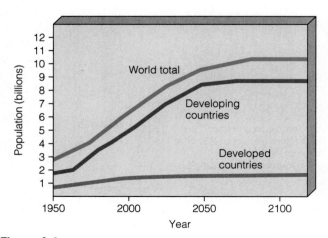

Figure A.1
Population in developed countries, developing countries and the total world population from 1950, projected into the 21st Century.

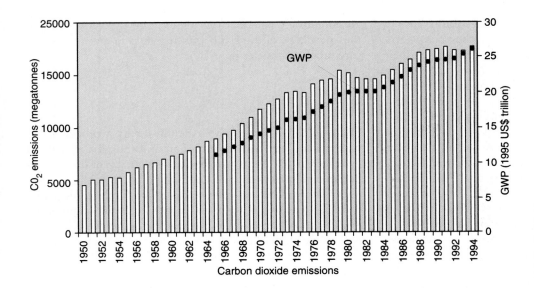

Figure A.2
Global carbon dioxide emissions compared to gross world product (GWP) during the latter part of the 20th Century.

The sense of urgency among politicians and scientists has produced a flurry of documents promoting sustainable development while also emphasizing the need for environmental stewardship. Neighboring countries which may have once viewed one another with suspicion or fear, must now collaborate because environmental problems do not respect international borders. In 1992, the Earth Summit in Rio de Janeiro produced treaties that proposed to curb carbon dioxide emissions, tackled the threatened extinction of endangered species, and provided an action plan for sustainable development in the 21st Century. A commitment to environmental stewardship was further indicated by the Kyoto Protocol in 1998, which was signed by 160 countries and set specific targets to reduce fossil fuel emissions.

In 1993, nearly 1700 scientists from 71 countries signed a declaration entitled "World Scientists' Warning to Humanity," which identified critical stresses in the atmosphere (ozone depletion, air pollution, acid rain), hydrosphere (exploitation of groundwater, heavy demands on surface water), biosphere (depletion of the world's fish stocks, destruction of the tropical rain forests, irreversible loss of species), and the solid Earth (loss of soil productivity). They drew attention to the fact that the Earth's ability to absorb wastes, provide food and energy, and sustain population growth is finite. They also noted that the developed nations are the largest polluters in the world today, due to overconsumption.

In order to rectify these problems, they concluded that: (1) environmentally damaging activities must be brought under control, (2) resources crucial to human welfare must be managed more efficiently, (3) population must be stabilized; a goal that is possible only if social and economic conditions improved and effective voluntary family planning is adopted, (4) poverty must be reduced and eventually eliminated, and (5) sexual equality must be ensured and women's control over reproductive decisions must be guaranteed.

Despite the severity of this prescription, the fact that we have recognized the enormous problems that confront us is grounds for optimism. Our expanding knowledge and understanding of the Earth means that we have the opportunity to decide our own fate. We have learned more about our planet in the last 40 years than we had in the previous 2 millennia. We have come to understand that the Earth is a dynamic planet, constantly renewing itself in ceaseless cycles of creation and destruction powered by plate tectonics. Even Alfred Wegener would marvel at the ways in which our modern understanding of plate motions have vindicated his hypothesis of continental drift.

We also know that processes within the solid Earth have provided us with the watery and gaseous envelopes that encircle the Earth and, in shaping our oceans and atmosphere, have provided us with an environment that can support life. We also understand the important role of the biosphere in purifying the air we breathe. We understand many of the processes that influence ocean chemistry and the forces that drive its circulation. Satellite imagery can track marine circulation patterns in both surface and bottom waters, and has shown their relationship to climate. We are now aware of the importance of groundwater resources. We have developed ingenious methods of finding these resources and sophisticated analytical techniques for monitoring water quality and identifying the slightest contamination. Ironically, some of the most tragic cases of contamination come from our efforts to provide food and energy to populated regions, either by application of fertilizers or by diverting the natural flow of surface waters. Hopefully, we have learned from the painful lessons of these tragedies.

We have identified the gaseous components in the atmosphere and have come to understand why its composition is unique within the Solar System, and how it has changed as the Earth evolved. Major advances in our understanding of weather and climate patterns have been achieved by examining the interaction between the hydrosphere and the atmosphere, and how both of these surface reservoirs are affected by radiant energy from the Sun.

This understanding of the modern Earth has also been applied to the evolution of the planet through geologic time. By examining the rock record, we have traced back Earth's history nearly 4 billion years and can recognize the earliest oceans and have identified traces of life in the rocks deposited along their continental margins. Developments in instrumentation now allow us to determine the age of the oldest rocks and minerals with amazing precision. These techniques, coupled with on-site field investigations, now provide surprisingly precise constraints on conditions in the primitive Earth and the timing of major events in the evolution of the Earth systems that have punctuated Earth history. We can recognize cyclicity in the creation and destruction of oceans as envisaged by Tuzo Wilson and expressed by the supercontinent cycle. We can also recognize the relationship between this cycle and our planet's long history of global change in sea level, evolution, and climate.

Our ceaseless quest to understand the world around us has led us into space. We have walked on the Moon and sent spacecraft to investigate celestial bodies in and beyond the Solar System. We have identified planets orbiting distant stars and plan to send humans to Mars in the early 21st Century. From these investigations, we have learned much about the cosmos and our place in it, and we have also come to appreciate the special uniqueness of planet Earth.

We know that our day-to-day existence utilizes the Earth's resources. But we now understand that mineral and fossil fuel deposits are commonly related to plate tectonic processes and formed over millions of years. Consequently, we consume these resources faster than nature can replenish them. Although sensitive instruments installed in aircraft and satellites can now complement on-the-ground field investigations in our efforts to find more resources, we recognize that they are nonrenewable and cannot be replaced.

One of the most important breakthroughs in our understanding of the Earth came with the recognition of the interconnected nature of the Earth systems, first popularized by Sir James Lovelock's Gaia hypothesis and now expressed in the emerging interdisciplinary science of biogeochemistry. Ironically, the Gaia hypothesis has been cited by some industrialists as evidence that the Earth will repair itself from all of our environmental tampering, so that we need not be concerned about environmental problems. However, careful analysis of the biogeochemical cycling of carbon, nitrogen, and phosphorus has clearly identified the detrimental long-term effects of human activities on the environment.

Finally, we have come to appreciate the unique perspective that Earth science brings to bear on environmental problems. By putting it all together, we can gain an appreciation for the Earth's natural beats and rhythms over its 4.6 billion year history and in doing so, we are able to distinguish between natural and human-induced change.

Future Research

Although we have made enormous strides in the last 40 years, much remains to be done in all aspects of Earth science. Never has this research been more important. You, today's students, are tomorrow's potential researchers, and by contributing to the knowledge of the Earth you may help to ensure the very survival of our species upon it.

Solid Earth

Studies of the solid Earth help identify the basic processes that have shaped our planet's evolution and the pace of natural change. Only 40 years ago, our understanding of the generation of earthquakes and volcanoes was crude at best. Plate tectonics provided the conceptual framework for the understanding and prediction of these events, as well as the origin of the world's mountain ranges and ocean basins. Yet we have huge gaps in our understanding of Precambrian geology compared to that of the Phanerozoic because the Earth's endless cycles of creation and destruction have destroyed much of the ancient rock record. Most of our understanding of the Precambrian Earth is based on geologic mapping, through studying the spatial distribution of surface rock formations and the contacts between them. More recently, this work has been complemented by sophisticated geologic analyses like those involved in the dating of rocks.

Important areas where Precambrian processes are preserved are natural laboratories for studying Earth processes, and precise age dating can determine the rates of natural change. For example, geologic field investigations in northern Canada have identified the oldest known remnants of oceanic crust, providing strong evidence that plate tectonics, in more or less its modern form, existed 2 billion years ago. Likewise, it was information from geologic maps in Greenland that stimulated the detective work that resulted in the discovery of carbon spheres that may represent the world's oldest fossil remains at 3.86 billion years. This discovery places life closer to the birth of the planet, and so further constrains its early evolution. Like a world record in athletics, every time-barrier that is broken has important implications.

But there is much ground to cover, and much remains to be done. Geologists working on regional studies are providing important pieces of the jigsaw puzzle. Questions concerning the earliest history of the Earth will continue to focus on the oldest rocks, and will test how far back the principle of actualism can be extended. Late Archean studies will address the role of plate tectonics and atmospheric conditions at that time. Rocks such as Banded Iron Formations will continue to attract special attention because of their relative abundance in a narrow period of geologic time, and because of their important role in the evolution of the atmosphere. Studies of intense mountain building activity in the Late Proterozoic will test proposals for the existence of Late Proterozoic supercontinents, and may reveal their configurations. By doing so we will find out whether the Late Paleozoic supercontinent, Pangea, was just an aberration, or whether supercontinent assembly and breakup is a common theme of Earth history. Studies of Late Proterozoic soft-bodied organisms will test theories of an evolutionary bridge to the next eon, the Phanerozoic.

For the great majority of geoscientists, plate tectonics forms the backdrop for studying the Earth's evolution in the

Phanerozoic. However, the processes responsible for plate motion remain as enigmatic as they were in the time of Wegener. We still have much to learn about the deep-seated processes that drive plate motion and the details of many of the consequences such as mountain building, magma compositions, oceanic ridge formation, and the evolution of active and passive continental margins. Major advances will continue to be derived from targeted field work (either on land or at sea) complemented by sophisticated laboratory analyses.

Although plate tectonics provides an explanation for the origin of geologic hazards, such as volcanoes and earthquakes, we need to continue to develop methods that will predict these events with a precision that can protect human life and can influence land development policies.

Hydrosphere

Present theories concerning the origin of water on the Earth are no more than educated speculations. We need to find ways to test these theories and to determine whether water volume has increased with time, either because of the burning of comets in the upper atmosphere or because of plate tectonics and igneous processes within the solid Earth. If not, we need to establish that the vast amount of water in our modern oceans was present 4 billion years ago and has it been recycling ever since. It is possible that minerals in ancient sedimentary rocks may trap portions of this ancient water. If so, what is the composition of this water and can trace elements and isotopes dissolved in the water help us determine its origin?

In the context of global warming, studies of sea level change and its possible relationship to climate are urgently needed. Is the melting of ice sheets the dominant short-term factor controlling this phenomenon, or are there other more important forces at work? For example, the melting of a continental ice sheet should cause the loaded crust beneath it to rebound as the weight is removed. This would contribute to a drop in sea level. But to what degree would this rebound outweigh the rise in sea level brought about by melting and an increase in ocean water volume? The relationship of sea level to plate tectonics and the ever-changing geometry of ocean basins also needs to be quantified. We need to know more about oceanic circulation patterns, the development of mesoscale eddies and their relationship to marine ecology, fish population, and climate change. How does an El Niño, the periodic warming of equatorial ocean waters in the Pacific Ocean, affect ocean circulation patterns and climate?

We now know that the chemistry of ocean water is relatively homogenous due to efficient circulation. This aids in the detection of chemical anomalies, such as those associated with pollution, both in coastal regions and in the open seas. But can geochemical techniques be sufficiently refined to identify the source of such pollution, and provide a chemical fingerprint that might, for example, link oil spills to the tankers from which they leaked? Although the source may be obvious in major spills, oil tankers routinely leak smaller portions of their cargo that generally go undetected.

Although the detailed chemistry of submarine hydrothermal systems is not fully documented, it is clear that in addition to chlorine and sulfates, they contain abundant precious metals such as gold. If mining techniques develop to the point where they can be exploited, will these systems be an exploration target in the future? If so, what will be the impact of exploration on the delicate ecosystems that depend on them? Access to fresh water for domestic and industrial purposes may be one of our greatest challenges of the 21st Century. Improved exploration techniques are essential as are ongoing studies into problems of subsidence related to groundwater extraction. Our search for water may lead us to exploit polar ice sheets. If so, what would be the effect of this exploitation on global ecosystems?

Atmosphere

Although we understand the major influences on the composition of our atmosphere, the global warming debate has taught us that we know little about the rates at which gases from either volcanoes or fossil fuel emissions migrate through the atmosphere. We know even less about the manner in which they are absorbed by the biosphere, hydrosphere, and solid Earth. Much quantitative work remains if we are to fully understand these processes so we can assess the influence of human activities.

There is also important fundamental and practical research to be done in the science of meteorology if we are to improve our understanding of weather and climate systematics. With the advent of satellites and remote sensing techniques, there have been many advances in this field, but much remains to be done. The most important practical advances would be those that might improve our ability to predict major storm systems. What, for example, influences their origin and defines their precise path? In the case of monsoon climates (on which much of Asia depends for its water), what causes the rains to fail in some years and be more intense in others? Can their varying nature be predicted ? If these changes could be anticipated it would greatly aid relief efforts and allow for preparations. What is the relationship between El Niños and the failure of the monsoon rains, and are the intensity and frequency of El Niños affected by global warming, or is this a separate phenomenon?

Longer-term studies are also needed to determine whether global warming is increasing the abundance and ferocity of hurricanes and typhoons, and why El Niños are accompanied by major disruptions in climate patterns.

Earth Resources

Projections that place the world's population at 10 billion by the middle of the 21st Century indicate that mineral and energy resources will still be in great demand. The challenge will be to develop responsible and environmentally sensitive ways to meet this need. New technologies, improved recycling, and the development of substitutes will all be needed if the supply of mineral commodities is to be maintained.

Similarly, a more efficient means of tapping the world's renewable energy resources will be needed if we are to reduce our dependence on fossil fuels, and so lessen the impact of our energy usage on the environment. There is certainly potential; heat flow maps show that the Earth's internal energy is escaping virtually untapped from the surface of the Earth (see Chapter 1). In tectonically active regions such as California and Iceland, high heat flow facilitates the harnessing of geothermal energy, which has reduced the dependency on fossil fuels. But geothermal energy may also be important in some tectonically stable regions where water circulation exploits ancient fault structures, or where water circulating within the crust is heated by the decay of radioactive elements. Ironically, the warming of the world's surface temperatures may also provide additional opportunities to harness wind or tidal power.

The time and money invested in the development of renewable resources pales in comparison to that invested in nonrenewable resources. For the foreseeable future, exploration for minerals and fossil fuels is certain to continue demanding ever more sophisticated exploration and exploitation techniques.

Astronomy

We are entering an exciting and challenging phase in our quest to understand our position in the cosmos. The recent controversy over life on Mars is likely to dominate the science for the foreseeable future. By the year 2005, NASA intends to send a robot mission to bring back samples from the Martian surface. These will be intensely studied by teams of planetary scientists and research students including geologists, physicists, chemists, and biologists. By analogy, samples taken from the lunar surface by Apollo astronauts changed not only our preconceived notions about the antiquity of the Moon, but had a profound impact on theories about the origin of the terrestrial planets.

Furthermore, given that life on Mars, if it exists, is most likely to reside beneath the Martian surface, techniques will have to be devised to obtain images of the subsurface. Who knows what surprises lie in store.

Space probes have greatly increased our understanding of our planetary neighbors and their moons. But if an equivalent fly-by was attempted for the Earth, how much would we learn? We would certainly notice its continents and oceans, but how much of the Earth's history would be revealed? Imagine how much more can be gained from detailed studies of our celestial neighbors!

The potentially exciting discoveries of new solar systems and ongoing revelations provided by the Hubble space telescope will provide impetus for research for many years to come. In fact, in the short time since writing Chapter 16 on the Sun and the Stars, planets orbiting distant stars are claimed to have been observed *directly*. (Recall from Chapter 16 that we stated the presence of planets orbiting distant stars was *inferred* from the wobble of the star; the planets themselves had not been observed). This exemplifies the dramatic and important revelations that continually add to our knowledge and understanding. Only a generation ago such revelations would have been impossible to fathom. The consequences for us and our position in the universe are enormous. As more "solar systems" are found, attention will turn to deducing atmospheric composition and surface temperature of their planets in order to identify those that might harbor life. This research will require an exciting collaboration of physicists, chemists, geologists, and evolutionary biologists who will need to combine their talents for a common purpose.

We also have much to learn about the Sun and how its radiation and that emitted from other stars and from supernovae may affect our planet.

Environment

The science of biogeochemistry is an emerging discipline that focuses our attention on important environmental issues. The carbon, nitrogen, and phosphorus cycles are first-order examples of the biogeochemical approach to the study of the Earth's reservoirs. In theory, a biogeochemical cycle could be derived for all chemical species. Within the framework of this new interdisciplinary science, however, it is important that the construction of individual biogeochemical cycles does not produce a new style of division and compartmentalization. As the interaction of carbon and calcium in the carbon cycle illustrates, the behavior of a particular chemical element and its biogeochemical pathway often depend on the availability of other species.

The differing behavior and mobility of chemical elements means that, like runners in a race, they are rarely bunched in an ideal configuration for the proliferation of life. However, over the course of the Earth's 4.6 billion year history, there must have been periods when this did occur. It is interesting to speculate, for example, whether the explosion of life at the beginning of the Cambrian Period might have been due to more optimal abundances of various essential chemical nutrients such as calcium and phosphorus at this time in the Earth's history when compared to other periods.

In Chapter 19, we treated the carbon, nitrogen, and phosphorus cycles independently. But how do alterations to one cycle affect another? How do excessive inputs of one chemical compound to an environment affect the fixation or release of another compound? For example, if an ecosystem whose development is limited by its nitrogen or phosphorus content artificially receives these nutrients, what other nutrients become the limiting factor to plant growth? And how does this affect the biogeochemical cycle? Clearly, much work remains to be done in documenting, understanding, and quantifying the migration of chemical species through the Earth's surface reservoirs. This research will require a willingness to break down the artificial barriers among the traditional sciences.

The relationships between ozone depletion and CFCs on one hand, and between fossil fuel emissions and the greenhouse effect on the other, are likely to remain controversial, and much basic interdisciplinary research remains to be done on both issues. Ozone is a fragile molecule, and even if the war waged against CFCs is won, it is conceivable that other artificially induced chemical substances may attack it. The greenhouse effect is tied to the carbon biogeochemical cycle and the behavior of elements such as calcium that extract carbon dioxide from the oceans. Although the general scheme has been worked out qualitatively, little is known about the rates at which carbon migrates from one "rest stop" to another along the biogeochemical highway, or the residence time of carbon at each stop.

As James Lovelock pointed out, we know that both positive and negative feedback systems are common in nature, and that they respectively amplify or dampen trends of global change. However, studies of these systems and their complex interactions are in their infancy. For example, if elements like calcium are so important, what conditions favor input of calcium into the sea? Is this input also self-regulated? Could the increased evaporation rates and higher water temperatures associated with global warming result in more efficient extraction of calcium from rocks and soils that would counteract the greenhouse effect? Such hypotheses may shed light on this important issue, but are as yet untested.

Final Words

In using this book to complete your course, we hope you have gained a new appreciation for our wondrous, but fragile, planet. A basic understanding of the Earth and the processes that have shaped its evolution are essential if humans are to survive on this planet. Let us hope that we leave our children the best legacy of all—a safe and healthy planet.

Appendix A

ENGLISH-METRIC CONVERSION CHART

ENGLISH UNIT	CONVERSION FACTOR	METRIC UNIT	CONVERSION FACTOR	ENGLISH UNIT
Length				
Inches (in.)	2.54	Centimeters (cm)	0.39	Inches (in.)
Feet (ft)	0.305	Meters (m)	3.28	Feet (ft)
Miles (mi)	1.61	Kilometers (km)	0.62	Miles (mi)
Area				
Square inches (in²)	6.45	Square centimeters (cm²)	0.16	Square inches (in²)
Square feet (ft²)	0.093	Square meters (m²)	10.8	Square feet (ft²)
Square miles (mi²)	2.59	Square kilometers (km²)	0.39	Square miles (mi²)
Acres (ac)	0.404	Hectares (Ha)	2.47	Acres (ac)
Volume				
Cubic inches (in³)	16.4	Cubic centimeters (cm³)	0.061	Cubic inches (in³)
Cubic feet (ft³)	0.028	Cubic meters (m³)	35.3	Cubic feet (ft³)
Cubic feet (ft³)	0.028	Cubic meters (m³)	264	Gallons (gal.)
Cubic yards (yd³)	0.8	Cubic meters (m³)	1.3	Cubic yards (yd³)
Gallons (gal.)	3.79	Liters (l)	0.264	Gallons (gal.)
Cubic miles (mi³)	4.17	Cubic kilometers (km³)	0.24	Cubic miles (mi³)
Flow				
Cubic feet/sec (cfs)	0.028	Cubic meters/sec (m³/s)	35.3	Cubic feet/sec (cfs) [= 448.8 gal/min]
Gallons/min (gpm)	3.79	Liters/min (lpm)	0.264	Gallons/min (gpm) [× 1440 gallons/day (gpd)]
Cubic feet/sec (cfs)	—	—	7.48	Gallons/sec (gps)
Weight				
Ounces (oz)	28.3	Grams (g)	0.035	Ounces (oz)
Pounds (lb)	0.45	Kilograms (kg)	2.20	Pounds (lb)
Short tons (st)	0.91	Metric tons (t)	1.10	Short tons (st)

Temperature °C = 5/9(°F–32).

To convert degrees Fahrenheit (°F) to degrees Celsius (°C):

Subtract 32 degrees from °F, then divide by 1.8.

To convert degrees Celsius (°C) to degrees Fahrenheit (°F):

Multiply °C by 1.8, then add 32 degrees.

Appendix B

PERIODIC TABLE OF THE ELEMENTS

Group

Atomic number → 11
Symbol → Na
Atomic mass → 22.99

Atomic masses are based on carbon-12. Numbers in parentheses are mass numbers of most stable or best known isotopes of radioactive elements.

Noble Gases (18)

← Transition Elements →

Period

IA(1)	IIA(2)	IIIB(3)	IVB(4)	VB(5)	VIB(6)	VIIB(7)	VIII (8)	VIII (9)	VIII (10)	IB(11)	IIB(12)	IIIA(13)	IVA(14)	VA(15)	VIA(16)	VIIA(17)	(18)
1 H 1.008																	2 He 4.003
3 Li 6.941	4 Be 9.012											5 B 10.81	6 C 12.01	7 N 14.01	8 O 16.00	9 F 19.00	10 Ne 20.18
11 Na 22.99	12 Mg 24.31											13 Al 26.98	14 Si 28.09	15 P 30.97	16 S 32.06	17 Cl 35.45	18 Ar 39.95
19 K 39.10	20 Ca 40.08	21 Sc 44.96	22 Ti 47.90	23 V 50.94	24 Cr 52.00	25 Mn 54.94	26 Fe 55.85	27 Co 58.93	28 Ni 58.7	29 Cu 63.55	30 Zn 65.38	31 Ga 69.72	32 Ge 72.59	33 As 74.92	34 Se 78.96	35 Br 79.90	36 Kr 83.80
37 Rb 85.47	38 Sr 87.62	39 Y 88.91	40 Zr 91.22	41 Nb 92.91	42 Mo 95.94	43 Tc 98.91	44 Ru 101.1	45 Rh 102.9	46 Pd 106.4	47 Ag 107.9	48 Cd 112.4	49 In 114.8	50 Sn 118.7	51 Sb 121.8	52 Te 127.6	53 I 126.9	54 Xe 131.3
55 Cs 132.9	56 Ba 137.3	57* La 138.9	72 Hf 178.5	73 Ta 180.9	74 W 183.9	75 Re 186.2	76 Os 190.2	77 Ir 192.2	78 Pt 195.1	79 Au 197.0	80 Hg 200.6	81 Tl 204.4	82 Pb 207.2	83 Bi 209.0	84 Po (210)	85 At (210)	86 Rn (222)
87 Fr (223)	88 Ra 226.0	89** Ac (227)	104 Unq (261)	105 Unp (262)	106 Unh (263)	107 Uns (262)	108 Uno (265)	109 Une (266)									

Inner Transition Elements

Lanthanide Series 6 *

58 Ce 140.1	59 Pr 140.9	60 Nd 144.2	61 Pm (145)	62 Sm 150.4	63 Eu 152.0	64 Gd 157.3	65 Tb 158.9	66 Dy 162.5	67 Ho 164.9	68 Er 167.3	69 Tm 168.9	70 Yb 173.0	71 Lu 175.0
90 Th 232.0	91 Pa 231.0	92 U 238.0	93 Np 237.0	94 Pu (244)	95 Am (243)	96 Cm (247)	97 Bk (247)	98 Cf (251)	99 Es (252)	100 Fm (257)	101 Md (258)	102 No (259)	103 Lr (260)

Actinide Series 7 **

Appendix C

MINERAL IDENTIFICATION CHART

MINERALS WITH GENERALLY METALLIC LUSTER. NOTE: MANY METALLIC MATERIALS COLLECTED BY NOVICE MINERAL COLLECTORS ARE ACTUALLY ARTIFICIAL, NOT MINERALS.

NAME	FORMULA	COLOR	DENSITY	HARDNESS	CLEAVAGE	GEOLOGIC SETTING	REMARKS
Gold	Au	Yellow	15–19	2.5–3	None—malleable	Low-temperature metamorphosed volcanic rocks, often old volcanic island arc terranes. Also gains in stream deposits.	Visible gold is extremely rare. Gold is much denser than lead, flattens and does not break if hit with a hammer (malleable), and is unaffected by most acids. If a mineral lacks any of these properties, *it is not gold.*
Copper	Cu	Red-orange	8.9	2.5–3	None—malleable	Volcanic rocks and conglomerate in Michigan, granite and metamorphic rocks of copper deposits.	Found in northern Michigan and some western copper deposits.
Galena	PbS	Silvery	7.6	2.5	Excellent cubic	In carbonate rocks in U.S. midwest.	Density and cubic cleavage are distinctive.
Pyrite	FeS$_2$	Yellowish	5.0	6–6.5	None	Very common, all types of rock.	Fool's gold. Not as dense as gold and not malleable. Often in good crystals. Not as yellow as gold.
Chalcopyrite	CuFeS$_2$	Yellow, often with greenish cast	4.2	3.5–4	None	In carbonate rocks in U.S. midwest and low-temperature metamorphic rocks.	Usually yellower than pyrite, often with a slight yellow-green cast.
Magnetite	Fe$_3$O$_4$	Dark gray	5.2	6	None	Igneous and metamorphic rocks.	Magnetic. No other tests needed.

MINERALS THAT CAN BE EITHER METALLIC OR NONMETALLIC IN LUSTER.

NAME	FORMULA	COLOR (METALLIC)	COLOR (NONMETALLIC)	DENSITY	HARDNESS	CLEAVAGE	GEOLOGIC SETTING	REMARKS
Hematite	Fe$_2$O$_3$	Gray	Red or brown	5.2 in massive forms	1 in earthy forms, up to 6 in metallic forms	Rarely visible	Sedimentary and metamorphic rocks.	Metallic: gray, flaky texture (specular hematite). Nonmetallic: massive or earthy, sometimes oolitic (small rounded pellets). Always streak.
Sphalerite	ZnS	Brown or gray	Yellow to red	4.0	3.5–4	Good, very complex, in six directions	In carbonate rocks in U.S. midwest and low-temperature metamorphic rocks.	Streak always lighter than mineral. Usually with galena and chalcopyrite.
Graphite	C	Dark gray	Dark gray	2.2	1	Rarely visible	Metamorphic carbon-rich rocks.	Very soft, leaves marks on paper, greasy feel.

MINERALS WITH NONMETALLIC LUSTER AND GOOD CLEAVAGE.

NAME	FORMULA	COLOR	DENSITY	HARDNESS	CLEAVAGE	GEOLOGIC SETTING	REMARKS
Gypsum	$CaSO_4H_2O$	Clear, white, pink	2.3	2	Excellent, thin sheets	Sedimentary rocks	Softness and lack of solubility distinctive.
Halite	$NaCl$	Clear, white	2.2	2.5	Excellent cubic	Sedimentary rocks	Salty taste. No other test needed.
Fluorite	CaF	Clear, white, green, violet	3.2	4	Excellent, four directions	Carbonate sedimentary rocks	Colors are often distinctive, cleavage complex, but distinctive.
Calcite	$CaCO_3$	Clear, white	2.7	3	Excellent, cleaves into skewed box shapes	Carbonate sedimentary rocks	Fizzes vigorously in acid.
Dolomite	$CaMg(CO_3)_2$	Clear, white, light colors	2.85	3.5–4	Excellent, cleaves into skewed box shapes	Carbonate sedimentary rocks	Looks almost identical to calcite, but slightly harder, fizzes weakly. Cleavage planes often slightly curved.
Kyanite	Al_2SiO_5	Blue, white	3.6	5 along crystal, 7 across	Cleaves along length of crystal	High-temperature metamorphic rocks; gneisses	Color, occurrence, and the hardness variation are distinctive.
Sillimanite	Al_2SiO_5	White, clear	3.2	7	One good cleavage direction	High-temperature metamorphic rocks; gneisses	Common, but inconspicuous. Often fibrous in appearance.
Amphibole	A group name. Complex, but Si/O ratio is 4/11.	Black, green, brown; some can be light or white	3–3.5	5–6	Cleavage planes parallel to length of crystal at 56 and 124 degree angles	Igneous and metamorphic rocks	In granites, more likely to occur than pyroxene.
Pyroxene	A group name. Complex, but Si/O ratio is 1/3.	Black, green; some can be light or white	3–3.5	5–6	Cleavage planes parallel to length of crystal at near right angles	Mafic igneous rocks, metamorphic rocks	Dark minerals in igneous rocks with long crystals are most likely pyroxene.
Biotite	Complex, K, Fe, Mg, and Si/O ratio of 2/5.	Black, brown, light when weathered	3	2.5–3	Perfect cleavage into thin, elastic sheets	Igneous and low-temperature metamorphic rocks	Color and cleavage distinctive. Weathered varieties are often mistaken for gold but are far less dense.
Muscovite	Complex, K, Al, and Si/O ratio of 2/5.	Clear, white, gray, silvery	2.8	2–2.5	Perfect cleavage into thin, elastic sheets	Granitic and low-temperature metamorphic rocks	Color and cleavage distinctive. No other minerals are similar to the micas.
Plagioclase	Mixture of anorthite $CaAl_2Si_2O_8$ and albite, $NaAlSi_3O_8$	White, gray	2.6–2.7	6	Good, two directions	Igneous and some metamorphic rocks	Large crystals may have a striated appearance. Some dark gray varieties show color flashes and are popular ornamental stones.
Potassium feldspar	$KAlSi_3O_8$	Pink, white	2.6	6	Good, two directions	Granitic and metamorphic rocks	Pink color is common but not universal.

NAME	FORMULA	COLOR	DENSITY	HARDNESS	GEOLOGIC SETTING	REMARKS
Tourmaline	Complex, with boron and Si_6O_{18} rings	Black, sometimes pink or green	3.1	7–7.5	Granitic rocks.	Most often as deep black accessory mineral in granite. Crystals have triangular cross-section.
Chlorite	Mg, Fe, with Si/O ratio 2/5	Green	2.6–2.9	2–2.5	Low-grade metamorphic rocks.	Crystals rarely visible. The principal coloring agent in low-grade schists and slates.
Serpentine	$Mg_3Si_2O_5(OH)_4$	Usually dark green or blue-green	2.3–2.7	2–5	Ophiolites, low-grade metamorphic rocks.	Crystals rarely visible. Sometimes occurs as the fibrous form asbestos.
Talc	$Mg_3Si_4O_{10}(OH)_2$	White or light green	2.7–2.8	1	Low-grade metamorphic rocks.	Softness and greasy feel distinctive.
Opal	SiO_2 plus water	Light colors	1.9–2.2	5–6	Weathered rocks.	A mineraloid. Usually has a waxy look, conchoidal fracture. Rarely shows a play of color.
Chert	SiO_2	White, gray, light colors, but can be any color	2–2.6	7	Sedimentary rocks.	Can be soft and chalky or massive. Microcrystalline quartz. Very common.
Quartz	SiO_2	Clear, white, light colors, can be any color	2.65	7	All rock types.	The most common mineral in the Earth's surface. If it scratches glass, suspect quartz.

Appendix D

TOPOGRAPHIC MAPS

Figure D1a shows part of a reservoir with a small island. The island is the top of a submerged hill. In (b) through (e), the water in the reservoir is lowered in steps of ten meters. As the water drops, more and more land is exposed. At each step, the shoreline marks a line of constant elevation. In each figure, the former shorelines are also marked. In the final figure (f) the reservoir is completely drained and the old shorelines make up a topographic map of the area.

Elevation is shown on topographic maps with contour lines, lines of equal elevation like the shorelines above. Contour lines follow two basic rules:

1. They never cross (why not? —answer follows).
2. They never end, through they may be cut off at the edge of the map. A contour just above sea level may wrap all the way around a continent before closing, but it does eventually close. In some places, contour lines are omitted to make the map more readable, but the actual line of constant elevation on the ground is continuous.

*Answer to question:*If contour lines crossed, the same point on the ground would simultaneously be at two different elevations. In theory, an overhanging cliff or natural arch could result in crossing contours, but in practice such features are almost never large enough to show on most maps.

Earth scientists can tell a lot about an area's geology just by reading topographic maps and interpreting the landforms. Figure D2 shows some common landforms on topographic maps.

a. Flat areas have no topographic contours.
b. Areas of gentle slope have widely spaced contours.
c. Areas of steep slope have closely spaced contours.
d. Hills are usually marked by sets of concentric contours increasing in elevation toward the center.
e. Ridges are like hills, but there may be a set of summits in a row, or a single, long-closed contour.
f. Valleys are marked by contours decreasing in elevation toward the center. Often there is a stream in the bottom.

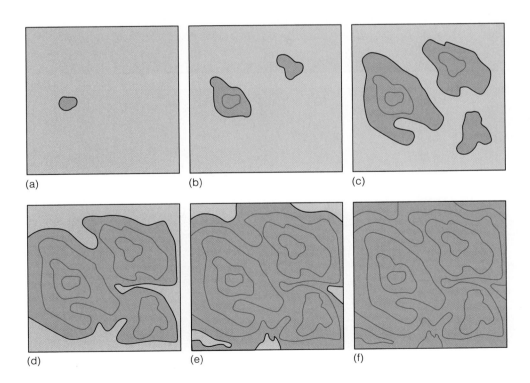

(a) (b) (c)
(d) (e) (f)

Figure D1

The contours make a V pointing upstream as they cross the valley bottom.

g. Closed depressions are rare landforms. They are usually shown by tick marks pointing downhill.

In areas of complex topography (Fig. D3) it may not be immediately obvious what a feature is. Nevertheless, by reading the map carefully, you can figure out the topography. Don't let the complexity intimidate you; proceed slowly and carefully.

If you don't know the contour interval, find two labeled contours and count the intervals between them (a). The contour interval is the difference between the labeled contours divided by the number of intervals. On this map, the interval is ten meters.

If you're not sure what a contour is, find a contour that is labeled and count up or down, using the contour interval of the map, to find the elevation. Find the nearest label and follow the contour as close to your location as you can. Contour (b) is at 380 meters.

Some elevation points like hilltops are also marked on topographic maps. You can count up or down from these points in the same way. Contour (c) is at 370 meters.

Is feature (d) a valley, or is (e)? Find contours of known elevation to determine the slope on either side of the feature. The land slopes on both sides away from (d), so (d) must be a ridge. It slopes toward (e) on both sides, so (e) must be a valley.

If (f) a ridge or a valley? Streams only flow in valleys, so (f) must be a valley.

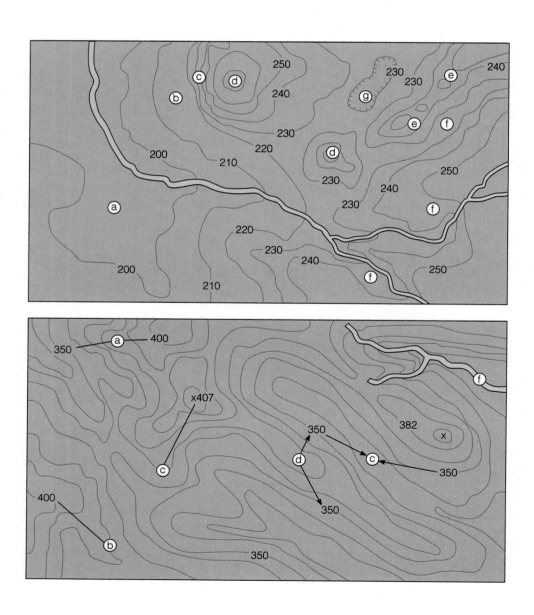

Figure D2

Appendix E

To use the star charts in this book, select the appropriate chart for the date and time. Hold it overhead, and turn it until the direction at the bottom of the chart is the same as the direction you are facing.

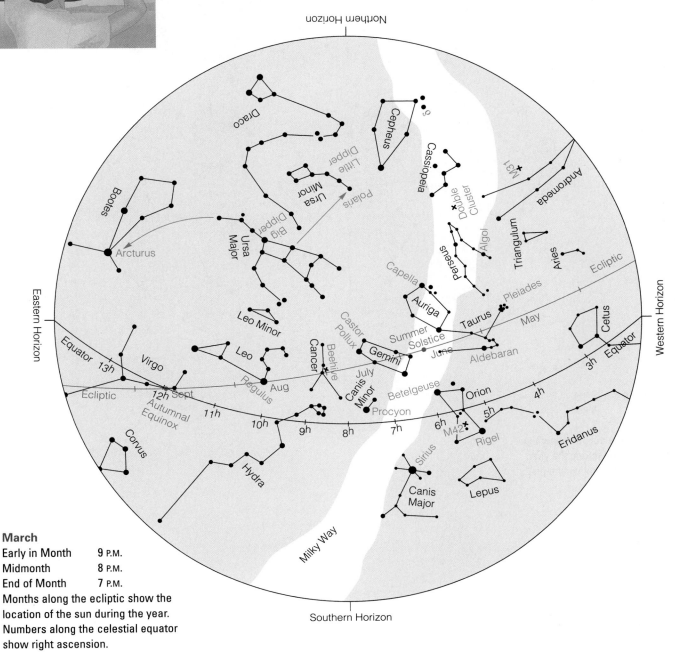

March

Early in Month   9 P.M.

Midmonth   8 P.M.

End of Month   7 P.M.

Months along the ecliptic show the location of the sun during the year. Numbers along the celestial equator show right ascension.

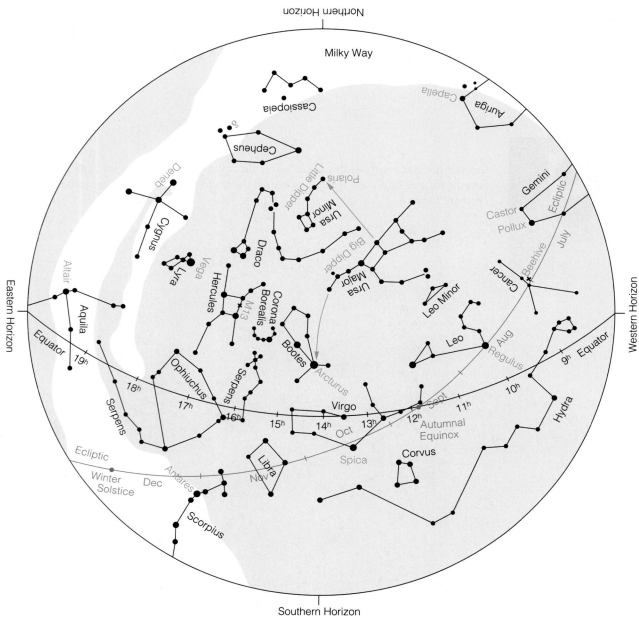

June

Early in Month	9 P.M.
Midmonth	8 P.M.
End of Month	7 P.M.

Months along the ecliptic show the
location of the sun during the year.
Numbers along the celestial equator
show right ascension.

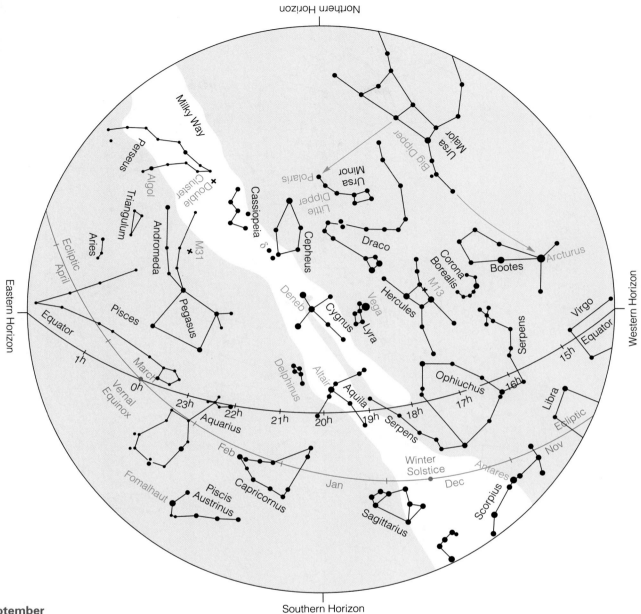

September

Early in Month	9 P.M.
Midmonth	8 P.M.
End of Month	7 P.M.

Months along the ecliptic show the
location of the sun during the year.
Numbers along the celestial equator
show right ascension.

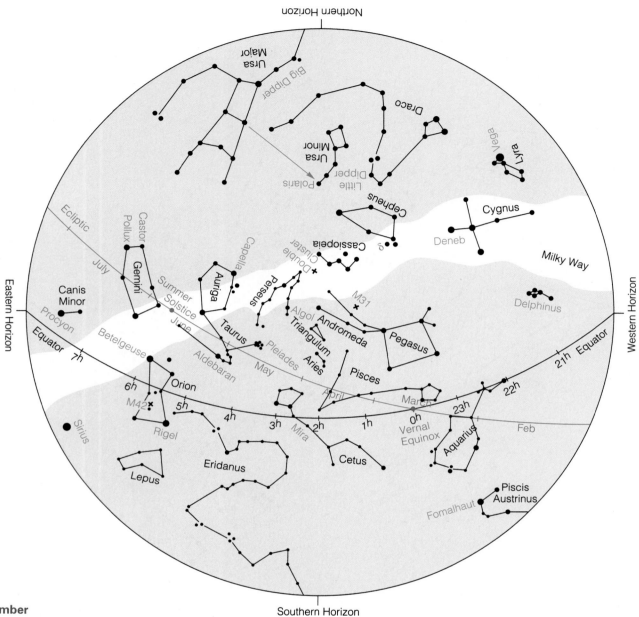

December

Early in Month	9 P.M.
Midmonth	8 P.M.
End of Month	7 P.M.

Months along the ecliptic show the location of the sun during the year. Numbers along the celestial equator show right ascension.

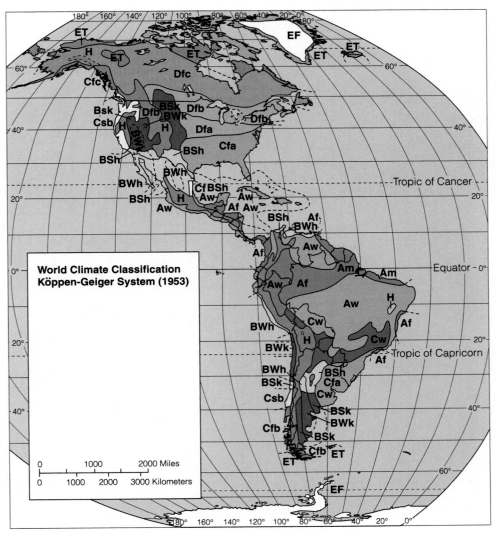

World Climate Classification Köppen-Geiger System (1953)

0   1000   2000 Miles
0 1000 2000 3000 Kilometers

Af: Tropical Rainforest climate
Am: Tropical Monsoon climate
Aw: Tropical Savanna climate
BSh: Tropical Steppe climate
BSk: Mid-latitude Steppe climate
BWh: Tropical Desert climate
Bwk: Mid-latitude Desert climate
Cfa: Humid Subtropical climate

Cfb, Cfc: Marine climate
Csa, Csb: Mediterranean climate
Cwa, Cwb: Subtropical Monsoon climate
Dfa, Dwa: Humid Continental—Warm Summer climate
Dfb, Dwb: Humid Continental—Cool Summer climate
Dfc, Dwc, Dfd, Dwd: Taiga (or Sub-arctic) climate
EF: Polar (or Icecap) climate
ET: Tundra climate

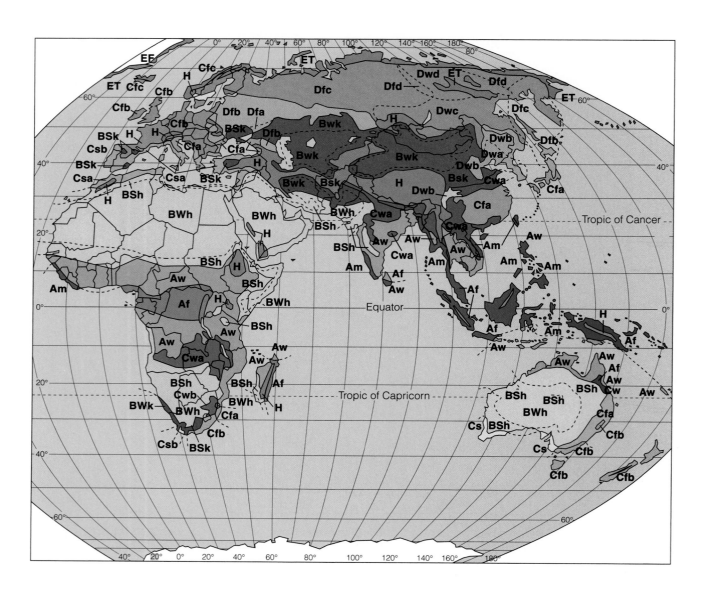

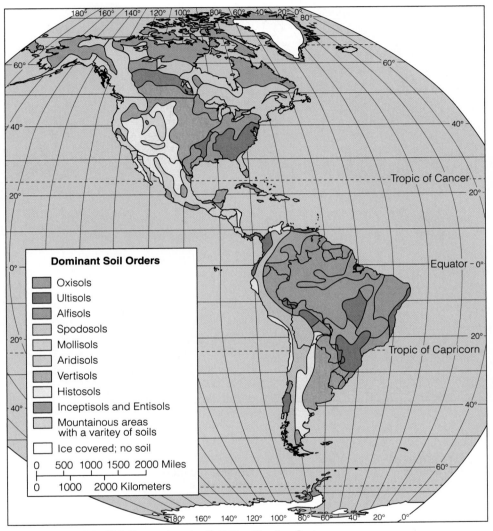

Dominant Soil Orders

- Oxisols
- Ultisols
- Alfisols
- Spodosols
- Mollisols
- Aridisols
- Vertisols
- Histosols
- Inceptisols and Entisols
- Mountainous areas with a varitey of soils
- Ice covered; no soil

0   500 1000 1500 2000 Miles

0   1000   2000 Kilometers

Global distribution of the soil orders of the U.S. Comprehensive Soil Classification System (the Soil Taxonomy).

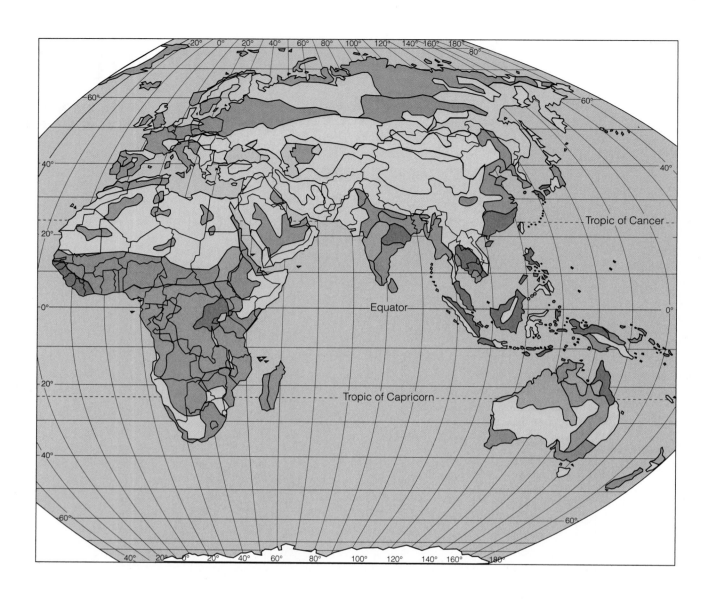

ORDER	EQUIVALENT SYSTEMS: 1938 GENETIC SYSTEM (CANADIAN SYSTEM)	GENERAL LOCATION AND CLIMATE	WORLD LAND AREA (%)	U.S. LAND AREA (%)
Oxisols	Latosols, lateritic soils	Warm, humid tropical regions	9.2	0.02
Ultisols	Red-yellow podzolic	Humid subtropical forest regions	8.5	12.9
Alfisols	Gray-brown podzolic, degraded chernozem (luvisol)	Humid temperate forest regions	14.7	13.4
Spodosols	Podzol, brown podzolic (podzol)	Cool, humid regions; needleleaf forest cover	5.4	5.1
Mollisols	Chernozem, chestnut (chernozemic)	Subhumid and semiarid temperate grasslands	9.0	24.6
Aridosols	Red desert soils, gray desert soils, sierozems	Tropical, subtropical, and temperate deserts	19.2	11.5
Vertisols	Grumusols, tropical black clays	Restricted semiarid grassland regions	2.1	1.0
Histosols	Peat, muck, bog (organic)	Poorly drained lowlands and depressions	0.8	0.5
Entisols	Azonal soils, tundra soils	Recently formed or disturbed soils in all climates	12.5	7.9
Inceptisols	Subarctic brown forest soils, lithosols (brunisol, cryosol, gleysol wet)	Immature soils mostly within mountain areas, floodplains, or tundra regions	15.8	18.2
Andisols		Active volcanic regions	<1.0	<1.0

Glossary

A

Absolute age: The numerical age giving the time of formation of geologic material, usually obtained by radiometric dating.

Absolute age dating: A method, such as radiometric dating, tree ring counting, and varve counting, that provides a numerical age for geologic materials (e.g. rocks, minerals, fossils).

Absorption spectrum: The spectrum emitted from a radiating body in which dark lines appear on a spectrum of visible light that represent the absorption of photons at particular wavelengths by the gas through which they travelled. (See *emission spectrum.*)

Abyssal plain: The relatively flat, extensive and deep region of the ocean floor located at the base of the continental rise having a thin layer of sediment overlying basaltic crust. The average depth of the abyssal plain is 4 to 5 km (about 3 mi) below sea level.

Accreted terrane: A definable package of crustal rocks (e.g. island arcs or microcontinents) that collided with a continental landmass. An accreted terrane of unknown or distant origin is known as *suspect* or *exotic terrane.*

Accretion (planetary): The process by which cooling nebular dust and gas may adhere to form planetesimals that coalesce together to form larger bodies, protoplanets, which may gain more material and become planets. (See *homogeneous accretion* and *inhomogeneous accretion.*)

Accretionary wedge: The material accumulated from slivers taken off the subducted slab during the subduction process. The material is collected in a wedge-shaped manner near the trench and typically is represented by a chaotic melange of rock that has undergone low temperature and high pressure metamorphism.

Active continental margin: The continental edge of a tectonic plate that undergoes subduction, collision or transcurrent movement, also termed Pacific-stage or *active margin.* (See *passive continental margin.*)

Aerobic: Pertains to an environment containing free oxygen (atmospheric or in water) or to the organisms (e.g. eukaryotes) that live in such an environment. (See *anaerobic.*)

Air mass: A relatively homogenous body of moving air, consisting of a mixture of atmospheric gases and minute particles of solids and liquids.

Air pressure: *See* atmospheric pressure.

Albedo: The ability of a surface to reflect rather than absorb incoming radiation.

Alleghanian orogenic belt: The Carboniferous to Permian orogenic event during which the Appalachian orogenic belt of eastern North America was formed from the closure of Iapetus Ocean during the formation of Pangea.

Alpha decay: One of the processes by which a radioactive atomic nucleus forms a more stable product: the emission of an alpha particle (composed of two neutrons and two protons) from the nucleus. (See *beta decay, electron capture.*)

Aluminum-26: A radioactive isotope of aluminum that decays to magnesium-26, that has a short half-life and is believed to have been a major contributor of radiogenic heat during the early formation of the Earth.

Amniotic egg: An egg in which the embryo, which contains yolk and waste sacs, develops in a liquid-filled cavity. The amniotic egg occurs in reptiles, birds, and egg-laying mammals.

Amphibians: Vertebrate animals that have gills and live in water in their early stages of life, and in later stages develop lungs and live on land.

Anaerobic: Pertains to an environment in which free oxygen is not available, or to the organisms (e.g. early prokaryotes) that live in such oxygen-deficient conditions. (See *aerobic.*)

Andean-type subduction: A shallowly dipping, subducted oceanic plate beneath continental crust forming a mountain chain with no associated back-arc spreading. This is exemplified by the Andean mountains situated along the west coast of South America. (See *Marianas-type subduction.*)

Andesite: A dark gray, fine-grained volcanic rock of intermediate composition that is common in mountain belts like the Andes. It is the volcanic equivalent of diorite and is similar to dacite but with more plagioclase feldspar and mafic minerals. It is common in continental and island arc environments.

Angle of repose: The maximum slope at which loose sediment can be stable on the Earth's surface without moving due to gravity. It is usually between 25° and 40°.

Angular unconformity: An unconformity, or erosional surface, which represents a significant gap in the geologic record and below which older beds dip at a different angle (usually steeper) than the overlying younger strata.

Anorthosite: An intrusive rock that contains predominantly one type of mineral: plagioclase feldspar. It formed predominantly in the Middle Proterozoic within the Precambrian shield; its origin is enigmatic.

Anthropogenic: Environmental changes that have been influenced by the activity of mankind.

Anticline: A folding of layered rocks which resembles an arch, such that each layer is highest near the center of the anticline and lower farther from the center. Sometimes an anticline forms a structural trap, which is important in forming a petroleum reservoir. (See *syncline.*)

Anticyclone: The circular movement of an air mass around a high pressure zone. This movement is clockwise in the northern hemisphere and counter-clockwise in the southern hemisphere. (See *cyclone.*)

Appalachian orogenic belt: An orogenic belt along the eastern margin of the North American continent extending from Georgia to Newfoundland, Canada. It formed from the closure of Iapetus Ocean which involved numerous mountain building events that culminated with the joining of Gondwana and Laurasia.

Aquifer: A layer of porous and permeable rock or unconsolidated sediment in the subsurface that permits the storage and movement of groundwater. An aquifer sandwiched between impermeable layers is called a *confined aquifer.* An *unconfined aquifer* is one where the upper boundary of the aquifer is the water table and the aquifer is open to groundwater infiltration from above. (See *artesian flow* and *potentiometric surface.*)

Arc-trench gap: The zone between the volcanic arc and the ocean trench, the distance of which indicates the angle or slope of the descending slab. A steeply dipping slab will result in a shorter distance between the volcanic arc and the trench, a more shallowly dipping slab will result in a larger gap.

Archean Eon: The geologic time which began about 4.0 billion years ago, and ended 2.5 billion years ago. It was during this eon that prokaryote cells (the first signs of life) and cratonic nuclei developed.

Artesian flow: The natural flow of water above an aquifer because the underlying aquifer is under hydraulic pressure. Water reaching the surface in this way forms a *natural spring*. If a well intersects below the potentiometric surface of the aquifer, the water will rise unaided into the well. (See *aquifer* and *potentiometric surface*.)

Aseismic ridge: A long, linear ridge or broad, plateau-like feature that rises as much as 3 km (2 mi) above the surrounding sea floor and that lacks seismic activity. These ridges are probably caused by the motion of oceanic crust above a mantle plume.

Assimilation: A process in which the magma in a magma chamber reacts with the pre-existing rock with which it comes into contact.

Asteroid: A small planetary body made up of rocky material that orbits the Sun. An asteroid belt lies between the orbit of Mars and Jupiter, and another lies beyond Pluto.

Asthenosphere: The plastic or "soft" region of iron- and magnesium-rich silicate rocks, lying below the lithosphere and above the lower mantle. A zone of low seismic velocity, approximately 100 km (62 mi) below the surface, defines the asthenosphere's upper boundary. The lower boundary is at about 700 km (about 430 mi) in depth. Numerous seismic velocity discontinuities occur between 400 and 700 km (250 to 430 mi), however, so that this part of the asthenosphere is referred to as the *transition zone*.

Astronomical unit (AU): One astronomical unit is equivalent to the average distance between the Earth and the Sun, 150 million kilometers (93×10^6 mi).

Atlantic-stage: The time in the plate tectonic cycle that is dominated by rift-related processes, where continental fragments drift apart and develop continental margins. (See *Pacific stage*.)

Atmosphere: The Earth's gaseous envelope and one of its four main surface reservoirs. It is made up of 78% nitrogen, 21% oxygen, 0.9% argon, 0.03% carbon dioxide and trace amounts of other gases including water vapor and ozone.

Atmospheric circulation: Refers to the horizontal and vertical movement of air in the atmosphere, driven by temperature and pressure contrasts. Persistent surface air circulation patterns over ocean bodies include doldrums, easterlies, horse latitudes, trade winds, and westerlies.

Atmospheric pressure: The force per unit area exerted by the weight of a column of air directly above it. Commonly called *air pressure*.

Atom: The fundamental unit of every chemical element. They are made up of a central nucleus containing positively charged protons and neutrons (neutral), and a cloud of orbiting negatively charged electrons; atoms are typically about 0.0000000001 meters (10^{-10} m) across. An atom can be characterized by its *atomic number* (the number of protons in an atom's nucleus; this determines the element to which the atom belongs) and its *atomic mass number* (the sum of protons and neutrons in an atom's nucleus).

Aulacogen: Continental rift system where spreading of the crust ceases, leaving a valley system that may retain seismic activity. Commonly, such areas are the site of continental drainage and sedimentation. Also known as a failed rift. (see *failed rift*).

Aurora: The display of colored light and sounds usually observed in the skies of polar regions due to the interplay of solar emissions and the Earth's electromagnetic field.

Autotrophic: Organisms, such as photosynthesizing bacteria and plants, that synthesize organic nutrients from inorganic raw materials.

B

Back-arc basin: A basin opened by extensional forces behind an island arc and where sea-floor spreading locally occurs.

Baltica: A Late Proterozoic continent (present day Europe) involved in the Paleozoic collision with Laurentia, forming Laurasia by the early Silurian.

Banded Iron Formation (BIF): An iron-rich sediment formed by the oxidation of dissolved iron in seawater and subsequent precipitation of iron-oxide minerals. It formed predominantly during the Archean in seawater.

Basalt: A dark volcanic mafic rock that formed by partial melting of mantle rocks and is common on the ocean floor, in continental rift settings, and above hotspots.

Base level: The lowest level to which a stream will erode, usually at the level of a large body of surface water. The *ultimate base level* for a stream is sea level. A temporary erosional level which a stream can not pass (e.g. a lake) is called a *local base level*.

Base map: A map generally containing topographic information, upon which geologic information such as rock type, rock distribution and structural data are superimposed.

Bathymetry: Measurement of the depth to the ocean floor and of ocean floor topography.

Beach: A narrow coastline area of unconsolidated sediment extending landward from low tide to a cliff face or where permanent vegetation begins.

Bed: An individual sedimentary layer over 1 cm thick that is distinguished from beds above and below by differences in composition, grain size, color or a combination of these. The upper and lower surfaces of the bed that separate it from the beds above and below are referred to as *bedding planes*. (See *strike* and *dip*.)

Beta decay: One process by which a radioactive nucleus of an atom forms a more stable product. A neutron in the nucleus breaks down, forming a proton and ejecting an electron (i.e. beta particle). Since the atom now has an additional proton, it has a higher atomic number than before and is therefore a different element. The atomic mass, however, remains unchanged. (See *alpha decay* and *electron capture*.)

Big Bang theory: A theory about the origin of the universe: that 12 to 15 billion years ago the universe existed as a single point of superdense, hot gas which exploded outward (the *big bang*) and began to expand and cool. This theory is supported by the presence of primordial background radiation and by the speed and direction that galaxies are travelling.

Big crunch: A model which proposes that the expanding universe will eventually reverse its direction and collapse upon itself.

Biochemical sedimentary rock: A nonclastic sedimentary rock, formed by chemical processes and activities of living organisms from the fossil remains of the organisms themselves (e.g. coal) or from the remains of their skeletons (e.g. chert, limestone).

Biogeochemical cycle: The cyclical behavior of chemical elements (e.g. carbon, nitrogen and phosphorous) that migrate between the biosphere, atmosphere, hydrosphere, and solid Earth.

Biomass energy: Surface organic matter such as trees and solid waste that is processed or combusted to create an energy supply.

Biosphere: The realm of life and one of Earth's four main surface reservoirs. It consists of all living animals and plants, and those in the process of decay.

Black smokers: See *hydrothermal vents*.

Blueschist facies: Metamorphic rocks formed at the very high pressure and relatively low temperature conditions under which the blue amphibole *glaucophane* is stable. The Blueschist facies is characteristic of the shallower parts of subduction zones.

Bonding (chemical): The process that enables atoms or ions to join to each other. Types of bonding include *covalent, ionic,* and *metallic bonds* and *van der Waals' force.*

Bowen's Reaction Series: A concept describing the sequence of reactions in a magma between the minerals which crystallize out and the melt which remains. The reactions cause mafic magmas to evolve toward more silica-rich compositions. The mineral series, whereby the mineral's composition (plagioclase) continuously changes in chemistry as it interacts with changing chemistry of the magma process, is termed a *continuous reaction series.* A *discontinuous reaction series* occurs whereby the crystallization of early-formed minerals reacts with the evolving magma to produce a different mineral phase (e.g. olivine to pyroxene).

Btu (British thermal unit): A measurement given in terms of the quantity of heat required to raise the temperature of a pound of water by 1°F.

C

Caldera: A crater-like volcanic depression; formed as a result of volcanic eruption, by collapse after the magma is expelled, or by explosion.

Calorie: A unit of measurement of the heat required to raise the temperature of 1 gram of water by 1°C.

Capillary fringe: The area immediately above the water table where the water molecules move upward by a few centimeters because of the attractive forces between liquid and solid material.

Carbon cycle: The movement of carbon through Earth's reservoirs from its origin through volcanic and solar sources, its exchange between living organisms, the hydrosphere and atmosphere, and involvement upon death, decay and burial of the organisms into the solid earth to be exhumed and follow the cycle again.

Carbon-14: The radiogenic product of cosmic bombardment of nitrogen-14 within Earth's atmosphere. Its incorporation into living organisms and their subsequent death and burial yield information on the rate of cosmic radiation during prehistoric times. Also, the decay of carbon-14 is useful as an age-dating method applied to buried organic material; useful ages can be determined to about 100,000 years ago.

Carbonate rock: A sedimentary rock of biochemical origin (e.g. limestone), or metamorphic origin (e.g. marble) that is predominantly made up of carbonate minerals such as calcite ($CaCO_3$) or dolomite ($CaMg[CO_3]_2$).

Cells: Discrete, fundamental units that contain the properties of life for an organism. These properties include the ability to reproduce and self-regulate.

Cenozoic: The most recent era of geologic time, dominated by mammals and spanning 66 million years to the present. Known as "Recent Life."

Chemical formula: A list of ions in a mineral as symbols according to the relative proportion in which they occur in the mineral structure. For example, quartz contains twice as many ions of oxygen (O) as silicon (Si) and so has the chemical formula SiO_2.

Chemical (sedimentary) rock: An inorganic sedimentary rock produced by precipitation, most commonly when a body of salt water evaporates. Chemical rocks tend to have a nonclastic texture and include rock salt, gypsum and anhydrite.

Chemical weathering: The breakdown of crustal material on or near the Earth's surface due to chemical reactions and dissolution of the parent material by interaction with water.

Chemosynthesis: Derivation of energy for life forms from chemical reactions involving submarine organisms near ocean vents. (See *hydrothermal vents.*)

Chlorofluorocarbons (CFC's): A gaseous molecule produced by industry that destroys ozone in the stratosphere.

Circum-Pacific belt: See *ring of fire.*

Clastic: A sedimentary rock texture comprising broken fragments of pre-existing rocks or minerals.

Cleavage: A mineral property describing the tendency of many minerals to break along preferred directions that reflect systematic weaknesses in their internal structure.

Climate: Description of regional weather patterns established over a period of time. (See *weather.*)

Clouds: A collection of water vapor droplets or ice crystals that commonly form in the troposphere between 2 to 14 km above the surface due to cooling and condensation in rising air masses. Basic cloud formations include: cirrus, stratus, cumulus and anvil.

Coal: A black, combustible biochemical sedimentary rock formed from highly compressed plant remains accumulated within an oxygen-deficient environment. Progressive burial initially forms peat followed by lignite, subbituminous coal and bituminous coal. Anthracite coal is associated with areas that have undergone regional metamorphism.

Co-evolution hypothesis: A model that looks at the progression of life and its surrounding environment as evolving simultaneously in which life has an influence on its environment, and vice versa.

Cold front: A front where a cold air mass advances, wedging itself under a warm air mass. This often produces cumulus-type clouds, giving rise to thunderstorms. An advancing cold air mass that catches up to a warm front and forces the warm air mass up is called an *occluded front.* (See *fronts, warm front.*)

Comets: Small bodies (less than a few kilometers diameter) composed of a mixture of dust and ice with the majority believed to originate from the Oort Cloud. They orbit the Sun and are noted for their long tails that point away from the Sun and reveal the presence of the solar wind. (See *solar wind.*)

Compression: Stress or pressure applied on a body, such as rocks, which results in volume reduction or shortening of the affected material.

Conduction: The transfer of heat through a solid body from molecule to molecule without mass movement.

Cone of ascension: Near coastal areas, if fresh groundwater is excessively used then underlying, denser salt water moves upward in a cone-shaped manner and contaminates the fresh groundwater.

Cone of depression: A cone shape centered about a well at which groundwater is depleted faster than it can be recharged by the surrounding water table.

Constellation: A named cluster or group of stars that have a fixed pattern when viewed in Earth's night sky. Also refers to a region in the sky, in which all stars located therein are recognized as part of that constellation. The stars in the cluster need not be related to one another.

Continental crust: Predominantly silicate material (average thickness of 35 km) that overlies the mantle and consists of a wide variety of igneous, sedimentary and metamorphic rocks. Due to isostasy, it is thickest in regions of high topography. It extends from the continental landmass beneath the oceans as far as the continental slope. A division into *upper* and *lower* crust is made on the basis of a seismic discontinuity and is believed to be characterized by a crust of granitic composition overlying a high grade metamorphic rock of dioritic composition. Density of 2.7 g/cm³, which is significantly lower than that of oceanic crust (2.9 g/cm³).

Continental drift: The theory that the continents move relative to one another proposed by Alfred Wegener in 1912. (See *Pangea.*)

Continental platform: The relatively flat portion of the continent having an elevation range between 1000 m and sea level.

Continental redbeds: Red clastic sediments that accumulate on the continent; red coloration is due to the oxidation of iron. First appeared about 1.8 billion years ago, indicating the presence of oxygen in the atmosphere.

Continental rise: Located at the base of the continental slope, where thick piles of sediment are deposited that are derived from the continental shelf at the outer edge of the continental landmass.

Convection: The mass movement of a body as a result of density differences producing whole-scale mixing. The differences may be produced by thermal (e.g. Hadley cell) or compositional contrasts (e.g. Earth's outer liquid core).

Convergent (destructive) plate boundaries: The zone where subduction and/or collision occur at the margin of two plates that move towards each other. Mountain belts, volcanic or island arcs, oceanic trenches and seismicity occur along or adjacent to these boundaries. It is referred to as a *continental-continental plate boundary* if two continental lithospheric plates collide (e.g. the collision of India with Asia).

Copernican (heliocentric) system: A model that described the Earth's position in the universe as centered about the Sun. (See *Ptolemaic system.*)

Cordilleran orogen: A mountain belt produced by episodes of tectonic activity along the western margin of North America from Late Permian to Early Cenozoic time; usually divided into four separate phases that have resulted in the westward growth of North American plate through subduction and the accretion of terranes (e.g. the *Sonoma, Nevadan, Sevier,* and *Laramide orogenies*).

Core (Earth): Central interior of the Earth beneath the mantle beginning at a depth of about 2,900 km. It is divided into a liquid outer and a solid inner core and is composed mostly of iron and nickel.

Coriolis effect: The apparent deviation of a body that follows a straight path as it moves above a rotating body. The Earth's rotation beneath moving objects deflects the objects to the right in the northern hemisphere and to the left in the southern hemisphere.

Correlation: The demonstration of the equivalency of rock layers in different areas from their physical properties, fossil content or age of formation.

Cosmic radiation: Detectable energy waves received on Earth from the cosmos that represent radiation from the Big Bang and stellar explosions within the universe.

Crater: A circular depression resulting from the extrusion of volcanic gases, pyroclastic materials, or lava, or as a result of a meteorite impact.

Craton: The stable core of crust in the continental interior established by the end of the Archean consisting of a nucleus of granite-gneiss complexes and greenstone belts surrounded by Proterozoic mountain belts and thick, continental shelf-type sedimentary sequences.

Creep: The solid state flow of material that results in slow movement of soil or rock.

Crust: The relatively thin layer of rigid material that forms the outer rocky surface of the Earth. It may be composed of either buoyant continental material (granitic) or dense, oceanic material (basaltic).

Crystal: The regular, external geometric form of a mineral which reflects the orderly internal arrangement of its atoms.

Crystal settling: The process whereby early-formed crystals in a cooling magma sink to the bottom of the magma chamber if their density is greater than that of the magma.

Crystallization: The process of solidification of a magma involving the growth of minerals as the magma cools.

Curie point: The temperature at which magnetism is retained within a cooling igneous body.

Cyclone: Refers to the movement of an air mass circulating about a low pressure zone. Due to the Coriolis effect, cyclones rotate counterclockwise in the northern hemisphere, and clockwise in the southern hemisphere. (See *anticyclone.*)

D

Delta: A triangular or fan-shaped accumulation of sediment formed at the mouth of a stream when the quantity of sediment the stream supplies is greater than that which can be removed by ocean waves.

Density: The mass of a body divided by its volume. Materials that have a higher density tend to sink beneath less dense material.

Depositional environment: The physical setting (e.g. floodplain, shallow marine or desert) in which sediments are deposited.

Desert: An area that receives minimal precipitation and typically has poorly developed soil and vegetation. Due to Hadley cell circulation, major deserts are found at latitudes 30° north and south. Deserts are also located in rain shadow regions.

Detrital sedimentary rock: Rocks (e.g. shale, siltstone, sandstone, conglomerate) made up of transported fragments of rock and mineral (detritus) produced by mechanical erosion of pre-existing rocks.

Dike: A sheet-like body of igneous rock that is *discordant* with (cuts through) the layering of the surrounding rock. See *rule of cross-cutting relationships*

Dip: A measure of the maximum inclination of a plane relative to the horizontal. It is used with the strike of a bed to define the orientation of a rock layer. (See *strike.*)

Discharge: The outflow of groundwater onto the Earth's surface producing a *spring*. This is also the term used for water flowing out of the mouth of a stream into the ocean.

Disconformity: A boundary that separates parallel sedimentary layers formed at distinctly different times. (See *unconformity.*)

Discontinuity (seismic): A marked change in the velocity of seismic waves within narrow zones in the Earth's interior, indicating a significant change in the composition of Earth materials or their properties. (See *asthenosphere* and *low velocity zone.*)

Divergent plate boundary: The boundary between two plates where oceanic crust is created by the upwelling of mantle material as the plates move away from each other. Basalt volcanism, seafloor spreading and shallow seismicity characterize diverging plate boundaries. Also known as a *constructive plate boundary*.

Divide: The separation of adjacent drainage basins by a topographical barrier.

DNA (deoxyribonucleic acid): The substance consisting of complex, helical molecules containing the genetic material of all organisms except bacteria.

Doldrums: Persistent but light surface winds near the equator associated with the a low pressure region known as the intertropical convergence zone.

Doppler effect: The apparent change in the wavelength of radiation or sound waves from a body as it moves away or draws closer to a stationary body, the Doppler effect is used to determine the movement of celestial bodies (e.g. stars, galaxies). A *blue shift* is an apparent shift of wavelength to the blue end of the electromagnetic spectrum; a *red shift* is an apparent shift of wavelength to the red end.

Downwelling: Downward movement of surface ocean waters due to density contrasts with underlying waters. Zones in which this occur are located in polar, subpolar regions, or along coastal waters where surface waters become denser than the underlying waters promoting the downward movement of ocean water. In polar and subpolar regions, it is the source of oceanic deep water circulation.

Drainage basin, outlet: See *stream*.

E

Earth system science: The study of the interactions between the Earth's surface reservoirs (the solid Earth, hydrosphere, atmosphere and biosphere).

Earthquake: Seismic vibrations caused by the sudden release of energy within the Earth. Usually rupture occurs with a major earthquake followed by less intense movements called *aftershocks*. The majority of earthquakes occur adjacent to plate boundaries along the subducted slab referred to as the *Benioff zone*.

Earthquake focus (hypocenter): The zone within the Earth's interior where an earthquake originates. The point vertically above the focus is called the *epicenter*.

Earth's tilt: The angle that the Earth's axis makes from a vertical position with respect to its orbit about the Sun. Presently 23.5°, but it varies within 1.5 degrees over a period of 40,000 years.

Eclipse: The positioning of the Moon and Earth with the Sun such that the light of the Sun is blocked. *Solar eclipses* occur where the Moon's position is between the Sun and the Earth. If a halo of light remains as a ring around the Moon, it is referred to as an *annular eclipse*. A *lunar eclipse* occurs when the alignments are such that the Earth is situated between the other two bodies and its shadow is cast upon the Moon.

Ediacara fauna: A collective name for all Late Proterozoic (670 to 550 m.y.a.) faunas containing soft-bodied metazoan animal fossils similar to those of the Ediacara Hills of Australia.

Ekman spiral: The changing direction of ocean surface layer motion with depth, tracing out a spiral path due to the interplay between prevailing wind direction, frictional forces, and the Coriolis effect. This results in an overall direction of water movement at a high angle to the prevailing wind.

El Niño: Spanish for "the boy", this refers to the periodic suppression of upwelling currents off the western coast of equatorial South America and the buildup of warm surface waters in the equatorial Pacific Ocean as a result of the waning trade winds.

Elastic rebound: The tendency of solid material that is stressed to return to its original shape when the stressing pressure is released or lessened.

Electromagnetic spectrum: A range of wavelengths of radiated energy from a hot body (e.g. the Sun) that may include radio waves, visible light and gamma rays.

Electron capture: One process by which a radioactive atomic nucleus forms a more stable product. The captured electron (beta particle) combines with a proton to form a neutron (e.g. Potassium-40 to Argon-40). This reduces the amount of protons, and hence the atomic number of the atom by one unit. (See *Alpha decay* and *Beta decay*.)

Element: There are 92 naturally occurring elements that are identified by their *atomic number* (i.e. the number of protons in their nucleus). Each element can be identified by their physical and chemical properties.

Emission spectrum: A collection of bright lines of colored light on a dark spectral background. The lines are characteristic of atoms in a gas which absorbed and then emitted photons from a radiating body and represent the wavelength characteristic of the elements that comprise the gas. (See *absorption spectrum*.)

Energy consumer: Organisms, like carnivores and plant eaters (herbivores), that use more energy from their environment than they replace.

Energy producers: Organisms, like plants, that interact with their environment and store energy as a byproduct of that interaction. (See *fossil fuels*.)

Energy resource: A supply of energy from a naturally occurring source (e.g. petroleum, coal, hydroelectric, tidal and wind power).

Equinoxes: The time when the Sun shines directly on the Earth's equator. In the modern world, they occur twice annually, March 20 and September 22. Due to precession, these dates progressively shift with time. (See *Milankovitch cycles*.)

Erosion: The breakdown of Earth's surface material and subsequent transportation by agents of erosion (i.e. water, wind or ice).

Escape velocity: The velocity required for a gas to escape the gravitational attraction of a planet.

Eukaryote: An aerobic, single-celled organism having some organized cell structure; capable of asexual and sexual reproduction; existed since the Middle Proterozoic.

Eutrophication: The natural or artificial process by which a body of water becomes enriched in plant nutrients.

Evaporite A mineral (such as halite or gypsum) or chemical sedimentary rock (such as rock salt and gypsum) produced by the evaporation of salt water.

Evolution: An explanation that accounts for changes through time. Within the biosphere organisms evolve from older life forms.

Extrusive rock: A volcanic rock formed on the Earth's surface (e.g. basalt, pumice, rhyolite) from solidifying lava.

F

Facies: See *Metamorphic facies* and *Sedimentary facies*

Failed rift: During continental breakup, the inactive arm of a triple point rift where crustal spreading no longer operates. Also known as an aulacogen. (See *aulacogen*.)

Fault: A fracture in the crust or upper mantle along which measurable movement has occurred. Rock containing angular fragments resulting from the crushing of pre-existing rock during movement along a fault is known as *fault breccia*. Slow, continuous movement along a fault allowing stress to be dissipated is known as *fault creep*. A fault that is not exposed on the surface of the Earth is known as a *blind fault*.

Fault movement: Described by the relative motion of the blocks on either side of the fault. For a dipping fault the area of rock that overlies the fault plane is known as a *hanging wall*, and the block that lies beneath, a *footwall*. (See *normal*, *reverse*, and *thrust faults*.)

Feedback: The amplification of a trend, either in a positive or negative way, by the natural response of an environment to the initial change.

Felsic: A descriptive term applied to all light-colored igneous rocks, such as granite and rhyolite, that are rich in feldspar and silica. The composition of such rocks and their corresponding magma would be relatively rich in silica, sodium, potassium, and aluminum, but poor in calcium, iron, and magnesium. (See *mafic* and *intermediate*.)

Ferromagnesian silicate: An iron- and/or magnesium-bearing silicate mineral (e.g. olivine or pyroxene) that typically occurs in mafic rocks; commonly darker and denser than nonferromagnesian silicates (feldspar and quartz). (See *mafic*.)

Fissures: Fractures in the Earth's crust that reach the surface; often emitting magma or steam (e.g. *fumaroles*). (See *hydrothermal vent*.)

Flood basalt: A thick layer of basaltic lavas formed by voluminous melting in the mantle that accumulate on continental landmasses. Also known as plateau basalts.

Floodplain: The flat area surrounding the banks of a stream that is periodically flooded and where sediment accumulate.

Fold: The deformation by buckling or bending of layered rocks due to applied stress or compression. Characterized by an axial plane, (an imaginary plane that divides a fold into two equal parts), which is perpendicular to the direction of stress forces, and by a fold axis.

Foliation: A metamorphic texture seen in slate, schist and gneiss, in which the segregation of minerals define a closely spaced planar layering.

Food chain: The movement of energy and nutrients when they enter the biosphere from producers (plants), to consumers (plant-eaters and carnivores).

Foreland basin: The trough-like structure landward of a mountain belt formed in front of thrust sheets; collects detritus shed from the mountain belt.

Formation: A sequence of igneous, sedimentary or metamorphic rocks that can be mapped throughout a region in which the rocks commonly share a similar origin.

Fossil: The markings or solid body part of prehistoric animals or plants preserved in sedimentary rocks. If the preserved imprint is a track, trail or burrow, it is called a *trace fossil*.

Fossil fuel: An energy source made up of the remains of buried organic material (e.g. natural gas, oil, and coal). (See *energy producers*.)

Fractionation: The progressive change in the composition of a magma during crystallization as a result of processes such as crystal settling.

Fracture: Any break in a mineral other than along the cleavage, or any crack in a rock (such as a joint or fault) caused by mechanical failure along which no appreciable movement has occurred. (See *fault*.)

Fraunhofer line: Absorption lines superimposed on the solar spectrum that are characteristic of the composition of atoms within the Sun. The lines represent absorption of photons with specific wavelengths within the solar atmosphere. (See *absorption spectrum*.)

Frictional resistance (drag): A force applied to a body that is opposite to the direction of motion of that body.

Front (weather): The zone that represents the collision between contrasting air masses, usually between warm and cold air masses. (See *cold front* and *warm front*.)

Frost action (heaving or wedging): The breakdown of surface rocks by repeated freezing and thawing of water in fractures.

G

Gabbro: A coarse-grained plutonic igneous rock of mafic composition containing calcium-rich plagioclase and pyroxene.

Gaia Hypothesis: The proposal that Earth has self-regulating mechanisms (*positive* and *negative feedbacks*) that counteract any changes that threaten its equilibrium.

Galaxy: A cluster of stars, dust and gas that are held together by mutual gravitational attraction.

Geologic time scale: A chronological chart divided into geologically significant units having the oldest part of geologic time (4.6 billion years ago) at the bottom, and progressively younger designations above. Geologic time may be subdivided into units of time, decreasing in size from eon to era, period and epoch.

Geology: The science that studies Earth by looking at its composition and the processes past and present that shaped it, both on the surface (including the atmosphere and hydrosphere) and within its interior.

Geosphere: Encompasses the region and all processes that affect solid Earth (both on the surface and in the interior) and includes also the hydrosphere and atmosphere.

Geothermal energy: A natural energy source obtained from extracting heated subsurface water; often associated with areas of recent volcanism.

Geysers: The cyclical and explosive eruption of superheated water and gases from hot springs.

Glacier: A long-lasting body of ice produced by the accumulation, compaction and recrystallization of snow on land, showing evidence of motion. Depending on the size and location of a glacier, they may classify it as either a *mountain glacier* or an *ice sheet*.

Gondwana: A Late Proterozoic continent formed by the amalgamation of microcontinents in the southern hemisphere that produced the Pan-African mountain belts located in present day South America and Africa. Gondwana began fragmenting during the Mesozoic.

Granite: A coarse-grained plutonic igneous rock of felsic composition, rich in quartz and potassium feldspar.

Granite-gneiss complex: Part of the Late Archean cratonic nuclei; associated with greenstone belts. They consist of a region of high-grade metamorphic gneiss intruded by bodies of granite.

Gravitational collapse: The inward movement of material within a large body (e.g. a solar nebula, or dying star), due to the force of gravity.

Greenhouse effect: The warming of the atmosphere linked to increase concentrations of greenhouse gases (e.g. carbon dioxide, water vapor), from the trapping of solar heat.

Greenhouse gases: Atmospheric gases, both artificially and naturally introduced (e.g. carbon dioxide, chlorofluorocarbons, methane, ozone, and water vapor), that trap incoming and outgoing solar radiation within the atmosphere.

Greenstone belt: A Late Archean or Proterozoic sequence, typically consisting of submarine ultramafic (at the base) through felsic volcanic rocks overlain by marine sedimentary rocks. These belts are located within stable craton.

Groundwater: Subsurface water that resides within pores or fractures in rocks or unconsolidated sediment.

Gyres: Large-scale, circulating surface water currents observed in ocean basins as a result of interaction between the surface layer, persistent air circulation patterns and the Coriolis effect. The pattern of movement spans the width of the ocean basins and in the northern hemisphere they move clockwise and in the southern hemisphere counterclockwise.

H

Hadean Eon: The geologic time spanning the formation of the Earth (4.6 b.y.a.), to the oldest known rocks (3.96 b.y.a.). Vertical processes such as the formation of the Earth's layers are believed to have dominated over horizontal plate-driven processes. The development of primitive oceans (*proto-oceans*) and crust (*proto-crust*) probably occurred during this time.

Hadley cell: The atmospheric convection cells located between the equator and the latitudes 30° north and 30° south related to the global transfer of heat from the equator to the poles.

Half-life: A measurement of the time it takes for one-half of a radioactive parent atom to decay to its stable daughter product. The radioactive isotope has a non-linear decay referred to as an exponential decay.

Hard water: A term used to describe groundwater that has a high concentration of dissolved ions such as calcium, iron and magnesium.

Heat: The total energy of motion within a given body. The amount of heat is related to the size of the body. (See *temperature*.)

Heat capacity: The quantity of energy required to raise or lower the temperature of a substance by 1°C.

Heat flow: A measure of the amount of heat that flows from the Earth's heated interior to its surface. The rate of change in temperature is known as the *geothermal gradient*, where temperature increases with depth in the Earth's interior.

Heat of formation: During the early stages of planet formation, the heat produced from accretion of solid particles during rapid growth.

High pressure zone (atmospheric): A region in the troposphere created by a descending circulating cold air mass where the direction of motion is influenced by the Coriolis effect.

Homogeneous accretion model: A model that proposes that planetary evolution occurred from growth of coalescing planetesimals with homogeneous chemistry. According to the model, the Earth's compositional layering of the planet occurred later.

Hominid: The family to which humans belong; includes bipedal primates *Australopithecus* and *Homo*.

Homo: The genus to which humans belong: includes *Homo erectus* and *Homo sapiens*.

Hotspot: Volcanic activity produced above a relatively stationary zone of melting believed to have originated deep in the mantle beneath the lithosphere. Due to the motion of the overlying plate, volcanic island chains may form in oceanic crust, whereas continental hotspot activity is characterized by areas of uplift followed by rift-basin structures. (See *mantle plume*.)

Hot springs: The surface discharge of heated or *geothermal water* from fractures or faults. A hot igneous body or radioactive rocks at depth may provide the source of heat for the groundwater or the water may have risen from great depths.

Hubble constant: A number that gives the rate of expansion of the universe. It was first determined by Edwin Hubble, but is continuously undergoing refinement.

Humus: Dark, decomposed plant or animal material within soil; it provides the nutrient base for much of the life on land.

Hurricane: An intense low pressure region (cyclone), usually originating over warm tropical oceanic water characterized by high winds which causes destruction when it reaches land. The central core is characterized by clear skies and calm weather and is known as the *eye of the hurricane*. Also known as a *tropical cyclone* or a *typhoon*.

Hydrologic cycle: The dynamic and continuous motion of water between the oceans, the atmosphere, and the continents such that seawater balance is maintained and there is no annual change in sea level. Water is exchanged between land and ocean by interchange of evaporated water with the atmosphere that then precipitates its load. The return of this water to the ocean is a result of groundwater discharge and surface runoff. Also, evaporation on land and transpiration by the biosphere to the atmosphere returns moisture to the oceans.

Hydrosphere: The Earth's watery envelope and one of its four main surface reservoirs. It includes the surface waters of oceans, streams and lakes, the water trapped in snow and ice, and all groundwater.

Hydrothermal vent: Fractures in the crustal surface through which heated or hydrothermal fluids travel. *Black smokers* may form from this venting of gases near oceanic ridges resulting in a chimney-like build up of metallic minerals. If cloudiness surrounds the sea floor vents, they are referred to as *white smokers*. (See *chemosynthesis*.)

Hypothesis: A tentative proposal put forward to explain the behavior of a natural phenomenon. It is tested to determine its validity and is less firmly founded than a theory.

Hypsometric curve: Graphical representation of the proportion of Earth's surface elevation above a given elevation.

Iapetus Ocean: A Paleozoic ocean that separated North America from Europe; the Iapetus Ocean began closing when North America, Europe and Africa started converging and it was eliminated when these continents collided during the Late Paleozoic.

Ice Age: A period of time in which glaciers (ice sheets) extended over a significant proportion of continental landmasses. The Ice Age which glaciated North America and Europe from the Pleistocene until 10,000 years ago is called the *Last Ice Age*.

Ice sheet: A very thick and extensive body of ice that flows outward in all directions from a source over land, and is not confined to valleys; usually more than 50,000 km^2 in area.

Igneous rock: Any rock formed by the crystallization of a molten magma.

Inclusions: Fragments of foreign material contained within a rock.

Indenter model: Describes the effect of a continent-continent collision where the boundary of the colliding plate is irregular. This plate, often a more rigid continent, produces major deformation in the other landmass, and in regions not directly under compression, extension occurs along major transcurrent faults.

Index fossils: Fossils which are useful in determining the age of the rock they are found in. They are useful because the fossils are easily identifiable and can be found in many areas but are not found in rocks from many different time periods.

Index mineral: A metamorphic mineral that forms only within a restricted range of temperature and pressure and may be used to define low-, intermediate-, or high-grade metamorphic zones.

Industrial minerals: See *nonmetallic minerals*.

Infiltration: The passage of water from the surface through soil or sediment.

Inhomogeneous accretion model: A model of planet formation which assumes that a compositionally inhomogeneous solar nebula cooled and accreted so that the denser iron-rich material accreted first, forming planetary cores, followed by less dense silicate material that mantled the core, followed by an outer accumulation of still less dense material.

Inner core (of the Earth): The solid core of the Earth whose upper surface is located at a depth of 5100 km (3162 mi) and is made up of pure iron or iron-nickel alloy. Temperature exceeds 5500 degrees Celsius.

Intermediate (igneous composition): A descriptive term applied to igneous rocks or magma, such as diorite and andesite, that have compositions that are intermediate between felsic and mafic compositions.

Intertropical convergence zone: The region of low pressure near the equator where the return flow from the northern and southern hemisphere Hadley cells converge.

Intrusive body: A molten igneous body that was emplaced and cooled within the Earth's crust. If the intrusive body is a relatively large, slowly cooled, coarsely crystalline body, it is referred to as a pluton. Intrusions that have been uplifted to the surface after their formation and which are over 100 km2 in area are called batholiths.

Ion: An atom that has a positive or negative electrical charge. Positively charged ions are called *cations*, negatively charged ions are called *anions*.

Ionosphere: The region of the atmosphere, extending from the upper mesosphere to the thermosphere, that contains particles ionized by interaction with the solar wind and solar radiation.

Iron catastrophe: The segregation of dense iron into the Earth's core in response to gravitational forces. This occurred within 20 million years of the Earth's formation.

Island arc: The arcuate line of volcanic islands formed on an overriding oceanic plate above a subducting slab.

Isobar: A line on a weather map which represents places of equal pressure in the atmosphere.

Isostasy: The balance reached by the Earth's crustal material floating on the denser, more plastic mantle. This balance in response to Earth's topographic variations is referred to as an *isostatic equilibrium*.

Isostatic rebound: An equilibrium that is reached whereby a region of the Earth's crust moves upward in response to erosion of crustal material or melting of an ice sheet.

Isotope: Atoms of the same element which have different numbers of neutrons within the nucleus. All isotopes of an element have the same chemical properties. However, due to differences in their mass they may behave differently in different physical environments.

J

Jet streams: Fast moving, eastward flowing currents of air located along zones of the tropopause which have rapid changes in altitude. Two such air currents are the *polar front jet* and the *subtropical jet*. The presence of jet streams has a significant influence on weather patterns because zones of low or high pressure are generated below them.

Jovian planets: Jupiter, Saturn, Uranus, and Neptune: the four outer planets of the Solar System (excluding Pluto). The Jovian planets are large and have low densities, indicating that they are composed mostly of lightweight gases, such as hydrogen and helium, and frozen compounds, such as ammonia and methane. Although it is located in the outer reaches of the Solar System, Pluto is not considered a Jovian planet, for several reasons.

Juvenile water: Water released from a molten igneous body that has not yet participated in the hydrologic cycle.

K

Karst topography: Concentrated regions of relatively large, circular depressions on the Earth's surface, caused by interaction of groundwater with soluble rocks such as gypsum and limestone. Karst topography is characterized by springs, sinkholes, disappearing streams, and caves, having features such as stalactites and stalagmites.

Kepler's Laws of Planetary Motion: Three laws, proposed by Johannes Kepler in the 17th century, that allow the motion of planets to be predicted: (1) The paths of the planets around the Sun are ellipses. (2) A planet moves about the Sun such that its radius vector sweeps out an equal area in equal times. The motion of the planet varies depending on its position within the ellipse. (3) The square of the time needed to make a revolution around the Sun is proportional to the cube of the planet's average distance from the Sun.

Kimberlite: An intrusive rock, often occurring in the shape of a pipe (called a *diamond pipe*), derived from deep within the mantle. The mineral diamond is often associated with this rock.

Kinetic energy: The energy of a moving object.

L

La Niña: Spanish for "the girl", this refers to the intensification of cooling conditions of oceanic circulation observed off the west coast of South America. (See *El Niño*.)

Latent heat: The energy absorbed or released when a substance changes state without a change in temperature or pressure.

Laterite: A soil formed under tropical conditions where chemical weathering is intense and leaching of soluble minerals is complete.

Laurasia: A Late Paleozoic, northern hemisphere continent which was composed of the present-day continents of North America, Greenland, Europe, and Asia.

Laurentia: The Early Proterozoic (1.8 b.y.a.) continent formed from the collision of numerous Archean cratons and Proterozoic microcontinents. Laurentia was the precursor to the North American continent.

Lava: Magma that reaches (and typically flows on) the Earth's surface. *Lava fountains*, especially common with basaltic lava, may be produced by the rapid expansion of trapped gas bubbles as the molten rock reaches the Earth's surface and is abruptly decompressed.

Law: A scientific edict describing the conclusively proven behavior of a natural phenomenon under given conditions.

Law of Gravity: The force of gravity between two separate objects is proportional to their mass and inversely proportional to the square of the distance between their two centers. Hence an object of large mass has a stronger gravitational attraction than an object of small mass, and increasing the distance between two objects greatly lessens the gravitational attraction they have for each other.

Layered mantle convection: Mass movement of material occurring in convection cells located within the upper and lower mantle.

Life forms: Organisms (including plants and animals) made up of one or more cells which are able to reproduce and self-regulate (viruses are not considered life forms).

Light year: The distance traveled through space by light in one year, approximately 10^{13} km.

Lightning: A bright electrical discharge between a thunderhead and an object of opposing charge, including the ground or another region of itself. Two types of lightning are *fork lightning* and *sheet lightning*.

Lineament: A term referring to the alignment of surface features on the Earth which may indicate the presence of faults or fractures.

Limestone: A chemical or biochemical sedimentary rock consisting largely of calcium carbonate, usually in the form of calcite.

Lithification: The conversion of unconsolidated sediment into solid rock, by such means as cementation or compaction.

Lithosphere: The Earth's rigid outer layer, which includes the crust and part of the upper mantle. It is broken up into *plates* and overlies the more pliable, or plastic, asthenosphere. From *Lithos*, the Greek word for rock.

Little Ice Age: The period between 1450 and 1850 a.d. when temperatures were significantly cooler in Europe, permitting a greater extension of glaciers on the continent. (See *Maunder Minimum*.)

Longshore current: The nearshore current resulting from waves breaking obliquely to the shoreline. The current parallels the shoreline transporting sediment as it moves. Sediments transported by this process accumulate as *longshore drift*.

Low pressure zone (atmospheric): A region of low pressure in the troposphere created by a warm air mass which ascends and circulates due to the influence of the Coriolis effect.

Low velocity zone: The region of the upper mantle, lying approximately 100 to 250 km below the Earth's surface, in which there is a noticeable decrease in velocity for earthquake-generated P and S waves, due to elevated temperatures and a small degree of partial melting. The zone closely parallels the boundary between the lithosphere and asthenosphere.

Lunar day: The period of time that a line of Earth's longitude directly under the Moon takes to regain its position directly under the Moon, after one revolution of the Earth: twenty-four hours and 50 minutes. This takes into account the Moon's movement about the Earth.

M

Mafic: A descriptive term applied to all rocks that are rich in magnesium- and iron-bearing minerals and typically occur in dark-colored igneous rocks, such as gabbro and basalt. These rocks and their molten counterparts are relatively poor in silica and alumina. (See *felsic* and *intermediate*.)

Magma: A body of molten rock in the Earth's interior. Magma may accumulate in a large underground cavern, called a *magma chamber*.

Magma mixing: The mixing of magmas having different compositions in the Earth's interior.

Magmatic segregation: The concentration of minerals through chemical and/or physical processes within a cooling magma chamber.

Magnetic field (of the Earth): The flow of energy from a positively charged pole to a negatively charged pole, due to the convection and rotation of electrically charged liquid in Earth's outer core.

Magnetic reversal: The phenomenon in which the orientation of the north and south magnetic poles is reversed. The reason for this is unknown.

Magnetosphere: The area around the Earth in which the Earth's magnetic field interacts with the solar wind.

Magnetotellurics: A technique by which Earth's subsurface rock and sediment layers are revealed through interpretation of their magnetic and electrical responses.

Magnitude (earthquake): The total amount of energy released by an earthquake at its source.

Mantle: The region below the Earth's crust, from approximately 100 km to 2900 km in depth, comprising dense ultramafic rocks and varying proportions of melted material. Believed to be chemically homogeneous, it is divided into a plastic, or pliable, *upper mantle* (the asthenosphere) and a rigid *lower mantle*, which is between the asthenosphere and the outer core.

Mantle drag: Refers to the mechanism by which the convecting asthenosphere moves the overlying plates.

Mantle plume: A column of magma that originates deep within the Earth's mantle and rises to the surface to form hotspot volcanoes or flood basalts. The column is thought to be more or less stationary relative to the Earth's plates which move above them.

Mantle wedge: A triangularly-shaped zone above the subducting slab where rock melts within the mantle.

Mare (plural *Maria*): Low-lying, dark areas on the Moon's surface, covered with lunar basalt flows.

Marianas-type subduction: The subduction of a cold, steeply dipping, dense oceanic slab beneath a less dense plate, in which subduction is relatively rapid, the subduction zone is steep, and extension occurs behind the arc, forming back-arc basins such as those seen in the western Pacific Ocean near the Marianas Islands. (See *Andean-type subduction*.)

Marine regression: The retreat of a sea, caused by falling sea level or uplift in coastal land. This is documented in the sedimentary record by the seaward migration of coastal depositional environments.

Marine transgression: The landward advance of a sea, caused by rising sea level or subsidence of land. This is documented in the sedimentary record by the shoreward migration of coastal depositional environments.

Marsupial: Pouched mammals such as opossums, kangaroos, and wombats. At present, marsupials are common only in Australia.

Mass extinction: A major global decline in Earth's biological population over a relatively short period of time. The end of the Paleozoic and Mesozoic eras are marked by such events.

Mass wasting: Downslope movement of material under the influence of gravity (e.g. *debris flow, landslide, mudflow*).

Maunder Minimum: The period of time from 1645 to 1715 when few sunspots were observed on the Sun, which resulted in a reduction in the solar wind. This period was within the time period called the Little Ice Age.

Mechanical weathering: The physical breakdown of bedrock at or near the surface. (See *physical weathering*.)

Medieval Optimum: A period of time in the 14th century characterized by a relatively warm climate in Europe.

Mesoscale eddies: Produced from exaggerated loops of major oceanic surface currents which detach from the main current, becoming relatively self-contained *rings* of water that may last for several months.

Mesozoic Era: Meaning "middle life," this is a major geologic division of time, encompassing the domination of reptiles on land and the breakup of the supercontinent Pangea. The era was brought to a close by a mass extinction, which is thought to be related to an extraterrestrial impact.

Metallic mineral: Minerals made up of chemical elements from the metal group that have physical and chemical properties which are characteristic of metals, such as being electrically conductive, soft, and malleable.

Metamorphic facies: An association of metamorphic rocks that were metamorphosed under similar conditions of pressure and temperature (e.g. *eclogite, greenschist*, and *granulite facies*).

Metamorphic grade: The intensity of metamorphism, which increases as the pressures and temperatures of metamorphism increase.

Metamorphic rock: Any rock that has undergone metamorphism (e.g. *marble, gneiss*, and *slate*).

Metamorphism: The process of change in the texture and chemical composition of rocks, caused by heat, pressure and chemically active fluids. The original rocks crystallize in different forms than they had originally, forming new minerals and textures. *Contact metamorphism* may occur in rocks where they meet a hot igneous body. *Regional metamorphism* may occur in the roots of mountain belts due to the elevated temperatures and pressures there. *Dynamic metamorphism* may occur along active fault zones in rocks subjected to high stresses.

Metazoan fossil: Multicelled animals showing complex cellular structure that first appeared in Late Proterozoic.

Meteorite: Extraterrestrial material, that has fallen to Earth's surface. Meteorites are classified as *iron meteorites* that are made up of iron and nickel and may represent the core of asteroids (fragmented planetesimals, *achondrite*) or *stony meteorites* composed predominantly of silicate minerals that resemble ultramafic rocks within Earth's mantle (*chondrules*).

Meteoroid: Small pieces of debris found within the Solar System, less than a few meters in size, and made up of rock or ice; may be either fragmented asteroids or comets. If small pieces of cosmic matter move into Earth's atmosphere and burn up, they are referred to as *meteors*, showering fine particles of dust over Earth's surface.

Microcontinents: Small, identifiable blocks of continental crust, often formed from volcanic island arc. They may reside in oceanic crust or may be accreted to an active continental margin.

Mid-latitudes: The regions on either side of the equator between latitudes 45° and 65° that are characterized by a temperate to cool climate with high annual precipitation.

Mid-oceanic ridges: The underwater mountain chain that follows the oceanic spreading centers where new oceanic crust is being formed.

Milankovitch cycle: Explanation of the cyclic variation in Earth's climate and the onset of ice ages as a result of cyclic variation in the eccentricity of the Earth's orbit, its tilt and precession. (See *Earth's tilt, Precession.*)

Milky Way Galaxy: The area in which a spiral-shaped collection of 100 billion stars exists, one of which is our Sun.

Mineral: A naturally occurring, inorganic crystalline solid. Each mineral is classified depending on its mineral properties.

Mineral properties: A mineral can be characterized by visible or physical properties such as color, cleavage, habit, hardness, luster, and density. These properties reflect the crystal's composition, structure, impurities and grain size.

Mineral reserve: See *reserve.*

Mineral resource: Minerals that can be mined and are considered to have value.

Mohorovicic Discontinuity (Moho): Characterized by an increase in the velocity of seismic waves, the moho is the transition area between the Earth's crust and mantle, and is on average 30 to 40 km below the surface.

Monsoon: Seasonal continuous rain due to changes in the position of persistent high and low pressure regions. This occurs primarily in Asia.

Moon: Earth's satellite body, in synchronous rotation with respect to the Earth. Its origin is attributed to an impact by a Mars-sized meteorite with the Earth. Discredited models for its formation include the *capture theory* and the *condensation hypothesis.*

Mountain belt: An extensive and linear zone of crustal rocks that have been uplifted through plate tectonic processes.

N

Natural gas: A gaseous mixture of hydrocarbons, (predominantly methane), produced from the burial of marine organisms in an oxygen-deficient environment at greater depths and higher temperatures than those producing liquid petroleum.

Natural selection: A mechanism proposed to account for evolution of life forms; as a result of natural selection, organisms best adapted to their environment are more likely to survive and reproduce.

Neap tide: Diminished tidal range during the first- and third-quarter phases of the moon, due to the nonalignment of the Sun, Earth and Moon during these periods. (See *spring tides.*)

Nemesis: The Sun's hypothetical companion star. Its existence was proposed because such a star might account for the disturbance of comets out of the Oort cloud and into the Solar System.

Neutrinos: Subatomic particles emitted from the Sun as a byproduct of thermonuclear fusion, that have no charge and negligible mass. They travel at the speed of light and are very difficult to detect because they rarely interact with matter.

Nitrogen cycle: This cycle traces the path of nitrogen from volcanic gas emanations to the atmosphere and into the biosphere and soils. Although there is very little interaction between nitrogen and these surface reservoirs and even less with the hydrosphere, the study of nitrogen is important because of its role as a nutrient to the biosphere.

Nonclastic: A sedimentary rock texture comprising interlocking crystals.

Nonconformity: An unconformity in which younger sedimentary rocks overlie an igneous intrusion or metamorphic rocks. (See *unconformity.*)

Nonfoliated metamorphic rock: Any metamorphic rock (such as marble and quartzite) that lacks a closely spaced planar layering or foliation. (See *foliation.*)

Nonmetallic (industrial) minerals: Minerals that do not have the characteristics of metallic minerals and that are often used in industries for such things as fertilizer and solvents or as materials in such things as roofing or buildings.

Nonrenewable resource: A commodity, such as groundwater, a fossil fuel, or a mineral resource, that is not replenished by nature as quickly as it is consumed due to the slow geologic process by which it was formed.

Nonsilicates: Minerals such as calcite and dolomite, in which metallic ions are bonded to elements other than silicon and oxygen.

Normal fault: A dipping fault in which the hanging wall has moved downward relative to the footwall. This is commonly the result of extensional forces and is most common in continental rifts and mid-oceanic ridges. (See *reverse fault.*)

Nuclear fission: The process by which atoms of radioactive elements are bombarded by subatomic particles (neutrons) to break down the atomic nucleus, thereby releasing a great amount of energy.

Nuclear fusion: The natural process by which elements in the universe form, through bombardment of a lower mass element with subatomic or atomic particles producing a higher mass element. For example, the nuclear fusion of hydrogen into helium is the source of fuel for the Sun.

Nuclear power: Electricity produced by harnessing the heat from slow, controlled nuclear fission reactions.

O

Oasis: A natural spring that occurs in the desert. These springs (plural *oases*) are usually fed by an artesian flow.

Obduction: The process in which, during plate convergence, a sliver of oceanic plate is detached from the subducting slab and transported onto the leading edge of the other plate. (See *ophiolite.*)

Ocean: A large body of salt water surrounding the continents that has accumulated within a topographically depressed basin floored by oceanic crust; an average depth of 4 km below sea level.

Oceanic circulation: Currents may be found within the ocean's surface, intermediate, or deep layers. The *surface layer* is a relatively thin but important zone where interaction with atmospheric circulation and solar radiation occurs. (See *gyres* and *mesoscale eddies.*) Below, the *intermediate layer* is a region of mixing that is characterized by rapid variations in temperature, salinity, and/or density (known as *thermocline, halocline* and *pycnocline* respectively). The *deep layer* usually occurs below 2000 m depth and is characterized by cold, relatively dense water. Two regions of the deep layer have been named the *North Atlantic deep water* and the *Antarctic Bottom water.*

Ocean-continent (plate) convergence: The site where oceanic plate is subducted beneath the buoyant continental plate at a convergent boundary (e.g. an *Andean-type subduction*).

Ocean-ocean (plate) convergence: A type of convergent plate boundary along which two oceanic lithospheric plates collide and the denser one (usually the older one) is subducted (e.g. a *Marianas-type subduction*).

Oceanic trench: The deep, narrow, trough-like crustal depression formed where the oceanic plate sinks and becomes subducted.

Oceanic crust: Originally formed at mid-oceanic ridges, it ranges from 5 to 10 km thick and predominantly consists of gabbro, overlain by basalt dikes and flows that are topped by a thin layer of marine sediment. Its average density is 2.9 g/cm³, and its average age is less than 200 m.y.a.

Oceanic ridge: An underwater mountain system found at constructive plate boundaries in all oceans; the site where new lithosphere is created. It is composed predominantly of basalt and displays features produced by rifting. Its elevation is related to mantle upwelling.

Oil: A liquid petroleum formed by heat and pressure within a buried oxygen-deficient environment from the remains of marine animals. Refining of the oil by cracking yields products including gasoline and fuel oil. Oil may also be contained within organic-rich mudstones (oil shale) that yield petroleum when heated.

Oil window: Refers to the narrow range of temperature and pressure in which liquid petroleum forms.

Oort cloud: A collection of comets, proposed to lie in the outer reaches of the Solar System.

Ophiolite: Uplifted fragments of oceanic crust emplaced on the edge of the overriding plate at a subduction zone. They have a stratigraphic sequence that is characteristic of oceanic crust, consisting of a thin cap of sedimentary rocks overlying pillow lavas, sheeted dikes, layered gabbro, and ultramafic rocks. (See *obduction*.)

Orbit: The elliptical path travelled by one body around another body having a much larger mass. Orbits of planets and other minor celestial bodies within the Solar System have elliptical paths about the Sun which can be described by their *eccentricity*. Comets have high eccentricity meaning that their paths are more elongate than circular.

Ore: Rocks that contain minerals which are in sufficient concentration to be extracted for economic benefit. A definable zone of ore grade material is called an *orebody*.

Ore grade: In coal or ore, the degree to which the mineable resource is free from contaminants that diminish its economic value.

Orogenesis: Refers to the processes involved in mountain building, which include the convergence of plates and their subsequent deformation, metamorphism, and magmatism through subduction and collision.

Orogeny: Mountain building produced by tectonic activity such as folding, faulting, and associated metamorphic and igneous activity. It usually takes place at convergent plate boundaries from subduction- or collision-related processes.

Outcrop: A body of rock that is exposed on Earth's surface.

Outer core: The region of the Earth beneath the mantle that is believed to be composed mostly of liquid iron and nickel, enveloping a solid inner core which is thought to be made of a purer iron-nickel. The boundary between mantle and core is at a depth of 2900 km, where the temperature is believed to reach 4800°C.

Oxygen isotope ratios: A proportion of stable ^{18}O to ^{16}O isotopes. These are useful when identified in marine fossils and glacially entrapped gas molecules. Based on the differing behavior of the stable isotopes when moving from the hydrosphere to the atmosphere and to the Earth's surface.

Ozone: A gas molecule (O_3) composed of three oxygen ions. It is very useful in the ozone layer, where it naturally occurs, but ozone near the ground (which only happens artificially) is a health hazard, as ozone is poisonous.

Ozone layer: A layer in the stratosphere approximately 35 km from the ground, in which ozone molecules are anomalously abundant. It is very useful to life on Earth, since it blocks a large proportion of the Sun's ultraviolet rays.

P

P wave (primary wave): A compressional body wave that is generated by earthquake activity and is transmitted within the Earth. It travels through both liquid and solid states of matter and moves faster than an S wave. (See *S wave*.)

Pacific Ring of Fire: See *Ring of Fire*

Pacific-stage: The period of time when the edges of ocean basins begin to subduct, similar to subduction observed in the Ring of Fire. (See *Atlantic-stage*.)

Paleoclimate: The climate that occurred on Earth in prehistoric times, which has been inferred by the use of proxy methods.

Paleomagnetism: The determination of past magnetic fields on Earth through analysis of ancient basalt flows. This is useful in determining the earlier positions and shapes of continents.

Paleontologist: A geologist who studies and documents ancient plant and animal life.

Paleozoic Era: Meaning "ancient life". The geologic era spanning the time from the Cambrian (545 m.y.a.) to the end of the Permian (245 m.y.a.). During this era, shelled organisms were prolific, and land colonization occurred by both plants and animals. The Paleozoic culminated with the amalgamation of the supercontinent Pangea.

Pangea: The Late Paleozoic supercontinent that amassed through continental collisions until the end of the Permian. It began splitting apart in the Jurassic, allowing the Atlantic Ocean to form between the pieces, which became the modern continents.

Partial melting: Incomplete melting of crust or mantle leaving a residue of solid material that is compositionally distinct from that of the melt.

Passive continental margin: Continental margin edges that are *not* bounded by subduction or collision-related processes (e.g. the eastern North American plate). (See *active continental margin*.)

Peat: A brown, water-rich accumulation of compressed plant material that can be used as an energy resource once desiccated. Lignite is formed from buried peat, and coal is similarly formed from lignite.

Permeability: The degree of connection between pore spaces in rocks or unconsolidated sediment in the subsurface. A highly permeable material is traversed easily by fluids. (See *porosity*.)

Petroleum: A fossil fuel such as oil or natural gas that is mostly made up of naturally occurring hydrocarbons from buried marine microorganisms. It can exist in a solid, liquid or gaseous state and is the major source of nonrenewable energy worldwide. Naturally occurring liquid and solid petroleum products are known respectively as bitumen and kerogen.

Pillow lavas: Submarine tubes of basaltic lava that form pillow shapes. These are common in basalt flows that make up the oceanic crust and can also form from hot spot activity within the oceanic crust.

Phanerozoic Eon: The geologic time span that covers the appearance of shelly life in the Cambrian Period, 545 million years ago, to the present.

Phase change (mineral): Changes in the internal structure of a mineral but not in its chemical composition. A phase change is thought to occur in olivine and pyroxene within the transition zone which divides the upper and lower mantle.

Phase (of the Moon): Refers to the proportion of the Moon that is illuminated at a specific time (as seen from the Earth). The Moon's phases are cyclical, from new moon (no illumination), to first-quarter moon (illuminated on the right side), to full moon (fully illuminated), to third-quarter moon (illuminated on the left-side), back to new moon. While the Moon's phases are between new moon and full moon (i.e. as the illumination increases), it is said to be waxing, and it is said to be waning between full moon and new moon (as the illumination decreases).

Phosphorus cycle: The movement of phosphorus from its origin in an igneous setting; its exhumation to enrich soils and ocean waters, allowing interaction with the biosphere. The atmosphere plays a relatively minor role in the recycling of phosphorous.

Photon: The transmission of energy as a particle of light. Photons with long wavelengths transmit lower energy (have a lower intensity) than photons with short wavelength (higher intensity).

Photosphere: The visible surface of the gaseous Sun from which we receive most of the Sun's light. It has an average temperature of 5800°C.

Photosynthesis: The process by which living organisms use the Sun's radiant energy to convert water and carbon dioxide into food (carbohydrates) and oxygen.

Placer deposit: A concentration of heavy minerals (such as gold or platinum) which accumulated by physical or mechanical enrichment during naturally occurring processes (such as falling to the bottom of streams).

Planet: A relatively small, cold body composed of a mixture of rock, liquid or gas, that orbits around a star.

Planetesimal: Small solid aggregations of relatively cool nebula dust and gas. Planetesimals may stick together to form protoplanets, which may in the same way grow to become planets.

Plankton: Microscopic plants and animals that float in marine waters. They are a very important part of the food chain.

Plate boundary: The margins or edges of lithospheric plates where tectonic activity predominates. (See convergent, divergent, and transform plate boundaries.)

Plate tectonic theory: The unifying theory that explains many of Earth's orogenic features resulting from the interaction between moving plates which cover the Earth's surface. At plate boundaries oceanic crust may be generated, be consumed by subduction or slide past one another depending on the relative motion of plates. Collision of plates may produce mountain ranges along their boundary whereas passive movement of plates may occur along transform faults.

Porosity: The availability of open cavities in rocks or unconsolidated sediment in the subsurface. A material with a high volume of pore spaces has a high porosity. (See *permeability*.)

Potentiometric surface: The highest elevation where recharge of an aquifer occurs, this may be at the Earth's surface or the water table. At any point below this surface, water will flow upwards freely. (See *aquifer* and *artesian well*.)

Precambrian: The interval of geologic time from the beginning of Earth (4.6 b.y.a.) to the appearance of shelled organisms of the Cambrian period (545 m.y.a.). Includes the three eons; Hadean, Archean and Proterozoic.

Precambrian shield: Stable continental crust formed before the Cambrian period from accretion of Archean greenstone and granite-gneiss complexes, and Proterozoic microcontinental rocks.

Precession: The "wobble" or change in the direction of the Earth's axis that traces out a cone-shaped path every 26,000 years. (See *Earth's tilt*.)

Pressure gradient (atmospheric): The difference in atmospheric pressure between adjacent regions. Air moves from high pressure to low pressure regions, causing winds.

Principle of actualism: The idea that geological processes which operate in the present were also active in the past. This encompasses gradual changes, such as erosion, and more abrupt changes, like volcanism and earthquakes.

Principle of fossil succession: The principle, based on the theory of evolution, that living organisms have evolved through time in a definite sequence and that these changes are reflected in the fossil record. Establishing the order in which different ancient organisms occurred allows relative age dating of sedimentary rocks: those rocks that contain fossils similar to more recent life forms are younger than those that contain older life forms.

Principle of gradualism: The belief that processes that shape the Earth make changes only gradually, over a long period of time.

Principle of original horizontality: The principle that sediment is deposited, compacted, and lithified in a horizontal position.

Hence, sedimentary layers that are currently inclined were deposited horizontally and tilted at a later time.

Principle of original lateral continuity: The principle that sediment is deposited in a laterally continuous manner such that a sedimentary layer can usually be traced in all directions until it becomes negligibly thick.

Principle of superposition: The principle that in any undeformed sequence of strata, younger layers will have formed on top of older layers, and will therefore still be above them.

Prograde metamorphism: Metamorphism that occurs under conditions of increasing pressure and temperature, such as that accompanying progressive burial.

Prokaryote cell: A single-celled organism that first appeared in the Archean in stromatolite-bearing rocks.

Prominence: Arcuate-shaped protrusions of hot gas from the Sun's surface within the chromosphere. Prominences are trapped by the magnetic field around sunspots.

Proterozoic Eon: The geologic time from 2.5 billion years ago to the beginning of the Cambrian Period in 545 m.y.a. The Proterozoic is characterized by the development of an oxygen-rich hydrosphere and atmosphere and by the development of horizontally-dominated plate movement.

Proxy methods: Determining ancient geological characteristics through indirect methods. A proxy method to determine the nature of the Paleolithic climate was by using oxygen isotope ratios of gas bubbles found in ice sheets.

Ptolemaic (geocentric) system: A discredited model that placed the Earth at the center of the universe.

Pull-apart basin: Narrow depressions of restricted size that occur as a result of extensional stresses along transform faults.

Pyroclastic eruption: An explosive volcanic eruption that jettisons fragments of hot magma and surrounding rock into the atmosphere. The turbulent movement of hot gas and ash down the flanks of the volcano is referred to as a *pyroclastic flow*. The settling out and cooling of the material yields a *pyroclastic deposit*.

Q

Quaternary: The geologic period that spans the last 1.6 million years and is characterized by a time of glacial and interglacial activity. It is subdivided into two epochs, the Pleistocene and the Holocene. The Holocene, which began 10,000 years ago, is the latest interglacial phase.

R

Radiation zone (of the Sun): Surrounding the Sun's core, a zone where thermal transfer of heat from thermonuclear reactions in the core occurs through outward or radiating movement of photons.

Radioactivity: The spontaneous breakdown of an unstable atom (the *parent isotope*) to form a new product (the *daughter isotope*) by ejecting or absorbing particles within the nucleus. (See *beta decay, alpha decay,* and *electron capture*.)

Rain shadow: A region of land on the leeward side of mountain ranges that receives little precipitation and consequently undergoes desert-like conditions.

Raised beach: A beach that has been stranded above sea level due to a drop in sea level or tectonic uplift of the coastline.

Rank: The evaluation of coal based on the amount of carbon and the amount of heat produced when combusted. In increasing order, coal is ranked as: *peat, lignite, subbituminous, bituminous, anthracite*.

Recharge: The replenishment of a groundwater aquifer through the downward percolation of precipitation.

Reflection: The rebound of a travelling seismic, light, or water wave as it meets a material.

Refraction: The change in direction of a wave as it travels from one material into a material with a contrasting character.

Relative age: The age relationship of one thing to another: one thing is younger than another, which is older than the first. Similarly, this applies to events: one happened before the other. (See *absolute age*.)

Relative age dating: The method by which the relative age of a rock unit or event (relative to other rock units or events) is established using stratigraphic principles. (See *rule of cross-cutting relationships* and *absolute age dating*.)

Relative humidity: The amount of water vapor in the air relative to the amount it can hold at a given temperature.

Remote sensing: Examination from a distance. It is possible to explore an area without being in that area to collect the information, by using satellite image technology and other means.

Renewable resource: A commodity that can replenish itself roughly as fast as it can be depleted (e.g. energy produced by tidal power, hydroelectric power).

Reserve: The known amount of an ore or resource that is available for economic extraction. The amount of the reserve which can be extracted and used is referred to as *recoverable reserves*.

Reservoir rocks: Rocks such as sandstone or limestone that have sufficient porosity and permeability to store petroleum.

Residence time: The average period of time an element remains in seawater, atmosphere or other Earth reservoir, from the point of its introduction to the time it leaves.

Residual deposits: The economic minerals which have concentrated near their source because of chemical weathering and erosion of crustal material.

Resource: A potential or known source or supply of a naturally occurring commodity, whether it is economical or not.

Reverse fault: Occurs in a compressional environment where the hanging wall is moved upward relative to the footwall on a dipping fault. (See *Thrust fault*.)

Revolution: The movement of a celestial body around another (e.g. Earth's orbit around the Sun). The time it takes for a body to complete a full revolution is its period of revolution.

Ridge push: The gravitational force associated with plate movement away from the buoyant, spreading ridge margin.

Rift valley: A narrow, fault-bounded valley that is created from extensional processes related to the splitting of continents and the formation of oceanic crust.

Ring of Fire: Refers to the numerous volcanic island arcs and mountain chains that lie on the perimeter of the Pacific Ocean basin due to subduction of oceanic plates.

Rock: An aggregate of one or more minerals either cemented together or grown together in an interlocking mosaic.

Rock cycle: The continuous process by which each of the three rock groups (igneous, sedimentary and metamorphic) is transformed into one of the others as the Earth's ever-changing surface goes through episodes of uplift, erosion, deposition, and burial.

Rock-forming mineral: The minerals which constitute the major components of igneous, metamorphic, or sedimentary rocks. The minerals may be silicates (e.g. feldspar, quartz, and mica) but, as in the case of chemical sedimentary rocks, may also contain carbonates, oxide, and halite groups of minerals.

Roll-back: Oceanward movement of the sinking, dense oceanic plate resulting in extension behind the arc and formation of back-arc basins in the overriding plate.

Rotation: The spinning movement of a body about its central axis. The Earth, for example, makes a full rotation every 24 hours.

Rule of cross-cutting relationships: A rule which states that any body or structure that displaces or truncates another occurred later than the body it cuts across. For example, a dike is younger than the rocks it cuts across. This is useful as a means of determining the relative age of a geologic event.

Rule of inclusions: A rule that states that any fragments contained within a rock body are older than the body. This is because any fragments from younger material (which formed later than the body) could not have entered the body as a fragment. This rule helps to establish relative ages of rocks.

S

S wave (secondary wave): A seismic body wave that moves at right angles to the direction of propagation and moves slower than the P wave. It cannot be transmitted through liquid. Also known as a shear wave. (See *P wave*.)

Salinity index: A measurement of the degree of *salinity* (the total amount of dissolved elements in water) in a specified volume of water. The salinity index is measured in parts per thousand. Common values of ocean water range from 33‰ to 37‰.

Salt domes: Diapirs or dome-shaped concentrations of rock salt in the subsurface. Rock salt moves upward within the Earth's crust because of density contrasts with the surrounding rocks.

Scale: The ratio that compares the distance between two points on a map to the same two points on the Earth's surface.

Scientific method: The reasoning followed by scientists in formulating and evaluating a hypothesis about some phenomena, usually involving observation, analysis, synthesis and deduction.

Sea level: The mean level of the sea surface, taken as the average point between high and low tide.

Sea-floor spreading: The divergent movement along plate margins associated with the production of oceanic crust that is equally distributed at the margins of each plate. The rate of growth of each plate is commonly a few centimeters per year.

Seamount: A submerged volcanic island within an ocean that was formed by hotspot activity. It may be isolated or part of a chain or group of submerged volcanic islands such as the Hawaiian-Emperor seamount chain. *Fringing reefs* or *atolls* may surround the structure.

Sediment: Unconsolidated material produced by weathering and erosion or by chemical or biochemical precipitation, which may be transported by glaciers, running water, wind, gravity, or waves.

Sediment load: Material carried by a stream. Very small particles can be carried in solution (a *dissolved load*), and larger ones may be pushed along the channel bed (called a *bed load*).

Sedimentary facies: A body of rock that is characterized on the basis of its compositional makeup, origin of formation or fossil association.

Sedimentary rock: Any rock formed by the burial and cementation of sediment. Classification is based on origin and texture of the rock. (See *detrital, chemical*, and *biochemical sedimentary rocks*.)

Seismic reflections: A geophysical technique, used in petroleum exploration, that looks at the Earth's subsurface from the response of seismic (shock) waves generated on the surface.

Seismic sea waves: Earthquake-generated ocean waves that move rapidly and forcefully across ocean basins. Destruction occurs when waves steepen as they enter shallow waters near coastlines and crash upon shorelines. Also known as *tsunami*.

Seismic tomography: Images produced of Earth's deep interior from interpretation of how seismic waves respond as they travel through different material.

Seismic waves: Shock waves that emanate from rupture along a fault or are produced artificially from explosions, that travel either near the Earth's surface or penetrate deeper like body waves. (See *P waves* and *S waves*.)

Shadow zone (seismic): P wave shadow zones are the regions between 103° and 143° from the earthquake epicenter where no direct P waves are recorded due to the refraction of P waves as they travel through the liquid outer core. The S *wave shadow zone* is the region beyond 103° from the earthquake epicenter, where no S waves are recorded, due to the inability of S waves to travel through the liquid outer core.

Sheeted dike: A layer of multiple, near-vertical fractures injected by basaltic magmatism within the oceanic crust. These are produced at regions of sea-floor spreading where magma is fed from the magma chamber to the surface. (See *ophiolite*.)

Silicates: Rock-forming minerals (such as feldspar and hornblende) in which silicon and oxygen are chemically bonded to a variety of metallic ions.

Silicate tetrahedron: The basic structural unit of all silicate minerals in which a central silicon cation is bonded to four oxygen anions in a tetrahedral or pyramid-like arrangement.

Sill: A sheet-like body of igneous rock that is *concordant* or parallel to the layering of the surrounding rock. (See *dike*.)

Slab-pull: A mechanism which explains the subduction of an old oceanic plate as a function of its density: at its leading edge this oceanic plate is denser than the underlying asthenosphere and will sink, pulling the rest of the plate with it.

Slate: Fine-grained, foliated metamorphic rock containing aligned flaky minerals such as muscovite that give the rock a preferred direction of splitting. Typical of low grade metamorphism. (See *foliation*.)

Soil: The loose sediment that occurs on the Earth's surface above bedrock that is able to support plant and animal life.

Soil profile: A vertical section of the loose sediment that includes all of the soil layers from the top surface to bedrock. The horizons are lettered from top to bottom: O, A, B and C, reflecting organic-rich layers near the top with humus and detrital grains, to fractured bedrock in C below.

Solar constant: The amount of solar radiation that reaches the Earth's surface: 1.36 kilowatts/m².

Solar energy: An energy resource that utilizes the Sun's radiant energy (e.g. photovoltaic cells).

Solar flare: A short-lived, flame-like disturbance within the Sun's surface linked to sunspots. A significant amount of energy is released from solar flares: this may affect Earth's magnetosphere and result in upper atmosphere aurora displays.

Solar nebula theory: The theory of the formation of the Sun, planets and other bodies in the Solar System from accretion of particles in a swirling cloud of dust and gas that collapsed and flattened into a disk.

Solar radiation: The radiation of light energy emitted from the Sun's surface. This encompasses wavelengths between 10^{-12} (Gamma rays) to 10^3 (Radio waves).

Solar spectrum: The band of light energy emitted from the Sun that is represented by wavelengths between 10^{-12} (Gamma rays) to 10^3 (Radio waves). The energy that we see (*visible light*), represents only a small portion of the spectrum of light energy.

Solar System: The nine planets which revolve about the Sun, as well as their satellite bodies and numerous asteroids. The Solar System formed around 4.6 billion years ago.

Solar wind: Outpouring of ionized gas from the Sun that travels through the Solar System. The solar wind transfers heat to the Earth's thermosphere and distorts its magnetosphere.

Solid Earth: The Earth beneath our feet and its principal reservoir. It comprises the Earth's crust, mantle and core.

Solstice: The two days of the year, June 21st and December 21st, when a hemisphere will receive the greatest amount of daily sunlight or the smallest. Which of these is naturally related to whether it is summer or winter in the hemisphere. For example, on December 21st it is winter in the northern hemisphere (which will then have the least daily sunlight of the year), but it is summer in the southern hemisphere (which will have the greatest daily sunlight of the year).

Sorting: The segregation of moving particles carried by water or wind, based on grain size and density.

Source rock (petroleum): A sedimentary rock layer capable of producing petroleum if it is buried under the proper conditions.

Southern Oscillation: The observation that persistent high and low pressure systems located in the equatorial Pacific Ocean interchange their positions, affecting ocean water circulation and contributing to the effects of El Niño or La Niña.

Specific gravity: The ratio of a mineral's weight to the weight of the same volume of water.

Spicules: Momentary, wave-like extrusions from the Sun's chromosphere above the photosphere that may extend into the corona.

Spring tides: The amplification of tidal range when the Sun, the Earth and the Moon are aligned.

Stable isotopes: Nonradiogenic isotopes of an element. (See *isotope*.)

Star: A huge body of swirling gas which radiates energy due to thermonuclear fusion in its core. Stars may be classified by their size and the amount of energy they give off. *Giant* stars are massive, producing enormous energy, while *dwarf* stars are relatively small and weak. There are over 100 billion in our Milky Way galaxy alone.

Stick-slip fault: Refers to the inhomogeneous movement along a strike-slip fault. During periods of quiescence, no movement is observed and stress builds up. When the fault reaches its maximum strain it ruptures, releasing enormous amounts of stored energy. Often observed within segments of continental transform faults, such as the San Andreas Fault.

Strain: The response of a body to an external force or stress. For a stressed body, the *strain rate* is the amount of change over time.

Strata: A term referring to layers of sedimentary rocks.

Stratigraphic sequence: Rocks which can be placed within a sequence based on their origin, and spatial and temporal relationship (e.g. greenstone belts).

Stratigraphic traps: Sedimentary layers or boundaries that act as barriers to the upward movement of oil and natural gas due to changes in lithology (e.g. an impermeable clay layer).

Stratigraphy: The study of layered rocks with respect to their compositions, origins, and their spatial and temporal relationships.

Stream: The surface flow of water in a channel. A meandering stream is a sinuous channel, while a braided stream is a numerous networking of channels. Streams are fed by groundwater, precipitation, and tributaries. The total land area encompassing the main stream channel, from its headwaters to its mouth, including all of its tributaries is referred to as a drainage basin. The discharge zone of a drainage basin, commonly a large lake or ocean, is called a drainage outlet.

Stream gradient: A measure of the average slope of a stream from its headwaters to its drainage outlet. (See *stream*.)

Stress: An external force that is applied to a body.

Strike: The direction of a horizontal line in a plane. The measurement of strike of a bedding plane indicates the direction of continuity of the rock unit.

Strike-slip fault: A steeply dipping fault where movement occurs parallel to the fault's strike and the two blocks slide past each other. (See *transform faults*.)

Stromatolite: The layered or dome-shaped build up of sediments by the action of blue-green algae or cyanobacteria in shallow marine environments. Stromatolites have been preserved in ancient rocks of the Archean.

Structural trap: The entrapment of oil and natural gas within folded or faulted rock layers.

Subduction zone: The area delineated by an oceanic trench in which dense oceanic crust bends down beneath the margin of a less dense plate (oceanic or continental), and pushes into the underlying asthenosphere, perhaps even reaching the lower mantle. Characterized by intermediate to deep focus earthquake activity, volcanic arc magmatism and granitic intrusion.

Subequatorial: The latitudes between 10° north and 10° south, which have a warm, moist climate.

Submarine canyons: Deeply incised ravines located on the continental shelf and slope through which turbidity currents often travel, depositing sediment at its mouth.

Subsidence: The lowering of a region of Earth's crust that may result from crustal deformation, glaciation, thermal cooling, or depletion of an aquifer.

Subtropical: The region between the tropics and the temperate zone (i.e. mid-latitudes), where the climate is warm and dry due to the descending Hadley cells. Subtropical climate is characterized by a net annual deficit of rainfall.

Subtropical high: The region of high pressure centers near 30° north and 30° south, caused by descending air of the Hadley cells.

Sun: A star formed 4.6 billion years ago and made up primarily of hydrogen and helium, having minor amounts of other heavier elements. Its large mass provides the gravitational attraction that keeps the planets of the Solar System in orbit. It has a diameter of 1.4 million km, is 5800 °C at its surface and approximately 15,000,000 °C at its core. Although it is gaseous, it is believed to have distinctive layers: a *core, radiation zone, convection zone* and *photosphere* (the Sun's surface). In addition, it is surrounded by a *chromosphere* and *outer corona*.

Sunspot cycle: The number of sunspots on the Sun's surface increases and decreases over an 11 year cycle. Recent peak activity occurred in 1991, the recent minima in 1997. A relationship has been proposed between anomolous sunspot minima and climate changes on the Earth. (See *Maunder Minimum.*)

Sunspots: Dark, relatively cold regions on the Sun's surface (photosphere) where there is an abnormally strong magnetic field. The presence of sunspots is linked with increased acceleration of particles in the solar wind. (See *solar flares, sunspot cycles.*)

Supercontinent Cycle: The cycle of supercontinent breakup, continent dispersal, subduction, and amalgamation that is believed to span some 400 million years. Supercontinent amalgamation is documented in the Late Proterozoic and Late Permian.

Supernova explosion: The explosion of a massive star. As massive stars run nearly out of energy, they collapse quickly, which increases pressures and temperatures one final time, causing a catastrophic explosion of the star's gases into the surrounding medium.

Surface water: Fresh or saline water on the Earth's surface that may be located in fresh and saline lakes or in streams (i.e. *surface runoff*).

Suspension: One of the means of transport of detrital sediment, in which a particle is suspended or floats within the moving water, as opposed to being dissolved in the water or bouncing along the stream bed. (See *sediment load.*)

Sustainable development: Activity and growth of society which does not deplete Earth's natural resources upon which future populations and economic growth depend.

Suture: The zone within a landmass where two continents once amalgamated: This is a major geologic boundary.

Syncline: A folding of layered rocks which resembles a trough, such that such that each layer is lowest near the center of the syncline and higher farther from the center. The opposite of an anticline. (See *anticline.*)

T

Tectonic activity: Pertaining to deformation events which may involve folding or faulting of rocks and associated igneous or metamorphic activity.

Temperate zone: The bands of latitude north and south of the equator between 45° and 65°. The temperate zone is so called because these latitudes have temperate, moderate climates.

Temperature: A measure of the average energy of motion within a body. (See *heat.*)

Terrestrial planets: The four inner planets of our Solar System: Mercury, Venus, Earth, and Mars, which are relatively small and made predominantly of dense, rocky material. (See *Jovian planets.*)

Tertiary Period: The geologic time beginning with the Cenozoic Era in which mammals became prolific and the Himalaya mountains formed. It ended 1.6 million years ago, when the last Ice Age began.

Theory: An hypothesis that has withstood repeated testing and application but has not been proven conclusively enough to be accepted as a law.

Theory of evolution: The theory that over long periods of time, life forms adapt to the environment in which they live, such that they do better in that environment. A history of these changes exists in the fossil record.

Thermonuclear fusion: A process in which particles collide and fuse, forming an element with a greater mass, and releasing energy. This process occurs naturally in stars and is also responsible for the force of hydrogen bombs.

Thrust fault: A fault that has a shallow dip, in which the hanging wall moves upward relative to the footwall. Often, thrust faults cut strata at a low angle to bedding. They are formed as a result of compressional forces.

Thunder: The loud noise created by the rapid expansion of air around lightning.

Tidal range: The vertical measurement of the difference between the low tide mark and the high tide mark along a coastline.

Tides: The ebb and flood of sea water along a coastline during a lunar day. Tides are caused by the gravitational attraction between the Earth, the Moon and the Sun (e.g. *diurnal, mixed,* and *semi-diurnal* tides).

Till: The general term given to sediment deposited from a melted glacier that is usually poorly sorted (e.g. *drumlin, moraine*).

Tornado: A funnel of fast-moving (up to 350 km/hr), swirling wind resulting from the presence of cold dense air above a layer of relatively warm air leading to upwelling and intensification of a low pressure zone over land. Tornadoes are also called *twisters.* A narrow belt in the southern United States has the world's highest concentration of tornadoes and is known as *Tornado Alley.*

Trade winds: The persistent easterly surface air movements from latitudes 30° north and 30° south that converge on the equator. The flow is from northeast to southwest in the northern hemisphere and southeast to northwest in the southern hemisphere.

Transform fault: A special type of strike-slip fault which terminates abruptly, occurring along the boundary of continental and/or oceanic plates where fault movement is horizontal. They are often associated with mid-oceanic ridges.

Transform (conservative) plate boundary: A major strike-slip fault between two plates where the plates slide horizontally along each other (e.g. the San Andreas Fault). Crust is neither formed nor destroyed at transform plate boundaries.

Trans-Hudson orogeny: An Early Proterozoic (2.0 to 1.8 b.y.a.) mountain building event in northern Quebec that is significant because it is believed to be the oldest known example of continental collision and closure of an ocean basin. (See *Wilson cycle*.)

Tree-ring dating: The process of determining ages by matching the annual growth rings in trees with those of other trees. This has allowed the determination of dates 7000 years before present.

Trench: See *oceanic trench*.

Triple point: Three main fractures that connect to form a "Y", originating from initial rifting of continental or oceanic crust. Also known as a *triple junction*.

Trophic level: A level within the food chain characterized by organisms which derive nourishment in the same manner.

Tropics: The region between latitudes 15° and 23.5° north and south surrounding the tropics of Cancer and Capricorn, respectively. Humid, high, temperatures and lush vegetation characterize the climate of this zone.

Tropic of Cancer: The latitude of 23.5° north, which is the northern boundary of the region where the Sun's rays shine vertically on the Earth.

Tropic of Capricorn: The latitude of 23.5° south, which is the southern boundary of the region where the Sun's rays shine vertically on the Earth.

Tropopause: The boundary of troposphere with the stratosphere. It is delineated by an isobar, and consequently it varies in altitude from 8 km to 18 km.

Troposphere: The layer of the atmosphere closest to the ground that encompasses most of Earth's weather and participates in the hydrologic cycle.

Tsunamis: See *seismic sea waves*.

Turbidity current: A dense, bottom-flowing current of sediment-laden water that moves rapidly downslope, throwing sand and mud into suspension.

U

Unconformity: An erosive boundary between older and younger rock units that represents a significant passage of time.

Uniformitarianism: see *principle of actualism*.

Universe: All known matter and space. The universe is thought to have formed by the Big Bang. (See *Big Bang*.)

Upwelling: Deep, cold currents that move upward. This occurs predominantly along coastlines and in the equatorial region, in zones where persistent air circulation patterns remove saline surface waters such that deeper, nutrient-rich waters travel upwards to replace it.

V

Varve: Alternating dark and light, finely-laminated sediments representing seasonal changes in glacial lakes. Annual sedimentation is represented by deposit of a dark layer in winter and a light layer in summer.

Viscosity: A measure of a fluid's resistance to flow. A geologic substance having high viscosity would be rhyolite lava, which moves much more slowly than basaltic lava.

Volcanic island: An island made predominantly of basaltic magma originating from underlying hotspot activity.

Volcanic rock: An ultramafic to felsic igneous rock formed by lava from volcanoes or fissures. The presence of ultramafic volcanic rock is restricted to the Archean and Early Proterozoic eras.

W

Warm front: The boundary between a warm air mass and a cold one where the warm air mass advances, rising above the cold air along a gentle slope. Widespread precipitation may follow. (See *fronts* and *cold front*.)

Wave: The means of transferring energy through a medium with no net movement of that medium (e.g. seismic waves, water waves). Defined by a sinuous or oscillatory path. Wave properties include: a *crest* and *trough* which is the wave's highest and lowest point, respectively; its *height* which is the difference in elevation between its crest and trough; the wave *amplitude* which is a measurement of one-half the wave height; and its wavelength. (See *wavelength*.)

Water table: The boundary between the zone of aeration and the underlying zone of saturation. (See *zone of aeration* and *zone of saturation*.)

Wave velocity: A wave's rate of movement. The velocity of water waves is affected by the depth of the water. The velocity of seismic waves is affected by the rigidity of the medium (the material the waves travel through). Faster waves have more energy.

Wavelength: The length of one wave cycle (e.g. from crest to crest, or trough to trough). Waves with short wavelengths travel faster than longer ones.

Weather: Describes the hourly or daily atmospheric conditions at a specific place. (See *climate*.)

Weathering: The disintegration and decomposition of rocks by physical and/or chemical processes as a result of their interaction with water, ice and air.

Wein's Law: Expresses the inverse relationship between radiation from a celestial body (wavelength) and temperature. A hotter body will emit shorter wavelength radiation and look bluer, whereas a cooler body will emit longer wavelength radiation and look redder.

Westerlies: Large-scale surface air masses between latitudes 30° and 60° in both the northern and southern hemisphere that move poleward from the west.

Whole-mantle convection: A model explaining movement of Earth's plates from large convection cells that dominate the mantle. If this model is correct it would account for the presence of detached subducted slabs in the lower mantle and constrain mantle changes to be a function of mineral phase changes with depth.

Wilson cycle: The repeated process by which continents break up, spread apart, and form oceans, culminating in continental collision before the cycle begins again. It differs from the supercontinent cycle in that it is focused on regional tectonic activity. (See *supercontinent cycle*.)

Wind: Movement of air caused by differences in atmospheric pressure. Air moves from areas of high pressure to low pressure areas.

Z

Zircon crystal: A zirconium-bearing silicate mineral formed predominantly as a minor constituent within igneous rocks of granitic composition. It contains minute amounts of the radioactive element uranium. This factor plus the mineral's resistance to chemical weathering makes it very useful for radioactive age dating.

Zone of aeration: The area in the subsurface where pore spaces are filled with a mixture of air and water. It lies above the zone of saturation.

Zone of saturation: The area in the subsurface where groundwater fills all of the pore spaces. It lies below the zone of aeration.

Credits

Credits are listed according to figure number.

Chapter 1
Opener Paul Kenward/Tony Stone Images
1.1 Heather Dutton
1.4 Damian Nance
1.06a Courtesy of the Department of Geology, University of Delaware
1.06b Donald Mullineaux, USGS
1.07a Krafft/Explorer Science Source/Photo Researchers
1.07b GECO/UK Science Photo Library/Photo Researchers

Chapter 2
Opener Robert de Gugliemo/Science Photo Library/Photo Researchers
2.5a Courtesy of the Smithsonian Institute
2.5b-c Courtesy of Sue Monroe
2.6a USGS
2.6b-c Paul Silverman/Fundamental Photographs
2.6d Photri-Microstock
2.9a-c Courtesy of Sue Monroe
2.10 Fundamental Photographs
2.11a,b Courtesy of Sue Monroe
2.12 Courtesy of the Smithsonian Institute
2.13 Biophoto Associates/Photo Researchers
2.14a Paul Silverman/Fundamental Photographs
2.15 Courtesy of Sue Monroe
2.16, top left Michael Dalton/Fundamental Photographs
2.16, top right and bottom left Courtesy of Sue Monroe
2.16, bottom right Paul Silverman/Fundamental Photographs
2.18a Soames Summerhays/Science Source/Photo Researchers
2.18b,c Courtesy of Sue Monroe
2.20a Ken M. Johns/Photo Researchers
2.20b WB Hamilton/USGS
2.21 Courtesy of Sue Monroe
2.24a-c Heather Dutton
2.26 Photri-Microstock
2.27a Courtesy of Keith Ronnholm
2.27b PhotoDisc
2.28 USGS
2.30 USGS
2.32 Damian Nance
2.33a,b Peter Scholle/AAPG
2.34a-c Courtesy of Sue Monroe
2.34d AJ Copley/Visuals Unlimited
2.34e E.R. Degginger
2.34f Courtesy of Sue Monroe
2.34g Joyce Photographics/Photo Researchers
2.34h Courtesy of Sue Monroe
2.34i Paul Silverman/Fundamental Photographs
2.34j-m Courtesy of Sue Monroe
2.38a-c Courtesy of Sue Monroe
2.38d Courtesy of Reed Wicander
2.41a-e Bernard Pipkin

2.41f Joyce Photographics/Photo Researchers
2.41g L.S. Stepanowicz/Visuals Unlimited
2.42, top Paul Silverman/Fundamental Photographs
2.42, bottom A.J. Copley/Visuals Unlimited

Chapter 3
Opener Damian Nance
3.1 Damian Nance
3.2 Natural History Museum, London
3.4 Natural History Museum, London
3.5b Courtesy of Dorothy L. Stout
3.7 Natural History Museum, London
3.10 Natural History Museum, London
3.14b Damian Nance
3.16 Reed Wicander

Chapter 4
Opener USGS EROS Data Center
4.14a National Geographic
4.15b W.A. Banaszewski/Visuals Unlimited

Chapter 5
Opener James Balog/Tony Stone Images
5.21 Reuters/Bettmann Archive
5.22b K. Staufel
5.22d Keith Ronnholm
5.22e Bernard Pipkin
5.30 NGDC
5.32 Courtesy of the USGS

Chapter 6
Opener Derke O'Hara/Tony Stone Images
6.1 Derke O'Hara/Tony Stone Images
6.5a-d Courtesy of Ross Griffiths
6.11 Bruce Hands/Tony Stone Images
6.20 Geospace/Science Photo Library/Photo Researchers
6.26 John Waldron
6.28a Courtesy of NOAA
6.28b Brendan Murphy
6.28c B. Bradley, University of Colorado
6.29a,b Brendan Murphy
6.29c Glen Stockmal
6.30b John Waldron
6.31a,b John Waldronz
6.32 Brendan Murphy
6.33a,b John Waldron
6.34 Damian Nance

Chapter 7
Opener Ken Lucas/Visuals Unlimited
7.1b U.S. Geological Survey
7.17 Fundamental Photographs
7.22 Courtesy of *American Scientist*
7.23 Courtesy of Robert van der Hilst

Chapter 8

Opener Warren Bolster/Tony Stone Images
8.3b Society of Sedimentary Geologists
8.13b Courtesy of U.S. Department of Navy
8.14 Courtesy of the Department of Geology, University of Delaware
8.18a VU/WHO1-D Foster
8.18b John Delaney, University of Washington
8.19b-d Woods Hole Oceanographic

Chapter 9

Opener Courtesy of NASA/JPL
9.1 Clint Farlinger/Visuals Unlimited
9.5a,b Courtesy of the USGS
9.6 Research by Philip L. F. Liv, Seung Nam Seo, and Sung Burn Yoon, Civil and Environmental Engineering, Devine, Cornell Theory Center
9.8 Peter K. Ziminski/Visuals Unlimited
9.10 Frank Hanna/Visuals Unlimited
9.12 Nova Scotia Board of Tourism
9.26 O. Brown, R. Evans, M. Carle, U. of Miami Rosenstiel School of Marine and Atmospheric Science
9.29b NOAA/UCAR. Gene Carl Feldman, NASA-GSFC, Otis Brown, U of Miami
9.31b NOAA/UCAR. Gene Carl Feldman, NASA-GSFC, Otis Brown, U of Miami
9.33b International Research Institute, Lamont-Doherty Earth Observatory
9.34 Courtesy of NOAA
9.35 Courtesy of NOAA

Chapter 10

Opener Michael Fairchild/Peter Arnold, Inc.
10.1 Damian Nance
10.2 Brendan Murphy
10.9 Charlie Plummer
10.15 Cindy Murphy
10.16a,b Cindy Murphy
10.17c Cindy Murphy
10.18 David Halpern/Photo Researchers
10.20a A.J. Copley/Visuals Unlimited
10.20b Dale Jackson/Visuals Unlimited
10.24b John Buckland-Nicks
10.25a,b Joe Martinez
10.26b Joe Martinez
10.26d Brendan Murphy
10.27 Jan Kopec/Tony Stone Images
10.30 Richard Thom/Visuals Unlimited
10.33e R.C. Scott, Introduction to Physical Geography, 2nd Edition, West Publishing Company 1996
10.34a Brain Brake/Photo Researchers
10.34b Garry D. McMichael/Photo Researchers
10.36 BP/NRSC/Science Photo Library/Photo Researchers
10.38a,b Courtesy of EMI
10.40a USGS
10.40b John Forsythe/Visuals Unlimited
10.40c Jalain/Explorer/Photo Researchers
10.42 Courtesy of EMI
10.44b Joe Martinez
10.44c Damian Nance

Chapter 11

Opener Tony Stone Images
11.1 Courtesy of the Department of Geology, University of Delaware
11.9a Explorer Berthoule/Photo Researchers
11.9b Bill Bachman/Photo Researchers
11.10b Malcolm Lockwood/Photri
11.11c John Buckland-Nicks
11.17b Patricia Peticolas/Fundamental Photographs

Chapter 12

Opener NASA/Peter Arnold
12.3 Tony Stone Images
12.9 Brendan Murphy
12.17b John Spragens, Jr./Photo Researchers
12.17c John Buckland-Nicks
12.17d Yves Momatiuk and John Eastcott/Photo Researchers
12.17e David and Doris Krumholz/Fundamental Photographs
12.26 AP/Wide World Photos
12.29 Kent Wood/Photo Researchers
12.32 Howard Bluestein/Photo Researchers
12.34b NOAA(National Weather Service)
12.35a,b NOAA National Hurricane Center, Courtesy of Ed Nuhfer and AIPG
12.36b NASA

Chapter 13

Opener Stephen Mojzsis/National Geographic
13.5a Stephen Mojzsis/National Geographic
13.5c Phillip E. Playford/Geological Survey of Western Australia
13.7 NASA
13.11b R.V. Dietrich
13.16a-g Geological Survey of Canada
13.18, top Stanley Awramik
13.18, bottom Andrew Knoll, Harvard University

Chapter 14

Opener Hank Williams
14.2b A.J. Copley/Visuals Unlimited
14.6c,d Hank Williams
14.7 NASA
14.14 Palisades Interstate Park Commission
14.15b John Shelton
14.32a,b E.R. Degginger/Photo Researchers
14.32c George Whiteley/Photo Researchers
14.33a Canadian Geological Survey
14.37 McCutcheon/Visuals Unlimited
14.38 T. Spencer/Colorific
14.40 Charlie Plummer
14.41 Brendan Murphy

Chapter 15

Opener NASA/JPL
15.4 Brendan Murphy
15.10 Lick Observatory
15.11 NASA
15.13 Lick Observatory
15.19b,c NASA
15.21 NASA
15.22 NASA
15.23 NASA/USGS Courtesy of Alfred S. McEwen

15.24 NASA/USGS
15.26 NASA/JPL
15.28a-d NASA/JPL
15.30 NASA/JPL
15.31a-d NASA
15.32 NASA and Lick Observatory
15.33 NASA
15.34 NASA
15.35 NASA/STScl
15.36a R.W. Doty
15.36b National Astronomical Observatory of Japan
15.37 NASA
15.40 USGS

Chapter 16
Opener NASA
16.2 Brendan Murphy
16.5 Deutsches Museum, Munich
16.9a,f National Optical Astronomy Observatories
16.9c Association of Universities for Research in Astronomy, Inc.
     Sacramento Peak Observatory
16.9d Grundy Observatory
16.9e NASA Skylab
16.9g William P. Stern, Jr.
16.10 National Optical Astronomy Observatories
16.12a Visuals Unlimited
16.12b NASA
16.13 NASA
16.14 National Optical Astronomy Observatories
16.16 Copyright Anglo-Australian Observatory
16.20b,c Copyright Anglo-Australian Observatory
16.21 NASA
16.22 NASA

Chapter 17
Opener Dan Kontak
17.2 D.D. Trent
17.3 George Whiteley/Photo Researchers
17.6 Janet Gordon
17.7 Sinclair Stammers/Science Photo Library/Photo Researchers
17.8 Damian Nance
17.9 Brendan Murphy
17.11 API/Explorer/Photo Researchers
17.12 Courtesy of De Beers Consolidated Mines
17.13 Brendan Murphy
17.14 Mark Burnett/Photo Researchers
17.17 Brendan Murphy
17.18 Dane S. Johnson/Visuals Unlimited
17.19 Damian Nance

17.22 College of Geographic Sciences
17.24 Science VU/Visuals Unlimited

Chapter 18
Opener Georg Gerster/Photo Researchers
18.7 Courtesy of AAPG
18.10a U.S. Department of Energy
18.21 Richard Keeler, Southern California Edison Company
18.22 D.D. Trent

Chapter 19
Opener John Buckland-Nicks
19.10 E.R. Degginger

Chapter 20
Opener NASA
20.10 Ingeborg Lippman/Peter Arnold
20.18 F.H. Kolwicz/Visuals Unlimited

Afterword
Closing shot Bruce Roberts

Index

Ice jams, 284–286
Ice sheets, 263–264, 266
 albedo of, 619–620
 extent of, in Quaternary Period, 427–432, 428f–429f
 as proxy for measuring global temperature, 613f–616f, 613–617, 624
Icebergs, 107, 108f, 427
Icehouse effect, 621, 621f
Iceland, 144
Icy comets, 226
Igneous rock, 15, 26, 28–36, 50
 classification of, 29–31, 31f, 50
 coarse-grained, 28, 31f
 evolution of
  chemical, 31–33, 34f, 50
  physical (volcanic eruption), 33–36
 felsic, 29–30, 31f
 fine-grained, 28, 31f
 formation of, 28–29
 intermediate, 29–31, 31f
 mafic, 29–30, 31f
 mineral deposits in, 530–531, 531f, 541
 plutonic or intrusive, 28–29, 36, 50, 147
 in rock cycle, 26–28, 27f
 silica content of, 29–31, 31f, 36
 ultramafic, 29, 31, 31f, 222
 volcanic or extrusive, 28, 29f, 36, 50
Ilmenite, 517
Inclusions, rule of, 62, 63f, 76
Indenter model, 154, 154f, 156b, 402, 404f, 418, 420f
Index fossils, 70
India
 climate of, 156b
 collision with Asia, 152–154, 153f–154f, 171, 418, 419f–420f
Indirect high tide, 239–240, 240f
Indochina, 154, 154f
Indus River, 156b–157b, 262
Industrial minerals, 514, 515t, 520–524, 540
Inertia, principle of, 446
Infrared radiation, 484
Inhomogenous model, of planetary accretion, 448f, 448–449
Inner core, 175, 183f, 184, 188
 composition of, 188
 general characteristics of, 182t
 melting point, 192f
 temperature of, 188
Insecticides, 292–293, 293f
Interglacial stage, 426, 429
Interior, Earth, 175–199, 183f
 circulation of, 188–192, 195–197
 general characteristics of, 182t
 and origin of water, 223–226
Interior mountains, 161, 161f
Interior oceans, 160, 162
Intermediate rock, 29–31, 31f
Intertropical convergence zone, 331f, 351, 352f
Intrusive rock, 28–29, 36, 50
Invertebrate(s)
 Mesozoic, 412
 Paleozoic, 396, 396f
  mass extinction of, 407, 407f
Io, 472, 472f, 478
Ion(s), definition of, 17
Ionic bonds, 16–17, 18f, 49
Ionosphere, 313–314, 314f
Iridium, 416f, 416–417, 418b, 520
Iron
 in inner core, 188

origin of, 503, 503f
 in outer core, 187–188
 in primitive oceans, 371f, 371–372, 602
Iron catastrophe, 450
Iron meteorites, 187, 477
Iron ore, 515, 515t, 516f–517f
 global distribution of, 528, 529f
Irrigation, 295–296, 296f
Island arcs. See Volcanic island arcs
Islands, volcanic. See Volcanic island(s)
Isobars, 312, 315–317, 318f–319f, 335, 335f
Isostasy, 106–109, 108f, 157b, 408, 428–429
Isostatic equilibrium, 109
Isostatic rebound, 109f, 109–110, 429, 429f
Isotope
 daughter, 71f, 71–74, 72b–73b, 73f, 76
 definition of, 71, 72b
 parent, 71f, 71–74, 72b–73b, 76

J

James Bay Project, 295, 295f, 564
Japan, 117, 117f, 121
Jeffreys, Harold, 86
Jet streams, 342–343, 355
 definition of, 312, 342
 El Niño and, 255, 344b–345b, 344f–345f
 locations of, 312f
 polar front, 312f, 342–343, 343f–345f, 344b–345b, 351–352, 355
 and seasons, 342–343, 343f
 subtropical, 342, 343f
Johnson, Donald, 424f
Johnstown flood, 286
Joly, John, 70
Jovian planets, 441–442, 468–474, 478
 composition of, 468
 density of, 458, 478
 internal structure of, 468–469, 468f–469f
Juan de Fuca plate, 125, 125f
Jupiter, 441–442, 468–469, 470f
 atmosphere of, 469
 color bands of, 469, 470f, 478
 comet collision with, 469, 471f, 471b
 composition of, 469, 469t
 current data for, 450t
 Great Red Spot of, 441–442, 469, 470f, 478
 internal structure of, 468–469, 468f–469f
 mass of, 469
 moons of, 442, 445, 450, 469–473, 472f, 478
 orbit of, 445t, 450t, 468
 rotation of, 469
Jupiter Serapias, 60, 62f
Jurassic Period, 69f, 70, 407–416, 409f, 413f
Juvenile water, 226

K

Kalahari Desert, 330
Kant, Immanuel, 446–447
Kaolinite, 20, 46, 275t
Karst topography, 275–277
Karsts, 275–277
Kelvin, Lord, 70
Kepler, Johannes, 10, 441, 444–445
Kepler's laws of planetary motion, 10, 444f, 444–445
Kerogen, 554
Kilauea, 34f, 132
Kimberlite, 188b–189b, 189f, 523, 524f
Kinetic energy, 317, 322
Kirchoff, Gustav, 486
Kuiper belt, 475
Kumme, Bernhard, 80–81

Kupferschiefer, 518
Kuroko-type massive sulfides, 535, 541
Kyanite, 45b, 45f

L

La Niña, 252, 253f
Lagoons, 41f, 41–42, 50
Lake Baikal, 153f, 154
Lake Bonneville, 38b, 38f, 432
Land bridges, 84, 212, 394, 423f, 423–424
Land of Midnight Sun, 328, 328f
Landfills, 292, 293f
Landslides, 431b–432b, 431f–432f
Laplace, Pierre-Simon, 446–447, 451
Laramide orogeny, 415b, 418–420, 421f
Latent heat, 310, 339, 355
 of fusion, 310f
 of vaporization, 310f
Laurasia, 409f
Laurentia, 384, 384f, 397f–398f
 and continental rifting, 382, 383f
 formation of, 378, 378f, 385f, 385b–387b
 and mountain building, 382
 ophiolites on, 402, 402f–403f
Lava
 dating of, 75, 75f
 and igneous rock, 28–29, 29f, 50
 in interaction of Earth reservoirs, 6–7, 7f
 magnetism of, 88f, 88–90, 100
  marine, 91f, 91–92
 pillow, 92, 115, 137, 224b, 373, 374f, 402f–403f
Lava fountains, 33–34, 34f, 121
Law, definition of, 10
Layered igneous complex, 534
Layered mantle convection, 192, 193f, 194, 197
Lead, 518, 519f
Leakey, Maeve, 424
Leakey, Richard, 424f
Leaning Tower of Pisa, 290, 290f–291f
Left lateral motion, of fault, 125
Legrange, Joseph, 451
Lehmann, Inge, 184
Lena River, 287t
Levee
 artificial, 284
 natural, 283–284, 285f
Life. See also Biosphere
 definition of, 368
 origin and early evolution of, 368–372, 372b–373b, 476
 visible, Phanerozoic, 393–394
Light year, 494, 495f
Lightning, 344–347, 345f
 conductors and rods, 345, 347f
 forked, 344–345, 346f–347f
 sheet, 344
Light-water reactors, 559–560, 560f
Lignite, 556, 557f, 568
Lime, 521–522
Limestone
 composition of, 26, 40f
 continental drift and, 84
 crushed, 521
 ecosystems, 219
 formation of, 37, 39
 metamorphism of, 48
 porosity and permeability of, 268, 268t
 solubility of, 275
 weathering of, 274, 275t
Limestone caves, 275–277, 276f
Limonite, 515

Lineaments, 538
Liquids, elastic behavior of, 178–180
Lithification, 27
Lithosphere, 6, 104f–105f, 105–106, 186, 186f
Little Ice Age, 492–493, 494f, 497f, 498, 506, 612, 612f
Local base level, 279, 281f–282f
    dams and, 294, 295f
Lodestone, 86–87
Loihi, 144
Loma Prieta earthquake, 128b, 129f
Lomonosov, Mikhail V., 460
London Basin, 272f, 272b
Longshore current, 237f, 237–238
Louanne Salt, 38b
Lovelock, Sir James, 591b–592b, 619
Lowell, Sir Percival, 441, 462, 464
Lower crust, 186
Lower mantle, 106, 175, 186–187
Low-velocity zone, 105, 105f, 186
Lunar day, definition of, 240
Lunar eclipse, 455b, 455f–456f
Lunar maria, 451f, 451–452, 452f–453f
Lunar Prospector, 456
Luster, 23, 24f
    metallic, 23, 24f
    nonmetallic, 23, 24f
Lyell, Charles, 60–62, 62t, 62f
Lystosaurus, 84, 85f

M

Mafic rock, 29–30, 31f
Magma
    definition of, 28
    viscosity of, 33–34, 36, 50
Magma chambers, 28, 35
Magma processes
    crystal settling, 32–33, 33f
    crystallization, 28, 31–33, 32f
        and mineral deposits, 530
        and release of water, 226, 227f
    fractionation, 32–33, 33f
    magma mixing, 33, 34f
    partial melting, 158b–159b
    segregation, 530, 534, 541
Magmatic deposits, 530–531, 541
Magnesite, 516
Magnesium, 219, 515t, 516f, 516–517
Magnetic anomalies
    marine, 90–92, 91f
    mineral deposits, 536, 536f
Magnetic epochs, 90
Magnetic events, 90
Magnetic field
    generation of, 187–189, 190b–191b, 191f
    strength of, and cosmic radiation, 495f, 497f, 497–498, 506
Magnetic inclination, 87–88, 87f–88f
Magnetic polarity
    normal, 90, 90f
    reversal, 90, 90f, 100
        marine, 91f, 91–92
Magnetism, 86–92, 87f
Magnetite, 86–88, 515, 523–524
Magnetometer, 88
Magnetosphere, 491–492, 492f
    strength of, and cosmic radiation, 495f, 497f, 497–498, 506
Magnetotellurics, 288–289, 289f
Main Sequence, of stars, 498, 502f
Malachite, 518
Mammals

age of, 422, 433
    Cenozoic, 422f–423f, 422–424, 433
    marsupial, 412, 422–424, 423f
    Mesozoic, 412, 414f, 433
    placental, 412, 414f, 422f–423f, 422–424
Manganese, global distribution of, 528, 529f
Manganese nodules, 539–540, 540f
Mantle, 175, 183f, 186–187, 450
    average density of, 366f
    circulation in, 188–192, 195–197
    composition of, 186–187, 197
    convection of, 189–192, 197
        layered, 192, 193f, 194, 197
        mantle plumes, 192, 193f, 197
        whole, 192, 193f, 194, 196b, 197
    depleted, 158b
    general characteristics of, 182t
    lower, 106, 175, 186–187
    melting in, 158b–159b, 186, 187f, 189–192, 192f
    and origin of water, 226
    Precambrian, 363, 366–367
    radioactive decay in, 189
    seismic tomography of, 192, 194–196, 195f–196f, 196b, 197
    upper, 104f, 106, 175, 186–187
Mantle drag, 136, 136f, 138
Mantle plumes, 130, 133f, 135, 137, 144, 144f–145f
    in mantle convection, 192, 193f, 197
Mantle wedge, 118
Marble, 47f, 48
Marcasite, 556
Marcy, Geoffrey, 442, 500b
Mare, lunar, 451f, 451–452, 452f–453f
Margulis, Lynn, 591b–592b
Mariana Trench, 117, 217
Marianas-type subduction, 121, 122f, 136–137, 145, 146f, 151
Marin Headlands, 149f, 149–150
Marine deposit, 39, 41, 41f, 50
Marine magnetic anomalies, 90–92, 91f
Marine regression, 42, 43f, 410f
Marine transgression, 42, 43f, 410f
Maritimes Basin, 402, 404f
Mars, 462–464
    atmosphere of, 307, 441, 463, 467
        comparison with Earth and Venus, 306, 306t, 309, 463t
    current data for, 450t
    density of, 449, 450t
    ice/water on, 206, 463–464, 464f–465f
    life on, possibility of, 372f, 372b–373b, 441–442, 462, 464, 466
    moons of, 462
    orbit of, 445, 445t, 462–463
    seasons of, 463
    surface of, 462f, 462–463
    surface temperature of, 206, 463–464
    volcanoes of, 462f, 463–464, 465f, 467
Marsupials, 412, 422–424, 423f
Martinique, volcanic eruption in, 34
Mass, conversion to energy, 483–484, 506
Mass extinction
    Cretaceous (dinosaurs), 394, 412, 416f, 416–417, 433
    extraterrestrial impact and, 416–417, 416f–417f, 418b, 433, 476
    Paleozoic, 394, 406–407, 407f, 433
    Phanerozoic, 394, 406–407, 407f, 433
    Precambrian, 364–365, 380–381, 389
Mass number, 71, 72b

Mass spectrometer, 72b
Mass wasting, 431b–432b, 431f–432f
Massive sulfide deposits, 518, 535, 541
Matthews, Drummond, 91–92
Mauna Kea, 132
Mauna Loa, 132
Mauna Loa Observatory, 609, 609f
Maunder Minimum, 492–493, 494f, 497f, 612f
Mayan calendar, 55
Meandering streams, 268–269, 279, 279f–280f, 282–283, 283f–284f, 297
Mechanical enrichment, 532
Medieval Optimum, 497f, 498
Mediterranean climate, 315
Mediterranean Sea, 38b, 418
Mekong River, 156b
Mélange, 415b, 415f
Melting
    in mantle, 158b–159b, 186, 187f, 189–192, 192f
    partial, 158b–159b
Melting points, of Earth's interior, 189–192, 192f
Mendocino fault, 125, 125f
Mercury (planet), 441, 459
    current data for, 450t
    density of, 449, 450t, 459
    internal layers of, 459, 460f
    orbit of, 445, 445t, 459
    rotation of, 459
    surface of, 459, 460f
    surface temperatures of, 459, 467
Mercury barometer, 315, 317f
Mesosaurus, 84, 85f
Mesoscale eddies, 249, 250f, 258
Mesosphere, 313
Mesozoic Era, 67, 69f, 70, 74, 393–394, 432
    atmosphere in, 412–416
    biosphere in, 394, 407, 412, 413f
    climate in, 412–416, 414f
    definition of, 393–394, 407
    Earth systems in, 407–416
    hydrosphere in, 394, 412
    mass extinction in, 394, 412, 416f, 416–417, 433
    solid Earth in, 394, 408–412, 409f–411f, 415b, 415f, 433
Metal(s)
    abundant, 514–517, 515t, 516f, 540
    enrichment factors for, 517, 518f
    native, 22, 518–520
    scarce, 514, 515t, 517–520, 519f, 539–540
Metal reserves, global distribution of, 528, 529f
Metallic bonds, 17, 49
Metallic luster, 23, 24f
Metallic minerals, 514–520
    classification of, 514, 515t
    definition of, 514
Metamorphic facies, 48–50
Metamorphic grade, 46, 50
Metamorphic rock, 15–16, 26, 43–50
    classification of, 46–48, 47f
    foliated, 46, 47f–48f, 50
    formation of, 43–44, 44f
    mineral deposits in, 533, 541
    nonfoliated, 46, 47f, 48, 50
    in rock cycle, 26–28, 27f
Metamorphism
    contact, 44
    definition of, 16, 43
    and deformation, 49–50
    dynamic, 44–45
    key minerals in, 45b, 45f

Volcanic eruption(s) *(continued)*
 Hawaiian, 33–34, 34*f*, 36*f*
 in interaction of Earth reservoirs, 6–7, 7*f*,
  307–308
 Martinique, 34
 Mount Piñatubo
  atmospheric effects of, 9, 9*f*, 306, 307*f*,
   326, 603, 610–611
  prediction of, 35
  as pyroclastic eruption, 119, 119*f*
 Mount St. Helens, 33–34, 35*f*, 119, 120*f*
 and origin of water, 226
 prediction of, 34–36, 36*f*
Volcanic glass, 28, 275*t*
Volcanic hotspots. *See* Hotspots
Volcanic island(s), 98, 117–118
Volcanic island arcs, 96, 117–118, 121–123, 137,
  145–146, 146*f*
 accretion of, 148–151, 148*f*–152*f*, 408–412,
  415*b*
 and greenstone belts, 373–375, 375*f*–376*f*
Volcanic island chains, 130*f*, 130–133, 144
Volcanic mountains, 98, 117–118, 121–123
Volcanic rock, 28–36, 29*f*, 50
 lunar, 451–452, 457
Volcanoes
 definition of, 28
 on Jupiter moon, 472, 472*f*, 478
 Martian, 462*f*, 463–464, 465*f*, 467
 and origin of life, 369–370
 Pacific Ring of Fire, 94, 96, 96*f*–97*f*, 119,
  145, 146*f*
 Precambrian, 366–367
 subduction zone, 96, 96*f*–97*f*, 99*f*, 117*f*,
  117–121, 145–147, 146*f*
Von Fraunhofer, Joseph, 486

W

Walcott, Charles D., 396
Walker, Gilbert, 255
Waning, of moon, 454*b*, 454*f*
Water. *See also* Fresh water; Ocean Water
 abundance on Earth, 206, 207*f*
 boiling temperature of, 209
 bonding and polarity of, 209*f*, 209–210
 density of, 209–210, 227
 heat in, 210, 227, 310*f*, 310–311, 317–319
 interaction with air, 310*f*, 310–311

 inventories of, 263–264, 264*f*, 264*t*
 juvenile, 226
 light penetration of, 210, 227
 location of, 206, 207*f*
 molecular structure of, 17–18, 209–210
 molecular weight of, 209
 natural states of, 206, 209
 orbital motion of, 232*f*–233*f*, 233
 as solvent, 210, 219, 220*f*, 227, 261, 273–277
 unusual properties of, 209–210, 227
Water budget, 263–264, 264*f*, 297, 311*f*
Water debt, 263
Water pressure, 271
Water table, 266, 266*f*–267*f*, 297
 changes in, 269*f*, 269–270
 elevation variations of, 270–271
 exploration for, 287
 and wells, 268, 269*f*
Water vapor
 in atmosphere, 306, 310–311
 and relative humidity, 319–321
Water witch, 287
Waterfalls, 279, 281*f*–282*f*
Wave(s), 232*f*–233*f*, 232–236, 258
 in coastal regions, 236–238, 236*f*–238*f*, 258
 orbital motion and, 232*f*–233*f*, 233
 seismic sea, 234*b*–235*b*, 234*f*–235*f*
 terminology, 232*f*, 232–233
Wave energy, 562
Wave height, 232–233
Wave velocity, 233
Wavelength, 233, 233*f*
Waxing, of moon, 454*b*, 454*f*
Weak force, 484*b*
Weather, 314–321
 definition of, 314, 322
 fronts and, 336–342, 339*f*–342*f*
 jet streams and, 342–343
 storms, 344–351
Weather forecasting, 314–315
Weathering, 27–28
 chemical, 274–275, 275*t*, 297
 freeze-thaw cycle, 210
 and mineral deposits, 532, 541
 moving water and, 280–282
Wegener, Alfred, 80–86, 99, 206
Well logging, 61*f*, 61*b*
Westerlies, 351

White dwarf, 502, 502*f*
White light, 484
White smoker, 225*b*
White, Tim, 424*f*
Whole-mantle convection, 192, 193*f*, 194, 196*b*,
  197
Wien, Wilheim, 484–485
Wien's Law, 484–485, 485*f*
Wilson cycle, 156–159, 157*f*, 171
 and North America, 386*b*, 387*f*
 Precambrian, 378, 379*f*, 380*b*–381*b*, 389
Wilson, J. Tuzo, 83, 130, 156–159, 171
Wind(s)
 atmospheric pressure and, 315, 319*f*, 322,
  335*f*, 335–336
 and El Niño, 252, 253*f*, 255*f*, 255–256, 258
 global patterns of, 351
 hurricane, 348–351
 and ocean currents, 249*f*, 249–252,
  251*f*–252*f*, 258
 and sea level, 212
 and waves, 232–236
Wind farms, 565, 565*f*
Wind power, 546, 565*f*, 565–568
Wind power density, 565–566
Wind tunnels, 342
Windchill factor, 317, 319*t*
Wrangellia, 149*f*, 151, 151*f*

Y

Yangtze River, 156*b*–157*b*, 262, 287*t*
Yellow dwarf, 482
Yellow River, 156*b*, 262
Yellowstone National Park
 caldera, 134, 135*f*
 geyser, 273, 275*f*
Yenisei River, 287*t*
Yucatan Peninsula, 417
Yukon, 150

Z

Zinc, 518, 519*f*
Zion National Park, 65–66, 68*f*, 70
Zircon, 73*b*, 74, 367, 368*f*
Zone of accumulation, 580
Zone of aeration, 266–267
Zone of saturation, 266, 297